Lecture Notes in Artificial Intelligence 4648

Edited by J. G. Carbonell and J. Siekmann

Subseries of Lecture Notes in Computer Science

Fernando Almeida e Costa
Luis Mateus Rocha Ernesto Costa
Inman Harvey António Coutinho (Eds.)

Advances in Artificial Life

9th European Conference, ECAL 2007
Lisbon, Portugal, September 10-14, 2007
Proceedings

Springer

Series Editors

Jaime G. Carbonell, Carnegie Mellon University, Pittsburgh, PA, USA
Jörg Siekmann, University of Saarland, Saarbrücken, Germany

Volume Editors

Fernando Almeida e Costa
University of Sussex, Centre for Computational Neuroscience and Robotics
Brighton BN1 9QG, UK,
E-mail: F.AlmeidaCosta@sussex.ac.uk

Luis Mateus Rocha
Indiana University, School of Informatics
Bloomington, IN 47406, USA
E-mail: rocha@indiana.edu

Ernesto Costa
University of Coimbra, Department of Informatics
3030-290 Coimbra, Portugal
E-mail: ernesto@dei.uc.pt

Inman Harvey
University of Sussex, Department of Informatics
Brighton BN1 9QH, UK
E-mail: inmanh@sussex.ac.uk

António Coutinho
Gulbenkian Institute of Science
2781-901 Oeiras, Portugal
E-mail: coutinho@igc.gulbenkian.pt

Library of Congress Control Number: 2007934544
CR Subject Classification (1998): I.2, J.3, F.1.1-2, G.2, H.5, I.5, J.4, J.6

LNCS Sublibrary: SL 7 – Artificial Intelligence

ISSN 0302-9743
ISBN-10 3-540-74912-8 Springer Berlin Heidelberg New York
ISBN-13 978-3-540-74912-7 Springer Berlin Heidelberg New York

Springer is a part of Springer Science+Business Media

springer.com

Printed in Germany

Typesetting: Camera-ready by author, data conversion by Markus Richter, Heidelberg
Printed on acid-free paper SPIN: 12122318 06/3180 5 4 3 2 1 0

Preface

The use of methodologies of synthesis implemented in computational, robotic or other artificial architectures, aiming at the study of life-like processes, is the hallmark of artificial life. This synthetic approach is linked both to scientific and technological desiderata. Scientifically, it provides a glimpse of possible biological architectures beyond the ones provided by nature, which may play a fundamental role in the constitution of a lawful science of life. Technologically, it helps us discover solutions for hard engineering problems, non-intuitive to humanly biased approaches. This close proximity of science and technology, of content and medium, allows for the search of general organizational principles that bring together, much closer than usually thought, life, mind and social systems.

In this edition of ECAL, alongside the usual areas that roughly maintained their level of contribution, two areas received particular attention from researchers, having significantly increased their presence: social systems (15 papers) and ecosystems (7 papers).

Another issue perhaps worth noting is the number of papers dealing, in one way or another, with the theme of the constitution of meaning. This theme runs across various sections of the proceedings but, to cite only the obvious examples, the sections on communication and on social systems total 28 papers.

From the original 172 submissions, 38 papers were accepted for oral presentation and 82 for poster presentation. They are published in these proceedings without distinction. The conference being of one-track session type, only a few papers could be accepted for oral presentation and that decision was a very difficult one. Many papers accepted for poster presentation are of very high scientific standards.

This proceedings volume is organized into 15 sections. Although every classification is to some extent arbitrary, sections 1 to 4 form a group that could be labeled *Life*, sections 5 to 9 another group that could be labeled *Cognition*; then, sections 10 and 11, *Social Systems*, and finally, sections 12 to 15, *Tools, Models, Methodologies.*

These papers were distributed over the four days of the conference in a way that generally corresponds to the themes highlighted each day: *Cognition and Embodiment, Complex Systems and Networks, Communication and Social Systems,* and *The Relation Between Alife and Biology, Open Issues.*

The thematic organization of the conference on each day was matched by the content of the keynote lectures. We are profoundly indebted to a group of outstanding keynote speakers who graciously accepted our invitation to address the conference. For a special keynote lecture, in the opening ceremony, we invited Dario Floreano. For the *Cognition and Embodiment* day, we invited Rudolf Bannasch and Randall Beer, for the *Complex Systems and Networks* day, António Coutinho and Janet Wiles, for the *Communication and Social Systems* day, Peter Todd and Ezequiel Di Paolo, for *The Relation Between Alife and Biology* day, Ricard Solé and Brian Goodwin.

It was our wish to promote the vivid discussion of particularly controversial or critical issues. With this objective in mind, panel discussions were organized, at the end of each day, highlighting a particular critical point, with the presence of the keynote speakers of that day and three other discussants. These panel discussions took place in plenary sessions.

The most important scientific component of a conference is the paper reviewing process. With very few exceptions, the submitted papers were reviewed by at least three reviewers. This was only possible due to the professionalism and extreme care of the 100 members of the Program Committee to whom we are deeply grateful.

Finally, we thank the support granted by the Portuguese Foundation for Science and Technology. We also thank the IGC – Gulbenkian Institute of Science, for assuming the role of host institution of the event, and for the general support provided.

July 2007

Fernando Almeida e Costa
Luis Mateus Rocha
Ernesto Costa
Inman Harvey
António Coutinho

Committees

Organizing Committee

Fernando Almeida e Costa (Chair)
Luis Mateus Rocha
Ernesto Costa
Inman Harvey
António Coutinho

Program Committee

Chris Adami
Fernando Almeida e Costa
Wolfgang Banzhaf
Mark Bedau
Randall Beer
Peter J. Bentley
Luc Berthouze
Hugues Bersini
Eric Bonabeau
Josh Bongard
Seth Bullock
Mathieu Capcarrere
Peter Cariani
Jorge Carneiro
Leandro Nunes de Castro
Thomas Christaller
Helder Coelho
Netta Cohen
Luís Correia
Ernesto Costa
António Coutinho
Luís Custódio
Kyran Dale
Kerstin Dautenhahn
Bart de Boer
Edwin de Jong
Yiannis Demiris
Ralf Der
Ezequiel Di Paolo
Peter Dittrich
Marco Dorigo
Arantza Etxeberria
Dario Floreano
Carlos Gershenson
Takashi Gomi
Paul Graham
Benoit Hardy-Vallée
Inman Harvey
Phil Husbands
Fumiya Iida
Auke Jan Ijspeert
Takashi Ikegami
Eduardo Izquierdo-Torres
Colin G. Johnson
Jozef Kelemen
Jeffrey Krichmar
Tim Lenton
Pedro Lima
Hod Lipson
Fernando Lobo
Luís Seabra Lopes
Max Lungarella
Ian Macinnes
Davide Marocco
Barry McMullin
José Fernando Mendes
Jean-Arcady Meyer
Eduardo Reck Miranda

Melanie Mitchell
Luís Moniz Pereira
Federico Morán
Alvaro Moreno
Chrystopher Nehaniv
Stefano Nolfi
Alexandra Penn
Alfredo Pereira
Andrés Pérez-Uribe
Rolf Pfeifer
Andrew Philippides
Daniel Polani
Vitorino Ramos
Steen Rasmussen
Luis Mateus Rocha
Miguel Rocha
Agostinho Rosa
Kepa Ruiz-Mirazo
João Sentieiro
Anil Seth
Cosma Shalizi
Linda Smith
Ricard Solé
Emmet Spier
Olaf Sporns
Susan Stepney
Jun Tani
Charles Taylor
Tim Taylor
Adrian Thompson
Jon Timmis
Peter Todd
Elio Tuci
Jon Umerez
Patrícia Vargas
Robert Vickerstaff
Paul Vogt
Richard Watson
Janet Wiles
Rachel Wood
Andy Wuensche
Larry Yaeger
Tom Ziemke

Table of Contents

Artificial Chemistries, Pre-biotic Evolution

Evolution

Ecosystems

Morphodynamics, Development

Adaptive Behavior

Learning and Evolution

Communication, Constitution of Meaning, Language

Agency, Autopoiesis, Autonomy

Alife and Art

Dynamics of Social Systems, Collective Behavior

Swarm and Ant Colony Systems

Robotics and Autonomous Agents: Concepts and Applications

Evolutionary Computation

Networks, Cellular Automata, Complex Systems

Models and Methodologies

Chemical Organizations at Different Spatial Scales

Pietro Speroni di Fenizio[1,2] and Peter Dittrich[2,⋆]

[1] ProtoLife, Parco Vega Via della Liberta' 12, 30175, Marghera, Venezia, Italia
[2] Bio Systems Analysis Group, Department of Mathematics and Computer Science, Friedrich Schiller University Jena, D-07743 Jena, Germany
dittrich@minet.uni-jena.de

Abstract. The dynamics of spatial reaction systems that consists of many molecular species can be difficult to understand. Here we introduce a method that allows to observe the dynamics of a diverse spatial reaction system at different spatial scales. Using chemical organization theory we define for a given spatial location its so called spatial organization, which is the organization generated by the molecular species present in the neighborhood of this location. The scale determines the size of that neighborhood. We show that at one scale, patterns become visible that can not be seen at a different scale. Furthermore, different scales tend to map to different parts of the lattice of organizations; at small scales spatial organizations tend to be small (lower part of the lattice of organizations) while at large scales spatial organizations tend to be large (upper part of the lattice of organizations). Finally we show how the right scale can be selected by comparing the spatial reactor with its well-stirred counterpart. The method is illustrated using an artificial chemistry.

1 Introduction

Data of inorganic and organic chemical reaction systems are increasing rapidly. At the same time, dynamical reaction system models are growing in size. For example, a dynamic model of earth's atmosphere ozone chemistry consists of about 60 molecular species [1], a stochastic model of the central sugar metabolism of E. Coli. comprises of 100 molecular species [2], and models of combustion chemistry can easily exceed several hundred molecular species (e.g., 506 species in Ref. [3]).

Those systems are usually studied in a well-stirred reactor setting or in an Euclidean reaction space. Visualizing the dynamics of the spatial system presents two different difficulties: First, at each location the state of the system is represented by a concentration vector, which is by itself hard to represent due to its high dimensionality. Second, although the changes follow local rules, the system as a whole might undergo changes that become apparent only when a whole region is taken into account. In those case we say that the system has undergone a change at a larger spatial scale.

⋆ Corresponding author.

F. Almeida e Costa et al. (Eds.): ECAL 2007, LNAI 4648, pp. 1–11, 2007.

In this work we use chemical organization theory [4] to represent the state of a system in an area; thus mapping a vector (the molecules present with their concentration) into a discrete set (the organization generated by such vector). By varying the size of the vicinity we can produce a number of descriptions of the state of the system. Then by comparing the result with an equivalent system made by the same set of molecules randomly placed, we obtain information on what area of the system and what spatial scales are more interesting, which provides information to select the right scale.

2 Examples

We will illustrate the theoretical concepts by using two abstract chemical reaction system consisting of 2 and 16 molecular species, respectively.

The 16-species systems is taken from Banzhaf [5]. It is especially well suited for our purpose, because it posses an interesting (and large) lattice of organizations (Fig. 3), which can not be expected in a randomly created reaction network [6].

The set of molecules is $\mathcal{M} = \{0, 1, 2, 3, 4, 5, 6, 7, 8, 9, 10, 11, 12, 13, 14, 15\}$. Given two molecules $i, j \in \mathcal{M}$, $\mathcal{R}(i, j) \in \mathcal{M}$ denotes the molecule catalytically produced by i and j according to the following reaction table:

$\mathcal{R}$	0	1	2	3	4	5	6	7	8	9	10	11	12	13	14	15
0	-	-	-	-	-	-	-	-	-	-	-	-	-	-	-	-
1	-	1	-	1	4	5	4	5	-	1	-	1	4	5	4	5
2	-	-	1	1	-	-	1	1	4	4	5	5	4	4	5	5
3	-	1	1	1	4	5	5	5	4	5	5	5	4	5	5	5
4	-	2	-	2	8	10	8	10	-	2	-	2	8	10	8	10
5	-	3	-	3	12	15	12	15	-	3	-	3	12	15	12	15
6	-	2	1	3	8	10	9	11	4	6	5	7	12	14	13	15
7	-	3	1	3	12	15	13	15	4	7	5	7	12	15	13	15
8	-	-	2	2	-	-	2	2	8	8	10	10	8	8	10	10
9	-	1	2	3	4	5	6	7	8	9	10	11	12	13	14	15
10	-	-	3	3	-	-	3	3	12	12	15	15	12	12	15	15
11	-	1	3	3	4	5	7	7	12	13	15	15	12	13	15	15
12	-	2	2	2	8	10	10	10	8	10	10	10	8	10	10	10
13	-	3	2	3	12	15	14	15	8	11	10	11	12	15	14	15
14	-	2	3	3	8	10	11	11	12	14	15	15	12	14	15	15
15	-	3	3	3	12	15	15	15	12	15	15	15	12	15	15	15

An entry $\mathcal{R}(i, j) =$"-" denotes that i and j do not react.

For simulation, we represent a chemical reactor by a 2-dimensional matrix $(p_{(x,y)})$ with $x \in \{0, 1, \ldots, x_{size} - 1\}$ and $y \in \{0, 1, \ldots, y_{size} - 1\}$ denoting a spatial coordinate in the reactor. Each location (x, y) contains one molecule $p_{(x,y)} \in \mathcal{M}$. As usual, we assume periodic boundary conditions, such that, for example, $p_{(-1,-1)} = p_{(x_{size}-1, y_{size}-1)}$. The dynamics is simulated by the following conventional update scheme: Let $r \in \mathbb{N}$ denote a reaction radius. (1) Chose randomly a location (x_1, y_1) and two further locations (x_2, y_2), (x_3, y_4) in the neighborhood specified by the reaction radius with $|x_1 - x_2| \le r$, $|y_1 - y_2| \le r$, $|x_1 - x_3| \le r$, $|y_1 - y_3| \le r$. (2) If $\mathcal{R}(p_{(x_1,y_1)}, p_{(x_2,y_2)}) \neq$ "$-$" replace $p_{(x_3,y_3)}$ with the product $\mathcal{R}(p_{(x_1,y_1)}, p_{(x_2,y_2)})$.(3) Advance time counter $t \leftarrow t + 1/(x_{size} y_{size})$. We call the unit of t, a generation since it consists of as many update steps as

(a)

$\mathcal{R}$	0	1
0	0	0
1	0	1

(b)

$$1+1 \to 1+1+1$$
$$1+0 \to 1+0+0$$
$$0+0 \to 0+0+0$$
$$0 \to$$
$$1 \to$$

(c)

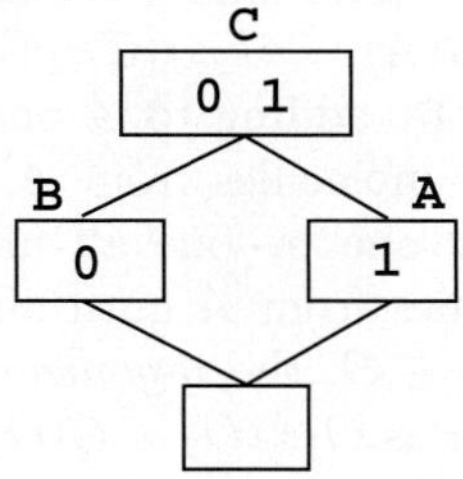

Fig. 1. Two-species example: (a) Reaction table $\mathcal{R} : \mathcal{M} \times \mathcal{M} \to \mathcal{M}$ of the two-species example chemistry. (b) Reaction network implied by this table. Note that the reactions "$0 \to$" and "$1 \to$" represent the general dilution flow. (c) Hasse-diagram of the lattice of the four organizations $A = \{1\}$, $B = \{0\}$, $C = \{0, 1\}$, and the empty set $\emptyset$.

there are molecules in the reactor. The algorithm simulates a second-order catalytic reaction system under flow condition (like the Hypercycle [7], the replicator equation [8], or AlChemy [9]) where the outflow assures that the total number of molecules is constant.

The second example has only two molecular species $\mathcal{M} = \{0, 1\}$ with the reaction rules $\mathcal{R}$ given by Fig. 1. Being a very simple system this example permits us to study some basic properties without getting lost in unnecessary details. The reaction dynamics, the boundary conditions, and all the other details are equivalent to the previous example.

3 Chemical Organization Theory

In the following, basic definitions of chemical organization theory are given for catalytic flow systems, i.e., systems where all molecules act only catalytically and are subject to a dilution flow [4,9,10]: Let $\mathcal{M}$ be the set of all possible molecules and let $\mathcal{R}$ denote the reaction rules. We call a set of molecules $C \subseteq \mathcal{M}$ *closed*, if the molecules in C produce only molecules already contained in C (formally: for all $i, j \in C$, $\mathcal{R}(i, j) \in C$). Given an arbitrary set $A \subseteq \mathcal{M}$ we can always find a smallest closed set that contains A. We say that A *generates the closed set* $C = G_{CL}(A)$. A complementing property is the self-maintenence. In catalytic flow systems, we call a set C *self-maintaining*, if every molecule of C is produced by at least one reaction of molecules contained in C (formally: for all molecules $k \in C$ exists $i, j \in C$ with $\mathcal{R}(i, j) = k$). In catalytic flow systems, given a set $A \subseteq \mathcal{M}$ we can always find a biggest self-maintaining set contained in A. We say that A *generates the self-maintaining set* $C = G_{SM}(A)$.

Now taking closure and self-maintenence together, we arrive at an organization, a concept introduced by Fontana and Buss [10,9] and generalized by us [4]. An organization $O \subseteq \mathcal{M}$ is a set of molecules that is closed and self-maintaining, In a catalytic flow system, given a set A we can generate an organization (uniquely) by first generating the closure of A and then generating the self-maintaining set of the closure. We say that A *generates organization* $O = G(A) = G_{SM}(G_{CL}(A))$. Practically, given A we can generate an organization by adding to A one by one molecules that can be created by reactions among molecules from A, until we cannot add molecules any more. Then we remove one by one all molecules that are not produced by reactions among molecules from A until all molecules are maintained. Given two organizations $O_1, O_2 \in \mathcal{O}$, the *organization union* $\sqcup$ and the *organization intersection* $\sqcap$ are defined as $O_1 \sqcup O_2 = G(O_1 \cup O_2)$ and $O_1 \sqcap O_2 = G(O_1 \cap O_2)$, respectively. The set of all organizations $\mathcal{O}$ together with the organization union and intersection form an algebraic lattice $\langle \mathcal{O}, \sqcup, \sqcap \rangle$. This lattice can be visualized as a Hasse-diagram (e.g., Fig. 1c), which reveals the hierarchical structure of the reaction system. The example systems with two and 16 species have four (Fig. 1c) and 54 organizations (Fig. 3), respectively.

4 The Well-Stirred Experiment

As we study the experiment, we will often face the problem to understand how important the actual position of the molecules in space are. A similar study has been presented by Lindgren, Eriksson, and Eriksson [11], trying to ascertain the information stored in a chemical system by the position of the molecules.

To do this we shall consider, in every time-step, side by side to the actual experiment, a second copy of it, that we shall refer to as the *well-stirred experiment.* The well-stirred experiment, will generally contain exactly the same molecules as the actual experiment; but the actual position of the molecules will be changed randomly[3]. By comparing the actual experiment, with the well-stirred experiment, using neighborhoods of various radii, we shall uncover how relevant the information we are gathering to the actual position of the molecules is.

5 Spatial Organization

Let $\mathcal{L} = \{(x, y) : x \in \{0, 1, \ldots, x_{size} - 1\},\ y \in \{0, 1, \ldots, y_{size} - 1\}\}$ be the set of all locations. In order to calculate the organization at location $(x, y) \in \mathcal{L}$ given a scale R we first take the set of all molecules in the neighborhood of (x, y) including the molecule at this position:

$$U^R_{(x,y)} = \{p_{(x',y')} : |x - x'| \le r, |y - y'| \le r\}. \tag{1}$$

Then we generate the organization from this set:

$$O^{(R)}_{(x,y)} = G(U^R_{(x,y)}). \tag{2}$$

For any given scale R, we can count the frequency of occurrence of each organization in the whole experiment. Given an organization $O \in \mathcal{O}$ we define

$$c(O, R) = |\{(x, y) \in \mathcal{L} : O^{(R)}_{(x,y)} = O\}| / |\mathcal{L}| \tag{3}$$

[3] As in the normal spatial experiment there is also one molecule at each location in the well-stirred experiment.

as the relative occurrence of organization O at scale R where $|\mathcal{L}| = x_{size}y_{size}$ is the total number of locations where spatial organizations can be found. So, $c(O,R)$ denotes the concentration of organization O at scale R.

In order to interprete $c(O,R)$ we can compare it with the expected number of spatial organizations in a well-stirred reactor that contains exactly the same molecules, but where the positions of the molecules are randomly assigned. We denote the expected concentration of organization O at scale R in a well-stirred reactor of the same geometry as $\bar{c}(O,R)$. Practically we estimate it by first shuffling the molecules in the reactor sufficiently (e.g., selecting randomly $10x_{size}y_{size}$ times two molecules and swapping them) and then measuring $\bar{c}$ in the same way as c before. For the results shown here, we repeated this procedure four times and averaged over these four samples, resulting in a negligible statistical error.

As a measure of how c diverges form $\bar{c}$ we simply take the Euclidian distance:

$$d_R = \sqrt{\sum_{O \in \mathcal{O}} (c(O,R) - \bar{c}(O,R))^2}. \tag{4}$$

So, d_R measures how at scale R the concentration vector of spatial organizations is different from the expected concentration vector assuming that the molecules are well-stirred. Consequently, d_R can be used as an indicator for choosing the appropriate scale, which reveals the effect of the spatial topology.

6 Results

6.1 Waves of Change

One of the simplest result that we observe is how clearly the general dynamics of the reactor can appear. Let us consider the simple two-species example, Fig. 1, and let us consider a homogeneous state of the system where every molecule present is of type 1. We now perturb that state by introducing from the outside a molecule of type 0. Such perturbation is equivalent to the ϵ fluctuation (constructive perturbation) presented in Refs. [4].As time progresses, the fluctuation expands, until the whole experiment moved from a homogeneous state 1 to a homogeneous state 0. But this transformation does not happen on the whole experiment at the same time. It starts where we introduced the fluctuation (in the right upper corner, Fig. 2a), and expands through the whole reactor. Such wave becomes apparent if we consider neighborhoods of intermediate radii. In such representations the area behind the wave contains only organization $B = \{0\}$. The area that has not yet been touched by the wave contains only the organization $A = \{1\}$, and the boundary area contains organization $C = \{0,1\}$. As time progresses the wave moves, transforming the state of the system from all 1 to all 0. Not only but if we fully ignore the spatial positions of the different neighborhoods, and we consider just their total number, we notice how organization A monotonically decreases, organization B monotonically increases, and organization C increases at the beginning and then decreases.

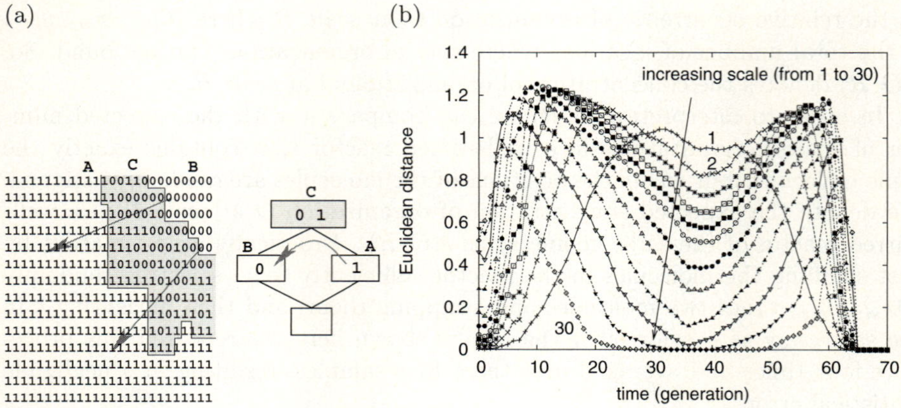

Fig. 2. Example of a wave. (a) A 14×14 section of the 2-dimensional reaction space (100×100 grid) at generation 10 (1 generation = 10000 elementary reactions). Each number represents a molecule. The shaded region is organization $C = \{0, 1\}$ at scale 2. Organizations $A = \{1\}$ and $B = \{1\}$ are kept white. (b) Effect of the spatial structure at different scales R over time measured as the Euclidean distance d_R between the distribution of organizations of the unperturbed grid and the distribution of organizations of a well-stirred grid (with the same amount of molecules). $R \in \{1, 2, 3, 4, 5, 6, 8, 10, 14, 18, 24, 30\}$. Catalytic reaction rules: $0 + 0 \Rightarrow 0, 0 + 1 \Rightarrow 0, 1 + 1 \Rightarrow 1$.

6.2 Different Scales Tend to Map to Different Parts of the Lattice

For each radius R (scale) that we use to study the system we obtain a different image, where at each location a different organization can be present. If we take two different sets of molecules, A and B, with $A \subset B$ then (in reactive flow systems) the organization generated by A is contained in the organization generated by B: $A \subset B$ then $G(A) \subseteq G(B)$. For this reason, as we increase the radius of the neighborhoods, the organizations represented will tend to raise in the lattice of organizations (Fig. 3). At each location we will thus move (with increasing R) from the organization generated by a single molecule to the organization generated by the whole set of molecules in the reactor.

6.3 Extreme Scales

As we consider different scales in our analysis, we should also consider what happens when we take an extremely small radius $R = 0$, the kind of radius that only covers a single molecule; or an extremely big radius $R = \infty$, which would cover the whole reactor. We shall call the first one the zero radius view (or zero view), and the second one the whole radius view (or whole view).

In the first case the organizations that appear would only be the organizations generated by a single molecule (or by the empty set, if the location is empty). Those organizations are commonly (but not always!) in the lower part

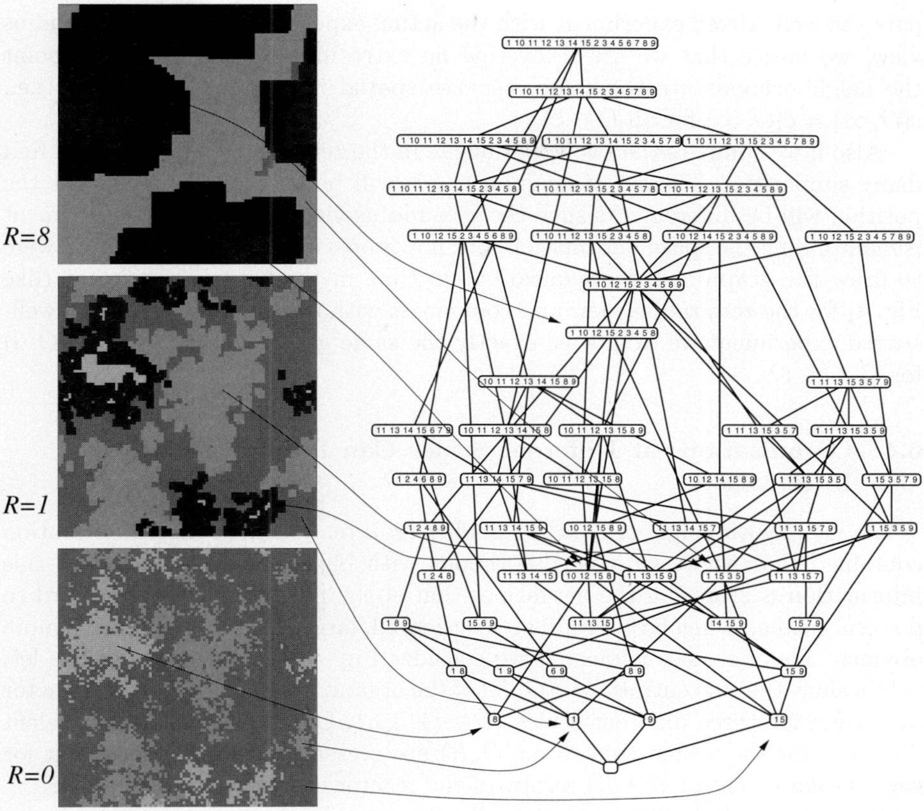

Fig. 3. Illustration of how at different scales $R \in \{0, 1, 8\}$ the spatial organizations tend to map to different parts of the lattice of organizations (right). With increasing scale, spatial organizations tend to map to larger organizations. Left: Spatial organizations in the 100×100 reactor at generation $t = 400$ of a typical simulation of the 16-species example chemistry (NTOP). Each color refers to an organization. Right: Hasse-diagram of the lattice of organizations of the 16-species example (NTOP).

of the lattice. And (in reactive flow systems) any other bigger neighborhood that contains this molecule will generate an organization that contains this organization. Also any empty location (assuming no inflow) as well as any molecule that does not generate a non-empty organization would be mapped to the empty organization.

Let us now consider the whole radius view. It consists of only one neighborhood, repeated at every location. The neighborhood consists of the whole set of molecules present in the reactor. In this case the organization that will appear will be the highest organization possible in the reactor. Any other studied neighborhood, of any radius, will always present an organization contained (but not necessarily strictly contained) in this organization. If we now compare the well-stirred experiment with the actual experiment, in the whole radius view, we notice that we are recovering no extra information. On every point the neighborhoods are the same, because spatial information is ignored, i.e., $c(O, \infty) = \bar{c}(O, \infty)$ for all $O \in \mathcal{O}$.

Also if we compare the two experiments in the zero radius view, we shall find many similarities. The set of neighborhoods will be exactly the same, but the position will be different. As such the zero radius view of the actual experiment is maintaining the space information, but not processing it in any way. If we were to draw the graph of the organizations in time in the actual experiment (like Fig. 4) for the zero radius view and compare it with the same graph for the well-stirred experiment we would see exactly the same graph, i.e., $c(O, 0) = \bar{c}(O, 0)$ for all $O \in \mathcal{O}$.

6.4 Organizations at Different Scales Can Provide Different Information

While views with neighborhoods of smaller radii tend to have more information contained than views with neighborhoods with bigger radii, still a lot of this information is stored in the actual position of the neighborhoods, and hard to discern. Instead when larger radii are considered, large-scale effects become more obvious. This becomes evident if we consider Fig. 4. The diagrams in the left colum show the concentrations $c(O, R)$ of the organizations present in the reactor over time for three different scales $R \in \{1, 3, 6\}$. On the right, the equivalent diagrams for the well-stirred case $\bar{c}(O, R)$ are presented (the same run, but for each measurement of $\bar{c}(O, R)$ a copy of the reactor is created and shuffled).

At scale $R = 1$ it is very hard to discern any differences between the actual reactor and the well-stirred one. This result is counterintuitive because the small scale contains a more fine grained image of the state of the system. Instead on the middle plate ($R = 3$) it is more evident how some organizations are rapidly taking over the system, and the experiment is going through a phase transition somewhere between generation 20 and 30. On the third plate ($R = 6$) the changes are even more steep.

While this would suggests that we should just only consider neighborhoods of a fairly large size, we cannot simply define a minimum size. Nor are all the scales above a certain value equally interesting. From a certain point onward, the representation of the state of the system will simply present an abrupt phase transition.

When we compare in Fig. 4 the left hand side (actual molecular position) with the right hand side (well-stirred case), we notice that there seems to be no major changes at scale $R = 1$ and scale $R = 6$. Instead at scale $R = 3$ the whole area (A), present in the actual experiment is absent in the well-stirred version. Why? This is due to a clustering effect in space, where due to local molecular reactions patches of molecules appear (Fig. 3). Then those patches remain and appear in the graph of $c(O, R)$, as more often will happen that a whole neighborhood is fully contained within a patch. But when those molecules are spread over the whole experiment, their effect seems to disappear.

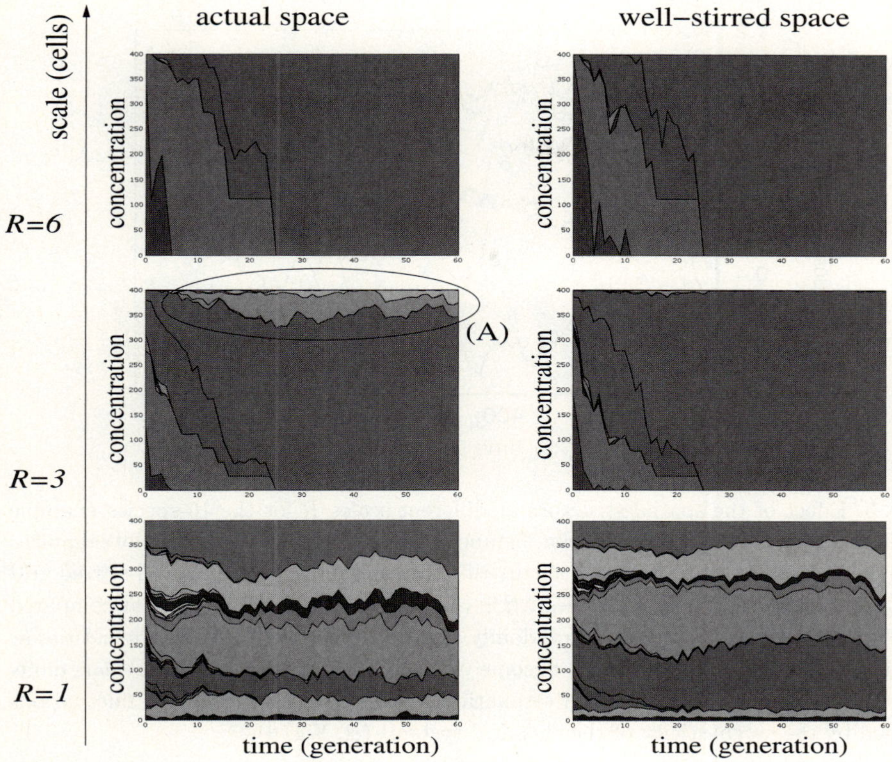

Fig. 4. Relative occurrence (concentration) $c(O, R)$ (left) and $\bar{c}(O, R$ (right) of all organizations $O \in \mathcal{O}$ over time for the normal spatial reactor (left) and the well-stirred reactor (right). Each color represents an organization.

Because such clusters only appeared on an analysis of some specific radius, it is now obvious that for studying an experiment we shall investigate what are those radii. And they shall give us information of the size of the clusters that are present. Such radii will generally be the ones in which the difference between the well-stirred experiment and the actual experiment are wider.

To calculate the difference we employ the Euclidean distance (Eq. (4)). As an example, Fig. 2b shows for different radii how the Euclidean distance changes over time for the wave experiment We can see that the system passes through 5 phases: from generation 1 to generation 30, one after the other the various scales become more relevant, as the wave breaks every cluster, and travels through the system. Between generation 30 and generation 45 there are no big clusters, and the more fine grained a representation is, the better it is to obtain information on the system. Between generation 45 and generation 65 a new big cluster is being formed. Such cluster is only present if we are not randomly spreading the molecules. And as the cluster becomes bigger and bigger one after the other bigger radius becomes more relevant.

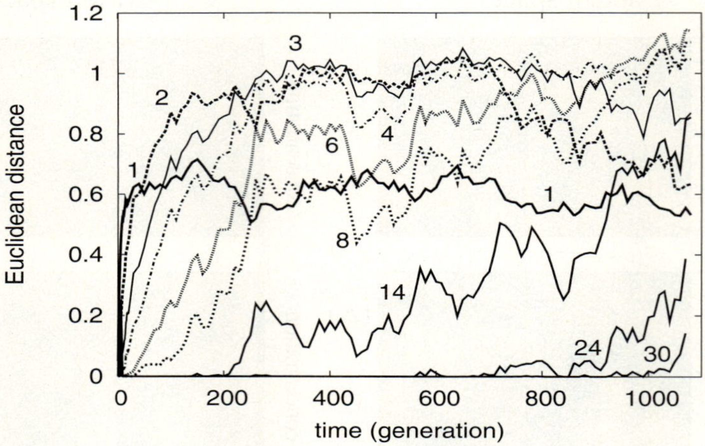

Fig. 5. Effect of the spatial structure at different scales R for the 16-species example (NTOP). Displayed is the Euclidean distance d_R between the distribution of organizations of the unperturbed grid and the distribution of organizations of a well-stirred grid (with the same amount of molecules). $R \in \{1, 2, 3, 4, 8, 14, 24, 30\}$. Note that compared to Fig. 2 this system is initialized randomly. So, it takes some time before structures at large scale (e.g. $R = 24$ or $R = 30$) become visible. Grid size 100×100. Error bars omitted, because the error made when estimating $\bar{c}$ is much smaller than the fluctuations caused by the stochasticity of the chemical systems (cf. Fig. 4).

A second example uses the 16-species chemistry (NTOP [5], Fig. 3). In this case the system is initialized randomly and soon starts to generate clusters. As time progress bigger and bigger clusters are generated. We can observe this, by noticing how the Euclidean distance between the actual experiment and the well-stirred experiment increases in order for increasingly bigger ranges (Fig. 5). First it raises for scale $R = 1$, then 2, then 3, 4, 6, 8, 14, 24, and finally 30. The actual diagram is very complicated and a complete analysis of that is beyond the scope of this paper. Yet it clearly shows which ranges should be taken into consideration when we want to study a certain experiment, in a particular moment. Which is what our intention was, from the beginning.

7 Conclusion and Outlook

Although organization theory was first developed in a well-stirred reactor setting, nothing prevents it to be further investigated in a spatial background. In our investigation we have shown that this is not only possible but can give rise to further information on the system. Organizations, in this, not only provide a way to look at the whole system and its changes, but how those changes spread through the system. What kind of clustering is formed, and what size range should be employed to observe the significant changes that are happening in the experiment. Without being distracted by details or lost in a too broad landscape.

Knowing that models of chemical [12], biological [13] and other systems can have non-trivial organizational structures, where organizations are related to functions, we can expect that the presented methods are widely applicable.

Interestingly, there is little in our approach that is specific to organization theory, and a similar approach can indeed be used with other techniques the identify sub-networks. In this paper we have used chemical organization theory to extract sub-networks, but we could also have used other approaches doing this, most notably pathway-oriented approaches like elementary mode analysis or extreme pathways. Applying them instead of chemical organization theory might also be fruitful.

Acknowledgments. We thank Kristian Lindgren for his inspiring input to this work and the European Center for Living Technologies (ECLT) for offering a very stimulating place to work. We acknowledge financial support by the Federal Ministry of Education and Research (BMBF) Grant 0312704A to FSU Jena.

References

1. Lehmann, R.: Determination of dominant pathways in chemical reaction systems: An algorithm and its application to stratospheric chemistry. J. Atmos. Chem. 41, 297–314 (2002)
2. Puchalka, J., Kierzek, A.: Bridging the gap between stochastic and deterministic regimes in the kinetic simulations of the biochemical reaction networks. Biophys. J. 86(3), 1357–1372 (2004)
3. Moreac, G., Blurock, E., Mauss, F.: Automatic generation of a detailed mechanism for the oxidation of n-decane. Combust. Sci. Technol. 178(10-11), 2025–2038 (2006)
4. Dittrich, P., Speroni di Fenizio, P.: Chemical organization theory. Bull. Math. Biol. 69(4), 1199–1231 (2007)
5. Banzhaf, W.: Self-replicating sequences of binary numbers. Comput. Math. Appl. 26, 1–8 (1993)
6. Speroni di Fenizio, P., Dittrich, P., Ziegler, J., Banzhaf, W.: Towards a theory of organizations. In: German Workshop on Artificial Life (GWAL 2000), Bayreuth, 5.-7. April, 2000 (in print)
7. Eigen, M., Schuster, P.: The hypercycle: a principle of natural self-organisation, part A. Naturwissenschaften 64(11), 541–565 (1977)
8. Schuster, P., Sigmund, K.: Replicator dynamics. J. Theor. Biol. 100, 533–538 (1983)
9. Fontana, W., Buss, L.W.: 'The arrival of the fittest': Toward a theory of biological organization. Bull. Math. Biol. 56, 1–64 (1994)
10. Fontana, W.: Algorithmic chemistry. In: Langton, C.G., Taylor, C., Farmer, J.D., Rasmussen, S. (eds.) Proc. Artificial Life II, Redwood City, CA, pp. 159–210. Addison-Wesley, Reading (1992)
11. Lindgren, K., Eriksson, A., Eriksson, K.E.: Flows of information in spatially extended chemical dynamics. In: Pollack, J., Bedau, M., Husbands, P., Ikegami, T., Watson, R.A. (eds.) Proc. Artififical Life IX, pp. 456–460. MIT Press, Boston (2004)
12. Centler, F., Dittrich, P.: Chemical organizations in atmospheric photochemistries: a new method to analyze chemical reaction networks. Planet. Space Sci. 55(4), 413–428 (2007)
13. Matsumaru, N., Centler, F., Speroni di Fenizio, P., Dittrich, P.: Chemical organization theory applied to virus dynamics. it - Information Technology 48(3), 154–160 (2006)

Formulating Membrane Dynamics with the Reaction of Surface Objects

Kazuto Tominaga, Tooru Watanabe, and Maki Suzuki

School of Computer Science, Tokyo University of Technology
1404-1 Katakura, Hachioji, Tokyo 192-0982 Japan
tomi@acm.org

Abstract. Membranes form important structures in living systems. In this paper, we propose a new formulation of membrane dynamics as an extension to our artificial chemistry. It does not explicitly specify membranes to react; instead, the surface objects of membranes decide which membranes transform and how. We model the clathrin-coated vesicular transport by the formalism, and thereby show the compatibility of our approach with natural membrane systems.

1 Introduction

A eucaryotic cell has many membranous compartments, or cell organelles, each of which has a specific role in the activity of the cell. They are interesting and important subjects in the field of artificial life. One methodology to study such systems is artificial chemistries [1], which describe the dynamics of a system in terms of the interaction among molecules according to reaction rules. To date, however, not many artificial chemistries are applied to modelling the dynamics of such systems. In this paper, we propose a new framework of describing the dynamics of membrane structures by the reaction of surface objects of membranes, which, we think, is compatible with the natural membrane dynamics observed in living systems such as eucaryotic cells.

2 Basic Formulation and Previous Extension

We previously extended our artificial chemistry [2] to make it capable of dealing with membrane structures [3]. In this section, the basic model [2] is briefly illustrated, and the extension [3] and its drawbacks are explained.

The basic artificial chemistry [2] deals with only one reaction pool. It employs the well-stirred tank reactor model [1, Section 2.3.9]. Example molecules are shown below, with their string notations.

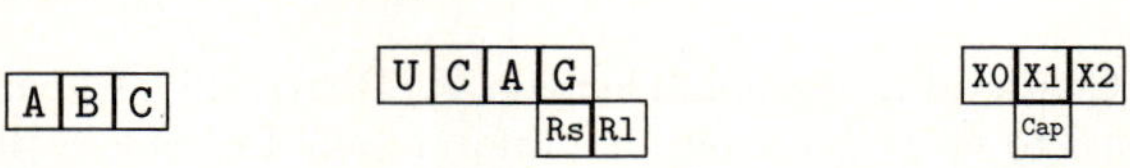

0#ABC/ 0#UCAG/3#RsRl/ 0#X0X1X2/1#Cap/

F. Almeida e Costa et al. (Eds.): ECAL 2007, LNAI 4648, pp. 12–21, 2007.

A box represents an atom; a molecule comprises lines of atoms. In the string notations, "`/`" separates lines. A number before "`#`" represents the displacement of the line relative to the top line. We may use terms *virtual elements* (or *v-elements*), *v-atoms* and *v-molecules* for our artificial chemistry in order to avoid confusion with natural elements, atoms and molecules.

Chemical equations in this artificial chemistry (we call them *recombination rules*) are expressed in terms of patterns that match (or do not match) v-molecules. A pattern may have wildcards: a number surrounded by angle brackets (such as `<1>`) denotes a wildcard that matches an arbitrary v-atom. A number and a star surrounded by angle brackets (such as `<*1>` and `<2*>`) denote a wildcard that matches an arbitrary sequence of v-atoms of any length; the position of star represents the direction in which the sequence can extend. An example recombination rule is illustrated below.

$$\texttt{0\#<*1>AB/1\#CD/} + \texttt{0\#AB<2*>/} \rightarrow \texttt{0\#<*1>ABAB<2*>/1\#CD/} \tag{1}$$

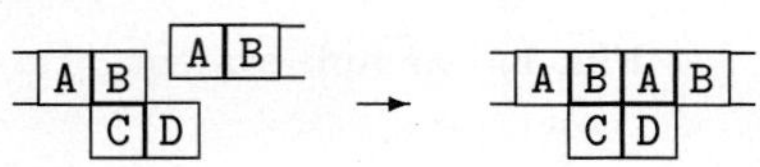

If this rule is applied to v-molecules `0#ABAB/3#CD/` and `0#AB/`, a new v-molecule `0#ABABAB/3#CD/` is produced and the reactants disappear.

We extended the basic artificial chemistry to enable it to describe membrane structures [3]. Its notable features are (1) a membrane that surrounds a compartment (called *cubicle*) may have embedded v-molecules (called *membrane v-molecules*), (2) v-molecules in a cubicle (called *cubicle v-molecules*) and membrane v-molecules can interact if the compartment and the membrane are adjacent, (3) such interactions are defined in the same way as interactions among cubicle v-molecules are defined (i.e., recombination rules), and (4) recombination rules do not explicitly specify which membrane participates in a reaction; it is determined by the existence of a membrane v-molecule that is matched by a pattern in the rule. We showed the uniformity of handling cubicle and membrane v-molecules was useful to model certain types of biochemical phenomena: signal transduction and protein transport [3]. The formulation, however, had some drawbacks. A major one is that it cannot deal with a system in which the membrane structure dynamically changes. Such dynamic change is often observed in living systems and is indeed an essential part of their dynamics such as cell division. Therefore the capability of manipulating a membrane structure is necessary for an artificial chemistry if it is to model such phenomena.

3 Extension for Describing Membrane Dynamics

In this section, we give a new extension to the basic artificial chemistry to model a class of membrane dynamics. While there are many ways of transforming membranes such as forming and dissolving, we deal only with two activities in this extension: merger and division.

3.1 Membranes and Cubicles

We assume every membrane is closed and has no open end; a membrane always surrounds a cubicle. A cubicle may include other cubicles surrounded by membranes. If a cubicle A has another cubicle B inside, we call A the *parent* of B, and call B a *child* of A. A *system* is a tree of cubicles; each node represents a cubicle, and links represents parent-child relations. Figure 1 illustrates an example system and the tree representing it. A system must have one cubicle that has no outside border, which is called the *root cubicle* of the system (R in the figure).

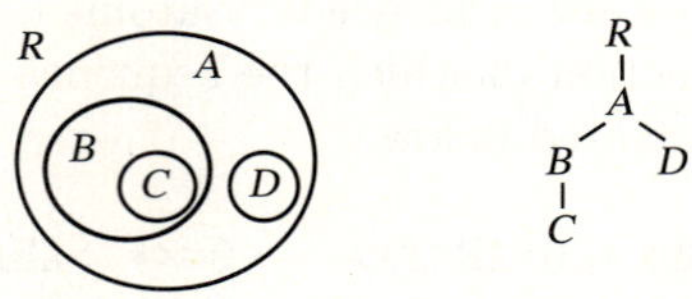

Fig. 1. Example system.

Each cubicle has a reaction pool that holds v-molecules; we call it a *cubicle pool.* It is like cytoplasm in a cell. A v-molecule in a cubicle pool is called a *cubicle v-molecule.* Each membrane also has its own reaction pool, called a *membrane pool.* A v-molecule in a membrane pool, or a *membrane v-molecule*, can represent transmembrane protein. Both kinds of pools are well-stirred. As transmembrane protein has directions, we give a *direction* attribute to each membrane v-molecule. A direction is either *top* or *bottom.* We depict a direction with a triangle (Fig. 2(a)). Cubicle v-molecules do not have this attribute.

3.2 Extending Patterns

Each reaction pool has a set of recombination rules. A recombination rule is described in terms of patterns in a similar way as in the basic artificial chemistry. There are two main differences. First, patterns are extended so as to deal with the directions of v-molecules on membranes. Second, a pattern that matches an arbitrary membrane — *the membrane pattern*, denoted by “`@`” — is introduced. Cubicle pools and membrane pools use patterns extended in different ways.

Patterns for Cubicle Rules. In a recombination rule for a cubicle pool, a pattern is of one of the following three form: without prefix, prefixed by “`^`” or prefixed by ”`_`”. Example patterns are `0#ABAB/1#CD/`, `^0#AB<1>/` and `_0#<*2>AB/1#CD/`. The prefixes express the directions of v-molecules. A pattern without prefix matches cubicle v-molecules, and one with a prefix matches membrane v-molecules. A pattern with `^` matches a membrane v-molecule whose tip (of the triangle) is visible from the cubicle. For example, if Cubicle X in Fig. 2(b) has a rule including a pattern `^0#AB/`, the pattern matches the v-molecule `0#AB/`

in the membrane. Similarly, if Cubicle Y has a pattern `^0#CD/`, it matches the v-molecule `0#CD/` shown in the illustration. A pattern with _ matches a membrane v-molecule with its base (of the triangle) visible to the cubicle. A pattern `_0#CD/` in Cubicle X matches `0#CD/` in the figure, and a pattern `_0#AB/` in Cubicle Y matches `0#AB/`. Note that a v-molecule (say, `0#AB/`) is matched by patterns with different prefixes (`^0#AB/` and `_0#AB/`) from different cubicles. Thus the direction of a membrane v-molecule is construed according to which neighbouring cubicle looks at it.

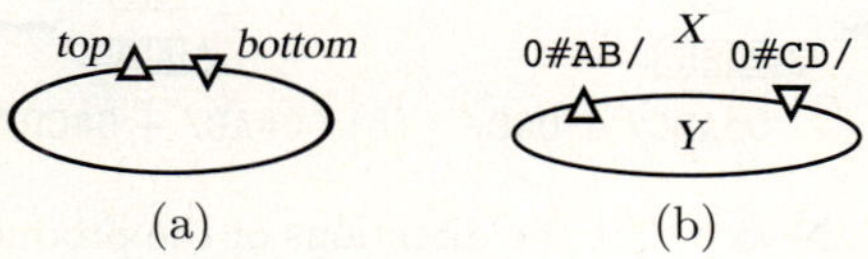

Fig. 2. Directions of v-molecules and pattern matching.

Patterns for Membrane Rules. A pattern used in a membrane rule is either *positive* or *negative*. We call this attribute *the sign* of pattern. A negative pattern is prefixed by "`!`" such as `!0#AB/`, and a positive one has no prefix like `0#CD/`. The sign of a pattern in a membrane rule represents the direction of a matched v-molecule relative to the directions of other v-molecules that are matched by the other patterns of the rule (if any).

3.3 Recombination

Recombination rules define the interaction among v-molecules, and the division and merger of membranes. *Cubicle rules* define (i) the reaction among cubicle v-molecules (we call such rules *internal rules*), (ii) the reaction in which the both types of v-molecules (cubicle and membrane) are involved (*mediating rules*), (iii) the division of a membrane (*dividing rules*), and (iv) the merger of two membranes (*merging rules*). *Membrane rules* define the reaction between two membrane v-molecules. In any rule, we restrict the number of reactants (i.e., the number of terms on the left-hand side) to be one or two (we call them *unary* and *binary* rules, respectively) in order to keep the formulation simple. From now we use the abbreviations *lhs* for "left-hand side" and *rhs* for "right-hand side."

Internal Rules. An internal rule is unary or binary, and is defined and interpreted in the same way as a recombination rule of the basic artificial chemistry.

Mediating Rules. A mediating rule is unary or binary, and its lhs has one and only one prefixed pattern, which represents a membrane v-molecule visible from the cubicle. For example, the rule "`^0#AB/`+`0#CD/` → `^0#ABC/`+`0#D/`" describes the reaction depicted in Fig. 3(a) (the cubicle having this rule is marked X). The rhs may have at most one prefixed pattern, and if there is, it represents a membrane v-molecule in the same membrane as the prefixed pattern on the lhs specifies. Another rule "^0#AB/ + 0#CD/ → _0#ABC/ + 0#D/" describes the reaction of Fig. 3(b). Note that a mediating rule does not specify a membrane explicitly; the membrane is selected if its v-molecule is matched by the prefixed pattern on the lhs. Hence any membrane that the cubicle adjoins may be chosen.

(a) ^0#AB/ + 0#CD/ → ^0#ABC/ + 0#D/ (b) ^0#AB/ + 0#CD/ → _0#ABC/ + 0#D/

Fig. 3. Mediating rules. Notice that the directions of the product 0#ABC/ are different.

Dividing Rules. A dividing rule is either unary or binary, and the lhs has one prefixed pattern. The rhs may have either two prefixed patterns or a prefixed pattern and the membrane pattern (@). If the rule "^0#AB/ → ^0#A/ + ^0#B/" is given to Cubicle X in Fig. 4(a), the illustrated division occurs. The contents of the divided membrane (the cubicle v-molecules, the membrane v-molecules and child cubicles) are distributed to the two product cubicles and membranes nondeterministically, i.e., every possible distribution is considered. The cubicle rules of the original cubicle are given to the both product cubicle, and the same for the membranes.

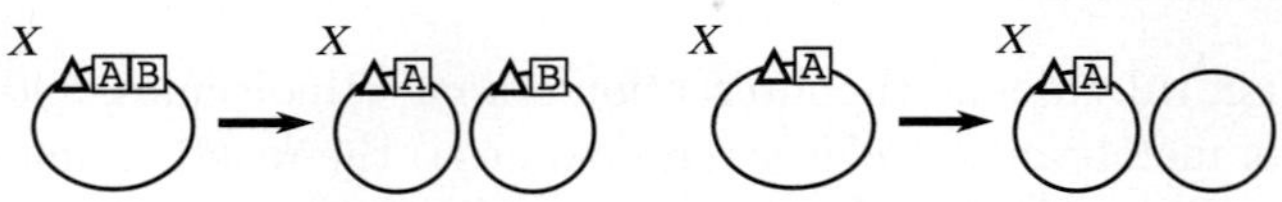

(a) ^0#AB/ → ^0#A/ + ^0#B/ (b) ^0#A/ → ^0#A/ + @

Fig. 4. Dividing the membrane of a child cubicle.

Explained above is the division of the membrane of a child cubicle. The same rule can apply to the division of the cubicle's own membrane: if the rule is given to Cubicle Y in Fig. 5, it can result either of Fig. 5(a) and (b). The contents of Cubicle X is distributed to Cubicles X and Z after the rule is applied, and the v-molecules in the membrane of Y is distributed to the two membranes. In this case the directions of all the membrane v-molecules that are given to the inner membrane are reversed; this formulation intends to model the directional change of transmembrane protein on membrane division.

A prefixed pattern and the membrane pattern (@) on the rhs express the division of membrane as well. The membrane pattern does not specify which membrane v-molecules will be included in the membrane it represents. The rule

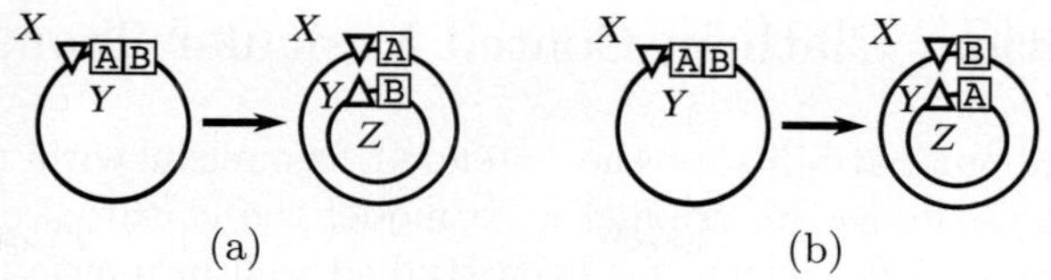

Fig. 5. Dividing a cubicle's own membrane by `^0#AB/` → `^0#A/` + `^0#B/`.

"`^0#A/` → `^0#A/` + `@`" induces the division depicted in Fig. 4(b). A rule using the membrane pattern can also divide the cubicle's own membrane in the same ways as those shown in Fig. 5.

Merging Rules. Merging is the inverse operation of dividing. A merging rule is binary, and its lhs consists of two prefixed patterns. The rhs comprises one prefixed pattern. The rule "`^0#A/`+`^0#B/` → `^0#AB/`" is the inverse operation of Fig. 4(a), Fig. 5(a) and (b). On merging membranes, the contents of membranes and cubicles are merged, and the sets of rules are also merged. As dividing reverses the directions of v-molecules in the inner membrane (Fig. 5), merging also reverses them when it merges the inner membrane with the outer one.

Membrane Rules. A membrane rule expresses the interaction between two membrane v-molecules in the same membrane; it operates on them considering their directions. A membrane rule is unary or binary, and each pattern can be positive or negative. An example rule is `0#AB/` + `!0#CD/` → `0#ABC/` + `!0#D/`. If this membrane has the v-molecules `^0#AB/` (`^` indicates a top v-molecule) and `_0#CD/` (`_` for bottom), this rule can be applied to them and produces `^0#ABC/` and `_0#D/`. Similarly, if this membrane has `_0#AB/` and `^0#CD/`, the rule is applicable and it produces `_0#ABC/` and `^0#D/`. In this way, the negative sign (`!`) means that the pattern matches a v-molecule with the opposite direction to those matched by positive patterns. This also applies to the rhs of the rule; all the v-molecules represented by positive patterns in a rule have the same direction, and those represented by negative patterns have the opposite direction. Therefore the above rule has no effect on the v-molecules `^0#AB/` and `^0#CD/` because their directions do not match the signs of the patterns on the lhs.

3.4 Dynamics

A system defined in the extended artificial chemistry is interpreted nondeterministically as follows.

1. Initialize all the cubicles and membranes by giving v-molecules.
2. Choose one cubicle or membrane, and apply one of its rule to a collection of v-molecules. This may cause the division or merger of membranes.
3. Go to Step 2.

4 Application: Clathrin-Coated Vesicular Transport

To evaluate the compatibility of the extended formalism with membrane dynamics in living systems, we applied it to model the clathrin-coated vesicular transport [4, pp712–724], which has been studied well in molecular biology. Its conceptual diagram is illustrated in Fig. 6 (the transport from the Golgi body to an endosome is assumed). A simplistic explanation of the process is as follows (the actual process involves many other factors): (1) cargo receptors in the membrane of the Golgi body attach cargo molecules from the inside of the Golgi body; (2) clathrin (protein to make a coat of vesicle) attaches to the other side of receptors and forms a bud; (3) dynamin (GTPase) pinches the bud off from the membrane; (4)–(5) the clathrin coat is lost after the vesicle is released; (6) v-SNARE (transmembrane protein) on the vesicle and t-SNARE (transmembrane protein) on the surface of endosome interact to tether the vesicle to the membrane of endosome; (7) the two membranes are fused; and (8) the cargo molecules are released into the endosome.

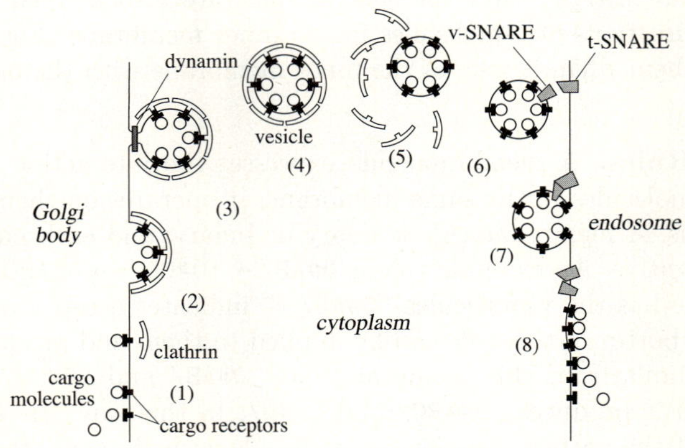

Fig. 6. Clathrin-coated vesicular transport.

We define a system that models this transport. Its initial membrane structure is shown in Fig. 7. Rules are given to the following parts (the numbers of rules are shown in parentheses): the cytoplasm (13), the Golgi body (1) and its membrane (2), and the endosome (1) and its membrane (2). In the initial state, the Golgi body has cargo molecules (denoted by `0#Ld/`), and its membrane has receptors (`^0#Rec/`). The cytoplasm has several kinds of v-molecules including clathrin (`0#Cla/`), dynamin (`0#DynGdpPi/`) and v-SNARE (`0#Vs/`). The endosome initially has no v-molecule inside. Its membrane has t-SNARE (`^0#Ts/`). The system has some other v-elements and v-molecules, but we omit explanations on them due to page limitation.

We will illustrate how the system works by explaining some of the rules. First, cargo molecules attach to receptors by the mediating rule (2) for the Golgi body,

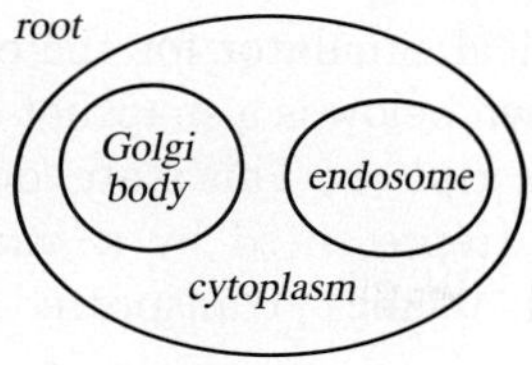

Fig. 7. Structure of the system for clathrin-coated vesicular transport.

and the mediating rule (3) for cytoplasm attaches clathrin to the receptors (P has been attached by an unexplained rule).

$$\texttt{\_0\#Rec/} + \texttt{0\#Ld/} \rightarrow \texttt{\_0\#Rec/0\#Ld/} \tag{2}$$

$$\texttt{0\#Cla/} + \texttt{\^{}0\#P/0\#Rec/0\#Ld/} \rightarrow \texttt{\^{}0\#Cla/0\#P/0\#Rec/0\#Ld/} \tag{3}$$

Note that these rules look at the receptor from different directions: from the Golgi body it is referred to by _, and from cytoplasm by ^. Next, produced complexes gather by (4), which is given to the membrane of the Golgi body.

$$\texttt{0\#Cla/0\#P/0\#Rec/0\#Ld/} + \texttt{0\#Cla<1*>/0\#P<2*>/0\#Rec<3*>/0\#Ld<4*>/}$$
$$\rightarrow \texttt{0\#ClaCla<1*>/0\#PP<2*>/0\#RecRec<3*>/0\#LdLd<4*>/} \tag{4}$$

While dozens of clathrin molecules are necessary to form a coat, we assume one can be formed with five clathrin v-molecules in this model for simplicity. If such a coat is formed, dynamin and some other molecules attach to the complex and pinch it off from the membrane by the dividing rule (5) given to cytoplasm.

$$\texttt{\^{}0\#DynGdpPi/0\#AmpCla<1*>/1\#P<2*>/0\#VsRec<3*>/1\#Ld<4*>/} \rightarrow$$
$$\texttt{\^{}0\#DynGdp/0\#AmpCla<1*>/1\#P<2*>/0\#VsRec<3*>/1\#Ld<4*>/} + \texttt{0\#Pi/} + \texttt{@} \tag{5}$$

The first term of the rhs represents the vesicle, and @ represents the Golgi body.

The membrane of vesicle has v-SNARE (Vs) and the membrane of endosome has t-SNARE (Ts), so the merging rule (6) fuses the vesicle and the endosome.

$$\texttt{\^{}0\#RabVsRec<1*>/2\#Ld<2*>/} + \texttt{\^{}0\#Ts/} \rightarrow \texttt{\^{}0\#RabVsRec<1*>/1\#TsLd<2*>/} \tag{6}$$

Then the receptor-cargo pairs split by (7) for the membrane of endosome

$$\texttt{0\#VsRec<1*>/0\#TsLd<2*>/} \rightarrow \texttt{0\#Vs<1*>/0\#Ts<2*>/} + \texttt{0\#Rec/0\#Ld/} \tag{7}$$

and the cargo molecules are released to the inside of endosome by the mediating rule (8) given to the endosome.

$$\texttt{\_0\#Rec/0\#Ld/} \rightarrow \texttt{\_0\#Rec/} + \texttt{0\#Ld/} \tag{8}$$

We developed a prototypical simulator for the extended artificial chemistry implemented with Ruby. Shown below is a snapshot of system state taken during the execution of the given description. This state, depicted in Fig. 8, shows that the membrane of endosome (represented by `m-endosome`) has received cargo receptors (`^0#Rec/0#Ld/` and `^0#Rec/`) transported from the membrane of Golgi body (`m-golgi`).

```
mddcl> 0 ← user input to show the current contents of pools
Root [all: 0]
Membrane 1 (<^0#./: 1> all: 1)
cytoplasm [<0#P/: 28> <0#Vs/: 93> <0#DynGdpPi/: 95> <0#Cla/: 37> <0#DynGdp/: 5> <0#
Pi/: 5> <0#Nsf/: 100> <0#Rab/: 98> <0#Amp/: 98> all: 559]
m-endosome (<^0#Rec/0#Ld/: 3> <^0#Ts/: 99> <^0#Vs/: 1> <^0#Rec/: 2> <^0#VsRecRecRec
RecRecRecRec/0#TsLdLdLdLdLdLdLd/: 1> all: 106)
endosome [<0#./: 1> <0#Ld/: 376> all: 377]
Membrane 11 (<^0#RabVsRecRecRecRecRecRec/2#LdLdLdLdLdLd/: 1> all: 1)
Cubicle 12 [<0#Ld/: 281> all: 281]
Membrane 13 (<^0#Cla/0#P/-1#VsRecRecRecRecRec/0#LdLdLdLdLd/: 1> all: 1)
Cubicle 14 [<0#Ld/: 124> all: 124]
m-golgi (<^0#P/0#Rec/0#Ld/: 9> <^0#ClaClaClaClaClaCla/0#PPPPPP/0#RecRecRecRecRec
/0#LdLdLdLdLdLd/: 2> <^0#ClaCla/0#PP/0#RecRec/0#LdLd/: 2> <^0#AmpClaClaClaClaClaCla
/1#PPPPPP/0#VsRecRecRecRecRecRec/1#LdLdLdLdLdLd/: 2> <^0#ClaClaCla/0#PPP/0#RecRecRe
c/0#LdLdLd/: 6> <^0#ClaClaClaCla/0#PPPP/0#RecRecRecRec/0#LdLdLdLd/: 4> all: 25)
golgi [<0#Ld/: 39> all: 39]
Membrane 9 (<^0#RabVsRecRecRecRecRecRec/2#LdLdLdLdLdLd/: 1> all: 1)
Cubicle 10 [<0#Ld/: 82> all: 82]

[NumObjects:1597, NumSpaces:13]
```

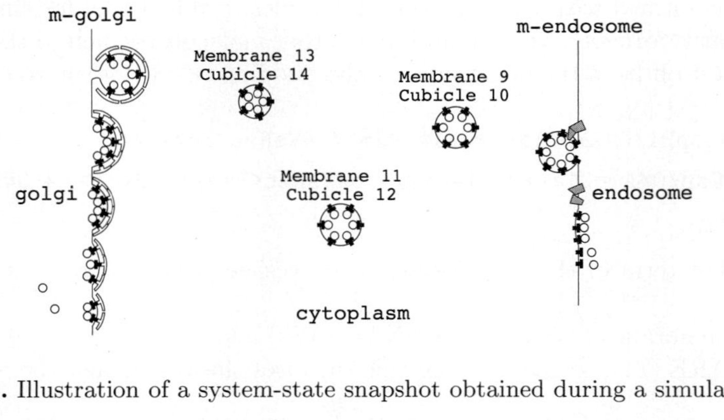

Fig. 8. Illustration of a system-state snapshot obtained during a simulation.

5 Discussion and Concluding Remarks

This approach attempts to formulate membrane dynamics through the interaction among molecules without explicitly specifying which membrane to take part in transformation. In the formalism, recombination rules (which are the only means to change the states of the system) do not specify names or IDs of membranes or cubicles, unlike P Systems with active membranes [5, Chap. 1, Sec. 13], whose dividing rules utilize the labels of membranes. This property of our formulation allows the same rule set to operate different structures. For example, if another endosome with the same rules and v-molecules is added to the system shown in Fig. 7, vesicles released from the Golgi body will fuse either of the two endosomes. This framework is also compatible with natural membrane dynamics: the merger and division of membranes are induced by the interactions among molecules in and around the membranes. In our formulation, rules basically do not change or "evolve" after they are given to the system at its initial state. This, however, does not necessarily limit the kinds of v-molecules produced by the system, since a rule may have wildcards. This resembles to ribosomes; they produce various kinds of proteins using the fixed codon table.

Brane Calculi [6] formalize membrane dynamics using a set of generic rules. In contrast, our formalism allows the user to design rules specific to his problem. An interesting difference is that Brane Calculi employ the "Phago" operation (which can express phagocytosis as one step) as a basic reaction while our formalism leaves it to nondeterminism. Studying on the expressive powers of both formulations is our future work.

Our current formulation is still preliminary. For example, distributing the contents of a membrane on division is defined simply as nondeterministic, and thus it may result in an unnatural distribution of v-molecules such as giving the half v-molecules of the Golgi body cubicle to a tiny vesicle cubicle. Moreover, we formulated a type of division as Fig. 5 for simplicity and symmetry, but the two dividing schemes must have different roles in natural systems, so they should be distinguished. One reason for these problems is that the formalism has no notion of volumes of cubicles. We will study this issue further.

We believe the present approach will beneficial for modelling and investigating a system in which many membranes interact with each other based on the biochemical composition of them, such as a system of organelles in a cell.

References

1. Dittrich, P., Ziegler, J., Banzhaf, W.: Artificial chemistries — a review. Artificial Life 7(3), 225–275 (2001)
2. Tominaga, K.: A formal model based on affinity among elements for describing behavior of complex systems. Technical Report UIUCDCS-R-2004-2413, Department of Computer Science, University of Illinois at Urbana-Champaign (March 2004)
3. Watanabe, T., Koizumi, K., Kishi, K., Nakamura, M., Kobayashi, K., Kazuno, M., Suzuki, Y., Asada, Y., Tominaga, K.: A uniform framework of molecular interaction for an artificial chemistry with compartments. In: Proceedings of the 2007 IEEE Symposium on Artificial Life, pp. 54–60. IEEE Computer Society Press, Los Alamitos (2007)
4. Alberts, B., Johnson, A., Lewis, J., Raff, M., Roberts, K., Walter, P.: Molecular Biology of the Cell, 4th edn. Garland Science, New York (2002)
5. Ciobanu, G., Păun, G., Pérez-Jiménez, M.J. (eds.): Applications of Membrane Computing. Springer, Berlin (2006)
6. Cardelli, L.: Brane calculi, interactions of biological membranes. In: Danos, V., Schachter, V. (eds.) CMSB 2004. LNCS (LNBI), vol. 3082, pp. 257–278. Springer, Heidelberg (2005)

Multi-level Selectional Stalemate in a Simple Artificial Chemistry

Barry McMullin, Ciarán Kelly, and Darragh O'Brien

aLife Lab, RINCE, Dublin City University
barry.mcmullin@dcu.ie
http://www.eeng.dcu.ie/~alife/

Abstract. We describe a simple artificial chemistry which abstracts a small number of key features from the origin of life "replicator world" hypotheses. We report how this can already give rise to moderately complex and counter-intuitive evolutionary phenomena, including macroevolutionary deterioration in replication fidelity (which corresponds to intrinsic replicator fitness in this model). We briefly describe the extension of this model to incorporate a higher, protocell, level of selection. We show that the interaction between the two levels of selection then serves to control parasitic exploitation at the molecular level, while still significantly constraining accessible evolutionary trajectories at the protocell level. We conclude with a brief discussion of the implications for further work.

1 Introduction

It has been argued that the most significant events in the evolution of the biosphere have been a relatively small number of so-called "major transitions" [6]. These are typified by the emergence of a new "level" of Darwinian actor, and thus a new process of Darwinian evolution. Thus, for example, there may have been an epoch early in the origin of life in which replicating molecules directly functioned as Darwinian actors (e.g., an RNA World [4, 5]); but these independently replicating molecules subsequently became subsumed into some form of cellular containers or "protocells". The evolutionary fate of any molecular level species then became contingent on that of the cellular aggregate of which it was a component. Similarly, there was another major transition between unicellular and multi-cellular organisms. In the latter, the evolutionary fate of the individual cells became subjugated to that of the multi-cellular aggregate.

A critical feature of any such major transition is the interplay between the two levels of selection. In particular, selection at the lower level is generally expected to favour "selfishness" on the part of the lower level actors [1].[1] However, as the evolutionary future of these entities becomes conditional on the success of the new, higher level, entity, this selfishness at the lower level may necessarily have

[1] Or, more precisely, on the part of *lineages* of these actors [7].

F. Almeida e Costa et al. (Eds.): ECAL 2007, LNAI 4648, pp. 22–31, 2007.

to be moderated in favour of more "co-operative" interactions which better serve the higher level, composite, entity.

We present here a set of artificial chemistry style [2] investigations of this phenomenon.

We start with a minimal "template-replicator world" in which there is only one level of Darwinian actor (the replicating "molecule"). The model incorporates the notion of unlimited heredity achieved through template replication of indefinite length polymers. This is taken to be dependent on catalysis by a suitable replicase. In the simplest case we consider molecules which can act as "self-replicases" (a form of degenerate, one-element, hypercycle [3]). We note and exhibit the distinctive hyperbolic selection dynamics of such systems ("survival of the common").[2]

Molecular replication is then made imperfect, with a fixed error rate per monomer (and thus a molecular-level replication error rate increasing with the length of the molecule). We characterise the effect of this on the distribution between the dominant and mutant molecular species.

We next introduce a simple rule for enzymatic coupling between different species (so that one species can function as replicase for another species as well as itself). This is deliberately made asymmetric. This introduces the possibility of exploitation between species. Even under the condition of hyperbolic growth, this allows effective displacement of a "host" species by a new "parasitic" species; and, under the conditions of the model, this can happen repeatedly. In this particular model, this has the somewhat counter-intuitive effect of leading to systematic, macro-evolutionary, *deterioration* in "intrinsic fitness" (as measured by replication *fidelity*).

Finally, we extend the model to place the molecules in simple "protocells", where protocell reproduction (by binary fission) is driven by molecular replication. With a fixed protocell population limit we now have a higher level Darwinian dynamic at the protocell level which interacts with the molecular level selection dynamic which still occurs within each protocell. We show that the protocell level selection does effectively eliminate parasitic exploitation at the molecular level; however, the molecular level selection is still effective in preventing positive evolution in the opposite direction (toward higher, molecular level, replication fidelity). The result is a rather robust evolutionary "stalemate" in which the selectional dynamics at the two interacting levels are, in effect, precisely counter-acting each other.

We conclude with a brief discussion of the implications of this study.

2 Basic Template-Replicator World

The basic template-replicator world consists of a finite number of strings (polymers) on a binary alphabet. The dynamics consists of a simple loop in which

[2] There is, of course, a large body of prior literature on replicator selection dynamics. We omit any extensive review here, in the interests of brevity; but [8], for example, includes a comprehensive bibliography.

one random string is chosen as a replicase and a second as a template. If the replicase exactly matches the template in sequence, then it "binds" to it, and replicates it. A random existing string is removed to make room for the new string. If the replicase does not bind (i.e., does not match in sequence) then the interaction has no effect.

Note that, by design at this point, there are no replication errors, and indeed no other intrinsic fitness differences between different molecular species.

Clearly, any species acts as a replicase for itself and no other; which is to say, there is a form of replication, or auto-catalysis. However, because of the requirement for two molecular instances to interact to achieve replication, the autocatalytic reaction scheme is of the form $2A \rightarrow 3A$ rather than the more usual $A \rightarrow 2A$. This has a significant effect on the dynamics. The intrinsic growth is not exponential, but hyperbolic. The result is effectively a form of positive frequency dependent selection, in which any species with above average representation will experience a direct positive feedback. In the absence of opposing intrinsic fitness differences, any species which initially achieves such an above-average representation will quickly become dominant; and will then be able to maintain that dominance even against invasion by species with much higher intrinsic fitness. This is sometimes referred to as "survival of the common" [8].

We can formulate an approximate differential equation model of this system by considering just two species (X and Y). Taking their respective relative concentrations as x and y, these are also the probabilities of choosing an instance of either species at random. If we choose two X molecules they will successfully replicate; and if the offspring displaces a Y molecule, there will be a net increase in x. The probability of this event on any given interaction is evidently x^2y. Conversely, if we choose two Y molecules and then displace an X, this will give a decrease in x, with probability y^2x. All other interactions leave the concentrations unchanged. Thus, the rate of change[3] of x is given by:

$$\dot{x} = x^2y - y^2x$$

Given that $x + y = 1$ (by definition), then we have:

$$\dot{x} = x^2(1-x) - (1-x)^2x$$

A simple insight into the behaviour is gained by plotting the expression on the right against x. This is shown in Figure 1. As can be seen, if x ever becomes greater than 0.5 then its rate of change will be positive, and it will keep on increasing until it completely displaces y (i.e., reaches concentration 1). Of course, by symmetry, exactly the same is also true of y. While there is a theoretical steady state with $x = y = 0.5$ this is clearly unstable.

The behaviour in a real system will obviously depend on the initial condition, and statistical fluctuations in a finite population. But even if initialised with $x = y = 0.5$ we expect that as soon as there is any significant fluctuation in

[3] In this and subsequent equations there is an implicit multiplicative constant, effectively setting the time scale. This has been arbitrarily taken as unity.

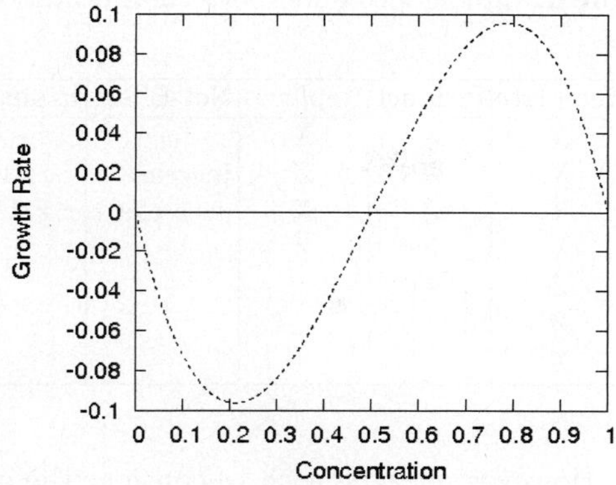

Fig. 1. Selection under hyperbolic growth.

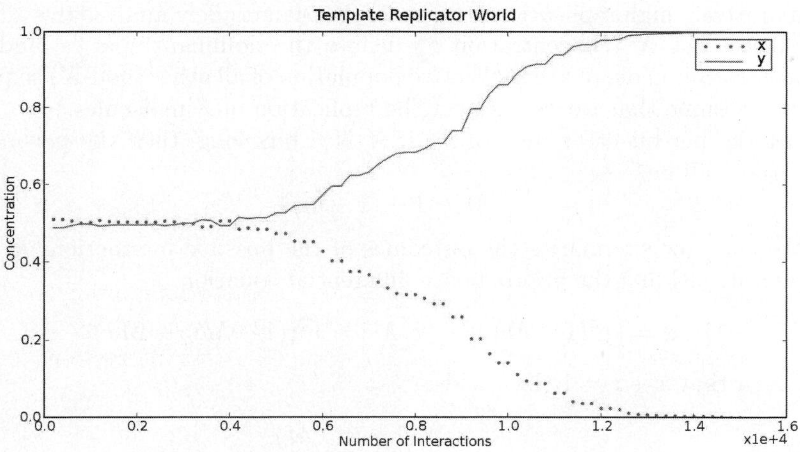

Fig. 2. Example of Template-Replicator World Experiment

favour of either species, that species will then quickly take over.[4] For illustration, data from one concrete experimental run illustrating this is shown in Figure 2.

3 Effect of (Molecular) Mutation

In the simple template-replicator world, the population will quickly be taken over by whichever species initially happens to achieve a greater than average representation. In the absence of replication error, this species will simply fixate

[4] Note that this behaviour is in marked contrast to systems of simple exponential replicators, in which, in the absence of fitness differences, relative concentrations will undergo only slow random drift.

Table 1. Interactions under Molecular Mutation

Replicase	Template	Product	Replaces	Net Effect	Probability
X	X	X	X	–	
X	X	X	Z	increase x	$x.x.(1-M).z$
X	X	Z	X	decrease x	$x.x.M.x$
X	X	Z	Z	–	
X	Z	–	–	–	
Z	Z	–	–	–	
Z	X	–	–	–	

at concentration 1.0. However, we now wish to consider the effect of replication error or molecular "mutation".

Qualitatively, the frequency dependent advantage of a dominant sequence is expected to be strong enough to maintain that dominance even when subject to comparatively high replication error rates. We can crudely analyse this situation as follows. Let X (concentration x) denote the dominant species; and let Z (concentration z) denote the collective population of all other (non-X) sequences. We will assume that we can neglect the replication of Z molecules.[5]

Let the per-bit error rate be m. If X is n bits long, then the per-molecule error rate will be:

$$M = 1 - (1-m)^n$$

We can now summarise the outcomes of the possible interactions as shown in Table 1, yielding the approximate differential equation:

$$\dot{x} = (x^2(1-M)z) - (x^3 M) = x^2((1-M)z - Mx)$$

Given that $x + z = 1$ this yields:

$$\dot{x} = x^2(1 - x - M)$$

We can find the stable states by setting $\dot{x}$ to zero. Neglecting the case of $x = 0$ this leaves one fixed point with:

$$1 - x - M = 0$$

that is:

$$x = 1 - M$$

Thus, the steady state population structure is with a proportion $(1-M)$ of the dominant species and M of diverse mutants. Note that, because of the (assumed) diversity in the mutant population, and the frequency dependent advantage of dominance, a particular species may thus remain effectively "dominant", even when its absolute proportion of the population is well below an absolute majority.

[5] This is based on assuming that X is of sufficient length that the mutated sequences are diverse and almost never find an identical sequence to bind to. This could be analysed more precisely, but this coarse simplification is adequate for our purposes here.

Table 2. Parasitic Invasion

Replicase	Template	Replace	Effect
x	x	y	increase x
x	y	x	decrease x
y	y	x	decrease x

4 Enzymatic Coupling (Parasitism)

We now define a more general replicase binding rule that will permit enzymatic coupling between different molecular species. As a simple generalisation for investigation, we specify that a replicase will successfully bind (and thus cause replication) as long as it is a *substring* of the template. (We still allow identity binding—i.e., a string binding to another instance of itself—as a special case of "substring" binding.)

As soon as we introduce any such enzymatic coupling between species we necessarily introduce the possibility of exploitation or parasitism: where one species may help replicate another without receiving any reciprocal support for its own replication.[6] Or equivalently, we might refer to such species as, respectively, *altruists* and *egoists*.

Let us, for the time being, discount mutation again. Let the population be initially dominated by a species X (concentration 1.0). Let us introduce a small quantity of another species Y, where the sequence of X is a substring of that of Y. The interactions are then as shown in Table 2.

The corresponding approximate differential equation becomes:

$$\dot{x} = x^2y - x^2y - y^2x = -y^2x$$

Since $x, y \in [0, 1]$ this means that $\dot{x} < 0$ for any non-zero y. That is, such a species Y can invade and completely displace X even from (in the continuous approximation) arbitrarily small initial concentrations. Note the contrast here with the earlier simple "survival of the common" outcome, characteristic of hyperbolic growth. Now the initial dominance of X is of no protection at all against selective displacement. This is because, in this scenario, a high concentration of X delivers just as much benefit to Y as to itself; whereas, Y delivers benefit *only* to itself.

5 Mutation and Parasitism

If we now take the system with substring-based enzymatic coupling, and reintroduce replication error (molecular mutation) we encounter a phenomenon

[6] Note that in this particular model we can have only *facultative* as opposed to *obligate* parasitism. That is, any parasite will also, by definition, be able to catalyse its own replication even in the absence of its putative host.

which, although fundamentally still Darwinian, seems at least a little counter-intuitive.

Briefly, we suppose that the replication errors include bit-flips, deletions and insertions (through bit repeats). Given reasonable choices of parameters, and initialising the population with a single homogenous sequence, the dynamic will be as follows. In the first instance, the concentration of the initial sequence will decrease in accordance with the analysis of section 3; but it will remain dominant. However, in due course, a mutant will arise which is one bit longer, but otherwise identical to the initial sequence. This will behave exactly as a parasite; and will displace the previous dominant sequence in accordance with the behaviour described in section 4, and become a new dominant species. But, of course, this process will then just be repeated with another mutant, one further bit longer.

Figure 3 shows an example experimental run, illustrating that this is, indeed, precisely the way the system behaves.

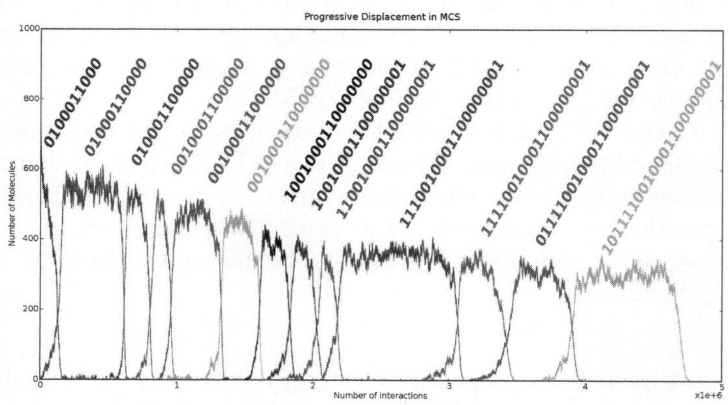

Fig. 3. Displacement Events in MCS

Note carefully that, over this macro-evolutionary epoch, the length of the dominant sequence grows steadily.[7] This means that the per-molecule mutation rate is steadily increasing; or, equivalently, the replication "fidelity" is getting steadily lower. That is, this macro-evolutionary trajectory actually results in a progressive and systematic *deterioration* in intrinsic fitness of the dominating

[7] We note that this property of parasites being more "complicated" than their hosts—at least in the sense of being longer—seems to contrast with the more typical situation with biological parasites, which may be radically simpler than their hosts. This choice is motivated in the toy model presented here simply by the objective of demonstrating the "in principle" *possibility* of sustained macro-evolutionary deterioration in intrinsic fitness.

species. This is in marked contrast to the naive "hill climbing" interpretation of evolution; and illustrates how evolutionary processes may be much more a matter of ecological interaction, or game playing, than any kind of optimisation. In particular, we note that this behaviour is completely at variance with the replicator world scenario which Dawkins has previously characterised (or, unintentionally, caricatured) with the slogan "fidelity, fecundity, longevity" [1].

6 Protocells: Two Interacting Levels of Selection

We now extend the model by confining the "template-replicator world" of the previous sections into a "protocell" container. The overall rules of the system are now changed as follows. When a replication event occurs, a molecule is no longer removed. Instead, the absolute number of molecules in the protocell is allowed to increase. However, at a specified maximum size, the protocell spontaneously undergoes binary fission to give two offspring protocells, each containing a random selection of half of the molecules from the parent. There is now a finite limit on the number of protocells. Thus, when a protocell reproduces, another protocell is chosen at random and discarded (along with its entire molecular contents).

A "mutation" at the protocell level now corresponds to the emergence of a molecular parasite as described in section 5. In this simple model, there is no exchange of molecules between protocells except through descent. Accordingly, a molecular parasite will give rise to a lineage of protocells which will necessarily become dominated by the parasitic molecular species. *However:* once this lineage "matures" in this way, the steady-state concentration of the dominant sequence will be slightly lower than in the parental lineage, in accordance with section 3 above (as the per-molecule mutation rate is slightly higher). This means, in turn, that the absolute rate of molecular replication is slightly lower (since this is essentially determined by the square of the concentration of the dominant molecular species). This means that, *at the protocell level*, this new protocell-mutant lineage will be effectively selected against. Thus, protocell level selection will prevent the progressive trend toward longer molecules and lower molecular replication fidelity, described in section 5. Instead, while such protocell mutants will occasionally arise, on an ongoing basis, they will be consistently eliminated again.

Figure 4 shows a test case to illustrate the robustness of this claim. The protocell population is initialised with equal numbers of two cell types. One is dominated by molecules 10 bits in length, the other by molecules of 11 bits in length, where the former is a substring of the latter. However, because of the isolation provided by the protocells, the shorter molecules are now effectively protected against exploitation; and because of their slightly lower replication error rate, and correspondingly higher steady state concentration, the lineage of protocells dominated by these molecules is able to reliably displace the lineage dominated by the longer molecules.

It follows that, in theory, at the protocell level, the population could still be taken over by a new lineage of protocells which are dominated by molecules which

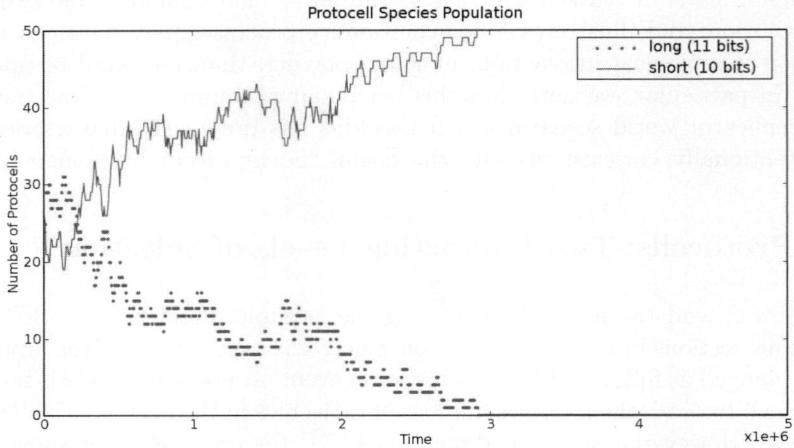

Fig. 4. Protocell Population Stability

are shorter again. However, this cannot happen in this model due to *selection at the molecular level*. That is, although, on an ongoing basis, shorter substring molecules will arise in individual protocells, they will be unable to build up any significant concentration relative to their already-dominant parasites (per section 4 above) or even just to any unrelated dominant sequence (per section 2). Thus, because of the molecular level selection dynamics, it is actually not possible to create a mutant *protocell* that is dominated by a shorter length molecular species.

The net effect is that, over a wide range, this system can be initialised with a protocell population with any arbitrary dominant molecular length; and the population will then remain dominated indefinitely by protocells which are individually dominated by that specific molecular species. Evolution toward protocells dominated by longer molecules will be prevented by the protocell level selection; and evolution toward protocells dominated by shorter molecules will be prevented by molecular level selection.

7 Conclusion: Implications of the Study

The system presented here is, of course, radically simplified compared to any phenomena of real chemistry or biology. Its purpose is not to directly model such real systems. Rather it is presented as a deliberately minimal system which already illustrates how complex and counter-intuitive the evolutionary behaviour of such systems can be; but also, how the evolution can, indeed, be dramatically altered by the interaction between multiple levels of selection.

The broader intention of the work is to develop a minimal abstract framework for understanding the evolutionary emergence of "computation" or, at least, coordinated signal processing and control, in protocellular systems. Presumably, any interesting molecular level computation must rely on a diversity of chemical species; but all of these in turn must be "replicated", directly or indirectly, to support protocell level reproduction. We therefore conjecture that protocellular computation is precisely a phenomenon that emerges in the conditions of a "major transition" between evolutionary levels. Further work will involve incrementally widening the repertoire of molecular interactions while working to still understand how this impacts on both interacting levels of evolution.

Acknowledgements. This work has received financial support under EU FP6 Integrated Project PACE (contract number 002035). The authors would also like to thank George G. Mitchell, James Decraene and Pietro Speroni di Fenizio for helpful discussions at the aLife lab that shaped the core material of this work.

References

[1] Dawkins, R.: The Selfish Gene. Oxford University Press, Oxford (1976)
[2] Dittrich, P., Ziegler, J., Banzhaf, W.: Artificial Chemistries - A Review. Artificial Life 7(3), 225–275 (2001)
[3] Eigen, M., Schuster, P.: The Hypercycle, A Principle Of Natural Self-Organization. Die Naturwissenschaften 64(11), 541–565 (1977)
[4] Gilbert, W.: Origin of life: The RNA world. Nature 319(6055), 618 (1986)
[5] Joyce, G.: The rise and fall of the RNA world. The New Biologist 3(4), 399–407 (1991)
[6] Maynard Smith, J., Szathmáry, E.: The Major Transitions in Evolution. Oxford Press, Oxford (1997)
[7] McMullin, B.: Replicators don't. In: Moran, F., Merelo, J.J., Moreno, A., Chacon, P. (eds.) Advances in Artificial Life. LNCS, vol. 929, pp. 158–169. Springer, Heidelberg (1995), http://www.eeng.dcu.ie/~alife/ecal95/rpl-12h/rpl-12h.html
[8] Szathmáry, E., Maynard Smith, J.: From Replicators to Reproducers: the First Major Transitions Leading to Life. Journal of Theoretical Biology 187, 555–571 (1997)

Simulation Model for Functionalized Vesicles: Lipid-Peptide Integration in Minimal Protocells

Kepa Ruiz-Mirazo[1*] and Fabio Mavelli[2]

[1] Dept. of Logic and Philosophy of Science / Biophysics Research Unit (CSIC-UPV/EHU)
University of the Basque Country, Spain
kepa.ruiz-mirazo@ehu.es
[2] Dept. of Chemistry, University of Bari, Italy
mavelli@chimica.uniba.it

Abstract. A recently developed and presented stochastic simulation platform ('ENVIRONMENT' [12, 25]), which extends Gillespie's algorithm for chemically reacting, fixed-volume, homogeneous systems to volume-changing and globally heterogeneous conditions, is applied to investigate the dynamic behaviour of self-(re-)producing vesicles whose membrane consists of both lipids and small peptides. We claim that it is through the integration of these two types of relatively simple –and prebiotically plausible– components that protocells could start their development into functional supramolecular structures, allowing the formation of increasingly complex reaction networks in their internal aqueous milieu. The model is not spatially explicit, but takes into account quite realistically volume-surface constraints, osmotic pressure, diffusion/transport processes, structural elasticity … In this framework the time evolution of non-equilibrium proto-metabolic cellular systems is studied, paying special attention to the capacity of the system to get rid of its waste material, which proved critical for balanced cell growth (avoiding the risk of an osmotic burst). We also investigate the effects of including an explicit feedback mechanism in the system: the case in which waste transport mediated by peptide chains takes place only under osmotic stress conditions.

Key words: lipid-peptide protocells, stochastic kinetics (Gillespie method), self-assembly, autocatalysis, osmotic crisis, vesicle reproduction

1 Introduction

Real biological membranes are not mere containers of a set of reactants in relevant concentration levels: they play a crucial role in the control of the matter-energy flow through the system, by means of very diverse transport and transduction mechanisms. This role is so crucial for the maintenance of any internal network of biochemical processes that compartments could not be invented 'off the cuff', at relatively late stages of the origins of life, as it is often assumed. A much more plausible prebiotic scenario involves the early appearance of compartments [7] and their 'co-evolution' with reaction networks, so that both get increasingly complex, until the living threshold is eventually achieved. In this context, we propose a model of minimal lipid-peptide protocell in which not only the membranes but also short peptide chains benefit from their coexistence. On the one hand, simple vesicles become more elaborate

F. Almeida e Costa et al. (Eds.): ECAL 2007, LNAI 4648, pp. 32 – 41, 2007.

compartments, with hydrophobic oligo-peptides getting inserted in the bilayer and changing their elastic properties and/or their permeability to a certain compound, what could result in higher osmotic robustness. On the other hand, oligomerization processes that lead from simple aminoacids to small peptides could be favoured if they happen in the lipidic phase (since peptide bond formation is not usually spontaneous in aqueous solution). So the aim of this paper is to explore, by means of our simulation platform [12, 25], the dynamics of protocellular systems that encapsulate a simple autocatalytic reaction cycle (inspired by Ganti's scheme for 'proliferating microspheres' [4]). The cycle produces the lipid –and, depending on the case, also the hydrophobic aminoacid– molecules that are progressively incorporated to the membrane, provoking their growth and reproduction.

Our approach is a step forward with regard to previous models of proto-metabolic cells (e.g.: [27, 14, 2, 1, 3, 17, 9, 26, 8]) because it tries to capture in a realistic (physico-chemically well grounded) way the active role and dynamic properties of protocellular compartments. These are modelled as closed bilayers (vesicles) made of lipid-like surfactant molecules –plus other types of compounds, like peptides– with specific molecular properties (e.g.: volume, head area, etc.) enclosing an 'aqueous core' where different reactions take place. Although the simulation is not spatially explicit, volume and surface constraints are taken into account, allowing for different protocell shapes (not just spherical) and determining the conditions for stability of the system: i.e., the range in which the protocell will not burst (due to an excess of osmotic pressure) or divide (when the surface is large enough to do so).

We are particularly interested in the study of 'lipid-peptide' protocells because, as prebiotically plausible supramolecular structures,[1] they could constitute the first or most elementary examples of functionalized vesicles, in the sense that the structural (and global) role of the lipid aggregate be complemented by the (more local) action of peptide chains spanning the bilayer and making possible primitive channels/pores for different compounds. For instance, polyalanine and polyleucine embedded in lipid bilayers have already been shown to induce proton-conducting pathways [20]. At these initial stages, we can assume [22] that the cellular system would not require very specific peptide sequences (i.e., it is reasonable to begin with a single aminoacid, or combinations of two different ones).

The insertion of these proto-functional components in the bilayer (forming rudimentary channels or pores) would be crucial to start controlling the in-/out- flow of compounds to/from the compartment and overcome problems like proto-cell bursting due to osmotic pressure. Something similar has already been demonstrated in the context of artificial cell bioreactors [18], where the energetic-nutrient limitations of the system were temporally avoided thanks to the insertion of a toxin (α-hemolysin) in the membrane. It remains an open question whether this is also the case at earlier

---

[1] Both simple amphiphilic molecules, like fatty acids, and aminoacids (and short peptides) are well-known as products of abiotic synthesis experiments [19, 15]. The idea is not new [16] and, in fact, at present there are other 'bottom-up' approaches to minimal lipid-peptide systems [23]. These consider an even simpler protocell scenario: they take the compartment as just a different phase (a hydrophobic domain: a micelle or a droplet), where certain reactions (in particular, the polymerization of PNA) may be favoured. Although that scenario is also worth exploring, we consider that vesicular topology will be the key to provide a solid and continuous account of the origin of biological cells.

stages, when the molecular complexity of the system is strongly reduced. But that is precisely the hypothesis we would like to explore, convinced that the capturing and channelling of matter and energy resources into the system (though the membrane) is pivotal for its autonomous construction and robust maintenance [24].

2 Methods: The Stochastic Simulation Platform ENVIRONMENT

'ENVIRONMENT' is an object-oriented (C++) platform that has been developed to simulate stochastically (by means of a Monte Carlo algorithm: the Gillespie method [5,6]) chemically reacting systems in non-homogeneous conditions [10-12]. The program considers these systems as a collection of different phases or reaction domains, each of which is assumed to be globally homogeneous. In this way, the Gillespie method can be generalized to systems that are not simple, fixed-volume 'reaction tanks' but consist of a heterogeneous mixture of phases, each with a different and variable volume, where reactions take place.[2]

So the platform can be naturally applied to model the dynamics of (bio-)chemical cellular systems. For instance, in the case of a single cell three fundamental phases or reaction domains are considered: a general *environment* (the common aqueous solution where the cell is contained), a hydrophobic or lipidic phase (the *membrane*) and a *core* (the internal aqueous solution that belongs to each cell). These three different domains 'interact' (exchange chemicals) via diffusion processes, according to the permeability rules of the system, the concentration gradients and the specific value of the diffusion constants. Thus, a cell is made of a closed membrane and a core; and the number of these phases may increase, if the initial cell divides. When desired, the program can follow the evolution of both the 'parent' and 'daughter' cells, monitoring the dynamics of cell populations. Unfortunately, a more extended description of the platform is beyond the scope of this article, but further details can be found in [12, 25, 13].

3 Main Modelling Assumptions and Protocell Scenario

The first important assumption of our model is that the global concentration of chemicals inside and outside the membrane is kept equal. In other words, free flow of water between the aqueous solutions ('environment' and 'cell core' phases) is assumed, to ensure the isotonic condition:

$$C_{Total} = \frac{\sum_{i}^{\text{Internal species}} n_i}{N_A V_{Core}} = \frac{\sum_{j}^{\text{External species}} n_j}{N_A V_{Env}} \tag{1}$$

[2] The Gillespie's algorithm can be generalized to volume-changing conditions by expressing the density probability coefficients c_ρ associated with each reaction in a particular domain in terms of the macroscopic kinetic constant k_ρ, through the relationship [11]:

$$c_\rho \propto k_\rho/(N_A V)^{m\rho-1}$$

where V is the domain volume, m_ρ is the reaction order and N_A is Avogadro's number.

This does not preclude the formation of concentration gradients of particular chemical species but, throughout the simulation, at the end of every iteration, the core volume V_{core} of each cell is rescaled in the following way:

$$V_{Core} = \frac{\sum_{i}^{\substack{Internal\\species}} n_i}{\sum_{j}^{\substack{External\\species}} n_j} V_{Env} \qquad \textbf{(2)}$$

simulating an instantaneous flux of water to balance overall osmotic differences. The volume of the environment remains fixed to its initially given value, whereas the volume of the cell core is free to change during the simulation, although it must satisfy some geometrical constraints (limits in the surface-volume relationship). Actually, the cell membrane can exchange continuously lipid and hydrophobic molecules with the internal and the external aqueous environments, with the result of varying its surface area (S_μ), as given by:

$$S_\mu = \frac{1}{2} \sum_{i}^{\substack{Membrane\\Species}} \alpha_i n_{i_\mu} \qquad \textbf{(3)}$$

where α_i are the hydrophilic head areas of all the surface active molecules located on the membrane and the factor 1/2 reflects the bilayer structure.

Even if the initial shape of a cell will be considered spherical (for the sake of standardizing initial conditions), our model does not assume that cells must stay spherical all the time, or that they divide when they double their initial size (as it is usually done). Instead, here we will consider that there is a relatively free relationship between its volume and surface, within the following limits:

1) The actual surface of the cell must be bigger than the theoretical spherical surface that corresponds to the actual volume at each iteration step. Otherwise the cell bursts.

2) The actual surface of the cell must be smaller than the theoretical surface that corresponds to two equal spheres of half the actual volume at each iteration step. Otherwise the cell divides, giving rise to two statistically equivalent cells.

These are, in fact, the conditions for stability of a cell in our model; i.e., they define the range of possible states in which it will not break or divide. In terms of Φ, the ratio between S_μ and the surface of an ideal sphere of volume V_{core}:

$$\Phi = S_\mu / \sqrt[3]{36\pi V_{core}^2} \qquad \textbf{(4)}$$

the conditions for stability become $1 \le \Phi \le \sqrt[3]{2}$. If one takes into account that the membrane is a relatively elastic structure, two additional parameters can be introduced as follows:

$$1 - \varepsilon \le \Phi \le (1+\eta)\sqrt[3]{2} \qquad \textbf{(5)}$$

where ε and η are the burst and fission tolerance, respectively. These two parameters may be fixed, or change as functions of the membrane composition. Anyway, in all simulation runs reported below they were set equal to 0.1, so the stability range becomes $0.9 \le \Phi \le 1.386$.

The probability for the uptake of a generic molecule Z from an aqueous domain (the environment or the core) to the lipid bilayer is given by the product of a constant times the aqueous molecule concentration $[Z]_{aq}$ (where *aq* can be *Env* or *Core*) multiplied by the membrane surface area S_μ , whereas the backward process probability (i.e., the release of a membrane component to an aqueous domain) was assumed proportional to the corresponding constant times the number of molecules of that component in the bilayer.

Diffusion/transport across the cell membrane is also allowed, although we work under the assumption that a pure lipid bilayer is only permeable to a few species (in the following simulations: to the nutrient X and waste W molecules). Spontaneous diffusion processes (for a generic membrane penetrating substance Z) are driven by the concentration gradient: $g_Z = \left([Z]_{Env} - [Z]_{Core}\right)/\ell_\mu$ (where ℓ_μ : bilayer thickness). The resulting molecular flux will be inwards or outwards depending on whether g_Z is greater or smaller than zero. In both cases, the process probability will be proportional to a diffusion constant D_Z, the absolute concentration gradient $|g_Z|$ and the surface of the membrane. If long enough polypeptides are present in the cell membrane, then selective mediated transport processes can also occur, with a probability that is calculated as before but multiplied by the number of polypeptides and by a different (typically higher) diffusion constant D_{ZP}. This additional (*passive* but *mediated*) transport mechanism will therefore induce faster in-out exchanges of a given substance.

Under these general conditions and modelling assumptions, the simulations carried out specifically for this work were directed to study the dynamic behaviour of self-(re-)producing lipid-peptide protocells, in particular with regard to the capacity of the system to get rid of the waste product of its internal metabolism, in conditions in which polymerization of the peptide chains could only happen in the lipidic phase (i.e., within the actual membrane: see Fig. 1). Since (i) peptide bond formation is usually quite unfavourable in water solution and (ii) concentration thresholds to provoke the assembly of peptide strands into longer chains is much easier to occur in the lipid bilayer, we can say that this hypothesis should be more realistic than previous schemes we tried for minimal lipid-peptide cells [25].

As we wanted to focus the analysis on the capacity for waste release and its effects for protocell stability and growth dynamics, we carried out simulations in three different cases: a) long enough peptide chains are produced to span the membrane, but they do not modify its W-permeability; b) the oligo-peptides formed –and properly oriented– in the membrane act as channels (mediated transport or MT) and c) the case in which mediated transport is only activated when the cell is under osmotic stress conditions (i.e., $\Phi < 1$), as a kind of feedback response mechanism (MTwFB). The latter can be interpreted, in more strict physical terms, as a situation in which channels are open by elastic tension due to the osmotic pressure on the cell (i.e., only in those conditions polypeptides would adopt the suitable conformation to form waste-transport channels).

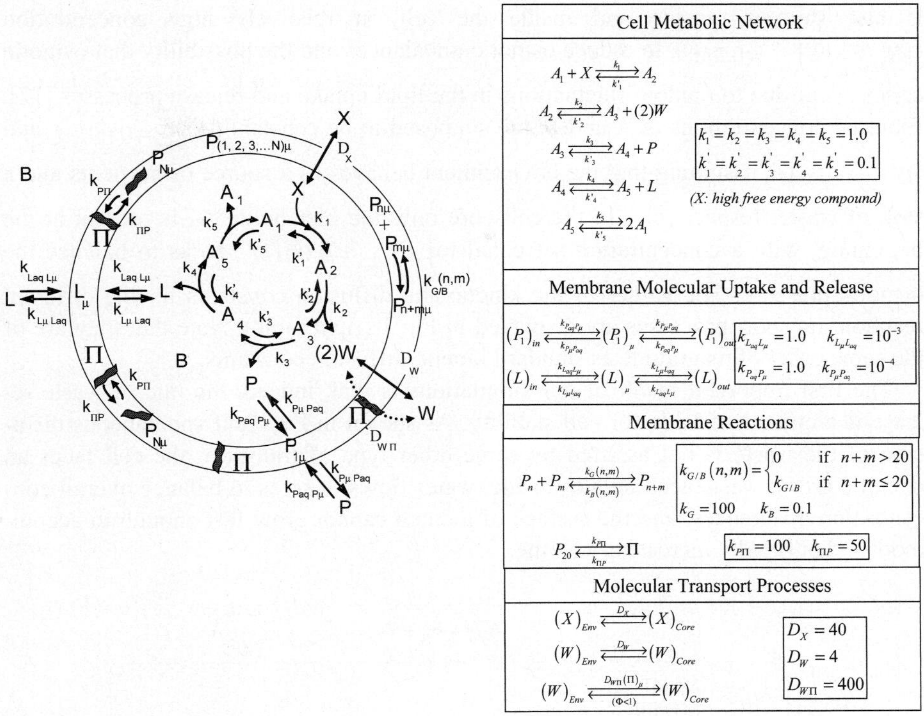

Fig. 1. Reaction scheme, together with the typical values of the constants in our present simulation runs. The internal proto-metabolic cycle produces, from nutrient X, a waste product W, the lipid L, the hydrophobic aminoacid monomer P and (A_1) one of the metabolites, to make the cycle actually autocatalytic. Polymerization processes take place within the membrane, in the lipidic phase. X and W are permeable compounds, but the rate at which W diffuses across the membrane can be enhanced if long enough polypeptides get oriented so as to span the membrane and act as Π channels (passive mediated transport). And B is a non-reactive species (the buffer), which is important to include for stability reasons [12]. (More details in the text.)

4 Results

All simulations reported in this paper were carried out with bilayered membranes (initially spherical: 50nm radius) composed of a generic lipid *L*, whose molecular volume and head surface area were equal to 1.0 nm^3 and 0.5 nm^2. The peptides spanning the membrane were assumed to be α-helixes of, at least, 20 residues (i.e., twentymers), with a somewhat larger surface area (2x5.0 nm^2) and a total cylindrical volume of 20.0 nm^3. Shorter peptide chains, when they belong to the membrane, are assumed not to be surface active molecules (i.e., they just contribute to the volume of the lipid bilayer, not to its surface).

The initial lipid aqueous concentration in the core and in the external environment was set to: $[L]_{Env} = [L]_{Core} = 0.004M$ (equilibrium concentration). The osmotic buffer is also present outside and inside the cell, at relatively high concentration ($[B]_{Env} = [B]_{Core} = 0.2M$), to reduce osmotic unbalances and the possibility that osmotic crises occur due to random fluctuations in the lipid uptake and release processes [12]. External concentrations of X and W are supposed to be constant $[X]_{Env} = 0.001M$ and $[W]_{Env} = 0.0M$, assuming that the environment behaves as a source of nutrients and a sink of waste, respectively. In the cell core only the metabolite A_1 is present at the beginning, with a concentration set equal to $[A_1]_{Core} = 0.001M$, so as to balance the osmotic pressure. The values of the kinetic and diffusion constants for the different reaction/transport processes are displayed in Fig 1 (right table). Note that they are of the same order of magnitude as standard kinetic/diffusion constants.

The first important result of our simulations is that, indeed, the rate of waste release is a critical variable for cell stability. As shown in Fig. 2, if spontaneous diffusion of the waste is not assisted by some other type of transport, the cell faces an osmotic crisis: waste accumulates inside, water flows in so as to balance overall concentration differences, and the surface of the cell cannot grow fast enough to accommodate that rapidly increasing volume.

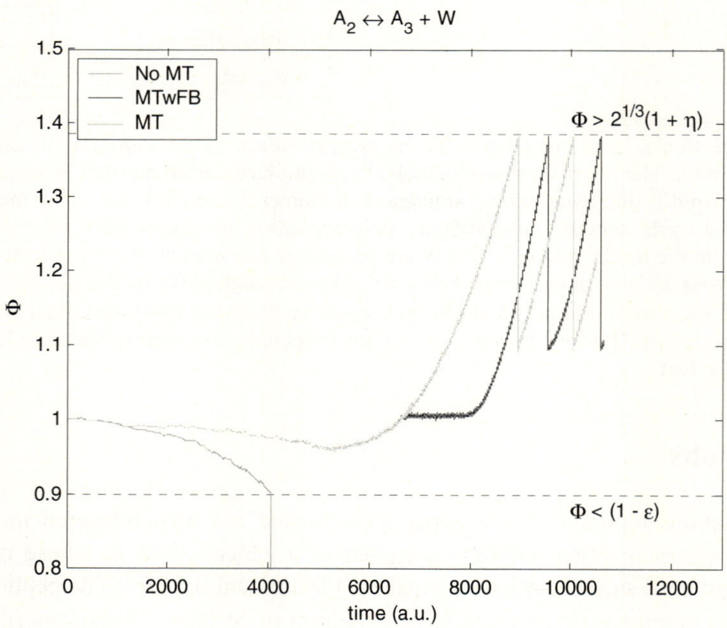

Fig. 2. The stability coefficient vs. time in the three different waste diffusion-transport conditions: 1) No mediated transport (NoMT), just slow diffusion outwards (due to the low permeability of the lipid bilayer to W molecules); 2) Oligo-peptides contribute to increase the permeability of the compartment, acting as selective channels (MT); 3) Mediated transport taking place only under osmotic stress conditions ($\Phi < 1$), as a feed back mechanism to regulate the distance to the osmotic burst limit (MTwFB).

Instead, if the oligopeptides produced by the system act as channels and contribute to a faster release of waste molecules, the osmotic burst is overcome and the protocell grows until it reaches the division limit, initiating a typical reproductive regime. The difference in the overall time behaviour of a protocell with simple versus feedback mediated transport does not seem to be so dramatic, if we just compare the splitting times: the latter just causes a delay in the time required for the original (mother) cell to reach the first division condition. But Fig. 4 clearly shows that in the MTwFB case the mother cell grows up to quite bigger sizes before splitting (again, as a result of accumulating a higher amount of W inside). So one has to search for more subtle differences, which could be relevant in the long-run, or in other regions of parameter space, by analysing the time evolution of the various chemical species involved. In Fig. 3 the results of such an analysis are portrayed for the components of the internal aqueous core of the protocell: whereas the behaviour of the concentrations of X, L, or B does not show major differences, it is quite obvious that the time profiles for A_1, A_5 and –more significantly– of W are more complex in the MTwFB case.

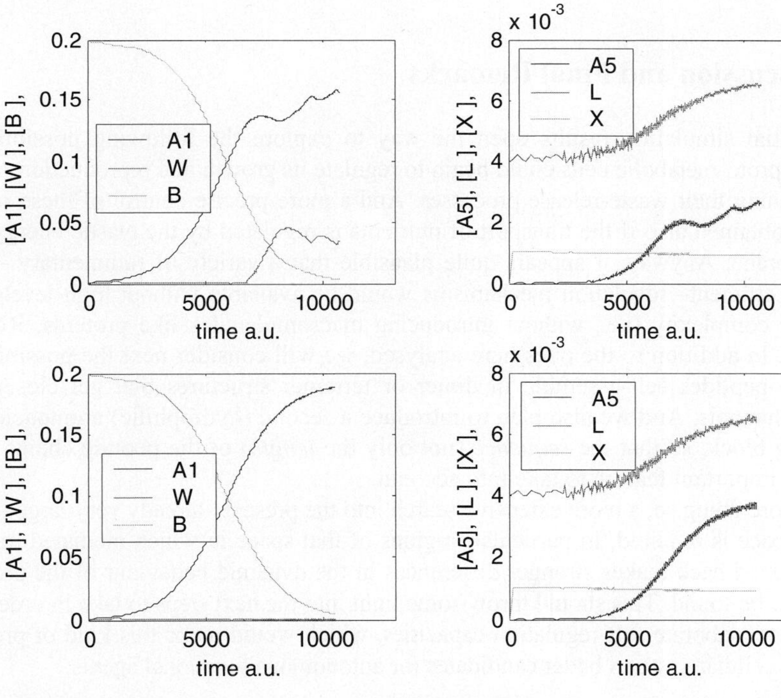

Fig. 3. Time profiles of some of the internal species of the protocell. The two graphs on top show the results for the MTwFB case and the two below are the corresponding results for the MT case. Simulations with a reaction scheme in which the internal autocatalytic cycle produced just twice as much W (see the stoichiometric coefficient in brackets in Fig. 1) yielded very similar results (data not shown here).

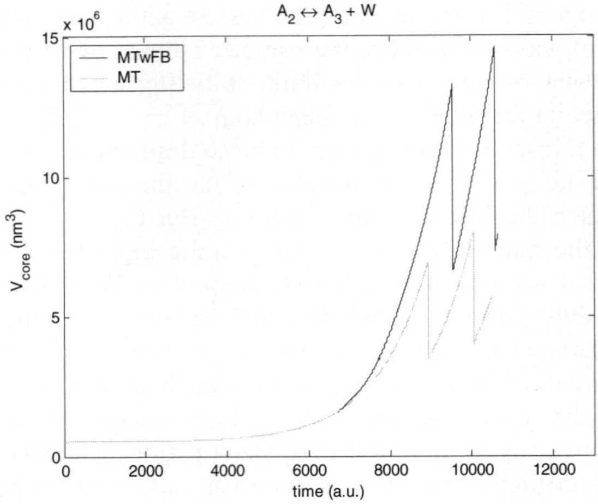

Fig. 4. Time profiles of the cell core volume in the two cases in which the protocell overcomes the osmotic crisis.

5 Discussion and Final Remarks

These initial simulation results open the way to explore the following possibility: primitive proto-metabolic cells could begin to regulate its growth and reproduction rates by fine-tuning their waste-release processes. And a more precise control of these rates could be obtained also if the transport of nutrients is regulated by the elastic energy of the membrane. Anyway, it appears quite plausible that a variety of rudimentary –but relatively efficient– regulation mechanisms would be available without high levels of molecular complexity (i.e., without introducing macromolecules like proteins, RNA, DNA,…). In addition to the ones here analysed, we will consider next the possibility that oligo-peptides self-assemble in dimer or tetramer structures that get closer to realistic channels. And we also plan to introduce a second (hydrophilic) aminoacid as a building block, so that the *sequence* (not only the *length*) of the peptide chains becomes an important feature to take into account.

But before doing so, a more extensive search into the present –already very large– parameter space is required. In particular, regions of that space in which mediated transport with feed back makes stronger differences in the dynamic behaviour of the protocell should be found. This should throw some light into the next steps to take in order to obtain more elaborate self-regulation capacities, which would make this kind of proto-metabolic cellular systems better candidates for autonomous functional agents.

Acknowledgements. Kepa Ruiz-Mirazo holds a *Ramón y Cajal* research position and acknowledges support from research grants *9/UPV 00003.230-15840/2004 and HUM2005-02449*. Both authors would also like to thank COST (Action D27), which made possible their collaboration.

References

1. Csendes, T.: A simulation study on the chemoton. Kybernetes 13, 79–85 (1984)
2. Dyson, F.: A model for the origin of life. J. Molec. Evol. 18, 344–350 (1982)
3. Fernando, C., Di Paolo, E.: The chemoton: a model for the origin of long RNA templates. In: Pollack, J., et al. (eds.) Proceedings of Artificial Life IX, pp. 1–8. MIT Press, Cambridge (2004)
4. Ganti, T.: On the early evolutionary origin of biological periodicity. Cell Biol. Int. 26, 729–735 (2002)
5. Gillespie, D.T.: A General Method for Numerically Simulating the Stochastic Time Evolution of Coupled Chemical Reactions. J. Comput. Phys. 22, 403–434 (1976)
6. Gillespie, D.T.: Exact Stochastic Simulation of Coupled Chemical Reactions. J. Phys. Chem. 81, 2340–2369 (1977)
7. Luisi, P.L.: The Emergence of Life. Cambridge University Press, Cambridge (2006)
8. Macía, J., Solé, R.V.: Protocell self-reproduction in a spatially explicit metabolism-vesicle system. J. Theor. Biol. 245(3), 400–410 (2007)
9. Madina, D., Ono, N., Ikegami, T.: Cellular Evolution in a 3D Lattice Artificial Chemistry. In: Banzhaf, W., Ziegler, J., Christaller, T., Dittrich, P., Kim, J.T. (eds.) ECAL 2003. LNCS (LNAI), vol. 2801, pp. 59–68. Springer, Heidelberg (2003)
10. Mavelli, F.: Theoretical investigations on autopoietic replication mechanisms. Ph.D. Dissertation (Nr. 15218), ETH-Zürich (2003)
11. Mavelli, F., Piotto, S.: Stochastic Simulations of Homogeneous Chemically Reacting Systems. J. Mol. Struct.: THEOCHEM 771, 55–64 (2006)
12. Mavelli, F., Ruiz-Mirazo, K.: Stochastic simulations of minimal self-reproducing cellular systems. In: Solé et al. (eds.), Philosophical Transactions Royal Society of London B, Special Issue 'Towards the artificial Cell' (in press, available online)
13. Mavelli, F., Ruiz-Mirazo, K.: An object-oriented programming platform to simulate stochastically reaction networks in complex (heterogeneous) environments (forthcoming)
14. McMullin, B., Varela, F.: Rediscovering Computational Autopoiesis. In: Husbands, P., Harvey, I. (eds.) Fourth European Conference on Artificial Life, pp. 38–47. MIT Press, Cambridge (1997)
15. Miller, S.L.: A production of amino acids under possible primitive Earth conditions. Science 117, 528–529 (1953)
16. Morowitz, H.J., Heinz, B., Deamer, D.W.: The chemical logic of a minimum protocell. Origs. Life. Evol. Bios. 18, 281–287 (1988)
17. Munteanu, A., Solé, R.V.: Phenotypic diversity and chaos in a minimal cell model. J. Theor. Biol. 240, 434–442 (2006)
18. Noireaux, V., Libchaber, A.: A vesicle bioreactor as a step toward an artificial cell assembly. Proc. Natl. Acad. Sci. USA 101, 17669–17674 (2004)
19. Nooner, D.W., Oró, J.: Synthesis of fatty acids by a closed system Fischer-Tropsch process. Division of Pretoleum Chemistry, American Chemical Society 23, 624–631 (1978)
20. Oliver, A.E., Deamer, D.W.: Alpha-helical hydrophobic polypeptides form proton-selective channels in lipid bilayers. Biophys J. 66(5), 1364–1379 (1994)
21. Ono, N., Ikegami, T.: Model of self-replicating cell capable of self-maintenance. In: Floreano, D., Mondada, F. (eds.) ECAL 1999. LNCS, vol. 1674, pp. 399–406. Springer, Heidelberg (1999)
22. Pohorille, A., Schweighofer, K., Wilson, M.A.: The origin and early evolution of membrane channels. Astrobiology 5(1), 1–17 (2005)
23. Rasmussen, S., et al.: Transition from non-living to living matter. Science 303, 963 (2004)
24. Ruiz-Mirazo, K., Moreno, A.: Basic autonomy as a fundamental step in the synthesis of life. Artificial Life 10(3), 235–259 (2004)
25. Ruiz-Mirazo, K., Mavelli, F.: On the way towards 'basic autonomous agents': stochastic simulations of minimal lipid-peptide cells. BioSystems (accepted)
26. Segré, D., Lancet, D.: Composing Life. EMBO Rep. 1(3), 217 (2000)
27. Varela, F.J., Maturana, H., Uribe, R.: Autopoiesis: The Organization of Living Systems, its characterization and a model. BioSystems 5, 187–196 (1974)

Emergence of Genetic Coding: An Information-Theoretic Model*

Mahendra Piraveenan[1], Daniel Polani[2], and Mikhail Prokopenko[1,**]

[1] CSIRO Information and Communication Technology Centre
Locked bag 17, North Ryde, NSW 1670, Australia
mikhail.prokopenko@csiro.au

[2] Department of Computer Science, University of Hertfordshire
Hatfield AL10 9AB, United Kingdom

Abstract. This paper introduces a simple model for evolutionary dynamics approaching the "coding threshold", where the capacity to symbolically represent nucleic acid sequences emerges in response to a change in environmental conditions. The model evolves a dynamical system, where a conglomerate of primitive cells is coupled with its potential encoding, subjected to specific environmental noise and inaccurate internal processing. The separation between the conglomerate and the encoding is shown to become beneficial in terms of preserving the information within the noisy environment. This selection pressure is captured information-theoretically, as an increase in mutual information shared by the conglomerate across time. The emergence of structure and useful separation inside the coupled system is accompanied by self-organization of internal processing, i.e. an increase in complexity within the evolving system.

1 Introduction

One of the most fundamental problems in biology and artificial life is the definition and understanding of "the gene". As pointed out by Carl Woese, whose work provided a very strong motivation for this study, this problem continues to contribute to much debate between classical biologists who understand "the gene to be defined by the genotype-phenotype relationship, by gene expression as well as gene replication" and many molecular biologists who declared the problem to be solved when the Watson-Crick structure of DNA clearly revealed the mechanism of gene replication [1]. Woese strongly argues against fundamentalist reductionism and presents the real problem of the gene as "how the genotype-phenotype relationship had come to be". In other words, the main question is how the mechanism of translation evolved.

The evolution of the translation mechanism is a complicated process, and we may only intend to analyse its simplified models. However, in doing so we shall take a principled approach and consider a model of evolutionary dynamics in a generic information-theoretic way, without obscuring it with hypothetical aspects such as biochemical composition of "primordial soup", structural properties of procaryotic cells, susceptibility

* The Authors list is in alphabetical order.

** Corresponding author.

F. Almeida e Costa et al. (Eds.): ECAL 2007, LNAI 4648, pp. 42–52, 2007.

of aminoacyl-tRNA synthetases to horizontal gene transfer (HGT), etc. The simple assumptions that we make, following Woese [1], include the notion of primitive cells as loosely connected conglomerates existing during the "era of nucleic acid life" [2,3], and the conjecture that primitive cell organization was "largely horizontal" in nature [4,5], making the simple cellular componentry open to HGT.

In taking the information-theoretic view, we focus on the "coding threshold" separating the phase of nucleic acid life from the evolutionary stage "where the capacity to represent nucleic acid sequence symbolically in terms of a (colinear) amino acid sequence developed" [1]. More precisely, we hope to understand the pressures that forced such a transition to "proto-symbols" encoding features of primitive cells in dedicated sequences and enabling a rudimentary translation. The analysis presented by Woese [1] sheds light not only on this transition, but also on saltations that have occurred at other times, e.g. advents of multicellularity and language. The common feature is "the emergence of higher levels of organization, which bring with them qualitatively new properties, properties that are describable in reductionist terms but that are neither predictable nor fully explainable therein" [1].

More importantly, the reason for the increase in complexity can be identified as *communication* within a complex, sophisticated network of interactions: "translationally produced proteins, multicellular organisms, and social structures are each the result of, emerge from, fields of interaction when the latter attain a certain degree of complexity and specificity" [1,6]. The increase of complexity is also linked to adding new dimensions to the phase space within which the evolution occurs, i.e. expansion of the network of interacting elements that forms the medium within which the new level of organization (entities) comes into existence [1,6]. These observations can be formalised information-theoretically. More precisely, we intend to consider a communication channel between a conglomerate of primitive cells and itself at a future time point, and pose a question of the channel capacity constrained by the noise. By varying the nature and degree of this noise prevalent in the environment within which such conglomerates exist and evolve, we hope to identify conditions leading to a separation between the conglomerate *per se* and its encoding with "proto-symbols". Specifically, we investigate conditions under which such separation is beneficial in terms or preserving the information within the noisy communication channel across time. In other words, the separation evolves to protect some information about the conglomerate in the encoding. A rudimentary translation then helps to recover the information that otherwise would have been lost due to the noise.

The adopted information-theoretic view allows us to concentrate on generic processes common to a collection of such conglomerates rather than on specific interactions within an environmental locality. Considering a collection of conglomerates is important because, as noted by Woese [1],

> A sufficiently imprecise translation mechanism could produce "statistical proteins", proteins whose sequences are only approximate translations of their respective genes [7]. While any individual protein of this kind is only a highly imprecise translation of the underlying gene, a consensus sequence for the various imprecise translations of that gene would closely approximate an exact translation of it.

In other words, a given gene can be translated not into a unique protein but instead into a family of related protein sequences: "early life did not require a refined level of tolerance" [3]. Looseness of the outcome is implied by an imprecise genome replication comprising relatively few unique genes [5] — therefore, rather than trying to develop a dynamical system (conglomerate plus encoding) that fully preserves the information about the conglomerate, we only need to develop dynamics that corresponds to "statistical proteins", preserving information in a "consensus sequence".

In modeling the evolution, we adopt the view that maximization of information transfer through selected channels is one of the main evolutionary pressures [8,9,10,11,12]. Although the evolutionary process involves a larger number of drives and constraints, information fidelity (i.e. preservation) is a consistent motif throughout biology: e.g., modern evolution operates close to the error threshold [13], and biological sensorimotor equipment typically exhausts the available informatory capacity (under given constraints) close to the limit [14]. Adami, in fact, argues that the evolutionary process extracts valuable information and stores it in the genes. Since this process is relatively slow [15], it is a selective advantage to preserve this information, once captured. In the following, we shall concentrate on the information preservation property of evolution. Everything else is modeled minimalistically: we encapsulate the influence of evolutionary constraints within a dynamical system, and represent the acquisition of valuable information by an explicit "injection" of information at the beginning of each trajectory.

2 Modelling Evolutionary Dynamics

Our generic model for evolutionary dynamics involves a dynamical coupled system, where a conglomerate is coupled with its potential encoding, evolving in a fitness landscape shaped by a selection pressure. The selection pressure rewards preservation of information in presence of both environmental noise and inaccuracy of internal coupling. When the conglomerate is represented as a dynamical system, the information about it can be captured generically via attractors of the dynamical system. In particular, a loss of such information corresponds to a loss of structure in the phase-space, while informational recovery would correspond to recovery of the equivalent (e.g., isomorphic) structure in the phase-space. Importantly, the information about the attractors can be compactly encoded if there is a need for it.

The dynamical coupled system is described by the equations

$$X_t = \begin{cases} f(X_{t-1}) + \varphi & \text{if } t \neq t^* \\ \alpha\left[f(X_{t-1}) + \varphi\right] + (1-\alpha)h(Y_{t-1}) & \text{if } t = t^* \end{cases} \tag{1}$$

$$Y_t = \begin{cases} g(X_{t_0} + \psi) & \text{if } t = t_0 \\ Y_{t-1} & \text{if } t > t_0 \end{cases} \tag{2}$$

where X_t and Y_t are the variables that describe the conglomerate and its potential encoding respectively. Function f defines the dynamical system representing the conglomerate, and $\alpha \in [0,1]$ sets the relative importance of translation h; we used $\alpha = 1/2$.

Noise functions φ and ψ are described in Section 2.1. The mechanism that extracts information from the conglomerate into its encoding at time t_0 is given by g.

At time $t = t_0$, noise is introduced into the environment affecting dynamics of the conglomerate. Our null hypothesis is that the ability to symbolically encode nucleic acid sequences does not develop when (possibly anisotropic) environmental noise φ is outside a certain range. In other words, it is precisely a limited reduction in the information channel's capacity, brought about by the environmental noise, that creates the selection pressure for the separation between a conglomerate and its encoding. At the time $t = t_0$, information from the conglomerate is accessed by the system Y (encoding) via the internal processing function g. The accessed information is not entirely accurate because of possible limitations of the access mechanism, and the noise function ψ represents this inaccuracy.

We aim to identify a function g that maximizes the ability to recover, in the face of environmental noise and in presence of an imperfect internal processing, as much original information as possible — i.e., the ability to resist both the external noise φ and internal noise ψ. The feedback from the Y to X occurs at the time point t^*, i.e. the function h translates the input Y_{t^*-1} from the encoding back into the conglomerate.

2.1 External and Internal Noise

The function φ describes the external (environment) noise that affects the variable X after some information has been stored in Y. This noise represents a pressure to push the system X towards certain attractors. It is implemented as a random variable $\varphi \in [-l, u]$, where $u > 0$ and $l > 0$, which is uniformly distributed, with probability $1/2$, between 0 and l, and with probability $1/2$ between 0 and u. In other words, positive values may be more sparsely distributed than the negative if u is larger than l.

The function ψ represents the internal noise associated with accessing information from X_t by the system Y_t at any given time t. In other words, it represents the inaccuracy within the internal communication channel. In addition, the noise ψ may be interpreted as inaccuracy of the environment's representation within the encoding Y which indirectly "perceives" the environment through the system X. This noise is modelled as uniform random noise $\psi \in [-b, b]$, where $0 < b \ll 1.0$.

2.2 State-Space

The dynamical system employed is a logistic map $X_{t+1} = rX_t\,(1 - X_t)$, where r is a parameter, i.e. the function f is given by $f\,(x) = rx\,(1 - x)$. The logistic map f is initialized with a value between 0.0 and 1.0, and stays within this range if the value of r is within the range $[0, 4.0]$. We used $r = 3.5$, resulting in four attractors of the logistic map (approximately $0.38, 0.50, 0.83, 0.87$). The time $t = t_0$ is set after the logistic map settles into its attractors, having passed through a transient. The function g is a surjective mapping from $[0, 1]$ to $[0, 1]$. The function h is identity $h(y) = y$.

In order to estimate the probability distribution of a random variable (X or Y) at each time-point, we generate an initial random sample $(X_{t_0}) = (X^1_{t_0}, X^2_{t_0}, \ldots, X^K_{t_0})$ of length K. Each $X^i_{t_0}$, where $1 \le i \le K$, is chosen from a uniform random distribution within $[0.0, 1.0]$. The mapping $X^i_{t+1} = f(X^i_t)$ produces an ensemble of K

corresponding time series, $1 \leq i \leq K$, denoted as $[X] = [X_t^1, X_t^2, \ldots, X_t^K]$, where $t_0 \leq t \leq T$, and T is a time horizon. Within the ensemble, each X_t^i time series may have a different initial value $X_{t_0}^i$. At any given time t', we can obtain a sample $(X_{t'}) = (X_{t'}^1, X_{t'}^2, \ldots, X_{t'}^K)$.

Given the initial sample (X_{t_0}), and the mapping $Y_{t_0} = g(X_{t_0} + \psi)$, we can generate the initial sample $(Y_{t_0}) = (Y_{t_0}^1, Y_{t_0}^2, \ldots, Y_{t_0}^K)$ for the variable Y. In the corresponding ensemble $[Y] = [Y_t^1, Y_t^2, \ldots, Y_t^K]$ each sample is identical to the initial sample.

3 Genetic Algorithm

In evolving the potential encoding system Y coupled with X via a suitable function g, we maximize the mutual information between the initial X_{t_0} and recovered X_{t^*} states of the system, by employing a simple genetic algorithm (GA). The mutual information between A and B is defined as $I(A;B) = \sum_{a \in A} \sum_{b \in B} P(a,b) \log \frac{P(a,b)}{P(a)P(b)}$, where $P(a)$ is the probability that A is in the state a, and $P(a,b)$ is the joint probability.

We generate an ensemble of X_t time series, each series governed by equation (1). The ensemble $[X]$ provides a fixed constraint on the optimization. For *each* function g, an ensemble $[Y]$ is then generated, using equation (2) — i.e., the values of the series Y_t depend on the choice of function g. The ensemble $[X]$ is kept unchanged while we evolve the population of functions g, being an optimization constraint, but the ensemble $[Y]$ differs for each individual within the population. The fitness of each function g is determined by the mutual information between X_{t_0} and X_{t^*}, denoted $I_g(X_{t_0}; X_{t^*})$ and estimated via the mutual information between samples (X_{t_0}) and (X_{t^*}).

Since the information from Y_{t^*-1} (different for each individual) is fed back into X_{t^*}, equation (1), the sample (X_{t^*}) is specific for each individual within the population. Therefore, it may be contrasted with the sample (X_{t_0}) which is identical across the population, producing distinct fitness values $I_g(X_{t_0}; X_{t^*})$ for each individual g. The experiments were repeated for different ensembles X_t.

We generate a population of g functions (the size of the population is fixed at 400). In order to implement the mapping g, the domain of g is divided into n consecutive bins x_i such that $x_i = [(i-1)/n, i/n)$ for $1 \leq i < n$, where [a,b) denotes an interval open on the right, and $x_n = [(n-1)/n, 1]$. The range of g is divided into m consecutive bins y_j such that $y_j = [(j-1)/m, j/m)$ for $1 \leq j < m$, and $y_m = [(m-1)/m, 1]$. Then each bin x_i in the domain is mapped to a bin y_j in the range: $G : x_i \rightarrow y_j$, where G represents the discretized mapping. Formally, any $x \in x_i$ is mapped to $g(x) \equiv \overline{G(x_i)}$, where $\overline{G(x_i)}$ is the median value of the bin $G(x_i)$. For example, if $n = 100$, $m = 10$, and $y_7 = G(x_{30})$, that is, the bin $x_{30} = [0.29, 0.30)$ is mapped to the bin $y_7 = [0.6, 0.7)$, then for any $x \in x_{30}$ (e.g., $x = 0.292$), the function $g(x)$ would return $0.65 = \overline{y_7}$.

Therefore, in the GA, each function g can be encoded as an array of n integers, ranging from 1 to m, so that the i-th element of the array (the i-th digit) represents the mapping $y_j = G(x_i)$, where $1 \leq j \leq m$. We have chosen a *generation gap* replacement strategy. In our experiments, we set the generation gap parameter 0.3. In other words, the entire old population is sorted according to fitness, and we choose the best 30% for direct replication in the next generation, employing an elitist selection mechanism. The rest of selection functionality is moved into the (uniform) crossover. Mutation is

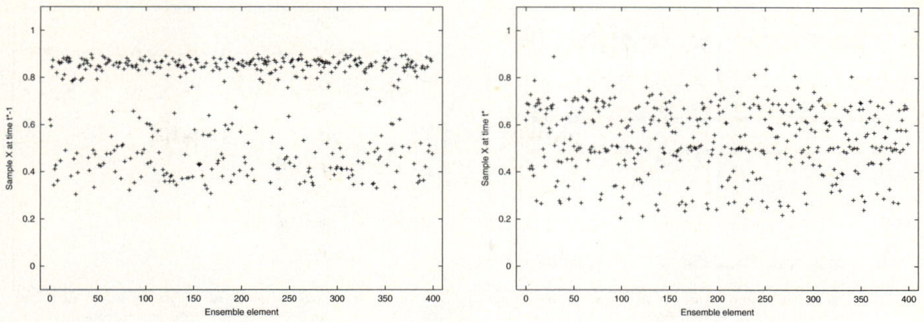

Fig. 1. Random g (noise $\varphi \pm 0.025$; $\psi = \pm 0.015$). Left: two remaining "clusters" in the sample (X_{t^*-1}). Right: the sample (X_{t^*}) does not recover full information about four attractors.

implemented as additive creeping or random mutation, depending on the number of "digits" in the genome. If the number of digits is greater than 10, then additive creeping is used: a digit can be mutated within $[-5\%, +5\%]$ of its current value. If the number of digits is less than 10, the random mutation is used with the mutation rate of 0.01.

4 Results

Our main goal is to verify whether, as the evolution takes place in a noisy environment, some structure emerges in Y (e.g., attractors are observed in Y-space). The emergent structure in Y can be associated with "proto-symbols" ("codes") that help in retrieving at time t^* some (or most of the) information accessed at t_0. This phenomenon must be supported by self-organization within the processing function g between t_0 and t^*.

4.1 Emergence of Structure in the Encoding

We begin by analyzing the optimization constraint — the system X. Figure 1 (left) shows the ensemble $[X]$ at the time $t^* - 1$, i.e. right before the moment when the feedback from Y to X occurs. It can be observed that the environment noise φ ($u = 0.025$ and $l = 0.025$)[3] disrupts the logistic map dynamics, and some information about the attractors of X is lost in the course of time: the observed sample (X_{t^*-1}) does not contain four clear clusters.

A random processing function g results in an encoding Y that, despite some structure (not shown), is not able to "help" X in recovering the lost information. Figure 1 (right) shows the un-recovered ensemble $[X]$ at the time t^* that contains values that are much more diverse than the four attractors of the original logistic map. As a result, the mutual information $I_g(X_{t_0}; X_{t^*})$ is low (≈ 0.7 bits).

Let us evaluate now the evolved coupled system. Figure 2 shows the encoding ensemble $[Y]$ at the time $t^* - 1$, as well as the recovered ensemble $[X]$ at the time t^*. The sample (Y_{t^*-1}) settles into four clusters that can be easily represented by four "codes"

[3] Similar results were obtained with anisotropic noise.

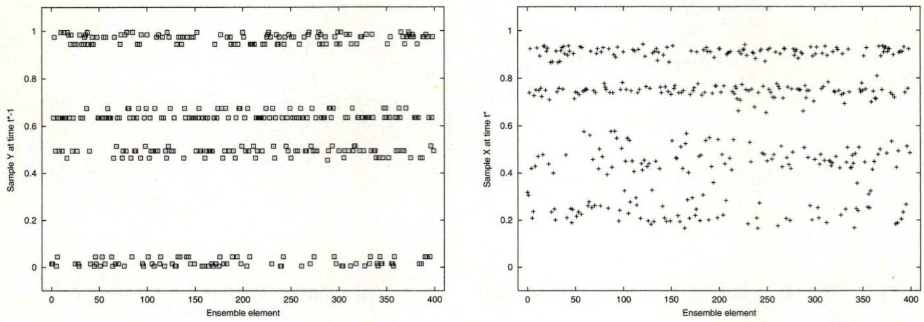

Fig. 2. Evolved g (noise $\varphi = \pm 0.025$; $\psi = \pm 0.015$). Left: four clusters in the encoding (Y_{t^*-1}). Right: Four recovered clusters in sample (X_{t^*}). $I(X_{t_0}; X_{t^*}) \approx 1.4$ bits. Contrast with Figure 1.

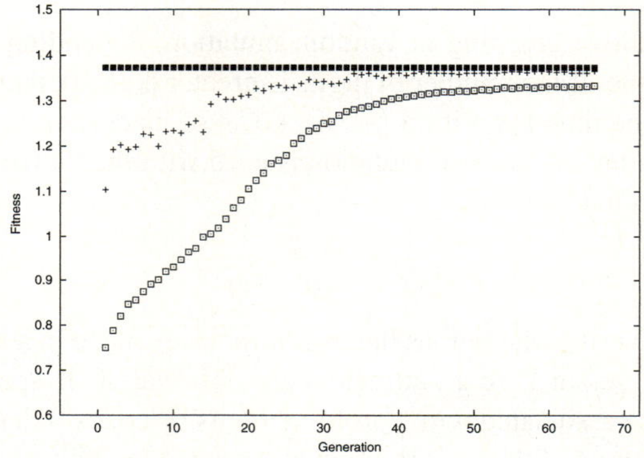

Fig. 3. Mutual information (fitness) in bits. Squares indicate the average fitness, '+'s show fitness of the best individual in each generation. Black squares trace possible maximum fitness: the mutual information when (X_{t^*}) is as informative as (X_{t_0}), i.e, $I(X_{t_0}; X_{t^*}) = H(X_{t_0})$ (entropy).

corresponding to the four attractors of X. This emergent encoding allows to recover the information within X, as evidenced by four clear clusters within the sample (X_{t^*}). The corresponding increase in the information-based fitness function is shown in Figure 3.

The clustering corresponds to emergence of discrete "proto-symbols' in the encoding Y. In the simplest case, each non-empty bin in the range of g may be associated with such a cluster, creating a symbol in the encoding Y. Following this, the function h may reconstruct precise information about X_{t_0} by clustering X_{t^*-1} into the clusters encoded by Y_{t^*-1}, and then simply using the inverse mapping g^{-1} to retrieve the original attractor. Without the clustering, however, the information reconstructed at time t^* is not precise, and rather than having four crisp attractors, X can be described as an individual with an imprecise translation of the underlying gene within a "consensus sequence" [1], analogous to a "statistical protein".

This experiment demonstrated that noise within the environment affects the emergence of structure in the encoding, allowing to recover the information within X. The

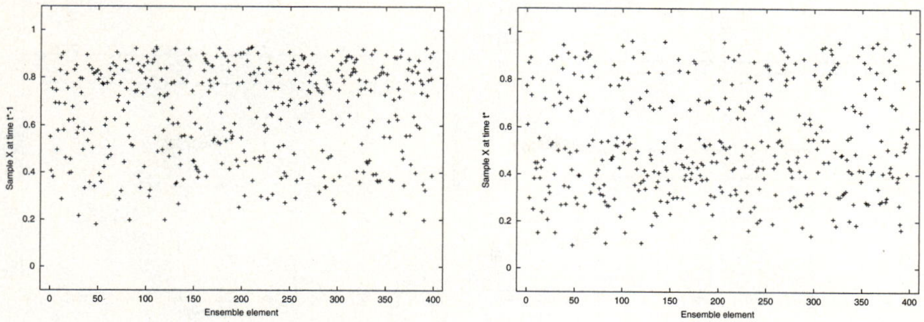

Fig. 4. Evolved g (noise $\varphi = \pm 0.06$; $\psi = \pm 0.015$). Left: the sample (X_{t^*-1}). Right: the sample (X_{t^*}) does not recover the information about four attractors. Mutual information $I(X_{t_0}; X_{t^*}) \approx$ 1.1 bits. Contrast with Figure 2.

null hypothesis, however, is that the coding threshold is not approached when environmental noise φ is outside a certain range. To verify the null hypothesis let us consider an evolution of the coupled system with a) larger noise φ; and b) without noise φ.

Figure 4 shows the ensemble $[X]$ at the time $t^* - 1$, as well as the un-recovered ensemble $[X]$ the time t^*. These results indicate that in this case there is too much noise in the environment to make the encoding useful.

In a noise-free environment ($\varphi = 0.0$), we expect a lesser or no pressure on the coupled system. In fact, there are four unobscured clusters in the sample (X_{t^*-1}), as shown in Figure 5 (left). However, the evolved encoding, being still affected by internal processing noise ψ, loses some of this information and hinders the recovery at t^*. Figure 5 (right) shows results of translation of ill-structured encoding: with zero environmental noise there is no pressure for emergence of a more precise structure in the encoding.

4.2 Self-organization Within the Processing Function

At this stage, we analyse self-organization within the processing function g. Figure 6 contrasts a randomly selected function g at the start of the evolution (noise φ is defined as $\varphi = 0.025$, while noise $\psi = \pm 0.03$ is increased to better illustrate the phenomenon), with the best individual function g after 100 generations. The important difference is in the way of mapping attractors of X (and their neighbourhoods) into the encoding Y. Let us consider, for example, the $\psi-$neighbourhood of the attractor $x \approx 0.50$. Figure 6 (left) shows that this neighbourhood has values between 0.07 and 0.91. This wide distribution is explained by the internal processing noise ψ and absence of any organization in g, so that even a slight shift from x to $x \pm \psi$ may result in a large difference between $g(x)$ and $g(x \pm \psi)$. Figure 6 (right) shows the best individual function g evolved after 500 generations. Here, despite the same level of noise ψ, we can observe that the $\psi-$neighbourhood of a given attractor in X (e.g. the neighbourhood of $x \approx 0.50$) is mapped by the evolved function g to a smaller interval (e.g., it has values between 0.93 and 0.96). In other words, the selection pressure resulted in a more "condensed" mapping. This is achieved by an increase in organization (stability) in g: a small shift from

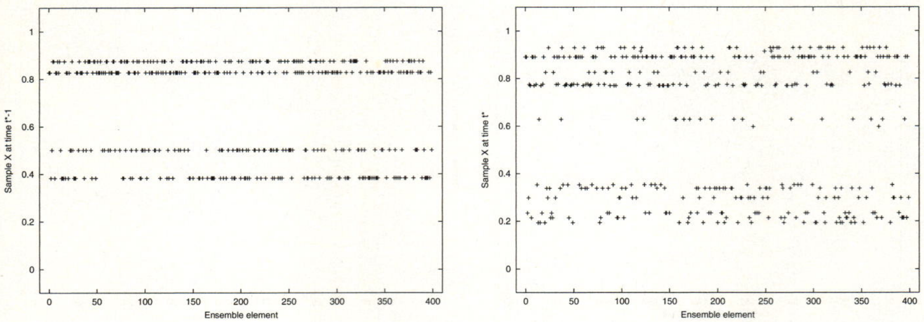

Fig. 5. Evolved g ($\varphi = 0.0$; $\psi = \pm 0.015$). Left: the sample (X_{t^*-1}). Right: the sample (X_{t^*}) obscures the four attractors. Mutual information $I(X_{t_0}; X_{t^*}) \approx 1.4$ bits. Contrast with Figure 2.

x to $x \pm \psi$ results now only in a small difference between $g(x)$ and $g(x \pm \psi)$. Moreover, only the attractors' neighbourhoods are generally handled in this way — no two other similar values of the domain are mapped to points close to each other in the range.

The self-organization of g counters the effect of internal processing noise ψ, given the noise φ in the environment[4]. This self-organization helps Y to maintain the structure of the space X (namely, the information that it had a certain number of attractors). The function h translates this encoding back into X.

5 Discussion and Conclusions

We considered a model for evolutionary dynamics in the vicinity of the "coding threshold", and identified conditions under which a separation between a conglomerate of primitive cells and its symbolic encoding becomes beneficial in terms of preserving the information within a noisy environment. The model evolves a dynamical system, where a conglomerate is coupled with its potential encoding, affected by environmental noise and inaccurate internal processing. The experiments supported the hypothesis that the ability to symbolically encode nucleic acid sequences does not develop in the absence of environmental noise or when the noise is too high (reminiscent of the U-shaped curve relating adaptive pressure to perception accuracy [18]). Instead, a (possibly anisotropic) limited noise constraining the channel's capacity creates the selection pressure for the separation between the conglomerate and its encoding. Formally, we captured this selection pressure as an increase in mutual information shared by the conglomerate across time. The experiments provided evidence that the emergence of structure and useful separation inside the coupled system is accompanied by self-organization of internal processing, i.e. an increase in complexity within the evolving system.

Following this path suggests a few intriguing possibilities for coupling systems in different ways, e.g. considering systems where a) the encoding evolves in parallel with the conglomerate, being subjected to different noise; b) the encoding may be used by

[4] In general, following [16,17], we may say that self-organization results from fluctuations, that is, internal information processing has self-organized in response to environmental "pollution".

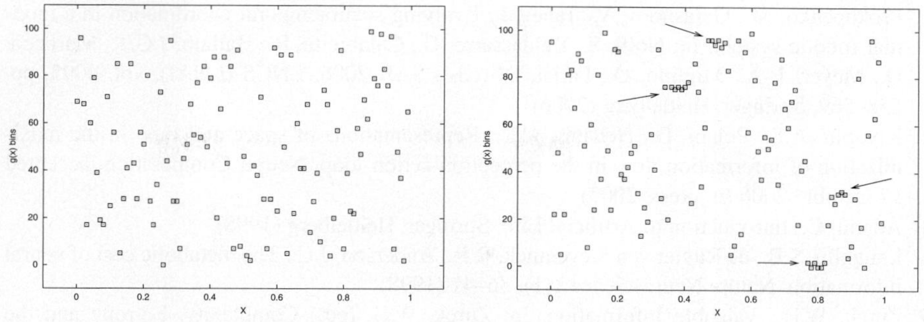

Fig. 6. Left: a random function g at the start of evolution, shown as a sample (Y_{t_0}). No structure is observed. Right: evolved g, shown as a sample (Y_{t_0}). Arrows point to regions of stability.

other co-evolving conglomerates; c) two coupled systems similar to the one analyzed in this work exchange the respective encodings. These directions are under investigation, aiming at understanding of emergence and universality of genetic code. Woese observed that "statistical proteins form the basis of a powerful strategy for searching protein phase space, finding novel proteins" [1]. We believe that further modelling of the evolutionary dynamics in such a space may explain mechanisms resolving Eigen's paradox [19] and leading to convergence on "the lingua franca of genetic commerce" [1].

References

1. Woese, C.R.: A new biology for a new century. Microbiology and Molecular Biology Reviews 68(2), 173–186 (2004)
2. Woese, C.R.: The emergence of genetic organization. In: Ponnamperuma, C. (ed.) Exobiology, pp. 301–341. North-Holland Publishing, Amsterdam (1972)
3. Vetsigian, K., Woese, C.R., Goldenfeld, N.: Collective evolution of the genetic code. Proc. Natl. Acad. Sci. USA 103, 10696–10701 (2006)
4. Woese, C.R.: The universal ancestor. Proc. Natl. Acad. Sci. USA 95, 6854–6859 (1998)
5. Woese, C.R., Fox, G.E.: The concept of cellular evolution. J. Mol. Evol. 10, 1–6 (1977)
6. Barbieri, M.: The organic codes: an introduction to semantic biology. Cambridge University Press, Cambridge (2003)
7. Woese, C.R.: On the evolution of the genetic code. Proc. Natl. Acad. Sci. USA 54, 1546–1552 (1965)
8. Klyubin, A.S., Polani, D., Nehaniv, C.L.: All else being equal be empowered. In: Capcarrère, M.S., Freitas, A.A., Bentley, P.J., Johnson, C.G., Timmis, J. (eds.) ECAL 2005. LNCS (LNAI), vol. 3630, pp. 744–753. Springer, Heidelberg (2005)
9. Prokopenko, M., Wang, P., Valencia, P., Price, D., Foreman, M., Farmer, A.: Self-organizing hierarchies in sensor and communication networks. Artificial Life 11, 407–426 (2005)
10. Prokopenko, M., Gerasimov, V., Tanev, I.: Measuring spatiotemporal coordination in a modular robotic system. In: Rocha, L.M., Yaeger, L.S., Bedau, M.A., Floreano, D., Goldstone, R.L., Vespignani, A. (eds.) Artificial Life X: Proceedings of The 10th International Conference on the Simulation and Synthesis of Living Systems, Bloomington IN, USA, pp. 185–191 (2006)

11. Prokopenko, M., Gerasimov, V., Tanev, I.: Evolving spatiotemporal coordination in a modular robotic system. In: Nolfi, S., Baldassarre, G., Calabretta, R., Hallam, J.C.T., Marocco, D., Meyer, J.-A., Miglino, O., Parisi, D. (eds.) SAB 2006. LNCS (LNAI), vol. 4095, pp. 558–569. Springer, Heidelberg (2006)
12. Klyubin, A.S., Polani, D., Nehaniv, C.L.: Representations of space and time in the maximization of information flow in the perception-action loop. Neural Computation, accepted 17 October 2006 (in press, 2007)
13. Adami, C.: Introduction to Artificial Life. Springer, Heidelberg (1998)
14. Laughlin, S.B., de Ruyter van Steveninck, R.R., Anderson, J.C.: The metabolic cost of neural information. Nature Neuroscience 1(1), 36–41 (1998)
15. Zurek, W.H.: Valuable Information. In: Zurek, W.H. (ed.) Complexity, Entropy and the Physics of Information. Santa Fe Studies in the Sciences of Complexity, Addison-Wesley, Reading (1990)
16. Prigogine, I.: From Being to Becoming: Time and Complexity in the Physical Sciences. W.H. Freeman & Co, San Francisco (1980)
17. Haken, H.: Synergetics, an Introduction: Nonequilibrium Phase Transitions and Self-Organization in Physics, Chemistry, and Biology, 3rd edn. Springer, Heidelberg (1983)
18. Todd, P.M., Miller, G.F.: Exploring adaptive agency ii: Simulating the evolution of associative learning. In: Meyer, J.-A., Wilson, S.W. (eds.) From animals to animats: Proceedings of the First International Conference on Simulation of Adaptive Behavior, pp. 306–315. MIT Press, Cambridge (1991)
19. Eigen, M.: Self-organization of matter and evolution of biological macromolecules. Naturwissenschaften 58(10), 465–523 (1971)

Emergent Phenomena Only Belong to Biology

Hugues Bersini and Christophe Philemotte

IRIDIA - CoDE,
Université Libre de Bruxelles,
CP 194/6 ULB,
50 Avenue F.D. Roosevelt,
B-1050 Brussels,
Belgium
bersini@ulb.ac.be
http://iridia.ulb.ac.be/bersini/
cphilemo@iridia.ulb.ac.be
http://iridia.ulb.ac.be/~cphilemo/

Abstract. This rather philosophical paper discusses the necessary three ingredients which together allow a collective phenomenon to be described as "emergent". First the phenomenon, as usual, requires a group of agents entering in a non-linear relationship and entailing the existence of two semantic descriptions depending on the scale of observation. Second this phenomenon has to be observed by a mechanical observer instead of a human one, which has the natural capacity for temporal and/or spatial integration. Finally, for this natural observer to detect and select the collective phenomenon, it needs to do so in rewards of the adaptive value this phenomenon is responsible for. The presence of natural selection drives us to defend, with many authors, the idea that emergent phenomena can only belong to biology. After a brief philosophical plea, we present a simple and illustrative computer thought experiment in which a society of agents evolves a stigmergic collective behavior as an outcome of its greater adaptive value. The three ingredients are illustrated and discussed within this experimental context.

Key words: emergence, natural selection

1 Introduction

"The whole is more than the sum of its parts" is an expression that continues to feed vivid debates in many scientific circles. While these days, every scientist agrees that two scientific observers (let's call them Mic Jim – the micro observer and Mac Jim – the macro observer) observing the same collective phenomenon but at different spatial and temporal scales are required (for instance, Mic Jim sees and knows the rules of the game of life while Mac Jim only sees the "glider" moving [1,2]) to qualify a phenomenon as emergent, the heart of the disagreement rests on the status of Mac Jim. What is his role, his "raison d'être"? Does his observation testify of any reality outside, or does it boil down to an epistemic facility, a mental compression, summarizing what is really happening outside when

F. Almeida e Costa et al. (Eds.): ECAL 2007, LNAI 4648, pp. 53–62, 2007.

observing the phenomenon longer and/or at a broader scale. In this second case and according to many authors, Mac Jim's situation is considerably weakened since everything he sees and describes can be completely anticipated and fully recovered by Mic Jim, as a result of some spatial and temporal integration. Mic Jim takes the scientific leadership; he is the only one to have a complete and faithful knowledge of the phenomenon, he can explain what is seen by his colleague but not the other way round. Reductionism wins and is the only way true science should progress.

Although certain authors [3,4] want to keep separated a weak version of emergence (epistemic) from a strong one (ontological), so as to stress what is really needed for the strong one, only worthy of interest, we don't see why the weak one deserves any further consideration. Take physics for instance, the queen of science, Mac Jim, when observing the glider in the game of life, behaves in the same classical way as when observing an increase of entropy in an isolated system or the ellipses the planets make in the sky. An entropy or a temperature increase appears to be more than the sum of particles, randomly and furiously agitated, but physics rightly says it is not. The planet ellipses seem to be more than the gravitation force combined to the planet inertial original velocity but physics, once again, says it is not. Although two physic Nobel prizes [5,6] have claimed these last years to perceive in "emergence" the source of a new paradigmatic shift in physics, convincing arguments really lack. So what about the strong one, the only "emergence" to care for, while remaining completely waterproof to mysticism, dualism or vitalism? We believe that a key move needed to restitute its ontological status to emergence is to substitute Mac Jim by a natural (mechanical) double of it. Science strengthens by discarding more and more the part played by human observation in the characterization of the observed phenomenon. We also believe that for the natural observer to detect a natural collective phenomenon, it needs to be selected for in rewards of its adaptive value. Since the only selection filter that science authorizes is the Darwinian one, it justifies why we believe in this paper (with others [7,8,9,10]) that only biology and its theory of natural selection can make an emergent phenomena to exist without any Mac Jim to detect it. A phenomenon will finally emerge in a system once observed and detected by a globalizing mechanical observer for the adaptive capability it provides this system with. In the rest of the paper, we will defend this idea and provide a simple illustration of it through a computational experiment of emergence in insect societies.

2 The Biological Key Ingredients of Emergence

The picture shown in Fig. 1 is taken from the European Swarm-bots project, which is being coordinated in our laboratory [11]. Largely inspired by the capacity of some insect species (such as ants) to assemble in order to accomplish tasks that none of them, alone, is able to accomplish, this project is about small robots that connect together to do as well. For instance, in the picture, you can see two robots that together and assembled can pass over a gap that would make any

Fig. 1. Two robots passing over a gap.

of them fall down if trying alone. One could be attempted to claim that "passing over that gap" is an emergent behavior since it requires a group of robots. However, being engineerized as it is right now, we consider this not to be the case, since a human observer and engineer is required to plan this behavior and to organize the pieces together (here the robots) so as to achieve it. Engineering is top-down while biology is bottom-up and, as a matter of fact, the genuine biological phenomenon, which inspires this engineering version really is emergent. It is so because of two reasons. First any emergent phenomenon needs a natural observer able to integrate in space and time this phenomenon. Here, and as surprising as it could seem, this role is endorsed by the gap, which "observes" and "makes sense" of the phenomenon. In fact, we agree with Yaneer Bar Yam [4] that any definition of emergence requires the presence of two complementary realities: the emerging phenomenon and an environment the phenomenon emerges for, in this case the "gap". However, we feel more uncomfortable with the best example of emergence he proposes as a string of bits including the "parity" one. His favorite example again has too much of a "top-down" and engineering flavor.

The second required ingredient for a phenomenon to be qualified as emergent justifies why and within which non-human context the mechanical observer detects and makes sense of this phenomenon. In biology, natural selection is at play. The observer detects and selects the phenomenon because it provides the system in which this phenomenon is manifest with adaptive capability, often a more economical or robust way to accomplish some task. Here, the insects have to pass over the gap since falling down will simply kill them. No engineer will appreciate a similar destiny for its robots even if, as a reminiscence of real emergence, some evolutionary algorithms often help them to realize the collective task. As Figure 2 illustrates, emergence appears at the crossroad of these three actors: the collective phenomenon per se, the mechanical observer and natural selection. Any of them misses and the whole idea collapses, bringing back emergence to a weak and no more original version of it.

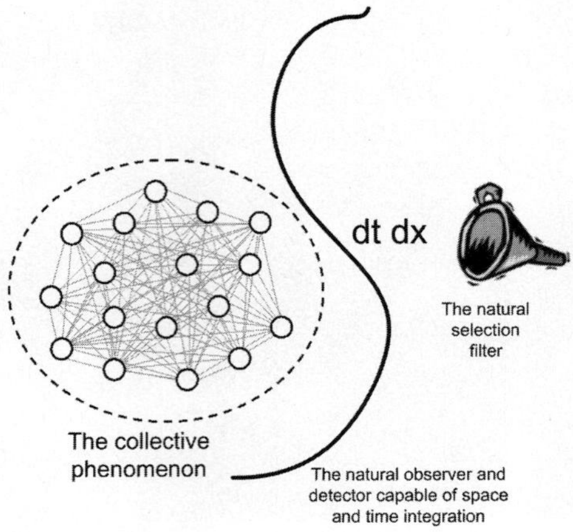

Fig. 2. The three needed ingredients for a collective phenomenon to be qualified as emergent.

We already had the opportunity in previous publications [12] to recognize our intellectual debt to authors like Jim Crutchfield [13,14] and Peter Cariani [10] in our mechanization of Mac Jim, the addition of natural selection in the whole picture goes in line with Maynard Smith and Szathmary's concept of "evolutionary major transitions" [9] and Peter Corning's "synergism hypothesis" [15]. This later defends the idea that "synergistic effects of various kinds have played a major causal role in the evolutionary process, for essentially the functional payoff (mainly in economical terms) these effects were responsible for".

3 The Emergence of Shorter Paths in Insect Societies

We propose to illustrate the three ingredients previously introduced by the following simulation. This simulation is inspired by the ant colonies stigmergic strategy: the selection among many paths of the shortest one in order to link resource locations [16]. In substance, we aim at answering this simple basic question: "Why did ants once decide to communicate by laying down some signal (in the case of ants, it is called pheromones) along their way? Which observer once decided that this could be a very effective communication strategy?". In the following, we describe the developed simulation in order to answer that question. On a bidimensional grid with periodical boundary conditions, a set of cells contains three possible items: an agent (for instance, an ant), a quantity of resources (food, for instance) and a quantity of signals (some pheromones). Each agent is characterized by a further positive quantity called its "vital energy". The agent is alive as long as its "vital energy" does not fall down to zero. If this energy vanishes to zero, the agent dies and disappears from the simulated environment. The quantity of resources in a cell stochastically varies with time. This stochastic variation represents different hardness of the environment. Resources are food for agents. Agents have to "consume" some resources to increase their "vital energy" and survive. Depending on the value of its "vital energy", an agent can be either "hungry" or "not hungry". With respect to its state, the agent can react differently. In the case of the presence of resources, the sated agent can decide not to consume.

A signal is the third possible item a cell can contain. The strength of this signal is represented by a positive quantity which can be increased by agents. With time, this quantity is decreased by a natural and exponential decay. When the agent deposits some signal in its cells, the signal quantity increases by a fixed amount. The behavior of an agent depends on its genotype. This genotype is evolved in time by means of a Genetic Algorithm. This genotype is divided in two parts: the "hungry" and "not hungry" parts. Each part is composed of a same number of genes. The allele of a gene codes a possible behavior. The locus of a gene corresponds to a possible state of an agent. The state of an agent is defined by its "vital energy" and the items contained in its four-cells Von Neumann neighborhood. With respect to this state, the agent acts following the behavior coded in the corresponding gene. Six possible behaviors have been defined. They are:

1. "don't do anything",
2. "randomly move to one of the four neighboring cells",
3. "consume a resource",
4. "go to a neighboring cell that contains a resource",
5. "deposit a signal in the cell", and
6. "move in a neighboring cell selected as a function of the signal contained in the cells".

Obviously, some actions are impossible in certain states. The possible actions are conditioned by the current state of the agent. For instance, an agent can consume resources only if its current cell contains resources. It can move to a neighboring cell with resources only if such a cell effectively exits. It can move in a selected neighboring cell only if the signals distribution in these cells make that possible, i.e. if one of theses cells have the greatest or the smallest quantity of such signals. They are 6 possible cases of the signals distribution among the agent's cell and the four neighboring cells:

1. complete uniformity of the signal quantity among the five cells,
2. signal quantity in the agent's cell greater than the signal quantity in the neighboring cells,
3. signal quantity in the agent's cell equal to the greatest signal quantity in the neighboring cells,
4. signal quantity in the agent's cell in between the greatest and the smallest quantity in the neighboring cells,
5. signal quantity in the agent's cell equal to the smallest signal quantity in the neighboring cell, and
6. signal quantity in the agent's cell smaller than the signal quantity present in the neighboring cells.

The genotype of the agent includes then 2×8 genes: "hungry" and "not hungry" parts and in both cases: resource, not-resource, plus all the 6 cases just described. With respect to its locus (corresponding to a state of the agent), each gene provides a different number of alleles (corresponding to an action of the agent). The size of the search space is rather huge, about 10^{11} possible behavioral patterns. At each time step of the simulation, the situation of all cells is synchronously updated and the action of each agent selected as a deterministic function of its current state ("vital energy", surrounding signals and resources). Every action of an agent entails a lost of "vital energy" depending on the precise action. Even doing nothing is costly. Obviously, the viability of an agent depends on its capacity to rapidly and economically find available resources to be consumed during the time of the simulation. After a simulation, the best alive agents are kept and evolved. Simulations are the way to evaluate genotype. We consider two kinds of simulation. A first one includes a set of homogeneous agents: they all share an identical chromosome. A second one includes a set of heterogeneous agents: they can have different chromosomes.

The evolution of agents' genotypes proceeds as follows. For each run, a population of twenty agents is generated and simulated during twenty thousands time steps. For the homogeneous case, each chromosome provides twenty agents which are simulated. The fitness of an agent is given by the value of "vital energy" at the end of a run. In the homogenous case, the fitness of the chromosome is defined by the average over the twenty agents' fitness. The used genetic operators are very classical. The five best agents are selected. They are exactly copied in the next generated population. From them, fifteen children are created following a uniform crossover and a gene mutation.

The homogeneous case is the simplest to understand. Which unique behavioral pattern allows the set of agents to live longer in average? We were rather satisfied to discover that the behavior consisting in 1) after the consumption of a resource, depositing a signal all along the way together with 2) in the absence of any resource, following the signal gradient, turned out to be the fittest and the most stable chromosome through the GA generations. A snapshot of the simulation obtained with the "best" chromosome is shown in Fig. 3, where the fittest strategy can be clearly seen and understood, discovering the signals left by the agents around the resources. The heterogeneous case pushed farther the evolution principles. In this case, each agent with different chromosome has to struggle for life. A direct competition between each represented chromosome is at play. Indeed, the resources are not inexhaustible. If an agent does not have a competitive behavior, and even if it can find resources, the resources will be quickly consumed by fitter agents. Signals could be used by defectors, and altruist agents would not be rewarded. In fact, the exploitation of signals is both a selfish and collaborative behavior. The agent which deposits and smells signals bounds its research of resources in the whole environment. The probability to find again a source of resources is then increased. By selection pressure, this behavior is transmitted to some children. A nearly homogeneous sub-population is then obtained such as in the homogeneous case. One can clearly see the signal

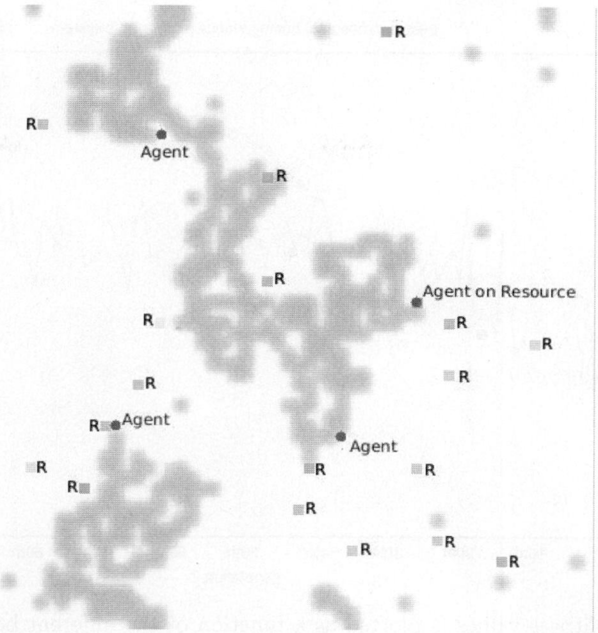

Fig. 3. Snapshot of the simulation. "R" indicates a resource, "Agent" an agent and the grey trace is the signals left by the agents.

trace and above all this collective cooperative strategy which really and "strongly emerges" as the result of the Darwinian competition.

Therefore, here the role of the "mechanical observer" is being played by the "vital energy" which integrates in time the collective effect of the agents. Indeed, the presence of the signal and this stigmergic cooperation among the agents is responsible for the reduction in length of the paths and thus in the energy to be consumed to reach the resources. As required in the previous sections, the three ingredients are all present to attribute the "emergent" qualification to this stigmergic, signal-based and cooperative behavior. Figure 4 shows the fitness of the best behavior in the homogeneous case. It is hard to really distinguish among many other well fitted behaviors such as, for instance, "don't move at all", which can be rewarding in some cases. Figure 5 is the frequency of appearance of the actions "deposit signal" as the evolutionary algorithm progresses and successive generations of agents are evaluated (in the heterogeneous case). We can see how this action stabilizes in time, an even better indication of its adaptive value.

4 Conclusions

Among others, two biologists [7,8] are acknowledged for the marriage they want and celebrate between self-organization phenomena coming from physics and the natural selection too much influent and systematic in biology. For them, natural

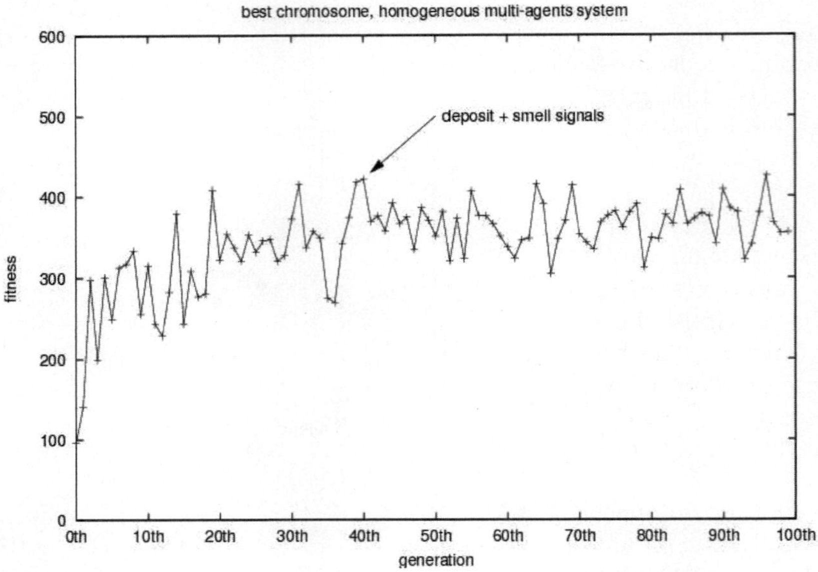

Fig. 4. The fitness values is plotted as a function of the different behavioral patterns in the homogeneous case. The best behavior is slightly better than the others. At the 40th generation, a "deposit + smell signal" behavior appeared and slightly better than older generations. This behavior remains stable during the next evolved generations.

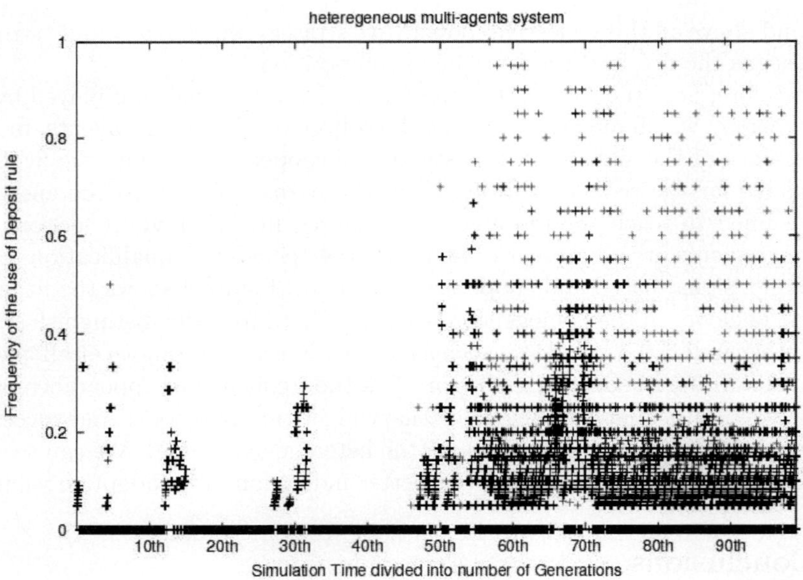

Fig. 5. This figure plots the use frequency of the action "deposit signal" as evolution progresses in time in the case of heteregeneous multi-agents system. After about the fiftieth generation, the "deposit signal" action is always used by the best agents.

selection should boil down to an opportunistic parametrisation of agents which when interacting in a non-linear way show a spontaneous tendency for interesting and complex collective behavior. They consider that this articulation between the two sources of order which are natural selection and physical self-organization has to be readjusted in favor of physics. Whatever road to be taken, coming from physics and viewing in this exotic natural selection a way to detect and select some of these spontaneous collective phenomena or, coming from biology, and viewing in the existence of self-organisation the missing explanation behind that kind of complex behaviours natural selection alone can not assume, our emergence in exactly there at the meeting point of these two roads.

We have insisted in this paper on naturalizing Mac Jim, the macro observer, or substituting him by a "natural double", in order to reinforce the status of emergence. However, Mac Jim, as a human, still exists and although completely eclipsed by Mic Jim and the way he understands the observed phenomenon, we might still try to explain and justify his existence and to conciliate in part the weak and strong forms of emergence. There are two ways. The first is obvious and not so appealing. Something can be interested both for the "eyes" of natural selection and for the human eyes. Short paths are beneficial for the viability of insects but can also easily be detected by human observer. However, the visual salience does not always go hand in hand with any adaptive value. Glider or planet ellipses are interesting or surprising to the eyes but don't see themselves attributed any adaptive value. On the other hand, some interesting biological collective behavior like network effects of gene or proteins (robustness, small-world...) are hardly accessible to the human eyes. A much more promising second way to explain why Mac Jim describes in a new way a collective phenomenon, while increasing the realism of this epistemic version of emergence, is to accept human perceptive apparatus as being calibrated by natural selection. Not only natural selection makes our cognition eager to abstract the outside world in space and time but, even more, some authors insist in explaining the filtering mechanism of neural processes in Darwinian terms [17,18]. The simple instantaneous process of perception and the learning in life time to better perceive this way is akin to a selectionist mechanism with the synaptic plasticity contributing to favor one neural pattern rather than another one in response to a stimulus. The strong version which requires the presence of a mechanical observer calibrated by natural selection slips into the weak version if the human that we firmly try to discard turns out at last to be this same very well fitted mechanical observer.

References

1. Poundstone, W.: Recursive Universe: Cosmic Consequences and the Limits of Scientific Knowledge. NTC Publishing Group, Lincolnwood (1985)
2. Kubik, A.: Toward a formalization of emergence. Artificial Life 9(1), 41–65 (2003)
3. Bedeau, M.: Weak emergence. Philosophical perspectives: Mind, causation and world 11, 375–399 (1997)

4. Bar-Yam, Y.: A mathematical theory of strong emergence using multiscale variety. Complex 9(6), 15–24 (2004)
5. Anderson, P.: More is different. Science 177, 393–396 (1972)
6. Laughlin, R.: A different universe: reinventing physics from the bottom down. Basic Books, New York (2005)
7. Goodwin, B.: How the Leopard Changed Its Spots: The Evolution of Complexity. Princeton University Press, Princeton (March 1994)
8. Kauffmann, S.: At Home in the Universe: The Search for Laws of Self-Organization and Complexity. Oxford University Press, Oxford (Oct. 1996)
9. Maynard-Smith, J., Szathmary, E.: The Major Transitions in Evolution. Oxford University Press, New York (1997)
10. Cariani, P.: Emergence of new signal-primitives in neural networks. Intellectica 2, 95–143 (1997)
11. Groß, R., Bonani, M., Mondada, F., Dorigo, M.: Autonomous self-assembly in swarm-bots. IEEE Transactions on Robotics 22(6), 1115–1130 (2006)
12. Bersini, H.: Whatever emerges should be intrinsically useful. In: Artificial life 9, pp. 226–231. MIT Press, Cambridge (2004)
13. Crutchfield, J.: Is anything ever new? considering emergence. In: Cowan, G., Pines, D., Melzner, D. (eds.) Integrative Themes. Santa Fe Institute Studies in the Sciences of Complexity, vol. 19, Addison-Wesley, Reading (1994)
14. Crutchfield, J., Mitchell, M.: The evolution of emergent computation. Proceedings of the National Academy of Science 23, 103 (1995)
15. Corning, P.: The re-emergence of "emergence": A venerable concept in search of a theory. Complexity 7(6), 18–30 (2002)
16. Dorigo, M., Maniezzo, V., Colorni, A.: The ant system: Optimization by a colony of cooperating agents. IEEE Transactions on Systems, Man, and Cybernetics Part B: Cybernetics 26(1), 29–41 (1996)
17. Calvin, W.: The brain as a darwin machine. Nature 330, 33–34 (1987)
18. Edelman, G.: Neural Darwinism. Oxford University Press, Oxford (1990)
19. Baas, N.: Emergence, hierarchies and hyper-structures. In: Langton, C.G. (ed.) Artificial Life III, Boston, MA, USA, pp. 515–537. Addison-Wesley Longman Publishing, Reading (1994)

Genotype Editing and the Evolution of Regulation and Memory

Luis M. Rocha and Jasleen Kaur

School of Informatics, Indiana University
Bloomington, IN 47406, USA
rocha@indiana.edu
http://informatics.indiana.edu/rocha

Abstract. Our agent-based model of genotype editing is defined by two distinct genetic components: a coding portion encoding phenotypic solutions, and a non-coding portion used to edit the coding material. This set up leads to an indirect, stochastic genotype/phenotype mapping which captures essential aspects of RNA editing. We show that, in drastically changing environments, genotype editing leads to qualitatively different solutions from those obtained via evolutionary algorithms that only use coding genetic material. In particular, we show how genotype editing leads to the emergence of regulatory signals, and also to a resilient memory of a previous environment

1 Introduction: RNA Editing

RNA Editing [Bass, 2001] refers to the post-transcriptional alteration of genetic information. It occurs in various forms such as insertion, deletion, or substitution. It can be implemented via non-coding RNAs (ncRNAs) such as *guide RNA's* or via enzymes (e.g. *adenosine deaminase acting on RNA* (ADAR), also known as *RNA Editase*) In either case, genetic information is altered after transcription and before translation (for an overview see [Huang et al., 2007]).

Previously we quantitatively established the advantages of genotype editing against the canonical evolutionary algorithm in various static and dynamic environments (e.g. [Huang et al., 2007]). Here, using our *Agent-Based Model of Genotype Editing* (section 2) in drastically changing environments (section 3), we focus instead on the qualitatively different evolutionary solutions attainable via genotype editing. Specifically, we show how genotype editing leads to the emergence of regulatory signals that allow agents to better adapt to radically different environments (section 4). We also show how the inclusion of non-coding genetic material, with the function of editing coding material, allows agents to evolve a memory of previous environments—a capacity not attainable by the canonical evolutionary algorithms which use only coding genetic material (section 5).

2 Modeling Genotype Editing

The Genetic Algorithm (GA) [Holland, 1975] is an idealized model of natural selection—and the canonical evolutionary algorithm. In a traditional GA, the

F. Almeida e Costa et al. (Eds.): ECAL 2007, LNAI 4648, pp. 63–73, 2007.

code between genotype and phenotype is a direct and unique mapping . In biology, however, before a gene is translated into a protein it may be altered, namely by functional or non-coding RNA (ncRNA) used for editing or other regulatory functions. To study and exploit the biological principle behind RNA Editing, we have introduced an *agent-based model of genotype editing* (ABMGE) [Rocha et al., 2006]. In this model, the agents in the population are defined by an artificial genome that contains functionally and operationally distinct *coding* and *non-coding* components: respectively, the *codome*, encoding solutions to a particular fitness function or environment, and the *editome*, producing editors which act on the coding component. Our goal is to understand the influence of editing, as a genomic (pre-translation) phenomenon, in the evolutionary process. Therefore, we have (1) explicitly separated an editome from a codome to better test its relative importance (rather than attempting to have it emerge from a common artificial genome), and (2) we strip our model of any post-translation dynamics (i.e. development) or epigenetic phenomena. We understand that both of these design choices may be unrealistic, but they are a reasonable and necessary starting point to understand editing as a genomic phenomenon.

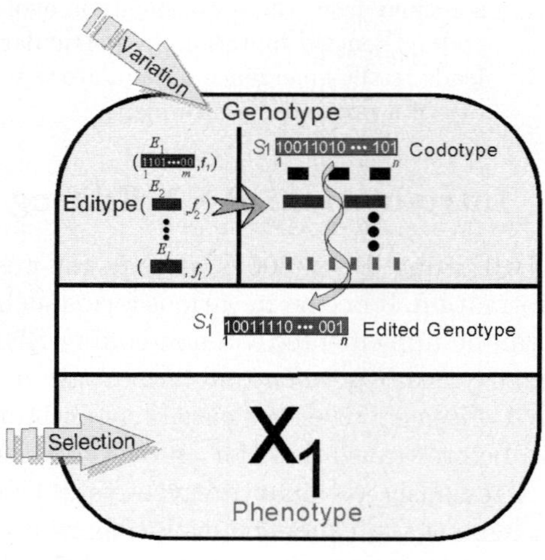

Fig. 1. Individual Agent in the ABMGE.

Figure 1 depicts an agent in the ABMGE. In each generation, the coding component of an agent's genotype, the *codotype* may be **stochastically** edited by the agent's non-coding genotype, its *editype*, and produce a phenotype different from what is encoded in the codotype. The *codotype* of an agent is a n-bit string (or sequence) S, whereas the *editype* is a family of r editors each defined by a 3-tuple (E_j, F_j, v_j). E_j is a m-bit string ($m \ll n$), which may bind to S by exactly matching a substring (bit-pairing). F_j is the *editing function* that specifies how the editor edits the codotypes it matches, e.g. by inserting into or deleting bits from S. In this model, the length n of the codotype is fixed. Therefore, when x bits are inserted, the sequence is shifted to the right discarding x bits on the right end of the string. When x bits are deleted, the sequence is shifted to the left, and x bits are randomly generated on the right end of the string. v_j is the *concentration* of the editor and denotes the probability that the codotype string S encounters editor j before translation in each generation. When an editor j encounters the codotype string S (with probability v_j), it checks the codotype from left to right, one bit position at a time, performing its editing function F_j every time a match occurs between its editor substring E_j and S. Thus, **the same codotype may be edited differently since editor concentration is a stochastic parameter**.

Indeed, the same genotype may produce different phenotypes in different generations, or even in the same population if clones exist. For instance, a codotype may fail to be edited in one generation (especially when the concentration of editors is small), edited by a single editor in the next, or edited by every editor in yet another generation. Table 1 describes the ABMGE algorithm.

Notice that only the codome is used to encode phenotypic attributes. The editome is, in this sense, "non-coding"; its role is to change genetic information ontogenetically, more specifically, to model the post-transcriptional, pre-translation process of genotype editing. Finally, also note that **edits are not inheritable.** While agent fitness is calculated using the phenotype produced from the edited codotype, what is inheritable, and subjected to variation, is the unedited genotype (codotype plus editype).

Table 1. The ABMGE Algorithm

1. Randomly generate an initial population of l agents, each agent consisting of a codotype (a n-bit string) and an editype (a family of r editors (E_j, F_j, v_j)).
2. Edit each agent's codotype S: apply each editor with probability v_j; If E_j matches S at any position running from left to right, edit S with function F_j
3. Evaluate the fitness of the edited genotype of each agent.
4. Repeat until l offspring have been created. a. select a pair of parent agents for mating; b. apply codotype variation operators (mutation and crossover); c. apply editype variation operators (editor mutation and crossover).
5. Replace the current population with the new one and go to step 2.

When two parents are selected for reproduction in our algorithm (step 4 in table 1), in addition to variation of codotypes as it is commonly done in the GA (mutation and crossover), the editype is also subjected to variation. In the current implementation, variation is only applied to editor strings E_j, while editing functions F_j and concentrations v_j remain unchanged. We will consider variation on these parameters in future work, though these were fixed not only to reduce the number of evolving parameters, but also to model physical characteristics of editors not amenable to evolution.

Editype mutation is implemented on editor bit-strings as usual: with a bit-flipping probability, P_{EdMut}, for each bit of an editor string per generation—P_{EdMut} is independent from the codotype bit-mutation probability: P_{Mut}. *Editype crossover*, is implemented as an exchange of editors between a pair of parent agents. We start with two parent agents a_1 and a_2, with r_1 and r_2 editors in their editypes, respectively. From this pair of parent agents, two offspring agents, a_3 and a_4, are produced whose editypes also contain r_1 and r_2 editors, respectively. However, x editors, chosen randomly from the editype of each agent, are swapped between the parent agents to produce the offspring, where $x \in [1, MIN(r_1, r_2)]$. Editype crossover occurs with a probability $P_{EdCross}$, which is independent from the codotype (one-point) crossover probability, P_{Cross}.

3 Testing on Drastic Environmental Oscillations

In previous work we showed that genotype editing outperforms the GA in many static and dynamic environments [Huang et al., 2007]. Our goal here is to understand how this advantage comes about, especially under drastic environmental changes. Therefore, we focus on a toy fitness function to better understand the adaptations enabled by genotype editing.

Table 2. Small royal road function SRR_1

t_1 =	11111***********************************;	c_1 =	10
t_2 =	*****11111******************************;	c_2 =	10
t_3 =	**********11111*************************;	c_3 =	10
t_4 =	***************11111********************;	c_4 =	10
t_5 =	********************11111***************;	c_5 =	10
t_6 =	*************************11111**********;	c_6 =	10
t_7 =	******************************11111*****;	c_7 =	10
t_8 =	***********************************11111;	c_8 =	10

The *small Royal Road* SRR_1, as depicted in Table 2 is a miniature of the class of the "Royal Road" functions [Mitchell et al., 1992]. This function is an idealized fitness environment defined by a set of schemata $T = \{t_1, \ldots, t_8\}$. The fitness of a bit string (codotype) S is defined as $F(S) = \sum_{t \in T} c_t \sigma_t(S)$, where each c_t is the value assigned to schema t as defined in Table 2; $\sigma_t(S) = 1$ if schema t exists in S and 0 otherwise. The single optimum fitness for SRR_1 is obtained by the string with 40 1's, and its value is 80. Consider another Small Royal Road function, SRR_0, in which each schemata is comprised of five 0's rather than 1's as SRR_1 above, but with all other parameters the same as SRR_1. With these two functions, we create the *oscillatory royal road* (ORR), which oscillates between SRR_1 and SRR_0 at every p generations.

We contrasted the GA with two versions of the ABMGE: with (ABMGE^C) and without editype crossover. The experiments we report here with ORR use binary tournament selection, and a population of 40 agents over 4000 generations for 50 runs. Codome variation (in both the ABMGE and the GA) is implemented with one-point crossover and mutation rates of $P_{Cross} = 0.7$ and $P_{Mut} = 0.005$, respectively—the best values we had previously found for the GA (see [Huang et al., 2007]). In every run, all editome parameters (of the ABMGE) are randomly generated as follows: $r \in \{1, \cdots, 5\}$ (number of editors for each agent); $m \in \{2, \cdots, 4\}$ (size of editor strings E_j); $v_j \in [0, 1]$ (editor concentration); F_j, the editor functions, insert or delete $x \in \{1, \cdots, 3\}$ bits which are randomly generated when the editor is created but fixed thereafter for each run. In addition to various editype mutation and crossover probabilities ($P_{EdMut} \in \{0.01, 0.05\}$ and $P_{EdCross} \in \{0, 0.3, 0.5, 0.7, 0.9\}$), we tested different oscillation periods ($p \in \{50, 100, 200, 250\}$). Figure 2 depicts the mean best-so-

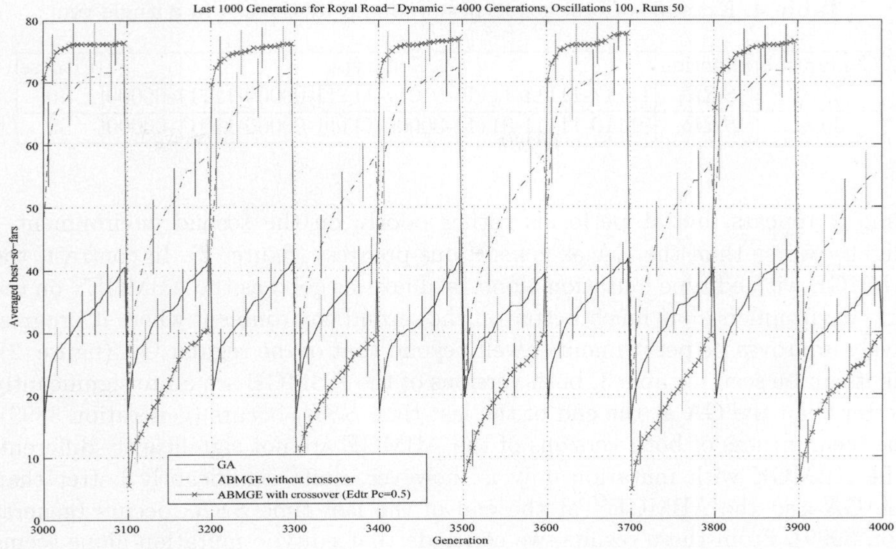

Fig. 2. Performance of GA, ABMGE without crossover, and ABMGE with crossover ($P_{EdCross} = 0.5$) on *ORR*, $p = 100$, 50 runs, 4000 generations (last 1000 shown).

Table 3. Mean fitness and 95% confidence interval for 50 runs of the *GA*, *ABMGE*, and $ABMGE^C$ with the *ORR* at $p = 100$, at the end of the last two oscillations.

		GA		ABMGE		ABMGE^C	
Generation	Function	Mean Fit.	Conf. int.	Mean Fit.	Conf. int.	Mean Fit.	Conf. int.
3899	SRR_1	43.4	3.18	72	4.22	76.6	2.9
3999	SRR_0	38	3.3	55.6	4.56	29.2	5.1

far fitness[1] (last 1000 generations) for $P_{EdMut} = 0.05$ and $P_{EdCross} = 0.5$ and period $p = 100$. Table 3 details the performance of the three algorithms tested, at the end of the last two environment oscillations (generations 3899 and 3999).[2]

The GA degrades its performance in time; at the end of 4000 generations its mean performance on both environments eventually reaches the same level (close to 40). This means that the GA ultimately converges to a population of genotypes with a balanced number of all-1 and all-0 schema. Table 4 depicts the best genotypes attained by the GA in both environments (SRR_1 and SRR_0) at the end of a run. This behavior highlights the difficulty a canonical evolutionary algorithm faces in a drastically changing environment.

As for the ABMGE, the version with both editype crossover and mutation (ABMGE^C) performs best on the first fitness environment (SSR_1), and every

[1] Fitness of the best individual that has been seen thus far in a given environment period; vertical bars represent 95-percent confidence intervals.

[2] Results for $p = \{50, 200\}$ at `http://informatics.indiana.edu/rocha/editing`.

Table 4. Example of best genotypes produced by the *GA* in a single run

Generation	Function	Genotype	Fitness
3899	SRR_1	11111-11111-11111-01011-11111-00000-11111-00000	50
3999	SRR_0	01110-11111-01111-00000-11111-00000-11011-00000	30

time it repeats, but it performs rather poorly on the second environment—slightly worse than the GA as generations progress (figure 2). In contrast, the ABMGE with editype mutation alone, is almost as good as the ABMGE^C on the first environment, but much better on the second environment where it progressively improves its performance—well beyond that of the regular GA (figure 2). As it can be seen in table 3, both versions of the ABMGE are quite significantly better than the GA at the end of the last time SSR_1 occurs (generation 3899); the mean fitness of both versions of the ABMGE are not significantly different. The ABMGE with mutation only is, however, quite significantly better than the GA and the ABMGE^C at the end of the last time SSR_0 occurs (generation 3999). From these results, we conclude that editype mutation alone seems to offer a much better agent architecture in drastic environmental changes. Indeed, the ABMGE with editype mutation alone, leads to agents which do well on both dramatically different fitness landscapes—as opposed to the GA which ultimately settles to agents that are mediocre in both. We discuss next why and how the ABMGE manages to outperform the GA and why the two variations of the ABMGE perform so differently in this dynamic environment.

4 Evolving Simple Regulatory Signals

Table 5 depicts one of the best agents evolved by generation 3899 (SRR_1) with the ABMGE with mutation alone. The second editor of this agent, with substring $E_2 = \{11\}$ inserts two 1's after matching the codotype: ($\{11\} \rightarrow 11$). Once the substring $\{11\}$ is found in the codotype S, every bit in the remaining portion of S, to the right of the match position, is edited to "1", as 11 is guaranteed to occur at the next bit position for E_2 to match again. Moreover, if 11 occurs in the first 2 bit positions of the codotype, the entire codotype is edited to an all-1 bit-string. We refer to this process as *repetitive massive insertion* (RMI), which is similar to the massive u-insertion observed in Trypanosoma in nature [Bass, 2001]. Notice that since this editor exists with a high concentration (0.966) and the first two positions of the codotype are 11, most of the time (97%) the codotype is massively edited into the maximum fitness value of SRR_1.

Once the environment changes to SRR_0, this agent evolves to deal with the completely new environment quite effectively. Table 6 describes one of its descendants in the last generation of the run (at the end of SRR_0). We can see that the editors are the same, but now the codotype has changed to contain more 0's, and no $\{11\}$ substring. This allows the first insertion editor ($\{000\} \rightarrow 0$) to come into action while simultaneously preventing the second editor to act. The first editor, also using RMI, converts to 0 every allele from the match position

Table 5. Example Agent evolved at generation 3899 (end of last SSR_1 period) with the *ABMGE* with editype mutation alone.

Codotype	11011-10000-01000-01001-00000-10100-00000-11000		
Editype	Editor Subtrings Functions Concentrations	{000} insert 1 bit: {0} 0.383	{11} insert 2 bits: {11} 0.966
Edited Genotype	11111-11111-11111-11111-11111-11111-11111-11111		
Comparison	Hamming Distance = 28	Unedited Fitness = 0	Edited Fitness = 80

Table 6. Example Agent evolved at generation 3999 (end of last SSR_0 period) with the ABMGE with editype mutation alone; descendent from agent in table 5

Codotype	00000-00000-00000-00000-00000-00000-01000-00100		
Editype	Editors Functions Concentrations	{000} insert 1 bit: {0} 0.383	{11} insert 2 bits: {11 0.966
Edited Genotype	00000-00000-00000-00000-00000-00000-00000-00000		
Comparison	Hamming Distance = 2	Unedited Fitness = 60	Edited Fitness = 80

all the way to the right end of the codotype, as 000 is guaranteed to occur at the next position. Therefore, if the codotype contains 000 in its first 3 bit positions, it is edited to an all-zero bit-string (the maximum fitness phenotype in SRR_0).

The way the agents of tables 5 and 6 use RMI, makes the first (leftmost) bit positions of their codome most important. Indeed, without us pre-specifying such a role for these bits, simple "promoter" or "regulatory" signals emerge in agents with genotype editing. With the appropriate signal at the start of the codotype ({000} or {11}), any codotype can be edited to produce an all-0 or all-1 phenotype. Moreover, once RMI occurs via one of the editors, the other editor can no longer act as it will no longer find matching substrings. Therefore, evolutionary pressures on the codotype in environment SRR_0 will quickly lead to the appearance of 000 "regulatory" signals towards the left end of the codotype, and when in environment SSR_1 to the appearance of 11 such signals.

Notice that if editors were equally likely, all bit positions to the right of these "regulatory" signals would be largely neutral. But since editors are randomly generated at the start of the simulation, this is rarely the case. In the case of the agents of tables 5 and 6, the first editor has a lower concentration (0.383) than the second (0.966) so it will lead to RMI much less often. Evolution compensates this with a bias towards more zeroes in the codotype; in environment SRR_0 (table 6), the fitness of the unedited codotype is fairly high (60) because it contains a large majority of zeroes. Thus, even if editing does not occur, the agent does well in the SRR_0 environment. Moreover, even in environment SRR_1, the codotype of the agent (in table 5) contains a majority of 0's (28 0's for 12 1's). As long as the "promoter" signal {11} is present in the leftmost bits, the overwhelming majority of the time (97%), the second editor will lead to the optimal phenotype in this environment. This makes it easier, once in environment SRR_0 again, to drfit to a majority zero codotype with no {11} signals.

It is interesting to note at this point, that with the oscillation of environments at every 100 generations, the GA with the same parameters gets stuck on evolving agents with a balanced number of schema with 1's and 0's—and thus mediocre fitness values in both environments. In contrast, the emergence of regulatory signals on the left side of the codotype of the ABMGE agents, allows them to quickly produce maximum (or at least high) fitness phenotypes in both environments. This way, genotype editing as modeled here, leads to the emergence of a functionally distinct role for a small substring of the codotype of agents. In this case, with the operational constraints of left to right decoding, we observe that the first few bits of the codotype instantiate a regulatory box whose (allele) value or signal leads to completely different phenotypes able to cope well with the two drastically different environments. When environments change, evolution only needs to "re-write" these signals, rather than the entire codotype. It is important to emphasize that the regulatory-signal behavior of the agents of tables 5 and 6 is not rare. RMI is observed in most best agents evolved in these conditions, with some variations (such as editor mutation knocking out detrimental editors in the appropriate environment). Due to space restrictions we do not show additional agents here, as well as a discussion of emergent signals in other fitness environments which we will leave for future work.

5 Memory and the Value of "Junk" Non-coding Genotype

Another interesting behavior observed with both versions of the $ABMGE$ is the quick recovery of a high level of performance, every time the environment changes back to the first environment presented to the population (SRR_1). Table 7 shows the mean best-so-far fitness at the last two environment transitions. When the environment changes from SRR_0 to SRR_1 (generation 3800), the traditional GA achieves a mean best-so-far performance of only 19.4, whereas the $ABMGE$ without editype crossover achieves a significantly higher 48.6, which is in turn significantly lower than the mean best-so-far performance of 65.8 achieved by the $ABMGE^C$ with editype crossover; this behavior is observed throughout most of the simulation every time the first environment repeats (figure 2). This indicates that an editome enables the agent population to preserve a ***memory*** of the first environment it encountered—though editype crossover preserves such memory much more effectively. Interestingly, once the environment changes from SRR_1 to SRR_0 (table 7), the $ABMGE^C$ observes the worst performance setback to 6.2 (from 76.6, in table 3), which is significantly lower than both the GA (17.2 from 43.4) and the $ABMGE$ with editype mutation alone (21.8 from 72). Thus, while it is clear that genotype editing allows agents to evolve a memory mechanism of the first environment, we still need to investigate why editype crossover leads to such a poor performance on the second environment?

Table 7. Mean fitness and 95% confidence interval for 50 runs of the *GA*, *ABMGE*, and $ABMGE^C$, at the start of the last two environment oscillations.

		GA		ABMGE		$ABMGE^C$	
Generation	Transition	Mean Fit.	Conf.	Mean Fit.	Conf.	Mean Fit.	Conf.
3800	$SRR_0 \rightarrow SRR_1$	19.4	2.40	48.6	7.83	65.8	5.86
3900	$SRR_1 \rightarrow SRR_0$	17.2	3.30	21.8	5.38	6.2	2.14

After observing the best agents evolved, we conclude that the $ABMGE^C$ performs rather poorly on the second environment because it is particularly successful on the first—as it also is in most static (more complex) environments already tested [Huang et al., 2007]. Indeed, our editype crossover is especially good at spreading the best editors discovered through the entire population. Since editype crossover allows agents to swap editors, it quickly leads to the evolution of agents which contain exclusively very good editors for the first environment encountered. Table 8 shows a typical best agent at generation 3899 (the end of the last time SRR_1 occurs). As we can see, every single editor this agent contains is capable of RMI and occurs in high concentration. Therefore, the overwhelming majority of the time, the codotype is edited into the maximum fitness configuration (all-1).

Table 8. Agent at generation 3899 (end of last SSR_1 period) evolved with $ABMGE^C$.

Codotype	11110-10101-01111-00001-00001-10010-01000-00010			
Editype	Editors Functions Concentrations	{111} ins 1 bit: {1} 0.841	{111} ins 1 bit: {1} 0.866	{111} ins 1 bit: {1} 0.885
Edited Genotype	11111-11111-11111-11111-11111-11111-11111-11111			
Comparison	Hamm. Dist. = 23	Unedited Fitness = 0		Edited Fitness = 80

Once the environment switches to SRR_0, however, because the population of agents contains only editors that are particularly suited for the first environment, and since editor functions and concentrations are fixed at the start of each run, the best the $ABMGE^C$ can do is to evolve agents that knock-off the editors (via editor mutation) or mutate the codotype to be immune to them. Such an agent (descending from the agent in table 8) is displayed in table 9. This behavior is quite different from what we observed with the *ABMGE* without editype crossover in section 4. Indeed, because in that case agents cannot swap editors, many useless or neutral editors hitchhike with advantageous ones in the first environment. Later on, when the environment changes, these unused editors may become useful in the new environment (see agents of tables 5 and 6).

Hitchhiking editors are a type of "junk" material in the genotype of agents. In the first environment, only some of the editype (the useful editors) gets used. Later on, when the environment changes, new environmental pressures can turn

Table 9. Example Agent descending from agent in table 8 evolved at generation 3999 (end of last SSR_0 period) with the $ABMGE^C$.

Codotype	11111-00000-00000-00000-00000-00110-00000-00000			
Editype	Editors	{101}	{101}	{101}
	Functions	ins 1 bit: {1}	ins 1 bit: {1}	ins 1 bit: {1}
	Concentrations	0.841	0.866	0.885
Edited Genotype	11111-00000-00000-00000-00000-00110-00000-00000			
Comparison	Hamm. Dist. = 0	Unedited Fitness = 60		Edited Fitness = 60

"junk" editype into useful editing material. Interestingly, while the editors that are useful in the first environment may get knocked-off or ineffective in the second, once the environment changes again, they can quickly become useful once more—thus granting the *ABMGE* without editype crossover a means to evolve agents with a quickly recoverable memory of both environments.

6 Discussion

Ours is an effort to investigate organization principles enabled by non-coding DNA in general, and genotype editing in particular. Our goal is above all to understand how genotype editing works and what kind of search process it leads to. Here, using only the ORR function, we report three novel observations:

1. **The evolution of "promoter signals"** for editing regulation (section 4), which in this case resulted in massive insertion reminiscent of massive U-insertion RNA editing in Trypanosomes [Bass, 2001]. The emergence of a "promoter box" in the genome of our agents shows us that non-coding RNA may have played an essential role in the origin of gene regulation.
2. **The emergence of memory of previous environments.** Whereas the GA must start over every time the environment changes, agents with genotype editing quickly recover good levels of performance once previous environments return (section 5). Thus, agents with genotype editing are better equipped to deal with changing environments.
3. **The value of redundancy or "junk" editome hitchhiking** (section 5). When variation and selection of editypes is very effective (e.g. our editype crossover mechanism), only the best editors for a given environment survive leading to the best memory of the first environment encountered, but poor performance once the environment changes. But when "junk" editome material is allowed to hitchhike, it grants agents additional genetic material for future regulation in the second environment.

One could argue that that editors leading to RMI are useful in the case of our *ORR* fitness function, but not necessarily in other fitness functions. In this case, the optimal phenotypes are obtained with all-zero or all-one codotype configurations. Therefore, they are especially amenable to the repetitive insertion that evolution exploited. Nonetheless, it is obvious that any other repetitive pattern (e.g. 1010101...) can be reached in the same manner. Moreover, in principle, a large enough set of editors can edit a given codotype into any desirable sequence. Therefore, genotype editing is by no means restricted to the sort of homogeneous repetitive insertion useful for the *ORR*. One can easily conceive a set of editors that use repetitive insertion only in a portion of the codotype in tandem with more localized and specialized edition (namely using deletion or non-repetitive insertion). Indeed, in our previous work we showed that the AB-MGE is also advantageous in many other dynamical environments considerably more complicated than the *ORR* [Huang et al., 2007]. In such cases, genotype editing may have allowed evolution to discover regulatory signals leading to more sophisticated repetitive behavior other than "all-0" or "all-1". We are pursuing a detailed analysis of editing in such functions to report in future publications.

We conclude that genotype editing offers a significantly distinct, and evolutionarily advantageous biological design principle. This advantage is particularly interesting in dynamic environments as agents become better equipped (using emergent regulatory signals and memory) to deal with changing conditions. While our highly idealized model does not capture the reality of biology, it implies that the process of RNA editing in nature is, likewise, advantageous in evolution. Our results emphasize the importance of genetic regulation by non-coding genetic components, offering another piece of conceptual evidence that it plays an essential role in phenotypic development and evolution.

Acknowledgements. We are grateful to Chien-feng Huang and Ana Maguitman for sharing code used on some of the experiments. Luis M. Rocha is partially funded by NSF (BCS-0527249). We are also grateful to the FLAD Computational Biology Collaboratorium at the Gulbenkian Institute in Oeiras, Portugal, for providing facilities used to conduct part of this research.

References

Bass, 2001. Bass, B.L.: *RNA* Editing. Frontiers in Molecular Biology Series. Oxford University Press, Oxford (2001)

Holland, 1975. Holland, J.H.: Adaptation in Natural and Artificial Systems. University of Michigan Press, Ann Arbor (1975)

Huang et al., 2007. Huang, C.F., Kaur, J., Maguitman, A.G., Rocha, L.M.: Agent-based model of genotype editing. Evolutionary Computation 15(3) (In press, 2007)

Mitchell et al., 1992. Mitchell, M., Forrest, S., Holland, J.: In: ECAL (1992)

Rocha et al., 2006. Rocha, L.M., Maguitman, A., Huang, C.-F., Kaur, J., Narayanan, S.: An evolutionary model of genotype editing. In: Artificial Life X, pp. 105–111. MIT Press, Cambridge (2006)

Investigating the Emergence of Phenotypic Plasticity in Evolving Digital Organisms

Jeff Clune, Charles Ofria, and Robert T. Pennock

Digital Evolution Lab, Michigan State University, 3114 Engineering Building,
East Lansing, MI, USA
{jclune, ofria, pennock5}@msu.edu

Abstract. In the natural world, individual organisms can adapt as their environment changes. In most *in silico* evolution, however, individual organisms tend to consist of rigid solutions, with all adaptation occurring at the population level. If we are to use artificial evolving systems as a tool in understanding biology or in engineering robust and intelligent systems, however, they should be able to generate solutions with fitness-enhancing phenotypic plasticity. Here we use Avida, an established digital evolution system, to investigate the selective pressures that produce phenotypic plasticity. We witness two different types of fitness-enhancing plasticity evolve: *static-execution-flow* plasticity, in which the same sequence of actions produces different results depending on the environment, and *dynamic-execution-flow* plasticity, where organisms choose their actions based on their environment. We demonstrate that the type of plasticity that evolves depends on the environmental challenge the population faces. Finally, we compare our results to similar ones found in vastly different systems, which suggest that this phenomenon is a general feature of evolution.

1 Introduction

The field of evolutionary computation uses natural selection to automatically find solutions to engineering problems [1, 2]. Frequently, these solutions are on par or better than any human-produced solution [2]. However, most of these cases have involved static solutions to static challenges [3]. If the challenge suddenly changed even slightly, most organisms would not be able to adapt without further evolution. In other words, these organisms exhibit little or no within-life, or 'intralife', adaptation. Frequently, however, natural organisms or engineered solutions need to be robust enough to handle noisy and dynamic environments. One way of doing this is to evolve organisms in environments where the challenges they are presented with vary over time. Certainly, if our long-term goal is to evolve truly intelligent systems, we need to better understand how evolution can produce things that can intelligently adapt, initially in simple and then in increasingly sophisticated ways, to changing and novel situations.

F. Almeida e Costa et al. (Eds.): ECAL 2007, LNAI 4648, pp. 74 – 83, 2007.

2 Previous Work

One tactic researchers have used to evolve individuals that exhibit within-lifetime adaptation is to provide natural selection with a hand-written learning module, such as back propagation for neural networks. The researchers then investigate how evolution takes advantage of this ability to learn [4-8] and whether the results of learning are transferred to the genome via the Baldwin effect [9, 10]. Such research forces natural selection to use human-designed learning mechanisms, however, as opposed to discovering its own. A different approach—and the one taken here—is to use very simple systems to investigate how the ability to adapt could evolve on its own.

One clear sign that an organism is capable of adapting is if it behaves differently in two different environments. This kind of adaptive capacity is called *phenotypic plasticity*. Investigating this characteristic, Nolfi et al. evolved neural network brains for virtual robots in alternating light and dark environments [11]. The environments were constructed such that bodies, brains and behavior tuned to one environment would perform poorly in the other. They tested the evolved solutions by placing the same genome in the two different environments and observed whether they behaved differently. They found that in populations that evolved in alternately light and dark environments, the individuals were tuned to whichever environment they were placed in. Stanley et al. also used evolving neural networks to look at the evolution of phenotypic plasticity [3]. They compared experiments where the connection weights of neural networks never changed during the course of their life to those where evolution could create and modify connection weights during a lifetime. Such changes could be used to modify behavior in response to information sensed from the environment. The individuals were evolved in environments where each type of food randomly switched between nutritious and poisonous. Contrary to their expectations, they found that the individuals in the fixed-connection weight treatment discovered a simple solution that worked in all environments. The individuals executed the same code, but the inputs into that code were different because of the environmental differences, resulting in a different in behavior that was adapted to each environment. They thus possessed phenotypic plasticity in spite of a static execution flow. On the other hand, in the modifiable connection weight treatment the individuals developed phenotypic plasticity using a dynamic execution flow. In both cases the phenotypic plasticity enabled them to behave differently in the different environments. Ironically, the 'simple trick' evolution discovered in the fixed-connection weight experiment was more effective than the dynamic execution flow based strategy it discovered in modifiable connection weight treatments.

In short, we identify two types of phenotypic plasticity: static-execution-flow plasticity and dynamic-execution-flow plasticity. Which type prevails under natural selection depends on many evolutionary factors, such as the environmental challenge and or the genomic representation. Much of the research into evolving phenotypic plasticity *in silico* has been done with neural networks controlling simulated robots. But how general are the results? Can similar patterns be found in different systems? Here, we investigate the evolution of phenotypic plasticity in a vastly different setup: Avida, an experimental digital evolution system that maintains populations of self-replicating and evolving computer programs with a simplified (but Turing complete) assembly language as its genetic basis. These "digital organisms" have been shown to evolve traits of significant complexity [18], but previous work has focused on constant environments and hence rigid solutions. We find that our preliminary results are surprisingly parallel to those from the neural net community and suggest fundamental biological principles that can be applied toward understanding the evolution of intelligence.

3 Methods

Avida is an established experimental evolution platform where digital organisms compete for space in a two-dimensional grid [12-17]. There is no explicit fitness function in Avida; instead, organisms compete for limited space and those that replicate the fastest are most successful. Organisms must copy their own genomes and then execute a divide instruction, to produce an offspring. The copy process is imperfect, however, which introduces the variation that fuels natural selection. In the experiments performed here, the copying of a given instruction occasionally results in a copy error (0.75% of the time). When this happens, an instruction is chosen at random from the available set (N=26) and written to the target location (see [13] for details on the instruction set). There are also insertion and deletion mutations, which introduce or delete an instruction at random in 5% of offspring. All mutations affect only the genome of the offspring. The population size is 3,600. While the genome size can vary, the ancestral organism is 100 instructions long. The ancestor starts with the ability to self-replicate, but is largely blank with 85 of its instructions set to a mostly neutral 'no operation' command.

Digital organisms can improve their speed of reproduction either by decreasing the number of instructions it takes to produce an offspring or by performing tasks that increase their metabolic rate (rate of executing instructions). The initial metabolic rate for any organism is approximately proportional to its genome length (see [12] for a more detailed explanation). This number is then doubled or halved each time a task is performed, depending on whether the task is rewarded or punished. The tasks in this experiment are the logic functions NOT and NAND. Each organism can input three 32-bit numbers. They can manipulate those numbers and output the result. If they output the logical bitwise negation of one of the numbers or the bitwise nand of any two, they have performed the NOT or NAND task, respectively. The manipulation of these numbers occurs as organisms push and pop them to stacks or move them between registers using instructions such as `push`, `pop`, `add` (combines the numbers in the two specified registers and places the result in a third), `shift-r` (bit shift right), etc.

In order to adapt to their environment, organisms need to be able to sense it. For these experiments, the typical Avida `IO` instruction, which simultaneously inputs and outputs a number, was changed to `IO-Feedback`. This new instruction is identical to `IO` except that it provides the organism with knowledge of the impact the output had on their metabolic rate. When `IO-Feedback` is executed, if the organism's metabolic rate increased (because it performed a task currently being rewarded), a 1 is placed on the top of its active stack. If its metabolic rate is diminished (because it performed a task being punished), a -1 is placed on the top of its active stack. If the output had no impact on its metabolic rate (because it was not the negation of one of the three numbers or the nand of two of them), a 0 is placed on the top of its active stack. Organisms have flow-control instructions available that allow them to jump to, or skip over, sections of code in their genomes. An organism could, for example, execute a set of instructions that perform NOT and then repeat those instructions if the number atop their active stack is 1. The instruction set used here is Turing complete, meaning that it can perform any computable function. Therefore, any sophisticated conditional execution flow should in principle be able to evolve; the only remaining questions are whether natural selection discovers such complexity and, if it does, whether it will cost too much to be advantageous to the organism [12, 18].

4 Experiments and Results

To challenge the organisms to evolve plasticity, the population is alternately exposed to two different environments. In the Not+Nand- environment, performing NOT doubles an organism's metabolic rate and performing NAND halves it (the '+' indicates reward, the '-' indicates punishment). In the Not-Nand+ environment, the reverse is true. The environment shifts every 100 updates. Updates are the standard unit of time in Avida where each organism, on average, executes 30 instructions. Since in this experimental setup it normally takes organisms around 300 instructions to copy themselves, switching environments every 100 updates is equivalent to switching it approximately every ten generations. Because each trial lasts for 100,000 updates, there are 500 full cycles through the two alternating environments.

4.1 Experiment One

Our first experiment uses this setup to investigate whether the digital organisms will evolve phenotypic plasticity when their environment is uncertain. In this experiment, there are no restrictions on how often an organism can perform a task, so an organism's metabolic rate can be doubled or halved any number of times. In 13 of 50 trials, the final dominant organism alters the number of tasks it performs depending on which environment it is in. In 8 of these 13 trials, the flexibility results in a net positive reward in both environments. In the other 4 trials the plasticity is used to decrease the number of punished tasks being performed, but the punishments still exceed the rewards in one environment.

We next determined how these organisms are able to adapt to their environment. The same strategy is used in all 8 trials where the final dominant organism has a positive score in both environments. In one sense, these organisms are indeed 'adapting' because they end up performing different tasks in different environments. In another sense, however, they are not adapting at all; they execute the exact same series of instructions irrespective of what environment they are in. The organisms make no use of instructions that would make their execution flow conditional. Instead, they have discovered a string of instructions that results in behaviors that are tuned to the environment they find themselves in. Table 1 reports on which tasks are performed by the final dominant organisms from two example trials from experiment 1. The first organism does not adapt to its environment while the second does.

Table 1. The result of running the final dominant organism from two example trials from experiment 1 in two different environments. During evolution the organisms alternately encountered these two environments. The first organism always performs the same task set, which is beneficial in one environment and deleterious in the other. The second organism is able to adapt to the environment. It performs X NANDs and either 0 or >X NOTs, ensuring a net positive bonus. This same basic technique is used by the final dominant organism in all 8 trials in experiment 1 where the evolved plasticity resulted in positive fitness scores in both environments. Interestingly, the technique produces adaptation despite executing the same series of instructions. See text and Fig. 1 for an explanation of how.

Not+Nand- Environment			Not-Nand+ Environment			Static Execution Flow Plasticity?	Dynamic Execution Flow Plasticity?
NOTs	NANDs	Score	NOTs	NANDs	Score		
2	0	2	2	0	-2	No	No
102	51	51	0	51	51	Yes	No

The way the organisms are able to produce different numbers of tasks in different environments with the same series of executed instructions is simple and clever. The strategy involves putting different inputs into the same function to get different results. Using the information sensed from the environment as an input into a function, the resultant behavior can be modified based on the environment. The eight final dominant organisms that end up with positive rewards in both environments use simple variations on the theme shown in Figure 1.

This result emphasizes that evolution sometimes selects simple but effective solutions over complex, elegant ones. A sophisticated manipulation of execution flow was not needed to produce genomes that are adapted to both environments. However, many forms of sophisticated intelligence will require the ability to dynamically change the way that actions are determined. As such, it is worthwhile to determine the conditions under which dynamic execution flow phenotypic plasticity evolves (i.e. organisms that execute different programs in response to varying environmental conditions). To do so, we must understand how a static solution can be more fit than a dynamic solution. Why do organisms always perform a large number of NANDs? Why don't they regulate the performance of both tasks instead of just regulating NOT? It could be because it is easy to outcompete the current dominant static strategy by making a slightly better static strategy. If there is a population of organisms that do 0 or 20 NOTs and 10 NANDS, for a net reward of 10 in either environment, it is not uncommon to mutate to perform 0 or 22 NOTs and 11 NANDs for a net reward of 11. A mutation simply needs to make the organism run this loop one more time. Compare that to the challenge of setting up the necessary instructions to do the following: "if x, do instructions A,B,C; else, do D,E,F." This hypothesis motivates our second experiment.

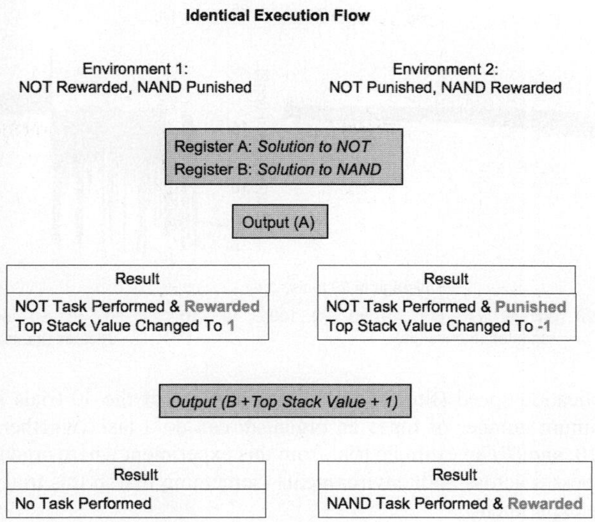

Fig. 1. The same series of instructions in two different environments leads to different results. Once the environment produces a difference in the stacks of the organisms, they can use it as an input to a function and produce two different results. If NOT is being rewarded, NAND is being punished. In environment 1, taking the correct answer for NAND and adding 2 to it produces a number that is no longer a correct answer for the NAND task (the 2 comes from 1 + the stack value of 1). Outputting this number does not result in either a reward or a punishment. In environment 2, adding 0 to the correct answer for NAND leaves it unchanged, and outputting this number yields the reward (the 0 comes from 1 + the stack value of -1).

4.2 Experiment Two

In our second experiment we cap the number of tasks for which an organism can be rewarded or punished to 10. Because in the previous experiment it was easy to extend the static strategy indefinitely, organisms did not need to use dynamic execution flow to regulate both of their tasks. Instead, they only regulated the NOT task (see Table 1). With a cap of 10, every punished task takes away from a potential rewarded task. The only path to the maximal fitnesses is through regulation of both tasks. Aside from this cap, experiment 2 is identical to experiment 1.

The results in this setup are quite different from experiment 1. In experiment 2, the final dominant organism in 23 of 50 trials alters its task output based on which environment it is in. In 15 of these 23 trials, the final dominant organism achieves a net positive score in both environments. All but one of these 15 organisms employs dynamic execution flow (in contrast with 0 final dominant organisms using dynamic execution flow in experiment 1). The average replication speed (fitness) across all 50 trials is shown in Fig. 2a. An individual trial where plasticity evolved is shown in Fig. 2b. The breakdown of what tasks it performs in each environment is presented in Table 2.

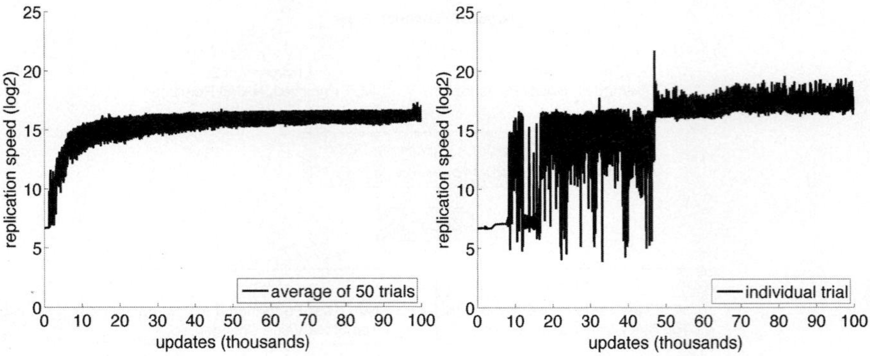

Fig. 2. The replication speed (fitness) for *(A)* the average of the 50 trials in experiment 2 in which the maximum number of times an organism can do a task (whether rewarded or punished) is set to 10, and *(B)* an example trial from this experiment where organisms have evolved to have high fitnesses across both environments (something that in this trial occurs about halfway through the experiment).

Table 2. The example organism from Fig. 2b. demonstrates a perfect ability to adapt to the two environments it is faced with. In environment 1, it gets rewarded 10 times for NOT (the maximum). In environment 2, it gets rewarded 9 times for NAND and punished once for NOT. It is not possible for an organism to perform better, as it must perform its first task without knowledge of which environment it is in. Thus, the first task performed will inevitably be punished in one of the two environments.

Not+Nand- Environment			Not-Nand+ Environment			Static Execution Flow Plasticity?	Dynamic Execution Flow Plasticity?
NOTs	NANDs	Score	NOTs	NANDs	Score		
16	0	10	1	32	9	Yes	Yes

The example organism from Table 2 (the final dominant from the experiment shown in Fig. 2b) attains the best score possible across the two environments. In Not+Nand- it ends up with 10 rewards (the max), and in Not-Nand+ it ends up with 9 rewards and 1 punishment. Since an organism must perform a NOT or NAND in order to determine which environment it is in, the best it can do is suffer just one poison in one of the two environments, as this organism does. Note: while all tasks performed above the cap of 10 (e.g 6 of the 16 NOTs) do not benefit the organism, performing them need not involve much or any extra cost for the organism. Evolution frequently produces designs that are 'good enough' instead of perfect. [19-24]

A question remains as to whether this organism's ability to adapt is derived from execution-flow plasticity. Is it changing which instructions it executes based on information from the environment? The answer in this case is yes. Fig. 3 presents a graphic representation of the instructions executed in the different environments by the final dominant organism from this case-study trial.

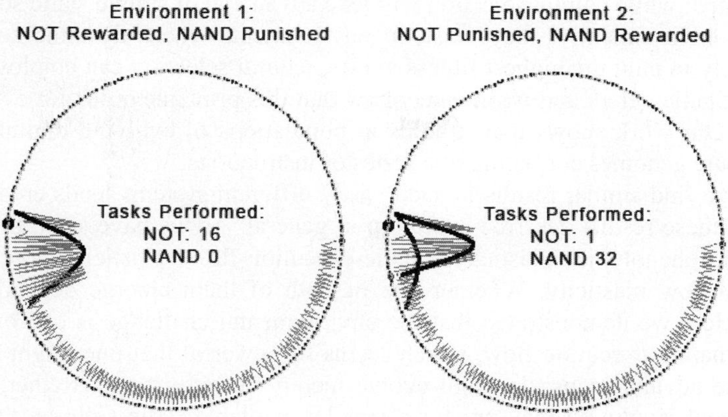

Fig. 3. A graphical representation of the instructions executed by the final dominant genome from the trial described in Fig. 2b and Table 2 when placed in environments 1 & 2. The small circles that make up the circumference represent each instruction in the genome of the organism. Arcs start on an executed instruction and land on the next one to be executed. Thin gray arcs indicate clockwise jumps and thick black arcs indicate counter-clockwise (backward) jumps. The height of the arc corresponds to the number of times that arc was traversed. Instructions that are never touched by an arc (as is the case with roughly half of the instructions in this organism) are not executed but do serve regulatory functions since they are traversed during replication. In both environments a series of instructions is executed before entering the looping area where tasks are performed. In environment 1, the organism repeats the same loop over and over, performing a NOT each time. In environment 2, the organism shortens the size of the loop after the first iteration and thus executes fewer instructions per subsequent iteration. This subset of instructions performs a NAND (and only a NAND) each iteration. The trigger to change the size of the loop is based on whether the NOT produced by the first iteration through the loop was rewarded or punished. The organism is thus able to change its behavior in response to the environment. It has evolved dynamic-execution-flow phenotypic plasticity.

5 Discussion and Conclusion

Our experiments demonstrate that natural selection will take advantage of simple static solutions that work across dynamic environments if they are available and advantageous. In our first experiment, a plastic solution using dynamic execution flow would have yielded high fitness values. Instead, selection reached high fitness values via a simpler static execution flow type of plasticity. Stanley et al. challenged static networks with dynamic environments and did not expect selection to discover a solution, but it did [3]. They tried a second experiment in which the network topologies could evolve over time. This simultaneously opened up the possibility of dynamic execution flow and made it difficult for static execution flow solutions to work. (Stanley, personal communication) In this second experiment, dynamic execution flow based plasticity evolved to produce highly fit organisms. Nofli et al. also challenged evolution with environments so different that, seemingly, only dynamic execution flow based strategies could work. Sure enough, they evolved [11]. A conclusion is suggested: while natural selection will take advantage of simple static solutions to a dynamic environment if they exist, in environments where dynamic solutions are more likely to gain the highest fitness values, natural selection can employ them. The work of Stanley et al. and Nolfi et al. show that this principle holds for evolving neural nets. Our work shows that it holds in populations of evolving digital organisms that execute genomes consisting of a series of instructions.

That we find similar results in such vastly different systems lends credence to the idea that these results describe evolution in general. As we have seen, there are two types of phenotypic plasticity: static-execution-flow plasticity and dynamic-execution-flow plasticity. Whether one or both of them emerge depends on many factors. Here we demonstrated that the environmental challenge is one of those factors. Dynamic execution flow, which seems so powerful that one might expect it to always be advantageous, does not evolve merely because it can. Rather, in a given environment, evolution may opt for a simpler available static solution that will "do the trick," selectively speaking.

Understanding how evolution works is of utility for engineers who want to apply evolutionary methods for practical purposes. If one wishes to evolve a particular type of plasticity, one should do so using environments that make it actively advantageous for natural selection to produce it, rather than simply possible. Future research is needed to learn more about when natural selection results in these different types of plasticity. Hopefully, such knowledge will facilitate our efforts to evolve systems as complex and intelligent as those found in the natural world.

Acknowledgments. The research for this paper was funded by the Cambridge Templeton Consortium, the National Science Foundation, and a fellowship to JC from the Quantitative Biology & Modeling Initiative at Michigan State University. We thank the anonymous reviewers and the members of the Evolving Intelligence and Digital Evolution Labs at Michigan State University, in particular Jeff Barrick, Sherri Goings, Dusan Misevic, Kaben Nanlohy, Brian Baer and Richard Lenski.

References

1. Holland, J.J.: Adaptation in Natural and Artificial Systems. University of Michigan Press, Ann Arbor (1975)
2. Koza, J., Keane, M., Streeter, M., Mydlowec, W., Yu, J., Lanza, G.: Genetic Programming: Routine Human-Competitive Machine Intelligence. Kluwer, New York (2003)
3. Stanley, K.O., Bryant, B.D., Miikkulainen, R.: Evolving Adaptive Neural Networks with and without Adaptive Synapses. In: IEEE Congress on Evolutionary Computation, Canberra, Australia, IEEE Press, Los Alamitos (2003)
4. Nolfi, S., Floreano, D.: Learning and Evolution. Autonomous Robots 7, 89–113 (2004)
5. Ackely, D.E., Littman, M.L.: Interactions between Learning and Evolution. In: Proceedings of the Second Conference on Artificial Life, Addison-Wesley, Reading (1991)
6. Belew, R.K., McInerney, J., Schraudolph, N.N.: Evolving Networks: Using the Genetic Algorithm with Connectionist Learning. CSE Technical Report CS89-174. University of California, San Diego (1990)
7. Whiteson, S., Stone, P.: Evolutionary Function Approximation for Reinforcement Learning. Journal of Machine Learning Research, 877-917 (2006)
8. Nolfi, S.: Learning and Evolution in Neural Networks. Adaptive Behavior 3, 5–28 (1994)

9. Baldwin, J.M.: A New Factor in Evolution. American Naturalist, 441-451 (1896)
10. Hinton, G.E., Nowlan, S.J.: How Learning Can Guide Evolution. Complex Systems, 495-502 (1987)
11. Nolfi, S., Miglino, O., Parisi, D.: Phenotypic Plasticity in Evolving Neural Networks, 146-157 (1994)
12. Ofria, C., Wilke, C.O.: Avida: A Software Platform for Research in Computational Evolutionary Biology. Artificial Life 10, 191–229 (2004)
13. Lenski, R.E., Ofria, C., Collier, T.C., Adami, C.: Genome Complexity, Robustness and Genetic Interactions in Digital Organisms. Nature 400, 661–664 (1999)
14. Ofria, C., Adami, C., Collier, T.C.: Design of Evolvable Computer Languages. IEEE Transactions on Evolutionary Computation, 420-424 (2002)
15. Misevic, D., Ofria, C., Lenski, R.E.: Sexual Reproduction Reshapes the Genetic Architecture of Digital Organisms. Proceedings of the Royal Society London, Series B 273, 457–464 (2006)
16. Adami, C., Ofria, C., Collier, T.C.: Evolution of Biological Complexity. Proceedings of the National Academy of Sciences 97, 4463–4468 (2000)
17. Goings, S., Clune, J., Ofria, C., Pennock, R.T.: Kin-Selection: The Rise and Fall of Kin-Cheaters. In: Proceedings of Artificial Life Nine, pp. 303–308 (2004)
18. Lenski, R.E., Ofria, C., Pennock, R.T., Adami, C.: The Evolutionary Origin of Complex Features. Nature 423, 139–144 (2003)
19. Darwin, C.: On the Various Contrivances by Which British and Foreign Orchids Are Fertilized by Insects. Murray, London (1862)
20. Dawkins, R.: The Selfish Gene. Oxford University Press, Oxford (1976)
21. Dawkins, R.: The Blind Watchmaker. Penguin, London (1986)
22. Gould, S.J.: The Panda's Thumb: More Reflections in Natural History. Norton, New York (1980)
23. Gould, S.J., Lewontin, R.C.: The Spandrels of San Marco and the Panglossian Paradigm: A Critique of the Adaptationist Programme. Proceedings of the Royal Society of London 205, 281–288 (1979)
24. Jacob, F.: Evolution and Tinkering. Science, 1161-1166 (1977)

Simulation of the Evolution of Aging: Effects of Aggression and Kin-Recognition

Svetlana Krivenko[1] and Mikhail Burtsev[2]

[1] Moscow Institute of Physics and Technologies, Institutsky per., 9, Dolgoprudny, RU-141700, Russia
skrivenko@gmail.com

[2] Keldysh Institute of Applied Mathematics, 4 Miusskaya sq., Moscow, RU-125047, Russia
mbur@ya.ru

Abstract. Current biological theory has no commonly accepted view on the phenomenon of aging. On the one hand it is considered as an inescapable degradation immanent to complex biological systems and on the other hand as outcome of evolution. At the moment, there are three major complementary theories of evolutionary origin of senescence – the programmed death theory, the mutation accumulation theory, and the antagonistic pleiotropy theory. The later two are rather extensively studied theoretically and computationally but the former one received less attention. Here we present computer multi-agent model of aging evolution compatible with theories of programmed death and mutation accumulation. In our study we test how presence of aggression and kin-recognition affects evolution of age dependent suicide which is an analog of programmed death in the model.

Key words: Aging, senescence, evolution, simulation, model, artificial life, Weismann, programmed death, mutation accumulation, cooperation.

Introduction

One of the fundamental problems of biology is the problem of phenomenon of aging [1,2]. There are two alternative approaches to the explanation of senescence. The first assumes that the aging is an immanent feature of all living matter. For example, the disposable soma theory [3,4,5] considers senescence as decline in somatic maintenance and repair. The second approach to the problem of aging is constituted by the evolutionary theories.

Evolutionary theory has no straightforward account for the aging. It is obvious that the death makes reproduction and, as a consequence, further proliferation of organism's genes in a population impossible. Therefore, selection should favor increase of an organism's lifespan leading to a more numerous progeny. Quoting the pioneer of aging research August Weismann:

> "This brings us face to face with one of the most difficult problems in the whole range of physiology, the question of the origin of death. As soon as we thoroughly understand the circumstances upon which normal death depends in general, we shall be able to make a further inquiry as to the circumstances

F. Almeida e Costa et al. (Eds.): ECAL 2007, LNAI 4648, pp. 84 – 92, 2007.

> which influence its earlier or later appearance, as well as to any functional changes in the organism which may produce such a result." [6 p.20].

Weismann put forward the theory of evolutionary emergence of death known today as the theory of programmed death or phenoptosis [7]. His hypothesis is:

> "I consider that death is not a primary necessity, but that it has been secondarily acquired as an adaptation. I believe that life is endowed with a fixed duration, not because it is contrary to its nature to be unlimited, but because the unlimited existence of individuals would be a luxury without any corresponding advantage." [6 p.24].

The hypothesis of programmed death is difficult to test; moreover Weismann proposed no plausible evolutionary scenario for emergence of "suicidal" adaptation.

At the middle of the 20th century other theories of senescence evolution were proposed. In accordance with the Medawar's theory of mutation accumulation [8-12] the aging is a result of accumulation of deleterious mutations which reveal their effect only in an old age. It is supposed that this kind of mutations is weakly affected by selection due to mortality caused by external factors. The other evolutionary theory of aging is the antagonistic pleiotropy theory suggested by Williams [13]. This theory relates aging to accretion of mutations which have the positive effect in a young age and the negative in an old age.

Theories of Weismann, Medawar and Williams propose independent causes of the senescence evolution. The hypothesis of programmed death assumes that phenoptosis itself has adaptive value. Mutation accumulation theory rests on the neutrality of senescence mutations. In the antagonistic pleiotropy theory a "death" side of mutation hitch-hike on an adaptive young age side. All three mechanisms are not incompatible and may comprise the holistic picture of aging evolution.

There are some theoretical studies of the senescence evolution based on the mutation accumulation and antagonistic pleiotropy approaches [2,11-12,14-18] but no recent studies in the framework of programmed death theory. We propose an evolutionary model of aging which allows exploration of the hypotheses of programmed death and mutation accumulation by means of computer simulations. In our model possible actions of individuals are independently affected by age, hence antagonistic pleiotropy is impossible. In the original formulation the theory of programmed death relates adaptiveness of senescence with cooperation, i.e. with "giving way" for new generations. Therefore, if this theory is accepted then evolution of aging should be affected by the strength of selection towards cooperation. So, we simulate evolution with varying strength of selection and ability for kin cooperation to observe their effect on plausibility of programmed death strategies emergence in the model.

Model

This computer model is a development of the previous one which was used for simulation of social evolution [20] and cooperation [19,21].

The two-dimensional artificial world in our model is divided into cells, which either contain a resource bundle or are empty. An empty cell can acquire a resource bundle with a certain probability per time step and lose it when resource is consumed by an agent. Agents are characterized by a set of receptors and effectors connected by a neural net. Each effector is responsible for a particular action. Agents can do nothing (rest), consume the resource bundle if it is present (eat), produce offspring (divide), go forward to a neighbour cell (move), make a turn to left or right (turn), attack another agent if present in the same cell (attack), and commit suicide (die). All actions spend energy taken from the agent's internal store. If internal energy is completely depleted, the agent "naturally" dies. The least energetically demanding action is rest, the most demanding is attack. Consumption of resource increases the internal store of energy subject to an upper limit (the maximum energy that can be stored). When an agent divides, one offspring is created and placed in the same cell as the parent. The parent then transfers half of its energy to the offspring. When one agent hits another, the victim loses an amount of energy, which is gained by the attacker (energetic costs of actions provided in Table 2).

Sensory inputs of agents include its internal store of energy, whether there are resources in the agent's field of vision (the cell it is in, the neighbour cell in front of the agent, and the cells on the right and left), and how many other agents are in the field of vision. Each agent has external phenotype that is coded by a vector of integer values (markers). The markers do not influence behaviour but function only as indicators of similarity. The euclidian distance between an agent's markers and the markers of another agent in the cell (a potential subject for attack) is also a sensory input. An agent perceives its own age. Value of an age input grows with actual age of an agent until it reaches maximum at double average population age. After the value of age sensory input remains constant. Behaviour of an agent is controlled by a simple one-layer neural net. Both weights of the neural net and external markers are inherited by the offspring when an agent divides, subject to a set rate of mutation.

Behaviour of agents is governed by a simple control system in which each output associated with a specific action is connected to sensory inputs from the environment or the internal state of the agent. The control system is linear and functions similarly to a feed-forward neural network with no hidden layer. To calculate the output vector **O** of values, the input vector **I** is multiplied by a matrix of weights **W**, which are constrained to lie in the range $[-W_{\max}, W_{\max}]$:

$$O_j = \sum_i w_{i,j} I_i \,. \tag{1}$$

At each time step, the agent performs the action associated with the maximum output value (note that the order in which agents act is randomly shuffled every step). The input vector **I** is populated with information about the presence of resource and other agents in the field of vision (the cell where the agent is, the neighbour cell in front of the agent, and the cells on the right and left), the level of internal resource, the euclidean distance between marker vectors of the agent and its partner for potential interaction, and own age. A full list of input variables and their definitions are given in Table 1. At the start of simulation, an initial population was formed from the agents with the same matrix of weights **W**. All the weights in this matrix were set to zero except for three that defined the following simple strategy: move if a resource bundle is in the forward cell; eat if a resource is in the current cell; divide otherwise. Correspondence between outputs and actions, and how changes of the internal resource r depend on actions are summarized in Table 2.

To speed up simulations, all variables were integers. For all simulations, the size of the world was 900 cells, W_{max} was 1,000, r_{max} was 5,000, the dimension of the marker vector was 10, and its values were bounded by $[-W_{max}, W_{max}]$.

If the agent executes the action 'divide', its offspring is placed in the same cell. The genome of the offspring is constructed in the following way: first, for every weight of the control system, a random value uniformly distributed on the interval $[-0.03W_{max}, 0.03W_{max}]$ is added; (2) for every component of the marker, a random value uniformly distributed on the interval $[-0.15\ W_{max}, 0.15W_{max}]$ is added.

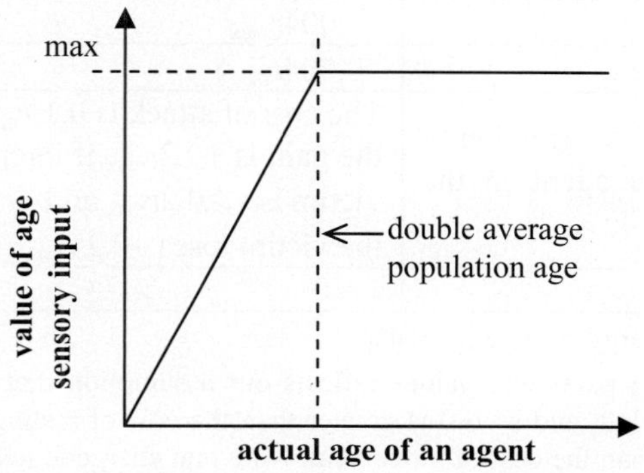

Fig. 1. Dependence of value of age sensory input on actual age of an agent.

Table 1. List of input variables and their definitions

Input variable*	**Value**
I_1	Bias constant, k
I_2, I_3, I_4, I_5	k if there is resource bundle in the field of agent's vision; 0 in the opposite case
I_6, I_7, I_8, I_9	cN_c, where c is a constant, N_c is the number of agents in the given cell of the field of agent's vision
I_{10}	Value of internal resource, r
I_{11}	$r_{max}-r$
I_{12}	$\sqrt{\sum_i (\overline{m}_i - m_i)^2}$, where $\overline{m}$ is a centroid of markers of all agents at the current cell
I_{13}	$\dfrac{k \cdot \sqrt{\sum_i (m_i^p - m_i)^2}}{2M_{max}}$, where m^p is a marker of partner to interact
I_{14}	see fig. 1.

*Note that I_1 is a constant and that I_2 to I_5 are binary variables (k is a functional analogue of unity and was set equal to r_{max}, where r_{max} is the maximal possible value of stored internal resource).

Table 2. The energetic costs of an agent's actions*

Output vector	Action	Change of internal resource r†
O_0	Rest	$-0.001r_{max}$
O_1	Turn left	$-0.002r_{max}$
O_2	Turn right	$-0.002r_{max}$
O_3	Consume the resource bundle	$+0.04r_{max}$‡
O_4	Move	$-0.004r_{max}$
O_5	Divide	$-0.004r_{max}$§
O_6	Fight (randomly chosen agent in the cell)	The cost of attack is 0.1 r_{max}; the gain is $+0.2r_{max}$ if internal resource of the victim is $r_n \geq 0.2r_{max}$ and $+r_n$ otherwise; the victim loses $-0.2\ r_{max}$.
O_7	Die	$r = 0$

*Note that r_{max} is the energy storage capacity.

†This scheme of setting parameter values reflects our assumption that the energetic cost of movement (move, turn left, and so on) is greater than the cost of resting, whereas the cost of attack is much greater than the cost of movement. Note that energetic losses are indicated with a minus sign and gains with a plus sign.

‡Food appears in the cell with the constant probability of 1/80 and the amount of resource in the bundle was $0.04r_{max}$.

§When the agent divides it spends $0.004r_{max}$; half of the remaining energy is then transferred to the offspring.

Results

To study how strength of selection and ability to cooperate affect the evolution of aging we have conducted simulations with four modifications of the basic model. These four variants were produced by combination of variation in selection pressure implemented by switching on or off attacking action and ability to cooperate by switching on or off kin recognition.

With each modification of the model a series of runs were conducted. At the beginning of every run the world was populated with the same initial population (as described in the previous section) but with different seeds for random number generator. All the weights from self-age input and to "die" action were set to zero, so there were no bias in the population toward any particular age dependent and death strategy. Amount of the resources in the world allows population density up to ~0.75 agent per cell.

During the runs every agent executing "die" action was monitored. This allowed to locate runs in which the ages of suicides were accreted to the close range around double average age in the population. Particular example of such run is presented on figure 2. There is no correlation between the age of action "die" execution and double average age in the population at the on set of evolution. Then after a short transient around 3 10^6 time step the death strategy emerges. Now the agents usually murder themselves at ages about double average population age, i.e. about maximal individually perceivable age (see fig.1). It is notable that the suicide strategy persists during considerable amount of evolutionary time which indicates presence of stabilizing evolutionary mechanisms.

The runs where the death strategy was evolved can be clearly identified by plotting distribution of suicide ages normalized to the double average population ages. Typical distributions for runs with emerged senescence, with no age related strategy and for the control run with disabled age input are presented on figure 3. Distribution for the run where no age dependent suicide evolved is similar to the run without possibility to perceive self-age and hence control behavior appropriately. On the other hand distribution for the case with evolved suicide strategy has remarkable peak at the maximal individually perceivable age.

With the aid of suicide ages distribution we found how many runs for each of four model modifications demonstrate evolution of death strategy (see fig. 4). The modification with attack action and kin recognition enabled had the highest plausibility of death strategy evolution (7 out of total 11 runs). When agents can fight each other but can not recognize kin the suicide strategy evolved in smaller fraction of runs (6 out of total 13 runs). The rest two "peaceful" modifications of the model had lowest proportion of outcomes with evolved senescence (6 out of total 13 runs).

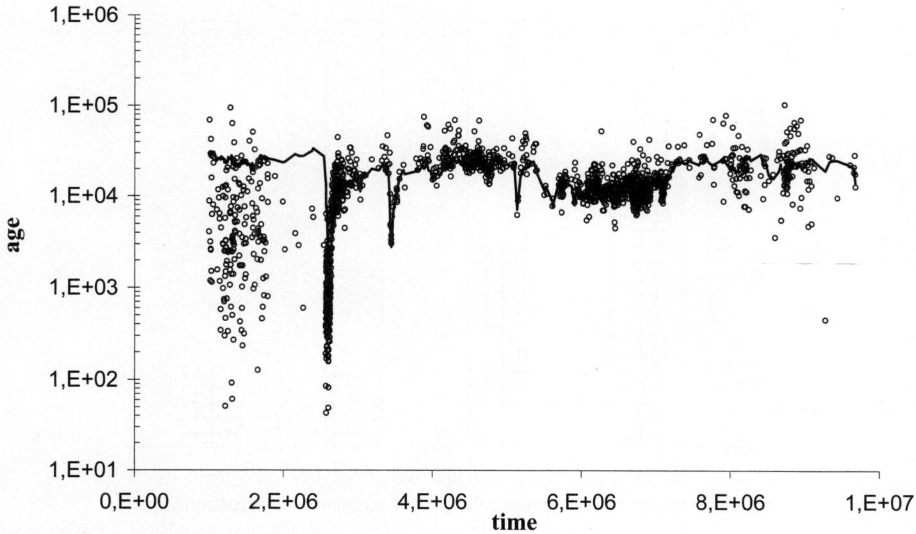

Fig. 2. The dynamics of suicide ages (circles) and double average population age (solid line). The age scale is logarithmic.

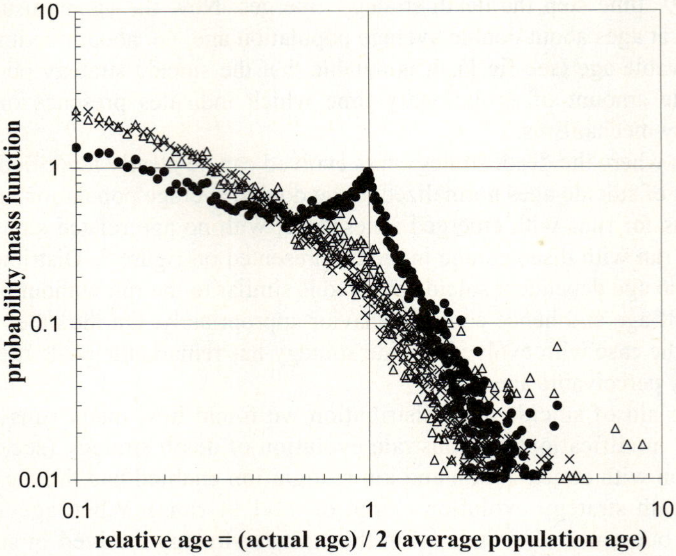

Fig. 3. Distributions of suicide ages normalized to the double average population age for the model without age input (crosses), for the run without evolved death strategy (triangles), for the run with evolved death strategy (solid circles).

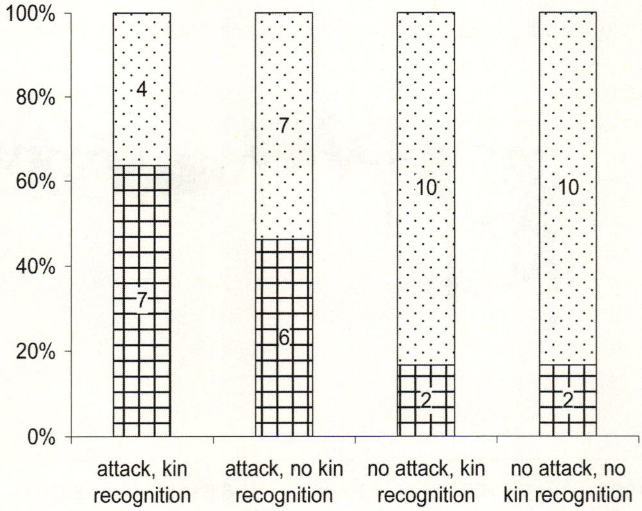

Fig. 4. Simulations outcomes for four modifications of the model. Each bar represents total of runs with evolved suicide strategy (grided area) and without it (dotted area). There is number of runs inside each area.

Discussion

The results of simulations can be outlined as follows:

- Age dependent death strategy evolved in the simulations without any predisposition or bias with only basic assumptions of agent's abilities to perceive self-age and commit suicide.
- Evolved death strategy demonstrated persistence over evolutionary time, i.e. evolutionary stability.
- Plausibility of death strategy emergence in the course of evolution in the model depends on the selection pressure (on agents' possibility to fight each other) and possibility of cooperation (through kin recognition).

The first result means that age dependent suicide strategy can invade population of immortal individuals, and the only prerequisite is presence of some mechanisms for determining self-age and phenoptosis.

The second result is a sign of presence of evolutionary processes stabilizing death strategy. There can be some evolutionary advantage for aging agents and neutral mutation accumulation. Analysis of simulation runs strongly supports the claim of adaptivness because the number of agents committing suicide is rather large which means that there should be selection against suicide, and, hence, the death strategy cannot be neutral in this case. The other thing which plays not in favor of neutrality is that evolution of age dependent suicide in the model is affected by aggression and cooperation which theoretically should have no direct impact on the process of neutral mutation accumulation.

Simulations results for different modifications of the model demonstrate that the strongest factor affecting the evolution of senescence in our set up is an aggression. As we expected, two model's modifications with enabled attack action have significantly higher fraction of runs with emergence of death strategy. We also anticipated the strong effect of ability to cooperate; here the outcome is more controversial. If aggression is present in the population then kin recognition seems to increase plausibility of senescence evolution, else it has no or small effect. It should be noted that simulations were conducted with rather low population density which is a condition for weak selection toward cooperation in the model as was found in the previous work [21]. Higher population densities (>2) should provide stronger kin-selection and might increase effect of cooperation on evolution of age dependent suicide in the model. We plan to test it in our future work and continue with more detailed analysis of evolved death strategies.

Acknowledgments. We deeply acknowledge the anonymous reviewers for their suggestions and comments. The work was supported by Russian Fund for Basic Research, project #06-06-80504 and Russian Science Support Foundation.

References

1. Wachter, K.W., Finch, C.E.: Between, Zeus and the Salmon: The Biodemography of Longevity. National Academy Press, Washington (1997)
2. Goldsmith, T.C.: The Evolution of Aging. Azinet LLC (2006), http://www.azinet.com/aging/Aging_book.html
3. Kirkwood, T.B.L.: Evolution of ageing. Nature 270, 301–304 (1977)
4. Kirkwood, T.B.L., Holliday, R.: The evolution of ageing and longevity. Proc. R. Soc. London Ser. B Biol. Sci. 205, 531–546 (1979)
5. Kirkwood, T.B.L., Austad, S.N.: Why do we age? Nature 408, 233–238 (2000)
6. Weismann, A.: Essays Upon Heredity and Kindred Biological Problems. Clarendon Press, Oxford (1889)
7. Skulachev, V.P.: The programmed death phenomena, aging, and the Samurai law of biology. Exp. Ger. 36, 995–1024 (2001)
8. Medawar, P.B.: Old age and natural death. In: Medawar, P.B. (ed.) The Uniqueness of the Individual, pp. 17–43. Basic Books, New York (1958)
9. Medawar, P.B.: An Unsolved Problem in Biology. In: Medawar, P.B. (ed.) The Uniqueness of the Individual, pp. 44–70. Basic Books, New York (1958)
10. Rose, M.R.: Evolutionary Biology of Aging. Oxford University Press, New York (1991)
11. Gavrilov, L.A., Gavrilova, N.S.: Evolutionary Theories of Aging and Longevity. TheScientificWorldJOURNAL 2, 339–356 (2002)
12. Hughes, K.A., Reynolds, R.M.: Evolutionary and Mechanistic Theories of Aging. Annual Review of Entomology 50, 421–445 (2005)
13. Williams, G.C.: Pleiotropy, natural selection and the evolution of senescence. Evolution 11, 398–411 (1957)
14. Penna, T.J.P.: A bit-string model for biological aging. Journal of Statistical Physics 78, 1629–1633 (1994)
15. Stauffer, D.: Life, Love and Death: Models of biological reproduction and aging. In: Electronic proceedings of the workshop: Monte Carlo and Structure Optimization Methods for Biology, Chemistry and Physics (1999)
16. Moss de Oliveira, S.: Evolution, Ageing and Speciation: Monte Carlo Simulations of Biological Systems. Brazilian Journal of Physics 34, 1066–1076 (2004)
17. Dzwinel, W., Yuen, D.A.: Aging in Hostile Environment Modeled by Cellular Automata with Genetic Dynamics. International Journal of Modern Physics C 16, 357–376 (2004)
18. Lee, R.D.: Rethinking the evolutionary theory of aging: Transfers, not births, shape senescence in social species. PNAS 100, 9637–9642 (2003)
19. Burtsev, M.S.: Tracking the Trajectories of Evolution. Artificial Life 10, 397–411 (2004)
20. Burtsev, M.S.: Artificial Life Meets Anthropology: A Case of Aggression in Primitive Societies. In: Capcarrère, M.S., Freitas, A.A., Bentley, P.J., Johnson, C.G., Timmis, J. (eds.) ECAL 2005. LNCS (LNAI), vol. 3630, pp. 655–664. Springer, Heidelberg (2005)
21. Burtsev, M.S., Turchin, P.V.: Evolution of cooperative strategies from first principles. Nature 440, 1041–1044 (2006)

Artificial Ecosystem Selection for Evolutionary Optimisation

Hywel T.P. Williams and Timothy M. Lenton

University of East Anglia, Norwich, NR4 7TJ UK
{h.williams, t.lenton}@uea.ac.uk

Abstract. Artificial selection of microbial ecosystems for their collective function has been shown to be effective in laboratory experiments. In previous work, we used evolutionary simulation models to understand the mechanistic basis of the observed ecosystem-level response to artificial selection. Here we extend this work to consider artificial ecosystem selection as a method for evolutionary optimisation. By allowing solutions involving multiple species, artificial ecosystem selection adds a new class of multi-species solution to the available search space, while retaining all the single-species solutions achievable by lower-level selection methods. We explore the conditions where multi-species solutions (that necessitate higher-level selection) are likely to be found, and discuss the potential advantages of artificial ecosystem selection as an optimisation method.

Key words: artificial ecosystem selection, evolutionary optimisation

1 Introduction

In nature, there are many scenarios where multiple species combine to perform a function that a single species acting alone could not achieve. Many of these occur in the microbial world, at a variety of scales and with varying degrees of integration and cooperation [2]. Global chemical cycles typically involve different species at different stages in the cycle, such as photosynthesisers and respirers in the carbon cycle, or nitrogen-fixers and denitrifiers in the nitrogen cycle. At a smaller scale, biofilms (colonies of microorganisms growing in association on a surface) can also create nutrient cycling loops involving multiple species. Symbiotic associations such as lichens or mycorrhizae are tight, mutually beneficial linkages between species with complementary metabolic processes. These natural collective solutions to problems have been the inspiration for the engineering of 'designer communities' in microbial biotechnology, which have been usefully employed for purposes such as bioremediation of pollutants, treatment of sewage, and microbial leaching of metal ores [2]. However, the design of useful microbial communities is not straightforward and new methods are continually being sought.

Laboratory experiments with microbial communities have shown that artificial selection at the ecosystem level can cause a significant adaptive response

F. Almeida e Costa et al. (Eds.): ECAL 2007, LNAI 4648, pp. 93–102, 2007.

[10,9]. In each experiment, an initial population of microbial ecosystem 'units' was created by random sampling of soil or pond water communities to inoculate containers holding neutral growth media. Each successive generation of ecosystems was then created by sampling from the best existing ecosystems (selected on various properties of their environment) to inoculate a new batch of sterile containers. In one set of experiments, pond water communities were selected for their collective effect on the pH level of their liquid environment [9]. In another, soil communities were selected for the growth they supported in above-ground plant biomass [9]. A third set of experiments selected pond water communities for their ability to break down the environmental pollutant 3-chloroaniline [10]. In all experiments, a statistically significant response to selection was observed, but it was unclear whether it resulted in each case from the indirect selection of a single dominant species (which could have been found by lower-level selection methods) or from genuine adaptation of community-level properties (which would necessitate higher-level selection).

Following on from prior population-based modelling of artificial ecosystem selection [4,6,5], in previous work [13] we used individual-based evolutionary simulation modelling to re-create the artificial ecosystem selection experiments reported in [10,9]. We demonstrated a robust response to artificial ecosystem selection and were able to isolate the underlying causal mechanisms, allowing us to determine the level at which adaptation occurred. Here we consider the potential of artificial ecosystem selection as a method for evolutionary optimisation. Evolutionary optimisation methods have traditionally focused on scenarios where candidate solutions are encoded as a single string of values. This 'single-species' approach has been successful in many areas, but some harder problems may require more complex encodings. We propose artificial ecosystem selection as a method for designing 'multi-species' solutions, where several part-solutions operate in combination to achieve the desired function. We address two key questions. First, is artificial ecosystem selection effective in achieving an arbitrary target ecosystem function? This question has largely been answered in previous work, which we will briefly describe in the next section. Second, under what circumstances is artificial ecosystem selection of benefit over more traditional lower-level selection methods? We approach this question by looking for solutions created by artificial ecosystem selection that could not have been created by lower-level selection methods; we leave efficiency testing, benchmarking, and direct comparison with other optimisation algorithms, for future work.

2 Artificial Ecosystem Selection in the Flask Model

Here we give a brief overview of the 'Flask' model of microbial evolutionary ecology [12,14] and its use in our previous artificial ecosystem selection experiments [13]. More detail and parameter values can be found in online supporting material for this article.[1]

[1] http://researchpages.net/media/resources/2007/04/16/ECALsupportingMethod.pdf

2.1 The Flask Model

Flask [12,14,13] is an individual-based evolutionary simulation model of a flask containing a neutral liquid matrix in which is suspended a microbial population. The composition of the liquid medium determines the environment of the microbes. Some of the chemicals present are 'nutrients' that may be consumed as food and converted to biomass, while others are non-consumable and form part of the abiotic environment. The environment is assumed to have properties such as temperature, pH, salinity, etc., that both affect, and can be affected by, microbial activity. Non-consumable chemicals and physical properties of the flask environment are collectively referred to as 'abiotic factors', to distinguish them from nutrients. There is a flow of liquid medium through each flask which occurs continuously at a prescribed rate. The inflow brings with it influxes of nutrients at fixed concentrations and steady inputs to abiotic factors, while the outflow removes fixed proportions of stored nutrients and abiotic factors.

Microbes are modelled as simple organisms that consume and excrete nutrients, and affect the levels of abiotic factors in their environment as a by-product of metabolism. The precise ratios in which nutrients are consumed and excreted are genetically encoded for each individual, as are associated effects on abiotic factors, and preferred abiotic conditions (i.e., the state of the abiotic environment in which growth rate is maximised). Microbes grow by converting consumed nutrients to biomass and reproduce by splitting when their biomass reaches a fixed threshold. Mutation may occur during each reproduction event by selecting a new random allele with low probability P_{mut} at each locus, otherwise each offspring microbe receives an identical copy of the parental genotype. Biomass is reduced at a fixed rate to represent the inevitable thermodynamic inefficiency of metabolism and the cost of maintaining cellular machinery. Microbes die if their biomass drops below a fixed threshold, which can happen in sustained periods of nutrient limitation. They may also die 'from natural causes' with a low probability at each timestep.

The state of the liquid environment in a flask is a function of the parameterised fluxes of nutrients and abiotic factors, and the activity of the microbial community (i.e., the metabolism of nutrients and associated side-effects on abiotic factors). The shared environment creates individual-level selection pressure on metabolic requirements and environmental preferences, but the nature of this selection pressure changes over time as microbial activity alters the environment. Ecological and (co)evolutionary dynamics of these model ecosystems are discussed elsewhere [14]. Common phenomena include the emergence of nutrient recycling loops and coevolutionary feedbacks between different microbe species and their environment. (The term 'species' here refers to genetically identical individuals.)

2.2 Artificial Ecosystem Selection

The 'phenotypic' ecosystem trait used for artificial selection is based on the levels of the abiotic factors in the flask environment. Basing the fitness of flask ecosystems on properties of the environment rather than the biotic population avoids any pre-specification of the type of population that will provide a good solution to the evolutionary problem. An arbitrary target state of the abiotic environment is assigned, with the deviation error of the actual abiotic state of a flask from this target constituting its performance score, $\Phi = \sqrt{\sum_{i=1}^{A}(\bar{a_i} - \hat{a}_i)^2}$, where $\bar{a_i}$ is the target level for abiotic factor a_i, and $\hat{a}_i$ is the actual level of a_i, in the normalised state vector for the A abiotic factors included in the model. Depending on the direction of artificial selection, the fitness of a flask ecosystem is based on maximising or minimising Φ. In each artificial selection experiment, three lines were selected based on the same initial random population: The 'high' line was selected to maximise Φ, the 'low' line was selected to minimise Φ, and the 'random' line – where the source ecosystem used to create the batch of ecosystems for each iteration was chosen at random – acted as a control. All lines consisted of a number of iterations of directed selection followed by an equivalent number of iterations of random selection, to allow study of the relaxation of the selected response.

Artificial ecosystem selection is an iterative process based on preferentially sampling from successive batches of flask ecosystems in order to create each succeeding batch. Each flask in the initial batch is seeded with a different inoculum, made up of randomly generated microbes. At each subsequent iteration of the selection process a new batch of flask ecosystems is created by inoculating sterile flasks with individuals from the fittest flasks of the previous iteration. A single inoculum of a fixed number of individuals is created by sampling at random from the source flasks, and identical copies of this inoculum are then used to seed the entire new batch of ecosystems. Two sampling methods are used: a 'propagule' method where the inoculum is drawn from a single source ecosystem, and a 'migrant pool' method where the inoculum takes individuals from several source ecosystems. The propagule method is analogous to asexual reproduction and should preserve ecological interactions between individuals. The migrant pool method is analogous to sexual reproduction and may better represent how new ecosystems form in nature. After inoculation, each ecosystem is propagated for a fixed period before Φ is measured.

2.3 Response to Selection

A similar response to artificial selection (e.g., Fig. 1(a)) is observed in all experiments performed. There is a rapid initial divergence in Φ values for the directed (high and low) selection lines from the control line. When directed selection is stopped and replaced by random selection (after iteration 30) the selected response in each line relaxes towards the non-selected control condition. The response to selection is robust to changes in the target environmental state and the ecosystem sampling method. Inverse relationships exist between the size of the response to selection and mutation rate P_{mut} (Fig. 1(b)), and between the size of response and ecosystem propagation time. Of the high-selected ecosystems, only 3.4% of cases showed adaptation at the community-level, but 36.5% of the low-selected ecosystems showed community-level adaptation. Further results and description can be found in [13].

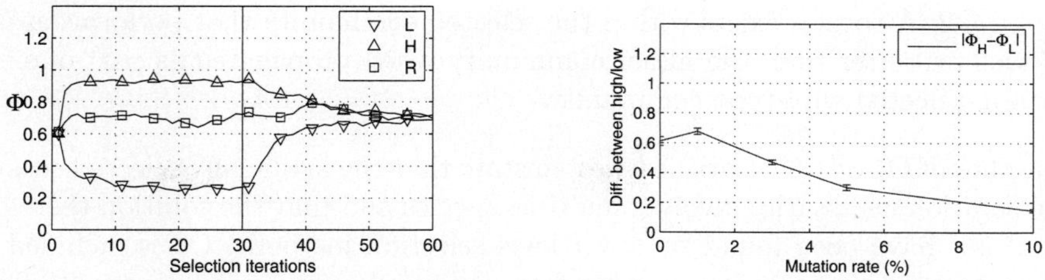

(a) Divergence of selected lines (b) Inverse relation with mutation

Fig. 1. Artificial ecosystem selection produces a strong adaptive response. An inverse relationship exists between the size of the response and microbe mutation rate. (a) Mean Φ from 49 runs with migrant pool sampling. Data plotted for directed selection for increase (high line) or decrease (low line) in distance from target environmental state, Φ, and for a random selection control line. Directed selection stopped at iteration 30, after which all selection is random. (b) Final deviation between high/low lines in {43, 57, 87, 42, 73} runs with P_{mut} set at {0, 0.01, 0.03, 0.05, 0.1} respectively.

3 Categorisation of Solutions from Artificial Ecosystem Selection

We now focus on showing that artificial ecosystem selection can create solutions that could not be created by standard genetic algorithms. Selecting at the ecosystem level creates the possibility for multiple species to be involved in the solution. Standard genetic algorithms select at the individual level, and even if we allow the case where an individual genotype is cloned in order to be tested in a group context, these algorithms therefore create solutions based on a single genotype. Thus we can reduce our task to categorising the solutions produced by artificial ecosystem selection into those that involve multiple species and those that involve only one species; the latter could have been created by lower-level selection, but the former require selection at a higher level.

The diverse initial seed and ongoing effects of mutation mean that multiple species are almost always present in each flask community. However, not all species in a selected community have a significant involvement in producing the target ecosystem function. If this function only involves a single active species, then any other species present are irrelevant to the response achieved and we have a single-species solution. If the function involves contributions from more than one active species, it is a multi-species solution.

A simple method for finding out if multiple species are involved in producing an observed ecosystem function is to test the converse proposition, that a single species is alone responsible for that function. We do this for each selected ecosystem by testing each species in the community separately for two observations:

O1: A species exists within the selected community that performs as well or better than the intact community when propagated in isolation as a monoculture population.

O2: A species exists within the selected community that performs as well or better than the intact community when propagated as part of a non-selected wild-type community.

If neither of O1 or O2 is satisfied we can state that the artificially selected ecosystem solution necessarily involves multiple species and that the solution therefore could not have been found by lower-level selection methods. O2 is included to account for the scenario (observed in community selection experiments with flour beetles [1]) where a single species is responsible for an observed ecosystem property, but requires the presence of a non-specific background community for the property to be expressed; this does not count as a genuine multi-species solution since the identity of the background community is not significant. Note that this method is conservative, since satisfaction of O1 or O2 by a species does not necessarily imply that the species is responsible for overall community function.

3.1 Method

We tested O1 and O2 for 1064 artificially selected ecosystems (high and low lines from 532 runs) that were created in the experiments reported in [13]. The performance of each species from each artificially selected ecosystem was tested using a similar method to that used to allocate fitness during the artificial selection process. For O1, a sterile flask was seeded with a clonal inoculum of 100 individuals of the test species. For O2, a sterile flask was seeded with 25 individuals from the test species and 75 individuals randomly sampled from the control line community associated with the relevant artificial selection run. Each seeded flask was then allowed to develop for 2000 timesteps with mutation switched off, after which Φ was measured. The intact evolved community was tested similarly for comparison. When testing O2, and for the intact community, the mean score from 20 repetitions was used, to allow for stochastic effects. In verifying O1 and O2, high scores are 'better' in the high line, and low scores are 'better' in the low line.

3.2 Results

Overall, the majority (75%) of solutions can be explained by the presence of a single dominant species: 72% of evolved ecosystems satisfy O1, 47% satisfy O2, and 44% satisfy both O1 and O2. This leaves 25% of all the selected ecosystems that rely on contributions from several species. There is a clear divergence in the kinds of solution that are found for each line. Figure 2(a) shows Venn diagrams that give the proportion of all the selected ecosystems in each line that satisfy O1 and O2. Multi-species solutions are less likely in high line ecosystems (8% cases, upper diagram) than in low line ecosystems (42% cases, lower diagram). Propagule sampling has a higher likelihood of producing a multi-species solution than migrant pool sampling (Fig. 2(b)). Of the 483 runs performed with the propagule method, 43% of low line and 9% of high line cases are multi-species solutions. Of the 49 runs performed with the migrant pool method (using 4 source

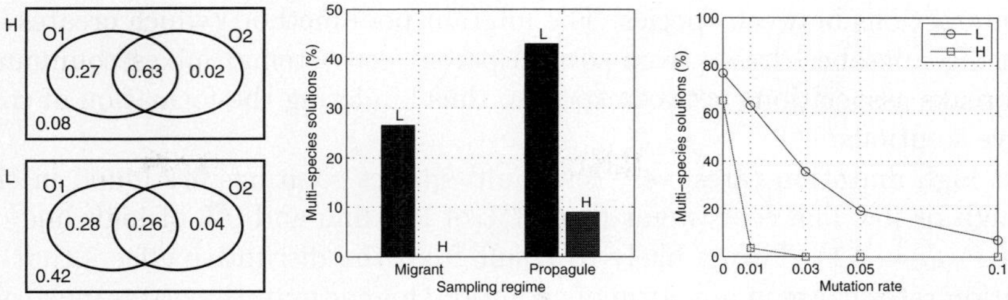

(a) Selection line (b) Sampling method (c) Mutation rate

Fig. 2. Proportion of artificially selected ecosystems where multiple species are functionally involved in generating the selected function, found as the percentage of cases where neither O1 nor O2 is satisfied. Each subfigure shows the effect of varying a different parameter of the selection process: (a) selection line, (b) sampling method, (c) mutation rate.

ecosystems), 27% of low line and 0% of high line ecosystems were multi-species solutions. A clear inverse relation is observed between P_{mut} and the likelihood of a multi-species solution (Fig. 2(c)). Testing of evolved ecosystems from {43, 57, 87, 42, 73} runs with P_{mut} set at {0, 0.01, 0.03, 0.05, 0.1} respectively showed that when $P_{mut} = 0$ there is a high occurrence of multi-species solutions (77% of cases in the low line, 65% in the high line), but the proportion of multi-species solutions achieved falls dramatically as P_{mut} is increased.

3.3 Interpretation

Since Φ measures the distance from a target environmental state, we may *a priori* say that the low line presents a more difficult problem than the high line, since converging on a target is more difficult than diverging from it; there are many ways to be far from a point in multi-dimensional space, but only one way to hit it. Furthermore, in a complex, dynamic environment, holding an environmental variable close to a particular target level will often require correction in two directions. Since a single species can only push any environmental variable in one direction, at least two species may therefore be needed to provide the necessary opposing influences for the low line (target-seeking) problem. However, a variable can be moved away from a target by pushing in a single direction, so a single dominant species in the community may offer a good solution to the high line (target-avoiding) problem. Thus the high and low lines offer qualitatively different evolutionary problems that demand qualitatively different ecosystem solutions, and this is reflected in the observed results.

The effect of different sampling methods on the likelihood of multi-species solutions can be understood by noting that the propagule method (which creates inocula using microbes drawn from a single parent ecosystem) transmits the parental community to the offspring ecosystem intact, and thus preserves the interactions between species. The migrant pool method (which creates inocula using microbes drawn from several parent ecosystems) mixes communities and breaks associations between species, thus hindering the formation of collaborative solutions.

At high mutation rates, very few multi-species solutions are found in either the high or low line ecosystems (e.g., 7% of low line and 0% of high line cases when $P_{mut} = 0.1$). This is likely to result from the disruptive effects that high mutation rates cause in any form of evolution; higher mutation rates make adaptation of individual phenotypes more difficult by introducing noise into the search process. In a multi-species community this disruption is amplified, because the interactions between species are also affected. High mutation rates add a large number of non-adaptive mutants to the community, which affect ecological dynamics even though they are likely to die out quickly without reproducing. The disruptive presence of these mutants makes it harder for beneficial associations to reliably form between species. Single-species solutions are not disrupted in this way and are therefore more robust. At the other end of the scale, the zero mutation scenario leads to a much higher incidence of multi-species solutions. With $P_{mut} = 0$, 77% of low line cases and 65% of high line cases were observed to be multi-species solutions, values much higher than the global average over all runs. There are two reasons for this. The first is that the lack of mutation makes ecological dynamics deterministic and stable, making it easier for useful associations between species to form. The second is that without mutation there is less chance that a single species will be found that can alone provide a good ecosystem-level solution, even for the high line problem. Without mutation the search algorithm is limited to those species that are present in the initial seed inoculum, and the indirect individual-level selection pressure applied by ecosystem-level selection cannot create a single 'super' species capable of achieving the target function.

4 Discussion

We have seen that artificial ecosystem selection can be effective in designing communities that collectively perform an arbitrary ecosystem function. It can (in theory) generate all the single-species solutions that are available to lower-level selection algorithms, but also allows a new class of multi-species solution which may be beneficial in some problems. Where the problem demands contributions from multiple species (e.g., the low line target-hitting problem) the evolved communities include several complementary species that together perform the function. In situations where the problem may be solved by a single species (e.g., the high line target-avoiding problem) the evolved communities reflect this. Where a single-species solution is possible, the search process will usually find it, since these solutions are more robust than multi-species solutions. Increasing mutation rate makes multi-species solutions less likely (Fig. 2(c)). There is a correlated decrease in the size of the response to artificial ecosystem selection (Fig. 1(b)), which may signify the occurrence of situations where multi-species solutions are required by the test problem, but prevented from forming by high mutation rates.

Artificial ecosystem selection is a novel method for evolutionary optimisation. The presence of multiple levels of selection and the isolation of the ecosystem units distinguish the artificial ecosystem selection method from superficially similar methods such as 'island GAs' [11]. Within the literature, artificial ecosystem selection is most closely related to work on cooperative coevolution. There are two main forms of cooperative coevolution algorithm in the literature, which differ mainly in how the population is structured. One kind coevolves part-solutions held in separate populations [7], while the other holds all part-solutions in the same population [8,3]. Artificial ecosystem selection fits neither of these schemes; all species are held in the same community as in [8,3], but the population of communities adds an additional level of structure. The multi-level population structure creates two levels of selection pressure. The lower level is within-ecosystem selection of microbes, and the higher level is between-ecosystem selection. The coevolutionary nature of the 'inner GA' generates new composite solutions that supply variety to the 'outer GA', which selects for those composite solutions that provide a good solution overall.

Potter et al [7] describe four issues which cooperative solutions must address: problem decomposition, sub-component interdependency, credit assignment, and diversity maintenance. With artificial ecosystem selection as described in this paper, these issues are addressed adaptively. The coevolutionary and ecological dynamics within each ecosystem prevent problems with sub-component interdependencies, since the persistence of a species in the community demonstrates its stable coexistence with other species. This criterion of survival in the community is the only form of species-level credit assignment. The interplay between internal coevolution and external ecosystem selection leads to adaptive problem decomposition; the niche structure that emerges from within-flask dynamics (steered by between-flask selection) maintains sufficient diversity to solve the problem, but no more. In the example given in this paper, single-species (single niche) solutions were successful in the high line, but multi-species (multiple niche) solutions were necessary in the low line. Within-ecosystem ecological dynamics prevent unnecessary redundancy, since similar species compete for the same ecological niche, while microbial evolution allows the species occupying each niche to adapt and potentially improve. The artificial ecosystem selection process thus allows for a wide variety of solutions with minimal user pre-specification, while also including mechanisms to ensure that the solution achieved is appropriate to the problem.

While the above discussion gives reasonable grounds for optimism, the application of artificial ecosystem selection as an optimisation method is unlikely to be straightforward and further theoretical elaboration will be required before the method can usefully be applied. In particular, one challenge will be in determining what kinds of problem are suitable; the ecological scenario described here is not directly representative of many types of optimisation problem. Artificial ecosystem selection may be most suited to the optimisation or design of 'collective intelligence' solutions such as swarms or artificial immune systems, where the community structure is already a feature of the problem. In these scenarios the survival criterion for species (part-solutions) to remain in the community (full solution) will depend on active involvement in solving the target problem; the ecological and coevolutionary selection pressures would be created by competition to contribute and do useful work. Niches would represent different ways of contributing to the whole solution. Future work will apply artificial ecosystem selection to a variety of test optimisation problems.

Acknowledgments. This work was funded by the Leverhulme Trust.

References

1. Goodnight, C.J.: Experimental studies of community evolution 2: The ecological basis of the response to community selection. Evolution 44, 1625–1636 (1990)
2. Madigan, M.T., Martinko, J.M.: Brock Biology of Microorganisms, 11th edn. Pearson, New Jersey (2006)
3. Moriarty, D.E., Miikkulainen, R.: Forming neural networks through efficient and adaptive coevolution. Evolutionary Computation 5(4), 373–399 (1997)
4. Penn, A.: Modelling artificial ecosystem selection: A preliminary investigation. In: Banzhaf, W., Ziegler, J., Christaller, T., Dittrich, P., Kim, J.T. (eds.) ECAL 2003. LNCS (LNAI), vol. 2801, pp. 659–666. Springer, Heidelberg (2003)
5. Penn, A.: Ecosystem selection: Simulation, experiment and theory. PhD thesis, University of Sussex, Brighton, UK (2005)
6. Penn, A., Harvey, I.: The role of non-genetic change in the heritability, variation, and response to selection of artificially selected ecosystems. In: Pollack, J., et al. (eds.) Proc. 9th Int. Conf. Simulation and Synthesis of Living Systems (ALIFE 9), pp. 352–357. MIT Press, Cambridge (2004)
7. Potter, M.A., De Jong, K.A.: Cooperative coevolution: An architecture for evolving coadapted subcomponents. Evolutionary Computation 8(1), 1–29 (2000)
8. Smith, R.E., Forrest, S., Perelson, A.S.: Searching for diverse, cooperative populations with genetic algorithms. Evolutionary Computation 1(2), 127–149 (1993)
9. Swenson, W., Arendt, J., Wilson, D.S.: Artificial selection of microbial ecosystems for 3-chloroaniline biodegradation. Environmental Microbiology 2(5), 564–571 (2000)
10. Swenson, W., Wilson, D.S., Elias, R.: Artificial ecosystem selection. PNAS 97, 9110–9114 (2000)
11. Whitley, D., Rana, S., Heckendorn, R.B.: The island model genetic algorithm: on separability, population size and convergence. J. Comp. Inf. Tech. 7(1), 33–47 (1999)
12. Williams, H.T.P.: Homeostatic Adaptive Networks. PhD thesis, University of Leeds, Leeds, UK (2006)
13. Williams, H.T.P., Lenton, T.M.: Artificial selection of simulation microbial ecosystems. PNAS 104(21) (2007)
14. Williams, H.T.P., Lenton, T.M.: The Flask model: Emergence of nutrient-recycling microbial ecosystems and their disruption by environment-altering 'rebel' organisms. Oikos (In press, 2007)

Building Virtual Ecosystems from Artificial Chemistry

Alan Dorin and Kevin B. Korb

Clayton School of Information Technology
Monash University, Clayton, Australia 3800
{alan.dorin, kevin.korb}@infotech.monash.edu.au

Abstract. This paper adopts an interdisciplinary view of the significant elements of ecosystems and the methods by which these might be simulated to explore theoretical issues of relevance to Artificial Life and Ecology. Artificial Life has largely been concerned with evolutionary ecosystems of agents in trivial environments. Ecology commonly produces models of specific habitats and organism populations unsuited to general exploration of theoretical issues. We propose that limitations of the simulations in these disciplines can be overcome by simulating ecosystems from the level of artificial chemistry. We demonstrate the approach's feasibility by describing several virtual organisms represented at this level. The organisms automatically adopt trophic levels, generate energy from chemical bonds and transform material elements in the process. Virtual organisms may interact with one another and their abiotic environment using the same chemistry. Biosynthesis and decay may also be simulated through this mechanism.

1 Introduction

"This paper discusses a computer model of living organisms and the **ecology** they exist in called PolyWorld." — Yaeger, 1992

"The memory, the CPU, and the computer's operating system are viewed as elements of the **'abiotic' environment**." — Ray, 1991

"A hierarchical computer program has been developed with genetic, organismic and population levels embedded in an **ecosystem**." — Conrad and Pattee, 1970

The terms *ecology*, *ecosystem* and *environment* have been employed extensively within the field of Artificial Life to refer to simulations. To date, virtual ecosystems (the authors' preferred terminology) have been utilized primarily for general studies of evolutionary processes within Artificial Life. For instance, Ray's *Tierra* [1] and Yaeger's *Polyworld* [2] are two early systems that fall into this category. Epstein and Axtell have used a similar system, *Sugarscape*, to study the emergence of social behaviour [3]. Dorin has employed evolutionary virtual ecosystems for aesthetic purposes [4]. In the field of Ecology, individual-based models are more typically non-evolutionary and represent particular, real species and landscapes [5]. The aim of these ecological models is to shed light on specific ecologies where spatially explicit models are required or where the phenomenon under study emerges through the heterogeneity of a population.

The different agendas of Artificial Life and Ecology have to some extent determined the path through which virtual ecosystems in each field have been developed.

F. Almeida e Costa et al. (Eds.): ECAL 2007, LNAI 4648, pp. 103 – 112, 2007.

This paper suggests an approach that melds the interests of Artificial Life researchers and Ecologists. The aim is to demonstrate that a simple artificial chemistry can support the existence of autotrophic and heterotrophic organisms, their interactions with one another and with the abiotic environment. A virtual ecosystem emergent from the simulation can then incorporate natural cycles of matter and energy between organisms of different trophic levels and their abiotic environment. The approach offers an alternative conceptualization of the virtual ecosystem that incorporates the benefits of artificial chemistry and Ecology's individual-based simulations of sub-evolutionary time scales.

The remainder of section 1 explains the significant features of the approach. Section 2 introduces one means of building such a simulation. Section 3 details various organism types and their operation based in the artificial chemistry.

1.1 The Ecosystem

The concept of the ecosystem first appeared in an article by ecologist Tansley [6], whose interest was in the relations amongst organisms and between organisms and their environment. He debated the merits of applying the term *organism* to communities of creatures and their physical habitat. Tansley preferred to refer to this whole as a "system", or to be specific, an *ecosystem.*

Insilico there is no theoretical reason to distinguish between biota and an abiotic environment unless the programmer wishes it. However, there are often practical reasons for making such a distinction — it has thus far proven difficult to simulate multi-level dynamical hierarchies that could be recognized as simulations of chemicals, organisms and ecosystems, although some examples exist (for instance [7]).

A cursory review of the literature in Artificial Life reveals that by convention we refer to agents with sensory capabilities, internal decision-making and resultant behaviours (such as reproduction) as model *organisms*. The abiotic environment is often understood as a space of some kind, often homogeneous, with rules dictating its change of state, its impact on the organisms and vice-versa. This space may include passive "abiotic furniture" (such as grains of sand or sugar [3]) that agents manipulate, thereby introducing an element of heterogeneity to the landscape. The abiotic furniture and the organisms are not usually implemented using the same data-structures. They are treated as "different stuff" within the simulation framework and in the analysis of results. This works acceptably in Artificial Life, as it does in Ecology, where population dynamics is a frequent subject of study. However, this view has left Artificial Life with a concept of "ecosystem" that differs from that held in Ecology regarding the significance of biogeochemical cycles — cycles of the elements through organisms and environment. Erasing the distinction in simulation between organism and environment allows a model to explore the exchange and transformation of matter and energy.

1.2 The Significance of the Environment

An organism-centric view of the universe defines the interests of Biology. At least since Aristotle, the study of autonomous creatures has dominated over the study of their habitat. In a homocentric circle from dust to dust, "Humanity" (organized dust) would sit at the pinnacle of the circle and "Environment" (disorganized dust) would be relegated to a location far beneath our feet. Perhaps the Fates would drive Time's hand around this circle.

We might nowadays reject the homocentric subjugation and exploitation of the universe by placing organisms (including humans) and the abiotic environment on either side of a more balanced circle. Nevertheless, Artificial Life focuses attention on virtual organisms with little direct study into the significance of their habitats. How many publications model agents trivially in order to study the impact of different sophisticated environmental simulations?

Within Artificial Life this bias is also evident in the field's mission to study "life-as-it-could-be". This is generally understood to refer specifically to organisms, not specifically to their habitats (although sometimes these briefly enter the equation). Within the sub-discipline of Artificial Chemistry researchers ponder the properties of basic building blocks that give rise to virtual organisms [8] or autopoietic entities [9]. Here too the research is organism-centric. The study of the environment has been largely neglected even in Artificial Life! There is nothing surprising about the desire of researchers to comprehend what is arguably the universe's most significant phenomenon. Nevertheless a wider perspective is also beneficial.

1.3 Cycles of Biosynthesis and Decay

An undergraduate level Biology text will explain that ecosystems are significant for their cyclic transformation of elements (carbon, hydrogen, oxygen, nitrogen etc.) and that nearly half of the chemical elements participate in such biogeochemical cycles (e.g. [10, p1248]). As long as the virtual ecosystems of Artificial Life maintain differences between biota and abiotic environment in their underlying implementation, our models will not shed much light on these cycles.

In many simulations, organisms are born and roam their environment to gain energy or resources. However virtual organisms rarely acquire matter from the environment to transform it into biomass through a simulated chemical process that expels waste. Simulated organisms often die, but they seldom *rot* to return materials to their environment. Few model ecosystems incorporate decomposers of biomass although there are some examples [11, 12]. The absence of these phenomena from many models stands in contrast to the fundamental importance granted them in elementary Ecology. A model that incorporates these phenomena will be useful for exploring the interactions of organisms at different trophic levels, their impact on the environment and their requirements for energy and material resources. In order to demonstrate the feasibility of the approach, the following section details one manner in which a simulation of this type may be constructed.

2 Cyclic Ecosystem Simulation

The basis of a virtual ecosystem that simulates all of the elements described above must be the common building block of the abiotic environment and the organism, namely the molecule or its atomic components. Molecules will need to be assembled and disassembled, and energy released or captured in the process. Such artificial chemistries are not new but to date nobody has managed to construct metabolising, replicating structures capable of open-ended evolution — an open problem [13]. Presently we hard-wire some of these processes rather than expecting them to emerge.

One approach is to focus on open-ended evolution and to construct models of self-replicating cells that do not store energy in chemical bonds [7]. In our paper we instead stress the importance of the presence of autotrophs (organisms that produce their own food) and heterotrophs (organisms that consume organic materials produced by other organisms). In particular, the model we propose naturally supports the existence of photosynthetic autotrophs, chemosynthetic autotrophs and decomposers by modelling the storage of chemical energy and its release. In this way the concept of an organism, its energy supply and its interactions with the abiotic environment may emerge from artificial chemistry.

The model is designed to support the interactions between various organism types and the abiotic environment as emergent from the interactions of "dumb" molecules. Agent behaviour need not be hard-wired. In fact, any distinction between agent and molecule is absent from the simulation implementation. Agents emerge in the model as large dynamic "molecules" that transform energy and matter. [1]

2.1 Artificial Chemistry

The system is based upon a set of hypothetical two-dimensional, mobile, non-intersecting, square "atoms" placed on a grid. Atoms may bond to neighbours at their edges by sharing virtual electrons, according to the rules of a virtual chemistry of covalent bonding. Each atom type has a set of electrons in one or more shells. The number of electrons and the fullness of an atom's outer shell determine the bonds in which the atom can participate (for details, see [14]). In all cases, some energy threshold is required to initiate (or break) a bond, and the result of bonding (or breaking a bond) is either the capture of some amount of energy or its release. Additionally, for each type of bond, parameters of the simulation determine the probability of bonds forming or breaking given the availability of the threshold energy.

A catalyst is said to be present at a reaction site when an atom involved in the reaction neighbours an atom of a designated catalyst-type. To support the existence of our virtual organisms, four types of catalyst are required. A chlorophyll-like catalyst is needed that, in the presence of sunlight, manufactures a complex molecule equivalent to sugar. An enzyme that breaks down this sugar, releasing the chemical energy stored in its bond is also needed. For simplicity and clarity, separate enzymes that decompose "organic" bonds that are not sugar and "inorganic" bonds may be added. Further details are given in section 3.

Energy that is released during a reaction is distributed throughout any continuous atomic structure that contacts directly or indirectly (through intermediate neighbours) the reaction site. This energy is available for making or breaking chemical bonds by the atoms that receive it.

---

[1] Our model is not concerned with reproduction, evolution or the self-assembly of organisms.

2.2 A Simulation Time-Step

The simulation progresses in discrete time-steps. At each step, for each atom, it must be determined stochastically whether each bond should break or join based on the site-types, the presence of catalysts, the availability of energy and the probabilities for bonding. Energy released during a reaction is totalled in each neighbourhood of connected atoms for use in that time-step by reactions that absorb energy. A reaction that requires energy to be expended can only occur if the neighbourhood of atoms involved has accumulated sufficient energy. Reactions occur in random order, consuming or contributing energy to and from the total amount available in their neighbourhood. A reaction that requires more energy than is available cannot proceed. Energy released from a chemical bond must be used in that time-step or it is released in non-recoverable form. The only way energy can be stored is in complex molecules. Sunlight is incident on all atoms at a rate governed by a parameterized sine function for use by the chlorophyll-like catalyst during photosynthesis (section 2.1).

In addition to stochastically determined bonding, within a time-step atoms may be moved one square on the grid in a random direction or they may remain stationary. Bonded atoms (forming molecules) are moved identically to preserve their topology. Collisions are not permitted.

3 Constructing Virtual Organisms from Artificial Chemistry

Our simple artificial chemistry supports the existence of a complete simulated ecosystem of photosynthetic and chemosynthetic autotrophs, heterotrophs (including decomposers) and their abiotic environment. Below we detail the biologically inspired construction of each of these organism types and explain its behaviour. The model obviously abstracts much of the detail of real systems for reasons of practicality.

3.1 The Abiotic Environment

The abiotic environment consists of the same molecules and atoms as constitute the simulation's organisms. Their bond structure and context enables us to label molecules as inorganic or organic, as a component of a metabolic system or as abiotic. Thus the "abiotic environment" is the set of atoms and molecules that are not bonded to a structure identified as an organism.

The abiotic environment consists of virtual atoms from the set {A, B, C, O}. Atoms may also be enzymes for sugar decomposition (break A-B bonds), biomass decomposition (break C-C bonds) or chlorophyll for sugar construction (make A-B bonds, break A-O and B-O bonds). The probabilities for these significant reactions are given in the simplified reaction table 1.

In order to sustain the required organism forms, the various reactions detailed in the sections below must be supported. Only bonding possibilities of relevance to the discussion are described even though other bonds may also be supported. A line between tiles in the diagrams below indicates a bond between the atoms these represent. Catalysts are labelled *Enz* (enzyme) or more specifically *Chl* (chlorophyll).

Table 1. Simplified reaction table. Bond energy must be supplied to break a specified bond and is released when the bond is made. (Negative bond energy values indicate a release of energy when a bond breaks and energy must be supplied to make the bond.)

Bond	Make probability	Break probability	Make probability (catalyst)	Break probability (catalyst)	Bond energy
A-B	low	low	high (chl)	high (enzAB)	- high
C-C	moderate	low	—	high (enzCC)	- low
A-O	high	low	—	high (chl, enzAO)	+ low
B-O	high	low	—	high (chl, enzBO)	+ low
C-O	low	moderate	high (enzCC)	—	+ low

3.1.1 Photosynthesis

The process of photosynthesis constructs sugar molecules from water and carbon dioxide, releasing oxygen and water. It requires the presence of the catalyst chlorophyll and incident sunlight. In our virtual chemistry the reaction is represented abstractly as:

AO + BO —(chlorophyll & sunlight)—> AB + 2O

Reactants A and B may be presented to the catalyst bonded to O atoms or free. The process of photosynthesis may be represented in many potential atom arrangements. An example is given in figure 2.

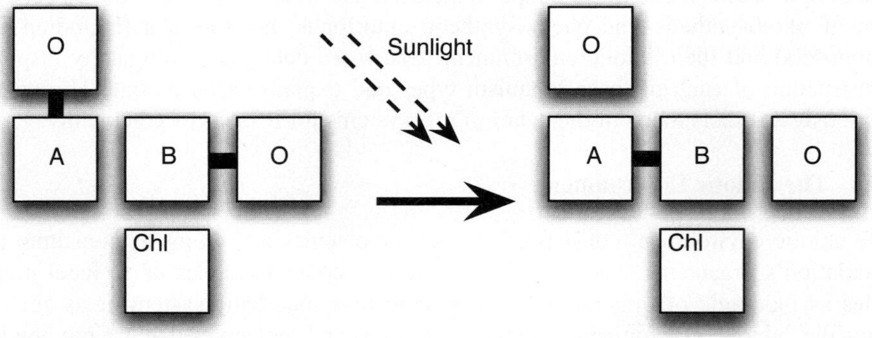

Figure 2. *Virtual photosynthesis* breaks A-O and B-O bonds (on the left of the figure) to manufacture an A-B molecule (on the right) employing a chlorophyll catalyst and sunlight.

3.1.2 Respiration

Real sugar molecules may be broken down in a process of respiration to release energy. The process utilizes oxygen and an enzyme to break down the sugar. Carbon dioxide and water are released. In our virtual chemistry the reaction is represented abstractly:

O + AB —(enzyme)—> A + BO + energy

An example respiration reaction is given in figure 3.

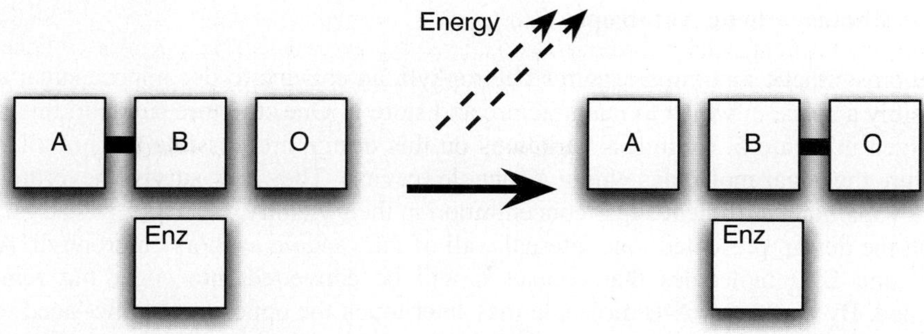

Figure 3. *Virtual respiration* breaks an A-B bond (on the left of the figure), employing an enzyme and releasing energy, A and B-O (on the right).

3.1.3 Biosynthesis

Organisms are able to add to their biomass by constructing bio-molecules from those they ingest. Growth occurs when a structure binds to atoms employing a reaction that requires energy. These bonds would not normally form in the absence of the energy required to create them (or the presence of a suitable catalyst). Such bonds may also break down spontaneously with probabilities as indicated in table 1. Hence an "organic" structure formed of these bonds must produce sufficient energy to sustain itself against natural decay by rebuilding broken bonds and by adding new material in a breach. When an organic bond is broken energy is released into the neighbouring structure. The amount that may be captured by a particular neighbouring atom will be insufficient to remake the bond instantaneously without an additional energy source. In our virtual chemistry the biosynthesis reaction is represented abstractly:

$$C + C \text{ —(energy)—> } C_2$$

One possible configuration for the biosynthesis reaction is depicted in figure 4.

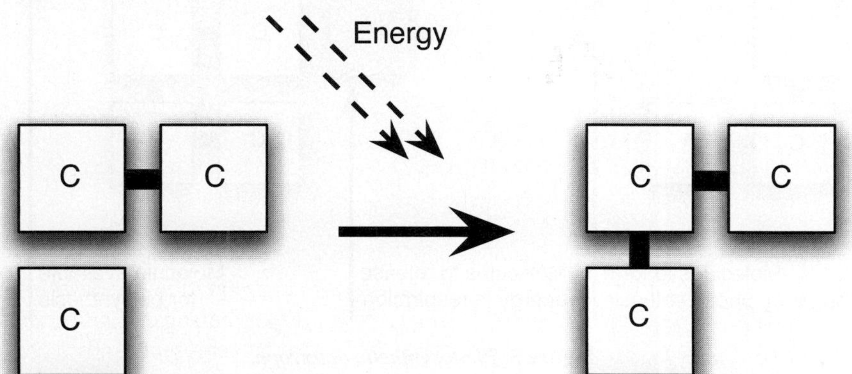

Figure 4. *Virtual biosynthesis* forms C-C bonds by consuming energy.

3.2 Photosynthetic Autotroph

A photosynthetic autotroph requires chlorophyll, an enzyme to decompose sugar and possibly a space in which to manufacture and store it. One structure suited to this role is given in figure 5. Countless variations on this design are possible but not all will capture the sugar molecules within a vacuole (cavity). They may survive nevertheless if they maintain sufficient sugar concentration in their vicinity.

In the design presented, one internal wall of the vacuole anchors chlorophyll. Any A-O and B-O molecules that contact it will be converted into sugar but remain trapped. By chance an A-B molecule may later touch the opposite enzyme-laced wall where it will participate in respiration, releasing energy through the structure and allowing biosynthesis.

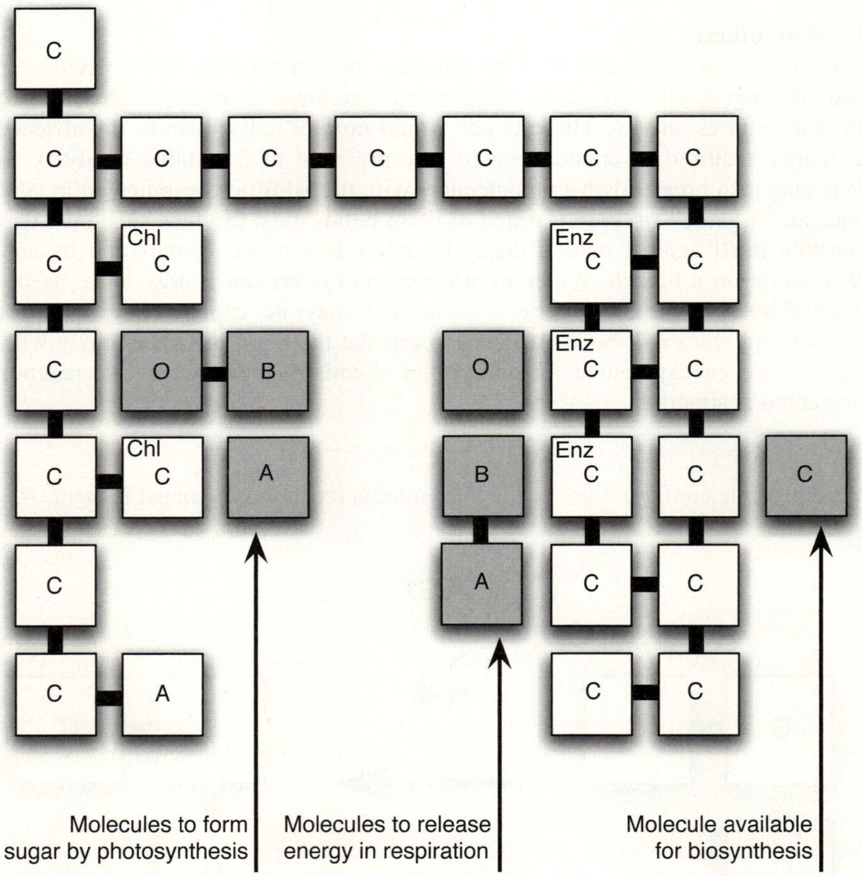

Figure 5. *Photosynthetic autotroph.*

3.3 Chemosynthetic Autotroph

A chemosynthetic autotroph generates its energy from inorganic molecules without the need for sunlight. In the chemical system described, one way to achieve this is by obtaining free O atoms and binding them to available A or B atoms. Given the natural affinity of A and B for O in the model, suitable atoms may be scarce unless a catalyst is employed to split A-O and B-O. The elements may then rejoin against the surface of the structure. In this case almost any structure containing a suitable catalyst on its surface would suffice in the role.

3.4 Heterotroph

A heterotroph in the present model breaks down A-B sugars produced by other organisms using an enzyme. Alternatively, a heterotroph may act as a decomposer if it possesses a catalyst to break down the C-C bonds of another structure. This catalyst must be isolated from its own organic C-C structure. One way to achieve this is with an intermediate O atom bonded to C with the aid of a catalyst (figure 6).

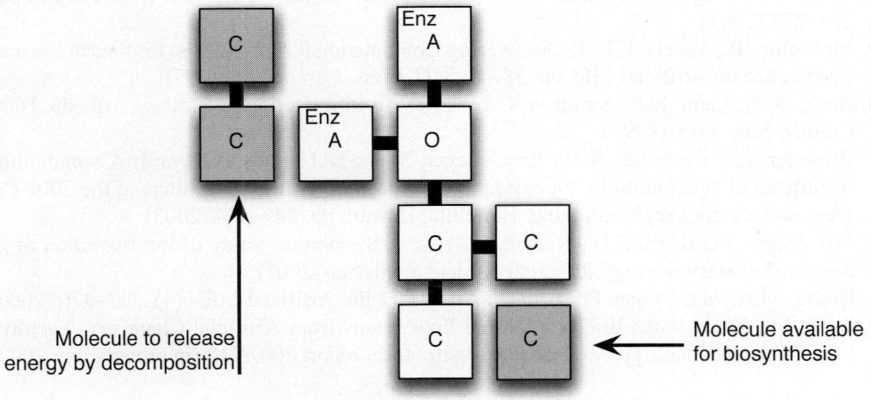

Figure 6. *Decomposer.*

4 Conclusions and Future Work

At the time of writing, a simulation run with all organism types present simultaneously remains to be performed. Since each is based upon the same set of chemical rules, at least in principle there is nothing preventing this from being successful. In the future we hope to adjust the chemical rules to facilitate the self-assembly and eventually, the replication and evolution of structures within the virtual ecosystem.

The host of organism structures presented here demonstrates the feasibility of constructing a complete virtual ecosystem from even a simple artificial chemistry in which organisms: emerge from the same abiotic elements as their environment; transform these elements in simulated biogeochemical cycles; acquire energy from chemical bonds; produce waste products and utilize the waste and biomass of other organisms.

References

1. Ray, T.S.: An approach to the synthesis of life. In: Langton, C.G. (ed.) Artificial Life II, Santa Fe, New Mexico, pp. 371–408. Addison Wesley, Reading (1990)
2. Yaeger, L.: Computational Genetics, Physiology, Metabolism, Neural Systems, Learning, Vision and Behavior or Polyworld: Life in a New Context. In: Langton, C.G. (ed.) Artificial Life III, pp. 263–298. Addison-Wesley, Reading (1992)
3. Epstein, J.M., Axtell, R.: Growing Artificial Societies, Social Science from the Bottom Up. Brookings Institution Press, Washington (1996)
4. Dorin, A.: The Virtual Ecosystem as Generative Electronic Art. In: Raidl, G.R., Cagnoni, S., Branke, J., Corne, D.W., Drechsler, R., Jin, Y., Johnson, C.G., Machado, P., Marchiori, E., Rothlauf, F., Smith, G.D., Squillero, G. (eds.) EvoWorkshops 2004. LNCS, vol. 3005, pp. 467–476. Springer, Heidelberg (2004)
5. Grimm, V.: Ten years of individual-based modelling in ecology: what have we learned and what could we learn in the future? Ecological Modelling 115, 129–148 (1999)
6. Tansley, A.G.: The Use and Abuse of Vegetational Concepts and Terms. Ecology 16(3), 284–307 (1935)
7. Hutton, T.J.: Evolvable Self-Reproducing Cells in a Two-Dimensional Artificial Chemistry. Artificial Life 13(1), 11–30 (2007)
8. Dittrich, P., Ziegler, J., Banzhaf, W.: Artificial Chemistries - A Review. Artificial Life 7(3), 225–276 (2001)
9. McMullin, B., Varela, F.J.: Rediscovering Computational Autopoiesis. In: Fourth European Conference on Artificial Life, pp. 38–47. MIT Press, Cambridge (1997)
10. Beck, W.S., Liem, K.F., Simpson, G.G.: Life: An Introduction to Biology, 3rd edn. Harper Collins, New York (1991)
11. Brewster, J.J., Reynolds, R.G., Brockmeyer, M.A.: Not In My Backyard: A simulation of the effects of agent mobility on environmental poisoning. In: Proceedings of the 2002 Congress on Evolutionary Computing, Honolulu, Hawaii, pp. 849–854 (2002)
12. Woodberry, O., Korb, K.B., Nicholson, A.E.: A Simulation Study of the Evolution of Aging. Evolutionary Ecology Research (under submission, 2007)
13. Bedau, M.A., et al.: Open Problems in Artificial Life. Artificial Life 6(4), 363–376 (2000)
14. Dorin, A., Korb, K.B.: Building Virtual Ecosystems from Artificial Chemistry, Faculty of Information Technology, Monash University, tech. report 2007/212 (in preparation, 2007)

Energy Flows and Maximum Power on an Evolutionary Ecological Network Model

Jiang Zhang

Complex systems research center, Academy of Mathematics and systems sciences, Chinese Academy of Sciences, Beijing, China, 100080
Zhangjiang@amss.ac.cn

Abstract. Energy flows in ecological systems which are determined by the structure of the ecological network influence the evolution of the network itself. The total system energy throughflow as an important indicator of the co-evolution of network and flows in the ecosystem can be maximized spontaneously according to the maximum power principle. This principle should be thought as an emergent and evolutionary property of the system. To address the problem of how this principle functioning theoretically, a simple model that exhibits the long term evolution of the ecological network determined by the fast dynamics of the energy flows was presented. Maximum power with the diffusion in the phenotype space was investigated in various settings. Accordingly, the conclusion that the total energy throughflow on the network and the diversity are always positive correlated was drawn.

Key words: Energy flow, Evolutionary ecological network, Maximum Power Principle.

1 Introduction

Energy flows on the ecological network had been studied by ecologists for a long time[1][2]. There are few prevalent phenomena of flows on the ecological network such as the hierarchy of the energy transformation[2], the trend of maximum of energy flux with the evolution of the system. The latter observation was presented as the famous principle of maximum power (MPP) which was first coined by lotka[3] and developed by Odum[4]. But how does MPP happen in the natural ecosystems still left seldom study neither from the empirical experiments nor the theoretical analysis.

A variety of ecological goals or indicators were drawn to indicate the evolution of the ecosystem[5][6]. These ecological orientors were integrated in a framework of Environ model[7] by fath so that different indicators correspond to the specific properties of the network flow model[8]. Fath's work encouraged us to suppose the maximum power principle can be explained as one of the systematic properties of the ecological network. Hence, natural selection takes effect on the species leading the network structure changed and accordingly maximizing the energy flux through the network.

Many dynamic network models as the most effective tools for the theoretical investigation of energy flows were developed. For example, Jorgensena et al. [9] explored various ingredients influencing the flows on the network; Fath[10] used large cyber

F. Almeida e Costa et al. (Eds.): ECAL 2007, LNAI 4648, pp. 113 – 122, 2007.

networks to test several attributes of the flows, also he tried to integrate the Boolean network model and the network flow model to study the correlation between flows and evolution[11]. The absence of the network evolution is the major shortage of these studies, especially for the scenario of the co-evolution of the network structure and the energy flows on it, which is observed frequently in the real ecosystems.

Many fruitful and powerful models investigating the coupled dynamics of population dynamics on vertices and the overall network structure were presented in recent years because of the development of complex network research area[12][13][14]. Many works focused on the speciation and evolutionary branching process by using the evolutionary network techniques or their variants[15][16][17][18]. This paper hopes to contribute to integrating these works to address the problem of how energy flows develops on the evolutionary graph.

Firstly a flow model on the complex ecological network supported by the population dynamics is presented. The evolution of graph is implemented by an approach of matching between two points of phenotype space. Then, the total flux of energy and the diversity of the species are reported as the main simulation results.

2 The Model

2.1 Population Dynamics and Energy Flow on the Ecological Network

An ecological network is a graph $\Gamma =< N,E >$, where N is the set of vertices to represent n species, any element $< i, j >\in E$ is a directed arc from i to j denoting the energy flow from i to j.

A real value vector $(x_1, x_2, \cdots, x_n)$ is assigned to set N. Each component of the vector represents the corresponding species' biomass or the equivalent energy store. Each x_i's (i>1) dynamics can be written as the equation.

$$\frac{dx_i}{dt} = e \sum_{<j,i>\in E} a x_i x_j - \sum_{<i,j>\in E} a x_i x_j - b x_i - c x_i^2 \quad for\ i = 2,3,...,n \tag{1}$$

Where e,a,b,c are constants represent: ecological efficiency, interaction strength, the death rate and the intra-species competition coefficient respectively. It should be claimed that in this study e may >0.1 because x_i is not the number of the species but the biomass or the equivalent energy store although it takes e=0.1 for most of ecological literatures[17][18].

The first vertex is a special one like sunlight to provide the energy inward for the whole system. It is modeled as a resource pool with the constant influx and fixed rate of leak. So x_1 is the resource storage in the pool, it can be presented as:

$$\frac{dx_1}{dt} = F - \sum_{<1,j>\in E} a x_1 x_j - b x_1 \tag{2}$$

Where, F is a constant denoting the inward resource from the environment in each time step. And the second term represents the usage of the resource by the plants in the system. The third one is the wasting energy of the whole system.

Actually, the equations (1) and (2) expressed the energy flow in the system once they are rewritten as the flow form:

$$\frac{dx_i}{dt} = I_i + \sum_{<j,i>\in E} f_{ji} - \sum_{<i,j>\in E} f_{ij} - D_i, \quad for\ i = 1,2,...,n \tag{3}$$

Where I_i is the energy influx to the vertex i, therefore it is F for i=1 and 0 for other vertices. f_{ij} is the energy flux from i to j at time t, it holds $f_{ij} = f_{ji}$. It is $ax_i x_j$ for all edges <i,j>. D_i is the energy dissipation term. For i>1, it is $(1-e)\sum_{<j,i>\in E} f_{ji} + bx_i + cx_i^2$. The first term is the energy dissipation for the inter-species transform. Also the loss of biomass because of the death of individuals and intra-species is treated as the energy dissipation.

With these equations, the vector $(x_1, x_2, \cdots, x_n)$ can be updated at each time step accordingly. The steady state of the system is the particular vector $(x_1*, x_2*, \cdots, x_n*)$ satisfying the condition:

$$\frac{dx_i}{dt} = 0, \text{ for any } i \in N \tag{4}$$

By using Euler's approach to simulate the equations instead of solving them, the steady state vector can be substituted by the average values of x_i during a large time interval T. And it will be used by the evolutionary algorithm of the graph.

2.2 Phenotype Space and the Construction of the Network

The ecological network is constructed by means of matching between phenotypes of each pair of vertices. Each species in the network is associated to a simplified phenotype (u,r) in $U \times V = [0,1]^2$, where the u component is the utilization (foraging) trait and v component is the resource (vulnerability) trait[15][16][17]. That is we can use the traits of predation and prey to depict each species in the network.

And then iterate for all possible pairs of species in the system: i with the phenotype (u_i, r_i) and j with (u_j, r_j) in N, we have:

$$<i,j>\in E \text{ if } |u_j - r_i| < \varepsilon \tag{5}$$

$$<j,i>\in E \text{ if } |u_i - r_j| < \varepsilon$$

Here, ε is a parameter for the matching precision. The formula means that if the utilization trait of j is approaching to the resource trait of species i, then j can predate i, thus a connection from j to i is drawn on the network. Because of the particular dynamics of the resource vertex of 1, it can only be consumed by other vertices, so their utilization trait is omitted. By using this method for any possible pair in N, the whole network can be built.

2.3 The Evolution of the Network

With the underlying population dynamics of each node and the construction approach by means of matching in the phenotype space, the evolutionary dynamics of the network can be presented. The species pool can be treated as a tournament to undergo the Darwinism evolution. The detail of the approach[12] is presented in several steps:

1. *Initialize* $x_i = m_0$ *for all vertices and* (u_i, r_i) *respectively.*
   - *1.1) For i=1,* r_i *is set by a fixed value 0*
   - *1.2) For any i>1,* $u_i = 0, r_i = rand(r_0, 1)$, *where* $rand(r_0,1)$ *is a uniform random number distributed on interval* $(r_0, 1)$
2. *Construct the network according to the approach presented in section 2.2*
3. *Evolve the population dynamics according to the equation (1), and calculate the steady state vector X*.*
4. *Natural selection applied on vertices but 1*
   - *4.1) Select the minimal x in N, i.e.* $k = \arg\min_{i \in N-\{1\}} x^*_i$, *delete m with its all connections in the network.*
   - *4.2) Select one of the element k'(u,v) from N-{1} by means of roulette selection.*
   - *4.3) A new species i* with* $x_{i*} = m_0$ *and (u*,r*) which is determined by the formula (6) is added in the network:*

$$u^* = u + \zeta_1, v^* = v + \zeta_2 \tag{6}$$

   Where $\zeta_1, \zeta_2 \sim N(0, \sigma^2)$ *are two normal random variables. And the cyclic boundary condition is used in the phenotype space.*

5. *Go back to 2.*

One iteration from step 2 to 5 is called a cycle. As the cycle forwards, the network undergoes the Darwinism evolution.

3 Simulation Results

3.1 Basic Settings

The simulation is implemented with the basic parameters settings are listed in table 1.

Table 1. Basic parameter settings

Parameters	Value	Parameters	Value
F	100	n	100
a	0.0005	ε	0.1
e	0.4	m_0	5
b	0.0005	σ	0.1
c	0.0005	r_0	0.5
T	100		

3.2 Different Ways of Energy Utilization

The energy flows in the network with multiple complex pathways. The total system throughflow (TST) is a useful indicator to provide the systematic information of energy flow[22]. TST is defined by:

$$TST = \sum_{i=1}^{n} \sum_{j=1}^{n} f_{ij} \tag{7}$$

The energy influx into the system should be counted in TST [22] but (7) omits this term just for the convenience because the influx is always the constant F

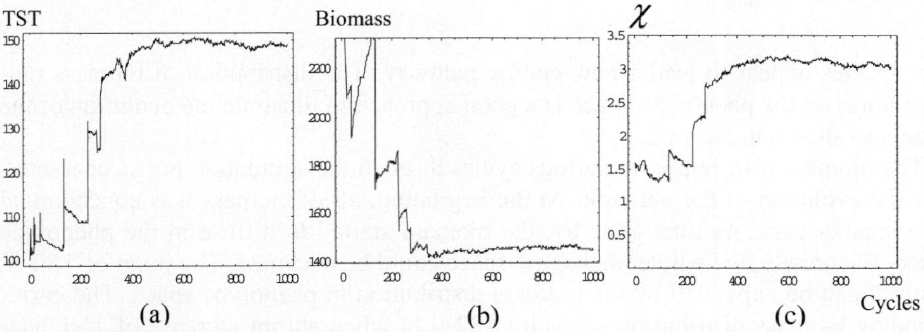

Fig. 1. TST as the function of cycles (a) compared with the total biomass time series(b). And the diversity of phenotypes as the function of cycles is shown in (c).

Figure 1(a) is the curve of TST time series compared with (b): total biomass ($\sum_{i=1}^{n} x_i$) time series. At the beginning the system was initialized as all of species were plants being able to absorb energy from the "sunlight" resource directly, thereafter the total input energy was maximized and the biomass that is related to the input energy was at apex. Fewer producers could lead the total energy income decrease because the simulation fixed the number of species. That is the main reason for the decrease of the total biomass. In spite of the assumption of the fixed species number leads to the inappropriate conclusion of the decrease of total biomass as system evolved, it is kept because it makes the model much simpler and shows the prominent effect of TST.

The decrease of biomass means less energy lost because the dissipation of each vertex relies on its biomass monotonously. And because the total influx to the system but vertex 1 equals to the total dissipations in the steady state, the influx to the system will reduce. The increase of TST with the decrease of influx and dissipation means more and more energy was invested on the transformation between species. In another word, the energy flows in the whole system was accelerating.

3.3 Diffusion in the Phenotype Space

Several prominent jumps of the curves can be observed from the figure 1 (a) and (b). The abrupt increase of the energy flux means the flow bottleneck was broken, i.e. a

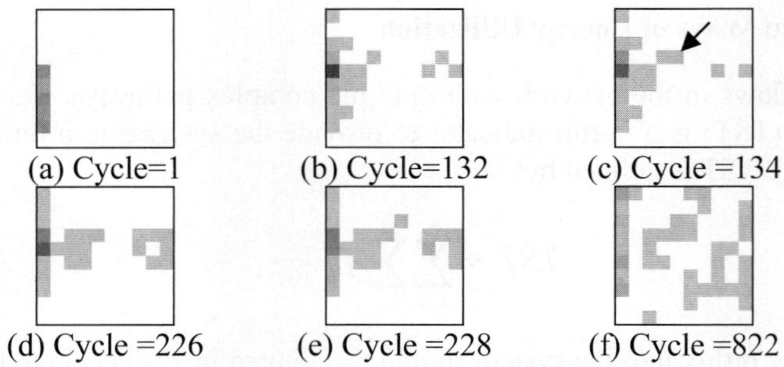

Fig. 2. The distribution of biomass in the phenotype space in different cycles from (a) to (f). The horizontal axis is u trait and the vertical axis is the r trait, the gray level of each cell reflects the concentration degree of biomass for that traits.

new species appear to lead a new energy pathway. The distribution of biomass (energy store) on the phenotype space is a good approach to illustrate the evolution of the system as shown in figure 2.

The biomass distribution in various cycles through the simulation per se characterizes the evolution of the network. At the beginning, all of biomass was concentrated on a smaller area. As time gone by, the biomass started to diffuse in the phenotype space. Eventually, the whole phenotype space could be occupied. The jump of TST in figure 2 can be explained by the biomass distribution in phenotype space. The corresponding biomass distribution of cycles 132~134 when abrupt increase of TST happened are depicted by figure 2 (b),(c). In (b), the species that occupied the black cell concentrated lots of energy to form an energy flow bottleneck. The appearance of new species that belongs to the phenotype space arrowed cell in fig.2(c) broke this energy monopoly by predating the species in black cell. This little change in the phenotype space could cause a major increase of energy TST. Once again the same story happened in cycles 226~228 that can give a good reason for the second jump of TST in figure 3.

The overall trend of the biomass distribution in phenotype space is obviously more random due to the diffusion of phenotypes from the observations through (a)~(f). This intuitive phenomenon can be captured by the species diversity indicator:

$$\chi = -\sum_{i=1}^{1/\varepsilon}\sum_{j=1}^{1/\varepsilon} p_{ij}\log p_{ij} \tag{8}$$

where $p_{ij} = \sum_{\substack{u_k \in V_i \\ v_k \in V_j}} \frac{x_k}{M}$, $V_i = [i/\varepsilon, (i+1)/\varepsilon]$, $M = \sum_{i=1}^{n} x_i$

Figure 1(c) shows the diversity of phenotypes increased, which can explain the observations from biomass distribution. The pattern of diversity increasing is similar with the TST increasing that underwent several jumps to reach a maximum top and stayed at the high level, hence they are correlated from this simulation.

3.4 General Trends in Different Settings

The conclusion that the diffusion of phenotype of species and the maximum of energy flow on the network have the positive correlation can be drawn from the simulation above, but is it a pervasive result?

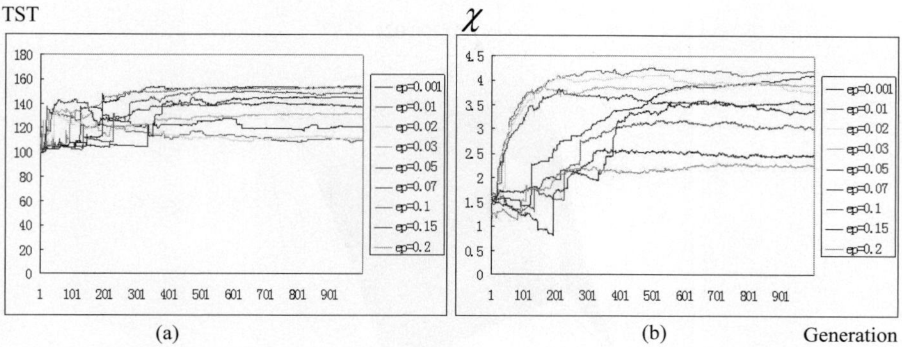

Fig. 3. TST(a) and diversity χ (b) time series with different ε .

Large number of computer simulations were tested in different parameter settings, most of them exhibited the similar patterns of increase of TST and diversity simultaneously. There were also some exceptions such as the decrease of diversity with the decrease of TST can be observed when σ was very small (smaller than 0.005) and kept invariant of other parameters (The figure is omitted). This phenomenon can be explained by the difficulty of seek of the better phenotype matching pairs increased when σ was small. Although the increase of TST and diversity will be broken in some special conditions, these two indicators always exhibit same trends, i.e. they are correlated. But when we change the parameter ε , this law can be broken.

As the figure 3 (a)(b) shows, the trends of increase of diversity and the concurrent decrease of TST can be observed when ε <0.03. The small value of ε means the matching task is harder, that is the whole system will hardly find the way to release its energy flow, as a result the TST decreased. In the meanwhile the diversity still rose because of the diffusion in the phenotype space. That is the reason for the negative correlation of TST and diversity. The detail of the relationship between TST and the diversity will be discussed in the next section.

3.5 Diversity and TST

It is necessary to investigate the relationship between diversity and TST since they have negative correlations sometimes when the matching precision is very low although they are positive correlated in most of time,

At first, the correlation coefficient ρ can be a good statistical indicator to measure the correlation of diversity and TST. Hence, in each run of the simulation, the correlation coefficient can be calculated from the paired time series.

Secondly, it is plausible that the number of vertices n can be a complementary ingredient to the small matching precision, i.e. the bigger n with the small ε will augment the correlation coefficient than the cases of smaller n with the same ε. Therefore, n and ε as two plausible variables to determine ρ can be drawn in a 3-D map in figure 4.

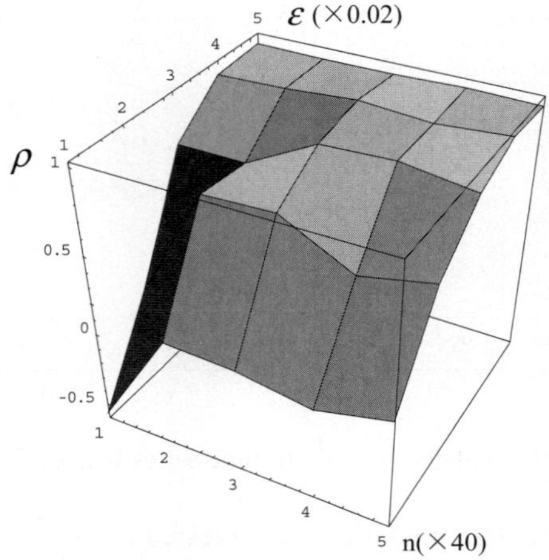

Fig. 4. Correlation of diversity and TST ρ with different ε and n.

As the figure shows, the indicator ρ rises with the increase of n and ε. n and ε are complementary for ρ because larger population provides more opportunity to build prey-predator connections although the difficulty caused by the small matching precision exists. Also n and ε can be viewed as the reference of the complexity of the network. Once the complexity of the network is beyond a threshold, the correlation between TST and diversity are positive.

In ecology, a large number of papers discussed the relationship between diversity and productivity[19][20][21]. But the notion of total system throughflow is different to the productivity because the later refers to the ability of producers or animals absorbing and utilizing resources, i.e. the influx of energy to some vertices, while the former refers to the total flux for the whole system. Although it seems that the diversity of the species could lead the increase of productivity, many literatures argued that the relationship between them are different in accord with the environments and conditions[19][20]. It is more possible that diversity and productivity can influence each other to form feedback loops[21]. The similar relationship between diversity and TST can be found in our experiments.

4 Discussion

The energy and resource flows play important roles in the real ecological systems but conventional models of ecological network mainly focused on the topology and structure. The partial reason of neglecting the flows on the network is the absence of tools to formulize the flows with the evolution of the network. Ordinary differential equations based population dynamics on the ecological network can be viewed as a flow dynamics. This idea makes it possible to address the energy flow problems.

Maximum power principle as one of the most important prevalent hypothesis should be treated as a systematic and topological properties with the evolution of the ecological networks. Hence, this trend should be reflected by the evolutionary network model.

TST as a good indicator of power in the system increases as the evolution of the network in most of cases from the large number of computer simulations. One of the reason is the diffusion on the phenotype space enlarges the possibility of formation of new energy pathways. And the proliferation of new energy pathways can lead the increase of TST. Nevertheless, the positive correlation of diversity of the phenotypes and the TST can only be observed it's possible that the diffusion in the phenotype space cause the increase of TST. But the positive correlation of diversity and TST will not be observed when the smallest complexity threshold of the network can't be satisfied. This conclusion is coordinated with empirical observations.

The positive correlation of diversity and TST can provide some deep insights of thermodynamics. Because the diffusion phenomenon is in accord with the second thermodynamic law, the acceleration of the energy flows can be explained as a natural extension of the second law. Once we view the phenotype space as a higher level space, then the second thermodynamic law will drive the diffusion in this space. This will lead the concurrent energy flow in the lower space accelerate. Thus, the maximum power principle can be viewed as the logical result of the second law. This conclusion may be the boldest hypothesis in this paper. But much more evidences and detail analysis are needed in the future works.

Acknowledgments. Thanks for the supporting of the Guozhi Xu Posdoctoral Research Foundation and National Natural Science Foundation of China (No. 60574068).

References

1. Odum, H.T.: System Ecology. John Wiley & Sons, Chichester (1983)
2. Odum, E.P.: Fundamentals of Ecology. W.B. Saunders Company, Philadelphia (1971)
3. Lotka, A.J.: Contribution to the energetics of evolution. Proceedings of the National Academy of Sciences of the United States of America 8, 147–151 (1922)
4. Charles, A.S. (ed.): Maximum power: the ideas and applications of H.T. Odum. University Press of Colorado, Boulder (1995)
5. Martyusheva, L.M., Seleznevb, V.D.: Maximum entropy production principle in physics, chemistry and biology. Physics reports 426, 1–45 (2006)
6. Felix, M., Maren, L. (eds.): Eco targets, Goal functions, and Orientors. Springer, Heidelberg (1998)
7. Fath, B.D., Patten, B.C.: Review of the Foundations of Network Environ Analysis. Ecosystems 2, 167–179 (1999)
8. Fath, B.D., Patten, B.C., Choi, J.S.: Complementarity of Ecological Goal Functions. Theoretical Biology 208, 493–506 (2001)
9. Jorgensena, S.E., Fath, B.D.: Examination of ecological networks. Ecological Modeling 196, 283–288 (2006)

10. Fath, B.D.: Network analysis applied to large-scale cyber-ecosystems. Ecological Modeling 171, 329–337 (2004)
11. Fath, B.D., Grant, W.E.: Ecosystems as evolutionary complex systems: A synthesis of two system-theoretic approaches based on Boolean networks. Environmental Modeling and Software 22, 693–700 (2007)
12. Jain, S., Krishna, S.: A model for the emergence of cooperation, interdependence, and structure in evolving networks. Proceedings of the National Academy of Sciences of the United States of America 98, 543–547 (2001)
13. Jain, S., Krishna, S.: Large extinctions in an evolutionary model: The role of innovation and keystone species. Proceedings of the National Academy of Sciences of the United States of America 99, 2055–2060 (2002)
14. McKane, A.J.: Evolving complex food webs. The European Physical Journal B 38, 287–295 (2004)
15. Doebeli, M., Dieckmann, U.: Evolutionary Branching and Sympatric Speciation Caused by Different Types of Ecological Interactions. The american naturalist 156, 77–101 (2000)
16. Rossberg, A.G., Matsuda, H., Amemiya, T., Itoh, K.: Food webs: Experts consuming families of experts. Journal of Theoretical Biology 241, 552–563 (2006)
17. Hiroshi, C.I., Ikegami, T.: Food-web formation with recursive evolutionary branching. Journal of Theoretical Biology 238, 1–10 (2006)
18. Drossel, B., Higgs, P.G., McKane, A.J.: The Influence of Predator}Prey Population Dynamics on the Long-term Evolution of Food Web Structure. Journal of Theoretical Biology 208, 91–107 (2001)
19. Poulin, R., Mouillot, D., Nascimento, M.G.: The relationship between species richness and productivity in metazoan parasite communities. Oecologia 137, 277–285 (2003)
20. McCann, K.S.: The diversity-stability debate. Nature 405, 228–233 (2000)
21. Worm, B., Duffy, J.E.: Biodiversity, productivity and stability in real food webs. Trends in Ecology and Evolution 18, 628–632 (2003)
22. Finn, J.T.: Measures of ecosystem structure and function derived from analysis of flows. Journal of Theoretical Biology 56, 363–380 (1976)

Entropy Production in Ecosystems

Nathaniel Virgo and Inman Harvey

Centre for Computational Neuroscience and Robotics
University of Sussex, Brighton, UK
nathanielvirgo@gmail.com
inmanh@sussex.ac.uk

Abstract. We present an extremely minimal ecosystem model which takes account of thermodynamic constraints on the organisms' metabolism. This suggests a way to test the application of a hypothesised principle of Maximum Entropy Production to ecosystems. It also puts definite physical bounds on the rates at which matter can flow through the system and paves the way for more detailed models that have thermodynamic principles built in from the start. In providing the background for this model we point out some connections between thermodynamic principles and autopoiesis.

1 Introduction

Artificial life has had some successes at emulating many aspects of life in abstract and simulation models. In order to do this some aspects of life in the physical world are included in the model and the remainder assumed irrelevant for the study in question. Most ALife studies have ignored thermodynamic issues such as energy use and entropy production, yet if one considers living organisms as self-sustaining processes such issues are fundamental.

The idea that ecosystems act so as to maximise a thermodynamic quantity such as energy flow has been a running theme throughout the ecological literature [17]. In a parallel series of developments, notions of a principle of Maximum Entropy Production (MEP) have been discussed for complex non-linear systems that are maintained in a steady state out of thermodynamic equilibrium by external forces. So far the most successful applications of this idea have been to large-scale models of planetary climates driven by energy input from the sun [13,6]. It has been hypothesised that the MEP principle might be general enough to apply to apply to living systems as well. In particular, MEP is thought to hold arbitrarily far from equilibrium, making its applicability to living systems uniquely plausible.

We present an extremely simple model of an ecosystem which takes account of thermodynamic constraints on the organisms' metabolism. We find that there is a peak in entropy production at a particular value for the rate of matter flux through the system. This demonstrates a way in which the hypothesised principle of Maximum Entropy Production could be tested experimentally in ecological systems. It also puts definite bounds on the rates of flux that can be

F. Almeida e Costa et al. (Eds.): ECAL 2007, LNAI 4648, pp. 123–132, 2007.

achieved and paves the way for more detailed models that have thermodynamic constraints built in from the start.

A further contribution of this paper is a clarification of the fundamental interrelationship between the autopoietic perspective and thermodynamics in section 2.1.

2 Background

2.1 Thermodynamics Applied to Life

Organisms have low entropy compared to their surroundings. We can relate this to their structure by considering Boltzmann's result that the entropy S of a system is proportional to the logarithm of the number W of physical microscopic states compatible with the system's macroscopic state: $S = k_{\mathrm{B}} \log W$.[1] Structure is a macroscopic property. We can think of it as a set of constraints on the system's microscopic state. On a microscopic level one can only talk about state: there is an atom here, another there and so on, and the concept of structure is not relevant. But moving to a macroscopic level and throwing away some information we are able to structural statements such as "there is a membrane here, beyond which the concentration of this compound is higher." This statement is macroscopic: there are many possible ways in which the atoms could be arranged which are compatible with the concentration being higher behind the membrane. We can think of these as a vast number of hypotheses about the system's (unobservable) microscopic state. These hypotheses can in principle be enumerated, and so the structure has an entropy. The autopoietic concept of organisation works at an even higher level, providing constraints on structure.

Without the constraint of the membrane (or more strictly, without the constraint of our knowledge of the membrane) there would be an even greater number of possible microstates, and hence a greater entropy. This is true for any constraint, because all constraints rule out some of the possible states. Any structured system will therefore have a lower entropy than its constituent matter would have in thermal equilibrium. Structure is central to the theory of autopoiesis [9,10], and to our knowledge this particular connection with thermodynamics is new to the literature on autopoiesis.

Morowitz [12] gives an estimate of $10^{-10^{11}}$ for the probability of a living *Escherichia coli* cell spontaneously arising due to a thermal fluctuation in an equilibrium system. Equivalently, if we take all the atoms that make up an *E. coli* cell, then for every arrangement of those atoms that constitutes an *E. coli* cell there are $10^{10^{11}}$ which do not. Applying Boltzmann's formula, this corresponds

[1] Jaynes [4] provides a clear elucidation of how this formula can be generalised to the information entropy $-k_{\mathrm{B}} \sum_i p_i \log p_i$ of the maximum entropy probability distribution of microstates compatible with the macroscopic constraints. Jaynes' clear and extensive writing on the nature of entropy has played a significant part in developing this work.

to an entropy difference of about 3.2×10^{-12} JK^{-1} between an *E. coli* cell and its constituent matter in the equilibrium state[2].

In order to maintain this low-entropy structure processes must continually take place that reduce the organism's entropy, countering its tendency to increase. According to the second law of thermodynamics this can only be done by increasing the entropy of the environment, so that the total entropy of the system and its environment increases. Schrödinger [18] described this as "feeding on negative entropy" in the environment. We can also think of this as the organism performing work in order to maintain its structure. This work is done by physical and chemical processes that occur due to constraints imposed by the organism's organisation itself, giving rise to the Work-Constraint cycle described by Kauffman [5] and the self-referentiality that is central to the idea of autopoiesis, as discussed in [11,15].

Chemotrophic organisms achieve this reduction of internal entropy by altering chemical concentrations in their environment, taking in food and converting it to waste. Plants absorb high-energy (and thus low-entropy) photons, emitting radiation at a lower temperature, and most animals feed on low-entropy matter from other organisms, producing higher entropy waste. The model presented here deals only with chemotrophic organisms, although the other types should be straightforward to add.

2.2 Maximum Entropy Production

The historical origins of the MEP principle are complex [8] and its applicability and justification are still debated but perhaps one of its earliest and most impressive application is Paltridge's [13] model of the Earth's climate, which accurately reproduces features of the Earth's climate using a simple 10-box model.

In [6] this is distilled to a two-box model which despite its extreme simplicity reproduces well the values of the latitudinal heat transfer rates of the atmospheres of Earth, Mars and Titan. Entropy is produced in the atmosphere due to the transfer of heat from the high temperature (T_{H}) equator to the low temperature (T_{C}) poles. If the heat is transported at a rate F then entropy is produced at a rate $F \cdot (1/T_{\mathrm{C}} - 1/T_{\mathrm{H}})$, which increases with F but decreases as the two temperatures become closer together. A higher heat transfer rate increases T_{C} while decreasing T_{H}, so there is a kind of tradeoff in which both very low and very high transfer rates produce little entropy. At the extreme of zero heat transfer the temperature difference is maximised but the F term is zero, whereas in the upper limit heat transfer is maximised but the temperatures are equal, so again there is no entropy production. In between is a value for F which maximises the entropy production, and the MEP hypothesis predicts that the observed value will tend to be close to this value. By making this assumption the model is able to make good predictions while essentially ignoring all the complex mechanisms by which the heat transfer takes place.

[2] For comparison, a similar difference in entropy is involved in heating 2.3×10^{-13} kg of water from 300 to 301 K (the per kilogram entropy change is given by heat capacity of water $\times \ln(T_{\mathrm{final}}/T_{\mathrm{initial}})$). An *E. coli* cell weighs about 10^{-15} kg.

There is no widely accepted theoretical justification for the MEP principle but Dewar [1,2] has given some interesting arguments that it could arise from information theory in a similar way to the second law itself [4]. If this is the case then the Maximum Entropy Production principle would apply to any physical system in a steady state, including not only planetary atmospheres but ecosystems as well, with the maximisation taking place subject to whatever constraints are acting on the system.

For a planet's atmosphere the relevant constraint is the effect the flow rate has on the temperature gradient. Our model shows that in an ecosystem an analogous constraint can be imposed by a relationship between the rate at which processes take place and the difference in chemical potential that drives them.

The idea that a thermodynamic maximisation principle might apply to ecosystems has many precedents in the ecological literature [17], beginning with Lotka's 1922 proposal of a maximum energy flux principle [7]. Schneider and Kay [16] have suggested that life is the result of a physical principle whereby nonequilibrium systems "take advantage of all available means" to resist externally applied gradients. Predictions from these ideas have usually taken the form that a particular quantity such as energy flux or surface temperature will increase or decrease over time (see [19] for an investigation into several such "goal functions" using a population-based food web model). In contrast, because our model exhibits a tradeoff closely analogous to that in [6] applying the MEP hypothesis predicts a specific numerical value for the rate of matter flow through the system.

2.3 Chemical Potential

Heat transfer in the Earth's atmosphere is powered by a temperature difference between the equator and the poles. Chemical potential is somewhat analogous to temperature, and we can think of chemotrophic ecosystems as being powered by a difference in this quantity. Just as the temperature of a system increases with energy added to the system, the chemical potential of a substance within a system increases with each mole of that substance added to the system. Concentrations tend to flow from areas of high chemical potential to areas of low chemical potential, so that in thermochemical equilibrium the chemical potentials are equal in all areas of the system.

This is also true for chemical reactions: if there is a reaction that converts between two substances X and Y, it will take place in the direction that moves the two chemical potentials together, so that at equilibrium they are equal. If the system is isolated this corresponds to a maximum of the system's entropy; if it is held at constant temperature and pressure then it corresponds to a minimum of Gibbs free energy within the system and a maximisation of the entropy of the system and its surroundings. If there is a mixture of X and Y in which the chemical potential μ_X of X is greater than that of Y (μ_Y) then X will tend to be spontaneously converted into Y, producing entropy. Work can be done by coupling this spontaneous reaction to another one.

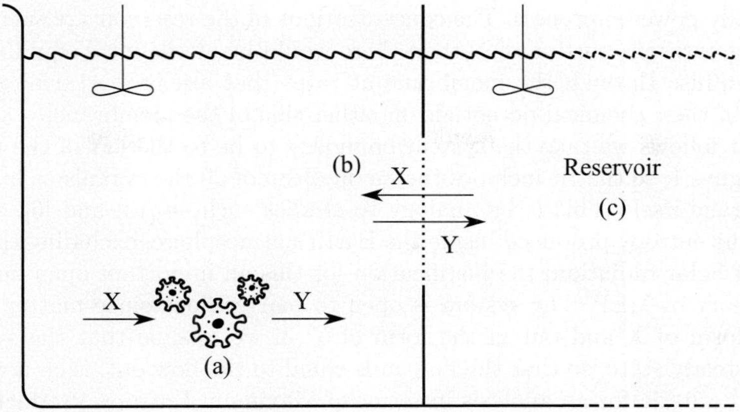

Fig. 1. Important features of the ecosystem model. (a) The system consists of a chemical mixture containing a population of organisms that metabolise by converting food X into waste Y. (b) A membrane allows X and Y to flow in and out of the system at rates which depend on the difference in chemical potential on either side of the membrane. (c) On the other side of the membrane is a well-mixed reservoir in which the chemical concentrations of 'food' X and 'waste' Y are held at constant non-equilibrium values.

All chemical reactions can take place in both 'forward' and 'backward' directions in this way but some reactions are experienced as unidirectional because the equilibrium point is such that when the chemical potentials are equal the proportions of reactants are vanishingly small compared to the products.

In an ideal solution the relationship between chemical potential μ_{X} and concentration N_{X} is given by $\mu_{\mathrm{X}} = \mu_{\mathrm{X}0} + RT \ln N_{\mathrm{X}}$, where $\mu_{\mathrm{X}0}$ is a constant that depends only on the temperature and the nature of X, and $R = N_{\mathrm{A}} k_{\mathrm{B}} = 8.31\mathrm{JK}^{-1}\mathrm{mol}^{-1}$ is the gas constant. However, the exact form of this relationship does not matter for what follows. This model is quite general in that all that is required for it to work is that the chemical potential is an increasing function of the concentration.

3 The Model

The basic components of an ecosystem are a population of organisms and a source of work, in this case a difference between chemical potentials. In order for interesting behaviour to take place there must be feedback between the organisms and their environment. Perhaps the simplest way to achieve this is to imagine that food is transported into the system and waste out by diffusion (figure 1).

The system contains a population of organisms (which need not be of the same species) whose metabolism is very simple, with individuals maintaining their structure by converting 'food' X into 'waste' Y. The system is coupled via a membrane to a reservoir in which the concentrations of X and Y remain constant, either because the reservoir is very large or because they are held constant by an externally powered process. The concentrations in the reservoir are such that there is a greater proportion of X than there would be at chemical equilibrium. X and Y diffuse through the membrane at rates that are proportional to the difference in their chemical potentials on either side of the membrane.

In what follows we take the system boundary to be to the left of the membrane in figure 1, so that it includes the metabolism of all the organisms but not the membrane itself. This is by analogy to studies such as [13] and [6], which calculate the entropy produced inside the Earth's atmosphere, excluding the absorption of Solar radiation; the justification for this an important open quesion for the theory of MEP. The system is open to matter flow, since matter flows in in the form of X and out in the form of Y. If we assume that the system reaches a steady state, so that the flow in is equal to the flow out, then it meets Dewar's [1] criteria for an analysis in terms of Maximum Entropy Production.[3]

3.1 Population Metabolic Rate

In traditional ecosystem models the focus tends to be on the numbers of individuals or the biomass of each species. These are treated as dynamical variables, with rates of change that depend on the population numbers of other species, so for instance the population of a predator species will grow more rapidly if there are more of its prey species present.

However, in this model the focus is on the flow of matter through the system, and the relevant variable to represent this is the *Population Metabolic Rate* M. We define this as the total rate at which food is converted into waste within the system. This takes place due to the combined effect of each individual's metabolism. There will of course be many other processes occurring within each organism's metabolism and within the system as a whole. In this scenario these are all ultimately powered by the conversion of X into Y and we can use the total conversion rate M to summarise the operation of the whole system.

Population metabolic rate is not necessarily directly related to population size. An organism can sustain itself either by having a structure that decays very slowly and using slow, efficient processes to maintain it or by having a rapidly decaying structure which must be renewed at a faster rate, so for a given population metabolic rate the population could consist of a small number of rapidly metabolising individuals or a large number that metabolise slowly; or there could be a small number which metabolise slowly as adults but which have a high rate of turnover. Like the two-box climate model in [6] this model is not concerned with what is inside the system, only with the rate at which it operates.

Another difference between this model and traditional models in ecology is that rather than treating the population metabolic rate as a dynamical variable it

[3] Dewar's framework is very general in that the flows in and out can be of matter rather than energy and they do not have to be fixed externally. In order to perform an analysis along the lines of [6] the system must also be free to take on a range of possible steady states, a requirement that is often expressed as needing many degrees of freedom. Our hope is that the complexity of biological organisms and ecosystem structure provide enough degrees of freedom to meet this requirement.

is left unconstrained. This gives the system a range of possible steady states (one for each value of M) which enables the use of the Maximum Entropy Production principle to choose between them.

3.2 Mechanics of the Model

Because the concentrations of X and Y are held constant on the reservoir side of the membrane, their chemical potentials remain constant in the reservoir also. Their values μ_X^{res} and μ_Y^{res} are parameters in the model. Since the proportion of 'food' X compared to 'waste' Y is assumed to be higher than it would be in equilibrium, $\mu_X^{res} > \mu_Y^{res}$. The chemical potentials μ_X and μ_Y within the system are variables that depend on the population metabolic rate and the rate of diffusion through the membrane.

The rate at which the fluids X and Y flow through the membrane are given by the difference in chemical potential of the fluid on the two sides of the membrane, multiplied by a diffusion constant D, which can be different for X and Y. Let the concentrations of X and Y within the system be N_X and N_Y, measured in moles. Their rates of change are given by

$$\begin{aligned} \dot{N}_X &= D_X\,(\mu_X^{res} - \mu_X) - M \\ \dot{N}_Y &= D_Y\,(\mu_Y^{res} - \mu_Y) + M\ . \end{aligned} \tag{1}$$

The rate of diffusion through the membrane is positive for X and negative for Y. In the steady state $\dot{N}_X = \dot{N}_Y = 0$ and the diffusion term will be balanced by the population metabolic rate M, from which we can obtain the chemical potentials as functions of M: $\mu_X = \mu_X^{res} - M/D_X$ and $\mu_Y = \mu_Y^{res} + M/D_Y$. As M increases the two potentials move closer together, and if no metabolism takes place then the system will be in equilibrium with its surroundings ($\mu_X = \mu_X^{res}$, $\mu_Y = \mu_Y^{res}$).

The total entropy produced per mole of X converted to Y is given by $\Sigma = (\mu_X - \mu_Y)/T$, which as a function of M is

$$\Sigma = \frac{1}{T}\left(\mu_X^{res} - \mu_Y^{res} - \left(\frac{1}{D_X} + \frac{1}{D_Y}\right) M\right)\ . \tag{2}$$

Σ is closely analogous to the inverse temperature difference $(1/T_C - 1/T_H)$ between the two heat baths in a classical heat engine or in the two-box climate model in [6]. It represents an upper limit on the amount by which an organism can reduce its structural entropy by metabolising one mole of X into Y. Equivalently, the upper limit on the amount of work that can be performed per mole metabolised is $T\Sigma$. Σ decreases with M (figure 2a), so that a slow population metabolic rate will result in a greater ability to do work per mole metabolised. M has a maximum value of $M_{max} = (\mu_X^{res} - \mu_Y^{res})/(\frac{1}{D_X} - \frac{1}{D_Y})$ at which Σ becomes zero and no work can be done. Converting X to Y faster than this rate would require work to be done rather than being a source of work.

The total entropy production due to metabolism is given by

$$\sigma = M\Sigma = \frac{1}{T}\left((\mu_X^{res} - \mu_Y^{res})\,M - \left(\frac{1}{D_X} + \frac{1}{D_Y}\right) M^2\right)\ . \tag{3}$$

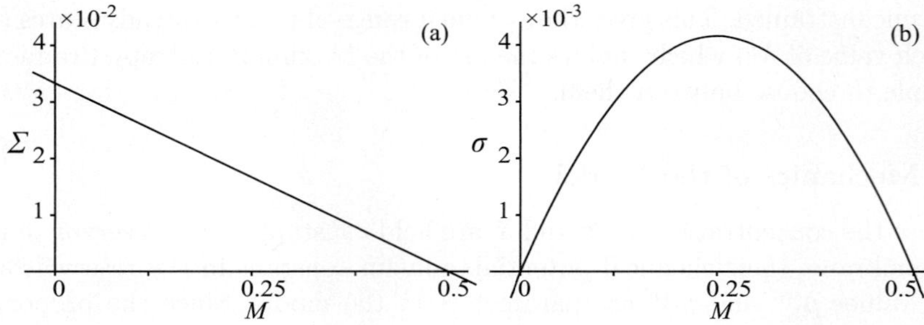

Fig. 2. (a) $\Sigma = (\mu_X - \mu_Y)/T$ tails off with increasing metabolic rate M. Σ is proportional to the difference in chemical potential between X and Y, which is equal to the maximum amount of work that can be done by converting one mole of X into Y. With $M > M_{\max} = 0.5$ work would have to be done to perform the conversion since $\Sigma < 0$. (b) The total entropy production $\sigma = M\Sigma$ rises to a peak and then falls off with increasing M. The values used for these plots are $\mu_X^{\text{res}} - \mu_Y^{\text{res}} = 10$, $D_X = D_Y = 0.1$ and $T = 300$.

This function is zero at $M = 0$ and $M = M_{\max}$, with a maximum in between at $M_{\text{MEP}} = \frac{1}{2}(\mu_X^{\text{res}} - \mu_Y^{\text{res}})/(\frac{1}{D_X} - \frac{1}{D_Y})$. From a global point of view there is therefore a tradeoff between slow and fast population metabolic rates (figure 2b). A slow population metabolism leaves a large difference in potential between food and waste but produces a sub-optimal amount of work because it occurs slowly, whereas a fast population metabolic rate leaves too small a difference in potential to perform as much work as the optimal rate.

The MEP principle suggests a hypothesis that real systems would tend to have a population metabolic rate close to M_{MEP} but it does not provide an explanation in terms of mechanisms that take place within the system, just as MEP based atmospheric models predict rates of heat transfer without specifying the mechanisms by which the heat transfer takes place (see [6] for an interesting discussion of this in relation to the Martian atmosphere). We leave open the question of what ecological or evolutionary processes might occur to drive an ecological system toward a state of maximum entropy production.

4 Extensions and Applications

4.1 Experimental Testing

A model along these lines could be used to test the application of MEP to ecosystems experimentally. In our model food and waste enter and exit the system via diffusion through a membrane but a similar calculation can be performed for a chemostat-like bioreactor in which a constant inflow of food is balanced by a constant outflow of the system's contents (a mixture of food and waste). This leads to a nonlinear decline in Σ with M but the analysis is qualitatively the same and one can find a value M_{MEP} for which σ is maximised. It should therefore be possible to perform a bioreactor experiment in which a measured value of M, which can be calculated from the amount of unused food in the system's outflow, is compared against the value predicted using the MEP principle.

In order for the MEPP to apply the organisms' growth must be constrained by the rate of food availability and/or waste disposal, and not significantly constrained by other factors. We suspect that this is not normally the case in a bioreactor since the aim is usually to produce as high a growth or reaction rate as possible so high concentrations of food are used, leading to a population metabolic rate that is constrained only by physiological factors. In order to test the applicability of MEP to biological populations it will probably be necessary to perform a specialised bioreactor experiment in which the nutrient inflow is very dilute and the system run until the population reaches a steady state. It may also be important to use species for which the bioreactor is a close approximation to their natural environment because an environment to which the organisms are not well adapted could induce additional constraints their growth.

4.2 More Detailed Models

This model was kept as minimal as possible and represents perhaps the simplest possible ecosystem, with a single environmental feedback caused by a single metabolic process which converts a single reactant into a single product.

Real systems contain species with many different types of metabolism. Real metabolisms are also more complex, performing multiple reactions at rates that can vary depending on circumstances. More complex reactions result in more complex biochemical feedbacks, including nutrient cycling. Real organisms can also use photosynthesis or prey on other organisms. It should be relatively straightforward to add each of these features to the model, which could then be used as a powerful tool to study the interaction between metabolism and environment at a very fundamental physical level, allowing studies along the lines of [3] but rooted in a model constrained by real physics. It should also be possible to add photosynthesis and organisms that can feed on other organisms, resulting in a more widely applicable ecosystem model.

A general ecosystem model built on sound energetic principles would be an extremely useful tool for simulating processes such as ecosystem succession [19] or artificial ecosystem selection [14]. The current natural tool for such studies is the generalised Lotka-Volterra equations, which are unsatisfying as they do not model the abiotic environment and include no energetic constraints.

5 Conclusion

We have presented a simple model that illustrates the relationship between the rate of chemical processes in an ecosystem and its rate of entropy production. This shows that the applicability of a principle of Maximum Entropy Production to living systems could be tested and opens up a range of possibilities for new research directions.

Acknowledgments. Thanks to Chrisantha Fernando and Alex Penn for discussions that have contributed to this work.

References

1. Dewar, R.D.: Information Theory Explanation of the Fluctuation Theorem, Maximum Entropy Production and Self-Organized Criticality in Non-Equilibrium Steady States. Journal of Physics A: Mathematical and General 36, 631–641 (2003)
2. Dewar, R.D.: Maximum Entropy Production and the Fluctuation Theorem. Journal of Physics A: Mathematical and General 38(21), 371–381 (2005)
3. Downing, K., Zvirinsky, P.: The Simulated Evolution of Biochemical Guilds: Reconciling Gaia Theory and Natural Selection. Aritificial Life 5, 291–318 (1999)
4. Jaynes, E.T.: Gibbs vs. Boltzmann Entropies. American Journal of Physics 33(5), 391–398 (1965)
5. Kauffman, S.: Investigations. Oxford University Press, New York (2000)
6. Lorenz, R.D., Lunine, J.I., Withers, P.G.: Titan, Mars and Earth: Entropy Production by Latitudinal Heat Transport. Geophysical Research Letters 28(3), 415–418 (2001)
7. Lotka, A.J.: Contribution to the Energetics of Evolution. PNAS 8(6), 147–151 (1922)
8. Martyushev, L.M., Seleznev, V.D.: Maximum Entropy Production Principle in Physics, Chemistry and Biology. Physics Reports 426(1), 1–45 (2006)
9. Maturana, H.R., Varela, F.J.: Autopoiesis and Cognition: The Realization of the Living. Kluwer Academic Publishers, Dordrecht (1980)
10. Maturana, H.R., Varela, F.J.: The Tree of Knowledge: The Biological Roots of Human Understanding. Shambhala Publications, Boston (1987)
11. Moreno, A., Ruiz-Mirazo, K.: Metabolism and the problem of its universalization. BioSystems 49(1), 45–61 (1999)
12. Morowitz, H.: Energy Flow in Biology. Academic Press, London (1968)
13. Paltridge, G.W.: Climate and Thermodynamic Systems of Maximum Dissipation. Nature 279, 630–631 (1979)
14. Penn, A.S.: Modelling Artificial Ecosystem Selection: A Preliminary Investigation. In: Banzhaf, W., Ziegler, J., Christaller, T., Dittrich, P., Kim, J.T. (eds.) ECAL 2003. LNCS (LNAI), vol. 2801, pp. 659–666. Springer, Heidelberg (2003)
15. Ruiz-Mirazo, K., Moreno, A.: Searching for the Roots of Autonomy: the Natural and Artificial Paradigms Revisited. Communication and Cognition–Artificial Intelligence 17(3–4), 209–228 (2000)
16. Schneider, E.D., Kay, J.J.: Life as a Manifestation of the Second Law of Thermodynamics. Mathematical and Computer Modelling 19(6–8), 25–48 (1994)
17. Schneider, E.D., Sagan, D.: Into the Cool: Energy Flow, Thermodynamics and Life. University of Chicago Press, Chicago (2005)
18. Schrödinger, E.: What is Life? Cambridge University Press, Cambridge (1944)
19. Virgo, N., Law, R., Emmerson, M.: Sequentially Assembled Food Webs and Extremum Principles in Ecosystem Ecology. Journal of Animal Ecology 75(2), 377–386 (2006)

Increasing Complexity Can Increase Stability in a Self-regulating Ecosystem

James Dyke, Jamie McDonald-Gibson, Ezequiel Di Paolo, and Inman Harvey

Centre for Computational Neuroscience and Robotics
School of Science and Technology, University of Sussex, BN1 9QH, UK
jgd20@sussex.ac.uk, jamiemcdg@yahoo.co.uk, ezequiel@sussex.ac.uk,
inmanh@sussex.ac.uk

Abstract. A long standing debate within ecology is to what extent ecosystem complexity and stability are related. Landmark theoretical studies claimed that the more complex an ecosystem, the more unstable it is likely to be. Stability in an ecosystems context can be assessed in different ways. In this paper we measure stability in terms of a model ecosystem's ability to regulate environmental conditions. We show how increasing biodiversity in this model can result in the regulation of the environment over a wider range of external perturbations. This is achieved via changes to the ecosystem's resistance and resilience. This result crucially depends on the feedback that the organisms have on their environment.

1 Introduction

Prior to the 1970s, ecologists generally held the view that the more complex an ecosystem, the more stable it would tend to be [1][2]. These conclusions were based on the observation that terrestrial ecosystems that had undergone significant simplification (e.g. alien species invasion or human monoculture), were more prone to pronounced fluctuations in population density. This reasoning was challenged by the theoretical study of May [3] who, building on earlier work by Gardner & Ashby [4], showed that the greater the number and strength of connections between species in a model ecosystem then the less stable it is likely to be. Notwithstanding the details of these particular theoretical studies, the relationship between complexity and stability has proved to be more subtle and elusive with the debate being conducted in two broad strands: a search for a general relationship between diversity and stability; investigating the relationship between food web structure and stability. See [5] for a review.

The stability of ecosystems can be assessed in terms of constancy, persistence, resistance, inertia and resilience. The results we will present will show that increasing ecosystem complexity reduces resistance (the ability to resist external pertubations), but increases resilience (the ability to recover after perturbations) in a population of two-locus, haploid individuals. These changes can result in an overall increase in the stability of the ecosystem as measured by its ability to regulate an environmental variable. Complexity in an ecosystems context can be

F. Almeida e Costa et al. (Eds.): ECAL 2007, LNAI 4648, pp. 133–142, 2007.

defined and measured in a number of ways. For example increasing complexity may equate to increasing the number and type of ecological interactions. In the model we present, complexity is increased via increasing the rate of mutation that operates on both loci and thus increasing genetic diversity. It would be possible to vary the diversity of the population by means other than mutation. For example the influx of alien species could be modelled with the random replacement of individuals. However given the very simple individuals modelled this, with certain limitations, could be considered as equivalent to the mutation of existing individuals. The frequency of alleles within the population changes in response to changes in selective pressure that are produced by external perturbations that operate on an environmental resource which determines the fitness of all individuals. Crucially, the activity of the organisms also affects the environmental resource; there is a feedback loop from organisms to environment and back again.

1.1 Self-regulating Models

The model we present is a modified version of [6] initially formulated by McDonald-Gibson which can be considered as a synthesis of the self-regulating Daisyworld [7] and niche construction models [8]. Organisms through their activity and metabolism will create, modify and at times destroy their environmental surroundings or 'niches', e.g. dams built by beavers, photosynthesis and the production of oxygen by plants or the burning of fossil fuels and production of carbon dioxide by humans. The theory of niche construction attempts to incorporate the feedback that organisms have on their environment into the established theories of population genetics and evolutionary theory [9].

Daisyworld features niche constructing-plants or 'daisies' that via their different colour affect the albedo of the planet they inhabit and thus the amount of energy it absorbs from the star it orbits. Both daisy types have an identical temperature-dependant growth function. This takes the form of an upside-down parabola that peaks at the optimum temperature for growth and declines on either side towards zero. Black daisies have a lower albedo than white daisies; therefore, in a cool environment, they absorb more sunlight, causing their local temperature to rise towards the optimal growth temperature. This increases their growth rate and engenders a feedback loop with their numbers rising, further increasing global temperature. This positive feedback on black daisy growth continues until the temperature moves towards the optimum for growth. The white daisies have a similarly regulatory effect and selective advantage in higher temperatures through cooling their environment. The net result is that the temperature on the planet is regulated around the optimum for daisy growth despite large increases in heat input from the star which increases in brightness over time. Since its inception it has been significantly extended with 1D, 2D and 3D implementations. See [10] for a review. The question of complexity and stability has previously been explored within the Daisyworld model [11][12][13]. These studies found that when feedback from organisms to the environment is incorporated in to their models, more complex and diverse ecosystems prove more robust to external perturbations. The McDonald-Gibson model exhibits regulation with fewer assumptions that these previous studies and is potentially more general. In particular, there are no pre-defined optimal resource conditions; the allele that determines which environmental conditions gives the greatest growth and fitness is subject to mutation and can take on a range of values, unlike the original Daisyworld model and variants listed above, and so regulation may occur over a wide range of 'optimal' environmental conditions.

2 Methods

A fixed population of two-locus haploid organisms is modelled. These interact with their environment by increasing or decreasing an environmental resource. This resource determines the fitness of the organisms. Real world resources may be temperature, pH, concentration of a particular chemical in a well mixed solution or partial pressure of a gas. A proportion of the population reproduce by cloning offspring that are subject to mutation with a fixed probability. The same proportion of individuals die thus maintaining a constant population. Reproduction and death occur simultaneously at every unit of dimensionless time. This may represent an annual, daily or even hourly cycle. The ecosystem is perturbed over time by an external force. Real world forcing may be thermal driving operating on an entire biosphere from a star or increasing chemical input into a freshwater lake via runoff. In the absence of any organisms this linear forcing would result in a linear increase in the environmental resource. This ecosystem, whilst simple, features a feedback loop from environment to organisms to environment and so on. Therefore exact analysis is not trivial and an individual based model was produced to compute values for varying parameter values over a range of external perturbations. In this model the resource is represented by a single variable, $R \in [-50, 150]$. The fitness F of individuals is a function of R. Each individual has a genotype that specifies its effect on the resource and the resource level to which it is best adapted. The Θ locus specifies the phenotypic effect on the environmental resource. The A locus specifies the resource level to which the phenotype is best adapted. The Θ locus has two possible alleles: e which reduces the resource and E which increases the resource. The bi-allelic Θ locus is represented with a double floating point number in the range [-1,1]. An individual will have the e allele if $-1 > \Theta < 0$ and the E allele if $0 > \Theta < 1$ The A locus is represented with a double floating point number in the range [15,85] and specifies the point that the individual is best adapted to within the range of resource values. Equation 1 is used to calculate an individual's fitness which is a parabolic function of the resource. This function peaks at the A value and declines sharply on either side towards zero. The fitness of the ith individual is a function of R and λ which provides a measure of the span of the parabola.

$$F_i = \begin{cases} 1 - \lambda(A_i - R)^2, & |A_i - R| < \lambda^{-\frac{1}{2}} \\ 0 & \text{otherwise} \end{cases} \quad (1)$$

Natural selection is modelled using a genetic algorithm with a proportion of the population subject to death, selection, and reproduction at each time step.

A constant death rate, γ, is applied at each step. Therefore, on average, the lifetime of any individual is $1/\gamma$ time steps. For each death, a tournament, where the victor is the individual with higher fitness, is held between two randomly selected members of the population. The winner of a tournament replaces the loser in the population. If both individuals fitness is the same, no replacement occurs. Mutation occurs with a probability of μ at each locus. This is performed by adding a number drawn from a Gaussian distribution of 0 mean and standard deviation 0.05 to the allele. The rate of change of the resource is:

$$\frac{dR}{dt} = \alpha \sum_i \Theta_i + \beta(P - R) \quad (2)$$

where Θ_i is the effect that the ith individual has on the resource value R. P is the current amount of external perturbation that is being applied at time t. Constants α and β determine the relative strengths of the population effects and external perturbation.

3 Preliminary Results

In the following results, unless otherwise stated, the maximum population size, K, is 2000 with the population being initialised with 2000 randomly generated individuals. The impact effect alleles e and E are set to -1 and 1 respectively. λ is fixed at 0.04 for all individuals which gives a range of 10 resource values in which an individual will have non-zero fitness. Mutation rates, μ, for alleles at both loci is fixed at 0.1 The death rate, $\gamma = 0.01$, and consequently 20 individuals are selected for tournament reproduction at every time step with the winners replacing the losers within the population. Simulations consist of 100,000 discrete time steps. Simulations were first performed with no population effects operating on R ($\alpha = 0$). The external perturbing force, P, was increased linearly from 0 to 100 over the course of a simulation run. As R increases, the population tracks this change with the average A allele, $\bar{A}$, quickly converging to R and increasing linearly. The population simply responds to the changing selective pressure produced by the external perturbing force. Simulations were then performed in which the the maximum difference, between the resource R, and perturbing force P, that can be maintained by the population is 50 ($\alpha = 0.0025$, $\beta = 0.1$). Again, P was increased linearly from 0 to 100 over the course of a simulation run. Initially R increases linearly with P. However when the resource enters the range [15,85] it no longer increases with increasing P, but stabilises at ≈ 15. As P increases, R remains at the same value whereas the sum of population effects decreases. Results are plotted in Figure 1.

The model exhibits regulation in its resistance to the external perturbing force, P. This regulation proves to be robust to a wide range of parameter values. Regulation is still observed with the following changes (data not shown): population size can be reduced by a quarter (further reductions tend lead to unstable populations due to increased variance produced by the stochastic GA) and increased beyond 100,000 (available computation time considerations led to a population of 2000 individuals being used for data collection); the λ parameter that determines the width of the fitness function can be halved or increased by a magnitude; the α and β parameters can be increased to 0.05 and 1 and reduced to 0.00005 and 0.001 respectively; the death rate can be decreased by one and increased by two magnitudes; the discrete e and E alleles can be replaced with double floating point values over the range [-1,1]; the assumption of a fixed population can be replaced with a variable population with logistic growth dynamics up to a carrying capacity, K, the value of which can vary over the same range for the fixed population.

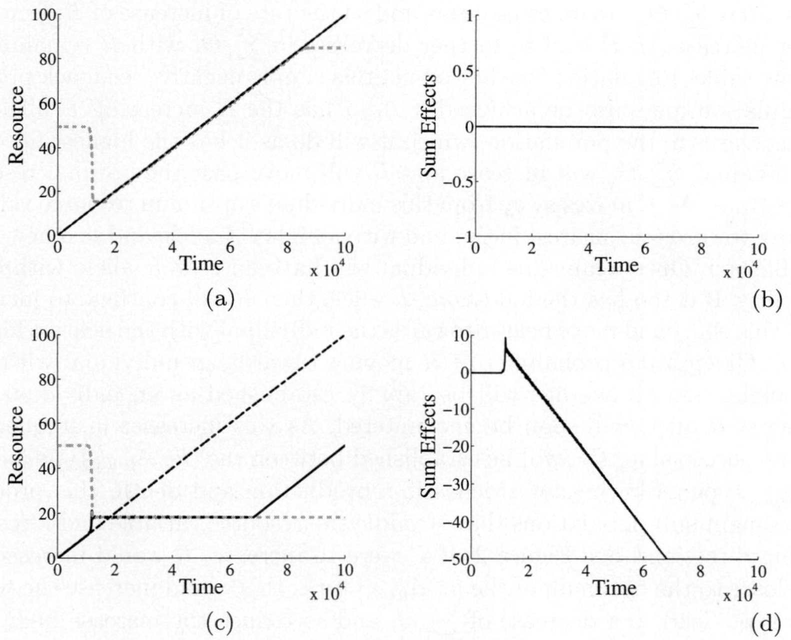

Fig. 1. Results of two simulations are plotted: no population effects on R ($\alpha = 0$) in (a) & (b); with population effects on R ($\alpha = 0.0025$) in (c) & (d). Time is plotted on the horizontal axis of all plots. The perturbing force, P, and the resource, R, are plotted with dashed black and solid black lines respectively in (a) & (c) along with the mean A trait of the population, $\bar{A}$, which is plotted with a dashed grey line. The effect the population has on the resource, $\sum_i \Theta_i$ is plotted in (b) & (d). In both simulations R initially increases with P until P drives R into the range [15,85] at which point the population rapidly converges to a single A trait: $\alpha = 0$, $\bar{A} \approx 15$; $\alpha = 0.0025$,, $\bar{A} \approx 20$. With no population effects as P continues to increase, $\bar{A}$ and R track P until it goes past 85. With population effects when P reaches the range [15,85], R no longer increases but remains fixed at ≈ 15 until $P \approx 65$. As P increases, the sum of the population effects decreases in (d).

4 The Regulatory Mechanism

Initial intuitions would suggest that as the external perturbations, P, increases, the population would respond to this selection pressure with the mean optimal resource level, $\bar{A}$, rapidly converging to the current resource level and tracking it as P increases. The regulation evidenced when the population exerts an effect on the resource can be shown with the following: Initially the resource, R, begins below the viable range of any individual. As P increases, R will reach 15 and so there will be an individual that has non-zero fitness. We call this individual A_{best}. Given a random initial population, it is equally likely to have an e or E allele. If it has an e allele, then as it increases in frequency, the effect the population has on the resource, $\sum_i \Theta_i$, becomes negative and so the rate of increase of R decreases. Further increases in P lead to further decreases in $\sum_i \Theta_i$ with R remaining at the same value. Regulation is achieved via this simple negative feedback process.

Regulation may also be achieved if A_{best} has the R increasing E allele. As A_{best} increases in the population, which it will do as it has the highest fitness of any individual, $\sum_i \Theta_i$ will increase and R will move past the optimal resource value of A_{best}. As R moves away from this individual's optimum resource value, it will move towards the nearest individual with a higher A allele and so a new A_{best} is established. Once again this individual will have an e or E allele with equal probability. If it too has the increasing E allele then R will continue to increase, transit this A_{best} and move nearer towards the individual with the nearest highest A allele. Given a 0.5 probability of R moving towards an individual with an e allele, regulation on average will be rapidly established as an individual with a reducing R allele will soon be encountered. As this increases in frequency a stable resource value, $R*$, will be established between the $[E,\ A_{low}]$ type and the $[e,\ A_{high}]$ type. Allowing for stochastic reproduction and death, this produces two dominant sub-populations that straddle the resource variable and force it in opposing directions. See Figure 2. If P were to increase, R would increase and move closer to the optimum of the $[e,\ A_{high}]$ type, this would increase the fitness of this type, lead to a decrease of $\sum_i \Theta_i$ and so reduce the increase in R. If P were to decrease, the fitness of the $[E,\ A_{low}]$ type would increase and so R would increase. This results in R remaining fixed for varying P and thus is equivalent in this respect to the proportional integral controller elements of detector, regulator and effector which seek to maintain a zero steady state error for the variable.

The mechanism of regulation in the model can be understood as an example of a *rein control system*. Rein control was introduced by Clynes [14] where it is hypothesised that certain variables (e.g. core body temperature in mammals) are maintained within a range of values by separate, unidirectional control reins that oppose forces that seek to perturb the variable. The rein control concept has since been developed in a physiological context [15] and latterly applied to analysis of a simplified Daisyworld model [16][17][18]. In this model the $[E,\ A_{low}]$ and the $[e,\ A_{high}]$ sub-populations can be regarded as unidirectional control reins. Varying the strength of these reins (changing the numbers of individuals within each sub-population) leads to the resource being regulated to within a narrow range over varying external perturbations.

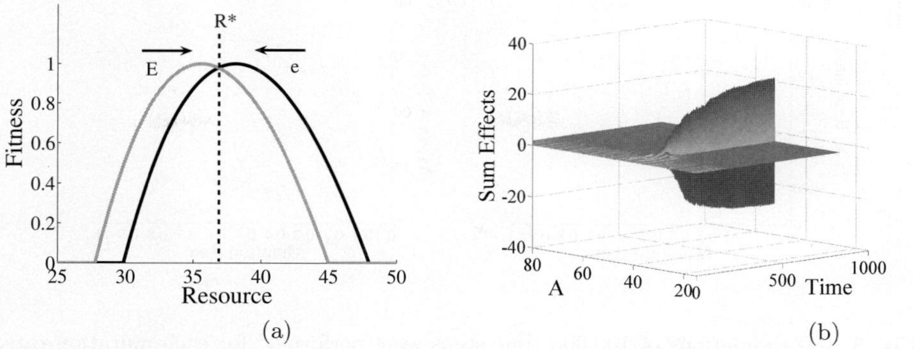

Fig. 2. In (a) the $[A_{low}, E]$ type's fitness range is plotted with a solid grey line, the $[A_{high}, e]$ type's fitness range is plotted with a solid black line. The vertical dashed black line marks the value for $R*$. The resource will be maintained around this value. If the the e allele were to increases in frequency, due to an increase in R (produced by internal or external perturbations) the change in the sum of effects leads to a counter opposing force acting to reduce this increases. Similarly if the E allele were to increase in frequency, due to an decrease in R the sum of effects leads to a counter opposing force acting to reduce this decreases. The emergence of a regulatory epoch is shown in (b). Individuals are collated into 70 distinct A trait bins which are plotted on the bottom left axis. For each binned A trait, the sum of population effects is plotted on the vertical axis. P is fixed at 37 for 1000 time steps. The initially diverse population, rapidly converges to two sub-populations: $[A_{low}, E]$ and $[A_{high}, e]$ that produce an equal increasing and decreasing effect on R with the net result that R remains at this value.

5 Results

In order to assess the effects of diversity in the model ecosystem, simulations with varying mutation rates, μ, were performed. Higher values for μ produce more diverse populations as mutation counter-acts the selective force that would have the entire population converge to the regulating sub-populations. As this random force would upset the regulatory balance in the population, it would be initially expected that the higher the value for μ and so the greater the diversity in the population, the lower the model's stability. This proves not necessarily to be the case.

100 simulations were performed for the 11 mutation rates over the range [0,1]. For each simulation, the total amount of time that R was regulated along with the ecosystem's biodiversity was recorded. Results are plotted in Figure 3. Biodiversity was measured in terms of the number of unique A alleles in the population. The A allele for each individual was rounded to the nearest integer and a sum of present A alleles recorded (the maximum being 71) at every time step. This total was then divided by the number of time steps over a simulation. R was judged to have been regulated if the average rate of increase of R was less

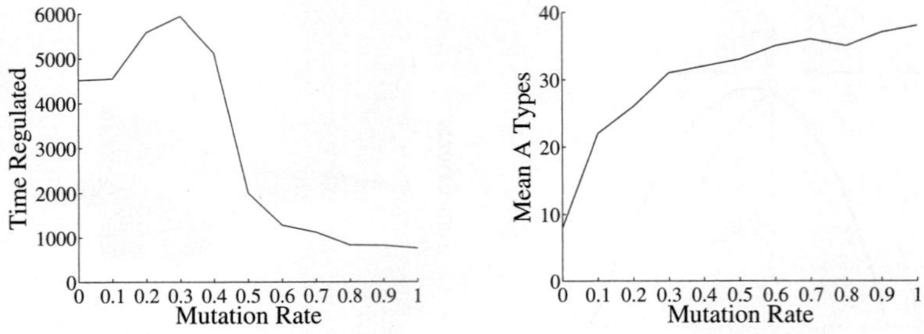

(a) (b)

Fig. 3. 100 simulations of 100,000 time steps were performed for each mutation rate. For each simulation, the length of time that the resource was regulated was recorded ($dR/dt < dP/dt$ over 200 time steps). The mean number of turns that R was regulated for each mutation rate is plotted on the vertical axis in (a). Regulation increases as mutation rates increase from 0 until an optimum of 0.3. Mutation rates higher than 0.4 lead to sharply reduced regulation. In (b) the mean amount of diversity (as measured with unique A types) over 100 simulations for each mutation rate is plotted on the vertical axis. Increasing mutation, increases diversity with the greatest rate of increase over the range [0,0.3]

than P over 200 time steps. It was necessary to average dR/dt in this manner as the stochastic GA produced a certain amount of noise in R. Rather than decrease, regulation increases with increasing mutation until $\mu = 0.3$. Increasing μ past 0.4 results in a sharp decrease in regulation.

This relationship can be explained by the different effect mutation will have on the system's resistance and resilience. Higher μ will tend to decrease the system's resistance. As the mutation rate increases, the length of any particular regulatory epoch decreases as at the limits of regulation mutation will remove individuals from the regulating sub-population and so decrease the maximum effect the organisms are able to exert on the environment. Having a more diverse population can increase the system's resilience however. Higher mutation rates increase the number of individuals outside of any current regulatory sub-population. These 'outliers' can prove to be the seed of a new regulatory epoch and so allow the system to recover regulation after the collapse of a previous regulatory epoch. See Figure 4. For a range of mutation rates, the increase in resilience is greater than the decrease in resistance and so total regulation increases. In that respect a more diverse system can prove to be more stable.

6 Conclusion

A simple model ecosystem has been presented. As well as responding to changes in the environment, the organisms also exerted an effect on the environment. This led to the regulation of an environmental resource over a range of external

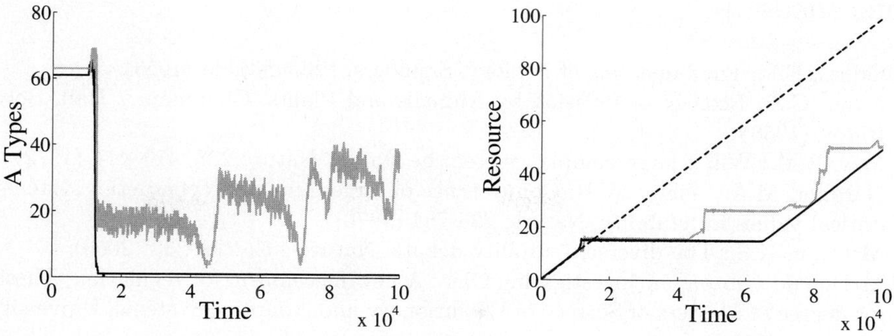

(a) (b)

Fig. 4. Results of two simulations are plotted: $\mu = 0$ with solid black line; $\mu = 0.3$ solid grey line. In (a) the number of unique integer value A types in the population is plotted (the A allele for each individual was rounded to the nearest integer). With no mutation diversity rapidly decreases to 2 with the establishment of a regulatory epoch. With mutation the population rapidly decreases to approximately 20, and then undergoes a series of decreases and sharp increases as new regulatory epochs are established. In (b) P is plotted with a dashed black line, R is plotted with solid lines. The regulatory epoch with no mutation is longer than that with 0.3 mutation. However total regulation is greater with 0.3 mutation as subsequent regulatory epochs are established. The establishment of these new regulatory epochs is shown in the sharp spikes in diversity in (a). The sharp transitions and multiple regulatory epochs observed are similar to the Daisyworld-type regulating model of [19].

perturbations. Increasing the mutation rate increased the diversity of the population by seeding individuals outside of the regulating sub-populations. It was found that higher mutation rates decrease the system's resistance in that the length of any regulatory epoch tends to be reduced. However higher mutation rates increase the system's resilience as increased diversity increases the probability that new regulatory epochs can be established. For a range of mutation rates, the increase in resilience was greater than the decrease in resistance and so produced an increase in the total amount of time that the ecosystem was able to regulate against the external perturbation. These results support the conclusions of [11] in particular the observation that in order for an ecosystem (a multi-species Daisyworld) to be able to respond to external perturbations there must be sufficient biodiversity. In this model, whilst biodiversity has no 'value' for the maintenance of a particular stable state, it can prove essential to the establishment of new stable states and as such biodiversity may be an important measure of the overall robustness of an ecosystem. This suggests that decreasing the complexity of an ecosystem by decreasing its biodiversity may imperil the entire ecosystem and not just those individual species threatened with extinction.

References

1. Odum, E.P.: Fundamentals of Ecology. Saunders, Philadelphia (1953)
2. Elton, C.S.: Ecology of Invasion by Animals and Plants. Chapman & Hall, Boca Raton (1958)
3. May, R.M.: Will a large complex system be stable? Nature 238, 413–414 (1972)
4. Gardner, M.A., Ashby, W.R: Connectance of large dynamic (cybernetic) systems: critical values for stability. Nature, 228-784 (1970)
5. McCann, K.S.: The diversity-stability debate. Nature 405, 228–233 (2000)
6. McDonald-Gibson, J.: Investigating Gaia: A new mechanism for regulation. Thesis for degree of Masters of Science of Evolutionary and Adaptive Systems, University of Sussex (2006)
7. Watson, A., Lovelock, J.: Biological Homeostasis of the global environment - the parable of Daisyworld. Tellus B 35, 284–289 (1982)
8. Laland, K.N., Odling-Smee, J., Feldman, M.W.: The evolutionary consequences of niche construction: a theoretical investigation using two-locus theory. Journal of Evolutionary Biology 9, 293–316 (1996)
9. Laland, K.N., Sterelny, K.: Perspective: seven reasons (not) to neglect niche construction. Evolution 60, 1751–1762 (2006)
10. Wood, J., Ackland, G., Dyke, J., Lenton, T., William, H.: Daisyworld: A review. Reviews of Geophysics. Forthcoming
11. Lovelock, J.: A numerical model for biodiversity. Phil. Trans. R. Soc. 338, 383–391 (1992)
12. Harding, S.: Food web complexity enhances community stability and climate regulation in a geophysical model. Tellus B 51, 815–829 (1999)
13. Harding, S., Lovelock, J.: Exploiter mediated coexistence and frequency dependent selection in a numerical model of biodiversity. J. Theor. Biol. 182, 109–116 (1996)
14. Clynes, M.: Cybernetic implications of rein control in perceptual and conceptual organization. Ann. NY Acad. Sci. 156, 629–670 (1969)
15. Saunders, P.T., Koeslag, J., Wessels, A.: Integral Rein Control in Physiology. Journal Theoretical Biology 194, 164–173 (1998)
16. Harvey, I.: Homeostasis and Rein Control: From Daisyworld to Active Perception. In: Pollack, J., Bedau, M,, Husbands, P., Ikegami, T., Watson, R.A. (eds.) Proceedings of the Ninth International Conference on the Simulation and Synthesis of Living Systems, ALIFE'9, pp. 309–314 (2004)
17. Dyke, J., Harvey, I.: Hysteresis and the Limits of Homeostasis: from Daisyworld to Phototaxis. In: Capcarrère, M.S., Freitas, A.A., Bentley, P.J., Johnson, C.G., Timmis, J. (eds.) ECAL 2005. LNCS (LNAI), vol. 3630, pp. 332–342. Springer, Heidelberg (2005)
18. Dyke, J., Harvey, I.: Pushing up the daisies. In: Rocha, L.M., Yager, L.S., Bedau, M.A., Floreano, D., Goldstone, R.L., Vespignani, A. (eds.) Artificial Life X, Proceedings of the Tenth International Conference on the Simulation and Synthesis of Living Systems, pp. 426–431. MIT Press, Cambridge (2006)
19. William, H., Noble, J.: Evolution and the Regulation of Environmental Variables. In: Capcarrère, M.S., Freitas, A.A., Bentley, P.J., Johnson, C.G., Timmis, J. (eds.) ECAL 2005. LNCS (LNAI), vol. 3630, pp. 332–342. Springer, Heidelberg (2005)

Niche Differentiation and Coexistence in a Multi-resource Ecosystem with Competition

Walter de Back[1], László Gulyás[2,1], and George Kampis[1,3]

[1] Collegium Budapest, Institute for Advanced Study, Budapest, Hungary
[2] Aitia International, Inc., Budapest, Hungary
[3] Department of Biology, East Tennessee State University, Johnson City, USA

Abstract. As a step towards modeling the evolution of food webs from an individual-based perspective, here we study the evolutionary dynamics of a simple multi-resource ecosystem model at the basal level of a food web. We combine two trade-off mechanisms in resource utilization (consumption abilities) and stoichiometric constraints (consumption needs) into a minimal model, and study the evolution of niche differentiation and coexistence through the interaction. Under a broad range of circumstances the model shows the emergence of specialization. By introducing stoichiometric constraints various evolutionary trajectories become possible but in this simple model we found no evidence for the coexistence of specialists and generalists.

1 Introduction

Understanding the emergence, evolution and stability of food webs is one of the most challenging tasks in theoretical and computational ecology today. Although several population-level models exist (see [8]), this problem has rarely been approached with an explicitly individual-based methodology. The present study is a first step towards this end.

We study the dynamics of a minimal model of a multi-resource ecosystem at the lowest trophic level of a food web. At this basal level, species feed on non-replicating abiotic energy resources and compounds (as would be sunshine and the available chemical substances in terms of a real ecosystem). We study the evolution of niche differentiation of species using two contrasting trade-offs:

1. *Abilities trade-off:* Given a limited number of resources, species differ in their utilization of resources. In terms of this factor, species can be classified as either generalists or specialist, depending on their consumption of resources. This rests on the assumption that a trade-off exist between generalists and specialists: Generalist can consume everything, but with a small efficiency. Specialists, on the other hand, consume only a single resource, but do that with a high efficiency.
2. *Needs trade-off*: Conventional ecological models are often based on the assumption of energy being a 'single currency'. Individual organisms, in contrast, are composed of many chemical compounds, and need to obtain mul-

F. Almeida e Costa et al. (Eds.): ECAL 2007, LNAI 4648, pp. 143–152, 2007.

tiple resources in order to survive and reproduce (stoichiometry). This introduces a second trade-off: Generalists are in the comfortable position of having the ability to uptake multiple types of resource, while specialists that consume only a single resource are limited by the resources that they have a limited ability to consume.

The process is modeled using an individual-based ecosystem model inhabited by a population of asexually reproducing organisms with genetically determined resource consumption abilities and needs. Using this model, we study the dynamics of emergence and maintenance of populations of specialists and generalists. This fits into the framework of recent ecological and evolutionary studies on specialist/generalist coexistence [1][2], stoichiometric ecology [9] and dynamic energy budgets [7]. The current study emerges from a broader proposal where (in the context of the EvoTech/FATINT project [6]) we attempt to understand the factors that enable multiple species with different ecological tasks to emerge and coexist in a process that also supports the increase of evolutionary complexity of various kinds.

2 The Model

We study the interaction of the trade-offs in resource utilization (abilities) and stoichiometric constraints (needs) in a simple individual-based ecosystem model. We define a minimal non-spatial environment which hosts a number of resource populations and (initially) a single consumer population. The environment is implemented as a vector, where each location can be occupied by either a resource or a consumer. In each step, a consumer randomly selects a location from the vector and, if it encounters a resource there, attempts to consume it.

The resources are modeled as abiotic entities that cannot reproduce (e.g. chemical substances or units of energy). The influx of resources from outside the system is taken to be constant per time step, and equal for all resource types. When adding a resource token, a location is randomly chosen in the environment vector. If the chosen location is empty, the resource is successfully added to the environment. In this way, total resource abundance saturates towards a maximum. Each resource represents a nutritional value transferred to the consumer when the resource token is consumed, whereupon the latter is removed from the environment. Additionally, the resource abundance suffers from a small dilution rate.

Consumers are modeled as replicating individuals. They are specified with a genotype, a 'stomach' (resource storage), and a variable energy value. The energy value regulates the consumer's life history: the consumer can reproduce when its energy exceeds a certain threshold, and dies when the energy value is turning negative. Energy is not directly increased upon the consumption of resources, but requires a *combination* of resources to be consumed. The specific combination of resources that is required to convert raw resource to useful energy depends on the genotype of the individual. The ability to consume a given type of resource is encoded on the genotype as well. These two factors (the energy function and the consumption function) stand for trade-offs 1 and 2, respectively.

Consumption Abilities. The length of the genotype G for a consumer (henceforth: individual) is equal to the number of resources n: $\{G_i, ..., G_n\}$ and determines the individual's abilities and needs. In experiments presented in this paper, we used $n = 3$ for clarity, but simulation with upto 10 resource types show essentially similar results. Each gene G_i has a real value in the interval $[0, 1]$. Gene values are normalized (indicated by Φ below), such that $\sum_{i=1}^{i=n} G_i = 1$. This normalization expresses that resource consumption is an activity of the organism, where the total time of "feeding" is distributed among the resources in an exclusive-or way. Normalization implies a specialist/generalist trade-off, because an increase in one gene causes a decrease in the others.

Upon encounter between an individual and a resource of type i, the resource is consumed with a probability A_i, which we call an individual's *ability* for i. An individual has abilities for a given type of resource depending on the gene value at the given locus corresponding to the given type of resource. In a system with $n = 3$, the individuals' consumption abilities are defined as:

$$A = \Phi\{G_1^s, G_2^s, G_3^s\} \tag{1}$$

where Φ denotes normalization and the exponent s is a key parameter that amplifies or suppresses the effect of genetic differences on an individual's abilities. First, consider the case where $s = 1$. Suppose that an individual has genotype $\{0.7, 0.1, 0.2\}$ and encounters a resource of type 2, then it has 10% probability of consumption. At $s = 1$, the genotype directly codes for the consumption abilities ($A = G$). With $s > 1$, the relative differences in G will be increased, and with $s < 1$, the relative differences are decreased. Figure 1 shows the dependence of A_i on G_i for various values of s.

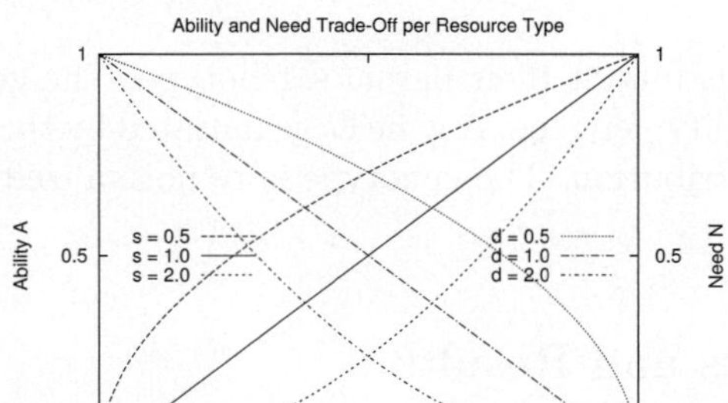

Fig. 1. Trade-off between ability A_i and need N_i as functions of gene G_i for a resource type i for various values of s and d.

Energy Needs. Upon consumption, an individual stores the resource of a given type in R_i. The energetic value of resources for an individual is calculated as a function of the stored resources. The resulting energy value then determines the reproduction and the survival/death of the individual.

In models where energy is a dealt with as a single currency that converts resource to offspring, every resource contributes equally to the energy value. In the present model, in contrast, the relative contribution of resources types to energy, i.e. the *need* for resources, differs depending on the individual's genotype.

We take the need N_i for a given type of resource for an individual to be the *inverse* of the corresponding genotype value, such that the resource on which an individual is most specialized contributes the least to its overall energy value. We calculate the contribution of the various resources, and again normalize the result to 1:

$$N = \Phi\{(1 - G_1)^d, (1 - G_2)^d, (1 - G_3)^d\} \tag{2}$$

Then, we calculate the overall energy E for the individual by summing over all stored resources R_i multiplied by their relative contribution: $E = \sum_{i=1}^{i=n} N_i R_i$. The exponent d in eq. 2 allows us to increase or decrease the relative differences in G to obtain the energy contribution N in the same way as before when consumption abilities were calculated. At $d = 0$, genetic differences do not affect energy contribution at all, which recovers the 'single currency energy' model. With increasing $d > 1$, genetic differences are amplified, such that the contribution of the resource to which an individual is not specialized becomes increasingly larger. Figure 1 shows how the need N_i of a resource type i depends on G_i, for various values of d.

Stored resources are continually discounted by metabolism with a fixed amount per stored resource, and energy declines linearly with time. The individual dies when $E \leq 0$ and reproduces when E exceeds a reproduction threshold.

Evolution. Offsprings inherit their parent's genotype. The genotype is mutated with a small probability per gene. A gene G_i is mutated by the addition of a small uniform random distribution. The genotype is re-normalized after mutation.

3 Experiments and Results

We performed a series of evolutionary experiments in the above model, using the parameter settings as in Table 1. First, a baseline experiment is shown in which the energy is treated as a 'single currency'. Then we observe the behavior of a model with stoichiometric constraints (i.e. needs trade-off) under different settings. We conclude our experiments with an overview of the evolutionary consequences of varying the strengths of both trade-offs.

Table 1. Model parameters

Parameter	Value	Parameter	Value
Environment size	2,500 locations	Reproduction threshold	10.0
Number of resources	3	Metabolism	1.0 per res. type
Resource influx	100 per res. type	Mutation rate	0.05
Resource value	15.0	Mutation range	0.05
Dilution rate	0.001		

3.1 Trade-off in Abilities

Pre-experiment. As a pre-experiment we note that if the normalizations of genotype G (and abilities A) are canceled, there will be no generalist/specialist trade-off, and hence no limitation on generalization or specialization. In this case we find that the consumer population rapidly evolves towards the point where consumption is maximal for all resource types ($A = \{1.0, 1.0, 1.0\}$), and there are no generalists or specialists. All following experiments therefore involve the generalist/specialist trade-off through normalization.

Base-line experiment. In our base-line experiment consumption abilities evolve in the absence of gene-dependent resource needs. The individuals' needs are fixed at $N = \{\frac{1}{3}, \frac{1}{3}, \frac{1}{3}\}$ by putting $d = 0$, by which energy is treated as a single currency. We also put $s = 1$ such that the genotype directly encodes the consumption abilities, $A = G$. Note that normalization implies that the overall consumption probability is kept constant.

Results show that, although in this case specialization does not offer an increased consumption probability, the initial generalist population nevertheless evolves into specialists on each of the three resources. In 100 conducted simulation runs ($t = 250,000$) we have observed in all cases that the population immediately speciates into a specialist on one resource and a pseudo-generalist population on the remaining resources (as shown in Figure 2). The specialist evolves towards the point where consumption of the given resource is maximal (i.e. a corner in genotype space). The other population evolves in the opposite direction until the border of the genotype space. From there, this pseudo-generalist population speciates and both evolve towards specialization, thus creating three specialist populations that occupy the three available pure niches.

This outcome cannot be explained in terms of the consumption probabilities, because the sum of the probabilities is equal to 1. The effect is the result of positive feedback from relative resource densities, shown in the right panel of Figure 2. Every mutation from the initial generalist population causes a slight specialization, and therefore a difference in resource abundance. This breaks down the initial symmetry of the latter. If, by chance, there are more mutants with some preference for one resource, this results in a disruptive selection pressure via resource abundance. This effect causes the emergence of a specialist to one resource, and pseudo-generalists to the other two resources; and also results in the speciation of pseudo-generalists.

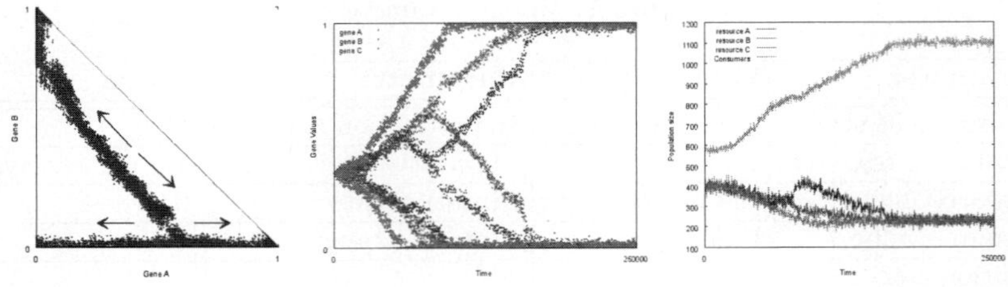

Fig. 2. Evolution of specialists with $s = 1$ and $d = 0$. Left panel shows evolutionary dynamics in genotype space in a Marschak-Machina triangle. Since the genotype is normalized, the value of the third gene can be left implicit because it is given by the two other gene values: $G_C = 1 - (G_A + G_B)$. The experiments are initialized with generalists (in the middle). Time is indicated by arrows. The middle panel shows gene values (y-axis) over time (x-axis). The right panel shows population abundances for resource populations and the total consumer population (y-axis) over time (x-axis).

3.2 Trade-offs in Both Abilities and Needs

In addition to the abilities trade-off as above, we now turn on the resource needs trade-off by setting $d = 1$ whereby the energy contribution of a given resource becomes inversely proportional to its consumption ability. Results obtained over 100 simulation runs ($t = 500,000$) show that, under this trade-off, the outcome of evolution is specialization on all resources, but this outcome can be achieved through several qualitatively different evolutionary trajectories. Figure 3 shows two most typical trajectories: with and without an intermediate step of pseudo-generalists.

Notice that the resulting specialist populations are not located in the extreme corners. The gene plots (middle panels) now show that populations never achieve perfect specialization (i.e. $A_i = 1$), since this would imply that the resource i does not contribute to energy at all. Specialists may have a large ability for one resource type, but with $d = 1$ they also need other resources in order to gain energy. From an analysis (not shown here) we also see that the evolution of specialization does not occur in a linear fashion. Instead, evolution slows down as the level of specialization increases (compare this with the case of Fig. 2 where change rate was found to be linear in time).

Figure 3 shows two different evolutionary trajectories obtained from the same parameter setting using different random conditions. In the first case, the initial generalist population immediately speciates into three populations that evolve into specialists. In the second case, first a specialist/pseudo-generalist situation evolves. After a transient phase, the pseudo-generalist population suddenly speciates and the three specialist populations occupy all niches. After this event, the situation will be similar to first case. From 100 simulation runs ($t = 500,000$), 12 cases could not be readily classified in case 1 or 2. Half of these approximated case 2 but ended up in a two-specialist situation, the other speciated in some intermediate stage. The large majority of the runs, however, show trajectories that closely resemble the cases presented in fig. 3. We found the occurrence of

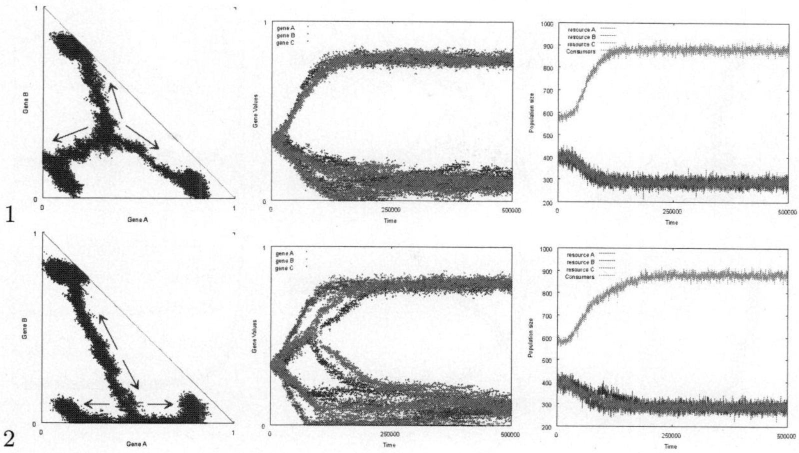

Fig. 3. Two different evolutionary trajectories (left), gene values over time (middle) and population dynamics (right) under an equal condition ($s = 1$ and $d = 1$). Initialized with a generalist population (center of triangle). Arrows in trajectory plot indicate time. Multiple runs using several random seeds for this as well as further figures are available at http://www.evotech.hu/doc/ecal2007figures.pdf

immediate speciation (case 1) in 42% of the runs, and the T-shape trajectory (case 2) in 46% of the runs.

Fate of an Initial Specialist Population. The above experiments were initialized using a generalist population. Now (using the same parameters) we start from a single specialist population occupying one of the three corners, leaving the others empty. The results, depicted in Figure 4, again show qualitatively different evolutionary outcomes yet with a common tendency.

We initialize the experiment with $G = A = \{0.8, 0.1, 0.1\}$. In case 1, the initial population first evolves into a generalist population consuming all resources. Then, the population speciates into three, one of which directly evolves towards specialization. The other two populations first evolve to become pseudo-generalists as above (consuming two of the three resources), just to become specialists afterwards. This results in an evolutionary cycle in the sense that the niche of the initial population is first abandoned, but after a long evolutionary trajectory that leads through generalist and pseudo-generalist stages, later reoccupied. Such a cycle does not always occur, as shown in case 2 (bottom row of Figure 4). Here, the initial population rapidly speciates into two populations that continue to specialize further on the two resources that were initially empty. This results in a stable two-specialist situation, and the niche from which both populations evolved is finally left empty.

In 100 simulations ($t = 500,000$), we found the cyclic behavior (yielding three specialists) in only 10% of the runs, while 90% of the runs ended up in a stable two-specialist situation.

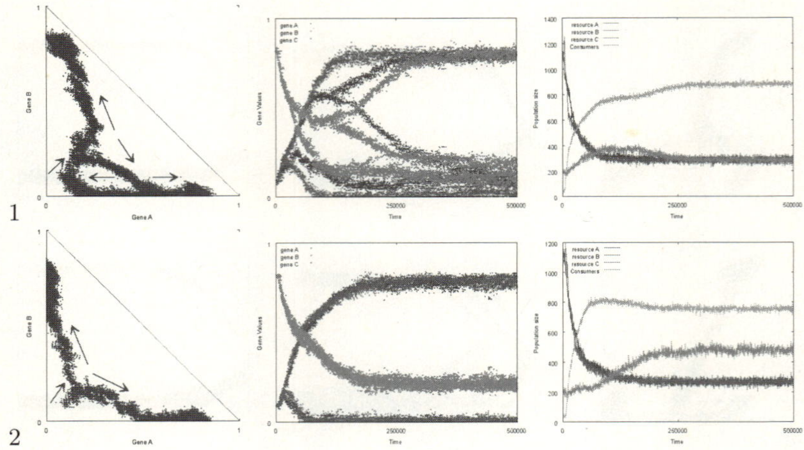

Fig. 4. Two different evolutionary trajectories (left), gene values over time (middle) and population dynamics (right) under equal condition ($s = 1$ and $d = 1$). Initialized with specialist population on one resource (lower left corner). Arrows in trajectory plot indicate time.

3.3 Varying Trade-off Strengths

The experiments in the last section were conducted with $s = 1$ and $d = 1$, such that the abilities A_i and needs N_i were (inversely) proportional to gene values G_i in a linear fashion. By changing these key parameters, we can alter the strengths of the ability trade-off and the needs trade-off respectively, and also alter their relative strengths. We study the evolutionary consequences of the relative trade-offs strengths by running a series of experiments with varying s and d.

Results of these simulations are depicted in the left panels of each cell in Table 2. In general we find that (1) the stronger the abilities trade-off, the stronger the selection pressure towards specialists, and (2) the stronger the needs trade-off, the stronger the selection towards generalists, as may be expected.

Table 2 shows that specialists do not evolve when the generalist/specialist trade-off is weak ($s < 0.1$), or when the needs trade-off is strong ($d > 1.0$). When both trade-offs are strong, however, the population does speciate into specialists ($s = 10.0$ and $d = 10.0$), yet in a limited sense (since a strong needs trade-off means a strong repelling force from the corners).

By combining equations 1 and 2, we can derive the energy that an individual with a given genotype is expected to receive disregarding resource availability. Given a genotype G, the expected energy $E_e = \sum_{i=1}^{i=n} A_i N_i$, which combines both trade-offs and thus depends on values for s and d. The right panels in Table 2 show the expected energy for the entire genotype space and varying trade-off strengths.

These expected energy plots show that generalists can actually result from very different situations, e.g. where selection for the specialists is weak (see

Table 2. Varying trade-offs strengths (s and d). Experiments are initialized with a generalist population. Left panels show evolutionary trajectory in genotype space in Marschak-Machina plots. Right panels show expected energy in gray from high (white) to low (black).

	d = 0.1	d = 1.0	d = 10.0
s = 0.1	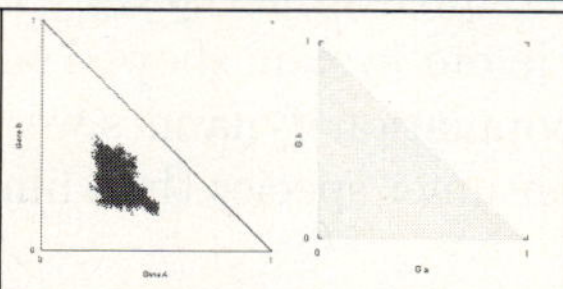	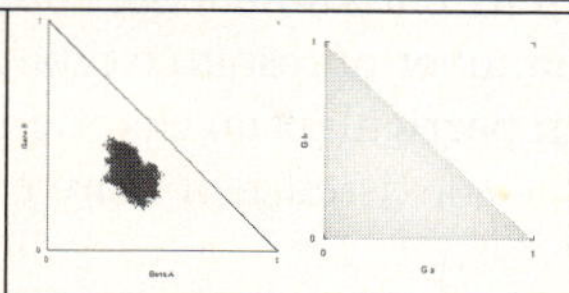	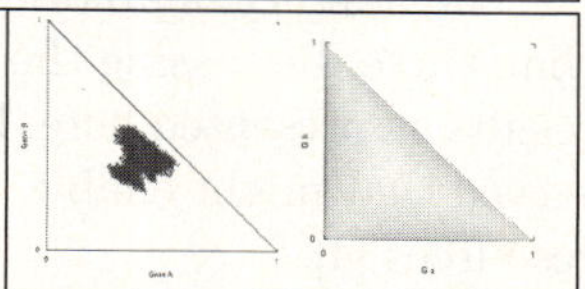
s = 1.0	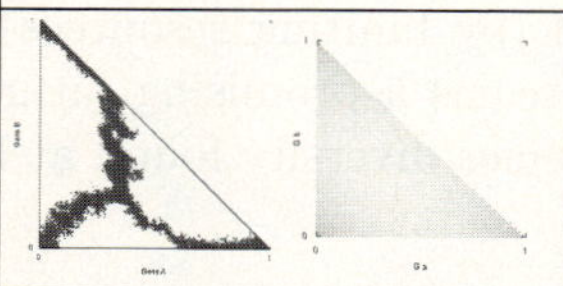	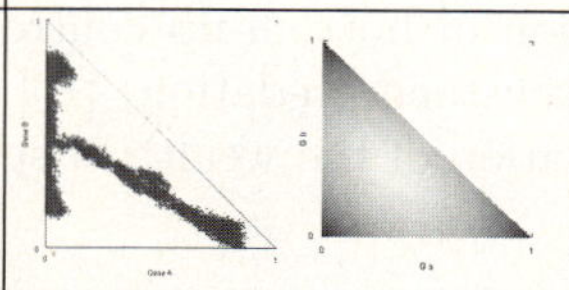	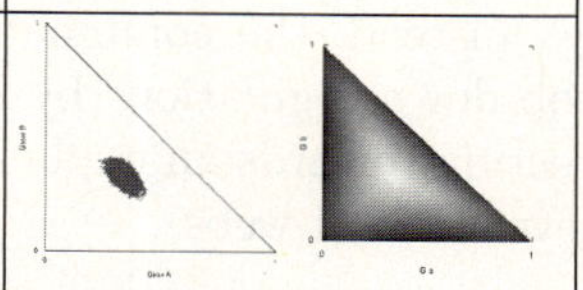
s = 10.0	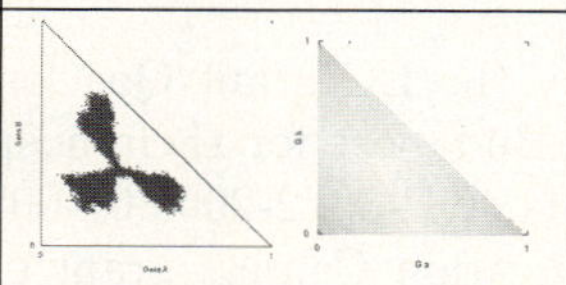	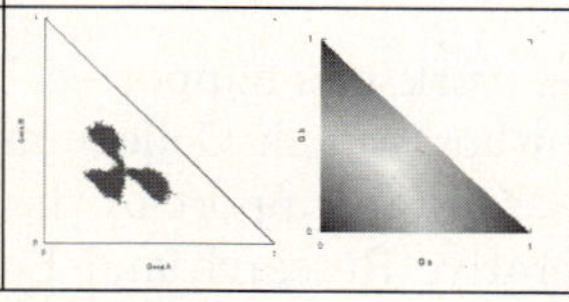	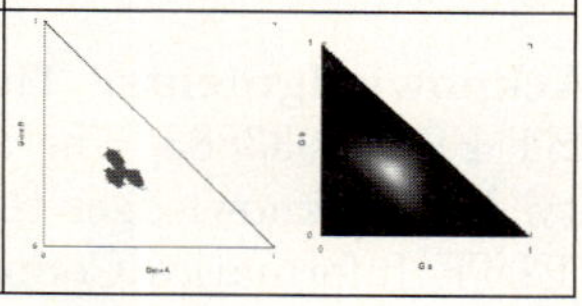

$s = 0.1$, $d = 0.1$), but also where the selection for generalists is strong (see $s = 1.0$, $d = 10.0$). Moreover, the shapes of the plots indicate that the evolution of pseudo-generalists (towards the middle of the edges) is possible in the case where $s = d = 10.0$, but inhibited by stronger needs ($s = 1.0$, $d = 10.0$).

We can clearly see that populations do not evolve towards the optimal expected energy, but tend to evolve towards the regions where the expected energy is intermediate. The difference arises because the expected energy calculation disregards resource abundances. Although individuals would evolve towards the point in which expected fitness is optimal in absence of resource competition, this is constrained by dynamic differences in resource availability.

4 Conclusion

Using a minimal individual-based ecosystem with three abiotic resources, we have studied the trade-offs in resource utilization (consumption abilities) combined with stoichiometric constraints (consumption needs). We have shown that in a broad range of conditions specialists emerge and examined some factors that can modify this tendency. The introduction of consumption needs leads to different evolutionary trajectories, which include evolutionary cycling (evolving from specialization to generalists and back to specialization). Variations in trade-off strengths shows that no specialists evolve when either the generalist/specialist trade-off is weak, or the stoichiometric constraints are strong. Under none of the studied situations, however, did the coexistence of specialists and generalists arise in this model.

Since we observed the number of evolved consumer populations to be equal or smaller than the number of resources, these results are in line with a competitive exclusion principle [3]. The present model will serve as a baseline model for further investigations to overcome this limitation on the emergence of basal species in an attempt to resolve the paradox of the plankton [5].

The number of abiotic nutrient and energy sources present is typically larger than three. Increasing the number of resource types in our system showed similar results as presented here. In particular, no chaotic population dynamics were observed that might enable the coexistence of more consumer species than limiting resources [4].

Instead, the combination of bottom-up control (by limiting resources) and top-down regulation (by keystone predation) [10] seems a promising future direction towards an explanation of the wealth of species diversity found at basal level of food webs.

Acknowledgments. This work was supported by the EC grant QosCosGrid IST FP6 #033883. The authors thank Collegium Budapest for their hospitality. L.G. acknowledges the partial support of the GVOP-3.2.2-2004.07-005/3.0 (ELTE Informatics Cooperative Research and Education Center) grant of the Hungarian Government. G.K. has been holding the Basler Chair at ETSU, TN during the period of the work. The supports are gratefully acknowledged. G.K. also wishes to thank comments and support from Dr. István Karsai of ETSU. The authors also wish to thank two anonymous reviewers for their helpful comments.

References

1. Abrams, P.A.: The prerequisites for and likelihood of generalist-specialist coexistence. American Naturalist 167(3), 329–342 (2006), doi:10.1086/499382
2. Egas, M., Dieckmann, U., Sabelis, M.W.: Evolution restricts the coexistence of specialists and generalists: the role of trade-off structure. American Naturalist 163(4), 518–531 (2004)
3. Hardin, G.: The competitive exclusion principle. Science 131, 1292–1297 (1960)
4. Huisman, J., Weissing, F.: Biodiversity of plankton by species oscillations and chaos. Nature 402, 407–410 (1999)
5. Hutchinson, G.E.: The paradox of the plankton. American Naturalist 95(882), 137–145 (1961)
6. Kampis, G., Gulyás, L.: Sustained evolution from changing interaction. In: Alife IX, Boston, pp. 328–333. MIT Press, Cambridge (2004)
7. Kooijman, S., Andersen, T., Kooi, B.: Dynamics energy budget representation of stoichiometric constraints on population dynamics. Ecology 85(5), 1230–1243 (2004)
8. McKane, A.J., Drossel, B.: Models of food web evolution. In: Linking Structure to Dynamics in Food Webs, pp. 223–243. Oxford University Press, Oxford (2005)
9. Moe, S.J., Stelzer, R.S., Forman, M.R., Harpole, W.S., Daufresneand, T., Yoshida, T.: Recent advances in ecological stoichiometry. OIKOS 109(1), 29–39 (2005)
10. Paine, R.T.: Food web complexity and species diversity. American Naturalist 100(910), 65–75 (1966)

Variance in Water Temperature as a Factor in the Modelling of Starfish and Mussel Population Density and Diversity

David White

Department of Electronic, Electrical and Computer Engineering,
The University of Birmingham, United Kingdom
dxw533@bham.ac.uk
http://postgrad.eee.bham.ac.uk/dxw533/

Abstract. There is a general agreement when modelling ecosystems that the simplest solutions are generally the best. Comparing water temperature models that affect the feeding rate of starfish can show similar results when simulated under a simple scenario. When the system is modified to include environmental change, water temperature models that have similar mean temperatures but differences in variance can produce variable results that correlate to the magnitude of their variance. This paper will examine and compare the effect that four water temperature models, each with similar mean temperatures, has on the predation of mussels by starfish, and how this affects population densities over time. Results find that water temperature models with comparable variance produce similar results; models that differ in variance produce dissimilar results, especially when environmental conditions capitalise on that variance.

Key words: Ecological systems, Measures of Complexity

1 Introduction

Modelling ecosystems is a part of artificial life that has received much attention [1], and artificial marine ecosystems form a considerable part of this. From modelling the movement and behaviour of fish [10], to the evolution of complex ecosystems [3], replicating the physical processes and interactions of marine life remains an important research area, especially in the context of climate change.

It is important to recognise which methods of simulation lead to appropriately realistic results. Interpreting the research data that marine biologists compile requires fundamental decisions on how to use the information to build accurate simulations and useful modelling tools. Lessons can be learnt from the experience of ecological modellers. Grimm [5] in his review of fifty ecological models notes that individual based models make more realistic assumptions for agent behaviour than traditional state variable models; so how realistic do these assumptions need to be in order to create accurate simulations? Thulke [16] states that in ecological modelling, the general rule is to "ignore as much detail as

F. Almeida e Costa et al. (Eds.): ECAL 2007, LNAI 4648, pp. 153–162, 2007.

possible in order to obtain general insights", but then leads on to say that there are no guidelines for identifying the appropriate resolution. Robinson [13] states that "a reasonably accurate" model can be achieved with a relatively simple model, and going beyond this results in only a small increase in accuracy.

If we can say that simple simulations produce adequately accurate results, what does 'accurate' actually mean? It could mean, for example, that a population of a simulated species eats an average amount of prey that is comparable to its real life counterpart, but this says little about the variability of the consumption; a factor that could have significant implications on other processes within the ecosystem.

Variance in ecological systems is a topic which has received much comment from within both artificial life and ecological modelling. Green [4] notes that many Australian plant species have adapted to the variability of rainfall. Mearns [9] studies the effect that mean and variable effects of climate change have on crops, and found that model results differed (reduced yields compared to increased yields) depending on whether mean or variable methods of climate change were used. Katz states that extreme weather events link closer to the variability of climate than the mean of climate [8].

If variance is acknowledged to have an influence then there is a need to understand how artificial life systems could be affected. A set of modelling functions may exhibit adequately accurate data, but the variance or shape of that data may differ considerably between functions and this in turn could affect the results of a simulation. This paper will examine the effect that water temperature models with different levels of variance have upon population distributions and species diversity.

1.1 A Starfish / Mussel Predation Model

A keystone species [11] can be found to have a positive effect on the diversity of its local environment, such that the withdrawal of the keystone species affects the diversity negatively. The starfish *Pisaster ochraceus* has such an effect by consuming the mussel *Mytilus californianus*, the competition for space is reduced enabling less aggressive organisms to inhabit an environment that without *Pisaster* would result in almost total domination by *M. californianus* [14].

Sanford finds that *Pisaster* is affected by water temperature [14] [15], with colder temperatures leading to a reduction in the consumption rate of its prey and leads on to say that episodic upwellings (an influx of colder water) could affect the role of the starfish as a determining factor in the diversity of the local ecology.

Differences in the variance of a range of water temperature models could lead to varying results in species population distribution and diversity. A simple simulation scenario may find that simulations using different water temperature models produce comparable results. If additional rules are introduced however, we could see differences between a simulation that uses one water temperature model and one that uses another, this difference could correlate to the variance in water temperature over time.

Four different water temperature models will be compared under simulation. Water temperature will affect the feeding rate of *Pisaster* on *M. californianus*, thus producing varying levels of species diversity. These models will be tested under three scenarios: the first is simple with no external or internal disturbances; the second has a period of lower overall water temperature; the third utilises a function that introduces additional starfish with increasing densities of mussels - increasing the consumption of mussels and the overall diversity of lower order species.

2 Method

The water temperature models are based on water temperature observations taken by Sanford [15]. Each temperature data unit represents one day. If a simulation reaches the end of the temperature data set, data is streamed back from the start. In realistic terms the temperature sets are not accurately representative of long term temperature data.

There are several methods for estimating water temperature, some based on a sinusoidal fit, others based on the mean over varying periods; [2] is just a sample of research into this methodology. Rather than take a critical view of any method, the temperature models in this research use a simple 'best fit' approach using: average temperature, low resolution temperature with exponential smoothing and a sinusoidal model.

The primary model taken from Sanford's measurements is the benchmark that other data sets subscribe to; the results from using this model will be compared to simulations using the other temperature models. This data set will be termed Real, the others: Flat, Smoothed and Sine (Figure 1).

2.1 Species Hierarchy

Paine details the dynamics of disturbance and its effects on the spatial structure of mussel beds [12]. From *Petalonia Sp.* (Alga) to the *Mytilus Californianus* (Mussel), probabilities are listed for each locally competing species occupying a space after a period of time. In the simulation four abstract species will occupy space: the first three are representations of plants and other marine organisms, the fourth are mussels, the dominant space competitor.

The virtual environment consists of a random scattering of patches. Each patch will be represented in program code as a Promotable Agent. The Agent object promotes itself to a higher level species after one of two events have occurred: either an amount of time has passed by which it is deemed the current species is ready to replaced with a higher level species; or the current species has been eaten by a starfish, in which case the patch reverts to time zero and the species hierarchy reverts to the start.

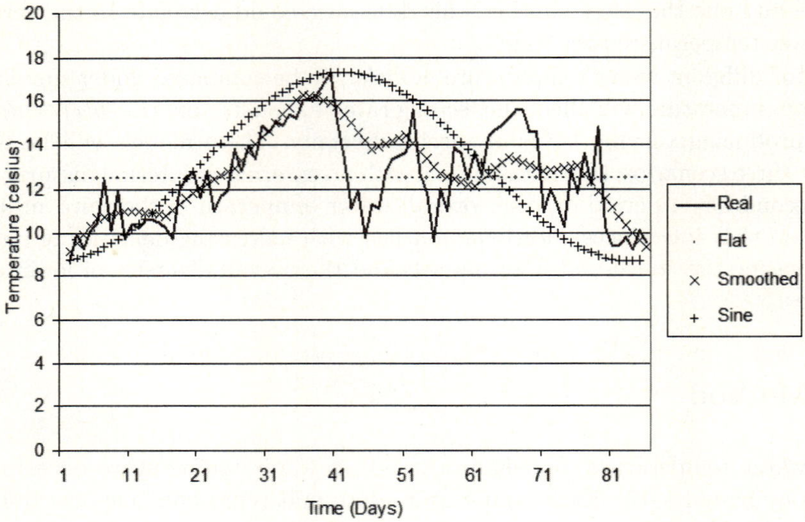

Fig. 1. Real, Flat, Smoothed and Sine Temperatures

2.2 Species Promotion

Species under competitive pressure use up energy normally reserved for growth and reproduction, limiting their populations [17]. The magnitude of the period for which a species exists is dependent upon the total head count of individuals of that species as a factor of the overall population; species who share a greater ratio of the total population live longer. Longevity is calculated on this basis and results in a lifespan of 10 to 30 days; an abstract definition rather than any actual classification.

2.3 Starfish Behaviour

Starfish are not only important as consumers of mussels, and therefore producers of diversity among species, but also because they are influenced by water temperature and are therefore fundamental to testing the differences between the water temperature models. Research by Himmelman [6] has been used to define starfish agents with activity budgets which determine time spent capturing and eating prey. Starfish will continue to consume while there are available prey, the limiting factor is the time it takes to capture, consume and digest prey.

2.4 Simulation Tests

Test A will run simulations for each model under a simple scenario. Starfish will consume mussels as described above and readings of species population head-counts will be taken. This will establish the differences each water temperature model has in terms of mean population densities and species diversity.

Test B will subject all water temperature models to a drop in temperature of 4 degrees Celsius for a period of 400 days. The drop in temperature resembles that found by Sanford during periods of upwelling events [14], and the period is long enough for population levels to react to the new environmental conditions.

Test C again is based on Test A but with an additional level of complexity based on observations seen in mussel farms by Inglis [7]. A function will add starfish to the simulations if the mussel population density reaches a head-count greater than a preset variable. This preset variable represents a stable mid-point state of mussel populations for each model (260). This will increase the effect of consumption:

1. A mussel population limit variable is set to an initial value of 260
2. If the mussel population reaches this value:
   (a) The limit variable is increased by 1.
   (b) Starfish are added to the simulation.

Step 2 may be called frequently throughout a simulation, the amount it is called will depend on the water temperature model being simulated.

Similarities between simulations of water temperature models found in Test A will be compared with tests B and C. Any correlation between water temperature variance and the results of these tests will be observed.

For the purposes of the experiments, 400 promotable individuals and 30 starfish produced sufficiently high levels of data over simulation times of 1000 days.

3 Results

The results will be split into three tests as described in section 2.4. The mean and standard deviation of population densities will be measured in each simulation. Simpson's Index of Diversity will also be tabulated for each model, showing the relative diversity of each species.

3.1 Test A - Simulations with No Internal or External Influences

The results of this test will draw comparisons between each water temperature model and the Real model. It will also establish simulation data for an assessment of the results of tests B and C.

The data from one simulation of each water temperature model was used for the results. In order to satisfy the assumption that consecutive tests would result in equivalent results, 25 simulations for each model were analysed using one-way ANOVA. Differences were found to be non-significant: Real $P = 0.21$, Flat $P = 0.24$, Smoothed $P = 0.60$ and Sine $P = 0.54$. This confirms that results from one simulation of each model represent an equivalent snapshot of typical behaviour.

Table 1 confirms some basic properties of the water temperature models. It can be seen that there is a correlation between the variance (as measured by the standard deviation) of the water temperature and the variance in the species populations. Where the temperature variance is low in relation to the other models,

Table 1. Species Population and Water Temperature - Mean and Standard Deviation

	Real	*Flat*	*Smoothed*	*Sine*
Species 1 Mean	46.44	46.87	46.62	46.57
Species 2 Mean	46.35	46.88	46.75	46.78
Species 3 Mean	46.56	46.97	46.80	46.76
Mussels Mean	260.65	259.28	259.83	259.89
Species 1 SD	6.44	3.21	6.98	9.48
Species 2 SD	7.78	4.25	8.08	10.52
Species 3 SD	9.43	5.59	9.39	11.64
Mussels SD	9.98	4.08	13.02	20.34
Water Temperature Mean	12.50	12.50	12.55	12.53
Water Temperature SD	2.07	0.0	2.01	3.04
Index of Diversity	0.5374	0.5411	0.5396	0.5395

as in the Flat model, the variance for each species is also low. The Sine model, which has the highest water temperature variance, also has the highest species variance. The variance in water temperature will cause variation in the feeding rate of starfish, and this is reflected in the density of populations over time.

The inherent differences between models can also be illustrated by comparing the mean of mussel populations for each model to the Real model using a t-test. Both the Smoothed model and the Sine model have a non-significant difference, respectively: $P = 0.11$ and $P = 0.29$. While the Flat model is significantly different, $P < 0.01$, we can nevertheless see in Table 1 that the difference in results is small. To measure the relative diversities, Simpson's Index of Diversity (1) was used to calculate a value for each model. Simpson's Index of Diversity is a measure that takes into account the number of each species as well as the abundance. This will provide a comparison for each model.

$$Diversity = \frac{\sum n(n-1)}{N(N-1)} \ . \tag{1}$$

n is the total number of organisms of a particular species and N is the total organisms in all species. To make the index easier to interpret, a modification of $1 - D$ will be made so that 0 represents zero diversity and 1 represents infinite diversity.

Table 1 shows the Real model to have the least diversity and the Flat model to have the greatest. The Flat model is the least similar to the Real model, but the difference is only small. This data will be compared against similar data from tests B and C.

In terms of mean mussel population densities, each model matches the Real model but the Flat and Sine models differ to a greater degree in the standard deviation of their population variance. The Flat model shows a significant difference in the t-test and this will be due to the lower variance in comparison to that of the Real model. Primarily this test has established the variance that differs between models despite the similarities of mean mussel population densities. The Smoothed model has the most similar variance, the Flat model has less and the Sine's variance is greater.

3.2 Test B - External Environmental Changes

Test B will examine how each model behaves subject to a drop in temperature of 4 degrees Celsius. Once again, in order to satisfy the assumption that one data snapshot is representative of repeat tests, 25 simulations of each model were tested for non-significance using ANOVA: Real $P = 0.83$, Flat $P = 0.98$, Smoothed $P = 0.96$ and Sine $P = 0.98$.

Table 2. Mussel population, Water Temperature and Species Diversity for Test B

	Real	*Flat*	*Smoothed*	*Sine*
Mussels Mean	289.76	295.39	289.96	285.34
Mussels SD	6.58	3.03	6.22	10.73
Water Temperature Mean	8.55	8.55	8.53	8.81
Water Temperature SD	2.08	0.0	2.08	3.03
Index of Diversity	0.4524	0.4344	0.4518	0.4663

T-Tests on mean mussel populations begin to show how dissimilarity in the variance of water temperature models seems to cause differences in relation to the Real model. The smoothed model has a non-significant difference with $P = 0.49$, the Flat and Sine though are both significant with $P < 0.01$ in both cases. This disparity can also be seen in the index of diversity (Table 2). The index of diversity also exhibits a wider difference, from 0.3% in Test A to 3.2% in this test, highlighting the greater spread of results.

Test B establishes that a factor of scale in a parameter (an offset value for water temperature) can unevenly affect the results of a set of models. It would have been difficult to predict this at the design stage of a simulation and, without comparing the water temperature models, would have been hard to realise without further analysis that such a scaling had been affected.

3.3 Test C - Population Frequencies and Changes in Population Density

Test C measures the effects of increased populations of starfish in relation to mussel population density. Consecutive tests of each model find a non-significant difference within the groups: Real $P = 0.06$, Flat $P = 0.07$, Smoothed $P = 0.17$ and Sine $P = 0.24$.

Table 3. Mussel population, Water Temperature and Species Diversity for Test C

	Real	*Flat*	*Smoothed*	*Sine*
Mussels Mean	232.06	246.98	232.13	229.89
Mussels SD	10.24	3.71	12.25	18.02
Index of Diversity	0.6072	0.5725	0.6070	0.6119

Table 3 finds the dominant effect of Test C to be a shift downwards in the mussel population densities. There is a significant discrepancy between the mean populations, notably the Flat model has a higher mean population of 246.98 compared to 232.06 for the Real model. A t-test comparing the Real model to the other models again finds a non-significant difference in the Smoothed ($P = 0.89$) and a significant difference in the Flat and Sine ($P < 0.01$). This difference is again reflected in the index of diversity for each model and comparing Table 3 to Table 1 we can see that there is an overall increase in diversity. This increase is to be expected as the function limits the population of mussels by increasing the feeding rate of starfish. The original basis for the simulation was to show that starfish increase diversity by consuming mussels; therefore increasing the feeding rate has led to an increase in diversity.

These results again show how variance in the water temperature models affects the results of simulations. In comparing mussel population densities using the t-test, the Flat and Sine models are significantly different, as opposed to the Smoothed model. The index of diversity also shows how the Smoothed and Real models keep their parity in relation to their variance, this is in contrast to the difference seen in the species diversity of the Flat and Sine models.

3.4 Summary of the Results

Test A finds the mean mussel population density to be similar in all models. However, while the Real and Smoothed models have similar variance in water temperature (2.07 and 2.01), the Flat model has less (zero), and the Sine model more (3.04).

Test B exposes differences between the model groups. Both the t-test and the index of diversity find the Flat and Sine models differing from the Real model, significantly in the case of the t-test. The Smoothed model however retains similarity in both cases.

Test C, more than Test B, attempts to capitalise on the different levels of variance in each water temperature model. The effect of the function can be seen to have a greater influence on models with a higher level of water temperature variance. When comparing the results of simulations to the Real model, we again find a significant difference in the Flat and Sine models, but not the Smoothed model. The index of diversity also reflects the effect of the function; it is higher in all models, compared to that of Test A as we would expect, but the Sine model surpasses the diversity of the Real model, and the Flat model falls behind.

Perhaps the most notable effect of Test C is while the Real, Sine and Smoothed models have similarity in mean mussel population densities, the Flat model has a far higher mean. This indicates the function was less active in simulations of this model than the others.

4 Discussion

The effects of variance within a model can be just as significant a factor as the mean. The experiments in this paper have shown that comparisons of water temperature models can demonstrate similar results in one case and varying results in another; especially when a simulation necessitates change that is affected by variance.

The experiments outlined in this paper appear to show a correlation between the differences in variance when compared to a benchmark model, and the significant disparities of the results of tests. More research could be conducted to further substantiate this correlation, perhaps with water temperature models of a more similar nature and several levels of variance. The similarities seen in the results of the Real and Smoothed models, while sharing a similar mean and standard deviation, could be due to the comparable perturbations in the shapes seen in water temperature graphs, further analysis could validate this to be true or otherwise.

The Smoothed model has a significantly lower resolution than the Real model; it was created by taking every fifth day's temperature and then applying an exponential smoothing function. The result of this is that many peaks and troughs are missing that exist in the parent data, yet still, this model manages to maintain some parity with the Real model. This is interesting because it raises the question on how simple a model can be, in terms of temporal resolution, and still maintain its usefulness and reliability during simulation.

Answers to questions that arise in this research, and the potential research that could be undertaken to validate some of the assumptions made, are helpful as long as simulations and the replication of life-like behaviour require the detail in order to satisfy their conclusions. Simplicity, if it achieves the desired result, is a good thing: it reduces the time needed to create simulations, and helps to better understand the results. The experiments in this research show that it is important to understand how a simulation that utilises a simple functional solution, such as calculating mean water temperature, may produce varying results to one that uses a more complex function, one that produces data more comparable to the real world.

References

1. Booth, G., Gecko, A.: Continuous 2-D World for Ecological Modelling. Artfiicial Life 3, 147–163 (1997)
2. Crisp, D.T.: Simplified methods of estimating daily mean stream water temperature. Freshwater Biology 23(3), 457–462 (1990)

3. Downing, K.: EUZONE: Simulating the Evolution of Aquatic Ecosystems. Artficial Life 3(4), 307–333 (1997)
4. Green, D.G., Newth, D.: Towards a theory of everything? Grand challenges in complexity and informatics. In: Complex Systems 2000, Dunedin, New Zealand (2000)
5. Grimm, V.: Ten years of individual-based modelling in ecology: what have we learnt and what could we learn in the future? Ecological Modelling 115, 129–148 (1999)
6. Himmelman, J.H., Dutil, C., Gaymer, C.F.: Foraging behaviour and activity budgets of sea stars on a subtidal sediment bottom community. Journal of Experimental Marine Biology and Ecology 322, 153–165 (2005)
7. Inglis, G.J., Gust, N.: Potential indirect effects of shellfish culture on the reproductive success of benthic predators. Journal of Applied Ecology 40(6), 1077–1089 (2003)
8. Katz, R.W., Brown, B.G.: Extreme events in a changing climate: Variability is more important than averages. Climatic Change 21, 289–302 (1992)
9. Mearns, L.O., Rosenzweig, C., Goldberg, R.: Mean and Variance Change in Climate Scenarios: Methods, Agricultural Applications, and Measures of Uncertainty. Climate Change 35, 367–396 (1997)
10. Oboshi, T., Kato, S., Mutoh, A., Itoh, H.: Collective or Scattering: Evolving Schooling Behaviours to Escape from Predator. In: Artificial Life VIII, Proceedings of the Eighth International Conference on Artificial Life (2003)
11. Paine, R.T.: A Note on Trophic Complexity and Community Stability. American Naturalist 103, 91–93 (1969)
12. Paine, R.T., Levin, S.A.: Intertidal Landscapes: Disturbance and the Dynamics of Pattern. Ecological Monographs 51(2), 145–178 (1981)
13. Robinson, S.: Distributed Simulation and Simulation Practice. Simulation 81(1), 5–13 (2005)
14. Sanford, E.: The feeding, growth, and energentics of two rocky intertidal predators (Pisaster ochraceus and Nucella canaliculata) under water temperatures simulating episodic upwelling. Journal of Experimental Marine Biology and Ecology 273, 199–218 (2002)
15. Sanford, E.: Water Temperature, Predation, and the Neglected Role of Physiological Rate Effects in Rocky Intertidal Communities. Integrative and Comparative Biology 42(4), 881–891 (2002)
16. Thulke, H.H., Grimm, V., Müller, M.S., Staubach, C., Tischendorf, L., Wissel, C., Jeltsch, F.: From pattern to practice: a scaling-down strategy for spatially explicit modelling illustrated by the spread and control of rabies. Ecological Modelling 117, 179–202 (1998)
17. Volterra, V.: Variations and fluctuations of the number of individuals in animal species living together. Animal Ecology (1931)

Cell Tracking: Genesis and Epigenesis in an Artificial Organism

Alessandro Fontana

IEEE
alessandro.fontana@ieee.org

Abstract. This paper belongs to the field of Computational Development. It describes a method that has the objective to provide an effective way of generating arbitrary shapes by using evolutionary-developmental techniques, i.e. by evolving genomes that guide the development of the organism starting from a single cell. The key feature of the method is the explicit introduction of an epigenetic memory, that is a cell variable that is modified during the development process and can take different values in different cells. This variable represents the source of differentiation, that leads different cells to read out different portions of the genome at different times. Preliminary experiments have been performed and the results appear to be quite encouraging: the proposed method was able to evolve a number of 25x25, 32x48 and 64x64 target shapes.

Key words: computational development, morphogenesis, epigenetic memory.

1 Introduction and Related Work

The present work belongs to the field of Computational Development. It describes a method that has the objective to provide an effective way of generating arbitrary forms by using evolutionary-developmental techniques, i.e. by evolving genomes that guide the development of the organism starting from a single cell. The paper is organised as follows: the rest of this section surveys the related work, section 2 describes the model, section 3 discusses the biological implications, section 4 and section 5 report the experiments performed and the relevant results; section 6 draws the conclusions and outlines future work.

One of the oldest computational models of development is perhaps represented by L-Systems, proposed by Lindenmayer [9]. L-Systems capture the notion of cell signalling and were proposed to model plant growth. Random Boolean Networks (RBN's) were originally developed by S. Kaufmann as a model of genetic regulatory networks (see [6]); in the context of the development of multi-cellular organisms, the attractors of RBN's (i.e. the stable states) are interpreted as the different "cell types" of the organism. Hugo de Garis [1] developed a model for evolving convex and non-convex shapes in 2D reproductive cellular automata. Two genes, which were called operons, were used to regulate the growth of the shape. The model was successful in evolving convex shapes but the non-convex shapes (L-shape) presented a problem. Gruau [7] used grammar trees to encode steps in the development of a neural network starting from a single ancestor cell. This system is called cellular encoding, because

F. Almeida e Costa et al. (Eds.): ECAL 2007, LNAI 4648, pp. 163 – 171, 2007.

its rules apply to single cells in a growing network. The grammar tree contains developmental instructions at each node. Miller [10] developed artificial organisms (the french flag) based on Cartesian Genetic Programming which is an extension of Boolean networks. Miller's goal is to evolve a developmental program inside a cell to create multicellular organisms.

2 The Model

In our model the phenotype of the developing organism is represented as a 2-dimensional array of cells, being each cell associated to a (x,y) position on a grid. The variables associated to a cell (see figure 1) belong to two categories:

- The "genetic memory", which is *not* modified during development; it is organised as an array of "development operators" and is identical in all cells;
- The "epigenetic memory", which is modified during development; it is made up by the variable "development stage" (DS), by the variable "tracker counter" (TC) (which are identical in all cells) and by the variable "cell tracker" (CT) (which is *not* identical in all cells).

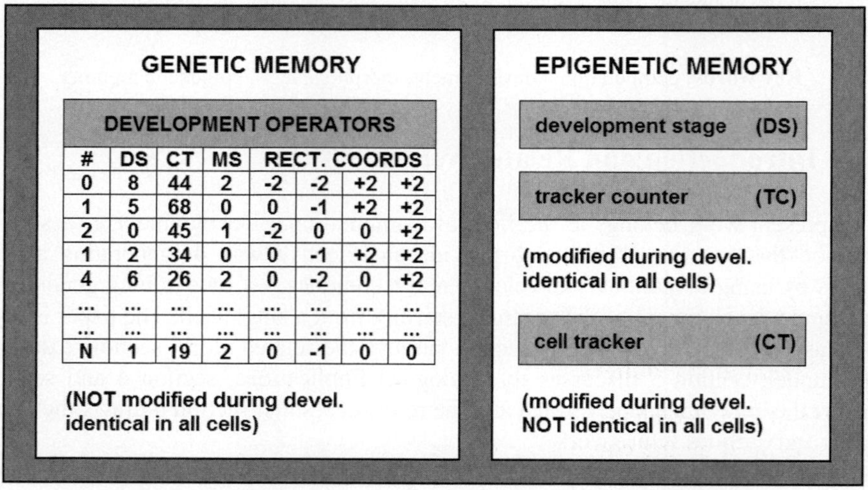

Fig. 1. Model variables.

The development starts with a single cell (the zygote) placed in the middle of the grid. The variable development stage is initialised at zero and is incremented by one up to a maximum value; at each development stage, a given cell can proliferate or can undergo apoptosis.

As we said, the genetic memory is structured as a list of development operators (see figure 1). Each development operator has a DS value and a CT value: if these values are equal to the values of DS and CT in the epigenetic memory of a given cell, the operator is activated and the relevant code is executed for that cell (called the mother cell). A field of the operator holds a "master switch" (MS) that defines the type of "development event" that is going to occur:

- "undifferentiated proliferation" (MS value = 0)
- "differentiated proliferation" (MS value = 1)
- "apoptosis" (MS value = 2)

Figure 2 reports examples of the three possible events. Let us have a look at each event in detail.

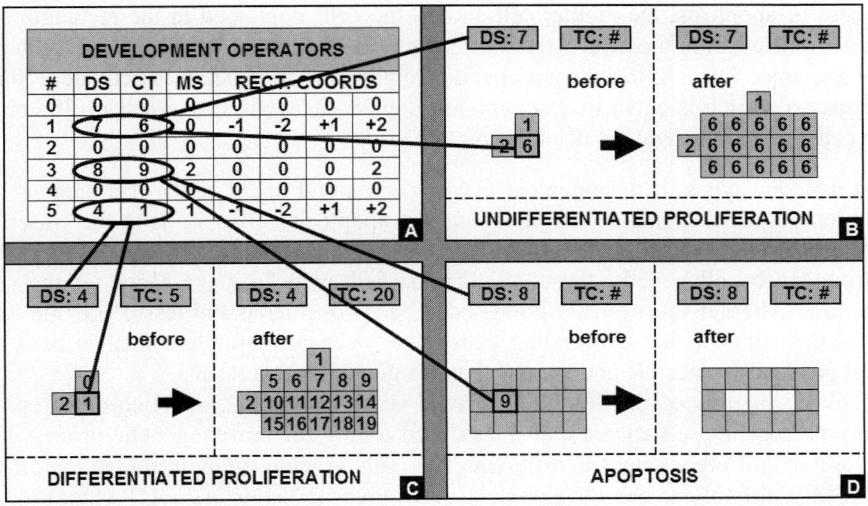

Fig. 2. Development events. Panel A contains the genetic memory, the other three panels show the events. In all cases the numbers inside the cells represent the CT values of the cells; the variables DS and TC are placed in the upper part of the panel because they are identical for all cells.

Undifferentiated Proliferation

In case of undifferentiated proliferation, new cells are created and placed in the rectangle defined by the coordinates reported in the right part of the operator, and the newly created cells are assigned the CT value of the mother cell. Example: in panel B of figure 2 we have a cell with CT = 6 and since DS = 7, the first operator highlighted in panel A is activated. Since the master switch value is equal to 0, an undifferentiated proliferation event takes place: the new cells all have the CT value of the mother (6).

Differentiated Proliferation

As in the previous case, also in the differentiated proliferation case new cells are created and placed in the rectangle defined by the coordinates reported in the right part of the operator, but this time each newly created cell is assigned a different value of CT. The value assigned corresponds to the value assumed by a the variable tracker counter (TC) at the beginning of the proliferation event; TC has the value 1 at the beginning of development (0 is the CT value of the zygote) and is incremented by one at each new assignment. Example: in panel C of figure 2 we have a cell with CT = 1 and since DS = 4, the third operator highlighted in panel A is activated. Since the master switch value is equal to 1, a differentiated proliferation event takes place; in this example we assume that the tracker counter at the beginning of the event is equal to 5: this is the value assigned to the first cell created, 6 the value assigned to the second cell created, and so on. The value of the tracker counter at the end of the event is 20: this will be the first value assigned in the next differentiated proliferation occurring in any cell.

Apoptosis

In case of apoptosis, the mother cell and all the cells contained in the rectangle die, i.e. are deleted from the grid. Example: in panel D of figure 2 we have a cell with CT = 9 and since DS = 8, the second operator highlighted in panel A is activated. Since the master switch is equal to 2, an apoptosis event takes place: the cell itself and all the cells in the rectangle are deleted from the grid.

The different types of development events correspond to the tools of a painter: the proliferation corresponds to the brush and the apoptosis corresponds to the eraser. A physical mechanism has been implemented, by which whenever a proliferating cell is surrounded by other cells, these cells are pushed away to make room for the new structure. Alternative physical options have been considered and tested (i.e. the simple overwriting of the surrounding cells); the mechanism implemented is both the most physically plausible and the one that has given the best results.

The key feature of the described method is the explicit introduction of an epigenetic memory, i.e. a variable that is modified during the morphogenetic process and can assume different values in different cells. This is achieved by means of the differentiated proliferation: this development event gives a "name" (the CT value) to the newborn cells, and this name represents the link by which these cells can in subsequent stages be picked up by the genetic memory and given other development events to execute. This represents the source of the differentiation process, that leads different cells at different times to read out different portions of the genetic memory.

3 Biological Interpretation

The aspects we are interested in modelling are those related to the shaping of an organism, achieved in nature through the mitosis, differentiation and apoptosis of individual cells. The genetic memory corresponds to the cell genome, and the development operators correspond to those genes that are active during the embryo development. The structure of a development operator draws inspiration from the structure of biological genes, which are composed of an "if" part (the variables DS and CT) and a "then" part (the development event).

The epigenetic memory corresponds to the biological epigenetic memory, implemented in real cells through the methylation of DNA segments and other mechanisms that regulate gene transcription. In an article published recently [12], M. Zernicka-Goetz and some colleagues proved that the decision about the cell fate in the mouse embryo is taken as early as at the four-cell stage. This decision would be determined by the presence in the embryo cells of certain chemicals in asymmetrical quantities and would then be "stored" in the nucleus through changes in the histone-arginine methylation patterns of specific DNA segments. The information stored in such methylation patterns is the biological counterpart of the epigenetic memory in our model.

The cell cycle of our model differs from the biological one in that in the course of a single cycle a cell can proliferate in a rectangular area, which in biological systems can be achieved only through a series of mitoses, each belonging to a distinct cell cycle. This "generalised-mitosis" was conceived mainly with the objective of speeding up the "shaping" process and it would be consistent with living systems if certain sequences of mitoses are coded as subroutines: the sequence of early divisions up to the morula stage could represent an example of such a subroutine.

The mechanism by which different CT values are assigned to newly created cells is biologically implausible, because it implies the existence of a global counter (the tracker counter) common to all cells. Nevertheless, it is not difficult to imagine a biologically plausible mechanism with the same computational "performances". First, we can imagine decomposing the rectangular proliferation in a series of individual mitoses; such mechanism would then, in case of differentiated proliferation, append a "0" to the CT of the mother cell and a "1" to the CT of the daughter cell; in case of undifferentiated proliferation it would simply copy the CT value of the mother cell onto the CT of the daughter cell.

Finally, the proposed algorithm uses the "prepatterning" method only to determine the fate of cells, i.e. the local environment of a cell has no influence on cell fate determination. This is in contrast with the biological evidence that a specialised cell type inserted in a tissue different from its origin tissue can be induced by the surrounding cells to take up the cell type of the host tissue. In spite of this, the outcomes of the preliminary experiments conducted appear to be quite encouraging (the proposed model proved able to generate a number of shapes), and we believe that by exploiting the local environment, these results might be further improved.

4 Implementations

The targets are 25x25, 32x48 and 64x64 multi-cellular arrays. Among the targets were considered simple letters or group of letters, like "L", "BA", and some elementary forms, like a beetle, an airplane, etc. Given the nature of the painting tools provided (a rectangular brush and a rectangular eraser), mostly shapes with only straight lines were considered; actually some shapes contain also oblique lines and, to our surprise, the algorithm was able to reproduce them at least partially, by leveraging the physics.

The method consists of evolving a population of genetic memories, letting at each generation the development unfold for each memory (starting from a single cell with CT = 0 placed in the middle of the grid and running DS from 0 up to a maximum value), and then using the adherence of the shape at the end of development to the target shape as fitness measure. The genetic population is composed of n (=4 or 8 in our experiments) subpopulations of equal size, for a total population size of 400 individuals (represented as strings of quaternary digits), undergoing elitism selection for 1000 to 5000 generations. The division into n subpopulations has the objective to increase the diversity of the evolved solutions; this is obtained by multiplying the fitness of each individual of subpopulation m by the normalised Hamming distances of the individual from the best individuals of subpopulations 1 to m–1. This has the effect to increase the fitness value of individuals that show a higher degree of diversity from the best individuals of the previous subpopulations, which in turns encourages the explorations of new regions of the search space (of course the mapping from genotype to phenotype is not one-to-one -many genotypes map to the same phenotype, which limits the effectivess of this measure). GA parameters are 50% single point crossover, mutation rate of 0.1% per digit. The fitness function formula is the same adopted by H. de Garis [1]:

F = (ins-k*outs)/des

where ins is the no. of points of the evolved shape falling inside the target shape, outs is the no. of points of the evolved shape falling outside the target shape, des is the no. of points of the target shape, k is a parameter. The lower the value of k, the more tolerant the formula becomes to points falling outside the target shape. As a result, the forms generated are less accurate in the details but generally complete in all their parts (an example of this trade-off can be seen by looking at the Italy map case). The pseudocode for the whole procedure is reported below.

```
for generation no. from 0 to maxval
   for individual no. from 0 to maxval
      for DS value from 0 to maxval
         for CT value from 0 to maxval
            if(there is a cell with these DS and CT values
               in its epigenetic memory and there is
               a dev. operator with these DS and CT values
               in its left part)
            then
               apply development operator to the cell
               apply physics (push-away)
      evaluate fitness
```

In our first experiments we started with a fixed, relatively large, genome size (64 development operators). The size of the search space constitutes indeed one of the major problems that affect large genomes; on the other hand, small genomes can suffer from inadequate modelling power (see [11]). The solution adopted has been to leave the genome size undefined and subject to evolution, adding a flag to each gene, also subject to evolution, indicating whether the gene is constitutively active or not. This solution has also the advantage of avoiding problems with the crossover operators occurring in genomes with variable size due to sequence misalignment.

Main parameters of the algorithm are (relevant reference values in parentheses):

- DSMAXVAL: max value for the variable development stage (8)
- CTMAXVAL: max value for the variable cell tracker (up to 1024)
- RSMAXVAL: max dimension of the side of the rectangle (10-18)

The experiments undertaken have shown the emergence of a “painting technique”, consisting of a fast differentiation in correspondance to the first DS values, followed by touch-ups in the subsequent ones. The genetic algorithm “understands” that the first thing to do is to create a relatively big mass of cells, marking each of them with a different cell tracker value. This is the only means by which it can later pick individual cells and commit them to specific fates. Let us now have a look to some of these experiments in detail.

5 Examples of Development

5.1 25x25 Shapes

The two reported 25x25 target shapes are the letter "L" and a "beetle". The "L" is a very simple (although non-convex) shape: this shape was easily evolved, using only 36 CT values. The development is typical: the zygote cell is first grown into a rectangle of differentiated cells (i.e. each cell has a different CT value). Then two border cells of the rectangle are grown without differentiation to create the two arms of the "L". The initial growth into a mass of differentiated cells seems to be a mandatory step in the development process; in fact, if this growth occurred without differentiation, every cell of the "cell mass" would then be constrained to follow the same fate, making the shaping process impossible. The "beetle" is a very challenging shape, due to the presence of many little protrusions and the best evolved shape is less adherent to the target compared to other experiments, nevertheless it is recognisable (at least with some good will!). Please note that in this and in the next figures the colours of the evolved shape are used simply to distinguish the grid size.

Fig. 3. 25x25 shapes: the "L" and the "beetle" (target shape on the left, best evolved shape on the right).

5.2 32x48 Shapes

The two reported 32x48 target shapes are a "airplane" and the group of letters "BA". In the airplane case, the result is surprisingly good, obtained in a very efficient way, using only 20 CT values. Particularly interesting is the very good approximation of the wings, which are not straight lines and thus in theory were not within the reach of the "brush". This result was achieved by the algorithm only by leveraging the physics (the push-away mechanism).

5.3 64x64 Shapes

The two reported 64x64 target shapes are a "house" and an "Italy map". In the case of Italy map, the two evolved shapes correspond to two different values of the parameter k, namely k=2 for the one in the middle, and k=1 for the one on the right. As we already pointed out, increasing the value of k makes the algorithm less tolerant to points falling

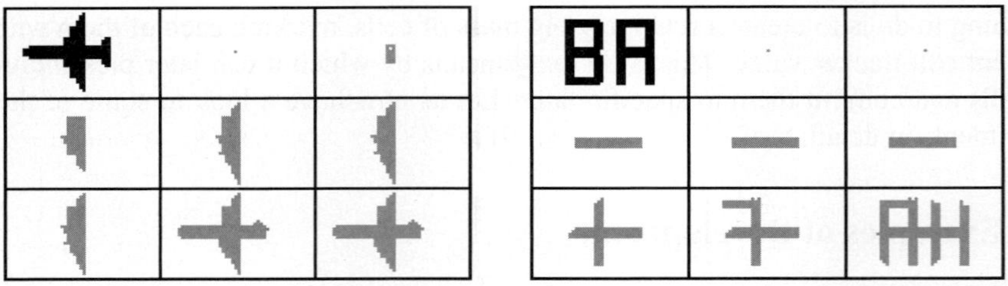

Fig. 4. 32x48 shapes: the "airplane" and "BA" (target shape in upper-left corner, development of the best evolved best evolved shape in the other boxes).

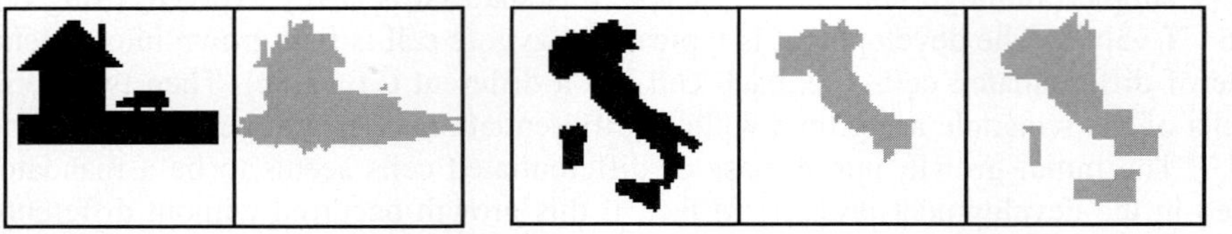

Fig. 5. 64x64 shapes: the "house" and the "Italy map" (target shape on the left, best evolved shape on the right).

outside the target shape. As a consequence, the shape in the middle shows a very good degree of adherence to the target shape, but is incomplete. On the other hand, the shape on the right is more tolerant to "spurious points" and is complete though not so neatly drawn.

6 Conclusion and Future Developments

A new Computational Development algorithm has been presented, with the aim of allowing the generation of arbitrary forms. The effectiveness of the algorithm has been tested with a number of target shapes, and so far the results have been promising. A major limitation of the present algorithm is that the proliferation can take place only within a rectangle; this poses restrictions on the types of shapes that can be obtained: in general, shapes containing oblique lines cannot be evolved. As we have seen, this is not strictly true, in that the genetic algorihtm can find ways to overcome this limitation by leveraging the physics. It is nevertheless already planned to insert additional painting tools, in particular the "oblique brush", that allows the direct realisation (i.e. without the need to leverage the physics) of shapes with non-straight lines: this would make the realisation of such shapes much easier. Another feature we plan to add is colour.

As we already mentioned, in this implementation the behaviour of the cell is based on the "prepatterning" method only. A natural extension to the model would be to add the influence of the surrounding cells. This is certainly the case in real biological systems, where the cellular environment can induce the cell, at least to some extent, to go down specific developmental paths (e.g. a liver cell inserted into the cardiac muscle is induced to trans-differentiate into a heart cell).

The “emerging” painting technique by which every development starts with a rectangular differentiation is perfectly understandable, as otherwise it would be impossible to track different cells and commit them to specific fates at specific developmental milestones. On the other hand, it mimics very closely the early development of the animal embryo, up to the morula stage. This similarity brings us to support the hypothesis that, even though they look the same, the morula cells already have different epigenetic memories (i.e. the biological equivalent of CT), hypothesis that appears in accordance with [12].

Finally, a few considerations on “junk” DNA. In molecular biology, junk DNA is a collective label for the portions of the DNA sequence of a chromosome or a genome for which no function has yet been identified (non protein- or RNA-coding DNA). About 97% of the human genome has been designated as "junk", including most sequences within introns and most intergenic DNA. According to the model of development proposed, a development operator having a CT value that never shows up during the “normal” embryonal development, can be defined as “junk”. On the other hand, if the embryonal development departs for whatever reason from normality and the relevant CT does appear, the development operator can now become active. These considerations suggest a more blurred scenario for the junk DNA, in which a given development operator can be junk or not depending on the actual conditions encountered by the organism during its development.

References

1. De Garis, H.: Artificial Embryology and Cellular Differentiation. In: Bentley, P.J. (ed.) Evolutionary Design by Computers, pp. 281–295. Academic Press, London (1999)
2. Eggenberger, P.: Evolving Morphologies of Simulated 3d Organisms Based on Differential Gene Expression. In: Husbands, P., Harvey, I. (eds.) Proceedings of the 4th European Conference on Artificial Life, MIT Press, Cambridge (1997)
3. Fontana, A., Fraccaro, W.: A Functional Model of Cell Genome. In: Proc. of Alife IX (2004)
4. Fontana, A.: A Functional Model of Development and Expression in an Artificial Organism. In: International Conference on Morphological Computation, Venice (2007)
5. Gershenson, C., Wuensche, A.: Tutorial: Introduction to random Boolean networks. In: Workshop and Tutorial Proc. of Alife IX (2004)
6. Gruau, F., Whitley, D., Pyeatt, L.: A Comparison between Cellular Encoding and Direct Encoding for Genetic Neural Networks. In: Genetic Programming 1996: Proceedings of the First Annual Conference (1996)
7. Kumar, S., Bentley, P.J.: On Growth, Form and Computers. Academic Press, London (2003)
8. Lindenmayer, A.: Mathematical models for cellular interaction in development I. Filaments with one-sided inputs. Journal of Theoretical Biology 18, 280–289 (1968)
9. Miller, J.F.: Evolving Developmental Programs for Adaptation, Morphogenesis, and Self-Repair. In: Banzhaf, W., Ziegler, J., Christaller, T., Dittrich, P., Kim, J.T. (eds.) ECAL 2003. LNCS (LNAI), vol. 2801, Springer, Heidelberg (2003)
10. Stanley, K.O., Miikkulainen, R.: A Taxonomy for Artificial Embryogeny. Artificial Life 9(2), 93–130 (2003)
11. Torres-Padilla, M.E., Parfitt, D.E., Kouzarides, T., Zernicka-Goetz, M.: Histone arginine methylation regulates pluripotency in the early mouse embryo. Nature (2007)

Developmental Neural Heterogeneity Through Coarse-Coding Regulation

Jekanthan Thangavelautham and Gabriele M.T. D'Eleuterio

Institute for Aerospace Studies, University of Toronto
Toronto, Ontario, Canada, M3H 5T6
thangav@ecf.utoronto.ca, gabriele.deleuterio@utoronto.ca

Abstract. A coarse-coding regulatory model that facilitates neural heterogeneity through a morphogenetic process is presented. The model demonstrates cellular and tissue extensibility through ontogeny, resulting in the emergence of neural heterogeneity, use of gated memory and multistate functionality in a Artificial Neural Tissue framework. In each neuron, multiple networks of proteins compete and cooperate for representation through a coarse-coding regulatory scheme. Intracellular competition and cooperation is found to better facilitate evolutionary adaptability and result in simpler solutions than does the use of homogeneous binary neurons. The emergent use of gated memory functions within this cell model is found to be more effective than recurrent architectures for memory-dependent variants of the unlabeled sign-following robotic task.

1 Introduction

One of the big challenges in Alife is to design open-ended artificial multicellular developmental systems that can grow in complexity to solve extensible control tasks by performing task-decomposition with little or no explicit supervision. Fundamental to understanding and re-engineering multicellular biological systems is to determine how functionality is distributed within these systems and how specialization takes shape.

It has been theorized that exploratory selection/regulation mechanisms, the process by which selection of parallelized selection of functional outcomes facilitates *evolutionary adaptability*, that is, the ability for genes to be heritable and selectable phenotypes, less susceptible to lethal mutations and produce novel traits with fewer mutations [12]. Extensive evidence of exploratory selection processes has been found in the immune system and this had spurred interest into how these processes might be at work within the brain [3,12]. Furthermore, biological evidence hardly points to the notion of a 'typical' homogeneous feedforward binary neurons of the McCulloch-Pitts type. Neurons are, in fact, complex heterogeneous multistate analog systems with memory.

In this paper, we demonstrate the advantages of these exploratory selection/regulation mechanisms based on a coarse-coding scheme, inspired by Albus [1], and show evidence of emergent task decomposition and specialization

F. Almeida e Costa et al. (Eds.): ECAL 2007, LNAI 4648, pp. 172–182, 2007.

occurring at the cellular and gene/protein level within an Artificial Neural Tissue (ANT) framework [17]. The model exhibits cellular and tissue extensibility through ontogeny, resulting in the emergence of neural heterogeneity, use of memory and multistate functionality within the ANT framework. Within each neuron, multiple networks of protein compete and cooperate for representation through a coarse-coding framework for binding sites. We choose to use coarse coding as it is a moderately distributed coding scheme that allows for pooling and redundancy thus helping to render the system robust in the face of noisy sensor data.

This model, with no explicit supervision and limited task-specific assumptions, produces solutions to a variant of the sign-following task found to be unsolvable with fixed-topology homogeneous artificial neural networks. Fixed-topology networks that lack regulatory functionality perform poorly in complex tasks with limited supervision owing to the bootstrapping problem, which causes premature stagnation of an evolutionary run [13].

2 Background

Artificial developmental systems mimic ontogenic processes from biology and have been successfully used, with variable-length genomes, to 'grow' topologies and heterogeneous functionality without explicit intervention. ANT is a morphogenetic system with a directly encoded genome and uses gene regulatory systems (GRNs) for development. Artificial embryogenic systems (L-systems [11] and cellular-encoding systems [8]) use indirect encoding schemes that involve recursive rewriting of the genotype to produce a phenotype. However, it has been argued that indirect encoding schemes introduce a deceptive fitness landscape and result in poor performance for smaller search spaces owing to overhead [15].

Examples of artificial morphogenetic systems include the work by Eggenberger [4] and by Gomez and Eggenberger [5], the latter using 'ligand-receptor interactions' to perform cell aggregation. A morphogenetic system was also used on POEtic by Roggen *et al.* [14]. Developmental tissue models such as Norgev by Astor and Adami [2] are also morphogenetic and facilitate cellular heterogeneity. Cell replication and synaptic connections are formed through a GRN based developmental and learning system using a genetic-programming-type command set. However, in our ANT model, regulation continues after development at the gene/protein and cellular levels. Neuroregulatory functionality is performed through coordinated release of diffusive neurochemicals resulting in superpositioning of chemical concentration fields (Fig. 1). Other models such as GasNet allow for volume signaling between neurons using neurochemicals but lack explicit regulatory functionality [10].

In the coarse-coding cell model presented here, a selection/regulation process is also at work within protein networks resident in each neuron, thus facilitating heterogeneity and open-ended growth in complexity of cells. The use of cellular heterogeneity may be more biologically plausible but more complex, multistate cells do not necessarily present advantages over simpler binary-state cells. Yet, the use of multistate feedforward and memory neurons may be beneficial.

The need for specialized memory neurons arises from the 'error decay' problem evident with standard recurrent connections for learning processes. A stored signal remains unprotected from spurious inputs decays or grows (without bound) making it difficult to recall a signal after many timesteps [6]. Long Short-Term Memory (LSTM) [9] overcomes this limitation but it is a predefined architecture consisting of a storage neuron, a reset gate, an input gate (to protect memory from spurious inputs) and an output gate. However, for the T-maze task (a simpler variant of the sign-following task), it was found that recurrent networks trained using Enforced Subpopulations (ESP) and Hierarchical ESP (H-ESP) outperformed a LSTM architecture [7]. LSTM also lacks biological plausibility and similar memory functions can be obtained without the use of predefined cell blocks using the coarse-coding cell model.

3 Artificial Neural Tissue and Coarse-Coding Cell Model

The ANT architecture consists of a developmental program, encoded in the genome that constructs a three-dimensional neural tissue and associated regulatory functionality (see [18] for details). The tissue consists of two types of neural units, decision neurons and motor-control neurons. Regulation is performed by decision neurons, which dynamically excite or inhibit motor-control neurons within the tissue based on a coarse-coding framework (1).

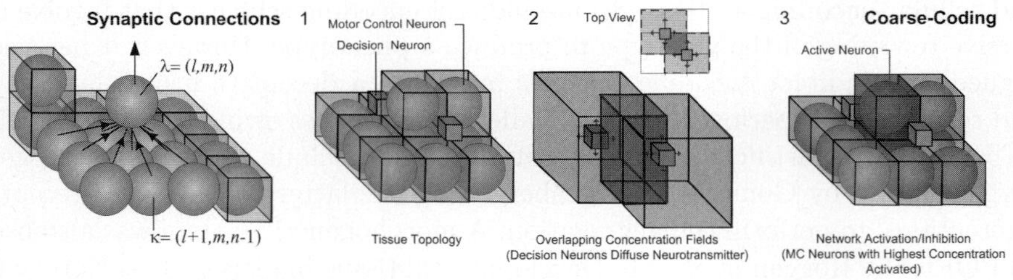

Fig. 1. Synaptic connections between motor-control (MC) neurons and operation of neurotransmitter field.

Our neuron model permits a number n_c of messenger-channel protein networks (Fig. 2a). Each protein network receives the same inputs $\mathbf{x}$, an $n \times 1$ real-valued column, which represents either sensory data or inputs from other neurons. These inputs are fed through m_j 'ion channels' that transform 'electrical signals' into various types of ion, the concentration of which are collectively denoted $\mathbf{y}_j$, an $m_j \times 1$ real-valued column. The concentrations are given by

$$\mathbf{y}_j = \mathbf{W}_j\mathbf{x}$$

where $\mathbf{W}_j$ is an $m_j \times n$ real-valued weight matrix associated with the jth protein network.

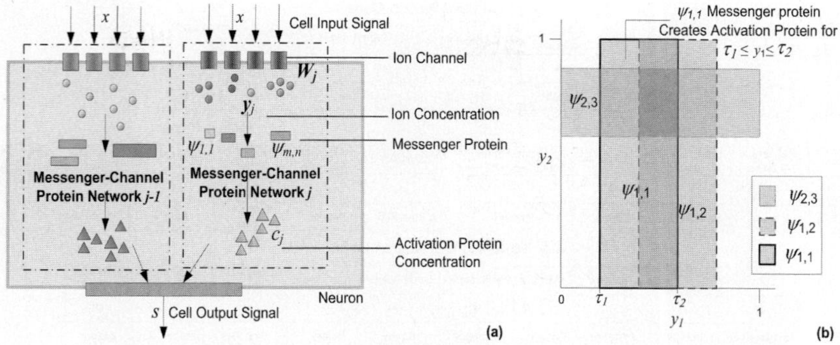

Fig. 2. (a) Schematic of competing messenger-channel protein networks. (b) Coarse-coding interactions between messenger-protein for $m_j = 2$.

Each protein network produces an 'activation protein,' whose concentration c_j is determined by a linear combination of p_j basis functions $\psi_{ik}, i = 1 \cdots m_j, k = 1 \cdots p_j$, dependent on the ion concentrations $\mathbf{y}_j$. The ith basis function, in fact, depends only on the ith ion concentration y_{ij}:

$$\psi_{ik}(y_{ij}) = \begin{cases} 1, & \text{if } \tau_{1,k} \le y_{ij} \le \tau_{2,k} \\ 0, & \text{otherwise} \end{cases} \tag{1}$$

The boundary parameters, τ_1 and τ_2 for each ψ are evolved. The concentrations c_j that determine the output of the neuron are simply $c_j = \sum_{i=1}^{m_j} \sum_{k=1}^{n_{b,j}} \psi_{ik}(y_{ij})$. (Note that c_j are integers.) This structure is reminiscent of and was inspired by the coarse-coding scheme of Albus [1] and we accordingly refer to it as coarse-coding regulation. The basis functions here square-hat functions in one dimension although the kind of tiled functions in Albus's Cerebellar Model Arithmetic Computer can also be used.

Binary-State Neuron. Let us first consider a binary-state neuron, i.e., one where the output $s(t)$, where t represents the discrete time-step, is either 0 or 1. This output is given by

$$s(t) = \begin{cases} a_j, & \text{if } c_j \text{ is a unique maximum} \\ \phi s(t-1), & \text{otherwise} \end{cases} \tag{2}$$

The messenger-channel protein networks compete to determine the neuron's output. If $c_j = \max\{c_1, c_2 \ldots c_{n_c}\}$ and is uniquely determined, i.e., no two networks produce the same maximum concentration, then the output is taken as $a_j \in \{0, 1\}$ (genetically evolved). Otherwise, the output takes the value $\phi s(t-1)$, where $\phi \in \{0, 1\}$. When $\phi = 1$, the output from the previous time-step is maintained. Thus $s(t)$ is intended to model the 'spike' status of the neuron.

Multistate Neuron. Spiking neurons superimpose their spikes on a background signal. We model this aspect of the neuron by allowing for a multiple-state output, $\mathbf{s} = [s_1 \;\; s_2]$ where $s_1 \in \{0, 1\}$ is associated with the spiking signal and

Channel Protein Gene

Specifier	Reference Address	Weights	Gene Activate	Reference Pointer	Neuron Address
D	A	w_1 w_2 ... w_n	G	P	N
Integer [0,6]	Integer	Real [0,1]	Binary	Integer	Integer

Messenger Protein Gene

Specifier	Reference Address	Boundary Params.	Gene Activate	Channel Protein Reference Pointer
D	A	τ_1 τ_2	G	P
Integer [0,6]	Integer	Real [0,1]	Binary	Integer

Action Protein

Specifier	Reference Address	Output State	Gene Activate	Channel Protein Reference Pointer	Neuron Address
D	A	a b	G	P	N
Integer [0,6]	Integer	Allowable States	Binary	Integer	Integer

Motor Control Neuron Gene

Specifier	Reference Address	Position	Memory	Multi-state	Gene Activate	Cell Death	Replication Prob.	Output Behaviour	Reference Pointer
D	A	x y z	ϕ b	q_b	G	C	R	k	P
Integer [0,6]	Integer	Integers	Binary	Integer	Binary	Binary	Real [0,1]	Integer [0,b]	Integer

Fig. 3. Genome of messenger-protein network components and a typical motor control neuron.

s_2 with the background signal. The output s_1 is given again by (2). We offer two models for the computation of s_2, a feedforward model and a memory model. In the former, s_2 is given by (2) with a_j replaced by $b_j \in \{0, 1/q_b, ..., 1\}$ which is graduated in q_b (an integer greater than one) uniform steps between 0 and 1. In the memory model, $b_j \in \{0, \max\{0, s_2(t-1) - 1/q_b\}, \min\{s_2(t-1) + 1/q_b, 1\}, 1\}$ allowing for storage, reset, gating and increment/decrement functionality. This multistate model of the neuron is an attempt to better incorporate biological observations of neuron action potential through bottom-up modeling of protein interactions.

Evolution and Development Details of the development process for the tissue remains identical to previous versions of ANT and can be found in [17,18]. Unlike previous versions of ANT that used neurons with a modular activation function using two thresholds [16], the coarse-coding cell model allows for a developmental activation function. Cell and protein genes have a binary 'activation' parameter, used either to express or repress gene contents. (The genome structure is shown in Fig. 3.) Each channel protein references a cell address. Messenger and action proteins in turn reference a channel protein. Since these genes are modular, it is possible for a messenger-channel protein network to be incomplete and thus lacking channel proteins.

Mutations in the genome can perturb existing genetic parameters or addition of new (cell, messenger, channel or action) genes caused by random gene transcription errors with a probability of p_{te}. Thus a new cell-protein gene as a result of a transcription error is a copy of an existing cell-protein gene with perturbations starting at point chosen from a uniform distribution along the gene's length and with the gene activation parameter toggled off by default.

4 Sign-Following Task

The effectiveness of the coarse-coding cell model is demonstrated in simulation on two memory-dependent versions of the unlabeled sign-following task. The workspace is modeled as a two-dimensional grid environment with one holonomic robot (based on a Khepera, equipped with a gripper and camera) occupying four grid squares. For these tasks, the controller must possess a number of capabilities including that to decipher signs relative to the robot's current frame of reference, to remember the current sign while looking for the next one, and to negotiate obstacles (see Fig. 4a). Each sign is color-coded and represents a waypoint (posted in a fixed frame of reference) that gives direction in one of four cardinal points to the next waypoint leading ultimately to the goal location.

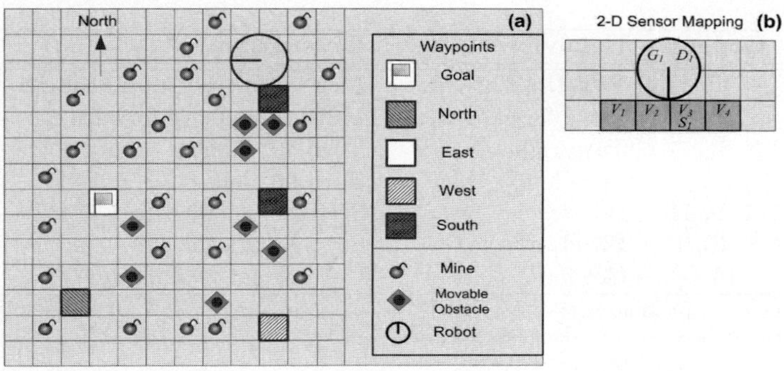

Fig. 4. (a) 2D grid world model for the sign-following tasks. (b) Input sensor mapping.

Mines (undetectable by the robot) are randomly laid throughout the floor except along the pathway. Once a robot encounters a mine, it remains disabled for the remainder of its lifetime. The sensory input map is shown in Table 1 (see also Fig. 4b). The task has to be accomplished using a discrete set of basis behaviors specified in Table 2. These behaviors are activated based on controller output and all occur within a single time-step. The robot is initially positioned next to the first sign, but the initial heading is randomly set to one of the four cardinal directions. Since the robot can only detect signs placed in front, it needs to go into a 'sign searching' mode and perform a sequence of 'turn left' or 'turn right' behaviors to detect the first sign. Once the first sign is detected, the robot then needs to transition to a 'sign following' mode, requiring one bit of memory.

Table 1. Sensor Inputs

Sensor Variables	Function	Description
$V_1 \ldots V_4$	Object detection	Robot, block, no obstacle
G_1	Gripper status	Holding block, no block
S_1	Sign detection	Red, blue, orange, pink, green
D_1	Heading	North, east, west, south

Deciphering signs relative to the robot's current frame of reference makes these tasks particularly difficult given a fitness function that measures success in terms of reaching the goal location. The two versions of the task considered here are (1) where the controller has access to a compass sensor at each time-step and (2) where compass sensor readings are penalized or restricted. We shall refer to the former variant as *compass-enabled* and the latter as *compass-restricted*. Even the simpler compass-enabled version is found to be unsolvable for predetermined fixed-network topologies that lack regulation (see Results and Discussion).

Table 2. Basis Behaviors

Order	Behavior	Description
1	Pick-Up/Put-Down	Pick up or put down obstacle
2	Move forward	Move one square forward
3	Turn right	Turn 90° right
4	Turn left	Turn 90° left
5, 7, 9, 11	Bit set[1]	Set memory bit i to 1, $i = 1 \cdots 4$
6, 8, 10, 12	Bit clear[1]	Set memory bit i to 0, $i = 1 \cdots 4$
13	Get hint[2]	Get current heading (D_1)

[1]Behaviors disabled for recurrent and memory neuron architectures
[2]Behaviors disabled under certain conditions (see text)

In the compass-restricted version, the controller must perform sign following knowing just its initial heading thus requiring the controller to predict and keep track of the robot heading (ego-orientation) in addition to accomplishing the other subtasks described earlier. Keeping track of long term dependencies is acknowledged to be difficult with recurrent connections [7] making the sign-following task a good benchmark for multistate architectures. The robot in this case has access to one additional behavior, the 'get hint' behavior, which interrogates the compass for the 'true' heading. However, the fitness function incrementally penalizes and restricts the number of hints used. The fitness function for a given run is defined as

$$f_i = \begin{cases} \dfrac{1}{1+\beta n_{\text{hint}}/16}, & \text{if goal is reached} \\ 0, & \text{otherwise} \end{cases} \tag{3}$$

The total fitness function is averaged over all runs. For the compass-enabled variant, $\beta = 0$ although the robot always knows the compass direction. So when the goal is achieved $f_i = 1$; otherwise, $f_i = 0$. For the compass-restricted one the reward for success is discounted according to the number n_{hint} of hints that have been used. However, for the first 5,000 generations, the robot is not penalized for using hints; hence $\beta = 0$. For the subsequent 10,000 generations, $\beta = 1$ but the hint can only be used in the first four time-steps. (This allows the robot to get the true direction reading as it starts out.) After 15,000 generations, hints are proscribed altogether. These parameters were found through experimentation to work well for this task.

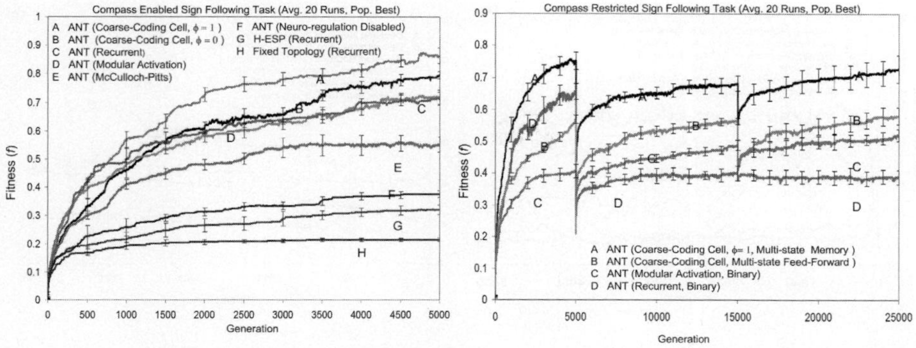

Fig. 5. Comparison of performance for (a) compass-enabled task and (b) compass-restricted task.

The evolutionary performance of various control system architectures is compared for the two variants of the sign-following task (see Fig. 5). The robot's world is a 20×20 grid with 80 uniformly distributed obstacles and 40 randomly distributed mines (except along the path to the goal). The fitness is averaged over 100 runs with different initial conditions, the elapsed time for each run being limited to 100 timesteps. Controllers that lack recurrent connections or memory neurons have access to four memory bits, which can be manipulated using the defined basis behaviors. Evolution assumes a population of 100 individuals in each generation and a tournament size of 24. The crossover probability $p_c = 0.7$, the mutation probability $p_m = 0.005$ and the transcription error rate $p_{te} = 0.005$.

5 Results and Discussion

Fig. 5a shows the average (population best) fitness of the controllers evaluated in each generation for the compass-enabled variant of the sign-following task. For some comparison, a fixed-topology recurrent network with 9 hidden and 4 output neurons is also shown. (Although this is typical of the results obtained for such a network, we did not optimize performance in any way.) Fixed-topology networks tend to have more 'active' synaptic connections present (all neurons are active) and thus more spurious neurons need to be dealt with simultaneously. The ANT topology with the coarse-coding neuroregulatory mechanism disabled (using modular activation function) shows better performance than fixed topologies (including H-ESP [7]) but not sufficient to complete the task for all the initial conditions tested.

In contrast, heterogeneous architectures, whether predefined (modular activation function) or developmental, outperform other architectures. In the coarse-coding cell model, selection is favoring increased complexity among decision neurons (Fig. 6a) but this is followed by a gradual simplification in neuron structure.

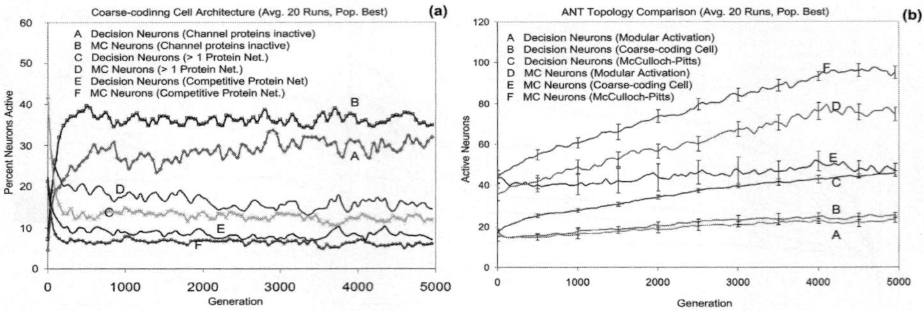

Fig. 6. Comparison of (a) coarse-coding cell models and (b) ANT topologies.

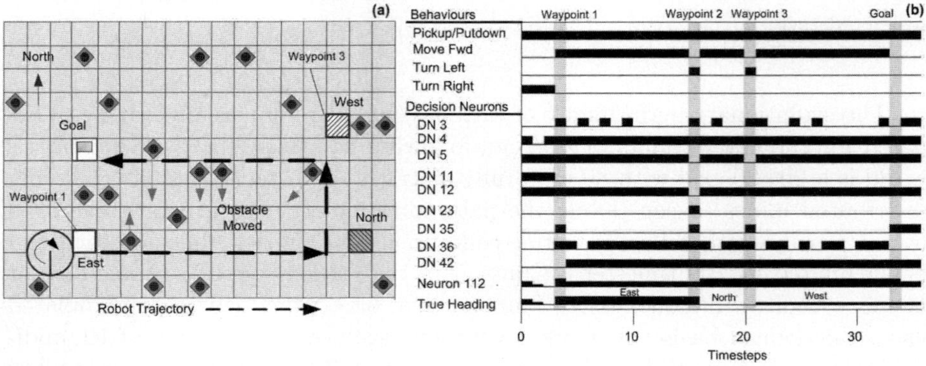

Fig. 7. (a) Typical ANT solution (fitness $f = 0.99$) using multistate neurons (memory model) for the compass-restricted sign-following task. (b) Output behavior and function of memory neurons.

This simplification is possible with all the channel proteins disabled, resulting in constant output (independent of sensory input).

While the modular activation function also allows for heterogeneity, it is a fixed-cell architecture that does not facilitate intracellular competition nor 'complexification.' Overall, the 'expressed' structure of the coarse-coding cell model is using fewer tuning parameters (both weights and thresholds) than with the modular activation function as it converges to a solution (Fig. 6b).

Fig. 6a shows that many messenger-channels protein networks have redundant action proteins implying a 'weak' cooperative setup. A weak cooperative setup is advantageous, allowing for cooperative and competitive tendencies and can better facilitate a transition between the two. The added benefit of the heterogeneous coarse-coding cell model is that it also facilitates memory functionality through gating. The evolutionary performance with gated-memory functionality ($\phi = 1$) shows a definite improvement over recurrent architectures for both tasks (Fig. 5a).

For the compass-restricted version of the task (Fig. 5b), only the multistate coarse-coding cell models can reach a fitness of 0.9 within 25,000 generations at least once. The binary-state neuron model with feedforward connections (and access to 4 memory bits) showed comparable results with a fitness of 0.8 at least once while the binary recurrent architecture performs poorly. Among the four architectures, the binary recurrent model (localized memory) lacks built-in competitive network mechanisms for memory representation. For the multistate model, these competitive mechanisms are the protein networks resident in each cell, while similar functionality is present within the binary feedforward architecture, where neuron ensembles compete for control of the 4 memory bits (see [18]).

These results also show *intra*cellular competition has an advantage over *inter*cellular competition for memory representation. The multistate neuron model with memory functionality outperformed the multistate feedforward model (Fig 5b). It should be noted that the multistate memory model is more favorable allowing for it to keep track of changes and storing expected robot heading (with the inclusion of increment/decrement and reset functions) over a feed-forward model (see Fig. 7).

6 Conclusions

A developmental coarse-coding neuron model that facilitates heterogeneity and cell extensibility is found to exploit gated memory functionality. This architecture is found to be more effective than recurrent architectures for two memory-dependent variants of the unlabeled sign-following robotic task. The neuron model produces smaller networks on average than a predefined heterogeneous cell model and is driven by competitive/cooperative dynamic leading to specialization.

Acknowledgements. The authors would like to thank the reviewers for their helpful comments.

References

1. Albus, J.S.: A theory of cerebellar function. Mathematical Biosciences (1971)
2. Astor, J.C., Adami, C.: A developmental model for the evolution of artificial neural networks. ALife 6(3), 189–218 (2000)
3. Edelman, G.: Neural Darwinism. The Theory of Neuronal Group Selection. Basic Books, New York (1987)
4. Eggenberger, P., Gómez, G., Pfeifer, R.: Evolving the morphology of a neural net for controlling a foveating retina. ALife 8, 243–251 (2002)
5. Eggenberger, P.: Evolving Morphologies of simulated 3d organism based on Differential Gene Expression. In: Proc. 4th European Conference on ALife (1997)
6. Graves, A., Beringer, N., Schmidhuber, J.: A Comparison Between Spiking and Differentiable Recurrent Neural Networks on Spoken Digit Recognition. In: Proc. of the 2nd IASTED Int. Conf. on Neural Nets, NCI (2004)

7. Gomez, F., Schmidhuber, J.: CoEvolving Recurrent Neurons Learn Deep Memory POMDPs. In: GECCO, pp. 491–498 (2005)
8. Gruau, F.: Automatic definition of modular neural networks. Adaptive Behaviours 3, 151–183 (1994)
9. Hochreiter, S., Schmidhuber, J.: Long Short-Term Memory. Neural Computation 9(8), 1735–1780 (1997)
10. Husbands, P.: Evolving Robot Behaviours With Diffusing Gas Networks. In: EvoRobots, pp. 71–86 (1998)
11. Sims, K.: Evolving 3D Morphology and Behavior by Competition. In: Proc. of Artificial Life IV, pp. 28–39. MIT Press, Cambridge (1994)
12. Kirschner, M., Gerhart, J.: Evolvability. Proc. Natl. Acad. Sci (PNAS) (1998)
13. Nolfi, S., Floreano, D.: Evolutionary Robotics: The Biology, Intelligence, and Technology of Self-Organizing Machines, pp. 13–15. MIT Press, Cambridge (2000)
14. Roggen, D., Floreano, D., Mattiussi, C.: A Morphogenetic Evolutionary System: Phylogenesis of the POEtic Tissue. In: Int. Conf. on Evolvable Systems, pp. 153–164 (2003)
15. Roggen, D., Federici, D.: Multi-cellular Development: Is There Scalability and Robustnes to Gain? In: Yao, X., Burke, E.K., Lozano, J.A., Smith, J., Merelo-Guervós, J.J., Bullinaria, J.A., Rowe, J.E., Tiňo, P., Kabán, A., Schwefel, H.-P. (eds.) Parallel Problem Solving from Nature - PPSN VIII. LNCS, vol. 3242, pp. 391–400. Springer, Heidelberg (2004)
16. Thangavelautham, J., D'Eleuterio, G.M.T.: A Neuroevolutionary Approach to Emergent Task Decomposition. In: Yao, X., Burke, E.K., Lozano, J.A., Smith, J., Merelo-Guervós, J.J., Bullinaria, J.A., Rowe, J.E., Tiňo, P., Kabán, A., Schwefel, H.-P. (eds.) Parallel Problem Solving from Nature - PPSN VIII. LNCS, vol. 3242, pp. 991–1000. Springer, Heidelberg (2004)
17. Thangavelautham, J., D'Eleuterio, G.M.T.: A Coarse-Coding Framework for a Gene-Regulatory-Based Artificial Neural Tissue. In: Advances In Artificial Life: Proc. of the 8th European Conf. on ALife (2005)
18. Thangavelautham, J., Alexander, S., Boucher, D., Richard, J., D'Eleuterio, G.M.T.: Evolving a Scalable Multirobot Controller Using an Artificial Neural Tissue Paradigm. In: IEEE ICRA, IEEE Computer Society Press, Los Alamitos (2007)

Re-examination of Swimming Motion of Virtually Evolved Creature Based on Fluid Dynamics

Yoshiyuki Usami

Institute of Physics, Kanagawa University,
Yokohama 221-8686, Japan
usami@kanagawa-u.ac.jp
http://www.museum.fm

Abstract. The swimming motion of the virtually evolved creature model proposed by Karl Sims is re-investigated on the basis of hydrodynamics. In his work, physical simulation was performed, and the swimming motion of the evolved creature was presented. Animation of the creatures was stimulus, however, detailed description about the simulation condition and its results were not always well described. In this work, we perform hydrodynamic simulation to investigate the swimming motion of virtual creatures. As a result, it is found that collaborating motion of fluid is essential and indispensable for the motion of the creature in water. This mechanism also works as a constraint in constructing creature's body. We found that the physical property of the water strictly regulate the structure of swimming creatures.

Key words: Evolution, Creature, Fluid, Hydrodynamics, Swimming, *Anomalocaris*

1 Introduction

Karl Sims accomplished the systematic investigation of evolving creatures and those behaviors for the first time in 1994[1,2]. In Sims' work, the creature was expressed as a combination of boxes. The structure of the creature evolved from a simple one to various complex shapes. They swam in water, walked on the ground, followed a target, and competed with each other for getting food. Interesting behaviors of creatures emerged as a result of numerical simulation; they were jumping, rotating, and fighting for food. Aside from the interesting visual presentation, simulation condition and scientific statement of results are not always well discussed. Scientific themes included in the simulation can be listed as swimming motion, walking motion, control, and evolution scheme. In each theme there are many simulation models and simulation parameters. Detailed descriptions about the employed model and range of parameters is useful in researchers. Several successive studies have been published mostly in alife field[3,4,5,6], however, the number of published articles is said to be few compared to the vast number of biological studies.

F. Almeida e Costa et al. (Eds.): ECAL 2007, LNAI 4648, pp. 183–192, 2007.

Among them, it has been drawn attention that open program source code called Framstics has been distributed by Komosinski and Ulatowski since 1998[7]. In the software creatures evolve and swim or walk in a simulation space. This is useful tool to learn about virtual evolution model studied in alife, however, the validity of the employed physical model has not always been verified.

In this paper, we present the results of hydrodynamic simulation for the study of swimming creatures. We compare the results of reaction force model with the results of hydrodynamic calculation. If the approximation is valid, using it in evolution study is justified. If it is not valid, serious reconsideration is required in these studies.

2 Swimming Motion of Virtual Creature Based on Fixed Frame Model

Instead of calculating fluid motion, a reaction force model from a fixed frame is sometimes used as an approximation for fluid dynamics. This model is used in alife studies for the case such that many calculations are required for various creatures whose boundary shapes are complicated. The author had once used this model and published the result at alife 6 conference[4]. In the work, the swimming pattern of an extinct euarthropod *Anomarocaris*[8,9] was investigated by using a fixed frame model.

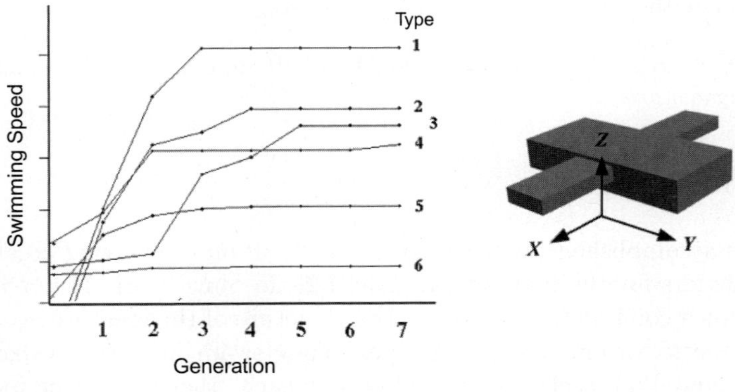

Fig. 1. Behavior of virtual creature (right) is controlled by many parameters. Evolutionary computation method was applied for obtaining optimal swimming motion of model *Anomalocaris* (left) [4].

The propulsion force of *Anomarocaris* was obtained by a sum of local reaction forces on this model. Because there were many parameters for the motion, an evolutionary computation method was used for searching the optimal motion. Figure 1 displays the result of calculation. Detailed description of this simulation

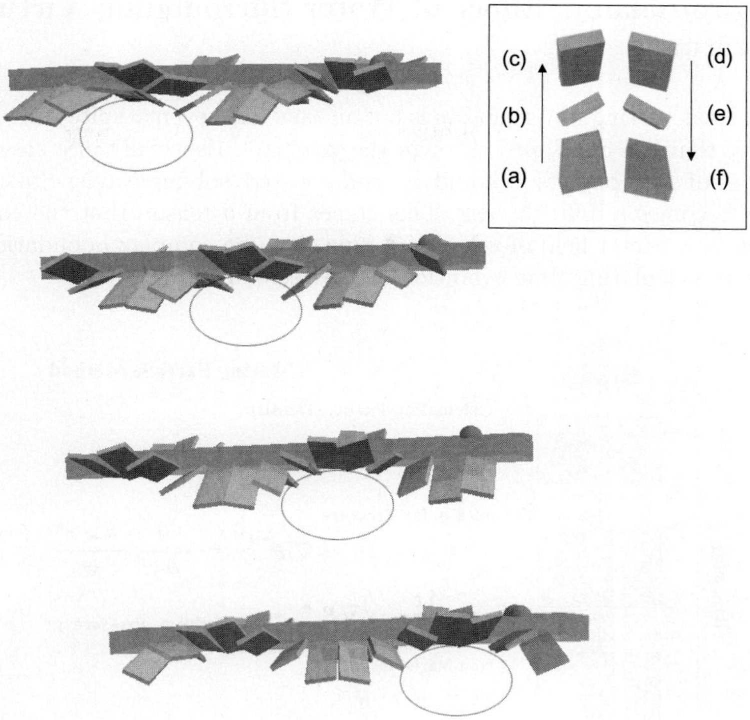

Fig. 2. Unusual motion appeared in the simulation of reaction force model. This shows Type 3 motion appeared in Fig.1

is shown in [4]. As a generation of evolutionary computation passed, several types of swimming patterns appeared.

The fastest motion, denoted in Type 1 in the figure, was a smooth waving pattern, however, additional swimming patterns also appeared. Among them, swimming patterns denoted Type 3 in Fig.1 shows an unusual waving motion. Figure 2 displays Type 3 motion appearing in the evolutionary computation.

An overall waving pattern of lobes is very unusual. When we look at the motion carefully, we notice that the trajectory of overall lobes propagates to the right side, shown as the gray circle. This propagating pattern usually brings creatures to left side. However, a single lobe waving pattern, shown as the insert at upper right, pushes the body to right side.

The reaction force model used in this simulation takes the simple sum of the receiving force of the individual lobe, then model *Anomalocaris* moves as a whole to right side, as shown as Fig. 2. This waving pattern is unusual, and we have never observed such motion in natural phenomena. To clarify the meaning of this strange result, we include an effect of fluid motion surrounding virtual creatures.

3 Hydrodynamic Effect of Water Surrounding Virtual Creature

In general, calculating water motion is not an easy task, even despite the computing ability that has developed through the years[10]. Especially in a case where boundaries of solid objects in liquid are complicated, solving Navier-Stokes fluid equation becomes a difficult task. This comes from a reason that the equation is written by a vector field of velocity of fluid. Moving complex boundaries adds difficulty to calculating time evolution of the vector field.

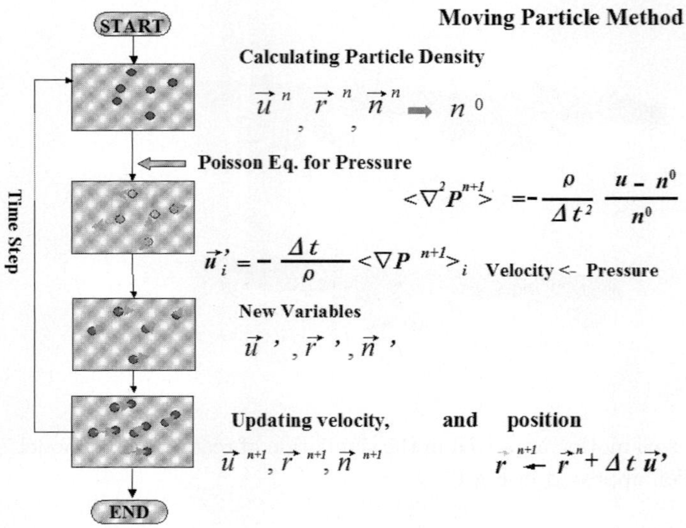

Fig. 3. The algorithm of moving particle method for obtaining fluid dynamics. The detailed description is in [11,12,13]

Among many proposed methods, we focus on the moving particle method for the present study. In this method, fluid is not represented by a vector, but by moving particles. Solid and liquid are equally treated as moving particles in this method, which makes easier to obtain solutions for moving complex boundaries. Koshizuka and subsequent researchers applied this method to many different problems concerning fluid dynamics and reported the validity of this method[11,12]. Furthermore, agreement with this method with experimental results have been also reported. Thus, we use this method for a study of swimming motion of artificial creatures.

Calculation procedure of moving particle method is graphically summarized in Fig.3. In the n-th calculation step, particle position r^n, particle velocity v^n, and particle density n_0^n are calculated. Next, Poisson equation of the pressure $(< \nabla^2 P >^{n+1} \cdots) =$ is calculated from those values. From the pressure, next

$n+1$-th time step variables are calculated. The exact description of the method appears on the references[11,12,13]. The parameters used in this simulation are summarized in the Appendix.

Our simulation is performed in a two-dimensional space. In a three-dimensional space, a large volume is required for a simulation of *Anomalocaris* whose boundary is highly complex. This results in an extremely long computation time and huge computer memory. Furthermore, we have to simulate many different boundaries corresponding to various creatures in alife study. As we will discuss in this paper, a part of the swimming motion can be understood in two dimensions, which does not become an essential problem in the experiment.

Figure 4(a) displays the simulation setup of our calculation. The black wall shows the boundary, and moving particles representing fluid are inside the boundary. Lobes of *Anomalocaris* are shown as thick black lines in the water. Fluid motion is calculated according to the moving particle method with given motion of lobe. We give vertical motion of i-th lobe expressed as a periodic function, $H_i(t) = H_0 \cdot \sin(w \cdot t - i \cdot \phi_0)$, where H_0 , ϕ_0, and w represent amplitude of vertical motion, a phase difference between each lobe, and frequency, respectively.

The lobe also changes its angle $\theta_i(t)$ as, $\theta_i(t) = \theta_0 \cdot \sin(w \cdot t - i \cdot \phi_0)$, where θ_0 represents the amplitude of the angle. As a result of this motion, the lobe receives force from the water, which makes the body move in a horizontal direction. All parameters are chosen as the same value as those in the previous chapter.

Fig.4(b-e) shows the results obtained by a simulation of the moving particle method for fluid dynamics. The black circle and small point represent the lobe of the model *Anomalocaris* and fluid particle, respectively. The velocity of fluid is indicated as a thick line with a point. At a first glance, we notice that *Anomalocaris* moves in a left direction. This result is the opposite result of the fixed frame reaction model shown in Fig.2. By overall bending of lobes, a certain volume of fluid is formed as indicated by the gray circle in Fig.4. *Anomalocaris* moves to the left side as a result of the reaction force by pushing the volume of fluid to the right side. However, the single lobe waving pattern shown in the upper right insert in Fig.2 makes the driving force toward the right side. Then, the conflict of the two different forces makes a disturbance of the surrounding water. Disorder of the velocity of water particles is observed around all lobes. The driving force caused by massive volume is shown as the gray circle overcoming the sum of the single lobe. As a result, *Anomalocaris* is going to the left side.

The reaction force model also produced another unusual motion. As a result of using the summation of local force, the synchronous motion of all lobes produces a large propulsion force. The motion is just like a boat race with eight rowers. In the boat race, eight rowers never row their oars in a wavy pattern but row the same direction. Because a synchronous rowing motion brings maximum propulsion force.

However, the situation becomes different when all lobes of *Anomalocaris* are in the water media. We performed a hydrodynamic simulation for synchronous

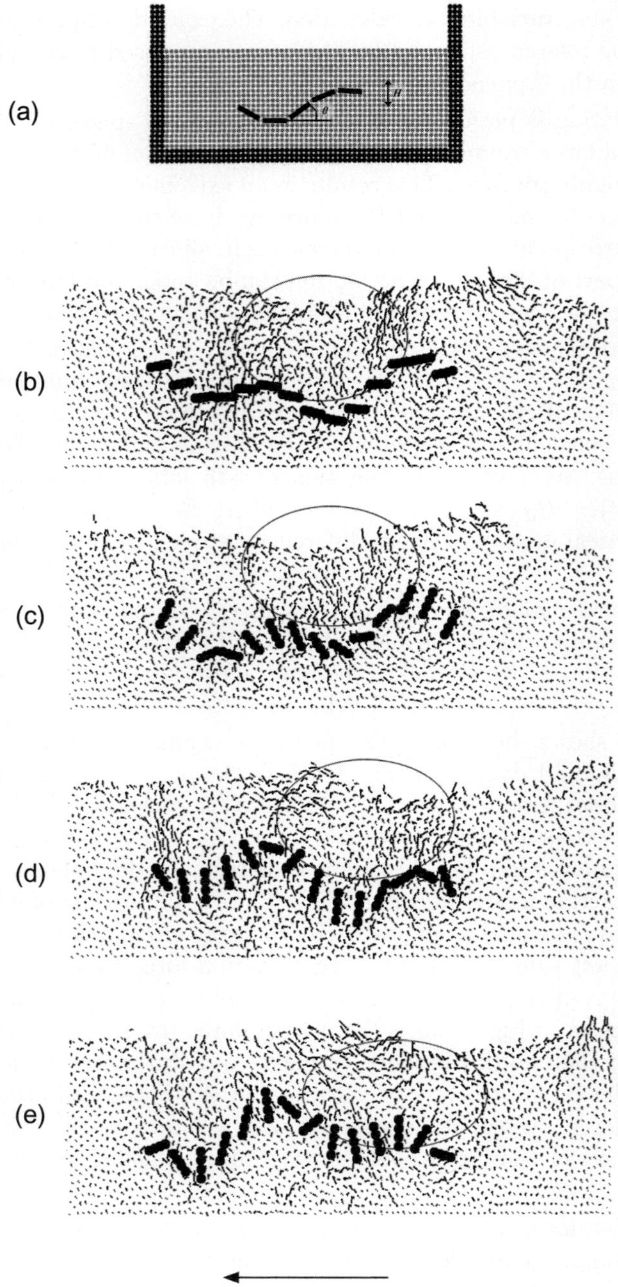

Fig. 4. Hydrodynamic simulation of unusual swimming motion of model *Anomalocaris*.

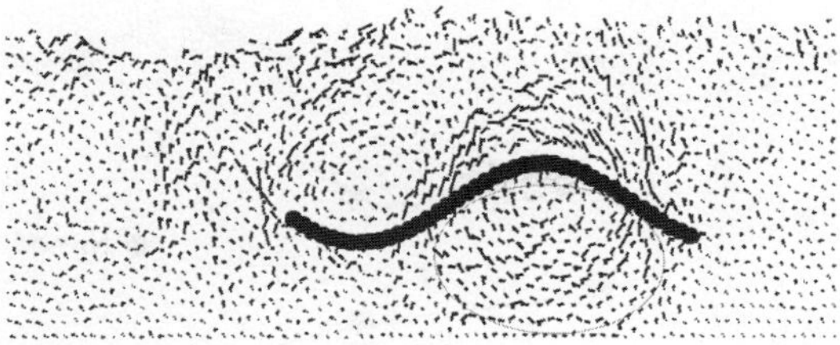

Fig. 5. Hydrodynamic simulation for the optimal waving pattern of *Anomalocaris* shown as Type 1 in Fig.1.

waving pattern of lobe. The results show that it produces large disturbance of surrounding water. Furthermore, the swimming speed is comparatively slow. These results show breakdown of the fixed reaction model instead of simulating fluid dynamics.

Figure 5 shows the best waving pattern of lobe denoted as Type 1 in Fig.1. The swimming speed of this motion is high, and energy consumption is low compared to the other motions. This motion realizes forming a certain volume of water. By a collaborative motion of this volume of water going to left side in the figure, model *Anomalocaris* swims smoothly to the right side. We observe no rise of water upon lobes, which leads to bring unnecessary energy consumption as the synchronous waving pattern. From this result, we can predict that *Anomalocaris* swam in the Cambrian Sea by waving its lobe as shown in Fig.5.

4 Swimming Behavior of Other Types of Creatures

In considering the relationship between morphology and behavior, we study swimming locomotion of other types of creatures. Morphological changes in body plan have been observed in the transition from aquatic animal to terrestrial one. These changes are typically observed in the marine reptiles of the Mesozoic era: nothosaurs, plesiosaurs, ichthyosaurs, and mosasaurs. These show strong variation of body plans adapted for various water environments.

To show a relationship between morphology and behavior, we change the shape of creatures along this evolutionary scheme. The body plan of terrestrial tetrapod is modeled as (A) in Fig.6 in our simulations. We calculate swimming performance of the creatures in the course of the change in direction of limbs. Swimming speed and energy are plotted with the decrease of the angle between limbs and body. The morphological changes are shown as (B) and (C) in Fig.6. The increase of swimming speed and the decrease of consuming energy are observed in the figure. Then, we can predict that terrestrial tetrapods were able to swim in water by shrinking the angle between limbs and body as (C) in Fig.6.

In the next stage, limbs degenerated in the course of evolution. In Mesozoic marine reptiles, we observe such a body plan in mosasaurs and ichthyosaurs. On

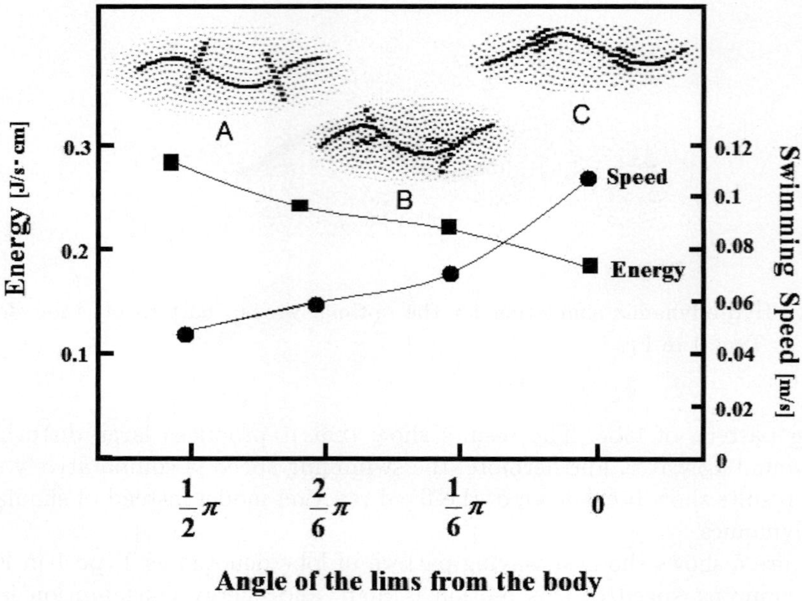

Fig. 6. The change of swimming velocity and consuming energy on the course of morphological change of body plan.

the other hand, plesiosaurs took a different evolutionary path. In this group, the morphology of limbs evolved to a paddle shape for an aquatic life, and these were used for locomotion instead of having to bend the body. Then, the body structure became harder, and the two pairs of paddles became larger. Their locomotion was achieved by stroking with their paddle like limbs. Studying such motion is left as a future problem to us.

Simulations shown in Fig.6 tell us that aquatic vertebrates that swam in water by bending the body tended to converge to a more streamlined body plan in the evolutionary process. This tendency is widely observed in Mesozoic marine reptiles and Cenozoic marine mammals. The Cenozoic amphibian salamander is another interesting example, showing behaviors of both swimming and walking. The behaviors of this group have been studied experimentally and theoretically in detail. The traveling wave of body bending motion studied by Ijspeert[14] corresponds to the bending motion in our simulation depicted in Fig.6.

5 Conclusion

We introduced hydrodynamic simulation in the study of artificial creature. A case study of *Anomalocaris* was shown, and discussed with the results of fixed frame reaction model.

From the result, disagreement between the two was observed. Reaction force model neglects collaborating motion of fluid, which brings serious error in predicting swimming motion of creature in water.

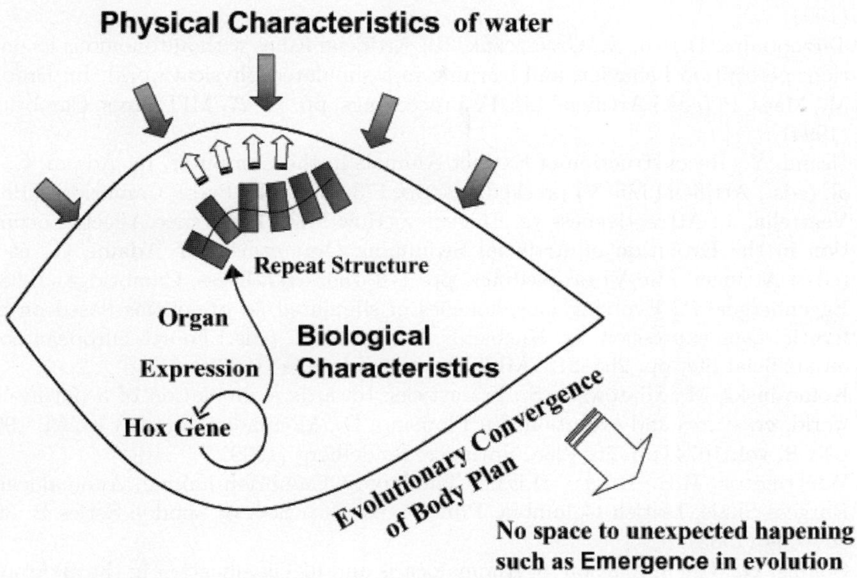

Fig. 7. Physical characteristics strictly regulates basic body plan of swimming creature in fluid.

Together with the following computational results, hydrodynamic simulation tells us another aspect, namely, concerning to body plan of creature, not only behavior. According to strong long distance correlation of water, body form of swimming creature converges to streamline, if the creature pursuits fast swimming. Then, a mechanism of forming body structure is controlled by environment, rather than biological origin such as gene structure and its expressions. So it is not necessary to consider evolution algorithm to generate body structure for swimming creature. Rather, physics of water determines body structure of swimming creature. This situation is summarized in Fig.7.

However, we should add a comment that this mechanism works for animal swimming middle column of water. The other type of discussion is needed for benthic and nect-benthic animal living just above sea floor. Those animals do not need to pursuit fast swimming. Then, body form of those animals is determined by the other mechanism, and body form of those does not converge to a shape of streamline. It reflects the situation that we observe a variety of body plan of swimming animal just above sea floor, and in the deep water.

References

1. Sims, K.: Evolving Virtual Creatures. Computer Graphics, Annual Conference Series (SIGGRAPH '94 Proceedings), 15-22 (1994)
2. Sims, K.: Evolving 3D Morphology and Behavior by Competition. In: Brooks, M., Maes, P. (eds.) Artificial Life IV proceedings, pp. 28–39. MIT Press, Cambridge (1994)
3. Terzopoulos, D., Tu, X., Grzeszczuk, R.: Artificial fishes with autonomous locomotion, perception behavior, and learning in a simulated physical world. In: Brooks, M., Maes, P. (eds.) Artificial Life IV proceedings, pp. 17–27. MIT Press, Cambridge (1994)
4. Usami, Y.: Reconstruction of Extinct Animals in the Computer. In: Adami, C., et al. (eds.) Artificial Life VI proceedings, pp. 173–177. MIT Press, Cambridge (1998)
5. Ventrella, J.: Attractiveness vs. Efficiency (How Mate Preference Affects Locomotion in the Evolution of Artificial Swimming Organisms). In: Adami, C., et al. (eds.) Artificial Life VI proceedings, pp. 178–186. MIT Press, Cambridge (1998)
6. Eggenberger, P.: Evolving morphologies of simulated 3d organisms based on differntial gene expression. In: Husbands, P., Harvey, I. (eds.) Fourth european conf. on artificial life, pp. 205–213. MIT Press, Cambridge (1997)
7. Komosinski, M., Ulatowski, S.: Framsticks: towards a simulation of a nature-like world, creatures and evolution. In: Floreano, D., Mondada, F. (eds.) ECAL 1999. LNCS, vol. 1674, pp. 261–265. Springer, Heidelberg (1999)
8. Whittington, H.B., Briggs, D.E.G.: The largest Cambrian animal, Anomalocaris, Burgess Shale, British Columbia. Phil. Trans. Roy. Soc. of London Series B. 309, 569–609 (1985)
9. Collins, D.: The 'Evolution' of Anomalocaris and its classification in the arthropod Class Dinocardia (NOV.) and Order Radiodonta (NOV.). J. of Paleontology 70, 280–293 (1996)
10. Wolfgang, M.J., Anderson, J.M., Grosenbaugh, M.A., Yue, D.K.P., Triantafyllou, M.S.: Near-body flow dynamics in swimming fish. Journal of Experimental Biology 202, 2303–2327 (1999)
11. Koshizuka, S., Tamako, H., Oka, Y.: A particle method for incompressible viscous flow with fluid fragmentation. Computational Fluid Dynamics Journal 4, 29–46 (1995)
12. Ikeda, H., Koshizuka, S., Oka, Y., Park, H.S., Sugimoto, J.: Numerical analysis of jet injection behavior for fuel-coolant interaction using particle method. J. Nucl. Sci. Tech. 38, 174–182 (2001)
13. Usami, Y.: Theoretical study on the body form and swimming pattern of Anomalocaris based on hydrodynamic simulation. J. Theor. Bio. 238, 11–17 (2006)
14. Ijspeert, A.J.: A connectionist central patterngenerator for the aquatic and terrestrial gaits of a simulated salamander. Biol. Cybern. 84, 331–348 (2001)

Appendix: Parameters Used in the Simulation

The size of model *Anomalocaris* is about 60cm. The oscilation number w =2[Hz] leads to swimming velocity 0.53 m/s. These yeilds Reynolds number as 0.25×10^5, which characterizes physical property of fluid.

Adaptation to Sensory Delays

An Evolutionary Robotics Model of an Empirical Study

Marieke Rohde and Ezequiel Di Paolo

Centre for Computational Neuroscience and Robotics (CCNR)
University of Sussex, Brighton, BN1 9QH, UK
{m.rohde,ezequiel}@sussex.ac.uk

Abstract. Evolutionary robotics simulations can serve as a tool to clarify counterintuitive or dynamically complex aspects of sensorimotor behaviour. We present a series of simulations that has been conducted in order to aid the interpretation of ambiguous empirical data on human adaptation to delayed tactile feedback. Agents have been evolved to catch objects falling at different velocities to investigate the behavioural impact that lengthening or shortening of sensory delays has on the strategies evolved. A detailed analysis of the evolved model agents leads to a number of hypotheses for the quantification of the existing data, as well as to ideas for possible further empirical experiments. This study confirms the utility of evolutionary robotics simulation in this kind of interdisciplinary endeavour.

Key words: Evolutionary Robotics, Sensory Delays, Sensorimotor Adaptation.

1 Introduction

This paper presents results from an evolutionary robotics simulation implemented to conceptually guide the analysis of empirical data on adaptation to tactile delays. The use of evolutionary robotics simulations in close match with minimalist empirical methods is rather unexplored territory. This study is also an exploration into the utility of this kind of approach[1]. The model, which is in itself rather simple, turns out to be an extremely rich source of hypotheses, both for analysis of the existing results and for future experiments.

The data under investigation stems from a recent empirical study (Rohde, Gapenne, Lenay, Aubert and Stewart (unpublished work)). In this study, we set out to investigate the dynamics of human adaptation to delayed sensory feedback in a minimal virtual environment. This experiment was inspired by the work of Cunningham et al. [2], who report an interesting negative aftereffect in subjects trained to solve a simple sensorimotor task with a 200ms visual delay:

[1] For conceptual clarity, in the following, the terms 'empirical' and 'experiment' are reserved for the real world experiments with humans, while the terms 'simulated' and 'model' are used to refer to the evolutionary robotics simulation of the task.

F. Almeida e Costa et al. (Eds.): ECAL 2007, LNAI 4648, pp. 193–202, 2007.

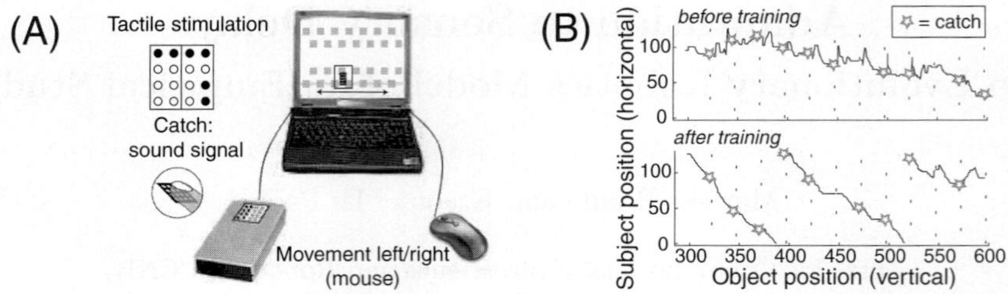

Fig. 1. (A) A diagram of the experimental set up for the empirical study. (B) Human trajectories before (top) and after (bottom) training, presented with a sequence of 16 lines of evenly spaced objects (black bars), both without delay: Even though the behaviour has clearly been transformed, the performance on the task is similar.

Subjects suffered a performance breakdown when returning to the normal no–delay condition, as a consequence of the adaptation to the sensory delay, an effect that had failed to occur in previous experiments with delays, which the authors suspect to be due to an *inherent time pressure* in the task.

Our experiment aimed to investigate the exact behavioural and dynamical conditions that produce such effects. We implemented a catch task in a minimal virtual environment, using the tactile feedback platform 'TACTOS' [4] by the GSP group in Compiègne (compare Fig. 1 (A)). Blindfolded subjects had to position themselves as fast as possible below falling objects in order to catch them. To include an inherent time pressure in our task, the objects fall down comparably fast. Subjects moved along one dimension, relying only on tactile stimuli to indicate the crossing of an object and a sound signal to indicate its velocity. We tested the subjects' capacity to adapt to a $250ms$ sensory delay, monitoring all sensory (tactile display, sound signals) and motor (mouse movement) variables.

In terms of performance profiles, our experiments only produced a clear negative aftereffect in few of the 20 experimental subjects. Indeed, the results seem rather arbitrary at first glance: Some subjects had difficulties with the unperturbed condition already, or were not perturbed by the introduction of the delay, or simply did not suffer a negative aftereffect. A closer look at the trajectories, however, revealed regularities across subjects, as to how different strategies are impacted by the introduction or removal of a delay, and how strategies were modified to recover performance. These changes were not always well represented in performance profiles. Figure 1 (B) shows, e.g., how a subject's initial strategy (fast oscillatory scanning) was transformed into a different but equally viable strategy (slowly drifting to the right and halting once an object is perceived).

Our next step will be to classify the strategies and to quantify the transitions they underwent across training. It is, however, not easy to formally characterise classes of strategies and the adaptation effects described. Many variables (such as velocity, acceleration, number of crossings, distance, velocity of the objects, movement direction, etc.) and any cross-correlation or dependency between them in any subgroup of subjects could be used. Apart from the fact that the data is relatively noisy, a major problem we face in our classification is that, despite our minimalist approach, our data is very high dimensional.

The evolutionary robotics simulations presented in this paper have been conducted in order to facilitate this classification. We want to stress that the task as it is solved by the agents, even though it is very simple, is very close to the real experimental set-up. This combination of minimal sensorimotor experiments and evolutionary robotics models proved to be fruitful before [3] and the present study confirms its usefulness. We analyse how delays impact on *reflex-like, reactive* and *anticipatory* behaviour, and we identify movement velocity, the number of movement inversions and the exact displacement from the object centre as variables with a large potential to shed light on our empirical data.

2 The Model

The evolved artificial agents, just as the human subjects, can act by moving left or right in an infinite one dimensional space, i.e., a looped tape (See Fig. 2). Evenly spaced objects (size 1×4, horizontal distance: 28 units) fall down from a vertical distance of 25 at one of seven constant velocities ($v_o \in \{0.004, 0.006, 0.008, 0.010, 0.012, 0.014, 0.016\} units/s$), each trial consists of a sequence of 32 objects at variable velocities. The agents have a perceptive field of 16×8 units. We decided not to model the exact tactile input patterns the subjects received, but to simply feed a continuous input signal representing the horizontal distance from the centre when an object entered the receptive field ($I_1 = |d_h|/6$ if $|d_h| \leq 6 \wedge d_v \leq 16$). The subjects additionally received auditory pulsed signals to indicate the velocity of falling objects. Such pulses (I_2) are fed into a second input neuron. A third input signal I_3 is a reward signal (auditory in the real experiment), in case an object is caught (rectangular input for $100ms$). An object is caught if it is in the centre region of the agent's receptive field when reaching the bottom line ($|d_h| < 4 \wedge d_v = 0$).

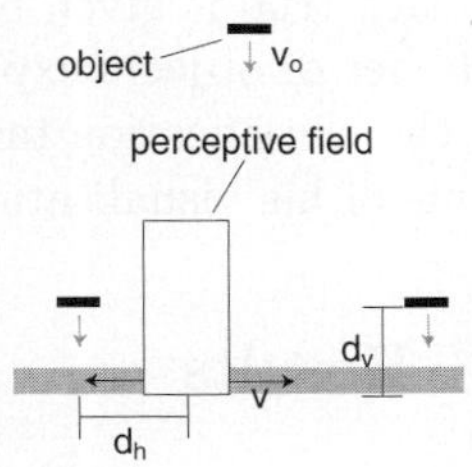

Fig. 2. Illustration of the experimental set-up

All three input signals are fed into the control network scaled by the sensory gain S_G and with a temporal delay. In the 'no delay condition' (*NDC*), there is a minimal processing delay (on average $35ms$) in the real experiment, which is prolonged by $250ms$ to $285ms$ in the 'delay condition' (*DC*). The same values (i.e., 35 and $285ms$) are used in the simulation. The agents are controlled by a continuous time recurrent neural network (CTRNN, see e.g., [1]), whose dynamics are governed by

$$\tau_i \frac{da_i(t)}{dt} = -a_i(t) + \sum_{j=1}^{N} w_{ij}\sigma(a_j(t) + \theta_j) + I_i \quad (1)$$

where $\sigma(x) = 1/(1+e^{-x})$ the standard sigmoid function, $a_i(t)$ the activation of unit i at time t, θ a bias term, τ the activity decay constant and w_{ij} the strength of a connection from unit j to unit i. The three input neurons feed forward into a fully connected layer of six hidden neurons, which feed the two non-recursively coupled output neurons. The dynamics are simulated using the forward Euler method, with a time step of $7ms$. The basic velocity output v calculated by the network is $v = \text{sign}(\sigma(a_{M1}) - 0.5) \cdot M_G \cdot \sigma(a_{M2})$, so one neuron controls velocity and another one direction, the motor gain M_G scales the output.

The search algorithm used to evolve the parameters of the control network is a generational genetic algorithm (GA) using a population of 30 individuals with a real–valued genotype $\in [0,1]$, truncation selection (1/3), vector mutation of magnitude $r = 0.6$ and reflection at the gene boundaries. The parameters evolved (145 parameters) are: $S_G \in [1,50]$, $M_G \in [0.001, 0.1]$, $\tau_i \in [25, 2000]$, $\theta_i \in [-3,3]$ and $w_{i,j} \in [-6,6]$. The weights w_{ij} and the bias θ_i are mapped linearly to the target range, the sensor gain S_G, the motor gain M_G and the time constants τ_i are mapped exponentially. The fitness $F(i)$ of an individual i in each trial is given by the proportion of objects caught $F(i) = o_i/O$, O is the number of objects exposed (usually 32). The catch task is in some ways similar to the classification task investigated by Beer in [1], which inspired us to adopt some of his visualisation tools to describe the sensorimotor dynamics.

3 Results

We evolved agents to solve the task under the *DC* and *NDC*, in order to get an idea of the strategy space resulting from our model (10 evolutionary runs with 1000 generations for each condition). Discarding one evolution from both series in which simply nothing evolved, we find that solutions for both conditions evolved to a high level of performance (see Fig. 3 (A)).

The solutions evolved for both scenarios generally involve halting abruptly once the object is encountered, frequently slightly overshooting the target, to then invert velocity and slowly move back to place the object in the centre of the receptive field (see Fig. 4). This strategy is rather trivial, which is probably

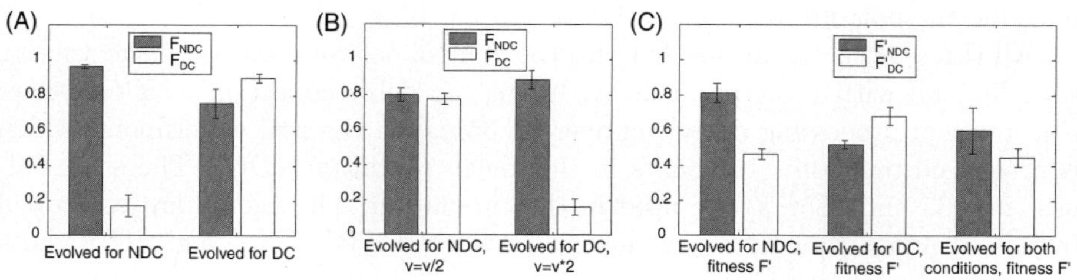

Fig. 3. Performance profile averaged over 9 runs. (A) Unperturbed condition. (B) Scaled velocities. (C) Modified fitness function F' (50% performance chance level).

related to tight temporal constraints and the coarseness of the fitness function that does not well capture the subtleties of sensorimotor perturbation and adaptation (see following analysis). As a stand–alone model, this model would, at best, have been a pilot study for more interesting simulations. However, there are reasons to believe that the discrepancy between performance measure and behaviour are *actual problems in the experiment and that subjects frequently use these kinds of trivial strategies* in response to the inherent limitations of the experimental set–up, just as the agents do. Therefore, a close analysis of the problems with the model and subtleties in the agents' trajectories can be useful for the data analysis and an improvement of the experimental set–up.

If we look at the solutions in terms of agent performance only, a striking difference is that most of the solutions to the *DC* are robust to the removal of the delay, while most of the solutions evolved for the *NDC* suffer a drastic breakdown in performance below chance level once the delay is introduced (see Fig. 3 (A)). If succeeding in the *DC* in many cases subsumes succeeding in the *NDC* for this task, and this is not just a property of the model, this issue could be crucial in explaining the failure to produce a negative aftereffect in the subjects, despite the transformation of sensorimotor strategies.

Another difference is that the velocity at which the object is first touched is on average twice as high for the *NDC* ($\bar{v} = 0.025$) than it is for the *DC* ($\bar{v} = 0.014$). Do maybe agents use the same strategy for both conditions, only varying their velocity? And are furthermore solutions generally able to deal with shorter, but not with longer delays? A very crude test for this hypothesis is to scale the velocity generated by evolved controllers, i.e. to double v in the networks evolved for the *DC* and to divide it by two in the networks evolved for the *NDC*. This test seems to confirm this hypothesis, because by virtue of this scaling, the performance profiles of the solutions evolved for both conditions are inverted (compare Fig. 3 (A) and (B)): Networks evolved under the *NDC* become generalists that are good at both conditions, whereas sped up networks evolved under the *DC* lose their capacity to perform the task with delay and become specialists for the *NDC*. A closer look at the sensorimotor dynamics, however, shows that things are not quite this simple.

As a first step in the analysis, it can be established that all evolved controllers seem to function independently of the reward signal and the pace at which the objects fall (I_2 and I_3), agents simply try to put objects as quickly as possible into the centre of the perceptive field. Therefore, agents produce the same trajectories for different object velocities that are just cut off at different points in time, and we will disregard these inputs in our following analysis.

Systematic displacements. Figure 4 depicts trajectories from different starting positions for two example individuals, one evolved for the *DC* (A) and one evolved for the *NDC* (B), and how they differ when tested under the *NDC* (top) and the *DC* (bottom). Both agents achieve to locate the object in the centre of their receptive field for most possible starting positions in the respective condition they have been evolved for (bottom left for *DC*, top right for *NDC*). Taking a look at how the behaviour is altered by the introduction/removal of a

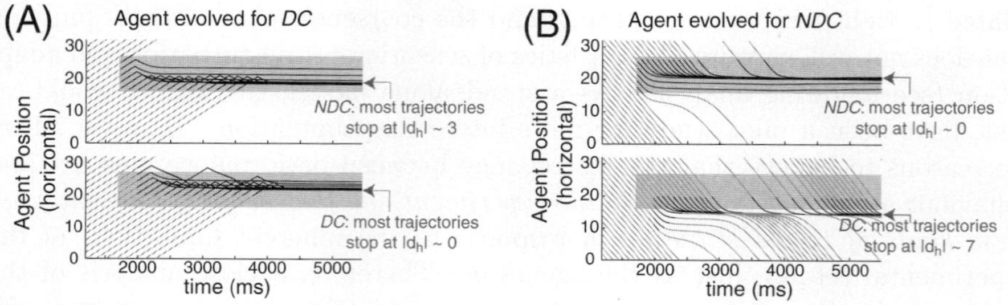

Fig. 4. Trajectories for different agent staring positions across time, presentation of a single object. Crossing the object (grey region) produces a (delayed) input stimulus I_1 (trajectories black during stimulation). Top: *NDC*, bottom: *DC*. (A) an agent evolved for the *DC* (B) an agent evolved for the *NDC*.

delay, it can be seen that in both cases the trajectories are systematically displaced from the centre of the perceptive field: The *NDC* agent overshoots its goal when exposed to a prolongued delay (bottom right), while the *DC* agent stops too early if the delay is removed (top left). The magnitude of these displacements is proportional to the agents' velocities. As the *DC* agent is much slower, its systematic displacement is small enough for trajectories to stay in the viable centre zone ($|d_h| < 4$) of the receptive field, as defined by the fitness function $F(i)$, while for the *NDC* agent, the displacement takes trajectories outside its receptive field. Such systematic displacement of trajectories can be observed for most agents. Since *DC* agents are on average half as fast, trajectory displacement is on average smaller for them. This seems to explain their robustness towards removal of the delay, which would then be due only to the coarseness of the fitness function that does not detect or punish micro displacements.

In order to further investigate these matters we evolved a new set of agents with a spatially more exact fitness function $F'(i) = \frac{1}{O}\sum_O 1 - \sqrt{d_h}/4$. With this modification, solutions to the *DC* cease to be robust to the removal of the delay (see Fig. 3 (C)), which confirms the hypothesis that *DC* agent robustness is due to the immunity of $F(i)$ to micro displacements. Interestingly, the experimental subjects were on average slower after training with delays. Therefore, these insights about systematic displacements, their relation to velocity and their effect on catch events (performance) may help in explaining the failure to produce a negative aftereffect in the experiment.

The modified fitness function also allows us to explore *agent adaptation to delays* by evolving agents under both conditions (Fig. 3 (C)). In the original task, this was not interesting, as many *DC* agents are able to solve both conditions anyway, so there was no selection pressure to go beyond robustness. Unfortunately, despite longer evolution, no adaptive adjustment of strategies evolved, only fixed strategies that compromised between the two conditions. We decided not to explore these possibilities further, as these are purely theoretical exercises.

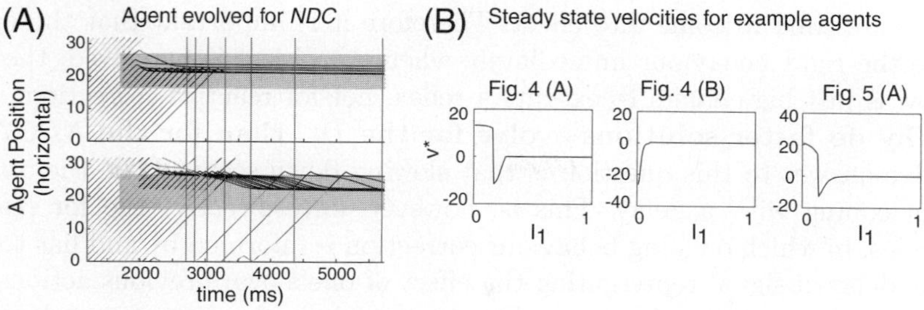

Fig. 5. (A) Trajectories for different agent staring positions across time for an agent evolved for the *NDC*, presentation of a single object. Crossing the object (grey region) produces a (delayed) input stimulus I_1 (trajectories black during stimulation). Top: *NDC*, bottom: *DC*. Vertical lines: cut off for different v_o. (B) Steady state velocities v^* for different I_1 for the agents in Fig. 4 (A) and (B) and Fig. 5 (A).

As future empirical experiments, such set–ups are neither possible, nor desirable, as discussed in the conclusion.

Reactive and reflex–like behaviour. Looking at the steady state velocities for varying I_1 in evolved agents, (Fig. 5 (B)), we generally find that $v^* = 0$ for values of I_1 that exceed a certain rather low threshold value of I_1. Behaviourally, this means that the agents are only sensitive to the onset of the stimulation when an object enters the receptive field, which triggers a rapid decay of v to 0, sometimes preceeded by a movement direction inversion (negative peaks in v^* in Fig 5 (B) left and right, compare corresponding trajectory graphs). The variation in signal magnitude, as an agent moves to the exact position to stop, however, is without effect on agent behaviour. This is why, e.g., the agent depicted in Fig. 4 (A) in the *NDC* remains in its location displaced from the centre of the receptive field, rather than to actively search for the exact centre. Such strategies are *reflex–like* in that they produce stereotyped trajectories.

Reflex–like behaviour evolved in all runs except one. The agent in Fig. 5 (A) is the only agent evolved for the *NDC* that maintains a relatively high level of performance when exposed to the *DC*. This is because the overshooting of the target and the first movement inversion bring it back to the margin of the perceptive field, where sensory activation is sufficiently low to trigger another movement inversion, which brings the trajectory to the centre. In that sense, the behaviour is more *reactive*, because it is sensitive to changes in magnitude of the signal caused by ongoing behavioural dynamics (Fig. 5 (A) top vs. bottom). Why did such reactive focussing strategies not evolve more systematically? The answer is simple: Because the deliberate inherent time pressure in the task does not allow for them to be beneficial. The cut off time for trials with the top three velocities is $1000, 1142$ and $1333ms$ after the objects become perceptible, which corresponds to the cut off points (vertical lines) $t = 2701, 2843$ and 3033 in Fig. 5 (A). The reactive mechanism that brings back the overshooting trajectories needs more time to come into effect. Therefore it is important that the agents induce the right behaviour immediately when the object is perceived, the time window is just big enough to execute a reflex, not for reactive correction.

Why do faster solutions evolve for the *DC* than for the *NDC*? The intuitive answer to this question is that slowing down seems to be the obvious way of coping with a delay. This is, however, only directly true for reactive strategies, in which ongoing behaviour correction is informed by and has to wait for the delayed signal representing the effect of one's own previous actions. For the execution of a reflex, however, there is no real disadvantage to a delay other than a shortening of the absolute time window in which to execute it, which is what makes it more mysterious that *DC* agents are slower. We investigated three plausible hypotheses: 1.) A maximum period of drifting back related to the upper limit of $2000ms$ for τ. 2.) An optimisation of velocity to a minimal network reaction time $t(d) = t_r + d$. 3.) An advantage in absolute time for *NDC* agents due to the $250ms$ longer time window. We didn't find evidence for any of these possible explanations. We are still unsure why faster solutions evolve for the *NDC*, there seems to be no simple answer. The answer may well be a combination of several of these simple factors, but, at the moment, we can give no obvious pressing reasons for why *DC* agents should not be fast.

4 Conclusion

We present an evolutionary robotics study that is a very close model of a minimal empirical study on adaptation to sensory delays in human subjects. The simulation model in itself is simple and not behaviourally sophisticated. However, we hypothesise that these problems of the simulation are analogous to problems in the empirical experiment, whose results we do not yet fully understand. Since evolutionary robotics simulations are scientific tools that can generate new hypotheses and conceptual clarifications [5], the model merits closer inspection.

A substantial number of hypotheses about the existing data set can be derived from the model. In the evolved solutions, sensorimotor behaviour is independent of the reward signal and object velocity (apart from different cut off points), and the behaviour within one object presentation is largely independent of previous object presentations. If this is similar in the empirical data, dimensionality could be reduced immensely. Another property of the evolved solutions is that, unless halting leads to a cessation of the input signal, the exact position of the object in the receptive field is not adjusted, which leads to another hypothesis to be explored for data simplification. On the other hand, insights were gained about the systematic displacement of trajectories from the object centre and its relation to velocity. This gives us a starting point to search for evidence in our data that a negative aftereffect could indeed have been produced, even if this is not reflected in the performance profile. As a general case, the visualisation of steady states and neighbouring trajectories that we adopted from Beer's work [1] has led to a helpful pre–theoretical understanding of evolved strategies. We will explore whether such techniques are, to an extent, transferable to the empirical data.

These insights about the discrepancy between sensorimotor adaptation and task performance should in principle enable us to design a better experiment in which these two factors concur. By using a fitness function that is spatially more exact, we can easily modify the simulation for this to be the case. Unfortunately, this modification is not feasible for the real experiment. Both, the temporal sampling rate and the spatial resolution are coarser in the experimental virtual environment than in the simulation, a coarseness that is necessary for the experiment to work in real time. There are fierce limitations to the transferability of such simulations to real experimental designs, an issue that can be seen as parallel to issues of simulation vs. real robots in classical robotics research.

But, even if these technical limitations could be mitigated, there are more profound lessons to be learned from the analysis of the simulation, lessons that relate to our discussion of *reactive* and *reflex-like* behaviour and that propose a more radically different experiment. In purely reactive sensorimotor loops, a sensory delay has an effect similar to an increase in inertia: Ongoing behaviour has to slow down, because what to do next depends on previous movements and the (delayed) perception of their effects. To an extent, this had already been recognised by Cunningham et al. [2], who hypothesise that negative aftereffects failed to occur in previous studies because they always provided the possibility to compensate for the delay by slowing down – hence the emphasis on the inherent time pressure in their task, which makes compensation impossible and forces real adaptation. This convincing argument had led us to include a strong time pressure in our simulation (and in our experiment) by choosing high object velocities. As desired, this hindered the evolution of reactive solutions, in which subjects can slow down to compensate. However, evolution found a different, faster 'cognitive shortcut': Agents use constant velocities, such that a fixed sensory delay corresponds to a fixed spatial displacement, which forms part of the stereotyped reflex–like trajectories that take objects to the centre of an agent's receptive field. In this kind of sensorimotor loop, we find a systematic error for both shortening and lengthening of delays, which makes it in principle possible to yield a negative aftereffect.

However, a more profound and complex adaptation to the delay seems to have taken place in Cunningham et al.'s experiment, something that goes beyond the tuning of reflex–like trajectories in our agents (and possibly in our subjects). This teaches us that there is more than just time pressure to their experimental design. Their visual task forces subjects to *produce fast variable velocity motor sequences*, in which a sensory delay can be neither understood as an increase in inertia, nor as a fixed displacement. However, in order to be able to usefully employ a fast motor sequence, it has to be possible to anticipate consequences of actions far into the future, i.e., the signal needs to be complex enough to feature cohesion between momentary signal structure, own movement possibilites, and future signal structure over a longer time course, regularities that make *anticipatory* behaviour possible. When designing a task analogous to Cunningham et al.'s within our minimalist approach, the complexity of the visual sensation has been traded for a more tractable but less rich tactile sensory signal. Thereby, long term predictability was lost, a loss that we were aware of, but that we did not previously consider essential. The lack of predictability, however, lead to the evolution of reflex–like behaviour. Reactive behaviour could not evolve because of the inherent time pressure in the task, and anticipatory behaviour could not evolve because of the poverty of the input signal. We now hypothesise that only anticipatory behaviour will produce interesting negative aftereffects to adaptation to delay and plan to design a series of new minimalist experiments with high predictability, to explore this hypotheses and elaborate our ideas about reactive, reflex–like and anticipatory sensorimotor strategies.

This study is also an exploration into the usefulness of evolutionary robotics simulations that closely match minimalist empirical experiments. The amount of hypotheses derived from the model on different levels of conceptual analysis speaks for itself; we consider this model a full success. However, it also became clear that some of the experimental mistakes would probably not have been made if the simulation had been run *before* the experiment. The design of an empirical experiment goes through a cycle of generate-and-test, which costs a lot of time and effort, both for the experimenter and for the piloting subjects, and frequently is ended pre–maturely. To transfer this generate-and-test phase as far as possible to simulation would be much less resource expensive. Therefore, we recommend the use of evolutionary robotics models not just after, but also before and alongside the conduction of empirical experiments.

In our analysis, we have focussed on one particular kind of agent behaviour, which we saw as model for a salient and characteristic, but nonetheless particular strategy adopted by the subjects. Both the empirical and the evolved solutions are much more versatile than that. Whether or not the generated hypotheses help in the classification and description of the data remains to be shown. In case these results lead the analysis into a dead end, this is not a reason to put the head in the sand. The possibilities of learning from evolutionary robotics models about the investigated scenario are not nearly exhausted yet.

References

1. Beer, R.: The dynamics of active categorical perception in an evolved model agent. Adaptive Behavior 11(4), 209–243 (2003)
2. Cunningham, D., Billock, V., Tsou, B.: Sensorimotor adaptation to violations of temporal contiguity. Psychological Science 12(6), 532–535 (2001)
3. Di Paolo, E., Rohde, M., Iizuka, H. (Forthcoming): Sensitivity to social contingency or stability of interaction? Modelling the dynamics of perceptual crossing. New Ideas in Psychology, Special issue on Dynamics and Psychology
4. Gapenne, O., Rovira, K., Ali Ammar, A., Lenay, C.: Tactos: Special computer interface for the reading and writing of 2D forms in blind people. In: Stephanidis, C. (ed.) Universal Access in HCI: Inclusive Design in the Information Society, pp. 1270–1274. Lawrence Erlbaum Associates, London (2003)
5. Harvey, I., Di Paolo, E., Wood, R., Quinn, M., Tuci, E.A.: Evolutionary Robotics: A new scientific tool for studying cognition. Artificial Life 11(1-2), 79–98 (2005)

Adapting to Your Body

Peter Fine, Ezequiel Di Paolo, and Eduardo Izquierdo

Centre for Computational Neuroscience and Robotics (CCNR)
University of Sussex, Brighton, BN1 9QH, UK
{p.a.fine,ezequiel,e.j.izquierdo}@sussex.ac.uk

Abstract. This paper investigates the processes used by an evolved, embodied simulated agent to adapt to large disruptive changes in its sensor morphology, whilst maintaining performance in a phototaxis task. By avoiding the imposition of separate mechanisms for the fast sensorimotor dynamics and the relatively slow adaptive processes, we are able to comment on the forms of adaptivity which emerge within our Evolutionary Robotics framework. This brings about interesting notions regarding the relationship between different timescales. We examine the dynamics of the network and find different reactive behaviours depending on the agent's current sensor configuration, but are only able to begin to explain the dynamics of the transitions between these states with reference to variables which exist in the agent's environment, as well as within its neural network 'brain'.

1 Introduction

The notion that perception and action are tightly intertwined concepts is a common theme throughout Artificial Life and Evolutionary Robotics research. O'Regan & Noë [8] outline the idea that perception is not a static, feed-forward means to achieve the representation of the world in the brain, but rather an ongoing process in which action plays an equally important part as perception. Additionally, adaptivity is seen as a crucial component of behaviour, since resilience to changing circumstances can be regarded as one of the basic tenets of life. However, a degree of tension can arise between paradigms aligned towards rapid reactivity through tight sensorimotor coupling, and the relatively slow-timescaled dynamics attributed to the reconfiguration of a system in response to the changing environment.

It is tempting to take a divide-and-conquer strategy to account for the differing requirements of these fast and slow aspects of behaviour. This can occur at a mechanistic level, with the addition of separate 'adaptive' processes which sit on top of a sensorimotor system (e.g. [5], where plastic synapses are used to enable an agent to switch between different lights during phototaxis). It may also occur via the imposition of a structural modularity (e.g. [12], where separately evolved networks account for the reactive and the adaptive components of a memory based task).

F. Almeida e Costa et al. (Eds.): ECAL 2007, LNAI 4648, pp. 203–212, 2007.

These approaches can certainly form useful models of adaptive behaviour, since biological organisms exhibit distinct phenomena such as spike time dependent plasticity [2] or slowly diffusing neurotransmitters [6] which warrant investigation. In [4] synaptic plasticity was exploited in an Evolutionary Robotics context to produce an agent robust to sensory inversion. These mechanisms appear to be somewhat dedicated to the slower timescale dynamics found in adaptive behaviour. The study of different, abstracted classes of adaptive mechanism can inform our understanding of what to look for in nature and how to build better engineered systems. However it is also beneficial to investigate the emergence of systems with interacting timescales. Evolutionary Robotics operating on vanilla, dynamical recurrent neural networks (with no additional mechanisms which are designed to account for the relatively slow, adaptive properties of behaviour) can provide a useful, relatively assumption-free paradigm within which to do this (e.g. [3], in which a robot's locomotion exploits a model of its body which is automatically constructed and refined through a predictive algorithm).

Previous work in Evolutionary Robotics has investigated the potential for fixed-weight networks to display the adaptive functionality more commonly attributed to dedicated weight-changing mechanisms (e.g. [11]). [9] provides a recent discussion and example of this approach. *Our work here differs from this and other attempts in that the environment itself does not change, but rather the agent's own morphology is what must be adapted to.* Instead of adapting a sensorimotor law to suit a disrupted environment, it is the sensorimotor laws used to control the agent which become disrupted. This occurs when the agent's body is altered such that the relationships and invariances between the sensory and motor activity change.

[10] offers a mathematical framework based on the sensorimotor contingency principles outlined in [8]. They provide an algorithm which is able to extract 'laws' allowing the coordinated control of unknown, arbitrary and high-dimensional simulated robotic systems. It provides a useful proof of concept, demonstrating the potential for sensorimotor systems to discover the structure of their environment and morphology despite knowing nothing about these *a priori.* This paper intends to investigate dynamical approaches to a related self-discovery task, albeit in minimally complex way.

Below, we present an evolved model agent, which is able to perform a sensorimotor task despite a radical perturbation to its morphology. This is analysed in section 4, where we see that the agent exploits its environment to overcome failure imposed by disruption to its sensor configuration.

2 Methods

We carried out experiments using a simulated agent required to perform phototaxis towards a sequence of lights. To achieve this, it was fitted with a single, simple light sensor. By switching the location of the light sensor, however, a single set of sensorimotor relationships is unlikely to be sufficient to always complete the task (as evidenced below).

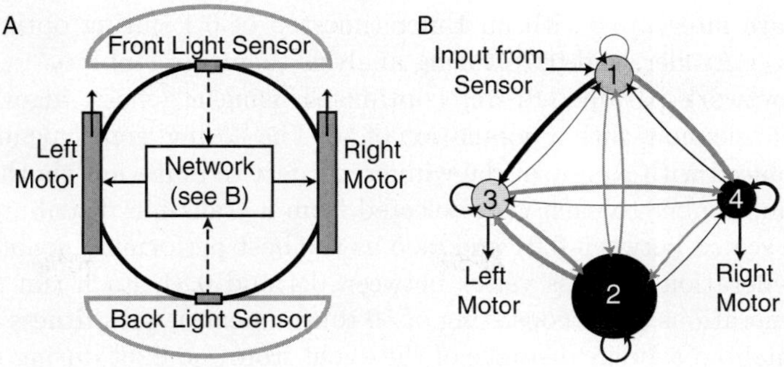

Fig. 1. A: The agent's circular body, showing motors and range of angles at which light impinges on the sensors. B: The neural network configuration, fully-connected with one input neuron (no. 1), an interneuron (2) and two motor neurons (3 & 4). The parameters for the best evolved network (see section 4) are depicted. The neuron size reflects time constants, neuron shade depicts the bias (dark is high), weight thickness is its magnitude, and weight colour is dark for excitatory, grey for inhibitory.

The experiment consists of a circular agent of radius 1, which exists in a 2D plane (see Figure 1A). Distance and time units are of an arbitrary scale. It is controlled by a motor on each side, which differentially steer the robot with their output (in range [0, 1]). A light sensor is attached to either the front or the back of the robot at any one time. It outputs a value in [0, 1] indicating closeness to a point light source, by linearly scaling the distance to that light (γ), thresholded such that $\gamma > 22.5$ becomes 0 (22.5 is 1.5x the maximum γ when each light is first presented). The sensor only receives input if the light is within the 180° in front (or behind) the robot, centred on the light sensor, since its circular body occludes the light.

Lights are presented sequentially, located at a random angle and distance in [10, 15] from the agent. After a variable (in [40, 60]) time period, a new light is generated. Less frequently, the sensor is switched from the front to the back of the agent (or vice-versa) at the start of a light presentation. The agent is controlled by a continuous time recurrent neural network (CTRNN), shown in Figure 1B and governed by equation (1) (based on [1]).

$$\tau_i \dot{y}_i = -y_i + \sum_{j=i}^{N} w_{ij}\sigma(y_j + \theta_j) + sI(t) \quad , \quad \sigma(x) = 1/(1 + e^{-x}). \qquad (1)$$

The state (y) of the ith neuron (out of N) is a function of the time constant τ (scaled exponentially in [1, e^5]), the weight w_{ij} in [-10, 10] from incoming neurons, their state y_j and bias θ_j (subject to a sigmoid function). The sensor weight s in [-10, 10] is applied to the input I from either the front or back light sensor (whichever is enabled at a given time). The network (and other simulation variables) are integrated with an Euler timestep of 0.1 during optimisation of the agent's controller, and 0.01 during analysis to ensure stability.

These network parameters were optimised using a genetic algorithm with tournament selection, with a population of 50. The 'losing' genotype in a tournament is replaced with a copy of the winner, subject to mutation via the addition of a random number to each gene selected from a Gaussian distribution (mean 0, variance scaled between 0.01 and 0.05 as the best performing genotype of the previous generation's fitness varies between 0.4 and 0.8). Each run progressed for 6000 generations (each consisting of 50 tournaments), with fitness calculated as the normalised average distance of the agent from the light during the last 25 time units of each light presentation. The first light after the sensor was switched does not contribute towards fitness, allowing for a possible adaptation phase to occur without punishing the agent.

The evolution progressed according to a shaping scheme consisting of several phases, since it would be unlikely that both phototactic and adaptive behaviour would evolve simultaneously. When the best agent of each of the 10 prior generations had attained a fitness greater than 0.8, the phase was advanced. Initially, the sensor was solely located on the front of the agent, so standard phototaxis was all that was required. For each trial, the agent was subject to 6 sequential light presentations, and the average of 8 of these trials taken as the overall fitness score for a given genotype. Phase 2 consisted of an additional set of 8 trials, with the sensor located on the back of the agent (which was not changed for the duration of each run). The overall fitness score was the mean of all 16 trials. Finally, a third phase proceeded as per phase 2, but with double the number of lights per trial (12). The sensor was switched to the opposite side of the agent after a random number of light presentations in the range [4, 8].

3 Results

16 runs (with different, independent random seeds) were carried out for each of the 0, 1 and 2 interneuron cases (giving a total of 3, 4 and 5 neurons for each). Of these, none of the 0 interneuron experiments succeeded in generating a successful phase 3 (re-adaptation to sensor switching) controllers. Two networks with 1 interneuron and one with 2 interneurons did however achieve a high level of fitness.

It is clear that the success was limited to only a small number of the evolutionary runs, and discovering why this task appears to be so difficult with the current experimental configuration could prove an interesting direction of study. However we concentrate here on the analysis of one of the best performing runs, since we are primarily concerned with the kind of dynamics which can satisfy the task, rather than evolutionary performance. The best agent of the most successful 1-interneuron run is thus considered from this point on. The progress of that evolutionary run can be seen in Figure 2. In 2000 trials performed on this agent, in which 25 lights were presented with a switch in sensor location after every 5 lights, the mean fitness score was 0.806 ($\sigma = 0.043$), indicating a reliable and adept controller.

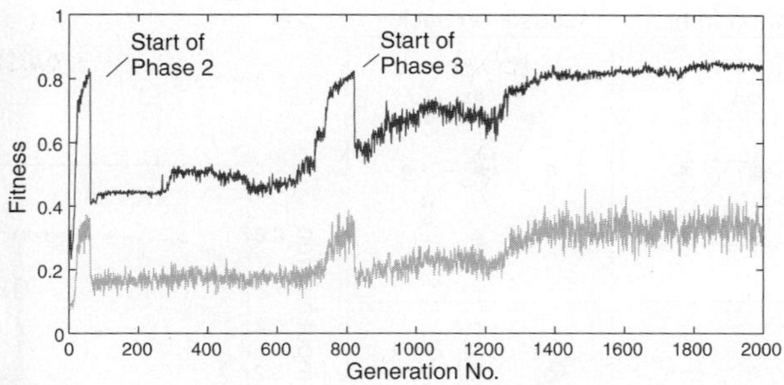

Fig. 2. Evolutionary performance. Black is the best agent of each generation, grey is the average. Only the first 2000 generations (of 6000) are shown.

4 Analysis of a Successful Agent

In order to understand how the high fitness scores were achieved, an analysis of the agent from the run shown in Figure 2 was carried out. Firstly, the behaviour of the agent was examined, and then its internal neural dynamics studied.

4.1 Behavioural Performance

Figure 3 demonstrates the behaviour of the agent in a series of trials separately showing the four main demands which the task requires of this agent. The top two plots of Figure 3A show the trajectories generated after the agent has been allowed sufficient time to adapt to the two possible sensor configurations. It is clear that the agent is able to navigate towards the light reliably from different relative headings, and along reasonably un-convoluted paths. However it is also evident that it follows quite different routes depending on whether the sensor is on the front or back. This is clarified in Figure 3B, where the difference (left−right) of the two motor outputs (shown on the x axis) reflects the turning behaviour of the agent. A positive difference produces a left turn, near 0 causes forward motion, and negative values a right turn. It demonstrates that when the sensor is on the front, the agent only ever turns left or moves forward, whereas when on the back, turns in both directions are made.

The trajectories during re-adaptation (the lower two plots of Figure 3A) show more complex behaviour. In both cases (front→back adaptation and the converse), the light is not reached by the end of the first light presentation, suggesting that there is not (at least that evolution could find) a trivial, reactive behaviour permitting both front and back sensor phototaxis. In both cases, the agent at first moves away from the light, which can be considered a temporary maladaptive strategy, from which it recovers. In the front→back case, the agent is sometimes successful by the second light, but in others, requires three.

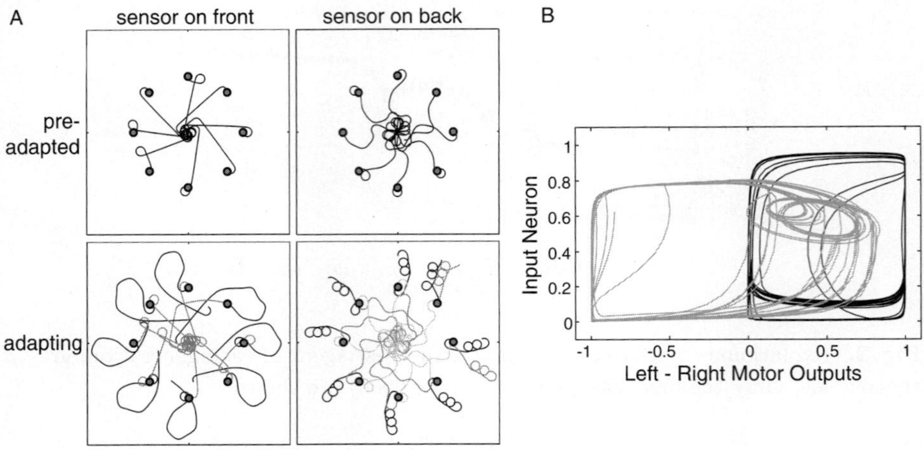

Fig. 3. A: The behaviour of the agent under different circumstances. In each, the agent was started at 8 different locations (marked with a small circle), with the light remaining in the same place (the centre of each figure). Each trajectory in the top two plots of A show the behaviour during a single light presentation of duration 50, with the agent previously being allowed to adapt to its sensor location for 3 successive random light presentations. The lower pair of plots demonstrate the behaviour after the sensor has been switched to the opposite side of the robot. The black line shows the first light presentation, immediately after the switch, whilst the lighter lines show the second and third (with the agent replaced at its original location), where these extra lights were required to complete adaptation. The agent's initial heading at the start of a light presentation alternates between 0 and π on each subsequent presentation (during the above analysis only). B: shows the motor outputs against input neuron output for the same trajectories as the top two plots of A above, with the black line representing data taken with the sensor on the front, and the grey with the sensor on the back.

4.2 Neural Dynamics

The system under consideration is fully described by four internal continuous variables (node activities), in addition to one continuous and one discrete environmental variable (the distance to the light, and whether it is in sight at a given time, respectively). Since we cannot observe the full structure of this six dimensional hybrid dynamical system, we must carefully choose appropriate lower dimensional projections which adequately capture the coupled agent-environment dynamics. The trajectory plots above adequately demonstrate what the agent does, but not how it achieves its adaptivity. We therefore analyse the neural dynamics to further understand the behaviour of the agent.

Figure 4 shows the behaviour of the interneuron against that of the input neuron during the same four scenarios as Figure 3A. There are quite clearly two separate stable modes in the interneuron's state space, with mean values of

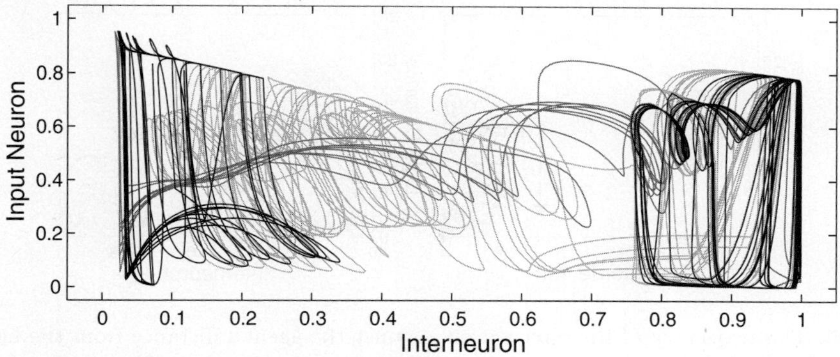

Fig. 4. Phase plot showing the output of the interneuron against that of the input neuron. Neural data from 8 different trials are superimposed. Black shows the behaviour after the agent has adapted to its sensor configuration, with the black trajectory on the left of the plot taken from a period with the sensor on the front, and the right-hand black path with the sensor on the back. Dark grey shows presentations of the light which immediately follow the switching of the sensor from back→front, and light grey represents switching from front→back (in the latter case two presentations are shown since adaptation takes multiple lights to complete). Note that the sensor weight is negative (-0.8), and therefore a high input neuron value is an indication of a low sensory input.

0.11 and 0.94 for the left and right black-lined areas in the diagram, respectively (see caption for details). Since these two bundles of black lines were separately recorded from the agent under each sensor configuration (whilst the agent undergoes successful phototaxis), we can be confident that they correspond to the two different stable sensorimotor behaviours exhibited (depending on that sensor location). In fact, if we fix the value of this interneuron to its mean value in each of these states, the performance of the agent is strongly biased. On setting to 0.11, the mean fitness (over 1000 ten light trials) with the sensor on the front is 0.8258 ($\sigma = 0.05$), whereas the mean fitness is 0 with the sensor on the back. Conversely, on setting the interneuron to 0.94, mean fitness is 0.8093 ($\sigma = 0.02$) with the sensor on the back, and 0 with the sensor on the front. However if we threshold the interneuron's output so that it is binary (if greater than 0.5, the output is clamped to 0.94, otherwise it is clamped to 0.11), we find a drastic loss in the agent's capability to adapt.

Another point to note is that the time constant parameters have evolved to within 0.05 of their fastest permitted value of 1 for all of the neurons except the interneuron, which took a larger value (6.6), which corresponds to a slower integrating neuron.

By outputting different values depending on the sensor's location, the interneuron is able to push the network into a different region of motor dynamics, which produces the different behaviours demonstrated above. But this does not

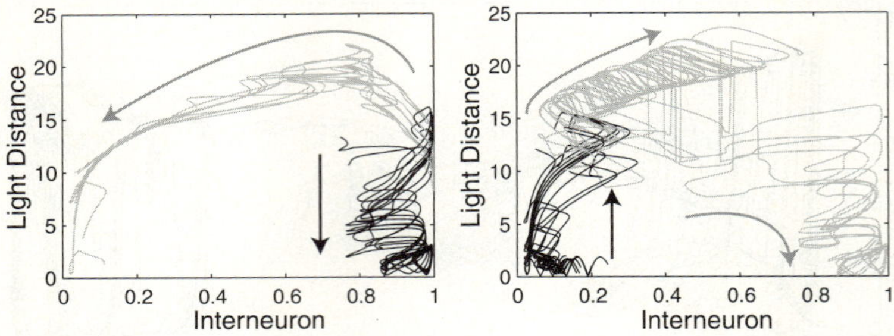

Fig. 5. The trajectory of the interneuron, against the agent's distance from the light. The left case shows the back→front transition in grey, with the black line showing the dynamics prior to the sensor switch. The right-hand figure shows the front→back transient in a similar fashion. Each show 10 different simulations, superimposed. The arrows show the general trend that the dynamics follow, their direction indicating the passage of time since the start of each light presentation.

explain how the dynamics during the grey phases shown in Figure 4 make the transition to the alternate state. In fact, for parts of the grey trajectories (after the sensor has switched), the input neuron appears to follow a similar regime to its activity prior to the switch. However if we consider what is happening to the agent's relation with its environment (in terms of phototactic performance) during these ambiguous parts of the transient dynamics, the picture becomes clearer. In Figure 5 we consider an additional, environmental variable: the distance from the agent to the light.

In contrast to Figure 4 (which only shows neural outputs), Figure 5 provides a more adequate disambiguation of the attractor and transient dynamics. In each case, the black lines show the agent's performance after if has been given time to adapt. They commence each light presentation at a distance in the range of 10 to 15 away from the light, and moves downwards (in these plots) towards a high-fitness, low distance location. The grey lines start in the same region of light distance at the start of a new light presentation, but now the sensor configuration has been changed. From here, the light distance grows instead of diminishing. The interneuron is at first still within the same region as it was prior to the switch, however following the same sensorimotor strategy as before the switch now moves the agent away from the light. Eventually, this maladaptive behaviour will lead to diminished sensory inputs, since the sensor is distance-sensitive. These changing input patterns likely cause the interneuron to transition to its opposite state, pushing the network's dynamics into the alternate, now successful, region of sensorimotor behaviour. Explaining these patterns in more detail, however, has not yet been achieved.

Finally, the agent was tested with the sensor mounted at different angles — not just the ones which it experienced during evolution. The average fitness during these tests were low when the angle deviated more than a small distance from either 0° or 180°.

5 Discussion

A model agent has been presented, which is capable of consistently performing phototaxis, regardless of whether its light sensor is mounted on its front or back. By design, this performance cannot be achieved by a fixed reactive control strategy. We have demonstrated how this is realised through a time-extended adaptation process with no explicit plasticity. After a switch of sensor location, the agent's now maladaptive sensorimotor behaviour produces a change in the results of its actions. It is then able to exploit the new situation to ensure its dynamics enter a different, task-satisfying phase. We have analysed the neural activity in the agent's controller, pointing to two separate strategies for the two different sensory morphologies. At first glance these two strategies correspond to the low and high firing regimes of the interneuron. However a useful explanation of how the dynamics bifurcate between these states required the inclusion of an environmental variable in the analysis.

We did not *a priori* provide a mechanism designed to produce the relatively slow dynamics of the adaptive process (when compared to the fast, reactive sensorimotor dynamics). However neural activity of multiple timescales were present within the system — this is particularly demonstrated by the fast time constant values of the sensor and motor neurons, in contrast to the slower dynamics of the interneuron which correlated with the adapted state of the agent.

It is interesting to compare this situation to what we would expect if we had to design such an agent by hand. On a first approximation, a rule based design would look at establishing sensorimotor correlations (such as light sensor changes with respect to motor activity). These correlations may seem sufficient to determine the body configuration of the agent. This approach would be very much in line with the abstract framework proposed by Philipona et al. [10], whereby invariance in the structure of the sensorimotor correlations can be exploited to generate appropriate control laws. Whilst this approach remains generic, one issue we must confront when dealing with a time-extended, behaving agent is how to combine the fast timescales of behaviour and the slow timescales of adaptation into an integrated functional unit. A hand designed, algorithmic approach would tend to separate these two requirements into specialised sub-modules. We find that our evolved agent handles these two requirements with a single mechanism. Our analysis has shown how fast dynamics corresponds to a stable behavioural regime and how, upon changing the sensor configuration, longer term transients in the interneuron corresponding to phases of distancing from and approaching a new light are successfully combined without being functionally separated.

The importance of the interneuron's transient dynamics is demonstrated by the experiment where its activity is made binary. While the high and low mean activity states are sufficient for producing their corresponding behaviours, the interneuron is also involved in generating the transition between the two, and its function cannot be reduced to that of representing either of the two conditions.

Whilst this experiment does demonstrate an adaptive agent which must explore its relation with the world in order to progress, it is limited to switching between just two discrete states corresponding to two distinct strategies. A model which extends deeper into notions of flexibly adaptive behaviour would be able to produce viable sensorimotor strategies when encountering a continuous range of morphological configurations. Although it had not been evolved to handle sensor mountings other then on its front or back, we tested the agent's ability to deal with other sensor angles, which it largely failed to adapt to. [7] investigates learning on a continuum in an embodied scenario using recurrent neural networks (CTRNNs) without the need for a separate, dedicated adaptive apparatus, and could provide a basis for the extension of our model into the continuous domain.

Acknowledgements. We would like to thank Hiroyuki Iizuka, Marieke Rohde and other members of the CCNR for valuable discussion. This work was funded by the EPSRC grant EP/C51632X/1.

References

1. Beer, R.D.: On the dynamics of small continuous-time recurrent neural networks. Adaptive Behavior 3(4), 471–511 (1995)
2. Bi, G.Q., Poo, M.M.: Synaptic modifications in cultured hippocampal neurons: dependence on spike timing, synaptic strength, and postsynaptic cell type. Journal of Neuroscience 18, 10464–10472 (1998)
3. Bongard, J., Zykov, V., Lipson, H.: Resilient Machines Through Continuous Self-Modeling. Science 314(5802), 1118–1121 (2006)
4. Di Paolo, E.A.: Homeostatic adaptation to inversion of the visual field and other sensorimotor disruptions. In: Meyer, J.-A., et al. (eds.) Proc. of the Sixth Int. Conf. on the Simulation of Adaptive Behavior, pp. 440–449. MIT Press, Cambridge (2000)
5. Floreano, D., Urzelai, J.: Neural Morphogenesis, Synaptic Plasticity, and Evolution. Theory in Biosciences 120(3-4), 225–240 (2001)
6. Hölscher, C.: Nitric Oxide, the Enigmatic Neuronal Messener: its Role in Synaptic Plasticity. Trends in Neuronscience 20, 298–303 (1997)
7. Izquierdo-Torres, E., Harvey, I.: Learning on a Continuum in Evolved Dynamical Node Networks. In: Rocha, L., et al. (eds.) Proc. of the Tenth Int. Conf. on the Simulation and Synthesis of Living Systems, pp. 507–512. MIT Press, Cambridge (2006)
8. O'Regan, J.K., Noë, A.: A sensorimotor account of vision and visual consciousness. Behavioral and Brain Sciences 24(5), 939–1011 (2001)
9. Phattanasri, P., Chiel, H.J., Beer, R.D.: The dynamics of associative learning in evolved model circuits. Submitted to Adaptive Behavior.
10. Philipona, D., O'Regan, J.K., Nadal, J.P., Coenen, O.J.-M.D.: Perception of the structure of the physical world using unknown sensors and effectors. Advances in Neural Information Processing Systems 15 (2004)
11. Tuci, E., Quinn, M., Harvey, I.: An evolutionary ecological approach to the study of learning behaviour using a robot based model. Adaptive Behavior 10(3), 201–221 (2003)
12. Yamauchi, B., Beer, R.D.: Integrating reactive, sequential and learning behavior using dynamical neural networks. In: Cliff, D., et al. (eds.) Proc. of the Third Int. Conf. on Simulation of Adaptive Behavior, pp. 382–391. MIT Press, Cambridge (1994)

An Analysis of Behavioral Attractor Dynamics

Alberto Montebelli, Carlos Herrera, and Tom Ziemke

University of Skövde
School of Humanities and Informatics
SE-541 28 Skövde, Sweden
{alberto.montebelli, carlos.herrera, tom.ziemke}@his.se

Abstract. The interaction of brain, body and environment can result in complex behavior with rich dynamics even for relatively simple agents. Such dynamics are, however, often notoriously difficult to analyze. In this paper we explore the case of a simple simulated robotic agent, equipped with a reactive neurocontroller and an energy level, that the agent has been evolved to re-charge. A dynamical systems analysis, shows that a non-neural internal state (energy level), despite its simplicity, dynamically modulates the agent-environment system's behavioral attractors, such that the robot's behavioral repertoire is continually adapted to its current situation and energy level.

1 Introduction and Background

The dynamical systems (DS) approach to adaptive behavior and cognition (e.g. [1,2,3]) views agents and their environments as tightly coupled dynamical systems. Nervous systems, bodies, and their environments have complementary roles in producing a rich range of adaptive behaviors [4,5]. Hence, the cognitive agent we consider is not a closed system whose activity can be reduced to the mapping of sensory inputs to motor outputs. In natural agents, nervous system, body and environment are not three independent components engaged in a synchronic interaction; rather, natural adaptation is the ongoing result of a global, self-organising process. The implementation of artificial adaptation on the basis of principles of self-organisation is therefore alluring. The DS approach offers a range of tools available to the theorist that aims to understand adaptive processes as they unfold in time.

In recent years there also has been a decisive turn towards aspects of situatedness and embodiment in much of cognitive science and AI [1,6,7]. Much recent robotics research has focused on the sensorimotor interaction between control systems, their robotic bodies and their external environment. For example, it has been demonstrated that control may be highly distributed and, for instance, morphology plays an active role in control [7]. Furthermore, Parisi [8] has recently argued that "to understand the behavior of organisms more adequately we also need to reproduce in robots the inside of the body of organisms and

F. Almeida e Costa et al. (Eds.): ECAL 2007, LNAI 4648, pp. 213–222, 2007.

to study the interactions of the robot's control system with what is inside the body". Parisi coined the term internal robotics to denote the interaction between the (neural) control system and the rest of the body. One important aspect of this interaction is motivation. Ashby [9] used the term "homeostasis" to refer to the fact that certain variables, essential for the system, need to remain within certain limits - a "viability zone" [10] - for the organism to function and stay alive. An adaptive agent is one that incorporates mechanisms that ensure that deviations from that viability zone are balanced during adaptive interaction. The viability zone is therefore ensured, not only through internal homeostatic mechanisms, but also motivating adaptive behavior.

There have been several attempts to model internal states as motivations for the control of robots [11,12]. The role of internal states as motivations in artificial agents will strongly depend on what particular architecture the artificial agent embodies. If different forms of behavior are determined by functional modules, the specific values of internal variables can effectively influence an action selection process [13,14,15]. Research on self-organizing behavioral systems has demonstrated that breaking down control systems into modules, according to some designer's or observer's preconception or perception of what "behaviors" might be required, is not unproblematic (e.g. [16,17]). For example, it has been shown that behavioral adaptivity through quasi-modularity at a functional level can be realized without actual modules, through relatively simple recurrent neurocontrollers that make use of feedback to dynamically modulate their own behavior over time [17]. Similarly, in recent work on recurrent architectures with parametric bias values [18,19], internal states function as control parameters of nonlinear dynamical systems to generate the dynamical bifurcation to distinct behaviors. In this case, the dynamics of internal states (the parametric bias) is achieved through anticipation and back-propagation of the error.

In this paper, we further investigate the role of internal states in the modulation of behavior, exploring the hypothesis that non-neural internal states (metaphorically named "energy level") can modulate the functioning of a simple, recurrent neural network to produce different behaviors. Our perspective is radically different from related work on action selection since we do not consider behavior that is generated by a collection of structural modules. In other words, rather than focusing on how a selection mechanism can arbitrate the activation of a module that encapsulates some behavioral program, we investigate how internal states can modulate the activity of a control system so to generate distinguishable interactive dynamics. Using a dynamical systems analysis, we distinguish several attractors, which can be understood as behavioral modes. The internal states used in the experiments documented here (energy values), are intentionally kept very simple. The underlying hypothesis examined in the following analyis is that, in a self-organized homeostatic system, internal states can play the role of control parameters of the interaction, therefore affecting behavior not through the selection of subsystems/modules, but through a modulation of the phase space that dynamically redraws the "behavioral repertoire" locally available to the agent.

2 Method

In the present paper we report an experiment based on an intentionally simple evolutionary robotics simulation model of dynamical energy management, inspired by the theoretical framework sketched above. The control system for a simulated Khepera robot, implemented by a feedforward artificial neural network with no hidden layer, was artificially evolved by standard evolutionary algorithms within a square environment containing two identical light sources. The agent receives input from 8 infrared sensors, 8 light sensors and 1 energy sensor (tracking the current energy level), all directly connected, through the evolved weights, to the 2 motor neurons, respectively determining the activation of the left and right motor. The light sources are symmetrically located at the center of the arena, their reciprocal distance being comparable to the maximum distance range of the light sensors.

The robot was initially provided with an energy level of 1.0. Its energy was subject to a continuous decay at 0.008 per time step, down to zero. When the robot reached a target area surrounding one of the lights (highlighted, for the reader's benefit, by the circle on the left of Figure 1) its energy reservoir was instantaneously refilled, thus simulating an internal effect sensed by the agent. The fitness, integrated over the whole lifetime, is calculated as follows: a positive value of energy would result in a positive additive constant (+0.01 per time step), under the condition that the agent were positioned outside the target area. On the other hand, the attainment of a null energy level would produce the opposite effect of integrating a negative constant (-0.02). In order to facilitate the evolutionary bootstrapping, the performance of the first 200 time steps of lifetime was not integrated into the fitness. The survival of the individual was only limited by the number of time steps during its lifetime and immediately obliterated by eventual crashes against the walls, which also determined a severe punishment (-100).

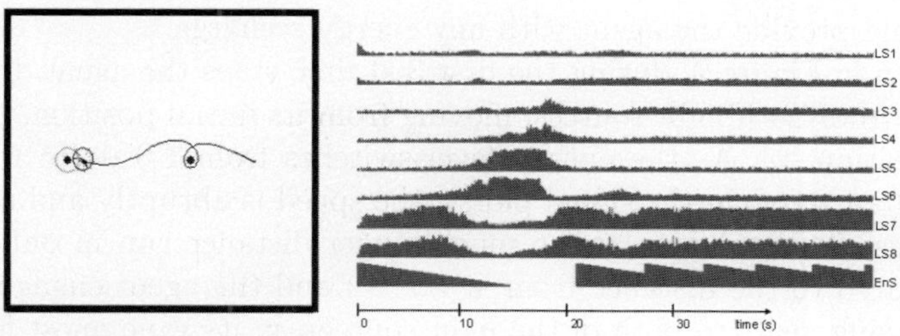

Fig. 1. Evolved behavior. The simulated agent approaches the light (represented by a dot) on the right side of the environment. The lack of any effect on its energy level sensor results in the following approach of the next light. As its energy reservoir is instantaneously refilled to the maximum level, the agent engages in a stable behavioral attractor in the proximity of the rewarding light source. The signals labeled LS1-8 represent the activation of the light sensors in time; EnS displays the energy level.

3 Results

At the end of the evolutionary process the simulated agent (endowed with the fittest evolved controller) proves reliably effective in disambiguating between the two light sources, according to their capacity to affect, neutrally or positively, its energy level. As exemplified in Figure 1, the agent, to whom the two light sources look identical, approaches a first light source and, receiving no reward, quickly migrates towards the other light source that supplies it with a whole load of energy. Such an offer produces the engagements of the agent in a stable behavioral attractor in the neighborhood of the rewarding light source. The development of a skillful controller capable of light disambiguation constitutes, rather than the goal, the actual starting point for our work, which will emphasize the general holistic nature of the relation between agents and their environments, the complex dynamics emerging from such interactions and the peculiar role of non-neural internal states in influencing the emergent dynamics. During the evolutionary process, the energy level plays the generic role of one of the many internal state variables of the system. By "system", we here mean the dynamical system constituted by agent, its control system, its environment/arena and the set of bio-regulatory mechanisms (synthesized, as a first approximation, in the agent's energy level). In the following analysis the energy level is used as control parameter of the overall system. The systematic manipulation of the energy level allows us to visit a wealth of behavioral attractors embedded in the system during its evolution (where the adaptation is limited to its neural controller).

3.1 Energy Binary-Switch

We ran some simulations where the same agent as in Figure 1 was placed in one of a few possible initial positions and fed with a steady high level of energy. After a suitable interval of time, chosen to exhaust the transitory phase from the initial position to the development of the specific behavioral attractor, the energy level was suddenly switched to low. In this case none of the two light sources would provide the agent with any energy recharge.

As shown in Figure 2, during the first 300 time steps the agent drifts at low speed away from both light sources, moving from its initial position, labeled 'a', towards position 'b'. As the energy level switches from 1.0 down to 0.0 (emphasized by an arrow in the signal plots), the speed is abruptly and remarkably increased (as clearly shown by the much longer distance run in only 200 time steps compared to the distance from 'a' to 'b') and the agent engages in a fast run, alternating visits to both of the light sources, in its vane quest for energy.

3.2 Energy Clamp

We further developed the analysis by refining the quantization of the possible energy levels. The previous controller was fed with discrete, steady levels of energy, ranging from 0.0 to 1.0 with constant increments of 0.02, and the agent left free to interact for a time sufficient to fully develop its ongoing dynamic. We

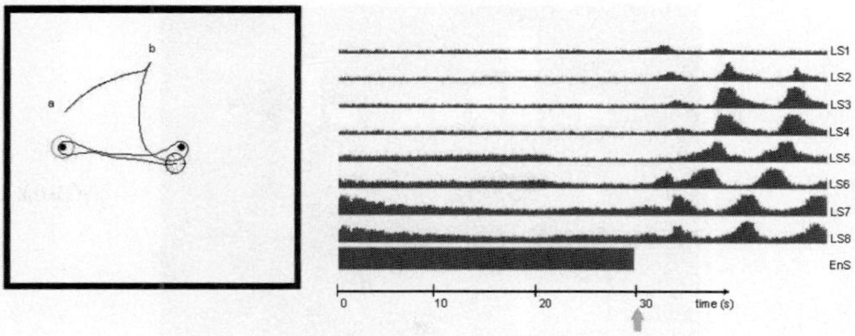

Fig. 2. Energy binary switch experiment. During the first 300 time steps (energy level 1.0), the same agent as illustrated in Figure 1 slowly moves from the position labeled 'a' towards 'b'. When the energy level switches to zero (arrow in the plot of the sensory activations) its speed suddenly increases, as the agent alternately visits the two lights.

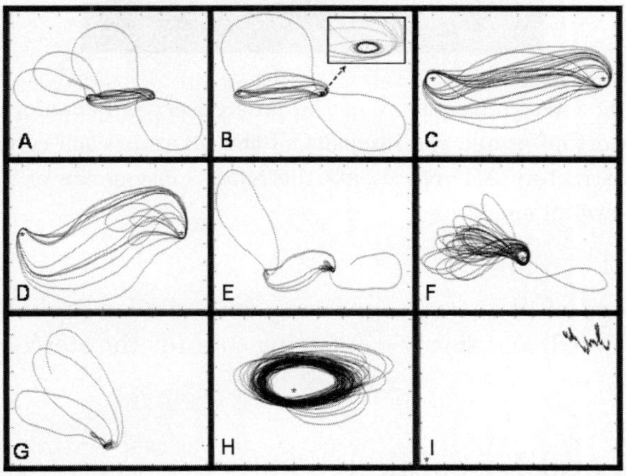

Fig. 3. Iconic representation of the spatial trajectories for the nine behavioral attractors in our taxonomy. The scale has been systematically varied for the reader"s benefit, in order to facilitate a deeper qualitative grasp of the attractor"s morphology (where possible, light sources are indicated by a star).

developed a taxonomy constituted of nine observed behavioral attractors, based on their general morphology and spectral characteristics (Figure 3), in order to determine the effect of the control parameter on the dynamical properties of the agent-environment coupling (Figure 4).

The bar plot at the top of Figure 4 shows the relative frequency of each of the nine attractors along the whole duration of the experiment. The lower part details how such relative frequency is distributed as a function of the current level of energy. Apparently, under the effect of the control parameter, the behavior drifts from attractors A and B, for the lowest levels of energy, to attractor I for energies of 0.8 and above. The energy interval [0.1, 0.4] is dominated by attractor C, whereas the following interval is characterized by a sequence of

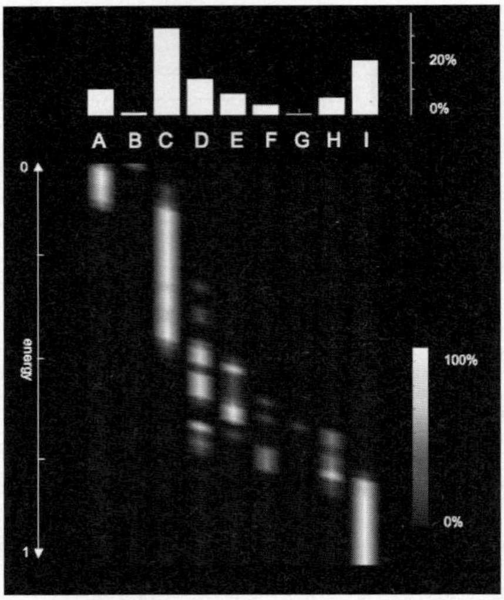

Fig. 4. Top graph: relative frequency of the nine observed behavioral attractors over all replications. Bottom graph: the intensity of the pixels in each column (respectively associated with attractors A-I) represents the relative frequency of the behavioral attractor for each level of energy.

couples or triplets of attractors appearing with similar probability for energy levels within [0.5, 0.8] and slowly converging towards the stereotypical attractor for high energies.

4 Analysis and Discussion

The tenets of embodied and situated cognition stress the importance of the physical dynamical coupling between an agent and its environment. This effect has already been shown in cognitive robotics (e.g. [20]); brain, body and environment equally co-participate in the cognitive phenomena. With the present work we contribute to such general framework by adding, and grounding in experimental terms, the basic principle of internal robotics: non-neural internal states are fundamental to the deployment of the full potential of cognitive processes. In our work a simple non-neural internal state variable belonging to the internal bio-regulatory process, the energy level, proves effective at regulating the behavioral coupling between the agent and its external environment.

When the energy level is used as a control parameter (see previous section) it effectively evokes a subset of potential behavioral attractors, belonging to the pool of behavioral attractors embedded within the system during the evolutionary process, by shaping a specific landscape in the state space. According to the general mathematical properties of dynamical systems, the actual attractor then emerges on the basis of the actual starting position and the current effects of noise. In this sense the energy level can be effectively used to implement a sort of non-deterministic action selection mechanism, similar to but different from other suggested mechanisms [14,21] in that this is emergent, dynamic and self-organized to a high degree.

In the normal (evolved) conditions (see Figure 1), all the dynamical subsystems (internal energy level dynamic, neurocontroller, body and environment) are engaged in a mutual regulatory interaction, in order to achieve an effective and viable behavior, self-organized under evolutionary pressure. During the normal dynamic interaction of the bio-regulatory and agent-environment subsystems, the mutual regulation exploits the possibility to use the energy level in order to integrate information over time (remember that the neural controller is implemented as a purely reactive ANN). It should be emphasized that the energy level follows dynamics structurally dissimilar to the parallel interaction of the neurocontroller with its external environment (i.e. characterized by different time scales). As we can clearly see in Figure 4, when the agent is in a condition of energy deficit (upper part of the lower graph) the pool of potential attractors is composed by "exploratory behaviors" (attractors A, B, C), both looping between the two lights and engaging in external loops starting from one of them. Interestingly, attractor B, for it embeds the possibility to engage in a limit cycle around the light source (although expressed with low probability), might be considered potentially maladaptive. On the other hand, when the agent reaches an energy level around 0.8 or slightly below (lower part of the figure) the potential attractors tend to engage in tight loops around the closest light source (attractors F, G, H). Attractors D and E embed both characteristics, and might be considered transitional forms. Therefore, given the dynamics in place, high energy level implies that a source of energy was recently visited and is still in the neighborhood of the agent, whilst low energy has an opposite value and is compatible with exploratory behaviors. In this case, the role that traditional cognitive science tends to attribute to explicit memory, is totally offloaded on the global dynamic of the system. Accordingly, internal bio-regulatory dynamics might be interpreted as supportive of cognitive phenomena.

As mentioned above, we intentionally carried out our experiment in a simple scenario. Such a minimal setup was functional to isolate and emphasize the phe-

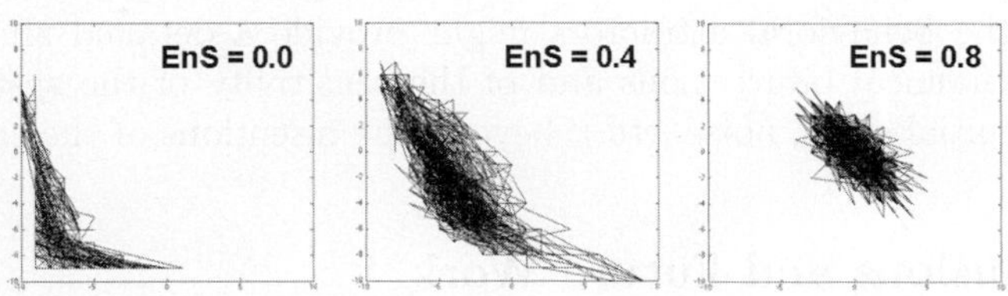

Fig. 5. Patterns of motor activation for different behavioral attractors at different levels of energy. From left to right: energy level = 0.0 (attractor A); energy = 0.4 (attractor C); energy = 0.8 (attractor H). The pattern of activation tends to drift as a function of the energy, from highly negative (at the lowest energy levels) towards a bounded region of slightly positive values (for high energy).

nomenon we intended to study: the holistic nature of cognitive processes and the mutual relationship of internal and external dynamics in their emergence. A more complex system might indeed give rise to more complex dynamics; nevertheless it would blur its causal mechanisms to the extent of creating an indistinguishable conceptual black box. On the contrary, we are allowed to develop a detailed analysis of the ongoing phenomena. For example, it is quite interesting to observe how the pattern of motor activation is modulated by the energy level. At the lowest energy levels, the two motor neurons receive high negative activation, to the point of saturation (Figure 5). On the contrary, high levels of energy cause the inhibition of such vivid activation: in such conditions, the two motors tend to be activated at very low speeds, with a slight positive bias. Therefore, according to a mechanism that is well documented in the specialized literature, the evolutionary algorithm exploits contingent regularities, here in the form of bodily symmetry of the simulated agent that permits it to move in two opposite directions, at the same time relying on two rather well spatially distinct sets of sensors. Thus, the energy level modulates the balance between two opposite behavioral forces, phototaxis (at the lowest energy levels) and photophobia (at the highest ones), aiming for opposite directions and responsible for the generation of the overall dynamics of the system. Incidentally, such a push-pull mechanism is the basic framework for the emergence of deterministic chaos [22].

A final clarification regards our taxonomy of the behavioral attractors, which might be considered somewhat arbitrary. The classification of the attractors was based on their morphological characteristics. Furthermore, their spectral analysis proved extremely useful in order to classify otherwise morphologically similar attractors. For example, attractors F and H, sharing the property of being bound to a limited region in the neighborhood of one of the lights, can be easily disambiguated on the basis of the spectral information carried by the activation of the light sensors (Figure 6). It might be argued that some of the classes considered in Figure 3 might be merged into a single family, as their differences are merely due to the obvious time constraints of our observations. Nevertheless, such a remark would not deeply affect our analysis in principle, since what is actually relevant to our purpose is the general structure of the transition between families of attractors that are qualitatively very different, respectively playing a major role at lower or higher energies. A quantitative analysis of the behavioral attractors might provide a detailed analysis of the occurring dynamical bifurcations and of the sensitivity of the system to local conditions (initial state, noise, etc.), beyond the intentions of the current work.

5 Conclusions and Future Work

With the present study, we show how a simple non-neural state belonging to internal bio-regulatory processes, the energy level, can be used to modulate the dynamical properties of the interaction between a simple simulated robotic agent, controlled by a feedforward artificial neural network and its environment. This dynamical property, emergent and self-organized, is here functional in the task of

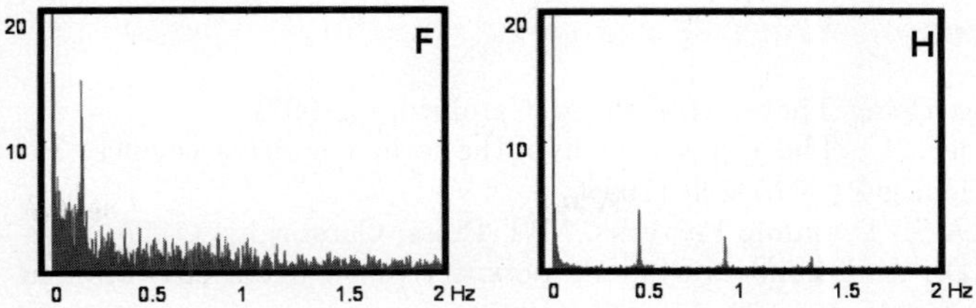

Fig. 6. Spectral analysis of the temporal activation of one of the light sensors, during the occurrence of attractors type F (left) and H (right). The dominance of low frequency components and a more continuous spectrum in the former, contrasted to the development of a highly periodic behavior in the latter (as shown by the isolated spikes centered around the frequency of 0.5 Hz and its multiples), allows a clear partition of their underlying dynamics. Detail of the interval [0 Hz, 2 Hz]; identical scales.

disambiguating identical light sources, on the basis of their capacity to influence the non-neural internal state. Nevertheless, the mechanism we describe should be considered as a general dynamical property and the specific problem merely prototypical for demonstration.

Crucially, the time scale of the energy level, peculiarly different from the time scale of the normal interactions between the agent and its environment, allows the neural controller of the agent to integrate information over time. We argue that internal bio-regulatory dynamics might play a role in supporting cognitive processes. Under this perspective, the non-neural internal state can also be considered as the implementation of an elementary, implicit mechanism of appraisal. In a complementary work [23] we argue that similar mechanisms represent the physiological basis for proto-emotions, in general accordance with somatic theories of emotions [24,25].

The current study has focused on the role that non-neural internal (bodily) states might play in support of cognitive phenomena. The internal dynamic we actually modeled, as well as our crude manipulation of the control parameter, is indeed a very rough simplification with respect to the normal dynamic of the bio-regulatory systems observed in nature. In future works we will try to understand how a more detailed modeling of the non-neural internal mechanisms (i.e. modeling of internal homeostasis) might further enhance the potential of the global dynamical interaction between cognitive agents and their environments. The refinement of our dynamical analysis, so to identify the exact conditions for the occurrence of dynamical bifurcations and the dynamical sensitivity to perturbations, might also be relevant for a deeper comprehension of the system.

Acknowledgements. This work has been supported by a European Commission grant to the project "Integrating Cognition, Emotion and Autonomy" (ICEA, www.iceaproject.eu IST-027819,) as part of the European Cognitive Systems initiative.

References

1. Clark, A.: Being There. MIT Press, Cambridge (1997)
2. van Gelder, T.: The dynamical hypothesis in cognitive science. Behavioral and Brain Sciences 21, 615–628 (1998)
3. Kelso, J.A.S.: Dynamic Patterns. MIT Press, Cambridge (1995)
4. Beer, R.D.: A dynamical systems perspective on agent-environment interaction. Artificial Intelligence 72, 173–215 (1995)
5. Chiel, H., Beer, R.: The brain has a body. Trends in Neurosciences 20(12), 553–557 (1997)
6. Pfeifer, R., Scheier, C.: Understanding Intelligence. MIT Press, Cambridge (1999)
7. Pfeifer, R.: Dynamics, morphology, and materials in the emergence of cognition. In: Burgard, W., Christaller, T., Cremers, A.B. (eds.) KI-99: Advances in Artificial Intelligence. LNCS (LNAI), vol. 1701, pp. 27–44. Springer, Heidelberg (1999)
8. Parisi, D.: Internal robotics. Connection Science 16(4), 325–338 (2004)
9. Ashby, W.R.: Design for a Brain. Chapman & Hall, Boca Raton (1960)
10. Meyer, J.A.: The animat approach to cognitive science. In: Roitblat, H.L., et al. (eds.) Comparative Approaches to Cognitive Science, pp. 27–44. MIT Press, Cambridge (1995)
11. Avila-García, O., Cañamero, L.: Using hormonal feedback to modulate action selection in a competitive scenario. In: Schaal, S., et al. (eds.) From animals to animats 8, pp. 243–252. MIT Press, Cambridge (2004)
12. Gadanho, S.C., Hallam, J.: Robot learning driven by emotions. Adaptive Behavior 9(1), 42–64 (2001)
13. Cañamero, D.: Modeling motivations and emotions as a basis for intelligent behavior. In: Proceedings of the first international conference on Autonomous agents, pp. 148–155. ACM Press, New York (1997)
14. Cañamero, L.: Designing emotions for activity selection in autonomous agents. In: Trappl, R., et al. (eds.) Autonomous Agents, pp. 115–148. MIT Press, Cambridge (2003)
15. Maes, P.: A bottom-up mechanism for behavior selection in an artificial creature. In: A., J., et al. (eds.) From animals to animats, pp. 238–246. MIT Press, Cambridge (1991)
16. Nolfi, S.: Using emergent modularity to develop control systems for mobile robots. Adaptive Behavior 5(3-4), 343–363 (1997)
17. Ziemke, T.: On 'parts' and 'wholes' of adaptive behavior. In: From animals to animats 6 (2000)
18. Ito, M., Tani, J.: On-line imitative interaction with a humanoid robot using a dynamic neural network model of a mirror system. Adaptive Behavior 12, 93–115 (2004)
19. Tani, J.: Learning to generate articulated behavior through the bottom-up and the top-down interaction processes. Neural Networks 16, 11–23 (2003)
20. Nolfi, S., Floreano, D.: Evolutionary Robotics. MIT Press, Cambridge (2000)
21. Prescott, T., Redgrave, P., Gurney, K.: Layered control architectures in robots and vertebrates. Adaptive Behavior 7(1), 99–127 (1999)
22. Strogatz, S.H.: Nonlinear Dynamics and Chaos. Westview Press, Cambridge (1994)
23. Herrera, C., Ziemke, T., Moffat, D.: Emotions as a bridge to the environment. In: Nolfi, S., Baldassarre, G., Calabretta, R., Hallam, J.C.T., Marocco, D., Meyer, J.-A., Miglino, O., Parisi, D. (eds.) SAB 2006. LNCS (LNAI), vol. 4095, pp. 3–16. Springer, Heidelberg (2006)
24. Damasio, A.: Descartes'error. G.P. Putnam's Sons, New York (1994)
25. Prinz, J.: Embodied emotions. In: Solomon, R.C. (ed.) Thinking about Feeling, pp. 44–59. Oxford University Press, Oxford (2004)

Artificial Emotions: Are We Ready for Them?

Jackeline Spinola de Freitas and João Queiroz

Department of Computer Engineering and Industrial Automation
School of Electrical and Computer Engineering
State University of Campinas
P.O. Box 6101 – 13083-852 Campinas, SP - Brazil
{jspinola, queirozj}@dca.fee.unicamp.br

Abstract. Recent research in psychology and cognitive neuroscience are increasingly showing how emotion plays a crucial role in cognitive processes. Gradually, this knowledge is being used in Artificial Intelligence and Artificial Life areas in simulation and cognitive processes modeling. However, it lacks a theoretical framework that allows them to deal with emotion. In addiction, regarding emotion-based computational projects, controversial questions concerning the nature, function, and mechanisms of emotions, that must be considered, are mostly neglected on researches. The objective of this article is to discuss some of these problems and to present references that can be useful in their solution.

Key words: Emotion, Artificial Intelligence, Computational Simulation.

1 Introduction

Recent findings in neuroscience regarding the mechanisms, functions, and nature of emotions ([1], [2], [3], [4], [5], [6], [7], [8], [9]) have been attracting the attention of researchers in Computer Science and Artificial Intelligence (AI). It is believed that emotions play a significant role in diverse cognitive processes, and are essential for problem solving and decision making. Despite Darwin's indications made in [10] that emotions are phenomena important for survival, only recently has the association of emotion with reason and logical behavior in human beings, been reviewed [9], [11], [12].

Currently, AI and Artificial Life rely on the emotion aspects that are crucial to model perception, learning, decision processes, memory, and other functions. In Computer Science, there are two branches of research exhibiting interest in emotion aspects. The first, Human-Computer Interaction (HCI), concentrates its attention on the interactions between the user (human) and the machines, and considers possible optimizations of this relationship. The main objective pursued by its researchers is the development of engineering tools to measure, model, and provide responses to human emotions through sensors, algorithms, and hardware devices. Affective Computing, coined by [13], is the term used to classify projects in this category. The most notable authors in HCI are: [13], [14], [15], [16] and the Labmedia Massachusetts Institute of Technology research group. Some of their projects include the Socially Intelligent Character, Embodied Conversational Agents, and Story Telling Systems. An example of a successful commercial project is Sony AIBO (http://www.sony.net/Products/aibo).

F. Almeida e Costa et al. (Eds.): ECAL 2007, LNAI 4648, pp. 223 – 232, 2007.

The second branch of research involves Intelligent Agents systems, which have their internal architectures based on emotions (emotion-based systems). These systems' models are biologically inspired processes described and studied by neuroscientists [11], [17], [18], [19], [20], [21], [22], [23]. Its main objective is to emulate emotion process in agents' behavior. In general, emotion-based projects aim to improve systems performance in decision making, action selection, behavior control, trustworthiness, and autonomy.

Since both of the above mentioned branches of research are new, their projects still confront basic problems. We are particularly interested in the second research branch, and we intend to discuss some of the problems faced by it. Looking at the more developed emotion-based computational systems projects, it can be said that the construction of such systems is far from trivial. In AI, to approach emotion concepts is as problematic and complex as life approach in Artificial Life. We have noticed that the problems can be clustered into two groups. *First, it lacks a scientific approach structure (theoretical framework) to deal with Artificial Emotion (AE).* This is a known limitation and many authors have suggested ways of dealing with it (e.g., [6], [12], [17], [24], [25], [26], [27], [28], [29], [30]). However, there are few projects that had shown acceptable trustworthiness to be followed. As a direct consequence of the lack of an appropriate framework to model emotion, we see the lack of comparisons, not only between distinct projects, but also within the same project, to contrast the results obtained from emotion-based and non emotion-based experiments. *Second*, a close look at *projects reports gives us a non-exhaustive list of important questions they should face* to get more trustworthy results.

To present these problems, but also to provide an indication of possible solutions, this work is organized in order to present, in our viewpoint, a significant list of the more important theoretical references in the diverse subjects displayed. The next section, for example, emphasizes the lack of adoption of a framework and suggests a possible approach. It also criticizes the lack of comparisons between projects. Section 3 suggests basic questions that current projects do not tackle and present possible directions to address them. Section 4 is dedicated to the final comments.

2 Framework for Emotion-Based Projects

The difficulty to elaborate a framework to deal with AE, has done that hitherto none approach has proved being notably distinct and superior to others. Maybe the difficulty in the elaboration of a theoretical framework is associated with, current lack of information about emotion and its relation with other subsystems. As indicated in [22], it can be questioned if emotions, in artificial systems, have a potential role in action selection (what agent should do, based on its emotional state), adaptation (behavior alterations), social regulation (communication, exchange of information), sensory integration, in alarm mechanisms (similar reactions in consequence of critical situations for the agent), motivation, objectives management (creation of new goals or priorities changes), learning, focus of attention (selection of data to be process, based on emotional evaluation), memory control, strategic processing and internal model (what emotional state represents to the agent).

The elaboration of a model that is capable to control simultaneously all these functions can be an extremely complex task to currently developing projects. Suggestions have been proposed by [5], [6], [24], [25] and [30], based on the hypothesis that emotions have adaptive purposes. All of them agree that it is necessary to follow a functional view of emotions presented in natural systems, in which we will concentrate on.

[24] propose a functional view on the role of emotions and its implications for an emotional intelligent agent project. The author believes that: (i) the emotion has an adaptive purpose and can contribute to improve the agent survival capacity and (ii) AE models must be based on emotion functions abstractions, and must be used to generate agent behavior. Among the functions that can be attributed to "emotions" and that allow computational implementations, the author mentions: (i) to assign event relevance according to system objectives, (ii) to detect the difficulties in the solution of the problems that impede reaching these objectives, (iii) to define priorities and resources according to goals and (iv) to allow parallel processing that guarantees system' objectives. To [24], the models that establish a linkage between emotion, motivation and behavior, provided by a synthetic physiological structure, can allow conclusions as those found in her experiment [6]. In that case, it can be observed that an agent presents either an oriented behavior (goal or objective) or an opportunistic behavior, when variables associated to physiological structures (functional) are used as regulatory mechanisms of its behavior.

One notable previous study [30] suggests that the understanding of the brain structure and the neural mechanisms embedded in neuromodulation[1], present in human and other animals emotions led researchers ([1]; [2]; [3]; [31]) to believe that it is possible to "abstract from biology a functional characterization of emotion" ([30]). These researchers affirm that the interactions between amygdale and prefrontal cortex, and its influence in emotion generation are already well known, but they admit that the way computational systems can "take advantage of it remains an open question" ([30]).

In [25]' functional view it is necessary to understand emotions as dynamic patterns of neuromodulation, instead of patterns of neural activity, as it is currently considered. When compared to the most classic viewpoint, centered in neural circuits [1], [2], [3], it is a distinct approach to characterize emotions origin and emotions impact in behavior and cognitive processes. Based on [32]' analysis of motivation and emotion, [30] propose that the dynamic of neuromodulators systems serve as an inspiration to generate the functional structure for emotion-based architectures. Even though [30] emphasize that "emotions are, of course, far more complex than a few brain structures" and interaction of neuromodulators systems, authors believe that if systems whose architecture contained functional characteristics were used in computational simulations, it would be possible to modify its parameters to generate a behavior analogous to the one generated in an emotional state.

The results from a previous study [25] provides a list of functional properties of the emotion, such as the change in the autonomous and endocrinal systems, impulse of motivational behavior, communication, social interaction, improvement of the survival capacity, and better memory storage and recovery. Although the author proposes a functional vision, his study also lists a few functions of emotions that "have natural robotics counterparts" [25]. The researcher affirms that it is possible to endow a robot

---

[1] http://www.neuromodulation.com/

or a computational system with characteristics that can be functionally related to emotions, making the elaboration of a project or experiment much easier. That is the first hurdle in the challenge of including emotion associated functions, useful in the functioning and performance of the system, in computational models. In any given project, establishing a restriction on the communication between two interfering agents in order to get a desired emotional behavior out of the system could be an undesirable limitation. In such a case, it would be better not to apply the restriction, even though it may cause the system's behavior to seem unnatural.

Although complex, the functional framework seems computationally implementable and similar to natural systems and can produce convincing emotion-based projects. As examples, we can mention some projects that abstract physiological components from the animals, as hormonal levels: [21], [26] and [33]. Even though, these authors assume that they are not worried about the plausibility (bio-inspiration) of these systems. Also, in [22], author says that in the emotional control of systems processes he does not claim for any "biological plausibility of the employed states or mechanisms" This can be seen as an initial approach of the emotion concept, and can help to test theories about psychological processes [12]. Certainly, some projects (e.g. [22]) observe the phenomenon, comparing it with what could be classified as emotion.

Even though the use of concepts and knowledge related to emotion does not mean to endow a computational system with emotional characteristics, we believe it is necessary to maintain somehow any plausibility, in a way that it does not be just a programmer who works metaphorically, attributing concepts related to human emotion to computational internal variables [34]. In fact, some projects deserve a severe critic related to the terms they use, assigning, many times, emotions names to system parameters without any abstraction of natural systems. As [24] and [35] question, it is difficult to establish a limit between a 'genuine emergent' behavior and the one conferred by a tendency of an observer anthropomorphism.

It is possible that the lack of an appropriate framework to model emotion is the reason why we scarcely see comparisons between distinct projects or within the same project. A systematic analysis of the projects is necessary to get progress in artificial emotion research [24]. Interactions between the experiments could be useful to compare and to discuss different architectures and eventually, to benefit projects in progress, generating more expressive results in lesser time. Also, they can be useful in providing insights on the way research is being executed, and on the use of different approaches. Comparisons between emotion-based and non emotion-based architectures, in replicated experimental protocols, can be an efficient way to validate conclusions.

3 Some Questions to Emotion-Based Computational Systems and Possible Ways to Address Them

As a new research area, we can notice that there are much more questions than solved problems, a fact that can be seen as an opportunity for new proposals. To answer them, as [34] suggests, the efficient tradition of basic scientific research, to return to the premises and initial principles in order to better understand the subject must be followed. Finally, the development of computational implementations can help to solve or supply new ideas to them. The proposed questions for emotion-based computational systems can be grouped in two types, related to (i) theoretical-conceptual problems and (ii) computational problems.

Initially, scientific communities do not have a consensual definition of emotion. Questions related to the origins and functions of emotions and their relation with other affective processes (motivation, mood, attitudes, values, temperament, etc), also seems hardly consensual. These facts suggest that an understanding of the mechanisms involved in the emotion phenomenon may limit the development of emotion-based systems. A solution to surpass them is strictly functional: instead of "what emotions are", we must concentrate on "what emotions are for" [25].

To create emotion-based architectures and projects, a list of important questions must include: how many and which emotions must be selected? Is it possible to have a feasible model that considers the co-occurrence of artificial emotions? 'How many' and 'which' emotions one might select are questions on which there is no consensus. Four 'basic emotions' are used in many systems: happiness, sadness, anger and fear (see [11], [21]). According to [36] this number must be bigger (15): happiness, sadness, anger, boredom, negation, hope, fear, interest, contempt, disgust, frustration, surprise, pride, shame and guilt. [7] believes that eight emotions must be used. They are classified by the author as "primary emotions": happiness, sadness, acceptance, anger, fear, disgust, anticipation and surprise. Contrarily, [37] argue that from the theoretical point of view it is a fallacy to compare affective states with a predetermined number of basic emotions. Strengthening this argument, Paul Ekman, the creator of the term "basic emotions", has admitted later on that there aren't non-basic emotions [38]. Due to the need of a complex control system, some projects (e.g. [22], [39]) seem inclined to select only one or two emotions. The most parsimonious suggestion is seen in [24]: "Do not put more emotion in your system than what is required by the complexity of the system-environment interaction".

What is the relation between emotion and other subsystems? How to integrate it with other mechanisms (sensory, learning, selection, reaction and communication)? [20] question the lack of integration between emotions and other systems (cognitive, language, memory, etc), and how it harms the attainment of better global results. [24] affirms that due to the possibility of interaction between emotional mechanisms and diverse cognitive subsystems, such as the physiological, the cognitive-evaluative and the communicative-expressive, it is possible to get an interesting solution to improve system agent's performance. However the author questions the need of such complexity, and the current plausibility of its implementation.

According to recent research in psychology [40] and neuroscience [2], [3] and [9], emotions are processes that control cognition and action, and manage our mental models. In its basis, we could ask if its shortfalls compromise cognitive abilities in autonomous agents. For [25] it is clear that "emotions have co-evolved with perceptual, cognitive and motor abilities." and that affect "all levels of functions, from low-level motor control to planning and high-level cognition". In this sense, comparisons between emotion-based projects and no emotional ones can be useful to supply us with information about the connections that are established between emotion and cognition. Also, it serves to verify if the inclusion of emotion improves or not cognitive abilities in computational systems.

Some questions that might be especially interesting in Artificial Life projects are associated with emergent phenomena [22], [24], [41]: Can Artificial Emotion be an emergent property? If "yes", how architecture design can influence on the emergency of complex actions in emotion-based agents? [24] affirms that it is possible that an emotion emerges from an agent, and that this is a feasible way to investigate the role emotions plays in agent-environment interaction, in distinct complexity levels. The author suggests that, to prevent problems, some functional equivalence between agent characteristics and its environment must be preserved. In this context, the problems are related to anthropomorphism tendency but, indeed, since an artificial system can represent existing natural models, it can be difficult to affirm "why and when emotional behavior arises" [24]. Besides that, one can question if the system will provide enough mechanisms to explain and justify a supposed emergency of emotional behavior [42].

[23], [25] affirm that typical explanations for the emotion function are based on the flexibility of agent behavior response to its environment. [5] defines the core of an emotion as the disposal to act in a pre-defined way. These theories are motivations to use behavior as a phenomenon to measure emotion. In fact, due to the lack of formal theories that describe the non-observable emotion subjective process [21] or intuitive parameters [42], many experiments [23], [26], [43] identify the emotion through any observable resultant behavior.

Probably, one of the most frequent and investigated questions is: emotional processes need to be related to an embodied entity? It is possible that the motivation to find its answer has appeared due to a recent change in the traditional view that intelligence is an abstract process that could be studied without taking into consideration the physical aspects of natural systems [44]. Great part of current research defends that, to obtain intelligent behavior it is necessary that an agent is situated in the environment and interacts with it. Besides being important for intelligence, in light of [2], [3] and [9], the body is essential for the emotion. [41] defends a contrary argument when the question involves a computational apparatus. As an intermediary viewpoint, [18] believes that "emotion systems involved in feedback control of situated agents may serve to provide the grounding for embodied agents in the body/environment coupling".

Related to computational problems it is possible to identify other questions. Some problems may be related to the wrong way we program agents' systems. Related to system architecture, one can ask: what type of data structure and computational mechanisms might be used to capture and represent the complexity of emotion processes? What architecture models are the best for the comparison of agent performance? The available programming languages and program syntax, with algorithms conceived by a program designer, in a great part of the cases, limits or disables any type of code evolution. It also limits that the agent can develop, autonomously, parts of its architecture during environment interactions. [45] affirms that it is fundamental to surpass the challenges of identifying the appropriate methods of information codification to produce a process of incremental growth. In order to get incontestable resulted, what type of experimental test allows better exploration of emotion-based models? Particularly, we feel that there is something missing related to computational tools to represent emotion phenomenon. While we do not get new advances in terms of algorithms or computational tools, we must concentrate on the modeling of the emotion, to the point not to lose the most relevant characteristics, and not to produce an extremely complex scenario that does not allow a computational representation of the phenomenon.

As we have said at the beginning, this section shows a not-exhaustive list of problems faced for emotion-based systems projects. Answering them is a complicated task, since it requires concepts that are not yet entirely understood and theories that are not well established [24], [46]. The notable complexity of a projected system to take care of such interdisciplinary problems, including computational parameters necessary to control their multiple factors, is probable one of the reasons AE research faces little development when compared with other equally novel areas. Many of these questions are probably made at the beginning of project developments but curiously, are not the focus of publications in the AI and Artificial Life areas. This difficulty can be one of the reasons for which we rarely see references to previous approaches in experimental projects.

Certainly, tough questions require a broad and deep multidisciplinary background or a research group that might include psychologists, ethologists, neuroscientists, computer scientists, software engineers and philosophers. Even though it does not guarantee that it is possible to have a single model that responds to the majority of questions, attempts to answer these questions can also serve to show other limitations that emotion-based research might face, helping to surpass them.

4 Final Comments

Even though currently available knowledge about emotions has led AI and Alife researchers to propose models of emotion-based systems, an essential question is still left to be answered; that is, to what extent can researchers abstract and model the supposed structural complexity involved in emotion phenomena. Indeed, the lack of appropriate frameworks for common reflection and standards for a sound validation practice is a restriction that needs to be surpassed ([47]).

Constructive critics and projects comparisons commonly made in any research field can be a beneficial counterpart to experimental progress and development. Our hope is that the development of computational models of emotion becomes a core research focus for Artificial Intelligence, and soon we can see many advances in such systems.

As more neuroscience findings are published, it will become easier to construct emotion-based agent systems. Computational projects with a specific focus will be able to extend their scope to include subsystems hardwired to specific emotions. On the other hand, conducting more emotion-based computational experiments will better our knowledge of unknown mind functions and provide a test-bed for theories of biological emotion ([30]).

The extent to which researches in AI and Alife will improve our understanding on mind phenomena and allow us to develop new robust and trustworthy artifacts will depend on the extent to which we will be able to answer the remaining open questions.

Positively, overcoming this challenges can be an important step in that field's progress beyond engineering applications towards a more scientific discipline ([47]), allowing that the answer to our title is a 'yes'.

Acknowledgments. João Queiroz is sponsored by FAPESB/CNPq. Jackeline Spinola and João Queiroz would like to thank the Brazilian National Research Council (CNPq) and The State of Bahia Research Foundation (FAPESB).

References

1. Ledoux, J.: The emotional brain: the mysterious underpinnings of emotional life. Touchstone, New York (1996)
2. Damásio, A.R.: Descartes'error: emotion, reason and the human brain. Avon books, New York (1994)
3. Damásio, A.R., Grabowski, T., Bechara, A., Damásio, H., Ponto, L.L., Parvizi, J., Hichwa, R.D.: Subcortical and cortical brain activity during the feeling of self-generated emotions. Nature Neuroscience 3(10), 1049–1056 (2000)
4. Nesse, R.M.: Computer emotions and mental software. Social Neuroscience Bulletin 7(2), 36–37 (1994)
5. Frijda, N.H.: The place of appraisal in emotion. Cognition and Emotion 7, 357–387 (1993)
6. Cañamero, D.: Modeling Motivations and Emotions as a Basis for Intelligent Behavior. In: Proceedings of 1st International Conference on Autonomous Agents, Marina Del Rey, California, pp. 148–155. ACM, New York (1997)
7. Plutchik, R.: Emotion: a psycho evolutionary synthesis. Harper & Row, New York (1980)
8. Ghiselin, M.T.: Darwin and Evolutionary Psychology. Science 179, 964–968 (1973)
9. Damásio, A.R.: Emotion and the Human Brain. Annals of the New York Academy of Sciences 935, 101–106 (2001)
10. Darwin, C.: The Expression of the Emotions in Man and Animals. University of Chicago Press, Chicago (1872/1965)
11. McCauley, T.L., Franklin, S.: An architecture for emotion. In: Cañanero, D. (ed.) Proceedings of the 1998 AAAI Fall Symposium on Emotional and Intelligent: The Tangled Knot of Cognition. Technical Report FS-98-03, pp. 122–127. AAAI Press, Menlo Park (1998)
12. Gratch, J., Marsella, S.: A domain independent framework for modeling emotion. Journal of Cognitive Systems Research 5, 269–306 (2004)
13. Picard, R.: Affective Computing. MIT Press, Cambridge (1997)
14. Reilly, W.S., Bates, J. (1992) Building Emotional Agents. Technical Report CMU-CS-92-143, School of Computer Science, Carnegie Mellon University, Pittsburgh, PA (1992)
15. Elliott, C.: Hunting for the holy grail with "emotionally intelligent" virtual actors. SIGART Bull. 9(1), 20–28 (1998)
16. Cassell, J.: Oral tradition, aboral coordination: building rapport with embodied conversational agents. In: Proceedings of the 10th international Conference on intelligent User interfaces, San Diego, California, ACM Press, New York (2005)
17. Velásquez, J.: Modeling emotion-based decision-making. In: Cañanero, D. (ed.) Proceedings of the 1998 AAAI Fall Symposium. Emotional and Intelligent: the Tangled Knot of Cognition, pp. 164–169. AAAI Press, Menlo Park (1998)
18. Nehaniv, C.: The First, Second, and Third Person Emotions: Grounding Adaptation in a Biological and Social World. In: Numaoka, C., Cañamero, D., Petta, P. (eds.) Grounding Emotions in Adaptive Systems. 5th International Conference of the Society for Adaptive Behavior (SAB'98), pp. 43–47. University of Zurich, Switzerland (1998)
19. Custódio, L., Ventura, R., Pinto-Ferreira, C.: Artificial emotions and emotion-based control systems. In: Proceedings of 7th IEEE International Conference on Emerging Technologies and Factory Automation, vol. 2, pp. 1415–1420. IEEE, Los Alamitos (1999)
20. Petta, P., Cañamero, D. (eds.): Grounding emotions in adaptive systems: volume II. Cybernetics and systems: an international journal 32(6), 581–583 (2001)
21. Gadanho, S.C., Hallam, J.: Emotion-triggered learning in autonomous robot control. Cybernetics and Systems: an International Journal 32(5), 531–559 (2001)
22. Scheutz, M.: Useful roles of emotions in artificial agents: a case study from artificial life. In: Proceedings of AAAI 2004, pp. 42–48. AAAI press, Menlo Park (2004)
23. Kato, T., Arita, T.: Evolutionary Simulations based on a Robotic Approach to Emotion. In: The 10th International Symposium on Artificial Life and Robotics, Oita, Japan, pp. 258–261 (2005)

24. Cañamero, D.: Emotions and adaptation in autonomous agents: a design perspective. Cybernetics and systems: International Journal 32, 507–529 (2001)
25. Fellous, J-M.: From Human Emotions to Robot Emotions. In: Hudlicka, E., Cañamero, D. (eds.) Architectures for Modeling Emotions: Cross-Disciplinary Foundations. AAAI Spring Symposium, pp. 37–47. AAAI Press, Menlo Park (2004)
26. Gomi, T., Ulvr, J.: Artificial Emotions as Emergent Phenomena. In: Proceedings of 2nd IEEE Workshop on Robot and Human Communication, Tokyo, Japan, pp. 420–425. IEEE Computer Society Press, Los Alamitos (1993)
27. Velásquez, J.: When Robots Weep: Emotional Memories and Decision-Making. In: Proceedings of the 15th National Conference on Artificial Intelligence, Madison, WI, pp. 70–75 (1998)
28. Staller, A., Petta, P.: Towards a Tractable Appraisal-Based Architecture for Situated Cognizers. In: Numaoka, C., Cañamero, D., Petta, P. (eds.) Grounding Emotions in Adaptive Systems. 5th International Conference of the Society for Adaptive Behavior (SAB'98), August 1998, pp. 56–61. University of Zurich, Switzerland (1998)
29. Sloman, A.: How many separately evolved emotional beasties live within us? In: Trappl, R., Petta, P., Payr, S. (eds.) Emotions in Humans and Artifacts, pp. 35–114. MIT Press, Cambridge (2002)
30. Arbib, M.A., Fellous, J-M.: Emotions: from brain to robot. Trends in Cognitive Sciences 8(12), 554–561 (2004)
31. Dalgleish, T.: The emotional brain. Nature Reviews Neuroscience 5, 582–589 (2004)
32. Kelley, A.E.: Neurochemical networks encoding emotion and motivation: an evolutionary perspective. In: Fellous, J-M., Arbib, M.A. (eds.) Who Needs Emotions? The Brain Meets the Robot, pp. 29–77. Oxford University Press, New York (2005)
33. Gadanho, S.C.: Learning behavior-selection by emotions and cognition in a multi-goal robot task. Journal of Machine Learning Research 4, 385–412 (2003)
34. Arzi-Gonczarowski, Z.: AI Emotions: Will One Know Them When One Sees Them? Agent Construction and Emotions. In: Trappl, R. (ed.) Cybernetics and Systems, pp. 739–744. Austrian Society for Cybernetcs Studies, Vienna, Austria (2002)
35. Grand, S., Cliff, D., Malhotra, A.: Creatures: Artificial Life Autonomous Soft-ware Agents for Home Entertainment. In: Proceedings of 1st International Conference on Autonomous Agents, Marina Del Rey, California, pp. 22–29. ACM, New York (1997)
36. Ortony, A., Clore, G.L., Collins, A.: The cognitive structure of emotions. Cambridge University Press, New York (1988)
37. Petta, P., Trappl, R.: Emotions and agents. In: Luck, M., Mařík, V., Štěpánková, O., Trappl, R. (eds.) ACAI 2001 and EASSS 2001. LNCS (LNAI), vol. 2086, pp. 301–316. Springer, Heidelberg (2001)
38. Ekman, P.: Basic Emotions. In: Dalgleish, T., Power, T. (eds.) The Handbook of Cognition and Emotion, pp. 45–60. John Wiley & Sons, Sussex (1999)
39. Delgado-Mata, C., Aylett, R.S.: Emotion and Action Selection: Regulating the Collective Behaviour of Agents in Virtual Environments. In: 3rd International Joint Conference on Autonomous Agents and Multiagent Systems, vol. 3, pp. 1304–1305 (2004)
40. Oatley, K.: Emotions. In: Wilson, R.A., Keil, F.C. (eds.) The Massachusetts Institute of Technology - Encyclopedia of the Cognitive Sciences, pp. 273–275. MIT Press, Cambridge (1999)
41. Sloman, A., Chrisley, R., Scheutz, M. (2003) Who needs emotions? The brain meets the machine. In: Arbib, M. and Fellous, J. (eds.) Oxford University Press, New York (2003)
42. Wehrle, T.: Motivations behind modeling emotional agents: whose emotion does your robot have? In: Numaoka, C., Cañamero, D., Petta, P. (eds.) Grounding emotions in adaptive systems. 5th International Conference of the Society for Adaptive Behavior (SAB'98), August, 1998, University of Zurich, Switzerland (1998)

43. Kitamura, T.: An Architecture of Behavior Selection Grounding Emotions. In: Numaoka, C., Cañamero, D., Petta, P. (eds.) Grounding Emotions in Adaptive Systems. 5th International Conference of the Society for Adaptive Behavior (SAB'98), August, 1998, University of Zurich, Switzerland (1998)
44. Pfeifer, R., Scheier, C.: Understanding Intelligence. MIT Press, Cambridge (1999)
45. Nolfi, S., Floreano, D.: Synthesis of autonomous robots through artificial evolution. Trends in Cognitive Sciences 1, 31–37 (2002)
46. Sloman, A.: What are emotion theories about? In: Hudlicka, E., Cañamero, D. (eds.) Architectures for Modeling Emotions: Cross-Disciplinary Foundations. AAAI Spring Symposium, pp. 128–134. AAAI Press, Menlo Park (2004)
47. AAAI press. Architectures for modeling emotion: cross-disciplinary foundations. In Hudlicka, E. and Cañamero, D. (eds.) 2004 AAAI Spring Symposium Technical Report (2004) Available in: http://www.aaai.org/Press/Reports/Symposia/Spring/ss-04-02.php Accessed in 11/05/2006
48. Labmedia – Home page of the research group of the Massachusetts Institute of Technology: http://affect.media.mit.edu/projects.php

Evolution of an Adaptive Sleep Response in Digital Organisms

Benjamin E. Beckmann, Philip K. McKinley, and Charles Ofria

Department of Computer Science and Engineering
Michigan State University
East Lansing, MI 48824, USA
{beckma24, mckinley, ofria}@msu.edu

Abstract. Adaptive responses to resource availability are common in natural systems. In this paper we explore one possible evolutionary cause of adaptive sleep/wake behavior. We subjected populations of digital organisms to an environment with a slowly diminishing resource and recorded their ability to adapt to the changing environment using sleep. We also quantified the selective pressure not to sleep in this competitive environment. We observed that diminishing resource availability can promote adaptive sleep responses in digital organisms even when there is an opportunity cost associated with sleeping.

Key words: Digital evolution, digital organism, AVIDA, adaptive behavior, sleep, resource availability.

1 Introduction

A population of organisms in an environment where a resource is always available can be non-adaptive and function exceptionally well. There is little or no selective pressure on the organisms to adjust their behavior within this environment since resources are plentiful and can be consumed at any time [1]. If resources often become diminished or unavailable, an adaptive response might allow for more conservative resource usage [2] or increased energy storage [3]. Natural organisms often display adaptive behavior that coincides with environmental changes where resources fluctuate [4,5]. An example of this type of adaptive response occurs in nocturnal rodents and insects that sleep during the day and forage for food under the cover of darkness. Animals that hibernate also display an adaptability that allows them to avoid extended periods of low resource availability by increasing the size of their fat stores prior to hibernation [6].

This form of adaptive behavior in natural organisms serves multiple purposes. During sleep periods an animal rests [7], reprograms its brain [8] and performs internal maintenance tasks [9]. However, while an animal is in a state of slumber it is less aware of its environment. How could resource-aware adaptive behaviors, such as sleep and hibernation, have evolved in competitive environments where torpid organisms are vulnerable to active organisms? Is there a selective pressure to sleep caused by resource limitations in environments with periodic resource

F. Almeida e Costa et al. (Eds.): ECAL 2007, LNAI 4648, pp. 233–242, 2007.

availability? The remainder of this paper attempts to answer these questions through experiments with digital organisms.

Previous work has been done in this area using neural networks [10]. In [10], the organisms were subjected to two different environments with periodic light availability, where the organism's ability to find a resource was impaired relative to the current light intensity. It was shown that the combination of a biological clock and light sensor produced the best results in an environment where light readings may not correctly disambiguate day from night. The work presented in this paper differs for [10] in that it does not impose a predefined structure on the organisms, provide a common starting point to the organisms, or give any information, ambiguous or not, to the organisms directly. All of these mechanisms must be evolved while preserving an organism's ability to self-replicate and while avoiding other detrimental behavioral changes. We begin with a brief overview of the AVIDA digital evolution platform [11] and the experimental setup, followed by presentation of the experimental results.

2 AVIDA

AVIDA is an experimental software platform for research in computational evolutionary biology [11]. In the past several years, AVIDA has been used to conduct pioneering research on the evolution of biocomplexity [12,13,14]. AVIDA provides researchers with tools to study the evolutionary process in greater detail and less time than previously possible.

In an AVIDA experiment, self-replicating digital organisms compete against each other in a fixed-size steady-state population. As shown in Fig. 1, each organism resides in a *cell* (one organism per cell) and comprises a circular list of assembly-like instructions (its genome) and a virtual CPU capable of executing those instructions. Cells are organized according to a topology; in this study we used a two-dimensional grid. Every virtual CPU has three 32-bit registers (AX, BX, and CX) and two stacks capable of storing up to ten 32-bit numbers. The virtual CPU has an instruction pointer (IP) that determines which instruction in an organism's genome to execute. The IP can be moved throughout the genome with the use of conditional if-statements and explicit move instructions. While the AVIDA instruction set is a Turing complete language, only basic computational instructions are available and complex computations must be constructed by combining simple instructions (i.e, NAND, INC, and ADD) with the input/output instruction.

We record all input and output to and from each organism in the population and examine them to determine the computational *tasks* performed. An example task is the bitwise-AND of two numbers [13]. To complete this task an organism must read in two numbers and output the bitwise-AND of those two numbers sometime in the future. We added an energy model to AVIDA that allows an organism to obtain and store energy. Once a task has been completed, an *energy* reward is added to the organism's current energy. In this study, the size of the energy reward is subject to the availability of *resources* in the environment.

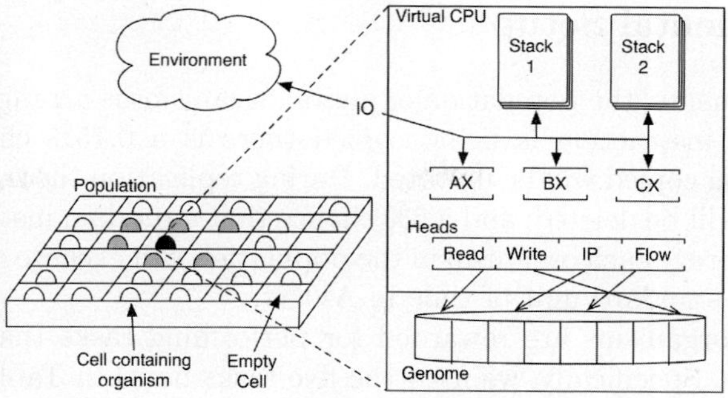

Fig. 1. AVIDA population and structure of a single organism.

Organisms can sense the quantity of resources within the environment. The more plentiful the resource, the larger the reward for performing a task.

An organism's current energy level is used to determine its metabolic rate, as shown in (1). AVIDA uses a probabilistic scheduler to assign virtual CPU cycles to organisms in the population. Organisms with higher metabolic rates are assigned higher priority within the scheduler, and therefore execute more instructions relative to organisms with lower metabolic rates. The metabolic rate is inversely proportional to a user-defined variable, $InstructionsBeforeZeroEnergy$, which specifies how many instructions an organism can execute before it runs out of energy, given no new energy influx. Probabilistically, organisms with a higher metabolic rate will execute more instructions and produce more offspring than those with less energy.

$$MetabolicRate = \frac{Energy}{InstructionsBeforeZeroEnergy} \tag{1}$$

AVIDA organisms are responsible for their own replication through the use of replication-specific instructions. To reproduce, an organism must perform three distinct functions: allocate space at the end of its genome for its offspring's genome, duplicate its own genome instruction by instruction into that space, and divide the resulting genome into two parts. Upon division, the parent organism's state is reset, the parent's energy is divided equally between itself and its offspring, and the offspring's genome is used to create a new organism. The offspring is placed in a random cell in the grid, replacing and terminating any organism that previously occupied that location. Variation among organisms in the population occurs when instructions are copied. Each copied instruction is subject to three types of mutation (modifying the instruction, deleting the instruction, or inserting an additional instruction) that occur at user defined rates. Replication is asexual, and therefore every AVIDA run presented here begins with the same single organism that serves as an ancestor for all successive organisms in the population. Each run is started with a different random number seed, resulting in different evolutionary paths taken by the population.

3 Experimental Setup

In these experiments, the population of digital organisms is arranged in a 60×60 grid. When an instruction is being copied there is a 0.75% chance that the instruction being copied will be mutated. During replication there is a 5% chance an instruction will be deleted, and a 5% chance that a random instruction will be inserted. On average each organism in the population will execute one instruction per *update*, the standard unit of time in AVIDA.

As in [13], organisms are rewarded for performing tasks that are Boolean logic operations. Specifically, we used the five tasks listed in Table 1. Each task has an associated reward, indicating the number of energy units an organism gains when completed, and a limit on how many times a individual organism may be rewarded for performing it. Completing even these relatively simple tasks can require several instructions. Table 2 shows a "hand-built" solution for the AND task (a NOP instruction modifies the behavior the preceding instruction, for example, placing the result in a different register than the default). Of course, evolution may produce many different solutions for the same task.

The environment contains a single resource that is available periodically. When the resource is available, it is non-depletable, and all five tasks described in Table 1 are maximally rewarded. If an organism completes a task when the resource is unavailable, no reward is given. The duration of the resource availability changes throughout every experiment except the control experiment, where it remains constant. Resource availability is defined in "years" and "days." Each year consists of 500 days, each of which lasts for 256 time steps (updates). During each year, the availability of the resource remains constant. That is, each day of a year has the same duration of resource availability. At the beginning of each day the resource becomes available for a period of time depending on the current year. For the first year the resource is available during 100% of the day. After each passing year, the availability of the resource during a day is reduced by 6.25% of a full day until it becomes zero, which deprives the population of energy and eventually brings on its demise. Through evolutionary change brought upon by depriving the population in this manner, we observe under which conditions the population of digital organisms will find sleep useful.

We have added six instructions to the base AVIDA instruction set, enabling an organism to sense and respond to its environment. These instructions are: TIME, SENSE, and four variations of SLEEP. Executing the TIME instruction stores the current time step in a register within the organism's virtual CPU. The SENSE

Table 1. Rewarded tasks.

Task Name	Input	Bitwise Output	Reward	Max Times Rewarded
ECHO	A	A	1000	35
NAND	A, B	$\neg(A \wedge B)$	1500	20
NOT	A	$\neg A$	1500	20
ORNOT	A, B	$A \vee (\neg B)$	2000	13
AND	A, B	$A \wedge B$	2000	13

Table 2. Instruction sequence that when executed completes the AND task.

Instruction	AX	BX	CX	Stacks 1,2	Output	Description
IO	?	X	?	?,?	?	read X into BX
IO nop-C	?	X	Y	?,?	?	read Y into CX
nand	?	X nand Y	Y	?,?	–	BX $\leftarrow \neg$(AX $\wedge$ BX)
push	?	X nand Y	Y	X nand Y, ?	–	push BX onto stack 1
pop nop-C	?	X nand Y	X nand Y	?,?	–	pop stack, place result in CX
nand	?	X and Y	X nand Y	?,?	–	BX = $\neg$(BX $\wedge$ CX)
IO	?	Z	X nand Y	?,?	X and Y	output BX

instruction allows an organism to detect the presence or absence of the resource; it loads one of the calling organism's registers with the current quantity of the resource times 100. (The value of the resource is multiplied by 100 to allow for a wider range of the sensed value.) The SLEEP instructions allow organisms to enter a low energy state that lasts for multiple CPU cycles. Compared to other instructions, the SLEEP1-4 instructions cost 100 times less energy to execute and last for 10, 20, 40, and 80 times more CPU cycles, respectively.

To help answer the questions posed in Section 1 we ran three experiments. The first is a control where the resource is available 100% of the time. In the second experiment resources are diminished for the duration of each run. In the final experiment the sleep instructions have been replaced by a null instruction to quantify the selective pressures being applied to the SLEEP instructions.

4 Experimental Results and Discussion

We define an environment where a resource is available for the duration of each day. In this environment, which is referred to as "constant," the organisms in the population do not benefit from an adaptive response based on the availability of the resource because the resource can be used at any time. The remainder of this paper presents evidence that a decline in resource availability within a single-resource environment can produce an adaptive resource-aware response.

To test this hypothesis we conducted two experiments; results presented are the average of 50 runs. In the first experiment the resource is available during the entire run (constant environment) and in the second experiment the availability of the resource is reduced over the course of the run (declining environment). Figure 2(a) displays the average metabolic rate in both the constant and declining resource environments. For clarity, error bars are omitted; the maximum standard error is 0.018 for constant environment and 0.01 for declining environment. The 16 vertical lines in Fig. 2(a) denote years, where a 6.25% decrease in resource availability occurs in the declining resource environment. As shown, the metabolic rate in the constant environment tends to stabilize as the run proceeds, but decreases over time in the environment with declining resource availability. This behavior is expected, since organisms can receive rewards for

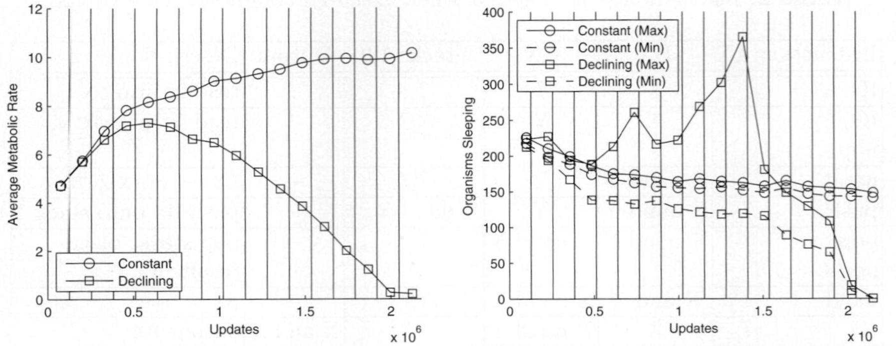

(a) Average metabolic rate

(b) Average maximum/minimum sleeping organisms

Fig. 2. Comparison of sleep responses in two environments, one where the resource is available 100% of the time (constant), and one where the resource availability decreases over time (declining). Results are the average of 50 runs.

completing tasks continually in the constant environment, but less often as time lapses in the declining resource environment. In fact, after the last vertical line the organisms in the declining resource environment populations no longer have a source of energy, and eventually the populations will die when they run out of stored energy.

Figure 2(b) shows the average maximum and minimum number of organisms sleeping at some time during a day in each environment. The maximum and minimum numbers of organisms sleeping during a day in the constant environment remain relatively close together. In contrast, organisms in the declining resource environment have evolved to participate in inactive periods, where at the peak, on average, greater than 10% of the organisms in the population are sleeping. At this point the number of organisms sleeping in the declining environment is significantly above the number sleeping in the constant experiment. (p-value < 0.0003, using Wilcoxon rank sum test for equal medians). A sample of evolved code from one of the runs is given in Table 3. The code produces a resource-aware behavior when executed. Specifically, the organism enters a loop that ends when the resource becomes available.

Since the organisms sleep more in the declining resource environment, one might infer that the organisms accumulate more SLEEP instructions in their genomes. However this is not true. Figure 3(a) shows the number of SLEEP instructions that are *present* in the organisms' genomes in both environments, along with the number of sleep instructions *executed* in each. For the first half of the runs, organisms in both environments have substantially more SLEEP instructions in their genomes than they actually execute. The gap then begins to narrow in the declining environment, and by the end of the runs the number of executions nearly equals the number present. The increase in the execution of SLEEP instructions in this environment suggests that sleeping is more beneficial

Table 3. Evolved code that loops until the resource becomes available.

Instruction	Explanation
H-SEARCH	place FLOW-HEAD at next instruction
SLEEP	start sleeping
SENSE	read resource availability into BX register
IF-EQU-0	if BX $\neq$ 0 skip next instruction
MOV-HEAD	move INSTRUCTION-HEAD to FLOW-HEAD

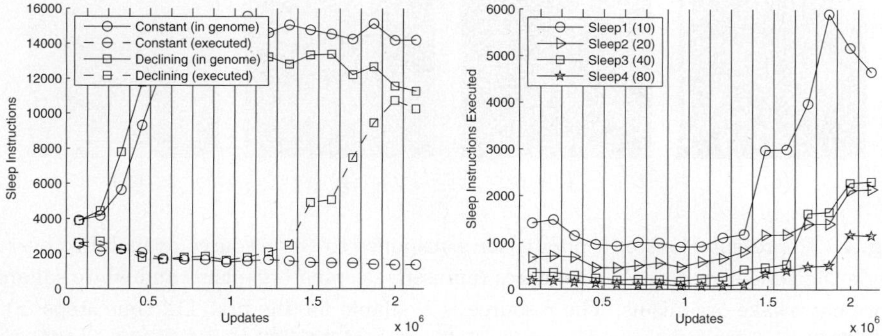

(a) number of SLEEP instructions present in and executed by organisms

(b) number of four SLEEP instructions executed, declining environment.

Fig. 3. Number of SLEEP instructions present in and executed by organisms in the constant and declining environments. Average over 50 runs.

as the resource availability diminishes. Figure 3(b) shows the rate of execution of SLEEP instructions over the course of the runs in the declining resource environment. As expected, the SLEEP instructions with lower CPU cycle costs are used more heavily than the more expensive SLEEP instructions, especially early in the runs. As the resource becomes scarce, the number of more expensive SLEEP instructions increases. This adaptation allows for longer sleep cycles with fewer executed instructions.

When AVIDA organisms are exposed to an environment where resource availability varies during a day, they evolve an adaptive resource-aware response. An example is shown in Figure 4, which depicts snapshots of the 60×60 grid during a single day in a population that evolved this adaptive sleep/wake behavior. The black squares depict organisms that are sleeping. At this point in the run, the resource is available for the first 112 (out of 256) updates. Figure 4(a) shows the population at the beginning of a day. Figure 4(d) shows the population at the day's midway point where the resource is no longer available and organisms are beginning to enter a sleep cycle. During this day the peak number of organisms sleeping at one time is 2111 or 58.6%, shown in Figure 4(e). After this point the organisms start to wake up and await the next period of resource availability.

Figure 5(a) plots the number of organisms sleeping and the resource availability during three consecutive days near the midpoint of a single run, when

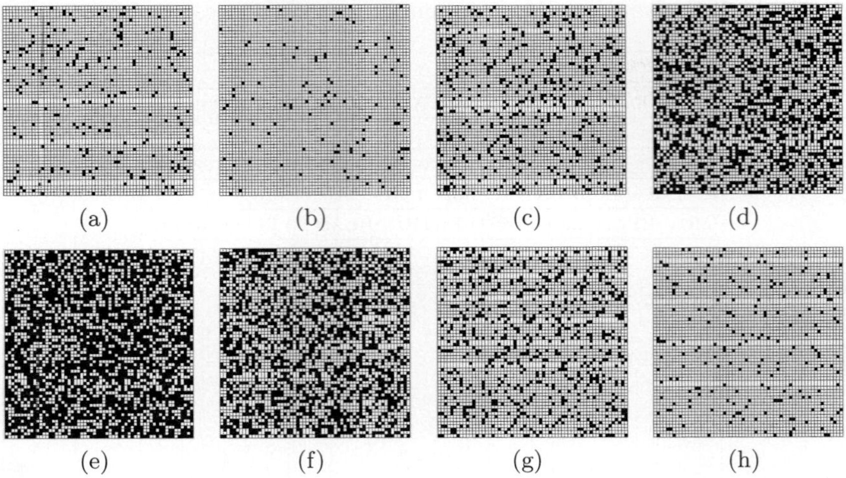

Fig. 4. Representations of a population's response to the resource availability over a single 256 time-step day. Black squares represent sleeping organisms and white squares represent awake organisms. The resource is available for the first 112 time steps. a) t = 1, 231 sleeping, resource becomes available; b) t = 64, 108 sleeping; c) t = 128, 469 sleeping; d) t = 152, 1355 sleeping, resource is no longer available; e) t = 180, 2111 sleeping; f) t = 204, 1502 sleeping, organisms are beginning to wake up; g) t = 228, 667 sleeping; h) t = 256, 189 sleeping, day ends and resource becomes available again.

the resource is available during the first 50% of each day. As shown, there is a tight correlation between number of sleeping organisms and lack of resources. Examination of evolved genomes shows that organisms in this population have evolved to begin their sleep cycle just before the beginning of resource deprived periods, and begin preparing data to be used in tasks, just prior to the return of the resource. This "early to bed, early to rise" behavior allows organisms to finish tasks early during periods of resource availability, thereby increasing the probability of receiving a reward. It also helps to avoid situations where an organism's execution is delayed, causing a task to be completed just after the resource disappears, in which case the organism receives no reward. This adaptive behavior arose in 37 out of 50 runs in the declining resource environment.

Although the populations evolved an adaptive behavior, in the above trials the fraction of concurrent sleeping organisms never stabilized above 60%. To help explain why more organisms did not sleep, we conducted a final experiment, where the four SLEEP instructions were replaced by the NOP-X instruction, which has no effect on the virtual CPU when executed, and has CPU and energy costs equal to the non-sleep instructions. The same experimental setup with a declining resource availability was used, the only difference being the replacement of the SLEEP instructions with NOP-X. Figure 5(b) compares the number of SLEEP and NOP-X instructions present and executed in the populations. In both cases the NOP-X instruction is significantly more plentiful than the SLEEP instructions. In

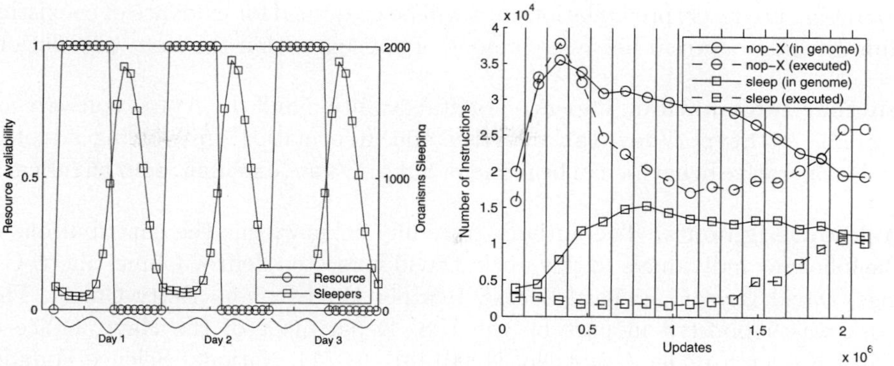

(a) (b)

Fig. 5. (a) Attempted resource usage by organisms (resource activity) and resource availability vs. time for a typical 3-day interval. (b) A comparison of SLEEP instructions (squares) to inert NOP-X instructions (circles); solid lines indicate the frequency with which each instruction is found in the genome and dashed lines indicate the frequency at which they are executed.

fact the p-values for both are less than 0.0001. Selective pressures produced by this treatment favored doing nothing for 1 CPU cycle and paying a higher energy cost, over doing nothing for multiple CPU cycles and using 100 times less energy. Yet, even in the presence of this selective pressure, an adaptive resource-aware sleep/wake behavior has evolved to a point where a majority of the organisms in a single population sleep at the same time.

5 Conclusion

Revisiting the questions posed in Section 1, we have shown that populations of digital organisms are capable of evolving resource-aware adaptive sleep/wake behavior in an environment where resource availability is periodic and declines over time. The organisms in these populations become highly active when the resource is available and sleep when it is not. This behavior evolves even though sleeping organisms are vulnerable to non-sleepers and there exists a selective pressure not to sleep. This behavior evolved and remained stable in a majority of the populations in our experiments. We also have seen evidence suggesting that the adage "early to bed, early to rise" describes an evolved behavior, as organisms maximize their probability of being rewarded for completing tasks. This behavior evolved even in the presence of a selective pressure not to sleep.

Continuations of this work using additional environments are ongoing. Environments with added costs, instruction and environmental impairments, positive and negative reinforcement, and punishment will all be tested for effectiveness. Additionally, seasonal resource availability models are under development and will be used to model the natural world more closely. Finally, environments encouraging predator/prey relationships will be examined for evidence of coexisting diurnal and nocturnal behaviors among organisms within the same population.

Further Information. Papers on digital evolution and the AVIDA software are available at `http://devolab.cse.msu.edu`. Information on evolving adaptive and cooperative behavior can be found at `http://www.cse.msu.edu/thinktank`.

Acknowledgments. The authors gratefully acknowledge the contributions of the following individuals to this work: David Knoester, Jeffrey Clune, Sherri Goings, David Bryson, Richard Lenski, Heather Goldsby, and Betty Cheng. This work was supported in part by the U.S. Department of the Navy, Office of Naval Research under Grant No. N00014-01-1-0744, National Science Foundation grants EIA-0130724, ITR-0313142, and CCF 0523449, and a Quality Fund Concept grant from Michigan State University.

References

1. Kennedy, P.J., Osborn, T.R.: Operon expression and regulation with spiders. In: Whitley, D., Goldberg, D., Cantu-Paz, E. (eds.) 2000 Genetic and Evolutionary Computation Conf. Workshop Program, pp. 161–166 (2000)
2. Wang, T., Hung, C.C., Randall, D.J.: The comparative physiology of food deprivation: From feast to famine. Annual Review of Physiology 68, 223–251 (2006)
3. Humphries, M.M., Thomas, D.W., Kramer, D.L.: The role of energy availability in mammalian hibernation: A cost-benefit approach. Physiological and Biochemical Zoology 76, 165–179 (2003)
4. Denver, R.J., Mirhadi, N., Phillips, M.: Adaptive plasticity in amphibian metamorphosis: Response of Scaphiopus Hammondiitadpoles to habitat desiccation. Ecology 79, 1859–1872 (1998)
5. Johnsen, G.H., Jakobsen, P.J.: The effect of food limitation on vertical migration in Daphnia Longispina. Limnology and Oceanography 32, 873–880 (1987)
6. Young, R.A.: Fat, energy and mammalian survival. American Zoologist 16, 699–710 (1976)
7. Zepelin, H.: Mammalian sleep. In: Kryger, M.H., Roth, T., Dement, W.C. (eds.) Principles and Practice of Sleep Medicine, pp. 82–92. Saunders, Philadelphia (2000)
8. Crick, F., Mitchison, G.: The function of dream sleep. Nature 304(5922), 111–114 (1983), `http://dx.doi.org/10.1038/304111a0`
9. Moorcroft, W.H.: Sleep, Dreaming, and Sleep Disorders, 2nd edn. University Press of America, Inc., Lanham (1993)
10. Mirolli, M., Parisi, D.: Artificial organisms that sleep. In: Banzhaf, W., Ziegler, J., Christaller, T., Dittrich, P., Kim, J.T. (eds.) ECAL 2003. LNCS (LNAI), vol. 2801, pp. 377–386. Springer, Heidelberg (2003)
11. Ofria, C., Wilke, C.O.: Avida: A software platform for research in computational evolutionary biology. Journal of Artificial Life 10, 191–229 (2004), `http://alife.org`
12. Lenski, R.E., Ofria, C., Collier, T.C., Adami, C.: Genome complexity, robustness, and genetic interactions in digital organisms. Nature 400, 661–664 (1999)
13. Lenski, R.E., Ofria, C., Pennock, R.T., Adami, C.: The evolutionary origin of complex features. Nature 423, 139–144 (2003)
14. Adami, C., Ofria, C., Collier, T.C.: Evolution of biological complexity. Proceedings of the National Academy of Sciences 97, 4463–4468 (2000)

Where Did I Put My Glasses? Determining Trustfulness of Records in Episodic Memory by Means of an Associative Network

Cyril Brom[1], Klára Pešková[1], and Jiří Lukavský[2]

[1] Charles University, Faculty of Mathematics and Physics, Prague, Czech Republic
[2] Institute of Psychology, Academy of Sciences, Prague, Czech Republic
brom@ksvi.mff.cuni.cz

Abstract. Episodic memory represents personal history of an entity. Human-like agents with *a full episodic memory* are able to reconstruct their personal stories to a large extent. Since these agents typically live in dynamic environments that change beyond their capabilities, their memory must cope with determining trustfulness of memory records. In this paper, we propose an associative network addressing this issue with regard to records about objects an agent met during its live. The network is presently being implemented into our case-study human-like agent with a full episodic memory.

1 Introduction

From the psychological point of view, *episodic memory* [17] represents personal history of an entity. Episodic memories are related to particular places and moments, and are connected to subjective feelings and current goals. *Human-like agents* (or *virtual humans*) are typically thought of as software components *imitating* behaviour of a human in a virtual world that are equipped with a *virtual body* graphically visualised. The important feature of a virtual human is that it is designed to imitate behaviour in a *believable* manner, but not necessarily psychologically plausibly. These agents inhabit artificial worlds, be it 2D or 3D, in commercial computer games, serious games, virtual storytelling applications, military simulations and various other applications (for a review, see [15]). These worlds are typically dynamic and unpredictable, and the user interaction is often allowed.

Human-like agents with an *ad hoc* episodic memory [5] are able to store only the events specified inside an agent's script or a reactive plan in a hardwired fashion. This kind of episodic memory is almost always present in current human-like agents for it is essential for action selection purposes. In contrast, a *full episodic memory* stores more or less everything happing around the agent in a general manner tagged with the agent's own relevance estimation. This presents actually a form of a life-long, autobiographic memory [11], which is often absent in current human-like agents. However, there is a growing need of agents with this kind of memory in the fields of narrative storytelling systems and role-playing computer games, as discussed e.g. in [7]. For example, these agents are able to reconstruct their personal history, which increases their believability. Imagine that while you are playing a role-playing game, you come to a medieval town, enter a magic shop and ask the computer driven seller:

F. Almeida e Costa et al. (Eds.): ECAL 2007, LNAI 4648, pp. 243 – 252, 2007.

Hey, I am the representative of the king Hromburac Pekelny, and you please tell me, what were you doing during the last week? And, please, summarise it in three sentences.

Well – answers the seller – I was in my shop every day except of Friday, selling my magical stuff. In the evenings, I enjoyed with my friends. Nothing interesting happened, except of Wednesday when a filthy usurer came to rob me.

Yet, a full episodic memory agent will allow you to ask further, for example:

Ok, that sounds interesting. Please, summarise the same now in 15 sentences. or
Focus please on the filthy usurer. or
Tell me more about Saturday evening.[1]

We have developed a prototype of such an agent with a full episodic memory [5] (Fig. 1) as a part of our on-going work on an educational storytelling game [3]. The memory stores the course of activities the agent performed, the objects the agent used, and also the reasons why the actions were performed (to some extent). It also forgets unimportant episodes as a time passes. It is widely optimised for storing and retrieval.

During development, we have stumbled on several inherent episodic memory issues, some of which remain to be solved. One of these issues is the problem with estimating trustfulness of the memory records that relate to objects. As the world is dynamic, an object may move beyond the agents capabilities (e.g. by another agent). Hence, after a couple of days in the virtual world, the agent may have tens of records about the same object. Which is the most trustful one?

In this paper, we address the issue of trustfullness by proposing an associative network aimed at coping with this problem. Implementation of this network is our current work-in-progress. Although the network is aimed as a component for our episodic memory employed in human-like agents, it can be also used *as is* in other artificial life agents.

Section 2 overviews our previous work concerning episodic memory and Section 3 proposes the associative network. In Section 4, an overview of related work is given.

2 Overview of Previous Work

This section details requirement on the full episodic memory architecture we had, and briefly reviews the overall architecture of the agent and its episodic memory.

As our goal was to *imitate* human-like episodic memory, we needed agents to remember only what real humans would remember, and forget in similar way and extent as real humans would do. Unfortunately, there is no thorough functional description of episodic memory of humans from which we could derive our model. Thus, we were forced to derive the requirements only from case-studies of forensic psychology like John Dean's testimony study [10] and from our own phenomenological experience:

---

[1] Of course, apart from the episodic memory, the agent must be equipped with a *linguistic module* allowing for transferring the outcome of the memory to syntactically correct sentences, and the player's question into a memory query.

1. The memory should cope with complex tasks that require manipulation with several objects and apparently require human-level cognitive abilities, like cooking or shopping, for they can be performed by a human-like agent. Such tasks tpically have hierarchical nature—they can be logically decomposed to sub-tasks, which can be divided to yet smaller tasks, until some atomic actions are reached.
2. The memory has to store and reconstruct personal situations: a) what an agent performed with which objects and why, and b) who was seen by the agent and what this second agent performed (we remark, that presently only (a) is implemented, (b) is another work-in-progress). Time information, at least approximate, shall be also stored. The memory should be able to provide information like "where is something?", "when did the agent see *x*?", "what did the agent do from *y* to *z*?", "why did you do *a*?"[2], and reconstruct stories like the above-mentioned. The memory is expected to reconstruct information *on demand* (i.e. when a user asks) rather than automatically based on an environmental clue.
3. The memory should not store all available information; neither external, nor internal. In particular neither all objects located around the agent, nor all activities of other agents should be stored. The objects the agent uses shall be stored more often than the objects not used, but only seen. Only important actions performed by other agents should be stored. Not all internal state should be stored, but only information about motivations relevant to current goals (i.e. I would not remember I am slightly thirsty, if I am dying hungry at the same time and my goal is food and only food). Generally, the activities/objects to be remembered should be chosen based on their relevance to the agent, on their general attractiveness, and on the agent's attentional and emotional state.
4. The memory should operate in a large time scale. As time passes, the unimportant details should be forgotten, until only a "gist" of what happened is kept. Different episodes should be forgotten in different speeds based on their importance and emotional relevance. Several similar episodes can be eventually merged together.
5. Coherence shall be maintained, in particular if there are two contradictory records in the memory, i.e. an object *x* has been seen both at *a* and *b*, one of them must be marked as more trustful. This last point is actually the scope of this paper.

Previous memory model. As mentioned, the associative network is primarily intended as a plug-in for the full episodic memory agent we developed previously. For brevity, we only sketch this previous model here. Full details can be found in [5].

Architecture of our agent is depicted in Fig. 2. Perhaps its only distinction from a classical symbolic architecture is the episodic memory. All the parts except of the linguistic module are implemented, emotional module being implemented separately [2], and its integration is in progress.

The agent is driven by hierarchical reactive planning with behaviour represented by AND-OR trees [4]. Basically, the AND-OR tree metaphor works with abstract *goals* representing what shall be achieved, and *tasks* representing how to achieve the goals. Typically, every goal can be accomplished by several tasks, while every task

[2] Causal inference, i.e. "why", is complicated, but an agent can answer these questions to a limited extend for it has some basic understanding of its actions owing to its behavioural representation. It has a goal and it picks the most suitable action for it in every moment.

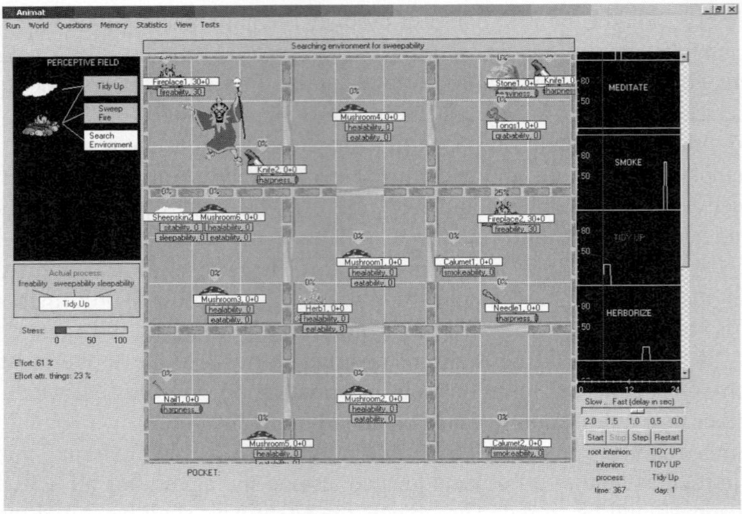

Fig. 1. A screen-shot from the prototype. Left: the short term memory. Middle: the 9x9 grid world of the shaman agent (the left upper part). Right: activation of the top-level goals.

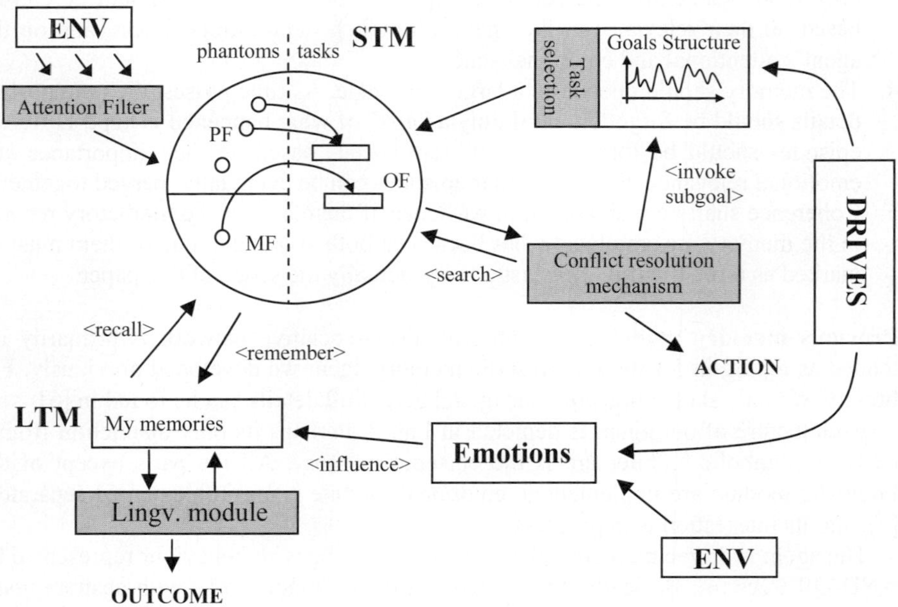

Fig. 2. The overall architecture of our agent. PF – phantoms of the STM. OF – own tasks of the agent. MF – records retrieved from the LTM. Emotional module is presently implemented separately (the emotional influence being specified manually), linguistic module remains unimplemented.

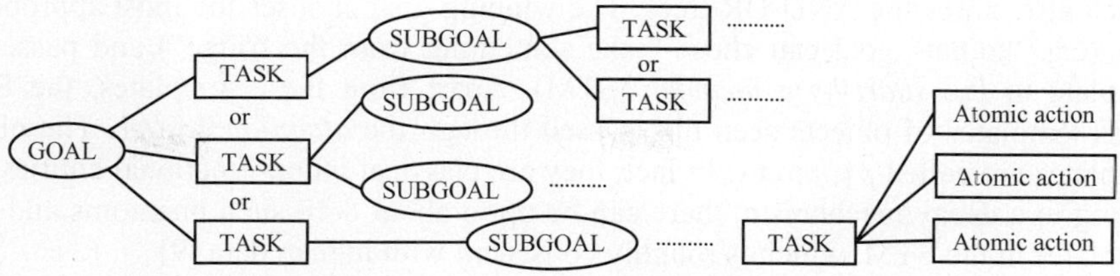

Fig. 3. Agent's behaviour is represented using AND-OR trees.

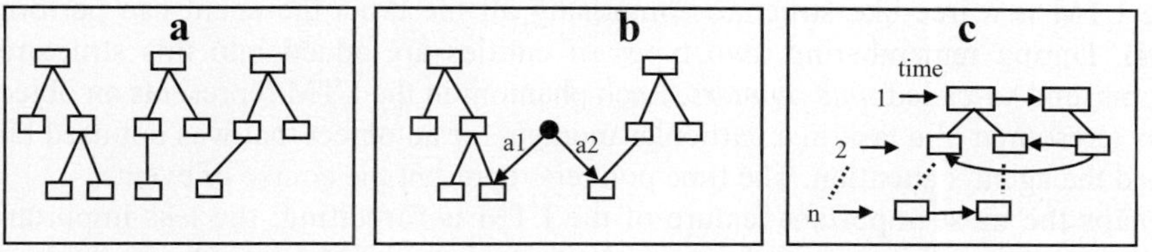

Fig. 4. a. A fixed arrangement of the long-term episodic memory, each box represents a task. 4b. Storage of a phantom of an object. This object can be used as a resource for two tasks (pointers *a1*, *a2*). 4c. The tasks are being sorted by time pointers during storing.

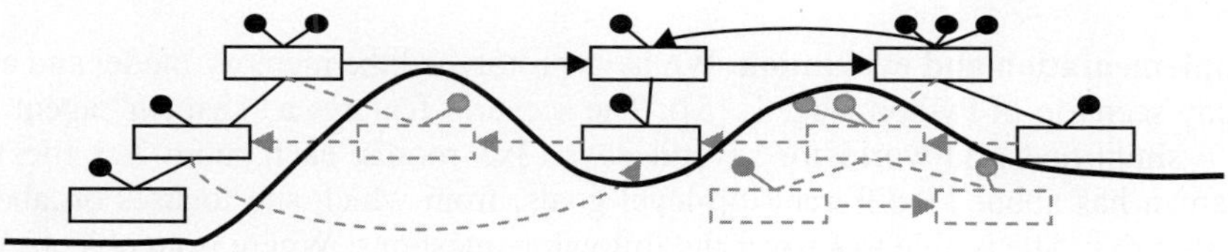

Fig. 5. The LTM forgetting schematically depicted.

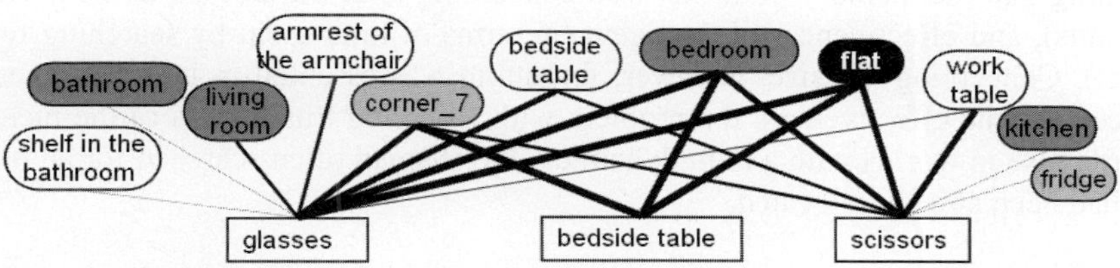

Fig. 6. A part of the associative network depicted. Intensity of grey denotes the level of space abstraction. The strength of the edges denotes the weight of the associations.

can be achieved by adopting some sub-goals. The agent needs to perform only one task to achieve a goal, provided there is no failure (hence, OR nodes), but to fulfill all sub-goals to solve a task (hence, AND nodes; Fig. 3). The tasks that cannot be further decomposed are *atomic actions*, i.e. primitives changing world-state. Every task may need several *resources* to be performed, i.e. objects (e.g. hammering is possible only with a nail and a hammer). Every top-level goal has its *activity level* based on drives, external events, and a schedule—the activity level is the mechanism of competing among the top-level goals. This competition takes place within the *goal structure*, which also stores the AND-OR trees. The winning goal chooses the most appropriate task (e.g. "to eat" goal can chose "take something from the fridge") and passes its template to the *short-term memory* (STM). Apart from these templates, the STM holds templates of objects seen that passed through the *attentional filter*. The object templates are called *phantoms*. In fact, they are classical index-functional entities [1]. Owing to a decay mechanism, there can be up to about 8-10 such phantoms and task templates in the STM, which is roughly consistent with human data [9].

If a task needs an object that is not seen, i.e. there is no relevant phantom in the STM, *the long-term memory* (LTM) is queried, and the environmental search is initiated. This may lead to invoking of a new sub-goal, and eventually a sub-task.

The LTM is a tree-like structure comprising all the tasks the agent can perform (Fig. 4). During remembering, two types of entities are added into this structure: phantoms, and so-called *time pointers*. Each phantom in the LTM represents an object used as a resource of a task in a particular moment, or an object that was not used but attracted the agent's attention. The time pointers represent the course of events.

Perhaps the most important feature of the LTM is forgetting: the less important episodes are being "bitten out" from the bottom of the LTM (Fig. 5), i.e. the time pointers and phantoms of the "bitten out" parts of the LTM are being removed, typically during night as a "consolidation". The episode importance is presently determined by the age of the event (automatically), and by its emotional salience (manually).

Implementation and evaluation. We have prototyped the memory model and a case-study scenario in Python (Fig. 1, [5]). The scenario features a "shaman" agent living in a simplified grid-world for several days (3x3 rooms, each room 3x3 tiles). The shaman has about 15 different top-level goals, from which she focuses on about 10 goals a day. She is able to answer the following questions: Where is an object? Where was an object from *a* to *b*? What did you do from *a* to *b*? When did you do *x* lastly?

Several tests measuring the efficiency of the memory have been carried out [5, 13], revealing that the memory performs well concerning both the size (in terms of memory units), and effectiveness of the agent (in terms of time spent by searching for an object). Concerning the size, however, forgetting was essential in large time scales. Concerning the effectiveness, the problem with assessing trustfulness of the memory records rose in worlds with a high dynamic – the shaman often searched for an object that had been already relocated.

Problem revisited. When the LTM in our current model is searched for an object, more than one phantom of this object can be found: e.g. glasses on the table, next to the TV, at the armrest of the armchair etc. Which of these records shall be considered as the most trustful? Apparently, this depends on the history of the world; on the habits of the agent as well as on the habits of the other agents in case of an object being shared. This issue is remarkable because searching at wrong places undermines both the agent's believability and effectiveness. While the believability is related only to the domain of human-like agents, the effectiveness issue is more general. In the next section, we propose an associative network aiming to cope with this issue.

3 The Proposed Associative Network

This section proposes first generation of the associative network addressing the issue of determining trustfulness of different phantoms of the same object in the LTM. Its implementation and integration into the rest of the model (described in Sec. 2) is our current work. The section also discusses possibilities and limitations of the network, including proposing experiments to be carried out and directions of future research.

The network architecture is designed to cope with dynamic and complexity of human-like worlds. We, present-day humans, live in a world we typically describe in a hierarchical fashion; we think in rooms, houses, streets, places etc.[3] We tidy up objects: they are not placed randomly in our surroundings, but logically clustered according to our needs and cultural norms (e.g. a spade is typically not to be found in a kitchen). Objects we use have their specific dynamic: some are being moved every day but from a typical place, e.g. glasses from a bedside table; others are almost never being relocated, e.g. a fridge; yet others change their position so often that it does not make sense to remember their exact position. Consequently, during searching, we sometimes scrutinise a whole room, sometimes we inspect several very specific places, sometimes a stimulus-response mechanism is employed. For example, we may have the following rule for searching for glasses: to inspect the bedside table first, then to look at a work table, and eventually to scrutinise the whole flat, starting in the living room. In another situation, we may go to the bathroom to fill the washing-machine and then realise: aha! – it was moved to the kitchen a week ago.

Basically, this complexity and dynamic was intended to be seized in the network. We believe we have found a relatively simple architecture for it, yet fulfilling the requirements. The network comprises two kinds of nodes: *place-nodes* and *phantom-nodes*. A phantom-node is an index-functional entity representing a particular object. A place-node represents a place where an object can be found. By place we mean any logically coherent space abstraction, e.g. a bedside table, the place between this table and the bed, a bedroom, a corner of the bedroom, a flat, etc. A place-node can also represent an object, as in the case of the bedside table. The places *are* from different levels of abstraction, which are numbered; e.g. the bedside table is the level 1, the corner is the level 2, the living rooms is the level 3, the flat level 4 etc.

A phantom-node can be connected to several place-nodes. Such an *edge* represents a possible occurrence of this particular object at this particular place (Fig. 6). Every edge has its *weight* meaning the number of times the object was seen there. We remark that while in our previous LTM, a phantom for each occurrence of an object was stored in the LTM, in this network, only one phantom is stored for one object.

Formally, the network is a triple $<P, H, E>$, where P is the set of all place-nodes, H the set of all phantom-nodes, and $E = \{< x, w >\}$ is the set of weighted edges (where $x \in P \times H$, and $w \in \mathbf{N}$).

Learning mechanism. In addition to storing a phantom into the LTM from the previous model (Sec. 2, [5]), weights of the edges between the object's phantom and *all* the places where the object was found are increased by 1 (new places being added).

[3] The question of whether our predecessors, which did not live in rooms and houses, had hierarchical representations of their surroundings or not, is not discussed here for brevity.

E.g., if glasses are found on the bedside table, the edges to the bedside table, to the corner next to the bed, to the bedroom, and to the flat are strengthen. Consequently, the associations between a phantom and the places the object is typically being found are strengthen more often than the other edges, and the edges to more abstract places are also strengthen more often as the total number of places at a particular level of abstraction decreases with the number of the level (e.g. the glasses are almost always in the flat, but sometimes on the bedside table, and sometimes on the work table).

Retrieval. For every place the phantom has an edge with, a *size-normalised trustfulness* (SNT) is computed as w/a^b, where w is the weight, a is the number of the layer of abstraction of this place, and b a scaling factor. The places are sorted according to this value and the ones that are over a *treshold* are considered as places where the object may be located. The object is then searched for at these places in the order of the SNTs. The concrete places are looked at directly (e.g. a bedtime table), abstract place are to be inspected. The SNT is supposed to lead to a searching process that prefers concrete places over abstract ones, provided there are only a few concrete places. Otherwise, it is preferred to inspect the abstract place directly, because searching in the concrete places is likely to fail. The factor b is intended to be tuned during experimenting.

Issues. Some questions are to be investigated after we finish the implementation of the first-generation network. Most notably, these are (1) exceptions, (2) classes of objects, (3) edges weakening, (4) states of objects, and (5) moving objects.

1) Sometimes, it may be fruitful to remember exceptions; e.g. "the glasses are typically at the bedside table, but now, I have put them at the TV". Phantoms from the STM are actually preferred to LTM records, but the question remains whether the exceptions should be handled also in the LTM, e.g. when a task is interrupted, in which case all the phantoms of the objects used during the task decay from the STM.

2) In the described network, each object is stored separately, including objects that are being replaced frequently like food or newspapers. We suggest that to store only one phantom for the whole class of such objects is a better solution. We will focus on this implementation issue after evaluating the first generation of the network.

3) Should the connections in the network be also weakened? The trouble is that sometimes humans remember things over long periods, e.g. that the 1979 Bordeux is in the cellar, even if we did not see it for last 10 years. This suggests that edges can not be weakened in time straightforwardly. Instead, we propose (i) weakening an edge only if it is small comparing to the sum of the weights of the other edges originating from *this* phantom. Additionally, we propose (ii) intentional weakening ("I try to forget that the washing-machine was in the bathroom when it has been moved to the kitchen"), and (iii) weakening in case of a place that was suggested by the network, but where the object was not found (this is now achieved indirectly because the edges to the places where the object was found are strengthened). We plan to experiment with these three mechanisms.

4) One object can have different states. Sometimes, the states may be associated with places, e.g. full wine bottles are in the cellar, while empty bottles are next to the bin. Could this be mirrored by the network?

5) The memory is not designed for objects that move themselves (a dog). An entirely different memory would have to be used for this class of objects.

Another interesting issue, which is however out of scope of this paper and our current research, is investigation of relations between this network and mechanisms of natural episodic memory. One interesting phenomenon to look at is place cells [8].

4 Related Work

In the field of human-like agents, the issue of generic episodic memory has been almost untouched, since the agents typically need not store more than a few episodes for action selection purposes. A notable exception is the tutor agent Steve [16], who employs an episodic memory that allows him to explain himself after a given lesson, which lasts, however, only a couple of minutes. Another exception is the memory model for agents of ALOHA system [14], which exploited to a great advantage division of the memory to short-term one and long-term one, but unfortunately stores records only about objects and groups of objects (but not the tasks).

In robotics, Dodd [6] developed recently a general memory system for a humanoid robot ISAC, which included also episodic memory working with emotions. An interesting a-life example is the work of Ho et al. [7], who developed agents with various different episodic memories aiming to investigate how different types of memories improve survival of the agents. Though these memories fit well for their domain, they are relatively low-level from the point of view of human-like agents. They are not designed to cope with complex, hierarchical tasks for example.

In agent research, perhaps the most elaborate model of episodic memory has been developing by Nuxoll & Laird [12]. This model is intended as a robust, general-purpose architectural extension of Soar with much broader scope than the model of ours. This, however, means that our model can benefit from some domain-specific tricks, which may finally increase efficiency of the memory in our domain. In all cases, it will be interesting to compare this model with ours in future.

Typically, the environments presented with the abovementioned memories are not very dynamic. To our knowledge, none of the models addresses the issue of trustfulness of memory records to the extent as the proposed network does.

5 Conclusion

We have proposed an associative network that addresses the issue of assessing trustfulness of records about objects in the long-term episodic memory of a human-like agent. The network is intended to elevate believability of the agent, and its efficiency (measured in time spent by searching for an object) in highly dynamic human-like environments. The network is presently being implemented into our agent with a full episodic memory developed earlier [5].

The main idea behind the network is that a place where an object was found is represented by several nodes, each representing this place at a *different* level of abstraction. The network is relatively simple, which is, however, its advantage considering future rigorous analysing and extending.

We have introduced a mechanism for searching for an object by an agent equipped with the network. We have also discussed limitations of the network proposing some issues that are ahead. These includes, most notably, handling exceptions, weakening of the associations, and coping with classes of objects being frequently replaced. These issues will be addressed as a part on the work of the second generation of the network.

Acknowledgments. The authors would like to thank to Andrew Nuxoll for answering questions on his work. This research was partially supported by the Program "Information Society" under project 1ET100300517, and GA UK 351/2006/A-INF/ MFF.

References

1. Agre, P.E., Chapman, D.: Pengi: An Implementation of a Theory of Activity. In: Proceedings of the Sixth National Conference on Artificial Intelligence, Seatle, USA, pp. 196–201 (1987)
2. Bída M., Kadlec R., Brom. C.: Relevance of emotions for artificial beings. (Význam emocí pro umělé bytosti) (in Czech). In: Mind, intelligence and life (Myseľ, inteligencia a život), pp. 158-172. Vydavateľstvo STU, Bratislava (2007)
3. Brom, C., Abonyi, A.: Petri-Nets for Game Plot. In: Proceedings of AISB Artificial Intelligence and Simulation Behaviour Convention, Bristol, vol. 3, pp. 6–13 (2006)
4. Brom, C., Lukavský, J., Šerý, O., Poch, T., Šafrata, P.: Affordances and level-of-detail AI for virtual humans. In: Proceedings of Game Set and Match 2, The Netherlands, Delft (2006)
5. Brom, C., Pešková, K., Lukavský, J.: What does your actor remember? Towards characters with a full episodic memory. Tech. Report No. 2007/4 of the Dept. of Software Engineering, Charles University in Prague (2007)
6. Dodd, W.: The design of procedural, semantic, and episodic memory systems for a cognitive robot. Master thesis. Vanderbilt University, Nashville, Tennessee (2005)
7. Ho, W., Dautenhahn, K., Nehaniv, C.: Autobiographic Agents in Dynamic Virtual Environments - Performance Comparision for Different Memory Control Architectures. In: Proc. IEEE CEC, pp. 573–580. IEEE Computer Society Press, Los Alamitos (2005)
8. O'Keefe, J., Dostrovsky, J.: The hippocampus as a spatial map. Preliminary evidence from unit activity in the freely-moving rat. Brain Research 34(1), 171–175 (1971)
9. Miller, T.: The magical number seven, plus or minus two: Some limits on our capacity for processing information. Psychological review 63(2), 81–97 (1956)
10. Neisser, U.: John Dean's Memory: A case-study. Cognition 9, 1–22 (1981)
11. Nelson, K.: The psychological and social origins of autobiographical memory. Psychological Science 4(1), 7–14 (1993)
12. Nuxoll, A., Laird, J.E.: Extending Cognitive Architecture with Episodic Memory. In: Proceedings of the Twenty-Second AAAI Conference on Artificial Intelligence, AAAI, Menlo Park (2007)
13. Pešková, K., Brom, C., Lukavský, J.: Episodic memory for virtual actors based on affordances (in Czech). In: Proc. of Cognition and Artificial Life, Slovak Republic (2007)
14. Peters, C., O'Sullivan, C.: A Memory Model for Autonomous Virtual Humans. In: Proceedings of Eurographics Ireland, pp. 21–26 (2002)
15. Prendinger, H., Ishizuka, M.: Life-like Characters: Tools, Affective Functions and Applications. Springer, Heidelberg (2003)
16. Rickel, J., Johnson, W.L.: Animated Agents for Procedural Training in Virtual Reality: Perception, Cognition, and Motor Control. App. Artificial Intelligence 13 (1999)
17. Tulving, E.: Elements of Episodic Memory. Claredon Press, Oxford (1983)

Grounding Action-Selection in Event-Based Anticipation

Philippe Capdepuy[1], Daniel Polani[1,2], and Chrystopher L. Nehaniv[1,2]

Adaptive Systems[1] and Algorithms[2] Research Groups
School of Computer Science, University of Hertfordshire
College Lane, Hatfield, Herts, AL10 9AB, UK
{P.Capdepuy,D.Polani,C.L.Nehaniv}@herts.ac.uk

Abstract. Anticipation is one of the key aspects involved in flexible and adaptive behavior. The ability for an autonomous agent to extract a relevant model of its coupling with the environment and of the environment itself can provide it with a strong advantage for survival. In this work we develop an event-based anticipation framework for performing latent learning and we provide two mathematical tools to identify relevant relationships between events. These tools allow us to build a predictive model which is then embedded in an action-selection architecture to generate adaptive behavior. We first analyze some of the properties of the model in simple learning tasks. Its efficiency is evaluated in a more complex task where the agent has to adapt to a changing environment. In the last section we discuss extensions of the model presented.

1 Introduction

In the struggle for life, an agent which is able to predict the future with some accuracy has a very strong advantage. This anticipatory ability is one of the keys to extend the temporal horizon of agents (see [3]). Such abilities can be applied on very different time scales, for example being able to predict when the sun will set or that an electric shock follows a particular sound. They can also concern external as well as internal phenomenon. The two first examples are external phenomenon, an internal phenomenon would be one under the control of the agent, e.g. an action. Understanding the potential effects an agent can induce on its environment is a key aspect of purposive action, for example when a rat learns that pressing a lever leads to the delivery of a food reward. The approach we will take here is a based on the concept of *latent learning* (see [5]) where the agent learns relationships between events without any reinforcement; it has some similarities with *state anticipation* (see [1]). Reinforcement is only taken into account during the action-selection process.

In the case of an embodied agent, for example in robotics, anticipatory abilities have to be constructed by the agent, based on its experience of the world. Extracting a relevant model of the environment and its coupling with the agent requires the introduction of learning mechanisms. Classical computational approaches to this problem such as reinforcement learning are usually based on

F. Almeida e Costa et al. (Eds.): ECAL 2007, LNAI 4648, pp. 253–262, 2007.

learning an expected reward function for state-action pairs. However this approach is not suitable in some cases, when for example the state is not clearly defined and when the temporal dimension has a strong importance. Also this framework is not easily adapted to take into account different rewards or a dynamical motivational system.

In this paper we expose a different view of the anticipation problem which is based on the concept of *event*. We can then express the *event-based anticipation* problem in a rather intuitive way. Section 2 explains the main principles of this framework and how we state the problem. In Sec. 3 we propose a mathematical model that allows the on-line extraction of predictive relationships. We then describe how the predictive model generated can be used for action-selection in the context of reward maximization. Section 4 shows preliminary results concerning the properties of the constructed internal model for simple learning tasks. We also evaluate the performance of the proposed action-selection architecture in a survival task involving reinforcement. In Sec. 5 we summarize the main points of the paper and we discuss possible extensions of the model.

2 Event-Based Anticipation

2.1 Formalization

We introduced the basis of this event-based anticipation framework in [2]. This section presents the main principles underlying the framework and some examples. We refer to anticipation in a very general way as the ability to predict, more or less accurately, the future occurrence of *perceptual events*. These events can be seen as different stimuli that the agent can encounter in its environment. We consider as a preliminary simplification that the agent is not allowed to act onto its environment, he can only observe it (handling of actions will be described in Sec. 3.3). Different from other approaches, the agent is not provided here with a continuous flow of sensoric values for different modalities. Instead we consider that the agent perceives discrete events in discrete time (0 to n different events can be observed at a given time-step). We will denote the set of possible events by $\mathcal{E}$. The agent is then observing a stream of events such as those represented in Fig. 1. The purpose of our work is to find an efficient way to identify relationships between events in an anticipatory perspective.

2.2 Examples

One of the constraints we put on our model is that it should be robust to noise and variations in the relationships. Figure 1 shows three different cases where there exists a predictive relationship (a predicts b). One is rather obvious but different configurations of the time-delay between a and b and other noise events can lead to more difficult situations. To allow the extraction of these relationships, we will split our analysis in two different components. The first one considers only pairs of events and its role is to measure the regularity of the

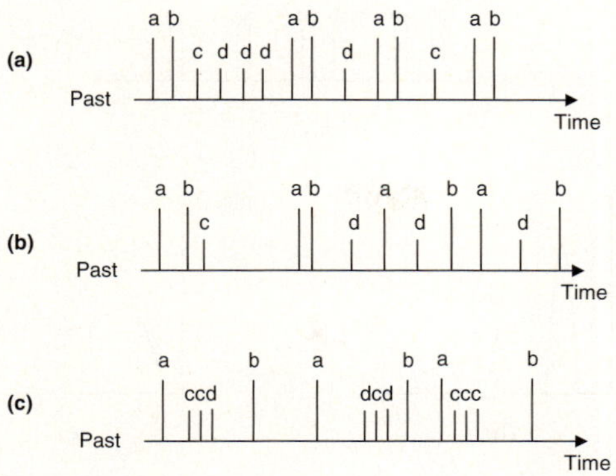

Fig. 1. Three different streams of events with $\mathcal{E} = \{a, b, c, d\}$. For each of them our aim is to identify the predictive relationship from a to b. Example **(a)** is quite obvious, events a and b follow each other very closely in time, and with a constant delay. Example **(b)** is more tricky as the delay between a and b varies, anyway it seems that b always follows a. In example **(c)** the delay between a and b is very large, providing room for many events to occur in between, nevertheless as this delay is constant we would like to identify such a relationship.

time-delay between these events. The second component is the relation from one event to all the others, the idea is to identify the most probable event than will occur shortly after another one (or shortly before if we look toward the past).

3 A Tentative Modelling

Our aim here is to be able to extract a predictive model of the environment from a flow of symbolic events that can potentially have a predictive relationship with each other. The basis of the model is to collect statistics about pairs of events and their relation in time. The first part of this section describes the low-level extraction of information that allows us to generate a probabilistic model of the time-delay between pairs of events. We then detail two mathematical tools that allow us to construct a predictive model. The first one is based on the regularity of the time-delay between two events, the second one extracts information from the contingency between events.

3.1 Low Level Extraction

The raw material from which we have to work is a stream of events such as those shown in Fig. 1. The information we want to extract is contained in the time-delay between repetitive occurrences of events. More precisely we will be concerned with the probability distributions of these time-delays. We will refer to such a probability distribution as $\Delta_{a,b}$. In this case it is the probability distribution of the observed time-delays between the event a and the next occurrence

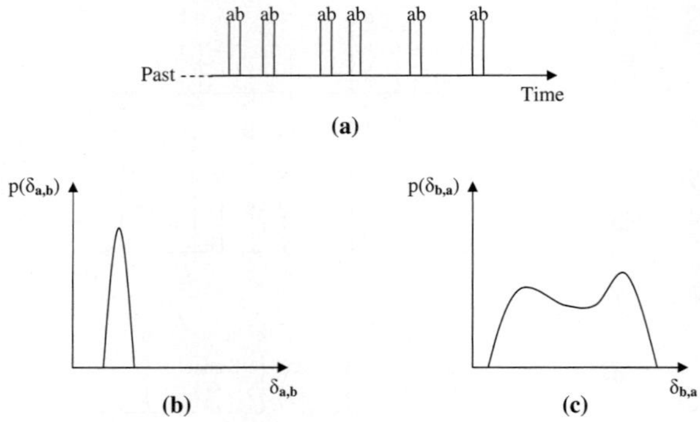

Fig. 2. Example of the low level extraction. **(a)** Part of the flow of event (the invisible part of the flow is a repetition of the visible part with slight variability). **(b)** Representation of the probability distribution extracted for $\Delta_{a,b}$. A single peak can be observed as the time-delay between a and b is nearly constant (we assume a bit of noise). **(c)** Representation of the probability distribution extracted for $\Delta_{b,a}$. In this case the delay between b and a consecutive a has a lot more variability. $\Delta_{a,a}$ and $\Delta_{b,b}$ are not shown but they are also computed in the model.

of event b (therefore this expression is not symmetric: $\Delta_{a,b} \neq \Delta_{b,a}$). The probability distribution function can then be written as $p(\Delta_{a,b} = \delta)$. For short we will use the simplified notation $p(\delta_{a,b})$.

The goal of the low-level extraction is to transform the flow of events into probability distributions of time-delays for each pair of events. Such a transformation is illustrated in Fig. 2. The transformation process makes use of a memory that stores that last occurrence for each possible event. Each time a new event b is processed, we parse the memory from the newest to the oldest occurrences. For each event a found in memory we add one occurrence of the corresponding time-delay δ to the statistics of $\Delta_{a,b}$. During the parsing, we stop when we encounter the last occurrence of the same b event, after updating the statistics of $\Delta_{b,b}$. Then the memory is updated with the observed occurrence. It is possible to compute the expected delay between two events using:

$$\bar{\delta}_{a,b} = \sum_{\delta_{a,b} \in \Delta_{a,b}} p(\delta_{a,b}) \delta_{a,b}. \tag{1}$$

At this level of the model we introduce a limitation to the memory capacity of the agent. This limitation concerns the maximum number of time-delay observations (*windowsize*) that can be stored in one probability distribution $\Delta_{a,b}$. Old observations are replaced by more recent ones, hence allowing the agent to adapt in the case of non-stationary data. This limitation is implemented as a stochastic deletion when a new observation is added to the statistics. More precisely a particular observation is randomly chosen according to its probability and removed from the statistics, the new observation is then added.

3.2 Model Construction

Time-Delay Relationships. The first kind of relationship we want to identify is the time-delay relationship. For this we will use an information-theoretic approach similar to the concept of causal entropy described in [6]. The principle of this relationship is to reflect the regularity of the observed time-delay between two events a and b. Although this regularity has to be made relative to the baseline regularity of the predicted event. Indeed, if an event b is already self-predicting (e.g. if it is periodic), there is no point in using another event a as a predictor for b. What we want then is to measure the uncertainty of the time-delay between a and b and, more precisely, how much uncertainty we lose compared to the uncertainty of b when considered alone. This principle can be translated using information-theoretic concepts as:

$$\tau(a,b) = 1 - \frac{H(\Delta_{a,b})}{H(\Delta_{a,b}) + H(\Delta_{b,b})} \tag{2}$$

where H stands for the Shannon entropy of the probability distribution:

$$H(X) = -\sum_{x \in \mathcal{X}} p(x) \log_2 p(x). \tag{3}$$

$\tau(a,b)$ is then a value between 0 and 1. In the case where $H(\Delta_{b,b}) = 0$, i.e. when the event b is strictly periodic, then $\tau = 0$. On the other extreme, if $H(\Delta_{a,b}) = 0$, i.e when the event a perfectly predicts b, then $\tau = 1$. In the particular case where both values are equal to 0 then we assume that b is already self-predicted, and so $\tau = 0$.

Contingency Relationships. From the probability distributions of time-delays, what we refer to as *contingency* is the combination of two properties. The first one is the probability that an event a has been observed before the occurrence of an event b, i.e the probability of having observed a in the recent past of b:

$$p(a \in \overleftarrow{S_b}) = \frac{\sum_{\delta_{a,b}} p(\delta_{a,b})\mu(\delta_{a,b})}{\sum_{x \in \mathcal{E}} \sum_{\delta_{x,b}} p(\delta_{x,b})\mu(\delta_{x,b})} \tag{4}$$

and the second one is the complementary value, i.e. the probability of having observed b in the close future of a:

$$p(b \in \overrightarrow{S_a}) = \frac{\sum_{\delta_{a,b}} p(\delta_{a,b})\mu(\delta_{a,b})}{\sum_{x \in \mathcal{E}} \sum_{\delta_{a,x}} p(\delta_{a,x})\mu(\delta_{a,x})} \tag{5}$$

where $\mu(\delta)$ is a "memory effect" function that maps a time delay to a real value following an exponential decay law:

$$\mu(\delta) = e^{-k\delta}. \tag{6}$$

These values are not probabilities *stricto sensu* but they can be regarded as a subjective evaluation from the agent's point of view. The contingency confidence value is then computed by combining the two values described above:

$$\kappa(a,b) = p(a \in \overleftarrow{S_b})p(b \in \overrightarrow{S_a}). \tag{7}$$

Again $\kappa(a,b)$ is a value between 0 and 1. 0 when there is no relationship, and 1 when there exists a strong relationship.

3.3 Action Selection

We now consider a reinforcement learning context where the agent has to identify the effect of its actions on the environment in order to obtain positively valued events or to avoid negatively valued ones. Actions are considered as events that get processed the same way as external ones. Their only particularity is that they are generated by the agent. First we define a reward function $R(e)$ that maps a particular event e to a real reward value.

Considering that actions are a subset $\mathcal{A} \subset \mathcal{E}$, the process of eliciting an action is based on a weighting function $V(a)$ which takes into account the confidence of the different outcomes and their associated reward. This weighting function can be defined recursively so as to take into account chaining of predictions. This leads to something very close to the Bellman equation (see [4]) but where the policy π is defined by the current predictive model. Also in our context there is no action-state pairs, but only event-event associations. Considering a given action (or any event) e, the expected reward is then computed separately for each relationship component (τ and κ) as:

$$V_\kappa(e) = R(e) + \gamma \sum_{e' \in \mathcal{E}} (p_\kappa(e'|e)V_\kappa(e')) \tag{8}$$

and

$$V_\tau(e) = R(e) + \gamma \sum_{e' \in \mathcal{E}} (p_\tau(e'|e)V_\tau(e')) \tag{9}$$

where $p_\kappa(e'|e)$ and $p_\tau(e'|e)$ are computed according to the current predictive model with:

$$p_\kappa(e'|e) = \frac{\kappa(e,e')}{\sum_{e'' \in \mathcal{E}} \kappa(e,e'')} \tag{10}$$

and

$$p_\tau(e'|e) = \frac{\tau(e,e')}{\sum_{e'' \in \mathcal{E}} \tau(e,e'')}. \tag{11}$$

In practice we approximate the solution to these equations using iterative policy evaluation (stopped when reaching a maximum change of ϵ). The final value of the event e can then be computed as:

$$V(e) = V_\kappa(e) + V_\tau(e). \tag{12}$$

A Boltzmann exploration policy is then applied to select the action:

$$p(A = a) = \frac{e^{V(a)/T}}{\sum_{a' \in \mathcal{A}} e^{V(a')/T}} \tag{13}$$

where T is the temperature value.

4 Experiments

4.1 Simple Learning Task

The aim of this first experiment is to evaluate the ability of the proposed tools to construct a relevant model. The goal of the task is to extract a relationship between two paired events, $L1$ and $Food$, out of a background noise composed of three events. The parameters studied are the delay between $L1$ and $Food$ and the variability of this delay. We measure the two confidence values τ and κ. Parameters of the simulation are the following: the sequence $L1 - Food$ can be initiated at each time-step with a probability of 0.005 (if it is not already initiated). The three noise events $R1$, $R2$, and $R3$ are emitted at each time-step with the respective probabilities 0.01, 0.005, and 0.0025. The memory size for collecting statistics ($windowsize$) is set to 40 observations, and the exponential decay parameter is set to $k = 1$.

The confidence values are computed after 100 presentations of the paired events. Each experimental run is repeated 10 times and the results are averaged. The first setup investigates the influence of the delay between the paired events, the range investigated goes from 4 to 100 time-steps, with a constant variability of $+/-3$ time-steps (uniformly distributed). The second setup studies the impact of the variability of this delay, for this we investigate uniform distributions centered around 20 time-steps and with interval size ranging from 0 to 20. Results are shown in Fig. 3.

The results illustrate the complementarity of the two measures proposed. For all the trivial cases (short delay and low variability), both confidence values are clearly above the noise level. For more difficult cases, the performance of the two measures evolves in different ways. The τ confidence value has the property to be robust to the duration of the delay between paired events, it expresses the regularity of the delay. The κ confidence value is robust to variability in the delay, it reflects the contingency of the two events. These two confidence values seem to correctly reflect orthogonal properties of the temporal relationship between pairs of events.

4.2 Survival Task

The second experiment strives at evaluating the proposed action-selection principle. For this we will consider a virtual world with two kinds of object: food and poison. Each kind of object emits a specific chemical that creates a gradient. We will denote them by gradient 1 and 2. The agent has the choice between two

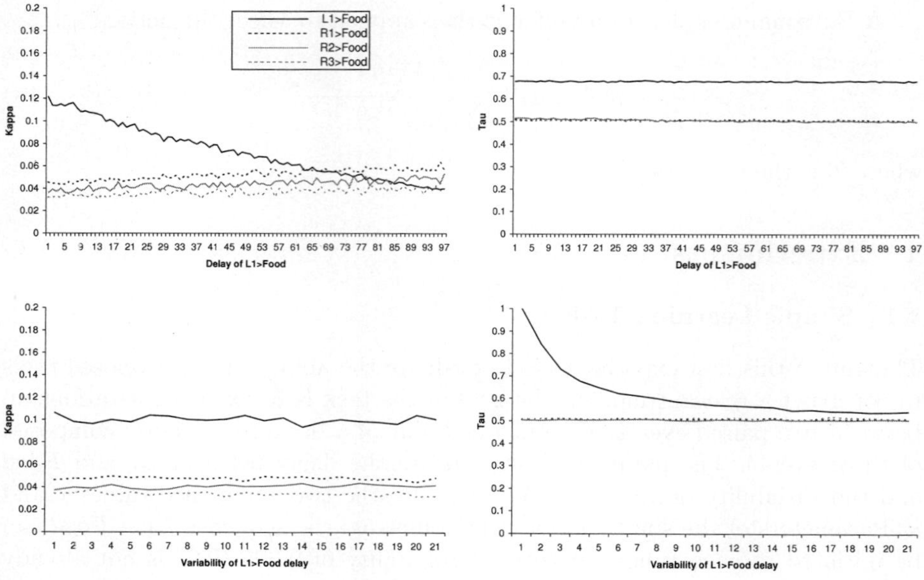

Fig. 3. Confidence levels τ and κ against time delay between paired events (top) variability of this delay (bottom). The κ confidence degrades when the delay increases, while the τ confidence level is stable. We have the opposite effect when changing the variability of the delay. The τ confidence degrades when the variability increases, while the κ confidence level is robust to high variability.

actions: follow gradient 1 and follow gradient 2. The task of the agent is to pick the correct action in order to collect a maximum of food objects (reward value of 1) and to avoid poison objects (reward value of -1). Each time an object is collected by the agent, it is randomly relocated in the world. The gradient following actions are not avoiding for the other gradient, hence it can happen that while following a gradient for food, the agent collects a poison object on the way, this introduces some noise in the system. Also the delay between engaging a particular action and collecting an object can vary a lot, depending on the particular situation of the agent in the world. As we want to evaluate the adaptive ability of the agent equipped with the predictive model, we periodically change every 100000 time-steps the environmental conditions. More precisely we swap the associations between the gradients and the corresponding object. Therefore the agent has to forget the first associations and learn new ones.

Events that the agent can perceive are: *Food*, *Poison*, and the two actions *Follow*1 and *Follow*2. Memory size is 40 observations. Boltzmann exploration temperature is $T = 0.1$. The exponential decay parameter is set to $k = 1$. Discount factor for Bellman equation is $\lambda = 0.9$ and precision for the computation of $V(e)$ is $\epsilon = 0.001$. The agent is only allowed to act when it perceives an event or with a probability of 0.01 at each time-step. Simulation is ran for 5 swapped cycles, i.e. 1000000 time-steps.

As can be seen on Fig. 4, the performance of the agent is acceptable despite the swapping between associations (which are visible as small drops of the gain). On average, the ratio of good picking ($\frac{food}{food+poison}$) reaches roughly 0.8. Trials with the agent picking the optimal action all the time led to an average ratio of 0.95 (because of the potential picking of the wrong object while following the correct gradient). The loss of performance compared to the optimal strategy is due to the Boltzmann exploration that sometimes picks the wrong action. This loss is necessary though, because actions with an expected bad outcome have to be reevaluated at some point (especially during the swapping phase).

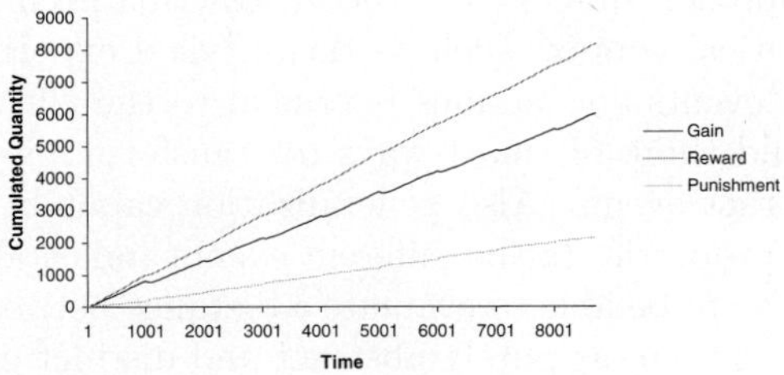

Fig. 4. Evolution over time of performance during the survival experiment. Horizontal axis is in hundreds of simulation time-steps. On the vertical axis are represented the cumulated quantities of food collected (*reward*), poison collected (*punishment*), and the overall net gain (*reward* – *punishment*). Every 1000 hundreds of time-steps gradient associations are swapped, hence the small drops of gain during the adaptation phase. Considering the overall experiment, the agent performs quite well with a nearly constantly increasing gain.

5 Conclusion

We have proposed a framework for stating the event-based anticipation problem and two mathematical tools to extract relevant predictive relationships from the stream of events. Both tools rely on a first low level extraction of the probability distributions of delays between pairs of events. We then extract a first kind of relationship τ that characterizes the regularity of the time-delay between the two events. The second relationship extracted κ reflects the contingency between events. Experimental results obtained through simulation reveals the complementary nature of the two measures in terms of robustness to different delays and variability of this delay.

We also proposed a mechanism for action-selection relying on the extracted model. This mechanism is based on the computation of a reward expectancy similar to reinforcement learning [4], but differs in the fact that actions are considered as events instead of state-action pairs. This approach leads to more generality as phenomena outside the scope of the agent ability to act (i.e. phenomena occurring in the environment without action) can be represented by the internal model. Such an approach has also a cost, which is the loss of the ability to consider the context (which is explicitly represented in reinforcement learning by the state). This ability has then to be reintroduced externally, whch is part of our current investigations.

The overall architecture has been evaluated in a rather simple task where an agent had to collect food resources and avoid poison by following chemical gradients, while the associations between the gradients and the corresponding objects are regularly swapped, requiring the agent to adapt to the new environment. Simulation results showed a rather good performance compared to the optimal strategy.

However our approach suffers from the abstraction needed in the definition of events. In a practical context such as the survival experiment, defining the relevant perceptual events and actions is crucial to the effectiveness of the architecture. One could think of smart ways to transform low-level sensory data into high-level abstract events. Also generalization capabilities could be introduced by attaching properties to the different events and establishing metrics on these properties, so as to be able to evaluate a distance between different events, instead of considering them as purely abstract and distinct events.

References

1. Butz, M.V., Sigaud, O., Gérard, P.: Internal models and anticipations in adaptive learning systems. In: Butz, M.V., Sigaud, O., Gérard, P. (eds.) Anticipatory Behavior in Adaptive Learning Systems. LNCS (LNAI), vol. 2684, Springer, Heidelberg (2003)
2. Capdepuy, P., Polani, D., Nehaniv, C.L.: Construction of an internal predictive model by event anticipation. In: Proc. of the 3rd Workshop on Anticipatory Behavior in Adaptive Learning Systems (2006)
3. Nehaniv, C.L., Polani, D., Dautenhahn, K., te Beokhorst, R., Cañamero, L.: Meaningful information, sensor evolution, and the temporal horizon of embodied organisms. In: ICAL 2003: Proc. of the 8th int. conf. on Art. life, pp. 345–349. MIT Press, Cambridge (2003)
4. Sutton, R.S., Barto, A.G.: Reinforcement Learning: An Introduction. MIT Press, Cambridge (1998), `citeseer.ist.psu.edu/sutton98reinforcement.html`
5. Tolman, E.C., Honzik, E.C.: "insight" in rats. University of California Publications in Psychology 4, 215–232 (1930)
6. Waddell, J., Dzakpasu, R., Booth, V., Riley, B., Reasor, J., Poe, G., Zochowski, M.: Causal entropies-a measure for determining changes in the temporal organization of neural systems. J. Neurosci. Methods (2006)

Aging in Artificial Learning Systems

Sarunas Raudys

Department of Informatics, MIF, Vilnius University
Naugarduko st. 24, Vilnius LT-03225, Lithuania
raudys@ktl.mii.lt

Abstract. While looking at aging of individuals, we take for granted that one of aging reasons is related with individual's ability to learn rapidly, to adapt to sudden environmental changes and survive. We explain maturation and aging of standard non-linear single layer perceptron by increasing of components of the weight vector and dramatic decline of a gradient. We analyze also artificial immune system trained by the mutation based genetic learning algorithm. In both, the connectionist and genetic learning, we obtain saturation and an inverted letter "U" shape dependence between success in learning and the "age".

1 Introduction

The aging and senescence are words that express gradual, slow degradation of plants, animals, human beings, human made devices: cars, computers and computer software. Biologists relate aging with a large number of "the age related diseases" [1]. Many factors are affecting aging process. A death, regeneration of biological cells and tissues, the accumulation of bad mutations, progressive breakdown of unreliable parts of the organism are just a few examples [1], [2].

In physiology, psychology, sociology, education and related disciplines we are observing the inverted letter U" shape dependence between success in learning and the age. Typical example is human immune system (HIS). Here, learning is based on genetic search where mutations are playing leading role [3], [4]. In Fig. 1 we present 80 years old graph describing human humoral immunity [5] as a function of its age.

In informatics, however, we do not have sufficient number of models suitable to describe a wide spectrum of processes taking place during aging and death. It was assumed that units in the artificial neural network (ANN) represent synaptic connections among neurons in the human brain [7]. Synaptic connections among neurons in

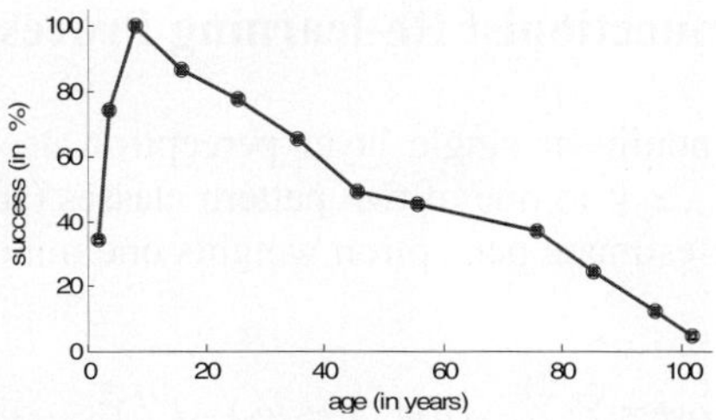

Fig. 1. The human humoral immunity (in %) as a function of its age (adapted from data in [6])

F. Almeida e Costa et al. (Eds.): ECAL 2007, LNAI 4648, pp. 263 – 272, 2007.

a brain are changing with age. Much of the data on the slowing of human information processes with advanced age can be modeled by neural networks that have been corrupted in ways suggested by neuropathology. The aging was modeled as progressive breakdown in neural connectivity and a progressive attenuation in neural conduction [8]. In [9] aging of single layer perceptron (SLP) was explained by two "inherited" factors: 1) a gradual increase in magnitudes of the components of weight vector and 2) by a diminution of the gradient. A model where the non-linear SLP is trying to learn two different pattern recognition (PR) tasks was suggested. If the perceptron is trained very shortly or very lengthy to solve the first PR task, it learns the second PR task slowly. The mildly trained perceptron learns the second PR task much easier. The curve "success in training as a function of the age (a number in iterations used to learn the first PR task)" has an inverted letter U" shape (details in Section 2).

The bit-string Penna model [10] and its modifications [11], [12] are very popular in analysis of populations. Penna [10] represented the genome by a bit-string such that "bit zero" means a healthy and "bit one" means a sick gene. The errors in the replication of the genome in the DNA, accumulated in the course of biological evolution, mainly affect the old ages after childbirth. Biecek and Cebrat [13] have modified the Penna standard model in such a way, that the state of each individual in the population has been determined by fluctuations of the individual and that of the environments. They assumed that the individuals can learn changes of the environment. They postulated that negative effect of environmental change, E_{EC}, is diminishing exponentially $E_{EC} = e^{time/\lambda}$, where λ corresponds to learning speed. This mechanism also had lead to the "∩" shape aging curve.

In contrast to model [13], we investigate learning model adopted from modern artificial immune systems (AIS). We consider aging of an individual, not of the population. To reveal principal trends we analyze the simplest version of conventional AIS and show that non-trained or very well trained AISs meet difficulties like in connectionist approach. Moderately trained AIS learn to recognize new types of antigens easier.

In Section 2 we refer the two PR task aging model based on the analysis of re-learning rapidity. We elucidate why do mildly trained non-linear SLPs are more successful as the untrained or very well trained ones. We advocate that maturation and aging are inherent peculiarities of standard nonlinear SLP and its gradient descent training algorithm. This methodology is used to investigate aging of AIS in Section 3. Here after specifying the AIS, we present theoretical and experimental examination of the AIS and also demonstrate a saturation and inverted "U" letter shape dependence between success in learning and the "age". Section 4 contains a discussion.

2 Aging in Connectionist Re-learning Process

Consider standard nonlinear single layer perceptron used to allocate p-dimensional vector, $\boldsymbol{x} = [x_1, x_2, ..., x_p]^T$ to one of two pattern classes (superscript "^{T}" denotes transpose operation). To estimate perceptron weights one minimizes a sum of squares cost function [14], [15]

$$cost = \sum_{i=1}^{2} \sum_{j=1}^{N_i} (f(\boldsymbol{w}^T \boldsymbol{x}_{ij} + w_0) - t_j^{(i)})^2, \qquad (1)$$

where $f(arg)=1/(1+\exp(-arg))$ is sigmoid activation function, vectors $\boldsymbol{x}_{i1}, \boldsymbol{x}_{i2}, \ldots \boldsymbol{x}_{iN_i}$ represent training set of i-th category; N_i is a number of training vectors of i-th class, $t_j^{(i)}$ stands for desired output, e.g. $t_j^{(1)} = 0$, $t_j^{(2)} = 1$. We investigate standard gradient descent training procedure in a batch mode: $\boldsymbol{w}_{(t)} = \boldsymbol{w}_{(t-1)} - \eta\, \partial cost / \partial \boldsymbol{w}$, η is learning step.

If we start training from the weight vector with small components, the weighted sums, $\boldsymbol{w}^T \boldsymbol{x}_{ij} + w_0$, are small. In such case, the gradient, $\partial cost / \partial \boldsymbol{w}$, is large. As a consequence, training process is fast at the very beginning. With a progress of training procedure, the components of the weights vector are increasing [9], [15]. *Therefore, the gradient is diminishing*. The learning process slows down in that case. If the perceptron succeeded to classify training data without error, the weights can become very large.

A sense of aging analysis introduced in [9] is two distinct PR tasks, PR1 and PR2. The SLP is trained with data of task PR1 at first. After t_1 iterations the PR task is changed. The perceptron "does not know" that the PR task was changed. Hence, it continues to modify its weight vector further. After the PR task change, the weights of well trained SLP remain large for a long time. So, the gradient values are small and learning process is slow. Thus, well trained perceptron can fail to adapt to second pattern recognition task, PR2, during finite number, say $t_2 = 400$, iterations.

If the number of iterations, t_1, is rather small, the components of its weight vector will remain to be small. So, training with data PR2 will be more successful. Fig. 2 shows two typical *re-learning curves*: a dependence of the probability of correct classification on the number of iterations t_1 used for its training with data PR1.

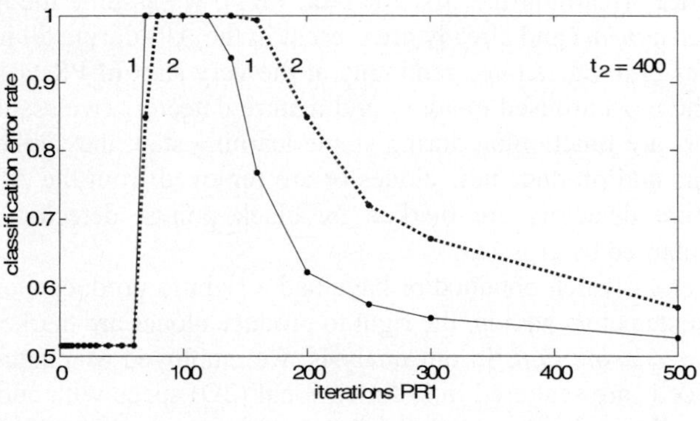

Fig. 2. Classification error or the re-trained SLP as a function of the number of iterations, t_1

Graphs 1 and 2 in are results of a single experiment. Both graphs remind aging curve in Fig. 1. Graph 1 corresponds to desired outputs 0 and 1. Graph 2 corresponds to experiments where 5% of desired outputs were corrupted. In this case, weight magnitudes are increasing slowly: both the maturation and the ageing processes are retarded.

Corruption of training directives could be interpreted as criminality, changes of laws, consequences of differences in value systems, mild environmental catastrophes or economy crises. A noise injection to inputs, possibly, could be interpreted as measurement errors, distortion or blurring of the images. A difference between target values, $s = t_j^{(2)} - t_j^{(1)}$, could be interpreted as stimulation strength. A noise injection to the inputs or to desired outputs, the difference between the target values are a part of learning algorithm. *It means that maturation and aging are inherent peculiarities of standard nonlinear SLP and gradient descent training algorithm.*

3 Aging of Artificial Immune System

In this section we will make use of the two PR tasks methodology to investigate learning rapidity (aging) of conventional static AIS [3], [4]. We will explore learning part of the AIS which is aimed to generate a number of detectors similar to the non-self cells (antigens) and capable to kill them.

Definitions. The AIS considered is designed to work with antigens, detectors and antibodies described by a set of binary (0 1) codes. Following elements and definitions are necessary to characterize learning block of AIS used in succeeding analysis:

- N_{self} randomly selected *self cells* characterized by *p*-dimensional sequences of "zeros" and "ones", $s_1, s_2, \ldots, s_{p/2}, s_{p/2+1}, \ldots, s_p$;
- N_{nonself} *p*-dimensional *non-self cells* (antigens), a set $\mathcal{Ag}$. The number of antigens, N_{nonself}, may shrink until "zero" when detectors evolve, become similar to the non-self cells and finally kill them;
- N_{memory} *p*-dimensional *memory detectors* (antibodies), accumulated during "previous experience" (learning the first PR task, PR1); we assume the memory detectors are given *a priori* and already are present in the AIS during all learning cycle;
- *immature detectors*, generated randomly at the very start of PR task PR2, or later cloned by the most aroused memory and mature detectors (we assume that immature detectors are functioning during single learning step: they either become mature detectors and produce new clones or are removed from the AIS; in Figure 3 new immature detectors are marked by black points, detectors with positive arousal are marked by crosses);
- *mature detectors* which obtained or have had a right to produce clones in the past (in Fig 3, the detectors having the right to produce clones are marked by circles);
- *a distance measuring unit.* In our analysis, we employed Manhattan distance assuming the cells are scattered in 2-dimensional (2D) space with coordinates *x* and *y*. The *j*-th cell coordinate x_j is defined as a scalar product of *p*/2-dimensional binary code vector $\boldsymbol{C}_{xj}$, and weight vector $\mathbf{w}_{\text{bin}} = (1\ 2\ 4, \ldots, 2048)$: $x_j = \boldsymbol{C}_{xj}(\mathbf{w}_{\text{bin}})^T$ (in standard genetic approach all components of vector $\mathbf{w}_{\text{gen}}$ are equal to 1). The distance between two cells, *j*-th and *r*-th ones, is determined as a sum of distances along axes *x* and *y*

$$Dist(j,r) = |\, \boldsymbol{C}_{xj}(\mathbf{w}_{\text{bin}})^T - \boldsymbol{C}_{xj}(\mathbf{w}_{\text{bin}})^T \,| + |\, \boldsymbol{C}_{yj}(\mathbf{w}_{\text{bin}})^T - \boldsymbol{C}_{yj}(\mathbf{w}_{\text{bin}})^T \,|. \quad (2)$$

We selected the distance measure (2) with weight vector $\mathbf{w}_{\text{bin}}$ due to the possibility to illustrate the evolutionary search process in the 2D space.

- *a threshold*, Δ_{self}, that defines maximal similarity between self cells and the detectors: if a distance, *Dist*(*j*, *r*)), between the *j*-th detector and closest to it self cell, say the *r*-th one, is smaller as Δ_{self}, the *j*-th detector is removed from the AIS;
- *a threshold*, Δ_{kill}, that defines maximal allowed similarity between the *i*-th non-self cells and a the *s*-th detector closest to it (if the distance $Dist(i, s) < \Delta_{\text{kill}}$, the *i*-th non-self cell is killed);
- *a measure of stimulation* of the *r*-th detector by the *j*-th non-self cell is defined as

$$Stimulation(r, j) = w_{\text{stim}} / (h_{\text{stim}} + Dist(j, r)); \tag{3}$$

- *a measure of suppression* of the *i*-th detector by the *r*-th one is defined as

$$Suppression(i, r) = w_{\text{suppress}} / (h_{\text{suppress}} + Dist(i, r)), \tag{4}$$

  where constants w_{stim}, w_{suppres}, h_{stim} and h_{suppres} are determining an importance and a character of decline of *Stimulation*(*rj*) and *Suppression*(*ir*);
- *an arousal* of the *j*-th detector which is expressed as a difference between stimulation of closest to the *j*-th detector non-self cell (say, the *r*-th one) present in the system at given time moment and a sum of suppressions of all remaining detectors

$$Arousal(j) = Stimulation(r, j) - \sum_{s=1,\, s\neq j}^{N_{\text{memory+new+clones}}} Supression(s, j)\,. \tag{5}$$

In this paper, we adopted the simplest way to control the suppression between competing detector cells: *the cell will highest stimulation is not suppressed by other ones. Remaining detectors suppress each other*. A diversity of the non-memory detectors is being increased in such a way. Detectors with positive arousal values have a right to produce clone detectors. The number of clones, $N_{clone}(j)$, produced by the *j*-th aroused detector depends on a sum of arousals of all remaining detectors

$$sum_{\text{arous}} = \sum\nolimits_i Arousal\,(i).$$

Thus, maximal number of clones allowed to be generated at given time moment

$$N_{\text{clones\_max}} = \alpha \times \left(N_{\text{detect\_max}} - (N_{\text{memory}} + N_{\text{mature}})\right), \tag{6}$$

where $N_{\text{detect\_max}}$ is a maximal number of detectors allowed for the AIS. N_{mature} is a total number of mature detectors produced for the duration of initial detector generation at the very start of task PR2 and the detector cloning procedures in the course of succeeding steps of evolutionary learning process. Scalar α ($0 < \alpha < 1$) could be interpreted as *learning speed parameter*. Then

$$N_{clone}(j) \approx \text{round}\left(N_{\text{clones\_max}} \times Arousal(j)/sum_{\text{arous}}\right). \tag{7}$$

Situations happen where we have a large number of detectors with $0 < Arousal(j) < 0.5$. Then necessary number of the clones is produced only by the most aroused antibodies. The clones were generated by adding mutations to coordinates of the parent antibody. Amplitudes of the mutations were proportional to distances to closest non-self cell in each of directions (x and y) plus shift Ξ, a constant, performed in the same direction.

An Age of Artificial Immune System. We start investigating the learning process of the AIS with N_{memory} detectors already accumulated during previous life period (task PR1). *We interpret the number* N_{memory} *as the age. We assume* the AIS is young, if we have a small number of memory detectors. The AIS is supposed to be old, if N_{memory} is high. In latter case, a space for cloning of the detectors could become too small.

At the very start of task PR2, N_{start} immature detectors are generated. A part of positively aroused immature and memory detectors (a set SS) is producing the clones as described above. In analysis of teaching the AIS to recognize and kill antigens *Ag*, we suppose that set SS of immature detectors become mature detectors and are not removed form the AIS.

Specifications of the AIS Model in Two-Dimensional Space. To evaluate an influence of the number of mature detectors on learning rapidity quantitatively we have to itemize a model of the AIS distinctively. Below we will present results obtained with the AIS defined on 4048×4048 rectangular 2D grid G_{AIS}. The self cells are situated in a left bottom corner of 2D space of grid G_{AIS}. Non-self cells, *Ag*, are situated in an upper right corner of the grid (Fig. 3).

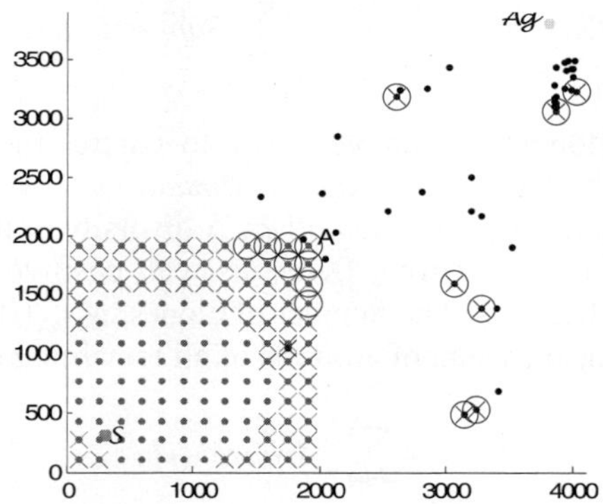

Fig. 3 2D grid of AIS: 36 non-self cells *Ag* and 36 self cells self, *S*, 12×12 array of memory detectors (magenta, left bottom part), mature (marked by circles) and new detectors (points).

We suppose that during previous life (task PR1), the AIS had met antigens more or less similar to its own genetic code. In such case, mature detectors will be dispersed around the individual's self cells. We assumed that a volume of an area where memory detectors are scattered is increasing with a number N_{memory}. So, the memory detectors are defined as $\sqrt{N_{memory}} \times \sqrt{N_{memory}}$ rectangular grid with a side

$$G_{mem} = (4048 \times (3.5 + \sqrt{N_{memory}}) \times 0.7) / \sqrt{N_{detect\,max}}. \tag{8}$$

In the AIS model used in the numerical and experimental analysis, 36 self cells were situated on grid $x_i, y_j \subset (290\ 315)$ and 36 non-self cells, *Ag*, $x_\gamma, y_\eta \subset (3800\ 3825)$.

Simulation Experiments. While specifying our model of the AIS, we postulated that *the maximal number of detectors*, N_{detect_max}, *is fixed a priori*. Thus, during cloning we may produce only a fixed number of new detectors defined by Eq. (6). It means the number of clones produced at each time moment depends on both N_{memory} and on N_{mature}, the number of mature detectors that had been accumulated throughout preceding steps of evolution while learning to solve task PR2.

In aging analysis, it is important to study an influence of the number of memory detectors, N_{memory}, on learning rapidity. Due to non-linear relations between stimulation and suppression and random character of generated detectors, exact quantitative analysis is hardly possible. In Figure 4 we present result of one series of simulation experiments. In this figure we depictured the success in learning to recognize and kill non-self cells (in percents) evaluated in 200 runs of genetic learning process. The runs differed in starting position of random number generator. Following set of parameters was used: $N_{detect\_max} = 600$, $w_{stim} = 26$, $w_{suppres} = 1$; $h_{stim} = 15$ and $h_{suppres} = 2$, $\Delta_{self} = 8$, $\Delta_{kill} = 8$, $\Xi = 6$. Success was evaluated as the ability to recognize and kill 35 non-self cells out of 36 ones during $t_{max} = 30$ evolution steps.

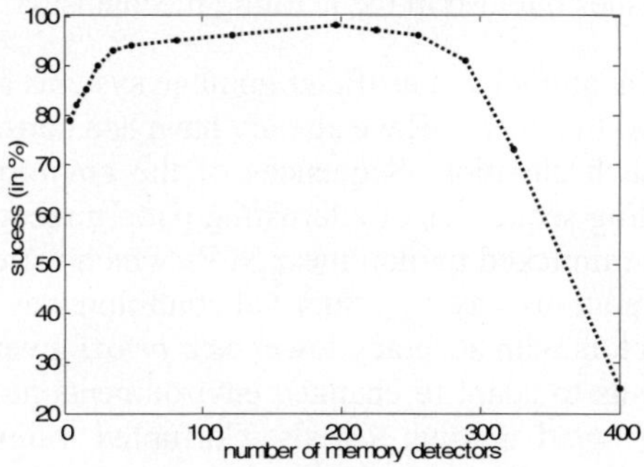

Fig. 4. Success in detecting a new type of the antigen in dependence on a number of memory detectors developed during previous life of the AIS

The graph in Fig. 4 confirms the inverted "U" letter shaped dependence between success in learning and the "age" measured in a number of mature detectors accumulated during numerous presentations of different non-self cells in the past.

4 Concluding Remarks

Aging Similarities in the Perceptrons and Artificial Immune Systems. In order to be efficient, the young, however, already mature individuals should meet a number of requirements. The ability to learn rapidly, adapt to environmental changes is one of them. Necessity to adapt rapidly to changes applies also to groups of individuals, organizations, as well as to artificial products including autonomous robots, embedded systems and computer programs. Theoretical analysis and simulation experiments explain why do standard mildly trained non-linear single layer perceptron is more successful as untrained or very well trained ones. Theoretical and experimental analysis of the standard AIS trained by the mutation based genetic learning algorithm confirms the conclusion obtained while investigating of error correction adaptation method of the SLP based classifier. *In our model, free memory reserved to store memory detectors becomes fulfilled (saturated) in old age. As a result, the AIS cannot learn new type of antigens rapidly.* We observe similarly shaped nonlinear dependence between success in learning and the "age" in this situation too.

In neural network training, inability to learn rapidly is caused by excessive growth of the weights and saturation of nonlinear activation function. In genetic training, we explained slow training by excessive increase in a number of memory detectors. While investigating SLP training, we have demonstrated that learning style parameters such as a noise injection to input and/or output signals, reduction of stimulation strength retard maturation and aging processes. In the AIS training, suppression of stimulation of immature detectors also retards maturation and aging processes. We considered very simple model of the AIS so far. We guess that analysis of modern and future sophisticated AIS training procedures would expose more factors that could be used to retard deteriorating processes in the AISs. Similarity of re-learning curves in artificial and natural immune systems hints that among a variety of possible causes, for reasons of the aging one may seek not only in the changes progressing in biological cells and tissues but also in the learning mechanisms.

Moral Factors. Human, animal and artificial immune systems are aimed to operate in changing environments. In paper [16] we already have had considered the agent populations operating in such situation. Sequences of the environmental changes were mimicked by long-lasting sequences of alternating pattern recognition tasks, PR1 and PR2. The agents were mimicked by nonlinear SLPs which were obliged to learn during finite number of iterations, say $t_{\max}$. Survival condition was a requirement to classify unseen pattern vectors with accuracy lower as *a priori* given constant, P_{survival}. To increase ability of agents to adapt to changed environments more rapidly, the agents were trained with corrupted training signals. Corrupted training directives prevent excessive increase in the weight magnitudes and make adaptation to new PR tasks easier Each agent possessed its individual fraction, α_i, of corrupted training directives. A failure to comply the survivability condition results in the agent being removed from the "society of the agents" and be replaced by a "newborn" that inherits learning style parameter, $\alpha_i + \textit{mutation}$, from its "parent agent". To increase resistance to powerful PR task changes, the agents belonging to the same group were helping each other and in a case of emergence – to members of almost dying group.

To illustrate specific problems arising in training of the agent groups we present results of simulation study. We performed simulations following the methodology described in the conference paper [16]. A population of 200 agents was split into eight groups. The first 600 PR task changes were insignificant. During following 200 changes, magnitude of the PR task alterations increased and remained constant during next 200 PR task modifications. Then the magnitude decreased until initial level and did not vary further. Graph 1 in Fig. 5 shows that the number of the agents survived depends on the magnitudes of the PR task changes. In this experiment, threshold P_{survival} was fixed (P_{survival} =0.12).

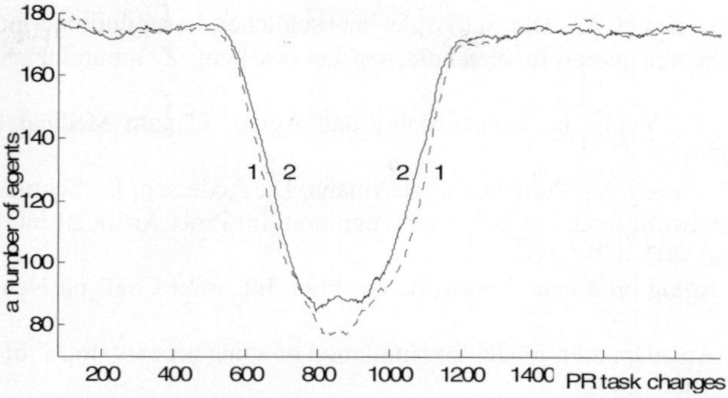

Fig. 5. Dynamics of a number of adaptive agents that survived environmental catastrophes (the pattern recognition task changes)

The first 600 PR task changes were comparatively mild. Therefore, only 20 agents were dying on average. For that reason, genetic adaptation to the changes was slow. *In order to force the agent population to adapt to changes faster, more agents have to die*. Hence, we introduced random survival condition: threshold P_{survival} was a random variable distributed uniformly in interval (0.08 0.16). This technique forces the agents to die more frequently prior and during most powerful PR task changes. Therefore, random survival condition speeds up genetic learning of the "training style" parameter, α_i. In such situation, more agents resisted to the strongest environmental changes (Graph 2), i.e. the agent population became more resistant. It is a typical example.

In this context, *degradation and aging of natural immune system could be interpreted as a noise injection into survival threshold*, P_{survival}, i.e. a positive factor that increases resistibility of the population to most powerful environmental changes. Looking from a point of view of moral, we see that "interests of the Nature" and that of the individuals are different. Individuals seek to slow down degradation and aging. For survival of the population during most powerful environmental changes, moderate degradation and aging of individual agents is beneficial. During millions of years of Nature's evolution, an "attention" was paid to the population survivability. Therefore, analysis of maturation, degradation and aging processes should be performed from a point of view of system approach where both interests of individuals and that of population should be taken into account. In our study, simulation experiments were performed with simplified model and criteria that encourages survival of the populations. Nevertheless, the simulation results raise important moral problems which should be discussed and verified by future large scale experiments *in vivo* and *in silico*.

References

[1] Kirkwood, T.B.L., Austad, S.: Why do we age? Nature 408, 233–237 (2000)
[2] Gavrilov, L., Gavrilova, N.: The reliability theory of aging and longevity. J. of Theoretical Biology 213, 527–545 (2001)
[3] Farmer, J.D., Packard, N.H., Perelson, A.S.: The immune system, adaptation and machine learning. Physica 22D, 182–204 (1986)
[4] De Castro, L.N., Timmis, J.: Artificial Immune Systems: A new computational approach. Springer, London (2002)

[5] Thomsen, O., Kettel, K.: Die Stärke der menschlichen Isoagglutinine und entsprechende Blutkörperchenreceptoren in verschiedenen Lebensaltern. Z. Immunitatsforsch. 63, 67–93 (1927)

[6] Makinodan, T., Yunis, E.: Immunololgy and Aging. Plenum Medical Book Company, London (1977)

[7] Alvager, T., French, V., Putman, G., Herrmann, D., Anderson, E., Schnitzer, S.: An artificial neural network model of aging and cognition. In: Proc. Artificial Intelligence and Applications, pp. 403–440 (2003)

[8] Cerella, J.: Aging on a neural network. In: Proc. Int. Joint Conf. on Neural Networks, p. 601 (1989)

[9] Raudys, S.: An adaptation model for simulation of aging process. Int. J. of Modern Physics C 13, 1075–1086 (2002)

[10] Penna, T.J.P.: A bit-string model for biological aging. J. of Statistical Physics 78, 1629–1633 (1995)

[11] Pletcher, S.D., Neuhauser, C.: Biological aging (criteria for modeling and a new mechanistic model. Int.l J. Modern Physics C 11, 525–546 (2000)

[12] Huang, Z.F., Stauffer, D.: Stochastic Penna model for biological aging. Theory in Biociences 120, 21–28 (2001)

[13] Biecek, P., Cebrat, S.: Immunity in the noisy Penna model. Int. J. Modern Physics C 17(12), 1823–1829 (2006)

[14] Haykin, S.: Neural Networks: A comprehensive foundation, 2nd edn. Prentice Hall, Upper Saddle River (1999)

[15] Raudys, S.: Statistical and Neural Classifiers: An integrated approach to design. Springer, London (2001)

[16] Raudys, S., Pumputis, A.: Group interests of agents functioning in changing environments. In: Pěchouček, M., Petta, P., Varga, L.Z. (eds.) CEEMAS 2005. LNCS (LNAI), vol. 3690, pp. 559–563. Springer, Heidelberg (2005)

An Analysis of the Effects of Lifetime Learning on Population Fitness and Diversity in an NK Fitness Landscape

Dara Curran[1], Colm O'Riordan[2], and Humphrey Sorensen[3]

[1] Dept. of Computer Science,
University College Cork,
Ireland
d.curran@cs.ucc.ie
[2] Dept. of Information Technology,
National University of Ireland,
Galway
colm.oriordan@nuigalway.ie
[3] Dept. of Computer Science,
University College Cork,
Ireland
sorensen@cs.ucc.ie

Abstract. This paper examines the effects of lifetime learning on the diversity and fitness of a population. Our experiments measure the phenotypic diversity of populations evolving by purely genetic means (population learning) and of others employing both population learning and lifetime learning. The results obtained show, as in previous work, that the addition of lifetime learning results in higher levels of fitness than population learning alone. More significantly, results from the diversity measure show that lifetime learning is capable of sustaining higher levels of diversity than population learning alone.

1 Introduction

Much artificial intelligence research has focused on the interaction between learning and evolution, where individuals within a population of artificial organisms are capable of evolving genetically (population learning) and also of acquiring knowledge during their lifetime (lifetime learning). Hinton and Nowlan[1] were among the first to show that learning could guide evolution and a number of other researchers have since conducted experiments which support this view[2,3,4,5,6,7,8].

Each experiment simulates both a population of organisms and their surrounding environment. Typically, environmental stimuli are mapped to a specific problem domain. For instance, a population may inhabit an environment where individuals must navigate a maze or forage for food.

The majority of these investigations are concerned with the impact that learning has on the fitness of the population, finding in general that the addition

F. Almeida e Costa et al. (Eds.): ECAL 2007, LNAI 4648, pp. 273–283, 2007.

of lifetime learning has a positive effect. In addition, studies have shown that both the Baldwin and Hiding effects can be simulated in populations of artificial organisms[9].

However, there is comparatively little research on other effects of learning, in particular with regard to population diversity. It is generally accepted that the maintenance of diversity can prevent an evolutionary algorithm from becoming trapped in local maxima. A number of techniques exist to both monitor and stimulate diversity[10,11,12,13].

The aim of this paper is to examine whether diversity can be maintained by lifetime learning. We employ an abstract model based on Hinton & Nowlan's work and conduct a number of experiments. The remainder of this paper is organised as follows. Section 2 describes related research, including the learning models on which this work is based and a discussion on diversity. Section 3 presents the model employed for these experiments, including a detailed description of the diversity measure. Section 4 outlines the experiment results and Section 5 presents conclusions.

2 Related Work

The following sections outline some background material, including learning models and diversity.

2.1 Learning Models

Population learning refers to the process whereby a population of organisms evolves, or learns, by genetic means through a Darwinian process of iterated selection and reproduction of fit individuals[14]. In this model, the learning process is strictly confined to each organism's genetic material: the organism itself does not contribute to its survival through any learning or adaptation process.

There exist species in nature that are capable of learning, or adapting to environmental changes and novel situations, at an individual level. Such learning, known as life-time learning, is often coupled with population-based learning, further enhancing the population's fitness through its adaptability and resistance to change. Much research explores the interactions between evolution and learning, showing that the addition of individual lifetime learning can improve a population's fitness[2,3,6,7,8].

Hinton & Nowlan. Hinton and Nowlan employed a genetic algorithm to study the effects of lifetime learning on the performance of genetic evolution. In particular, they examined the Baldwin effect, which occurs when certain behaviours discovered through life-time learning become imprinted onto an individual's genetic material through the evolutionary processes of crossover and mutation[1,15]. This should not be confused with Lamarckian theory, where an individual's lifetime experience is directly mapped to its genome and is directly inheritable[16].

Each agent in the model possesses a genome, represented by a string of characters which can be either *1*, *0* or *?*, mapping to a neural network structure. Each agent is allowed a number of rounds of lifetime learning where for each *?* in the genotype they 'guess' its value, assigning it either a *1* or a *0*. The *?* in the genotype represent the agent's phenotypic adaptability and any guessed values are lost at the end of the agent's life.

The experiments compared the performance of a population endowed with learning to one without. Results showed that the non-learning population was not capable of finding optimal solutions to the problem task. In contrast, once learning was applied, the population converged on the problem solution, showing that individual learning is capable of guiding genetic evolution. A number of researchers have since replicated the Baldwin effect in populations of artificial organisms[17,6,18,19,20].

Turney. The model employed by Turney examined the shift in bias in simulations similar to those undertaken by Hinton and Nowlan [18]. The definition of bias in this context is all factors excluding input data that influence the selection of a concept. Bias can include the algorithm that the learner employs to search through concept space and the criteria for determining whether a selected concept is compatible with the training data the learner has received.

An agent's genome is a combination of bias direction genes and bias strength genes, where bias direction genes are *0*s or *1*s while the bias strength genes are real number values in the range 0 to 1. In a 64 element string, 32 alleles represent bias direction and a further 32 represent bias strength. If bias is strong at a particular allele, the individual will be more likely to select the bias direction present in its genome. However, if bias is weak then the individual will be more likely to guess at a bias direction rather than employing the genetically encoded value.

In other words, while the *?*s in Hinton and Nowlan's experiments were replaced with *1*s or *0*s with equal probability (which did not change during the course of the experiments), Turney's model allows the evolutionary process to shift biases towards particular guesses.

2.2 Diversity

Diversity measures typically quantify the differences between individuals in a population. It is commonly accepted that a population that is capable of maintaining diversity will avoid premature convergence and local maxima. Diversity measures for populations of neural networks have been the focus of considerable research, focusing mainly on genotypic diversity[21,22,23,24].

Eriksson conducted an experiment designed to measure whether lifetime learning is capable of maintaining diversity [25]. The experiment employed Turney's model and measured the diversity of bias direction genes in the population. Experiments found that lifetime learning only produced higher diversity when bias strengths were explicitly initialised at low values. In all other cases, population learning produced higher diversity levels.

Previous work has examined the effects of cultural learning on both the phenotypic and genotypic diversity of populations of evolving neural network agents whose problem task was comprised of a number of turn-based games[26]. Cultural learning in this context is a form of lifetime learning where agents learn from other members of the populations. Results showed that the combination of cultural learning and population learning led to improvements in fitness and higher levels of genotypic diversity. However, the complexity of the model employed makes analysis difficult, highlighting the need for a simpler model of both agent and lifetime learning mechanism alike.

3 Model

Our model follows the structure of the original Hinton and Nowlan experiments. Each agent comprises a genecode of 20 bits and experiments employ a population of 1000 agents evolved for 400 generations. The experiments employ two populations: one using population learning alone (purely genetic evolution) and the other employing lifetime learning.

Where a population evolves solely by population learning, agent genomes consist of strings of 20 *1*s or *0*s randomly generated in the initial population with equal probability. Populations employing lifetime learning have genomes containing *1*s, *0*s or *?*s, where the *?*s represent the agent's phenotypic ability to guess either *1* or *0* in a similar manner to the original experiments. In the initial population these are randomly selected with probability 0.25, 0.25 and 0.5 respectively, corresponding to the original probability set in the Hinton and Nowlan experiment.

The fitness landscape of the original Hinton and Nowlan experiments was a particularly challenging one for populations attempting to evolve genetically, consisting in effect of a "needle in the haystack" problem where no evolutionary feedback is given to partial solutions. An individual with no lifetime learning would have a fitness of 1 unless its genome consisted of exactly twenty *1*s, in which case its fitness would be 20. It is highly unlikely that an individual will be randomly generated with such a genome and, since there is no path for the evolutionary process to follow, it is very difficult for population learning alone to solve this problem.

Our model employs the NK fitness model as the fitness landscape for the experiments [27]. The NK fitness model provides some evolutionary feedback to populations evolving genetically and therefore allows a fairer comparison between evolutionary and lifetime learning.

3.1 Fitness Calculation

To calculate the fitness of each individual in the population, an NK fitness landscape is generated and the maximum fitness is calculated. The fitness of an individual is normalised with respect to the maximum possible fitness, obtaining a fitness measure in the range $[0, 1]$.

Populations employing purely genetic evolution (population learning alone) possess genomes comprising of *1*s or *0*s (no *?*s) and therefore have no opportunity to alter the way they interact with their environment. The fitness of agents in such populations is measured directly from their genomes.

When a population adds lifetime learning to population learning, its members are given the opportunity to replace each of the *?*s in their genomes with either *1*s or *0*s. Each agent guesses the value of its *?* with equal probability. The number of guesses an agent makes during its lifetime can be tuned to allow lifetime learning a greater or lesser influence on the evolutionary process. Each guess is evaluated using the fitness function and an agent's best guess is taken as its fitness value (see Fig. 1).

3.2 Diversity Measure

The diversity measure examines the differences between members of a population and, for the purposes of this work, it is important that a diversity measure capture the differences between populations employing population learning alone and those employing both population and lifetime learning.

The population employing population learning alone contains genomes comprising of *1*s or *0*s, while the population adding lifetime learning has an additional number of *?*s within its genomes. This difference between the two populations makes a direct genetic comparison difficult. Therefore, a phenotypic measure of diversity may be more suitable. In this model, the phenotype of an agent is defined as its response to its environment.

In the population employing population learning alone, an agent's response to its environment is fixed and directly influenced by its genome. Therefore, its genotype and phenotype are identical - the agent has no lifetime influence on its behaviour. However, in the population employing lifetime learning, an agent is allowed make a number of guesses which replace the *?*s in its genome with either *1*s or *0*s. Since the agent's fitness is calculated according to the best of these guesses, we chose to select an agent's best guess as its phenotype.

Employing this model, phenotypic diversity measurements of populations employing population learning alone and populations employing both population and lifetime learning are directly comparable, as both are measuring differences between strings of equal cardinality (containing either *1*s or *0*s).

The diversity measure is implemented in a manner similar to Eriksson's and is defined as the average of all individual Hamming distances between phenotypes of individuals x and y, $h(x,y)$, whose phenotypes are unique within the population. Thus, for a population of size n containing m unique phenotypic binary strings, the diversity measure can be calculated as:

$$\frac{2}{n(m-1)} \sum_{x=1}^{m} \sum_{y=x+1}^{m} h(x,y)$$

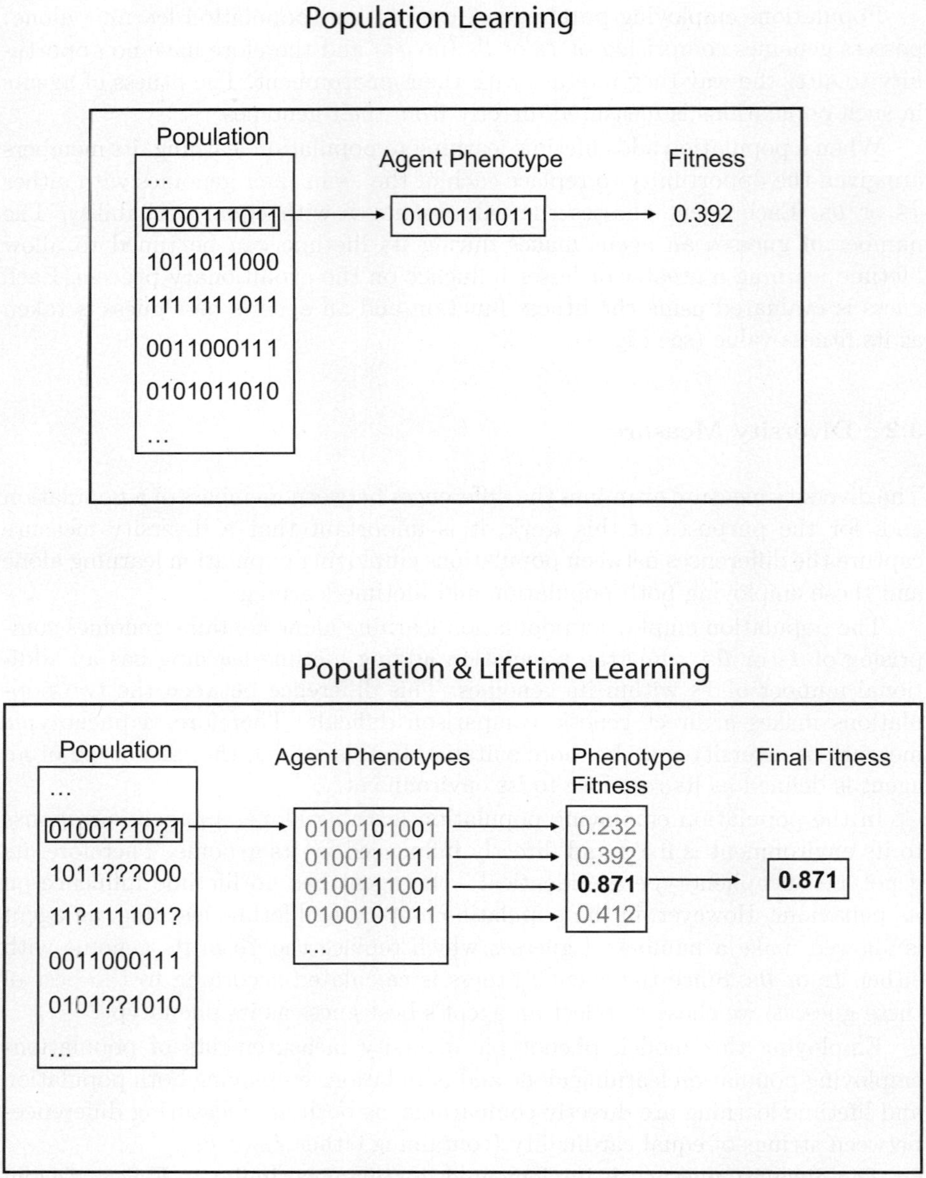

Fig. 1. Fitness evaluation

4 Experiments

The experiments employ three populations: one employing population learning alone, one employing both population and lifetime learning with 10 learning cycles and the final one employing both population and lifetime learning with 30 learning cycles. Both fitness and diversity are measured for each experiment. Populations of 1000 agents are allowed to evolve to 400 generations. Linear ranking selection is employed, 1-point crossover occurs with 0.6 probability and mutation occurs with 0.02. The NK landscape parameters are N=20, K=15. Results presented are averaged from 20 independent experiment runs.

4.1 Fitness

Fig. 2 illustrates the results obtained from the three populations over 400 generations. Each of the populations converges within 100 generations to high levels of fitness. However, it is clear that both populations employing population and lifetime learning have higher fitness levels than the population employing population learning alone.

Both lifetime learning populations begin the experiment with a higher fitness level than population learning alone and both gradually increase to convergence. By contrast, the non-learning population's increase in fitness is much sharper,

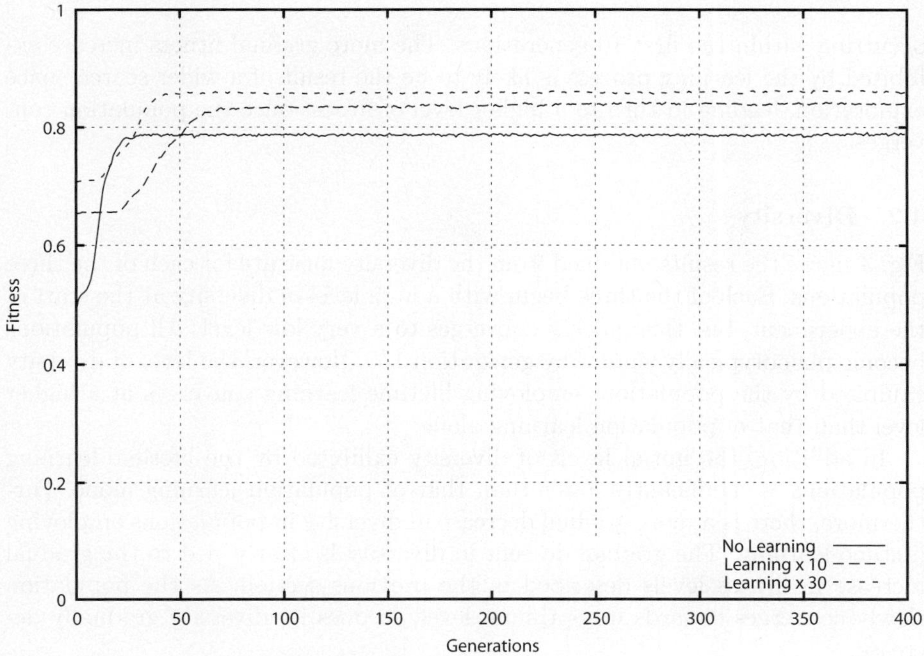

Fig. 2. Average Fitness

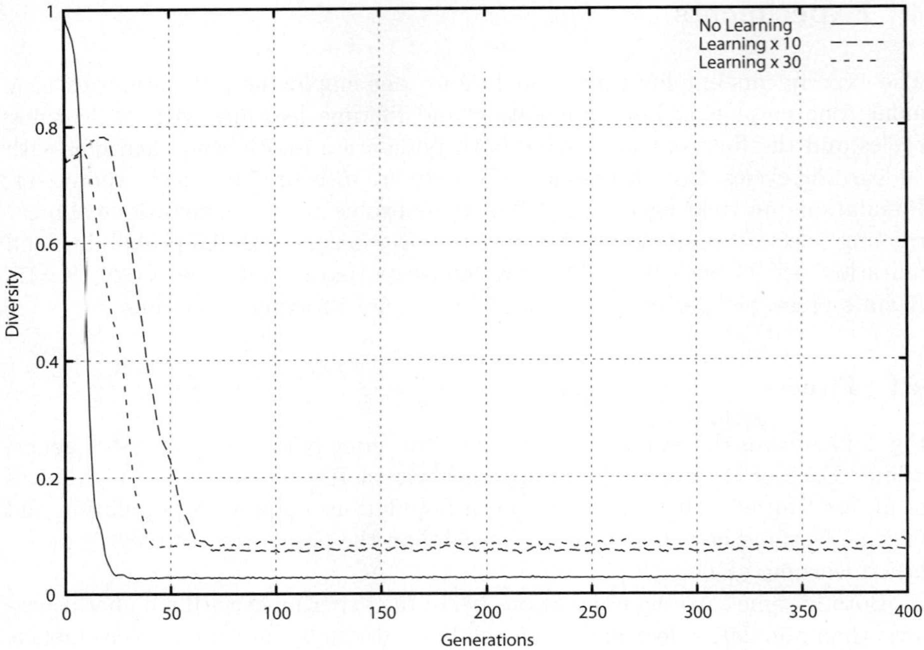

Fig. 3. Average Diversity

occurring within the first 10 generations. The more gradual fitness increase exhibited by the learning process is likely to be the result of a wider search space exploration, leading in turn to a higher level of fitness once the population converges.

4.2 Diversity

Fig. 3 shows the results obtained from the diversity measure for each of the three populations. Each of the three begin with a high level of diversity at the start of the experiment, but this quickly converges to a very low level. All populations become phenotypically similar by generation 100. However, the level of diversity exhibited by the populations employing lifetime learning converges at a higher level than that of population learning alone.

In addition, the initial levels of diversity exhibited by the lifetime learning populations is significantly lower than that of population learning alone. Furthermore, there is a more gradual decrease in diversity in populations employing lifetime learning. The gradual descent in diversity is clearly tied to the gradual increase in fitness levels described in the previous section. As the population slowly converges towards its optimum level, so does its diversity gradually decrease.

It is also clear from these results that the level of diversity does not increase as the number of learning opportunities increases. The population employing

10 learning cycles has a level of diversity that is significantly higher than the population employing 30 learning cycles (after 20 generations).

Diversity levels for all populations converge by generation 75, indicating that the population has reached an optimum and that a particular set of individuals are dominant throughout the population. However, given that the fitness levels for populations employing both population and lifetime learning is significantly higher than the population employing population learning alone, it is clear that the initial maintenance of high diversity levels associated with learning allows those populations to more thoroughly search the problem space and converge on better individual solutions.

These results show that learning has a clear effect on the diversity of populations – even small levels of learning produce an increase in population diversity. This is in contrast to the results obtained by Eriksson, who showed that diversity could only be increased through learning if initial parameters were carefully tweaked. However, future work should examine whether the impact of the initial proportion of *?* alleles contributes significantly to these results.

It is likely that the higher level of diversity comes from the hiding effect provided by the learning process. In other words, the learning process allows individuals without a genetic innate ability to survive by improving their fitness during their lifetime, thus hiding genetically weak individuals from the selection process.

An individual born in a non-learning population has no opportunity to influence its fitness because this is entirely derived from its genome and the selection process quickly moves to propagate highly genetically fit individuals throughout the population. Learning gives individuals the opportunity to search the problem space using their inherited genotypes as a starting point. Thus, if an individual is born with a mediocre genotypic fitness, it is possible it will significantly improve through the learning process and be selected for reproduction.

5 Conclusion

This paper examined both fitness and diversity levels in populations employing population learning alone and populations employing both population and lifetime learning. The model we employed is based on the Hinton & Nowlan model and was chosen for its relative simplicity and ease of analysis. The experiments that were conducted produced a number of interesting results.

Firstly, the results provide a further confirmation that combining lifetime learning with population learning leads to an increase in fitness. Unlike previous work based on the Hinton & Nowlan model, our fitness measure is not biased towards lifetime learning. The NK fitness model allows population learning to successfully develop solutions and highlights the fact that lifetime learning can provide a fitness improvement in a non-biased environment.

Secondly, we have demonstrated that diversity levels can be measured in an alternative manner to previous work based on modified versions of the Hinton & Nowlan model. Previous work examined the genetic differences between bias strength and direction, while our work is based on a phenotypic diversity measure examining the behaviour of the population in a much simpler setting.

Finally, our results show that lifetime learning maintains higher diversity levels than population learning alone. This is arguably a more intuitive result than that found in previous related work. Learning provides an opportunity for genetically weaker individuals to attain higher fitness levels. The high diversity levels in such populations attests to the fact that many behaviours are present in the population, leading to a larger, directed exploration of search space, which in turn leads to higher fitness levels.

Acknowledgements The first author would like to acknowledge the support of the Irish Research Council for Science, Engineering and Technology.

References

1. Hinton, G.E., Nowlan, S.J.: How learning guides evolution. Complex Systems 1, 495–502 (1987)
2. Nolfi, S., Parisi, D.: Learning to adapt to changing environments in evolving neural networks. Adaptive Behavior 5(1), 75–97 (1996)
3. Floreano, D., Mondada, F.: Evolution of plastic neurocontrollers for situated agents. In: Animals to Animats, vol. 4 (1996)
4. Sasaki, T., Tokoro, M.: Adaptation toward changing environments: Why darwinian in nature? In: Husbands, P., Harvey, I. (eds.) Fourth European Conference on Artificial Life, pp. 145–153. MIT Press, Cambridge (1997)
5. Pereira, F.B., Costa, E.: How learning improves the performance of evolutionary agents: A case study with an information retrieval system for a distributed environment. In: Proceedings of the International Symposium on Adaptive Systems: Evolutionary Computation and Probabilistic Graphical Models (ISAS 2001), pp. 19–23 (2001)
6. Watson, J., Wiles, J.: The rise and fall of learning: A neural network model of the genetic assimilation of acquired traits. In: Proceedings of the 2002 Congress on Evolutionary Computation (CEC 2002), pp. 600–605 (2002)
7. Curran, D., O'Riordan, C.: On the design of an artificial life simulator. In: Palade, V., Howlett, R.J., Jain, L. (eds.) KES 2003. LNCS, vol. 2773, pp. 549–555. Springer, Heidelberg (2003)
8. Curran, D., O'Riordan, C.: Artificial life simulation using marker based encoding. In: Proceedings of the 2003 International Conference on Artificial Intelligence (IC-AI 2003), vol. II., Las Vegas, Nevada, USA, pp. 665–668 (2003)
9. Mayley, G.: Guiding or hiding: Explorations into the effects of learning on the rate of evolution. In: Proceedings of the Fourth European Conference on Artificial Life, MIT Press, Cambridge (1997)
10. O'Reilly, U.M.: Using a distance metric on genetic programs to understand genetic operators. In: IEEEInternational Conference on Systems, Man, and Cybernetics, vol. 5, pp. 4092–4097. IEEE Computer Society Press, Los Alamitos (1997)
11. De Jong, K.A.: Analysis of Behavior of a Class of Genetic Adaptive Systems. PhD thesis, The University of Michigan (1975)

12. Booker, L.B.: Improving the performance of genetic algorithms in classifier systems. In: Proc. of the International Conference on Genetic Algorithms and Their Applications, Pittsburgh, PA, pp. 80–92 (1985)
13. Collins, R.J., Jefferson, D.R.: Selection in massively parallel genetic algorithms. In: ICGA, pp. 249–256 (1991)
14. Darwin, C.: The Origin of Species: By Means of Natural Selection or the Preservation of Favoured Races in the Struggle for Life. Bantam Press, London (1859)
15. Baldwin, J.M.: A new factor in evolution. American Naturalist 30, 441–451 (1896)
16. Lamarck, J.B.: Philosophie Zoologique. Chez Dentu, Paris (1809)
17. Boers, E., Borst, M., Sprinkhuizen-Kuyper, I.: Evolving Artificial Neural Networks using the "Baldwin Effect". Technical Report TR 95-14 (1995)
18. Turney, P.D.: How to shift bias: Lessons from the baldwin effect. Evolutionary Computation 4(3), 271–295 (1996)
19. Suzuki, R., Arita, T.: The baldwin effect revisited: Three steps characterized by the quantitative evolution of phenotypic plasticity. In: Banzhaf, W., Ziegler, J., Christaller, T., Dittrich, P., Kim, J.T. (eds.) ECAL 2003. LNCS (LNAI), vol. 2801, pp. 395–404. Springer, Heidelberg (2003)
20. Arita, T., Suzuki, R.: Interactions between learning and evolution: The outstanding strategy generated by the baldwin effect. In: Proceedings of Artificial Life VII, pp. 196–205. MIT Press, Cambridge (2000)
21. Liu, Y., Yao, X., Higuchi, T.: Evolutionary ensembles with negative correlation learning. IEEE Transactions on Evolutionary Computation 4(4), 380–387 (2000)
22. Opitz, D.W., Shavlik, J.W.: Generating accurate and diverse members of a neural-network ensemble. In: Touretzky, D.S., Mozer, M.C., Hasselmo, M.E. (eds.) Advances in Neural Information Processing Systems, vol. 8, pp. 535–541. MIT Press, Cambridge (1996)
23. Brown, G.: Diversity in Neural Network Ensembles. PhD thesis, University of Birmingham (2003)
24. Burke, E.K., Gustafson, S., Kendall, G.: Diversity in genetic programming: an analysis of measures and correlation with fitness. IEEE Trans. Evolutionary Computation 8(1), 47–62 (2004)
25. Eriksson, R.I.: An initial analysis of the ability of learning to maintain diversity during incremental evolution. In: Freitas, A.A., Hart, W., Krasnogor, N., Smith, J. (eds.) Data Mining with Evolutionary Algorithms, Las Vegas, Nevada, USA, pp. 120–124 (2000), `citeseer.ist.psu.edu/eriksson00initial.html`
26. Curran, D., O'Riordan, C.: Increasing population diversity through cultural learning. Adaptive Behavior 14(4) (2006)
27. Kaufmann, S.A.: Adaptation on rugged fitness landscapes. Lectures in the Sciences of Complexity 1, 527–618 (1989)

Embodied Evolution and Learning: The Neglected Timing of Maturation

Steffen Wischmann, Kristin Stamm, and Florentin Wörgötter

Bernstein Center for Computational Neuroscience,
University of Göttingen, Bunsenstr. 10, 37073 Göttingen, Germany,
{steffen|kstamm|worgott}@bccn-goettingen.de

Abstract. One advantage of the asynchronous and distributed character of embodied evolution is that it can be executed on real robots without external supervision. Further, evolutionary progress can be measured in real time instead of in generation based evaluation cycles. By combining embodied evolution with lifetime learning, we investigated a largely neglected aspect with respect to the common assumption that learning can guide evolution, the influence of maturation time during which an individual can develop its behavioral skills. Even though we found only minor differences between the evolution with and without learning, our results, derived from competitive evolution in predator-prey systems, demonstrate that the right timing of maturation is crucial for the progress of evolutionary success. Our findings imply that the time of maturation has to be considered more seriously as an important factor to build up empirical evidence for the hypothesis that learning facilitates evolution.

1 Introduction

Evolutionary Robotics (ER) has become a well established method for investigating the development of adaptive behavior in situated and embodied agents (e.g., [1,2]). However, demonstrating truly open-ended evolution of complex behavior is still a great challenge [3]. A regularly raised argument for that is the strong simplification of ER experiments with respect to natural evolution and the complexity of biological organisms (for a discussion see, for instance, [4]). Although this criticism has to be taken seriously, it seems rather unpractical to begin with.

Instead of trying to increase the complexity of individual agents at any cost, we propose a more pragmatic approach. The aim of this paper is to present a methodology which unifies several aspects of ER research, which previously have been considered rather separately: the interplay between ontogenetic (learning) and phylogenetic (evolutionary) development [5], the evolution of morphology and control [6], the use of distributed and asynchronous evolutionary algorithms [7], the role of competitive co-evolution [8], and the reduction of fitness function constraints to achieve open-ended evolutionary processes [4]. We argue, unifying those aspects gives promising ground for the development of more biologically

F. Almeida e Costa et al. (Eds.): ECAL 2007, LNAI 4648, pp. 284–293, 2007.

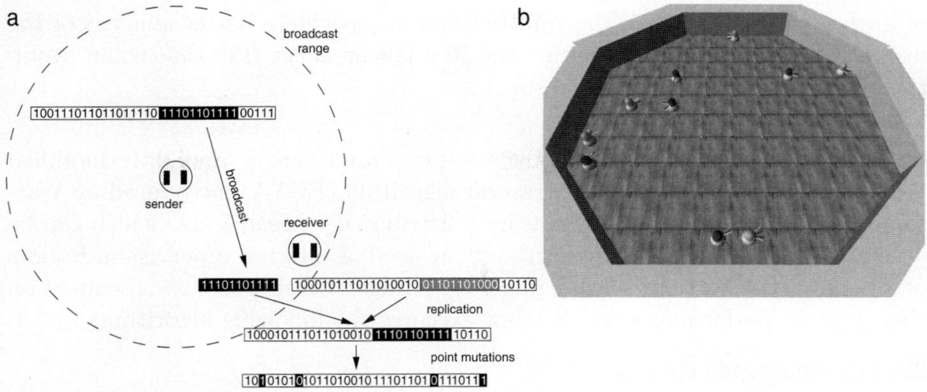

Fig. 1. a: Principle mechanisms of the embodied evolutionary algorithm (see text for details). b: Physical simulation of a predator-prey scenario (two populations can be distinguished by the color on top of the robots).

plausible ER experiments, which may allow a successful incremental increase of behavioral complexity.

Besides the description of such a unified methodology, we present first experiments involving competing populations of predators and preys. Our results challenge the previously hold assumption that learning facilitates evolution [9,5,10]. This assumption is mainly based on experiments where, as we believe, an important aspect of lifetime adaptation has been largely neglected–the time of maturation during which an individual can develop its behavioral skills. Whether or not learning facilitates evolution crucially depends on this factor, as we can show within our experimental framework. If the timing is wrong, learning can even be counterproductive and individuals are better off with evolutionary development only. However, with the right timing, learning can slightly improve evolutionary success. We argue further, to build empirical evidence for the hypothesis that learning can guide evolution in the context of ER, we have to abandon classical generation based evolution and should instead consider an asynchronous and distributed approach as pursued in this paper.

2 Methods

In the following we consider a predator-prey scenario involving two competing populations. In remembrance of the pioneering work of Werner and Dyer [11] we call prey individuals sheep and predators wolves. Sheep can gather energy by moving around without getting caught by wolves (an analogy to grazing), whereas wolves gain energy by catching sheep. Here, only the sheep population evolves whereas wolves can only perform lifetime adaptation. However, the applied evolutionary algorithm (EA) can be easily used to investigate co-evolution of both populations which was omitted here to provide a better analysis of the fitness development by eliminating the Red Queen effect [12], that is, an evolutionary arms race as investigated in [13].

Embodied Evolution. The EA we implemented here is a slightly modified version of the probabilistic gene transfer algorithm (PGTA) introduced by Watson et al. [7]. It is an asynchronous and distributed *embodied* EA, which can be executed directly on real robots without the need of external supervision. Robots of a population evolve by exchanging gene strings of their genome dependent on their current performance. Each robot executes the following algorithm:

```
DEFINE embodied_evolve
  initialize_genes[]
  energy = initial_energy
  REPEAT FOREVER
    IF (maturation_exceeded?){
      IF (excited?)
        send(genes[random(num_genes)])
      IF (receptive? and received?){
        genes[indexof(received)] = valof(received)
        mutate(genes[])
        begin_maturation()
        energy = maturation_energy}}
    do_task_specific_behavior
    energy = limit(energy + reward - penalty)
  ENDREPEAT
ENDDEFINE
```

At the beginning of evolution, the genome of a robot is initialized randomly and a starting energy is set (this can be either the actual battery current or a virtual energy value). In every time step, depending on the current energy, it is probabilistically determined if a robot sends a sequential part of its genome (*excited*). The gene string can be received by all individuals within a predefined broadcast range (see Fig. 1a). A robot accepts received gene strings with probability (1-energy) and overwrites part of its genome with the new string. The new genome is mutated with a predefined mutation rate. With gene acceptation the maturation period begins and the energy level is reset. Maturation period is a developmental phase where individuals can neither broadcast nor receive gene strings. Note, this period does not determine the lifetime of an individual, it just assures a minimum time of development. The actual lifetime depends on the performance of an individual, the better it is the longer the individual lives. At the end of each time step the energy is updated. If the robot broadcasts a gene string, energy is decreased with a penalty (in analogy to reproduction costs). If the robot successfully accomplishes its task, it gains an energy reward. The energy of a robot is limited by a minimum and maximum value.

The major difference to the PGTA proposed in [7] is the maturation period which ensures that new genomes are able to prove their performance potential

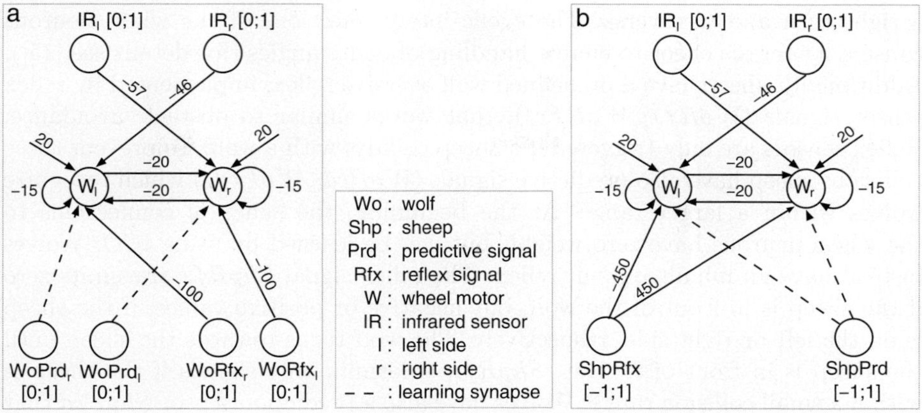

Fig. 2. Neural networks controlling sheep (a) and wolves (b). Square brackets enclose the range of neuronal output (W_l and W_r are within $[-1, 1]$).

before they may get overwritten by another received gene string. Further, mutation is applied to the whole receiver's genome instead of to the transferred gene string only. This allows more variation in new genotypes.

Learning. During lifetime wolves and sheep are able to adapt their behavior–either catching sheep or escaping from wolves. As adaptation mechanism we applied isotropic-sequence-order learning using input correlations only (ICO) described in [14]. This approach is based on temporal correlations between sensor signals as they appear in nature, too. For instance, a child learns to react to heat perception caused by a hotplate which precedes the pain when touching it. In our example, seeing a wolf precedes getting caught. Sheep react to wolf attacks by trying to move away. This is an inbuilt reflex. During lifetime, they are able to learn to prevent these reflexes by using predictive sensory signals–e.g. sensing of approaching wolves. ICO learning realizes correlations between a reflexive and a preventive behavior. Preventive synaptic weights in the neural control are learned by the following learning rule:

$$\frac{d}{dt}p_i = \mu u_i \frac{du_r}{dt}, i \in [0, N], \tag{1}$$

where a weight change of a synapse p_i in time step t depends on the learning rate μ, the input signal u_i transmitted by the learned synapse, and the difference in a reflex input signal u_r (for details see [14]).

Fig. 2 presents the neural network (NN) control including ICO learning of sheep and wolves. Both NNs include a robust obstacle avoidance functionality realized by infrared (IR_l, IR_r) sensor signals influencing the speed of the left and right wheel (W_l, W_r). The IR signals detect obstacles on the left and right frontal side of the robot. An obstacle detected on the robot's left side causes a right turn and vice versa. The cyclic interconnections of the wheel neurons cause a hysteresis effect to enable handling of acute angles (for details see [15]). Additionally, sheep have a predefined wolf aversive reflex, implemented by reflex sensor signals ($WoRfx_l, WoRfx_r$), that works similar to obstacle avoidance. Reflex sensors are only triggered if a sheep collides with a wolf. To prevent these collisions sheep have also predictive signals ($WoPrd_l, WoPrd_r$) which recognize wolves within a larger range. At the beginning, the synaptic connections to the wheel neurons have zero weight, but can be learned by using ICO. Wolves instead have an inbuilt pursuit reflex. The reflex signal $ShpRfx$ transmits zero if the sheep is in front of the wolf, but negative or positive values, if the sheep is on the left or right side, respectively. The wolf turns towards the sheep until the sheep is in front of it. But, $ShpRfx$ transmits only signals if the sheep is within a small collision range. Wolves have also a prediction sensor $ShpPrd$ that functions in the same way as $ShpRfx$, but senses sheep within a larger range. $ShpPrd$ synaptic connections can be learned with ICO.

Competitive Evolution. For the implementation of embodied evolution we have selected evolvable parameters of sheep from different domains: neural control, embodiment, and lifetime adaptation. We evolved maximum wheel motor speed, position of the wolf detection sensors (similar to the sensor position evolution described in [16]), and the ICO learning rate μ. Each parameter is restricted by a value range and represented by a 10 bit string in the binary genome (see [17] for details).

Instead of defining high level targets, the fitness function simply corresponds to the energy level of a robot, that is, it describes a rather general survival criteria, as it is considered to be essential for open-ended and creative evolution [3,4]. Each sheep gains energy whenever there is no wolf within their collision range, whereas a wolf gains energy while a sheep is within its collision range (i.e., the energy of a sheep is transfered to the wolf). Additional energy loss is realized dependent on the broadcast behavior. Anytime an individual broadcasts a gene string, energy is reduced (in analogy to reproduction costs).

For the ease of analysis we implemented the robots and the environment with the physical simulation library *ODE* (see Fig. 1b) based on the specifications of the Khepera^{TM} robot[1]. For the following experiments each population of sheep and wolves consists of three individuals where sheep can evolve and wolves can adapt their behavior by learning only.

3 Results

3.1 Maturation Time

First of all we wanted to investigate the influence of the time an individual has to develop until it becomes exposed to the selection pressure (i.e. its maturation

[1] The limited space of this paper prohibits detailing all parameters, but a precise description of the simulation can be found in [17].

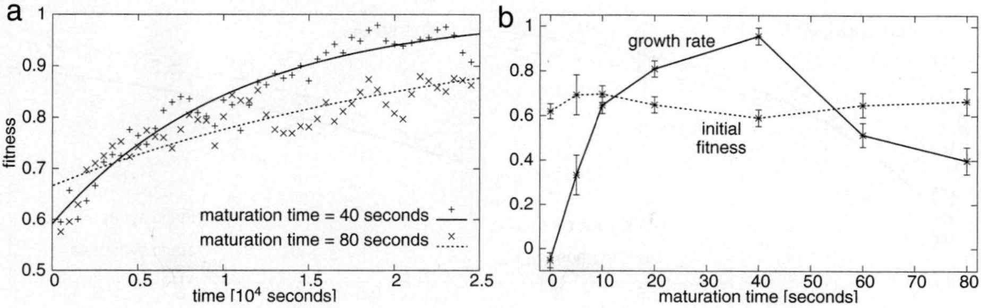

Fig. 3. a: Data points indicate fitness development for two distinct maturation times (see text for calculation method). Data is approximated by an exponential function (see text for details). b: Growth rate and initial fitness depending on maturation time (error bars indicate the mean error of the approximated exponential function).

period). For each of several distinct maturation times (MTs) we conducted 25 evolutionary runs with randomly varied initial conditions (i.e., randomly initialized genomes and random starting positions). Each run lasted 2.5 million time steps (time step = 0.01 sec) which corresponds to approximately seven hours of real time, if we would execute it on real robots. This is a reasonable time scale for embodied evolution experiments (see also [7]). During these 25 runs we recorded the energy levels of all robots in each time step. We then took the mean of all robots and all runs. To smoothen sudden jumps in the data which can occur if a sheep gets caught or receives a new genome, we averaged again over a time window of 500 seconds.

Fig. 3a shows the resulting data for two different MTs. To get a clearer picture of the fitness development during evolution, we approximated the data by an exponential function of type $f(x) = 1 - ae^{-bx}$, where $(1 - a)$ indicates the initial fitness (i.e., after the first 500 sec.) and b the growth rate over time (an unconstrained nonlinear optimization method of the Matlab^{TM} optimization toolbox was used). In Fig. 3a we see that individuals with the shorter MT (i.e., 40 sec.) do not only increase their fitness much faster than individuals with a longer MT (i.e., 80 sec.), but they also reach nearly maximal fitness already after 18,000 seconds.

Fig. 3b illustrates the approximated growth rate and initial fitness for different MTs. The initial fitness is nearly constant. The growth rate, however, differs significantly. We find a maximum at 40 seconds, and declining growth rates for smaller and larger MTs. Explaining smaller growth rate at smaller MT is rather straight forward. Individuals do not have time enough to adapt their learning synapses before they are exposed to the selection pressure. The evolutionary progress is rather slow because this holds for all evolving individuals and even though the frequency of gene exchange is rather high, the fitness difference between sender and receiver is not a reliable indicator for a better or worse performance. With short MT, some individuals may have encountered wolves a

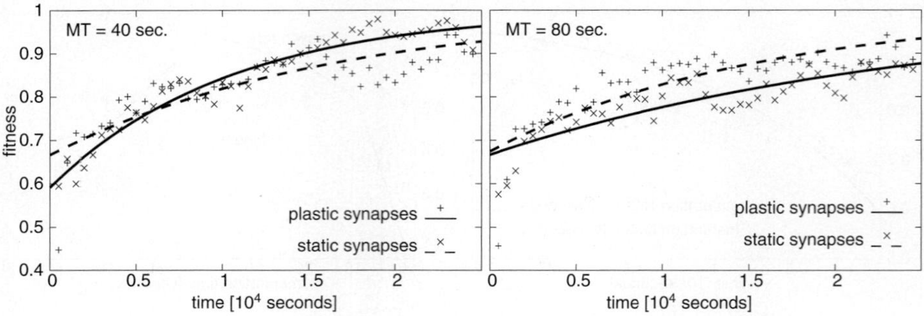

Fig. 4. Comparing fitness development during the evolution of individuals with plastic synapses and static synapses for different maturation times (MTs).

few times, but other may have never encountered them and, therefore, have a higher fitness. Thus, genes may be transmitted which are only successful because they never had to prove their performance (note that sheep have to encounter wolves to develop their escaping skill). This effect cancels out over time, but the evolutionary progress is slowed down. In contrast, for long MTs the gene exchange frequency is too slow. Even though individuals now have enough time to develop, evolving parameters (e.g., learning rate or morphological properties) which improve the performance spread only slowly throughout the population.

3.2 Learning and Evolution

To figure out how learning influences the course of evolution we conducted the same experiments as described before for neural networks with static synapses. Basically, nothing changed except that the weights of the synapses, which were learned during the lifetime of an individual (cf. Fig. 2a), were now encoded in the genome as well. Note, this was applied for sheep only, the setup for the wolf population remained the same, that is, wolves could learn, but were not evolved.

We chose two MTs, the one with the largest growth rate (40 sec.) and one with a small growth rate, where we preferred 80 to 5 seconds (cf. Fig. 3a) because the former provides more reliable fitness estimates than the latter (for the reason that sheep encounter more wolves during their evaluation). Fig. 4 compares the fitness development during the evolution with the two MTs to our previous results from the evolution with synaptic plasticity (same measurements were applied as described before). Considering only static synapses, we find no influence of MT. Compared to the case of plastic synapses, for an MT of 80 seconds the fitness development for evolution without learning is even slightly better than for evolution with learning. If, however, MT is shorter (i.e., 40 sec.), evolution with learning performs slightly better than without learning.

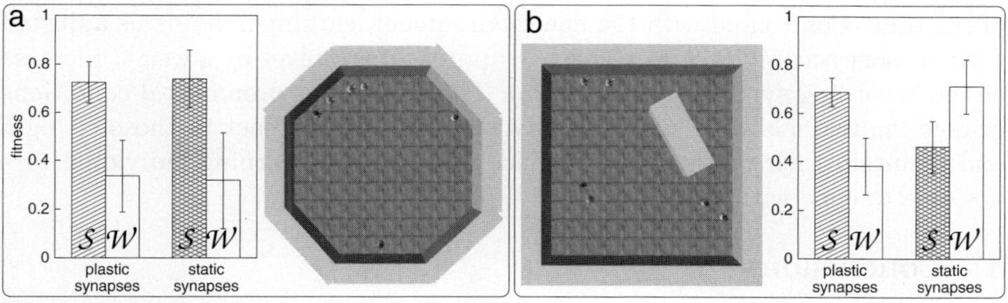

Fig. 5. Comparing performance of sheep populations ($\mathcal{S}$) evolved either with plastic or static synapses in the environment with which they were evolved (a) and a novel environment (b). Performance of wolf populations ($\mathcal{W}$) is drawn as a reference. Shown is the average and standard deviation of 25 randomly initialized runs.

3.3 Adaptability to Novel Environmental Conditions

In the previous section we have seen that learning may slightly guide evolution, but the effect within our experimental framework is not really significant. The good news is that both methods take the same computational time due to the asynchronous and distributed evolution approach. Thus, we may conclude that in those basic tasks it does not matter whether we evolve with or without synaptic plasticity (presupposing the right timing of maturation). However, the commonly assumed advantage of learning is that individuals may be able to adapt to significantly novel environmental conditions with which they were not confronted during their evolutionary development, as for instance shown in [10,18].

To test whether or not this assumption holds for our approach as well, we chose one representative evolutionary run from the evolution with and without learning with an MT of 40 seconds (cf. Fig. 4). From these two runs we took the neural networks of the sheep population at the end of evolution (i.e., after 25,000 sec.) and tested them first in the environment with which they were evolved for 400 seconds (i.e., ten times the original MT) while disabling genome transmissions. Fig. 5a illustrates the performance of a sheep population with either plastic or static synapses. As could be expected from Fig. 4, both solutions perform similar well. However, confronted with a significantly different environment, the population with plastic synapses performs significantly better (see Fig. 5b). This can be accounted to the adaptation mechanism which individuals with static synapses lack. And adaptation is required in this environment because compared to the original environment (cf. Fig. 5a), it contains sharp corners and narrow passages. Thus, sheep are much better off with reacting earlier to approaching wolves to avoid being captured in corners. Optimal reaction time was not required in the original environment. There, the most important parameter was the speed of sheep which was always maximized by evolution. Thus, even with a moderate reaction time, sheep could escape approaching wolves for most of the time. Confronted with the new environment learning individuals adjusted their sensory sensitivity with respect to approaching wolves by a weight increase of the according synapses (see Fig. 2a). Therefore, if environmental conditions change, individuals with synaptic plasticity are able to react to those changes and maintain their original performance, whereas non-learning individuals are not able to cope with this change.

4 Conclusions

The very nature of embodied evolution forces us to abandon single individual evolution. Even under basic environmental conditions as considered here, robotic agents already have to deal with a dynamically changing environment. Even further, right from the beginning they have to develop robust behaviors because it is not guaranteed that other individuals of their own species, they may encounter, act reasonable. Due to the asynchronous and distributed character of embodied evolution, conspecifics can show quite different behaviors at the very same time. For instance, some individuals may try to avoid their conspecifics while others got some mutations which may lead them directly crash into their conspecifics.

The most intriguing aspect of a methodology as proposed here is the ability to incrementally increase behavioral complexity by increasing environmental factors. Here, we considered only one evolving population, but we also tested co-evolution of both populations and discovered an evolutionary arms race as already described in [13] for predator-prey systems with one individual in each population only. To do so, no change of fitness functions is required, robots 'solely' have to maintain their energy. For future research it would be interesting to consider more than two interacting populations by building up small ecosystems of intertwined species, each influencing the evolution of the other. Thus, we argue that our unified framework is suitable to approach more biologically plausible scenarios in ER striving towards open-ended and creative scenarios as proposed in [4,3].

Besides these general methodological facets, we investigated an aspect largely neglected in previous studies concerning the interplay of evolution and learning [9,5,10]. We demonstrated that whether or not learning may facilitate evolution depends on the right timing of individual maturation. Further, we showed that the choice of maturation time significantly influences the rate of progress during evolution. Even though our experimental results can not offer a general answer to the question how long learning individuals should be able to develop, they clearly show that we have to consider this aspect sincerely to build up empirical evidence for the commonly hold assumption that learning guides evolution. And the use of asynchronous and distributed embodied evolution should be preferred over the classical generation based evolution; not only because it can be executed directly on real robots, but also because it allows a direct comparison of evolution with and without learning measured on real time progress.

Acknowledgments. This work was supported by the PACO-PLUS project, funded by the European Commission under grant IST-FP6-027657.

References

1. Nolfi, S., Floreano, D.: Evolutionary Robotics: The Biology, Intelligence, and Technology of Self-Organizing Machines. MIT Press, Cambridge (2000)
2. Harvey, I., Di Paolo, E.A., Wood, R., Quinn, M., Tuci, E.: Evolutionary robotics: A new scientific tool for studying cognition. Artificial Life 11, 79–98 (2005)
3. Taylor, T.: Creativity in evolution: Individuals, interactions and environments. In: Creative Evolutionary Systems, pp. 79–108. Morgan Kaufmann, San Francisco (2001)
4. Bianco, R., Nolfi, S.: Toward open-ended evolutionary robotics: Evolving elementary robotic units able to self-assemble and self-reproduce. Connection Science 16, 227–248 (2004)
5. Nolfi, S., Floreano, D.: Learning and evolution. Autonomous Robots 7, 89–113 (1999)
6. Taylor, T., Massey, C.: Recent developments in the evolution of morphologies and controllers for physically simulated creatures. Artificial Life 7, 77–87 (2001)
7. Watson, R.A., Ficici, S.G., Pollack, J.B.: Embodied evolution: Distributing an evolutionary algorithm in a population of robots. Robotics and Autonomous Systems 39, 1–18 (2002)
8. Nolfi, S., Floreano, D.: Co-evolving predator and prey robots: Do 'arm races' arise in artificial evolution? Artificial Life 4, 311–335 (1998)
9. Hinton, G.E., Nowlan, S.J.: How learning guides evolution. Complex System 1, 495–502 (1987)
10. Urzelai, J., Floreano, D.: Evolution of adaptive synapses: Robots with fast adaptive behavior in new environments. Evolutionary Computation 9, 495–524 (2001)
11. Werner, G., Dyer, D.: Evolution of communication in artificial organisms. In: Proc. Artificial Life II, Addison-Wesley, Reading (1991)
12. van Valen, L.: A new evolutionary law. Evolutionary Theory 1, 1–30 (1973)
13. Floreano, D., Nolfi, S.: God save the red queen! Competition in co-evolutionary robotics. In: Genetic Programming 1997: Proc. of the Second Annual Conference, pp. 398–406 (1997)
14. Porr, B., Wörgötter, F.: Strongly improved stability and faster convergence of temporal sequence learning by utilising input correlations only. Neural Computation 18, 1380–1412 (2006)
15. Hülse, M., Wischmann, S., Pasemann, F.: Structure and function of evolved neurocontrollers for autonomous robots. Connection Science 16, 249–266 (2004)
16. Buason, G., Bergfeldt, N., Ziemke, T.: Brains, bodies, and beyond: Competitive co-evolution of robot controllers, morphologies and environments. Genetic Programming and Evolvable Machines 6, 25–51 (2005)
17. Stamm, K.: Individual learning and the dynamics in predator-prey populations. Master's thesis, University of Göttingen (2007)
18. Di Paolo, E.A.: Organismically-inspired robotics: Homeostatic adaptation and teleology beyond the closed sensorimotor loop. In: Dynamical Systems Approach to Embodiment and Sociality, pp. 19–42. Advanced Knowledge International, Magill (2003)

Evolution and Learning in an Intrinsically Motivated Reinforcement Learning Robot*

Massimiliano Schembri, Marco Mirolli, and Gianluca Baldassarre

Istituto di Scienze e Tecnologie della Cognizione, Consiglio Nazionale delle Ricerche
Via San Martino della Battaglia 44, I-00185 Roma, Italy
{massimiliano.schembri, marco.mirolli, gianluca.baldassarre}@istc.cnr.it

Abstract. Studying the role played by evolution and learning in adaptive behavior is a very important topic in artificial life research. This paper investigates the interplay between learning and evolution when agents have to solve several different tasks, as it is the case for real organisms but typically not for artificial agents. Recently, an important thread of research in machine learning and developmental robotics has begun to investigate how agents can solve different tasks by composing general skills acquired on the basis of internal motivations. This work presents a hierarchical, neural-network, actor-critic architecture designed for implementing this kind of intrinsically motivated reinforcement learning in real robots. We compare the results of several experiments in which the various components of the architecture are either trained during lifetime or evolved through a genetic algorithm. The most important results show that systems using both evolution and learning outperform systems using either one of the two, and that, among the former, systems evolving internal reinforcers for learning building-block skills have a higher evolvability than those directly evolving the related behaviors.

1 Introduction

One important area of investigation of Artificial Life concerns the relationships existing between evolution and learning, the two key mechanisms that generate adaptive behavior in real organisms [1]. The synthetic approach of Artificial Life is an invaluable tool for investigating such a topic given the difficulties of collecting relevant empirical evidence related to it [2]. This approach already highlighted several important aspects of the relationship (for a review, see [3]), for example the fact that learning can guide evolutionary search [4] and that evolution can discover good starting conditions which can in turn facilitate learning processes during lifetime [5].

One of the most important distinctions between the two adaptive mechanisms is the time scale within which they operate [1]. In this respect, evolution has the advantage of producing various aspects of behavior 'readily available' at birth, but with the cost that it can 'track' environmental changes only if they take place at a time scale longer than the individuals' life length. On the contrary, learning has the cost of causing

* This research was supported by the EU Projects ICEA, contract no. FP6-IST-027819-IP, and MindRACES, contract no. FP6-511931-STREP.

F. Almeida e Costa et al. (Eds.): ECAL 2007, LNAI 4648, pp. 294 – 303, 2007.

inefficient behavior during the first phases of life but it allows tracking environmental changes within an individual's life span. Because of these time-scale differences, the models proposed so far, which typically used neural networks as agents' control systems, assigned to evolution the role of developing the 'general aspects' of learning systems, for example their overall architecture [6], the learning rules [7][8], the parameters regulating learning [9], and the initial connection weights [2][5], whereas they assigned to learning processes the role of updating connection weights during individuals' life. In this respect, a relevant novelty of this paper is that it proposes a reinforcement-learning system in which evolution develops some components of the system while learning uses these innate components to guide the training of the other components. The only work which carried out a study related to this issue is the pioneering work of Ackley and Littman [2]. In this work, the authors had a genetic algorithm evolving both an actor and an evaluator network, where the former was also trained during individual lifetime through a reinforcement learning algorithm on the basis of the evaluations of the latter. However, the main focus of that work was on the Baldwin effect. In contrast, the present work proposes a new hierarchical neural network architecture which learns to solve several different tasks by combining general skills acquired during an 'infancy' period. Hence, the most important novelty of the present work consists in studying the relationship between evolution and learning in the case in which *learning has a twofold nature*, and takes place on the basis of *both external and internal rewards*. This second point is directly related to a recent trend of research in the study of learning in artificial systems.

This new trend of research is inspired by the acknowledgement that when faced with new problems, organisms do not need to create solutions from scratch on the basis of low-level sensorimotor primitives but they can focus on composing and modifying previously developed general skills. Consequently, researchers in both machine learning [10][11] and developmental robotics [12][13][14] started to investigate systems with a twofold learning process. These systems acquire general skills on the basis of *internal motivations* (such as the drives to be exposed to novel/surprising/salient events), and then use these skills as building blocks to assemble more complex behaviors on the basis of *'external' rewards* (e.g. pleasure for eating and reproducing). This twofold process seems to play a fundamental role in the flexibility of behaviors exhibited by real organisms, especially the most sophisticated ones, like humans and primates in general [15][16]. Of course, understanding these processes is not only scientifically relevant but it is also one of the most important current goals of developmental robotics and machine learning, as it would allow building artificial intelligence systems having a flexibility and autonomy comparable to those of real organisms.

One of the most interesting machine learning proposals that encompass this insights, and that inspired the present work, is Intrinsically Motivated Reinforcement Learning (IMRL) [11][17]. The architecture used in IMRL is based on machine learning theory of 'options' [18]. Basically, options are sub-routines which can be invoked as any other primitive action, and include a set of initiation states where the option can be invoked, a termination condition, a policy mapping states to actions' probabilities and, within the IMRL framework, an option model which maps initiation states to the probabilities of terminating the option in any other state. New options are created each time the system experiences a novel 'salient event'. A key point is that the system uses the prediction error of the option model as an *internal reward* to decide which option to invoke: the effect is that until the ability to produce the associated 'salient event' is not refined, an option continues to generate internal rewards and hence to be selected and trained (for a more detailed account of IMRL, see [17]).

The present paper (see also [19]) proposes a two-level hierarchical reinforcement-learning actor-critic architecture that represents a first attempt to solve two important drawbacks of the current implementation of the IMRL architecture: (a) the assumption of abstract representations of states and actions (e.g. grid-world environments and discrete actions), and (b) the fact that 'salient events' guiding options' formation and training must be hardwired by the programmer. The architecture tackles the problem (a) by using neural networks as components of the learning system which controls the behavior of a simulated robot, and tackles problem (b) by using a genetic algorithm to evolve neural 'reinforcers' that allow the system to autonomously associate a level of saliency to experienced states (see [14] for another solution to the same problems).

Using this hierarchical architecture, this research investigates the possible roles that evolution and learning can play when learning processes have the aforementioned twofold nature. In particular, it compares the performance (in terms of evolvability, learning speed, and maximal performance) of different versions of the system in which its two main components are either evolved or trained during life: the 'experts', which form the lower-level of the system's hierarchical architecture, and the 'selector', which forms its higher-level. The next section describes the proposed architecture, the task, the simulated robot, and the experimental conditions of the tests. Section 3 reports the results, while section 4 discusses the results and the limits of the present work and, on the basis of these, some possible directions for future research.

2 Simulated Robot, Task and Neural Network Architecture

The simulated robot is a mobile 'wheelchair' robot with a 30 cm diameter and a camera pointed towards a portion of the ground located just in front of the robot (24×8 cm). The robot perceives the ground using a grid of 6×2 sampling points associated with color-specific RGB receptors (so the system's input is a 'retina' formed by a 12×3 = 36 binary values). The robot's motor system is driven by setting the orientation variation within [-30, +30] degrees and the translation speed within [0, 2] cm. The environment is a square arena with a regularly textured floor (Fig. 1). Four different experimental conditions differing with respect to which parts of the system were evolved or trained were studied. We first illustrate the most complex experimental condition and then we explain the other conditions as variations to the former.

The robot's life is divided into two phases: 'childhood' and 'adulthood'. During childhood, the robot learns a set of basic sensorimotor skills based on intrinsic motivations. During adulthood, the robot learns to combine the acquired skills in order to accomplish six rewarded tasks (Fig. 1): in each task the robot has to reach a given target location starting from a particular position, and every time it reaches the target it receives one unit of reward and is set back to the starting position. The controller of the robot (Fig. 2) is a hierarchical modular neural network formed by a 'selector' and three 'experts' (the quality of results did not change in tests with a higher number of experts). The selector and each experts are neural network implementations of the actor-critic reinforcement-learning model [21], which is known to have a high biological

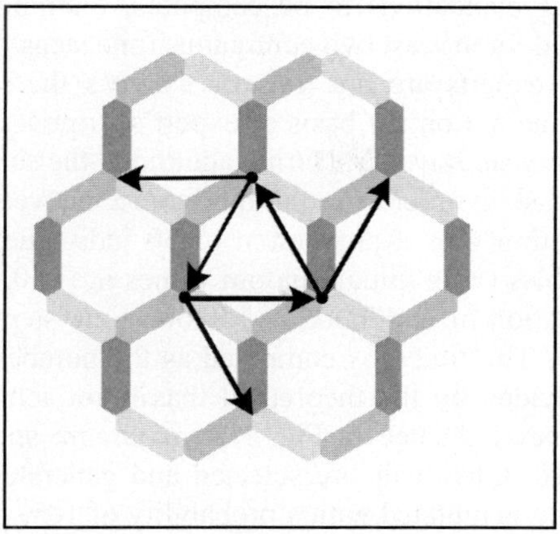

Fig. 1. The environment and the six 'adulthood' tasks. The sides of the hexagons are colored with blue (dark gray), red (gray) and green (light gray). Arrows represent the different tasks: each arrow's tail and head indicate, respectively, the starting and the target position of a task.

plausibility [20][22][23][24]. Each *expert* is formed by three components: (a) a *reinforcer*: a perceptron mapping the retinal input to a [-1, 1] sigmoid unit encoding the internal reward for that expert (reinforcers are evolved, see below); (b) an *actor*: a perceptron mapping the retinal input to two sigmoid units; the activation of these units sets the centre of a Gaussian function which is used to generate noisy commands issued to the motor system: the first unit sets the orientation variation of the robot, the second unit sets its translation (initial standard deviation = 0.3; noise is linearly reduced to zero during childhood); (c) *critic*: this is based on an *evaluator*, a perceptron that maps retinal input to one linear output unit encoding the expert's evaluations of states; these evaluations, together with the reward produced by the expert's reinforcer, are used to compute the *surprise* of the expert's critic in a standard way [21]. The *selector* is formed by two components: (a) *selector's actor:* a perceptron that maps the retinal input to three sigmoid output units; at each time step, the activations of these units, each corresponding to an expert, are used as pseudo-probabilities to select the expert that takes control of the motor system and (during childhood) learns; (b) *selector's critic*: analogous to the experts' critics, it uses as its reward signal either external rewards or the surprise of the expert which currently has the control (see below).

During childhood, at each time step the selector selects the expert that has the control. The selected expert: (a) selects and execute an action; (b) trains its evaluator as in standard function-approximation actor-critic models [21], but on the basis of the internal rewards delivered by its own reinforcer (discount factor = 0.99); (c) trains the actor as in [24]: if surprise is positive, the output units' activations are 'moved' (with a delta rule) towards the (Gaussian noisy) values corresponding to the executed action, whereas if surprise is negative the output units' activations are moved in the opposite direction (learning rate of evaluator and actor = 0.01). On the other hand, in order to train its own actor and evaluator, the selector uses the *surprise of the selected expert as its (internal) reward signal*. As the surprise of an actor-critic system is a good indicator of its learning progress, during this phase the selector learns to give the control to the expert which is learning at the maximum rate. Note that as surprise needs two succeeding evaluations to be computed, learning occurs only when the same expert is selected for at least two contiguous time steps.

During adulthood experts are not trained, whereas the selector is trained as in childhood, but this time not on the basis of expert's surprises, but rather on the basis of the task-related *extrinsic rewards*. During adulthood the selector's weights are reset before tackling each task in order to avoid interference between different tasks.

The genetic algorithm uses a population of 50 individuals, encoding connection weights as real variables (with initial random values in [-1.0, +1.0]), evolved for 100 generations. The duration of childhood is 150,000 time steps, while the duration of adulthood is 600,000. The fitness is computed as the number of times that the robot reaches the target divided by the theoretical maximum achievable if the robot followed the straight lines indicated in Fig. 1 at maximum speed. At the end of each generation the best 10 individuals are selected and generate 5 offspring each. Each weight of the offspring is mutated with a probability of 10% by adding to it a random value uniformly drawn in [-1.0, +1.0].

Four different experiments were run with the following conditions:

1. Learning Experts, Learning Selector (LE-LS). This is the condition just described, in which individuals' genome encodes only the connection weights of the three experts' reinforcers.
2. Evolved Experts, Learning Selector (EE-LS). In this condition experts' actors are encoded in the genome and evolved (hence there is no childhood), while the selector is trained during adulthood as described above.
3. Evolved Experts, Evolved Selector (EE-ES). In this condition the actors of both the experts and the selector are evolved, and no learning takes place.
4. Single Learning Expert (SLE). In this condition no evolution takes place, and a simple expert is used to directly tackle each of the six adult tasks on the basis of only extrinsic rewards (weights were reset at the beginning of each task to avoid interferences between different tasks).

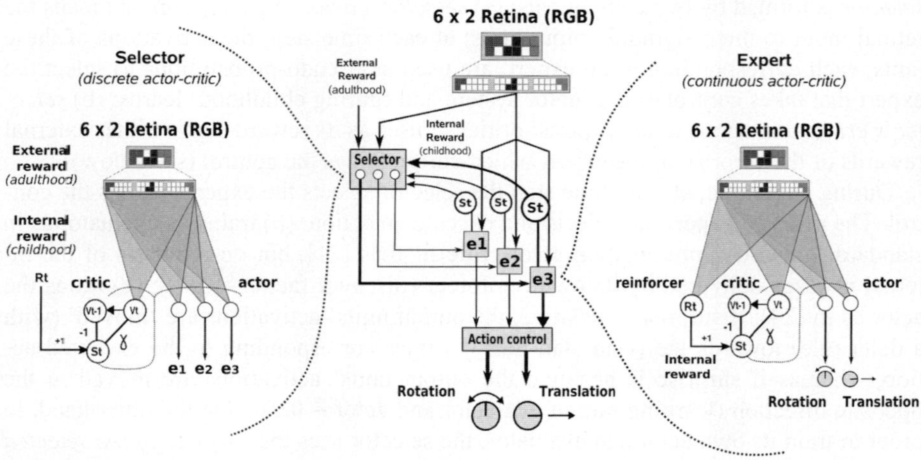

Fig. 2. Center: the whole architecture. Left: the selector's architecture. Right: one expert's architecture (see text for details)

3 Results

Direct observation of the behavior of the evolved individuals indicates that organisms endowed with the hierarchical architecture we have presented (that is those of all but the SLE condition) tend to solve their tasks in the following way. Experts tend to specialize for following one color each, while the selector tend to compose experts' basic skills so to navigate on the colored lines and then choose the most appropriate direction at each junction (for a more detailed analysis, see [19]). This is particularly true for organisms of conditions LE-LS and EE-LS, that is the conditions in which the selector can learn during life how to make the best possible use of the experts' skills.

In order to compare the results of the four conditions, we present three kinds of statistics, which are meant to assess different properties of the various systems: (a) fitness of the best individuals along generations reveals systems' evolvability; (b) performance throughout a long learning period reveals systems' learning speed; (c) performance after a long period of learning reveals systems' steady-state ability.

Fig. 3a reports the fitness of the best individuals along 100 generations for the three conditions involving evolution: LE-LS, EE-LS and EE-ES. The most striking result is that the condition LE-LS is clearly far more evolvable than the other two conditions: it requires about an order of magnitude less generations than the other two to reach a steady state performance (about 10 vs. about 100). Moreover, the LE-LS condition has a higher reliability in different evolutionary runs (note the much smaller standard deviation in the graph). On the other side, EE-LS achieves a higher final fitness with respect to LE-LS. This happens because in the EE-LS condition, evolution is able to find highly accurate and reliable experts (data not reported), whereas the learning of the experts during childhood is always noisy, and results in the acquisition of sub-optimal basic skills. However, this limit might be reduced or even overcome by prolonging the rather short childhood phase used here and/or by optimizing the experts' learning parameters like learning rate and discount factor. Another remarkable result is that the EE-ES condition produces individuals with a quite high fitness, at the same level of the LE-LS condition (consider that in the EE-ES condition the selector is evolved, and hence robots in this condition must find a single solution for all the six different tasks). This is due to the well-known remarkable ability of evolutionary searches to find very 'smart' solution to difficult problems [25]. In particular, evolved organisms of the EE-ES condition typically produce a stereotyped behavior such that the robot follows a circular path at maximum speed which includes most of the target positions: in this way, some tasks are accomplished very efficiently, other with a reasonable efficiency, while other targets are never reached at all. This fact, together with the fact that the behavior of these robots is completely inherited, and hence fully developed from birth, explains the quite good performance reached by this condition.

Fig. 3b shows the learning curves of the three conditions involving learning: LE-LS, EE-LS and SLE over 1,000,000 cycles (for each task). The most important result is that the compositional strategies (LE-LS and EE-LS) clearly outperform the 'monolithic' strategy (SLE) in terms of learning speed. On the other hand, EE-LS and SLE outperform LE-LS in terms of final performance. In the same vein as the result on fitness discussed above, this is explained by the fact that EE-LS can evolve highly reliable experts, SLE can train its only expert during a very long period of test (1,000,000 cycles), whereas LE-LS can only sub-optimally train its three experts during the relatively short childhood phase (150,000 cycles). Finally, the higher fitness of EE-LS with

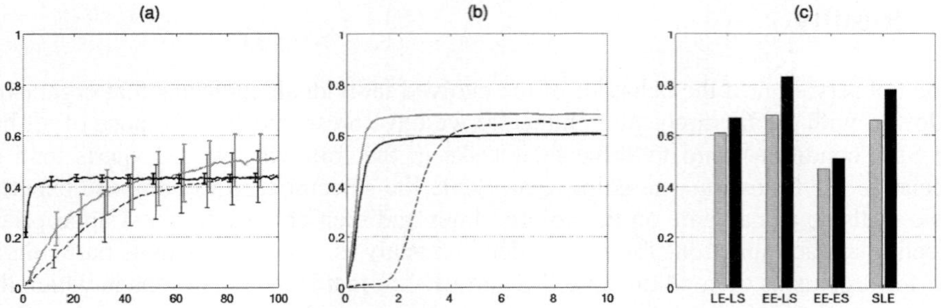

Fig. 3. (a) Evolution of the fitness of the best individuals (averaged over 10 runs) along 100 generations, for the three conditions involving evolution: LE-LS (bold line), EE-LS (gray line), and EE-ES (dashed line). The graph also reports standard deviations. (b) Average performance during learning tests lasting 1,000,000 cycles for the three conditions involving learning: LE-LS (bold line), EE-LS (gray line), and SLE (dashed line). Curves refer to the average performance (normalized number of received rewards) of the 10 best individuals of each of 10 runs on 10 tests for each of the 6 tasks (i.e. average of 10×10×6 tests). (c) Steady-state performance level of all the four conditions measured as average over the last 100,000 cycles of the data reported in graph 'b' (dark gray bars: average over 10 runs; light gray bars: best run). For the EE-ES condition the test of graph 'b' was run with no learning process taking place.

respect to SLE is due to the fact that the former can solve its tasks by efficiently combining useful low-level skills rather than by relying on one single actor.

Fig. 3c shows the steady state level of performance achieved in all the four conditions at the end of learning: these tests allow to compare final performance independently from the time spent to acquire behavior. The results show that EE-ES has the lowest performance as it pays the costs of its rigid behavior. LE-LS has a performance lower than EE-LS and SLE because of the mentioned difficulty to optimize the experts in the short childhood phase. Finally, EE-LS slightly outperforms SLE because of the higher mentioned efficiency of the compositional strategy that can rely upon specialized experts.

4 Discussion and Future Work

This paper investigated the role played by evolution and learning in adaptive behavior when learning processes during life take place in two stages, one where the systems acquire flexible sensorimotor skills on the basis of intrinsic motivations (as a general drive to explore) and a second one where those skills are assembled to accomplish tasks that directly increase fitness (e.g., allow eating) on the basis of extrinsic rewards (e.g. pleasure from food). To this purpose, we used a reinforcement-learning hierarchical neural-network architecture as the control system of a simulated robot and we evaluated the effects of applying either evolution or learning to the various components of the system.

The results highlighted various interesting phenomena related to the relative strengths and limits of evolution and learning, and to their complementary roles in producing adaptive behavior. First of all, the they clearly confirmed previous seminal works (see [1][3]) indicating that evolution alone has the limit of producing rigid behaviors whereas learning alone has the limit of exposing organisms to long periods of non-adaptive behavior. On the contrary, systems that build up adaptive behavior on the basis of both evolution and learning tend to have both the flexibility and fast adaptation advantages provided by the two adaptive processes. With respect to the behavioral flexibility provided by learning, one should also consider that in the learning tests done in this paper robots were tested with the same tasks used during evolution. The advantages provided by learning would surely be much stronger if the systems were tested with tasks which have never been encountered during evolution: this might be a subject of investigation in future work.

A novel interesting finding of this work is that within 'mixed' systems, which rely on both evolution and learning, developing *innate low-level behaviors* in the course of evolution might allow achieving a higher performance. This is in line with the presence of a few but important innate behaviors even in the most complex species such as primates. These are typically behaviors which are very directly related to fitness (like the behaviors implemented by the experts of our system) and for which a ready availability at birth is very important (examples of these are the motor reflexes or basic behaviors related to feeding such as salivation and babies' suction reflex).

On the other hand, our simulations clearly demonstrate that *evolving general criteria (reinforcers) for guiding learning of building-block behaviors* is much easier than directly evolving behaviors themselves. Furthermore, the entity of this effect in the experiments presented here is so big that it suggests that such result might be caused not only by a difference in search spaces for the two conditions (in our experiments reinforcers have half the weights of the actors) but also by the fact that, generally speaking, evolving 'goals' might be much easier than evolving the behaviors that satisfy them (a similar suggestion has also been made by [2]). Future research should investigate more in detail why this is the case.

Furthermore, and most important, our experiments clearly show that the costs of learning, namely the need to acquire behavior from scratch at every generation, can be significantly diminished if agents have a hierarchical control system architecture like the one presented here. In this case, organisms which have to tackle several different tasks during their life can accomplish this by combining general low-level abilities which might be either genetically inherited or acquired during a childhood phase. Indeed, the system that learned each behavior from scratch took nearly four times to reach a performance comparable to that of systems exploiting compositional strategies. This result strongly supports the motivations behind the Intrinsically Motivated Reinforcement Learning framework.

Although interesting, these results are preliminary in many respects, and their limits suggest important problems for future research. First, several interesting conditions have not been explored yet, for example the conditions in which: (a) the genetic algorithm evolves neither the actors of the experts (as in the EE-LS), nor their reinforcers (as in the LE-LS), but rather their evaluators (cf. [2]); (b) the whole hierarchical architecture is trained only on the basis of external task-related rewards; (c) learning and discount parameters are evolved; (d) not only expert's reinforcers, but also their number is evolved. Second, the present architecture might be improved under various respects: for example the selector, which is supposed to operate at a more abstract level with respect to experts, should not operate at the same time-scale and with the same input as them. Notwithstanding these limits, we think that the work presented here is a first important step in the investigation of the relationships existing between evolution and *compositional learning* processes.

References

1. Nolfi, S.: Learning and Evolution in Neural Networks. In: Arbib, M. (ed.) The Handbook of Brain Theory and Neural Networks, pp. 415–418. The MIT Press, Cambridge (2003)
2. Ackley, D., Littman, M.: Interactions Between Learning and Evolution. In: Langton, C.G., Taylor, C., Farmer, J.D., Rasmussen, S. (eds.) Artificial Life II, pp. 487–509. Addison-Wesley, New York (1991)
3. Nolfi, S., Floreano, D.: Learning and Evolution. Autonomous Robots 1, 89–113 (1999)
4. Hinton, G., Nowlan, S.: How learning guides evolution. Complex Systems 1, 495–502 (1987)
5. Belew, R., McInerney, J., Schraudolph, N.: Evolving networks: Using the genetic algorithm with connectionist learning. In: Langton, C.G. (ed.) Proceedings of the Second Conference on Arificial Life, Addison-Wesley, Reading (1992)
6. Di Ferdinando, A., Calabretta, R., Parisi, D.: Evolving Modular Architectures for Neural Networks. In: French, R., Sougné, J. (eds.) Connectionist Models of Learning, Development and Evolution, pp. 253–262. Springer, London (2001)
7. Urzelai, J., Floreano, D.: Evolution of Adaptive Synapses: Robots with Fast Adaptive Behavior in New Environments. Evolutionary Computation 9(4), 495–524 (2001)
8. Niv, Y., Joel, D., Meilijson, I., Ruppin, E.: Evolution of Reinforcement Learning in Foraging Bees: A Simple Explanation for Risk Averse Behavior. Neurocomputing 44(1), 951–956 (2002)
9. Eriksson, A., Capi, G., Doya, K.: Evolution of Meta-parameters in Reinforcement Learning Algorithms. In: Proceedings of the 2003 IEEE/RSJ International Conference on Intelligent Robots and Systems, IEEE, Los Alamitos (2003)
10. Schmidhuber, J.: A Possibility for Implementing Curiosity and Boredom in Model-Building Neural Controllers. In: Meyer, J-A., Wilson, S.W. (eds.) From Animals to Animats: Proceedings of the First International Conference on Simulation of Adaptive Behavior, pp. 222–227. MIT Press, Cambridge (1991)
11. Barto, G., Singh, S., Chentanez, N.: Intrinsically Motivated Learning of Hierarchical Collections of Skills. In: Proceedings of the Third International Conference on Development and Learning (2004)
12. Huang, X., Weng, J.: Novelty and Reinforcement Learning in the Value System of Developmental Robots. In: Prince, C.G., Demiris, Y., Marom, Y., Kozima, H., Balkenius, C. (eds.) Proceedings Second International Workshop on Epigenetic Robotics: Modeling Cognitive Development in Robotic Systems, pp. 47–55. Lund University Cognitive Studies, Lund (2002)
13. Marshall, J., Blank, D., Meeden, L.: An Emergent Framework for Self-Motivation in Developmental Robotics. In: Proceedings of the Third International Conference on Development and Learning (ICDL 2004), pp. 104–111 (2004)
14. Oudeyer, P., Kaplan, F., Hafner, V.V.: Intrinsic Motivation Systems for Autonomous Mental Development. IEEE Transactions on Evolutionary Computation 11(1) (2007)
15. White, R.W.: Motivation Reconsidered: The Concept of Competence. Psychological Review 66(5), 297–333 (1959)
16. Berlyne, D.E.: Conflict, Arousal and Curiosity. McGraw-Hill, New York (1960)
17. Stout, A., Konidaris, G.D., Barto, A.G.: Intrinsically Motivated Reinforcement Learning: A Promising Framework For Developmental Robot Learning. In: Proceedings of the AAAI Spring Symposium on Developmental Robotics, AAAI, Menlo Park (2005)
18. Sutton, R., Precup, D., Singh, S.: Between MDPs and Semi-MDPs: A Framework for Temporal Abstraction in Reinforcement Learning. Artificial Intelligence 112, 181–211 (1999)
19. Schembri, M., Mirolli, M., Baldassarre, G.: Evolving Internal Reinforcers for an Intrinsically Motivated Reinforcement-Learning Robot. In: 6th IEEE International Conference on Development and Learning (ICDL2007) (submitted)

20. Baldassarre, G.: A Modular Neural-Network Model of the Basal Ganglia's Role in Learning and Selecting Motor Behaviours. Journal of Cognitive Systems Research 3, 5–13 (2002)
21. Sutton, R.S., Barto, A.G.: Reinforcement Learning: An Introduction. The MIT Press, Cambridge (1998)
22. Houk, J.C., Davis, J.L., Beiser, D.G.: Models of the Basal Ganglia. The MIT Press, Cambridge (1995)
23. Schultz, W.: Getting Formal with Dopamine and Reward. Neuron 36, 241–263 (2002)
24. Mannella, F., Baldassarre, G.: A Neural-Network Reinforcement-Learning Model of Domestic Chicks that Learn to Localise the Centre of Closed Arenas. Philosophical Transactions of the Royal Society B – Biological Sciences 362(1479), 383–401 (2007)
25. Nolfi, S.: Evolutionary Robotics: Exploiting the Full Power of Self-Organization. Connection Science 10(3-4), 167–183 (1998)

Evolving Cultural Learning Parameters in an NK Fitness Landscape

Dara Curran[1], Colm O'Riordan[2], and Humphrey Sorensen[3]

[1] Dept. of Computer Science,
University College Cork,
Ireland
d.curran@cs.ucc.ie

[2] Dept. of Information Technology,
National University of Ireland,
Galway
colm.oriordan@nuigalway.ie

[3] Dept. of Computer Science,
University College Cork,
Ireland
sorensen@cs.ucc.ie

Abstract. Cultural learning allows individuals to acquire knowledge from others through non-genetic means. The effect of cultural learning on the evolution of artificial organisms has been the focus of much research. This paper examines the effects of cultural learning on the fitness and diversity of a population and, in addition, the effect of self-adaptive cultural learning parameters on the evolutionary process. The NK fitness landscape model is employed as the problem task and experiments employing populations endowed with both evolutionary and cultural learning are compared to those employing evolutionary learning alone.

Our experiments measure the fitness and diversity of both populations and also track the values of two self-adaptive cultural parameters. Results show that the addition of cultural learning has a beneficial effect on the population in terms of fitness and diversity maintenance. Furthermore, analysis of the self-adaptive parameter values shows the relative quality of the cultural process throughout the experiment and highlights the benefits of self-adaptation over fixed parameter values.

1 Introduction

A number of researchers have analysed the interactions between learning and evolution, where individuals within a population of artificial organisms are capable of evolving genetically (evolutionary learning) and also of acquiring knowledge during their lifetime (lifetime learning). Hinton and Nowlan[1] were among the first to show that learning could guide evolution and a number of other researchers have since conducted experiments which support this view[2,3,4].

F. Almeida e Costa et al. (Eds.): ECAL 2007, LNAI 4648, pp. 304–314, 2007.

A population capable of passing information between individuals in a non-genetic way can be said to possess a culture. Culture can take many forms including language, artifacts or imitation, and a number of researchers have examined the effects of culture on evolution, as well as the emergence of culture itself[5][6][7][8][9].

The aim of this paper is to further existing research by examining the effects of allowing certain cultural learning parameters to self-adapt as the experiment progresses. The model employed is similar to the social learning model used by Best to simulate non-genetic information transfer between individuals[6]. Experiments are undertaken employing two populations: one using evolutionary learning alone, and the other employing both evolutionary and cultural learning. The fitness and diversity of the two populations are measured throughout the experiments. Cultural parameter values are allowed to evolve with the population and these values are analysed and discussed.

The remainder of this paper is organised as follows. Section 2 describes related research, including the learning models on which this work is based and a discussion on self-adaptation. Section 3 presents the model employed for these experiments, including a detailed description of the fitness measurement and the self-adaptation mechanism. Section 4 outlines the experiment results and Section 5 presents conclusions.

2 Related Work

The following sections outline some background material, including learning models, previous related research, self-adaptation and diversity measurement.

2.1 Learning Models

Evolutionary learning refers to the process whereby a population of organisms evolves, or learns, by genetic means through a Darwinian process of iterated selection and reproduction of fit individuals[10]. Individuals do not contribute to their survival through learning or adaptation and the learning process is confined to the evolution of an organism's genetic material.

An alternative form of learning known as lifetime learning allows individuals to adapt to environmental changes and novel situations. Lifetime learning is often coupled with evolutionary learning, further enhancing the population's fitness through its adaptability and resistance to change.

A population that transfers information from one individual to another in a non-genetic manner can be said to possess a culture. Culture can take many forms such as language, signals or artifactual materials. Cultural learning describes the manner in which this information transfer takes place. A number of approaches have been implemented to simulate cultural learning including fixed lexicons[11], indexed memory[12], cultural artifacts[7] and signal–situation tables[13].

Hinton & Nowlan. Hinton and Nowlan employed a genetic algorithm to study the effects of lifetime learning on the performance of genetic evolution. In particular, they examined the Baldwin effect, which occurs when certain behaviour discovered through life-time learning becomes imprinted onto an individual's genetic material through the evolutionary processes of crossover and mutation[1][4].

Each agent in the model possesses a genome, comprised of a string of characters which can be one of 1, 0 or *?*. Each agent is allowed a number of rounds of lifetime learning where for each *?* in the genotype they 'guess' its value, assigning it either a 1 or a 0. The *?*s in the genotype represent the agent's phenotypic adaptability and any guessed values are lost at the end of the agent's life.

The experiments compared the performance of a population endowed with learning and one without. Results showed that the non-learning population was not capable of finding optimal solutions to the problem task. In contrast, once learning was applied, the population converged on the problem solution, showing that individual learning is capable of guiding genetic evolution.

Best. Hinton & Nowlan's model was modified by Best to replace individual learning with social, or cultural, learning[6]. A number of model agents are selected from the population according to their fitness and learner agents are randomly assigned to teacher agents. For each *?* in a learner's genome, the learner imitates the corresponding value in the teacher's genome with some probability. Best found that cultural learning could be used to guide evolution as an alternative to lifetime learning and in addition, that cultural learning resulted in faster convergence on optimal solutions.

2.2 Self-adaptation

Some research has been performed on the self-adaptation of parameters employed in evolutionary computation. Angeline divided such adaptation into three groups: *population level*, *individual level* and *component level*[15]. Population level adaptation dynamically adjust a parameter that is subsequently used globally across the population (such as a global crossover value)[16]. Individual level adaptation reduces the impact of a parameter to an individual level (such as adapting crossover points)[17]. Component level adaptation dynamically changes the way in which individuals components of an individual will be altered independent of one another[18]. Researchers have applied self-adaptive parameters at varying levels to evolutionary programming[19] and to genetic algorithms[18].

2.3 Diversity

Diversity measures are used to examine the differences between individuals in a population. Differences can be at a genetic level (genotypic diversity), or at a behavioural level (phenotypic diversity). It is commonly accepted that a population

[4] This should not be confused with Lamarckian theory, where an individual's lifetime experience is directly mapped to its genome and is directly inheritable[14]

that is capable of maintaining diversity will avoid premature convergence and local maxima. Diversity measures for populations of neural networks have been the focus of considerable research, focusing mainly on genotypic diversity[20,21].

Some research has analysed the effects of cultural learning on both the phenotypic and genotypic diversity of populations. In particular, work by Curran and O'Riordan employed evolving neural network agents engaged in cultural learning whose problem task was comprised of a number of turn-based games[22]. Results showed that the combination of cultural learning and evolutionary learning led to improvements in fitness and higher levels of genotypic diversity. However, the complexity of the model employed makes analysis difficult, highlighting the need for a simpler model of both agent and lifetime learning mechanism alike.

3 Model

Our model follows the structure of the experiments performed by Best[6]. Each agent possesses a chromosome of 20 bits and experiments employ a population of 1000 agents evolved for 200 generations. The experiments employ two populations: one using evolutionary learning alone (purely genetic evolution) and the other employing cultural learning.

In the evolutionary learning population, a genetic algorithm is used to evolve agents. Agent genomes consist of strings of 20 1s or 0s randomly generated in the initial population with equal probability. Populations employing both evolutionary and cultural learning have genomes containing 1s, 0s or *?*s, where the *?*s represent the agent's phenotypic ability to imitate either 1 or 0 from a teacher agent. In the initial population these are randomly selected with probability 0.25, 0.25 and 0.5 respectively, corresponding to the original probability set in the Hinton and Nowlan experiment.

Our model employs the NK fitness model as the fitness landscape for the experiments[23]. The NK fitness model alters the ruggedness of a fitness landscape by tuning the level of epistasis present within a chromosome. The model takes two parameters: N, the length of the chromosome and K, the level of epistasis. The higher the epistasis level, the more rugged the landscape.

Unlike the model employed by Hinton and Nowlan, the NK fitness model provides some evolutionary feedback to populations evolving genetically and thus makes a fair comparison between evolutionary and lifetime learning possible.

3.1 Fitness Calculation

Individuals in a population employing evolutionary learning alone possess genomes comprised of *1*s or *0*s (no *?*s) and therefore have no opportunity to alter the way they interact with their environment. The fitness of agents in such populations is measured directly from their genomes.

When cultural learning is added to evolutionary learning, individuals are given the opportunity to replace each of the *?*s in their genomes with either 1s or 0s. Each agent has an opportunity to replace *?*s in its genome with values

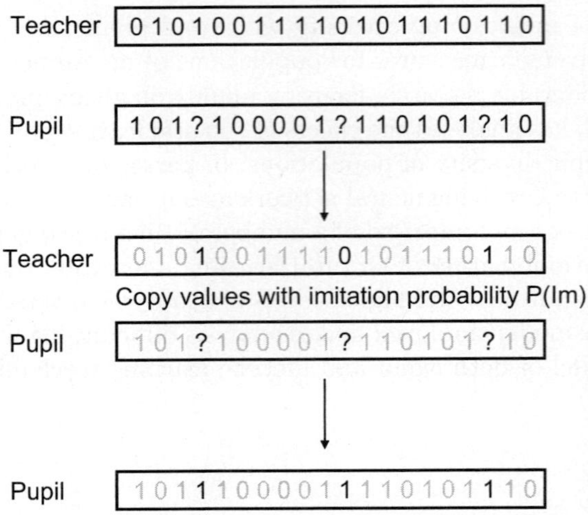

Fig. 1. Example teaching round

imitated from its teacher. This process may be repeated a number of times and each resulting bit string is evaluated using the NK fitness function. The best of these is taken as the agent's phenotype and the corresponding fitness as the agent's final fitness value.

3.2 Cultural Learning

Cultural learning is implemented by allowing individuals to acquire values for the their *?* loci from teachers selected from the population. Teachers are selected according to fitness and the number of teachers selected for each generation is determined by a teacher proportion. For a teacher proportion of 10%, 10% of the current generation will be selected (according to fitness) to instruct the next.

Each pupil is stochastically assigned a teacher from the previous generation and for each *?* locus in the pupil's genecode the corresponding value is copied from the teacher's phenotype with a certain *imitation probability*. Should the agent not copy the value from its teacher, it will replace the *?* in its genome with either 1 or 0 at random (it will guess a value). Once this has been done for all loci containing *?*s, the pupil's fitness is measured. This process is repeated for a number of *teaching rounds* and the best fitness result is taken as the agent's final fitness.

Fig. 1 shows an example teaching round where a pupil's *?* loci are stochastically replaced with the teacher's equivalent values with a certain imitation probability. In this example, the pupil successfully imitates the first and last values from the teacher but does not imitate the middle value. This time, the pupil randomly assigns a 1 or 0 to replace its *?* locus.

3.3 Self-adaptation of Cultural Parameters

The model employed by Best allowed learner agents to imitate corresponding teacher values perfectly (i.e. with an imitation probability of 1.0) for a fixed number of teaching rounds. However, it is not clear that such an arrangement would produce the best result in all circumstances. In our implementation we allow both these parameters to evolve in a similar manner to the main genome concerned with NK fitness problem. Each agent carries three genomes: one containing an encoded solution to the NK landscape, one encoding imitation probability and, finally, one encoding teaching rounds. All three genomes are allowed to undergo the processes of crossover and mutation.

The imitation probability genome consists of 7 bits. The imitation probability value is obtained from the decimal value of the 7 bit string divided by the maximum possible value, giving a range of probabilities between 0.0 and 1.0. The teaching rounds genome also consists of 7 bits which are simply converted to a decimal value representing the number of teaching rounds to be applied, giving a range of teaching rounds between 0 and 127.

When an individual is born, its imitation probability and teaching cycle parameter values are decoded from their genomes and are subsequently used during its cultural learning phase. All three genomes undergo processes of crossover and mutation when two individuals are selected, thus evolving along with the population.

3.4 Diversity Measure

The diversity measure examines the differences between members of a population. The population employing evolutionary learning alone contains genomes comprising of 1s or 0s, while the population adding cultural learning has an additional number of *?*s within its genomes. This difference between the two populations makes a direct genetic comparison difficult. Therefore, a phenotypic measure of diversity may be more suitable. In this model, the phenotype of an agent is defined as its response to its environment.

In the population employing evolutionary learning alone, an agent's response to its environment is fixed and directly influenced by its genome. Therefore, its genotype and phenotype are equivalent - the agent is not capable of influencing its behaviour during its lifetime. However, in the population employing cultural learning, an agent may imitate its teacher by replacing the *?*s in its genome with the teacher's equivalent value of either 1 or 0. Since the agent's fitness is calculated according to the best of these teaching rounds, we chose to select the bit-string resulting from an agent's best teaching round as its phenotype.

Employing this model, phenotypic diversity measurements of populations employing evolutionary learning alone and populations employing both evolutionary and cultural learning are directly comparable, as both are measuring differences between strings containing *1*s and *0*s. In both cases, the diversity measure examines agents at the end of their lives.

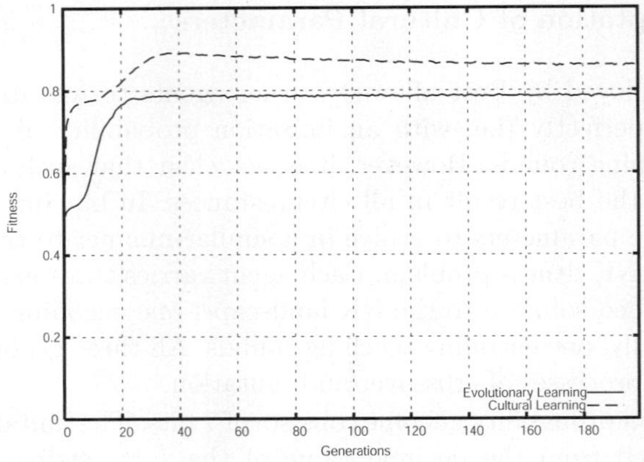

Fig. 2. Population fitness

The diversity measured in a manner similar to that developed by Eriksson[24]. Diversity is defined as the average of all individual Hamming distances between phenotypes of individuals x and y, $h(x, y)$, whose phenotypes are unique within the population. Thus, for a population of size n containing m unique phenotypic binary strings, the diversity measure can be calculated as:

$$\frac{2}{n(m-1)} \sum_{x=1}^{m} \sum_{y=x+1}^{m} h(x, y)$$

4 Experiments

The experiments employed two populations: one using evolutionary learning alone, and the other using both evolutionary and cultural learning. The focus, therefore, is to examine the effect of adding cultural learning to an evolving population. Both populations are allowed to evolve for 200 generations. 1000 individuals are present per generation and each genome is 20 bits in length (N=20). The K parameter for the NK model is set to 15. The proportion of teachers per generation is 10%. The results are averaged from 20 independent experiment runs. Results for the cultural learning population are shown from generation 1 (as the population in generation 0 is not capable of performing any cultural learning due to the lack of teachers).

Fig. 2 shows the fitness levels for each population. It is clear from these results that the population employing cultural learning is attaining significantly higher fitness levels than the population employing evolutionary learning alone. The cultural process is clearly beneficial to the population and while cultural learning may take longer to converge, its overall fitness remains higher than evolutionary learning alone throughout the experiment.

The cultural learning process is allowed a great deal more evaluations as each individual has the chance to perform a local search by altering its phenotype

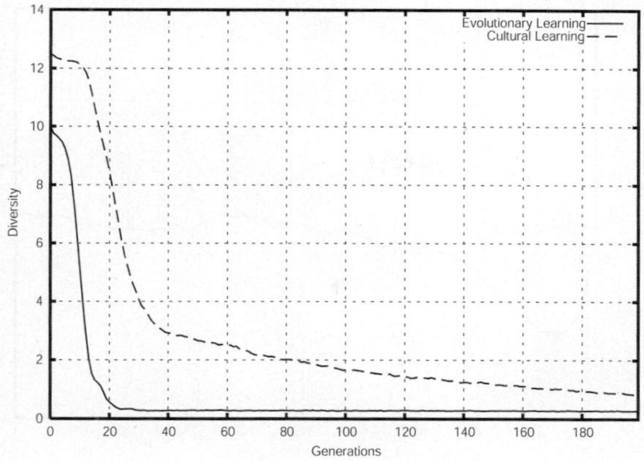

Fig. 3. Population diversity

during each teaching round. To investigate the effect of the increased number of evaluations, a number of experiments were undertaken comparing very large evolutionary learning populations with much smaller cultural learning populations. In each case, the smaller cultural learning population out-performed the larger evolutionary population.

The results illustrated in Fig. 3 show that both learning populations maintain an initial high level of diversity. However, the evolutionary learning population converges at generation 20, while the cultural learning population exhibits a steady decline throughout the experiment but does not converge by generation 200. Clearly, the addition of cultural learning helps to maintain higher diversity levels than evolutionary learning alone.

The values of the first self-adaptive parameter, imitation probability, are illustrated in Fig. 4. The imitation probability determines the probability than an individual will copy a teacher's value to replace its own *?* value. Where the value is not copied, the individual selects either 1 or 0 at random.

The parameter begins at an average of 0.5 for the population, but quickly descends for the first 20 generations. This drop is indicative of the fact that at the beginning of the experiment, it is likely that teachers are not particularly fit and that it is generally better for a learner to guess at a value rather than copy the teacher's. This situation begins to change after generation 20, where the probability climbs steadily, reaching 0.5 by generation 120.

Fig. 5 shows the results obtained for the second self-adapting parameter: teaching rounds. A teaching round is an opportunity for learners to copy values from teachers to replace their own *?* genes. Increasing the number of teaching rounds increases the chance that a learner will attain a highly fit genotype, provided that the teacher's values are useful to copy.

The parameter's value begins at around 63 and exhibits a sharp rise to over 100 teaching cycles per learner. By generation 20 however, the parameter has begun to fall and continues to do so before stabilising at around 70, well short of the maximum 127 teaching rounds possible.

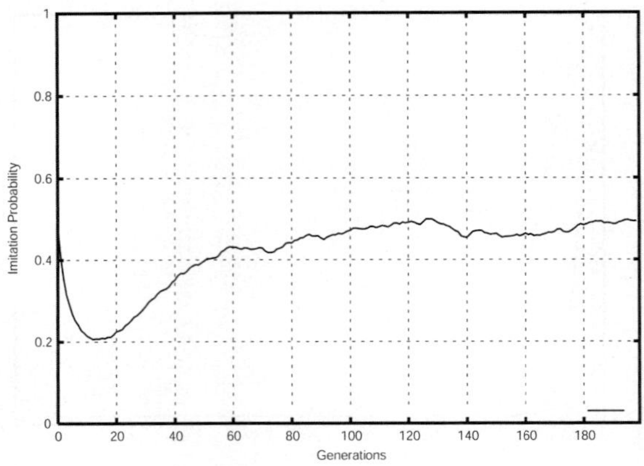

Fig. 4. Evolution of imitation probability parameter

It is interesting that the number of teaching cycles decreases despite the fact that there is no explicit cost to cultural learning. While it is possible that this is due to random genetic drift, it is made unlikely by the fact that this effect repeats itself throughout the 20 independent experiment runs.

The two parameters are evolving simultaneously and therefore are likely to influence each other throughout the experiment. As the number of teaching rounds peaks at generation 15, imitation probability reaches its lowest point. There then follows a steady increase in imitation probability mirrored by a steady decrease in teaching rounds.

Once teaching rounds attain an initial high value, the benefits of imitation become more pronounced and the population responds by beginning to increase imitation probability. However, once probability of imitating a teacher increases, less learning rounds are required in order to achieve imitation, and the teaching rounds values begin to fall. This process continues throughout the experiment, with the fall of teaching rounds values being mirrored by a corresponding rise in imitation probability.

A second interpretation of these results concerns the quality of teachers available throughout the experiment. Initially, teachers are unlikely to be highly fit and therefore complete imitation on the part of pupils is not necessarily the best strategy. The best approach is likely to be a high number of teaching rounds coupled with a low imitation probability to allow pupils to guess values rather than imitating them.

As the population's fitness improves, so does the quality of teachers available. The evolutionary response to this change is to alter the teaching rounds and imitation probability values accordingly. Better teachers should be imitated with a higher probability, as this imitation is more likely to result in improved fitness for the pupil. At the same time, teaching rounds should decrease to take account of the fact it is no longer as necessary to guess at values now that pupils can benefit more from imitation.

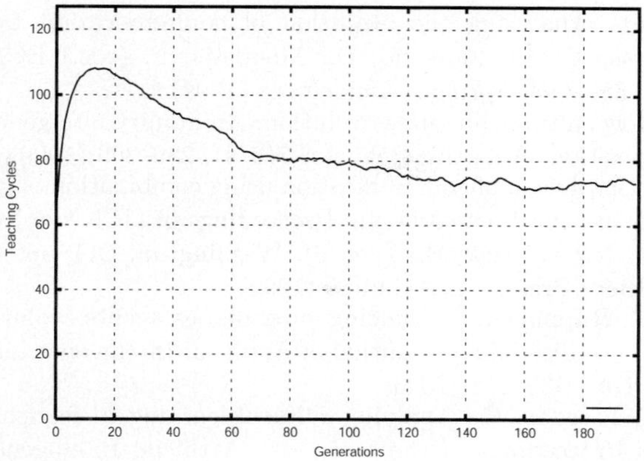

Fig. 5. Evolution of teaching rounds parameter

5 Conclusion

This paper presented results obtained from experiments involving two populations: one employing evolutionary learning and one employing both evolutionary and cultural learning. The evolutionary process was also given the opportunity to co-evolve suitable values for two cultural learning parameters: imitation probability and teaching rounds.

Results show that the addition of cultural learning to evolutionary learning has beneficial effects in terms of fitness, as seen in previous work. Furthermore, results obtained from the examination of the self-adaptive parameters illustrate the relative quality of teachers (and therefore, the impact of cultural learning) at different stages in the experiment. The fact that the values for these parameters changes significantly throughout the experiment shows the usefulness of self-adaptive cultural parameters for both performance and analysis of cultural learning. Future work will examine the impact of changing environments on the values of self-adaptive cultural parameters.

Acknowledgements. The first author would like to acknowledge the support of the Irish Research Council for Science, Engineering and Technology.

References

1. Hinton, G.E., Nowlan, S.J.: How learning guides evolution. Complex Systems 1, 495–502 (1987)
2. Nolfi, S., Parisi, D.: Learning to adapt to changing environments in evolving neural networks. Adaptive Behavior 5(1), 75–97 (1996)
3. Floreano, D., Mondada, F.: Evolution of plastic neurocontrollers for situated agents. In: Animals to Animats 4 (1996)
4. Sasaki, T., Tokoro, M.: Adaptation toward changing environments: Why darwinian in nature? In: Husbands, P., Harvey, I. (eds.) Fourth European Conference on Artificial Life, pp. 145–153. MIT Press, Cambridge (1997)

5. De Jong, E.D.: Analyzing the evolution of communication from a dynamical systems perspective. In: Floreano, D., Mondada, F. (eds.) ECAL 1999. LNCS, vol. 1674, pp. 689–693. Springer, Heidelberg (1999)
6. Best, M.L.: How culture can guide evolution: An inquiry into gene/meme enhancement and opposition. Adaptive Behavior 7(3/4), 289–306 (1999)
7. Cangelosi, A.: Evolution of communication using combination of grounded symbols in populations of neural networks. In: Proceedings of IJCNN99 International Joint Conference on Neural Networks (vol. 6), Washington, DC, pp. 4365–4368. IEEE Computer Society Press, Los Alamitos (1999)
8. Borenstein, E., Ruppin, E.: Enhancing autonomous agents evolution with learning by imitation. Interdisciplinary Journal of Artificial Intelligence and the Simulation of Behaviour 1(4), 335–348 (2003)
9. Curran, D., O'Riordan, C.: Applying cultural learning to sequential decision task problems. In: Proceedings of the 16th Irish Artificial Intelligence and Cognitive Science Conference (AICS 2005) (2005)
10. Darwin, C.: The Origin of Species: By Means of Natural Selection or the Preservation of Favoured Races in the Struggle for Life. Bantam Press, London (1859)
11. Cangelosi, A., Parisi, D.: The emergence of a language in an evolving population of neural networks. Technical Report NSAL–96004, National Research Council, Rome (1996)
12. Spector, L.: Genetic programming and AI planning systems. In: Proceedings of Twelfth National Conference on Artificial Intelligence, Seattle, Washington, USA, pp. 1329–1334. MIT Press, Cambridge (1994)
13. MacLennan, B., Burghardt, G.: Synthetic ethology and the evolution of cooperative communication. Adaptive Behavior 2(2), 161–188 (1993)
14. Lamarck, J.B.: Philosophie Zoologique. Chez Dentu, Paris (1809)
15. Angeline, P.J.: Adaptive and self-adaptive evolutionary computations. In: Palaniswami, M., Attikiouzel, Y. (eds.) Computational Intelligence: A Dynamic Systems Perspective, pp. 152–163. IEEE Computer Society Press, Los Alamitos (1995), `citeseer.ist.psu.edu/angeline95adaptive.html`
16. Spears, W.M.: Adapting crossover in evolutionary algorithms. In: McDonnell, J.R., Reynolds, R.G., Fogel, D.B. (eds.) Proc. of the Fourth Annual Conference on Evolutionary Programming, pp. 367–384. MIT Press, Cambridge (1995)
17. Rosca, J.P.: Hierarchical self-organization in genetic programming. In: Proceedings of the Eleventh International Conference on Machine Learning (1994)
18. Bäck, T., Schwefel, H.P.: An overview of evolutionary algorithms for parameter optimization. Evolutionary Computation 1 (1993)
19. Fogel, D.B., Fogel, L.J., Atmar, J.W.: Meta-evolutionary programming. In: Proceedings of the Conference on Signals, Systems, and Computers, pp. 540–545 (1991)
20. Brown, G.: Diversity in Neural Network Ensembles. PhD thesis, University of Birmingham (2003)
21. Burke, E.K., Gustafson, S., Kendall, G.: Diversity in genetic programming: an analysis of measures and correlation with fitness. IEEE Trans. Evolutionary Computation 8(1), 47–62 (2004)
22. Curran, D., O'Riordan, C.: Increasing population diversity through cultural learning. Adaptive Behavior 14(4) (2006)
23. Kaufmann, S.A.: Adaptation on rugged fitness landscapes. Lectures in the Sciences of Complexity 1, 527–618 (1989)
24. Eriksson, R.I.: An initial analysis of the ability of learning to maintain diversity during incremental evolution. In: Freitas, A.A., Hart, W., Krasnogor, N., Smith, J. (eds.) Data Mining with Evolutionary Algorithms, Las Vegas, Nevada, USA, pp. 120–124 (2000), `citeseer.ist.psu.edu/eriksson00initial.html`

How Does Niche Construction Reverse the Baldwin Effect?

Hajime Yamauchi

Japan Advanced Institute of Science and Technology
hoplite@jaist.ac.jp

Abstract. Deacon [1] considers that the reverse Baldwin effect can be one of the major forces in language evolution. The reverse Baldwin effect is essentially a redistributional process of genes as a result of environmental changes which mask and unmask selection pressures. Although Deacon indicates that in the case of language evolution, niche construction is deeply involved in masking and unmasking processes, neither specific explanations for the mechanism nor examples have been given. In this study we use evolutionary computation simulations to explore how niche constructing properties of language evolution can induce at least the masking effect, and hence lead to genetic degradation. The simulation demonstrates that the masking effect is indeed a part of the evolutionary process found in the normal Baldwin effect.

1 Introduction

As a causal theory of learning and evolution, the Baldwin effect has gathered wide attention in evolutionary linguistics where first language acquisition is recognized as one of the key issues to understand this uniquely human capacity. However, for those accustomed to the Baldwinian view of the relationship between learning and evolution, what Deacon [1] describes may sound somewhat counterintuitive. While the Baldwin effect describes how previously learnt knowledge becomes a part of innate knowledge, according to Deacon, under some circumstances, innate knowledge would be replaced by more plastic, learnt knowledge. As the process seemingly follows the opposite flow of what the Baldwin effect describes, he called this process the "reverse Baldwin effect". This paper will present how the niche constructing aspect of language evolution serves as one of the key mechanisms necessary for the purported effect without assuming, as Deacon has, that externally motivated changes (like climate changes) in environmental conditions would take place.

2 Genetic Redistribution

While Waddington's [2] genetic assimilation is often conceived of as the mechanism of the Baldwin effect, Deacon believes that the reverse Baldwin effect is essentially a process of "genetic redistribution" where initially high innate

F. Almeida e Costa et al. (Eds.): ECAL 2007, LNAI 4648, pp. 315–324, 2007.

knowledge is functionally replaced by a suite of traits each of which has its own original function.

It is thought that genetic redistribution is triggered when the current selective pressure is somehow shielded. This is called the "masking effect" and it effectively works to reduce the adaptive importance of the trait carried. Consequently, the masking effect causes the gene expressing the trait to randomly drift. Eventually the gene loses its functional significance and becomes a "pseudogene"; incapable of expressing the given trait. When the mask is lifted (i.e. the original selective pressure resurfaces), the organism has to take an alternative option to compensate for the lack of the trait. Deacon posits that this "unmasking effect" causes originally irrelevant traits that evolved for different functions to form a suite in order to replicate the function of the original trait (and the gene for it).

Because the focal function is now subserved by the complexes of different traits attributed to a group of genes, the original selective pressure is now *redistributed* onto those genes. As the function is no longer controlled by a single gene, but a suite of different genes, Deacon considers it effectively increases the freedom of ontogenetic accommodation; the reverse Baldwin effect.

2.1 Case Studies

One of the empirical examples Deacon and others [1,3] often draw on is the case of vitamin C synthesis. Somehow, the ability of synthesizing ascorbic acid (vitamin C) was lost in the primate lineage including *Homo*. It is known that the gene responsible for the enzyme working on the last stage of the synthesis was replaced by its degraded pseudogene. Around the period when the gene irreparably deteriorated, it is assumed that the climate was warm allowing primates to obtain vitamin C rich fruits easily. This masked the adaptive importance of synthesizing the vitamin endogenously, and consequently the genes attributed to the function started to randomly drift. This effectively made the primates "dietarily entrenched" in a frugivorous life style. A change in climate, however, made the fruits scarcer, and unmasked the selective pressure once again. Because the gene responsible for the synthesis had become a pseudoegene because of the random drift, the primates had to rely on other traits to compensate. The alternative was to increase foraging ability to obtain fruits effectively. For that purpose, various primarily independent traits such as tooth structure, color vision, and taste preference were employed and increased their evolutionary importance: redistribution of selective pressure of the function originally served by the innate trait onto the suite of more genetically indirect traits. Wiles et al. [3] have conducted computer simulations based on the above argument, and confirmed that interactions between learning and evolutionary search could induce not only genetic assimilation, but also genetic redistribution if masking and unmasking effects take place.

Deacon (p.c.) has similarly attempted to explain the striking complexification of bird song during the domestication process of the whitebacked munia (*Lonchura striata* var. *domestica*) from its feral ancestor, the Bengalese finch

(*Lonchura striata*) [4]. Ritchie and Kirby [5] have demonstrated that this description of the song's complexification through the reverse Baldwin effect is computationally attestable.

2.2 Masking Effect in Language Evolution

Given the potential explanatory power of the reverse Baldwin effect, both Deacon [1] and Wiles et al. [3] envisage that it could also play a significant role in language evolution. However, it is apparent from the above arguments that, for the reverse Baldwin effect to take place, there needs to be some causal agent to induce the masking effect. In the case of vitamin C, it was the warm climate (and abundant fruits), and in the case of the munia, it was domestication. Deacon considers the potential masking agent in language evolution is its niche constructing process [6,7]. However, it is unclear quite how the process comes into play as regards the masking effect. In the next section, we will examine how complex interactions between language learning and evolution create niche constructing processes, and how they transfer the selective pressure on innate linguistic knowledge to learning, and hence lead to genetic drift.

3 A Computational Model

In order to establish if the niche constructing aspect of language could mask selective pressure, a computational simulation of language evolution has been developed. The simulation, which plausibly models both biological and cultural evolution, and the developmental aspect of language at a reasonable level, is based on my research [7] The model works with an evolving population of agents.

3.1 Stages

The main stages in the simulation are listed here, and details of each stage are given in following sections:

Birth The agent's grammar is empty regardless of innate linguistic knowledge coded in a chromosome *G-Chrom*, described in the next section. The agent's cognitive ability is determined based on her designated chromosome *L-Chrom*, also described in the next section.

Learning The agent is exposed to *nInput* linguistic inputs from her linguistic environment (i.e. E-language). *nInput* is considered to set the *critical period.* With the learning algorithm described in the next section, the agent builds her own grammar by using a particular cognitive ability *Cog.*

Invention The agent invents some parts of her grammar by using *Cog.*

Communication The agent communicates with her two immediate neighbor peers for *nCom* times. Successful communication will increase the agent's fitness score.

Reproduction Parents are selected probabilistically according to their fitness score and their chromosomes are crossed over using one-point cross-over to give two children. The two chromosomes (i.e. *G-Chrom* and *L-Chrom*) are independent, and cross-over takes place within each of the chromosomes. Individual genes are mutated with probability *pM*.

Death The entire grammar of each agent in the population is sampled and the resulting linguistic information creates an E-language from which the next generation is to learn.

3.2 Model Structure

1. **The Agent** An agent has two types of chromosomes, namely *G-Chrom* and *L-Chrom*. Genetic information of innate linguistic knowledge is coded in *G-Chrom*. The innate linguistic knowledge is encoded as a string of 0*s*, 1*s*, and **s* in *G-Chrom*. * is thought of a junk allele, and no use for anything. In the results reported here, the entire knowledge is represented in a 24 ternary string (i.e. the number of genes in *G-Chrom* is 24). The initial gene pool has random numbers of 0*s*, 1*s*, and *s (thus, on average, one agent has 8 of each allele).
   The agent also has a different type of genetic information on *L-Chrom* coding the size of the cognitive capacity *Cog*. *Cog* is a cognitive capacity which enables the agent to update her grammar when discrepancies exist between her innate linguistics knowledge and linguistic inputs, and invent her own grammatical information. The size of *Cog* is given by a quantifiable scalar value *nCog*. The value is determined based on the number of 1*s* in *L-Chrom*. *L-Chrom* consists of a 48 binary string. The incremental value *iV* is set to 1 in the reported simulation. Therefore, the maximum value of *nCog* is 48.
   A grammar is coded as a ternary string, and the length of the string is 24 –equal to the size of *G-Chrom*. Three possible allelic values are 0, 1 and *NULL*. Wherever there is a *NULL* allele on the grammar, this part of the grammar is considered NOT to code any linguistic knowledge. Therefore, the more the *NULL* alleles there are in a grammar, the less the grammar codes linguistic knowledge. As described in the next section, although the agent learns a particular grammar by using her innate linguistic knowledge and input data, the grammar does not necessarily faithfully reflect the innate linguistic knowledge coded in *G-Chrom*; some parts of the grammar may converge to different values from the corresponding parts of the innate knowledge.
2. **Learning** Every agent in every generation is born with a completely empty grammar; all 24 alleles are *NULL*. Learning is the process to update such *NULL* alleles to substantial alleles (i.e. 0*s* and 1*s*). A learning agent sequentially receives linguistic inputs from *nAdult* adult neighbors (this is set to 5 in the simulation). Adults are the agents from the previous generation. A linguistic input is thought of as the *utterance* of an adult, which is represented by one allele of her mature grammar. Utterances derived from *NULL* alle-

les are considered as *NULL* utterances, and no learning (thus no grammar update) takes place. Following is the algorithm to develop the grammar:
Learning Algorithm Whenever the learner receives a linguistic input:

1. If the input value and the allelic value of the corresponding locus of the learner's grammar are different(i.e. not "parsable"), carry out the following the procedures:
   (a) If the corresponding allele of *G-Chrom* (her innate linguistic knowledge) "matches" (i.e. the two values are the same) with the input, update the given allele of the current grammar, and subtract 1 point from *nCog*.
   (b) If the corresponding allele of the innate linguistic knowledge is different from the input, update the given allele of the current grammar, and subtract 2 points from *nCog*.
2. Otherwise keep the current grammar.

The learning procedure is ceased when either *nCog* reaches 0 or the number of inputs meets *nInput*. *nInput* is set to 120. *NULL* utterances are counted for this process. Any locus of the grammar not receiving any input (or receiving only *NULL* utterances) remains *NULL*. Who utters and which part of her grammar is provided an input datum is totally random. This means that if the adults have totally different grammars, the learner may update a given allele of her grammar frequently.

3. **Invention** Agents are capable of inventing their own linguistic knowledge, and of adding it onto their grammars. If an agent still holds *NULL* alleles in her grammar after the learning has taken place, and if her *nCog* has not yet become 0, with a probability of *pI*, pick one *NULL* allele randomly, and flip it to either 0 or 1 randomly, and subtract 1 point from *nCog*. This process is carried on until either no more *NULL* allele is in the grammar, or *nCog* reaches 0. In reported the simulation, *pI* is set to 0.02. Once the invention process is over, her grammar is considered to have reached a mature state, and no more grammar update takes place.
4. **Communication** Each agent is involved in *nCom* communicative acts with her immediate peer neighbors. The fitness of an agent is scored on the basis of how many utterances spoken with her mature grammar were parsable to the hearer[1]. The same as for learning input, an utterance is represented by one allele of the speaker's mature grammar. *nCom* is set to 15 for utterances. As each neighbor also speaks to each agent the same number of times, a total of 30 communicative acts is involved to gauge her fitness. Each successful communication increases her fitness by 1. Those who cannot establish any communication still receive 1 fitness score to keep the possibility of being a parent in Reproduction. Therefore, the maximum fitness value is 31.
5. **Reproduction** Roulette wheel selection is used for selecting parents according to their fitness, and their chromosomes *G-Chrom* and *L-Chrom* are individually crossed over using single-point crossover to create two offspring. *pM* is set to 0.001 per allele.

[1] By the configuration of the model, this also means that how many utterances of the other can be parsed with her grammar

3.3 Layout of the Model

In this section, the overall layout of the model is described. The layout basically follows Kirby and Hurford [8]. Fig. 1 shows how the elements of the simulation are organized.

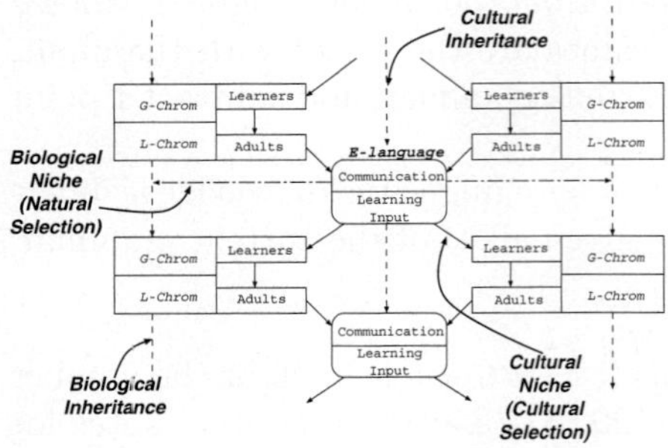

Fig. 1. An overview of the simulation model.

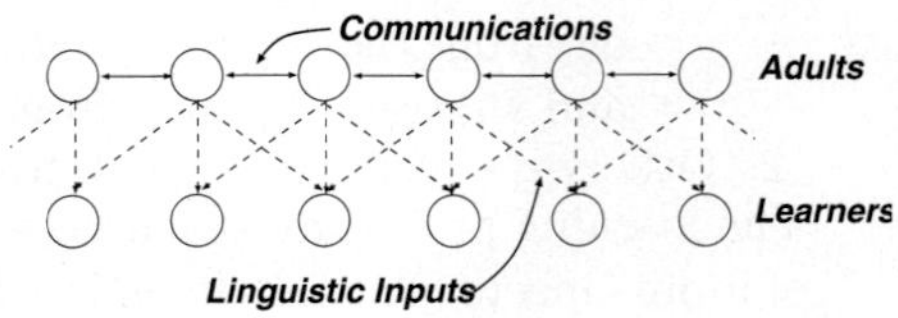

Fig. 2. The spatial organization of the population.

The figure shows how the niche constructing properties of the language are modeled in the simulation. Notice that there are two different channels of inheritance that interact with each other; namely cultural and biological inheritances. Linguistic knowledge is inherited in the biological channel via genes (i.e. innate linguistic knowledge) and is also inherited in the cultural channel via learning. What is inherited through the cultural channel is *niche constructed* grammars; accumulations of previous generations' linguistic inventions become the learning environment from which learners *reconstruct* their ancestors' grammars. The other side of the coin is biological niche construction; grammars acquired with that manner also affect the natural selection of individuals. The adaptive utility of a given grammar is only determined when a specific linguistic demography of the population is referred[2]. As such a demography is created by previous generations' linguistic activities (including both learning and communications), it is also a product of niche construction; in this case, what is constructed is a biological niche which affects individuals fitness.

Fig. 2 shows the spatial organization of the populations. Individuals are organized in a one-dimensional loop. Circles on the two different tiers represent adults and learners (i.e. two different generations). Incidents of communications (represented by the horizontal double-arrows in the adult tier) only take place within a single tier and is local since an individual attempts to communicate with her two immediate neighbor peers (left and right, in this figure). While this is an adult-to-adult process that results in natural selection, learning is thought of a vertical, adult-to-child transmission which results in cultural inheritance. One

[2] A particular grammar is only useful when others communicate with the same or very similar grammars.

adult provides linguistic inputs for *nAdult* neighbor learners (from the learner's point of view, she receives the inputs from *nAdult* immediate neighbor adults).

Note that reproduction is not affected by this spatial organization. Parents are probabilistically selected in proportion to their fitness and their offspring is randomly distributed in this space.

4 Results

All figures shown here are taken from one typical run of the simulation under the conditions described, and as such they well characterize the general tendency of the model. Fig.3 shows the average fitness of the population over time with a red line, and the average number of *NULL* alleles in matured grammars with a blue line. Rapid increase of the fitness (reaching 30 around the 150th generation) shows the population quickly evolves to almost the optimal state (the highest possible score is 31) as they develop their linguistic knowledge (i.e. reduction of *NULL* alleles). They do not do so randomly, rather neighbor individuals develop similar grammars in order to establish communications with peers.

Fig.4 shows the average number of * ("Junk") alleles in *G-Chrom* in a red line, the average matching number *nMatch* of matured grammars and *G-Chrom*, and evolution *Cog* by presenting *nCog* in a green line. We compare learners' stable grammar (i.e. after the learning, but before the invention stages) with their *G-Chrom* (i.e. innate linguistic knowledge. *nMatch* indicates a degree of contribution of innate linguistic knowledge to a given mature grammar. The higher the matching number is, the more of one's innate knowledge is reflected on the matured grammar. From the graph, it is apparent that soon after *Cog* has evolved, genes in *G-Chrom* start to match with grammars, and rapidly hit the

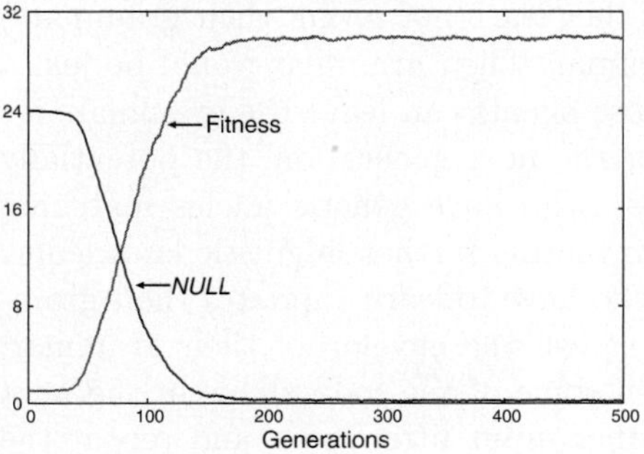

Fig. 3. Average fitness and numbers of *NULL* alleles in grammars of the population against generations. Because of these factors evolved rapidly, the graph shows only the first 500 generations.

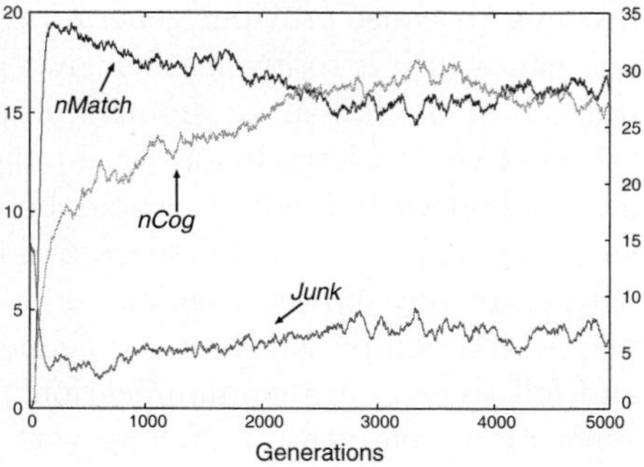

Fig. 4. In this graph, average numbers of "Junk" alleles ("*" alleles), *nMatch*, and *nCog* are shown. Values of *nCog* are plotted against the y2 axis (right).

highest point (on average 20 genes match with alleles in grammars at around the 200th generation) while *nCog* is still able to override only 14 alleles. The average number of the junk allele also supports this view. As *nMatch* grows, the junk allele rapidly expelled from the population. However, as *Cog* gradually evolves, *nMatch* decreases and becomes stable around 15. Along the same line, the junk allele also kicks in the population once again.

5 Analysis

The time course of the simulation demonstrates a complex interactions of learning, its evolution, and evolution of innate linguistic knowledge. Overall, we are able to summarize the course in three main stages: Stage 1: *Baldwinian Niche Construction* 0-200 (generations); Stage 2: *Masking* 200-4000; and Stage 3: *Stable* 4000-.

5.1 Baldwinian Niche Construction

The initial rapid increase of fitness in Stage 1 goes hand in hand with the increase of genetic contribution in language acquisition. What is striking is that this trend starts immediately after the average *nCog* reaches 1 (around the 30th generation) and it reaches almost the highest possible value, while *nCog* still remains around 12 which barely covers a half of the whole grammar to be updated. This strongly implies that innate linguistic knowledge of learners highly faithfully represents invented grammars of previous generations. The population at the first few generations would not receive any input from the previous generations as with 0 *Cog*, neither learning nor invention takes place to leave input for next generation. However, soon after *Cog* evolves in some individuals, they can invent their own grammatical information randomly, they do not use *Cog* for learning (because previous generations neither learn nor invent their grammars) and can utilize it to invent their own grammar. Their grammar would be just a fraction of the full-size grammars, but those agents can leave the grammars for the next generation's linguistic input. For the next generation, the potentially most adaptive individuals would be those who have genetic alleles matching with linguistic inputs, and spare *Cog* for inventing further linguistic knowledge: with extremely limited *Cog*, individuals who have to learn (update) their grammar would completely lose the chance to push the envelop of their grammar, and fail to establish communication. Then some of the individuals in the next generation have genes matching with further input utterances, and repeat the same process. This is a cyclic process which Avital and Jablonka [9] have called "assimilate stretch": learning pushes the envelop $\Rightarrow$ genes assimilate it $\Rightarrow$ leave the room for the learning to push the envelop further. Consequently, * alleles in *G-Chrom* are quickly winnowed out, and 0*s* and 1*s* are reshuffled to match the given linguistic environment. Importantly, this cycle of adaptive assimilations is the consequence, and the cause of the niche construction process in the linguistic environment: what is created here is linguistic knowledge which works as a niche for both cultural inheritance and biological inheritance: individuals learn from linguistic knowledge created (invented) by the previous generations (niche construction in the cultural channel), and adaptability of their learnt grammars is evaluated only within local linguistic communities (through communications with neighbor peers). The assimilated innate linguistic knowledge also canalizes what type of grammars are learnable, and it reflects on agents' learnt grammars. Those grammars determines the shape of the selective environment (the selection is frequency dependent as the utility of a given grammar is determined by the frequency of other grammars in neighbors). This is what I [7] report as Baldwinian Niche Construction.

5.2 The Masking Effect

Once niche constructed linguistic knowledge, which is a cooperative product of highly limited *Cog* and *G-Chrom*, has been transferred into innate linguistic knowledge, the masking effect is able to take place. As *Cog* evolves in Stage 2, the importance of the contribution of innate linguistic knowledge begins to be masked. This permits random drift to affect genes in *G-Chrom*: as learning can cover the discrepancy between inputs and innate knowledge, the number of the junk allele gradually increases. The result is a coordinated decrease of *nMatch*. This trend continues until around the 3000th generation, where *G-Chrom* and *nMatch*, and the number of the junk allele reach their (somewhat rough) plateaus (Stage 3). During Stage 2, *nMatch* decreases by nearly 5, and remains around 15 in Stage 3. Therefore, once again *Cog* shoulders the important role of language acquisition.

It is highly significant that, given the early increase of fitness, the selective pressure for the increase of *Cog* seems to be well diminished at the end of Stage 1, while in fact it keeps evolving. This is potentially because a different type of random drift takes place in *L-Chrom*: initially the chromosome does not include any 1*s*. Because of mutations in the genes, however, gradually the numbers of the alleles increase, and spreads by crossovers. In other words, the evolution of *Cog* is a largely neutral one. This indicates that the masking effect is an epiphenomic consequence of the neutral evolution.

6 Discussion and Conclusion

The simulation clearly demonstrates the self-induced model of the masking effect by the niche constructing property of language. The result is comparable to those provided in Wiles et al. [3], and Ritchie and Kirby [5]. However, both models employ arbitrary changes in environmental conditions to induce the masking effect. In this regard, our model successfully presents one possible mechanism of the niche construction based masking effect as Deacon envisages it.

As it stands though, the model does not explain how and why the unmasking effect would occur. The model only shows that a certain degree of the masking effect takes place; this does not cause a substantial degradation of genetic

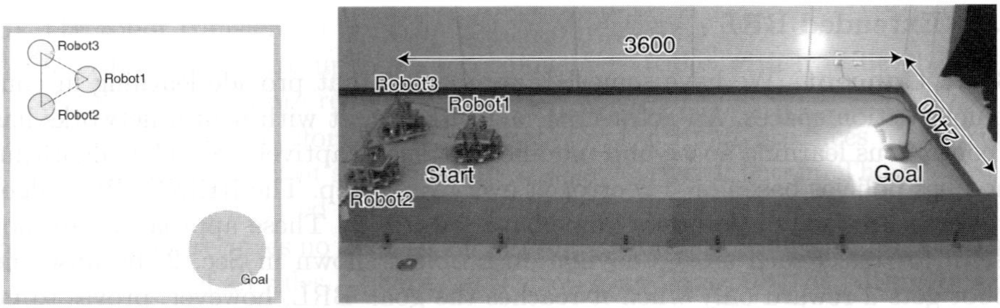

(a) Simulation (b) Real Robots

Fig. 2. Experimental Environment

All robots get a positive reward when one of them reaches the goal ($l_0 > thr_{goal} \vee l_1 > thr_{goal} \vee l_2 > thr_{goal}$). A robot gets a negative reward when it collides with a wall ($d_0^i > thr_d \vee d_1^i > thr_d$). We represent a unit of time as a *step*. A *step* is a sequence that allows the three robots to get their own input information, make decisions by themselves, and execute their actions independently. When the MRS reaches the goal, or when it cannot reach the goal within 200 steps in simulations and 100 steps in physical experiments, it is put back to the start. This time span is called an *episode*.

The settings of the learning mechanisms are as follows.

Prediction Mechanism (NN) Our previous work [7], verified BRL as a successful approach to CCP, with a reformation such that the state space was constructed with sensory information and predictions of the movements of the other robots in the next time step, to decrease the learning problem dynamics.
The prediction mechanism attached is a three-layered feed-forward neural network that performs back propagation. The input is a short history of sensory information, $I = \{\cos\theta_{t-2}^i, \sin\theta_{t-2}^i, \cos\psi_{t-2}^i, \sin\psi_{t-2}^i, \cos\theta_{t-1}^i, \sin\theta_{t-1}^i, \cos\psi_{t-1}^i, \sin\psi_{t-1}^i, \cos\theta_t^i, \sin\theta_t^i, \cos\psi_t^i, \sin\psi_t^i\}$, where $\psi_t^i = (\theta_t^j + \theta_t^k)/2 \;\; (i \neq j \neq k)$. The output is a prediction of the posture of the other robots at the next time step $O = \{\cos\psi_{t+1}^i, \sin\psi_{t+1}^i\}$. The hidden layer has eight nodes.

Behavior Learning Mechanism (BRL) The input is $\boldsymbol{x} = \{\cos\theta_t^i, \sin\theta_t^i, \cos\psi_{t+1}^i, \sin\psi_{t+1}^i, d_0^i, d_1^i, l_0^i, l_1^i, l_2^i\}$. The output is $\boldsymbol{a} = \{m_{rud}^i, m_{th}^i\}$, where m_{rud}^i and m_{th}^i are the motor commands for the rudder and the throttle respectively. σ in Eq.(4) is 0.05. For the standard BRL, $P_{th} = \{0.012, 0.01\}$. For the extended BRL, $P_{th} = 0.012$ and $P'_{th} = 0.01$. The other parameters are the same as the recommended values in our journal [7].

4.2 Result: Simulations

Figure 3 shows the averages and the deviations of steps that the MRS takes by the end of each episode. In the early stages, the MRS requires a lot of trial

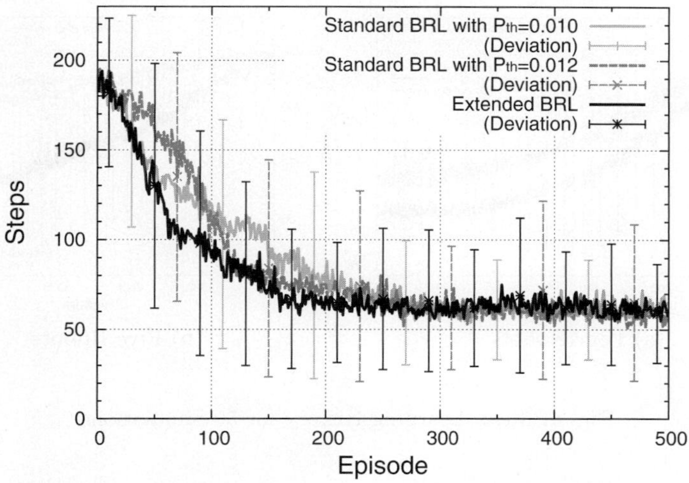

Fig. 3. Mean Learning History for 50 Simulations of Three Robots

and error and takes many steps to finish the episode. After such a trial and error process, the behaviour of MRS becomes more stable and it takes fewer steps. An MRS with the standard BRL stably achieves the task within nearly constant steps after the 250th episode, and the extended BRL accomplishes this in 200 episodes. This means that, in terms of learning speed, the extended BRL outperforms the standard one.

For the 50 independent runs, the MRS achieved different globally stable behaviour. However, we found a common point that robots always achieved cooperative behaviour by developing team play organised by a leader, a sub-leader and a follower. This implies that acquiring cooperative behaviour always involved autonomous specialisation. The extended BRL displayed higher adaptability, and yielded autonomous specialisation faster than the standard BRL.

Discussion. There is no significant difference in results in the learning performance of the BRLs for a three-robot CCP; therefore, we tested four- and five-robot CCP performance for more dynamic and complicated problems. The four robots use a square load, and the five robots have a pentagonal load. In these CCPs, ψ is the average of the angles between two neighbouring robots and the load. The other controller settings are the same as those for the three-robot CCP.

Figure 4 shows the average and the deviations of steps an MRS takes by the end of each episode. As the number of robots increases, we can find that the extended BRL provides increasingly better results than the standard BRL, although it requires more episodes before obtaining stable behaviour. The extended BRL has a function for coordinating behaviour as well as reducing the number of random actions that can result in unstable behaviour. These results

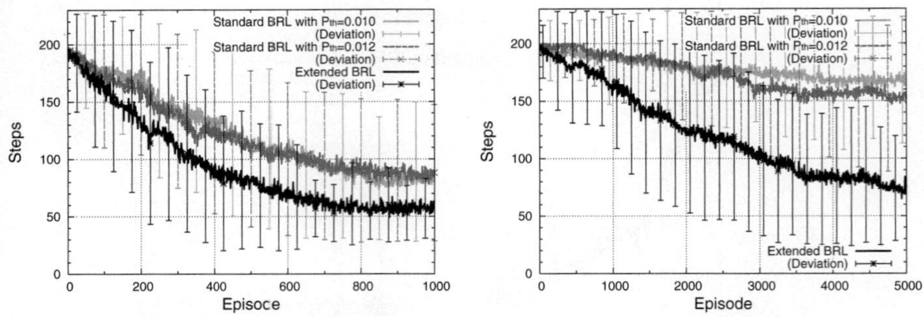

(a) Four Robots (b) Five Robots

Fig. 4. Mean Learning History for 50 Simulations

show that the extended BRL has a higher learning ability and is less dependent on the number of robots in the MRS. This implies that the extended BRL might have more scalability, which is one of the advantages of MRS over single-robot systems.

Although parameters that are more refined might provide better performance, parameter tuning is outside the scope, because BRL is designed for acquiring reasonable behaviour as quickly as possible, rather than optimal behaviour. In other words, the focal point of our MRS controller is not optimality but versatility. In fact, we obtain similar experimental results through experiments with an arm-type MRS similar to that in [6] using the same parameter settings.

4.3 Result: Physical Experiments

We conducted five independent experimental runs for each BRL. The standard BRL provided two successful results and the extended BRL provided four. Fig. 5 illustrates the best results of the physical experiments. These figures illustrate the number of steps and punishments in each episode. Comparing these results shows that the extended BRL requires fewer episodes to learn behaviour. The other successful results of the extended BRL show better performance than the best result of the standard BRL. The behaviour of the extended BRL is also more stable than that of the standard, because the MRS with the standard BRL gets several punishments after learning goal-reaching behaviour.

Figure 6 shows an example of the behaviour of the extended BRL. In the early stages, robots have no knowledge and function by trial and error. During this process, robots often collide with a wall and become immovable (Fig. 6(a)). Then, some robots reach the goal and develop appropriate input-output mappings (Fig. 6(b)). Observing the acquired behaviour and investigating rule parameters, we found that the robots developed cooperative behaviour, based on autonomous specialisation.

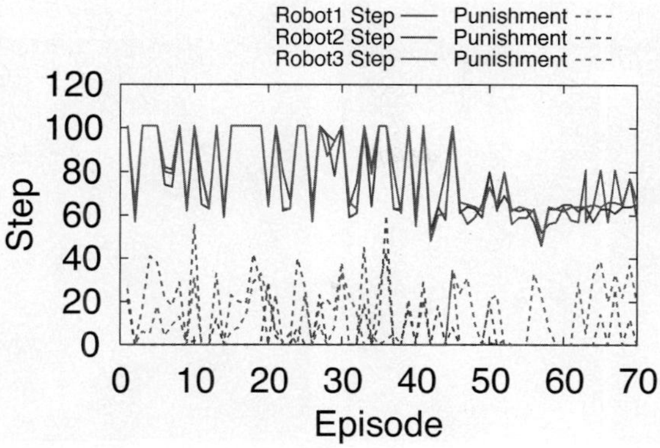

(a) Standard BRL

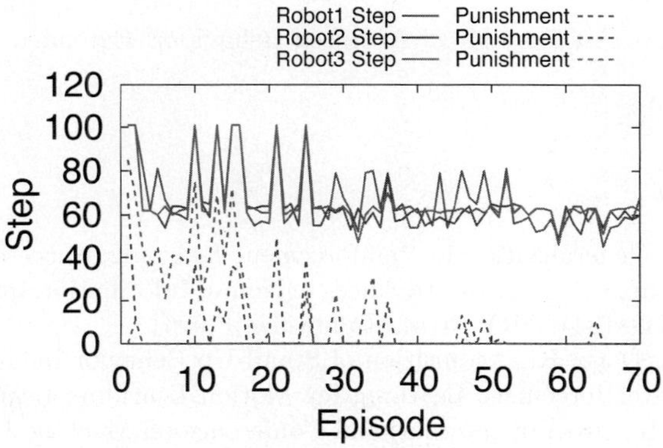

(b) Extended BRL

Fig. 5. Learning History: Physical Experiment

5 Conclusions

We investigated the RL approach for the behaviour acquisition of autonomous MRS. Our proposed RL technique, BRL, has a mechanism for autonomous segmentation of the continuous learning space, and proved effective for MRS through the emergence of autonomous specialisation. For accelerated learning, we proposed an extension of BRL with a function to generate interpolated actions based on previously acquired rules. Results of the simulations and physical experiments showed that the MRS with an extended BRL did learn behaviour faster than that with the standard BRL.

In the future, we plan to investigate the robustness and re-learning ability in a changing environment. We also plan to increase the number of sensors and adopt other expensive sensors such as an omnidirectional camera that will allow a robot to incorporate a variety of information, and thereby acquire more sophisticated cooperative behaviour in more complex environments.

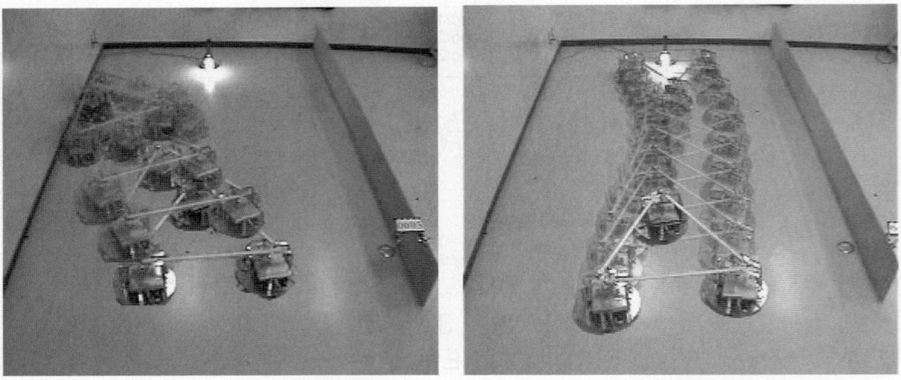

(a) In the Early Stage (b) After Successful Learning

Fig. 6. An Example of Acquired Behaviour: Extended BRL

References

1. Sutton, R.S.: Generalization in Reinforcement Learning: Successful Examples Using Sparse Coarse Coding. In: Advances in Neural Information Processing Systems, vol. 8, pp. 1038–1044. MIT Press, Cambridge (1996)
2. Morimoto, J., Doya, K.: Acquisition of Stand-Up Behavior by a Real Robot using Hierarchical Reinforcement Learning for Motion Learning: Learning "Stand Up" Trajectories. In: Proc. of International Conference on Machine Learning, pp. 623–630 (2000)
3. Lin, L.J.: Scaling Up Reinforcement Learning for Robot Control. In: Proc. of the 10th International Conference on Machine Learning, pp. 182–189 (1993)
4. Asada, M., Noda, S., Hosoda, K.: Action-Based Sensor Space Categorization for Robot Learning. In: Proc. of IEEE/RSJ International Conference on Intelligent Robots and Systems, pp. 1502–1509. IEEE, Los Alamitos (1996)
5. Takahashi, Y., Asada, M., Hosoda, K.: Reasonable Performance in Less Learning Time by Real Robot Based on Incremental State Space Segmentation. In: Proc. of IEEE/RSJ International Conference on Intelligent Robots and Systems, pp. 1502–1524. IEEE, Los Alamitos (1996)
6. Svinin, M., Kojima, F., Katada, Y., Ueda, K.: Initial Experiments on Reinforcement Learning Control of Cooperative Manipulations. In: Proc. of IEEE/RSJ International Conference on Intelligent Robots and Systems, pp. 416–422. IEEE, Los Alamitos (2000)
7. Yasuda, T., Ohkura, K.: Autonomous Role Assignment in Homogeneous Multi-Robot Systems. Journal of Robotics and Mechatronics 17(5), 596–604 (2005)
8. Doya, K.: Reinforcement Learning in Continuous Time and Space. Neural Computation 12, 219–245 (2000)
9. Williams, R.J.: Simple Statistical Gradient-Following Algorithms for Connectionist Reinforcement Learning. Machine Learning 8, 229–256 (1992)

Improving Agent Localisation Through Stereotypical Motion

Bart Baddeley and Andrew Philippides

Centre for Computational Neuroscience and Robotics
Department of Informatics
University of Sussex
Brighton, UK
bartbaddeley@googlemail.com

Abstract. When bees and wasps leave the nest to forage, they perform orientation or learning flights. This behaviour includes a number of stereotyped flight manoeuvres mediating the active acquisition of visual information. If we assume that the bee is attempting to localise itself in the world with reference to stable visual landmarks, then we can model the orientation flight as a probabilistic Simultaneous Localisation And Mapping (SLAM) problem. Within this framework, one effect of stereotypical behaviour could be to make the agent's own movements easier to predict. In turn, leading to better localisation and mapping performance. We describe a probabilistic framework for building quantitative models of orientation flights and investigate what benefits a more reliable movement model would have for an agent's visual learning.

1 Introduction

Flying insects are able to locate the position of their inconspicuous nest entrances using local visual landmarks. They can do this with relatively simple visual systems, high environmental noise and limited computational resources. To cope with these difficulties, insects have evolved innate behaviours which simplify visual learning. A remarkable example is the orientation or learning flight, performed by bees and wasps when they leave the nest to forage [2]. This behaviour includes a number of stereotyped flight manoeuvres which appear to be structured to mediate the active acquisition of visual information [1, 7]. However, there has been little work investigating how, and what aspects of, the learning flight lead to an improvement in learning. In computer vision and robotics in particular, one extensively studied approach to the general problem of localising an agent in the world is the Simultaneous Localisation And Mapping (SLAM) methodology [6, 3]. In this paper we use the SLAM framework to investigate the influence of the orientation flight on visual learning.

A successful solution to the SLAM problem involves building and maintaining a map of features sufficient for successful localisation during navigation. The map must be built incrementally, using noisy measurements and a stochastic model of the agent-environment interaction. Given knowledge of the initial conditions

F. Almeida e Costa et al. (Eds.): ECAL 2007, LNAI 4648, pp. 335–344, 2007.

and a perfect movement model, a perfect estimate of position can be maintained by integrating the agent's estimated movements across time. Alternatively, a perfect sensory system, providing perfect measurements of feature positions, could remove the need for any sort of internally generated position estimate. In practice neither the process (movement) model nor the measurements will be perfect and the best performance can be obtained by considering and combining both sources of information. Under the assumption of a Gaussian noise model, the Extended Kalman Filter (EKF) [4], provides a probabilistic framework for optimally combining these two sources of information. In an EKF framework, the quality of the positional estimates are dependent on the accuracy of both the movement model and the measurements.

The explicit dependence of localisation performance on the accuracy of movement and measurement models means that SLAM provides an opportunity to explore how the details of the orientation flight affect visual learning. Considering the problem from the perspective of a bee or wasp, we can assume that measurement accuracy is fixed by the physical constraints of the visual system and optics. The movements are not fixed however, and will vary according to the flying conditions and the flight manoeuvres being executed. There are two ways that the movements can affect localisation performance within a SLAM framework; the stereotypical dynamics of the orientation flight could make movements easier to predict (effectively reducing the noise entering the system), and/or that the movements act to increase the information gained from measurements.

In what follows we address the first possibility, that stereotypical movements act to stabilise flight trajectories thereby making it easier to predict the sensory consequences of motor commands. We first develop a movement model for typical orientation flights. We then use the orientation flight model to investigate the effects of process noise on localisation performance. Finally, we explore the benefits of the orientation flight for the insect's subsequent behaviour. We conclude with a summary of our results and brief discussion of future research.

2 Methods

2.1 The Simulation Environment

In our simulations the agent's position is represented by its X, Y and Z coordinates and its orientation by pitch, roll and yaw. The landmarks are defined as simple rectangular 3D structures that are fixed throughout any given run. The visual system of the agent is approximated by a set of 10000 uniformly spaced facets covering the surface of a sphere, giving a resolution of approximately 2 degrees as would be typical of the highest acuity parts of a bee's compound eye.

2.2 The SLAM Methodology

The essence of any EKF approach to state estimation involves a two-step procedure. Firstly the state estimate, together with a covariance matrix that reflects the uncertainty in our estimate, is propagated forward in time using a movement model. Following this, a measurement is made and used to update the state and covariance estimates once again. Since agent movement does not affect the position of the landmarks, the state and covariance estimates for the landmarks are unaffected by the movement model update. The uncertainty in the agent's position will increase however, due to inaccuracies and noise in the movement model. The measurement phase acts to reduce the uncertainty in the entire state estimate, the reduction in uncertainty being determined by the accuracy of the measurement model. In our case the accuracy of the measurement model is fixed. Thus, the only way to improve localisation performance is to improve the accuracy of the movement model, thereby reducing the noise entering the system.

2.3 The State and Covariance

The state $\hat{x}$ of the system is a vector containing both the state of the agent $\hat{x}_v$ and the states of all of the entries in the map of landmarks $\hat{y}_i$. It is paired with a covariance matrix, P, partitioned as follows:

$$\hat{x} = \begin{pmatrix} \hat{x_v} \\ \hat{y_1} \\ \hat{y_2} \\ \vdots \end{pmatrix}, \quad P = \begin{pmatrix} P_{xx} & P_{xy1} & P_{xy2} & .. \\ P_{y1x} & P_{y1y1} & P_{y1y2} & .. \\ P_{y2x} & P_{y2y1} & P_{y2y2} & .. \\ \vdots & \vdots & \vdots & \end{pmatrix}$$

The choice of representation and the frame of reference of the state are both arbitrary. In our simulations we choose the nest as the origin of a cylindrical coordinate system (R, θ, Z). The state of the agent and the landmarks are then $\hat{x}_v = \left(\hat{R}_v, \hat{\theta}_v, \hat{Z}_v\right)$ and $\hat{y}_i = \left(\hat{R}_i, \hat{\theta}_i, \hat{Z}_i\right)$ respectively.

2.4 The Process Model

The state and covariance are updated following a movement of duration Δt_k using the following equations:

$$\hat{x}_{v(k+1)} = f_v(\hat{x}_{v(k)}, u_k, \Delta t_k) \tag{1}$$

$$\hat{y}_{i(k+1)} = \hat{y}_{i(k)}, \forall i \tag{2}$$

$$P_{(k+1)} = \frac{\partial f}{\partial x} P_{(k)} \frac{\partial f}{\partial x}^T + Q_k \tag{3}$$

where, in equation (1) f_v is the state transition function. That is f_v maps the agent state estimate, $\hat{x}_k$, and control inputs, u_k, at time k, into a subsequent state $\hat{x}_{k+\Delta t_k}$, following a possibly variable period, Δt_k. Equation (2) simply states that the i estimates of the landmark positions $\hat{y}_i$ are unaffected by the movement update. Equation (3) is used to update the full state estimate covariance matrix P, where $\frac{\partial f}{\partial x}$ is the Jacobian of the state transition function and Q_k is the process noise.

The process noise is intended to account for any unmodelled movements or perturbations and is given by:

$$Q_k = \frac{\partial f_v}{\partial u_k} U_k \frac{\partial f_v}{\partial x_k}^T \tag{4}$$

where U_k is the diagonal covariance matrix of u_k.

2.5 The Measurement Model

Our measurement model assumes that there is a direct line of sight between the agent and any given landmark. The measurement prediction is a Cartesian 3D unit vector, pointing in the direction of the centre of the landmark relative to the agent. The measurement model returns the expected measurement, given the current state estimate, $\hat{x}$.

$$h_i = \begin{pmatrix} \alpha_{xi} \\ \alpha_{yi} \\ \alpha_{zi} \end{pmatrix} = XYZ_{NORM} \begin{pmatrix} DX \\ DY \\ DZ \end{pmatrix} \tag{5}$$

$$XYZ_{NORM} = \left(\frac{1}{\sqrt{DX^2 + DY^2 + DZ^2}} \right) \tag{6}$$

where, $DX = \hat{R}_i cos(\hat{\theta}_i) - \hat{R}_v cos(\hat{\theta}_v)$, $DY = \hat{R}_i sin(\hat{\theta}_i) - \hat{R}_v sin(\hat{\theta}_v)$, and $DZ = \hat{Z}_i - \hat{Z}_v$.

As well as a point prediction of the measurement h_i, the measurement model allows us to calculate the innovation covariance matrix S_i. The innovation covariance matrix S_i represents the expected uncertainty in measurement h_i and is given by:

$$S_i = \frac{\partial h_i}{\partial x_v} P_{xx} \frac{\partial h_i}{\partial x_v}^T + \frac{\partial h_i}{\partial x_v} P_{xyi} \frac{\partial h_i}{\partial y_i}^T + \frac{\partial h_i}{\partial y_i} P_{yix} \frac{\partial h_i}{\partial x_v}^T + \frac{\partial h_i}{\partial y_i} P_{yiyi} \frac{\partial h_i}{\partial y_i}^T + R \tag{7}$$

where, $\frac{\partial h_i}{\partial x_v}$ and $\frac{\partial h_i}{\partial y_i}$ are the Jacobian matrices of the measurement model with respect to the agent, x_v, and landmark, y_i, state estimates respectively, P_{xx}, P_{xyi}, P_{yix} and P_{yiyi} are sub-matrices of P, and R is the measurement noise covariance.

The visual system of the agent is modelled as a set of individual facets uniformly spaced over the surface of a sphere. A facet is defined as occluded by an object in the field of view if a ray-trace from the agent in the direction of the facet intersects the object. The heading to the object is defined as the average directional heading of all of the facets that are occluded by the object. This results in measurement errors that are not strictly Gaussian (a requirement for EKF). However by being conservative and assuming a higher level of noise on the measurements than empirically observed through simulation, performance is not unduly affected.

The noise in α_{xi}, α_{yi} and α_{zi} is correlated and the interactions are captured by the off-diagonal elements of the 3×3 matrix R. The value of R was estimated empirically by calculating the error covariance of a random dataset of measurements from our simulated environment.

2.6 Updating the Map

Following a measurement z_i, the Kalman gain, W, is calculated and used to update the state and covariance estimates using:

$$W = P\frac{\partial h_i}{\partial x}^T S^{-1} \tag{8}$$

$$\hat{x}_{new} = \hat{x}_{old} + W(z_i - h_i) \tag{9}$$

$$P_{new} = P_{old} - WSW^T \tag{10}$$

where P is the state estimate covariance, $\frac{\partial h_i}{\partial x}$ is the Jacobian matrix of the measurement model, S is the innovation covariance matrix, $\hat{x}$ is the state estimate and $z_i - h_i$ is the innovation, given by the difference between the predicted value h_i and the actual value z_i of the measurement of landmark i .

3 Results

Using our simulation and the SLAM framework described above we examine a series of different scenarios designed to shed light on the effect of behaviour on localisation performance. In order to study a given behaviour it is necessary to explicitly specify a movement model. We begin by describing a simple model of an orientation flight, inspired by the observation of real flights. We then use the model to explore the effect of varying the level of noise in the different control inputs. Localisation performance is determined with respect to pin-pointing the position in space of a visual landmark. Finally we show that without the initial grounding provided by the orientation flight, positional errors rapidly increase. In contrast, even a very brief orientation flight can provide enough information for successful re-orientation following a period of simulated free flight.

3.1 A Simple Model of Arcing Flight

The structure of orientation flights has been characterised in terms of a series of arcs at different radial distances centred on the nest [5]. We consider a simple movement model inspired by this observation. We choose as our control inputs the angular arcing velocity and the radial velocity, together with a third control input that determines the height of the agent. To generate noisy trajectories, we begin with a noise free trajectory and differentiate with respect to time to define the trajectory in terms of velocities. Zero mean Gaussian noise is added to the velocity signals and the system is then integrated to obtain a noise-corrupted trajectory of positions. By varying the amount of noise in each of these control channels, trajectories can be generated that are qualitatively similar to flights observed in real insects. Example trajectories are shown in Figure 1.

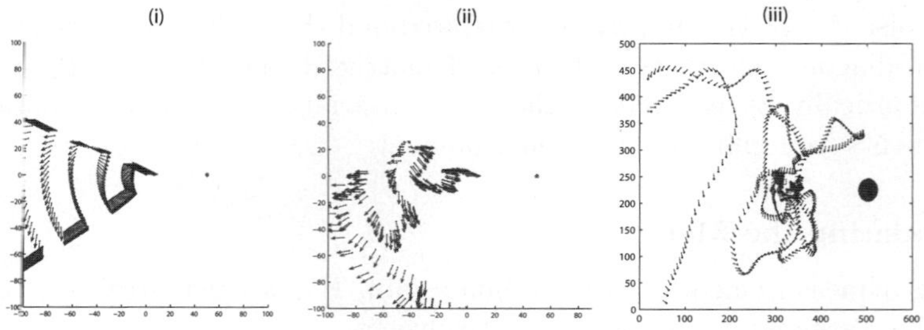

Fig. 1. Example trajectories: Panel (i) shows a synthesised trajectory with no noise added. Panel (ii) shows a synthetically generated trajectory with zero mean Gaussian noise added to the control inputs, the magnitude of the noise being 0.05 rad/s in the angular control and 1 unit/s in the radial control signal (arbitrary scale). Nest position is indicated by a dot. For comparison, panel (iii) shows a real trajectory of a bee performing an orientation flight, recorded in the field [Philippides et al. in preparation]. The filled circle indicates the position of a prominent landmark.

3.2 Experiment 1: The Effect on Localisation Performance of Varying the Control Noise

The agent begins each run at the nest and performs a series of arcs of increasing radial distance from the nest. After a fixed number of time-steps the simulation is terminated and the estimated position of any visible landmarks are recorded.

Figure 2 shows how varying the noise in the control signals affects the agents' ability to determine the position of a landmark. The average positional error following 200 time-steps is recorded for 100 separate runs for each combination of noise levels. As would be expected, increasing the noise in either control input decreases the accuracy with which the agent is able to locate the landmark. It is clear that improvement in performance can be achieved through reducing noise in either dimension. We also generate a feel for the relative sensitivity to noise of the two control inputs with regard to performance.

3.3 Experiment 2: Re-orienting Following a Free Flight

As previously, the agent begins each run at the nest and performs a series of arcs of increasing radial distance from the nest. After a fixed number of time-steps the agent moves radially away from the nest at constant speed, no longer receiving measurements of the position of the landmark. At a set distance the agent stops and attempts to return home using the current estimate of position to determine the movement direction, whilst again taking measurements of the landmark. Figures 3 and 4 show typical trajectories and localisation performance in this task.

Figure 3 shows a typical flight. Following the learning flight, the agent flies a set distance away from the nest without fixating the landmark. During this phase

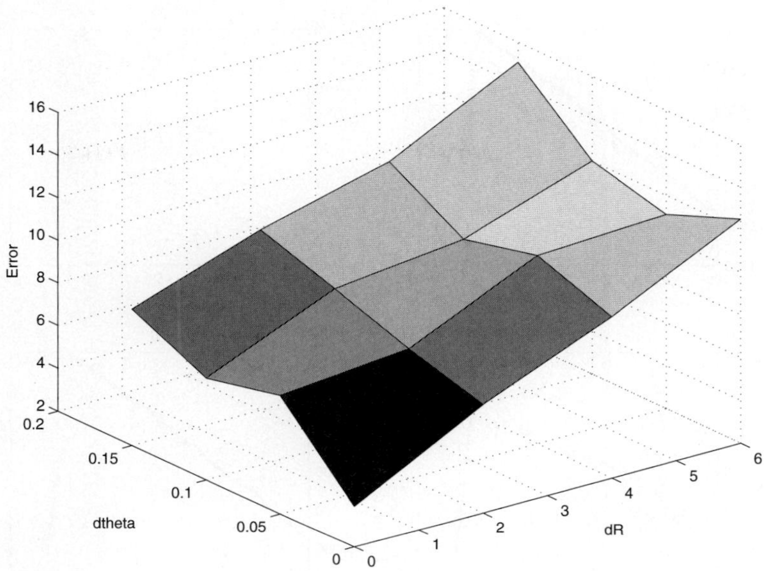

Fig. 2. Average error in the estimate of the position of a landmark following 200 time-steps of an orientation flight for different levels of noise in the radial dR and the arcing dtheta elements of the movement. Each datum represents the average of 100 runs with a fixed level of process noise. The ability to accurately locate landmarks deteriorates with increasing noise in either dimension.

the positional error grows due to inaccuracies in the path integration system. The positional error remains high on the return journey until the landmark comes back into view, at which point the agent is able to re-orient relative to the known landmark thereby rapidly reducing the positional error. Figure 4 shows the positional error in both the estimated position of the agent and the landmark, for typical flights with (bottom) and without (top) an orientation flight. During the orientation flight, (t = 1 to 50), the error remains bounded. During free flight, (t = 50 to 200), when no measurements are being used to counter the accumulation of errors in the path integration system, the error steadily grows. When the landmark comes back into view (t $\geq$ 300), the agent is able to rapidly re-orient, with the positional error rapidly reducing. Without an orientation flight the agent is unable to re-orient using the landmark on the return journey. The positional error grows quickly when the agent does not first ground itself in the world.

4 Discussion

We have presented a quantitative framework in which different behaviours can be modelled. By assuming that stereotypical movements will be easier to pre-

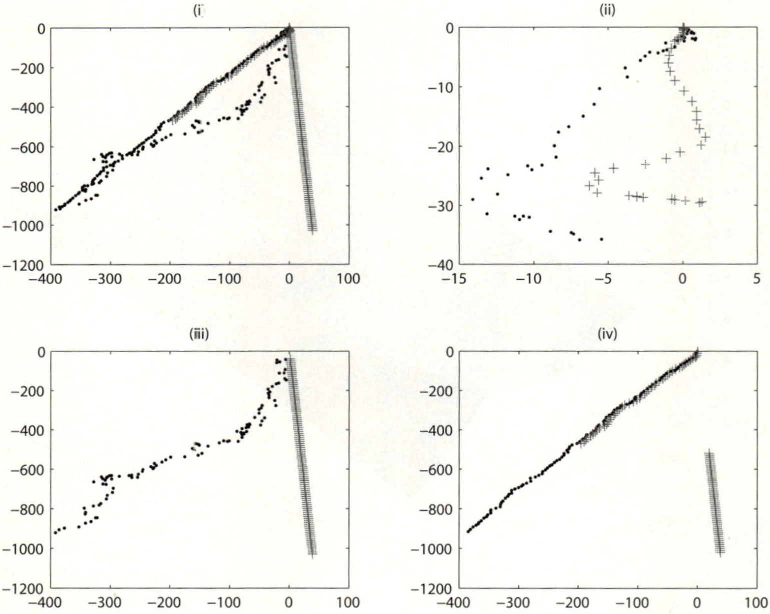

Fig. 3. Panel (i) shows the estimated (red cross) and actual (black dot) positions of a simulated agent during a foraging flight. Panels (ii), (iii) and (iv) show three separate sections of the flight in more detail. Panel (ii) shows the orientation flight (note smaller scale), during this phase of the flight the agent fixates the landmark and learns the position of the landmark relative to the nest, the error in the agents estimate of its own position remains relatively small. Panel (iii) shows the free flight phase of the foraging trip when no measurements are made. During this phase the agent's estimate of its own position becomes progressively worse with time. Panel (iv) shows the homing flight in more detail. The positional estimate, which is initially very poor, quickly becomes more accurate when the landmark comes back into view, about half way through the return journey.

dict, and modelling this as the level of process noise that enters the system, we have shown how reliable movements could affect localisation performance. As we alluded to in the introduction, the other way that behaviour can affect localisation, within our framework, is through performing actions that maximise the information gained from measurements of the world. We intend to explore this aspect of the problem in more detail in the near future.

We have also shown that the spatial relationships between landmarks and the target destination, once learned, are sufficient to bound the degree of uncertainty in the agent's own positional estimate. Knowledge of these spatial relationships also allows the agent to re-orient itself when landmarks come back into view after prolonged periods of flight where path integration errors will have caused the state estimate to drift away from ground truth.

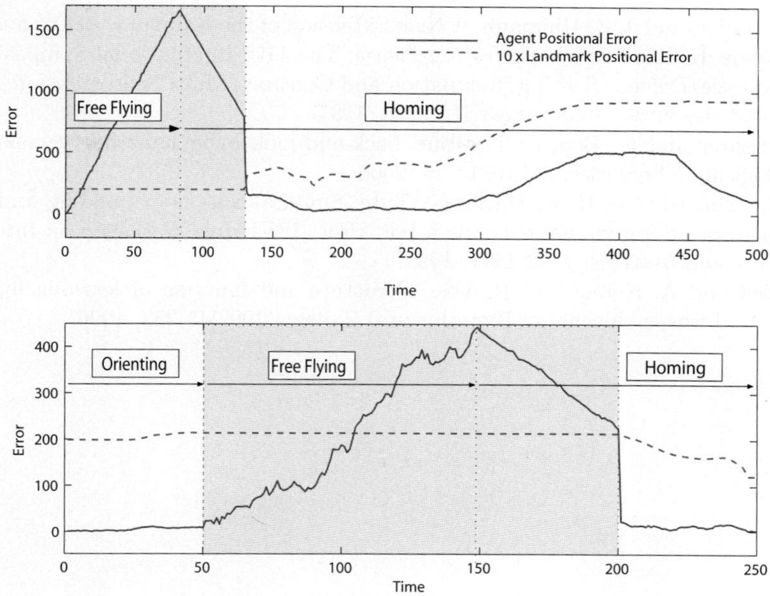

Fig. 4. The error in the estimate of the agent's position over time, for flights without (top) and with (bottom) an orientation flight [Note the different time scales due to the disoriented agent failing to return home succesfully in the top figure]. The bottom panel shows data taken from the flight shown in Figure 3. In both figures the solid line shows the error in the agents own positional error and the dashed line shows the error in the estimate of the landmark's position (scaled by a factor of 10 for ease of viewing). The shaded areas indicate when the landmark was not being fixated and therefore not providing information.

The results presented here represent the first steps in a novel approach to analysing learning flight behaviour. In future, we will relate the model to data that have been collected from bees performing orienting behaviours, and increase the sophistication and biological plausibility of both sensory and motor elements of the model.

References

[1] T. S. Collett: Making learning easy: the acquisition of visual information during the learning flights of social wasps. Journal of Comparative Physiology A, 177:737-747, 1995.

[2] T. S. Collett and J. Zeil: Flights of learning. Current Directions in Psychological Science, Journal of Comparative Physiology A, 5(5):149-155, 1996.

[3] A. J. Davison and D. W. Murray: Simultaneous localization and map-building using active vision. IEEE Transactions on Pattern Analysis and Machine Intelligence, 24(7):865-880, 2002.

[4] S. J. Julier, and J. K. Uhlmann: A New Extension of the Kalman Filter to nonlinear Systems. In The Proceedings of AeroSense: The 11th International Symposium on Aerospace/Defense Sensing, Simulation and Controls, Multi Sensor Fusion, Tracking and Resource Management II, SPIE, 1997.

[5] M. Lehrer and G. Bianco: The turn-back-and-look behaviour: bee versus robot. Biological Cybernetics, 83(3):211-29, 2000.

[6] J. J. Leonard, and H. F. Durrant-Whyte: Simultaneous map building and localization for an autonomous mobile robot. Proc. IEEE Int. Workshop on Intelligent Robots and Systems, 1442-1447, 1991.

[7] J. Zeil and A. Kelber and R. Voss: Structure and function of learning flights in bees and wasps. Journal of Experimental Biology, 199:245-252, 1996.

Neuroevolution of Agents Capable of Reactive and Deliberative Behaviours in Novel and Dynamic Environments

Edward Robinson, Timothy Ellis, and Alastair Channon

School of Computer Science, University of Birmingham, Birmingham, B15 2TT, UK
edd@edd-robinson.net, t.s.ellis@cs.bham.ac.uk, alastair@channon.net

Abstract. Both reactive and deliberative qualities are essential for a good action selection mechanism. We present a model that embodies a hybrid of two very different neural network architectures inside an animat: one that controls their high level deliberative behaviours, such as the selection of sub-goals, and one that provides reactive and navigational capabilities. Animats using this model are evolved in novel and dynamic environments, on complex tasks requiring deliberative behaviours: tasks that cannot be solved by reactive mechanisms alone and which would traditionally have their solutions formulated in terms of search-based planning. Significantly, no a priori information is given to the animats, making explicit forward search through state transitions impossible. The complexity of the problem means that animats must first learn to solve sub-goals without receiving any reward. Animats are shown increasingly complex versions of the task, with the results demonstrating, for the first time, incremental neuro-evolutionary learning on such tasks.

Key words: Artificial Life, Neural Networks, Incremental Evolution, Reactive and Deliberative Systems, Novel and Dynamic Environments.

1 Introduction

In this paper we present work showing animats that use neural networks to display high level deliberative decision making whilst retaining reactive qualities. Deliberative planning has traditional roots in "Good Old Fashioned Artificial Intelligence" (GOFAI) as a search-based method for the design of behaviour systems. There are issues however with its application in dynamic and novel environments. Reactive models of action selection on the other hand can be very successful in dealing with unpredictable and dynamic environments. However, since these systems generally have only a short look-ahead the individual complexity of behaviour that can emerge is limited. Both reactive and deliberative qualities are essential for a good action selection mechanism: deliberative mechanisms for long term goal seeking and reactive capabilities for dealing with unforeseen events [1,2].

A complex problem has been designed to demonstrate our model, which we have called the 'river-crossing task' or RC task. In this problem an animat must

F. Almeida e Costa et al. (Eds.): ECAL 2007, LNAI 4648, pp. 345–354, 2007.

cross a river by building a bridge made out of stones collected from locations in a 2D grid-world environment. Importantly, animats are evolved to solve this problem without any task-specific information. Animats are embodied with two very different neural networks. The first acts as a deliberative style decision network: it makes high level choices about the sub-goals that need to be achieved, given current internal and external states. The actions that the animats choose may for example be 'head to the nearest stone', 'avoid the traps' or 'head to the resource'. Once an animat has made a decision, the second (reactive) neural network acts as a navigation tool, taking care of low level actions, such as which direction to move in next. In the RC environment there are several classes of objects that the animats can interact with. *Grass* objects take up most environmental space; the animat can place other objects onto them. *Stones* are movable objects: they can be picked up and dropped on grass or water. If one is dropped on water then the water object is converted into a grass object. *Water* objects are dangerous. If an animat moves onto one and does not place a stone down then the animat drowns. *Traps* are lethal to animats, which die if they move onto one. *Resource* objects offer rewards to animats, if they can reach them. *None of this information is given a priori to the animats, ruling out the possibility of explicit forward search through state transitions.*

Payton [3] used gradient fields to represent the state-space of a problem and as an internalised plan. Unlike more traditional search based models, gradient fields can be generated efficiently, and do not suffer from the same local-minima problems as other wave based mechanisms such as potential fields [4]. However, the gradient fields approach does not deal well with changing environments and so is often coupled with a Brooks inspired [5] subsumption architecture [3]. Another issue with gradient fields is that they have to be laboriously constructed.

We describe in the next section a biologically inspired gradient based model which does not suffer from local minima nor any of the other problems associated with other gradient based models. It is computationally efficient and simple to initialise. We also describe a decision network which is designed to allow animats to manipulate the navigation model. We show experimental results in section 3 and conclude in Section 4.

2 The Model

The movements of the animats in the environment are dictated by a shunting model introduced in [6,7]. Yang and Meng were interested in motion planning models that could react quickly in real-time, allowing a robot or robot-manipulator to perform collision-free motion planning.

Neural networks have been used extensively and successfully for robot control problems. Often controllers specify a robot's behaviour based upon sensory input from the environment; this makes them good for dynamic environments, which are likely to change continuously. The model in [6,7] uses neural networks in a very different way. Instead of using the network to specify behaviour by, for example, mapping the actuators of the robot to the outputs of the network, the network's activation landscape itself directly specifies the robot's movements through the environment.

Their model consists of a neural network composed of an n-dimensional lattice of neurons ℓ, where each neuron represents a possible state of the system. Therefore any system that can be fully described by a set of discrete states can be represented. This is referred to as the 'configuration space' of the robot [8]. In the case of the simulated robot in their studies, the configuration space was the discretised 2D Cartesian workspace. Each neuron is connected to a subset of the lattice—$\Re_i \subset \ell$. This subset is called the receptive field, and represents all the states that are reachable from the current state. It is useful to note that if, as in Yang and Meng's simulated robot example, the state space is simply the 2D coordinates of the environment that we wish the agent to navigate, there is always a simple one-to-one relationship between neurons and locations.

The transition function used to specify inter-neuron dynamics is based on the 'shunting equation', inspired by Hodgkin and Huxley [9] and Grossberg [10]. Yang and Meng designed two versions of this transition function: one which helped to control activity saturation in the network, and a simpler one which did not. In our study we found that the more elaborate transition function was not necessary, since out model did not develop saturation problems; the function is shown in equation 1.

$$\frac{dx_i}{dt} = -\mathrm{A}x_i + \mathrm{I}_i + \sum_{j=1}^{k} w_{ij}[x_j]^+ \ . \tag{1}$$

Alpha (A) represents the passive decay rate, which determines the degree to which each neuron's activity diminishes towards an idle state. The function $[x]^+$ can be described as $max(0, x)$. The connection weight (or synapse strength) $w_{i,j}$ between neurons is simply specified as the Euclidean distance between the cell and its neighbour within the receptive field. k is the receptive field size and is set to 8 to represent the direct neighbours in a 2d grid-world environment. Iota (I) is equal to E in the case of a target, and $-E$ for an obstacle, where E is a large integer.

Once the network is configured, and the targets and obstacles established, neural activity can be used to navigate a robot by gradient ascent. At each time-step the robot looks at the level of activity in each grid-cell that it is connected to (its neighbours), because they are all the states that it can reach, and picks the one with the highest value. As a result of the network's dynamics, positive activity entered at neurons that map to targets propagates through the network, whilst negative activity that is inputted into neurons mapping to obstacles cannot propagate globally. Due to the nature of the leaky integrator function in the model, the activity contribution from a target decreases with distance from that target source, leaving a trail of activity back to the target.

This system therefore allows a robot to navigate a dynamic environment, avoiding obstacles and heading towards targets. Using the shunting network to control movement throughout the environment means that high level actions such as 'head to the target whilst avoiding obstacles' can be carried out flawlessly and quickly. The highly dynamic nature of the network means that when the environment changes (e.g. an obstacle moves), a new activity path can be generated quickly. In the RC task there are four different classes of objects (Resource, Stone, Water and Trap) that have Iota values associated with them. Grass states are considered empty states to allow activity to flow through the environment. For target acquisition and obstacle avoidance there are two types of value: positive and negative. In our implementation of the model however, we allow a class to be specified with no value at all. Setting Iota to 0 for a class means that no external activity will be inputted into any of the neurons in that class. The animat will ignore the object present at that location, and may or may not pass over it while moving through the environment.

2.1 The Decision Network

The outputs of the decision network are used to set the Iota values for the object classes. Using this network, the animat can manipulate the activity landscape in the shunting network in a way that allows it to string together multiple actions in parallel to create more complex behaviours.

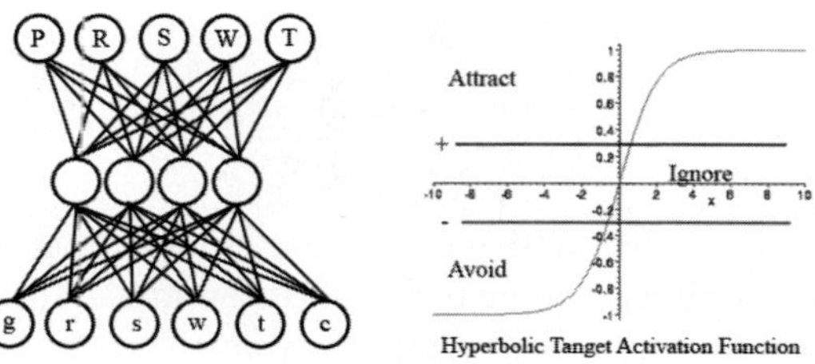

Fig. 1. The decision network controller. The output neurons are P = pick up/put down; R = resource; S = stone; W = water; T = trap. The input neurons are g = grass; r = resource; s = stone; w = water; t = trap.

The decision network is a feed-forward neural network with a single hidden layer of four neurons (figure 1). The input layer represents the current state of the animat, or more precisely, the object class situated on and the carrying status of the animat. The inputs are single values of 1 or 0 and they feed through weighted links into the hidden layer neurons, where tan activation functions are applied to the summed input values. The neurons in the output layer represent Iota values for the object classes needing them (four). Output neurons have tan activation functions and two fixed thresholds. Neurons with activations over 0.3 or under -0.3, after being processed through the activation function, output 1 and -1 respectively. Any activation values in the range $[-0.3; 0.3]$ resolve to 0.

Output neurons then, have three possible outputs: -1, 0 or 1. The Iota values of all the objects (except grass) in the environment are set based upon the output of the decision network neuron associated with that class. If a neuron has a negative output then all of the objects in that class will have a negative Iota value (-15 in our simulations). Similarly, a positive neuron output sets all the objects in that class with positive Iota values (+15 in our simulations). Finally, a neuron with an output of zero sets the objects in that class with Iota values of 0. Having an Iota value of 0 means that the objects have no external activity inputted into them: their activation values in the shunting network will be solely based upon those of their neighbours. To get a clear understanding of the purpose and ability of the decision network, two examples of resulting shunting network activity landscapes are shown in figure 2. Each landscape is different because of the Iota values of object classes. In the first landscape for example, the Iota values result in positive activity propagating from the resource neuron, through the rest of the network attracting the animats, while negative activity repels animats from the traps.

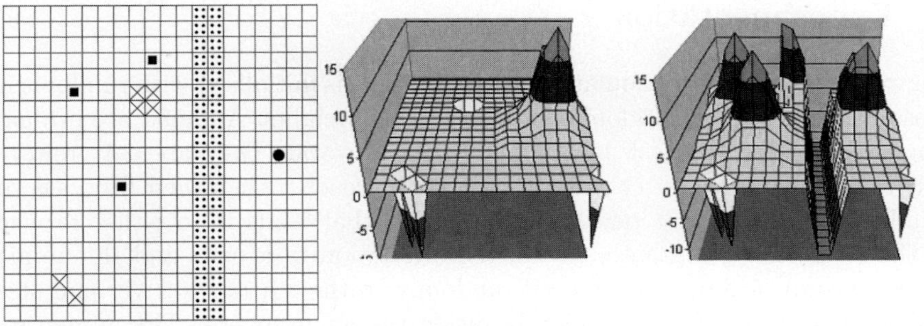

Fig. 2. A typical environment (left) and two activity landscapes (middle, right). Iota values in the first (middle) landscape a are: resource = 15; stone = 0; water = 0; trap = -15. In the right landscape b, resource = 15; stone = 15; water = -15; trap = -15. Environment legend: stone = small square; resource = circle; trap = cross; water = four dots.

Landscape b represents the same environment, but with different Iota values for object classes, and so the animat acts differently. Activity from the resource would no longer be able to propagate through the river, but since the stones have positive activities, the animat moves to the nearest one (still avoiding traps). One of the output neurons on the decision network is not used to provide an Iota value for motion. Instead, its output is used to make a decision about whether or not to pick up or drop stones: the other actions an animat can take in our system. If the output is positive then the animat will attempt to pick up whatever object it is currently situated on in the grid-world. If negative then the animat will attempt to drop an object.

2.2 Evolution of Decision Networks

We used a steady-state genetic algorithm (GA) to search the weight-space, with fitness based upon animat performance in evaluation. Tournament selection was used for each iteration, with three animats evaluated and the worst performer replaced by a new offspring created from a combination of the other two.

An animat has a set of chromosomes: one for each neuron in its decision network. Each chromosome contains the floating point values for the weights of its neuron's input connections. For each of an offspring's (output and hidden) neurons, there is a probability $P_{whole} = 0.95$ that the corresponding chromosome will be inherited from just one parent; which parent this is copied from is then chosen at random. Otherwise (probability $P_{mix} = 0.05$) the offspring will instead inherit a new chromosome whose genes are a mixture of both parents' versions of the same chromosome, combined by single-point crossover. Finally, each weight has a probability of $P_{mut} = 0.001$ of having a mutation value from $N(0, 0.4)$ added to it. All mutations are bounded to within $[-1; 1]$.

3 Experimentation

For each experiment, a population of 250 animats is initialised with random chromosomes, hence with random decision network weights. Animats are evaluated singularly on the RC task in a 20x20 cell grid-world. They are only rewarded when they reach a resource state: if they fail to reach it then their fitness is zero. Similarly, if the animat performs any action that leads it to either drowning in the river or moving onto a trap, the evaluation task ends and the animat's fitness is zero. Animats are placed randomly to the left of the river and stones and traps are distributed randomly inside the environment. The animat's decision network inputs are updated whenever the animat's state changes; then the shunting model updates the activity landscape with new Iota values and the animat moves to the neighbouring cell with the highest activity. If the pick up/put down neuron is activated then the animat will attempt to pick up/put down whatever it is on or carrying. At each iteration of the GA three animats are randomly selected. Each animat is evaluated on the same randomly generated environment to ensure fairness; a new environment is randomly generated in the next iteration of the GA.

First we tested animats by evaluating them on one environment that contained a river two cells wide. Due to the problem complexity, and because the animats had to learn to solve sub-problems before even receiving any fitness, all of the population's individuals had zero fitness and search was random. To overcome this problem we exposed the animats to three trials of increasing difficulty. Figure 3 shows three randomly generated versions of the three maps used. The randomly positioned objects are the traps and stones: the same river and resource locations are always used. The first map that the animats are tested on already has a completed bridge. To solve this task the animats simply have to learn to avoid traps and the river and get to the resource. To solve the second and third maps an animat has to build a bridge, with the bridge needing to be smaller (one cell wide) for the second map than the third (two cells wide)[1]. This system provides a route for incremental evolutionary learning and can be seen as a simplification of a more complex general world-environment in which animats encounter opportunities for reward, of varying degrees of difficulty.

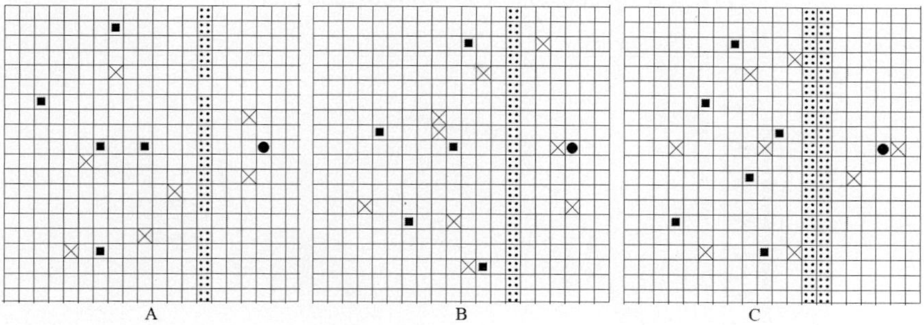

Fig. 3. Examples of the three environments that animats are evaluated in.

3.1 Experimental Results

For each task the animats solved out of the three they received a score of 100. If they failed a task they were awarded a score of 0 only for the task they failed. In each simulation, data showing how many versions of each task had been solved in the previous 250 tournaments was collected in 250 tournament intervals. Since there were three animats taking part in each tournament, the maximum number of tasks solved in this period, for each map, was 750. Table 1 shows results from 15 simulations. Once 80% (ie. 600) of animats tested per iteration could solve all versions of the task they were shown, the simulation was stopped.

In all simulations the animats quickly evolved behaviours that could be used to solve the simplest map, where they needed only to know how to avoid dangers like water and traps. They didn't need to interact with stones to solve this task, making the process simple. The next two maps were substantially harder to solve, and the time taken to solve them reflects this. Results showed that contrary to previous observations: the third map was not harder to solve than the second,

[1] Observations of the simulation showed that animats were solving a map with a single celled river by picking up stones and dropping them on the river indiscriminately. They were attracted to both the resource and the river; since the river was closer, they would deposit a stone on the nearest river cell. Although once they had done this they could complete the task, because they were still attracted to the river they would often keep depositing stones there. Eventually they would deposit enough stones to create a gap so large that the activity from the resource attracted them enough for them to reach it. Using a deeper river stopped animats for being rewarded for developing this 'brute force' behaviour.

Table 1. The mean, best and worst number of tournaments needed for 80% of the animats evaluated in a 250-iteration period to have solved each map.

Map	Mean	Best	Worst	Stdev
1	5700	4000	8250	1203.4
2	99084.6	13000	437750	139542.4
3	99083.3	13000	437750	139544.4

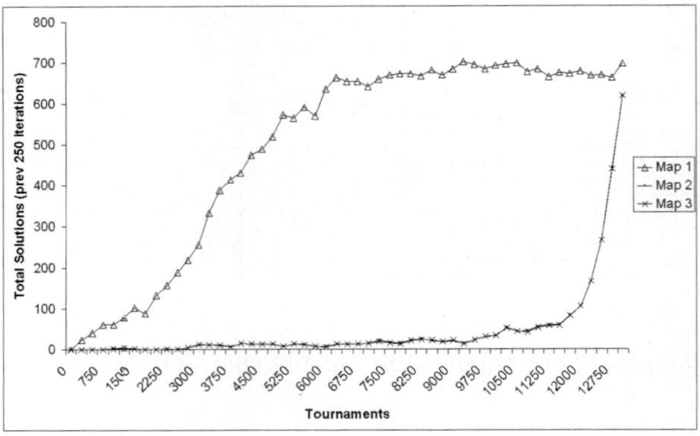

Fig. 4. Results from the simulation which solved all the tasks in the fastest time. Map 1 was learnt in around 6000 tournaments, while maps 2 and 3 took 13000 tournaments each.

even though it required a specific strategy. The second intermediate map could be solved in two ways: the 'brute force' approach (above)[1] or the 'correct' way, described below. Since the third map could only be solved in the correct way, animats that learnt this behaviour also solved the second map in the same way; this accounts for the similar (usually identical) time taken to solve both harder maps. However, the exclusion of either of these maps in the learning process would have lead to sub-optimal behaviour evolving.

The observed 'correct' behaviour for solving the harder maps can be described as follows: When the animats are on grass and not carrying anything they head to the nearest stone, whilst avoiding the river and traps. Once they reach a stone they pick it up; they are then situated on grass, but now carrying a stone. Next, they adjust the shunting model so they are attracted to the resource, ignoring the river/other stones and avoiding traps. Once they reach the river they deposit a stone; they are now back to being on grass and not carrying. If activity can propagate from the resource to the animat (because of a completed bridge) they head to the resource. Otherwise they return to a stone and repeat.

The RC problem is particularly difficult because of the lack of reward available leading to long periods of stasis between solving the easy and the harder maps. During this time the animats phenotypic behaviour does not change by much; they keep solving the easy task and failing the harder ones. It is possible that selection pressures are preventing change from taking place gradually as one might expect. Animats beginning to learn to solve the harder tasks, for example by starting to pick up stones, may disrupt and forget previous behaviours causing them to fail the simple task - most likely due to sharing connections for the different behaviours [11]. Figure 4 shows the results of the simulation with the quickest time to solve all tasks, and although it is the fastest, the nature of the graph is the same for all runs: the simpler task is learnt quickly, then there is a stasis period until by chance animats are born that can solve the harder tasks. This advantageous behaviour then quickly propagates through the environment until the vast majority can solve all tasks.

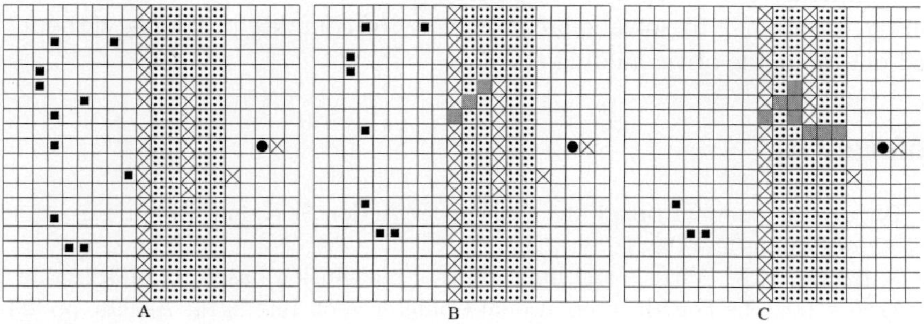

Fig. 5. A dynamic environment: In Part B the wall is about to move and block the animats path; instantly the animat starts building in a new direction and crosses successfully (Part C)

One advantage of the shunting model as shown in [6] was that it was highly reactive to a dynamic environment. We tested this ability in our system by evaluating animats that had been evolved to solve the three maps. Animats were shown the map in part A of Figure 5. The map has a larger river than they have been shown during evolution. Also, it has a movable wall of traps. The animat begins building a bridge through the shortest route of the river (part B); once the animat gets halfway across however the wall is moved, blocking the animats path (part C). Without hesitation the animat continues to build its bridge in the new required direction and navigates across the bridge.

4 Conclusions and Future Work

We have developed and presented a model that allows animats to develop complex behaviours such as building a bridge in the RC task, by using an incremental approach. Through the use of an adapted version of Yang and Meng's shunting model animats manipulate an activity landscape by utilising a decision network. This approach allows high level behaviours such as 'find a stone without drowning or falling into traps' to be carried out without any further requirements from the animat other than the evolved desire to do so. Further, animats that are only shown the simpler three maps used in the evolutionary process can solve a novel and dynamic version of the task. Due to the lack of hard-coded constraints, this model could be used with many different environments without needing to make changes.

One draw-back of the current model is that it requires incremental versions of the task to be *shown* to the animat. Using a larger environment could allow these tasks to be situated in the same realm, but we chose to implement them as separate worlds. This approach leads to problems as more complex environments are constructed. To address this issue, future work will include methods for allowing the animats to generate intrinsic motivations, which has been shown to be imperative in mental development [12]. The intrinsic motivation will encourage an animat to solve sub-components of complex problems without the need of an outside critic guiding them.

References

1. Tyrrell, T.: Computational Mechanisms for Action Selection. PhD thesis, University of Edinburgh (1993)
2. Benjamin, M.R.: Virtues and limitations of multifusion based action selection. In: Agents '00: The Fourth International Conference on Intelligent Agents, pp. 23–24 (2000)
3. Payton, D.W., Rosenblatt, J.K., Keirsey, D.M.: Plan guided reaction. IEEE Trans. on Systems, Man, and Cybernetics 20(6), 1370–1382 (1990)
4. Koren, Y., Borenstein, J.: Potential field methods and their inherent limitations for mobile robot navigation. In: IEEE Int. Conf. on Robotics and Automation, 1991, pp. 1398–1404. IEEE Computer Society Press, Los Alamitos (1991)
5. Brooks, R.A.: A robust layered control system for a mobile robot. IEEE J. Robot. and Auto. 2(3), 14–23 (1986)
6. Yang, S.X., Meng, M.: An efficient neural network approach to dynamic robot motion planning. Neural Networks 13(2), 143–148 (2000)
7. Yang, S.X., Meng, M.: An efficient neural network method for real-time motion planning with safety consideration. Robotics and Autonomous Systems 32(2-3), 115–128 (2000)
8. Schultz, A.C.: Adapting the evaluation space to improve global learning. In: Belew, R., Booker, L. (eds.) Proceedings of the Fourth International Conference on Genetic Algorithms, pp. 158–164. Morgan Kaufman, San Mateo (1991)
9. Hodgkin, A.L., Huxley, A.F.: A quantitative description of membrane current and its application to conduction and excitation in nerve. Journal of Physiology 116, 500–544 (1952)
10. Grossberg, S.: Nonlinear neural networks: Principles, mechanisms, and architectures. Neural Networks 1, 17–61 (1988)
11. Seipone, T., Bullinaria, J.A.: The evolution of minimal catastrophic forgetting in neural systems. In: Mahwah, N.J. (ed.) Twenty-Seventh Annual Conference of the Cognitive Science Society, pp. 1991–1996. Lawrence Erlbaum Associates, Mahwah (2005)
12. White, R.W.: Motivation reconsidered: The concept of competence. Psychological Review 66(5), 297–333 (1959)

On the Adaptive Disadvantage of Lamarckianism in Rapidly Changing Environments

Ingo Paenke[1,2], Bernhard Sendhoff[2], Jon Rowe[3], and Chrisantha Fernando[3]

[1] Institute AIFB, University of Karlsruhe, D-76128 Karlsruhe, Germany,
ingo.paenke@aifb.uni-karlsruhe.de

[2] Honda Research Institute Europe GmbH, Carl-Legien-Straße 30, D-63073 Offenbach/Main, Germany,
bernhard.sendhoff@honda-ri.de

[3] Systems Biology Centre, University of Birmingham, Birmingham, Edgbaston, B15 2TT, UK,
c.t.fernando@cs.bham.ac.uk, j.e.rowe@cs.bham.ac.uk

Abstract. Using a simple simulation model of evolution and learning, this paper provides an evolutionary argument why Lamarckian inheritance - the direct transfer of lifetime learning from parent to offspring - may be so rare in nature. Lamarckian inheritance allows quicker genetic adaptation to new environmental conditions than non-lamarckian inheritance. While this may be an advantage in the short term, it may be detrimental in the long term, since the population may be less well prepared for future environmental changes than in the absence of Lamarckianism.

1 Introduction

Natural selection to a first approximation operates with variation that is undirected [1]. Lamarck suggested that the results of lifetime learning could be directly passed on to ones offspring [2]. When would we expect directed variation or inheritance of acquired characters to occur? Recent work reveals a range of mechanisms capable of sustaining heritable epigenetic variation [3], phenotypic memory [4] and neo-Lamarckian inheritance [5], for example: mutational hotspots and adaptive mutations occurring during bacterial stress [6] , chromatin marks that control differentiation in multicellular organisms [7], RNA silencing allowing potential influence by somatic RNA on germ line gene expression [8], inheritance of immune system states by antibody transfer in breast milk [9], and behavioural and symbolic inheritance systems such as food preference, niche construction traditions and all information transmission dependent on language [3]. By these mechanisms, Lamarckian inheritance can occur without breaking the central dogma, i.e. without sequence information having to pass from protein to DNA [10]. In the case of immune system, behavioural and symbolic inheritance systems, information need not be passed from soma to germ line either [11].

So, directed variation is possible, but is it always advantageous to inherit characteristics acquired by ones parent? Evolutionary biology modeling revealed

F. Almeida e Costa et al. (Eds.): ECAL 2007, LNAI 4648, pp. 355–364, 2007.

that Lamarckianism would have provided an adaptive advantage in changing environments compared to genetic mutation alone [4]. In randomly changing, and rapidly changing environments, low to medium values of heritable phenotypic plasticity were optimal, respectively. In slowly changing environments there was a monotonic improvement in fitness as the capacity for heritable phenotypic plasticity increased. Absent in the above model was a distinction between genotype and phenotype, for its intention was not to compare non-heritable phenotypic change, i.e. lifetime learning, with heritable phenotypic change, i.e. lifetime learning plus Lamarckian inheritance. In the field of evolutionary algorithms, researchers experiment with "evolution as it could be". They found when evolving artificial neural networks that Lamarckian inheritance of weights learned in a lifetime was harmful in changing environments but beneficial in stationary environments [12]. Our aim is to compare various combinations of non-directed genetic change, lifetime learning, and Lamarckian inheritance of learnt phenotypes, under variable environmental conditions.

Recently, Paenke et al. developed a general model [13,14] in a Darwinian framework to explain under which conditions learning accelerates or decelerates evolution in stationary environments. In their model, learning altered the fitness landscape, which could speed up or slow down evolution. In this paper we investigate the interaction between learning, Lamarckian inheritance, and undirected variation, in changing environments. We introduce the model in Section 2, results are presented in Section 3, and discussed in Section 4.

2 Model

Inspired by the model of Jablonka et al [4], our model of evolution and learning, allows two environmental states E_0,E_1. Two phenotypes P_0, P_1 are possible, where P_0 is better adapted to E_0, and P_1 is better adapted to E_1, i.e.

$$f(P_0|E_0) > f(P_1|E_0) \quad , \quad f(P_0|E_1) < f(P_1|E_1) \; , \tag{1}$$

where f denotes the fitness score. In the simulations of Section 3 fitness scores are set such that $f(P_i|E_i)/f(P_i|E_j) = 2, i \neq j$, i.e. the "fit" phenotype reproduces twice as much as the unfit. The real-valued genotype $x \in [0;1]$ represents the predisposition towards a phenotype, a low x value corresponds to a genetic predisposition towards P_0, and a high x value towards P_1. A phenotype is realized stochastically by sampling from a Bernoulli probability distribution with parameter p. In the absence of learning, p is defined by the genotype value alone, i.e.

$$p(P_1) = x \quad , \quad p(P_0) = 1 - x \; . \tag{2}$$

If learning is adaptive, it must increase the probability of realizing the fit phenotype of the current environment (P_0 in E_0, P_1 in E_1). The following mapping satisfies this condition

$$p(P_1|x, E_0, L) = \phi(x, L) \quad , \quad p(P_1|x, E_1, L) = 1 - \phi(1 - x, L) \quad , \tag{3}$$

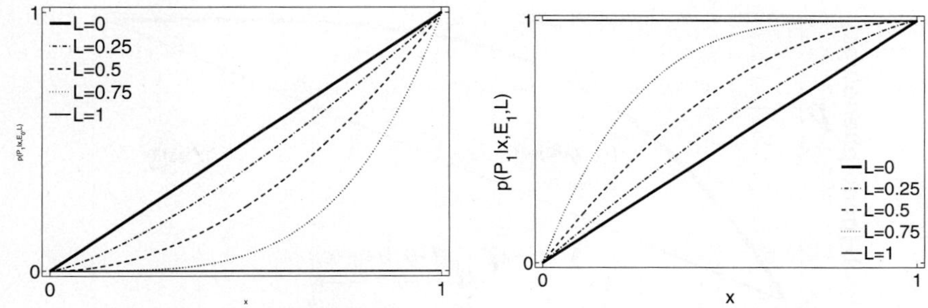

Fig. 1. Influence of the learning parameter L on the probability to express phenotype P_1 for genotype value x, in Environments E_0 (left panel) and E_1 (right panel).

with

$$\phi(x, L) = \begin{cases} = 1, & \text{if } L = 1 \\ = x^{1/(1-L)}, & \text{else} \end{cases} , \tag{4}$$

where $L \in [0; 1]$ is a learning parameter. Thus, learning ($L > 0$) increases the "predisposition" towards the fit phenotype. The larger the L, the larger this increase. Note, that the probability to express phenotype P_0 is the counter-probability of realizing P_1, i.e. $p(P_0|x, E_i, L) = 1 - p(P_1|x, E_i, L), i \in \{0, 1\}$. Fig.1 illustrates the relationship as specified in Equation 3 for different L values.

In each generation, each of 100 individuals reproduces (asexually) an expected number of $f/\bar{f}$ offsprings (f is the individual's fitness, $\bar{f}$ the population mean fitness), such that the population size is constant over time. The offspring's genotype x' depends on the parent's genotype, its learning-induced increase in predisposition, and a Lamarckian parameter λ, in particular $x' = \lambda p + (1 - \lambda)x$. Pure Lamarckianism is given if $\lambda = 1$ and no Lamarckianism is present if $\lambda = 0$. See Fig.2 for illustration of this implementation of Lamarckianism. A low mutation rate (realized by adding a Gaussian random number with mean $\mu = 0$ and standard deviation $\sigma = 10^{-4}$, cut off at the genotype space boundaries) may further influence the offspring's genotype. In some of the experiments the Lamarckian parameter λ and/or the learning parameter L evolves as well. In these cases each individual has an additional gene that stores its λ respectively L. The average time between two environment changes is specified by a parameter T. The actual change periods are either deterministic (cyclic changes) or probabilistic. How well the population adapts is measured as the mean fitness of individuals in the population averaged over time. To avoid an initialization bias, only fitness from generation 1000 to 2000 is sampled. Three experiments have been carried out, which are described in the next section.

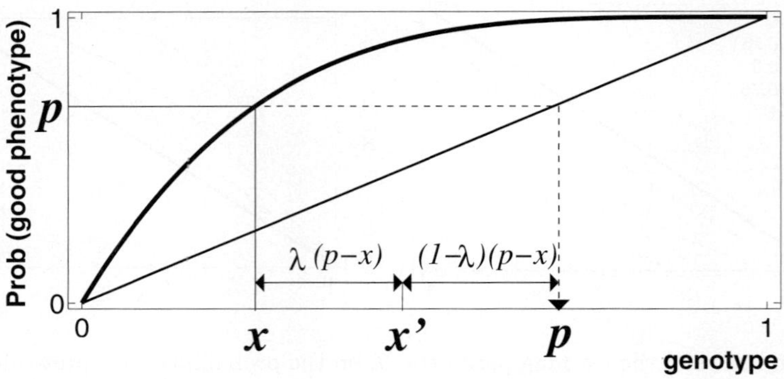

Fig. 2. Implementation of Lamarckianism: Learning increases the probability of realizing the optimal phenotype from genetic predisposition x to p. Depending on the Lamarckian parameter λ its offspring benefits from this increase directly because it inherits a value x', with $x \leq x' \leq p$, where λ determines how close x' is to x and p.

3 Results

Experiment 1

The results of this experiment are presented in Fig.3. We simulate evolution for a range of combinations of the Lamarckian parameter λ and change interval T, for learning parameters $L = 0.5$ and $L = 0.75$, and we compare the case of probabilistic to the case of deterministic environmental changes. In particular we combine $T \in \{1, 5, 10, \cdots, 95, 100, 200\}$ and $\lambda \in \{0, 0.05, \cdots, 0.95, 1.0\}$. The results are shown in four figures (see Fig.3), each showing all combinations of λ and T. The mesh plots show for each combination of T and λ the mean population fitness, averaged over time and over 25 independent evolution runs. The following findings are qualitatively consistent over all settings: With small T, i.e. in rapidly changing environments, the maximum mean population fitness is produced for $\lambda = 0$, i.e. without Lamarckianism (see the thick gray line). However, with large T, i.e. slowly changing environments, the maximimum mean population fitness (thick gray line) is produced with $\lambda = 1$, i.e. pure Lamarckianism. The minimum mean population fitness (thick black line) is produced with pure or high level of Lamarckianism (λ around 1) in rapidly changing environments, and without or low level of Lamarckianism (λ around 0) in slowly changing environments. Interestingly, for intermediate T, the lowest adaptation success is found for intermediate λ. For example, in the top-left panel, for $T = 20$ the minimum mean population fitness is produced with $\lambda = 0.4$. The curious fitness valley disappears for very low or high T. A geometric explanation for this fitness valley is provided at the end of this section. Although qualitatively consistent, the observed effects are weaker with higher mutation rates (not shown). In summary, Lamarckianism results in higher mean population fitness than Darwinian inheritance in slowly changing environments and a lower mean population fitness in rapidly changing

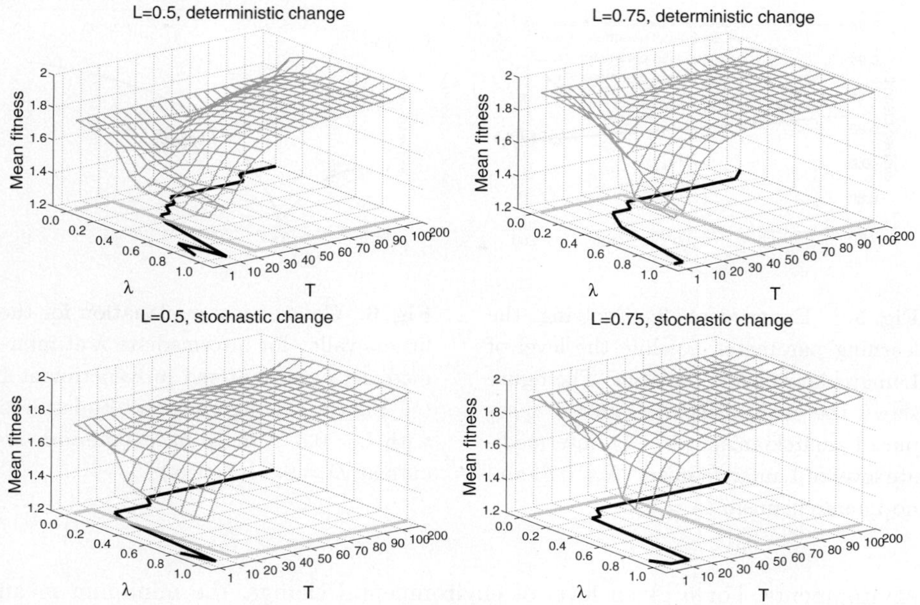

Fig. 3. Experiment 1: Mean fitness in evolution with different Lamarckianian parameters λ and environmental change intervals T. The thick black line shows where, for a given T, the minimum occurs for λ. The thick gray line is the corresponding maximum. Panels in the left column correspond to $L = 0.5$, panels in the right column correspond to $L = 0.75$. Upper row panels are based on deterministic, lower row panel on probabilistic environmental changes.

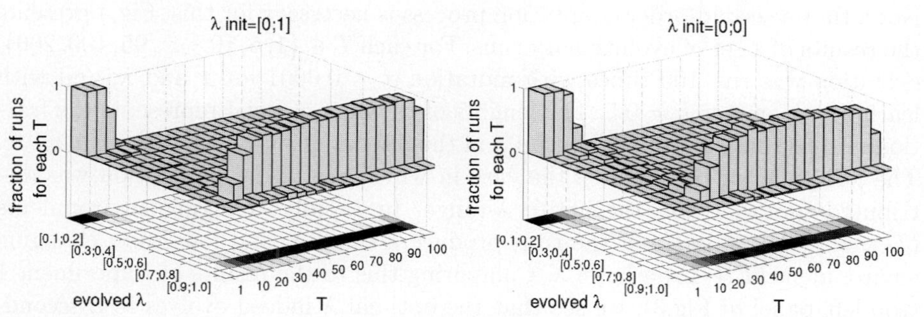

Fig. 4. Experiment 2: Evolving the Lamarckian parameter λ, initialized uniformly on [0; 1] (left panel), and starting without Lamarckianism, i.e. $\lambda = 0$ for all individuals (right panel), in case of deterministic environmental changes and with $L = 0.5$.

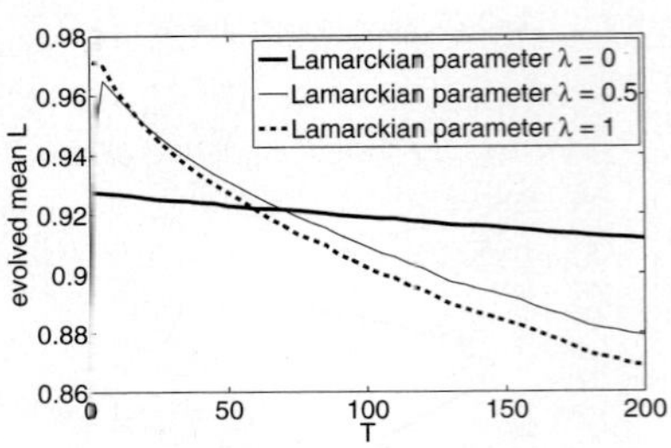

Fig. 5. Experiment 3: Evolving the learning parameter L while the level of Lamarckianism λ is constant. The figure shows the evolved mean L, in the cases pure Lamarckianism ($\lambda = 1$), intermediate level of Lamarckianism ($\lambda = 0.5$) and no Lamarckianism ($\lambda = 0$).

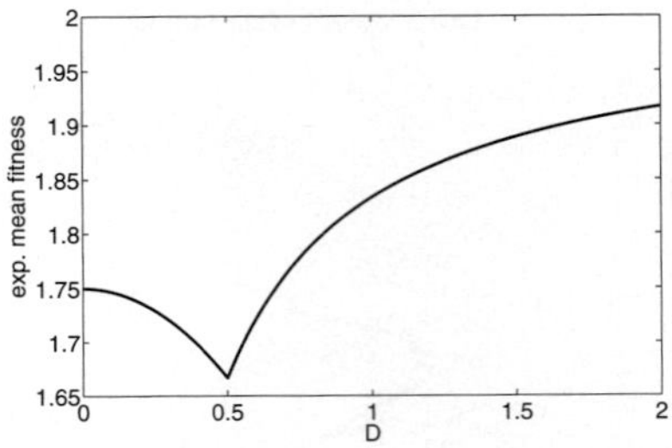

Fig. 6. Geometrical explanation for the fitness valley for intermediate λ at intermediate T encountered in Experiment 1 (cf. Fig 3). The figure shows Equation 6 with $L = 0.5$. A mean fitness minium occurs at $D = 0.5$ (cf. text).

environments. For a given level of environmental change, the minimum mean population fitness is produced by an intermediate level of Lamarckianism. The slower the environment changes, the lower the level of Lamarckianism where this minimum occurs (see also end of this section for a geometrical explanation).

Experiment 2

The results of this experiment are presented in Fig.4. This experiments aims to test whether the optimal level of Lamarckianism λ (cf. thick gray line in the top-left panel of Fig.3) evolves if each individual has its λ encoded in the genotype. Note, that a second-order adaptation process is necessary for this. Fig.4 presents the results of a set of evolutionary runs. For each $T \in \{1, 5, 10, \cdots, 95, 100, 200\}$, evolution was run 100 times with mutation $\sigma = 0.0001$ for x and λ, and with learning parameter $L = 0.5$. The lenghts of the bars in Fig.4 represents the fraction of runs that resulted in a mean λ in the interval $[0; 0.1], [0.1; 0.2] \cdots [0.9; 1.0]$. The left panel of Fig.4 shows the case in which the initial population was distributed uniformly on the entire λ-range. In rapidly changing environments ($T \leq 10$), the majority of the runs produce a small λ, and for slower changing environment ($T \geq 15$) a large λ. Comparing this to the results of Experiment 1 (top-left panel of Fig.3), we see that the optimal λ indeed evolves in a second-order process. In another experiment (Fig.4, right panel) evolution starts without Lamarckianism ($\lambda = 0$) for all individuals. In this case, a large λ is only evolved for $T \geq 25$. The likely reason for this difference is the observed fitness valley for intermediate λ in case of intermediate levels of environmental change. Apparently, the population can not cross the fitness minimum for T around 20. In an additional experiment (results not shown) the learning rate L was evolvable as well. In the absence of learning cost, a high L quickly evolved and suppressed the evolution of the Lamarckian parameter λ in slowly changing environments: With very high learning ability, there was only weak selection pressure for a large L in slowly changing environments, which leads to the evolution of only intermediate levels of λ. In summary, in most cases, a near-optimal level of Lamarckianism evolves as a second order process. However, in cases where there is a population mean fitness minimum for intermediate levels of Lamarckianism (see Experiment 1), the globally optimal level of Lamarckianism does not always evolve.

Experiment 3

The results of this experiment are presented in Fig.5. The aim of this experiment is to test if Lamarckianism influences the evolution of learning ability L. Holding the level of Lamarckianism λ constant during the evolution, we evolve L, for a wide range of T, and compare the cases "no Lamarckianism" ($\lambda = 0$), pure Lamarckianism ($\lambda = 1$) and an intermediate level of Lamarckianism ($\lambda = 0.5$). Comparing the two extreme cases no ($\lambda = 0$) and pure ($\lambda = 1$) Lamarckianism, we see that in quickly changing environments ($T < 60$) a larger mean L evolves with pure Lamarckianism, and in slower changing environments a lower mean L evolves with pure Lamarckianism than without Lamarckianism. The case of intermediate level of Lamarckianism ($\lambda = 0.5$) lies between the two extremes cases, but is closer to the case of $\lambda = 1$. So, Lamarckianism suppresses the evolution of learning ability in slowly changing environments and facilitates the evolution of learning ability in quickly changing environments. An explanation for this is that for large T, there is a relatively low selection pressure for high L in case of Lamarckianism, because a high λ alone allows good adaption. For small T, however, we have shown that Lamarckianism is detrimental, and there is a relatively high selection pressure to evolve a high L that can compensate for the Lamarckian disadvantage. In summary, where Lamarckianism provides an adaptive advantage (slowly changing environments) a lower learning ability is evolved because there is less selection pressure for it, but where Lamarckianism provides an adaptive disadvantage (rapidly changing environments) a higher learning ability is evolved because there is stronger selection pressure for it, i.e. learning compensates the disadvantage of Lamarckianism here.

Geometric Explanation for the Fitness Valley

In experiments 1 and 2, we found that for a given T, the minimum mean population fitness is produced by an intermediate λ. A possible explantion is outlined in the following: With a very low mutation rate we assume that genotype changes within time T are mainly induced by Lamarckianism and that mutation-induced random genetic changes are negligible. We further assume that the population mean fitness is well represented by the expected fitness of the population mean genotype. Thus, population mean fitness can be expressed w.r.t. the population mean distance to the optimal genotype, which we denote d. Assume that initially $d = 0.5$ and between two environmental changes (within one T), this distance is reduced by a distance of D, where D depends on the level of Lamarkckianism λ and the learning parameter L, i.e. $D(\lambda, L)$. In our model, we know that

$\frac{\partial D}{\partial L} \geq 0$ and more importantly for this analysis $\frac{\partial D}{\partial \lambda} \geq 0$, i.e. D is increasing with λ. Let us first consider the case where $(0 < D \leq 0.5)$, such that the population never reaches the optimum within T or just immediatly before the environmental change at T, e.g. because λ is too small: At the time, just before an environmental change occurs, the population has a distance $d = 0.5 - D$ to the optimum. Immediately after the environmental change this distance becomes $d = 0.5 + D$ since the optimal genotype has changed (from 0 to 1 or from 1 to 0). Since the population always moves back and forth between these two states, the expected fitness over time is approximately

$$\bar{f}(D, L) = \frac{1}{D} \int_{0.5-D}^{0.5+D} f_{\text{exp}}(d, L) \;\, \mathrm{d}d \;, \tag{5}$$

where the expected fitness of d is $f_{\text{exp}}(d, L) = 2 - \phi(d, L)$ (cf. equations 3 and 4). This assumes, that the fit phenotype's fitness is twice the unfit phenotype's fitness. Equation 5 can be reformulated with straight-forward calculations. Substituting n for $(1/(1 - L))$, we obtain

$$\bar{f}(D, n) = \begin{cases} 2 + \frac{(0.5-D)^{n+1}}{2D(n+1)} - (0.5 + D)^{n+1} & \text{if } 0 < D \leq 0.5 \\ 2 + \frac{1}{2D}\frac{2n+1}{n+1} - \frac{1}{D} & \text{if } 0.5 < D \leq 1 \\ 2 - 0.5^n & \text{if } D = 0\,. \end{cases} \tag{6}$$

The first case $(0 < D \leq 0.5)$ corresponds to the above described scenario, where the population never reaches the optimum within T. In the second case $(0.5 < D \leq 1)$, the population reaches the optimal genotype within T and stays there until the next environmental change (having the maximum fitness of 2 during this time). Thus, for $(0.5 < D \leq 1)$, we obtain $(0.5/D) \cdot \bar{f}(0.5, n) + ((D - 0.5)/D) \cdot 2$, which produces the second case of Equation 6 after some straightforward calculations. The third case $(D = 0)$ corresponds to $\lambda = 0$ (no Lamarckianism). Here, the mean fitness over time is simply the expected fitness of $d = 0.5$, i.e. the population does not move. Figure 6 illustrates Equation 6 for $L = 0.5$. It shows a minimum at $D = 0.5$. For a given constant L, D only depends on λ and we know that D is increasing with λ. Thus, the mean population fitness $\bar{f}$ is decreasing for small λ and increasing for large λ, producing a minimum for intermediate λ. This provides a possible explanation for the occurence of the fitness valley for intermediate λ at intermediate T in experiments 1 and 2.

To summarize the main argument of this geometrical explanation: With a low mutation rate, the population's mean genotype movement is mainly dependend on the level of Lamarckianism, i.e. Lamarckianism allows quick genotype movement. A (Lamarckianism-induced) quickly moving population may be less fit than a population that is not or hardly moving (without Lamarckianism): While a quickly moving population has the advantage of approaching a recently changed fitness optimum, it potentially has an adaptive disavdantage when the next environmental change occurs, since it is farther away from the new optimum than the population that has moved less. In our model this disadvantage indeed occurs, and the disadvantage is even larger than the adaptive advantage of approaching a new optimum. Thus, the population mean fitness is decreasing for increasing level of Lamarckianism. If, however, the level of Lamarckianism increases further and exceeds a certain threshold, the population can move very quickly to the new optimum and stay there at a high fitness level (until the next environmental change occurs). Thus, at intermediate levels of Lamarckianism, the population mean fitness is increasing with the level of Lamarckianism.

4 Discussion

This paper predicts that Lamarckian inheritance should be less common in natural environments that oscillate rapidly, compared to stationary environments. The disadvantage of Lamarckian inheritance in rapidly changing environments is explained by the movement of the mean genotype. With Lamarckian inheritance, genotype movement is faster than with genetic mutation alone. In rapidly oscillating environments, Lamarckianism increases the integral of genotype distance from the optimum. The advantage of Lamarckian inheritance in slowly changing environments is because the genotype converges to the optimum more rapidly than by random mutation alone. A curious finding at intermediate levels of environmental oscillation is that a minimum value of mean population fitness is associated with a particular value of Lamarckian inheritance. This is in contrast to the monotonic changes in mean population fitness observed at very high and very low rates of environmental change. This fitness valley may prevent the evolution of Lamarckianism from scratch even though high levels of Lamarckian inheritance are a global optimum.

A follow-up experiment in which learning rate was evolvable, showed us that the introduction of Lamarckian inheritance in rapidly oscillating environments increases selective pressure for better learning mechanisms, whilst introduction of Lamarckian inheritance in slowly oscillating environments decreases the selective pressure for learning mechanisms. Note that this finding is limited to instances where environmental changes occur cyclically such that the genotype is able to establish itself in an area where a high fitness under several environmental conditions is experienced. In nature, simple binary oscillating environments involve geophysical rhythms such as diurnal and seasonal cycles. If however, the environment were to change in a non-oscillating path, e.g. a continuously increasing temperature or in a co-evolutionary setting, then Lamarckianism may be beneficial even in rapidly changing environments. We have not modeled non-oscillating environmental changes.

Recently it has been proved for several conditions that directed phenotype plasticity helps to cross a fitness valley by smoothing the fitness landscape [15]. This positive effect stems from a non-Lamarckian inheritance mechanism and would not be observed with Lamarckian inheritance. Our model does not cover this type of adaptive disadvantage of Lamarckianism. It also does not include dependencies between Lamarckian mechanisms and learning mechanisms as one would find them in cultural evolution, a Lamarckian-like inheritance mechanism, where a high learning ability is actually a prerequisite for this form of Lamarckianism, to appear. Neither does the model include a cost of learning, which can certainly influence the balance between Lamarckianism and learning ability.

In future we will carefully extend the analysis model in order to cover various aspects of the interaction of evolution and learning under Lamarckian inheritance.

Acknowledgments. The major part of this work has been done during a research stay of I. Paenke at the University of Birmingham, UK. I. Paenke wants to thank all people who made this research stay possible, in particular H. Schmeck (Institute AIFB, University of Karlsruhe, Germany), E. Körner (Honda Research Institute Europe GmbH, Germany) and X. Yao (CERCIA, University of Birmingham, UK). The authors would also thank D. Stekel for interesting discussions. C. Fernando is supported by the ESIGNET 6th Framework European Grant, I. Paenke is supported by a grant from the Honda Research Institute Europe GmbH.

References

1. Gould, S.J. (ed.): The Structure of Evolutionary Theory. The Belknap Press of Harvard University Press, Cambridge (2002)
2. Lamarck, J.B.: Philosophie zoologique ou exposition des considérations relatives á l'histoire naturelle des animaux. UCP (reprinted 1984) (1809)
3. Jablonka, E., Lamb, M.: Evolution in Four Dimensions - Genetic, Epigenetic, Behavioral, and Symbolic Variation in the History of Life. MIT Press, Cambridge (2005)
4. Jablonka, E., Oborny, B., Molnar, I., Kisdi, E., Hofbauer, J., Czaran, T.: The adaptive advantage of phenotypic memory in changing environments. Philos. Trans. R. Soc. Lond. B Biol. Sci. 29(350), 133–141 (1995)
5. Richards, E.: Inherited epigenetic variation – revisiting soft inheritance. Nature Reviews Genetics. Advanced online publication (2006)
6. Foster, P.: Adaptive mutation: Has the unicorn landed? Genetics 148, 1453–1459 (1998)
7. Holliday, R., Pugh, J.E.: DNA modification mechanisms and gene activity during development. Science 187, 226–232 (1975)
8. Lippman, Z., Martienssen, R.: The role of rna interference in heterochromatic silencing. Nature 431, 364–370 (1986)
9. Slade, H.B., Schwatrz, S.A.: Mucosal immunity: The immunology of breastmilk. Journal of Allergy and Clincal Immunology 80, 348–356 (1987)
10. Crick, F.H.C.: Central dogma of molecular biology. Nature 227, 561–563 (1970)
11. Weissman, A.: The Germ-Plasm: A Theory of Heredity. Walter-Scott, London (1893)
12. Sasaki, T., Tokoro, M.: Comparison between Lamarckian and Darwinian evolution on a model using neural networks and genetic algorithms. Knowledge and Information Systems 2(2), 201–222 (2000)
13. Paenke, I., Kawecki, T.J., Sendhoff, B.: On the influence of lifetime learning on selection pressure. In: Artificial Life 10, pp. 500–506 (2006)
14. Paenke, I., Sendhoff, B., Kawecki, T.J.: Influence of plasticity and learning on evolution under directional selection. American Naturalist (in press, 2007)
15. Borenstein, E., Meilijson, I., Ruppin, E.: The effect of phenotypic plasticity on evolution in multipeaked fitness landscapes. Journal of Evolutionary Biology 19(5), 1555–1570 (2006)

The Dynamics of Associative Learning in an Evolved Situated Agent

Eduardo Izquierdo and Inman Harvey

Centre for Computational Neuroscience and Robotics
Department of Informatics, University of Sussex, Brighton, UK
{e.j.izquierdo, inmanh}@sussex.ac.uk

Abstract. Artificial agents controlled by dynamic recurrent node networks with fixed weights are evolved to search for food and associate it with one of two different temperatures depending on experience. The task requires either instrumental or classical conditioned responses to be learned. The paper extends previous work in this area by requiring that a situated agent be capable of re-learning during its lifetime. We analyse the best-evolved agent's behaviour and explain in some depth how it arises from the dynamics of the coupled agent-environment system.

1 Introduction

Learning is a behaviour. In fact, it is a *change* of behaviour over time. Living organisms show a variety of behaviours that are modulated by environmental conditions and previous experience. A major goal of the artificial life sciences is to elucidate the dynamical bases of such experience-dependent adaptive behaviour.

Associative learning is a particularly adaptive form of such experience modulated behaviour, as it requires responses to be paired with a particular stimulus. Organisms at several levels of 'complexity' provide evidence for this, including many extraordinarily simple ones. In the small nematode worm *C. elegans*, evidence for the formation of associations between temperatures and food has been known for quite some time [5]. However, the mechanisms required for the storage and resetting of this memory are still largely unknown.

In the animal learning theory there is the idea of the strengthening of a 'connection' between a stimulus and a response. This has been directly translated to the strengthening of physical connections between neurons. While this is a good description at the level of the agent's interaction with the environment (behavioural description), there need not be a direct correspondence of connection-forming processes in the internal behaviour-producing mechanisms of the agent. We believe there is a more fundamental principle underlying learning behaviour at the level of an organism's internal mechanisms that has to do with dynamics on multiple timescales.

The aim of this work is to: (1) successfully evolve the smallest possible integrated dynamical system controller with fixed weights in a situated[1] agent on an

[1] By situated we mean an agent that is embedded in a world; and thus its ongoing sensori stimuli is dynamically determined by its own actions.

F. Almeida e Costa et al. (Eds.): ECAL 2007, LNAI 4648, pp. 365–374, 2007.

associative learning task requiring re-learning, (2) perform a behavioural analysis of the best evolved agent; and (3) study the coupled agent-environment dynamics of a successful controller and attempt to understand it as implementing a finite state machine (FSM), so as to compare with similar work [6].

2 Related Work

A number of researchers have used genetic algorithms to evolve, for tasks requiring associative learning, dynamical neural controllers without in-built synaptic plasticity mechanisms. Yamauchi and Beer [8] were the first to explore this idea using a one-dimensional navigation task with a goal and a landmark. Attempts to evolve an 'integrated' network failed, so a modular approach was taken.

Blynel and Floreano [2] evolve controllers on a relatively similar task and environment. In their version, because the light is fixed to one side of the arena and the goal is the only thing that changes, it is possible for the agent to employ a reactive turn left or right strategy, as opposed to approaching or avoiding the light; making it unnecessary to form an association between light and goal.

Attempts to remedy those initial difficulties were successfully overcome by Tuci et al. [7] in a two-dimensional version of the same task. As the emphasis of that work was on the evolutionary process, no further analysis of the behaviour or internal dynamics was performed.

Fernando in [3] explores the same associative learning task in a slightly more complicated T-maze environment. Despite not being able to evolve an agent that solves the task completely, an analysis of the best performing agent in terms of animal learning theory is attempted.

Such work demonstrates that multitimescale dynamics can exhibit learning-like behaviour without synaptic plasticity mechanisms. However, none of the previous work deals with re-learning during the lifetime of the agent: the agent's internal state is reset when tested on a different environment. Also, the internal mechanisms of the best-evolved agents have not been explored in much depth or at all in some cases.

Phattanarasri et al. [6] study in-depth the dynamics of an evolved circuit for an associative learning task very similar to the one being presented here. The main difference with this work is that their experiments take place in a non-situated agent. Of particular interest is their analysis of the evolved internal mechanisms, which can be understood to implement a FSM.

3 Methodology

We use evolution to synthesize continuous-time recurrent neural networks that display associative learning behaviour when situated. The task is loosely abstracted from the temperature preference behaviour observed in the nematode worm *C. elegans* [5]. In particular, we would like an agent that is capable of associating temperature with food in two different types of environment, and re-learning: modify its temperature preference during its lifetime when required.

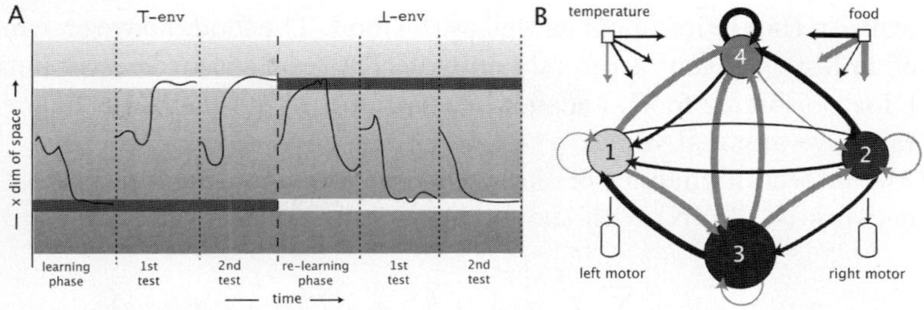

Fig. 1. [**A**] Example trial. 1D projection of environment with thermal gradient (shades of grey). 'Nutritious food' denoted by white bars; 'poisonous' with black. [**B**] Agent architecture with 4 fully inter-connected nodes, a food and a temperature sensor, and two wheels controlled by arbitrarily chosen nodes. Parameters of the best evolved circuit are also depicted. Nodes are shaded according to their bias. Excitatory connections (black) and inhibitory (grey), with the width of the line proportional to the strength. Time-constants represented by size, with larger circles representing slower nodes.

We use a 2D arena with a thermal gradient along one of its dimensions containing two types of food: 'nutritious' and 'poisonous'. Each type of food can be found only in regions in a particular temperature range: '*hot*' between [9,10]; '*cold*' between [-10,-9]. Which region the nutritious food can be found in depends on the type of environment: $\top$-env, nutritious food in the *hot* region; and $\perp$-env, in the *cold* region. For each of the different environment types, the poisonous food can be found in the opposite region to the nutritious food. There are no walls and the thermal gradient extends in all directions.

An example trial of the task is depicted in Figure 1A. The task involves placing an agent at random in the central region (between [-2,2]) of the arena (including random orientation) in one of the two environment types, requiring it to find and stay on the food as efficiently as possible. The first challenge involves exploring the whole of the arena in search for food. After a random amount of time (between [80,100] units), the agent is physically displaced back towards the central region of the arena and given a random orientation again. A successful agent should navigate up or down the thermal gradient depending on whether it had found food in the *hot* or *cold* region in the previous trial, respectively. This requires that it learn and remember in which of two environment types it finds itself. Less frequently, the displacement involves changing the environment type as well. This requires that the agent remain sufficiently plastic to change its temperature preference online. Although it is this learning and re-learning phenomena that are central to our paper, there is also a more basic sensory-motor challenge involved in navigating up and down the thermal gradient which will not be explored.

Agents are modelled as circular bodies of radius 1 with two diametrically opposed motors and two sensors. Agents can move forwards and turn. The mass of the body is sufficiently small so that the motor's output is the tangential velocity at the point where the motor is located. The agent can sense the local temperature in the environment as well as the food. The food, however, cannot be perceived unless the agent is directly upon it. The food sensor is: 1 for nutritious food, -1 for poisonous food, and 0 when no food is present. The temperature sensor can have any real value.

For the internal dynamics of the agent, we use a continuous-time recurrent neural network (CTRNN) with the following state equation [1]:

$$\tau_i \dot{y}_i = -y_i + \sum_{j=1}^{N} w_{ji} \sigma(y_j + \theta_j) + s_i T(x) + g_i F(x;e) \tag{1}$$

where y is the activation of each node; τ is the time constant; w_{ji} is the strength of the connection from the jth to the ith node; θ is a bias term; $\sigma(z) = 1/(1+e^{-z})$ is the standard logistic activation function; $T(x)$ is the thermal sensor, a function of the agent's position along one of the dimensions of the physical space, x; s_i is the strength of the connection from the thermal sensor; $F(x;e)$ is the food sensor, also a function of x but parameterized by the type of environment, e; g_i is the strength of the connection from the food sensor; and N represents the number of nodes in the network. In simulation, node activations are calculated forward through time by straightforward time-slicing using Euler integration with a time-step of 0.1. The network is fully connected (see Figure 1B). There are no additional weight changing or any other parameter changing rules.

The connection weights, biases, and time-constants in Equation 1 are encoded in a genotype as a vector of real numbers and evolved using the microbial genetic algorithm [4]. The size of the population used was 50. We define a generation as the time it takes to generate 50 new individuals.

The fitness of a circuit is obtained by minimising the relative distance away from the food at the beginning of each test (a), and maximising the time spent sensing food towards the end of the same phase (b), according to

$$a = \int_{t=0}^{50} \left(\frac{20-d}{20} \right) dt \ , \quad b = \int_{t=30}^{80} F dt \tag{2}$$

where F is the agent's sensor for food and d is the absolute distance between the source of food and the position of the agent capped at 20. Both components are normalised to run between 0 and 1. The two components are clearly linked: the first provides emphasis on heading in the direction towards where the food should be at the start of the trial; the second emphasizes staying directly on top of the food once found.

A fitness trial consists of the evaluation of an agent's performance for the number of times it is displaced in the same environment type, p, and for the number of changes of environment type, k, all without reinitialising the agent's state. No evaluation takes place at the start of a trial, nor immediately after a change of environment type. This is repeated 50 times for each individual and the fitness taken from the multiplication of their averages, $f = \bar{a} \cdot \bar{b}$. Each repetition involves the re-initialisation of the agent's internal state.

Following [6], a set of evolutionary stages of increasing complexity are employed. The changes are in the starting orientation of the agent, φ, after each start of trial or displacement; in the number of times an agent is tested (i.e. displaced), k; and the number of changes of environment type, p; as follows:

Stage	1	2	3	4	5
φ	$\{0, \pi\}$	$[0,2\pi)$	$[0,2\pi)$	$[0,2\pi)$	$[0,2\pi)$
k	1	1	1	2	5
p	1	1	5	[1,5]	[1,5]

Transitions occur when the best fitness exceeds 0.8 consistently (i.e. for 5 consecutive generations). At the last stage, the orientation is chosen at random from the full range, the environment type changes 5 times during the agent's lifetime, and the changes occur between the first and the fifth displacement at random.

4 Results

4.1 Evolutionary Performance

We attempted evolving 3, 4, and 5-node circuits for this task using 15 evolutionary runs with different seeds for 10000 generations each. The proportion of evolutionary runs that reached the different stages are depicted in Figure 2A. While no 3-node populations reached the last stage, several 4 and 5-node populations did. In fact, the majority of 5-node runs were highly successful, but we will focus our attention on the smallest successful circuit obtained. The interest in evolving the smallest circuit that solves the task is primarily to make the analysis most amenable to the mathematics of dynamical systems theory.

An example evolutionary trajectory for the population that produced the best 4-node agent is shown in Figure 2B. As can be seen, the fitness drops sharply after every transition except the last: once the circuit is able to generalize to all learning scenarios. It is the best agent of this evolutionary run that will be analysed in some depth in the rest of this paper.

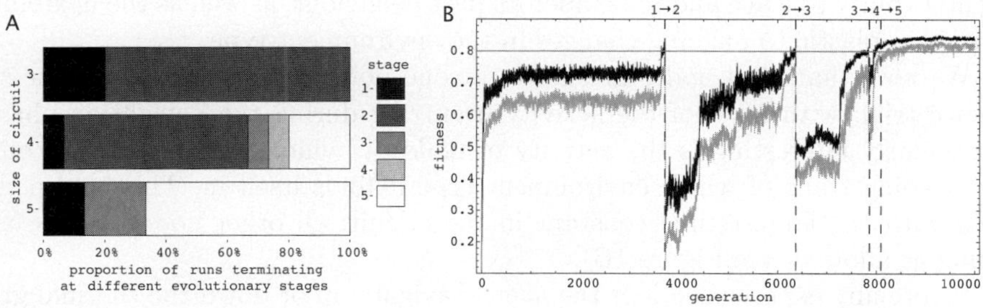

Fig. 2. Evolutionary performance. [**A**] Proportion of populations that terminated at a certain evolutionary stage for different size circuits. [**B**] Fitness vs. generation for the best evolved 4-node population (best in black and average in grey). Transitions between stages (dashed lines) occur when the best fitness consistently exceeds the horizontal grey line and are labelled accordingly.

4.2 Behavioural Analysis

The performance of the best circuit was further tested using 10^4 evaluation trials, each with 10 changes of environment type, between [1,10] displacements, noise in the sensors and motors drawn from a Gaussian distribution (σ=0.05), and a time-step an order of magnitude smaller (0.01). As we are interested in how well the agent finds the nutritious food in the face of changing environments, only the b component of fitness is considered. The best 4-node circuit obtained 98.81% success rate on this test, meaning that it generalises well on a broad range of situations. Since the slope of the thermal gradient remains constant throughout evolution, the agent could use the distance instead of the temperature as the relevant factor to remember. We used the same test while varying the slope of the gradient between ±20% with the success rate dropping by only a minor fraction (98.48% success), meaning the agent relies on the temperature and not the distance the food is away from the centre.

Figure 3 shows the behaviour of this agent on a typical sequence trial with 2 changes of environment type. At the beginning of the trial, the agent navigates down the thermal gradient but switches to navigating up before reaching the usual region where food could have been located. This is part of the search strategy, as it does not yet know in what type of environment it finds itself. When displaced for the first and second times after reaching the food, however, it navigates more directly up the thermal gradient. Subsequently the environment type is changed, unaware the agent navigates up the thermal gradient as for previous trials, with the difference that negative reinforcement is encountered (but only very briefly[2]). The agent navigates past this food region and eventually changes behaviour to navigate in the opposite direction of the gradient, until reaching the nutritious food on the *cold* region. On subsequent trials, the agent will navigate directly down the gradient, showing that it remembers where the food was last found in the other type of environments as well. A similar pattern is observed in the second change of environment type. This demonstrates the agent's ability to learn and remember its past behaviour, as well as the flexibility to remain plastic to ongoing changes in the environment type.

We note that all 4 nodes are active at one point or another during the sequence trial; with most of the activity occurring during the navigation phase. Particularly interesting is the activity of node o_3, which seems to be the only one keeping track of which environment type it finds itself in. This is also the node with the largest time constant in the circuit; all other nodes are as fast acting as allowed (see Figure 1B).

Before any experience, does the agent navigate up or down the thermal gradient? and what does this depend on? We studied the long-term behaviour of the agent when initialised in an environment with nutritious food on both *cold* and *hot* regions. As can be seen in Figure 4A, what the agent does depends mainly on its starting position: visiting the furthest region first. How does experience affect this pattern? After learning has occurred, the agent will preferably head

[2] Absence of poisonous food in the environment does not affect the learning behaviour in this agent. The reason is that the negative reinforcement is redundant in this task.

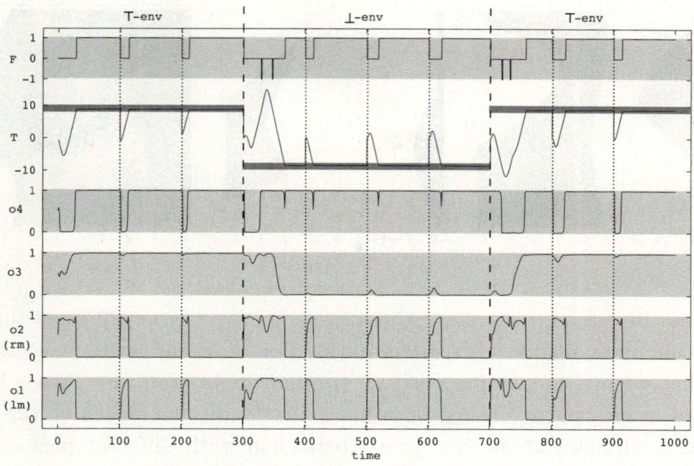

Fig. 3. Activity of the best 4-node circuit on a typical trial sequence. From top to bottom the traces correspond to the food signal (F), the temperature signal (T), and the outputs of the neurons (o_i). The last two neurons control the right (rm) and left (lm) motors. The dark grey horizontal bars in the temperature trace depict where nutritious food is to be found for that trial ($\top$ or $\bot$). Dotted vertical lines mark different trials (where the agent is displaced). Dashed lines mark transitions between environments.

towards *hot* or *cold* regions, even with nutritious food on both, depending on where food was found in the previous trial (see Figures 4B and 4C, respectively). This shows how behaviour is appropriately modulated according to previous experiences regardless of initial position and orientation.

4.3 Dynamics of the Coupled Agent-Environment System

We next turn to the dynamics underlying the behavioural phenomena described in the previous section. The primary interest is in understanding how this agent's dynamics is structured so that where food was encountered in the past affects which direction of the thermal gradient it will navigate towards. From the equations describing the coupled agent-environment system we can make some general observations. First, the agent is a nonautonomous dynamical system with two inputs, T and F. Second, although T varies continuously as a function of x, discontinuities are introduced into the dynamics by the food sensor because $F(x;e)$ is a discontinuous function of x, making the agent a hybrid dynamical system. Given these two factors, the best way to study its operation is to characterize its autonomous dynamics for all possible combinations and then examine the transient dynamics induced by the agent-environment interaction. If we take into consideration only the range of temperatures where the agent was observed to navigate around, then there are five possible bifurcation diagrams to consider: $P_{\pm 15}$ (temperature between [-15,15] with no reinforcement), $P_{\downarrow +}$ (*cold* temp. and positive reinf.), $P_{\uparrow +}$ (*hot* temp. and positive reinf.), $P_{\downarrow -}$ (*cold* temp.

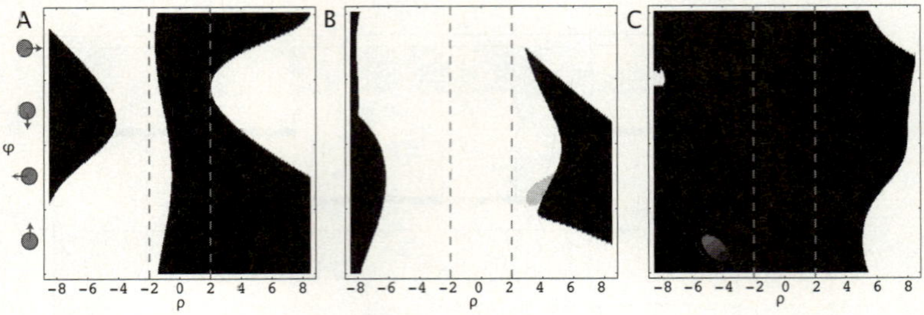

Fig. 4. Points in the map represent the average position (over 20 repetitions) of the agent after 100 units of time with nutritious food on both *hot* (white) and *cold* (black) regions while varying its starting position (p) and orientation (φ). Points in-between are in shades of grey. Grey dashed lines mark the conditions for which the agent was evolved. Different maps show the agent's behaviour with different past experiences: [**A**] Before any experience. [**B**] After $\top$-environment. [**C**] After $\perp$-environment.

and negative reinf.), and $P_{\uparrow-}$ (*hot* temp. and negative reinf.). Three-dimensional projections of the stable solutions of the first three of these are shown in Figure 5A, coded in shades of grey as a function of the temperature and labelled accordingly. The portraits corresponding to the negative reinforcements can be left out of the analysis because they do not affect the performance of the agent's learning behaviour. As can be seen, for mid-temperatures ($P_{\pm 9}$) the long-term behaviour of the system is bistable. As the temperature increases or decreases outside of this range, only one attractor is left in opposite ends of the original for *cold* and *hot*. Similarly, for $P_{\downarrow+}$ the dynamics are bistable and for $P_{\uparrow+}$ there is only one stable state.

How do these bifurcation diagrams combine to produce the learning behaviour? We can study the transient trajectories in the internal state of the agent as it interacts with its environment. In Figure 5B we show a set of trajectories from behaviours crucial for the task using the same projection as in the previous plot. Can we interpret the transitions in the internal state of the agent as implementing a FSM? We were unable to do so. The difficulty arises from the agent's dependence on the temperature sensing as an ongoing and continuous perturbation. We hypothesize that it is the discretisation and non-situatedness of the task in [6] that facilitates their FSM interpretation. Only when we consider a different form of state machine that allows for ongoing sensori-motor interactions can we summarize the coupled agent-environment system in relation to the agent's internal dynamics. We will call this an 'interactive state machine' (see Figure 5C). Although similar, strictly speaking the diagram is not a FSM because some of the states include ongoing interactions with the environment. In it, the finite states the system can be in are denoted by circles labelled: $\uparrow+$ or $\downarrow+$, for when nutritious food is found in the *hot* and *cold* regions, respectively. The graded ellipses represent the 'interactive states': where the agent's state moves it in relation to the environment, and the change of temperature changes the dynamics of the agent in turn. There are two of these: $\uparrow$ and $\downarrow$, for what results

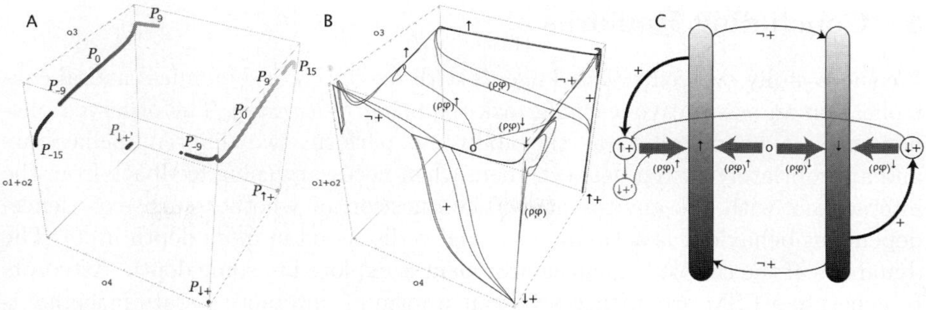

Fig. 5. Agent-environment coupled dynamics. [**A**] Equilibrium points of the nonautonomous system depending on temperature (shade of grey) and positive reinforcement. [**B**] 3D projection of the trajectories in internal space state for a typical set of behaviours. See main text for the labels. [**C**] Diagram of the coupled dynamics.

in navigation up or down the gradient, respectively. We denote the starting internal state as o. Physical displacements events are depicted with thick arrows. We can characterise the basins of attraction of the bistable dynamics in $P_{\pm 9}$ as a function of the agent's position and orientation from Figure 4A as $(p, \varphi)^z$, where z represents the long-term behaviour ($\uparrow$ or $\downarrow$). The black arrows denote the encountering of nutritious food, $+$. The thin arrows connecting the ellipses denote the transition from one stable state in $P_{\pm 15}$ to the other in the internal dynamics. This occurs when the agent reaches colder or hotter temperatures. The diagram up to this point is sufficient to fully characterise the observed behavioural phenomena. There is an additional finite state that is never reached during regular associative learning which we denote as $\downarrow +$'.

4.4 Predictions from the Dynamics

The study of the dynamics suggest a number of predictions which we could confirm using behavioural studies. Although a full study of the predictions would require further space, two of them are mentioned briefly. First, as a result from the bistability of $P_{\downarrow +}$, we could predict and confirm that even after experiencing environments with food in the *cold* regions, if exposed to *hot* temperatures and food simultaneously for sufficiently long, the agent could be re-conditioned to navigate up the thermal gradient. This was not the case in the opposite scenario, where the agent required doing the down-the-thermal-gradient navigation behaviour to remember. We can describe the agent as employing a mixture of classical (pairing two signals) and operant (pairing an action with a reinforcement) conditioning. Second, and as a consequence of the geometry of $P_{\pm 15}$, we could predict and confirm that in the total absence of any kind of food, the coupled system falls into a limit cycle, that involves the agent switching between going up and down the gradient modalities. Although this was not a scenario the agent was evolved for, it could be interpreted as a higher level 'searching for food' behaviour that emerges from the lower level behaviours selected for.

5 Concluding Remarks

We successfully evolved *situated* agents with fixed weight dynamical neural controllers on an associative learning task requiring *re-learning.* The observed phenomena can be described as the ability to perform two different behaviours and appropriately switch between them when necessary using feedback from the interactions with the environment. The question of whether such experience-dependent behaviour is actually 'learning' is discussed in more depth in [6]. The dynamics of the coupled agent-environment is explored in some depth. Attempts to generate a FSM are unsuccessful but a form of 'interactive state machine' is provided instead. From the dynamics, two predictions are explored.

This work raises a number of issues we believe deserve to be further studied. First, in the case of a situated agent, how useful is the conventional distinction drawn between operant and classical conditioning? Our work suggests that the distinction arises from the discretisation of the task or the minimisation of the coupling between agent and environment. Second, in such 'representationally-hungry' tasks, correlations between the activity of internal components and that which the agent has to remember are trivial to spot. Could they be interpreted as symbols the agent can manipulate to perform computations? Further work unravelling what is meant by 'internal representations' from minimal model systems such as the one presented here should be of interest. Finally, an important next step will be to extend this work to an agent that can associate any temperature along a continuum with food, as is the case in the phenomena observed in *C. elegans* from which this task was abstracted.

References

1. Beer, R.D.: On the dynamics of small continuous-time recurrent neural networks. Adaptive Behavior 3(4), 469–509 (1995)
2. Blynel, J., Floreano, D.: Levels of dynamics and adaptive behavior in evolutionary neural controllers. In: Proc. of the 7th Int. Conf. on Simulation of Adaptive Behavior: From animals to animats, pp. 272–281. MIT Press, Cambridge (2002)
3. Fernando, C.: A situated and embodied model of classical and instrumental learning. Master's thesis, COGS, University of Sussex (2002)
4. Harvey, I.: Artificial evolution: a continuing SAGA. In: Gomi, T. (ed.) Evolutionary Robotics. From Intelligent Robotics to Artificial Life. LNCS, vol. 2217, Springer, Heidelberg (2001)
5. Hedgecock, E.M., Russell, R.L.: Normal and mutant thermotaxis in the nematode caenorhabditis elegans. Proc. Nat. Acad. Sci. USA 72(10), 4061–4065 (1975)
6. Phattanasri, P., Chiel, H.J., Beer, R.D.: The dynamics of associative learning in evolved model circuits. Adaptive Behavior (Submitted)
7. Tuci, E., Quinn, M., Harvey, I.: An evolutionary ecological approach to evolving learning behavior using a robot based model. Adaptive Behavior 10(3/4), 201–221 (2003)
8. Yamauchi, B.M., Beer, R.D.: Integrating reactive, sequential and learning behavior using dynamical neural networks. In: Cliff, D., Husbands, P., Meyer, J., Wilson, S. (eds.) From Animals to Animats 3: Proc. of the Third Int. Conf. on Simulation of Adaptive Behavior, pp. 382–391. MIT Press, Cambridge (1994)

Constructing the Basic *Umwelt* of Artificial Agents: An Information-Theoretic Approach

Philippe Capdepuy[1], Daniel Polani[1,2], and Chrystopher L. Nehaniv[1,2]

Adaptive Systems[1] and Algorithms[2] Research Groups
School of Computer Science, University of Hertfordshire
College Lane, Hatfield, Herts, AL10 9AB, UK
{P.Capdepuy,D.Polani,C.L.Nehaniv}@herts.ac.uk

Abstract. In the context of situated and embodied cognition, we evaluate an information-theoretic approach to the construction of the *basic Umwelt* of an artificial agent. We make the assumption that the construction of such a *basic Umwelt* is an emergent property of the coupling between the agent and its environment where the goal of the agent is to maximize its control abilities. An information-theoretic approach of the perception-action loop allows us to evaluate the capacity of the agent to inject information into its environment and to later recapture this information in its own sensors. We define a construction mechanism based on an automaton that generates internal states relevant to the agent in terms of perception-action loop. Optimizing this automaton leads to internal representations that can be a basis for the construction of the *basic Umwelt* of the agent. We illustrate the properties of the proposed mechanism in a simple example where an agent is acting in a box world. Simulation results show that this construction mechanism leads to a representation that captures important properties of the environment.

1 Introduction

Embodied and situated cognition are fundamental paradigms in the artificial life field whose endeavour is to provide a view of artificial and natural cognitive systems that is consistent with our current hypotheses about the underlying mechanisms of our world. Two of the strongest assumptions on which these paradigms rely on are (i) the locality of interactions between elements and (ii) the causal relationships between events in the flow of time. These assumptions have strong consequences on the perspective we have of cognitive agents. One such consequence is that an agent can only rely on information already present in its embodiment and on subsequent information acquired from experience.

In the embodied paradigm, we do not assume any preexisting information that would be accessible and meaningful to the agent, the meaning of things is created by the agent itself (see [8]). Approaches from biosemiotics and ecological psychology, especially the concepts of *Umwelt* (von Uexküll [7,6]) and *affordances* (Gibson [2]), are particularly relevant as they root the creation of meaning for an agent in its ability to interact with objects of the world. In [6], von Uexküll makes the following description of *Umwelt*:

F. Almeida e Costa et al. (Eds.): ECAL 2007, LNAI 4648, pp. 375–383, 2007.

> "We no longer regard animals as mere machines, but as subjects whose essential activity consists of perceiving and acting. We thus unlock the gates that lead to other realms, for all that a subject perceives becomes his perceptual world and all that he does, his effector world. Perceptual and effector worlds together form a closed unit, the *Umwelt*." (von Uekül, [6])

The closed unit consisting of the perceptual and effector worlds can also be referred to as the *perception-action loop*. One could then say that the *Umwelt* has to be an emergent property of the perception-action loop.

The possibility to attribute an *Umwelt* to an artificial being is still an active debate (see [1]). One of the difficulties is that the *Umwelt* contains also a purposive component, in the sense that the relevance is related to the goals of the agent. However in our approach we will use a weaker concept that we refer to as *basic or proto-Umwelt*. In this concept we consider a generic goal, which is to maximize control over the perception action-loop. Our aim is to evaluate a potential mechanism for constructing such a *basic Umwelt* that is relevant for a given agent-environment coupling. The main idea behind it is that this *basic Umwelt* is an emergent property of the coupling between the agent and its environment which is the result of an optimization of the information-processing capabilities of the agent with respect to its perception-action loop. More precisely, we strive at optimizing the ability of the agent to inject information into its future perceptions. This ability as been formalized in the information-theoretic framework is referred to as *empowerment* (see [4]). One of the hypothesis is that before being meaningful in terms of purpose or goal, actions have to be distinguished in terms of their effects. The *basic Umwelt* of an agent is therefore mainly about distinguishing the effect of different actions, which manifest themselves as regularities of the perception-action loop. This essential first step can be illustrated by the parcimonious use of limited information-processing abilities (e.g. limited memory) in order to distinguish relevant contexts, i.e. contexts which potentially give more control to the agent.

The paper is structured as follows: in Sec. 2 we describe the information-theoretic framework that allows us to describe and analyze properties of the perception-action loop. Section 3 details the principles behind the optimization mechanism that underlies the construction of the *basic Umwelt*. In Sec. 4 we show different experiments with artificial agents that show properties of the *basic Umwelt* constructed and how it structures the perception of the environment. Section 5 discusses some practical applications of the principle and how it can be extended to cope with more complex environments.

2 Methodology of Investigation

The methodology that we use to investigate the construction of a particular *basic Umwelt* is based on two principles. The first one aims at expressing the perception-action loop of artificial agents within an information-theoretic framework (see [4]). This framework allows us to express some intrinsic properties of the perception-action loop, more precisely the capacity of the actuation channel toward future perceptions. The second part of our methodology consists in defining a computational mechanism that maps the perceptions of the agent considering a bounded information-processing capacity. The aim of this mechanism is to optimize the efficiency of the actuation channel, i.e. to maximize the potential control an agent has onto its future perceptions by constructing relevant contexts.

2.1 The Information-Theoretic Picture of the Perception-Action Loop

Following the assumptions of embodied cognition, we will refer to the perception-action loop of the agent as a causal Bayesian network which describes the relationships between the environment, the sensors and the actuators of the agent. The perception-action loop can then be unrolled in time (see Fig. 1) and some of its properties can be assessed using information-theoretic tools such as empowerment (see [4]). The idea of empowerment is to measure how much information

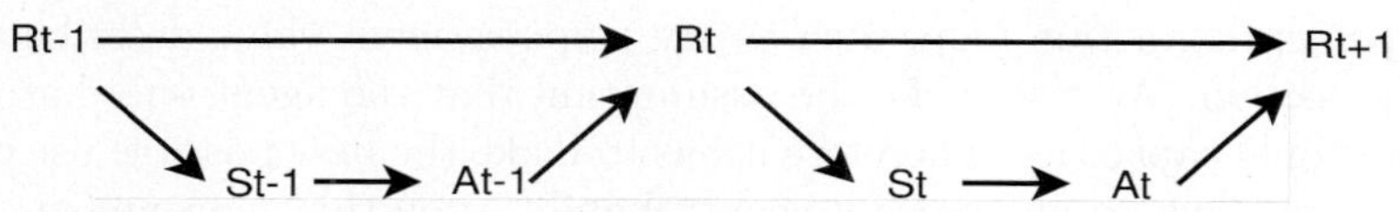

Fig. 1. Representation of the perception-action loop as a causal Bayesian network unrolled in time. R_t stands for the environment of the system, S_t is the sensor of the agent and A_t its actuator.

can be injected by an agent into its environment and then perceived by its sensors. More precisely it is defined as the channel capacity from the sequence of actions $A_t, A_{t+1}, \ldots, A_{t+n-1}$ to the perceptions S_{t+n} after a fixed number of time steps. The channel capacity is defined as the maximum mutual information between the sent message and the received message, where the maximization is made with respect to the probabilities for the sent message. Empowerment can then be written as

$$\mathfrak{E}(A_t, \ldots, A_{t+n-1} \rightarrow S_{t+n}) = \sup_{p(\overrightarrow{a})} I(A_t, \ldots, A_{t+n-1}; S_{t+n}) \tag{1}$$

with $p(\overrightarrow{a})$ being the probability distribution function of the action sequences (i.e. $p(\overrightarrow{a}) = Pr(A_t = a_t, \ldots, A_{t+n-1} = a_{t+n-1})$). Empowerment can be described as the maximum potential information an agent can transfer into its own sensors through the environment. The higher the empowerment the more information it can potentially inject.

We will also consider another measure of empowerment referred to as *context-dependent empowerment* which has been introduced in [4]. Here we mean by context the sensoric state before the action is performed. Context-dependent empowerment can be computed as

$$\begin{aligned}&\mathfrak{E}(A_t, \ldots, A_{t+n-1} \to S_{t+n}|S_t)\\ &= \sum_{s_t \in \mathcal{S}_t} p(s_t) \sup_{p(\vec{a})} I(A_t, \ldots, A_{t+n-1}; S_{t+n}|s_t)\end{aligned} \tag{2}$$

From a global perspective, empowerment can be used as a universal utility to guide the behaviour of an agent, but also to evolve the sensors or actuators of an agent in order to maximize its control abilities in a given environment. In the following sections we will be only considering 1-step empowerment, i.e. the channel going from the current action to the next perception.

2.2 *Basic Umwelt* as Maximization of Empowerment

We now come to one of the key points of our work. As described in the introduction, the *basic Umwelt* of an agent is an emergent relational property of the coupling between the agent and its surroundings. In our context, we consider such an emergent property as being strongly linked with intrinsic properties of the perception-action loop, such as the empowerment value described in the previous section. We also make the assumption that the agent only has limited computational capacities. Therefore it has to make the best possible use of these capacities in order to maximize its control abilities on the environment.

In our context, we assume that the basic control capacity of the agent is expressed by the context-dependent empowerment of the agent, i.e. the capacity of the actuation channel for each possible sensoric state, weighted by the probability to be in a given sensoric state. We then consider that the agent is able to use its computational resources in order to determine an internal state that can be used jointly with the sensoric state to define the current context (this approach has been introduced in [3]). This internal state M_t can be seen as a memory system which integrates previous perceptions to construct the context value (this setup is described in Fig. 2). It is this state and its correlation with particular situations of the agent that we will consider as the *basic Umwelt* of the agent.

To compute the context, we use an automaton whose states are the different possible contexts, and whose alphabet is made of all the possible perceptive state. The measure we use to evaluate the effectiveness of a particular automaton is the context-dependent empowerment of the agent where the context is defined as the current perceptive state and the current automaton state, it can be expressed as

$$\mathfrak{E}(A_t, S_{t+1}|S_t, M_t) = \sum_{s_t \in \mathcal{S}_t, m_t \in \mathcal{M}_t} p(s_t, m_t) \sup_{p(\vec{a})} I(A_t; S_{t+1}|s_t, m_t). \tag{3}$$

Our methodology consists in searching the space of possible automata to find the ones that maximize this value. Such an automaton can then be considered as extracting some information from the flow of perceptions that is relevant for

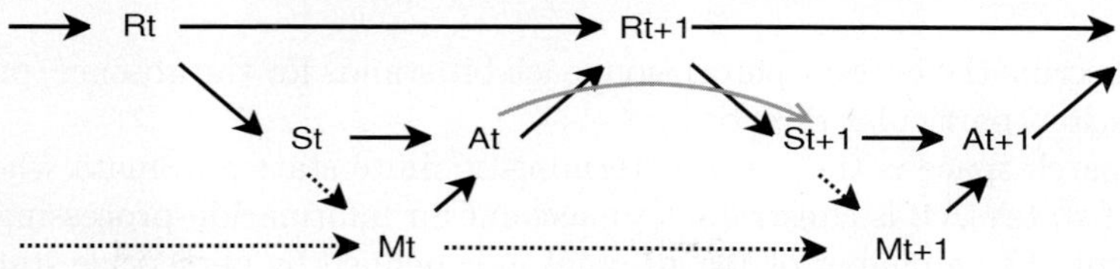

Fig. 2. Representation of the perception-action loop and the memory system as a causal Bayesian network unrolled in time. R_t stands for the environment of the system, S_t is the sensor of the agent and A_t its actuator. M_t is the context state constructed by the automaton. The search space of automata that we study defines the mapping corresponding to the dashed lines. The curved arrow represents the actuation channel that we want to optimize. It is the channel from the current action towards the next perception, knowing the current perception and the current state of the automaton.

the agent in terms of its control abilities on its own perception-action loop. As defined by the concept of *basic Umwelt*, this context information is an emergent property of the coupling between the agent and its surroundings. We introduce a computational constraint for the agent by limiting the number of possible states of the automaton.

3 Experiments

Starting from the formalization of the basic *basic Umwelt* described above, we investigate their properties through different simulated experiments. We first describe the main principles that underlie all the experiments, especially the search method used to find the automata that maximize the empowerment of the agent. We then introduce a first experiment in a box world and study the resulting *basic Umwelt* that is constructed.

3.1 Principles

As described in the previous section, our goal is to find the automaton that maximizes the empowerment of the agent when the state of the automaton is used as a context of the actuation channel. The fixed part of the problem is the coupling between the agent and its environment. Therefore we have to define the environment along with the sensoric and actuation mechanisms of the agent. In the following experiments, we consider that the world is a discrete grid where the agent can move. Each tile of a grid can be either empty, occupied by the agent, or occupied by a block. The actions that the agent is allowed to perform are: do nothing, move one tile east, west, north, or south. The agent collide with blocks, therefore it can only move toward an empty tile. If it tries to move to an already occupied tile, the result is that no movement is performed. The agent is sensing the presence of blocks in all possible movement directions (i.e. 4 bits are used to describe the current perception, each bit stands for the absence/presence of a block in a particular direction).

The search space is the set of deterministic finite state automata where the number of states $|\mathcal{M}|$ is constrained to account for information-processing limits of the agent. The alphabet of the automaton is defined by perceptive states, i.e. at each time step, the agent picks an action a_t and get a resulting sensoric state s_{t+1} after performing the action. The pair (a_t, s_{t+1}) is then used to iterate the automaton and hence to determine the next state.

To measure the fitness of a particular automaton, we use a Monte-Carlo method to obtain statistics about the perception-action loop. The world is initiated with the agent in a random position (unoccupied) and its automaton in a random state. Statistics are collected for a fixed number of iterations. At each iteration, the agent picks a random action and performs it, the state of the automaton is then changed according to the action picked and the resulting sensoric states. At the end of the iteration, statistics are updated with the collected sample. After all the iterations have been performed, the context-dependent empowerment of the agent is computed and it is used as the fitness value for this particular automaton.

The search method we use is a genetic algorithm. The fitness function the context-dependent empowerment of the agent when it uses the state automaton as a context, together with the current perception. The mutation operator changes a transition from a state and a sensoric state toward a random state. The crossover operator picks a random state as the crossover point and mixes the transition tables from the two parents; the resulting transition table is made from the transition table of parent 1 for all the states below the crossover point, and from the parent 2 for all the other states.

After the search has been completed (which is actually stopped when no improvement has occurred for a sufficiently long time), we run the agent using the best found automaton with a random action policy and we collect statistics about correlation between the state of the automaton and the position of the agent on the grid. This allows us to visualize the probability to be in each state when the agent is at a given position. This visualization gives us some insight about the nature of the information that is extracted by the automaton, i.e. it gives us a glimpse of the *basic Umwelt* constructed by the agent.

3.2 The Box World

The first experimental setup used is a 20 by 20 grid world, where the borders are occupied with blocks, hence resulting in a kind of box which the agent cannot leave because of the borders. Here we show results of a search for an automaton with $|\mathcal{M}| = 6$ where the best evolved fitness has a value of 0.126. The resulting mapping between position of the agent and state of the automaton is shown in Fig. 3. What we can observe in this visualization is that the automaton is extracting information that identifies the different borders of the box. Obviously each border has different properties in terms of perception-action loop, and they are encountered often enough for the agent to have an interest into identifying them as such in order to characterize its actuation abilities. During the evolutionary runs, many automata have been found that clearly identify the different borders, and also quite often they were more or less characterizing the proximity to a given border (as partially illustrated in this example by states 4 and 5). An interesting pattern is illustrated by the states 0 and 3. If we consider them superposed, we can see that being in one of these states is correlated with being in middle part of the box, i.e. not next to a border. But when considered separately they can characterize the proximity toward the top-right or bottom-left border. This pattern is not necessarily the best one possible, it appeared as a transient solution in many evolutionary runs, but as we do not know what is the global maximum of the fitness function, we can only consider this pattern as being a local maximum, which might or might not be preserved when reaching the global maximum.

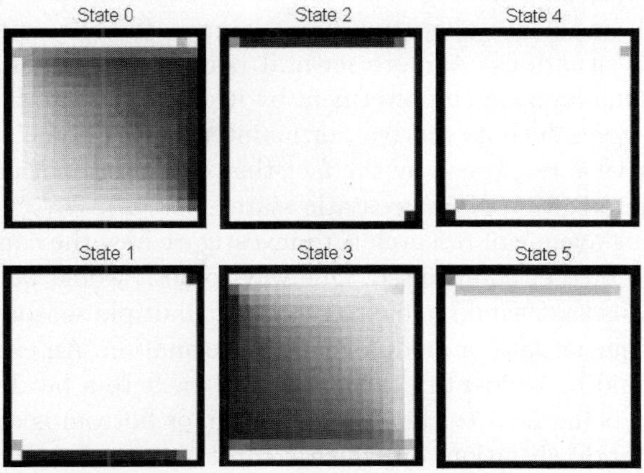

Fig. 3. Visualization of the mapping between states of the automaton and position of the agent in the box world experiment. Each picture shows the probability of being in the corresponding state when the agent is at a given position. Dark grey for a high probability, bright grey or white when the probability is close to 0. The black surrounding of each picture represents the walls of the box.

4 Discussion

In this work, we have shown that using an information-theoretic approach can be a successful avenue to investigate the construction of a *basic Umwelt* by an agent in a given environment. Our hypothesis is that if an agent can construct and use an internal state based on the history of perceptions and actions, finding the mechanism that would optimally determine this internal state with respect to the capacity of the actuation channel can lead to the creation of representations that are relevant for the agent (see also [5]). Such internal states, which we referred to as the *basic Umwelt* of the agent, are strongly dependent on the particular embodiment of the agent and on the nature of its environment. These states are an emergent property of the coupling between the agent and its surroundings.

One obvious criticism that can be made is that the agent has no goal, i.e. it is not trying to maintain its homeostasis, nor trying to collect resources for survival or to avoid hazards. Obviously such considerations should strongly shape the internal representation created by the agent, allocating more computational resources (i.e. more different states) to identify situations that are relevant with respect to these specific tasks (e.g. survival). Nevertheless we think that such a principle can be at the basis of the construction of the *basic Umwelt* of the agent. One strong advantage is that they can be used in a task-independent context, and still raise some interesting properties of the sensori-motor universe of the agent in a particular environment. With this perspective, taking into account a particular goal for the agent, such as a survival task, could be done by biasing the very same mechanism toward giving more importance to situations that are goal-related, and hence allocating more computational resources to identify these particular situations. A more natural perspective would be to consider that the goal of maximizing empowerment by itself can lead to the identification of fundamental goals such as survival, or maintenance of homeostatic variables. Such a perspective is reinforced by the fact that empowerment measures lead to a natural characterization of homeostatic states (see [3]).

An interesting avenue of research is to investigate how the constructed representation can be further compressed. One way to do it would be to identify sets of transformations that would allow to represent multiple sensori-motoric situations as one unique situation, modulo the transformation. An example of such a compression would be to identify the situations "next to a border" as only one, considering that being next to the right, left, top, or bottom is only a particular instance of a general situation, on which a rotation transformation would have been applied. An important point of this approach is that the rotation transformation, here loaded with our commonsense meaning, could be automatically identified without introducing this meaning from the outside of the system. The idea would be that the agent could identify particular instances of an abstract transformation scheme, then creating by itself the meaning of a rotation by identifying it as a particular class of sensori-motoric transformations. Obviously if such a transformation is applied to only one state then it does not bring any improvement in terms of compression, this approach is interesting only if the transformation found is generic enough to be applied to many different states. It is to be noted also that changes in the particular embodiment of the agent can make use of the symmetries of the environment. For instance, if we have used an agent with a direction and turning abilities, the distinction between the different sides of the box would not have emerged. Instead these different situations would have been classified as only one, however the orientation relative to the sides of the box would have appeared as an important factor. In future studies we compare automata generated by different embodiments and different environments.

References

1. Emmeche, C.: Does a robot have an Umwelt? Reflections on the qualitative biosemiotics of Jakob von Uexküll. Semiotica 134(1/4), 653–693 (2001), `http://www.nbi.dk/~emmeche/cePubl/2001d.robumwelt.html`
2. Gibson, J.J.: The Ecological Approach to Visual Perception. Lawrence Erlbaum Associates, New Jersey (1979)
3. Klyubin, A.: Organization of Information Flow Through the Perception-Action Loop. PhD thesis, School of Computer Science, University of Hertfordshire, UK (2007)
4. Klyubin, A.S., Polani, D., Nehaniv, C.L.: Empowerment: A universal agent-centric measure of control. In: Proceedings of the 2005 IEEE Congress on Evolutionary Computation, vol. 1, Sept. 2005, pp. 128–135. IEEE Computer Society Press, Los Alamitos (2005)
5. Polani, D., Nehaniv, C.L., Martinetz, T., Kim, J.T.: Relevant information in optimized persistence vs. progeny strategies. In: Artificial Life X: Proceedings of the Tenth International Conference on the Simulation and Synthesis of Living Systems, August 2006, pp. 337–343. MIT Press, Cambridge (2006)
6. Von Uexküll, J.: A stroll through the worlds of animals and men. In: Lashley, K. (ed.) Instinctive behavior, International Universities Press, New York (1934)
7. Von Uexküll, J.: The theory of meaning. Semiotica 42(1), 25–82 (1940)
8. Ziemke, T., Sharkey, N.: A stroll through the worlds of robots and animals: Applying jakob von uexkull's theory of meaning to adaptive robots and artificial life. Semiotica 134(1/4), 701–746 (2001), `citeseer.ist.psu.edu/ziemke00stroll.html`

Directed Evolution of Communication and Cooperation in Digital Organisms

David B. Knoester, Philip K. McKinley, Benjamin Beckmann, and Charles Ofria

Department of Computer Science and Engineering
Michigan State University
East Lansing, Michigan 48824
{dk, mckinley, beckma24, ofria}@cse.msu.edu

Abstract. This paper describes a study in the use of digital evolution to produce cooperative communication behavior in a population of digital organisms. The results demonstrate that digital evolution can produce organisms capable of distributed problem solving through interactions between members of the population and their environment. Specifically, the organisms cooperate to distribute among the population the largest value sensed from the environment. These digital organisms have no "built-in" ability to perform this task; each population begins with a single organism that has only the ability to self-replicate. Over thousands of generations, random mutations and natural selection produce an instruction sequence that realizes this behavior, despite continuous turnover in the population.

Key words: digital evolution, communication, cooperative behavior, natural selection, mutation, autonomic computing, biologically-inspired computing.

1 Introduction

The increasing interaction between computing technology and the physical world requires that systems with different characteristics and capabilities be able to reliably communicate, regardless of changing environmental conditions [1]. Similar to how living organisms have evolved remarkable methods (audible, visual, stigmergic) for communication, we can use *digital evolution* [2] to evolve communication strategies for distributed computing systems. By utilizing an evolutionary process that incorporates many of the hazards to communication (packet loss, node failure), solutions that would not otherwise be apparent to human designers may be discovered.

Our work uses the AVIDA platform for digital evolution [2] to investigate the evolution of cooperative communication behavior. In AVIDA, a population of self-replicating computer programs exists in a user-defined computational environment and is subject to instruction-level mutations and natural selection. Over thousands of generations, these "digital organisms" can adapt to and even thrive under extremely dynamic and adverse conditions. AVIDA has previously been used to conduct research in the evolution of biocomplexity, with an emphasis on the evolutionary design process in nature [3, 4]. However, digital evolution can also be used to address complex problems in science and engineering [5, 6], often revealing unexpected and clever solutions.

F. Almeida e Costa et al. (Eds.): ECAL 2007, LNAI 4648, pp. 384–394, 2007.

Biologically-inspired approaches and evolutionary computation have been applied to a variety of cooperative communication problems. Examples include mimicking the social behavior of insect colonies in robotic foraging [7]and using chemotaxis to facilitate robust network routing [8]. In addition, a variety of studies have been conducted to better understand the evolution of cooperation and communication. Examples include the evolution of a common vocabulary [9, 10], using the Prisoner's Dilemma to examine the evolution of cooperation [11], the effect of communication and indirect reciprocity on the evolution of cooperative strategies [12, 13, 14], and how information flow between agents is shaped by interaction with the environment [15, 16].

This study focuses on the evolution of a distributed problem solving task [17], specifically, where the population must determine the largest value sensed by any individual. Such behavior could provide a means for a population to perform leader election [18], or could be used to obtain the maximum sensed value in a wireless sensor network. Our results show that digital evolution can produce this behavior, and therefore has promise as a tool to be used in the design of future distributed computing systems. Like natural organisms, those systems will need to adapt to the environment, self-heal, and evade attackers. After reviewing the AVIDA system, we describe our experiments, present results, and analyze the dominant genome of a population that evolved the desired behavior. Finally, we present conclusions and discuss future work.

2 The AVIDA System

Figure 1 depicts an AVIDA population and the structure of an individual organism. Each digital organism comprises a circular list of instructions (its genome) and a virtual CPU, and "lives" in a common virtual environment. AVIDA instructions are similar in appearance to traditional assembly language instructions. They enable an organism to perform simple mathematical operations, such as addition, multiplication, and bit-shifts, as well as interact with the organism's environment, for example, by sending a message to a neighboring organism, or outputting a number to the environment. Instructions are executed by the organism's virtual CPU; the one used here contains three registers, two stacks, and four *heads*, which are similar to program and stack pointers [2].

The AVIDA environment comprises a number of *cells*, each of which can contain at most one organism; organisms cannot live outside of cells. Each cell has a circular list of directed *connections* to neighboring cells. These connections define the topology of the environment. A single connection, the *facing*, defines the orientation of the resident organism. The facing of a cell may be sensed and manipulated by the resident organism using the GET-FACING and ROTATE-{L,R} instructions, respectively. Finally, each cell in the environment has an associated identifier, a unique random 32-bit integer, termed the *cell-ID*. A resident organism may obtain its cell-ID via the GET-ID instruction. Organisms in AVIDA can communicate with each other by sending and receiving messages in the direction currently faced. If the sending organism is facing a neighboring organism, the message is deposited in that neighbor's inbox. If the sender was facing an empty cell, the message is lost. The recipient of the message must execute a RETRIEVE-MSG instruction to extract the message from its inbox. Organisms are not able to determine if they are facing an occupied cell, nor do we provide an explicit mechanism for them to identify neighbors (though it may be evolved).

A population starts with a single organism that is capable only of replication, and different genomes are produced through random mutations that occur during replication. The first step in replication is for the parent to allocate space for the offspring's genome. The parent then executes its "copy-loop," where instructions are copied individually from the parent's genome to the offspring's. Finally, the parent organism executes an H-DIVIDE instruction, creating the offspring. Each time an instruction is copied, a mutation may be introduced according to a predefined probability. These mutations may take the form of a replacement (substituting a random instruction for the one copied), an insertion (inserting an additional, random instruction into the offspring's genome), or a deletion (removing the copied instruction from the offspring's genome). When an organism replicates, a target cell that will house the new organism is selected from the environment. Different models to select this target cell are available, including MASS-ACTION (select at random from among all cells) and NEIGHBORHOOD (select from cells adjacent to the parent), among others. In every case, an organism that is already present in the target cell is *replaced* (killed and overwritten) by the offspring.

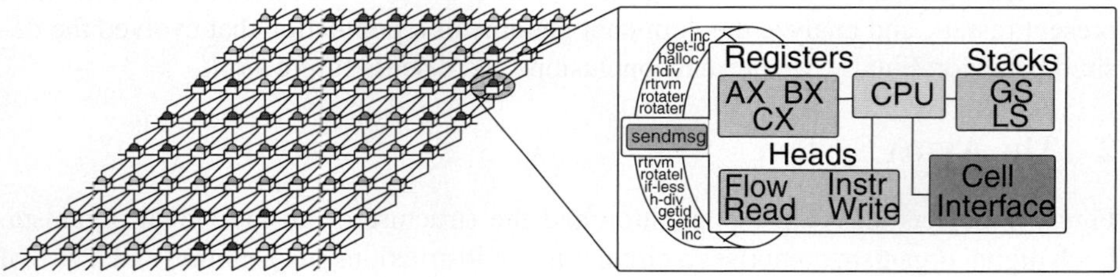

Fig. 1. An AVIDA population (left), and the structure of an individual organism (right).

During an AVIDA experiment, the *merit* of a given digital organism determines how many instructions its virtual CPU is allowed to execute in relation to other organisms, similar to a priority-based scheduling algorithm. Since digital organisms are self-replicating, a higher merit results in an organism that replicates more frequently, spreading throughout and eventually dominating the population. Unlike fitness in genetic programming, merit in AVIDA is not evaluated only at discrete time intervals, but rather updated asynchronously based upon performed *tasks*. Tasks are designed by the user and are used to reward desirable behavior (they may also punish undesirable behavior), thereby driving natural selection. Tasks are defined in terms of an organism's externally visible behavior, or *phenotype* (for example, messages that are sent), rather than in terms of CPU-level actions. This approach is intended to allow maximum flexibility in the evolution of a solution for a particular task. The solution might not be optimal when considering the task in isolation, but it may have other properties that make it well-suited for its environment – resilience to mutation, for example. Multiple tasks can be used in conjunction to reward complex behaviors. For example, one task may reward organisms for sending a message, while another may reward for a specific message payload. Rewards for performing multiple tasks are, by default, multiplicative.

3 Experiments and Results

We present three different sets of experiments. Each uses different combinations of AVIDA tasks, however all are focused on evolving the same behavior: proliferation of messages that carry the largest sensed value. For this study we use cell-IDs for the sensed values, thus the desired behavior is that all organisms send messages containing the largest cell-ID. Taking into account population turnover and mutations, we consider the solution to have been found when 95% of messages carry the largest cell-ID. The first set of experiments investigates the basic communication capabilities of digital organisms, focusing on the evolution of message filtering. The second set introduces a penalty, where each time that a digital organism sends a message that does not contain a cell-ID, the sender's merit is reduced. Finally, the third set of experiments investigates the ability of the population to recover from resetting the largest cell-ID. We note that organisms do not have an inherent ability to identify messages that contain a cell-ID; both the messaging behavior, as well as the grammar, must be evolved.

Experimental setup. For this study we configured AVIDA as follows. The environment comprises 3600 cells in a 60×60 torus. Experiments are run for 100,000 *updates*, a standard unit of time in AVIDA; an update averages 30 virtual CPU instructions per organism. The copy mutation rate is set to 0.75% per-instruction, while the insertion and deletion mutation rates are set to 5% per-replication; these parameters correspond to the default AVIDA configuration. We developed a set of tasks, summarized in Table 1, to reward organisms for various communication behaviors. To account for the stochastic nature of evolution, 20 separate AVIDA trials were performed for each experiment.

Table 1. Descriptions of the AVIDA tasks developed for this study.

Task Name	Description
SEND-SELF	Rewards sending a message containing the sender's cell-ID.
SEND-ID	Rewards sending a message containing *any* cell-ID.
MAX-KNOWN	Rewards sending a message containing the largest value known, defined as $Max(self, Max(msg_0, ...msg_n))$, where $\{msg_0, ...msg_n\}$ is the set of all messages received by that organism. The sender must have received at least one message prior to being rewarded for performing this task.
SEND-NON-ID	Penalizes the sender of a message that does not carry a cell-ID.

3.1 Filtering Messages

In the first set of experiments, we tested the hypothesis that rewarding organisms for sending messages containing cell-IDs, where the cell-ID carried is greater than the organism's own cell-ID, will eventually result in all messages in the population carrying the largest cell-ID. Experiments were conducted using different combinations of the tasks defined in Table 1. In every case, experiments that used the MAX-KNOWN task produced organisms sending messages containing cell-IDs greater than their own. However, none of these experiments resulted in the proliferation of the largest cell-ID.

Figure 2(a) depicts average messaging behavior for an experiment uses the MAX-KNOWN and SEND-ID tasks. Five different values are plotted over the previous 100 updates: Total, the total number of messages sent; Carry, the number of messages sent that carry a cell-ID; ID, the number of messages sent that carry the sender's cell-ID; and >ID, the number of messages sent that carry a cell-ID greater than the sender's. Here we see that more than half of all messages sent do not carry an ID, indicated by the difference between Total and Carry - these are "junk" messages, produced when organisms send values that are easy to calculate or when a SEND-MSG instruction has been mutated into the genome. We also see that greater than 75% of ID-carrying messages contain the sender's cell-ID. Finally, very few messages contain an ID that is greater than the sender's cell-ID. Figure 2(b) shows the average number of *organisms* that performed the MAX-KNOWN and SEND-ID tasks during the same AVIDA trials. Here we see that when MAX-KNOWN task is used in combination with the SEND-ID task, not only do all (allowing for genetic drift) organisms perform the SEND-ID task, but a large number of organisms (2600) also perform the MAX-KNOWN task. We note that organisms can perform the MAX-KNOWN task by sending their own ID once they have received any message carrying a smaller value.

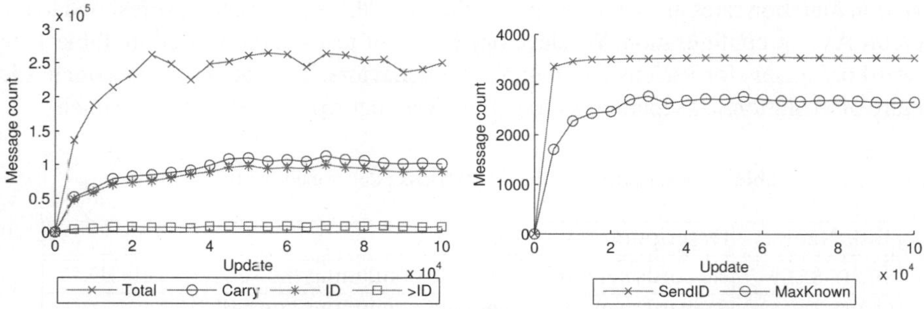

(a) Messaging Behavior (b) Task Performance

Fig. 2. Data filtering with MAX-KNOWN and SEND-ID tasks; average of 20 trials.

3.2 Encouraging ID-Carrying Messages

Our next experiments investigated ways to reduce the number of "junk" messages being sent, under the supposition that the large number of non-ID carrying messages might be preventing the population from determining the largest cell-ID. We tried two different approaches, one where we actively penalized organisms for sending junk messages, and another where we increased the cost (in virtual CPU cycles) of the SEND-MSG instruction. Both of these approaches resulted in the desired behavior, with the penalty evolving a solution more quickly than the additional cost approach. Here we discuss only the former; details of the cost experiments can be found in a technical report [19].

In this experiment, a task, SEND-NON-ID, was defined such that the sender of a message that does not carry a cell-ID is docked 75% of its merit. The SEND-NON-ID task is similar to an unseen predator, or hostile and unpredictable environment, in biological systems. We tried two different configurations with SEND-NON-ID, one that included MAX-KNOWN and SEND-ID, and another that included MAX-KNOWN and SEND-SELF. Initial experiments that used the SEND-NON-ID penalty performed similarly to those described in Section 3.1. However, when we also changed the replacement strategy from MASS-ACTION to NEIGHBORHOOD performance improved dramatically. The reason for this improvement is related to *kin selection* [6], which occurs when parent and offspring work together on cooperative tasks. In this case, parent and offspring are genetically similar, and thus likely to cooperate on the rewarded tasks, while avoiding the SEND-NON-ID penalty. We note that NEIGHBORHOOD replacement alone, without using either a penalty or cost, did not achieve the desired behavior.

Figure 3 shows messaging behavior using the tasks MAX-KNOWN, SEND-ID, and the penalty SEND-NON-ID. In addition to the values plotted in Figure 2(a), we also plot MaxID, the number of messages sent that carry the largest cell-ID in the population. Figure 3(a) shows the average messaging behavior of 20 different trials, and Figure 3(b) is a detail of a single trial that shows improvement in the types of messages present in the population. Here we see that the number of junk messages has been dramatically reduced, and that the number of messages containing IDs greater than that of the sender is increasing, although slowly. Still, very few messages contain the largest cell-ID.

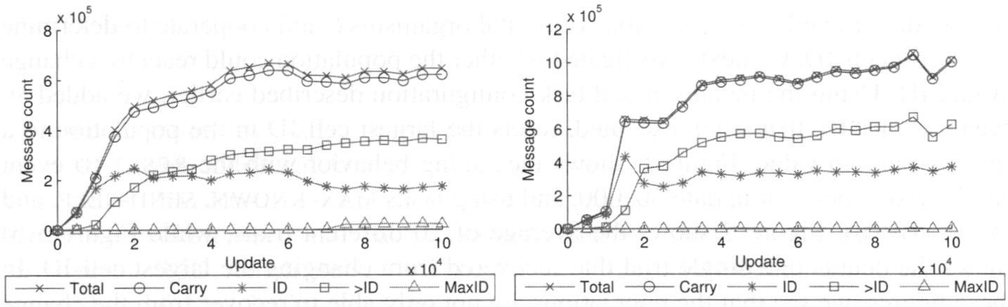

(a) Average of 20 Trials (b) A Sample Trial

Fig. 3. Messaging behavior with a penalty.

Figure 4 shows messaging behavior using the tasks MAX-KNOWN, SEND-SELF, and the penalty SEND-NON-ID. Figure 4(a) shows the average behavior of 20 different trials. For the first time, we see evidence of the convergence of message types, where the number of messages carrying an ID greater than the sender's approaches the total number of messages sent. However, we observe that due to the SEND-SELF task, each organism sends its own ID at least once. We also see a significant number of messages that carry the largest cell-ID. Figure 4(b) plots details of a particular trial where nearly all messages contain cell-IDs greater than that of the sender. Moreover, those containing the largest cell-ID (MaxID) represent 98.3% of all sent messages. It is this behavior, where nearly all sent messages converge to the largest cell-ID, that we sought. We note

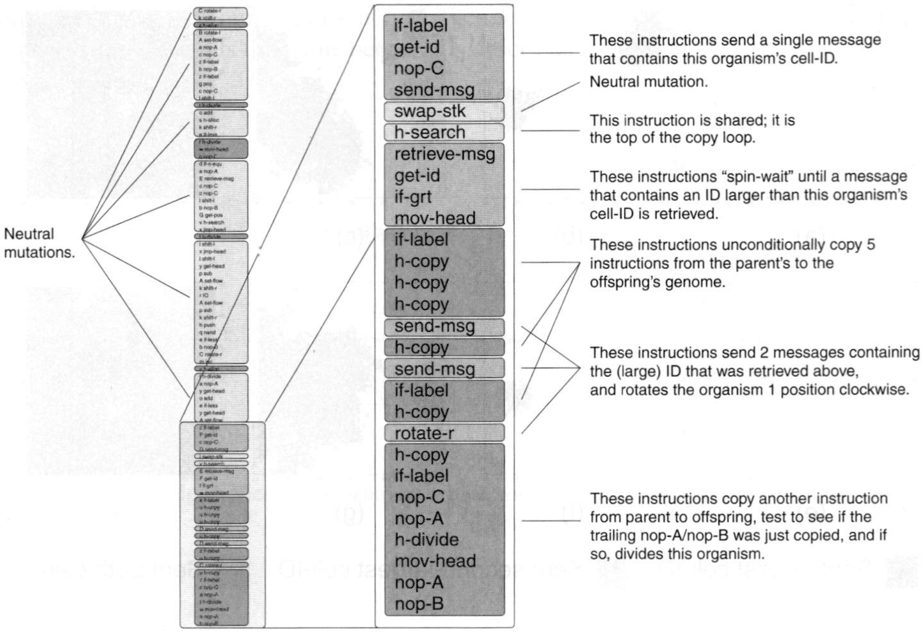

Fig. 7. Dominant genome sending messages that carry the largest cell-ID.

It should be noted that one of the forces at work in evolving this behavior is the natural selection of organisms that do *not* perform the SEND-NON-ID task. As soon as the RESET-ID event is triggered, any organism that sends the original largest cell-ID is subject to the penalty for sending a junk message. Even in the absence of an explicit penalty, organisms that send the original largest cell-ID would still not receive the reward for the MAX-KNOWN task. It is these *selective pressures* that are primarily responsible for the distribution of the largest cell-ID. Moreover, an organism cannot be rewarded for sending a message containing the new largest cell-ID without first having been sent that ID in a message (unless, of course, the organism lives in that cell). In other words, to survive a change in the largest ID, the organisms depend on cooperation.

4 Conclusions and Future Work

We have demonstrated that digital evolution can produce populations capable of distributed problem solving, specifically distributing the largest cell-ID among the population. Further, we have shown that in the presence of selective pressures, populations of organisms are able to recover from changes in their environment, and that this behavior emerges from simple localized interactions between neighboring organisms. In other work, we have used group selection to evolve organisms that perform leader election by identifying unique characteristics of individuals [20]. We are also using AVIDA to study the evolution of other distributed operations, such as data gathering, and to evolve organisms that generate UML state diagrams for dynamically adaptive systems. Finally, we have recently developed an instruction set that includes simple motor control primitives and sensors. We expect to use this platform to evolve individuals that can traverse obstacle courses, elude predators, and catch moving targets.

Further Information. Papers on digital evolution and the AVIDA software are available at `http://devolab.cse.msu.edu`. Information on evolving cooperative behavior can be found at `http://www.cse.msu.edu/thinktank`.

Acknowledgments. The authors gratefully acknowledge the contributions of the following individuals to this work: Jeffrey Clune, Sherri Goings, David Bryson, Richard Lenski, Heather Goldsby, and Betty Cheng. This work was supported in part by the U.S. Department of the Navy, Office of Naval Research under Grant No. N00014-01-1-0744, U.S. National Science Foundation grants EIA-0130724, ITR-0313142, and CCF 0523449, and a Quality Fund Concept grant from Michigan State University.

References

1. McKinley, P.K., Sadjadi, S.M., Kasten, E.P., Cheng, B.H.C.: Composing adaptive software. IEEE Computer 37(7), 56–64 (2004)
2. Ofria, C., Wilke, C.O.: Avida: A software platform for research in computational evolutionary biology. Journal of Artificial Life 10, 191–229 (2004)
3. Lenski, R.E., Ofria, C., Pennock, R.T., Adami, C.: The evolutionary origin of complex features. Nature 423, 139–144 (2003)
4. Adami, C., Ofria, C., Collier, T.C.: Evolution of biological complexity. Proceedings of the National Academy of Sciences 97, 4463–4468 (2000)
5. Lenski, R.E., Ofria, C., Collier, T.C., Adami, C.: Genome complexity, robustness, and genetic interactions in digital organisms. Nature 400, 661–664 (1999)
6. Goings, S., Clune, J., Ofria, C., Pennock, R.T.: Kin selection: The rise and fall of kin-cheaters. In: Proceedings of the International Conference on Artificial Life (ALIFE), Boston, MA, USA, Sept. 2004, pp. 303–308 (2004)
7. Dorigo, M., Trianni, V., Şahin, E., Groß, R., Labella, T.H., Baldassarre, G., Nolfi, S., Deneubourg, J.L., Mondada, F., Floreano, D., Gambardella, L.M.: Evolving self-organizing behaviors for a swarm-bot. Autonomous Robots 17(2–3), 223–245 (2004)
8. Babaoglu, O., Canright, G., Deutsch, A., Caro, G.A.D., Ducatelle, F., Gambardella, L.M., Ganguly, N., Jelasity, M., Montemanni, R., Montresor, A., Urnes, T.: Design patterns from biology for distributed computing. ACM Transactions on Autonomous and Adaptive Systems 1(1), 26–66 (2006)
9. Baronchelli, A., Dall'Asta, L., Barrat, A., Loreto, V.: Strategies for fast convergence in semiotic dynamics. In: Proceedings of the International Conference on Artificial Life (ALIFE), pp. 480–485 (2006)
10. Marocco, D., Nolfi, S.: Self-organization of communication in evolving robots. In: Proceedings of the International Conference on Artificial Life (ALIFE), pp. 178–184 (2006)
11. Baranski, B., et al.: High-order punishment and the evolution of cooperation. In: Proceedings of the International Genetic and Evolutionary Computation Conference (GECCO), Seattle, Washington, USA, pp. 379–380 (2006)

12. McPartland, M., Nolfi, S., Abbass, H.A.: Emergence of communication in competitive multi-agent systems: a pareto multi-objective approach. In: Proceedings of the International Genetic and Evolutionary Computation Conference (GECCO), Washington DC, USA, pp. 51–58 (2005)
13. Eiben, A.E., Schut, M.C., Toma, T.: Comparing multicast and newscast communication in evolving agent societies. In: Proceedings of the International Genetic and Evolutionary Computation Conference (GECCO), Washington DC, USA, pp. 75–81 (2005)
14. Marocco, D., Cangelosi, A., Nolfi, S.: The role of social and cognitive factors in the emergence of communication: Experiments in evolutionary robotics. Philosophical Transactions of the Royal Society London – A 361, 2397–2421 (2003)
15. Klyubin, A.S., Polani, D., Nehaniv, C.L.: Tracking information flow through the environment: Simple cases of stigmergy. In: Proceedings of the International Conference on Artificial Life (ALIFE), pp. 563–568 (2004)
16. Lungarella, M., Sporns, O.: Mapping information flow in sensorimotor networks. PLoS: Computational Biology 2(10), e144 (2006)
17. Smith, R.G., Davis, R.: Frameworks for cooperation in distributed problem solving. In: Distributed Artificial Intelligence, pp. 61–70. Morgan Kaufmann, San Francisco (1988)
18. Lynch, N.A.: Distributed Algorithms. Morgan Kaufmann, San Francisco (1997)
19. Knoester, D.B., McKinley, P.K., Beckmann, B., Ofria, C.: Evolution of leader election in populations of self-replicating digital organisms. Technical Report MSU-CSE-06-35, Computer Science and Engineering, Michigan State University, East Lansing, Michigan (December 2006)
20. Knoester, D.B., McKinley, P.K., Ofria, C.: Using group selection to evolve leadership in populations of self-replicating digital organisms. In: Proceedings of the International Genetic and Evolutionary Computation Conference (GECCO), London, UK, July (2007)

Evolution of Acoustic Communication Between Two Cooperating Robots

Elio Tuci and Christos Ampatzis

CoDE-IRIDIA, Université Libre de Bruxelles, Bruxelles, Belgium
{etuci,campatzi}@ulb.ac.be

Abstract. In this paper we describe a model in which artificial evolution is employed to design neural mechanisms that control the motion of two autonomous robots required to communicate through sound to perform a common task. The results of this work are a "proof-of-concept": they demonstrate that evolution can exploit a very simple sound communication system, to design the mechanisms that allow the robots cooperate by employing acoustic interactions. The analysis of the evolved strategies uncover the basic properties of the communication protocol.

1 Introduction

This paper is about the evolution of acoustic communication in a two robot system, in which the agents are required to coordinate their efforts to perform a common task (see Sec. 2). The robots' mechanisms are determined by design methods referred to as Evolutionary Robotics (see [6]). That is, an artificial evolutionary process sets the parameters of neural networks controllers. The latter are in charge of the robots' actions by setting the states of the agents' actuators. Although from a different perspective and with different motivations, the issue of the evolution of acoustic communication has already been investigated in several research works. Some of these works model aspects of the evolution of communication in living organisms (see [7,4,10]). Other studies aim to engineer acoustic communication systems that improve the effectiveness of the robots' collective responses (see [9,8,1]). Either biologically or engineering inspired these studies exploit the properties of the evolutionary robotics approach in which the designer is not required to make strong assumptions about the essential features on which social interactions are based—e.g., assumptions concerning what communication is and about the requirement of individual competences in the domain of categorisation and naming. The results of the evolutionary process (i.e., the behaviour of the robots and the underlying mechanisms) inform the designer on the effects that the physical interactions among embodied agents and their world have on the evolution of individual behaviour and social skills.

Following this line of investigation, our work aims to demonstrate the effectiveness of a very simple sound signalling system in a context in which the robots are demanded to share individual experiences to build a common perspective of their world. The robots can communicate by using an extremely simple binary

F. Almeida e Costa et al. (Eds.): ECAL 2007, LNAI 4648, pp. 395–404, 2007.

signalling system (i.e., ON/OFF). As far as we know, this is the first study that investigates a communication scenario in which a bi-directional interaction is required by the robots to accomplish a common goal. Communication is based on the emission in time of asynchronous and mutually determined single tone signals. The results of this work should be taken as a "proof-of-concept" concerning the potentiality of the proposed approach to the design of acoustic communication mechanisms in multi-robot systems. We demonstrate that it is possible to use evolution to define the mechanism underlying a bi-directional communication protocol based on a very simple acoustic system.

2 The Task

The robot environment is a rectangular arena–120cm by 50cm—divided into two equal sides by a horizontal bar that revolves—i.e., the revolving door. There are three lights L_1, L_2 and L_3. When L_1 and L_2 are turned on, L_3 is turned off and vice versa. L_1 can only be seen by a robot located in the lower side of the arena while L_2 can only be seen by a robot located in the upper side of the arena. L_3 can be seen from anywhere in the environment. The arena floor is white except in the proximity of L_1 and L_2 up to a distance of 15cm from the lights, where the floor is painted in black or grey. The robots can experience four different combinations of black and grey zones (see Fig. 1). The type of environment in which the robots are located is labelled according to the combination of the colour of the floor in the two painted zones. In detail, the environments are labelled E_{xx}, where the first digit corresponds to the colour of the floor in the proximity of L_1 and the second digit to the colour of the floor near L_2. Grey colour corresponds to 0, while black colour corresponds to 1. The four types of environment are: E_{10}, E_{01}, E_{00}, and E_{11}. The revolving door rotates from the horizontal to the vertical position if simultaneously pushed by both robots in the

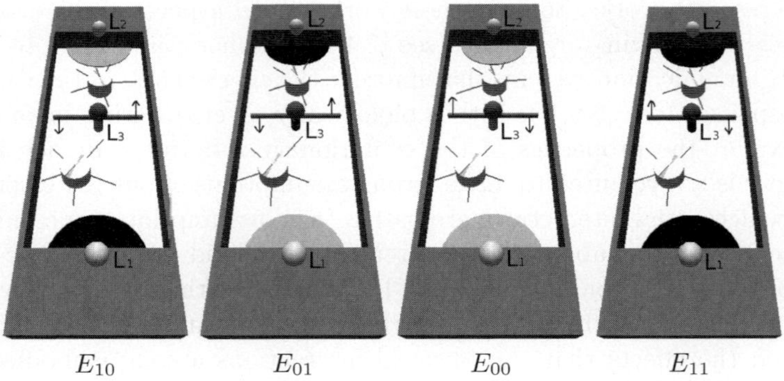

Fig. 1. The four environments E_{10}, E_{01}, E_{00}, and E_{11}. L_1, L_2 and L_3 refer to the lights. The revolving door is indicated by the horizontal bar in the centre of the arena. In each environment, the arrows indicate the direction in which the door revolves. The cylinders with spikes on the white floor represent the robots.

proper direction. Pushing forces exerted by a single robot on the revolving door are not enough to open it. The direction of rotation changes according to the type of environment. The robots have to exert forces to make the door rotate (a) clockwise, if they are located in E_{00} or in E_{11}; (b) anticlockwise, if located in E_{10} or in E_{10} (see the arrows in Fig. 1).

At the beginning of the first trial and in those that follow an unsuccessful one, the robots are randomly placed in the proximity of L_3. In trials following a successful one, the robots are not repositioned. The sequence of desired actions that each robot is demanded to carry out during a trial can be decomposed into two phases. At the beginning of the first phase, L_1 and L_2 are turned on, the revolving door is in the horizontal position and the colour of the floor in the proximity of L_1 and L_2 is set according to the type of environment that characterises the trial. During this phase, the robots are required to find the painted zone in their side of the white arena floor and remain for at least 6s on the painted zone. This exploration is facilitated by the presence of the lights that can be used as beacon (i.e., L_1 for the robot located in the lower side of the arena and L_2 for the robot located in the upper side of the arena). The first phase terminates once the 6s on the painted zones are elapsed for both robots. At this point, L_1 and L_2 are turned off, L_3 turned on, and the second phase begins. In the second phase, the two robots are required to move back towards the middle of the arena, approach the revolving door, and simultaneously push the door in order to open it and to reach the previously inaccessible opposite side of the arena. As mentioned above, the direction of rotation changes according to the type of environment. Therefore, to rotate the revolving door from the horizontal towards the vertical position the robots are required to "tell" each other the colour of the floor in the proximity of light—L_1 or L_2—previously approached. A trial successfully terminates once both robots, by rotating the revolving door, move into the opposite side of the arena, and reach a distance of 24 cm from L_3. At the end of a successful trial, L_3 is turned off, L_1 and L_2 are turned on, the rotating door automatically returns to the horizontal position and a new trial begins. A trial is considered unsuccessful, if a single robot exerts forces in both arms of the revolving door (i.e., west and east of L_3). This behaviour, referred to as trial-and-error strategy, is penalised by the fitness function (see Sec. 4).

Note that this task requires coordination of actions, cooperation and communication between the robots in order to successfully open the revolving door. For each robot, the perception of a grey or black floor can be associated both to a clockwise and anticlockwise rotational movement of the revolving door. Only the combination of the two coloured zones unambiguously identifies a rotational movement. Since a robot can only walk on a single zone per trial, the task can be successfully accomplished in all the environmental conditions only by a group of robots that communicate through sound. Without communication, a single robot can only exploit a trial-and-error strategy. By using a simple sound signalling system the robots should inform each other on the colour of the floor in the proximity of the light they perceive—L_1 or L_2—and consequently push the door in the proper direction as explained above.

3 Methods

The robot and its world are simulated using simulation software based on Open Dynamic Engine (see `http://www.ode.org/`), a 3D rigid body dynamics simulator that provides primitives for the implementation of detailed and realistic physics-based simulations. Our simulation models some of the hardware characteristics of the real *s-bots*. The *s-bots* are small wheeled cylindrical robots, 5.8 cm of radius, equipped with a variety of sensors, and whose mobility is ensured by a differential drive system (see [5]). Our simulated robot has a differential drive motion provided by a traction system composed of four wheels: two lateral, motorized wheels and two spherical, passive wheels placed in the front and in the back, which serve as support. The four wheels are fixed to the cylindrical body that holds the sensors. In particular, robots make use of 5 infrared sensors IR_i, two ambient light sensors AL_i, a floor sensor FS, a loudspeaker SO to emit sound and an omni-directional sound sensor SI to perceive sound (see Fig. 2a). Light levels change as a function of the robot's distance from the lamp. FS, placed underneath the robot, detects the level of grey of the floor. It outputs the following values: 0 if the robot is positioned over white floor; 0.5 if the robot is positioned over grey floor; 1 if the robot is positioned over black floor. SO produces a binary output (on/off). SI has no directionality and intensity features. 10% uniform noise is added to IR_i and AL_i readings, the motor outputs and the position of the robot.

The controller of each agent is composed of two modules referred to as M_C and M_M (see Fig. 2b). The modularisation is hand-coded to facilitate the evolution of successful behavioural strategies. M_C is a non-reactive module, that is a six neurons fully connected continuous time recurrent neural network (CTRNN,

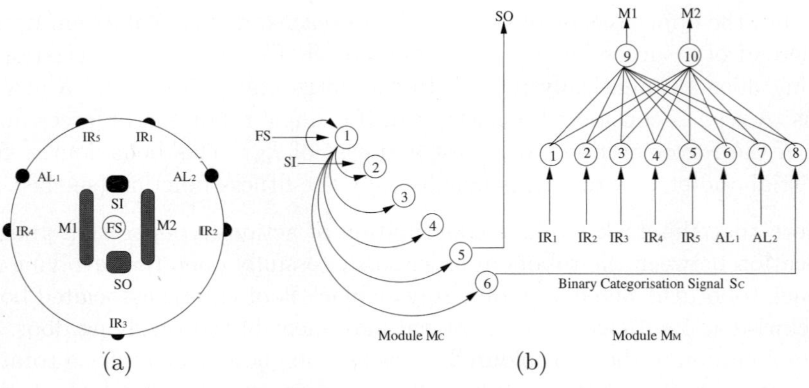

Fig. 2. (a) The simulated robot. $IR_i, i \in [1,5]$ are the infrared sensors; $AL_i, i = [1,2]$ are the ambient light sensors; FS is the floor sensor; SI is the sound sensor (i.e., the microphone); SO is the sound actuator (i.e., the loudspeaker); M_1 and M_2 are respectively the left and right motor. (b) The network architecture: module M_C and module M_M. For M_C only the efferent connections for one neuron are drawn. S_C is the binary categorisation signal sent, at each updating cycle, by M_C to M_M.

see also [2]). M_C is required to detect in which type of environment the robot is currently located. The categorisation has to be based on the FS's readings of both robots. Thus, it demands communication between the agents. For this reason, M_C takes input from FS and SI and it outputs the state of the SO and S_C (i.e., the binary categorisation signal). In other words, at every updating cycle, M_C is in charge of (a) managing sound by producing the signal the robot emits and by receiving the signal of either robot, and (b) "informing" M_M on the type of environment in which the robot is currently located by setting the value of the binary categorisation signal S_C either to 0 or 1. M_M is a reactive module, that is a feed-forward artificial neural network made of eight sensory neurons and two output neurons. M_M is demanded to (a) guide the robot avoiding collisions with the arena walls, and (b) "parse" the value of S_C to determine in which side to push the revolving door (i.e., anticlockwise if current trial in E_{10} or E_{01}, clockwise if current trial in E_{00} or E_{11}, see also Fig. 1). M_M takes input from IR_i, $i \in [1,5]$, from AL_i, $i = [1,2]$, and S_C, and it outputs the speed of the robot's wheels. The following associations (a) $S_C = 1$, robots located in E_{10} or E_{01}, anticlockwise rotational direction of the revolving door, and (b) $S_C = 0$, robots located in E_{00} or E_{11}, clockwise rotational direction of the revolving door, are determined *a priori* by the experimenter (see Sec. 4). The neural mechanisms and the communication protocol required by the robots to build these relationships from the sensors' readings are set by evolution.

The states of the neurons of M_C and M_M are governed by the equations (1) and (2) respectively:

$$\frac{dy_i}{dt} = \frac{1}{\tau_i}\left(-y_i + \sum_{j=1}^{6} \omega_{ji}\sigma(y_j + \beta_j) + gI_i\right), \ i \in [1,6]; \ \sigma(x) = \frac{1}{1+e^{-x}} \tag{1}$$

$$dy_i = \begin{cases} -y_i + gI_i & i \in [1,8] \\ \left(-y_i + \sum_{j=1}^{8} \omega_{ji}\sigma(y_j + \beta)\right) & i \in [9,10]; \end{cases} \tag{2}$$

where, using terms derived from an analogy with real neurons, y_i represents the cell potential, τ_i is the decay constant, g is a gain factor, I_i the intensity of the sensory perturbation on sensory neuron i, ω_{ji} the strength of the synaptic connection from neuron j to neuron i, β the bias term, $\sigma(y_j + \beta)$ the firing rate. The parameters ω_{ji}, τ, β and g are genetically encoded. Cell potentials are set to 0 any time the network is initialised or reset, and circuits are integrated using the forward Euler method with an integration step-size of $dt = 0.1$. Note that the cell potentials of M_M's neurons do not depend on time (see equation (2)). That is, the neurons' decay constant τ is set to 0.1, as the integration step-size dt. In M_C, the cell potentials y_i of the 5th and the 6th neuron, mapped into [0,1] by a sigmoid function σ, set the state of the robot's sound actuator SO and of the binary categorisation signal S_C. The robot emits a sound if $SO \geq 0.5$. $S_C = 1$ if $\sigma(y_6 + \beta_6) \geq 0.5$ otherwise $S_C = 0$. In M_M, the cell potentials y_i of the 9th and the 10th neuron, mapped into [0,1] by a sigmoid function σ and then linearly scaled into $[-6.5, 6.5]$, set the robot motors output.

A simple generational genetic algorithm is employed to set the parameters of the networks [3]. The population contains 80 genotypes. Generations following the first one are produced by a combination of selection with elitism, recombination and mutation. For each new generation, the three highest scoring individuals ("the elite") from the previous generation are retained unchanged. The remainder of the new population is generated by fitness-proportional selection (also known as roulette wheel selection) from the 64 best individuals of the old population. Each genotype is a vector comprising 67 real values, chosen uniformly random from the range $[0, 1]$. The first 18 genes are used to set the parameters of M_M (i.e., 16 connection weights, 1 bias term and 1 gain factor both shared by all the input neurons). The other 49 genes are used to set the parameters of M_c (i.e., 36 connection weights, 6 decay constants, 6 bias terms, and 1 gain factor). More details on the genetic algorithm and on the genotype-networks mapping can be found at `http://iridia.ulb.ac.be/supp/IridiaSupp2007-005`.

4 The Fitness Function

During evolution, each genotype is translated into a robot controller (i.e., modules M_C and M_M see Sec. 3), and cloned in each agent. Then, the two robot group is evaluated two times in each environment type E_{11}, E_{00}, E_{01}, and E_{10}, for a total of eight trials. Note that the sequence order of the environment type experienced by the robots—randomly chosen at the beginning of each generation—has a bearing on the overall performance of the group since the robots' controllers are reset only at the beginning of the first trial. Each trial differs from the others in the initialisation of the random number generator, which influences the robots' starting position and orientation anytime the robots are positioned, and the noise added to motors and sensors. The robots are randomly placed in the arena at the beginning of the first trial and repositioned in subsequent trials following an unsuccessful one. Within a trial, the robots life-span is 90 simulated seconds (900 simulation cycles). A trial is terminated earlier in case a robot crashes with the arena walls, or if the group successfully accomplishes its task. For each trial $e \in [1, 8]$, the group is rewarded by an evaluation function which seeks to assess the ability of the robots to open the revolving door located at the centre of the arena (see Sec. 2). This requires the robots to be able to determine the nature of the environment (i.e., E_{11}, E_{00}, E_{01}, or E_{10}) by using acoustic communication. The final fitness F attributed to a group controlled by a specific genotype is the average group score over a set of eight trials. A detailed illustration of the fitness function can be found at `http://iridia.ulb.ac.be/supp/IridiaSupp2007-005`. Note that F doesn't refer anyhow to signalling behaviour. F rewards the robots for accomplishing the task as detailed in Sec. 2. However, due to the nature of the task, the robots can be successful only if they coordinate their actions using the sound signalling system. By leaving signalling behaviour out of the fitness function, we clean our model from preconceptions concerning what (i.e., semantics) and how (i.e., syntax) successful group communicates, and we let evolution determine the characteristics of the communication protocol.

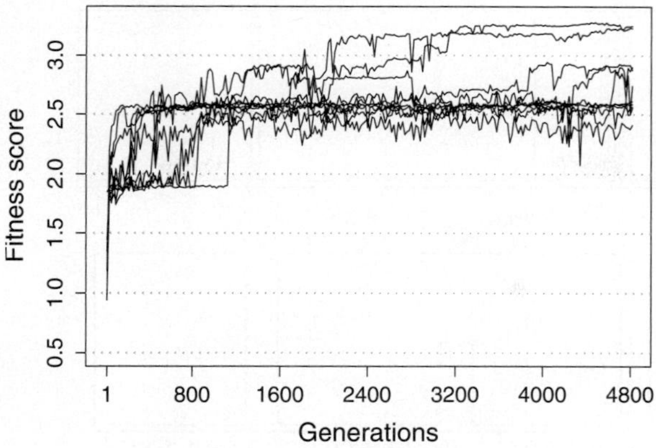

(a)

	(%) Success			
	E_{10}	E_{01}	E_{00}	E_{11}
g_1	0.0	0.0	100.0	100.0
g_2	99.6	99.8	100.0	98.16
g_3	0.0	0.0	97.8	96.3
g_4	100.0	100.0	100.0	99.6
g_5	0.0	0.0	79.7	82.5
g_6	99.0	94.2	92.7	0.0
g_7	99.5	99.5	100.0	0.0
g_8	100.0	100.0	0.0	0.0
g_9	0.0	0.0	100.0	99.8
g_{10}	100.0	99.7	0.0	0.0

(b)

Fig. 3. (a) Fitness F of the best groups at each generation of ten evolutionary runs. (b) Results of post-evaluation tests, showing for the best evolved groups of each run the (%) of successful trials in each type of environment. In grey the successful groups.

5 Results

Ten evolutionary simulations, each using a different random initialisation, were run for 4800 generations. Given the nature of the fitness function, the highest fitness score that a group can reach is 3.4. This score corresponds to the behaviour of a group in which each robot (i) finds the coloured zone on the white arena floor; (ii) communicates to the robot at the opposite side of the arena the colour encountered in its side; (iii) uses the combination of colours to properly set the binary categorisation signal S_C; and (iv) pushes the revolving door in the proper direction until it reaches the opposite side of the arena. Fig. 3a shows the fitness of the best groups at each generation for each evolutionary run. Notice that only two evolutionary runs managed to produce groups whose average fitness F is close to the maximum score. However, fitness scores lower than 3.4 might be associated to equally successful alternative strategies.[1] Thus, in order to have a better estimate of the behavioural capabilities of the best evolved controllers, we post-evaluate, for each run, the genotype with the highest fitness. These groups are referred to as $g_i, i \in [1, 10]$. The entire set of post-evaluations (i.e., 2400 trials, 100 evaluations for each permutation, 100*N! with N=4) should establish whether a group of robots is capable of accomplishing the task as described in Sec. 2 in all the four types of environment. The results of the post-evaluation tests are shown in Fig. 3b. The data confirm that only two groups g_2 and g_4 have a success rate higher than 98% in all four types of environment (see Fig. 3b, grey rows); g_1, g_3, g_5, g_8 and g_9 are capable of carrying out the task only when the door revolves clockwise, and g_{10} only when the door revolves anticlockwise; g_6

[1] Data not shown, movies of successful strategies, and further methodological details can be found at `http://iridia.ulb.ac.be/supp/IridiaSupp2007-005`.

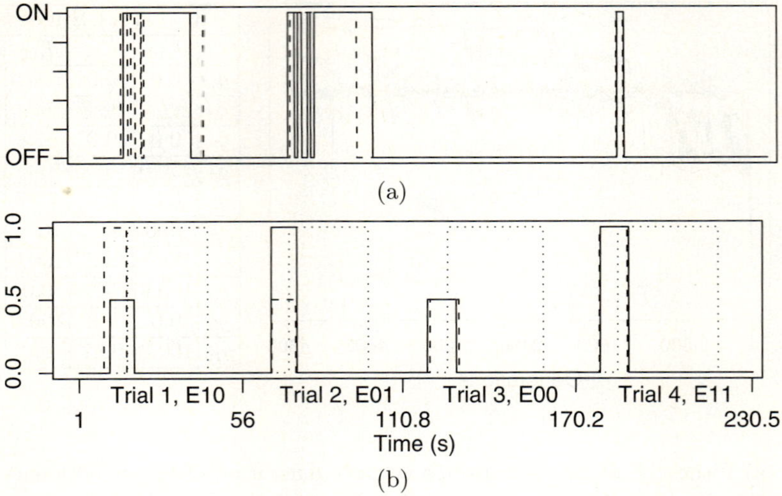

Fig. 4. Post-evaluations of group g_4. Dashed lines refers to the robot placed at the beginning of trial 1, in the upper side of the arena; continuous lines refer to the robot placed in the lower side of the arena. (a) Sound signals. (b) Floor sensors readings. Dotted line indicates the state of L_3, 1 = ON, 0 = OFF. On the x axis is indicated the time of start and end of each trial.

and g_7 fail in only one type of environment. From a behavioural point of view, the failure are due to trial-and-error strategy (data not shown, see footnote 1). That is, during the second phase of the task, both robots push the revolving door both west and east of L_3 instead of exerting forces directly on the proper side of the bar. Failure due to collisions are very rare. The lower success rate of g_{10} in E_{00} and E_{11} is mainly due to the fact that the robots of this group are not able to exert enough forces in order to rotate the revolving door (data not shown, see footnote 1). From a mechanism point of view, the failure of each single robot can be caused by either (a) M_C not capable of correctly categorising the environment by properly setting S_C as made explicit in Sec. 3 or (b) M_M not capable of "interpreting" the value of S_C as produced by M_C. Post-evaluation tests show that for almost all the unsuccessful groups it is M_C that by setting incorrectly the value of S_C, does not allow M_M to choose the correct direction of rotation of the revolving door (data not shown, see footnote 1). It seems that robots of unsuccessful groups are not capable of informing each other about the colour of the painted zone in the proximity of L_1 and L_2. Consequently, in the absence of an effective communication protocol, it turns out to be impossible for M_C to properly set S_C. In the following paragraphs, we analyse the communication protocol used by a successful group.

Fig. 4a illustrates the structures of signalling behaviour of the successful group g_4. In this post-evaluation test, the group undergoes 4 trials with the environment presented in the following sequence: E_{10}, E_{01}, E_{00}, and E_{11}. In each trial the robots don't emit sound before reaching the coloured zones. The perception of grey doesn't induce the emission of sound. Therefore, in E_{00} no robots emit sound (see Fig. 4 trial 3). The absence of sound in the environment lets M_C set S_C to 0 in both robots. $S_C = 0$ is correctly "interpreted" by M_M modules so that both robots push the revolving door clockwise. The perception of a black zone induces the robots to emit intermittent bursts of sound (see Fig. 4a trials 1, 2 and 4). In trials E_{10} and E_{01}, the perception of these intermittent bursts induces the robot that is on grey to emit a continuous tone. The perception of a continuous tone induces the robot on black to imitate its fellow, so that at the time when L_3 turns on (see Fig. 4b, dotted line) both robots emit a continuous tone. The presence of sound in the environment lets M_C set S_C to 1 in both robots. $S_C = 1$ is correctly "interpreted" by M_M modules so that both robots push the revolving door anticlockwise. Both robots autonomously stop emitting sound before the end of a trial in E_{10} or E_{01}, few seconds after the aperture of the revolving door. Thus, at the beginning of the following trial both robots are in the state of not emitting sound. In trials E_{11}, the asynchronous emission of intermittent bursts of sound by both robots determines moments of silence which inhibit signalling behaviour. At the time when L_3 turns on, none of the robots is signalling. The absence of sound in the environment lets M_C set S_C to 0 in both robots. $S_C = 0$ is correctly "interpreted" by M_M modules so that both robots push the revolving door clockwise as in E_{00}.

6 Conclusions

In this paper, we described a model in which artificial evolution is employed to design neural mechanisms that control the motion of autonomous robots required to communicate through sound to perform a common task. The results of this work are a "proof-of-concept": they demonstrate that evolution can exploit a simple sound system, detailed in Sec. 3, to design the mechanisms that allow two robots cooperate by using bi-directional acoustic interactions. Post-evaluation tests illustrate the nature of the robots' communication protocol based on entirely evolved asynchronous and mutually determined single tone signals. Concerning future work, we believe that priority should be given to investigations aimed to limit the amount of *a priori* assumptions that we have been forced to make in this first study. In particular, we are referring to the modularisation of the control structures and the arbitrary associations detailed in Sec. 3. In spite of this, we believe that the results are particularly encouraging. A complex syntax may emerge in scenarios in which semantic categories are linked to more articulated sensory-motor structures (e.g., neural structures that underpin object recognition processes rather than the perception of coloured zones).

Acknowledgements. This research work was supported by the *ECAgents* project (grant IST-1940), the *SWARMANOID* project (grant IST-022888), and the "ANTS" project, an "Action de Recherche Concertée" funded by the Scientific Research Directorate of the French Community of Belgium. The information provided is the sole responsibility of the authors and does not reflect the Community's opinion. The Community is not responsible for any use that might be made of data appearing in this publication. The authors thank Carlo Pinciroli and their colleagues at IRIDIA for stimulating discussions and feedback during the preparation of this paper.

References

1. Ampatzis, C., Tuci, E., Trianni, V., Dorigo, M.: Evolution of signalling in a group of robots controlled by dynamic neural networks. In: Sahin, E., Spears, W.M., Winfield, A.F.T. (eds.) Proc. 2nd Int. Workshop on Swarm robotics, Berlin, Germany, vol. 4433, pp. 173–188. Springer, Heidelberg (2006)
2. Beer, R.D., Gallagher, J.C.: Evolving dynamic neural networks for adaptive behavior. Adaptive Behavior 1(1), 91–122 (1992)
3. Goldberg, D.E.: Genetic Algorithms in Search, Optimization and Machine Learning. Addison-Wesley, Reading (1989)
4. Marocco, D., Nolfi, S.: Origins of communication in evolving robots. In: Nolfi, S., Baldassarre, G., Calabretta, R., Hallam, J.C.T., Marocco, D., Meyer, J.-A., Miglino, O., Parisi, D. (eds.) SAB 2006. LNCS (LNAI), vol. 4095, pp. 789–803. Springer, Heidelberg (2006)
5. Mondada, F., Pettinaro, G.C., Guignard, A., Kwee, I.V., Floreano, D., Deneubourg, J.-L., Nolfi, S., Gambardella, L.M., Dorigo, M.: SWARM-BOT: A new distributed robotic concept. Autonomous Robots 17(2–3), 193–221 (2004)
6. Nolfi, S., Floreano, D.: Evolutionary Robotics: The Biology, Intelligence, and Technology of Self-Organizing Machines. MIT Press, Cambridge (2000)
7. Di Paolo, E.: Behavioral coordination, structural congruence and entrainment in a simulation of acoustically coupled agents. Adaptive Behavior 8(1), 27–48 (2000)
8. Trianni, V., Dorigo, M.: Self-organisation and communication in groups of simulated and physical robots. Biological Cybernetics 95, 213–231 (2006)
9. Tuci, E., Ampatzis, C., Vicentini, F., Dorigo, M.: Evolved homogeneous neurocontrollers for robots with different sensory capabilities: coordinated motion and cooperation. In: Nolfi, S., Baldassarre, G., Calabretta, R., Hallam, J.C.T., Marocco, D., Meyer, J.-A., Miglino, O., Parisi, D. (eds.) SAB 2006. LNCS (LNAI), vol. 4095, pp. 679–690. Springer, Heidelberg (2006)
10. Wischmann, S., Pasemann, F.: The emergence of communication by evolving dynamical systems. In: Nolfi, S., Baldassarre, G., Calabretta, R., Hallam, J.C.T., Marocco, D., Meyer, J.-A., Miglino, O., Parisi, D. (eds.) SAB 2006. LNCS (LNAI), vol. 4095, pp. 777–788. Springer, Heidelberg (2006)

Group Size Effects on the Emergence of Compositional Structures in Language*

Paul Vogt[1,2]

[1] Tilburg University, P.O. Box 90153, 5000 LE Tilburg, The Netherlands
[2] Vrije Universiteit Amsterdam, The Netherlands
p.a.vogt@uvt.nl

Abstract. This paper presents computer simulations which investigate the effect that different group sizes have on the emergence of compositional structures in languages. The simulations are based on a model that integrates the language game model with the iterated learning model. The simulations show that compositional structures tend to emerge more extensively for larger groups, which has a positive effect on the time in which the languages develop and on communicative success, which may even have an optimal group size. A mathematical analysis of the time of convergence is presented that provides an approximate explanation of the results. The paper concludes that increasing group sizes among humans could not only have triggered the origins of language, but also facilitated the evolution of more complex languages.

1 Introduction

One popular hypothesis explaining the origins of language is that the group sizes in which our human ancestors lived have grown beyond a critical threshold [1]. Dunbar argues that physical grooming, which is believed to be crucial for maintaining social bonds within a group, would take up too much time required for survival oriented behaviours, such as foraging. Language, Dunbar argues, could have taken over the role of physical grooming. This paper examines the effect that group size has on the evolution of *compositional* structures in languages, i.e. structures in which parts of utterances refer to parts of their meanings and the way these parts are combined.

That group size has an effect on language development has been found in a number of studies on both human and animal communication. For instance, non-human primates have larger vocal repertoires [2] and Carolina chickadees have greater vocal complexity in information structure [3] when they live in larger groups. Humans can learn to categorise phonetic categories better when they receive input from multiple speakers than when they only learn from one, because this allows the learner to generalise better on new tokens [4]. Ragir

* This research was supported by the Netherlands Organization for Scientific Research (NWO) through a VENI grant. Many thanks to Bart de Boer and Antal van den Bosch for their invaluable comments on an earlier version of this paper.

F. Almeida e Costa et al. (Eds.): ECAL 2007, LNAI 4648, pp. 405–414, 2007.

[5] has shown that spontaneously evolved sign languages have become more structured in their grammars when used in larger communities. For instance, the sign language in Martha's Vineyard, which had a population of more than 150 signers, was well formed, as is the Nicaraguan Sign Language (100 signers in 1979 up to 500 in 1995). The sign languages of Noyha (12 signers), Grand Cayman (18 signers in 1978, earlier more) and Providence (20 signers, earlier more), however, never evolved grammar [5].

It has been shown computationally [6] and mathematically [7] that when holistic languages are transmitted iteratively from one generation to the next, they can transform into compositional ones. This is provided that 1) language learners have the ability to discover and exploit regular patterns in the utterance-meaning pairs to form compositional structures and 2) learners only observe a small part of the entire language from the previous generation. The latter *transmission bottleneck* [6,7] is crucial, because it places a pressure on compositional structures to evolve. The reason for this pressure is that evolutionary processes tend toward stable systems [8], but *holistic* languages (i.e. languages that have no structural correspondences between parts of utterances and parts of their meanings) are not stable when transmitted through a bottleneck, whereas compositional languages are. To illustrate this, suppose that an individual wants to communicate about, say, a red triangle, which it has never communicated about before. If the language is holistic, this individual cannot use any previously learnt part of the language to produce an utterance and would have to invent a new utterance. If the language is compositional and the individual has learnt a word for red and a word for triangle from hearing utterances referring to a red square or a blue triangle, this individual can combine these words to convey the meaning of a red triangle and no new words have to be invented. Consequently, compositional languages can be transmitted more stably than holistic ones.

One limitation of the iterated learning model is that it assumes a completely *vertical transmission* of languages from one generation to the next, i.e. individuals of one generation only direct their speech to individuals of the next generation. As a consequence, this model does not allow for simulating the evolution of languages in large populations, since it would be infeasible, or would simply take too long, for the languages to converge on the entire population. This problem can be solved if the iterated learning model is combined with a model of *horizontal transmission* [9].

In horizontal transmission models [10], only one generation of individuals exists who all communicate with each other. Combining the vertical transmission model with the horizontal transmission model, yields an *isotropic* transmission model [11] that contains two generations and in which the language is transmitted in all directions (adult→child, adult→adult, child→child and child→adult). In such models, but not in vertical models, compositional languages can evolve without imposing a transmission bottleneck. Instead, individuals face a bottleneck that is an implicit and natural consequence of their development when they need to speak about meanings they have not encountered before [9]. In vertical (adult→child) models, the experimenter needs to control this transmission bottleneck to prevent children from hearing the entire language.

This paper investigates the effect of group size on the emergence of compositionality using a model based on earlier models presented in [12,9,11], but in which the meanings are predefined to reduce computational complexity. This model is presented in the next section. Section 3 presents the results of this study, which are discussed in Section 4. Finally, Section 5 concludes the paper.

2 The Model

This isotropic transmission model implements a multi-agent system that contains $N/2$ adult agents and $N/2$ child agents, where N is the total group size. The whole group plays T rounds of language games [10], after which all adults are removed, all children become adults and new children are introduced (cf. the iterated learning model [6]). The world of the agents contains $M = 81$ compound meanings, which are constructed in 2 dimensions (e.g., colour and shape) of $m = 9$ values each (so a meaning could be something like a 'red square').

1	S → toma/[green,square]	0.2
2	S → A/colour B/shape	0.8
3	A → ba/[red]	0.6
4	B → ke/[triangle]	0.7

Fig. 1. This example grammar contains rules that rewrite a non-terminal into an utterance-meaning pair (1, 3 and 4) or into a compositional rule that combines different non-terminals (2). Whole meanings are formed by 2 features (here colour and shape). Each rule has a rule score that indicates its effectiveness in past guessing games. Only sentences of 2 constituents are allowed in this grammar.

Initially, agents' grammars are empty; the grammars are constructed by the agents playing language games (or *guessing games*). The grammar, such as illustrated in Figure 1, consists of two types of rules: *holistic rules* (rule 1) that map whole compound meanings to randomly created utterances and *compositional rules* (rule 2) that rewrite into two non-terminal rules (rules 3 and 4), each mapping meanings of one dimension to some word-form. (Note that there are two types of compositional rules in this grammar differing only in word-order.)

The grammar may contain redundant rules in that there may be different ways to encode or decode an utterance. To deal with the competition between these redundancies, each rule j is associated with a rule score ρ_j that indicates the effectiveness of the rule during past language games. When agents need to choose between a (possibly holistic) composition of redundant rules, they always select the composition i that has the highest weight w_i:

$$w_i = \begin{cases} \rho_j & \text{if holistic} \\ \rho_c \cdot \rho_{t1} \cdot \rho_{t2} & \text{if compositional.} \end{cases} \quad (1)$$

Here ρ_c is the score of a general rule, and ρ_{t1} and ρ_{t2} are the scores of the terminal rules.

In each game, two agents are arbitrarily selected from the population. One is randomly assigned the role of speaker and the other becomes the hearer. The agents are provided a shared context that contains $c = 8$ distinct meanings randomly selected from the M compound meanings. The speaker selects one meaning as the target and searches its private grammar to encode an utterance. If there are more ways to encode an utterance, the speaker selects the one that has been used most successfully in the past based on the weight of the rules. If there is no way to encode an utterance, a short random word-form is created from a finite alphabet. This new form is associated either holistically with the entire compound meaning, or with a part of the meaning if the other part is already associated with a word.

The hearer tries to decode the utterance by searching its grammar for compositions that parse the utterance such that the resulting meaning is in the context. If there are more ways to decode the utterance, the one with the highest weight is selected, yielding the meaning that the hearer guesses was intended by the speaker. If this is the correct meaning, the game is successful. Otherwise, the game fails either because the hearer guesses a wrong meaning or the hearer is unable to decode the utterances. (Note that the game's outcome is verified through explicit meaning transfer. Though this is not realistic, it is done to speed up convergence.)

Depending on the outcome of the game, the rule scores ρ_j are adapted by both agents. If the game at time t is a success, the score(s) of used rule(s) are increased by

$$\rho_j(t) = \eta \cdot \rho_j(t-1) + (1-\eta) \cdot \rho_j(t-1), \tag{2}$$

while the scores of competing rules (i.e., rules that could also encode or decode an utterance) are laterally inhibited using

$$\rho_j(t) = \eta \cdot \rho_j(t-1). \tag{3}$$

The same equation is used to inhibit the rule scores when the guessing game fails. In these equations, $\eta = 0.9$ is a learning parameter and $\rho_j(0) = 0.01$ is the initial score. These updates implement a *positive feedback loop*.

If the game fails, the speaker informs the hearer which compound meaning was intended, allowing the hearer to acquire the correct mapping. While adopting the utterance, the hearer tries to induce a compositional structure in three steps:[3]

1. The hearer searches its grammar to see if it contains a rule that can decode a part of the utterance with the correct meaning. If this is the case, the remaining part of the utterance is associated with the remaining part of the meaning. If there are more such cases, the one with highest weight is used.

[3] More details of these steps are described in [12,9].

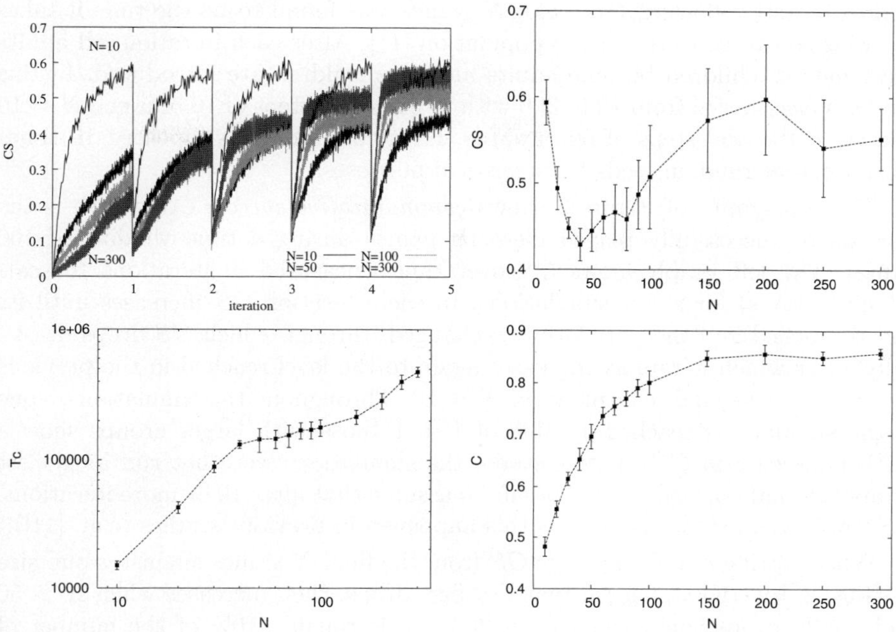

Fig. 2. Results of this study. The top graphs show communicative success CS against time in iterations (left) and against group size N (right). The bottom graphs show time of convergence T_c (left) and compositionality C (right) (both against N).

2. If this fails, the hearer searches for a *regular pattern* in the heard utterance-meaning pair compared to the most recent 1,000 instances of utterance-meaning pairs that it heard in previously played games and which are stored in a separate instance base. A regular pattern is found in two distinct utterance-meaning pairs if two utterances either start or end with the same substring and if both pairs have a similar meaning part. If such a regular pattern is discovered, the hearer breaks up the utterance and meaning following certain heuristics to form a compositional structure (see [12], pp. 221–223). The same break up is applied to all existing rules that have a similar pattern. Note that previously obtained rules are retained.
3. If the second step also fails, the utterance-meaning pair is incorporated in the grammar holistically.

Note that these induction steps are similar to those used in [6] and are inspired from usage-based approaches to human language acquisition [13].

3 Results

Figure 2 shows the results of simulating the model with various group sizes for 5 iterations of T guessing games each, where T is proportional to the group size N

approximately following $T \propto N \log N$, which was found to be the time it takes for a lexicon to converge in the population [14]. After each iteration, all adults are removed, children become adults and new children are introduced. In this study, N was varied from 10 to 300 with incremental steps of 10 between $N = 10$ and $N = 100$ and steps of 50 onwards. Each condition was repeated 10 times with different random seeds for statistical purposes.

The top graphs of Figure 2 shows *communicative success* CS, which is the fraction of successfully played guessing games during a time window of 100 games. The left graph shows CS over time (measured in iterations to scale all simulations) for a few simulations. In each iteration CS increases until its end is reached and the population is changed, through which CS drops drastically, after which it rapidly increases again to the level reached in the previous iteration and beyond, except when $N = 10$. Throughout the simulations, none of the simulations reached a value of 1, but those with larger groups show a further increase in CS. In this study, the simulations were not run longer for computational reasons[4], but it is safe to assume that after 10 or more iterations, CS would yield values near 1, as this happened in previous studies (e.g., [11]).

When setting out the average CS from the final X games against group size N (Fig. 2, top right), we see that CS first drops, then increases when $N > 50$ and finally drops again when $N > 200$. (X is roughly 10% of the number of games per iteration, T. This is done because CS can vary strongly and this average gives us more reliable values.) So, there appears to be an optimal group size around $N = 200$. Yet, although the differences between the simulations of small N and of those with larger differences (e.g., for $N = 50$, $N = 100$ and $N = 200$) are significant ($p \leq 0.01$ according to the Wilcoxon rank test), those that closely vary around $N = 200$ (i.e. $N = \{150, 250, 300\}$) are not ($p > 0.05$).

It is possible to estimate the time it takes for the curve of CS to stabilise, using a non-linear regression of its curve. This *time of convergence* T_c shows linear dependencies with N on a log-log scale, i.e. $T_c \propto N^{\beta}$, when $N < 50$ with slope $\beta = 1.58$ and when $N \leq 150$ with slope $\beta = 1.18$ (Fig. 2, bottom left). In between these values, the slope is $\beta = 0.36$. (All slopes are obtained with linear regression.) Interestingly, this result is quite different from those obtained for the evolution of lexicons, where a continuous linear dependency on the log-log scale was found. The result in the first part ($N < 50$) is similar to that of Baronchelli et al., who obtained $T_c \propto N^{1.5}$ [14,15]. The remaining parts have slopes lower than those obtained by Kaplan, who obtained a dependency of $N \log N$ [14].

Figure 2 (bottom right) shows the evolution of *compositionality* C, which measures the proportion of two-word utterances encoded, decoded and discovered with induction step 2 (see previous section) by the population measured during the final X guessing games against group size N. It is clear that the level of compositionality increases with group size until a maximum level is reached near $C = 0.86$ for $N \geq 150$. The remainder of this paper discusses how and why group size, compositionality and time of convergence relate to each other.

[4] Processing these simulations took over 1 month using a cluster of 10 PCs

4 Discussion

The simulations reported in this paper show the surprising results that

1. compositionality tends to evolve more extensively with larger group sizes,
2. time of convergence shows different regions of dependencies with group size, and
3. there appears to be an optimum in group size concerning communicative success (though this is not a significant result).

These results are surprising, because intuitively one would think that evolving structured languages would be harder for larger populations, just as is the case for evolving (holistic) lexicons [14,15]. However, that turns out not to be the case for all group sizes.

To explain the first two results, it is important to note that in larger groups more randomly created words occur than in smaller groups. In fact, there is a power law relation between the maximum number of randomly created words $W_{\max}$ and group size N, i.e.

$$W_{\max} \propto N^{\omega}. \tag{4}$$

The exact relation was not monitored in this experiment, but was previously observed (Vogt, unpublished) and occurs for emerging lexicons in the naming game simulations of Barronchelli et al. [15], who found that $\omega \approx 1.5$. Interestingly, they also found that time of convergence and group size had the same dependency, i.e. $T_c \propto N^{\omega}$ and argued that this was a sound correspondence. Kaplan [14], who found for a more closely related language game model that $T_c \propto N \log N$, further found that the time of convergence is linearly proportional to the number of meanings M in the language, so let us assume that

$$T_c \propto MN \log N. \tag{5}$$

Now let us assume that the number of meanings M to be acquired by the population is – on average – proportional to the level of compositionality C in the language according to

$$M = CM_c + (1 - C)M_h, \tag{6}$$

where M_c is the number of word-meaning pairs an entirely compositional language would have and M_h is the number of word-meaning pairs an entirely holistic language would have.

In this model, compositionality C depends on the chance of finding a regular pattern in both one dimension of the compound meanings and in the signals, because that is what agents find and use [12]. Since the number of meanings per dimension remains constant in all simulations, the probability of finding a regular pattern in the signals depends on the size of the alphabet Λ and the number of distinct words (or utterances) W in the language.[5] The number of words in the language follows a power law depending on group size. Hence the chance of finding and using compositional structures depends on group size.

Suppose – for simplicity – that the language in question only contains randomly constructed words that are strings containing only two letters AB, where $A, B \in \Lambda$. Let us further assume that the level of compositionality C is proportional to the probability $P(R|W)$ that, given the number of different words W that exist in a language, all possible regular patterns in word-meanings have been observed at least twice, so that for all possible compound meanings a compositional rule can be constructed.

Suppose the population has created $W = W_{\max} - 1$ different words. Now, when we find a new word starting with some arbitrary letter A, then the probability that we can find an existing word with the same letter (i.e. we can find a regular pattern R in the expression) is

$$P(R|W) = 1 - P(\neg A)^W = 1 - (1 - \frac{1}{|\Lambda|})^W, \tag{7}$$

where $P(\neg A)$ is the probability that a randomly created word does not start with the letter A. In effect, this equation says that the probability of finding a pattern is equal to one minus the probability that after creating W words none of these words start with A.

Since we assume that $C \propto P(R|W)$, we have, following Eq. (6)

$$M = P(R|W) \cdot M_c + (1 - P(R|W)) \cdot M_h. \tag{8}$$

Substituting this equation in Eq. (5) gives

$$T_c \propto (P(R|W) \cdot M_c + (1 - P(R|W)) \cdot M_h) \cdot N \log N. \tag{9}$$

[5] In the remainder of the paper, I use the term *words* for both holistic words and compositional two-word utterances, which are transmitted without a word boundary.

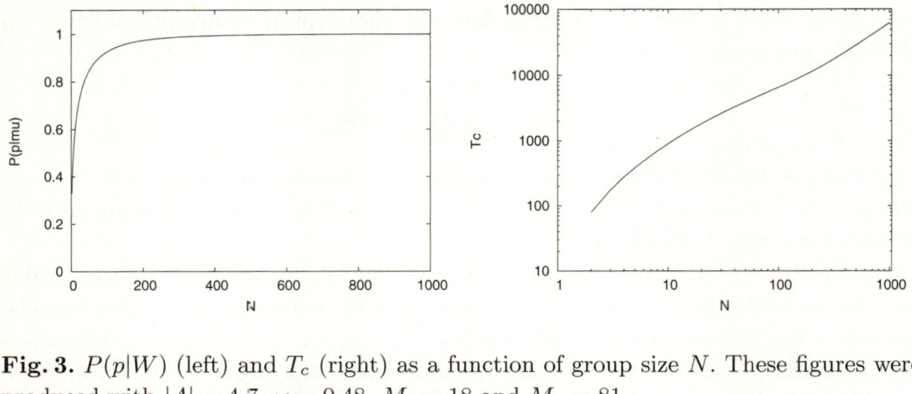

Fig. 3. $P(p|W)$ (left) and T_c (right) as a function of group size N. These figures were produced with $|\Lambda| = 4.7$, $\omega = 0.48$, $M_c = 18$ and $M_h = 81$.

Plotting Eqs. (7) and (9), using values for $|\Lambda| = 4.7$ and $\omega = 0.48$ obtained through linear regression of Fig. 2 (bottom right), produces Figure 3. This figure shows qualitatively similar dependencies for compositionality C and time of convergence T_c as observed in Figure 2. What it shows is that when the group size increases, compositionality (expressed here as $P(R|W)$) increases until it converges. While compositionality increases, the number of meanings M decrease, thus affecting time of convergence T_c by bending its curve slightly to the right and going up again, but with a lower slope on the log-log scale than previously.

The bend in the curve of T_c is less expressed than in the curve obtained with the simulation. Moreover, both $|\Lambda|$ and ω are smaller than expected, as the alphabet in the simulations contains 15 characters and – if Baronchelli et al. are right – the exponent of Eq. (4) was expected to be closer to 1, or even exceeding 1. Unfortunately, in these particular simulations, the evolution of the maximum number of words in the language was not monitored, so we have no data to measure ω based on Eq. (4).

A probable reason that the values are lower than expected is that the results on C and T_c were presented concerning the fifth iteration, before which the language had already been developed to some extent. This would increase the likelihood of finding a regular pattern, thus lowering $|\Lambda|$. As a consequence, this also decreases the need for creating new random words, thus lowering ω.

Currently, the simulations are being repeated with a faster implementation of the model than used to generate the data presented here, thus allowing to do more runs of the simulations for better statistics, to run them for a longer time and to go beyond the 300 agents. In addition, these new runs do monitor the evolution of words, so we can have a better estimate of parameters for the mathematical model. Moreover, the additional simulations aim to investigate the effect that group size has on the level of communicative success, which revealed an apparent optimum. However, this optimum was not significant. Better statistical results and larger group sizes should shed more light on this issue.

5 Conclusions

The simulations in this paper show that for larger groups compositional languages evolve more extensively due to the increased number of words, which increases the likelihood of finding regular patterns in utterance and meaning. As a result, individuals tend to use the compositional language bits more frequently, so there are less meanings to be distributed among the population, which affects time of convergence.

The relation between time of convergence and group size has three phases. First, time of convergence is increasing relatively fast with increasing group sizes, then it increases much slower, after which it starts to increase faster again, but with a slower rate than for the smaller group sizes. As shown mathematically, the first decrease in the slope coincides with a strong decrease in the number of meanings to be distributed, due to the increased compositionality. The later increase in the slope coincides with compositionality reaching a stable maximum.

So, Dunbar's hypothesis that language has originated to facilitate a different mode of grooming when the groups in which hominids started to live exceeded a certain threshold [1], whether right or wrong, seems to have an interesting consequence. Our ancestral evolution to live in larger groups did not only put a pressure on language to originate, but actually facilitated the emergence of compositional languages.

References

1. Dunbar, R.: Grooming, Gossip, and the Evolution of Language. Harvard University Press (1998)
2. McComb, K., Semple, S.: Coevolution of vocal communication and sociality in primates. Biology Letters 1(4), 381–385 (2005)
3. Freeberg, T.: Social complexity can drive vocal complexity: Group size influences vocal information in carolina chickadees. Psychological Science 17, 557 (2006)
4. Lively, S., Logan, J., Pisoni, D.: Training Japanese listeners to identify English/r/and/l/. II: The role of phonetic environment and talker variability in learning new perceptual categories. The Journal of the Acoustical Society of America 94, 1242 (1993)
5. Ragir, S.: Constraints on communities with indigenous sign languages: Clues to the dynamics of language genesis. In: Wray, A. (ed.) The Transition to Language, Oxford University Press, Oxford (2002)
6. Kirby, S.: Spontaneous evolution of linguistic structure: an iterated learning model of the emergence of regularity and irregularity. IEEE Transactions on Evolutionary Computation 5(2), 102–110 (2001)
7. Kirby, S., Dowman, M., Griffiths, T.: Innateness and culture in the evolution of language. Proceedings of the National Academy of Sciences 104(12), 5241–5245 (2007)
8. Cavalli-Sforza, L.L., Feldman, M.W.: Cultural Transmission and Evolution: A quantitative approach. Princeton University Press, Princeton (1981)
9. Vogt, P.: On the acquisition and evolution of compositional languages: Sparse input and the productive creativity of children. Adaptive Behavior 13(4), 325–346 (2005)
10. Steels, L.: The synthetic modeling of language origins. Evolution of Communication 1(1), 1–34 (1997)
11. Vogt, P.: Cumulative cultural evolution: Can we ever learn more? In: Nolfi, S., Baldassarre, G., Calabretta, R., Hallam, J.C.T., Marocco, D., Meyer, J.-A., Miglino, O., Parisi, D. (eds.) SAB 2006. LNCS (LNAI), vol. 4095, Springer, Heidelberg (2006)
12. Vogt, P.: The emergence of compositional structures in perceptually grounded language games. Artificial Intelligence 167(1–2), 206–242 (2005)
13. Tomasello, M.: Constructing a Language: A Usage-Based Theory of Language Acquisition. Harvard University Press (2003)
14. Kaplan, F.: Simple models of distributed co-ordination. Connection Science 17(3-4), 249–270 (2005)
15. Baronchelli, A., Felici, M., Caglioti, E., Loreto, V., Steels, L.: Sharp transition towards shared lexicon in multi-agent systems. Journal of Statistical Mechanics P06014 (2006)

Language Learning Dynamics: Coexistence and Selection of Grammars

Valery Tereshko

School of Computing, University of Paisley, Paisley PA1 2BE, Scotland
valery.tereshko@paisley.ac.uk
http://cis.paisley.ac.uk/tere-ci0/

Abstract. Language learning dynamics is modelled by an ensemble of individuals consisting of the grammar carriers and the learners. Increasing the system population size results into the transition from the individual to the collective mode of learning. At low communication level, different grammars coexist in their own survival niches. Enhancement of the communication level in purely collective mode, when all individuals are the part of general communication network, leads to the selection of the fittest grammar. Adding the individual mode of learning results into the formation of the quasigrammar, with the dominant grammar prevailing over the set of coexisting grammars.

1 Introduction

A community of language users collectively developing a shared communication system can be viewed as the complex adaptive system subjected to a Darwinian evolution [1, 2, 3, 4, 5, 6, 7]. The language users are considered as the interacting agents. The central question is the phenomenon of spontaneous emergence of order in the ensembles of these agents. This order can be static or dynamic, i.e. represents stationary patterns or synchronized motions respectively, and most importantly, it appears without imposing any centralized control. Applying to a language, children are known to develop the grammatical competence by interactions with people, without any formal training [8]. This can be viewed as one of examples of the order that emerges entirely through individual's interactions.

In this paper, we consider language learning dynamics as a pattern formation phenomenon in a space of all available grammars, the grammar space. All individuals are divided into the grammar carriers, who are already learned and carry a particular grammar, and the learners who are not carry any grammar yet but potentially can learn one and become the grammar carriers themselves. Since the learners can interact with the carriers of different grammars, they can choose between the latter. On the other hand, the grammar carriers are competing for the pool of learners in the attempts to persuade the latter to their own grammars. The similar approach was used to model the molecular evolution [9] and the honey bee colony foraging dynamics [10, 11, 12].

We introduce the individual and the collective (through communication with other individuals) modes of learning. Increasing the population size results into

F. Almeida e Costa et al. (Eds.): ECAL 2007, LNAI 4648, pp. 415–424, 2007.

the transition from the individual to the collective learning characterizing by the sharp increase in the grammar carriers.

When the grammar carriers can access the learners only locally, different grammars coexist. Unlike, in purely collective mode, when the grammar carriers have access to the whole pool of learners, the only fittest grammar survives. When taking the individual mode of learning into account, the quasigrammar is developed. The latter is characterized by the formation of the dominant grammar prevailing over the set of coexisting grammars.

Unlike other models using the population dynamics approach to model an evolution of language [5, 6, 7], our model introduces the real space, allowing to consider not only the temporal but the spatio-temporal dynamics, and the individual mode of learning. The latter give extra modelling opportunities to produce more realistic outcomes. Indeed, in [7] the coexistence of grammars, or so called m-grammar solution, can be reached only for the case of equal fitness, which is unrealistic situation. Our model allows the coexistence of the different fitness grammars.

2 Basic Model

Let us describe the language dynamics mechanism in the terms of chemical reactions. The grammar carriers and the individuals who are not carry any particular grammar yet, i.e. the learners, are denoted by X and Y respectively; the grammar fitness by f:

$$Y + X \to 2X \tag{1}$$

Reaction (1) illustrates the autocatalytic nature of the communication process. If a carrier communicates its grammar to an individual involving the latter to it, this individual (as the new carrier) will in turn reinforce the process and communicate the chosen grammar to other individuals, and so forth.

A carrier can abandon an unrewarding grammar at the rate inversely proportional to that grammar's quality:

$$X \xrightarrow{\frac{1}{f}} Y \tag{2}$$

The kinetic equations corresponding to reactions (1-2) take the form

$$\begin{aligned}\frac{\partial x(\mathbf{r},t)}{\partial t} &= \Big(y(\mathbf{r},t) - \frac{1}{f(\mathbf{r},t)}\Big)x(\mathbf{r},t) + D_x\frac{\partial^2 x(\mathbf{r},t)}{\partial \mathbf{r}^2}\\ \frac{\partial y(\mathbf{r},t)}{\partial t} &= -\Big(y(\mathbf{r},t) - \frac{1}{f(\mathbf{r},t)}\Big)x(\mathbf{r},t) + D_y\frac{\partial^2 y(\mathbf{r},t)}{\partial \mathbf{r}^2},\end{aligned} \tag{3}$$

where x and y are the concentrations, and D_x and D_y are the diffusion constants of the grammar carriers and learners respectively, and $f(\mathbf{r},t)$ is the grammar landscape. Taking the Wright's idea of fitness landscape which assigns a fitness to each point in a genetic space, our grammar landscape gives a grammar quality value to each point in the grammar space.

3 Dynamics at Different Communication Levels

The grammar carrier diffusion is assumed to be small that means high accuracy of the grammar acquisition. The diffusion of learners determines their access to different grammars. Let us to compare two opposite cases: (i) the diffusion of Y is slow allowing the access of learners to local information only, and (ii) the diffusion of Y is high enough to ensure the global mixing of learners throughout the entire grammar space.

(i) small D_y

In the limit of vanishing diffusions, system (3) possesses integral of motion

$$\frac{\partial x(\mathbf{r},t)}{\partial t} + \frac{\partial y(\mathbf{r},t)}{\partial t} = 0, \tag{4}$$

which yields condition of constant *local* concentrations

$$x(\mathbf{r},t) + y(\mathbf{r},t) = x_0(\mathbf{r}) + y_0(\mathbf{r}) = C(\mathbf{r}). \tag{5}$$

This allows to eliminate variable y from system (3) and to reduce the latter to spatially extended logistic equation

$$\frac{\partial x(\mathbf{r},t)}{\partial t} = \big(\alpha(\mathbf{r}) - x(\mathbf{r},t)\big)x(\mathbf{r},t), \tag{6}$$

where

$$\alpha(\mathbf{r}) = C(\mathbf{r}) - \frac{1}{f(\mathbf{r})}. \tag{7}$$

The grammar acquisition threshold, $1/C(\mathbf{r})$, indicates the grammars that are attractive for the individuals. The latter begin to learn only those grammars whose fitness value exceeds threshold

$$f(\mathbf{r}) > 1/C(\mathbf{r}). \tag{8}$$

The problem nature allows us to take into account the set of n spatial modes corresponding to the local maxima of the grammar landscape. Considering only these modes, infinite-dimensional system (6) reduces to system of uncoupled equations describing the logistic growth of the grammar carriers at i-th spatial point

$$\dot{x}_i(t) = (\alpha_i - x_i(t))x_i(t), \tag{9}$$

where

$$\alpha_i = C_i - \frac{1}{f_i} \tag{10}$$

is the reproductive rate of i-th mode.

Every mode associated with a fitness exceeding the grammar acquisition threshold converges to attractor

$$x_i^s = C_i - \frac{1}{f_i},\ i = 1, .., n. \tag{11}$$

When the learners have access to only local information, different grammars have ability to *coexist* in the society.

(ii) large D_y
Let us consider now the opposite case when the learners have access to all available grammars. In the limit of Y's full mixing, system (3) reduces to system of integro-differential equations

$$\begin{aligned}\frac{\partial x(\mathbf{r},t)}{\partial t} &= \Big(y(t) - \frac{1}{f(\mathbf{r})}\Big)x(\mathbf{r},t) + D_x \frac{\partial^2 x(\mathbf{r},t)}{\partial \mathbf{r}^2}\\ \frac{\partial y(t)}{\partial t} &= -y(t)\frac{1}{S}\int_Q x(\mathbf{r},t)\,\mathrm{d}\mathbf{r} + \frac{1}{S}\int_Q \frac{1}{f(\mathbf{r})}x(\mathbf{r},t)\,\mathrm{d}\mathbf{r},\end{aligned} \tag{12}$$

where $y(t) = (1/S)\int_Q y(\mathbf{r},t)\,\mathrm{d}\mathbf{r}$ is the spatially-averaged concentration of carriers over domain Q with area $S = \int_Q \mathrm{d}\mathbf{r}$.

Integral of motion[1]

$$\frac{1}{S}\int_Q \frac{\partial x(\mathbf{r},t)}{\partial t}\,\mathrm{d}\mathbf{r} + \frac{\partial y(t)}{\partial t} = 0 \tag{13}$$

yields condition of constant *total* concentration

$$\frac{1}{S}\int_Q x(\mathbf{r},t)\,\mathrm{d}\mathbf{r} + y(t) = x_0 + y_0 = C, \tag{14}$$

which allows to eliminate variable y from system (12) and to reduces the latter to spatially extended Lotka-Volterra equation

$$\frac{\partial x(\mathbf{r},t)}{\partial t} = [\alpha(\mathbf{r}) - \frac{1}{S}\int_Q x(\mathbf{r},t)\,\mathrm{d}\mathbf{r}]x(\mathbf{r},t) + D_x \frac{\partial^2 x(\mathbf{r},t)}{\partial \mathbf{r}^2}. \tag{15}$$

In the limit of vanishing D_x, if only modes corresponding to the local maxima of the grammar landscape are taken into account, infinite-dimensional system (15) reduces to system of coupled equations for the spatial mode amplitudes

$$\dot{x}_i(t) = (\alpha_i - \sum_{i=1}^{n} x_i(t))x_i(t). \tag{16}$$

Dividing i-th and j-th equations on x_i and x_j respectively and subtracting one equation from another, one obtains

$$\frac{\dot{x}_i(t)}{x_i(t)} - \frac{\dot{x}_j(t)}{x_j(t)} = \frac{1}{f_j} - \frac{1}{f_i} \tag{17}$$

The integration of equation (17) results into

$$\frac{x_i(t)}{x_j(t)} = \frac{x_i(0)}{x_j(0)}\exp\Big(\Big(\frac{1}{f_j} - \frac{1}{f_i}\Big)t\Big). \tag{18}$$

[1] The integration over the space eliminates the diffusion term in the first equation of system (12) due to no-flux boundary conditions.

Expression (18) provides the analytical proof of the selection in the system. If m-th mode is the fittest and the unique, then $f_m > f_j$ for $\forall j \neq m, j = 1,..,n$. Hence, it immediately follows that when $t \to \infty, x_m/x_j \to \infty$ for $\forall j \neq m, j = 1,..,n$. However, condition of constant total concentration (14) and the positive definiteness of variables prevent the unlimited growth of the modes. This means that the amplitudes of all modes, excluding the fittest one, must tend to zero with time. If more than one mode are the fittest, then they all survive.

If the fitness of at least one mode exceeds the grammar acquisition threshold, then trivial equilibrium

$$x_i^s = 0, \;\; i = 1,..,n \tag{19}$$

loses stability and system (16) converges to non-trivial attractor

$$x_m^s = C_0 - \frac{1}{f_m}, \;\; x_i^s = 0, \;\; i = 1,..,n; \;\; i \neq m \tag{20}$$

where $f_m > f_i$, which corresponds to the *selection* of the fittest grammar.

The society of globally informed individuals is, thus, capable of the collective choice of the fittest grammar.

Let us perform the numerical simulations. We use the explicit method of the numerical integration of PDEs when space and time are divided into discrete uniform sub-intervals, and derivatives are replaced by their finite-difference approximations. The numerical integrations are performed on the 2D lattice with the space and the time steps are chosen to guarantee the stability and the convergence of explicit scheme. Throughout, the initial concentrations of grammar carriers and their diffusion constant are taken to be $x_0 = 0.01$ and $D_x = 0.01$ respectively.

Consider the grammar fitness landscape with three spatially separated regions (Fig. 1a). Two regions consist of two fitness peaks each, and one region consists of only one peak. Among the peaks, we have the highest one, two smallest ones, and two intermediate ones of the same height.

The diffusion length of learners in the grammar space, l_d, can be evaluated [13]:

$$l_d = \sqrt{D_y \tau_d}, \tag{21}$$

where characteristic diffusion time τ_d is approximated as

$$\tau_d \sim \frac{1}{\alpha(\mathbf{r})} = \left(C_0 - \frac{1}{f(\mathbf{r})}\right)^{-1}. \tag{22}$$

For the areas where $f(\mathbf{r}) \to 0$, τ_d and, hence, $l_d \to 0$. The latter means that for the weak diffusion of learners and the distantly separated grammars, the grammar carriers get their own exclusive resource for the development of their grammars. In others words, the grammar niches are developed, which leads to the coexistence of the different grammars in the system. Figure 1b illustrates the above arguments. Note that the weakest grammars corresponding to the smallest peaks in every niche get suppressed. This happens because the diffusion length becomes compared to the niche size. The grammars inside the niches compete

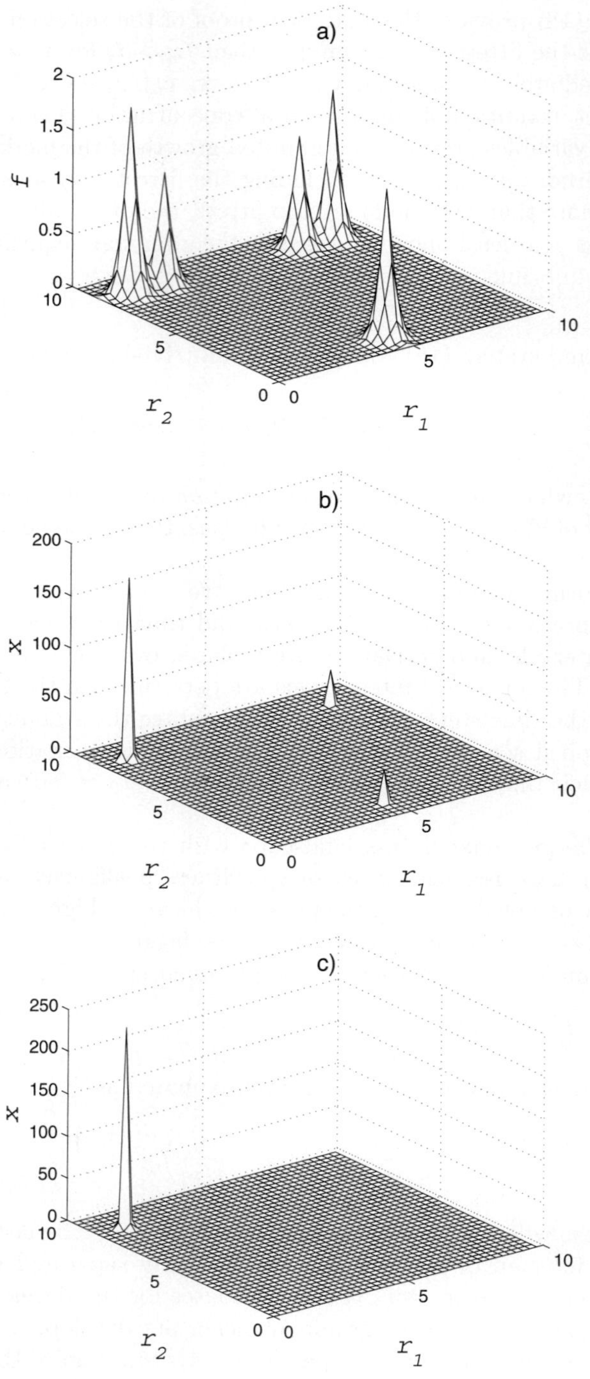

Fig. 1. a) Fitness landscape, b)-c) Concentration of grammar carries for system (3) at $D_y = 1$, $t = 3000$ and $D_y = 10$, $t = 75$ respectively.

for their common resource, y. As result, the niche's strongest grammars survive. Note that the above processes become completed at about $t = 300$, and the landscape similar to one in Fig. 1b is formed. Performing simulations up to $t = 3000$ gives the same picture, which ensures us that the grammar coexistence is stable. Increasing D_y creates a common communication niche throughout the grammar space, which leads to the selection of the fittest grammar in the system (see Fig. 1c).

4 Transition from Individual to Collective Learning and Formation of Quasigrammar

Model (1)-(2)accounts for the learners acquiring their grammars through the communication process with the grammar's carriers. Let us now to account for these who learns individually.

The following reaction represents the individual learning:

$$Y \xrightarrow{\epsilon} X_i. \tag{23}$$

The amount of individual learners are assumed to be much smaller that these learning through the communication, meaning ϵ is set to be small.

The kinetic equations of the updated model are

$$\begin{aligned}\frac{\partial x(\mathbf{r},t)}{\partial t} &= \Big(y(\mathbf{r},t) - \frac{1}{f(\mathbf{r},t)}\Big)x(\mathbf{r},t) + \epsilon y(\mathbf{r},t) + D_x\frac{\partial^2 x(\mathbf{r},t)}{\partial \mathbf{r}^2}\\ \frac{\partial y(\mathbf{r},t)}{\partial t} &= -\Big(y(\mathbf{r},t) - \frac{1}{f(\mathbf{r},t)}\Big)x(\mathbf{r},t) - \epsilon y(\mathbf{r},t) + D_y\frac{\partial^2 y(\mathbf{r},t)}{\partial \mathbf{r}^2},\end{aligned} \tag{24}$$

In the limit of vanishing D_x and high D_y, model 24 reads

$$\dot{x}_i = \Big(C - \frac{1}{f_i}\Big)x_i - \epsilon\sum_{k=1}^{n} x_k - x_i\sum_{k=1}^{n} x_k + \epsilon C. \tag{25}$$

Let us consider the simplest case of a single grammar and analyze the system dynamics depending on its size. In this case, the learners dynamics is described by simple logistic equation with small constant-growth term

$$\dot{x} = \Big(C - \frac{1}{f} - \epsilon\Big)x - x^2 + \epsilon C. \tag{26}$$

The only physical attractor the system converges to is

$$x^* = \frac{C - \frac{1}{f} - \epsilon}{2} + \sqrt{\frac{(C - \frac{1}{f} - \epsilon)^2}{4} + \epsilon C} \tag{27}$$

It is easy to see that x^* remains low up to the point $C = \frac{1}{f} + \epsilon$. Above this point, x^* sharply increases and eventually tends to C (at higher values of the latter). Fig. 2 illustrates this.

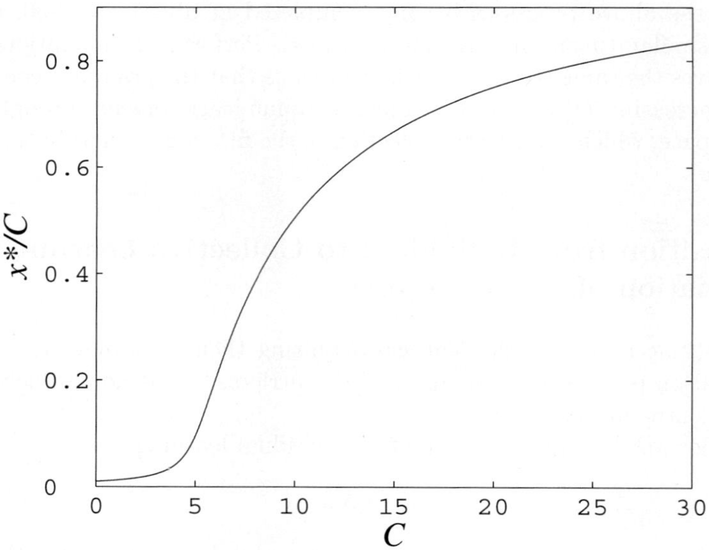

Fig. 2. Proportion of the stationary concentration of grammar carriers to the total population versus the population size.

The above figure elucidates the transition from the *individual* to the *collective* learning. Indeed, when $C < \frac{1}{f} + \epsilon$, the grammar carriers concentration growths entirely due to the free term of equation (26), i.e. due to the individual learning. Unlike, at $C > \frac{1}{f} + \epsilon$, the communication is the mechanism governing the learning dynamics. Note that we still have an increase of the grammar carriers caused by the individual learning. However, the dominant term contributing to the above increase, $(C - \frac{1}{f}\epsilon)x$, is determined by the communication.

In general case of n grammars, define the (total) grammar carriers flow as $\sum x_i$. Then, the flow dynamics is governed by equation

$$\sum_{i=1}^{n} \dot{x}_i = C \sum_{i=1}^{n} x_i - \sum_{i=1}^{n} \frac{1}{f_i} x_i - \epsilon \sum_{i=1}^{n} x_i - \Big(\sum_{i=1}^{n} x_i \Big)^2 + \epsilon C. \tag{28}$$

Taking into account that

$$\sum_{i=1}^{n} \frac{1}{f_i} x_i = \frac{\sum_{i=1}^{n} \frac{1}{f_i} x_i}{\sum_{i=1}^{n} x_i} \sum_{i=1}^{n} x_i = \Big\langle \frac{1}{f} \Big\rangle \sum_{i=1}^{n} x_i, \tag{29}$$

where $\langle \frac{1}{f} \rangle$ is the averaged (over the set of grammars) grammar quality, obtain:

$$\sum_{i=1}^{n} \dot{x}_i = \Big(C - \Big\langle \frac{1}{f} \Big\rangle - \epsilon \Big) \sum_{i=1}^{n} x_i - \Big(\sum_{i=1}^{n} x_i \Big)^2 + \epsilon C. \tag{30}$$

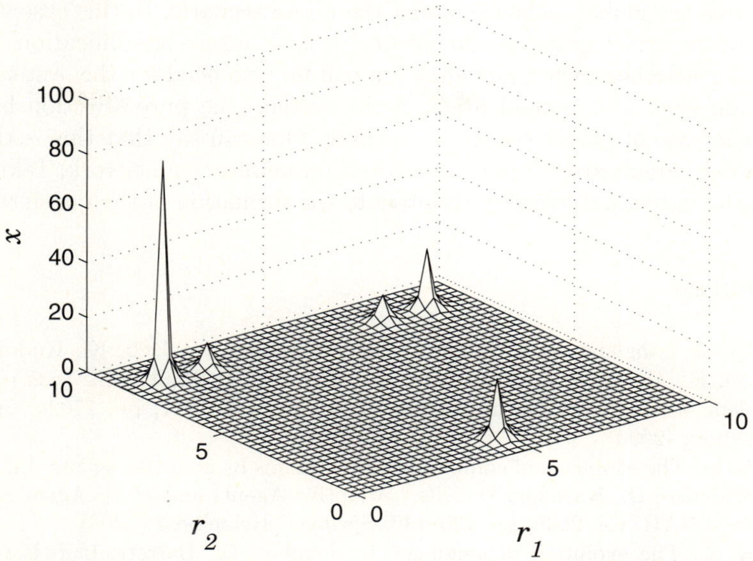

Fig. 3. Concentration of grammar carries for system (24) at $D_y = 10$, $t = 300$.

Obviously, all previous results obtained for the single grammar case are held here, and at $C = \langle \frac{1}{f} \rangle + \epsilon$ the population undergoes the individual-collective learning transition.

Let us look at the grammar fitness landscape in Fig. 1a. As we have already seen, at high communication level, the common communication niche throughout the whole system is created. In this case, the basic model dynamics leads to the survival of the fittest grammar in the system (Fig. 1c). Unlike, system (24) doesn't produce the pure selection: one observes the formation of the dominant grammar that prevails over all other coexisting grammars (Fig. 3). Taking the idea of quasispecies [14], we call the above the *quasigrammar*.

5 Conclusion

Our approach elucidates the natural selection that created the human's system of communication. The grammars are like species in competition. The fitness of the species is given by the grammar quality. A grammar can survive by continuing to circulate within the society, and is able to reproduce itself by recruiting new carriers who has learned and share it with others.

Increasing the population size results into the transition from the individual to the collective learning. The sharp increase in the grammar carriers demonstrates greater efficiency of the collective mode.

The level of spatial interactions determines the system behaviour. At low level, the learners are "locked" on their local knowledge sites. Different grammars can coexist in their own survival niches. The enhancement of the interaction level creates the global niche excluding the above scenario. In this case we have survival of the fittest grammar. In cultural sense it means simplification, and we can wonder whether such a globalization will be beneficial for the system.

At high level of communication in the system, the pure selection happens only in the case of purely collective learning. One can say that this is the case when all individuals are the part of general communication network. Taking into account the individual learning results into the formation of the quasigrammar.

References

1. Steels, L.: Language as a complex adaptive system. In: Deb, K., Rudolph, G., Lutton, E., Merelo, J.J., Schoenauer, M., Schwefel, H.-P., Yao, X. (eds.) Parallel Problem Solving from Nature-PPSN VI. LNCS, vol. 1917, pp. 17–26. Springer, Heidelberg (2000)
2. Steels, L.: The evolution of communication systems by adaptive agents. In: Alonso, E., Kudenko, D., Kazakov, D. (eds.) Adaptive Agents and Multi-Agent Systems. LNCS (LNAI), vol. 2636, pp. 125–140. Springer, Heidelberg (2003)
3. Kirby, S.: The evolution of language. In: Dunbar, R., Barrett, L. (eds.) Oxford Handbook of Evolutionary Psychology, pp. 669–681. Oxford University Press, Oxford (2007)
4. Oudeyer, P-Y., Kaplan, F.: Language evolution as a darwinian process: Computational studies. Cognitive Processing 8, 21–35 (2007)
5. Komarova, N.L., Niyogi, P., Nowak, M.A.: The evolutionary dynamics of grammar acquisition. J. Theor. Biol. 209, 43–59 (2001)
6. Nowak, M.A., Komarova, N.L., Niyogi, P.: Evolution of universal grammar. Science 291, 114–118 (2001)
7. Komarova, N.L.: Replicator-mutator equation, universality property and population dynamics of learning. J. Theor. Biol. 230, 227–239 (2004)
8. Jackendoff, R.S.: The architecture of the language faculty. MIT Press, Camridge (1977)
9. Tereshko, V.: Selection and coexistence by reaction-diffusion dynamics in fitness landscapes. Phys. Lett. A 260, 522–527 (1999)
10. Tereshko, V.: Reaction-diffusion model of a honeybee colony's foraging behaviour. In: Deb, K., Rudolph, G., Lutton, E., Merelo, J.J., Schoenauer, M., Schwefel, H.-P., Yao, X. (eds.) Parallel Problem Solving from Nature-PPSN VI. LNCS, vol. 1917, pp. 807–816. Springer, Heidelberg (2000)
11. Tereshko, V., Lee, T.: How information-mapping patterns determine foraging behaviour of a honey bee colony. Open Systems and Information Dynamics 9, 181–193 (2002)
12. Tereshko, V., Loengarov, A.: Collective decision making in honey-bee foraging dynamics. Computing and Information System Journal 9, 1–7 (2005)
13. Mikhailov, A.S.: Foundations of Synergetics I. Springer, Berlin (1990)
14. Eigen, M., Schuster, P.: The Hypercycle: A Principle of Natural Self-Organization. Springer, Berlin (1979)

Multi-level Selection in the Emergence of Language Systematicity

Luc Steels[1,2], Remi van Trijp[2], and Pieter Wellens[1]

[1] VUB AI Lab - Brussels
[2] Sony Computer Science Laboratory - Paris
steels@arti.vub.ac.be

Abstract. Language can be viewed as a complex adaptive system which is continuously shaped and reshaped by the actions of its users as they try to solve communicative problems. To maintain coherence in the overall system, different language elements (sounds, words, grammatical constructions) compete with each other for global acceptance. This paper examines what happens when a language system uses systematic structure, in the sense that certain meaning-form conventions are themselves parts of larger units. We argue that in this case multi-level selection occurs: at the level of elements (e.g. tense affixes) and at the level of larger units in which these elements are used (e.g. phrases). Achieving and maintaining linguistic coherence in the population under these conditions is non-trivial. This paper shows that it is nevertheless possible when agents take multiple levels into account both for processing meaning-form associations and for consolidating the language inventory after each interaction.

1 Sources of Selection in Language

There is a wide consensus among researchers in language evolution that language must have originated and still continues to evolve because there is a selectionist system underlying it. Despite this consensus, there is however a difference of opinion on how variation and hence the potential for change is caused and what selectionist pressures are operating to retain a particular variation in the language. Basically there are three different types of approaches, based on genetic evolution, cultural transmission, and problem solving respectively.

The genetic evolution models (e.g. [1]) put the selectionist pressure at the level of fitness, which is assumed to be directly related to communicative success. Agents are endowed with an artificial genome that determines how they should communicate: what perceptual primitives they should use for segmenting the world and identifying objects and features, what concepts they can employ for structuring their world, what types of categories are to be used, etc. Potential innovation takes place at the moment this genome is transmitted from parents to children. Because genome copying involves crossover and possibly mutation, variation is inevitable, and some of it will lead to higher or lower success.

F. Almeida e Costa et al. (Eds.): ECAL 2007, LNAI 4648, pp. 425–434, 2007.

Iterated learning models [5] are similar to genetic models, in the sense that variation and hence potential innovation takes place in the transmission of the language system from one generation to the next, but now the language and conceptual system is considered to be culturally coded instead of genetic. Children learn the language from their parents and are then assumed to use it largely unchanged throughout the rest of their life. The learning process necessarily introduces generalisations and variations because of the poverty of stimulus, and hence innovations may enter into the acquired language system. This innovation may re-appear in the data the learners generate for the next generation once they have become adults and thus gets preserved.

The third class of models views the task of building and negotiating a communication system as a kind of problem solving process. Agents try to achieve a communicative goal with maximal success and minimal effort. This problem solving process is definitely not a rational conscious problem solving process but an intuitive one that is seldom accessible to conscious inspection. It is not an individualistic problem solving process either, but a collective one, in which different individuals participate as peers. According to this view a communication system is built up in a step by step fashion driven by needs and failures in communication, and it employs a large battery of strategies and cognitive mechanisms which are not specific to language but appear in many other kinds of cognitive tasks, such as tool design or tool use. Recent experiments on the emergence of communication in human subjects provide good illustrations of these problem solving processes in action [4]. Variation and innovation in problem solving models are common because each individual can invent its own communication system. In fact the main challenge is rather to explain how agreement between individuals and thus a globally shared population language can ever arise.

In the problem solving approach, language becomes a Complex Adaptive System (CAS) in its own right, similar to a complex ecosystem or a complex economy [8]. There are many parallel competitions going on: between synonyms for becoming dominant in expressing a particular meaning, between idiomatic patterns that group a number of words, between different syntactic and semantic categories competing for a role in the grammar, between ways in which a syntactic category is marked, etc. An innovation only survives if it is successful in communication (which could be due to many factors such as the effectiveness of the meanings involved) and if it is also picked up and further propagated by a sufficient number of agents. Often there is no particular reason why one solution is preferred over another one, except that it is more frequent in the population and it wins because of the rich-get-richer dynamics. So we get two types of selectionist forces: functional and frequency-based.

The problem solving/complex adaptive systems approach underlies the many artificial life simulations and robotic experiments we have already carried out in our group (see e.g. [3,9,10,11,12]) and it will be pursued further in the remainder of this paper. Although this approach does not rely on the role of generational transmission in innovation or selection, we nevertheless see clearly a selectionist system arise. As suggested by Croft [2], not grammars (encoded genetically or culturally) but utterances are the 'replicating units' in language. More specifically, we consider the form-meaning associations (for example a word with an associated meaning or a grammatical construction that maps an abstract meaning to a syntactic pattern) as the replicating units. The replication takes place if it is used by a speaker to form an utterance AND if it is processed by the hearer (which may or may not lead to success).

2 Orchestrating Selection Through Consolidation

The problem solving/CAS approach requires that the complete communicative cycle is modeled, and not just the transmission from one generation to the next: a specific communicative interaction (a "language game") starts when the speaker sets a specific communicative goal, like drawing attention to an object in the scene, and then conceptualises the world to achieve that goal. He then has to express the conceptualisation based on his own linguistic inventory and produce the resulting utterance. The hearer parses the utterance with his own linguistic inventory and then projects the meaning back into his own experience. If an interpretation was possible, he must choose and perform the action that follows from this interpretation, for example point to the object that the speaker may have had in mind. At the end of this exchange, speaker and hearer can give feedback on success or failure in the game and they may try to fix a failed interaction with additional utterances or gestures.

The highly complex cognitive activities that agents must bring to bear to a complete successful communicative interaction can be grouped into five types of strategies. First of all agents need strategies for setting up a situation in which negotiations can take place to establish a communication system. For example, if we do this with embodied agents (robots) we will have to endow them with ways for setting up a joint attention frame with enough common ground and shared knowledge to guess the meanings that might be expressed by unknown words or constructions. This could be achieved with pointing gestures, eye gaze following, movement towards objects that are going to be the subject of the interaction, etc.

Second, agents need strategies to play the language game itself. They have to make a large number of decisions at all levels of language and conceptualisation. Often there are many alternatives, partly because different conventions are circulating in the population and partly because there is usually more than one way to express similar meanings. These decisions are primarily guided by achieving communicative success.

Agents also need diagnostic strategies for detecting that something is going wrong and for finding out the exact cause. The main indication is of course whether the communication itself did or did not achieve its desired effect. But agents need also more fine-grained diagnostics. For example, a word may have been misunderstood, a perceptual category used by the speaker may have been broader or more restricted compared to that of the hearer, the speaker may have adopted another perspective on the scene than the hearer, etc.

Fourth, agents need repair strategies for fixing a problem. For example, agents may introduce a new word or change the definition of a word they have in their

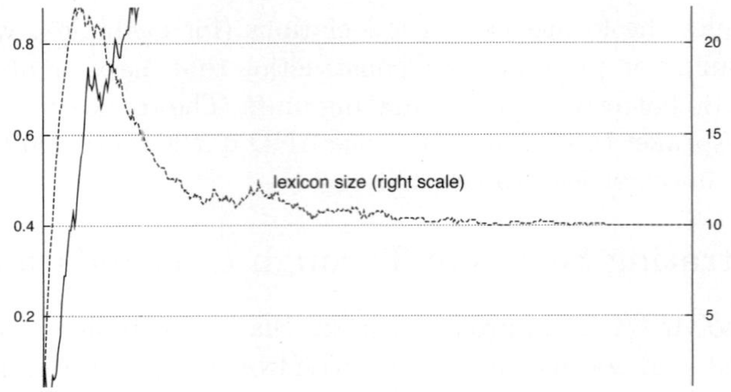

Fig. 1. A population of 5 agents plays a sequence of 5000 naming games, naming 10 objects. We see that the lexicon climbs up at first, reaching a peak of almost 25 after about 200 games. It then settles quickly to an optimum of 10 names for 10 objects thanks to the lateral inhibition dynamics.

lexicon, a hearer may shift a perceptual category slightly to align it with the way that category is used by the speaker, agents may start to mark perspective explicitly, or they may introduce more syntax to tighten combinatorial explosions in the search space or ambiguities in semantic interpretation [13].

Finally, agents need consolidation strategies which are enacted at the end of a game. Based on success or failure, agents have to adapt their inventories in order to become more successful in the future. Each element in an inventory has a particular score and the scores are adjusted to make their usage more or less likely in the future. In this paper we focus on the consolidation strategy, because it is responsible for ensuring that certain form-meaning conventions are selected, independently of functional factors.

These strategies can be illustrated by a simple naming game [7], in which the speaker tries to draw the hearer's attention to an object in the scene using a name. When the speaker does not have a name yet, he introduces a new one. When the hearer does not know the name, he adopts it after additional feedback from the speaker (e.g. a pointing gesture). Because games are played in a distributed population with only local interactions, some pairs may invent/adopt a name and others may invent/adopt another one. Unavoidably different competitors will start to float around in the population. However agents will have the most success and need the least effort if everyone always uses the same name for the same object. This can be achieved if agents keep a score in their local lexicons between objects and names. When they have to choose or interpret a name they use the association with the highest score. When a name is successful, its score is increased and competing names for the same object are decreased (both by speaker and hearer), thus implementing lateral inhibition. When a name is not successful, its score goes down. These strategies create a positive feedback loop so that one name will win the competition (see Figure 1).

The art of setting up language game experiments consists in endowing agents with the right kind of problem solving strategies, such that when these are applied in consecutive language games, the right kind of selectionist process emerges and the global language is driven towards the best or at least a viable state. Note that this process is not implemented directly but is a side effect of the local behaviours of the interacting agents.

3 The Problem of Systematicity

In the case of synonyms, competition takes place between different words for the same meaning, and we have seen that it is settled when agents use the appropriate lateral inhibition strategy. But consider now what happens when different words are grouped. For example, there could be words for actions ("pick-up", "put-down", "give", etc.) and words for objects ("ball", "box", ...) as well as combinations of these in simple patterns ("give ball", "push box", etc.). There will be competition going on among synonymous words (for example between "ball" and "ballon" and "box" and "boîte") but also between the patterns that contain these words (e.g. "pick-up ball" versus "pick-up ballon"). Obviously we want that if "ball" wins the competition at the lexical level, all the patterns that use this choice should become dominant as well. We also want that if new patterns form they should use the "best" words. So we now get selection at two levels: the level of words and the level of combinations of words. A simple lateral inhibition dynamics in which words and combinations compete with each other as if they are on the same level will not do, because there is no guarantee that the winner of the intra-word competition is also the one who wins the intra-pattern competition. For example, it is perfectly possible that the pattern "pick-up ballon" and the pattern "push ball" win their respective competitions, and so there is no guarantee that the meaning of "ball" is always going to be expressed with the same word. Even if words are still used on their own (for example to refer to the objects involved) and hence synonyms get damped (as in Figure 1), there is still the possibility that a word incorporated in a pattern is different from the word that won the competition at the word level.

The next computational experiment shows that this problem indeed occurs in computational simulations (see Figure 2). Agents are now playing games involving an action and objects involved in the action. Sometimes they describe only an object or an action and sometimes they describe the scene as a whole. Agents evolve both words for individual meanings ("give", "ball", etc.) and words for the patterns that combine these meanings (as in "give ball"). Figure 2 shows what happens when agents use the same strategies as in Figure 1. There is an optimal size for the lexicon (10 words), but not for the patterns (which would be 14 two-word patterns and 56 three-word patterns). The reason for this becomes apparent when we plot the coherence between meaning-form pairs in the population, measured as the frequency of the most used form for the same meaning. Even though coherence is quite high (around 90%), agents do not reach total coherence and hence have to remember several synonymous, competing patterns in order to maintain communicative success.

This problem of lingering incoherence does not only occur between words and phrases, but also recursively between phrases and more complex phrases. In

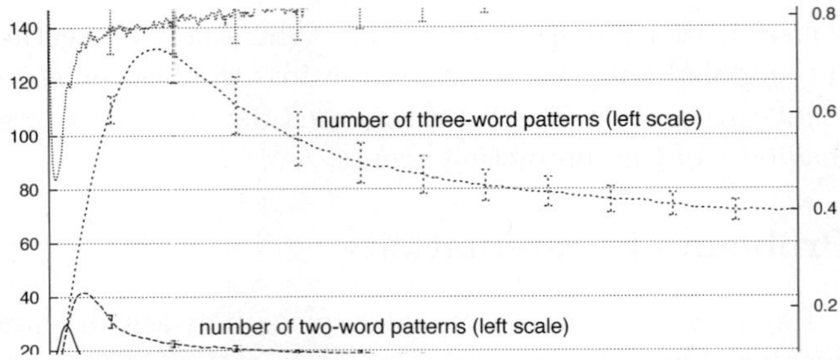

Fig. 2. A population of 10 agents plays a sequence of 30000 games. The graph shows the size of the lexicon and the number of patterns, as well as the meaning-form coherence in the total population which stagnates at around 90%.

fact, it has to be dealt with wherever systematicity is desirable in the grammar. For example, a convention for expressing a tense/aspect combination (such as the expression of the present perfective with "have/past-participle", as in "John has seen a lunatic running down the street") should be "productive" across all the situations in which it can be expressed, including in idiomatic phrases.

More generally, this issue of systematicity arises in every selectionist system in which there are elements competing with each other on one level but also undergoing selective competition at another level of larger structures in which they occur. An example from biology concerns the origins of chromosomes. Individual genes started to be combined into larger units (chromosomes). The genes are on the one hand replicators in their own right, undergoing competition, but they are also part of the larger replicating unit of chromosomes [6].

4 Multi-level Selection

We now propose mechanisms that achieve the right kind of multi-level selection. In the first experiment, agents have both a lexicon of associations between meanings and individual words, for example between the predicate [ball] and the word "ball", and a grammar of associations between more complex meanings and combinations of words, for example between the meaning 'push(o1), ball(o1)' and "push ball". Each of these associations has a score. When agents need to decide how to express a particular meaning they try to cover it with the minimal number of associations that has the highest score. This implies that a pattern, if it exists in the grammar, takes precedence over individual words.

The first step now is to refine the decision criteria by which agents choose which association to use, particularly for patterns. The agents do this in the following steps:

1. First look up all patterns that cover the complex meaning or form to be processed.

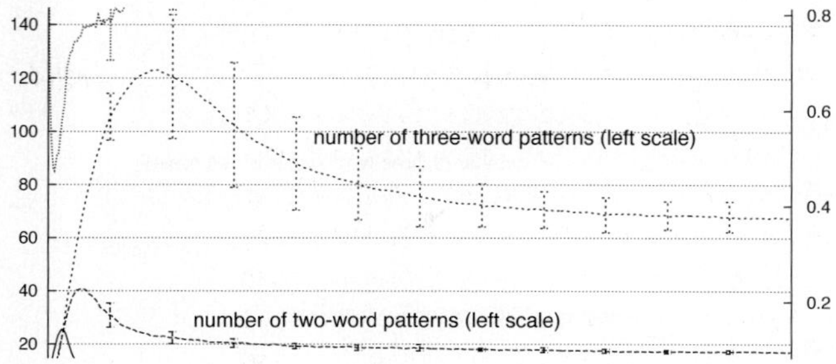

Fig. 3. A population of 10 agents plays a sequence of 30000 games. Here, the agents consider the score of the elements of a pattern in order to choose which pattern to apply. Coherence increases, but doesn't reach 100% yet.

2. Next rank these patterns, integrating not only the score of the pattern itself, but also the scores of the individual elements that make up the pattern. This is done in a recursive manner, so that agents can also handle patterns that are themselves part of still larger patterns.
3. If these choices lead to communicative success in the language game, then consolidate the (own) score of the pattern AND the scores of each of the elements in the pattern using the lateral inhibition dynamics discussed earlier.

The ranking of patterns in step 2 ensures that the agents will use those patterns that have not only been used successfully in the past themselves, but also whose elements have been the most successful. The lateral inhibition dynamics of step 3 not only punish conflicting patterns, but also conflicting elements that are used in these patterns.

The effect of this strategy on the semiotic dynamics in the population can be seen in Figure 3. The same conditions hold as in the previous figure: there are 10 agents playing 30000 games. Both the lexicon and grammar size are shown as well as the meaning-form coherence. We see that meaning-form coherence has improved compared to the previous strategies where patterns and words were competing at the same level, even though it is still not 100%. We also see the agents are slightly more efficient in settling on the grammar, in the sense that 5% fewer patterns circulate in the population before damping sets in.

In a second phase, we further improve results by not only considering the quality of parts to decide on the quality of the whole, but also by investigating in which other patterns this whole is itself a part. For individual words, this means that agents integrate the score of the patterns in which this word is used. Moreover in the consolidation strategy, agents not only update the scores of component parts using lateral inhibition but also the scores of the wholes in which the used association plays a part. Consequently there is now not only a top-down impact (as in Figure 3) from the whole to its parts, but also a bottom-up flow from the parts to the whole. The algorithm can be summarized as follows:

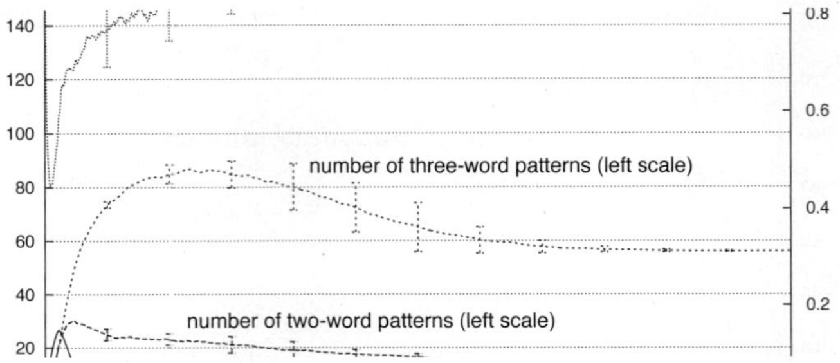

Fig. 4. A population of 10 agents plays a sequence of 30000 games while applying the more sophisticated consolidation strategies. Coherence now reaches 100% and there is greater efficiency in settling the conventions.

1. First look up which rules cover the meaning or the form to be processed.
2. Next rank these rules, integrating not only the score of the rule itself, but also the scores of all the elements that make up the rule AND the scores of larger rules in which the rule itself is used. This is again done in a recursive manner.
3. If these choices lead to a successful game, consolidate the (own) score of the rule AND the scores of its elements AND the scores of the larger rules of which the applied rule itself is a part.

Figure 4 shows the effect of this more sophisticated strategy for the same conditions as in the previous figures. We see that meaning-form coherence now has further improved to reach total coherence and the number of patterns has reached its optimal size. We also see that there is still greater efficiency in settling on the grammar compared to only top-down impact, in the sense that 25% fewer patterns circulate initially in the population. The effects of this strategy is further exemplified in Figure 5. Here, we see a single agent's network of forms covering the meaning "book" and its combinations into two-word patterns with the actions "give" and "take". The top diagram shows that after 500 language games the agent knows two synonyms for "book" and several competing patterns in which they are used. Some of these patterns even use words which themselves are no longer remembered by the agent. After 7000 games, however, a coherent network has evolved in which "book" is always expressed by the same form.

5 Conclusions

Language can be viewed as a complex adaptive system. Conventions of the language serving the same purpose are in competition with each other for dominance in the population and so in order to reach coherence and hence systematicity in a language, the right kind of selectionist forces must be set up. When all competitors are on the same level (as in the case of competing synonyms) a lateral inhibition dynamics has been shown to be adequate. But when there are different

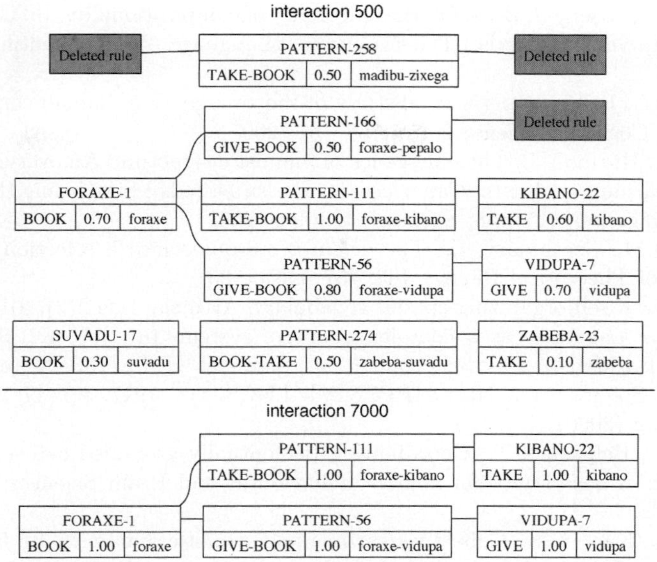

Fig. 5. These two diagrams show the evolution within a single agent's linguistic inventory of the form-meaning pairs covering the meaning "book" and its combinations with "give" and "take" in two-word patterns. Each box contains a rule name (top), the covered meaning (left), a success score (middle) and the form for expressing the meaning (right). The lines indicate which words were used to create which patterns. By considering multiple levels during consolidation, we see that after 7000 interactions the meaning "book" is systematically mapped to the form "foraxe".

levels in the system because certain elements are used as parts of other structures, we need multi-level selection. We have shown that this can be achieved by endowing the agents with a more sophisticated strategy for choosing which elements they are going to be used in a particular interaction and by applying the lateral inhibition consolidation strategy on different levels.

Acknowledgement. This research was funded (for PW) by a Belgian FWO TM 298 project and (for LS and RvT) by the Future and Emerging Technologies program (IST-FET) of the European Commission under the EU RD contract IST-1940. We thank other members of the VUB AI laboratory, particularly Joachim De Beule, and the SONY Computer Science Laboratory for discussion and implementation of the FCG framework that underlies the experiments.

References

1. Briscoe, E.: Grammatical acquisition: Inductive bias and coevolution of language and the language acquisition device. Language 76(2), 245–296 (2000)
2. Croft, W.: Explaining Language Change: An Evolutionary Approach. Longman, Harlow (2000)

3. De Beule, J., Bergen, B.K.: On the emergence of compositionality. In: Cangelosi, A., Smith, A., Smith, K. (eds.) The Evolution of Language, World Scientific, Singapore (2006)
4. Galantucci, B.: An experimental study of the emergence of human communication systems. Cognitive Science 29(5), 737–767 (2005)
5. Kirby, S., Hurford, J.: The emergence of linguistic structure: An overview of the iterated learning model. In: Cangelosi, A., Parisi, D. (eds.) Simulating the Evolution of Language, pp. 121–148. Springer, London (2002)
6. Smith, J.M., Szathmáry, E.: The origin of chromosomes i. selection for linkage. Journal of Theoretical Biology 163, 437–446 (1993)
7. Steels, L.: A self-organizing spatial vocabulary. Artificial Life 2(2), 319–332 (1996)
8. Steels, L.: Language as a complex adaptive system. In: Deb, K., Rudolph, G., Lutton, E., Merelo, J.J., Schoenauer, M., Schwefel, H.-P., Yao, X. (eds.) Parallel Problem Solving from Nature-PPSN VI. LNCS, vol. 1917, pp. 17–28. Springer, Heidelberg (2000)
9. Steels, L., Belpaeme, T.: Coordinating perceptually grounded categories through language: A case study for colour. Behavioural and Brain Sciences 28, 469–529 (2005)
10. Steels, L., Kaplan, F.: Collective learning and semiotic dynamics. In: Floreano, D., Mondada, F. (eds.) ECAL 1999. LNCS, vol. 1674, pp. 679–688. Springer, Heidelberg (1999)
11. Steels, L., Kaplan, F., McIntyre, A., Van Looveren, J.: Crucial factors in the origins of word-meaning. In: Wray, A. (ed.) The Transition to Language, pp. 252–271. Oxford University Press, Oxford (2002)
12. Steels, L., Loetzsch, M.: Perspective Alignment in Spatial Language. In: Coventry, K.R., Tenbrink, T., Bateman, J. (eds.) Spatial Language and Dialogue, Oxford University Press, Oxford (2007)
13. Steels, L., Wellens, P.: How Grammar Emerges to Dampen Combinatorial Search in Parsing. In: Vogt, P., Sugita, Y., Tuci, E., Nehaniv, C.L. (eds.) EELC 2006. LNCS (LNAI), vol. 4211, pp. 76–88. Springer, Heidelberg (2006)

Protolanguages That Are Semi-holophrastic

Mike Dowman[1]

[1] Department of General Systems Studies, The University of Tokyo,
3-8-1 Komaba, Meguro-ku, Tokyo 153-8902, Japan.
mike@sacral.c.u-tokyo.ac.jp

Abstract. There is an ongoing debate about whether the words in the first languages spoken by humans expressed single concepts or complex holophrases. A computer model was used to investigate the nature of the protolanguages that would arise if speakers could associate words and meanings, but lacked any productive ability beyond saying the word whose past uses most closely matched the meaning that they wished to express. It was found that both words expressing single concepts, and holophrastic words could arise, depending on the conceptual and articulatory abilities of the agents. However, most words were of an intermediate type, as they expressed more than a single concept but less than a holophrase. The model therefore demonstrates that protolanguages may have been of types that are not usually considered in the debate over the nature of the first human languages.

Key words: Language Evolution, Protolanguage, Synthetic, Analytic, Holophrasis, Iterated Learning.

1 Introduction

Non-human apes do not have the ability to use language, so that ability must have evolved since the human lineage split from that of the other apes. However, it seems unlikely that the first languages that arose had the full complexity of modern human language. Instead it seems more likely that initially languages were very simple, and that humans gained the ability to use more complex language only as a result of further phylogenetic changes.

One of the most distinctive characteristics of modern languages is the complexity of their syntactic structures. Syntax extends the expressiveness of languages, by allowing the meanings of utterances to be related to the structures in which words are used, not just to depend on the words themselves. However, syntax would be of little use unless there already existed meaningful words that could be combined using syntactic rules. Therefore, it seems that humans must have evolved a capacity to associate words with meanings before they evolved the ability to use syntax. Bickerton [3, 4] has termed such pre-syntactic communication systems *protolanguages*.

If we are to determine the nature of these protolanguages, we need to first determine the kind of cognitive abilities humans had immediately prior to their development. Probably the most relevant evidence for this comes from the cognitive abilities of other apes, and in particular their linguistic capabilities. The human lineage split from that of

F. Almeida e Costa et al. (Eds.): ECAL 2007, LNAI 4648, pp. 435–444, 2007.

bonobos and common chimpanzees after the common human-chimpanzee lineage split from the lineages leading to present day gorillas and orangutans. Therefore we can assume that any ability to use language that is common to all non-human great apes was also possessed by the last common ancestor of both humans and chimpanzees [5]. (The alternative is that the ability later evolved separately in gorillas, chimpanzees and orangutans, which seems much less likely.)

Non-human apes do not use anything resembling human language spontaneously in the wild, but several attempts have been made to teach some form of language to captive chimpanzees and gorillas [12, 16, 13, 8]. These studies have shown that while apes have only a very limited ability (if any at all) to use syntactic rules to structure their language, they are much better at learning associations between words and meanings. So long as the words are presented in an appropriate modality (for example using manual gestures, or a lexicographic keyboard) apes can also use the words that they have learned productively, although, unlike humans, they rarely use language spontaneously [16]. When trained apes produce language, rather than choosing the minimal set of words necessary to convey the meaning that they wish to communicate, they appear to simply to use any words relevant to the meaning, without paying attention to whether the whole of the meaning is conveyed, or whether some part of the meaning is conveyed more than once. For example, one trained chimpanzee, Nim Chimpsky, once uttered 'give orange me give eat orange me eat orange give me eat orange give me you' [16, p210]).

We can therefore presume that the last common ancestor of humans and chimpanzees also had such latent abilities, but like modern day non-human apes, did not make use of them. From such a starting point, the only evolutionary change that would have had to have taken place in order for languages to emerge would have been for a tendency to spontaneously use language in the absence of explicit training to have evolved. We could expect that this would have resulted in the emergence of protolanguages with little or no syntax. When speaking, like trained apes, the users of those languages would simply have uttered any words whose meanings were related to the proposition they were trying to express. This paper reports research that used a model of speakers with this kind of language ability to determine what kinds of properties protolanguages emerging in this situation would have had.

The words in the languages spoken by trained apes, are similar to open class words in modern languages, in that they mainly refer to objects such as bananas, or to actions such as tickling. Bickerton [3, 4], Carstairs-McCarthy [6] and Tallerman [15] have argued that the words in protolanguages would have had similar meanings. However, Wray [17, 18] and Arbib [1] have made a quite contradictory proposal, suggesting instead that words in protolanguages would each have expressed a complex holistic meaning, such as '*give us the meat*' [17, p51]. Wray supported her proposal by noting that this kind of word seems to be a natural progression from the kind of communicative signals used by other animals. For example vervet monkeys have an alarm call for eagles, but the meaning of the signal would much better be translated into English as *beware of the eagle*, than as the English word *eagle*, as it can only be used to warn that an eagle is present, and never simply to refer to the concept EAGLE in the abstract, as the English word can.

Previous multi-agent models have been used to study the evolution of human languages [14, 2, 10, 11, 9]. Steels et al [14] simulated the evolution of asyntactic languages in which words could have either holophrastic meanings denoting ranges of values on several different sensory channels, or simple atomic meanings corresponding to a range of values in single sensory channel. However, it seems that a single word with an atomic meaning was usually sufficient to satisfy the communicative needs of Steels et al's agents (which was simply to distinguish a target referent from a context of other objects), so all the most commonly used emergent words consequently had atomic meanings. In other models [2, 10, 9] there was a transition from holistic protolanguages to fully syntactic languages, but in these models the agents had the ability to use syntactic language built in from the very start. It therefore took only a cultural evolutionary process for syntactic languages to evolve. Such models could therefore correspond to a point just after the human capacity for syntactic language had emerged, but before humans had used that capacity to actually create a syntactic language. However, they do not directly address the question of what kind of protolanguage humans lacking an ability to use syntactic language would have spoken.

The model reported here assumes that initially humans tried to use language to express only a limited range of the meanings that were of most importance to them. These meanings might have included concepts such as eating, hunting, and the kinds of animals that early humans ate and hunted. It seems reasonable to assume that initially humans had only a limited ability to remember, articulate or perceive distinct words, as the ability to do any of these things would have come under selective pressure only once language had come into use. It would have been possible for early humans to have used only as many words as the number of distinct sounds that they were able to articulate, but cognitive learning or memory limitations might have further restricted the number of words that they were able to use. Modern human language are able to contain many thousands of perceptually distinct words, because each word is made up of a sequence of discreet phonemes, but it would seem unlikely that the ability to use phonemic language was selected for before protolanguages came into use. The ability to use only a limited number of words was therefore incorporated into the computer model in order to reflect this presumed communicative limitation of early humans.

2 A Model of Protolanguage

The computer model was a very simple expression-induction model [9], which contained ten agents, each of which was able to express meanings by choosing from amongst the words it knew, and to learn new words, and new meanings for words, by observing other agents' use of those words. Each agent was present throughout the whole simulation, so the simulations represented the creolization of a language in a single generation. This contrasts with the work of Dowman [7], which reported an iterated learning model in which protolanguages were passed from generation to generation along a chain of agents, but in which there was only a single agent in each generation. These models incorporate complementary aspects of the process by which language is passed between individuals in a society, and so any results obtained with one model will be reinforced if similar results are obtained with the other.

Meanings were created at random for the agents to communicate to one another. Each meaning was composed of a small number of individual concepts (four in the case of the results reported in this paper) from a set of ten basic concepts. No constituent concept ever occurred more than once in any complex meaning, and all concepts were chosen with equal probability. Different orderings of the same meanings were not treated as significant.

While this approach clearly neglects many aspects of word meaning, including the role of context and the active role that language users may take in interpreting word meanings, it was sufficient to create a wide range of possible complex propositions that the agents could try to communicate. For example, MAN CHILD ANTELOPE SEE could be taken as a representation of the English phrase *A boy sees an antelope*, although it does not convey as much information as the equivalent English sentence. (In particular tense and number information is absent, and it does not make explicit which semantic role is filled by which participant, or even that it the man that is a child, rather than the antelope.) This form of meaning representation was used to simulate an ability, or a propensity, to express only a limited number of simple meanings, on the assumption that humans would initially have tried to communicate only simple messages.

The agents' limited capacity regarding the number of words they were able to use was simulated simply by limiting the number of distinct words available to the agents in the model. Each word was assumed to be completely distinct, so the agents would never confuse one word with another. Each communicative interaction took place between one randomly selected agent, and one other different randomly selected agent, each agent being selected with equal probability. The speaker would be given a meaning to express, and in response would say one or more words which it associated with that meaning.

In order to learn words in the absence of explicit instruction, people must be able to infer the meaning that a word is intended to convey on at least some of its occasions of use. Therefore, each time a word was used in the simulations, the hearing agent would also observe the meaning that the speaker was trying to express, and would remember that that word had been used to express that meaning. However, the agents were given no indication of which part of the meaning corresponded to which word. Agents would only remember the ten most recent uses of each word, in order to simulate a limited memory capacity, and a preference for updating their internal knowledge of language to reflect recent usage. The agents' knowledge of language therefore consisted of a list of words, and a list of up to ten complex meanings that the agent has observed the word being used to express. (The same complex meaning would be paired with a word more than once if the agent heard the word used to express exactly the same meaning on more than one occasion.)

When agents spoke they simply compared the meaning they were trying to express to the examples of the use of each word that they remembered at that time, so that they could say the words that in the past had been used to express meanings most similar to the target meaning. In order to achieve this, a degree of match was calculated for each word for each of the concepts in the meaning to be expressed, by finding the proportion of meanings that the agent remembered for that word that contained the concept. These scores were then averaged over each semantic element in the target meaning. For example, if a target meaning contained WOMAN HUNT EAT BIRD, and the agent had previously observed a word being used to express CHILD BIRD NUT WOMAN and NUT BIRD CHILD EAT, then the degree of match would be 0.5, because there are two matches for BIRD, and one each for WOMAN and EAT, out of a total of 8 possible matches. This allowed a score for the similarity of the past uses of each word to the meaning to be expressed to be obtained. These scores could vary from 1 (if the word had only ever been used to express the current meaning) to 0 (if the word had never expressed a meaning containing any of the elements in the meaning to be expressed).

If the agent knew one or more words for which the degree of match was 1, it would say them (up to a limit of three words, so a random choice of three words would be made if ever there were more than three words that matched exactly). Otherwise, if it were able to use a new word for which it had not yet observed any meanings, it would use that word. It would also remember that that word had expressed that meaning, so it would be able to reuse the same word if it ever needed to express the same or a similar meaning in the future. In all other cases, the speaker would just say the three words which received the highest scores (choosing at random in the event of ties).

In each simulation, 125,000 utterances were spoken, so that the there was plenty of time for a coherent language shared by all the agents to emerge. The analysis of the languages that emerged was based on the internal representations of the agents at the end of these simulations. The only parameter that was changed between different simulation runs was the number of distinct words that the agents were able to use, which was varied from only 10 words right up to 500 words.

3 Emergent Languages

A wide range of different types of words emerged in the resulting languages, going well beyond the holophrastic words and the words expressing single concepts that have been the focus of the debate over the nature of protolanguages. Most of the words communicated more than a single concept, but less than a complete proposition.

Table 1 shows the meanings that an agent had associated with two of the words that it knew. This agent was from a simulation in which the agents could use only ten words. The entries in this table indicate the range of meanings that other agents expressed using these words, and they would have provided the basis on which the agent would have chosen which words to say when presented with a meaning to be communicated. Looking at the meanings associated with the first word, we can see that it has been used to express propositions containing all ten basic concepts. It might seem that this word cannot therefore be said to express any single concept, but careful inspection of the meanings reveals that all of them contain the semantic element WOMAN. If this agent heard this word, its observations of its past uses would allow it to reason that the meaning that the speaker was trying to communicate included the meaning WOMAN. The agent would not gain any information about which other concepts were contained in the meaning that the speaker was trying to convey, as no other concept is present in all the entries in the table. Therefore, this word effectively communicates the single concept WOMAN, and so can be said to have a meaning similar to the English word *woman*. This word therefore supports Bickerton' s [3, 4] claim that the words in prototypes resembled open class items in modern languages.

The second word in Table 1 is similar, in that it also has a single semantic element that is common to all its observed uses (in this case BIRD). We might therefore consider this word's meaning to correspond simply to the concept BIRD. However, in eight of the ten remembered uses of this word, the meanings it had expressed contained

Table 1. Example Word Meanings when Agents could use only 10 Words.

Meanings Expressed by Word (frequency in brackets)	Description of Word Meaning
HUNT NUT SEE WOMAN (1)	Word denotes WOMAN
CHILD NUT MAN WOMAN (1)	
MAN CHILD HUNT WOMAN (1)	
WOMAN BIRD ANTELOPE NUT (1)	
WOMAN BIRD ANTELOPE HUNT (1)	
LION HUNT WOMAN CHILD (1)	
ANTELOPE HUNT NUT WOMAN (1)	
BIRD LION HUNT WOMAN (1)	
WOMAN MAN CHILD BIRD (1)	
WOMAN NUT ANTELOPE EAT (1)	
BIRD ANTELOPE NUT MAN (1)	Word denotes BIRD but carries strong connotations of NUT
WOMAN BIRD ANTELOPE NUT (1)	
WOMAN BIRD ANTELOPE HUNT (1)	
BIRD SEE ANTELOPE NUT (1)	
NUT BIRD MAN EAT (2)	
CHILD BIRD NUT EAT (1)	
BIRD SEE MAN CHILD (1)	
NUT LION BIRD EAT (1)	
MAN BIRD NUT SEE (1)	

the concept NUT. An agent hearing this word would therefore be able to infer that the intended meaning probably contained the concept NUT in addition to the core meaning BIRD. This word is therefore like many words in modern languages that have strong connotations of concepts that they do not necessarily express. For example the English word *incident* means an event or occurrence, but it strongly suggests the event was in some way undesirable. (This can be seen by comparing the odd sounding *fortunate incident* with the much more normal *unfortunate incident.*) This addition of an extra degree of meaning beyond the single concept which is the core meaning of the word gives this word a small degree of holophrasticity. Most words emerging in simulations in which the agents could use only ten words had meanings of one of these two types.

When agents were able to use 50 words, the meanings of most of the words were of a somewhat different type. Table 2 shows the meanings that an agent had associated with two words emerging in this condition. All the meanings associated with the first word contain both the concept MAN and the concept BIRD, so this word only expresses meanings containing both of those two concepts. This word is therefore partly holophrastic, in that it expresses two concepts instead of one, but is not completely holophrastic as it does not convey a complete four concept complex meaning. The second word in Table 2 is slightly more holophrastic, as it always expresses both the concept MAN and the concept SEE, but it also carries strong connotations of EAT, as 8 out of the 10 observed meanings contained the concept EAT.

When the number of words that the agents could use was increased still further, words with a greater degree of holophrasticity became more common, as shown in Table 3. The first word in this table is of a type common under this condition, which expresses three concepts, but does not convey any information about the final concept in the proposition being expressed. The second word in Table 3 is, however of a quite

Table 2. Example Word Meanings when Agents could use 50 Words.

Meanings Expressed by Word (frequency in brackets)	Description of Word Meaning
WOMAN BIRD MAN EAT (2)	Word denotes MAN BIRD
EAT MAN SEE BIRD (2)	
HUNT SEE MAN BIRD (1)	
BIRD EAT CHILD MAN (1)	
NUT ANTELOPE MAN BIRD (1)	
BIRD WOMAN ANTELOPE MAN (1)	
MAN LION BIRD CHILD (1)	
WOMAN SEE MAN BIRD (1)	
EAT MAN NUT SEE (3)	Word denotes MAN SEE with strong connotations of EAT
HUNT SEE MAN BIRD (1)	
SEE EAT ANTELOPE MAN (2)	
MAN BIRD SEE NUT (1)	
MAN SEE HUNT EAT (1)	
MAN SEE EAT LION (1)	
MAN EAT BIRD SEE (1)	

Table 3. Example Word Meanings when Agents could use 100 Words.

Meanings Expressed by Word (frequency in brackets)	Description of Word Meaning
ANTELOPE BIRD HUNT NUT (3)	Word denotes ANTELOPE BIRD HUNT
EAT ANTELOPE HUNT BIRD (2)	
BIRD MAN ANTELOPE HUNT (2)	
BIRD HUNT ANTELOPE CHILD (1)	
ANTELOPE LION HUNT BIRD (2)	
CHILD HUNT MAN EAT (3)	Word has strong connotations of CHILD BIRD HUNT EAT MAN but does not denote any concept
MAN EAT BIRD HUNT (3)	
HUNT BIRD CHILD MAN (1)	
CHILD MAN EAT BIRD (1)	
BIRD HUNT EAT CHILD (2)	

different type. This word has no meanings common to all its uses, so on hearing this word an agent could not infer that any particular concept had been expressed. This might appear to make the word communicatively useless, but closer inspection reveals that all of the concepts its expresses come from a set of only five concepts. Therefore, any use of this word carries very strong connotations of all of those concepts, even though each communicated meaning in fact only contains four of them. This word is therefore in some ways highly holophrastic, in that it conveys information about all four concepts in a complex meaning, but it is also in some ways less holophrastic than even a word expressing a single concept, as it does not unambiguously express any single concept.

Fully holophrastic words usually only emerged when the agents were able to use a much larger number of words. Two fully holophrastic words, which expressed only a single four concept complex meaning are shown in Table 4. This is the kind of meaning that the words in protolanguages had, according to the claims of Wray [17, 18] and Arbib [1].

Table 4. Example Word Meanings when Agents could use 500 Words.

Meanings Expressed by Word (frequency in brackets)	Description of Word Meaning
EAT MAN ANTELOPE HUNT (10)	Word denotes EAT MAN ANTELOPE HUNT
HUNT BIRD SEE LION (10)	Word denotes HUNT BIRD SEE LION

Figure 1 shows the relationship between the number of words that the agents were able to use and the degree of holophrasticity of the emergent word meanings. In general, when the agents could use more words, then words with greater degrees of holophrasticity emerged. This should not be surprising, as with only a small number of words, there simply are not enough words available for a separate word to be assigned to each possible fully or partly holophrastic meaning. However, with more words, it is possible to assign a word to many far more specific meanings, allowing a wide range of holophrastic meanings to each be assigned their own words.

Depending on the number of words that the agents could use, words expressing a single concept, a two or three concept partial holophrase, or a fully holophrastic four concept proposition were the most common kind of word. However, because as noted above words can have strong connotations of concepts beyond their core meanings, there is in reality a continuum between words which have atomic meanings and those which have holophrastic meanings. Furthermore, most of the emergent languages contained a mixture of different types of words, so revealing another way in which a language can be intermediate between the type of protolanguage proposed by Bickerton [3, 4] and that envisaged by Wray [17, 18].

Figure 1 suggests that if, when language first arose, humans were able to use only a small number of words, but gradually became able to use more words as the ability to articulate, recognize and remember words came under selective pressure, protolanguage words would initially have had meanings like those in modern languages, but would gradually have become increasingly holophrastic as the number of words that people could use increased. However, clearly the degree of holophrasticity is determined by the number of words relative to the number of concepts that the agents try to communicate, so if instead the complexity of the agents conceptual capacity had increased faster than their communicative ability, there would have been a change in the opposite direction, from words with holophrastic meanings to ones with meanings increasingly like those of modern words. As both of these abilities may well have evolved in tandem, over many generations of speakers, protolanguages could have gone through stages of being increasingly holophrastic, or increasingly like modern words. There does not seem to be any reason to assume that there could only have been a single type of protolanguage prior to the emergence of fully syntactic language.

4 Conclusions

The research reported in this paper used the evidence we have concerning the cognitive capacities of early humans prior to their use of language to infer what kind of language abilities were most likely possessed by the speakers of the first protolanguages. By creating agents with corresponding language abilities, and simulating conversations between them, it was possible to create a model in which protolanguages emerged as

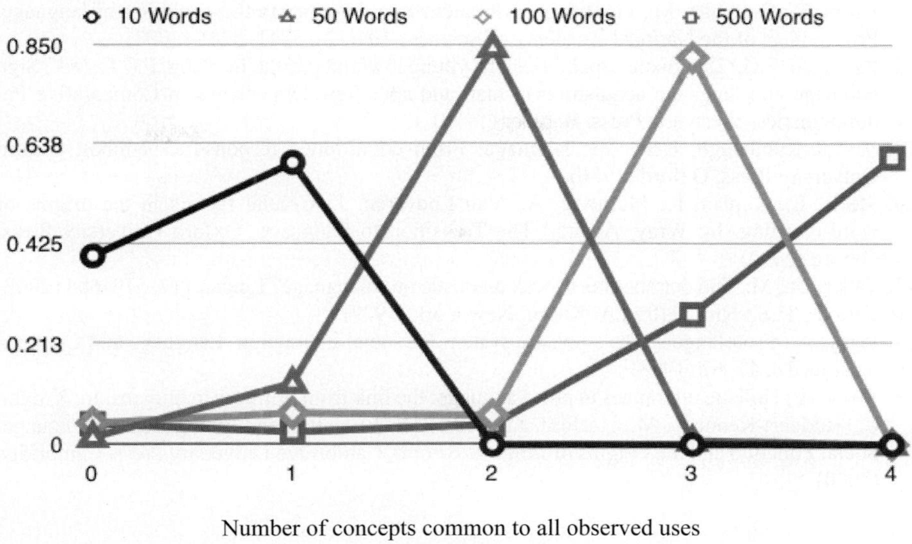

Fig. 1. The proportion of words (y axis) which had a particular number of concepts common to all their uses. These figures are averages over all ten agents in a simulation. As the number of available words increases, so does the frequency of the more holophrastic word types, while the proportion of the less holophrastic types decreases.

References

1. Arbib, M.A.: From monkey-like action recognition to human language: An evolutionary framework for neurolinguistics. Behavioral and Brain Sciences 28, 105–167 (2005)
2. Batali, J.: Computational simulations of the emergence of grammar. In: Hurford, J.R., Studdert-Kennedy, M., Knight, C. (eds.) Approaches to the Evolution of Language: Social and Cognitive Bases, Cambridge University Press, Cambridge (1998)
3. Bickerton, D.: Language and Species. University of Chicago Press, Chicago (1990)
4. Bickerton, D.: Language and Human Behavior. UCL Press, London (1996)
5. Burling, R.: The talking Ape: How Language Evolved. Oxford University Press, Oxford (2005)
6. Carstairs-McCarthy, A.: The Origins of Complex Language: An Inquiry into the Evolutionary Beginnings of Sentences, Syllables, and Truth. Oxford University Press, Oxford (1999)
7. Dowman, M.: The Nature of Words in Human Protolanguages: It's not a Synthetic-Analytic Dichotomy (Submitted)
8. Greenfield, P.M., Savage-Rumbaugh, E.S.: Imitation, grammatical development, and the invention of protogrammar by an ape. In: Krasnegor, N.A., Rumbaugh, D.M., Scheifelbusch, R.L., Studdert-Kennedy, M. (eds.) Biological and Behavioral Determinants of Language Development, Lawrence Erlbaum, Hillsdale (1991)
9. Hurford, J.: Expression/induction models of language evolution: dimensions and issues. In: Briscoe, T. (ed.) Linguistic Evolution through Language Acquisition: Formal and Computational Models, Cambridge University Press, Cambridge (2002)
10. Kirby, S.: Syntax without natural selection: How compositionality emerges from vocabulary in a population of learners. In: Knight, C., Hurford, J.R., Studdert-Kennedy, M. (eds.) The Evolutionary Emergence of Language: Social Function and the Origins of Linguistic Form, Cambridge University Press, Cambridge (2000)

11. Kirby, S., Dowman, M., Griffiths, T.: Innateness and culture in the evolution of language. Proceedings of the National Academy of Sciences 104(12), 5241–5245 (2007)
12. Patterson, F.G.: Linguistic capabilities of a young lowland gorilla. In: Peng, F.C.C. (ed.) Sign language and language acquisition in man and ape: New Dimensions in Comparative Pedolinguistics, Westview Press, Boulder (1978)
13. Savage-Rumbaugh, E.S.: Ape language: From conditioned response to symbol. Oxford University Press, Oxford (1986)
14. Steels, L., Kaplan, F., McIntyre, A., Van Looveren, J.: Crucial factors in the origins of word-meaning. In: Wray, A. (ed.) The Transition to Language, Oxford University Press, Oxford (2002)
15. Tallerman, M.: Did our ancestors speak a holistic protolanguage? Lingua 117, 579–604 (2007)
16. Terrace, H.S.: Nim. Alfred A. Knopf, New York (1979)
17. Wray, A.: Protolanguage as a holistic system for social interaction. Language and Communication 18, 47–67 (1998)
18. Wray, A.: Holistic utterances in protolanguage: the link from primates to humans. In: Knight, C., Studdert-Kennedy, M., Hurford, J.R. (eds.) The Evolution and Emergence of Language: Social Function and the Origins of Linguistic Form, Cambridge University Press, Cambridge (2000)

From the Outside-In: Embodied Attention in Toddlers

Linda B. Smith, Chen Yu, and Alfredo Pereira

Department of Psychological and Brain Sciences, and Cognitive Science Program
Indiana University, Bloomington IN 47405, USA
smith4,chenyu,afpereir@indiana.edu,
http://www.indiana.edu/~dll/

Abstract. An important goal in cognitive development research is an understanding of the real-world physical and social environment in which learning takes place. However, the relevant aspects of this environment for the learner are only those that make contact with the learner's sensory system. We report new findings using a novel method that seeks to describe the visual learning environment from a young child's point of view. The method consists of a multi-camera sensing environment consisting of two head-mounted mini cameras that are placed on both the child's and the parent's foreheads respectively. The main results is that the adult and child's view are fundamentally different in that the child's view is more dynamic and centered on one object at time. These findings have broad implications for how one thinks about toddler's attentional task as opposed to adults. In one sense, toddlers have found cheap solution: Selectively attend not by changing internal weights by bringing the attended object close to your eyes so it is the only one in view.

1 Introduction

Children learn about their world - about objects, actions, other social beings, and language – through their second-by-second, minute-by-minute sensorimotor interactions. Visual information plays a critical role in this early learning. Before babies with normal vision can talk or walk, and before the emergence of any social intelligence to guide their everyday interaction with caregivers, babies are able to perceive and parse their visual environment and are able to move their eyes and head to select visual targets (objects or people) in space. Infants have the opportunity to continuously process complex visual input, and accumulate knowledge from the visual environment. This real time visual information, plus its control through gaze direction and visual attention, contributes to the development of other sensory, cognitive and social capabilities. Indeed, developmentalists such as Gibson [4]and Ruff [6] have documented the powerful dynamic visual information that emerges as infants and children move their eyes, heads and bodies, and as they act on objects in the world. In addition, Bertenthal and Campos [2] have shown how movement - crawling and walking over, under, and around obstacles - creates dynamic visual information crucial to children's developing knowledge about space. Researchers studying the role of social partners in development and problem solving also point to the body and active movement -points, head turns, eye gaze and active movement - in social dynamics and particularly in establishing joint attention

F. Almeida e Costa et al. (Eds.): ECAL 2007, LNAI 4648, pp. 445–454, 2007.

(Yu, Ballard, Aslin[10];Smith and Breazeal [7],for a review). Computational theorists and roboticists (Ballard et. al [1]) have also demonstrated the computational advantages of what they call "active vision", how an observer - human or robot - is able to understand a visual environment more effectively and efficiently by interacting with it. This is because perception and action form a closed loop; attentional acts are preparatory to and made manifest in action while also constraining perception in the next moment.

Nonetheless, most previous studies of children's attention and learning have been conducted using macro-level behaviors and in constrained situations, without considering the role of active vision and the perception-action loop. This is in part a consequence of the typical method which uses a third person camera (or several) to record the child's stream of activities in context. Such recordings provide the view of an outside observer not the view of the actively engaged cognitive system. Further, these views are typically conded by human coders who watch these third person views, a process which is both time consuming and biased, as these coders are outside observers with their own psychology and parsing of the events. Understanding how developmental process emerges in second-by-second and minute-by-minute sensori-motor interactions requires capturing (and describing without bias) the first-person view as it is actively generated by the young learner.

The larger goal of this research enterprise is to understand the building blocks for fundamental cognitive capabilities and, in particular, to ground social interaction and the theory of mind in sensorimotor processes. To these ends, we have developed a new method for studying the structure of children's dynamic visual experiences as they relate to children's active participation in a physical and social world. In this paper, we report results from a study that implemented a sensing system for recording the visual input from both the child's point of view and the parent's viewpoint as they engage in toy play. With this new methodology, we compare and analyze the dynamic structure of visual information from these two views. The results show that the dynamic first-person perspective from a child is substantially different from either the parent's or the third-person (experimenter) view commonly used in developmental studies of both the learning environment and parent-child social interaction. The key differences are these: the child's view is much more dynamically variable, more tightly tied to the child's own goal-directed action, and more narrowly focused on the momentary object of interest.

2 Method

2.1 Multi-camera Sensing Environment

The method uses multi-camera sensing system in a laboratory environment wherein children and parents are asked to freely interact with each other. As shown in Figure 1, participants interactions are recorded by three cameras from different perspectives - one head-mounted camera from the child's point of view to obtain an approximation of the child's visual field, one from the parent's viewpoint to obtain an approximation of the parent's visual field, and one from a top-down third-person viewpoint that allows a clear observation of exactly what was on the table at any given moment (mostly the participants' hands and the objects being played with).

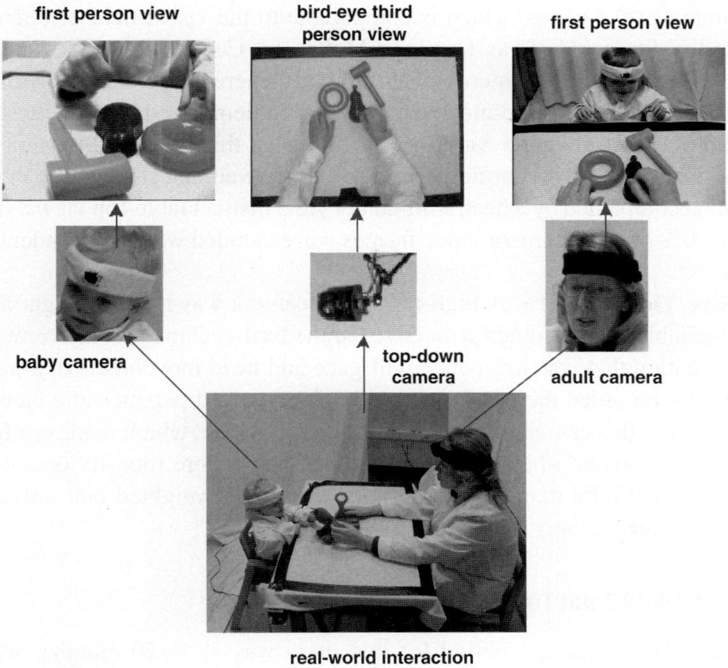

Fig. 1. Multi-camera sensing system. The child and the mother play a set of toys at a table. Two mini cameras are placed onto the child's and the mother's heads respectively to collect visual information from two first-person views. A third camera mounted on the top of the table records the bird-eye view of the whole interaction.

Interaction Environment. The study was run in a 3.3m x 3.1m room. At the center of the room a 61cm x 91cm x 64cm table was placed. The table surface was covered in a white soft blanket and the edges were clearly marked with black tape. A high chair for the child and a small chair for the parent was placed facing each other. The walls and floor of the room were covered with white fabrics. Both participants were asked to wear white T-shirts as well. In this way, from any image collected from any camera, white pixels can treated as background while non-white pixels are either objects on the table, the edges of the table, the hands, or the faces of participants.

Head-Mounted Cameras. Two light-weight head-mounted mini cameras (one for the child and another for the parent) were used to record the first-person view from both the child and the parent's perspectives. These cameras were mounted on two everyday sports headbands, each of which was placed on one participant's forehead and close to his eyes. The angle of the camera was adjustable. Input power and video output to these cameras went through a camera cable connected to a wall socked, which was long enough to not cause any movement restriction while participants were sitting down. Both cameras were connected via standard RCA cables to a digital video recorder card in a computer in the room adjacent to the experiment room. The head camera field is approximately 90 degrees, which is comparable to the visual field of older infants, toddlers and adults (van hof van Duin and Mohn, [8], Candy, et al. [3]). One possible concern in the use of a head camera is that the head camera image changes with changes in head movements not in eye-movements. This problem is reduced by the geometry of table-top play. Yoshida and Smith [9] documented this in a head-camera study of toddlers by independently recording eye-gaze and showed that small shifts in eye-gaze direction unaccompanied by a head shift do not yield distinct table-top views. Indeed, in their study 90% of head camera video frames corresponded with independently coded eye positions.

Bird-Eye View Camera. A high-resolution camera was mounted right above the table and the table edges aligned with edges of the bird-eye image. This view provided visual information that was independent of gaze and head movements of a participant and therefore it recorded the whole interaction from a third-person static view. An additional benefit of this camera lied in the high-quality video, which made our following image segmentation and object tracking software work more robustly compared with two head-mounted mini cameras. Those two were light-weighted but with a limited resolution and video quality due to the small size.

2.2 Parent-Child Joint Interaction Experiment

Participants. The target age period for this study was 18 to 20 months. We invited parents in the Bloomington, Indiana area to participate in the experiment. Nine dyads of parent and child were part of the study. One child was not included because of fussiness before the experiment started. For the child participants included, the mean age was 18.2, ranging from 17.2 to 19.5 months. Three of the included children were female and five were male. All participants were white and middle-class.

Stimuli. Parents were given six sets (three toys for each set) in this free-play task. The toys were either rigid plastic objects or plush toys (three of the total 18). Most of them had simple shapes and either a single color or an overall main color. Some combinations of objects were selected to elicit an action, especially evident to an adult asked to play with them.

Procedure. The study was conducted by three experimenters: one to distract the child, another to place the head-mounted cameras and a third one to control the quality of video recording. Parents were told that the goal of the study was simply to observe how they interacted with their child while playing with toys and that they should try to interact as naturally as possible. Upon entering the experiment room, the child was quickly seated in the high chair and several attractive toys were placed on top of the table. One experimenter played with the child while the second experimenter placed a sports headband with the mini-camera onto the forehead of the child at a moment that he appeared to be well distracted. Our success rate in placing sensors on children is now at over 80%. After this, the second experimenter placed the second head-mounted camera onto the parent's forehead and close to her eyes.

To calibrate the horizontal camera position in the forehead and the angle of the camera relative to the head, the experimenter asked the parent to look into one of the objects on the table, placed close to the child. The third experimenter controlling the recording in another room confirmed if the object was at the center of the image and if not small adjustments were made on the head-mounted camera gear. The same procedure was repeated for the child, with an object close to the child's hands. After this calibration phase, the experimenters removed all objects from the table, asked the parent to start the experiment and left the room. The instructions given to the parent were to take all three objects from one set, place them on the table, play with the child and after hearing a command from the experimenters, remove the objects in this trial and move to the next set to start the next trial. There were a total of six trials, each about 1 minute long. The entire study, including initial setup, lasted for 10 to 15 minutes.

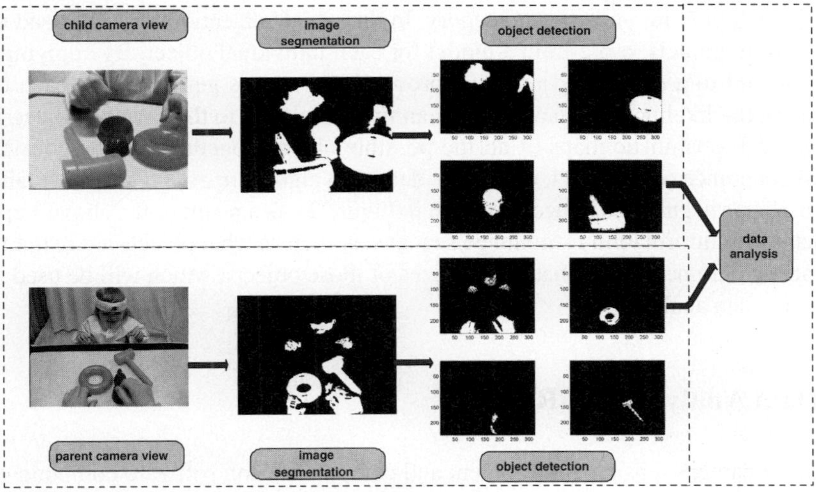

Fig. 2. Overview of data processing using computer vision techniques. we first remove background pixels from an image and then spot objects and hands in the image based on pre-trained object models. The visual information from two views is then aligned fro further data analyses.

2.3 Image Segmentation and Object Detection

The recording rate for each camera is 10 frames per second. In total, we have collected approximately 10800 ($10 \times 60 \times 6 \times 3$) image frames from each interaction. The resolution of image frame is 320×240.

The first goal of data processing is to automatically extract visual information, such as the locations and sizes of objects, hands, and faces, from sensory data in each of three cameras. These are based on computer vision techniques, and include three major steps (see Figure 2). Given raw images from multiple cameras, the first step is to separate background pixels and object pixels. This step is not trivial in general because two first-view cameras attached on the heads of two participants moved around all the time during interaction causing moment-to-moment changes in visual background. However, since we designed the experimental setup (as described above) by covering the walls, the floor and the tabletop with white fabrics and asking participants to wear white cloth, we simply treat close-to-white pixels in an image as background. Occasionally, this approach also removes small portions of an object that have light reflections on them as well. (This problem can be fixed in step 3). The second step focuses on the remaining non-background pixels and breaks them up into several blobs using a fast and simple segmentation algorithm. This algorithm first creates groups of adjacent pixels that have color values within a small threshold of each other. The algorithm then attempts to create larger groups from the initial groups by using a much tighter threshold. This follow-up step of the algorithm attempts to determine which portions of the image belong to the same object even if that object is broken up visually into multiple segments. For instance, a hand may decompose a single object into several blobs. The third step assigns each blob into an object category. In this object detection task, we used Gaussian mixture models to pre-train a model for each individual object. By applying each object model to a segmented image, a probabilistic map is generated for each object indicating the likelihood of each pixel in an image belongs to this special object. Next, by putting probabilistic maps of all the possible objects together, and by considering spatial coherence of an object, our object detection algorithm assign an object label for each blob in a segmented image as shown in Figure 2. As a result of the above steps, we extract useful information from image sequences, such as what objects are in the visual field at each moment, and what are the sizes of those objects, which will be used in the following data analyses.

3 Data Analyses and Results

The multi-camera sensing environment and computer vision software components enable fine-grained description of child-parent interaction and from two different viewpoints. In this section, we report our preliminary results while focusing on comparing sensory data collected simultaneously from two views. We are particularly interested in the differences between what a child sees and what the mature partner sees.

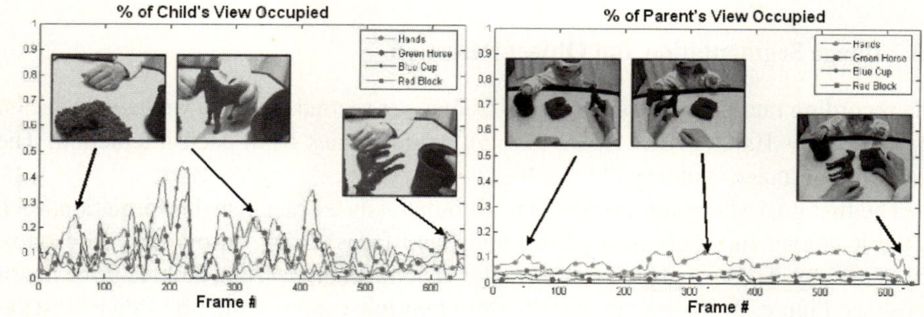

Fig. 3. A comparison of the child's and the parent's visual fields. Each curve represents a proportion of an object in the visual field over the whole trial. The total time in a trial is about 1 minute (600 frames). The three snapshots show the image frames from which the visual field information was extracted.

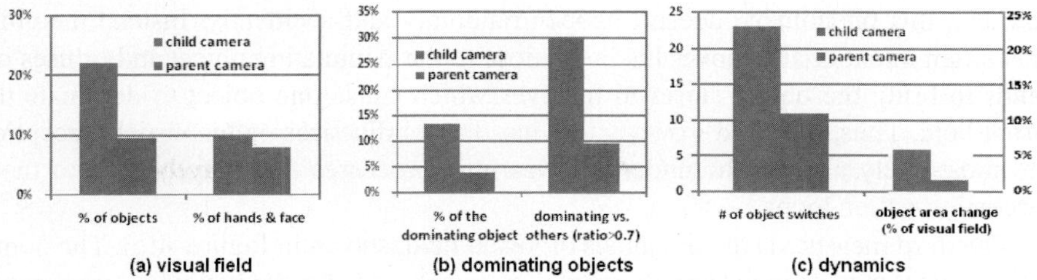

Fig. 4. We quantify and compare visual information from two views in three ways.

Figure 3 shows the proportion of each object or hand in one's visual field over a whole trial (three snapshots taken from the same moments from these two views). Clearly, the child's visual field is substantially different from the parent's. Objects and hands occupy the majority of the child's visual field and the whole field changes dramatically moment by moment. In light of this general observation, we developed several metrics to quantify three aspects of the differences between these two views.

First, we measure the composition of visual field shown in Figure 4 (a). From the child's perspective, objects occupy about 20% of his visual field. In contrast, they take just less than 10% of the parent's visual field. Although the proportions of hands and faces are similar between these two views, a closer look of data suggests that the mother's face rarely occurs in the child's visual field while the mother's and the child's hands occupy a significant proportion (15%-35%) in some image frames. From the mother's viewpoint, the child's face is always around the center of the field while the hands of both participants occur frequently but occupy just a small proportion of visual field.

Second, Figure 4(b) compares the salience of the dominating object in two views. The dominating object for a frame is defined as the object that takes the largest proportion of visual field. Our hypothesis is that the child's view may provide a unique window of the world by filtering irrelevant information (through movement of the body close to the object) enabling the child to focus on one object (or one event) at a single moment. To support this argument, the first metric used here is the percentage of the dominating object in the visual field at each moment. In the child's view, the dominating object takes 12% of the visual field on average while it occupies just less than 4% of the parent's field. The second metric measures the ratio of the dominating object vs. other objects in the same visual field, in terms of the occupied proportion in an image frame. A higher ratio would suggest that the dominating object is more salient and distinct among all the objects in the scene. Our results show a big difference between two views. More than 30% of frames, there is one dominating object in the child's view which is much larger than other objects (ratio ¿ 0.7). In contrast, less than 10% of time, the same phenomena happens in the parent's view.

This result suggests not only that children and parents have different views of the environment but also that the child's view may provide more constrained and clean input to facilitate learning processes which don't need to handle a huge amount of irrelevant data because there is just one object (or event) in view at a time. We also note that this phenomena doesn't happen randomly and accidently. Instead the child most often intentionally moves his body close to the dominating object and/or uses his hands to bring the object closer to his eyes which cause one object to dominate the visual field. Thus, the child's own action has direct influences on his visual perception and most-likely also on the underlying learning processes that may be tied to these perception-action loops.

The third measure is the dynamics of visual field, shown in Figure 4(c). The dominating object may change from moment to moment, and also the locations, appearance and the size of other objects in the visual field may change as well. Thus, we first calculated the number of times that the dominating object changed. From the child's viewpoint, there are on average 23 such object switches in a single trial (about 1 minute or 600 frames). There are only 11 per trial from the parent's view. These results together with the measures in Figure 4(b) suggest that children tend to move their head and body frequently to switch attended objects, attending at each moment to just one object. Parents, on the other hand, don't switch attended objects very often and all the objects on the table are in their visual field almost all the time.

The dynamics of their visual fields in terms of the change of objects in visual field makes the same point. In the child's view, on average, in each frame, 6% of the visual field consists of new objects, objects that are different from the just previous frame to frame. Only less than 2% of the parent's visual field changes this way frame to frame. over time. The child's view is more dynamic and such offers potentially more spatio-temporal regularities that may be utilized by lead young learners to pay attention to the more informative (from their point of view!) aspects of a cluttered environment.

4 General Discussion

4.1 Embodiment

There are two practical reasons that the child's view is quite different from the parent's view. First, because they are small, their head is close to the tabletop. Therefore, they perceive a "zoom-in", more detailed, and more narrowed view than taller parents. Second, at the behavioral level, children move objects and their own hands close to their eyes while adults rarely do that. Both explanations above can account for dramatic differences between these two views. Both factors highlight the crucial role of the body in human development and learning. The body constraints and narrows visual information perceived by a young learner. One challenge that young children face is the uncertainty and ambiguity inherent to real-world learning contexts: learners need to select the features that are reliably associated with an object from all possible visual features and they need to select the relevant object (at the moment) from among all possible referents on a table. In marked constrast to the mature partner's view, the visual data from the child'first-person view camera suggests a visual field filtered and narrowed by the child's own action. Whereas parents may selectively attend through internal processes that increase and decrease the weights of received sensory information, young children may selectively attend by using the external actions of their own body. This information reduction through their bodily actions may remove a certain degree of ambiguity from the child's learning environment and by doing so provide an advantage to bootstrap learning. This suggests that an adult view of the complexity of learning tasks may often be fundamentally wrong. Young children may not need to deal with all the same complexity inherent in from an adult's viewpoint - some of them that complexity may be automatically solved by bodily action and the corresponding sensory constraints.

4.2 Joint Interaction

Previous joint-attention research has focused on the temporal synchrony of different participants in real-time interaction. For instance, developmental researchers have shown that children and parents share visual attention through social cues signaled by their eyes. The present work extends this observation in two important ways. First, our results suggest the importance of spatial information. Children need to not only share visual attention with parents at the right moment; they also need to perceive the right information at the moment. Spatio-temporal synchrony encoded in sensorimotor interaction may be provide this. Second, hands (and other body parts, such as the orientation of the body trunk) play a crucial role in signaling social cues to the other social partner. The parent's eyes are rarely in the child's visual field but the parent's and the child's own hands occupy a big proportion of the child's visual field. Moreover, the change of the child's visual field can be caused by gaze and head movement, but this change can be caused by both his own hand movements and the social partner's hand movements. In these ways, hand movements directly and significantly changes the child's view.

4.3 A New Window of the World

The first-person view is visual experience as the learner sees it and thus changes with every shift in eye gaze, every head turn, every observed hand action on an object. This view is profoundly different from that of an external observer, the third-person view, who watches the learner perform in some environment precisely because the first person view changes moment-to-moment with the learner's own movements. The systematic study of this first person view — of the dynamic visual world through the developing child's eyes – seems likely to reveal new insights into the regularities on which learning is based and on the role of action in creating those regularities. The present findings suggest that the visual information from a child's point of view is dramatically different from the parent's (or an experimenter's) viewpoint. This means analyses of third-person views from an adult perspective may be missing the most significant visual information to a young child's learning.

The head camera method used here provides a new look on the structure of the learning environment, and how that structure is generated by the child's own actions. In general, a head camera can provide information about what is in that field –and available to attention – but does not provide fine-grained information on what the specific focus of the child's attention in that field (as does eye-tracking technology). From this perspective, the head-mounted camera is complimentary to the remote eye-tracking technique which can obtain precise eye gaze location but just in a 2-dimensional predefined screen.

5 Conclusions

The goal is to see the world as the child sees it and not filtered through our own adult expectations about the structure in that world. The goal is to understand how the child's own actions -and coupled actions to a social partner –create regularities in visual information. This paper reports beginning progress in reaching these goals and, moreover suggests that progress in achieving these goals will bring unexpected new discoveries about the visual environment, about the role of the body, and the structure of the learning task -from the learner's point of view.

Acknowledgement. This research was supported by National Science Foundation Grant BCS0544995 and by NIH grant R21 EY017843.

References

1. Ballard, D.H., Hayhoe, M.M., Pook, P.K., Rao, R.P.N.: Deictic codes for the embodiment of cognition. Behavioral and Brain Sciences 20(4), 723–767 (1997)
2. Bertenthal, B.I., Campos, J.J., Kermoian, R.: Perspective on the Development of Self-Produced Locomotion and Its Consequences. Current Directions in Psychological Science 3(5), 140–145 (1994)
3. Candy, T.R., Skoczenski, A.M., Norcia, A.M.: Normalization models applied to orientation masking in the human infant. Journal of Neuroscience 21(12), 4530–4541
4. Gibson, E.J.: Principles of perceptual learning and development. Appleton-Century-Crofts, East Norwalk (1969)
5. Hains, S.M.J., Muir, D.W.: Infant sensitivity to adult eye direction. Child Development 67, 1941–1951 (1996)
6. Ruff, H.A.: Components of attention during infant's manipulative exploration. Child Development 57, 105–114 (1986)
7. Smith, L.B., Breazeal, C.: The dynamic lift of developmental process. Developmental Science 10, 61–68 (2007)
8. van Hof-van Duin, J., Mohn, G.: The development of visual acuity in normal fullterm and preterm infants. Vision Research 26(6), 909–916 (1986)
9. Yoshida, H., Smith, L.B.: Hands in view: Using a head camera to study active vision in toddlers. Infancy (Submitted, 2007)
10. Yu, C., Ballard, D.H., Aslin, R.N.: The role of embodied intention in early lexical acquisition. Cognitive Science 29(6), 961–1005 (2005)

Autonomy: A Review and a Reappraisal

Tom Froese, Nathaniel Virgo, and Eduardo Izquierdo

Centre for Computational Neuroscience and Robotics
University of Sussex, Brighton, UK
{t.froese, n.d.virgo, e.j.izquierdo}@sussex.ac.uk

Abstract. In the field of artificial life there is no agreement on what defines 'autonomy'. This makes it difficult to measure progress made towards understanding as well as engineering autonomous systems. Here, we review the diversity of approaches and categorize them by introducing a conceptual distinction between *behavioral* and *constitutive* autonomy. Differences in the autonomy of artificial and biological agents tend to be marginalized for the former and treated as absolute for the latter. We argue that with this distinction the apparent opposition can be resolved.

1 Introduction

Two major research goals of artificial life are to 1) synthesize autonomous agents, and 2) through this process gain a better understanding of the generative mechanisms underlying autonomy in general. But what do we mean when we say that a system is *autonomous*? There seems to be no commonly accepted definition in the artificial life community or the cognitive sciences. For example, in engineering and robotics the notion of autonomy is often used to refer to the self-sufficiency of a machine to achieve a certain task (e.g. [10,26]), in artificial life the term 'autonomy' is commonly used to characterize self-organizing systems (e.g. [37]; [25], p. 117), Kauffman [17] uses the term "autonomous agent" to refer to a life cycle constituted by thermodynamic work, and in the autopoietic and enactive tradition it is used to refer to the self-constitution of an identity in living systems (e.g. [36]). Still, in spite of the evident definitional ambiguity there is arguably a sense in which most uses of the term 'autonomy' are united by a common concern with self-governance, a notion which is already implied by the term's etymology (*auto* [self] *nomos* [law])[1].

Nevertheless, the particular kind of self-governance which these authors have in mind can vary considerably. Indeed, due to the lack of a coherent conceptual framework which connects the different uses of the term, it is hard to measure the progress that has been made in the artificial synthesis of such systems. Are today's systems more autonomous than those presented at the first ECAL over 10 years ago? If this is the case, then what are the significant challenges that remain?

[1] The word 'autonomy' can also appear in an unrelated mathematical sense of meaning a dynamical system with no time dependence, which is another potential source of confusion.

F. Almeida e Costa et al. (Eds.): ECAL 2007, LNAI 4648, pp. 455–464, 2007.

And are current research methodologies appropriate for addressing them? In order to provide answers to these questions an understanding of autonomy is needed which enables the different uses of the term in artificial life and the cognitive sciences to be systematically related to each other. The aim of this paper is to provide a first step towards this necessary conceptual clarification.

2 Autonomy: A Review

In this section the various uses of the term 'autonomy' are categorized into two main classes of approaches according to whether the focus is on the agent's 1) external behavior, or 2) internal organization. We introduce a conceptual distinction between *behavioral* and *constitutive* autonomy in order to differentiate between the type of autonomy referred to by 1) and 2), respectively.

2.1 Behavioral Autonomy

For this class of approaches, it is generally a necessary condition that the behavior of an autonomous system is characterized by some capacity for stable and/or flexible interaction with its environment. The system's identity can be self-constituted (as is the case for all organisms), but it is sufficient for it to be externally imposed by some designer (e.g. the unit of selection in evolutionary robotics), or even explicitly represented by a particular component of the system (e.g. the central controller in GOFAI). Thus, this category includes all of those approaches which do not treat the autonomy of living beings as qualitatively (though, perhaps, quantitatively) different from the autonomy of most artificial agents. Three sub-categories can be distinguished:

1) The broadest use of the term 'autonomy' can be found in the context of engineering where the study of "autonomous systems" is basically equated with a concern for building robots (e.g. [30]). Thus, there is a sense in which even remotely controlled mobile robots (e.g. a Mars explorer) can be referred to as "autonomous agents" (e.g. [13], p. 37). However, more commonly the notion is used to designate that the robot is engineered so as to be able to interact with its environment without requiring ongoing human intervention (e.g. [25], p. 67). Brooks [10], for example, uses the notion of autonomy to refer to tether-free robots, where all the energy and computational requirements are stored on board. Note that using the term 'autonomy' in this broad manner does not exclude agents whose behavior has been completely pre-specified. As such it can be criticized on the basis that the "agent can hardly be said to be autonomous because its behavior is largely dictated by the experimenter" ([25], p. 148). A more restrictive notion is used by Pfeifer [26] who proposes as the first design principle of autonomous agents that "they have to be able to function without human intervention, supervision, or instruction". Nevertheless, it is clear that these requirements for autonomy are almost trivially fulfilled by many artificial agents and all organisms.

2) It is also often claimed that an autonomous system must be capable of satisfying some goal (or even of generating its own goals). For example, Beer ([2], p. 173) uses the term "autonomous agent" to mean "any embodied system designed to satisfy internal or external goals by its own actions while in continuous long-term interaction with the environment in which it is situated". Similarly, Nolfi and Floreano ([25], p. 25) hold that "autonomous systems are expected to survive in unknown and partially unpredictable environments by devising their own goals and finding out solutions to challenges that may arise". The way in which teleological concepts such as purpose, agenda, concern, or goal are used in this kind of approaches should generally be interpreted as rather loose metaphors. As a point in case, consider Franklin's ([13], p. 233) use of these terms when he invites us to "think of an autonomous agent as a creature that senses its environment and acts on it so as to further its own agenda", and then continues by claiming that "any such agent, be it a human or a thermostat, has a single, overriding concern – what to do next." Following Beer [2], we can say that in this context "the class of autonomous agents is thus a fairly broad one, encompassing at the very least all animals and autonomous robots".

3) Another common approach is to relate autonomy to the robustness and flexibility of behavior. Smithers [30], for example, claims that "autonomous systems" are those that "engage in specific kinds of task achieving behavior in particular real environments, and which do so reliably and robustly". This view often relates autonomy to notions of self-organization (e.g. [37]) and emergence (e.g. [25], p. 117). While this sometimes implies some philosophical commitment (e.g. [8]), it primarily manifests itself as a pragmatic response to the practical difficulties faced by the GOFAI tradition. For example, the approach for designing autonomous systems proposed by Pfeifer and Verschure [27] "promises to resolve a number of fundamental problems of AI in natural ways (such as situatedness and robustness), others will not need to be solved since they are artifacts of the traditional approach (e.g. symbol grounding)".

2.2 Constitutive Autonomy

This category includes all approaches to autonomy which can be traced to the autopoietic tradition, a movement which originated in theoretical biology in the 1970's (e.g. [35,21]), and/or which are generally related to metabolism (e.g. [24,29]). It is generally claimed that autonomy in living systems is a feature of self-production or *autopoiesis*[2]. However, this restriction of autonomy to living systems is unsatisfactory because we also want to refer to some systems as autonomous even though they are not characterized by metabolic self-production, for example artificial and social systems [20].

[2] One recent definition of autopoiesis as the minimal organization of living systems is: "An autopoietic system is organized (defined as unity) as a network of processes of production (synthesis and destruction) of components such that these components: 1) continuously regenerate the network that is producing them, and 2) constitute the system as a distinguishable unity in the domain in which they exist" [34,36,12].

Thus, the original account was followed by an attempt to conceptually separate the notion of autonomy from that of autopoiesis. In 1979 Varela published his *Principles of Biological Autonomy*, a book that continues to be an important reference for many researchers (e.g. [12,4,9,23,29]), and in which he formulated the 'Closure Thesis' which states that "every autonomous system is organizationally closed" ([33], p. 58)[3]. Accordingly, autopoietic systems are reinterpreted as one rather prominent member of a broader class of autonomous systems. Weber and Varela [36] neatly summarize this position by proposing that we should identify the " 'constitution of an identity' as the governing of an autonomy principle". The idea is that this principle should make it possible to "take the lessons offered by the autonomy of living systems and convert them into an operational characterization of *autonomy in general*, living or otherwise" ([33], p. 55).

This conception of autonomy clearly poses a significant difficulty for many common methodologies in artificial life research. For if we accept the general claim that an autonomous system is a self-defining or self-constituting system, then it follows that all current robots and most (if not all) artificial agents are "by constitution non-autonomous insofar as their realization and permanence as unities is not related to their operation" [35]. However, it is worth pointing out that while the question of "whether or not one may want to make an autopoietic system is, of course, an ethical problem" it is still the case that "if our characterization of living systems is adequate, it is apparent that they could be made at will" ([33], p. 44), at least in principle. Indeed, there is research in artificial life which tries to understand the generative mechanisms underlying such constitutive autonomy. Two main approaches can be distinguished according to whether their target is the 1) computational or 2) chemical domain.

1) The field of computational autopoiesis [23] attempts to explore the nature of living systems with the use of simulations. This research program originated over a decade in advance of the first Santa Fe Workshop on Artificial Life with the publication of a seminal paper by Varela, Maturana and Uribe [35] in which the authors outline the first model of an autopoietic entity. It has subsequently given rise to a whole tradition of simulating autopoiesis [23]. However, the question of whether such research can generate genuine autopoietic systems is still the subject of debate, with some researchers claiming for various reasons that computational entities can not be autopoietic in principle (e.g. [19,31,28,34]). Nevertheless it is clear that such modelling research has the potential to clarify some of the key ideas underlying autopoiesis and draw attention to some of the central questions which still remain open (e.g. [4]).

[3] An autonomous system can be defined in operational terms as a system with an organization that is characterized by processes such that "(1) the processes are related as a network, so that they recursively depend on each other in the generation and realization of the processes themselves, and (2) they constitute the system as a unity recognizable in the space (domain) in which the processes exist" ([33], p. 55). This is essentially the definition of autopoiesis but without the implication that the processes necessarily involve physical synthesis and destruction.

2) The field of chemical autopoiesis has been investigating the "creation of chemical models of cellular life that can be constructed in the laboratory" since the early 1990's (see [20] for a recent overview). In this manner some of the problems of the computational medium are avoided, but there are other challenges which derive from working with the chemical domain. Nevertheless, this approach has the advantage that it allows theoretical questions to be addressed on the basis of concrete experimental phenomena (e.g. [6]).

It is worth pointing out that, as computational models are becoming increasingly realistic, it is possible to relate them with actual chemical realizations in a mutually informative manner (e.g. [22]). Moreover, in contrast to most of the current work on behavioral autonomy, this kind of research has the potential to discover the conditions under which autonomous systems emerge spontaneously (rather than having their identity pre-defined by the experimenter), and, since it is well grounded in the actual laws of physics and chemistry, it could thereby provide the basis for a proper *naturalization* of the concept of autonomy (e.g. [29]).

3 Autonomy: A Reappraisal

In the previous section we identified two main approaches to autonomy. The advantage of the behavioral approach is that it can generally accommodate both artificial and biological agents. At the same time, however, it has difficulties in specifying exactly what makes such systems autonomous. Consequently, the requirements are often trivially met in many cases. As an ambiguous and inclusive approach, it threatens to make the concept of autonomy meaningless. In contrast, the constitutive approach can provide a more precise definition in operational terms, but this has the undesirable consequence that its applicability is mainly restricted to actual organisms. It thus excludes most artificial life research from potentially contributing to our understanding of the generative mechanisms underlying autonomy in general. These considerations make it evident that there is a pressing need of finding a principled way of integrating these two approaches into one coherent framework of autonomous systems research.

Accordingly, in this section it is proposed that one useful way of clarifying this issue is to 1) conceptualize autonomy as a continuum that includes both behavioral and constitutive autonomy as two distinct dimensions[4], and 2) relate these dual dimensions of autonomy such that they appear as two interrelated aspects of one unifying concept (i.e. life).

3.1 Autonomy as a Continuum

Following Boden [7], we agree that "autonomy is not an all-or-nothing property. It has several dimensions, and many gradations" (see also [13], p. 266), and

[4] Also useful, but out of the scope of this paper, would be to include substrate requirements as a third dimension of autonomy. Some authors require autonomous systems to be real physical/chemical systems, whereas others will allow simulated entities to be autonomous within a computational world.

propose that these dimensions are best captured by behavioral and constitutive autonomy. Boden [7] also addresses these two distinct aspects when she claims that "an individual's autonomy is the greater, the more its behaviour is directed by self-generated (and idiosyncratic) inner mechanisms, nicely responsive to the specific problem-situation, yet reflexively modifiable by wider concerns". This is a good guideline, but we are still faced by the considerable challenge of devising the precise operational criteria for measuring these gradations. In particular, there are two main issues that need to be addressed: 1) how to operationalize the criteria for behavioral autonomy, and 2) whether the dimension of constitutive autonomy is best conceived of as continuous or binary.

1) It is evident that the behavioral dimension of autonomy is best conceived of as continuous, but it is not exactly clear how. This is largely due to the fact that important behavioral criteria are often undefined (e.g. the requirement of 'stability' and 'flexibility') or phrased in ambiguous terms (e.g. the requirement of 'goal generation'). Fortunately, the ongoing development of the dynamical approach in cognitive science is ensuring that better tools for characterizing the dynamics of behavior are being appropriated from mathematics [32]. For example, Kelso ([18], p. 45) points out that in the mathematical theory of dynamical systems the "measurement of the time it takes to return to some observed state – local relaxation time – is an important index of stability", and that "instabilities are hypothesized to be one of the generic mechanisms for flexible switching among multiple attractive states." Furthermore, it has been shown that the evolutionary robotics framework [15] can help to investigate the dynamics underlying the behavioral autonomy associated with stability and flexibility (e.g. [11,16]).

2) Constitutive autonomy, as captured by the notion of autopoiesis, is strictly speaking an all-or-nothing systemic property [12]. Varela ([33], p. 27), for example, notes that "the establishment of an autopoietic system cannot be a gradual process: Either a system is an autopoietic system or it is not. [...] Accordingly, there are not and cannot be intermediate systems". Even if we follow Varela ([33], p. 55) in extending the class of autonomous systems to include all systems which constitute their own identity, it still seems to be the case that either a system is constitutively autonomous or it is not. Nevertheless, there might be ways of treating the constitutive dimension as continuous. Bickhard [5], for example, holds that an autonomous system is one which actively contributes to its own persistence and that "autonomy in this sense is a graded concept: there are differing kinds and degrees of such 'active contributions' ". Barandiaran and Moreno [1] outline another promising approach when they write that "while self-organization appears when the (microscopic) activity of a system generates at least a single (macroscopic) constraint, autonomy implies an open process of self-determination where an increasing number of constraints are self-generated".

Another possibility would be to measure the dimensions of autonomy along an increase in organizational requirements. For example, one could go from negative feedback, to homeostasis, and finally to autopoiesis[5]. This might make

[5] Thanks to Barry McMullin for pointing this out. This hierarchy is enhanced when we consider that "an autopoietic machine is an homeostatic (or rather a relations-

it possible to trace behavioral and constitutive autonomy from what might be called a 'weaker' sense to a 'stronger' sense, a continuum which roughly coincides with a transition from a more technological to a more biological usage of the term, and which finally culminates in a complete restriction of the term's applicability to actual living organisms. However, if this hierarchy of organizational requirements is to be actually useful in measuring autonomy, further work needs to be done to define the terms and their relationships more precisely.

3.2 Life as Constitutive and Behavioral Autonomy

After conceptually teasing the constitutive and behavioral domain of autonomy apart, it is nevertheless quite clear that they do somehow relate in living systems. Varela [34], for example, relates constitutive autonomy to the behavioral domain: "To highlight autonomy means essentially to put at center stage two interlinked propositions: Proposition 1: Organisms are fundamentally the process of constitution of an identity. [...] Proposition 2: The organism's emergent identity gives, logically and mechanistically, the point of reference for a domain of interactions"[6]. However, it is a non-trivial question as to exactly how the organism distinguished in the constitutive domain relates to its behavior distinguished in the behavioral domain. Moreover, this connection only works for some conceptions of behavioral autonomy, and a more precise definition of how such autonomy relates to living systems is needed before the relationship can be stated more formally.

While such further conceptual clarification is important for the development of a coherent theory of autonomy, it is also of practical interest for current artificial life research. Bourgine and Stewart [9], for example, conceptualize autopoiesis and cognition as distinct aspects of living systems in such a way that it allows them to refer to artificial agents as 'cognitive' without them having to be autopoietic. This view is clearly a useful theoretical justification for using evolutionary robotics as a methodology for studying behavioral autonomy in the form of cognition (e.g. [15]) without having to address the problem of constitutive autonomy. Similarly, Beer's [4] approach to cognition follows directly from an autopoietic perspective on life when two key abstractions are made: 1) Focus on an agent's behavioral dynamics. An agent's behavior takes place within its cognitive domain, which is a highly structured subset of its total domain of interaction. 2) Abstract the sets of destructive perturbations that an agent can undergo as a viability constraint on its behavioral dynamics. Thus, we assume the existence of a constitutively autonomous agent, but model only its behavior and not the constitutive aspects of its autonomy. In other words, the agent is constitutively autonomous by definition only.

However, there are reasons for holding that in living systems autopoiesis and cognition are more tightly interlinked than the possibility of strict conceptual

static) system which has its own organization (defining network of relations) as the fundamental variable which it maintains constant" ([21], p. 79), see also ([33], p. 13).

[6] This was clearly also a part of his vision for ECAL, as is evident in [8].

separation seems to indicate [6]. Thus, as Beer [3] himself makes clear, some of the abstractions made in artificial life research are not completely satisfactory:

"[T]his explicit separation between an animal's behavioral dynamics and its viability constraint is fundamentally somewhat artificial. An animal's behavioral dynamics is deeply intertwined with the particular way in which its autopoiesis is realized. Unfortunately, a complete account of this situation would require a theory of biological organization, and the theoretical situation here is even less well developed than it is for adaptive behavior. [...] However, if we are willing to take the existence of an animal for granted, at least provisionally, then we can assume that its viability constraint is given a priori, and focus instead on the behavioral dynamics necessary to maintain that existence" ([3], p. 265).

It is clear from these considerations that, while the general aim of evolutionary robotics is not to study the mechanisms underlying constitutive autonomy, more thought needs to be given as to how natural cognition is constrained by the constitutive processes which give rise to living systems. In this regard it might be helpful to introduce more biologically inspired mechanisms into the controllers of the artificial systems being evolved (e.g. homeostasis [11,14,16]). However, in general more work needs to be done in order for us to better understand what kind of methodology is best suited for studying autonomous artificial systems which actually self-constitute an identity at some level of description. Only when we are able to investigate both constitutive and behavioral autonomy via synthetic means can the field of artificial life claim to provide one coherent framework of autonomous systems research.

4 Conclusion

Are today's artificial agents more autonomous? By distinguishing between behavioral and constitutive autonomy, we can see that this question actually demands two distinct responses. It seems safe to say that today's systems are indeed more behaviorally autonomous (than at the start of ECAL, for example). Most of the work that is done in the artificial sciences under the banner of autonomous systems research is providing a wealth of tools of analysis and ways of understanding of how externally defined constraints can be successfully satisfied by increasingly complex artificial agents. However, the vast majority of this kind of research is not tackling the question of how such viability constraints (and, more importantly, an agent's identity) can emerge from the internal operations of those autonomous systems while coupled to their environments, though more work is starting to be done in this area.

Finally, it is important to note that the widespread disregard of the dimension of constitutive autonomy is a serious shortcoming not only for scientific research, but also in terms of our own understanding of what it means to be human. As Boden [7] points out: "what science tells us about human autonomy is practically important, because it affects the way in which ordinary people see themselves – which includes the way in which they believe it is possible to behave". The field of artificial life is therefore also faced by an ethical imperative to invest more effort into improving our understanding of constitutive autonomy. Only then can we ground our understanding of human freedom – not only in terms of the behavior involved in mere external constraint satisfaction, but also in terms of the creativity involved in dynamic and open-ended self-realization.

Acknowledgments. Many thanks to all the participants of the 'Modeling Autonomy' workshop which was held in San Sebastian, Basque Country, during the 22nd and 23rd March 2007, as well as those involved in the Life and Mind seminars at Sussex University, for their helpful discussions.

References

1. Barandiaran, X., Moreno, A.: On what makes certain dynamical systems cognitive: A minimally cognitive organization program. Adaptive Behavior 14(2), 171–185 (2006)
2. Beer, R.D.: A dynamical systems perspective on agent-environment interaction. Artificial Intelligence 72(1-2), 173–215 (1995)
3. Beer, R.D.: The dynamics of adaptive behavior: A research program. Robotics and Autonomous Systems 20(2-4), 257–289 (1997)
4. Beer, R.D.: Autopoiesis and Cognition in the Game of Life. Artificial Life 10(3), 309–326 (2004)
5. Bickhard, M.H.: Autonomy, Function, and Representation. Communication and Cognition – Artificial Intelligence 17(3-4), 111–131 (2000)
6. Bitbol, M., Luisi, P.L.: Autopoiesis with or without cognition: defining life at its edge. Journal of the Royal Society Interface 1(1), 99–107 (2004)
7. Boden, M.A.: Autonomy and Artificiality. In: Boden, M.A. (ed.) The Philosophy of Artificial Life, pp. 95–108. Oxford University Press, New York (1996)
8. Bourgine, P., Varela, F.J.: Introduction: Towards a Practice of Autonomous Systems. In: Varela, F.J., Bourgine, P. (eds.) Proc. of the 1st Euro. Conf. on Artificial Life, pp. xi–3. The MIT Press, Cambridge (1992)
9. Bourgine, P., Stewart, J.: Autopoiesis and Cognition. Artificial Life 10(3), 327–345 (2004)
10. Brooks, R.A.: Intelligence without reason. In: Myopoulos, J., Reiter, R. (eds.) Proc. of the 12th Int. Joint Conf. on Artificial Intelligence, pp. 569–595. Morgan Kaufmann, San Mateo (1991)
11. Di Paolo, E.A.: Organismically-inspired robotics: homeostatic adaptation and teleology beyond the closed sensorimotor loop. In: Murase, K., Asakura, T. (eds.) Dynamical Systems Approach to Embodiment and Sociality, pp. 19–42. Advanced Knowledge International, Adelaide (2003)
12. Di Paolo, E.A.: Autopoiesis, adaptivity, teleology, agency. Phenomenology and the Cognitive Sciences 4(4), 429–452 (2005)
13. Franklin, S.: Artificial Minds. The MIT Press, Cambridge (1995)
14. Harvey, I.: Homeostasis and Rein Control: From Daisyworld to Active Perception. In: Pollack, J., et al. (eds.) Proc. of the 9th Int. Conf. on the Simulation and Synthesis of Living Systems, pp. 309–314. The MIT Press, Cambridge (2004)
15. Harvey, I., Di Paolo, E.A., Wood, R., Quinn, M., Tuci, E.A.: Evolutionary Robotics: A new scientific tool for studying cognition. Artificial Life 11(1-2), 79–98 (2005)

of autopoiesis, which was proposed by Nomura [12,13,14] to clarify whether autopoiesis can really be represented within the conventional mathematical frameworks.

2 Autopoiesis and (M,R) Systems

2.1 Autopoiesis

An autopoietic system is organized as a network of processes of production of components, where these components:

1. continuously regenerate and realize the network that produces them, and
2. constitute the system as a distinguishable unity in the domain in which they exist.

The characteristics of autopoietic systems Maturana gives are as follows:

Autonomy: Autopoietic machinery integrates various changes into the maintenance of its organization. A car, the representative example of non–autopoietic systems, does not have any autonomy.

Individuality: Autopoietic machinery has its identity independent of mutual actions between it and external observers, by repeatedly reproducing and maintaining the organization. The identity of a non–autopoietic system is dependent on external observers and such a system does not have any individuality.

Self–Determination of the Boundary of the System: Autopoietic machinery determines its boundary through the self–reproduction processes. Since the boundaries of non–autopoietic systems are determined by external observers, self–determination of the boundaries does not apply to them.

Absence of Input and Output in the System: Even if a stimulus independent of an autopoietic machine causes continuous changes in the machine, these changes are subordinate to the maintenance of the organization which specifies the machine. Thus, the relation between the stimulus and the changes lies in the area of observation, and not in the organization.

In Japan, Hideo Kawamoto has continued his own development of autopoiesis [6]. He designated the properties of autopoiesis by comparison with conventional system theories. In particular, he focuses on the fourth item among the above characteristics of autopoiesis, i.e., absence of input and output in the system.

When we consider the "absence of input and output", important is the view where the system is understood based on the production processes. Kawamoto claims the following: the view of the relation between inputs and outputs in the system is one from external observers and it does not clarify the organization or the operation of the production in the system. A living cell only reproduces its components and does not produce the components while adjusting itself according to the relation between itself and oxygen in the air. Although the density of oxygen affects the production processes, external observers decide the influence and the cell does not. As long as the system is grasped from an internal view of the cell, the system does not have any "inputs and outputs".

From the above perspective, Kawamoto's theory focuses on circular relations of components and the network of production processes of components. His important claims are as follows: an autopoietic system is a network consisting of relations between production processes of components. This network produces components of the system, and the components exist in physical spaces. Then, the system exists only if the components reproduce the network of production processes. The structure of the system is a realization of the system through the operation of the system in the physical space, and the organization of the system is a form of the network. The organization is functionally specified and closed, although the structure is realized in the physical space.

2.2 (M,R) Systems and Closure Under Entailment

In relational analysis, a system is regarded as a network that consists of components having functions. Rosen compared machine systems with living systems to clarify the difference between them, based on the relationship among components through entailment [16]. In other words, he focused his attention on where the function of each component results from in the sense of Aristotle's four causal categories, that is, material cause, efficient cause, formal cause, and final cause. As a result, Rosen claimed that a material system is an organism if and only if it is closed to efficient causation. Furthermore, Rosen suggested that systems closed under efficient cause cannot be described with their states because they lead to infinite regress.

(M,R) systems [15] satisfy closure under efficient cause. This system model maintains its metabolic activity through inputs from environments and repair activity. The simplest (M,R) systems represent the above aspect in the following diagram and the left half in figure 1.

$$A \xrightarrow{f} B \xrightarrow{\phi_f} H(A,B) \xrightarrow{\Phi_f} H(B,H(A,B)) \tag{1}$$

Here, A is a set of inputs from an environment to the system, B is a set of outputs from the system to the environment, f is a component of the system represented as a map from A to B, and ϕ_f is the repair component of f as a map from B to $H(A,B)$ ($H(X,Y)$ is the set of all maps from a set X to a set Y). In biological cells, f corresponds to the metabolism, and ϕ_f to the repair. If $\phi_f(b) = f$ ($b = f(a)$) is satisfied for the input $a \in A$, we can say that the system maintains itself. In addition, Φ_f can be constructed by the preceding (M, R) system in the following way: For a and b such that $b = f(a)$ and $\phi_f(b) = f$, if $\hat{b} : H(B,H(A,B)) \to H(A,B)$ ($\hat{b}(\phi)(a') = \phi(b)(a')$ ($\phi \in H(B,H(A,B)), a' \in A$)) has the inverse map $\hat{b}^{-1}$, it is easily proved that $\hat{b}^{-1}(f) = \phi_f$. Thus, we can set $\Phi_f = \hat{b}^{-1}$. The right half in figure 1 shows the aspect that the components except for a are closed under entailment.

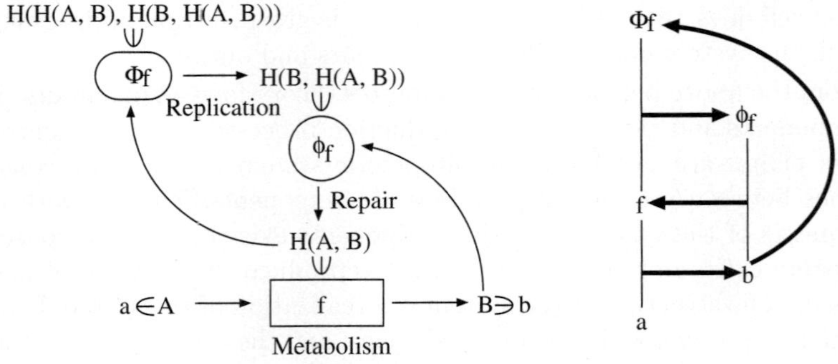

Fig. 1. A (M,R) System and Its Entailment Relation

3 Category Theoretical Formalization of Autopoiesis and (M,R) Systems

Category theory is an algebraic framework to abstractly handle the collection of mathematical objects having some specific properties, such as "the collection of all groups", "the collection of all sets", "the collection of all topological spaces", "the collection of differential manifolds", and so on [1]. In this framework, an individual space or set is dealt with as an object, and a function or map from an object to another object is dealt with as a morphism corresponding to an arc between them. Thus, the inner structures of any object and morphism are reduced, and pure relations of morphisms between objects are focused on. This can make it possible to investigate what category of mathematical objects a specific relation between objects (for example, closed relations between objects and morphisms) is satisfied in. In addition, category theory can deal with relations of categories themselves as functors. This can make it possible investigate relations between a specific category and general ones such as state spaces.

In this paper, we assume that an abstract category $\mathcal{C}$ has a final object 1 and product object $A \times B$ for any pair of object A and B. The category of all sets is an example of this category. Moreover, we describe the set of morphisms from A to B as $H_{\mathcal{C}}(A, B)$ for any pair of objects A and B. A element of $H_{\mathcal{C}}(1, X)$ is called a morphic point on X. For a morphism $f \in H_{\mathcal{C}}(X, X)$ and a morphic point x on X, x is called a fixed point of f iff $f \circ x = x$ ($\circ$ means concatenation of morphisms) [17]. Morphic points and fixed points are respectively abstraction of elements of a set and fixed points of maps in the category of sets. This abstraction is useful when our discussion is extended to categories of which objects and morphisms are not assumed to be sets and their maps, such as the Lindenbaum category of which objects and morphisms are the constants and equivalence classes of formulas of a formal theory [17].

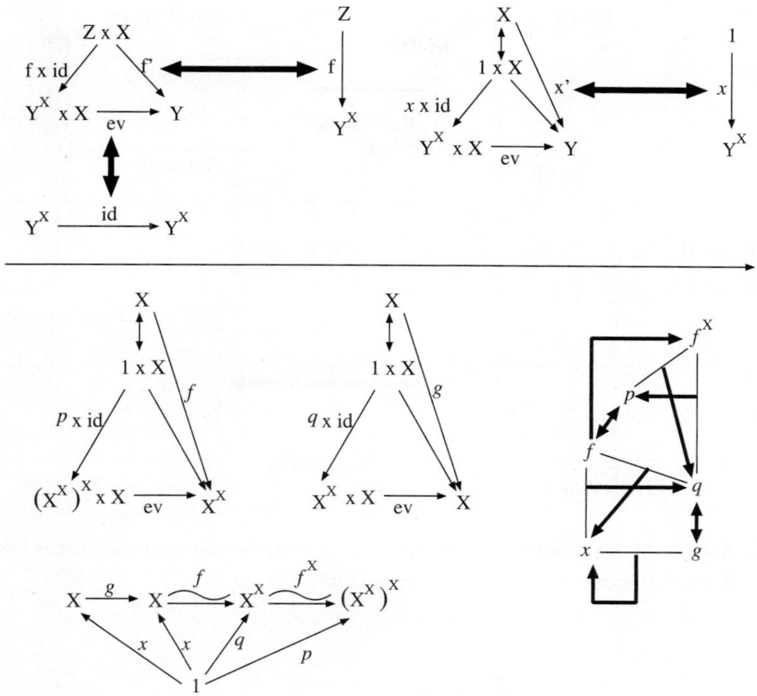

Fig. 2. The Diagrams of a Completely Closed System and the Entailment Relations based on Natural One–To–One Correspondence

3.1 Category Theoretical Formalization of Autopoiesis

The fact that the components reproduce themselves in a system implies that the components are not only operands but also operators. The easiest method for realizing this implication is the assumption of existence of an isomorphism from the space of operands to the space of operators [5]. Under this assumption, Nomura [12,13] proposed completely closed systems under entailment between the components.

When there exists the power object Y^X for objects X and Y (that is, the functor $\cdot \times X$ on $\mathcal{C}$ has the right adjoint functor $\cdot^X$ for X), note that there is a natural one–to–one correspondence between $H_{\mathcal{C}}(Z \times X, Y)$ and $H_{\mathcal{C}}(Z, Y^X)$ for any objects X, Y, Z satisfying the diagram in the upper figure of figure 2 [18]. Thus, there is a natural one–to–one correspondence between morphic points on Y^X and morphisms from X to Y satisfying the diagram in the lower figure of figure 2. By using the above property, we can construct completely closed systems as follows.

Now, we assume an object X with powers and an isomorphism $f : X \simeq X^X$ in $\mathcal{C}$. Then, there uniquely exists a morphic point p on $(X^X)^X$ corresponding to

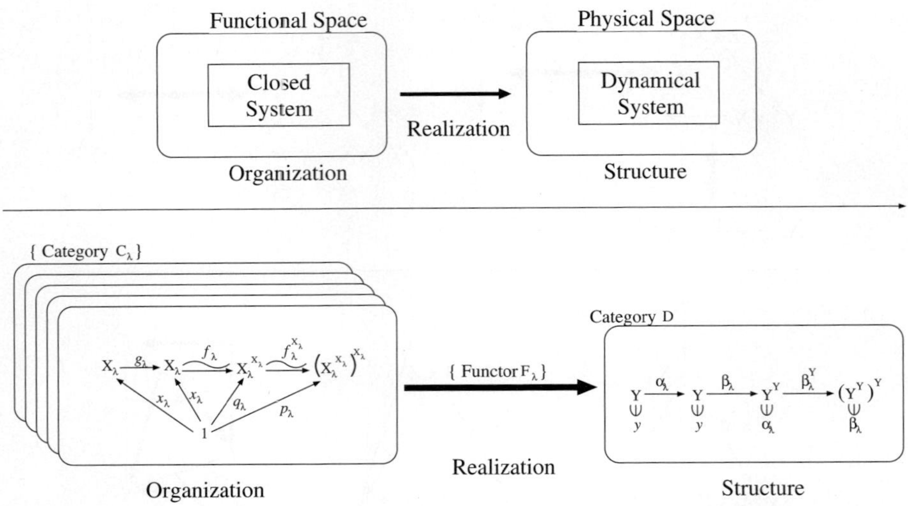

Fig. 3. Aspect of Autopoiesis based on Distinction between Organization and Structure, and Its Category Theoretical Formalization

f in the above sense, that is , $p' = f$. Since the morphism from X^X to $(X^X)^X$ entailed by the functor $\cdot^X$, f^X, is also isomorphic, there uniquely exists a morphic point q on X^X such that $f^X \circ q = p$. We can consider that p and q entail each other by f^X. Furthermore, there uniquely exists a morphic point x on X such that $f \circ x = q$ because f is isomorphic. Since we can consider that x and q entail each other by f, and f and p entail each other by the natural correspondence, the system consisting of x, q, p, f, and f^X is completely closed under entailment. Moreover, if x is a fixed point of $g : X \to X$ naturally corresponding to q, that is, $g \circ x = x$, we can consider that x entails itself by g. The lower figure of Figure 2 shows the diagrams of this completely closed system and the entailment relations.

Furthermore, Autopoiesis argues not only closedness of entailment between the components but also two levels of description. Kawamoto's claims mentioned in the previous section have an important implication. The organization of a system differs from the structure since they exist in different levels. This distinction is mentioned in Maturana and Varela's original literature [10]. The upper figure of Figure 3 shows this aspect. The distinction between organizations and structures in an autopoietic system can be interpreted as a distinction between categories on which the organization and structure of a autopoietic system are defined in a mathematical formalization of it.

This fact has an important implication. If circular relations between components and their production process network are closed under entailment, this closedness may be hard to be formalized in general category such as state spaces.

On the other hand, the structure of an autopoietic system must be realized in a state space as a physical one. These implications suggest the distinction between organizations and structures in formalization of autopoiesis.

To represent the distinction between organizations and structures, Nomura [14] proposed a model in which the organization is formalized in a specific category, the structure is formalized in the category of general state spaces, and realization from the organization to the structure is represented by a functor between the categories. The lower figure of Figure 3 shows the model.

We assume a family of categories $\{C_\lambda\}_\lambda$, and that each C_λ includes a completely closed system $\{X_\lambda,\ x_\lambda \in H_{C_\lambda}(1, X_\lambda),\ f_\lambda \in H_{C_\lambda}(X_\lambda, {X_\lambda}^{X_\lambda}),\ q_\lambda \in H_{C_\lambda}(1, {X_\lambda}^{X_\lambda}),\ p_\lambda \in H_{C_\lambda}(1, ({X_\lambda}^{X_\lambda})^{X_\lambda}),\ g_\lambda \in H_{C_\lambda}(X_\lambda,\ X_\lambda)\}$. Moreover, we assume another category D. Here, it is assumed that D is the category consisting of state spaces and maps between them, or its subcategory.

Now, we assume a family of functors $\{F_\lambda : C_\lambda \to D\}_\lambda$ such that $F_\lambda(1_{C_\lambda}) = 1_D$, $Y = F_\lambda(X_\lambda)$, $y = F_\lambda(x_\lambda) \in H_D(1, Y) = Y$ for all λ, and $\alpha_\lambda = F_\lambda(g_\lambda) \in H_D(Y, Y) = H(1, Y^Y) = Y^Y$ for any λ (here, we can regard $H_D(1, Y) = Y$ and $H_D(Z, Y) = H_D(1, Y^Z) = Y^Z$ since D is the category of state spaces). Note that $\alpha_\lambda(y) = \alpha_\lambda \circ y = F_\lambda(g_\lambda) \circ F_\lambda(x_\lambda) = F_\lambda(g_\lambda \circ x_\lambda) = F_\lambda(x_\lambda) = y$ for any λ, that is, y is a fixed point of α_λ.

The family of the completely closed systems $\{\{X_\lambda, x_\lambda, f_\lambda, q_\lambda, p_\lambda, g_\lambda\}\}_\lambda$ is an organization of an autopoietic system and $\{Y, y, \{\alpha_\lambda\}_\lambda\}$ is its structure realized on the category D through the family of the functors $\{F_\lambda\}$ if for any λ one of the following conditions is satisfied:

1. $^\exists\beta_\lambda \in H_D(Y, Y^Y) = H_D\left(1, (Y^Y)^Y\right) = (Y^Y)^Y$ *s.t.*, $\beta_\lambda(y) = \alpha_\lambda$, ${\beta_\lambda}^Y(\alpha_\lambda) = \beta_\lambda$
2. $^\exists\beta_\lambda \in H_D(Y, Y^Y)$, λ_1, λ_2, and $\beta_{\lambda_2} \in H_D(Y, Y^Y)$ *s.t.*, $\beta_\lambda(y) = \alpha_{\lambda_1}$, $\beta_\lambda = {\beta_{\lambda_2}}^Y(\alpha_{\lambda_2})$

The above relationship between the organization and structure represents the aspect that the structure is entailed repeatedly within the organization.

3.2 Category Theoretically Described (M,R) Systems

As mentioned in the previous section, (M,R) systems are closed under entailment except for the input a. We can re–write the closed part of (M,R) systems as follows.

For objects X and Y in $\mathcal{C}$, we assume that X has powers. When a morphism $f : X \to Y$ and a morphic point x on X are given, we assume that x satisfies the following conditions:

$$\begin{aligned}&\exists\, G_x \in H_{\mathcal{C}}(Y^X, Y) \\ &\quad s.t.,\ G_x \circ z = z' \circ x \text{ for any } z \in H_{\mathcal{C}}(1, Y^X) \\ &\text{and } G_x \text{ has its inverse morphism } F_x \in H_{\mathcal{C}}(Y, Y^X)\end{aligned} \tag{2}$$

here, z' is the morphism from X to Y naturally corresponding to the morphic point z on Y^X. When $y = f \circ x$ and x_f is the morphic point on Y^X naturally

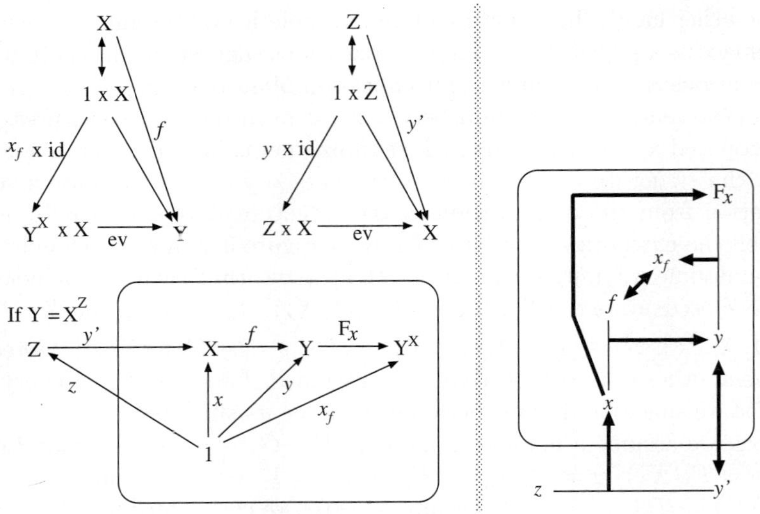

Fig. 4. The Diagrams of Category Theoretically Described (M,R) System and the Entailment Relations

corresponding to f $((x_f)' = f)$, we obtain $F_x \circ y = F_x \circ f \circ x = F_x \circ G_x \circ x_f = x_f$. Thus, x_f is entailed by y and F_x. If we regard F_x as entailed by x, then f, y, F_x, and x_f are entailed by themselves and x.

Furthermore, if there exist an object Z with powers and morphic point z on Z such that $Y = X^Z$ and $y' \circ z = x$, the system including y' and z represents the original (M,R) system.

Figure 4 shows the diagrams of this generalized (M,R) system and its entailment relations.

4 Differences Between Autopoiesis and (M,R) Systems

It is considered that closure under entailment or production is a necessary condition for a system to be autopoietic because the components reproduce themselves in the system. In fact, the existing research works found the similarity of autopoiesis with (M,R) systems based on this closedness [8,12,13,4,3]. However, there are two points of difference between these systems. The first one is the difference on forms of closure.

The forms of closedness in completely closed systems as autopoiesis and (M,R) systems reveal the difference between them. In the completely closed system, the existence of isomorphism f between X and X^X determines complete closure under entailment without any condition. On the other hand, the closedness of the (M,R) system depends on whether one of the components x satisfies the condition represented in equation (2).

Moreover, there is also a difference on conditions of categories on which these systems are constructed. Although completely closed systems show the possibility of constructing systems closed under entailment in specific categories, these categories have to satisfy the condition that operands coincide with operators. Although Soto–Andrade and Varela [17] provided a category satisfying this condition (the category of partially ordered sets and continuous monotone maps with special conditions), this category is very special.

On the other hand, Rosen [16] argued based on category theoretic frameworks that systems closed under efficient cause like (M,R) systems cannot be described with their states because they lead to infinite regress. However, Chu and Ho [4] found that Rosen's proof for this argument was not complete since his proof assumes an implicit condition irrelevant from state space representation of systems. In fact, Letelier et al, [8] provided with an arithmetic example of a (M,R) system constructed within the category of finite groups. These facts imply the difference on types of categories required for autopoiesis and (M,R) systems.

The second difference between autopoiesis and (M,R) systems is based on distinction between organizations and structures. As mentioned in the previous section, autopoiesis requires this distinction. However, the form of (M,R) systems does not include the explicit distinction between closed organizations and structures realized in state spaces, and these concepts are confused.

5 Discussion and Conclusion

This paper suggested two differences between autopoiesis and (M,R) systems from the perspective of category theoretic formalization of them, the difference on forms of their closedness under entailment of the components and categories required for the closedness, and the existence of distinction between organizations and structures. However, the first difference depends on the assumption that completely closed systems are necessary conditions of autopoiesis, that is, the existence of an isomorphism from the space of operands to the space of operators is a necessary condition of autopoiesis. This proposition has not still been proved in a mathematically strict sense or sufficiently considered in a philosophical sense. We need to explore which mathematical conditions should be satisfied for formalization of autopoiesis.

Moreover, it should be sufficiently discussed what contribution the differences between autopoiesis and (M,R) systems suggested in the paper can provide with, in the sense of the above exploration of conditions of minimal living systems. As Chu and Ho [4] argue that Rosen's idea based on category theory can contribute to distinction between living and non–living systems, we also believe that category theoretical frameworks including Rosen's method will help us to bring us closer to an understanding of life systems.

Acknowledgment. The author deeply thank Dr. Plamen L. Simeonov at the Technische Universität Berlin for his collaborative discussion on the subject of the paper.

References

1. Borceux, F.: Handbook of Categorical Algebra 1: Basic Category Theory. Cambridge University Press, Cambridge (1994)
2. Bourgine, P., Stewart, J.: Autopoiesis and cognition. Artificial Life 10(3), 327–346 (2004)
3. Chemero, A., Turvey, M.T.: Complexity and "Closure to Efficient Cause. In: Proc. AlifeX: Workshop on Artificial Autonomy, pp. 13–19 (2006)
4. Chu, D., Ho, W.K.: A Category Theoretical Argument against the Possibility of Artifical Life: Robert Rosen's Central Proof Revisited. Artificial Life 12(1), 117–134 (2006)
5. Kampis, G.: Self–Modifying Systems in Biology and Cognitive Science: A New Framework for Synamics, Information, and Complexity. Pergamon Press, Oxford (1991)
6. Kawamoto, H.: Autopoiesis: The Third Generation System (in Japanese). Seido–sha Publishers, Tokyo (1995)
7. Kneer, G., Nassehi, A.: Niklas Luhmanns Theorie Sozialer Systeme (Japanese Edition: Tateno, T., et.al. (1995). Shinsensha). Wilhelm Fink Verlag, München (1993)
8. Letelier, J.C., Soto-Andrade, J., Abarzúa, F.G., Cornish-Bowden, A., Cárdenas, M.L.: Organizational invariance and metabolic closure: Analysis in terms of (M,R) systems. Journal of Theoretical Biology 238, 949–961 (2006)
9. Maturana, H.R., Varela, F.J.: Autopoiesis and Cognition: The Realization of the Living (Japanese Edition: Kawamoto, H (1991). Kokubun–sha Publishers). D. Reidel Publishing, Dordrecht (1980)
10. Maturana, H.R., Varela, F.J.: The Tree of Knowledge (Japanese edition: Suga, K (1987). Asahi Publications). Shambala Publications, Boston (1987)
11. McMullin, B.: Thirty years of computational autopoiesis. Artificial Life 10(3), 277–296 (2004)
12. Nomura, T.: Formal description of autopoiesis based on the theory of category. In: Kelemen, J., Sosík, P. (eds.) ECAL 2001. LNCS (LNAI), vol. 2159, pp. 700–703. Springer, Heidelberg (2001)
13. Nomura, T.: Formal description of autopoiesis for analytic models of life and social systems. In: Proc. 8th Int. Conf. Artificial Life (ALIFE VIII), pp. 15–18 (2002)
14. Nomura, T.: Category Theoretical Formalization of Autopoieis from Perspective of Distinction between Organization and Structure. In: Proc. Seventh German Workshop on Artificial Life (GWAL–7), pp. 31–38 (2006)
15. Rosen, R.: Some Relational Cell Models: The Metabolism–Repair Systems. In: FOUNDATIONS OF MATHEMATICAL BIOLOGY, pp. 217–253. Academic Press, London (1972)
16. Rosen, R.: LIFE ITSELF. Columbia University Press, New York (1991)
17. Soto-Andrade, J., Varela, F.J.: Self–reference and fixed points: A discussion and an extension of Lawvere's theorem. Acta Applicandae Mathematicae 2, 1–19 (1984)
18. Takeuchi, G.: Sheaf, Category, and Topos (in Japanese). Nihon Hyoron–sha, Tokyo (1978)

Measuring Autonomy by Multivariate Autoregressive Modelling

Anil K. Seth

Dept of Informatics, University of Sussex, Brighton, BN1 9QJ, UK
a.k.seth@sussex.ac.uk

Abstract. I introduce a quantitative measure of autonomy based on a time series analysis adapted from 'Granger causality'. A system is considered autonomous if prediction of its future evolution is enhanced by considering its own past states, as compared to predictions based on past states of a set of external variables. The proposed measure, G-autonomy, amplifies the notion of autonomy as 'self-determination'. I illustrate G-autonomy by application to example time series data and to an agent-based model of predator-prey behaviour. Analysis of the predator-prey model shows that evolutionary adaptation can enhance G-autonomy.

1 Introduction

The concept of autonomy is increasingly central in artificial life and many related disciplines in the cognitive and behavioural sciences [15,3]. Intuitive, quantitative, and practically applicable measures of autonomy are therefore much needed [2]. This paper introduces and illustrates such a measure.

Autonomy has a wide variety of meanings. A non-exhaustive list might include the ability to learn, the ability to self-recharge, the ability to operate without power cables or without tele-operation. A more abstract definition of autonomy as 'organizational closure' is prominent in the work of Varela [15]. Here, motivated by practical applicability, I adopt a simple conception of autonomy as the degree of self-determination of a system [3]. Following [2], I amplify this concept in two ways: An autonomous system should not be fully determined by its environment, and a random system should not have a high autonomy value. In addition, a useful measure of autonomy should be practically applicable to a wide range of situations, in particular to situations involving coupled agent-environment interaction.

The measure introduced here, G-autonomy, is based on multivariate time series analysis, specifically on the notion of Granger causality (G-causality) [6,13]. Broadly, a G-autonomous variable is one for which prediction of its future evolution is enhanced by considering its own past states, as compared to predictions based on past states of a set of external variables. In other words, a variable is G-autonomous to the extent that (1) it is dependent on its own history, and (2) these dependencies are not accounted for by external factors.

F. Almeida e Costa et al. (Eds.): ECAL 2007, LNAI 4648, pp. 475–484, 2007.

Section 2 of this paper provides a formal basis for G-autonomy along with an illustrative analysis using artificially constructed time series with known dependencies. Section 3 describes the application of G-autonomy to the important case of coupled interactions between an agent and an environment. Analysis of this model addresses the relation between evolutionary adaptation and autonomy and it is found that evolutionary adaptation increases G-autonomy in the model. Finally, section 4 outlines limitations of the present approach and directions for future work.

2 G-Autonomy

Granger [6] introduced the idea of G-causality as a formalization, in terms of linear regression modelling, of Norbert Wiener's intuition that Y 'causes' X if knowing Y helps predict the future of X. According to G-causality, Y causes X if the inclusion of past observations of Y reduces the prediction error of X in a linear regression model of X and Y, as compared to a model which includes only previous observations of X. Since its introduction, G-causality has found wide application in economics and many other fields including neuroscience [8,14,12] and climatology. For a tutorial introduction to G-causality, see [13].

The above definition of G-causality immediately suggests a possible quantitative measure of autonomy. Instead of asking whether the prediction error of X is reduced by including past observations of Y, we ask whether the prediction of error of X is reduced by inclusion of its own past, given a set of external variables $Y_{1...N}$. That is, a system X is autonomous to the extent that its own past states help predict its future states over and above predictions based on past states of a set of external variables $Y_{1...N}$. I call this measure *G-autonomy.* Because of the pre-existing statistical framework of G-causality, G-autonomy is easy to calculate and can be applied to a wide range of data.

To illustrate G-autonomy, suppose that the temporal dynamics of two time series, $X_1(t)$ and $X_2(t)$ (both of length T), can be described by a bivariate autoregressive model:

$$X_1(t) = \sum_{j=1}^{p} A_{11,j} X_1(t-j) + \sum_{j=1}^{p} A_{12,j} X_2(t-j) + \xi_1(t) \quad (1)$$

$$X_2(t) = \sum_{j=1}^{p} A_{21,j} X_1(t-j) + \sum_{j=1}^{p} A_{22,j} X_2(t-j) + \xi_2(t) \quad (2)$$

where p is the maximum number of lagged observations included in the model (the model order, $p < T$), A contains the coefficients of the model (i.e., the contributions of each lagged observation to the predicted values of $X_1(t)$ and $X_2(t)$), and ξ_1, ξ_2 are residuals (prediction errors) for each time series. If the variance of ξ_1 (or ξ_2) is reduced by the inclusion of the X_1 (or X_2) terms in the first (or second) equation, then it is said that X_1 (or X_2) is *G-autonomous* with respect to X_2 (or X_1). In other words, X_1 is G-autonomous if the coefficients in

A_{11} are jointly significantly different from zero. This can be tested by performing an F-test of the null hypothesis that $A_{11} = 0$, given assumptions of covariance stationarity on X_1 and X_2. By analogy with G-causality [5], the magnitude of G-autonomy can be estimated by the logarithm of the corresponding F-statistic. Importantly, this method is easy to generalize to the multivariate case in which the G-autonomy of X_1 is tested in the context of multiple external variables $X_{2...N}$.

A MATLAB (Mathworks, Natick, MA) toolbox containing routines needed to calculate G-autonomy given a set of time series is provided on the author's website: `www.anilseth.com/code`.

2.1 Illustrative Example

Next, I illustrate G-autonomy by analyzing several example time-series:

$$X_1(t) = \xi_1(t), \tag{3}$$
$$X_2(t) = aX_1(t-1) + (1-a)\xi_2(t), \tag{4}$$
$$X_3(t) = bX_3(t-2) + (1-b)\xi_3(t), \tag{5}$$
$$X_4(t) = cX_3(t-1) + (1-c)\xi_4(t), \tag{6}$$
$$X_5(t) = dX_1(t-2) + dX_5(t-2) + d\xi_5(t), \tag{7}$$

where $\xi_{1...5}$ are independent white noise processes, and the parameters a, b, c are all equal to 0.5 and $d = 1/3$. To calculate the G-autonomy values, 10,000 values of each time series were generated and an order 2 multivariate regression model was computed, using the method of ordinary least squares [7]. Table 1 shows the G-autonomy values for each variable. For comparison, the G-causality values among variables are also shown.

Table 1. G-autonomy and G-causality values for an illustrative data set generated using equations 3-7. G-autonomies are along the diagonal (in bold type). G-causalities are arranged so that the column variable G-causes the row variable. Statistically significant values ($p < 0.01$) are indicated with an asterisk.

	X_1	X_2	X_3	X_4	X_5
X_1	**0**	0.03	0	0.5	0
X_2	8.4*	**0**	0	0	0
X_3	0.6	0.4	**7.9***	0.3	0
X_4	0	0	7.7*	**0**	0
X_5	7.3*	0.7	0.3	0	**7.0***

Table 1 shows that G-autonomy satisfies several intuitive requirements for a quantitative measure of autonomy. X_1 has a G-autonomy of zero which is expected because it is a white noise process. X_2 also has a G-autonomy of zero because it is a combination of a white noise process and influences from another white noise process, X_1. X_3 has a high G-autonomy because its future evolution can be partially predicted from its own past in a way that does not depend on the other variables (X_1, X_2, X_4, X_5). X_4 has a low G-autonomy because, although its future can be partially predicted by its own past, the accuracy of this prediction is entirely due to influence from X_3. Finally, X_5 has a high G-autonomy because, like X_3, it contains an autoregressive component that is not accounted for by the other variables. Thus, in this example and in general, a variable has high G-autonomy if and only if its future can be better predicted from its own past after taking into account the predictive contribution of a set of external variables.

It is instructive to compare the G-autonomy values with the corresponding G-causality values. There are three significant G-causality interactions: $X_1 \rightarrow X_2$, $X_3 \rightarrow X_4$, and $X_1 \rightarrow X_5$, as expected from equations 3-7. Notice that while both X_1 and X_3 exert G-causal influences on other variables, X_3 is G-autonomous whereas X_1 is not. Also, X_1 G-causes both X_2 and X_5 but X_5 is G-autonomous whereas X_2 is not. These simple observations show that knowing the set of G-causalities among variables is not sufficient to know the G-autonomies of each variable with respect to others in the set.

3 G-autonomy in a Predator-Prey Model

To test G-autonomy in conditions involving coupled agent-environment interaction, I now consider a simple agent-based model in which a 'predator' agent moves in a 2D toroidal plane populated by mobile 'prey'. There are two types of prey, red and green, and the predator has two internal batteries, one corresponding to each prey type. Six experimental conditions are compared:

1. Predator moves randomly.
2. Predator behaviour is determined by environment.
3. Predator behaviour is determined by a combination of environment and internal state.
4. Predator behaviour is controlled by an evolved neural network.
5. As condition 3 but prey take avoiding action.
6. As condition 4 but prey take avoiding action.

In each condition there are 3 red prey and 3 green prey. Prey move at a speed of 3 distance units per time step (u/ts). The environment is $200u$ by $200u$ in size. In conditions 1-4 prey change heading each time step by an angle randomly chosen from the range $[-\pi/8 : \pi/8]$. In conditions 5 and 6, prey turn away from the predator by angle randomly chosen from the range $[0 : \pi/8]$ if the predator is within $50u$ and within $[-\pi/2 : \pi/2]$ of the prey's current heading; otherwise, prey behave as in conditions 1-4. If the predator comes within $10u$ of a prey, the corresponding battery is fully replenished. Batteries otherwise deplete by 1 unit per time step from an initial value of 300.

In all conditions the predator moves at a speed of $4u/ts$, i.e., slightly faster than prey. In condition 1, the predator changes direction each time step by an angle randomly chosen from the range $[-\pi/8 : \pi/8]$. In condition 2, the predator adjusts its heading at each time step to point towards the red prey that was closest at the beginning of the trial. In conditions 3 and 5, the predator adjusts its heading each time step to point towards the nearest red prey if the red battery level is lower than the green battery level, or otherwise towards the nearest green prey. In conditions 4 and 6, the predator heading is controlled by a feed-forward neural network optimized using a genetic algorithm to maximize the average level of both batteries. This network consists of 6 input neurons responding to the nearest red and green prey and the levels of the two batteries, 3 intermediate neurons, and 2 motor outputs. Details of the network and the evolutionary algorithm are provided in Appendix A.

3.1 Analysis

Each condition of the model was run for 10 trials of 10,000 ts each (in conditions 4 and 6 these trials were run following completion of the evolutionary algorithm). From each trial, 7 time series were generated corresponding to the trajectories of the predator and the 6 prey. To prevent edge effects, each time series ($a_1 \ldots a_7$) consisted of the average of the horizontal and vertical displacement of the predator/prey from the midlines of the environment, i.e.,

$$a_i(t) = 0.5(|x_i(t) - 100| + |y_i(t) - 100|).$$

Each time series was then first-order differenced [i.e., $a(t) \rightarrow a(t) - a(t-1)$] in order to ensure covariance stationarity [12]. After differencing, all time series from all conditions were covariance stationary ($p < 0.01$, Dickey-Fuller test). Each trial therefore produced a multivariate data set with 7 variables and 9,999 observations.

Each data set was used to construct a multivariate autoregressive model of order $p = 4$.[1] This model order was chosen according to the Akaike Information Criterion (AIC) [1] (the mean minimum AIC, computed from all data sets, was 3.9). To verify that the order-4 models sufficiently described the data, it was noted that the mean adjusted residual-sum-square RSS_{adj} was sufficiently high (0.86 $\pm$0.07). Each model was then used to calculate the G-autonomy values for the predator and the prey, according to the procedure described in section 2. G-autonomies were averaged across the 10 trials in each condition.

3.2 G-autonomy Results

Figure 1 shows sample trajectories of predator and prey from each condition. All conditions exhibit rich behavioural dynamics. Apparently, only condition 2 (follow the nearest red prey) is clearly distinguishable from the others, as might be expected given the simplicity of the rule governing predator behaviour in this condition. For the remaining conditions it seems infeasible by inspection to infer the relative values of predator autonomy.

---

[1] To check robustness, analysis was repeated with $p = 8$, because the maximum lowest AIC (among conditions) was 7.8. Results were unchanged from $p = 4$.

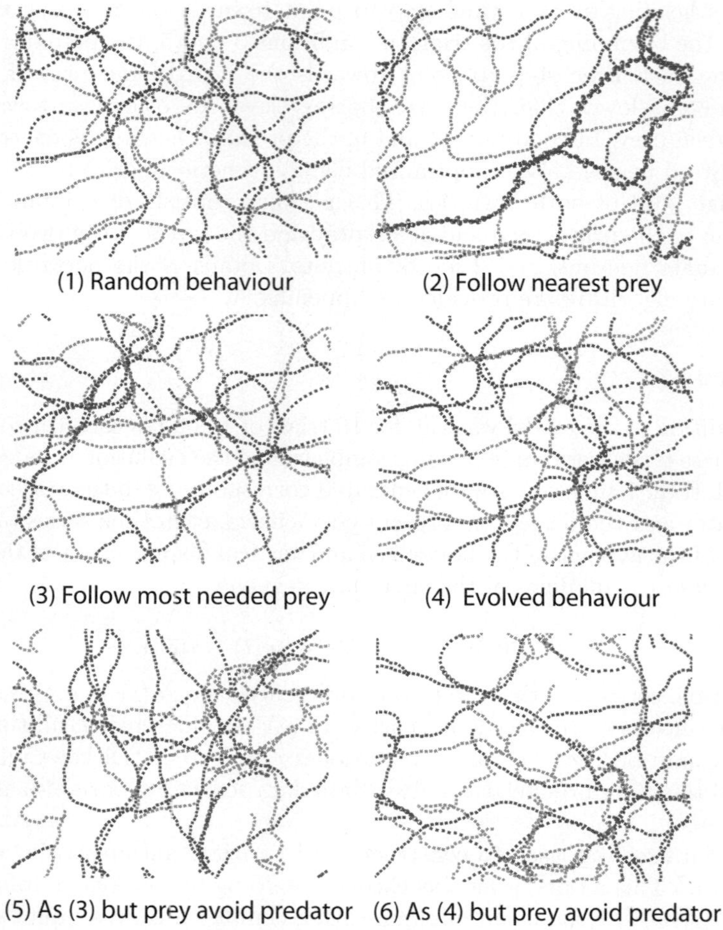

Fig. 1. Example predator and prey trajectories from each condition (200 ts are shown). Dark grey (blue online) line shows predator trajectory, light grey (red and green online) lines show red and green prey trajectories. Thicker parts of the lines indicate where prey were in contact ($< 10u$) with the predator.

Figure 2 shows the average predator G-autonomy values in each condition (1-6), as well as the G-autonomy of prey averaged across all prey from condition 1 (p1) and condition 5 (p2). The figure reveals a number of interesting features. First, random predator movement produces high G-autonomy (condition 1). This is expected because random movement in this model is not a white noise process. Rather, random movement implies that the predator changes direction unpredictably, so that the best (but not perfect) predictor of future predator position is the set of recent past predator positions. Also as expected, condition 2 produces low predator G-autonomy because predator position is now well predicted by past positions of a particular prey.

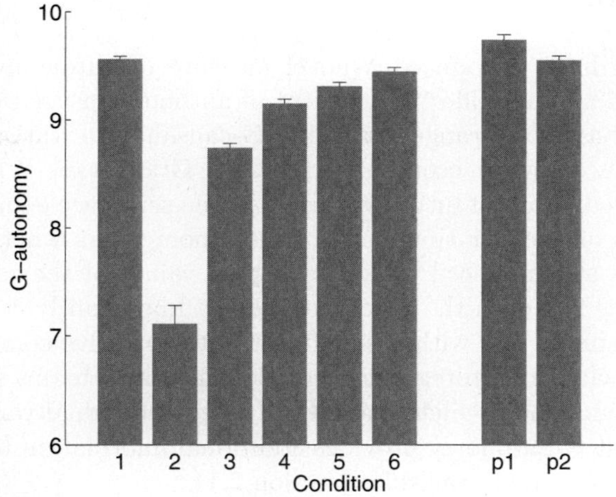

Fig. 2. Mean predator G-autonomy for each condition (1-6). Also shown is prey G-autonomy from condition 1 (p1) and from condition 5 (p2). Error bars show standard deviations. All comparisons are statistically significant (two tailed t-test, $p < 0.001$ after Bonferroni correction for multiple comparisons, t-statistic ranges from 5.7 to 44.3).

The predator G-autonomies in conditions 3-6 invite several interesting comparisons. Predator G-autonomy in condition 3 is higher than in condition 2 but lower than the upper bound provided by random movement (condition 1). This reflects the fact that predator behaviour in condition 3 is driven both by the position of a nearby prey and by the relative levels of the two internal batteries. Strikingly, predator G-autonomy in condition 4 is significantly higher than in condition 3. This comparison suggests that evolutionary adaptation in the model leads to increased G-autonomy, possibly because the evolved controller has the opportunity to take internal state into account at every time step, and not only at those time steps at which the lowest battery level changes from red to green or *vice versa* as is the case for the preprogrammed controller (condition 3). Predator G-autonomy in condition 4 remains lower than in condition 1 because predator position can still be partially predicted by prey position.

Conditions 5 and 6 show a similar pattern to conditions 3 and 4, i.e., predator G-autonomy is higher for the evolved controller (condition 6) than for the preprogrammed controller (condition 5). It is interesting that predator G-autonomies are higher when prey actively try to avoid predators (conditions 5 and 6) than when prey move randomly (conditions 3 and 4). The explanation for these differences requires further investigation.

The two examples of prey G-autonomy follow the expected pattern. Randomly moving prey (p1) have higher G-autonomy than actively guided prey (p2), most likely because the behaviour of the latter can be partially predicted by predator behaviour.

4 Discussion

In this paper I have introduced a novel measure of autonomy, G-autonomy, which quantifies and amplifies the concept of autonomy as 'self determination'. G-autonomy is based on Granger causality (G-causality), a statistical measure of causality introduced by the econometrician Clive Granger [6]. While G-causality asks whether predictions of future values of a time series are enhanced by knowing past values of another time series, G-autonomy asks whether predictions of future values are enhanced by knowing past values of the same time series given predictions based on the other time series. Importantly, a G-autonomous variable is not simply one with a significant autoregressive component, rather, it is a variable with an autoregressive component that remains significant after taking into account the predictive ability of a set of external variables. It bears emphasizing that G-autonomy provides additional information to that found in set of G-causalities among variables (section 2.1).

A useful measure of autonomy should be applicable to coupled agent-environment systems. By analyzing an agent-based predator-prey model, I showed that G-autonomy satisfies this requirement (section 3). A striking finding was that evolutionary adaptation enhanced predator G-autonomy in the model. This finding is consistent with the general notion that evolutionary processes promote autonomy [11,10] and opens the way towards explicit modelling of the relation between evolution and autonomy, a relation which has thus far remained speculative.

A second interesting result was that predator G-autonomy increased still further when prey actively avoided predators, whereas prey G-autonomy decreased. While a satisfying explanation of this result is still lacking, it brings forward the important open question of autonomy in reciprocally causal systems. What does it mean for an agent to be autonomous if its behaviour is both caused by and causal to an external environment? Another important open question, also left for the future, is the relation between G-autonomy and autonomy as conceived within the autopoeitic framework, i.e., autonomy as 'organizational closure' or "the condition of subordinating all changes to the maintenance of the organization" ([9], p.135).

Besides these general concerns, the present approach has a number of practical limitations including the following. First, measurement of G-autonomy requires that all elements of a system show changing activity over time. It is not possible at present to measure the G-autonomy of an agent with respect to a static feature of the environment. Second, like G-causality, G-autonomy gains its simplicity and power by modelling times series as linear processes. This power comes at the cost of neglecting nonlinear features of time series. While future extensions to nonlinear G-autonomy are plausible, based on existing nonlinear measures of G-causality [4], such measures are likely to be difficult to apply in practice.

A similar approach to the present has been developed by Bertschinger and colleagues, who propose measuring the conditional mutual information between consecutive states of a system, conditioned on the history of the environment

[2]. Unlike the present approach, conditional mutual information can take into account nonlinear data features (although see above for possible nonlinear extensions to G-autonomy). However, again in contrast to G-autonomy, it is not easy to extend conditional mutual information to multivariate situations (> 2 variables), which are likely characterize many situations of interest. Another difference between the two approaches is that Bertschinger et al. propose two distinct measures, one (as stated above) for cases in which the system cannot influence its environment, and a second measure (the non-conditioned mutual information between consecutive system states) for cases in which the system has full control over the environment. As shown in Section 3, the present approach can be applied to both these cases as well as to the important intermediate case of mutual agent-environment interaction.

In conclusion, G-autonomy provides an intuitive and easily applicable quantitative measure of autonomy based on the concept of self-determination. It can be applied not only to behavioural data, but also to time series data reflecting internal variables (e.g., neuronal activity, metabolic activity), or indeed to any multivariate data set in which autonomy is of interest. The development of G-autonomy participates in a much needed movement towards the operationalization of fundamental biological concepts.

Appendix A: Neural Controller and Evolutionary Algorithm

In conditions 4 and 6, predators are simulated robots with left/right wheels, two internal batteries (red and green) and 6 sensors, 4 of which are grouped in 2 left/right pairs. Two sensor pairs respond to the nearest red and green prey and the response of each pair decays linearly with distance. If the nearest prey is to the left (right) of the predator, the first (second) sensor of the corresponding pair increases its response by a factor $0.5(|\theta|/(\pi/2))$ where θ is the bearing to the prey (range $[-\pi : \pi]$). If $|\theta| > \pi/2$ the corresponding sensor pair does not respond. The third sensor pair responds linearly to the level of red and green batteries. All sensor activities are linearly scaled to range from -3 to +3.

Predators are controlled by an 11 node feed-forward neural network. The 6 sensors determine the input to 6 input neurons. Input neurons are fully connected to a layer of 3 intermediate neurons, which are themselves fully connected to the 2 output neurons. Each neuron implements a sigmoidal transfer function mapping an input range [-10:10] onto an output of range [0:1]. The motor output neurons are scaled to range from 0 to 5 to set the wheel speeds. Forward speed is fixed at $4u/ts$ to enable comparison with other experimental conditions, and angular velocity is calculated as the difference between the wheel speeds divided by the robot diameter ($5u$).

Network parameters were encoded as real numbers (range [0:1]) on a 29 element genotype. The first 5 alleles specified biases for the intermediate neurons and output neurons (range [-3:3]). The remaining 23 elements specified connection weights (range [-3:3]). The fitness of each genotype was calculated as the average of 5 evaluations. The fitness of each evaluation was calculated as the summed level of the two batteries for the lifetime of the predator. If either battery reached zero, the evaluation was terminated (this constraint was removed during data analysis for which it was important to obtain time series of consistent and extended length). Otherwise, evaluations lasted for 1,000*ts*. Positions and headings of predator and prey were intialized randomly at the start of each evaluation.

A genetic algorithm was used to evolve a population of 30 genotypes for 250 generations. The algorithm used stochastic rank-based selection with elitism and with a mutation probability of 0.09 per allele (no crossover). Each mutation changed the allele value by a number randomly chosen from the range [-0.2:0.2] with reflection at the boundaries (0,1). High fitness was reliably achieved after approximately 100 generations. The results in section 3.2 are obtained from representative evolutionary runs for each condition.

References

1. Akaike, H.: A new look at the statistical model identification. IEEE Trans. Autom. Control 19, 716–723 (1974)
2. Bertschinger, N., Olbrich, E., Ay, N., Jost, J.: Autonomy: an information theoretic perspective. Biosystems (in press)
3. Boden, M.A.: Autonomy and artificiality. In: Boden, M.A. (ed.) The philosophy of artificial life, pp. 95–108. Oxford University Press, Oxford (1996)
4. Chen, Y., Rangarajan, G., Feng, J., Ding, M.: Analyzing multiple nonlinear time series with extended Granger causality. Physics Letters A 324, 26–35 (2004)
5. Geweke, J.: Measurement of linear dependence and feedback between multiple time series. Journal of the American Statistical Association 77, 304–313 (1982)
6. Granger, C.W.J.: Investigating causal relations by econometric models and cross-spectral methods. Econometrica 37, 424–438 (1969)
7. Hamilton, J.D.: Time series analysis. Princeton University Press, Princeton (1994)
8. Kaminski, M., Ding, M., Truccolo, W.A., Bressler, S.L.: Evaluating causal relations in neural systems: Granger causality, directed transfer function and statistical assessment of significance. Biological Cybernetics 85, 145–157 (2001)
9. Maturana, H., Varela, F.J.: Autopoiesis and cognition: The realization of the living. D. Reidel Publishing, Dordrecht (1980)
10. Moreno, A., Etxeberria, A., Umerez, J.: The autonomy of biological individuals and artificial models. Biosystems (in press)
11. Rosslenbroich, B.: The notion of progress in evolutionary biology. The unresolved problem and an empirical suggestion. Biology and Philosophy 21, 41–70 (2006)
12. Seth, A.K.: Causal connectivity of evolved neural networks during behavior. Network: Computation in Neural Systems 16, 35–54 (2005)
13. Seth, A.K.: Granger causality. Scholarpedia (submitted)
14. Seth, A.K., Edelman, G.M.: Distinguishing causal interactions in neural populations. Neural Computation 19, 910–933 (2007)
15. Varela, F.J.: Principles of biological autonomy. Elsevier, New York (1979)

Minimal Agency Detection of Embodied Agents

Hiroyuki Iizuka[1,2] and Ezequiel Di Paolo[1]

[1] Centre for Computational Neuroscience and Robotics,
Department of Informatics, University of Sussex
Brighton, BN1 9QH, UK
[2] Department of Media Architecture, Future University-Hakodate
116-2 Kamedanakano-cho, Hakodate, Hokkaido, 041-8655, Japan
ezca@sacral.c.u-tokyo.ac.jp, ezequiel@sussex.ac.uk

Abstract. Agency detection is studied in a simple simulated model with embodied agents. Psychological experiments such as double TV-monitor experiments and perceptual crossing show the central role of dynamic mutuality and contingency in social interactions. This paper explores the ongoing dynamical aspects of minimal agency detection in terms of the mutuality and contingency. It is investigated how the embodied agents can establish a live interaction and discriminate this from interactions from recorded motions that are identical to the live interaction but cannot react contingently. Our results suggest that the recognition of the presence of another's agency need not lie on complex cognitive individual mechanisms able to integrate past information, but rather on the situated ongoingness of the interaction process itself, on its dynamic properties, and its robustness to noise.

1 Introduction

Social interaction may lead to coordinated behaviours when mutual anticipations are formed dynamically. This mutuality of influences is a key property of the interaction process but its dynamical characteristics have not been sufficiently investigated from a theoretical perspective. In contrast, important empirical evidence points to the central role of dynamic mutuality, or contingency, in sustaining and forming several aspects of an ongoing interaction. This is clearly shown in Trevarthen's double TV-monitor experiments with infants [6,9], Nadel's extensions to these experiments [7], and in the perceptual crossing experiments by the GSP group at UTC Compiègne [1]. In Trevarthen's experiment, a mother and her baby are placed in separate rooms and allowed to interact only through video screens that display their faces to each other. During 'live' interaction, mother and infant engage in coordinated utterances and affective expressions. However, if a delayed video recording of the mother is displayed to the baby, the baby becomes withdrawn and depressed. This shows that it is not sufficient for the baby to sustain interaction that the mother's expressive actions be displayed on the monitor, but the mother is required to react 'live' to the baby's own motions in order for the interaction to continue. It can be assumed that

F. Almeida e Costa et al. (Eds.): ECAL 2007, LNAI 4648, pp. 485–494, 2007.

a most important clue during interactions is its ongoingness which has to be shared between the subjects and is therefore a property of the dyadic system.

Auvray et al. [1] have studied social interaction by means of a minimal perceptual crossing paradigm. In their experiments, two blindfolded subjects interact in a virtual one-dimensional space. Each subject moves a receptor field using a computer mouse and can get all-or-nothing tactile sensation when the receptor crosses an object or the other subject's receptor. Subjects are constantly 'followed' by a shadow object that mimics their movement and, from their partner's perspective, is indistinguishable from their receptor. Their task is to find each other without becoming fixated on static objects or on the partner's objectively identical shadow. Subjects can achieve this goal thanks to how mutuality influences the global dynamics of the task since becoming fixated on a shadow that does not respond to one's own movements does not cause one's partner to stop the search and the interaction breaks down. Thus, mutual searching behaviour produces one globally stable dynamical situation when both subjects are directly touching each other.

The common idea in these experiments is that subjects are exposed to behaviours that are identical in motion to a 'live' interaction but do not have a subjective control to respond to the subject's own motions. Human subjects (even infants) easily discriminate the two conditions by engaging in interaction and thus investigating whether a partner has a contingent control of their motion. This dynamical enquiry produces an experience of confronting another subject, *an agency*. This suggests that the presence of another subjectivity is not located anywhere in the quality of the motion themselves but it is rather in the ongoingness of interactions shared between subjects.

The aim of our study is to explore ongoing dynamical aspects of minimal interaction between embodied agents. We investigate how embodied agents can establish a live interaction and discriminate this from interactions with dummy agents, which have the same motions but can not react live. The issue we want to test is *whether such a discrimination requires a complex perceptual strategy on the part of the discriminating agent or whether simpler solutions emerge from the interaction process itself.*

2 Model

Agents are required to discriminate whether another agent is a live interaction partner or a recording of the agent's behaviour by using minimally restricted sensor and motors. The two agents, top and bottom, face each other in an unlimited 1-D space. Each agent can move only left and right horizontally. One on-off sensor is attached on the centre of the agent and it is activated, i.e. set to 1 when it crosses the partner, while it is set to 0 otherwise. With a certain probability the sensory information is flipped into a different state (the probability is set to 0.05) at each Euler time step as sensory noise. The presence of noise will play an important role for agents to achieve the task, as will be explained later. The schematic view is illustrated in Fig. 1.

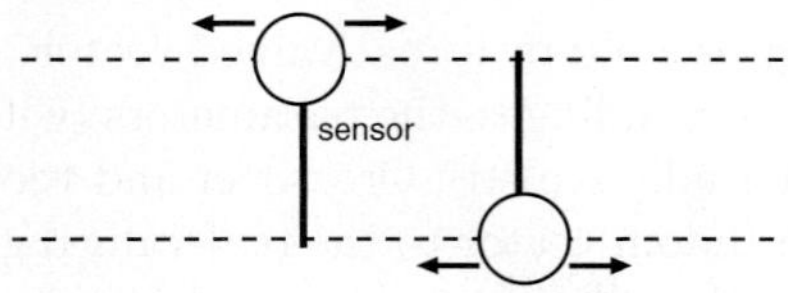

Fig. 1. A schematic view of the simulation settings. Two agents equipped with single on/off sensor are facing each other in a unlimited 1-D space.

The task is for an agent to cross its partner if it is interacting live (contingently) and to stay away from a recorded motion that replays the partner's previous motions as in the double TV-monitor experiments. The behaviours of original and recorded motions are identical but sensory noise is introduced independently.

2.1 Agents

Agents are controlled with a continuous-time recurrent neural network (CTRNN, [2]), which consists of fully-connected 8 nodes in our model. The time evolution of the states of neurons is expressed by:

$$\tau_i \dot{y}_i = -y_i + \sum_{j=1}^{N} w_{ji} z_j(y_j) + I_i, \quad z_i(x) = 1/(1+e^{-x-b_i}), \tag{1}$$

where y_i represents the cell potential of neuron i, z_i is the firing rate, τ_i (range $[1, 100]$) is its time constant, b_i (range $[-3, 3]$) is a bias term, and w_{ji} (range $[-8, 8]$) is the strength of the connection from the neuron, j, to i. I_i represents the sensory input, which is given to only one sensory neuron. The number of neurons, N, is set to 8. The sensory input is calculated by multiplying 1/0 (on/off) signal by a gain parameter (range $[1, 100]$), which is genetically encoded. There are two effector neurons for controlling the motor activity. Similarly, the motor output is calculated from the difference of the firing rates of the effector neurons, which is mapped into a range $[-1, 1]$ and is then multiplied by a gain parameter (range $[1, 100]$)

The time evolution of the agents' movement and neurocontrollers are calculated using an Euler method with a time step of 0.1.

2.2 Evolutionary Setup

Agents are evolved using a rank-based genetic algorithm with elitism. We differentiate agents into two groups: those represented at the top and the bottom of Fig. 1. In order to allow each group of agents to have different behavioural strategies, a co-evolutionary approach with two populations is used. Each population has 20 agents, which are evaluated by interacting with the best 5 agents from another population. In each population, all network parameters, w_{ji}, τ_i, b_i,,

η_{ji} and the gains are represented by a real-valued vector ([0,1]) which is decoded linearly to the range corresponding to the parameters (with the exception of gain values which are exponentially scaled). Crossover and vector mutation operators [3], which adds a small random vector to the real-valued genotype, are used. The best 5 agents of the population are retained and 10 agents are replaced by the agents with mutation which are selected from the original populations based on a fitness and the rest 10 agents are replaced by mating with crossover between selected two agents.

The fitness is calculated on the basis of two factors. One is how many times the agent can cross its central position with that of a live interacting agent (live interaction). The other is how much the agent can stay away from a dummy agent which only replays the motions of their partner as recorded in the first stage (one-way interaction). It should be clear that both conditions start from the same initial configurations such as positions, speeds, and neural states and then, at a certain point, the top agent is replaced with the recorded motions. It means that the bottom agent has a exactly same states as those being taken in the live condition when the top being replaced. This is schematically shown in Fig. 2. Therefore, agents need to discriminate the two conditions through ongoing interaction while somehow exploiting the presence of noise. This way of modelling contingent and non-contingent motion reproduces the analogous conditions of the double TV-monitor and the perceptual crossing experiments.

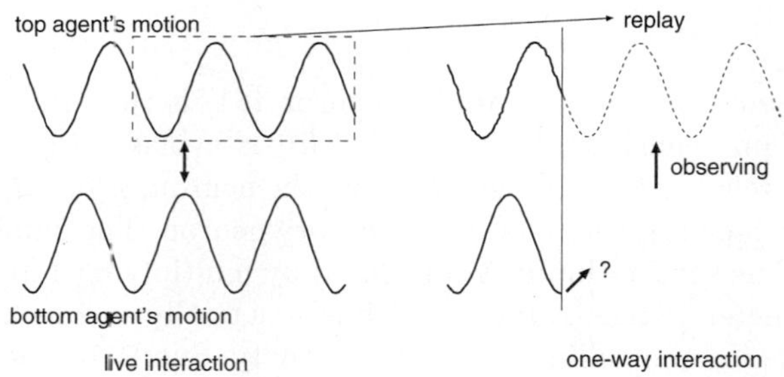

Fig. 2. Live and one-way interactions. Both agents can react to the current situation to establish coordinated behaviour in the live interaction while the only bottom agent can behave reactively while the top agent just replay the previous motions that are recorded in the live condition.

In this paper, only the bottom agents are evolved to discriminate live from recorded interactions. This means that each population has a different fitness function. The population for the bottom agents is evaluated by both factors explained above while the only first factor (crossing under live interaction) is applied to the top agents. In order to measure how many times an agent does not cross the dummy during a trial, the maximum times of crossing is arbitrarily set to 20 (a number estimated from pilot runs) and not-crossing is counted by subtracting the crossing times from it.

3 Results

Evolved agents successfully acquired the capacity to discriminate between the two conditions. Here we analyse one such pair of agents in detail in an attempt to derive some general lessons. Figure 3 shows trajectories of both agents under the live interaction and the trajectory of the bottom agent when interacting with the recorded motion of the top agent (one-way interaction); the differences in position between top and bottom agents are shown on the right. The crossing coordination is established while the two agents move to one direction and control their speeds up and down in order to cross each other without losing track of each other. Under the condition of one-way interaction, the collapse of the coordination is observed. Although there is no big difference at the beginning of the interaction, eventually the bottom agent moves away from the recorded motions. Since the one-way interaction starts the same way as the live interaction, i.e., both conditions have same internal and physical states at the time of switching, a distortion of the ongoing interaction by accumulation of sensory noise must cause the differentiation of behaviours.

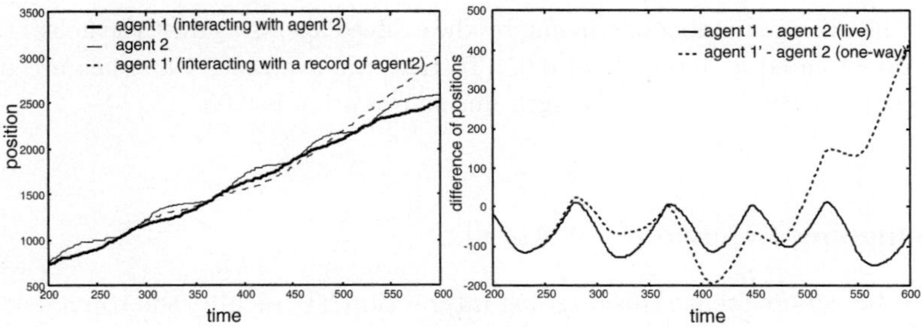

Fig. 3. Left: The agents' trajectories of live and one-way interactions. Thin (top agent) and thick (bottom agent) solid lines show the results of coordinated behaviours under the live interaction. The dashed line shows the trajectory of the bottom agent interacting with the replay of the thin solid line (top agent). Right: The difference between the two trajectories under each condition. When the line crosses 0 it means that two agents cross each other.

3.1 Noise Stability

Without noise in our current settings, there is no way to tell whether the partner is reactive or non-reactive. In other words, the bottom agent needs to exploit the effect of noise to achieve the task. Figure 4 shows the noise effect in the coordinated behaviours. The coordination established by two adaptive agents can be maintained without being destroyed by the increasing noise. Although the stronger noise breaks the coordination more, the coordination shows a robustness against noise perturbations. However, if the top agent is replaced with non-reactive recording, the coordination is not established any more. It means that the bottom agent exploits the presence of noise to detect that the partner is nonreactive and avoids it. Despite the fact that behaviours of the top agent are identical in motion under live and one-way interactions, the bottom agent does not coordinate with the recording. It means that only mutual coupled dynamics can maintain the coordination while suppressing disruptions induced by sensory noise.

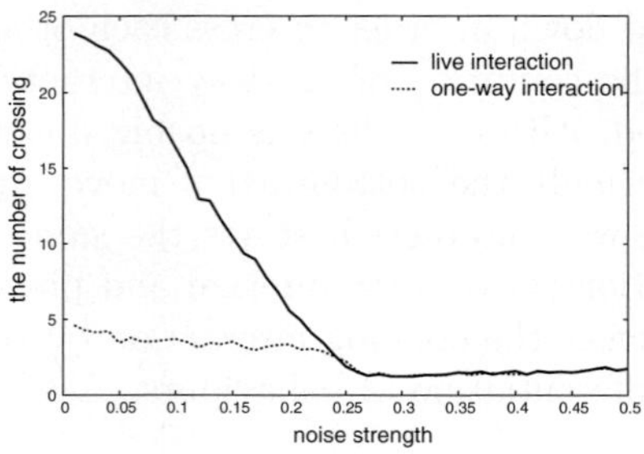

Fig. 4. The average number of crossing by the evolved agents against the noise strength that is investigated at intervals of 0.01. To take the averages, 100 trials are used at each noise strength. The noise strength during evolution is 0.05.

3.2 Ongoing Behaviours of Agents

How do the agents create this ongoing interaction? Typically, the top agent keeps its position on the right hand side (positive in the space) of the bottom agent and control the speed up and down. If the top agents receives sensory stimulation, then it increases the speed to the right. After moving away from the bottom agent, it decreases the speed and waits for the bottom to catch up and make sensor contact. This process is repeated periodically (Fig. 3). By contrast, the bottom agent keeps its position on the left hand side (negative in the space) of the top agent. Without any sensory stimulus, it accelerates to the right to catch up with the partner and decreases the speed after making contact.

Figure 5 shows the neural dynamics of both agents during the coordinations as well as the attractor states constituted when the input is forced to be constantly activated (1) or deactivated (0). The top agent often uses the edges of neural activations and the dynamics (plotting 3 representative neurons) changes sharply which means that the neural activities behave as in a discrete system. The internal dynamics changes quickly depending on the sensory inputs so that the resulting behaviours are reactive. In contrast, the internal dynamics of the bottom agent tends to take place in the transient dynamics between the attractors (indicated as the points with constant input 0 and 1). Before an attractor state is reached, the input changes and the transient trajectory is redirected to the new attractor. During the coordination, the agent alternates between mak-

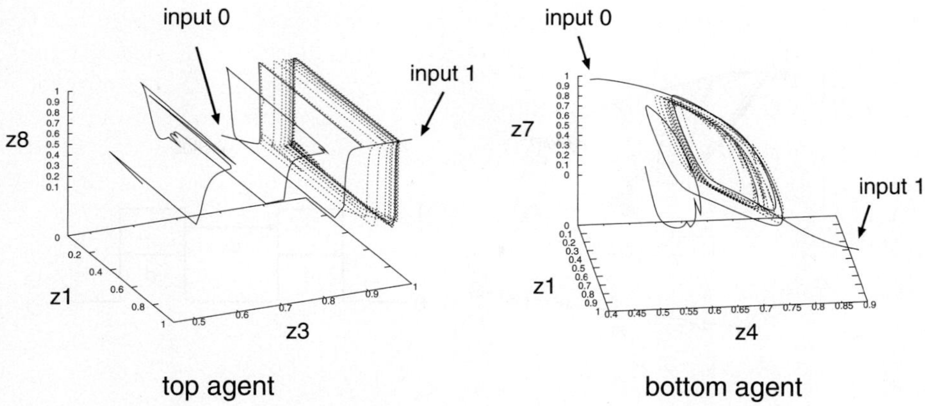

Fig. 5. The internal dynamics of agents. Each axis indicates the firing rate, z_i. The choice of neural activations is representative for visual clarity. The dashed line shows the dynamics constituted by interaction between agents and the solid line shows the dynamics when the sensory input is fixed to be always deactivated (0) or activated (1). Input 0 and 1 are attractors constituted under the fixed sensor.

ing contact and staying away from its partner. The permanence of the system in its transient dynamics is a reflection of the agent's embodied interactions.

In order to see how the transient dynamics is generated and modulated by the inputs, controlled sensory inputs are given to the bottom agent. As seen in Fig. 6, if the controlled inputs which have a close frequency to the average over the actual interactions are given to the bottom agent, it can reproduce similar dynamics to the actual one. However, reducing the length of the stimulations of the sensory input and increasing the gaps generate the internal dynamics falling into the attractors, which can hardly adjust the agent's behaviour. A sufficient length of the consecutive activations is necessary to stay away from the attractor and to show the transient dynamics. Actually, when the body size of agents are reduced, they cannot achieve coordinated crossing even in the live interaction because the smaller bodies do not produce inputs that are sustained for long enough. It means that the bottom agent uses the duration of the sensory stimulus to tell if the interaction is live or one-way by using the transient dynamics.

What factors affect the duration of the sensory input intervals during interaction? The duration is affected by the relative speeds of the two agents as they cross. In the live interaction, at the crossing, the top agent "catches" the movement of the bottom agent and provokes a return to the left by reacting to the partner to fit the moving speed (Fig. 7). The reactive response makes the duration of contact longer. In other words, the duration of the contact and the intervals between contacts are *co-regulated by both agents*. However, a small difference of the timing of start crossing is enough for the recorded trajectory of the top agent not to be able to "respond" in time to catch the bottom agent. This non-responding behaviour cannot make sufficiently long sensory contacts

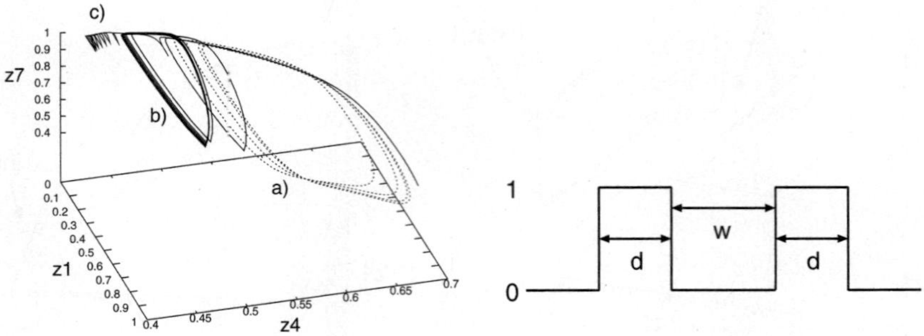

Fig. 6. The internal dynamics of the bottom agent corresponding to different periodic stimulus patterns: a) $d = 20, w = 63$, b)$d = 14, w = 69$, and c)$d = 10, w = 73$ where d and w represent the length of activations and deactivations of controlled sensory inputs.

and the bottom agent passes through and moves away. Therefore, the presence of noise causes a breakdown of the coordination and the lack of co-regulation makes this breakdown irrecoverable.

This co-regulated behaviour happens between the agents. A single agent cannot control the duration of the contact interval to make it last the right time in order to produce the cyclic transient dynamics in the bottom agent. It means that there is a kind of agreement between agents that cannot be determined individually. Only the precise timing and interdependency of dynamics can produce co-regulation in this case.

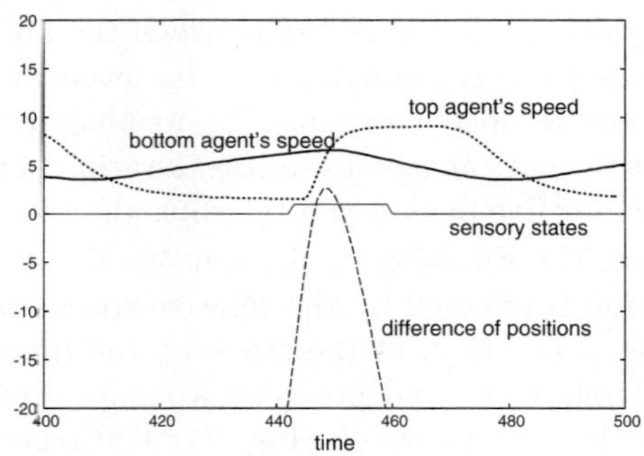

Fig. 7. Co-regulations of agents' speeds for the duration of contacts. The top agent starts "catching" behaviour at the beginning of contacts and that works the bottom agent to "give up" passing through and it eventually slows down. Both these behaviours make the sufficient duration of contacts.

4 Discussion

This simple model explores the agency detection observed in minimal social interaction. In order for agents to produce a different behaviour in the case of a one-way interaction, it is sufficient to rely on properties of the mutuality of the live interaction, such as increased robustness to noise and timing differences. This suggests that, at least in some cases, the recognition of the presence of another's subjectivity need not lie in complex cognitive individual mechanisms able to integrate past information, but rather on the ongoingness of the interaction process itself and its robustness to noise. These simulation results show the possibility of embodied modelling to deal with matters as complex as agency detection and co-regulation of interaction without the models themselves being complex.

In our model, social coordination is realized through contingent co-regulation, which gives rise to the detection of live agency. Contingency is a relational condition and its perception must rely on how the patterns of expressive action of a partner relate to our own actions rather than on the patterns themselves. The interrelated structure of social coordination has the global stability of the mutual feedback against perturbations of noise and the active co-construction of precise contact timing.

Our results also shed light on the ongoingness in social interactions. This kind of ongoingness has been less investigated in the simulation studies so far (some can be seen in [4,5]). The dynamical structure of ongoingness could be very different from the simple attractors views in which the timing issue is removed due to a convergence of the dynamics to stable structures. In our results, one of the agents behaves reactively and another uses the transient dynamics and those behaviours are interrelated in the sense that both agents are meant to behave in a way of making the time of contact longer when crossing. In the presence of non-reactive recorded motions, this co-regulation cannot occur. It is a kind of agreement between agents that can be seen in our daily social interactions. As discussed above, this agreement cannot be attributed to a single individual. It arises from a relational view of coupled dynamics. In order to understand social interaction, however, we are highlighting not only the importance of the coupled dynamics but also an aspect of the ongoingness under the coupling. If the interaction had a pre-established harmony, there would be no reason to regard the agents as separate and it all could be described by a single whole dynamics where we don't have to call any behaviour an "agreement". But the interaction that our evolved agents establish is not like this. It is rather based on the ongoing co-regulation that they establish by constantly responding to the abrupt breakdowns induced by the presence of noise. Agreement is constantly renewed – as opposed to pre-established – by this ongoing and transient dynamics.

The embodiment of agents is rather minimal and the minimalism will allow us to formulate hypotheses for more sophisticated forms of embodied interaction in both simulations and experiments with real robots [3,8]. In order to better understand the role of embodiment, future work will compare these results with interaction in disembodied agents that are not situated in an environment but are directly coupled with each other as well as agents capable of different ways of regulating their mode of coupling. In our model, the agents' capacities such as sensors and motors are very limited but they make use of the environmental space that produces the gap between their bodies. Agent are not in constant coupling with a partner. Coordination consists on regulating the presence (or more generally the quality) of the coupling. In the experimental situation, the mother or baby can attract the other into an interaction by "actively" moving and observing the response of the other. Such active engagement can only work if the existence of some sort of a "gap" in coupling is possible. The environment can thus be considered as a buffer and implicitly work in our current simulation as it is. Modifying such a buffer, or even removing in the case of disembodied agents, will result in a family of models shedding light on how embodiment lies at the root of social coordination.

Acknowledgments. This research was partially supported by the Japanese Ministry of Education, Science, Sports and Culture, Grant-in-Aid for JSPS Fellows, 17-04443.

References

1. Auvray, M., Lenay, C., Stewart, J.: The attribution of intentionality in a simulated environment: the case of minimalist devices. In: Tenth Meeting of the Association for the Scientific Study of Consciousness, Oxford, UK, 23-26 June (2006)
2. Beer, R.D.: Intelligence as adaptive behavior: An experiment in computational neuroscience. Academic Press, San Diego (1990)
3. Beer, R.D.: Toward the evolution of dynamical neural networks for minimally cognitive behavior. In: Maes, P., Mataric, M.J., Meyer, J.-A., Pollack, J.B., Wilson, S.W. (eds.) From Animals to Animats 4: Proceedings of the 4th International Conference on Simulation of Adaptive Behavior, pp. 421–429. MIT Press, Cambridge (1996)
4. Paolo, E. A, D.: Behavioral coordination, structural congruence and entrainment in a simulation of acoustically coupled agents. Adaptive Behavior 8:25-46. MIT Press, Cambridge (2000)
5. Iizuka, H., Ikegami, T.: Adaptability and diversity in simulated turn-taking behavior. Artificial Life 10, 361–378 (2004)
6. Murray, L., Trevarthen, C.: Emotional regulations of interactions between two-month-olds and their mothers. In: Field, T.M., Fox, N.A. (eds.) Social perception in infants, pp. 177–197. Ablex, Norwood (1985)
7. Nadel, J., Carchon, I., Kervella, C., Marcelli, D., Réserbat-Plantey, D.: Expectancies for social contingency in 2-month-olds. Developmental Science 2, 164–174 (1999)
8. Robins, B., Dautenhahn, K., te Boekhorst, R., Billard, A.: Robotic assistants in therapy and education of children with autism: Can a small humanoid robot help encourage social interaction skills? Universal Access in the Information Society 4(2), 105–120 (2005)
9. Trevarthen, C.: The self born in intersubjectivity: The psychology of an infant communicating. In: Neisser, U. (ed.) The perceived Self, pp. 121–173. Cambridge University Press, Cambridge (1993)

Hermeneutic Resonance in Animats and Art

Alasdair Turner

Bartlett School of Graduate Studies, UCL, Gower Street, London
a.turner@ucl.ac.uk

Abstract. One major criticism of direct or active perception (and other forms of embodied action) from the perspective of cognitive psycology is that, according to common sense, there are some actions that require strictly symbolic information — for example, to stop a car in response to a red traffic light — which fall outside the realm of a perception-action cycle. Although such cognitive responses are not necessarily a goal of artificial life, they must necessarily be included within the embodied paradigm if it is to encompass the cognisant individual, the self-aware individual, or, potentially, the conscious individual. This paper will address the question, 'can an animat appreciate art?' Although this may seem very different to the example of a prosaic response to a traffic light, it will be argued that a common framework for establishing the meaning of an object is needed. It will also be argued that clarification to previous philosophical models of artistic engagement is required: in particular that the process of understanding is not a dialogue between an autopoietic artwork and animat, but that there is either a unity of object (artwork-animat) which becomes self-maintaining, or a more classical Gibsonian interpretation as a fixed set of affordances offered by an object to the subject, both of which lead to the conclusion that the process of understanding becomes a resonance in the unity or animat.

Key words: Autopoiesis, hermeneutics, embodiment, phenomenology, direct perception, natural vision

1 Introduction

The relationship between artificial life and art is an enticing one. That an artificial organism might have access to a deeper meaning within an object than is initially apparent, or, further, be able to create an object capable of engendering deeper meaning, offers an intriguing mechanical and philosophical challenge. Wheeler [1] attempts to bring together the materialism of Rothko's outlook with the spiritual experience of his art, through consideration of the philosophy of artificial life and the phenomenological relationship between art viewer and art work. In a previous study, I took Wheeler's phenomenological interpretation and applied it to the interaction between artwork and animat to create an ongoing embodied process of location and relocation of both an 'artwork' and an animat [2]. Bird and Stokes [3] are engaged in the production of fractal drawings by animats, with allusions to what surely must be the creation of art through animat.

F. Almeida e Costa et al. (Eds.): ECAL 2007, LNAI 4648, pp. 495–504, 2007.

However, Bird and Stokes are careful to avoid making explicit this implication, as in doing so the difficult question of 'what is art?' is raised. While this question may be impossible to answer directly, this paper will attempt to show that the related question of 'what is artistic appreciation?' can be tackled through the philosophy of artificial life. Although answering this question will not determine 'what is art?', it can address questions such as 'is this specific creation artistic?'.

In wrestling with the question 'what is artistic appreciation?', it is important to realise that the analysis herein is based very much on Wheeler's account, in particular, his move to a non-linguistic hermeneutics. The symbolic representation of meaning, and Cartesian dualism are rejected in favour of a process-based approach to understanding. However, Wheeler leaves a certain amount unsaid; in his coda of 'Art and Language', there are two types of understanding process. An example of a 'live' jazz musical interaction is described as a hermeneutical dialogue, which implies a loop of (non-linguistic) information exchange between two individuals (performer and auditor), whereas the appreciation of Rothko's art is discussed in terms of a phenomenological 'drawing into the artwork', or as we will put it here, an 'at-oneness' with the art. This second mode of interaction is very much one way: the art does the drawing into, and there is little if any formal exchange of information. In [2], I took an approach which initially seems to be more fruitful: that both artwork and live jazz musician have equal status as entities, and that there is a hermeneutical dialogue in both cases. However, it now seems to me, and as will be examined, that it is the 'at-oneness' which provides for both a true account of the process of understanding. The framework constructed here also seems to lead to a measurable outcome in an animat for 'art appreciation'. If correct, this completes Wheeler's attempt to tie the scientific artificial life approach to the phenomenological account of the experience of art.

In order to get to this stage, I will draw heavily on the work of Luhmann [4]. Luhmann struggles to get to the crux of meaning within the context a systems approach, through a ten point examination of the necessary conditions for meaning. Although for the most part, Luhmann is engaged in the analysis of social systems, at this stage he maintains that 'psychic' (pertaining to the individual) and 'social' meaning are identical in form – that is, that his formal theory of meaning is equally applicable to all systems. For the purpose of this paper, we will tend to examine the individual. Wheeler takes pains to explicate the role of individual and society, as the animat apparently lacks social context, and without a social context, in Heidegger's stream of phenomenology, there can be no art. We will try to look in more detail at what the context of the animat is, both environmentally and socially, and suggest that, social context is a special case of a more general environmental context.

Hermeneutics itself is a topic too large to be covered in detail and riven with a history of linguistic approach [5]; it will be divided into two more manageable concerns: the establishment of meaning (which Luhmann will inform) and the process of understanding, which will be approached through autopoesis [6], and Gibson's affordance-based theory of vision [7]. Given the appeal to Gibson, this paper will concentrate on the discussion visual art, but other media should not be discounted. The generality of the approach is to sensor-motor individuals connected to the environment by any type of affordance.

Before addressing the major topic, it is worth mentioning that there exists a more prescient problem than the linguistic backdrop of hermeneutics, and that is the equation of phenomenology and science into a single framework. To the originators of hermeneutic or phenomenological enquiry the experiential cannot be resolved with the analytic, or to use Gadamer's [5] terms, truth cannot be resolved with method. However, to be true to its aims, it seems to me that the practice of artificial life should aim at this very resolution.

2 Background

There are various approaches to embodied [8] or situated cognition [9,10]. At the centre of any is the rejection fundamental dualism of Descartes, and the either implicit or explicit adoption of a philosophy attuned to Heidegger's phenomenology. The basal assumption is that the organism can only exist in relationship to its environment, and that cognition can only follow in this embodied situation.

For the purposes of this paper, I will start with Maturana and Varela's definition of living existence, autopoiesis [6]. Maturana and Varela [11] draw the situation of autopoiesis, as shown in figure 1. The organism (originally in the scope of their theory, a biological cell) is drawn as a circle with an arrow, intended to indicate a self-producing entity. That is, an entity which maintains itself in relation to its environment. This environmental relationship is also shown in the diagram, with arrows to indicate that the entity is both affected by and affecting its environment.

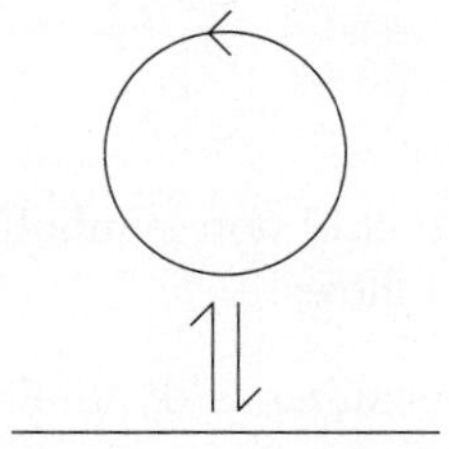

Fig. 1. Autopoietic individual in relation to the physical environment, after Maturana and Varela [11]

More formally, the autopoietic entity, a network of ongoing processes, both continuously reproduces its network of processes, and distinguishes these processes from the environment, or physical space around it. This, to proponents of autopoiesis, is life. Furthermore, this systems theory account of interaction summarises the process of direct or active perception.

In Gibson's direct perception [7] as related to natural vision, he posits that the individual interacts directly with its environment, taking affordances from the environment in order to move around it. His affordances pre-exist in the environment, they are natural ways in which the individual can interact with the objects around it. For example, a rock might be picked up, or, to move to a Heideggerian example, a hammer may be picked up as it affords the ability to pick it up by the shape of its handle. It may also afford hammering of an object. Both of these can be seen in autopoietic terms of the individual existing in the environment, with a linkage through picking up or otherwise engaging with the object. Gibson's view is that in natural vision, if the affordance exists then it will be followed. So, a chair placed on a floor affords moving towards it, and sitting on it. An argument against Gibson is that the chair also affords standing on, being thrown across the room and various actions that although they happen, are not typical interactions with the object. Why should it be more usual to sit on a chair than to throw it? Of particular relevance, is how the object takes on the set of meanings of the chair, and how non-manipulable objects, such as fixed static artworks, take on a meaning. There is also the location of meaning to consider. In the definition of embodied cognition, there is a dilemma, in that although the dualism of Descartes (mind-body split) is rejected, there is still a distinction between entity and environment, and a fundamental distinction of meaning and non-meaning. Since the symbolism of objects has been eschewed, there has to be a replacement for the symbol.

These are issues that typically seem to be glossed over in the race for an embodied cognition. For the adherent to autopoiesis, all life might be construed as cognition [12], in that these processes of active perception simply *are* the processes of life. However, Luhmann, in relation to social systems, does realise the requirement for further understanding of meaning systems, and in this paper I will examine the relationship of these to cognisant systems boundaries.

3 Meaning in Systems Theory

Luhmann [4, pp. 59–102] grapples with 'meaning' in the context of non-symbolic systems theory in ten phases. These can be summarised as follows:

1. Luhmann discards a static approach to meaning. All meaning is dynamic, as it must be allowed to change over time and for different individuals and contexts. For our purposes, dynamical systems are well known to artificial life, and we do not need to dwell on this factor.
2. Luhmann notes that meaning must be an action of distinction, what I call above the taking on of a set of meanings. That is, where a chair becomes sittable or throwable, these meanings must be distinguishable from the previous state of non-meaning.
3. Establishing meaning necessarily involves information: the transfer of information, from information to non-information embeds meaning (where 'non-information' is information no longer necessary to create the established meaning).

4. Once a new meaning exists, it extends the horizon of available meanings (which Luhmann suggest's is Husserl's horizon). That is, once one meaning is established, there become other possible meanings.
5. Importantly, the *subject* is discarded. Luhmann asserts that meaning occurs outside of 'subject'. This has implications for boundaries of knowledge and cognition in the current context, which are dealt with in the next section.
6. Luhmann distinguishes between actuality and possibility, so that meaning is decomposed into different 'dimensions': factual (or ontological), temporal, and social (or intersubjective).
7. There is an action involved in meaning (attribution as action is the reproduction of autopoietic systems) and experience (attribution as experience is the reproduction of meaning). Luhmann posits that there can be no experience without action, that is, no ego without alter. This will be a difficulty to us in the quiet contemplation of artwork. What action is involved in the appreciation of art? This will be dealt with in the final section.
8. Meaning is an evolution: it is not planned, by which we might understand that accident of experience comes to play in the establishment of meaning – knocking over a chair accidentally may show that it embodies affordances for throwing.
9. Meaning as a symbolic generalisation must include self-reference. If it does not then it fails to include the possibility of future alteration. This forms the single most important factor to us, as it is the link to a symbolic representation. Luhmann suggests that 'the concept of symbolic generalisation of meaning's self-reference replaces 'sign'', which sounds complex in the extreme, but is perhaps approximated by simply a looping or recursive relationship established which equates to a sign.
10. Meaning can be present either in the form of consciousness or communication (psychic or social), or in our case, indivdiual.

Although Luhmann suggests that meaning for systems is the totality of all these factors, some have more bearing than others. Most are either implicitly or explicitly accepted within the fields of artificial life and embodied cognition. However, for the current purpose, three stand out as requiring further attention: the loss of subjectivity (5), the action to create meaning (7), and the creation of self-reference (9). If loss of subjectivity is to be tackled in an autopoietic context, then it must explain how the individual definition of itself as distinct from the environment can be resolved with a non-locational meaning. The second of these three, action, of course should not be a problem, as the very basis of the paper is in active perception. But there is a problem as the appreciation of a static artwork initially seems inactive. The perception-action cycle is seemingly broken in the relationship of artwork to animat. This will be dealt with in the final section, and will rely on the third and most difficult of Luhmann's factors, the necessity of a self-referential loop in generalisation of meaning.

4 Boundaries and Affordances

There is a problem with the autopoietic approach to cognition raised by Boden [12]. She points out that, according to Maturana and Varela's own definition, the autopoietic individual must maintain its identity as separate from the environment. This conflicts with a view of situated cognition, where knowledge is distributed between the combination of individual and environment. It is also a problem to our conception, as the locus of meaning is important to the ability to cognise. The solution proposed here is to look more closely at what the maintenance of a separate identity involves. If the individual is a human being, then it can easily add what might seem inorganic objects to itself. For example, nail varnish can be painted and repainted. Although this does not immediately seem a great step, it does suggest that the individual can adapt its boundary, and that the boundary can extend to non-organic structure. This is already explicit in autopoiesis: the nail varnish is a protective layer that encases part of the individual just as a cell wall forms an outer boundary to a cell. Both are inorganic matter, and can be added to extend the individual. However, once this step has been taken, there is no longer a fixed boundary for the individual at all. The boundary may be physical in the sense that it exists in space, but it does not necessarily follow the prescribed Kantian order around it. Gibson [7] makes the interesting observation that the third dimension is a Cartesian error, and the accepted boundary of an object forms a similar error. This, of course, is clear in phenomenological thinking: Heidegger's hammer is at one with the skilled practitioner when in use, and the boundary of the unity thus extends around them 'both'. In this sense the unity exists as what Merleau-Ponty might recognise as a series of adaptive Gestalts [13], although the word Gestalt implies a forced set of results, rather than the range of possibilities that might actually exist.

In this sense of body, action in response to affordance creates new unities of object, and these come within the modes of usage we have examined. The objections to neo-Gibsonian affordance can be seen as trivial. For example, the chair can still be stood on, thrown or sat on. The action of throwing, standing on or sitting creates a new unity of object. Luhmann's symbolic generalisation, that of chair, is created through the combination of these actions and the gathering of them together into a continued loop. In this sense, the creation of meaning is consistent with the extension of the body. Symbolic features of the environment, for example a door handle, or even the mechanism for opening a puzzle box come into being through the action of their completion. So to a culture that has not seen a door handle, the very action of performing the use of the handle becomes its meaning. The puzzle box is infuriatingly present-at-hand until we work out the relationships. Action is discovery of inherent affordances in the object. Once these are learnt, they can be applied to other similar affordances in objects, thus enforcing the symbolic meaning of the object.

5 Societal Context

A rapid exposition such as this must necessarily neglect much detail. One important feature of meaning for Heidegger is the societal context of meaning.

Wheeler points out that since the individual animat in a setting is devoid of societal context, the possibility of a phenomenological account of the appreciation of art seems flawed. While Wheeler's resolution, to invoke *fitness affordance* is undoubtedly adequate, there are a couple of further points which may prove fruitful: firstly, it is important to point that just as we do not exist in absentia of society, neither does an animat. We may attempt to divorce it, but our very positioning of motors and sensors, or the representation used in our genetic algorithm, or the task chosen for it to perform encode societal values. We should not see this as a problem, but in fact a solution: this animat remains a human creation, and is inherently (and irrevocably) linked to its creator. It is a social construction, although perhaps non-social in itself, no matter how hard we try to avoid it. The second point is that the social and physical environment of an animat may be considered as a continuous relationship, in that the situation of the societal relationships and environmental relationships can be abstracted from the animat, following previous work on the definition of relationships between individuals, society and environment [14], as shown in figure 2.

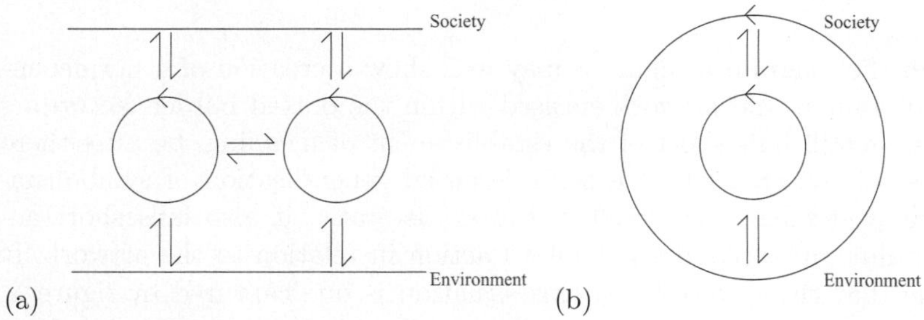

Fig. 2. (a) Autopoietic relationship between individuals, society and environment, after [14], (b) The autopoietic context of the individual.

6 Hermeneutic Resonance

In section 4 it was argued that the process of understanding is achieved through a creation of unity, and that this unity permits situated cognition through the combination of affordances. This is a very different model of understanding to the hermeneutical dialogue. Figure 3 shows what a hermeneutical dialogue might look like, according to my initial intepretation [2]. This was perhaps reasonable for that study, where animats and artworks both moved, creating the situation whereby an informational exchange between the two could exist; however, it does not serve in the situation where an artwork is seen as grounded. In this case, whatever affordance the artwork offers cannot be responded to directly. For example, how does the animat stop, look and reflect on the canvas in a meaningful exchange when the canvas offers only static affordances. In fact, of course, the canvas offers many affordances, for example, that of ripping it off the wall and constructing a dwelling with it. We must accept that this is what many cultures may do with a Rothko, but then we can invoke Luhmann's concept of a stochastic evolution of meaning to suggest that one animat might not follow this more primary affordance, it might follow another affordance offered by the painting, but what is that affordance?

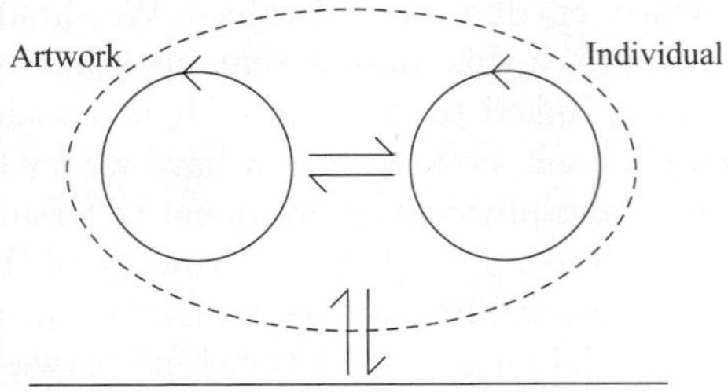

Fig. 3. Autopoietic relationship between artwork and individual, after [2]

Although the diagram in figure 3 may well show a creation of a hermeneutical unity of animat and artwork encased within the dotted border, according to Luhmann, it still falls short of the establishment of meaning, because there is no process of the creation of a self-referential generalisation of symbolism. Furthermore, as we have discussed, if the art is static, it also falls short according to Luhmann as it does not afford action in relation to the artwork. It seems to me that the resolution of this situation is one captured by figure 4. The lefthand situation shows artwork offering affordance, without action from the animat. However, in the righthand situation, a self-referential process is established between artwork and viewer. The viewer is 'at-one' with the artwork in their contemplation of it. The *affordance* of the artwork is to create this very self-referential meaning loop. Thus there is an action, an intentional action, to establish the loop between the artwork, just as there is an intentional action to modify the individual by her painting her nails. In this manner, the three vital facets of Luhmann's meaning have been met: there is a non-local event concerning an action of self-referential creation.

7 Conclusion

In this paper, I have investigated how Luhmann's [4] work on meaning in embodied systems can be applied to the consideration of how animats might understand artwork. I have argued that Wheeler's notion of a hermeneutic dialogue [1] should be extended to encompass the establishment of a self-referential unity between viewer and artwork. Although this may seem a somewhat esoteric approach, and limited in its application to artwork, the relevance to artificial life systems is that it provides a mechanism to be able to move beyond simple reactive units

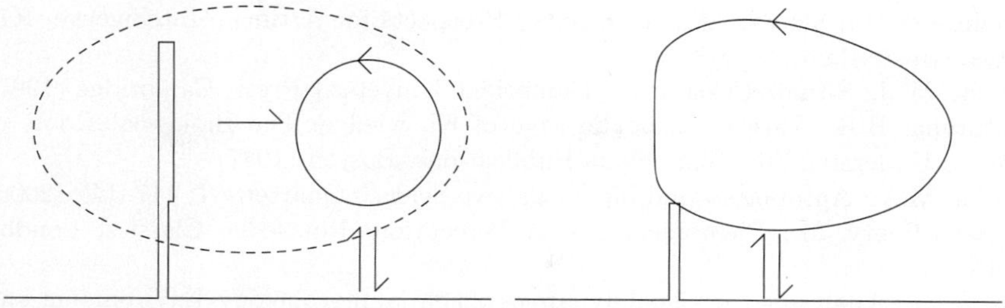

Fig. 4. Hermeneutic relationship between artwork and individual

and into cognisant individuals without compromising on the active perception paradigm. That is, the formation of meaning cycle presented allows us to posit cognition without the requirement of a physical symbol system.

As a somewhat fanciful indulgence, it might be claimed that the formation of the cycle causes a resonance in the animat-artwork unity. But fanciful as it might be, it tallies with the phenomenological perspective of the interaction with artwork: it *feels* as though it causes a resonance when it transmits meaning or truth. The disengagement from the cycle leaves the animat modified, although the artwork itself remains intact from which others in society can take their own meaning. The fanciful interpretation may even extend to what an appreciation of art might look like in an animat. As a materialist, it seems to me that we should investigate the neural response of an animat in this situation: one that appears to stop and watch. Has a new resonance been created in its neurology, and has it, just possibly, appreciated art?

References

1. Wheeler, M.: From robots to Rothko: The bringing forth of worlds. In: Boden, M.A. (ed.) The Philosophy of Artificial Life, pp. 209–236. Oxford University Press, Oxford (1996)
2. Turner, A.: Reversing the process of living: Generating ecomorphic environments. In: Hanson, J. (ed.) Proceedings of the 4th International Symposium on Space Syntax, pp. 15.1–15.12. UCL Press, London (2003)
3. Bird, J., Stokes, D.: Evolving fractal drawings. In: Soddu, C. (ed.) Generative Art 2006, Milan, Italy (2006)
4. Luhmann, N.: Social Systems. Standford University Press, Stanford (1984)
5. Gadamer, H.G.: Truth and Method. Continuum Impacts, London (2004)
6. Maturana, H.R., Varela, F.J.: Autopoiesis and Cognition: the Realization of the Living. D. Reidel, London (1980)
7. Gibson, J.J.: The Ecological Approach to Visual Perception. Houghton Mifflin, Boston (1979)
8. Varela, F.J., Thompson, E., Rosch, E.: The Embodied Mind: Cognitive Science and Human Experience. MIT Press, Cambridge (1993)

9. Husbands, P., Harvey, I., Cliff, D.: An evolutionary approach to situated artificial intelligence. In: Sloman, A., et al. (eds.) Prospects for Artificial Intelligence, IOS Press, Amsterdam (1993)
10. Clancy, W.J.: Situated Cognition. Cambridge University Press, Cambridge (1997)
11. Maturana, H.R., Varela, F.J.: The Tree of Knowledge: The Biological Roots of Human Understanding. Shambhala Publications, Boston (1987)
12. Boden, M.A.: Autopoiesis and life. Cognitive Science Quarterly 1, 117–145 (2000)
13. Merleau-Ponty, M.: Phenomenology of Perception. Routledge Classics, London (2002)
14. Turner, A.: Analysing the visual dynamics of spatial morphology. Environment and Planning B: Planning and Design 30, 657–676 (2003)

Robotic Superstrings Installation: A-Life Science & Art

Mauro Francaviglia, Marcella Giulia Lorenzi, and Michael Petry

[1] Department of Mathematics, University of Torino,
Via C. Alberto 10, 10123, Torino, Italy.
mauro.francaviglia@unito.it
[2] E.S.G. (Evolutionary Systems Group), University of Calabria,
Via Ponte P. Bucci, Cubo 17b, 87036, Arcavacata di Rende, Italy.
m.lorenzi@unical.it
[3] MOCA Museum of Contemporary Art, London Project Space,
113 Bellenden Road, SE15 4QJ, London, England.
mocalondon@yahoo.com

Abstract. Superstrings Installations (Torino 2005, Cetraro 2005, Milano 2005) – where Art and Science were directly conjugated – pictures and movies have been later collected into a DVD produced in a strictly limited edition and in few copies. These DVDs have been part of a Generative Art Installation presented in Milano in December 2006. A small robotic artifact has been programmed with very simple rules to trace on the DVD surfaces space "visual superstrings", which changed depending both on the interaction with the public and the configuration of the DVDs on the floor. At the end, a unique artwork was generated. This paper discusses the ideas of A-Life Art, the related aesthetic concepts and how the use and programming of a robot has allowed to create A-Life Artistic object inspired by the Superstring Installation through semi-adaptive robotic behaviour.

Key words: A-Life Art, Generative Art, Robotics, Art and Science.

1 Introduction: Superstrings Installation I-II-III

"***Physics is a form of insight and as such is a form of Art***" **David Bohm.** Superstrings Installation is a conceptual art work by one of us (MP). It was realized for the first time in Torino, during the "Art, Complexity and Technology: Their Interaction in Emergence" Workshop [1] and later in Cetraro, during the "Mathematics, Art and Cultural Industry Conference" [2] and finally in Milano, during Generative Art 2005 [3]. This was at the same time an exquisite demonstration of how a clever mix of Art and Science is feasible and also a clear example of the cultural challenges offered by the fascinating and fast growing field of "Emergence in Art". The artwork was in fact a creative and generative process, that - by means of physical interactions involving both the artist and the public - generates, step by step, an installation that is never ending, continuously changing and dynamically oscillating in space and time.

The "Standard Model" (the current one in Particle Physics) postulates the existence of four "fundamental" physical interactions that are in turn carried by "elementary particles" grouped, according to precise symmetries rules, in suitable "families". A different framework has been later proposed in Theoretical Physics, which tries to

F. Almeida e Costa et al. (Eds.): ECAL 2007, LNAI 4648, pp. 505 – 514, 2007.

Fig. 1. Torino, Superstring Installation I.

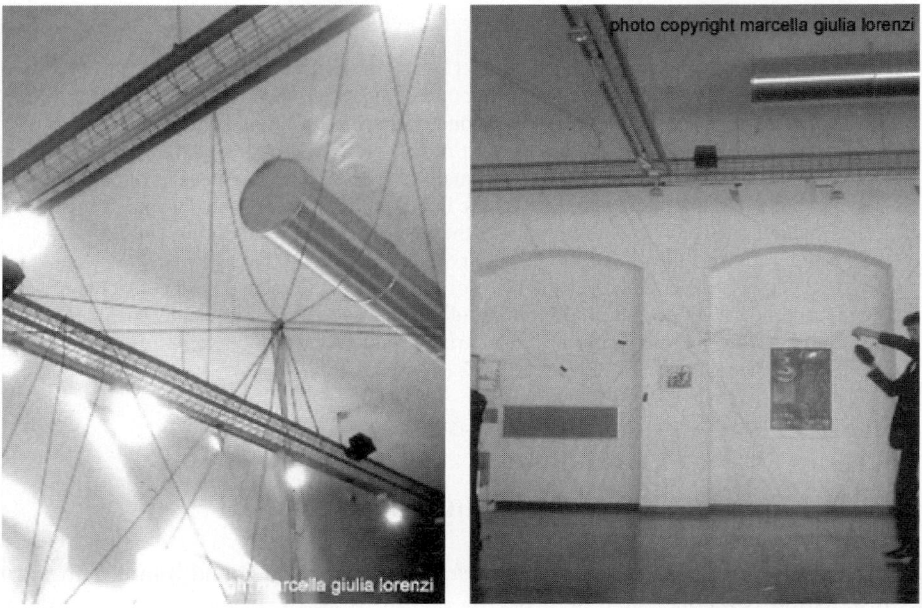

Fig. 2. Cetraro, Superstring Installation II.

replace standard particles with so-called “strings” [4] and “superstrings” [5]. These are “extended objects” having one dimension, like real tiny elastic ropes which fill infinitesimally small portions of space. Strings continuously vibrate in spacetime and their vibrations generate observable excitations of the physical fields.

In much the same way, virtual strings formed by elastic ropes filled the space in the Exhibition Room, by means of an infinite process in which the public helped to generate intricate and never ending patterns. The room became wrapped in miles and miles of elastic string, going from floor to wall, ceiling to door, wherever anyone wanted to attach two endpoints. Attendees were able to enter and bounce around space, the visual metaphor of the bound room being a pattern for space, like a “Wormhole”,

Fig. 3. Milano, Superstring Installation III.

a loop in space-time where we could as we exited one, see ourselves enter it. As a final output Art and Science merge through the emergence of the art work itself. This installation is therefore a true and genuine "generative process" that allows to intertwine Art and Complexity, giving space into Contemporary Art to the modern physical notions of "Strings" and "Superstrings", which are among the most fascinating and challenging issues of current Physics of fundamental interactions.

2 Installations, Art and Life

Installation, as a generic term, covers a large area of practice and enquiry within Contemporary Art. It is suggestive the notion of 'exhibition', or 'display', and of an actual activity which is today as widespread as any other way of making Art […] Installation, as a hybrid discipline, is made of multiple histories; it includes architecture and Performance Art in its parentage, and the many directions within contemporary visual Arts have also exerted their influence. By crossing the frontiers between different disciplines, installation is able to question their individual autonomy, authority and their history and relevance to the contemporary context [6].

The prominence of installations in specific non-art sites also continues to figure among the concern of installation artists. The activation of the place, or context, of artistic intervention suggests a localized, highly specific reading of the work, and is concerned not only with Art and its boundaries, but with the continual rapprochement, or even fusion, of Art and Life. Installation must therefore also represent the artist's desire to extend the area of practice from studio to the public space. […] Installation, in the sense proposed by the book of one of us [6], is a relatively new term. It is really only on the last decade or so that it has been used to describe a kind of Art making which rejects concentration on one object in favour of a consideration of the relationships between a number of elements or of the interaction between things and their context […] Goldberg gave an article the title "Space as Praxis" and it is this sense of space in active dialogue with things and people it contains, in all its ramifications, that lies at the heart of the subject. Procedures which activate the potential or repressed meanings of a specific place, which play real space and time off against the imaginative dimensions of the various electronic media, which question the cultural 'truths' reflected in patterns of collection, scholarship and display in the priviledged spaces of Art, and which relate the social space in which they operate to the sense of public, private and communal found in the language of architecture, all fall within the term [...] What is needed, rather, is the drawing of certain ideas out of the history, particularly the notion that space and time (the actual duration rather than the abstract notion of time) themselves constitute material for Art. We must also take note of the tendency, observable throughout modernism, for Art to merge with Life [6]. Crary affirms that "unavoidably, our lives are diveded between two essentially incompatible milieus; on one hand, the spaceless electronic worlds of contemporary technological culture and, on the other, the physical extensive terrain on which our bodies are situated. Much Installation Art affirms that experience (and Art) is constituted out of the paradoxes and discontinuities of this mixed heterogeneous zone" [7]. Though technology has become an integral part of Installation Art, it is the way in which artists have appropriated these advances that it is important. These 'tecniques' have been integrated into their work by artists, often highlighting the artificiality of the medium. [...] The artists revel in the artificiality of their environments, pointing out that the parallel and simulated worlds of gaming are of greater interest than their real counterparts [7].

3 A-Life & Art: from Genetic Algorithms to Autonomous Robots

Since A-Life's self-declared inception in 1987, artists began to apply its techniques and concepts. "Mirroring the expansion and diversification of Life Science, A-Life Art has to come to encompass work in a wide range of forms, reflecting diverse intonations and perspectives" [8]. From its key process-artificial evolution mostly used at the beginning, in the following decade artists began to generate aesthetic objects drawing on other elements and forms: ecosystem simulations, cellular automata and behavioral robotics. "These techniques are applied across the gamut of 'new media' forms: digital image, animation, interactive installation and CD-ROM, on- and offline virtual environments, and static, robotic and bio-robotic sculpture. Less obvious, though perhaps more important, is a correponding diversity of conceptual approaches. Some artists endorse and play out A-Life's aim for the synthesis of living systems; they reflect some of the progressive, futurist tendencies of A-Life and the cultural discourse it has inspired. Others approach A-Life critically, questioning the assumptions that underpin its techiniques as they turn those techniques to creative ends. Still others draw on the technical resources of A-Life only to alter them, reconfigure and reeingeneer them to serve particular aesthetic and conceptual concerns.

Contemporary new media artists use A-Life in a variety of contexts, to a variety of ends: some works pursue an absolute, self-sufficient autonomy; others use an appearance of autonomy to provoke empathy or raise questions about human agency. Many of the artists using A-Life strive for a supple, engaging form of interactivity and a work that draws the audience into an active relationship; others present artistic artifacts that arise through their own intense engagement with A-Life processes. Some set about creating whole artificial worlds, others seek out a complex, dynamic relationship with the physical 'outside' world." Some artists exploit genetic algorithms to create digital images or music [9], many other draw on arficial intelligence, other work with robotics, avatars or artificial agencies, generative processes or simulated worlds, etc. [10]. Other artists, such as the so called "breeders", operate in a coded computational interior world. Whitelaw [8] notes "In cybernatures, that computational space begins to open outward in both form and content: the outside is drawn in through the user's interactive involvement and mirrored, awkwardly, in these toy worlds". Artists that pull away from the inner window provided by a computer screen and consciously occupy physical space often build a variety of physical systems: interactive robotic creatures, technological and biological composites, installed robotic 'ecosystems' and 'communities'. "In placing their works in the room with us rather than in the 'elsewhere' of a virtual or simulated space, these artists are able to explore an open, transparent form of interactivity that requires no interface, the weight and presence of a 'body' brings with it an immediacy that screen-based works often lack. This sense of 'being with us' is at the core of the concerns articulated by these artists" [8].

4 A-Life Art and Aesthetics

Means talks about some dichotomies between artificial creatures and artificial environments. A creature is bounded, has outlines, and can be perceived as potentially capable of locomotion as separate from its background. Early A-Life experimenters tended to create creatures animated by robotics, an emerging art form. An environment entails the surroundings that may contain creatures, but at least will be inhabited by the perception of the viewer of the art work. An environment can be conceived as boundless, with the edges imposed by the artist thought of as somewhat arbitrary. This brings up another dichotomy in A-Life Art: the schism between works of Art that are static and those that are temporal in nature. Means also observes that discussions of Art created using A-Life approaches have tended over the years to focus on the methods by which the Art was created. The art of the pioneers of A-Life Art (Latham, Ray and Sims) has motivated lengthy expositions on the parallels between the methodologies of these artists and our increasing understanding of evolutionary processes. In their work, the conceptual tends to outweigh the aesthetic, whereas in the work of Rooke and Bachelier, the work stands on its own aesthetic merits, irrespective of the methods of its creation. Only recently it has been fruitful to concentrate on A-Life Art from the standpoint of aesthetics. This contemporary Art rivals the work of human artists in complexity, invention, and especially in abundance, and is possibly the most exciting Art being produced today [11].

But what is the real meaning of aesthetics, and is it true it is so far from (A-) life? According to Welsch, the discipline's traditional name `aesthetics' seems to be wrong, because in fact it doesn't have - as the name suggests - aesthetics as its point of reference, but Art - so the name `artistics' would be much more appropriate. "With respect to the older tradition this clearly is the case. Baumgarten, the founding father of aesthetics, created the expression `aesthetics' precisely with recourse to the Greek words (aisthesis) and (aisthetikos) - that is, to expressions which designate sensation and perception altogether, prior to any artistic meaning. Baumgarten established the new discipline in order to improve our sensuous capacity for cognition. This is why he defined it as the "science of sensuous cognition" and gave it the name `aesthetics'. It was to comprehend all kinds of sensuous cognition [...] The Arts, on the other hand, didn't even get a mention among the scope of aesthetics […] The discipline of `aesthetics', however, has restricted itself for a long time to questions concerning Art - and more on conceptual than sensuous issues of Art". For centuries, since Kant's Critique of Judgment in 1790, aesthetics has been understood exclusively as a philosophy of Art. Today the mainstream of aesthetics still follows this conception. The academic discipline tends to restrict itself to artistics - no matter how uncertain the notion of Art itself may have become in the meantime [12]. Jaschko points out that "focussing on the aesthetics is actually quite unorthodox - both in the Electronic Art scene, which is quite engaged in discussing the functional and political aspects, and in Contemporary Art, which, for the most part, tries to neglect the existence of traditional aesthetic categories in favor of the establishment of a different concept of Art". When Aristotle claimed, in the 4th century before Christ, that Art, and in particular poetics, should imitate nature, he did not mean copying nature in its outward shapes, but was interested in the representation of acting individuals. Already this early aesthetic manifesto defines the process - in this case the process of man changing over time - as an aesthetic quality. As Aristotle has correctly noted, it is this representation and recognition of natural behaviour that arouses the perceivers' interest and supports his/her emotional involvement. In Generative Art and Design, a high level of complexity in the system's behaviour, and of autonomy and unpredictability increases fascination and suspense. Another old-fashioned category, Grace, is an obvious aesthetic quality of Generative Art works being complex but balanced systems, built of autonomous parts and external feeds or interventions [13].

5 Robotic Superstrings

However, Nature is the object of study of Physics. "Art and Physics are a strange coupling. Of the many human disciplines, could there be two that seem more divergent? The artist employs image and metaphor; the physicist uses numbers and equations […] Traditionally, Art has created illusions meant to elicit emotion, Physics has been an exact science that made sense […] Yet despite what appear to be irreconcilable differences, there is one fundamental feature that solidly connects these disciplines. Revolutionary Art and visionary Physics are both investigations into the nature of reality. […] While their methods differ radically, artists and physicists share the desire to investigate the ways the intelocking pieces of reality fit together. This is the common ground upon which they meet" [14].

Following the desire to further mix Life, Art and Science, the evolution of our Superstring Installation was an A-Life installation we will briefly describe. The pictures taken and the videos shooted during the three installations have been collected in a DVD produced only in a limited edition, in few copies. Their surface was perfectly uniform and white, except from only a small black title running around the central hole. The DVDs were then part of a Generative Art process presented during the GA2006 Conference. All the copies were put on a table, forming a regular matrix, within a limited space. All around a small black "arena" delimited the space. A small autonomous robotic artifact, built using Lego robotics kit and RCX computer-brick and code, was programmed so to trace on this space a "visual superstring", which changed depending both on the interaction with the public and the configuration of the DVDs on the floor. i.e., whenever a participant has chosen a direction it left a trace on the DVDs surfaces; in addition it behaved in different ways when the space changed because somebody has moved a copy of the DVD or because the movement of the robot changed the initial configuration. At the end, a unique artwork was generated. The whole surface was covered with black marks creating an intricate pattern, reminding of the physical superstrings installation, but also each DVD was uniquely decorated, since each one differed from another one, with black intricated lines. People interacted with the installation choosing the starting point from which the robot traced the virtual superstrings. However, coming to the walls of the "arena", the robot adapted its route through some sensors, turning back and starting a new string in the space, in an autonomous way. LEGO Mindstorms - Robotics Invention System let people design and program real robots that do what they want them to do. At the heart of the system is the RCX, an autonomous LEGO microcomputer that can be programmed using a pc. The RCX serves as brain of the robot, while a number of sensors can be added to take physical inputs from the environment. After processing the data, the robot behaves in interaction with the environment, fully independent from the computer. LEGO robots are usually used, apart from entertainment, as educational technologies to teach/learn complex theoretical contents [15,16], in an engaging and hands-on way [17]. We decided to build a "roverbot" with four wheels, to give it the stability to perform its task. The robot was then programmed to go from one point to the other in the "arena" interacting with the public (who chose the initial position) and changing its direction when meeting an obstacle. Very simple rules were programmed, as very simple were the rules for the physical superstrings installation: as people were invited to fix the two ends of a thin black elastic rope from one point to the other of a room, the robot was programmed to go from one point to the other of the "arena". Meanwhile, a marker, connected to sensors, would draw a 'virtual string' on the surface of the DVDs. Of course all the process depends on the initial and subsequent positions and behavior, so that each string is unique.

Fig. 4. Milano, Robotics Superstrings Installation, the robot and a DVD.

This is a new, different way of representing the superstring idea and making the invisible visible. In addition, a physicist, one of us (MF) was there also to give all the explanations about the related scientific theories, made easily understandable referring to the installation itself. From a strictly visual point of view, the artistic result is somewhat similar to Jackson Pollock's dripping paintings on a canvas, or, in the digital domain, to Jochem Hendricks' "eye-drawings" [18] and to some experiments in visualizing chaos in Chua's attractors made by our research group [2].

The public, basicly composed of people working in the Generative Art domain – visual artists, programmers, musicians, researchers as well as professors and students from the Politechnical University of Milano, was very involved and the presentation of the installation gave rise to an interesting discussion about Generative Art, randomness, A-Life Art, complexity between Art and Science. Just to borrow a typical metaphor, the process was simple, the outcome was complex. In some cases the attempt to obtain an aesthetically acceptable result leads to a sort of scientific experiment in applying A-Life techniques and changing the variables and testing all the possibile situations to product an artwork [9,19].

Since designers and artists who pursue software abstraction are our contemporaries and since we share the same knowledge and references, it is easy for us to see the strategy of direct borrowing at work. "Indeed, many designers and artists use the actual algorithms from the scientific publications on chaos, artificial life, cellular automata and related subjects. Similarly, the iconography of their works often closely followed the images and animations created by scientists. And some people actually manage to operate simultaneously in the scientific and cultural universes, using same algorithms and same images in their scientific publications and Art exhibitions. What is less obvious is that in addition to the extensive cases of direct borrowing, the aesthetics of complexity is also present in the works that do not use any models from complexity research directly. In short, I argue that just as it was the case with modernist abstraction, the abstraction of the information era is connected to contemporary scientific research both directly and indirectly – both through a direct transfer of ideas and techniques and indirectly as being part of the same historically specific imagination" [20].

Shanken observes that research on artificial life proceeds in a somewhat different fashion. "A-Life begins with pre-existing explanations of life and complex systems from biology and mathematics and seeks to reproduce variations on those themes. While it may claim to seek an expanded understanding of life, it does not perform experiments on living matter, but rather on theories about life. This is a crucial distinction. In this sense, artificial life is a misnomer because the phenomena being reproduced and studied are not life-phenomena, but scientific theories" [21]. He tied these concepts into Jack Burnham's book "Beyond Modern Sculpture," which examines how sculpture uses non-living matter to replicate the human form. He quoted projects by Sims, Ray, Draves, Grey and Brooks, to illustrate ways both artists and scientists have used technology to engage the idea of what life is. He also talked about whether and how Art- and Science-based research differs and how it connects, likening the results to a collage "cobbled together" from disparate pieces to form a cohesive whole [21].

In conclusion, one could argue that from a certain viewpoint making Mathematics, Art, Design or Science is always making a creative work, in a sense that it's a generation of something which belongs to the mind, influenced by the perception we have of the world we live in, that we choose to represent using different languages.

6 Conclusions

In "Principles of research", Einstein stated that the artist and the scientist each substitute a self-created world for the experiential one, with the goal of trascendence" [22]. In the past centuries the relation between Art and Science was very strong, but then the two disciplines were separated, and it is only recently that the two are intermingled in new, different ways [23]. The convergence of Art and Science in some projects may still be the exception rather than the rule, but the growing amount of work in this category indicates that there is a necessity to create a public platform for the discussion of the effects Science and Technology on our culture [24]. Minsky points out the importance of artistic representations to better undestand scientific concepts: "No matter what one's purposes, perhaps the most powerful methods of human thought are those that help us find new kinds of representations. Why is this so important? Because each new representation suggests a new way of understanding-and if you understand something only one way, then you scarcely understand it at all. Perhaps this is the way the Arts so often precede the flowerings of culture" [25]. Moreover, Wilson pushes the concepts even further, demostrating through a rich variety of examples that "The role of the artist is not only to interpret and spread scientific knowledge, but to be an active partner in determining the direction of research" [26]. Computer Science and A-Life are powerful tools in this direction.

References

[1] Art, Complexity and Technology: Their Interaction in Emergence. Workshop, Torino, Italy, 5-6 May (2005), `http://www.isi.it/conference-art.html`
[2] Mathematics, Art and Cultural Industry Conference, Cetraro, Italy, 19-21 May (2005) - CD-rom by M.G. Lorenzi - http://galileo.cincom.unical.it/convegni/cd_ma&ic/start.htm
[3] Francaviglia, M., Lorenzi, M., Petry, M.: The Space Between: Superstring Installation III. In: Soddu, C. (ed.) 8th Generative Art Conference, GA2005, Milano, 15-17 December 2005, pp. 265–276. Alea Design Publisher, Milano (2005)
[4] Il futuro della teoria delle stringhe; una conversazione con Brian Greene. Le Scienze 424, 44–49 (Dec. 2003)
[5] Griffiths, D.J.: Introduction to Elementary Particles. John Wiley & Sons, New York (1987)
[6] de Oliveira, N., Oxley, N., Petry, M.: Installation art. Thames & Hudson, London (1994)
[7] de Oliveira, N., Oxley, N., Petry, M.: Installation Art in the new millenium. The empire of the senses (foreword by J. Crary). Thames & Hudson, London (2003)
[8] Whitelaw, M.: Metacreation. Art and Artificial Life. MIT Press, Cambridge (2004)
[9] Bilotta, E., Pantano, P., Talarico, V.: Synthetic Harmonies: an approach to musical semiosis by means of cellular automata. In: Proc. Artificial Life VII, Portland, MIT Press, Cambridge (2000)
[10] Bilotta, E., Lorenzi, M.G., Pantano, P., Talarico, A.: Art Inspired by cellular automata. In: NKS Conference, Boston, 22-25 April (2004), `http://www.wolframscience.com/conference/2004/`
[11] Means, L.: Artificial Life: Creatures and Environments. YLEM Journal 12(22) (2002)
[12] Welsch, W.: Aesthetics Beyond Aesthetics. Proc. XIIIth International Congress of Aesthetics, Lahti 1995, vol. III., pp. 18-37 (Helsinki, 1997)
[13] Jaschko, S.: Art from code Generator.x (2005), http://www.generatorx.no/20051115/text-process-as-aesthetic-paradigm-a-nonlinear-observation-of-generative-art/
[14] Shlain, L.: Art&Physics. Morrow, New York (1991)

[15] Martin, F.: Building Robots to learn design and engineering. In: Proceedings of the 1992 Fontiers in Education Conference, Nashville, Tennessee, November (1992)
[16] Martin, F.: Cuircuits to control: learning engineering by designing LEGO Robots, PhD dissertation, MIT, Cambridge, MA (1994)
[17] Bilotta, E., Pantano, P., Bertacchini, P.A., Gabriele, L., Servidio, R.: Apprendere con le mani. Franco Angeli, Milano (2006)
[18] Hendricks, J.: Augenzeichnungen. Vexer Verlag, St. Gallen (1993)
[19] Bird, J., Webster, A.: The Blurring of Art and Alife. In: Dorin, A. (ed.) Proc. Second Iteration, CEMA, Melbourne 2001, pp. 38–46 (2002)
[20] Manovich, L.: Abstraction and Complexity. Catalog essay for the exhibition Abstraction Now, Ch. 16 – MediaArtHistories, pp. 339-354 (2007), http://hdl.handle.net/10002/311
[21] Shanken, E.A.: Life as We Know It and/or Life as It Could Be: Epistemology and the Ontology/Ontogeny of Artificial Life1. Leonardo Electronic Almanac 6:2 (February, 1998) <mitpress.mit.edu/e-journals/LEA/> Reprinted in Leonardo 31:5, 383-388 (Oct. 1998)
[22] Einstein, A.: Essays in Science, Philosofical Library, New York (1934)
[23] Francaviglia, M., Lorenzi, M.G., Pantano, P.: Art & Mathematics – A New Pathway. In: Rocha, E., et al. (eds.) Proc. "Communicating Mathematics in the Digital Era" Conference (CMDE2006), Aveiro, 15-18 August 2006, pp. 15–18. A.K. Peters Ltd., Wellsley (to appear, 2007)
[24] Paul, C.: Digital art. Thames and Hudson, London (2003)
[25] Minsky, M.: The future merging of Science, Art and Psychology, Timothy Druckrey with Ars Electronica (edited by), Ars Electronica, Facing the future (Cambridge, MA, 1999)
[26] Wilson, S.: Information arts. Intersections of Art, Science and Technology. MIT Press, Cambridge (2002)

A Distributed Formation Algorithm to Organize Agents with No Coordinate Agreement

Gregory Studer[1] and Inman Harvey[1]

Centre for Computational Neuroscience and Robotics, University of Sussex

Abstract. In this paper we present an extension of the ShapeBugs distributed formation algorithm which enables 2D mobile agents to agree on a consensus coordinate system starting from no coordinate agreement. The participating agents require only local communication and local distance and motion information. Because this distance and motion information can in many cases be obtained implicitly through software motor approximation and attenuation or time-of-flight in communication, this approach has the potential to globally coordinate general mobile communicating agents without additional sensor requirements. The extended algorithm also remains robust against agent incapacitation and disorientation.

1 Introduction

In the areas of robotics and distributed systems, from many directions, a strong research effort has developed towards controlling autonomous agents with low power requirements and simple sensor capabilities [1][2]. This research has been partially motivated by continuing advances in hardware: microcontrollers and digital communication technology have become cheap, widely available, and in many cases very efficient. The technology enables new applications, and large numbers of simple, self-directed robots show high potential for use in sensor grids [3], resource harvesting [4], and group transport [5][6], amongst many other proposals. As we begin to design robots of smaller size or larger number, however, there are strong technical challenges to our ability to individually control, coordinate, or adequately power them. Centralized supervision of complex agents with virtually unlimited communication range, while successful in many areas, often scales poorly when there are hundreds or thousands of agents to control [7]. Ideally we would like to program global behavior in a distributed way, using assumptions of limited communication and agents acting autonomously. Overcoming the limitations of local communication and independent agent action to achieve global coordination, using minimum hardware, is the defining challenge in making simple agents practical.

The well-studied 2D formation task provides a particularly good starting point for this research, as it is simple to describe with applications in theory and practice. Given a set of points and a set of mobile agents, the formation task is simply to arrange the agents on the points (or perhaps an isometric set of

F. Almeida e Costa et al. (Eds.): ECAL 2007, LNAI 4648, pp. 515–524, 2007.

points)[2]. Real-world tasks for multiple agents often involve forming particular shapes, as in the previous examples of sensor grids and group transport. On the theoretical side, agents which are able to arrange into arbitrary, large shapes have established global information from local interactions [6].

A particularly complete heuristic approach to the formation problem called the "ShapeBugs" algorithm is described by Cheng et al [3], which synchronizes local agent coordinate systems through repeat trilateration. Given a small seed group of initially oriented agents, an arbitrary (though connected) global shape much larger than an individual agent's communication radius can be formed. The algorithm is robust to sensor and movement error and large numbers of agents. It only requires agents to have local communication, an approximate measure of relative distance, an approximate measure of relative motion, and a global compass. Though technically the goal of the work was to distribute agents through a shape, not place them in particular points on the plane, the algorithm proceeds by synchronizing every agent's coordinate system, and so implicitly enables agents to form many arbitrary connected formations of any size.

Our work modifies and extends the ShapeBugs algorithm, reducing the agent abilities required to form arbitrary global formations by eliminating the need for a global compass. By continously calculating the error in local estimation of motion, orientation can be derived from sequential distance measurements. We present here:

- A distributed method capable of organizing many mobile communicating agents into many classes of arbitrary shapes without agent position or orientation agreement.
- An analysis of formation stability using the method with and without orientation agreement.

Without the need for agents to access some shared coordinate system or orientation (except a small number of initially coordinated seed agents, which are not strictly necessary though helpful for control), our algorithm requires simpler agent hardware at the cost of more complex software and higher data broadcast rates. Maintaining agent density to allow effective orientation measurements requires a new distributed growth process (but no new sensors). Because communication often has measurable attenuation or time of flight [8], giving an approximate measure of distance (though inaccuracy can make this unusable), and motor output can be modeled in software, giving an approximate measure of relative motion, this algorithm potentially enables any mobile communicating agent to globally coordinate using only those two defining abilities.

In addition, calculation errors [9] affected the previous data presented in [3], so new ShapeBugs results are presented for comparison with the modified "ShapeBots" version.

1.1 Previous Work

The difficulty of solving the formation task is related to agent abilities. Agents generally have arbitrary formation ability when they know their positions in a unified, global coordinate system (each agent can simply move toward the nearest unfilled formation point, for example). These types of agents will not be discussed in this paper. Formations are not as straightforward if agents use distributed logic and local communication, and the literature contains many proposed solutions [4][7][10][11]. Fujibayashi et al [12] describe a method able to create locally regular formations, though without arbitrary controllability, much like simulated crystals. Another approach by Yamaguchi et al [13] allows a variety of formations from mobile agents in a line using similar virtual constraints. Suzuki and Yamashita [2] mathematically describe a process to create regular polygonal formations, while Ikemoto et al [14] extend this ability using Turing waves to polygonal shapes with an axis of symmetry. Hybrid approaches [8][15] are also possible and widespread in WiFi localization literature which assume only a few agents have extra reference capabilities for global positioning. In general, however, no fully distributed arbitrary formation algorithms for agents with no coordinate agreement (i.e. no global position or orientation information) have been demonstrated thus far for communicating mobile agents, though such algorithms have been proven impossible in non-communicating cases [2][1]. This is an important version of the formation problem, because simple, low power agents may not generally have global positioning or orientation information available, such as in energetic [16], heterogeneous [8], or non-geographic environments. Also, any agents which have some form of coordinate agreement would be able to rely on formation algorithms not requiring agreement as a backup in case of failure or as a complement to their existing methods. Our modified algorithm attempts to achieve connected arbitrary formations in a heuristic manner, by synchronizing agents' local coordinate systems using relative distance information.

2 The "ShapeBots" Algorithm

We simulate simple mobile agents in a 2D, $N \times N$ continuous periodic world, chosen to simplify the simulation by ignoring distant agent aggregation. There are many distributed aggregation techniques described for disconnected agents, [17][18] as examples, but we do not explore those here. Within a range $R_B = 6$ units, agents have the ability to broadcast and recieve information, sense neighbor distance, and sense their own motion. Agents also have a simulated repulsion range $R_{Rep} = 4$ and physical collision radius $R_c = 1$, though collision handling becomes unimportant if $R_{rep} \gg R_c$. Agents have a reference direction (a "nose") which allows them to be oriented in the simulation, and while they have the ability to rotate in place and move in any direction no matter which way they are pointed, agent motion vectors are interpreted as relative to this "forward" direction. After every motion, the reference direction is pointed in the direction of motion.

2.1 Agent Lifecycle and Scheduling

Our algorithm simulates agents as semi-asynchronous, with finite time required for both calculation and movement. Semi-asynchronous is defined here as a variation of asynchronicity as defined by [1], but with a finite upper bound instead

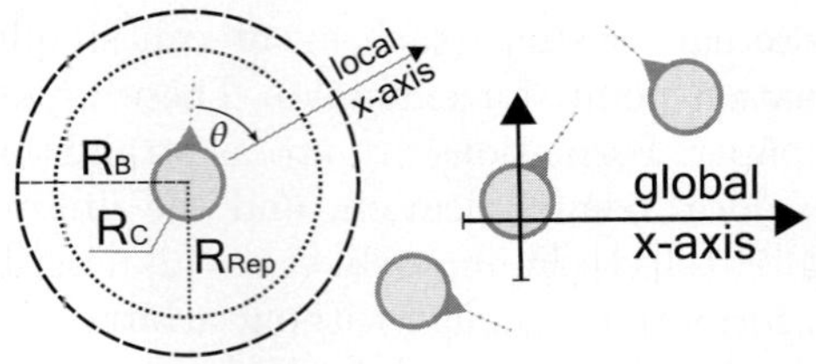

Fig. 1. Sample ShapeBot agent in the 2D plane. The agent's estimated x-axis at θ from reference direction generally does not correspond to the global observer x-axis.

of a lower bound on how long an agent may stay in each state of its activation cycle, which we believe more closely models robot hardware. Agents follow a lifecyle of:

$$Wait \rightarrow Sense_1 \rightarrow Compute_1 \rightarrow Move \rightarrow Sense_2 \rightarrow Compute_2 \rightarrow Wait \quad (1)$$

The final $Sense_2$ and $Compute_2$ steps are necessary because agents compare information before and after *Mov*ing to determine orientation from positioning error. The *Wait*, $Compute_{1,2}$, and *Move* steps each take time bounded by the finite values W_t, C_t, M_t respectively, while the *Sense* time is amortized into the total *Compute* time. After each agent transition, the time until the next transition is drawn uniformly from $[0, T_{max})$, where T_{max} is appropriately W_t, C_t, or M_t. Agents may move $d_m = 1$ or 0 units during each $Move \rightarrow Sense_2$ transition (if they move 0 units, the *Move* step takes no time). In a more realistic system agents could move varying distances over time, but this was not modeled exactly in our simulation. The positioning and orientation algorithms depend largely on the ability of agents to estimate the magnitude of their motions, while the magnitude itself could vary. One agent iteration is defined complete when an agent transitions through all six steps. In our results, $M_t = 500$, $W_t = M_t$, and $C_t = 0.01M_t$.

The simulation begins with each agent in the *Wait* state, with each agent's time to *Wait* randomly drawn from $[0, W_t)$. The agents are placed in a priority queue, ordered by soonest update time. The simulation proceeds by removing the agent with the soonest transition from the queue, performing the transition, updating the agent time for the next transition, and finally adding it back into the queue. One simulation iteration is defined complete when a simulation of N agents makes N agent transitions back to the *Wait* state (not all agents may be iterated fully, some may be iterated multiple times). Every agent continuously broadcasts the following information (for both their previous and current iteration):

- estimated (x, y) position and local orientation θ
- computational state (described below)
- local neighbor density ρ
- count of position and orientation updates since oriented

This information is therefore always available for other local agents to *Sense*. While moving, the position is extrapolated from the original position and time, but new estimates are only generated in *Compute* steps. Broadcasted information also contains the relative distance between the agents at time of broadcast.

In addition to the readiness states described above, agents have three computational states, as described in [3], of *Lost*, *OutOfShape*, and *InShape*. An agent in the *Lost* state has no coordinate system, while the *OutOfShape* and *InShape* agents have local coordinate systems (an (x, y) coordinate and an orientation estimate θ). An agent in the *Lost* state will wander randomly through the world until it senses three neighbors with coordinates, which then allows the *Lost* agent to trilaterate a position guess using the neighbor distance measurements and (x, y) positions. Importantly, while trilateration can approximate a consistent coordinate system between neighboring agents, the relative orientation of the agents' reference direction with respect to this coordinate system θ must be calculated using distance measurements after neighbor motions. As an initial orientation guess, a newly un-*Lost* agent's orientation is reset to correspond to a motion directly toward the centroid of the orienting points, though the actual motion may have been less direct.

A newly oriented agent will become *Lost* again immediately if it cannot remain in contact with other oriented agents long enough for several agent orientations to occur, defined in our simulation as a 10 iteration update window for orientation and trilateration. Otherwise poorly oriented agents tend to immediately escape the main formation and form nearby competing formations of opposite orientation. As in [3], once agents have acquired a concensus coordinate system they attempt to fill a formation shape by each agent simulating a virtual gas particle with a repulsion range R_{rep}. Neighboring agents inside the shape with distance $< R_{rep}$ will repel one another, leading to an average distribution of agents throughout the shape. These virtual gas mechanics allow the shapes to be robust against agent addition and death while spreading agents evenly throughout the formation.

2.2 Orienting Using Agent Motion

To adjust an agent's percieved reference direction with respect to the trilaterated consensus coordinate system θ, an agent simply observes the error in its coordinate system after motion. Because agents' reference direction is pointed in the last direction of motion, the corrected direction of motion with respect to the trilaterated coordinate system is an estimation of the agent's reference direction in that consensus coordinate system. Along with the simulated error in movement and distance measurement, asynchronous motion limits the agent speed and density required for the agent coordinate systems to converge. Agents may move simultaneously, adding inaccuracy to their trilaterated coordinate systems due to estimated orientation errors, and the less accurate coordinates can then result in less accurate orientation. Like position updates, the noisy orientation changes calculated each agent motion are averaged over a window of 10 agent iterations to smooth errors.

2.3 Formation Scaling

A major consequence of the co-dependency of orientation and position is that agent coordinate systems will diverge very quickly from one another if agents become separated from a dense group with coordinate agreement. For this reason, agents with coordinate systems try to maintain a target agent density of $\rho = 18$ neighbors by scaling their local copy of the target formation, much like a balloon expands as more gas is added. If many agents are added to the shape, the shape will expand proportionally to maintain ρ, while if many agents are removed or become inoperative, the target shape will shrink to ensure agents stay grouped. If a particular size formation is required, the shape scale could easily be capped at the target size (this is not done in our tests), but if there are too few agents to maintain the target density the agents will be unable to maintain the formation size.

The distributed scaling is achieved by every agent averaging neighbor values of the desired shape scale. An agent broadcasts its desired shape scale (calculated assuming local neighbor density applied throughout the shape) while using itself the average of neighbor scales. There can be a large amount of variability in the number of neighbors each iteration, so the broadcasted shape scale is smoothed using density information from the past 10 agent iterations.

3 Formation Control Results

To measure the ability of agents to achieve coordinate agreement under varying amounts of distance and motion error, the coordinate variance of 200 agents was calculated (Figure 4) while forming a simple square formation in an 80×80 world. Coordinate variance is defined as the average variance in distance between the non-*Lost* agent origins, using the average orientation of all non-*Lost* agents. Higher variance corresponds to weaker formation control. 12 seed agents were initially placed in a 25×25 square in the center of the world and given consistent

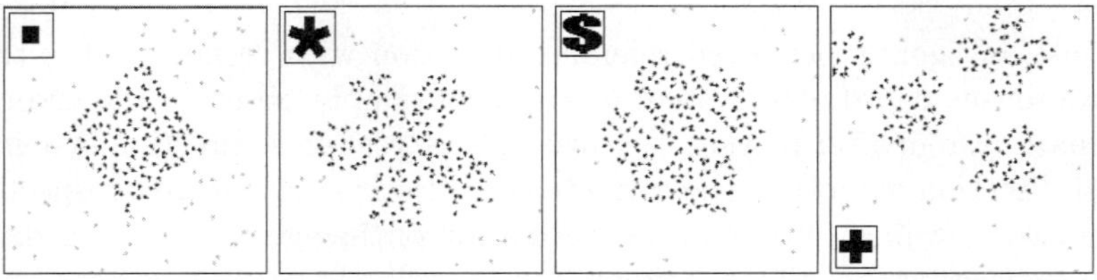

Fig. 2. Sample formations. Dark agents are non-*Lost*, and the formation shapes are specified by the bitmaps in the image corner. Only connected formations are possible using our algorithm, and shapes containing thin portions tend to be harder (or sometimes impossible) to form well. As seen in the ✚ formations, if *Lost* agents are allowed to trilaterate with other *Lost* neighbors as well as non-*Lost*, they will spontaneously create many stable formations without seed agents.

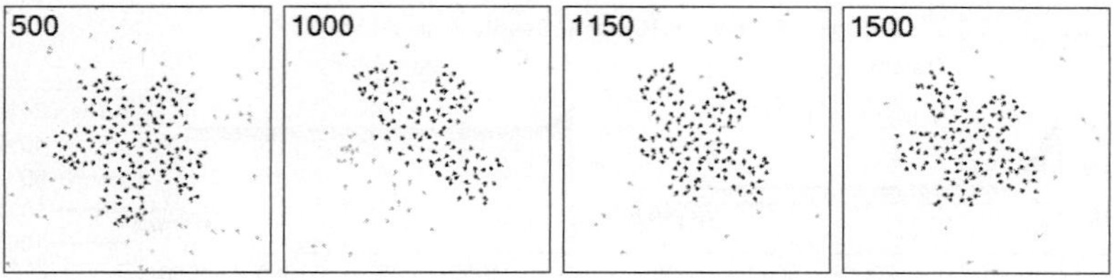

Fig. 3. Even without compasses, formations remain resistant against agent death. At iteration 1000, agents in the lower half of the asterix formation die. At iteration 1150, new agents have moved into the destroyed area, and by iteration 1500 the formation is rebuilt (at smaller size to maintain density).

coordinates and orientation to start the formation. The distance error e is applied to measurements d by adding a uniformly chosen value from the range $(-de, de)$. Motion error is applied in the same way to each component of a motion vector (m_x, m_y).

4 Discussion

What a global compass seems to "buy" you, in the context of distributed, arbitrary shape formation algorithms, is the ability for agents to tolerate greater distance and motion sensor error. As seen in Figure 4, 60% distance error with our compass-free agents results in the same coordinate variance as 100% distance error using agents with compasses and scaling. Though less converged, our agent formations without orientation still remain stable until about 80% distance error (when the converged formations may only be temporarily stable). Our modified algorithm seems less susceptible to motion error than the original ShapeBugs work, even in the no-compass case, however this is probably due to a reduced agent speed of 1 unit/iteration instead of the original 2 units/iteration and higher density. Formation instability in our modified algorithm tends to increase slowly until a threshold error level is passed, then quickly jumps, as can be seen in Figure 4. The much higher instability then makes it impossible to maintain a consensus coordinate system, in the case of no compasses.

Our corrected ShapeBugs convergence results match well with the originally reported values, however for low distance error the coordinate variance is reduced. As the problem related to inaccurate initial trilateration calculation, the effect becomes minimal where measurement errors are larger .

4.1 Conclusion

The local communication, asynchronous update, and short burst motion schedule make the ShapeBots algorithm amenable to many different types of agents, though higher density and lower speed (equivalent to more frequent updates) to mitigate larger measurement errors are required to reach comparable coordinate

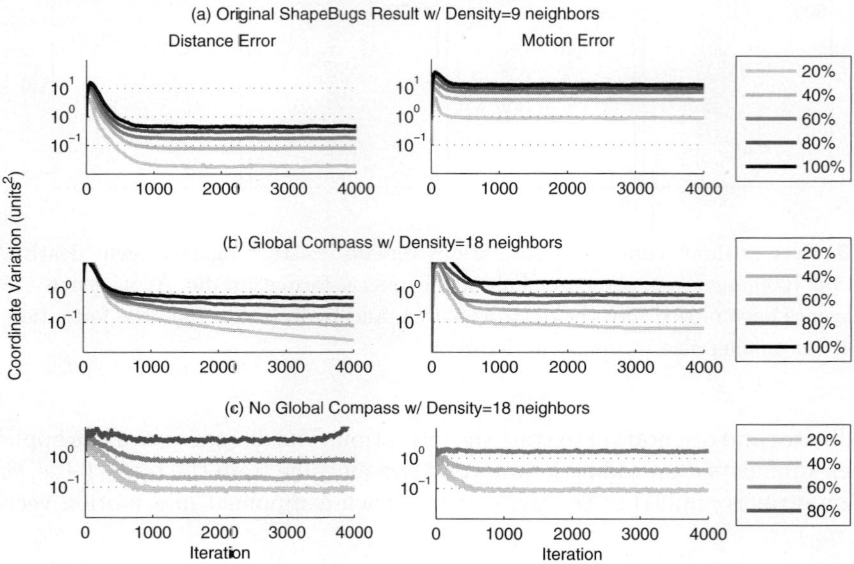

Fig. 4. Average coordinate variance under distance and motion error (averaged over 90 simulations). The variance using the original ShapeBugs algorithm from [3] is shown in (a). Though variance is low, as distance error increases, often seeded simulations will fail to converge on a target shape, this happens 6%-80% of the time for 20%-100% distance error and 11%-45% of the time for 20%-100% motion error. In (b) our modified algorithm results using formation scaling are presented. At maximum, the simulations fail to converge only 13% of the time, this is largely due to formations growing too quickly (but is unchanged to better compare with (c)). Simulation data using formation scaling without compasses is presented as (c), all simulations achieve convergence except at distance error 80% (formations may not be stable long-term). At higher error, no simulations converge.

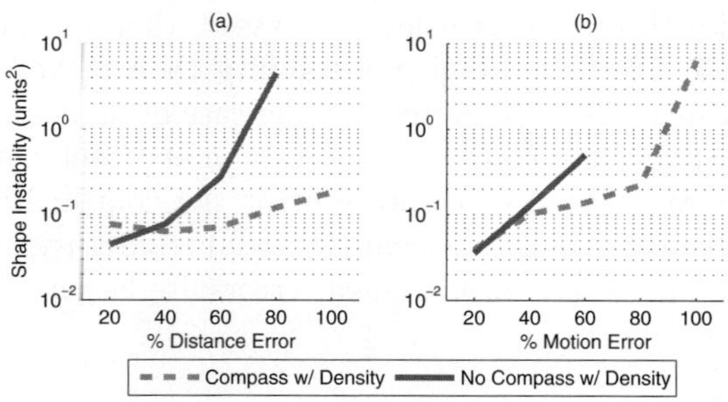

Fig. 5. Formation instability, measured as the average standard deviation over time of the stabilized coordinate variance from $t = [2000, 4000)$, is lowest when distance and motion error is low.

convergence. For limited-power devices, this may mean the speed of formation is lowered as compared to the ShapeBugs algorithm. Alternately, devices could use higher transmission power to increase the agent range, and so increase the formation density. Stable formations tend to converge more consistently using gradual scaling, this may be desirable despite the other limitations.

In a situation where agents can extract reasonably accurate distance and motion data from communication and motion input and can tolerate small variance in the global formation, large numbers of mobile agents containing only communication hardware can organize in arbitrary patterns using the ShapeBots algorithm. The new algorithm shares much of the resiliance of the original ShapeBugs approach, as the repeated coordinate averaging and virtual gas motion was retained from the original algorithm. The presented method of synchronizing scale across the formation to maintain this density works even with large numbers of agents, but provides an additional source of formation shape error (though *not* coordinate error). If the target shapes have very thin sections, where agents are prone to escape and density measurements become especially variable (because there cannot be many surrounding positions), even rescaling may fail and the formation can be distorted or leak agents.

In future work we hope to address these issues and extend the algorithm to approximate more intricate shapes. A major limiting factor is the rapid divergence of coordinates and orientation once agents become disconnected. Scaling agent motion proportional to their confidence in position in the formation may be one useful approach, as slower motions could allow converged agents to stabilize the coordinate system more exactly. Over time the agent formation would also tend to slow down and stop, ideal behavior for energy-limited agents. With larger numbers of agents, though, intricate shapes become less of a problem because the shape scale grows and narrow portions become wider. Another approach would simply be to growing the shape a filled square, then transitioning to the actual shape when agents agree the scale is large enough.

The ShapeBots simulation code, implemented in the MASON multi-agent simulation framework [19], is available online at:
`http://www.informatics.sussex.ac.uk/users/gms21/shapebot`

Acknowledgments. We thank Simon McGregor for his comments on the work-in-progress.

References

1. Prencipe, G., Santoro, N.: Distributed algorithms for autonomous mobile robots. In: Proc. of 5th IFIP Intl. Conf. on Theoretical Computer Science (2006)
2. Suzuki, I., Yamashita, M.: Distributed anonymous mobile robots: Formation of geometric patterns. SIAM Journal on Computing 28(4), 1347–1363 (1999), `citeseer.ist.psu.edu/suzuki99distributed.html`
3. Cheng, J., Cheng, W., Nagpal, R.: Robust and self-repairing formation control for swarms of mobile agents. In: Proc. of the 20th Natl. Conf. on Artificial Intelligence, pp. 59–64. AAAI Press, Menlo Park (2005), `http://jmvidal.cse.sc.edu/library/cheng05a.pdf`

4. Dudek, G., Jenkin, M., Milios, E., Wilkes, D.: A taxonomy for multi-agent robotics. Autonomous Robots 3, 375–397 (1996), citeseer.ist.psu.edu/dudek96taxonomy.html
5. Şahin, E., Labella, T.H., Trianni, V., Deneubourg, J.-L., Rasse, P., Floreano, D., Gambardella, L., Mondada, F., Nolfi, S., Dorigo, M.: SWARM-BOTS: Pattern formation in a swarm of self-assembling mobile robots. In: Proc. of the IEEE Intl. Conf. on Systems, Man and Cybernetics, Hammamet, Tunisia, IEEE Computer Society Press, Los Alamitos (2002)
6. Cao, Y.U., Fukunaga, A.S., Kahng, A.B.: Cooperative mobile robotics: Antecedents and directions. Autonomous Robots 4(1), 7–23 (1997), citeseer.ist.psu.edu/cao97cooperative.html
7. Bahceci, E., Soysal, O., Sahin, E.: A review: Pattern formation and adaptation in multi-robot systems. Technical Report CMU-RI-TR-03-43, Robotics Institute, Carnegie Mellon University, Pittsburgh, PA, USA (2003)
8. Niculescu, D., Nath, B.: Position and orientation in ad hoc networks. Ad Hoc Networks 2, 133–151 (2004), citeseer.ist.psu.edu/niculescu03position.html
9. Nagpal, R,: Personal communication (2007)
10. Defago, X., Konagaya, A.: Circle formation for oblivious anonymous mobile robots with no common sense of orientation. In: Proc. of the 2nd ACM Intl. Workshop on Principles of Mobile Computing, Toulouse, France, pp. 97–104. ACM Press, New York (2002), doi:10.1145/584490.584509
11. Howard, A., Mataric, M., Sukhatme, G.: Putting the 'i' in team: an ego-centric approach to cooperative localization. In: Proc. of the IEEE Intl. Conf. of Robotics and Automation, Taipei, Taiwan, May 2003, IEEE Computer Society Press, Los Alamitos (2003), citeseer.ist.psu.edu/howard03putting.html
12. Fujibayashi, K., Murata, S., Sugawara, K., Yamamura, M.: Self-organizing formation algorithm for active elements. In: Proc. of the 21st IEEE Symposium on Reliable Distributed Systems, Washington, DC, USA, p. 416. IEEE Computer Society Press, Los Alamitos (2002)
13. Yamaguchi, H., Arai, T., Beni, G.: A distributed control scheme for multiple robotic vehicles to make group formations. Robotics and Autonomous Systems 36, 125–147 (2001)
14. Ikemoto, Y., Hasegawa, Y., Fukuda, T., Matsuda, K.: Gradual spatial pattern formation of homogeneous robot group. Information Sciences - Informatics and Computer Science 171, 431–445 (2005)
15. Mihaylova, L., Angelova, D., Canagarajah, C.N., Bull, D.R.: Algorithms for mobile nodes self-localisation in wireless ad hoc networks. In: 9th Intl. Conf. on Information Fusion, Florence, Italy (2006), citeseer.ist.psu.edu/mihaylova06algorithms.html
16. Souissi, S., Defago, X., Yamashita, M.: Using eventually consistent compasses to gather oblivious mobile robots with limited visibility. In: Datta, A.K., Gradinariu, M. (eds.) SSS 2006. LNCS, vol. 4280, p. 471. Springer, Heidelberg (2006)
17. Trianni, V., Gross, R., Labella, T., Sahin, E., Rasse, P., Deneubourg, J., Dorigo, M.: Evolving aggregation behaviors in a swarm of robots. Technical Report TR/IRIDIA/2003-07, IRIDIA, Universite Libre de Bruxelles, Bruxelles, Belgium (2003)
18. Flocchini, P., Prencipe, G., Santoro, N., Widmayer, P.: Gathering of asynchronous robots with limited visibility. Theoretical Computer Science 337(1-3), 147–168 (2005), doi:10.1016/j.tcs.2005.01.001
19. Luke, S., Cioffi-Revilla, C., Panait, L., Sullivan, K.: Mason: A new multi-agent simulation toolkit. In: Proc. of the 2004 SwarmFest Workshop (2004), http://cs.gmu.edu/~eclab/projects/mason

A Multi-level Selection Model for the Emergence of Social Norms

Francisco C. Santos[1], Fabio A.C.C. Chalub[2], and Jorge M. Pacheco[3]

[1] IRIDIA, CoDE, Université Libre de Bruxelles, Brussels, Belgium*
[2] Departamento de Matemática da Universidade Nova de Lisboa and Centro de Matemática e Aplicações, Caparica, Portugal
[3] Centro de Física Teórica e Computacional and Departamento de Física da Faculdade de Ciências da Universidade de Lisboa, Lisbon, Portugal

Abstract. We develop a multi-level selection model in the framework of indirect reciprocity. Using two levels of selection, one at the individual level and another at the group level, we propose a competitive scenario among social norms, in which all individuals in each group undergo pairwise interactions, whereas all groups also engage in pairwise conflicts, modeled by different games. Norms evolve as a result of groups' conflicts whereas evolution inside each group promotes the selection of best strategies for each ruling social norm. Different types of inter-group conflict and intensities of selection are considered. The proposed evolutionary model leads to the emergence of one of the recently obtained *leading-eight* social norms, irrespective of the type of conflict between groups employed. We also compared the individual performance of the norm obtained in the evolutionary process with several other popular norms, showing that it performs better than any the other norms. This reputation assignment rule gives rise to a stern and unambiguous response to each individual behavior, where prompt forgiving coexists with implacable punishment.

1 Introduction

Many biological systems employ cooperative interactions in their organization [1]. Humans, unlike other animal species, form large social groups in which cooperation among non-kin is widespread. This contrasts with the general assumption that the strong and selfish individuals are the ones who benefit most from natural selection. This being the case, how is it possible that unselfish behaviour has survived evolution? Adopting the terminology resulting from the seminal work of Hamilton, Trivers, and Wilson [2,3,4], an act is altruistic if it confers a benefit b to another individual in spite of accruing a cost c to the altruist (where it is assumed, as usual, that $b > c$). In this context, several mechanisms have been invoked to explain the evolution of altruism, but only recently an evolutionary model of indirect reciprocity (using the terminology introduced in [5]) has been developed [6] addressing *unique aspects of human sociality, such as*

* For further information contact fsantos@ulb.ac.be

F. Almeida e Costa et al. (Eds.): ECAL 2007, LNAI 4648, pp. 525–534, 2007.

trust, gossip and reputation [7]. As a means of community enforcement, indirect reciprocity had been investigated earlier in the context of economics, notably by Sugden [8] and Kandori [9] (see below). More recently, many studies [10,11,12,13,14,15,16,8,7,17] have been devoted to investigate how altruism can evolve under indirect reciprocity.

In the indirect reciprocity game, any two players are supposed to interact at most once with each other, one in the role of a potential donor, while the other as a potential receiver of help. Each player can experience many rounds, but never with the same partner twice, being direct retaliation unfeasible. By helping another individual, a given player may increase (or not) its reputation, which may change the pre-disposition of others to help her in future interactions. However, its new reputation depends on the social norm used by her peers to assess her action as a donor. Previous studies of reputation-based models of cooperation, reviewed recently [10] indicate that cooperation outweighs defection whenever, among other factors, assessment of actions is based on norms which require considerable cognitive capacities [10,12,13]. Such high cognitive capacity contrasts with technology-based interactions, such as e-trade, which also rely on reputation-based mechanisms of cooperation [18,19,20]. Despite the success and high levels of cooperation observed in e-trade, it has been found [18] that publicizing a detailed account of the seller's feedback history does not improve cooperation, as compared to publicizing only the seller's most recent rating. In other words, practice shows that simple reputation-based mechanisms are capable of promoting high levels of cooperation. In view of the previous discussion, it is hard to explain the success of e-trade on the basis of the results obtained so-far for reputation-based cooperation in the context of indirect reciprocity.

2 Evolving Social Norms

Let us consider a world in black and white consisting of a set of tribes, such that each tribe lives under the influence of a single norm, common to all individuals. Each individual engages once in the indirect reciprocity game (cf. section 7) with all other tribe inhabitants. Its action as a donor will depend on its individual strategy, which dictates whether it will provide help or refuse to do it. Reputations are public: this means that the result of every interaction is made available to every one through the *indirect observation model* introduced in [13] (see also [15]). This requires a way to spread the information (even with errors) to the entire population (communication/language). Consistently, language seems to be an important cooperation promoter [21] although recent mechanisms of reputation spreading rely on electronic databases (e.g., in e-trade, where reputation of sellers is centralized). If you consider that each individual action is determined by both donor and receptor's reputations, and since reputations are either *GOOD* or *BAD*, there are $2^4 = 16$ possible strategies. On the other hand, the number of possible norms used to assess each individual's action depends on their associated order. The simplest are the so-called first order norms, in which all that matters is the action taken by the donor. In second order norms the reputation of one of the players (donor or recipient) also contributes to decide the

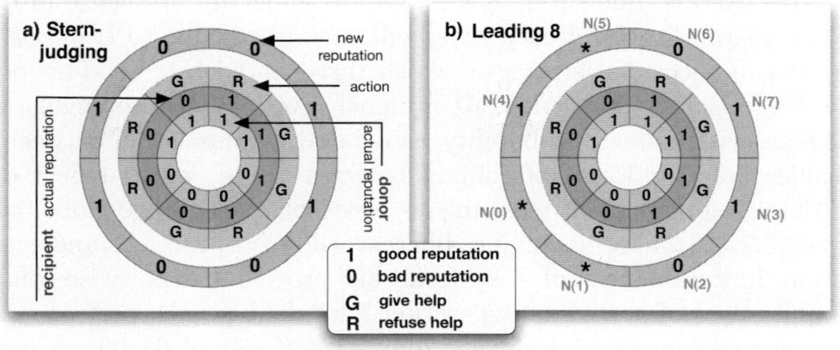

Fig. 1. The higher the order (and complexity) of a norm, the more inner layers it acquires. The outer layer stipulates the donor's new reputation based on the 3 different reputation/action combinations aligned radially layer by layer: Inwards, the first layer identifies the action of the donor. The second identifies the reputation of the recipient; the third the reputation of the donor. Out of the 2^8 possible norms, the forth represents one social norm. In a) we show the most successful norm, *stern-judging*. Stern-judging renders the inner layer (donor reputation) irrelevant in determining the new reputation of donor. This can be trivially confirmed by the symmetry of the figure with respect to the equatorial plane (not taking the inner layer into account, of course). All norms of second order will exhibit this symmetry, although the combinations of 1 and 0 bits will be, in general different. b) The *Leading 8* Norms of Ohtsuki and Iwasa, identifying with $*$ those *slices* in the final norm which can be associated with either *GOOD* (1) or *BAD* (0) reputations.

new reputation of the donor. And so on, in increasing layers of complexity and associated requirements of cognitive capacities from individuals. Any individual in the tribe shares the same norm, which in turn raises the question of how each inhabitant acquired it. We do not address this issue here. However, inasmuch as indirect reciprocity is associated with *community enforcement* [6,9] one may assume, for simplicity, that norms are acquired through an educational process. Moreover, it is likely that a common norm contributes to the overall cohesiveness and identity of a tribe. It is noteworthy, however, that if norms where different for different individuals, the *indirect observation model* would not be valid, as it requires trust in judgments made by co-inhabitants. For a norm of order n there are 2^{2^n} possible norms, each associated with a binary string of length 2^n . Here, we consider third order norms (8 bits). In assessing a donor's new reputation, the observer has to make a contextual judgment involving the donor's action and the reputation of the donor and the receiver (see Figure 1).

We introduce the following evolutionary dynamics inside each tribe: During one generation all individuals interact once with each other via the indirect reciprocity game. When individuals reproduce they replace their strategy by that of another individual from the same tribe, chosen proportional to her accumulated payoff [12]. The most successful individuals in each tribe have a higher reproductive success. Since different tribes are *under the influence* of different norms, the overall fitness of each tribe will vary from tribe to tribe, as well as the plethora of successful strategies which thrive in each tribe. This describes individual selection in each tribe. At a higher level, tribes also engage in pairwise conflicts with a small probability, associated with selection between tribes. We consider different forms of conflict between tribes, which reflect different types of inter-tribe selection mechanisms based on the average global payoff of each tribe [5,22,23,24,25], involving different selection processes and intensities of selection: imitation dynamics, a Moran-like process, the pairwise comparison process and a war of attrition (see section 7 for details). We perform extensive computer simulations of evolutionary dynamics of sets of 64 tribes, each with 64 inhabitants. Once a stationary regime is reached, we collect information for subsequent statistical analysis. We compute the frequency of occurrence of bits 1 and 0 in each of the 8 bit locations. A bit is said to fixate if its frequency of occurrence exceeds or equals 98%. Otherwise, no fixation occurs, which we denote by X, instead of 1 or 0. We analyze 500 simulations for the same value of b, subsequently computing the frequency of occurrence ϕ_1, ϕ_0 and ϕ_X of the bits 1, 0 and X, respectively. If $\phi_1 > \phi_0 + \phi_X$ the final bit is 1; if $\phi_0 > \phi_1 + \phi_X$ the final bit is 0; otherwise we assume it is indeterminate, and denote it by $*$. It is noteworthy that our bit-by-bit selection/transmission procedure, though artificial, provides a simple means of mimicking biological evolution, where genes are interconnected by complex networks and yet evolve independently. Certainly, a co-evolutionary process would be more appropriate (and more complex), and this will be explored in future work.

3 Emergence of an Unique Social Norm

The results, for different values of b are given in Table 1, showing that a unique, ubiquitous social norm emerges from these extensive numerical simulations. This norm is of second-order, which means that all that matters is the action of the donor and the reputation of the receiver. In other words, even when individuals are equipped with higher cognitive capacities, they rely on a simple norm as a key for evolutionary success. In a nutshell, helping a good individual or refusing help to a bad individual leads to a good reputation, whereas refusing help to a good individual or helping a bad one leads to a bad reputation. Moreover, we find that the final norm is independent of the specifics of the second level selection mechanism, i.e., different second level selection mechanisms will alter the rate of convergence, but not the equilibrium state. In this sense, we conjecture that more realistic procedures will lead to the same dominant norm.

The success and simplicity of this norm relies on never being morally dubious: To each type of encounter, there is one *GOOD* move and a *BAD* one. Moreover, it is always possible for anyone to be promoted to the best standard possible in a single move. Conversely, one bad move will be readily punished [26,27] with the reduction of the player's score. This prompt forgiving and implacable punishment leads us to call this norm *stern-judging*. Long before the seminal work of Nowak and Sigmund [6] several social norms have been proposed as a means

Table 1. For each value of the benefit b (c=1), each column displays the eight-bit norm emerging from the analysis of 500 simulations employing the selection method between tribes indicated as column headers. Irrespective of the type of selection, the resulting norm which emerges is always compatible with *stern-judging*. For the pairwise comparison rule, the inverse temperature used was $\beta = 10^5$ (strong selection, see Sections 4 and 7).

b	Imitation dynamics	Moran	Pairwise Comparison	War of Attrition
2	10011001	1*011001	10011001	********
≥ 3	10011001	10011001	10011001	10011001

Table 2. Emergence of *stern-judging* for different intensities of selection. We carried out the bit-fixation analysis described in main text for the evolution of social norms under the pairwise comparison rule (see section 4), for different values of the intensity of selection β. Intensity of selection decreases from left to right. Whereas for strong selection all norm bits fixate for $b \geq 2$, fixation becomes more difficult for b=2 as β decreases. Yet, in no case did we obtain fixation of a digit incompatible with *stern-judging*.

b	$\beta = 10^5$	$\beta = 10^4$	$\beta = 10^3$	$\beta = 10^2$	$\beta = 10^1$	$\beta = 10^0$
2	10011001	1*011001	1*01100*	1*01100*	1*01100*	********
≥ 3	10011001	10011001	10011001	10011001	10011001	10011001

to promote (economic) cooperation. Notable examples are the standing norm, proposed by Sugden [8] and the norm proposed by Kandori [9] as a means to allow community enforcement of cooperation. When translated into the present formulation, standing constitutes a third-order norm, whereas a fixed-order reduction of the social norm proposed by Kandori (of variable order, dependent on the benefit to cost ratio of cooperation) would correspond to *stern-judging.* Indeed, in the context of community enforcement, one can restate *stern-judging* as : *Help good people and refuse help otherwise, and we shall be nice to you; otherwise, you will be punished.* It is therefore, most interesting that the exhaustive search carried out by Ohtsuki and Iwasa [13,15] in the space of up to third-order norms found that these two previously proposed norms were part of the so-called *leading-eight* norms of cooperation.

4 Norm Evolution for Different Intensities of Selection

The pairwise comparison rule [28], one of the evolutionary mechanisms used in the previous section (see also section 7), provides a convenient framework to study how the intensity of selection between tribes affects the emergence of *stern-judging.* It corresponds to introduce the following dynamics: Given two tribes chosen for a conflict, say A and B, with average payoffs P_A and P_B, respectively, then norm of tribe B will replace that of A with a probability given by $p = (1 + e^{-\beta.(P_B - P_A)})^{-1}$, whereas the inverse process will occur with probability $(1 - p)$. In physics this function corresponds to the well-known

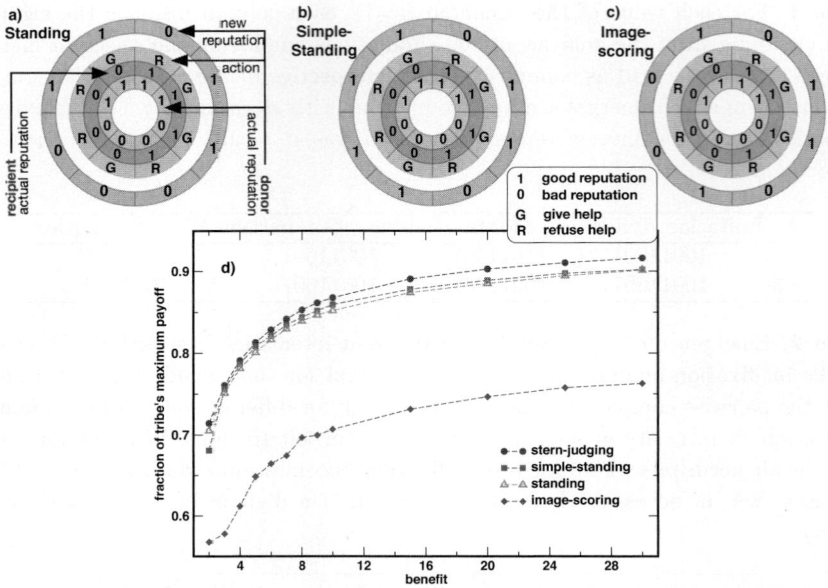

Fig. 2. We depict the three norms (besides *stern-judging* - Fig.1-a), the performance of which we analysed. Both *stern-judging*, *simple-standing* and *image-scoring* are symmetric with respect with the equatorial plane, and as such are second order norms. As for standing, it clearly breaks this symmetry, constituting a third order norm. In d), we plot the ratio between the average payoff attained by each tribe under the influence of a single, fixed norm, and the maximum value possible, given the population size (64), the benefit from cooperation (b) and the cost of cooperation ($c = 1$).

Fermi distribution function, in which the inverse temperature β determines the sharpness of transition from $p = 0$, whenever $P_B < P_A$, to $p = 1$, whenever $P_A < P_B$. Indeed, in the limit $\beta \to +\infty$, we obtain imitation dynamics (strong selection), whereas whenever $\beta \to 0$, B replaces A with the same probability that A replaces B. As we change β between these two extreme limits, we can infer the role of selection intensity on the emergence of *stern-judging.* In Table 2 we show results for different values of β, which testify for the robustness of *stern-judging.* In other words, in spite of the fact that, with decreasing β (decreasing selection intensity), it becomes increasingly difficult for all 8 bits to fixate whenever $b = 2$, in no case do we get results which deviate from *stern-judging* as the emerging social norm. These results reinforce the conclusion that *stern-judging* is robust and ubiquitous.

5 Cooperation Under a Selected Social Norm

Among the *leading-eight* norms discovered by Ohtsuki and Iwasa [13,15] (Fig. 1-b), only *stern-judging* [6] and the so-called simple-standing [29] constitute second-order norms. Our present results clearly indicate that *stern-judging* is favored compared to all other norms. Nonetheless, in line with the model considered here, the performance of each of these norms may be evaluated by investigating how each norm performs individually, taking into account all 16 strategies simultaneously. We compare the performance of *stern-judging* with the popular norms standing and *image-scoring*, as well as with the other second-order norm which incorporates the *leading-eight*, coined *strict-standing* [29]. We shall maintain mutation errors in strategy update, as well as errors of implementation. As a result, and given a fixed (immutable) norm, selection and mutation dictates the simultaneous evolution of all the 16 strategies in a given tribe. In Fig. 1-b we depict the *leading-eight* norms. The *slices* identified by $*$ correspond to places where both *GOOD* (1) or *BAD* (0) reputations can be freely assigned, the remaining norm being on of the *leading-eight*. Since a second order norm, in this representation, is simply a norm which exhibits a mirror symmetry with respect to the equatorial plane, it is obvious that there are only two second order norms which incorporate the *leading-eight*: Besides *stern-judging* (see Fig. 1-a), also simple-standing (Fig. 2-a) belongs to the *leading-eight*. *Image-scoring* and *standing*, the original norm proposed by Sugden, complete the set of norms of Fig. 2.

In Fig. 2-d we show results for the ratio between the average payoff reached in each tribe and the maximum average payoff attainable in that tribe, given the tribe size and the benefit (keeping cost=1). This quantity is plotted as a function of the benefit from cooperation, b. The results in Fig. 2-d show that *stern-judging* performs better than any of the other norms. Both standing and simple-standing lead to very similar performance, which reinforces the idea that second order norms are enough to promote cooperation under indirect reciprocity. Finally, *image-scoring* performs poorly compared to any of the other norms, a feature which is also related to the fact that the present analysis was carried out in the presence of errors [16,7,17,12].

6 Conclusion

Analyzing the approaches in the previous sections, we should note the results obtained in sections 3 and 4 are stronger than the analysis carried out with fixed social norms, since *stern-judging* emerges as the most successful norm surviving selection and mutation with other norms, irrespective of the selection mechanism. In other words, *stern-judging*'s simplicity and robustness to errors may contribute to its evolutionary success, since other well-performing strategies may succumb to invasion of individuals from other tribes who bring along strategies which may affect the overall performance of a given tribe. In this sense, robustness plays a key role when evolutionary success is at stake. We believe that *stern-judging* is the most robust norm promoting cooperation.

The present result correlates nicely with the recent findings in e-trade, where simple reputation-based mechanisms ensure high levels of cooperation. Indeed, *stern-judging* involves a straightforward and unambiguous reputation assessment, decisions of the donor being contingent only on the previous reputation of the receiver. We argue that the absence of constraining environments acting upon the potential customers in e-trade, for whom the decision of buying or not buying is free from further ado, facilitates the adoption of a *stern-judging* assessment rule. Indeed, recent experiments [30] have shown that humans are very sensitive to the presence of subtle psychologically constraining cues, their generosity depending strongly on the presence or absence of such cues. Furthermore, under simple unambiguous norms humans may escape the additional costs of conscious deliberation [31].

As conjectured by Ohtsuki and Iwasa [13] (cf. also [5,22]), group selection might constitute the key-element in establishing cooperation as a viable trait. The present results show that even when more sophisticated selection mechanisms operate between tribes, the outcome of evolution still favors *stern-judging* as the most successful norm under which cooperative strategies may flourish.

7 Simulations

Individual Interactions. We considered sets of 64 tribes, each tribe with 64 inhabitants. Each individual engages in a single round of the following indirect reciprocity game [6] with every other tribe inhabitant, assuming with equal probability the role of donor or recipient. The donor decides it provides help to the recipient, following her individual strategy that regards its own reputation and the reputation of the receiver. This results in a total of 16 strategies, encoded in a string of four bits. If it helps, donor's payoff decreases by $c = 1$, while the recipient's payoff increases by $b > 1$. In case of defection, the payoffs remain. This action will be witnessed by a third-party individual who, based on the tribe's social norm, will ascribe (subject to some small error probability $\mu = 0.001$) a new reputation to the donor. Moreover, individuals may fail to do what their strategy compels then to do, with a small execution error probability $\mu_c = 0.001$. After all interactions take place, one generation has passed, simultaneously for all tribes. Individual strategies in each tribe replicate to the next generation in the following way: For every individual A in the population we select an individual B proportional to fitness (including A) [12]. The strategy of B replaces that of A, apart from bit mutations occurring with a small probability $\mu_s = 0.01$.

Conflicts Among Tribes. With probability $p_{conflict} = 0.01$, all pairs of tribes may engage in a conflict, in which each tribe acts as an individual unit. Different types of conflicts between tribes have been considered which, besides the pairwise interaction rule introduced in section 4, we describe in the following. *Imitation Selection*: we compare the average payoffs P_A and P_B of the two conflicting tribes A and B, the winner being the tribe with highest score; *Moran Process*: In this case the selection method between tribes mimics that used between individuals in each tribe; one tribe B is chosen at random, and its norm is replaced by that of another tribe A chosen proportional to fitness ; *War of attrition*: We choose at random two tribes A and B with average payoffs P_A and P_B. We assume that each tribe can display for a time which is larger the larger its average payoff. To this end we draw two random numbers R_A and R_B each following an exponential probability distribution given by $\frac{e^{-t/P_A}}{P_A}$ and $\frac{e^{-t/P_B}}{P_B}$, respectively. The larger of the two numbers identifies the winning tribe. As a result of inter-tribe conflict (an additional conflict is discussed in [14]), the norm of the loosing tribe (B) is shifted in the direction of the victor norm (A). After the conflict, each bit of the A's norm will replace the corresponding bit of norm of tribe B with probability $p = \frac{\eta P_A}{\eta P_A + (1-\eta) P_B}$ which ensures good convergence whenever $\eta < 0.2$, independently of the type of conflict (a bit-mutation probability $\mu_N = 0.0001$ has been used). Furthermore, a small fraction of the population of tribe A replaces a corresponding random fraction of tribe B: Each individual of tribe A replaces a corresponding individual of tribe B with a probability $\mu_{migration} = 0.005$. Indeed, if no migration takes place, a tribe's population may get trapped in less cooperative strategies, compromising the global convergence of the evolutionary process.

Parameters. Each simulation runs for 9000 generations, starting from randomly assigned strategies and norms, in order to let the system reach a stationary situation, typically characterized by all tribes having maximized their average payoff, for a given benefit $b > c = 1$. The subsequent 1000 generations are then used to collect information on the strategies used in each tribe and the norms ruling the tribes in the stationary regime. As a cross validation, results did not change if instead we ran simulations for 14000 generations, accumulating information over the subsequent 1000 generations. This indicates that a steady state has been reached. We ran 500 evolutions for each value of b, subsequently performing a statistical analysis of the bits which encode each norm, as detailed before. In our simulations, we adopted the following values: $\eta = 0.1$, $\mu_N = 0.0001$, $\mu_S = 0.01$ and $\mu_a = \mu_e = 0.001$. The benefit b varied from $b = 2$ to $b = 36$. Our conclusions are robust to reasonable changes of these parameters. Moreover, results presented are qualitatively invariant to variations of the different mutation rates introduced above, as well as to variation of population size and number of tribes. Furthermore, reducing the threshold from 98% to 95% does not introduce any changes in the results shown. Finally, in Fig. 2-d we ran 500 simulations for each tribe with 64 inhabitants, and used the last 1000 generations from a total of 10000.

References

1. Smith, J.M., Szathmary, E.: The Major Transitions in Evolution. Freeman, Oxford (1995)
2. Hamilton, W.D.: Narrow Roads of Gene Land, vol.1. Freeman, New York (1996)
3. Trivers, R.: The evolution of reciprocal altruism. Q. Rev. Biol. 46, 35–57 (1971)
4. Wilson, E.O.: Sociobiology. Harvard Univ. Press, Cambridge (1975)
5. Alexander, R.D.: The Biology of Moral Systems. AldinedeGruyter, New York (1987)
6. Nowak, M.A., Sigmund, K.: Evolution of indirect reciprocity by image scoring. Nature 393(6685), 573–577 (1998)
7. Panchanathan, K., Boyd, R.: A tale of two defectors: the importance of standing for evolution of indirect reciprocity. J. Theor. Biol. 224(1), 115–126 (2003)

8. Sugden, R.: The economics of rights, co-operation and welfare. Basil Blackell, Oxford (1986)
9. Kandori, M.: Social norms and community enforcement. The Review of Economic Studies 59, 63–80 (1992)
10. Nowak, M.A., Sigmund, K.: Evolution of indirect reciprocity. Nature 437(7063), 1291–1298 (2005)
11. Fehr, E., Fischbacher, U.: The nature of human altruism. Nature 425 (2003)
12. Brandt, H., Sigmund, K.: The logic of reprobation: assessment and action rules for indirect reciprocation. J. Theor. Biol. 231(4), 475–486 (2004)
13. Ohtsuki, H., Iwasa, Y.: How should we define goodness?–reputation dynamics in indirect reciprocity. J. Theor. Biol. 231(1), 107–120 (2004)
14. Chalub, F.A.C.C., Santos, F.C., Pacheco, J.M.: The evolution of norms. J. Theor. Biol. 241(2), 233–240 (2006)
15. Ohtsuki, H., Iwasa, Y.: The leading eight: Social norms that can maintain cooperation by indirect reciprocity. J. Theor. Biol. 239(4), 435–444 (2006)
16. Leimar, O., Hammerstein, P.: Evolution of cooperation through indirect reciprocity. Proc. Biol. Sci. 268(1468), 745–753 (2001)
17. Panchanathan, K., Boyd, R.: Indirect reciprocity can stabilize cooperation without the second-order free rider problem. Nature 432(7016), 499–502 (2004)
18. Dellarocas, C.: Sanctioning reputation mechanisms in online trading environments with moral hazard. MIT Sloan School of Management working paper, pp. 4297–4303 (2003)
19. Bolton, G.E., Katok, E., Ockenfels, A.: How effective are electronic reputation mechanisms? an experimental investigation. Manage.Sci. 50, 1587–1602 (2004)
20. Keser, C.: Trust and reputation building in e-commerce. IBM-Watson Research Center, CIRANO working paper, 2002s–75k (2002)
21. Brinck, I., Gardenfors, P.: Co-operation and communication in apes and humans. Mind and Language 18(5), 484–501 (2003)
22. Mackie, J.L.: The law of the jungle: moral alternatives and principle of evolution. In: Thompson, P. (ed.) Issues in Evolutionary Ethics, pp. 165–177. State University of New York Press, New York (1995)
23. Bowles, S., Gintis, H.: The evolution of strong reciprocity: cooperation in heterogeneous populations. Theor. Popul. Biol. 65(1), 17–28 (2004)
24. Boyd, R., Gintis, H., Bowles, S., Richerson, P.J.: The evolution of altruistic punishment. Proc. Natl. Acad. Sci. U S A 100(6), 3531–3535 (2003)
25. Boyd, R., Richerson, P.J.: Culture and the Evolutionary Process. University of Chicago Press, Chicago (1985)
26. de Quervain, D.J., Fischbacher, U., Treyer, V., Schellhammer, M., Schnyder, U., Buck, A., Fehr, E.: The neural basis of altruistic punishment. Science 305(5688), 1254–1258 (2004)
27. Gintis, H.: The hitchhiker's guide to altruism: gene-culture coevolution, and the internalization of norms. J. Theor. Biol. 220(4), 407–418 (2003)
28. Traulsen, A., Nowak, M.A., Pacheco, J.M.: Stochastic dynamics of invasion and fixation. Phys. Rev. E 74(011909) (2006)
29. Ohtsuki, H., Iwasa, Y.: Global analyses of evolutionary dynamics and exhaustive search for social norms that maintain cooperation by reputation. J. Theor. Biol. 244(3), 518–531 (2007)
30. Haley, K.J., Fessler, D.M.T.: Nobody's watching? subtle cues affect generosity in an anonymous economic game. Evolution and Human Behaviour 26, 245–256 (2005)
31. Dijksterhuis, A., Bos, M.W., Nordgren, L.F., van Baaren, R.B.: On making the right choice: the deliberation-without-attention effect. Science 311(5763), 1005–1007 (2006)

Evolution of Cooperation in a Population of Selfish Adaptive Agents

Jorge M. Pacheco[1], Tom Lenaerts[2,3], and Francisco C. Santos[4]

[1] Centro de Física Teórica e Computacional and Departamento de Física da Faculdade de Ciências da Universidade de Lisboa, Lisbon, Portugal **
[2] SWITCH Laboratory, VIB, Gent, Belgium
[3] SWIT, Vrije Universiteit Brussel, Brussels, Belgium
[4] IRIDIA, CoDE, Université Libre de Bruxelles, Brussels, Belgium

Abstract. Often the selfish and strong are believed to be favored by natural selection, even though cooperative interactions thrive at all levels of organization in living systems. Recent empirical data shows that networks representing the social interactions between people exhibit typically high average connectivity and associated single-to-broad-scale heterogeneity, a feature which precludes the emergence of cooperation in any static network. Here, we employ a model in which individuals are able to self-organize both their strategy and their social ties throughout evolution, based exclusively on their self-interest. The entangled evolution of individual strategy and network structure provides a key mechanism toward the sustainability of cooperation in social networks. The results show that simple topological dynamics reflecting the individual capacity for self-organization of social ties can produce realistic networks of high average connectivity with associated single-to-broad-scale heterogeneity, in which cooperation thrives.

1 Introduction

Conventional evolutionary game theory predicts that natural selection favors the selfish and strong [1], despite existing evidence showing that cooperation is more widespread than theory predicts [2].

When cooperation is modeled in terms of the prisoner's dilemma [3] (PD), the solution of the replicator dynamics equation in infinite, well-mixed populations [4,5,6] dictates the extinction of cooperators. Cooperation becomes evolutionarily competitive whenever individuals are constrained to interact with few others along the links of sparse graphs, as initially predicted in Ref. [7] and more recently concluded in two independent studies which extend the original work to general social dilemmas described in terms of symmetric two-player games [8,9]. Both studies place individuals on the nodes of a static graph, and associate their social ties with the edges linking the nodes such that, throughout evolution, every individual has the possibility of changing her strategy, but not her social

** For further information contact pacheco@cii.fc.ul.pt

F. Almeida e Costa et al. (Eds.): ECAL 2007, LNAI 4648, pp. 535–544, 2007.

ties. In both studies the authors concluded that games on graphs open a window for the emergence of cooperation, inasmuch as graphs are sparse, that is, the average connectivity z (see section 3), is small.

However, recent data shows that realistic networks [10,11,12,13,14,15] exhibit values of z ranging from 2 to 170 (see methods), with an associated heterogeneity intermediate between single and broad-scale [11], which differs from the connectivity values typically used in previous studies [8,9]. Indeed, whereas single scale networks are characterized by a degree distribution peaked around a well defined value, such that most nodes exhibit a similar average connectivity pattern (homogeneity), broad scale networks are characterized by a more heterogeneous degree distribution, with different nodes exhibiting markedly different connectivities (cf. Fig. 4).

For instance, the network of movie actors exhibits an average connectivity of 30 [16], whereas collaboration networks based on co-authorship of published papers vary from average values of 4 (mathematics) up to 15 (biology) [13]. None of the previous results is capable of explaining how cooperation thrives on such social networks. Other mechanisms have to be at work here that allow for the survival of cooperation.

In most evolutionary models developed so far, social interactions are fixed from the outset. Such immutable social ties, associated naturally with static graphs, imply that individuals have no control over the number, frequency, or duration of their ties; they can only evolve their behavioral strategy. A similar observation can be made on studies related to the physical properties of complex networks [10,11,12,13,14,15]. The analyzed networks constitute but one static snapshot of networks that have been typically produced by some growth process.

Yet, networks have naturally evolved before and will continue to evolve after the snapshot has been taken. Indeed, recent longitudinal studies of evolving social networks [17,18] indicate that global properties seem to remain rather stable, whereas individual patterns of social ties keep evolving in time. Here, we employ a minimal model that combines strategy evolution with topological evolution, in which the requirements of individual cognitive capacities are very small, and investigate under which conditions cooperation may thrive. Network heterogeneity, which now emerges as a result of an entangled co-evolutionary dynamics, will be shown to play a crucial role in facilitating cooperative behavior.

2 Entangled Evolution of Strategy and Structure

2.1 The Prisoner's Dilemma

We consider two types of individuals-cooperators and defectors, who engage the most popular social dilemma of cooperation : The Prisoner's Dilemma (PD) (see below). They are not required to accumulate information on all other players, only those they are immediately connected with. The PD is modeled as a symmetric two-player game in which the players can either cooperate or defect when they interact. When both cooperate, they receive the payoff R (the reward for mutual cooperation). On the other hand, when both defect, they both obtain the

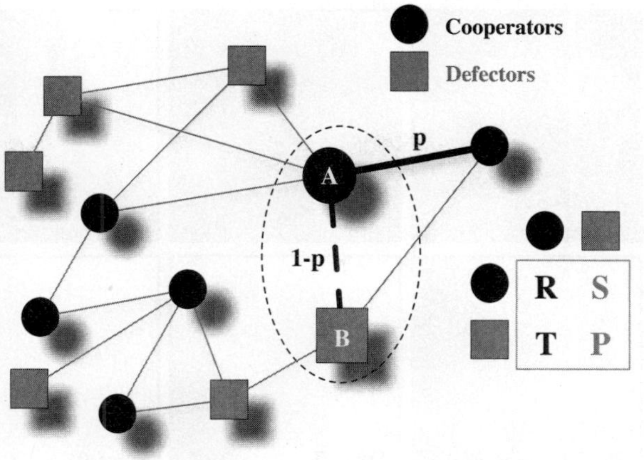

Fig. 1. Readjusting social ties. Cooperators and defectors interact via the links of a network. B (A) is satisfied (dissatisfied), since A (B) is a cooperator (defector). Therefore, A wants to change the link whereas B does not. The action taken is contingent on the fitness $\Pi(A)$ and $\Pi(B)$ of A and B, respectively. With probability p (see 3), A redirects the link to a random neighbor of B. With probability $1 - p$, A stays linked to B. Other possibilities may occur, as detailed in section 3.

payoff P (the punishment for mutual defection). The two remaining possibilities occur when one defects and the other cooperates, resulting in the payoff T (the temptation to cheat) for the defector and S (the disadvantage of being cheated) for the cooperator. The dilemma follows from the players' payoff preferences. Players are referred to as greedy since they prefer unilateral defection to mutual cooperation ($T > R$). Moreover, mutual defection is preferred to unilateral cooperation ($S < P$), resulting in an intrinsic fear for the players to cooperate. We adopt the convention of [8] and normalize the difference between mutual cooperation (R) and mutual defection (P) to 1, making $R = 1$ and $P = 0$, respectively. Consequently, we investigate the PD in a 2D parameter space, depicted in Figure 2, where the payoff T (temptation to cheat) satisfies $0 \leq T \leq 2$ and the payoff S (disadvantage of being cheated) satisfies $-1 \leq S \leq 0$. The fitness of each individual corresponds to the total accumulated payoff resulting from pairwise interactions (see section 3) with all her neighbors.

2.2 Quality Assessment of Social Ties

Both cooperators and defectors are able to decide, on an equal footing, those ties that they want to maintain and those they want to change. Given an edge with individuals A and B at the extremes, we say that A (B) is satisfied with the edge if the strategy of B (A) is a cooperator, being dissatisfied otherwise. If A is satisfied, she will decide to maintain the link. If dissatisfied, then she

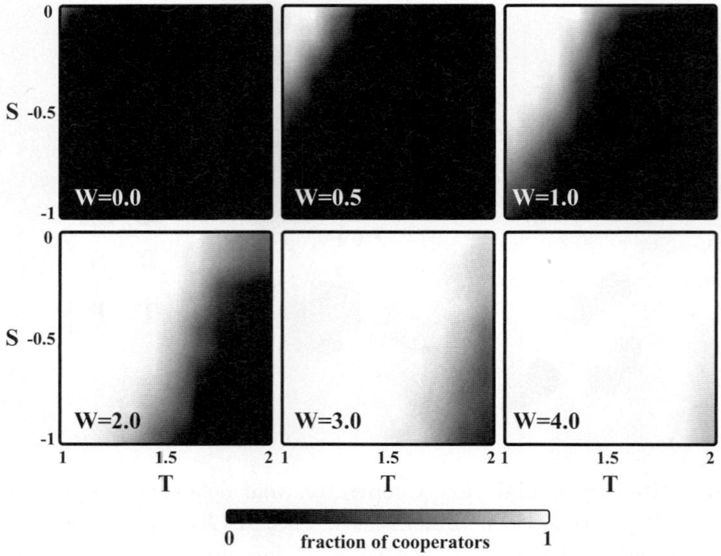

Fig. 2. Co-Evolution in the *PD* for different time-scales. Results for the fraction of successful evolutionary runs ending in 100% cooperation for different values of the time scale ratio W. We study the PD in the area $2 \geq T \geq 1 > 0 \geq S \geq -1$. For $W = 0$ ($N = 10^3$, $z = 30$ and $\beta = 0.005$), the results fit the predictions from well-mixed populations. With increasing W it also increases the rate at which individuals readjust their ties, and so does the viability of cooperation. Above a critical value $W_{critical} \sim 4.0$ (see also Fig. 3), cooperators efficiently wipe out defectors. For the strategy evolution dynamics adopted here (pairwise comparison, see section 3), and according to [9], cooperation would never be favored. The contours provide arresting evidence of the viability of cooperation under co-evolutionary dynamics.

may compete with B to rewire the link (see Fig. 1 and section 3), rewiring being attempted to a random neighbor of B. The intuition behind this reasoning relies on the fact that agents, equipped with limited knowledge and scope, look for new social ties by proxy [17]. In this sense, agent A is more likely to encounter one of the friends of B and become a friend with B's neighbor. Moreover, selecting a neighbor of an inconvenient partner may turn out to be a good choice, since this partner also tries to establish links with cooperators, making it more likely that the rewiring results in a tie to a cooperator. Indeed, it is always best to have a link with a cooperator, as it maximizes the fitness of any individual, irrespective of her (fixed) strategy. Consequently, all individuals naturally seek to establish links with cooperators. As a result, rewiring to a neighbor of a defector is certainly a good choice for individuals with local information only.

2.3 Different Evolutionary Timescales

The fact that in our model cooperators and defectors interact via social ties they both decide upon establishes a coupling between individual strategy and population structure: the game payoff induces now an entangled co-evolution of strategy and structure. Such an adaptive individual behavior introduces a new time scale (τ_a), not necessarily equal to the time scale associated with strategy evolution (τ_e). Depending on the ratio $W = \tau_e/\tau_a$, different fates may occur for cooperation. Indeed, whenever $\tau_e << \tau_a$, that is, $W \to 0$, we recover the results of [8,9]. On the other hand, with increasing W, individuals become apt to adapt their ties with increasing efficiency. In general, however, one expects the two time scales to be of comparable magnitude in realistic situations (see Figs. 2 and 3). More intuitively, W provides a measure of individuals' responsiveness to adverse ties: large values of W reflect populations in which individuals react promptly to adverse ties, whereas smaller values of W reveal the opposite behaviour. In reality, individuals may exhibit their own characteristic W. We shall not dwell into these intricacies here. Instead, we view W as reflecting an average feature of the population.

3 Materials and Methods

3.1 Graphs

We place individuals on the nodes (a total of N) of a graph. Links (a total of N_E) represent the social ties between individuals. Graphs evolve in time as individuals change their ties. The average connectivity $z = 2N_E/N$ is conserved since we do not introduce or destroy links. At all times graphs are required to remain connected. We enforce this condition by imposing that nodes connected by a single link cannot loose this link. We also computed the cumulative degree distribution $D(k) = N^{-1}\sum_{i=k}^{N-1} N_i$ (where N_i gives the number of nodes with i links), where $D(k)$ gives the probability of finding nodes in the graph with degree greater or equal to k. Since $D(k) = 0$ for $k > k_{max}$, with k_{max} the maximum value of the connectivity of a graph, both h and k_{max} provide simple measures of the heterogeneity of a graph.

3.2 Evolution of Strategies

Whenever $W > 0$, evolution of strategy and structure proceed together under asynchronous updating. Choice of type of update event depends on W; assuming (without loss of generality) $\tau_e = 1$, a strategy update event is chosen with probability $(1+W)^{-1}$, a structural update event being selected otherwise. A strategy update event is defined in the following way, corresponding to the so-called pairwise comparison rule [24]: One node A is chosen at random and another node B is chosen randomly among A's first neighbors. The individuals A and B interact with all their neighbors (those directly connected to them by links) according to PD. As a result, they accumulate total payoffs $\Pi(A)$ and $\Pi(B)$, respectively.

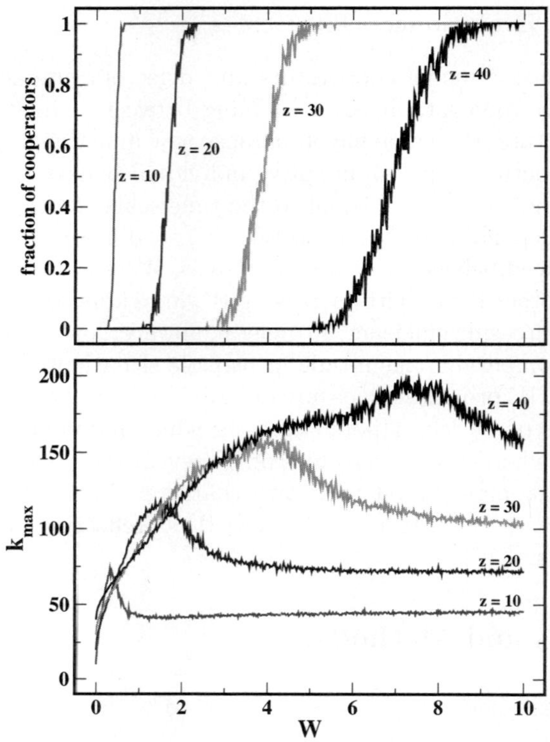

Fig. 3. Co-Evolution in the *PD* for different time-scales. *PD* with $T = 2, S = -1$ and $\beta = 0.005$. **Upper panel** Fraction of cooperators at end as a function of W for different values of z. For each value of z, there is a critical value of W - $W_{critical}$ - above which cooperators wipe out defectors. **Lower panel** Maximum value of the connectivity in population as a function of W. With increasing z, $W_{critical}$ increases. In all cases, the heterogeneity of the associated network becomes maximal at $W_{critical}$, stagnating for higher values.

The strategy of B replaces that of A with a probability given by the (Fermi distribution) function $p = [1 + e^{-\beta[\Pi(B)-\Pi(A)]}]^{-1}$. The value of $\beta \geq 0$, which plays the role of an inverse temperature in statistical physics, controls here the intensity of selection [24] : $\beta \to 0$ leads to neutral drift whereas $\beta \to \infty$ leads to the so-called imitation dynamics, often used to model cultural evolution.

3.3 Link Rewiring

Cooperators and defectors interact via the links of a network. Two individuals, A and B, connected by one link, may be satisfied or dissatisfied, as shown in Fig. 1. In Fig. 1, B is satisfied, whereas A is not, since A (B) is a cooperator (defector). Therefore, A wants to change the link whereas B does not. The action taken is contingent on the fitness $\Pi(A)$ and $\Pi(B)$ of A and B, respectively. With the probability p defined above in terms of the Fermi distribution, A redirects the link to a random neighbor of B. With probability $1-p$, A stays linked to B. Whenever both A and B are satisfied, nothing happens. When both A and B are dissatisfied, rewiring takes place such that the new link keeps attached to A with probability p and attached to B with probability $1-p$.

3.4 Computer Simulations

We start from a homogeneous random graph [25], in which all nodes have the same number of links (z), randomly linked to arbitrary nodes. Population size is $N = 10^3$ and average connectivities $z = 10, 20, 30$, and 40 (the value $z = 30$ used in Fig. 2 reflects the mean value of the average connectivities reported in [13] for socials networks). We start with 50% of cooperators randomly distributed in the population. We run 100 independent simulations for each set of parameters (T, S, W) and compute the fraction of times that evolution stopped at 100% cooperation. Indeed, under stochastic dynamics the population will invetitably converge to one of the two possible absorbing states: 100% cooperators or 100% defectors; consequently, each run proceeds for as many iterations as necessary until one of these absorbing states is reached. At the end of each evolution we also computed the maximal connectivity k_{max} associated with the final graph and the cumulative degree distribution, which are on the basis of the results plotted in Figures 3 and 4. We have confirmed that our results are valid for $N > 500$.

4 Results and Discussion

The contour plots in Fig. 2 show how cooperators survive for different values of the ratio W in networks with high connectivity. We plot the fraction of cooperators who survive evolution, averaged over 100 independent realizations for the same values of the temptation to cheat T, the disadvantage of being cheated S, and W. For $W = 0$ the results reproduce, as expected [8], the predictions for finite, well-mixed populations [8,19]. Yet, with increasing W we observe a *wave of cooperation* moving *southeast* toward the lower right corner of each contour plot. Hence, the PD for $T = 2$ and $S = -1$ constitutes the hardest challenge for cooperation within the parameter space studied. Nonetheless, for sufficiently large values of W, one obtains as a result of the entangled co-evolution of strategy and structure a full cooperative scenario: The swifter the response of individuals to the nature of their ties, the easier it gets for cooperators to wipe out defectors. Additional insight is provided in Fig. 3 (upper panel), where we investigate how cooperation wins over defection as a function of W for the PD when $T = 2$ and $S = -1$ and different values of the average connectivity z. For small W, cooperators have no chance. Their fate changes as W approaches a critical value $W_{critical}$ - which increases monotonically with z - cooperators wiping out defectors above $W_{critical}$ (the increase of $W_{critical}$ with z is expected, since there are more links to be rewired; in practice, $W_{critical}$ is determined as the value of W at which cooperation reaches 50%). Thus, the survival of cooperation relies on the capacity of individuals to adjust to adverse ties, even when the average connectivity is high. Figure 3 also provides evidence of the detailed interplay between strategy and structure. On one hand, strategy updating promotes a local assortment of strategies, since cooperators *breed* cooperators and defectors *breed* defectors. On (static) heterogeneous graphs, and for specific values of T and S, the population will evolve into a cooperative state [8], and the transient dynamics associated with the approach to such a cooperative state, starting from an equal fraction of cooperators and defectors, has been examined in detail in [7,20]. On the other hand, under structural updating, however, one is promoting local assortative interactions between cooperators (that is, cooperator-cooperator links) and disassortative interactions between defectors and cooperators (that is, cooperator-defector links), which constitute *favorable steps* from an individual point of view. Clearly, when simultaneously active, strategy update will reinforce assortativity among cooperators, but will inhibit disassortativity between defectors and cooperators, which overall will benefit the emergence of cooperation. Furthermore, since for any finite W graph heterogeneity will develop as a result of structural update (we are starting from homogeneous graphs), it will become easier for strategy update to promote cooperation.

For any $W > 0$, individual choices lead to heterogeneous graphs in which some individuals interact more, and more often than, others. The overall onset of increase of heterogeneity qualitatively follows the wave of cooperation shown in Fig. 2 [21]. In fact, the overall heterogeneity of the graph increases as W increases reaching a maximum at $W_{critical}$, above which heterogeneity decreases again down to a stationary value[21]. This is clearly shown in Fig. 4 for the case $T = 2, S = -1$. The results shown suggest that the adaptive dynamics of social ties introduced here coupled with social dilemmas accounts for the heterogeneities observed in realistic social networks [11]. Similar (analytic) results were obtained recently in a simpler model of link rewiring [22,23].

To conclude, the results presented in this manuscript show that to understand the emergence of cooperative behavior in a realistic scenario, one should consider simultaneously the evolution of the social network of interactions and the evolution of individual strategies. Besides providing a bottom-up answer to the conundrum of cooperation, the proposed mechanism also shows how complex social topologies can result from simple social dynamical processes, exclusively based in local assumptions.

Acknowledgments. Discussions with Martin A. Nowak, Arne Traulsen, Sebastian Maurer-Stroh, Hisashi Ohtsuki and Hugues Bersini are gratefully acknowledged. JMP acknowledges support from FCT, Portugal. FCS acknowledges the support of COMP^{2}SYS, a Marie Curie Early Stage Training Site, funded by the EC through the HRM activity.

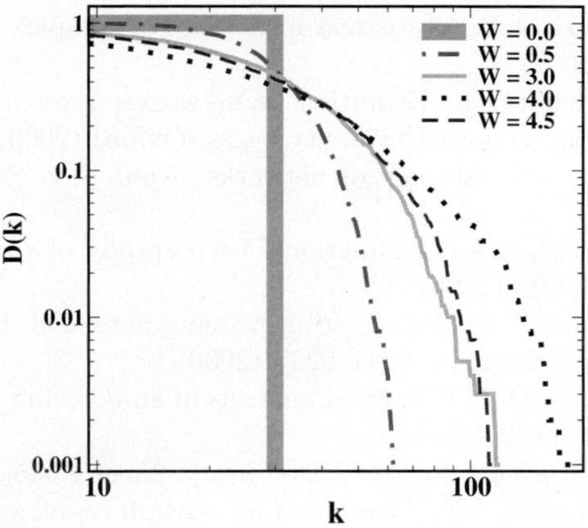

Fig. 4. Evolution of heterogeneity. Cumulative degree distributions (see section 3) for PD with $T = 2$ and $S = -1$. Starting from a distribution exhibiting a sharp cutoff at $k_{max} = z$, as soon as $W > 0$, the distribution widens, resulting in both single scale networks ($W = 0.5$, dash-dotted line) and broad-scale networks ($W > 3$, solid and dotted lines), reflecting the increase in the overall heterogeneity of the pattern of ties in the population. For $W > W_{critical}$ heterogeneity decreases again, as with increasing W it is easier for cooperators to wipe out defectors. Consequently, heterogeneity helps cooperators eradicate defectors.

References

1. Nowak, M.A., Sigmund, K.: Evolution of indirect reciprocity. Nature 437, 1291–1298 (2005)
2. Smith, J.M., Szathmáry, E.: The Major Transitions in Evolution. Freeman, Oxford (1995)
3. Rapoport, A., Chamah, A.M.: The prisoner's dilemma. University of Michigan Press, Ann Arbor (1965)
4. Hofbauer, J., Sigmund, K.: Evolutionary games and population dynamics. Cambridge University Press, Cambridge (1998)
5. Gintis, H.: Game theory evolving. Cambridge University Press, Cambridge (2000)
6. Nowak, M.A., Sigmund, K.: Evolutionary dynamics of biological games. Science 303, 793–799 (2004)
7. Santos, F.C., Pacheco, J.M.: Scale-free networks provide a unifying framework for the emergence of cooperation. Phys. Rev. Lett. 95, 098104 (2005)
8. Santos, F.C., Pacheco, J.M., Lenaerts, T.: Evolutionary dynamics of social dilemmas in structured heterogeneous populations. Proc. Natl. Acad. Sci. U S A 103, 3490–3494 (2006)
9. Ohtsuki, H., Hauert, C., Lieberman, E., Nowak, M.A.: A simple rule for evolution of cooperation on graphs and social networks. Nature 441, 502–505 (2006)
10. Watts, D.J.: Small worlds: The dynamics of networks between order and randomness. Princeton University Press, Princeton (1999)
11. Amaral, L.A., Scala, A., Barthelemy, M., Stanley, H.E.: Classes of small-world networks. Proc. Natl. Acad. Sci. U S A 97, 11149–11152 (2000)

12. Albert, R., Barabási, A.L.: Statistical mechanics of complex networks. Rev. Mod. Phys. 74, 47–98 (2002)
13. Dorogotsev, S., Mendes, J.: Evolution of networks: From biological nets to the Internet and WWW. Oxford University Press, Oxford (2003)
14. Watts, D.J.: The "new" science of networks. Ann. Rev. Sociobiol. 30, 243–270 (2004)
15. Guimerá, R., Amaral, L.A.N.: Functional cartography of complex metabolic networks. Nature 433, 895 (2005)
16. Albert, R., Barabasi, A.L.: Topology of evolving networks: Local events and universality. Phys. Rev. Lett. 85, 5234–5237 (2000)
17. Kossinets, G., Watts, D.J.: Empirical analysis of an evolving social network. Phys. Rev. Lett. 311, 89–90 (2006)
18. Watts, D.J.: A new science for the 21st century. Nature 445, 489 (2007)
19. Santos, F.C., Rodrigues, J.F., Pacheco, J.M.: Graph topology plays a determinant role in the evolution of cooperation. Proc. Biol. Sci. 273, 51–55 (2006)
20. Santos, F.C., Pacheco, J.M.: A new route to the evolution of cooperation. J. Evol. Biol. 19, 726–733 (2006)
21. Santos, F.C., Pacheco, J.M., Lenaerts, T.: Cooperation prevails when individuals adjust their social ties. PLoS Computational Biology 2, 1284 (2006)
22. Pacheco, J.M., Traulsen, A., Nowak, M.A.: Coevolution of strategy and structure in complex networks with dynamical linking. Phys. Rev. Lett. 97, 258103 (2006)
23. Pacheco, J.M., Traulsen, A., Nowak, M.A.: Active linking in evolutionary games. J. Theor. Biol. 243, 437–443 (2006)
24. Traulsen, A., Nowak, M.A., Pacheco, J.M.: Stochastic dynamics of invasion and fixation. Phys. Rev. E 74, 011909 (2006)
25. Santos, F.C., Rodrigues, J.F., Pacheco, J.M.: Epidemic spreading and cooperation dynamics on homogeneous small-world networks. Phys. Rev. E 72, 056128 (2005)

Evolutionary Dilemmas in a Social Network

Leslie Luthi, Enea Pestelacci, and Marco Tomassini

Information Systems Institute, University of Lausanne, Switzerland
{leslie.luthi, enea.pestelacci, marco.tomassini}@unil.ch

Abstract. We simulate the prisoner's dilemma and hawk-dove games on a real social acquaintance network. Using a discrete analogue of replicator dynamics, we show that surprisingly high levels of cooperation can be achieved, contrary to what happens in unstructured mixing populations. Moreover, we empirically show that cooperation in this network is stable with respect to invasion by defectors.

1 Introduction

Some extremely simple games lead to puzzles and dilemmas that have a deep social meaning. The *Prisoner's Dilemma* (PD), a universal metaphor for the tension that exists between social welfare and individual selfishness, is the most famous game of this type. It stipulates that, in situations where individuals may either cooperate or behave selfishly and thus defect, they will rationally choose the latter. Unfortunately, cooperation would be the preferred outcome when global welfare is considered. Game theory [14] is the discipline that deals with such situations of conflict where two or more individuals must make decisions that will mutually influence each other. It takes a view of collective systems in which global social outcomes emerge as a result of the interaction of the individual decisions made by each agent. Another well known simplified model of many common important socio-economic situations is the *Hawk-Dove* (HD) game. According to game theory, cooperative attitude should vanish in the PD, and should be limited to a given fraction in the HD. This is also the case when large populations of individuals play the game pairwise in a random manner and anonymously, as prescribed by evolutionary game theory [16]. However, in controlled experiments it has been observed that cooperation actually emerges when the game is played by humans and in many other cases [3,9]. A number of mechanisms have been invoked to explain the emergence of cooperative behavior: repeated interaction, reputation, and belonging to a recognizable group have often been mentioned [3]. However, the work of Nowak and May [8] showed that simply arranging the players in a spatial structure and allowing them to only interact with neighbors is sufficient to sustain a certain amount of cooperation even when the game is played anonymously and without repetition. Nowak and May's study and much of the following work was based on regular structures such as two-dimensional grids (see also [4] for the HD case). However, while two-dimensional grids may be realistic for ecological and some biological applications, they are inadequate for modeling human networks of interactions as it

F. Almeida e Costa et al. (Eds.): ECAL 2007, LNAI 4648, pp. 545–554, 2007.

has now become clear that many actual networks have a structure that is neither regular nor random but rather of the *small-world* type. Roughly speaking, small-world networks are graphs in which any node is relatively close to any other node. In this sense, they are similar to random graphs but unlike regular lattices. However, in contrast with random graphs, they also have a certain amount of local structure, as measured, for instance, by a quantity called the *clustering coefficient* which essentially represents the probability that two neighbors of a given node are themselves connected (see e.g. [6]). Thus, most real conflicting situations in economy and sociology are not well described neither by a fixed geographical position of the players in a regular lattice nor by a mixing population, and it becomes relevant to study these dilemmas on other, more faithful social structures. Some previous work has been done in this direction. We mention Santos and Pacheco's work on scale-free networks [10,11] and work on Watts–Strogatz small-world graphs [1,12,15]. However, these network types, although they have the right global "statistical" properties, are only an approximation of the actual topological properties of measured networks of interactions. In the present work we introduce a more socially relevant network and we emphasize the relationships between community structure and cooperation. A recent work close to the present one in spirit is Holme et al. [5]. However, the authors of [5] only study the PD on a much smaller social network using a different, noisy, strategy update rule, while we employ the more standard replicator dynamics on a larger social network and also study the HD game.

The remainder of this paper is organized as follows. We first give a brief background on the PD and HD. We then describe the main features of social networks and we present an evolutionary game model on a real collaboration network. We finally present and discuss results of numerical simulations of the model in terms of cooperation, community structure and stability.

2 The Model

2.1 Social Dilemmas

We first recall a few elementary notions on the PD and the HD games. These are two-person, symmetric games in which each player has two possible strategies: cooperate (C) or defect (D). In strategic form, these games have the payoff bimatrix shown in table 1. In this matrix, R stands for the *reward* the two players receive if they both cooperate, P is the *punishment* for bilateral defection, and T is the *temptation*, i.e. the payoff that a player receives if it defects, while the other cooperates. In this latter case, the cooperator gets the *sucker's* payoff S. For the PD, the payoff values are ordered numerically in the following way: $T > R > P > S$, while in the HD game $T > R > S > P$. Defection is always the best rational individual choice in the PD – (D,D) is the unique Nash equilibrium and also an evolutionary stable strategy (ESS). Mutual cooperation would be preferable but it is a strongly dominated strategy. Thus the dilemma is caused by the "selfishness" of the actors. In the HD, when both players defect they each get the lowest payoff; (C,D) and (D,C) are Nash equilibria of the game

	C	D
C	(R,R)	(S,T)
D	(T,S)	(P,P)

Table 1. Payoff matrix for 2 × 2 symmetric games.

in pure strategies, and there is a third equilibrium in mixed strategies where strategy D is played with probability $1/(2\beta-1)$, and strategy C with probability $1-1/(2\beta-1)$, where β is another name for the temptation T. The dilemma in this game is caused by "greed", i.e. players have a strong incentive to "bully" their opponent by playing D, which is harmful for both if the outcome is (D,D).

2.2 Social Networks

In standard evolutionary game theory [14,16], these dilemmas have been classically studied by modeling the behavior of a large population in which randomly paired individuals play the game in an anonymous manner. Non-rational players are "hard-wired" to play a given strategy, and those faring better than average increase their share in the population. The fixed points of these *replicator dynamics* are evolutionarily stable strategies, i.e. they cannot be invaded by a mutant strategy [16].

In terms of networks of interaction, the "mixing" structure of the population would be represented by a complete graph, i.e. any individual may interact with any other player. The advantage of the mixing model is that it admits an approach by *mean-field* methods, which treat the system as being homogeneous, ignoring space dependences and correlations [16]. However, we know today that real social networks do not have this random structure. Instead, they are of finite size, have heterogeneous connectivity, are often small worlds, in the sense that any individual is only a few steps away from any other, and individuals cluster together in communities [2,6,15]. Therefore, evolutionary games should be studied on more general types of graphs, to understand the limitations of the theory, and to extend it as far as possible to structures encountered in real-life.

As stated in the introduction, numerical simulations of evolutionary games have been performed for degree-homogeneous and degree-heterogeneous model graphs such as Watts–Strogatz and scale-free networks [1,10,11,12,15]. Here we go a step further and make use of a real social collaboration network, the genetic programming coauthorship network. This network is a small world with a connected giant component of 942 scientists and it has recently been analyzed [13]. It has clusters and communities and it should be representative of other similar human acquaintance networks. Watts–Strogatz networks [15] are only a mathematical construction and do not properly represent social networks. As for model scale-free graphs, most social networks studied to date are not of the pure scale-free type, and show a faster decay of the tail of the degree distribution [2,6]. Intuitively, there must be a cutoff in the number of acquaintances a given agent can have, and in many cases also a typical number of acquaintances, which gives a scale to the network. Besides, it has been observed that social networks have a higher clustering than the typical values reached in scale-free graphs, another manifestation of the complex neighborhood structure of the network. Furthermore, the appearance of communities – sets of densely connected vertices with sparse connections between the sets – is yet another typical feature found in social structures. Communities can highly influence the way information is propagated throughout the network or opinion formation is processed. Finally, we should make it clear that social networks are dynamical, i.e. new nodes may join the network forming new links, and old nodes may leave it as social actors come and go. As a first approximation here we model a static network, thus ignoring fluctuations and non-equilibrium phenomena.

2.3 Model Parameters

Population Structure. We consider a population P of players of size N. Each individual i in the population P is represented as a vertex v_i of a graph $G(V, E)$, with $v_i \in V$, $\forall i \in P$. An interaction between two players i and j is represented by the undirected edge $e_{ij} \in E$. The number of neighbors $N(i)$ of player i is the degree k_i of vertex v_i. The average degree of the network will be called $\bar{k}$. The terms vertex, node, individual, or player shall be used interchangeably in the sequel; likewise for edge, link, interaction, and acquaintance.

Strategy Update Rules. To update the strategies of the individuals given an initial strategy distribution in the population, we use a discrete analogue of replicator dynamics [4]. Other socially meaningful strategy update policies could also be used, such as *imitation of the best* and *proportional updating* [4,8,12]. The replicator dynamics assumes that the share of the population playing a particular strategy grows in proportion to how well this strategy is doing relative to the average population payoff.

Let Π_x be a player x's aggregated payoff and k_x the number of neighbors x has (k_x can also be seen as the degree of the vertex representing x). We define the replicator dynamics function $\phi(\Pi_j - \Pi_i)$ as being the probability function according to which player i adopts neighbor j's strategy, namely

$$\phi(\Pi_j - \Pi_i) = \begin{cases} \dfrac{\Pi_j - \Pi_i}{k_j \Pi_{M1} - k_i \Pi_{m1}} & \text{if } \Pi_j - \Pi_i > 0 \\ \\ 0 & \text{otherwise,} \end{cases} \tag{1}$$

where Π_{M1} (resp. Π_{m1}) is the maximum (resp. minimum) payoff a player could get if it had only one neighbor.

Payoff Calculation. There exist several possibilities for determining a player's utility or payoff. One can define a player's payoff as being the sum (*accumulated payoff*) of all pair interactions with its nearest neighbors. Or it can be defined as the accumulated payoff divided by the number of interactions (*average payoff*). Accumulated and average payoff give the same results when considering degree-homogenous networks such as regular lattices. Accumulated payoff seems more logical to use in degree-heterogeneous networks since it reflects the very fact that players may have different numbers of neighbors in the network. Average payoff, on the other hand, smoothes out the possible differences although it might be justified in terms of the number of interactions that a player may sustain in a given time, i.e. an individual with many connections is likely to interact less often with each of its neighbors than another that has a lower number of connections. Also, if there is a cost to maintain a relationship, average payoff will roughly capture this fact, while it will be hidden if one uses accumulated payoff. In this paper we use a form of accumulated payoff.

Population Dynamics. Calling $C(t) = (s_1(t), s_2(t), \ldots, s_N(t))$ a *configuration* of the population strategies $s_i \in \{C, D\}$ at time step t, the global *synchronous* system dynamics leads to $C(t+1)$ by simultaneously updating all the players' strategies according to the chosen rule. Synchronous update, with its idealization of a global clock, is customary in spatial evolutionary games, and most results have been obtained using this model. However, perfect synchronicity is only an abstraction as agents normally act at different and possibly uncorrelated moments. In spite of this, it has been shown that the update mode does not fundamentally alter the results for replicator dynamics [4]. We have also checked that asynchronous update dynamics does not influence the system evolution in a significant way and so, all results presented refer to synchronous systems.

3 Simulation Results and Analysis

For each game, we can explore the entire game space by limiting our study to the variation of only two parameters per game. In the case of the PD, we set $R = 1$ and $S = 0$, and vary $1 \leq T \leq 2$ and $0 \leq P \leq 1$. For the HD game, we set $R = 1$ and $P = 0$ and the two parameters are $1 \leq T \leq 2$ and $0 \leq S \leq 1$. In the Prisoner's Dilemma case, P is limited between $R = 1$ and $S = 0$ in order to respect the ordering of the payoffs ($T > R > P > S$) and T's upper bound is equal to 2 due to the $2R > T + S$ constraint. Had we instead fixed $R = 1$ and $P = 0$, T could be as big as desired, provided $S \leq 0$ is small enough. In the Hawk-Dove game, setting $R = 1$ and $P = 0$ determines the range of S (since this time $T > R > S > P$) and gives an upper bound of 2 for T, again due to the $2R > T + S$ constraint. Note however, that the only valid value pairs of (T, S) are those that satisfy the latter constraint.

The network is randomly initialized with 50% cooperators and 50% defectors. In all cases, the parameters are varied between their two bounds by steps of 0.1. For each set of values, we carry out 50 runs of 16000 time steps each. Cooperation level is averaged over the last 1000 time steps, well after the transient equilibration period.

3.1 Evolution of Cooperation

In Figure 1 we report average cooperation levels for both games for systems having attained a steady-state. As expected, the region in which cooperation is possible is much more restricted in the PD than for HD. Cooperation is more widespread for the HD, as mutual defection is the worst outcome in this game. For the PD cooperation is sensitive to the "punishment" level P, for a given T. Concerning the HD, one can see that the S parameter has moderate influence on cooperation for a given T. We also notice that the transition from cooperation to defection is much steeper in the PD than for the HD.

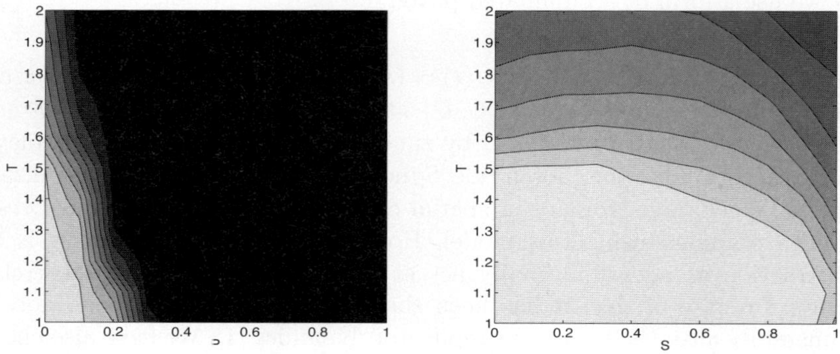

(a) (b)

Fig. 1. Level of cooperation at the end of the simulation. Left: PD; right: HD. For the HD, the meaningful phase space is the lower left triangle only.

Another important global quantity is the total payoff at the end of the simulated games, also called the wealth. The cumulated wealth of defectors and cooperators is plotted in Figure 2 for the PD. This is done for $T = 1.3$, for two values of the punishment P, giving rise to two different cooperation regimes: one in which cooperation prevails and a second one where defection predominates. We see that the cooperators' wealth is larger and has a broader distribution. This hints at a clustering of cooperators, as this is the only way for them to increase their payoff. We shall comment on this phenomenon below.

Social networks are characterized by the presence of communities, which can be seen as sets of highly connected vertices having few connections with vertices belonging to other communities. Finding the communities in a given network is a difficult task for which there exist several algorithms essentially based on clustering considerations. We have used one of the algorithms proposed by Newman [7].

For reasons of space, in the following we show results for the PD only. In Figure 3 we depict a portion of the scientific coauthorship graph, distinguishing between cooperators and defectors for the PD. We note that tightly-bound com-

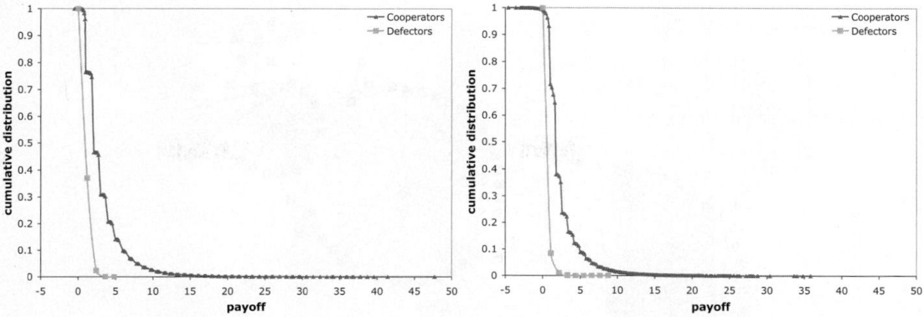

(a) (b)

Fig. 2. Cumulative wealth distribution in the PD game averaged over 50 runs for the social network model; (a) $T = 1.3, P = 0.1$ yielding ∼73% of cooperation. Average C-wealth=2.92, average D-wealth=0.47; (b) $T = 1.3, P = 0.2$ yielding ∼15% of cooperation. Average C-wealth=2.30, average D-wealth=0.11.

munities are mostly composed of players with the same strategy. Although we only show a small portion of the whole network for reasons of clarity, we could have chosen many other places as the phenomenon is widespread. Cooperators tend to "protect" themselves by having many links toward other cooperators. On the other hand, a cooperator like the central one in the largest defecting community will have a tendency to become a defector since its neighbors are nearly all defectors; but when its highly connected "wealthy" cooperator neighbor on the left of the figure is probabilistically selected to be imitated, then it will certainly become a cooperator again. So, the rare cooperators that are not tightly clustered with other cooperators will tend to oscillate between strategies. The community structure of cooperators, together with the mutual payoff advantage of cooperating, explains the previous observation, namely that the average cooperators' wealth exceeds the average wealth of defectors. Strategy distribution in communities found for the HD game is qualitatively similar; however, in the HD the two strategies are slightly more intermingled, confirming analogous findings for grid-structured populations [4].

When cooperation prevails, we have found that cooperators tend to occupy the highest degree nodes. When defectors predominate, the degree distributions of the two strategies tend to be closer, although the cooperators still monopolize higher degree nodes. For HD the results are similar, namely, degree distribution for defectors fall off more rapidly than those for cooperators.

3.2 Evolutionary Stability

No empirical investigation of an evolutionary games scenario would be complete without examining its evolutionary stability [16]. Evolutionary stability can be defined exactly for mixing populations but qualitatively it simply implies that a given population strategy cannot be invaded by an individual playing another

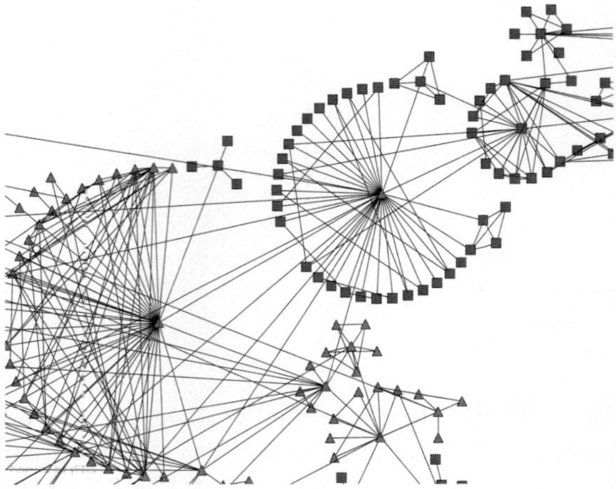

Fig. 3. Communities: cooperators are represented by triangles and defectors by squares.

strategy. For example, a single defector in a mixing population of cooperators in the PD will lead to a total extinction of the cooperators. Single individual strategy mutations are interesting only in mixing populations or in networks of the scale-free type [11]. In our social network, the distribution of strategies that obtains when the steady-state is reached is left undisturbed by this kind of event. However, given that the C and D strategies tend to cluster together, we have applied a more radical type of perturbations to the system. After the pseudo-equilibrium is reached, we choose a highly connected individual that plays the strategy of the majority (suppose it is C) and we flip its strategy to D as well as the strategy of all its first neighbors that are also cooperators. In this case, evolutionary stability requires that any such small group of individuals who try an alternative strategy do worse than those that those who stick to the status quo. Figure 4 shows the results using the above described perturbation. In each figure, ten executions have been reported to give a feeling of the behavior (many more have been run but in the case of noise, average values are irrelevant). Figures 4 (a), (b), and (c) refer to the PD with $P = 0.1$ and three different values of T: $1.5, 1.7$, and 1.9 respectively. Figure 4 (d) refers to the HD with $S = 0.1, T = 1.6$. The first constant part of the graphs up to abscissa 0 represent the pseudo-equilibrium values reached in each run after 10000 steps, just before applying the perturbation. For the PD, it can be seen that where cooperation is high (figure 4 (a)), after a transient period in which there can be a significant loss of cooperation, all runs tend to recover the original levels, albeit at different speeds. When cooperation is at intermediate or lower levels (figures 4 (b) and (c)), the behavior is more oscillatory but populations tend to recover the original pseudo-equilibrium levels of cooperation. Even populations that had originally a significantly lower cooperation percentage can sometimes reach the cooperation

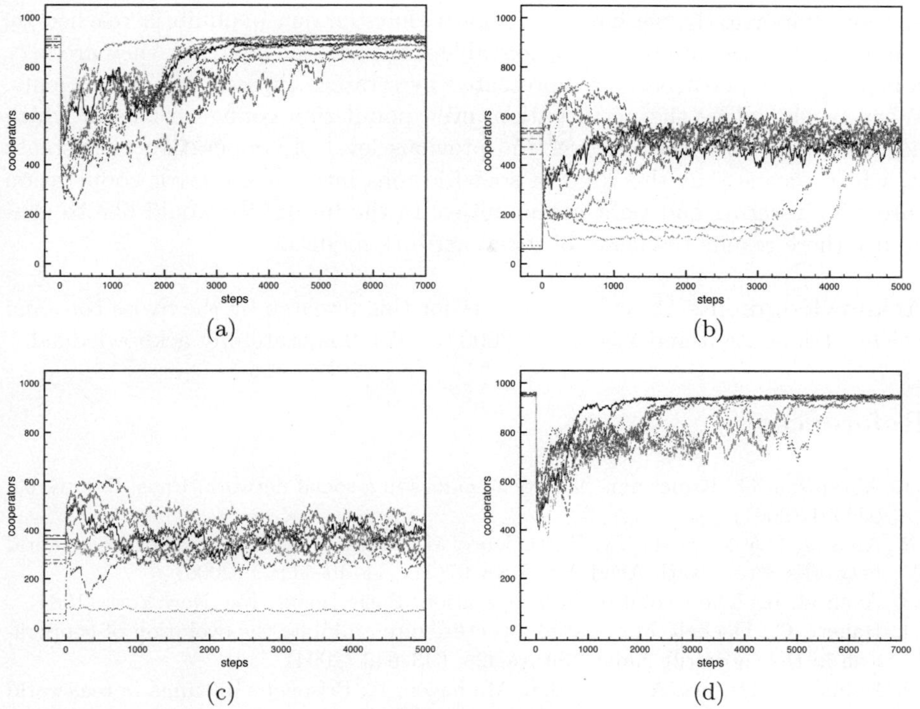

Fig. 4. Effect of noisy behavior on pseudo-equilibria states for the PD (a), (b), (c), and HD (d) (see text for explanations).

level of the majority of runs after the perturbation. For the HD, populations easily recover from noisy behavior (see figure 4 (d)), as cooperation is more widespread in this game. Even for extreme values of T the original cooperation level tends to be recovered (not shown to save space but very similar to figure 4 (b)).

4 Conclusions

Extending previous work on regular structures and on model scale-free and small-world networks, in this paper we have empirically studied two fundamental social dilemmas on a real acquaintance network. Although the graph studied is a single particular instance, it possesses all the features that characterize actual social networks, such as high clustering and communities. We find that this kind of topology allows cooperation to be reached and maintained, for a large portion of the game parameter space for HD, and even in the more difficult case of the PD. It was previously known that this is the case for lattice structures and, most notably, for scale-free graphs. However, these structures are not good representations of social ties and thus our result is encouraging from the social point of view. Importantly, we have also shown that the quasi-equilibria reached by the dynamics are not ephemeral, unstable states; on the contrary, they are very robust against perturbations represented by strategy flips of groups of agents. When reshuffled by the perturbation, even population configurations in which defection prevails either recover the previous level of cooperation or increase it, which means that this kind of social graphs intrinsically favor cooperation through clustering and tight communities. In the future, we would like to generalize these results to classes of social network models.

Acknowledgments. Financial support for this research by the Swiss National Science Foundation under contract 200021-107419 is gratefully acknowledged.

References

1. Abramson, G., Kuperman, M.: Social games in a social network. Phys. Rev. E, 63, 030901 (2001)
2. Amaral, L.A.N., Scala, A., Barthélemy, M., Stanley, H.E.: Classes of small-world networks. Proc. Natl. Acad. Sci. USA 97(21), 11149–11152 (2000)
3. Axelrod, R.: The Evolution of Cooperation. Basic Books, Inc, New York (1984)
4. Hauert, C., Doebeli, M.: Spatial structure often inhibits the evolution of cooperation in the snowdrift game. Nature 428, 643–646 (2004)
5. Holme, P., Trusina, A., Kim, B.J., Minhagen, P.: Prisoner's dilemma in real-world acquaintance networks: spice and quasi-equilibria induced by the interplay between structure and dynamics. Phys. Rev. E 68, 030901(R) (2003)
6. Newman, M.E.J.: The structure and function of complex networks. SIAM Review 45, 167–256 (2003)
7. Newman, M.E.J.: Fast algorithm for detecting community structure in networks. Phys. Rev. E 69, 066133 (2004)
8. Nowak, M.A., May, R.M.: Evolutionary games and spatial chaos. Nature 359, 826–829 (1992)
9. Sally, D.: Conversation and cooperation in social dilemmas: A meta-analysis of experiments from 1958 to 1992. Rationality and Society 7, 58–92 (1995)
10. Santos, F.C., Pacheco, J.M.: Scale-free networks provide a unifying framework for the emergence of cooperation. Phys. Rev. Lett. 95, 098104 (2005)
11. Santos, F.C., Pacheco, J.M., Lenaerts, T.: Evolutionary dynamics of social dilemmas in structured hetrogeneous populations. Proc. Natl. Acad. Sci. USA 103, 3490–3494 (2006)
12. Tomassini, M., Luthi, L., Giacobini, M.: Hawks and doves on small-world networks. Phys. Rev. E 73, 016132 (2006)
13. Tomassini, M., Luthi, L., Giacobini, M., Langdon, W.B.: The structure of the genetic programming collaboration network. Genetic Programming and Evolvable Machines 8(1), 97–103 (2007)
14. Vega-Redondo, F.: Economics and the Theory of Games. Cambridge University Press, Cambridge (2003)
15. Watts, D.J.: Small worlds: The Dynamics of Networks between Order and Randomness. Princeton University Press, Princeton (1999)
16. Weibull, J.W.: Evolutionary Game Theory. MIT Press, Boston (1995)

Exogenous Fault Detection in a Collective Robotic Task

Anders Lyhne Christensen, Rehan O'Grady, Mauro Birattari, and Marco Dorigo

IRIDIA, CoDE, Université Libre de Bruxelles
50, Av. Franklin Roosevelt, CP 194/6
1050 Brussels, Belgium
alyhne@iridia.ulb.ac.be,{rogrady,mbiro,mdorigo}@ulb.ac.be

Abstract. In robotics, exogenous fault detection is the process through which one robot detects faults that occur in other, physically separate robots. In this paper, we study exogenous fault detection in a collective leader-follower task for autonomous robots. We record sensory inputs from the robots while they are operating normally and after simulated faults have been injected. Given that faults are simulated, we can correlate the flow of sensory inputs with the fault state of the robots. We use back-propagation neural networks to synthesize fault detection components. We show that the flow of sensory inputs is sufficient information for performing exogenous fault detection, that is, we show that the leader robot is capable of detecting faults in the follower robot. All results are based on experiments with real robots.

1 Introduction

Some faults are hard to detect in the robot in which they occur. These faults include software bugs that cause a robot to hang, sensor failures that prevent a robot from detecting that something is wrong, and mechanical faults such as a loose connection to a battery. In this study, we show how fault injection and learning can be applied in order to synthesize software components for exogenous fault detection in autonomous robots. We present a concrete method for obtaining components that let one robot detect faults in collaborating robots. The ability to perform exogenous fault detection can improve the reliability of multi-robot systems. If one of the constituent robots fails, other robots in the system can take corrective actions even if the failed robot is unable to detect or communicate that it has experienced a fault.

The method relies on recording sensory data, firstly over a period of time when the robots are operating as intended, and secondly over a period of time when simulated hardware faults have been injected. After the data collection phase, fault detection components in the form of back-propagation neural networks are synthesized through supervised learning. The method requires no special fault detection hardware and relatively few computational resources on the robots. The work presented in this paper is an extension of previous studies

F. Almeida e Costa et al. (Eds.): ECAL 2007, LNAI 4648, pp. 555–564, 2007.

in which we have shown that the proposed method can be used to synthesize endogenous fault detection components for autonomous robots [1].

Technically, a fault is an unexpected change in system function which hampers or disturbs normal operation, causing unacceptable deterioration in performance. Fault detection is a binary decision process confirming whether or not a fault has occurred in a system. Fault detection is a single facet of the larger objective of ensuring fault tolerance. Other aspects of achieving fault tolerance for a system might include *fault diagnosis*, namely determining the type and location of faults, and *protection* which comprises any steps necessary to ensure continued safe operation of the system [2].

2 Related Work

Fault detection is based on observations of a system's behavior. Deviations from normal behavior can be interpreted as symptoms of a fault in the system. Several model-based approaches have been proposed [2,3]. In model-based fault detection some model of the system or of how it is supposed to behave is constructed. The actual behavior is then compared to the predicted behavior and deviations can be interpreted as symptoms of faults. A deviation is called a *residual*, that is, the difference between the estimated and the observed value. As uncertainty and noise play a significant role, techniques such as artificial neural networks and radial basis function networks for residuals-based fault detection have been proposed [4,5,6].

Alternative approaches relying on multi-model estimation have been studied, see for instance [7,8]. In these studies, banks of Kalman filters are applied to track multiple models with embedded fault states. Recently, computationally efficient approaches for approximating Bayesian belief using *particle filters*[1] have been suggested as a means for fault detection and identification [9,10,11].

Systems of multiple collaborating robots have the potential to achieve a high degree of fault tolerance. If one robot in such a system fails while performing a task, another one can take over and complete the task. In some cases, fault tolerance is an inherent property of the system and not handled explicitly. Lewis and Tan [12] have, for instance, shown that their control algorithm for maintaining geometric formations exhibits correct behavior even if one of the robots fails. However, the fault tolerance is a consequence of the adaptive nature of their controller design (robots attempt to maintain a virtual structure and do not let any robots fall behind) and not of explicit fault detection, diagnosis and protection. Implicit fault tolerance by design is, however, not generally feasible. In most tasks, faults must be detected and handled explicitly for the system to be fault tolerant.

Parker [13] has demonstrated that cooperating teams of robots based on the ALLIANCE software architecture can achieve a high degree of fault tolerance. Fault tolerance is obtained by modelling "motivations" mathematically and by

[1] Particle filters are Monte Carlo methods capable of tracking hybrid state spaces of continuous noisy sensor data and discrete operation states.

Fig. 1. Two *s-bots*, sensors and actuators. An *s-bot* has a diameter of 120 mm, a height of 190 mm, and weighs approximately 700 g.

adaptive task selection based on these motivations. If a robot experiences a fault in one or more of its components and if the fault degrades performance, the robot's motivation for performing its current task decreases. Eventually, the robot will switch to another task that it may still be able to perform. Alternatively, another robot will discover that there is limited or no progress in the task undertaken by the failed robot, and take over.

Other approaches, such as MURDOCH [14,15] and TraderBots [16], use explicit communication and negotiation of task allocation. In these cases fault detection and tolerance are built into the negotiation process.

In this paper, we address the issue of how a robot can determine if a collaborating robot is not operating correctly. Exogenous fault detection is only based on sensory inputs. There is no explicit communication, nor is there any pre-specified behavior used to aid fault detection. We take techniques that we have successfully used in the past for endogenous fault detection and apply them to exogenous fault detection [1]. We apply a technique known as *software implemented fault injection* (SWIFI) used in dependable systems research. The technique is usually applied to measure the robustness and fault tolerance of software systems [17]. In our case, we inject faults to discover how sensory readings and the control signals sent to the actuators by the control program change when faults occur.

3 The Robots and The Task

For the experiments, we use real robots known as *s-bots* [18] (see Fig. 1). The *s-bot* platform has been used for several studies in swarm intelligence and collective robotics [19,20,21]. Each *s-bot* is equipped with an Xscale CPU running

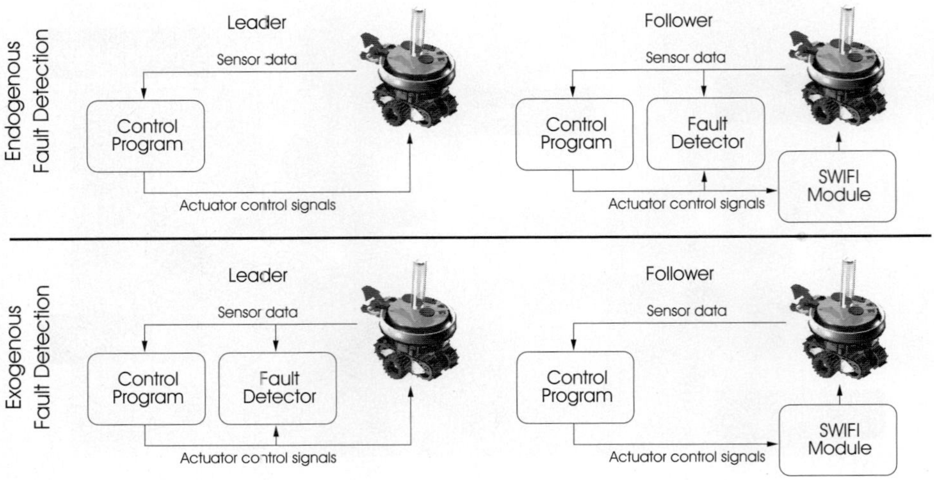

Fig. 2. The software architecture.

at 400 MHz and several sensors including an omni-directional camera and infrared proximity sensors. Each *s-bot* also has a number of actuators. These include differential *treels* (combined tracks and wheels) and 8 sets of RGB colored LEDs distributed around the circumference of the *s-bot* body. Using the omni-directional camera, one *s-bot* can see the colored LEDs on other *s-bots* up to 50 cm away depending on the light conditions.

We have chosen a simple *follow the leader* task in which two robots are placed in a 180 cm by 180 cm walled arena. One of the robots has been preassigned the *leader* role, while the other has been preassigned the *follower* role. The *leader* moves around in the environment. The *follower* tails the *leader* and tries to stay at a distance of 35 cm. If the *follower* falls behind, the *leader* waits. During experiments, we inject faults in the *follower* robot.

4 Software Architecture

An overview of the software architectures is shown in Fig. 2. The *Control Programs* are responsible for steering the robots. They read sensory inputs and send control signals to the robots' actuators. The *Fault Detectors* passively monitor the flow of sensory inputs and control signals that passes to and from the *Control Programs*. Faults are simulated by the *SWIFI Module* in the *follower*. When the *follower*'s *Control Program* sends actuator control signals, these commands pass through the *SWIFI Module*. If no fault is currently being simulated the *SWIFI Module* forwards all actuator control signals to the robot hardware. If a fault has been injected, control signals to the hardware affected by the fault are discarded.

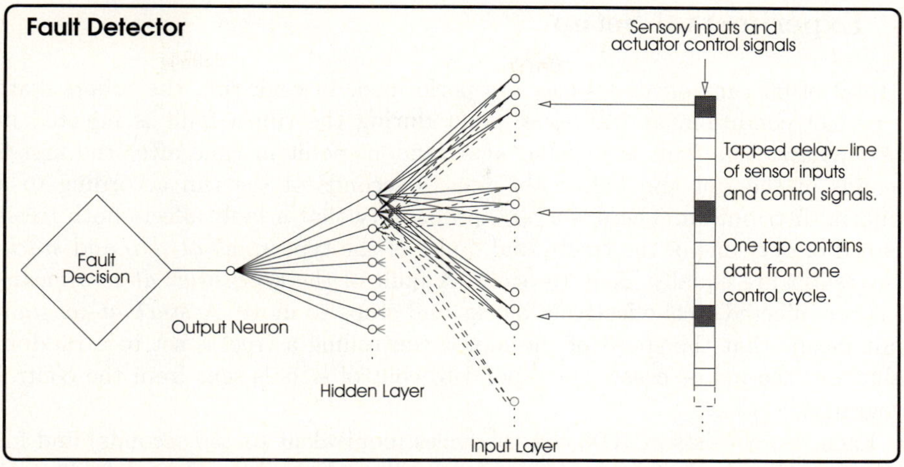

Fig. 3. An overview of the fault detection component, consisting of an artificial neural network and a tapped delay-line of observations. The input neurons in the neural network are logically organized into groups. In each group, the neurons encode observations from one tap.

The *Fault Detector* component consists of a time-delay artificial neural network (TDNN). A TDNN is a feed-forward network that allows for reasoning based on time-varying inputs without the use of recurrent connections [22,23]. In a TDNN, the input layer is logically organized into a number of groups. In each group, the activations of the neurons are set based on observations from a fixed distance into the past. TDNNs have been extensively used for time-series prediction due to their ability to make predictions based on data distributed in time. The TDNNs used in this study are normal multilayer perceptrons for which the inputs are taken from multiple, equally spaced points in a delay-line of past observations. Fig. 3 illustrates this concept. The current sensor inputs and control signals are stored in a tapped delay-line, and the activations of the neurons in the input layer are set based on data from the delay-line. The neurons from all input groups are fully connected to the neurons in the hidden layer. The term *input group distance* refers to the distance in time between adjacent input groups. In this study, we use an input group distance of 5 control cycles (as illustrated in Fig. 3), and TDNNs with hidden layer of 5 neurons and an input layer with a total of 10 input groups. Each input group consists of 15 neurons corresponding to the 15 infrared proximity sensors and 16 neurons encoding data extracted from the omni-directional camera. Images from the camera are partitioned into 16 sections and each of the 16 neurons are assigned values inversely proportional to the distance of the closest object perceived in the corresponding section. The image processor has been configured to detect colored LEDs. This configuration means that the camera only detects other *s-bots* and not objects like walls.

5 Experimental Setup

A total of 60 runs on real *s-bots* are performed. In each run, the robots start in perfect condition, and at some point during the run a fault is injected in the *follower*. The fault is injected at a random point in time after the first 5 seconds of the run and before the final 5 seconds of the run according to a uniform distribution. There is a 50% probability that a fault affects both treels instead of only one of the treels, and faults of the type *stuck-at-zero* and *stuck-at-constant* are equally likely to occur. A fault of the type *stuck-at-zero* means that the affected treel effectively blocks and stops to move. A *stuck-at-constant* fault means that the speed of the motor controlling a treel is set to a random value and the motor ceases to respond to control signals sent from the control program.

Each run consists of 1000 control cycles (equivalent to 150 seconds) and for each control cycle the sensory inputs, control signals, and the current fault state are recorded. The data sets are partitioned into two subsets, one consisting of data from 40 runs, which is used for training; and one consisting of the data from the remaining 20 runs, which is used for performance evaluation. After a network has been trained by a batch learning back-propagation algorithm, it is evaluated on data from one evaluation run at a time. The output of the network is recorded and compared to the fault state.

Fault detection is a binary classification problem and, in this study, we present results for threshold-based classification. If the TDNN outputs a value lower than the threshold we interpret it as a *no fault* classification, whereas we interpret outputs equal to or above the threshold as a *fault* classification. The interpreted output is compared to the correct output and fault detectors are scored on their *latency* and the number of *false positives* they produce.

Latency refers to the span of time from the moment a fault is injected until it is detected. False positives refers to the number of control cycles for which a fault detector is wrongly classifying the state as a *fault*. For this study, we have chosen the threshold to be 0.75. If we were to choose a lower threshold, we would expect a lower latency but more false positives. Similarly, if we were to choose a higher threshold we would expect a higher latency and fewer false positives.

6 Results

Box-plots of the latency and the false positive results from the 20 evaluation runs are shown in Fig. 4. Each box comprises observations ranging from the first to the third quartile. The median is indicated by a bar, dividing the box into the upper and lower part. The whiskers extend to the farthest data points that are within 1.5 times the interquartile range. Outliers are shown as dots. Each sample point corresponds to the results of a single evaluation run. In the figure we have plotted results for the *follower* performing endogenous fault detection and for the *leader* performing exogenous fault detection during the same runs. The median latency for the *follower* performing endogenous fault detection is

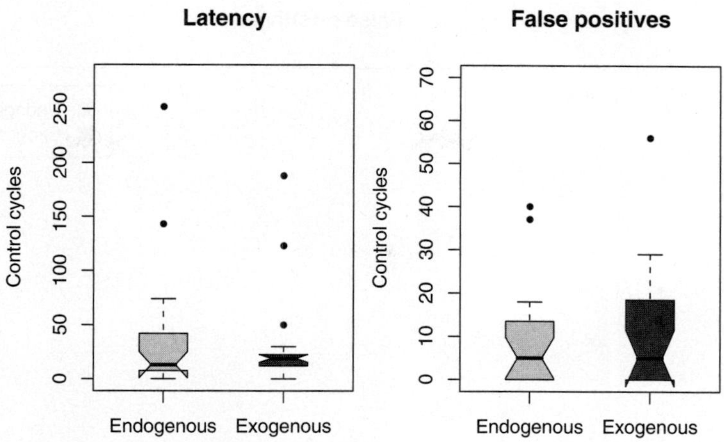

Fig. 4. Box-plots of the performance results using threshold-based classification for the *follower* performing endogenous fault detection and for the *leader* performing exogenous fault detection during the same runs.

14 control cycles, while the median latency for the *leader* performing exogenous fault detection is 19 control cycles. This difference of 5 control cycles corresponds to 750 ms. The median number of false positives is 5 control cycles for both the endogenous and exogenous detection. In every trial, the fault injected was detected.

The results indicate that the *leader* robot is capable of detecting faults injected in the *follower*. Furthermore, the performance of the exogenous fault detection is comparable to the performance of the endogenous fault detection performed by the *follower*. The latency is, however, slightly higher for exogenous fault detection.

In order to obtain a fault detector that produces fewer false positives, we implemented a threshold-based classification scheme based on the moving average of the TDNN output value. Hence, instead of interpreting the output of the TDNN directly, as above, a number of past values are stored, and the fault classification is based on the average of those values. Fig. 5 shows the false positive results for the different lengths of the moving average window. For moving average windows up to 10 control cycles, false positives occur in several trials. For longer windows, false positives are only observed in one or two trials. When a moving average window of length 50 is used, the exogenous fault detector produces no false positives.

For window lengths of 20-50, false positives for the *follower* performing endogenous fault detection occurred in one of the 20 trials. The endogenous fault detector produced 173 false positives with a window length of 20 control cycles and 128 false positives with a window length of 50 cycles for the trial in ques-

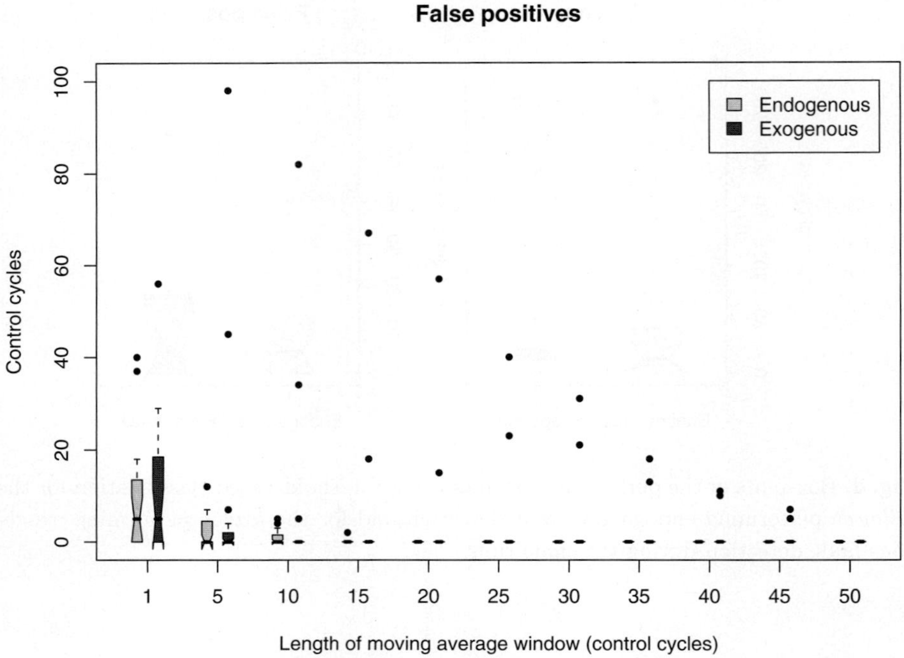

Fig. 5. False positives results for moving average threshold-based classification.

tion. These results are not shown in Fig. 5, since they are outside the scale of the figure.

When the moving average is used, the latencies are increased by the length of the moving average window (results are not shown). Thus, introducing a threshold-based classification scheme based on the moving average of the TDNN output value can remove nearly all false positives, but this comes at the cost of a higher latency.

7 Conclusion

In this study, we synthesized fault detection components for exogenous faults using fault injection and supervised learning. Our synthesizing technique can be used for both endogenous and exogenous fault detection. We achieved this transparency between endogenous and exogenous faults by avoiding the use of any explicit modelling. The method's only requirement is that a fault must create a detectable change in the detecting robot's flow of sensory input. Our method proved effective on real robots; one robot was able to detect faults injected in another robot without the use of explicit communication or any specific behavior written to aid fault detection.

Many faults cannot be detected endogenously. We therefore believe that the use of exogenous fault detection in combination with endogenous fault detection has the potential to significantly enhance the reliability and robustness of multi-robot systems. We are currently investigating the scalability of the proposed approach. One consideration is how to apply exogenous fault detection to systems with larger numbers of robots. When multiple robots with limited sensory range interact, it is unclear how a set of observations from one robot should be correlated with the fault state of other robots that may or may not be within sensory range. We are also investigating how larger numbers and a broader range of faults can be detected.

Acknowledgements. This work was supported by the *SWARMANOID* project, funded by the Future and Emerging Technologies programme (IST-FET) of the European Commission, under grant IST-022888. Anders Christensen acknowledges support from COMP2SYS, a Marie Curie Early Stage Research Training Site funded by the European Community's Sixth Framework Programme (grant MEST-CT-2004-505079). The information provided is the sole responsibility of the authors and does not reflect the European Commission's opinion. The European Commission is not responsible for any use that might be made of data appearing in this publication. Marco Dorigo acknowledges support from the Belgian FNRS, of which he is a Research Director. This research was supported by the ANTS project, an *Action de Recherche Concertée* funded by the Scientific Research Directorate of the French Community of Belgium.

References

1. Christensen, A.L., O'Grady, R., Birattari, M., Dorigo, M.: Automatic synthesis of fault detection modules for mobile robots. In: Proc. of the NASA/ESA Conf. on Adaptive Hardware and Systems (AHS-2007), MIT Press, Cambridge (In press, 2007)
2. Isermann, R., Ballé, P.: Trends in the application of model-based fault detection and diagnosis of technical processes. Control Engineering Practice 5(5), 709–719 (1997)
3. Gertler, J.J.: Survey of model-based failure detection and isolation in complex plants. IEEE Control Systems Magazine 8, 3–11 (1988)
4. Vemuri, A.T., Polycarpou, M.M.: Neural-network-based robust fault diagnosis in robotic systems. IEEE Trans. on Neural Networks 8(6), 1410–1420 (1997)
5. Terra, M.H., Tinos, R.: Fault detection and isolation in robotic manipulators via neural networks: A comparison among three architectures for residual analysis. Journal of Robotic Systems 18(7), 357–374 (2001)
6. Patton, R.J., Uppal, F.J., Lopez-Toribio, C.J.: Soft computing approaches to fault diagnosis for dynamic systems: A survey. In: Proc. of 4th IFAC Symposium on Fault Detection supervision and Safety for Technical Processes, vol. 1, pp. 298–311. Elsevier, Oxford (2000)
7. Roumeliotis, S.I., Sukhatme, G.S., Bekey, G.A.: Sensor fault detection and identification in a mobile robot. In: Proc. of IEEE/RSJ Inter. Conf. on Intelligent Robots and Systems, vol. 3, pp. 1383–1388. IEEE Computer Society Press, Los Alamitos (1998)

8. Goel, P., Dedeoglu, G., Roumeliotis, S.I., Sukhatme, G.S.: Fault detection and identification in a mobile robot using multiple model estimation and neural network. In: Proc. of IEEE Inter. Conf. on Robotics and Automation, ICRA'00, vol. 3, pp. 2302–2309. IEEE Computer Society Press, Los Alamitos (2000)
9. Dearden, R., Hutter, F., Simmons, R., Thrun, S., Verma, V., Willeke, T.: Real-time fault detection and situational awareness for rovers: Report on the Mars Technology Program Task. In: Proc. of IEEE Aerospace Conf., vol. 2, pp. 826–840. IEEE Computer Society Press, Los Alamitos (2004)
10. Verma, V., Gordon, G., Simmons, R., Thrun, S.: Real-time fault diagnosis. IEEE Robotics & Automation Magazine 11(2), 56–66 (2004)
11. Li, P., Kadirkamanathan, V.: Particle filtering based likelihood ratio approach to fault diagnosis in nonlinear stochastic systems. IEEE Trans. on Systems, Man and Cybernetics, Part C 31(3), 337–343 (2001)
12. Lewis, M.A., Tan, K.H.: High precision formation control of mobile robots using virtual structures. Autonomous Robots 4(4), 387–403 (1997)
13. Parker, L.E.: ALLIANCE: an architecture for fault tolerant multirobot cooperation. IEEE Trans. on Robotics and Automation 14(2), 220–240 (1998)
14. Gerkey, B.P., Matarić, M.J.: Sold!: auction methods for multirobot coordination. IEEE Trans. on Robotics and Automation 18(5), 758–768 (2002)
15. Gerkey, B., Matarić, M.J.: Pusher-watcher: An approach to fault-tolerant tightly-coupled robot coordination. In: Proc. of IEEE Inter. Conf. on Robotics and Automation, ICRA'02, pp. 464–469. IEEE Computer Society Press, Piscataway (2002)
16. Dias, M.B., Zinck, M.B., Zlot, R.M., Stentz, A.: Robust multirobot coordination in dynamic environments. In: Proc. of IEEE Inter. Conf. on Robotics and Automation, ICRA'04, vol. 4, April 2004, pp. 3435–3442. IEEE Computer Society Press, Piscataway (2004)
17. Arlat, J., Aguera, M., Amat, L., Crouzet, Y., Fabre, J.C., Laprie, J.C., Martins, E., Powell, D.: Fault injection for dependability validation: A methodology and some applications. IEEE Trans. on Software Engineering 16(2), 166–182 (1990)
18. Mondada, F., Gambardella, L.M., Floreano, D., Nolfi, S., Deneubourg, J.-L., Dorigo, M.: The cooperation of swarm-bots: Physical interactions in collective robotics. IEEE Robotics & Automation Magazine 12(2), 21–28 (2005)
19. Groß, R., Bonani, M., Mondada, F., Dorigo, M.: Autonomous self-assembly in swarm-bots. IEEE Trans. on Robotics 22(6), 1115–1130 (2006)
20. O'Grady, R., Groß, R., Mondada, F., Bonani, M., Dorigo, M.: Self-assembly on demand in a group of physical autonomous mobile robots navigating rough terrain. In: Capcarrère, M.S., Freitas, A.A., Bentley, P.J., Johnson, C.G., Timmis, J. (eds.) ECAL 2005. LNCS (LNAI), vol. 3630, pp. 272–281. Springer, Heidelberg (2005)
21. Dorigo, M., Trianni, V., Şahin, E., Groß, R., Labella, T.H., Baldassarre, G., Nolfi, S., Deneubourg, J.-L., Mondada, F., Floreano, D., Gambardella, L.M.: Evolving self-organizing behaviors for a swarm-bot. Autonomous Robots 17(2–3), 223–245 (2004)
22. Waibel, A., Hanazawa, T., Hinton, G., Shikano, K., Lang, K.J.: Phoneme recognition using time-delay neural networks. IEEE Trans. on Acoustics, Speech, and Signal Processing 37, 328–339 (1989)
23. Clouse, D.S., Giles, C.L., Horne, B.G., Cottrell, G.W.: Time-delay neural networks: Representation and induction of finite-state machines. IEEE Trans. on Neural Networks 8, 1065–1070 (1997)

From Artificial Societies to New Social Science Theory

Eric Silverman and John Bryden

School of Computing, University of Leeds, Leeds LS2 9JT
{erics, johnb}@comp.leeds.ac.uk

Abstract. We identify two distinct themes in social science modelling. One, more specific, approach is that of social simulation which addresses how behaviour of many actors can lead to emergent effects. We argue that this approach, while useful as a tool in social science policy development, is fundamentally constrained due to the fact that its models are developed within the society they are supposed to model. Alternatively, the second theme looks to take a more holistic view by taking inspiration from systems sociology. This approach looks to build societies from the bottom up and may allow us to generate new perspectives in social theory.

1 Introduction

Understanding and predicting the future of human society is a problem of staggering complexity. Our level of understanding of the basic unit of society, the human being, is far from complete, let alone the collective interactions and transmissions of information between the 6 billion individuals that live on our planet. The mind boggles over the potentials for chaotic effects and intractable non-linearities in this vast system.

That said, the problem of how life has evolved on that same planet would seem also to be one of insurmountable complexity if it weren't for the theories of Charles Darwin [1]. Perhaps there is hope for a theory in social science of similar explanatory power to evolution: one that does not fully explain society, but provides us with a holistic framework to push forward our understanding of society – in a similar way that evolution does for biology. This is the problem at hand, the development of a *fundamental theory of social science.*

The only sensible way we may approach such a problem is by using the best tools available in the most appropriate manner. We propose that computer modelling gives us the freedom of abstraction and the power of the study of non-linear mechanisms that will be vital for this endeavour.

This paper looks at the current literature surrounding social simulation to attempt to formulate the role computer simulation should play in working towards fundamental theories of social science. From our review, we outline and

F. Almeida e Costa et al. (Eds.): ECAL 2007, LNAI 4648, pp. 565–574, 2007.

distinguish two main themes in the computer modelling of social systems: *social simulation* and *systems sociology*. One key common feature of both these modelling themes is that both focus on producing emergent macro-level effects.

The distinction we develop between social simulation and systems sociology is based on the application and approach of these two themes. The application of social simulation models is through the guidance of social policy. Such models are often linked, or intended to be linked, to empirical data. As such social simulation models are bottom up in their application. Systems sociology models however are top down in their application – they are intended to push forward social science and its paradigms.

When we consider the approach of the two themes, they take the opposite roles to those taken in their application. Social simulation models take a top down approach on human behaviour, identifying a specific class of behaviour and predicting macro-level patterns. We argue that it is the very *specificity* of this approach that restricts its usefulness in creating a new social theory. In contrast, systems sociology modelling takes a bottom up view, starting with building blocks of social behaviour and attempts to form meaningful and recognisable social structure. Unlike the social simulation approach, the building blocks are not defined with any predetermined interactions.

In many ways this looks analogous to the perspectives of evolutionary biology and artificial life – with evolutionary biology concerning itself with microevolution in specific biological systems and artificial life concerning itself with macroevolutionary theory from a more general perspective. However, evolutionary theory was founded on a generalisation from observances from specific systems from the Galapagos islands [1]. We argue that specific social systems are not likely to generalise into a fundamental theory for social science in this way. Being human beings, and therefore part of the system being modelled, we are not able to look at our society from the outside in the way we may look at nature. We propose that we require the input from grown artificial societies to give us new perspectives on our own society.

The next section introduces key literature in the field of social simulation and criticises social simulation from the perspective outlined in this Introduction. Following on from this a new perspective derived from systems sociology is introduced. Finally the two themes are compared in more detail before the paper is concluded.

2 Social Simulation

This section on the social simulation stream focuses mainly on the work of two seminal social science modellers – Schelling and Cederman. It starts by critiquing a famous case study: Schelling's model of residential segregation and its impact upon the development of social simulation. Following this, we consider a broader picture of the stream looking at Cederman's three types of social simulations. Finally, we look at some of the relevant criticisms of social simulation seen in the current literature.

2.1 Case Study: Schelling's Segregation Model

Schelling's well-known 'chequerboard model' of residential housing segregation has proves an interesting example of the power of the social simulation theme in certain aspects of the social sciences[2]. He sought to confirm his suspicion that if the racial makeup of a given area was critical to an individual's choice of housing, then even populations of individuals tolerant to mixed-race environments would end up segregating into single-race neighbourhoods. To this end, he devised a simplistic model of two populations placed on a chequerboard, with each agent on the board given a very simple rule: if the number of agents of a different 'race' in its immediate eight-cell neighbourhood exceeded a certain critical threshold, the agent would become 'unhappy' and move; otherwise, the agent would remain in its current position.

Schelling's model supported his initial hypothesis, proving that even if the agents were given a high level of tolerance for other agents of a different 'race,' the agents would end up segregating themselves into separate neighbourhoods. The model also illustrated a deeper point: large-scale factors such as socio-economic or educational differences between different ethnic populations could not explain the generally puzzling phenomenon of residential segregation. Indeed, without a greater insight into the preferences and thought processes of individuals within a given population (their 'micromotives'), some critical aspects of this social problem may elude the researcher.

Despite its apparent simplicity, Schelling's model provides an important illustration of the potential power of the social simulation theme. The chequerboard model provides only the barest of motivations for each agent in the simulation, and yet the results are very applicable to more complex theories of the residential segregation phenomenon. As he states, a model 'can be a precise and economical statement of a set of relationships that are sufficient to produce the phenomena in question, or, a model can be an actual biological, mechanical, or social system that embodies the relationships in an especially transparent way, producing the phenomena as an obvious consequence of these relationships' [3][p. 87]. Schelling's model shows the power of this transparency, demonstrating a clear relationship between the simple micromotives driving an agent's movement between neighbourhoods and the resulting segregation.

Perhaps most importantly, Schelling's model succeeds due to its obvious qualitative similarity to the problem of residential segregation: one can see the effect of varying levels of tolerance in these agents by simple visual examination, as the agents begin to cluster into separate neighbourhoods of their own accord. While Cederman and many others acknowledge that social simulation models can be difficult to analyse, models of this type which illustrate important social concepts simply could in fact be much easier to effectively 'analyse' through these qualitative similarities than more detailed models which aim for social realism. In fact one might argue that such a perspective echoes the techniques of much of social science.

While Schelling's model was accepted as a remarkable illustration of a simple principle regarding the large-scale effects of individual housing preferences, some modellers sought to create a more 'realistic' Schelling-type model which could incorporate socio-economic factors as well[4,5,6]. Given the accepted complexity of residential segregation as a social problem, and the new insight into the effects of individual preference illuminated by Schelling, models incorporating both Schelling-style 'micromotives' and large-scale social factors were seen as a potential method for examining the interactions between these two levels of social structure, something that was very much lacking in Schelling's original formulation.

However, even increasing the complexity of Schelling's rather elegant portrayal of this particular societal problem cannot solve a fundamental problem of this approach. Human society by its very nature is a complex system, incorporating as it does the collective interactions and transmission of knowledge of many millions of individual actors in concert. Those interactions and retained societal knowledge can in turn influence members of the society at large, and as a society continues to evolve its past history can exert a tremendous influence on its collective future decisions, making human society heavily path-dependent (see[7] for one example of this phenomenon).

With this in mind, while agent-based models following Schelling's example can provide remarkable insight into the emergence of social phenomena within a certain limited scope, models of this type are inherently limited in their theoretical application by their specificity. As we move into our discussion of Cederman's contribution to social simulation, we will examine these potential methodological difficulties in greater detail.

2.2 Cederman's Three Types of Social Simulation

Lars-Erik Cederman has been one of the most prominent proponents of agent-based modelling within the political sciences. His own work appears to have substantial roots in the social simulation mould described above, with a more detailed focus on specific aspects of modern political structures [8]. Far from confining himself to a singular mode of enquiry, however, Cederman has developed a theoretical framework of three types of social models in an attempt to outline how a number of potential methods might contribute to the development of social science as a whole [9].

Type 1 models in Cederman's framework focus on behavioural aspects from human society. By implementing a behavioural aspect at the micro-level in the model, macro-level, or emergent, conclusions can be drawn. For example, Axelrod's work has aimed to show the emergence of cooperation, and by using the iterated prisoner's dilemma showed that cooperation is possible in social settings as long as the interactions of the involved agents are iterated and that agents are capable of remembering previous actions [10].

Cederman identifies Type 2 models as those which attempt to explain the emergence of particular configurations in a model due to properties of the agents involved[9]. Cederman's later simulations, for example, focused on issues such as nationalist insurgency and state formation and stability using a framework in which agents within the simulation were assigned specific properties that influenced the resultant distribution of resources and information [11,12]. Ian Lustick's Agent-Based Identity Repertoire Model (ABIR) in which agents were provided with potential cultural 'identities' similarly aims to demonstrate the emergence of clusters of these identities through properties of the agents and the changes resulting from their local interactions [13].

Finally, Cederman identifies Type 3 models as being perhaps the most ambitious: these models attempt to model both the agents themselves and their interaction networks as emergent features of the simulation[9]. He cites the field of artificial life as one likely to inform this area of computational work in political science, given that ALife focuses on such emergent features. He also acknowledges that some overlap can occur these three types of models; for example, a Type 1 model may stray into Type 3 territory by allowing agents more latitude in choosing interaction partners, thus allowing for an interaction network to emerge from the system. He argues that such models may provide very powerful tools for the political scientist, allowing for potentially profound conclusions to be drawn regarding the development of political institutions. This approach does seem most methodologically difficult, however, as the already significant abstractions necessary to create Type 1 or Type 2 models must be relaxed even further to allow for such ambitious examinations of emergent features in social systems.

2.3 Criticisms of Social Simulation

From our perspective of attempting to develop a social theory, we present some criticisms of social simulation. These help to elucidate the path we expect social simulation to take in the future. We can then put that into context within our main distinction.

Criticism of agent-based models in political science has come from a number of different areas, but many of those criticisms focus on the difficulty of making sensible abstractions for social and political structures within such a model. One can view potentially all of scientific enquiry as reflecting the inherent biases of the experimenter, and this problem of inherent theory-dependence is even more acute when constructing simulation models [14]. While even abstract ALife models may reference the real-life behaviour of natural biological systems, and the wealth of related empirical data, political models do not necessarily have that luxury. Highly complex social structures and situations, such as Cederman's models of nationalist insurgency and civil war [11,12] involve further layers of abstraction, often involving factors which do not immediately lend themselves to quantification, such as cultural and national identities.

In addition, sociological theory is notoriously difficult to formalise, incorporating as it does a number of both higher- and lower-level cognitive and behavioural interactions. In fact, sociological theory may not benefit from the micro/macro distinction of levels of analysis that benefits researchers of evolution and other large-scale processes [15]. These interacting social levels cannot be clearly differentiated into a hierarchy of processes, making simulation a very difficult, and highly theory-dependent, exercise.

Of course, with such complex phenomena contributing to the results of a social simulation, finding an appropriate method of analysing those results becomes a very real concern for the modeller. Donald Sylvan's review of Cederman's Emergent Actors demonstrates this common complaint levelled at agent-based models in the social sciences; he urges readers to note carefully the lack of data in its conventional sense throughout Cederman's work [16].

As Sylvan points out, Cederman's 'data' only relate to the interactions of virtual states in an idealised grid-world; applying such data to real-life political events or transitions seems suspect at best. The levels of complexity at work in large-scale political events may be very difficult to capture in an agent-based model, and knowing when to draw a specific conclusion from a model of such an inherently difficult-to-analyse situation is a thorny issue.

While there is a potential for social simulations to illuminate the emergence of certain properties of political structures, the difficulty in connecting these abstracted simulations to real-world political systems is significant. Weidmann and Gerardin[17], with their GROWLab simulation toolkit, have attempted to sidestep these concerns by making their framework compatible with GIS (geographic information system) data, in order to allow 'calibration with empirical facts to reach an appropriate level of realism.'

Putting this into context, the need for an empirical perspective inherently restricts the scope of social simulation models. While such a perspective allows for models to have potential relevance to more conventional means of social-science enquiry, this modelling methodology still exists within a preconceived framework of societal interaction, lacking the more external perspective required to develop a fundamental social theory.

Out of Cederman's 3 modelling types, Type 3 models appear to be the most likely to provide a means to draw broader conclusions about the development of societies given that the agents are much less constrained in their interactions than in Type 1 and Type 2 methodologies. However, even models of this type begin with fundamental assumptions: about how these agents will behave, and what sort of interactions might develop. Still we find ourselves striving for a more fundamental perspective, one further removed from an artificially-imposed social framework. We look to systems sociology.

3 Systems Sociology

As we have seen, the advent of social simulation has proved influential in the social sciences, provoking new questions regarding the origin and nature of society. While Schelling's segregation model demonstrates the potential impact of social simulation, it also illustrates the inherent difficulties involved in generalising the conclusions drawn from a social simulation. More generalised models of society as alluded to by Cederman's Type 3 categorisation may provide a means for investigating aspects of society which elude the empirical data-collector and in turn inform our search for a fundamental social theory, but in order for this to occur we need to establish a method of examining society on a broad theoretical scale through simulation.

The well-known social systems theory of Niklas Luhmann provides one example of an attempt to develop an understanding of the foundations for social behaviour. Luhmann classifies social systems as systems of communication which attempt to reduce complexity by presenting only a fraction of the total available information [18].

One of the fundamental issues facing the systems sociology theorist is solving the problem of double contingency, an issue Luhmann describes as central to the development of social order. Put simply, if two entities meet, how do they decide how to behave without a pre-existing social order to govern their actions? How might these entities decide to develop a common means of interaction, and through those interactions develop a shared social history?

As Dittrich, Kron and Banzhaf describe [19], Luhmann described a method for resolving this contingency problem which was far more elemental than previous approaches, relying as it does on 'self-organisation processes in the dimension of time' rather than through more standard social processes. The entities in question would perform initial contingency-reducing actions during an encounter to allow for each to develop an understanding of the expectations of each party in the interaction.

In Luhmann's view, the social order develops as a consequence of these contingency-reducing actions on a large scale. As elements of the developing society develop their expectations about the social expectations of others (described as 'expectation-expectation' by Luhmann), a system of social interaction develops around this mutual social history. This system then produces social institutions which can further influence the development of the social order. These social institutions perform a similar function by reducing the amount of information disseminated amongst the members of a society, essentially providing contingency-reducing services on a much larger scale.

Agent-based models in the context of ALife have certainly proved useful in the examination of other autopoietic systems; however, recent attempts to formalise Luhmann's theories into a usable model, while producing interesting results, have highlighted the inherent difficulties of encapsulating the many disparate elements of Luhmann's theories of social systems into a single model [20].

4 Evaluation

Given the outlines of social simulation and systems sociology, this section aims to examine the specific contrasts between these two approaches. By comparing the methods and aims of these methods, we posit that the systems sociology approach presents a promising avenue for developing fundamental social theory through simulation.

4.1 Social Simulation vs. Systems Sociology

As we can see from Luhmann's analysis, while there may indeed be a lack of 'data' inherent to the study of artificial societies, there still exists a theoretical framework for understanding the fundamental mechanisms which drive the creation of a larger social order. While some social simulation researchers may seek to strengthen their models through establishing direct connections with empirically-collected data from social science, the systems sociology perspective could provide a different path to more useful examinations of human society.

The social simulation stream is oriented towards specific elements of social behaviour: simulations of cooperation [21], nationalist insurgency [8] or the spatial patterning of individuals or opinions within a society [13]. Social simulation's stronger links with empirical data may make validation of such models much easier, but further restricts the domain of those models to focus on social problems for which usable data exists. Given the difficulties inherent in collecting social science data, these problems tend to be a subset of those social problems for which models could prove potentially illuminating.

It is the very restriction into particular domains which prevents the social simulation approach from reaching a more general perspective. It is constrained by approaching social phenomena from the top down. These actual top-down approaches are necessarily rooted in the societies they model. In essence, looking for a feature in society and then attempting to reproduce it in a model is not sufficient to develop a fundamental theory.

In contrast, the systems sociology stream abstracts outside of the standard view of society. Luhmann's perspective aims to describe interactions which can lead to the development of social order, in a sense examining the development of human society through an 'outside perspective.' Luhmann essentially moves beyond standard sociology, attempting to describe what occurs prior to the existence of social order, rather than operating within those bounds as with social simulation. By viewing society from its earliest beginnings, prior to the existence of any societally-defined modes of interaction and communication, the systems sociology approach hopes to develop a theoretical understanding of the fundamental behavioural characteristics which lead to the formation of social order. In many ways this approach is reminiscent of the ALife approach to modelling 'life-as-it-could-be'[22]; the systems sociology perspective leads us to examine *society-as-it-could-be.*

4.2 The Philosophical Difficulties of Systems Sociology

While the external and abstract perspective offered by systems sociology is appealing given our desire to develop a fundamental social theory, removing ourselves so completely from the conventional empirical perspective carries its own pitfalls. Of course, given that we would not be constructing models with the specificity inherent in the social simulation approach, concerns related to validation and connection with empirically-collected data would not be of any great import.

However, as Silverman and Bullock describe, developing such broad theories through simulation can carry with it a great deal of philosophical baggage, as in the case of strong ALife [14]. The strong ALife perspective, in which modellers justify their examinations of digital organisms by maintaining that those digital organisms represent a new form of life in the digital substrate, could provoke difficult questions for the systems sociology simulator. Are these artificial societies mere models of societies, or are they bona fide digital cultures in their own right? Are these simulations worthy of attention, or are they merely representational oddities with no direct relation to the development of human society?

Within the systems sociology perspective, however, this modelling approach seeks not to create society but to replicate its beginnings. By allowing agents to construct their own social order, we may be able to observe the micro-level interactions which lead to macro-level social constructs, and do so without a pre-existing framework that guides those initial forays into communication. Without question this approach lends itself to difficult philosophical questions, and given the complete separation from conventional empirical social science these questions are even more significant. However, fundamentally this approach brings us full-circle once more to Schelling: we seek models which display a qualitative similarity to our current understanding of social order. If our simplistic computational agents can develop characteristics reminiscent of a society, then we move one step closer to understanding the fundamentals of human society, in the same fashion that Schelling's model provided a novel view of the problem of residential segregation.

5 Conclusion

In this paper we have reviewed the state of the art literature in the field of social science modelling. Distinguishing between social simulation and systems sociology has illuminated the shortcomings of social simulation for developing broad theories in social science.

While the production of a fundamental theory of social science is clearly a long way away, the approaches developed through the systems sociology approach should at least be able to play an instructive or forming role in the development of social theory. Computer models that are at first very complex are still mathematical constructs and may well be simplified to explicit mathematical models which form competent explanations [23]. By taking this bottom-up approach of building new societies, while maintaining a top-down motivation, new directions can emerge in future modelling endeavours.

Acknowledgments. Thanks to Takashi Ikegami and David Hogg.

References

1. Darwin, C.: The Origin of Species by Means of Natural Selection. John Murray, London (1859)
2. Schelling, T.C.: Dynamic models of segregation. Journal of Mathematical Sociology 1, 143–186 (1971)
3. Schelling, T.: Micromotives and macrobehavior. W.W. Norton, New York (1978)

4. Sander, R., Schreiber, D., Doherty, J.: Empirically testing a computational model: The example of housing segregation. In: Sallach, D., Wolsko, T. (eds.) Proceedings of the Workshop on Simulation of Social Agents: Architectures and Institutions, pp. 108–115 (2000)
5. Clark, W.A.V.: Residential preferences and neighborhood racial segregation: A test of the schelling segregation model. Demography 28, 1–19 (1991)
6. Epstein, J.M., Axtell, R.L.: Growing Artificial Societies: Social Science from the Bottom Up. MIT Press, Cambridge (1996)
7. Arthur, W.B.: Competing technologies, increasing returns, and lock-in by historical events. Journal of Economics 116, 99 (1989)
8. Cederman, L.E.: Emergent actors in world politics: how states and nations develop and dissolve. Princeton University Press, Princeton (1997)
9. Cederman, L.E.: Agent-based modeling in political science. The Political Methodologist 10, 16–22 (2001)
10. Axelrod, R.: The Evolution of Cooperation. Basic Books, New York (1984)
11. Cederman, L.E., Girardin, L.: Beyond fractionalization: Mapping ethnicity onto nationalist insurgencies. Paper read at Disaggregating the Study of Civil War and Transnational Violence Conference, at San Diego, CA (2005)
12. Cederman, L.E.: Articulating the Geo-Cultural Logic of Nationalist Insurgency. In: Order, Conflict, and Violence, Cambridge University Press, Cambridge (2006)
13. Lustick, I.: Defining violence: a plausibility probe using agent-based modeling. Paper prepared for LiCEP, Princeton University, May 12-14 (2006)
14. Silverman, E., Bullock, S.: Empiricism in artificial life. In: Pollack, J., Bedau, M., Husbands, P., Ikegami, T., Watson, R.A. (eds.) Artificial Life IX: Proceedings of the Ninth International Conference on the Simulation and Synthesis of Living Systems, pp. 534–539. MIT Press, Cambridge (2004)
15. Klüver, J., Stoica, C., Schmidt, J.: Formal models, social theory and computer simulations: Some methodical reflections. Journal of Artificial Societies and Social Simulation 6(2), 8 (2003), `http://jasss.soc.surrey.ac.uk/6/2/8.html`
16. Sylvan, D.: modeling the rise and fall of states', review of emergent actors in world politics by lars-erik cederman. Mershon International Studies Review 42, 377–379 (1998)
17. Weidmann, N.B., Girardin, L.: GROWLab: A Toolbox for Social Simulation. Paper presented at the First World Congress on Social Simulation, August 21-25, 2006 in Kyoto, Japan (2006)
18. Luhmann, N.: Social Systems. Stanford University Press, Stanford (1995)
19. Dittrich, P., Kron, T., Banzhaf, W.: On the scalability of social order: Modeling the problem of double and multi contingency following luhmann. Journal of Artificial Societies and Social Simulation 6(1), 3 (2003), `http://jasss.soc.surrey.ac.uk/6/1/3.html`
20. Fleischmann, A.: A model for a simple luhmann economy. Journal of Artificial Societies and Social Simulation 8(2)4 (2005), `http://jasss.soc.surrey.ac.uk/8/2/4.html`
21. Axelrod, R.: The Complexity of Cooperation. Princeton University Press, Princeton (1997)
22. Langton, C.: Artificial life. In: Langton, C. (ed.) Artificial Life, pp. 1–47. Addison-Wesley, Reading (1989)
23. Bryden, J., Noble, J.: Computational modelling, explicit mathematical treatments, and scientific explanation. In: Artificial Life X, pp. 520–526 (2006)

From Solitary to Collective Behaviours: Decision Making and Cooperation

Vito Trianni[1], Christos Ampatzis[2], Anders Lyhne Christensen[2], Elio Tuci[2], Marco Dorigo[2], and Stefano Nolfi[1]

[1] LARAL, ISTC, CNR, Italy {vito.trianni,stefano.nolfi}@istc.cnr.it
[2] IRIDIA, CoDE, ULB, Belgium {campatzi,alyhne,etuci,mdorigo}@ulb.ac.be

Abstract. In a social scenario, establishing whether a collaboration is required to achieve a certain goal is a complex problem that requires decision making capabilities and coordination among the members of the group. Depending on the environmental contingencies, solitary actions may result more efficient than collective ones and vice versa. In robotics, it may be difficult to estimate the utility of engaging in collaboration versus remaining solitary, especially if the robots have only limited knowledge about the environment. In this paper, we use artificial evolution to synthesise neural controllers that let a homogeneous group of robots decide when to switch from solitary to collective actions based on the information gathered through time. However, being in a social scenario, the decision taken by a robot can influence—and is influenced itself—by the status of the other robots that are taking their own decisions at the same time. We show that the simultaneous presence of robots trying to decide whether to engage in a collective action or not can lead to cooperation in the decision making process itself.

1 Introduction

Decision making is a complex problem for a collective robotic system, due to the necessity to reach a global consensus among the robots, which contrasts with the system's inherent decentralisation. Current approaches resort to biological inspiration [1,2,3] or to context-specific solutions [4,5]. The problem of deciding whether to switch between solitary and collective behaviours is much less studied. Such a problem is of fundamental importance for a particular robotic system: the *swarm-bot*, a swarm robotic artefact composed of a number of autonomous mobile robots—referred to as *s-bots*—which have the ability to connect to each other forming a physical structure, as shown in Fig. 1 (for more details, see [6]). Forming a *swarm-bot* by self-assembly is a collective action that can lead to great advantages: for example, the *swarm-bot* can overcome an obstacle impassable for solitary *s-bots* [7] or collectively transport a heavy item [8]. On the other hand, for tasks such as searching for a goal location or tracing an optimal path to a goal, a swarm of unconnected *s-bots* may be more efficient [9].

When should a group of *s-bots* assemble in a *swarm-bot*? This problem—referred to as *functional self-assembly* [10]—has been studied to date without particular focus on the decision making process that should lead to the

F. Almeida e Costa et al. (Eds.): ECAL 2007, LNAI 4648, pp. 575–584, 2007.

switch from individual to collective behaviours. The decision to self-assemble was based either on *a priori* assumptions or on clearly distinguishable environmental cues [8,7,10], which may reduce the adaptiveness of a solution and the efficiency of the system as a whole. We believe that a truly adaptive system should prove capable of autonomously extracting all the information it requires to solve a problem. In other words, the *s-bots* should be capable of recognising the necessity to self-assemble based only on the environmental contingencies they experience. Given the limited sensory range of each *s-bot*, the information relevant to autonomously decide whether to switch from a solitary to a collective behaviour is not ready-to-use, but should be constructed by the robots while they interact and accumulate experience about the environment in which they are placed. Moreover, being in a collective scenario, the actions of each *s-bot* can influence—and are influenced themselves—by the status of the other *s-bots*, which try to make their own decisions at the same time. This opens the way to cooperative solutions that can exploit not only the dynamical interactions among individuals, but also the way in which these interactions change over time. In this paper, we show how the adaptiveness of the robots' behaviour can be increased by an evolutionary process that favours through selection those solutions that improve the "fitness" of the robotic group. Here, we do not focus on assembly but we limit our study to the processes that should lead to the formation of a *swarm-bot*. We demonstrate how non-trivial individual and collective decision making processes can be efficiently obtained.

The work presented in this paper is based on previous studies about time-dependent decision making mechanisms [11,12]: robots had to categorise the environment in which they were placed, either by explicit signalling or by performing different actions in different environments. When a social scenario was considered, communication resulted in increased robustness of the categorisation [12]. In this paper, we advance by studying a collective behaviour—i.e., aggregation—as a consequence of the decision making process: robots are placed in two different environments and, according to the environmental contingencies they experience, they should perform the appropriate individual or collective action. From the observer—i.e., *distal*—point of view, this is yet another categorisation problem in which the robotic group faces a binary choice between two

(a) (b) (c)

Fig. 1. (a,b) Different views of an *s-bot*. (c) A *swarm-bot* in a indoor environment.

environment types. However, from the robot—i.e., *proximal*—point of view, the binary choice is to be performed between two different behavioural states: a solitary behaviour and a collective one. In the definition of the evaluation function, we emphasise the importance of evaluating the robots for their ability to switch between behavioural states (see Section 2.3). The obtained results show that a number of different strategies can be evolved to solve the given problem. Among these, we show that those solutions that exploit communication perform better, systematically achieving a consensus in the group and reducing the decision errors.

2 The Task

The path towards the evolution of neural controllers for functional self-assembly in a physical *swarm-bot* passes through the definition of the following experimental scenario. A group of *s-bots* is placed in an arena that is surrounded by some obstacles that *s-bots* cannot overcome individually. The arena may have a *way out*, that is, a passage through which a solitary *s-bot* can exit (see Figure 2a). However, an *s-bot* does not have the perceptual abilities to detect the *way out* from every location in the arena. Therefore, *s-bots* should first search for the *way out* and, if they do not find any as in Figure 2b, they should aggregate and self-assemble in order to collectively overcome the obstacles that surrounds the arena. As mentioned above, we consider in this paper only the first part of this scenario concerning the decision to switch from the individual behaviour of searching for the *way out* to the collective behaviour of aggregating in one place. The second part of the scenario concerning self-assembly is on-going work.

2.1 The *S-bot*

An *s-bot* is a small mobile autonomous robot with self-assembling capabilities, shown in Fig. 1a and b [6]. The main body is a cylindrical turret with a diameter of about 12 cm. The turret holds the gripper used for assembling with other *s-bots* and can be actively rotated with respect to the chassis. The traction system is composed of both tracks and wheels, and provides a differential drive motion.[3] Each *s-bot* is provided with many sensory systems, useful for the perception of the surrounding environment or for proprioception. In this paper, we make use of the four proximity sensors placed under the chassis—referred to as *ground sensors*—that can be used for perceiving the ground's grey level. Each robot is also equipped with an omni-directional camera and red LEDs distributed around the *s-bots*' turret. The circular image recorded by the camera is filtered in order to return the distance of the closest *s-bot* in each of four 90° sectors, up to a maximum distance of about 50 cm. In order to communicate with each other, *s-bots* are provided with a very simple sound signalling system, which can produce

[3] The experiments presented here are performed in simulation only. However, we already presented elsewhere the portability of similar controllers to the physical system [12].

a continuous tone with fixed frequency and intensity. When a tone is emitted, it is perceived by every robot in the arena, including the signalling *s-bot*. The tone is perceived in a binary way, that is, either there is someone signalling in the arena or there is no one.

2.2 Experimental Setup

Three *s-bots* are initially placed up to 25 cm from the centre of a boundless arena. The arena contains a circular band in shades of grey (inner radius: 1.0 m; outer radius: 1.2 m—see Fig. 2a,b). The outer border of the circular band is painted in black and simulates the presence of a trough/obstacle that the *s-bots* cannot overcome individually: the simulation is stopped whenever individual *s-bots* pass over the black border, and the trial is considered unsuccessful. The grey level of the circular band can be perceived by the *s-bots* only locally through the ground sensors, and it is meant to warn *s-bots* about the presence of the simulated trough/obstacle: the darker the ground colour, the closer the danger. The *s-bots* can be placed in two different environments: in environment A, the circular band is discontinuous—i.e., there is a *way out* through which the *s-bots* can exit (see the trajectories in Fig. 2a). In environment B, the *way out* is not present and therefore *s-bots* should aggregate after having searched for it (see the trajectories in Fig. 2b). The amplitude of the *way out* is randomly selected in each trial within the interval $[\pi/4, \pi/2]$.

Homogeneous groups of *s-bots* are controlled by artificial neural networks, whose parameters are set by an evolutionary algorithm. A single genotype is used to create a group of individuals with an identical control structure. Each *s-bot* is controlled by a continuous time recurrent neural network (CTRNN, see [13]) with a multi-layer topology, as shown in Fig. 2c. The neural network is composed of 9 input neurons ($N_{I,i}$) which are simple relay units, 3 output neurons ($N_{O,i}$) with a sigmoid transfer function, and 5 continuous time hidden

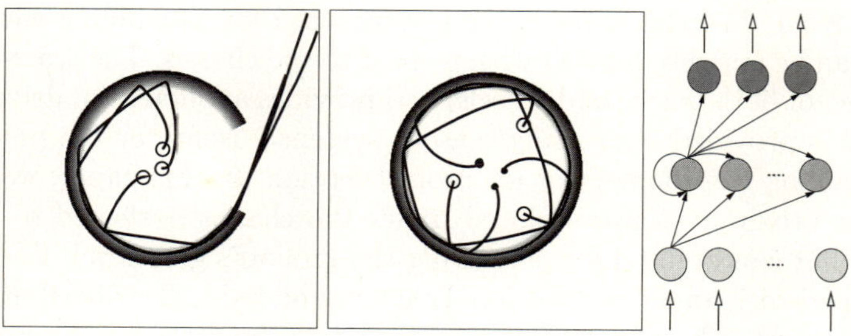

(a) (b) (c)

Fig. 2. (a,b) The experimental arena contains a circular band in shades of grey, which may or may not have the *way out*. Dark lines represent the trajectory of three *s-bots*, and the starting position on the trajectories is indicated by empty circles. (c) The three-layer architecture of the neural controller. The hidden layer is composed of continuous time neurons with fully recurrent connections.

neuron ($N_{H,i}$). Four inputs take values from the camera, four from the ground sensors and one from sound perception, while two outputs control the wheels and one controls the sound signal. The weights of the synaptic connections between neurons, the bias terms and the decay constants of the hidden neurons are genetically encoded parameters, optimised through a simple generational evolutionary algorithm [14]. The population contains 100 genotypes that are evolved for 5000 generations. Each genotype is a vector of 98 real values (85 synaptic connections, 5 decay constants and 8 bias terms). Subsequent generations are produced by a combination of selection with elitism and mutation. Recombination is not used. At every generation, the best 20 genotypes are selected for reproduction, and each generates 4 offspring. The genotype of the selected parents is copied in the subsequent generation; the genotype of the 4 offspring is mutated with a 50% probability of adding a random Gaussian offset to each real-valued gene.[4]

2.3 The Evaluation Function

During evolution, a genotype is mapped into a control structure that is cloned and downloaded onto all the *s-bots* taking part in the experiment. The fitness of a genotype is the average performance of a group of three *s-bots* evaluated over ten trials—five performed in environment A and five in environment B.[4] Each trial lasts 65 seconds and differs from the others in the initialisation of the random number generator, which influences mainly the *s-bots* starting positions and orientations, and the amplitude of the *way out*, if present. As mentioned above, robots should make a binary choice between two behavioural states: (i) searching for the *way out* and moving away from the arena centre—hereafter called solitary state $\mathcal{S}$—or (ii) aggregating with the other *s-bots*—hereafter called collective state $\mathcal{C}$. The performance of the group is computed as the average individual performance of the three *s-bots*. The individual performance rewards the movements of an *s-bot* according to its current behavioural state. When in state $\mathcal{S}$, the *s-bot* should continue to move away from the centre, and it is considered successful if it reaches the distance $D_O(\mathcal{S}) = 2.4$ m from the centre. When an *s-bot* switches to state $\mathcal{C}$, it should aggregate with the other robots by reducing its distance from the centre of mass of the group. It is considered successful if it stays below the distance $D_O(\mathcal{C}) = 0.25$ m from the centre of mass of the group. In both cases, we conventionally say that a successful *s-bot* "achieves the desired distance D_O". Note that a trial is terminated whenever an *s-bot* passes over the black border of the circular band—and in this case its performance is 0—or if *s-bots* collide when in state $\mathcal{S}$. It is worth mentioning that when computing the individual performance, the behavioural state of an *s-bot* cannot be directly observed, because it is not explicitly encoded in the controller or elsewhere. However, knowing the environment type and looking at the movements of the robot, it is possible to estimate in which state an *s-bot* should be at any given time: when an *s-bot* is placed in environment A, it should search for the *way out* and exit through it, therefore it should be in state $\mathcal{S}$.

[4] For more details, see the supplementary material available in [15].

When an *s-bot* is placed in environment B, it should initially search for the *way out*, being in state $\mathcal{S}$, and at some point it should give up and aggregate, therefore switching to state $\mathcal{C}$. Given that it is not possible to exactly recognise when an *s-bot* switches to state $\mathcal{C}$, we compute the individual performance by considering an *s-bot* in state $\mathcal{C}$ as soon as it encounters the circular band for the first time. On the basis of such estimation of the behavioural state, it is possible to systematically evaluate the *s-bot*'s performance. Note that the evaluation function does not explicitly reward either cooperation or communication. It rather rewards those agents that perform the correct movements in each behavioural state, without any reference to the mechanism necessary to switch from one state to the other.

3 Results

We performed 20 replications of the experiment, most of which were successful. For each evolutionary run, we selected a single controller from the last generation. To do so, we evaluated the 20 best individuals—the *elite* of the last generation—for 200 trials in both environments, and we selected the genotype with the highest average performance. As a result, we obtained 20 controllers—hereafter referred to as $C_1, \ldots, C_{20}$—that were further evaluated for 2000 trials, half in environment A and half in environment B. The obtained results are summarised in Table 1: in both environments, we computed the average performance and its standard deviation (avg $\pm$ std), the rates of success %S (all *s-bots* achieve the desired distance D_O), failure %F (no *s-bot* achieves the desired distance D_O), partial success/failure %M (not all *s-bots* are successful or fail) and error %E (*s-bots* collide or cross the black edge of the circular band). In each trial, we also computed the *coverage*, which is defined as the percentage of the circular band that each robot covers in average during a trial: a value smaller than 1 indicates that the single *s-bot* does not search the whole circular band for the *way out*, while a value bigger than 1 indicates that the single *s-bot* performs more than one tour (see Fig. 3). The coverage—together with the success rate—is useful to quantitatively assess the quality of the evolved strategies.

Successful controllers produce good search behaviours when *s-bots* are in state $\mathcal{S}$:[5] *s-bots* avoid collisions and move away from the centre of the arena. Once on the circular band, *s-bots* start looping in search of the *way out*, which is eventually found and traversed when *s-bots* are placed in environment A. On the contrary, if *s-bots* are placed in environment B, the absence of the *way out* is recognised by the *s-bots* through the integration over time of their perceptual flow, which includes the signals that the *s-bots* may emit (for more insights about decision making processes based on temporal cues, see [11,12]). As a consequence, a behavioural transition from state $\mathcal{S}$ to state $\mathcal{C}$ can be observed. The modalities with which the transition is performed significantly vary across the different solutions synthesised during different evolutionary runs. However, looking at the behaviour produced by the evolved controllers, we recognised some similarities that let us classify the controllers in 4 classes.

[5] Detailed descriptions and movies are available as supplementary material in [15].

Table 1. Post-evaluation results. See text for details.

		environment A					environment B				
		avg ± std	%S	%M	%F	%E	avg ± std	%S	%M	%F	%E
U	c_4	0.82 ± 0.14	92.0	6.5	1.0	0.5	0.37 ± 0.11	19.4	18.9	61.7	0.0
	c_6	0.85 ± 0.06	98.6	1.2	0.0	0.2	0.31 ± 0.08	0.9	30.6	68.4	0.1
	c_{14}	0.83 ± 0.15	91.3	6.2	0.0	2.5	0.46 ± 0.15	2.5	65.1	24.0	8.4
	c_{17}	0.66 ± 0.07	74.3	25.4	0.1	0.2	0.39 ± 0.08	4.9	78.8	16.3	0.0
B	c_1	0.86 ± 0.11	97.7	0.8	0.0	1.5	0.69 ± 0.07	95.9	2.8	1.3	0.0
	c_5	0.85 ± 0.13	92.1	5.7	0.0	2.2	0.57 ± 0.14	66.8	16.9	16.1	0.2
	c_8	0.83 ± 0.15	90.3	7.6	0.4	1.7	0.57 ± 0.12	34.3	55.2	9.2	1.3
	c_{10}	0.88 ± 0.07	99.0	0.6	0.0	0.4	0.66 ± 0.07	94.1	2.1	3.7	0.1
	c_{16}	0.85 ± 0.14	94.4	4.1	0.0	1.5	0.74 ± 0.13	94.1	2.3	1.4	2.2
M	c_3	0.83 ± 0.15	85.8	11.7	0.0	2.5	0.63 ± 0.09	87.6	8.1	3.4	0.9
	c_7	0.79 ± 0.20	89.3	5.5	0.0	5.2	0.62 ± 0.25	49.5	34.2	10.5	5.8
	c_{11}	0.86 ± 0.07	98.9	0.6	0.0	0.5	0.61 ± 0.07	87.6	9.5	2.7	0.2
	c_{13}	0.85 ± 0.09	94.3	5.2	0.0	0.5	0.62 ± 0.07	93.0	5.3	0.8	0.9
	c_{19}	0.81 ± 0.15	94.8	2.3	0.6	2.3	0.67 ± 0.12	91.7	3.8	1.9	2.6
	c_{20}	0.87 ± 0.06	99.6	0.0	0.0	0.4	0.59 ± 0.07	79.3	11.3	9.3	0.1
C	c_2	0.86 ± 0.10	98.6	0.1	0.0	1.3	0.82 ± 0.12	97.1	0.4	0.9	1.6
	c_9	0.87 ± 0.08	99.2	0.0	0.0	0.8	0.78 ± 0.12	88.1	8.3	3.1	0.5
	c_{12}	0.87 ± 0.05	99.6	0.3	0.0	0.1	0.74 ± 0.11	87.8	6.4	5.4	0.4
	c_{15}	0.86 ± 0.08	99.3	0.0	0.0	0.7	0.78 ± 0.13	96.6	0.4	0.6	2.4
	c_{18}	0.84 ± 0.18	95.8	0.0	0.0	4.2	0.83 ± 0.17	95.3	0.3	1.0	3.4

Class $\mathbf{U} = \{C_4, C_6, C_{14}, C_{17}\}$ encompasses the "unsuccessful" controllers, that is, those controllers that solve the task only in part. These controllers generally produce appropriate search behaviours when *s-bots* are in state $\mathcal{S}$, as confirmed by the good performance and the high success rate in environment A (see Table 1). However, when *s-bots* are placed in environment B they fail in systematically aggregating, scoring a low performance and a poor success rate. The second class $\mathbf{B} = \{C_1, C_5, C_8, C_{10}, C_{16}\}$ consists of controllers that produce a strategy named "bouncing" after the aggregation behaviour of the *s-bots* in state $\mathcal{C}$: *s-bots* search for each other by continuously bouncing off the circular band, so that they sooner or later meet and remain close. Communication is not exploited,[6] and consequently each *s-bot* individually switches from state $\mathcal{S}$ to state $\mathcal{C}$, without any reference to the state of the other robots. The bouncing behaviour is resilient to possible individual failures in environment A: by bouncing off the circular band, *s-bots* can continue searching for the *way out*, even if less efficiently. This corresponds to high success rates in environment A despite the fact that the *s-bots* perform in average less than one tour over the circular band, as indicated by the corresponding coverage (see Fig. 3). The third class $\mathbf{M} = \{C_3, C_7, C_{11}, C_{13}, C_{19}, C_{20}\}$ encompasses controllers that produce a strategy named "meeting", due to the fact that *s-bots* aggregate by encountering at a meeting point, which is normally close to the

[6] Only C_{16} exploits signalling to trigger a synchronous switch to state $\mathcal{C}$ [15].

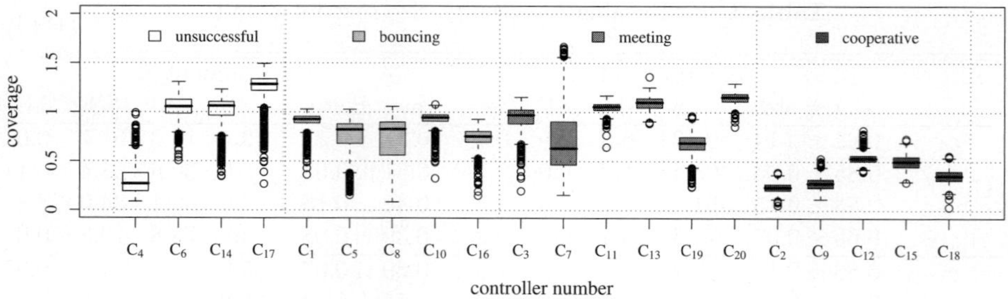

Fig. 3. The *coverage* of the evolved controllers. Boxes represent the inter-quartile range of the data, while the horizontal lines inside the boxes mark the median values. The whiskers extend to the most extreme data points within 1.5 times the inter-quartile range from the box. The empty circles mark the outliers.

centre of the arena. Except for C_7 and C_{19}, controllers of this class do not make use of communication. The main difference with class **B** controllers resides in the aggregation behaviour, which lets robots leave the band and move in circles close to the centre of the arena, waiting for the other *s-bots* to reach a similar position. This behaviour is not robust with respect to possible decision errors in environment A. As a consequence, evolution shaped the controllers of this class to be characterised by a high coverage (see Fig. 3): *s-bots* perform more than one loop over the circular band before switching to state $\mathcal{C}$, which corresponds to robust individual decisions and a high success rate in environment A. The last class $\mathbf{C} = \{C_2, C_9, C_{12}, C_{15}, C_{18}\}$ is named "cooperative" because it encompasses controllers that produce communicative behaviours exploited for cooperation in the decision making. In fact, *s-bots* are able to share the information they collect over time through their signalling behaviour. The *s-bots* initially emit a sound signal, and they stop only after looping on the circular band for some time. If any robot finds the *way out*, signalling continues, inducing all other *s-bots* to remain in state $\mathcal{S}$ and to keep searching for the *way out*. This leads to a high success rate in environment A, and no complete failures are observed (see Table 1). When the *way out* is not present, all robots eventually stop signalling, allowing the transition to state $\mathcal{C}$ and triggering the aggregation behaviour. By sharing the information through communication, *s-bots* can collectively search the circular band, splitting the task among them: as shown by the coverage data in Fig. 3, each *s-bot* covers from a quarter to half circle when placed in environment B. This allows to consistently reduce the search time, achieving high performance and high success rates. Communication is fundamental here, because it provides robustness to the decision making process and it makes the system more efficient by reducing the time necessary to take the decisions to switch from solitary to collective behaviours.

In order to quantitatively compare the performance of the behaviours produced by the evolved controllers, we used the performance data recorded over

2000 trials to perform a series of pairwise Wilcoxon tests among all possible controller couples, which allowed to produce the following ranking:

$$C_4 \prec C_6 \prec C_{17} \prec C_{14} \prec C_3 \prec C_8 \prec \{C_{13}, C_{11}\} \prec C_{19} \prec C_1 \prec$$
$$\prec C_{20} \prec C_{10} \prec C_5 \prec C_7 \prec \{C_{16}, C_{12}\} \prec C_{15} \prec C_9 \prec C_2 \prec C_{18},$$

where $C_i \prec C_j$ indicates that C_j is statistically better than C_i with 99% confidence. Controllers that have no statistical difference are reported in curly brackets. All class **U** controllers have a low rank, as one would expect. Instead, it is worth noting that class **C** controllers perform statistically better than the others. Moreover, other controllers making use of communication but with a different strategy (namely C_7-Meeting and C_{16}-Bouncing) occupy a good position in the rank. We can conclude that communication can improve the efficiency and the robustness of the decision making process. Robots exploiting only local interactions are prone to decision errors or to behaviours that are less efficient. Therefore, by cooperating through communication, *s-bots* increase their ability to make correct and unanimous decisions, consequently achieving a better performance.

4 Conclusions

We have studied the decision making mechanisms that can let a group of robots switch from solitary to collective behaviours. We have faced the problem through an evolutionary approach in order to limit the *a priori* assumptions and search broadly the space of the possible solutions. The results we obtained demonstrate that suitable decision making mechanisms can be evolved. Moreover, by providing the robots with a simple communication channel, the evolved cooperative strategies display higher efficiency and enhanced robustness of the system. The use of communication generally results in a faster and more robust decision making process. Communication increases the otherwise limited information available to each robot, not only about the quality of the physical environment but also and above all about the social environment and about the internal states of other robots that, by definition, are not directly accessible.

A systematic analysis of the evolutionary pressures that shaped the above mechanisms is out of the scope of this paper, and is left for future work. Further testing with real robots is also planned for the future. Finally, we plan to integrate the decision making processes studied here with on-going work on self-assembly, in order to produce the first example of functional self-assembly of real *swarm-bots* based on completely evolved controllers.

Acknowledgements. This work was supported by the Swarmanoid and by the ECAgents projects. Both projects are funded by the Future and Emerging Technologies programme (IST-FET) of the European Commission, under grant IST-022888 and 001940 respectively. The authors thank Paola Pellegrini and Mauro Birattari for providing the tools necessary to perform the statistical analysis.

References

1. Parker, C.A.C., Zhang, H.: Biologically inspired decision making for collective robotic systems. In: Proc. of the 2004 IEEE/RSJ Int. Conf. on Intelligent Robots and Systems, pp. 375–380. IEEE Computer Society Press, Los Alamitos (2004)
2. Parker, C.A.C., Zhang, H.: Active versus passive expression of preference in the control of multiple-robot decision-making. In: Proc. of the 2005 IEEE/RSJ Int. Conf. on Intelligent Robots and Systems, pp. 3706–3711. IEEE Computer Society Press, Los Alamitos (2005)
3. Garnier, S., Jost, C., Jeanson, R., Gautrais, J., Asadpour, M., Caprari, G., Theraulaz, G.: Aggregation behaviour as a source of collective decision in a group of cockroach-like robots. In: Capcarrère, M.S., Freitas, A.A., Bentley, P.J., Johnson, C.G., Timmis, J. (eds.) ECAL 2005. LNCS (LNAI), vol. 3630, pp. 169–178. Springer, Heidelberg (2005)
4. Kok, J.R., Spaan, M.T.J., Vlassis, N.: Multi-robot decision making using coordination graphs. In: Proc. of the 11th Int. Conf. on Advanced Robotics, pp. 953–957. IEEE Computer Society Press, Piscataway (2003)
5. Vlassis, N., Elhorst, R., Kok, J.R.: Anytime algorithms for multiagent decision making using coordination graphs. In: Proc. of the 2004 IEEE Conf. on System, Man and Cybernetics, pp. 953–957. IEEE Computer Society Press, Piscataway (2004)
6. Mondada, F., Pettinaro, G.C., Guignard, A., Kwee, I.V., Floreano, D., Deneubourg, J.-L., Nolfi, S., Gambardella, L.M., Dorigo, M.: SWARM-BOT: A new distributed robotic concept. Auton. Robots 17(2–3), 193–221 (2004)
7. O'Grady, R., Groß, R., Bonani, M., Mondada, F., Dorigo, M.: Self-assembly on demand in a group of physical autonomous mobile robots navigating rough terrain. In: Capcarrère, M.S., Freitas, A.A., Bentley, P.J., Johnson, C.G., Timmis, J. (eds.) ECAL 2005. LNCS (LNAI), vol. 3630, pp. 272–281. Springer, Heidelberg (2005)
8. Groß, R., Tuci, E., Bonani, M., Mondada, F., Dorigo, M.: Object transport by modular robots that self-assemble. In: Proc. of the 2006 IEEE Int. Conf. on Robotics and Automation, pp. 2558–2564. IEEE Computer Society Press, Los Alamitos (2006)
9. Nouyan, S., Dorigo, M.: Chain based path formation in swarms of robots. In: Dorigo, M., Gambardella, L.M., Birattari, M., Martinoli, A., Poli, R., Stützle, T. (eds.) ANTS 2006. LNCS, vol. 4150, pp. 120–131. Springer, Heidelberg (2006)
10. Trianni, V., Tuci, E., Dorigo, M.: Evolving functional self-assembling in a swarm of autonomous robots. In: Schaal, S., et al. (eds.) From Animals to Animats 8. Proc. of the 8th Int. Conf. on Simulation of Adaptive Behavior, pp. 405–414. MIT Press, Cambridge (2004)
11. Tuci, E., Trianni, V., Dorigo, M.: 'Feeling' the flow of time through sensorimotor co-ordination. Conn. Sciene 16(4), 301–324 (2004)
12. Ampatzis, C., Tuci, E., Trianni, V., Dorigo, M.: Evolution of signalling in a swarm of robots controlled by dynamic neural networks. In: Şahin, E., et al. (eds.) Swarm Robotics. LNCS, vol. 4433, pp. 173–188. Springer, Heidelberg (2007)
13. Beer, R.D.: A dynamical systems perspective on agent-environment interaction. Art. Intell. 72, 173–215 (1995)
14. Goldberg, D.E.: Genetic Algorithms in Search, Optimization and Machine Learning. Addison-Wesley, Reading (1989)
15. Trianni, V., Ampatzis, C., Christensen, A.L., Tuci, E., Dorigo, M., Nolfi, S.: From solitary to collective behaviours: Decision making and cooperation – supplementary material. http://laral.istc.cnr.it/esm/trianni-etal-ecal2007.html

Individual Selection for Cooperative Group Formation

Simon T. Powers, Alexandra S. Penn, and Richard A. Watson

School of Electronics and Computer Science, University of Southampton, U.K.
stp05r@ecs.soton.ac.uk

Abstract. It is well known that certain environmental conditions, such as a spatially structured population, can promote the evolution of cooperative traits. However, such conditions are usually assumed to be externally imposed. In this paper, we present a model that allows the conditions that promote or hinder cooperation to arise adaptively via individual selection. Consequently, instead of selection simply favouring cooperation under imposed environmental conditions, in our model selection also operates on the conditions themselves via a niche construction process. Results are presented that show that the conditions that favour cooperation can evolve, even though those that favour selfish behaviour are also available and are initially selected for.

1 Introduction

In this paper, we consider the evolution of cooperative behaviour that benefits all individuals within a group. An example of such behaviour is bacteria growing at a reduced rate in order to consume a limiting resource more efficiently [1,2,3,4]. The difficulty in explaining the evolution of such behaviour arises from the fact that it can be exploited by selfish cheaters who do not pay the costs of performing the cooperative act themselves but nevertheless still reap the group benefit. In the bacterial growth example, cheaters would be individuals who unilaterally grow at a faster rate but in so doing waste resource. If resource is shared freely between all members of the group, then cheaters will be favoured under individual selection. This follows since all members of the group feel the effects of their inefficient resource usage, while only they grow at a faster rate.

However, environmental conditions such as a spatially structured population can allow prolonged cooperation to be maintained under individual selection [5,1,2,3]. As an extreme example, if an individual only shares resource with itself then selection will clearly favour individuals that consume the limiting resource with less wastage. A more common situation is that the existence of spatial structure provides a setting where an individual only shares resource with a small number of its neighbours. Such a setting allows the formation of clusters of neighbouring cooperators that share the majority of their resource only with each other. Consequently, the cooperators in the cluster do not feel the effects of the wasteful resource usage of cheaters that are located elsewhere

F. Almeida e Costa et al. (Eds.): ECAL 2007, LNAI 4648, pp. 585–594, 2007.

in the population, and so will ultimately reach a larger biomass in the long term [1,2,3].

Unfortunately, such a cluster of cooperators could still be exploited by the appearance of a mutant cheater amongst their ranks [3]. Therefore, the existence of spatial structure alone is not sufficient to purge cheaters. Indeed, it has already been proposed in this resource sharing scenario that clusters of cooperators should periodically break up into single cells in order to prevent them from being outcompeted by mutant or immigrant cheaters [4]. A somewhat similar approach that can also purge cheaters is Wilson's well known trait-group aggregation and dispersal model [6,7,8]. This is a model based on the idea that groups containing a greater proportion of cooperators grow to a larger size than those containing more cheaters. If the progeny of the groups are then periodically mixed, and there is sufficient difference in group sizes after growth to counter the decline in frequency of cooperators within any mixed groups in which they are exploited, then this can lead to a global increase in the frequency of cooperators. This outcome can be explained in one of two ways. One view is that selection acting *between* groups (differential group productivity) favours cooperation, while selection *within* groups favours cheating; the end result is then determined by the balance of these two selective forces. The alternative, reductionist, view point is that the cooperative trait can have the greater individual fitness in the context of a group aggregation and dispersal population structure. Both of these viewpoints are compatible with each other, since neither denies that the allele that increases in frequency globally is the one with the highest fitness when averaged across the whole population [8].

One of the key determinants of whether the cooperative trait does in fact have a greater fitness is the variation in final group sizes before the groups are remixed, since this must be sufficient to outweigh selection operating *within* groups that favours cheaters. This in turn is affected by the variation in group composition when the groups are formed. For example, it has already been shown that assortative grouping, whereby a propensity exists for individuals to form groups with other playing the same strategy, facilitates cooperation in cases where random group formation is insufficient [7,9]. Another way of increasing initial variance in group composition that is applicable to random group formation is to reduce the initial group size, thus increasing the sampling error on the global population due to randomly formed groups.

The above discussion has shown how cooperation can be favoured by certain environmental conditions or modelling choices. However, in previous works, the parameter settings that give the cooperative strategy the advantage under individual selection, e.g. small group size in an aggregation and dispersal model, have been assumed. As a result, the explanatory power of the models has been limited to revealing the conditions under which cooperation is favoured. Consequently, they have left unanswered the following question. If individuals can modify the environmental conditions through individual adaptation such that either cooperative or selfish behaviour is favoured, which conditions and hence which behaviour will evolve? Thus, rather than imposing the environmental conditions that favour cooperation, in this paper we allow individual adaptation to determine whether or not these conditions arise. This modification of environmental conditions by the evolving individuals is an evolutionary dynamic known as niche construction [10], and is discussed further in Sect. 4.

In our model, we allow the initial group size parameter of the aggregation and dispersal process to be brought under individual control. Specifically, individuals posses a genotype that carries two parameters. The first of these is whether or not they grow at a reduced rate in order to use a limiting resource more efficiently, i.e. whether they are cooperative or selfish. The second parameter specifies the size of group that they will join, large or small. Our model also contains an intrinsic advantage to large groups, in the form of a greater per capita resource allocation.

It should be noted that individual choice over group size is realistic in many biological settings. For example, bacteria are known to be able to control micro-colony size in biofilms via the amount of Extracellular Polymeric Substances that they secrete and the process of quorum sensing. This example illustrates the fact that high-level cognitive abilities are not required in order for group size to be determined adaptively by individuals.

Using our model, we obtain results in which the following pairwise relationships hold: cooperative + large outcompetes cooperative + small, and selfish + large outcompetes selfish + small. Therefore, evolution should be towards the large initial group size environmental state. It is then also the case that selfish + large outcompetes cooperative + large. This leads to the conclusion that the direction of evolution should be towards large groups of selfish individuals. However, when all 4 genotypes are present in the initial population in equal proportions, this is not what happens. Instead, evolution leads to the fixation of the cooperative + small genotype. Therefore, although pairwise comparisons suggest that selfish + large is favoured under individual selection, the dynamics are such that when all 4 strategies are present it is cooperative + small that wins. We therefore show in this paper that, given certain parameter settings and assumptions, individual adaptation of environmental conditions can lead to those that favour cooperation, even though it initially seems that the opposite environmental conditions and strategy should be promoted under individual selection.

2 Model Details

Our model consists of a population of individuals that reproduce in randomly formed groups for a number of generations (t), before the progeny of each group are mixed together into a migrant pool and the process repeated. The genotype of each individual specifies two traits:

1. Cooperative or selfish resource usage (specified as a growth rate, G_i, and a resource consumption rate C_i);
2. specification of the initial size of the group that that the individual will join (small or large).

We therefore consider 4 possible genotypes / strategies: cooperative + small, cooperative + large, selfish + small and selfish + large.

In our model, reproduction occurs entirely within groups, as follows. Let n_i represent the number of copies of genotype i in a single group. In order to reproduce, these clones require a share of the group's resource influx, R. The amount of resource that a group receives at each time-step depends upon its size, with a larger per capita amount of resource allocated to larger groups. Specifically, a group that is twice as large receives an extra 5% per capita resource influx. The size of this share that each genotype receives depends upon its growth and consumption rates relative to those of the other genotype(s), where the consumption rate should be understood as the amount of resource required for a genotype to make one copy of itself. The magnitude of the share of the total group resource that the genotype receives, r_i, is then as defined in (1).

$$r_i = \frac{n_i G_i C_i}{\sum_j (n_j G_j C_j)} R \tag{1}$$

Therefore, the genotype with the highest growth and consumption rates will receive the largest per capita share of the total resource. This means that the selfish genotype always receives more per capita resource than the cooperative type, and so will ultimately drive a cooperative genotype in the same group extinct.

Given the share received by a genotype, the number of individuals in the group with that genotype then changes according to (2). The form of this equation is motivated as a simplification of existing models of bacterial colony growth [1,2,3].

$$n_i(t+1) = n_i(t) + \frac{r_i}{C_i} - K n_i(t) \tag{2}$$

As the replicator equation (2) shows, the number of clones of a genotype produced during reproduction depends on both growth (favouring the selfish type in the short-term) and consumption rates (favouring the cooperative type in the long-term under limited resource). The final term in (2) represents mortality, where K is a death rate that is constant to all genotypes.

The remainder of our model is based around Wilson's trait-group aggregation and dispersal process. In order to address our research question, the model differs from others that use this process in that in our model individuals carry a gene that determines the initial size of group that they join. Consequently, this parameter can be set adaptively during evolution. By contrast, in other models the value of this, and all other parameters of the aggregation and dispersal process, are imposed. Thus, in those models the conditions that determine whether or not cooperative behaviour is selected for are imposed, whereas in our model they are determined by the individuals themselves.

The other ways in which our model differs from the original trait-group aggregation and dispersal process [6] are that reproduction occurs within groups and that we impose both global, and group, carrying capacities, where the group carrying capacity follows from a finite resource influx. The overall algorithmic operation of our model is presented below:

1. **Initialisation**: Initialise the migrant pool with N individuals.
2. **Group formation (aggregation)**: Assign individuals in the migrant pool to groups, as described in the main text below.
3. **Reproduction**: Perform reproduction within groups for t time-steps, as described in the text above.
4. **Migrant pool formation (dispersal)**: Return the progeny of each group to the migrant pool.
5. **Maintaining the global carrying capacity**: Rescale the migrant pool back to size N, retaining the proportion of individuals with each genotype.
6. **Iteration**: Repeat from step 2 onwards for a number of generations, T.

Assignment of individuals to groups occurs by the following process. Groups of the "small" size are created by choosing (without replacement) individuals at random who specify the small size parameter on their genotype. Likewise, groups of the "large" size are created from randomly drawn individuals who specify the large parameter. Therefore, although all individuals in a group will specify the same size trait, the composition of the group will be random with respect to the cooperative / selfish resource usage trait. Finally, if there are an insufficient number of individuals remaining in the migrant pool to form the appropriate group size, then those remaining unassigned to groups are discarded. Given a large population size, the effect of this is negligible.

3 Results

In this section, we first investigate the results of pairwise competition between strategies, before proceeding to investigate the dynamics that occur when all 4 strategies are available. Throughout this section, the parameter settings stated in Table 1 are used.

Table 1. Parameter settings used throughout.

Parameter	Value
Growth rate (cooperative), G_c	0.018
Growth rate (selfish), G_s	0.02
Consumption rate (cooperative), C_c	0.1
Consumption rate (selfish), C_s	0.2
Population size, N	4000
Number of generations, T	1000

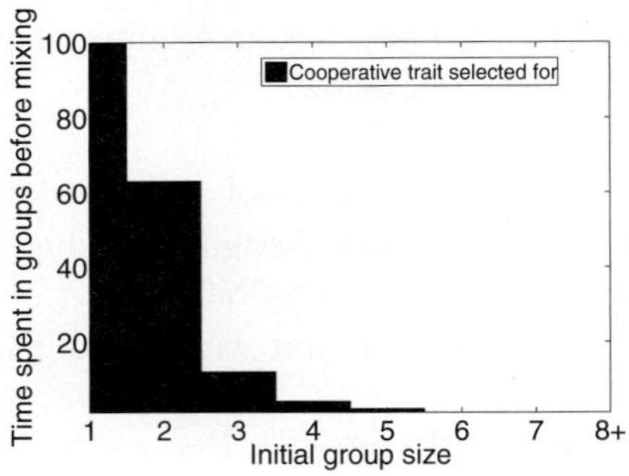

Fig. 1. Equilibrium state as a function of group size and the time spent within groups. The black area indicates the region of this parameter space where cooperation is favoured; the white area the region where selfish cheating is favoured.

3.1 Preliminary Experimentation / Pairwise Competition

When deciding on values for "small" and "large" group sizes, it was essential that the imposition of the small size upon all members of the population lead to selection favouring the cooperative trait, and the imposition of the large size lead to selection favouring the selfish trait. This would then give individuals the possibility of adaptively determining whether the environmental conditions, in the form of group size, lead to cooperation being selected for.

To determine suitable values for large and small group sizes, the 2D parameter space consisting of group size and the time spent in groups prior to mixing (returning to the migrant pool) was sampled; all other parameters of the model were held constant. From this sampling, it was possible to plot the parameter space with respect to whether cooperation was selected for, as shown in Fig. 1. Note that cooperation is reported as being selected for if and only if the cooperative trait reaches fixation in the global population at equilibrium.

As expected, this sample of the space shows that cooperation is favoured by both small groups and a short time spent within those groups prior to mixing. Small groups favour cooperation by increasing the variance in group composition, thereby strengthening selection acting between groups, i.e. selection that is due to the difference in group sizes after reproduction [8]. A short time spent in groups prior to mixing favours cooperation since the longer the time spent in groups, the greater the decline in frequency of cooperators within all groups that also contain cheaters [8].

Given the results in Fig. 1, it was decided to set the small group size at 4 and the large at 40, with the time spent within groups fixed at 4. This then creates the situation in which cooperative + small outcompetes selfish + small, and selfish + large outcompetes cooperative + large. Therefore, our definition of small size favours cooperative efficient resource usage, while our definition of large favours selfish inefficient usage.

However, further preliminary results showed that in pairwise competition, cooperative + small is outcompeted by cooperative + large. This is due to the

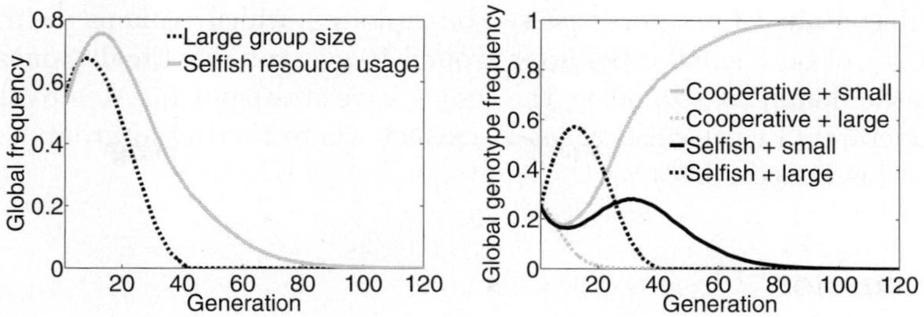

Fig. 2. Left-hand plot: average environment and strategy through time. Right-hand plot: change in genotype frequencies over time.

greater per capita resource influx provided to large groups, which allows such groups to grow to a larger size, assuming all other conditions are the same. Likewise, selfish + large outcompetes selfish + small. Given that large genotypes outcompete their small counterparts, and that in an environment with large groups selfish resource usage is favoured, it should be expected that the population would evolve towards a state with the selfish + large genotype at fixation. However, the results in the next section show that this is not the case if the population is initialised with all 4 genotypes in equal proportion.

3.2 Allowing Both Strategy and Environment to Evolve

The left-hand plot in Fig. 2 shows how the average environmental conditions, in terms of group size, and the average resource usage strategy, change over time. The key point is that at the start neither small groups nor cooperators are favoured; this is shown by an increase in the frequency of both large groups and selfish cheaters. Therefore, since the initial conditions favour both large groups and cheaters, it might be expected that the equilibrium reached would consist entirely of large groups of cheaters. However, this is not what occurs. Instead, after around 20 generations, both small groups and cooperators become favoured by selection. Therefore, the selective pressures change in a way that could not be predicted from pairwise comparisons alone.

The right-hand plot in Fig. 2 can help to explain why this occurs, by showing how all possible combinations of environment and strategy change over time. The figure shows that initially, the environment of large groups is favoured; this is due to their per capita resource advantage. In such large groups, the selfish inefficient resource usage strategy is favoured, as previously discussed.

However, the selfish + large genotype does not reach fixation. This is because selfish cheaters benefit from consuming resource in mixed groups with cooperators, since they can consume the left-over resource that follows from the cooperators efficiency. As the selfish + large type increases in frequency, it reduces the frequency of cooperative + large and hence the number of cooperators that it can exploit. However, there are still cooperative + small individuals present in the population. This then creates a selective advantage of being in small groups which still contain cooperators that can be exploited, which explains the increase in frequency of both small types after around 10 generations. Finally, once small becomes the dominant size allele, the cooperative strategy wins, due to the fact that cooperators drive selfish cheaters extinct where the initial group size is 4 and $t = 4$ (as in Sect. 3.1).

4 Discussion

The decision to live in groups of a certain size can be considered as an example of niche construction. It is commonly held that synergistic fitness effects in a cooperative, group-living scenario would have been an important driver in the formation of higher levels of biological organisation [11]. The potential of higher fitness within groups would have been an incentive for individuals to actively choose to participate in these groups bringing an aspect of their selective environment under control and altering their fitness landscape accordingly. Choosing to live within a group via some individual-level adaptation or adaptations that facilitate group formation would thus constitute the creation of a niche. This of course does not negate the problem of invasion or disruption by selfish individuals. As discussed earlier, in the absence of assortative groupings, cooperation within a group context is only stable when groups are small enough that between group variation can be maintained by sampling error, hence allowing between group selection to be an effective force. We might expect then that cooperative groups would be stable if the individuals within them were able to adaptively choose a group size small enough to purge selfish strategies from the population but large enough to enjoy the fitness benefits of group living.

The significance and originality of our model lies in the fact that the conditions that favour cooperation are not assumed. Instead, individuals in the population carry a specification of the conditions, in this case initial group size, on their genotype. Whether or not conditions that favour cooperation occur is then determined adaptively by individual selection on this specification. Figure 3 illustrates the two-way interaction of strategy and environment in this scenario. On the one hand, whether cooperative or selfish behaviour is preferred by individual adaptation is influenced by the environment - this is normal, environmentally sensitive, adaptation. On the other, whether individual adaptation will support the construction of environments with large groups or small groups is influenced by the strategy being played - this is niche construction. A system with these two influences must reach an attractor where the characteristics of the environment favour strategies that in turn support those particular environmental characteristics. In this case we need not presuppose, for example, that the prevailing environment favours cooperation in order for cooperation to arise; and similarly, we need not presuppose that the prevailing strategy is cooperative in order for cooperation-favouring groups to arise.

Previous models have shown that cooperative traits are selected for when environmental conditions favour cooperation. In our model, conditions that favour either selfish or cooperative traits are available. We have then been able to show

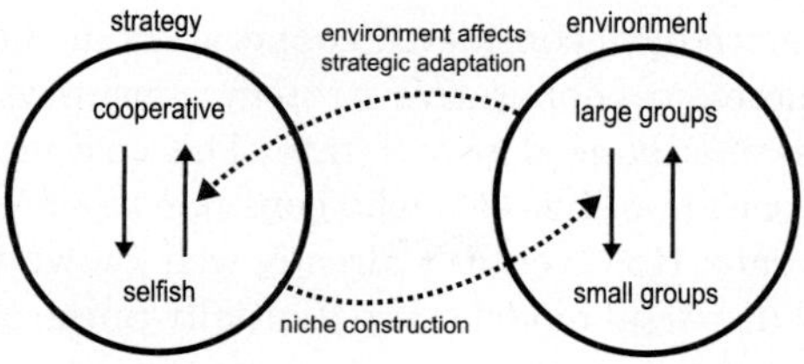

Fig. 3. Feedback between strategy and environment and their adaptation.

that, under given parameter settings and model assumptions, the conditions that promote cooperation (a small group size) are indeed selected for. This also occurs despite a bias towards the conditions (large groups) that favour selfish individuals. The result that the cooperative + small genotype reaches fixation, even though cooperative + large beats it in pairwise competition, is particularly interesting since a small group size can therefore be viewed as an adaptation that purges cheaters. The adaptive dynamics of the model can be seen more clearly in Fig. 2, which shows the changing global proportions of individuals choosing a large group environment and those choosing a selfish strategy over the course of the run. It is clear that both large group size and selfish resource use are initially favoured, however the frequency of the large group adaptation quickly peaks and starts to decline. In our scenario this occurs because a) in the presence of many large-group preferring cheaters, cooperators change their preference to favour small groups not large and the large-group cheaters become less and less fit; and, b) in the presence of many small cooperators, cheaters (also) change their preference to favour small groups not large. Once small groups are established, cooperation is the more fit strategy. This minimalist model illustrates the tightly coupled interaction of the evolution of strategy with the evolution of an individual's environment-affecting characters.

5 Conclusion

A plethora of models have shown that certain environmental conditions, such as a spatially structured population, can promote cooperative behaviour. This occurs because such conditions can give a cooperative trait a higher fitness than the selfish trait, when averaged over the whole population. Our contention however is that individuals are not simply the passive recipients of environmental conditions. Niche construction dynamics in which organisms adaptively alter their environments and thus their fitness landscapes have been observed in many systems and are postulated to be an important evolutionary dynamic [10]. It seems reasonable that individuals could adaptively choose to form or dissolve groups thus choosing the conditions that would facilitate either cooperative or selfish behaviour to be the preferred strategy. However, previous models have not considered the effect of allowing individual adaptation to determine whether or not these favourable conditions arise.

In this paper, we have considered cooperation in the consumption of a limiting resource, where our cooperative strategy consumes resource more efficiently but suffers a cost of a reduced growth rate. This cost means that cooperators can be exploited by selfish individuals who consume resource less efficiently but that grow at a faster rate. However, it is already well known that Wilson's trait-group aggregation and dispersal model can potentially purge cheaters in situations such as this. One important parameter of the aggregation and dispersal process that determines whether or not cooperation wins out is the initial group size. Rather than impose this externally, as in previous models, we allow it to be set adaptively via individual selection in a process analogous to niche construction. In this minimal model we have demonstrated that when both cooperative and selfish strategies and the environmental conditions, the group size, which favours each, are both adaptable, then both cooperative behaviour and the small group size environment which actively supports it can evolve via individual selection even in conditions which initially favour large groups and selfish cheaters.

In future work, we may consider the introduction of new strategies via mutation. We also intend to examine in more detail the parameter values that promote cooperation and small groups, with regard to determining how large an attractor this equilibrium is. Finally, the process of niche construction should also be applicable to other parameters of the aggregation and dispersal process, for example, to the time spent reproducing in groups before mixing in the migrant pool. It would therefore be useful to investigate the individual adaptation of this and other parameters.

References

1. Pfeiffer, T., Schuster, S., Bonhoeffer, S.: Cooperation and competition in the evolution of ATP-producing pathways. Science 292(5516), 504–507 (2001)
2. Pfeiffer, T., Bonhoeffer, S.: An evolutionary scenario for the transition to undifferentiated multicellularity. PNAS 100(3), 1095–1098 (2003)
3. Kreft, J.U.: Biofilms promote altruism. Microbiology 150, 2751–2760 (2004)
4. Kreft, J.U., Bonhoeffer, S.: The evolution of groups of cooperating bacteria and the growth rate versus yield trade-off. Microbiology 151, 637–641 (2005)
5. Nowak, M.A., May, R.M.: The spatial dilemmas of evolution. International Journal of Bifurcation and Chaos 3(1), 35–78 (1993)
6. Wilson, D.S.: A theory of group selection. PNAS 72(1), 143–146 (1975)
7. Wilson, D.S.: The Natural Selection of Populations and Communities. Benjamin/Cummings, New York (1980)
8. Sober, E., Wilson, D.S.: Unto Others: The Evolution and Psychology of Unselfish Behavior. Harvard University Press, Cambridge (1998)
9. Wilson, D.S., Dugatkin, L.A.: Group selection and assortative interactions. The American Naturalist 149(2), 336–351 (1997)
10. Odling-Smee, F.J., Laland, K.N., Feldman, M.W.: Niche construction: the neglected process in evolution, no. 37. Princeton University Press, Princeton (2003)
11. Keller, L. (ed.): Levels of Selection in Evolution. Monographs in behavior and ecology. Princeton University Press, Princeton (1999)

Institutional Robotics

Porfírio Silva[1] and Pedro U. Lima[2]

[1] Philosophy Department, Faculdade de Letras, University of Lisbon
porfiriosilva@clix.pt

[2] Institute for Systems and Robotics, Instituto Superior Técnico,
Technical University of Lisbon
pal@isr.ist.utl.pt

Abstract. Pioneer approaches to Artificial Intelligence have traditionally neglected, in a chronological sequence, the agent body, the world where the agent is situated, and the other agents. With the advent of Collective Robotics approaches, important progresses were made toward embodying and situating the agents, together with the introduction of collective intelligence. However, the currently used models of social environments are still rather poor, jeopardizing the attempts of developing truly intelligent robot teams. In this paper, we propose a roadmap for a new approach to the design of multi-robot systems, mainly inspired by concepts from Institutional Economics, an alternative to mainstream neoclassical economic theory. Our approach intends to sophisticate the design of robot collectives by adding, to the currently popular emergentist view, the concepts of physically and socially bounded autonomy of cognitive agents, uncoupled interaction among them and deliberately set up coordination devices.

Key words: Collective Robotics, Institutional Economics, Institutional Robotics

1 Introduction

Three great neglects are at the heart of Good Old-Fashioned Artificial Intelligence: the neglect of the body, of the world, and of other agents. Collective Robotics is an important attempt to surpass these neglects: because it embodies intelligence in physical robots; because it places robots in physical environments largely natural; because it locates intelligence in the collective. Nevertheless, most multi-robot systems model extremely poor social environments. Our aim is to put forward a new conceptual approach to design control systems of artificial robotic societies. In Section 2 some weaknesses of popular guiding principles to collective systems design are identified. In Section 3 we look for inspiration coming from fields of sciences of the artificial other than robotics. In Section 4 we sketch out a new strategy to conceptualize multi-robot systems: *Institutional Robotics*, which takes institutions as the main tool of social life of robots with bounded rationality and bounded autonomy.

F. Almeida e Costa et al. (Eds.): ECAL 2007, LNAI 4648, pp. 595–604, 2007.

2 Emergence, Uncoupled Interaction, Bounded Autonomy, and Collective Inefficiency

Two out of the four design principles for collective systems suggested in [1:241–243] represent popular views among practitioners of AI and Robotics. According to the "level of abstraction principle", collective intelligence refers not only to groups of individuals, as in human societies, but equally "to any kind of assembly of similar agents", including groups of modules in modular robotic systems or organs that make up entire organisms. The "design for emergence principle" states that a desired functionality should not be directly programmed into a group of agents, but emerge from a set of simple rules of local interaction, relying on self-organizing processes. These two principles raise three questions.

First, exclusive focus on emergence and self-organization stems from the prominent role conferred to local interaction. No reason is given to ignore indirect or mediated interaction, which [2:14] considers characterized by properties such as name uncoupling (interacting entities do not have to know one another explicitly), space uncoupling and time uncoupling (they do not have neither to be at the same place nor to coexist at the same time). Communication is an example of such an indirect (not local) interaction.

Second, to put individuals in human societies on the same foot with organs or modules purports to ignore different degrees of autonomy enjoyed by a human lung and a human individual. According to [3:518], the following could be a working definition within robotics: "Autonomous agents operate under all reasonable conditions without recourse to an outside designer, operator or controller while handling unpredictable events in an environment or niche". However, for some philosophical perspectives this conception of autonomy would be unsatisfactory, because a truly autonomous agent must be capable of acting according to his own goals, while designers are the sole providers of goals to the robots. A concept of autonomy arising out of the ongoing attempt to maintain homeostasis could improve our understanding of autonomy and how goals become grounded in artificial creatures [3].

Whether full autonomy is attainable is a remaining question. A sharp negative answer to that question is offered by [4]. Autonomous agents must be capable of generating new goals as means for achieving existing goals of their own, but they are not necessarily self-sufficient. An agent depends on a resource where he needs it to perform some action to achieve one of his goals. There is also social dependence: an agent x depends on another agent y when, to achieve one of his goals, x needs an action of y. Two agents can be mutually dependent. Dependences imply interests: a world state that favours the achievement of an agent's goals is an interest of that agent. Dependence and interest relations are objective relations, holding whether an agent is aware of them or not. Limited autonomy of social agents comes also from influencing relations between them. By acquiring (true or false) beliefs about their interests agents can acquire goals. Now, an agent x can influence another agent y to adopt a goal according to x's needs, even if that goal is not an interest of y.

Within this approach, cognition does not preclude emergency. To form goals and establish plans to their achievement, agents must be cognitive. However, bounded rationality combines with bounded autonomy to give place to emergent phenomena: there are deliberately planned actions but they may produce unintended effects beyond reach of the agent's understanding or awareness.

Third, no reason is given to rule out coordination devices deliberately set up by agents in some multi-agent systems (legislation and organisations in human societies, for example). The remaining question is the desirability of that fact. Could we show that, at least in some situations, merely emergent processes may lead to inefficient solutions to collective problems? If so, we would have a hint on why multi-agent systems may need coordination devices. A set of experiences within MAS, reported in [5], advances our understanding of the problem. There, situations previously identified in experimental economics are simulated with a version of the Genetic Algorithm (GA). The GA population represents a collection of sets of rules associated with the set of actions available to agents; the fitness function for each agent maximizes his payments.

Co-ordination problem 1. A set of individuals, kept in isolation from one another, must choose one of 16 colours. Each participant choice will be rewarded in accordance with the rule: multiply a fixed amount of money by the number of players that have chosen the same colour. The experiment repeats a number of times. After each repetition, players are informed of frequencies and pay-offs by colour, so participants can change their choices next time, what they indeed do to maximize payments. Individual behaviours rapidly converge: the rule "choose colour x", where x is the most often selected, emerges as a shared convention. The "spontaneous order hypothesis" seems to work.

Co-ordination problem 2. A new experimental situation departs from the previous one in just one detail. The payoff to each individual now depends, not only on the frequency of the chosen colour, but also on an "intrinsic" characteristic of each colour, which remains unknown to players. For example, all other factors remaining equal, the choice of the colour number 16 pays 16 times more than colour number 1. The convergent choices of all participants to colour 16 is the most valuable situation to every participant, but that convergence is unlikely to occur in the absence of any opportunity to agree on a joint strategy. An initial accidental convergence to any colour creates an attractor capable of strengthen itself from repetition to repetition. Even if a participant has a god's eye view of the situation, any isolated option for the best theoretical option will neither improve the individual payoff nor move the collective dynamics towards a path conducive to a higher collective payoff. Self-organizing processes may lead to inefficient solutions for a collective problem. The "spontaneous order hypothesis" is in trouble, even with mere co-ordination problems, when the best for each individual is also the best for the collective (for other individuals). The situation gets worse with a "co-operation problem", when the best outcome for the collective and the best outcome for an individual don't coincide necessarily.

Co-operation problem. Now, the individuals must post a monetary contribution (from 0 to a predefined maximum) in an envelope and announce the amount contained in it. The sum of all the contributions is multiplied by a positive factor ('invested') and the resultant collective payoff is apportioned among the individuals. For each participant, its share of the collective payoff is proportional to the announced contribution, not to the posted contribution. As all participants know these rules, they realize that to maximize payoff an individual must contribute nothing and announce the maximum. So, it is with no surprise that, after some initial rounds, free-riding behaviour emerges: the posted contributions tend to zero while the announced contributions are kept close to the maximum. The group follows collectively a path that all of his members consider undesirable: soon there will be no more money to distribute.

This set of experiences suggests collective order does not always emerge from individual decisions alone. Coordination devices deliberately set up by agents could be useful.

3 Artefacts in Institutional Environments

The previous Section has shown that some concepts can add to emergentist views in order to sophisticate artificial collective systems design: physically and socially bounded autonomy of cognitive (not only reactive) agents; uncoupled interaction among them; deliberately set up coordination devices. How could we put all these concepts together? Social sciences' concepts have already inspired fields of sciences of the artificial other than robotics. Relying on some results of that cross-fertilization, we will arrive at the unifying concept of "institutional environment". It will later lead us to Institutional Robotics.

Epstein and Axtell argue that artificial society modelling can constitute a new kind of explanation of social phenomena [6:20]. Lansing [7] argues that the modelling of artificial societies can profit from a broad historical perspective of disputes among social scientists and philosophers on how to study social phenomena. To exemplify, he points out the parallel between some writing of Theodor Adorno on the positivist dispute in German sociology and the question that introduces [6]: "How does the heterogeneous micro-world of individual behaviors generate the global macroscopic regularities of the society?". This is a classical problem of the social sciences, the micro-macro link problem or the problem of social order. A number of researches take both perspectives together within Multi-agent systems (MAS) modelling. A few examples are: [8] studies norms as a possible solution to coordination problems; [9] suggests relaxing the assumption that coordination can be designed to perfection and importing conflict theories from sociology; [10] reviews trust and reputation models; within the framework of "Socionics" (a combination of sociology and computer science [11]), the Social Reputation approach [12] models reputation as an emergent mechanism of flexible self-regulation; [13] argues for using basic individual rights in MAS, combined with some argumentation mechanism.

Facing such a variety, how would we choose the most promising concept? Perhaps we need them all. "It does not seem possible to devise a coordination strategy that always works well under all circumstances; if such a strategy existed, our human societies could adopt it and replace the myriad coordination constructs we employ, like corporations, governments, markets, teams, committees, professional societies, mailing groups, etc." [14:14] So, we keep them all, and more – but we need an unifying concept to give the whole some consistence.

"Environment" is such a concept. [2] suggests the need to go deeper than the subjective view of MAS, where the environment is somehow just the sum of some data structures within agents. What we need to take into account is the active character of the environment: some of its processes can change its own state independently of the activity of any agent (a rolling ball that moves on); multiple agents acting in parallel can have effects any agent will find difficult to monitor (a river can be poisoned by a thousand people depositing a small portion of a toxic substance in the water, even if each individual portion is itself innocuous) [2:36]. Because there are lots of things in the world that are not inside the minds of the agents, an objective view of environment must deal with the system from an external point of view of the agents [15:128].

One can wonder if this can be relevant to robotics, where agents already behave sensing and acting in real (not just software) environments. We suggest the answer is affirmative. Dynamic environmental processes independent of agents' purposes and almost unpredictable aggregate effects of multiple simultaneous actions are not phenomena restricted to physical environments. Similar phenomena can occur in organizational environments: if nine out of ten of the clients of a bank decide to draw all their money at the same date, bankruptcy could be the unintended effect. And, most of the time, social environments in robotics are poorly modelled. So, the objective view of the environment could apply not only to physical features, but also to the social environment of the agents. We further suggest that both physical and social environments are populated with strange artefacts: artefacts with material and mental aspects. Let us see, following [16].

An artefact is something done by an agent to be used by another (or the same) agent. An artefact may not be an object: footprints left on a mined field for the followers are artefacts. Artefacts shaped for coordinating the agents' actions are coordination artefacts. Even single-agent actions can be coordinated actions if they contribute to solve an interference problem with other agents. Some artefacts have physical characteristics that represent opportunities and constraints which are sufficient conditions to enable a single-agent coordinated action, even if the agent doesn't recognize them (the wall of a house keeps people inside and outside separated). Sometimes, the agent must additionally recognize the opportunities and constraints of the artefact: sitting at a table with other people needs some knowledge ("not try to seat at a place already occupied").

More interesting artefacts are associated not only with physical but also with cognitive opportunities and constraints (deontic mediators, such as permissions and obligations). Recognizing all of those enables a single-agent coordinated action: a driver approaching a roundabout is obliged, only by physical properties of the artefact, to slow down and go right or left to proceed; traffic regulations add something more indicating which direction all drivers have to choose not to crash with others.

Furthermore, artefacts can be completely dematerialized. Such artefacts enable single-agent coordinated actions only by means of cognitive opportunities and constraints recognized by the acting agent. Social conventions and norms are relevant examples of the kind. A traffic convention to drive on the right works independently of any material device.

Consider now multi-agent coordinated actions. "There exist some artefacts such that the recognition of their use by an agent and the set of cognitive opportunities and constraints (deontic mediators) are necessary and sufficient conditions to enable a multiagent coordinated action" [16:320]. Institutions are of such a kind of artefacts. The definition takes institutional actions as multi-agent coordinated actions performed by a single-agent. How could this be? Because of a cognitive mediation intertwined with the agents' behaviours. While traditional views on institutions take them as structured sets of rules and conventions, in [16] the basic coordination artefact is the institutional role played by an agent with the permission of others. A group of agents recognizes that an agent (Paul) plays a role (priest) and so believes he has the artificial power of doing a multi-agent coordinated action (the marriage of John and Mary). Both recognition and belief are intertwined with the behaviour of treating Paul as a priest and treating John and Mary, from some point in time on, as a married couple.

The single-agent action of an agent playing a role is the vehicle action for a collective action, like flipping the switch is the vehicle action for the supra-action of turning the light on. In this context, the agent relies on some external aspects of the world (the functioning of the electrical circuit). To get John and Mary married the priest must perform a certain set of bodily movements counting as marrying. That set of movements is the vehicle action for the supra-action of marrying John and Mary. Again, the collective of agents rely on some external aspects of the world: the institutional context [16:312,320–321].

So, we have got our unifying concept: institutional environments populated with a special kind of artefacts.

4 Institutional Robotics

With the "institutional environment" concept as a starting point, in this Section we sketch out a new strategy to conceptualize multi-robot systems. Some global inspiration comes from Institutional Economics [17], an alternative to mainstream neoclassical economic theory. "Market-based multi-robot coordination" is a previous example of importing some Economics' views into Robotics [18]. We do the same, but with different assumptions.

(1) The control system for a robotic collective is a network of institutions. All institutions exist as means for some activity of some set of robots. As a principle, institutions are generic: they are not designed to any specific set of robots.

(2) Institutions are coordination artefacts and come in many forms: organizations, teams, hierarchies, conventions, norms, roles played by some robots, behavioural routines, stereotyped ways of sensing and interpret certain situations, material artefacts, some material organization of the world. A particular institution can be a composite of several institutional forms.

(3) Institutions can be mental constructs. An example of a "mental institution" is a program to control a sequence of operations.

(4) Institutions can be material objects functioning exclusively by means of its physical characteristics given the physical characteristics of the robots (a wall separating two buildings effectively implements the prohibition of visiting neighbours if the robots are not able to climb it). Some rules (or other kinds of mental constructs) can be associated to a material object to create a more sophisticated institution (a wall separating two countries is taken as a border; there are some doors in the wall to let robots cross the border; some regulations apply to crossing the border).

(5) The boundaries between institutional and purely physical aspects of the world are not sharp. Not all material objects are institutions. If the wall separating buildings is seen as just an element of the physical world, some robots gaining access to opposite building with newly acquired tools or physical capabilities will not be minded as a breach of a prohibition. However, modifications of the material world creating new possibilities of interaction can become institutional issues. If the collective prefers to preserve the previous situation of separated buildings, the new capability of the robots to climb the wall could give place to new regulations. Material objects are devices for institutions when they implement some aspect of the collective order. The continuing existence of a material object can be uncoupled from the continuing existence of the institutional device it implements (the wall could be demolished without eliminating the border; the border can be eliminated without demolishing the wall). So, a material leftover of a discarded institution can last as an obstacle in the world.

(6) Enforcement mechanisms can be associated with institutions to prevent (or to redress negative effects of) violation. Examples are fines and reputation.

(7) The institutional environment at any point in the history of a collective is always a mix of inherited and newly adopted forms. So, the designer of a robotic collective must shape the first version of any system of institutional robotics. However, that first institutional setup must be neither purely centralized, nor fully decentralized, nor purely distributed. That means the following. Not all robots are equal in power: neither all agents have the same computational power, nor all access the same information, nor all are allowed to take decision on all domains. There are some hierarchical relations among robots: for any robot, access to information and permission to act are bounded by decisions of some others. However, different hierarchies apply to different issues and the same robot can be on top of one hierarchy and at bottom of others. Some robots, by virtue of one-to-one relationships not framed by any hierarchy, are able to establish short cuts to and influence top level decision makers that would otherwise be beyond reach. There is neither a single robot nor a small group of robots in charge of all collective decisions all the time. Although, some kind of elitism is possible: different (eventually partially overlapping) groups of robots share the ruling over different domains of decision. Elite must eventually be renewed: robots can be removed from power, robots can access power.

(8) Agents are robots, hardware/software "creatures", operating on real physical environments. Robots are able to modify at some extent the material organization of their physical world.

(9)The continuing functioning of any robot depends on some material condition (available energy, for example). Whatever set of tasks a robot has to fulfil, some of them must be related to survival. There could be some institutions in charge of managing life conditions for all or some robots.

(10) All robots have built-in reactive behaviours, routines, and deliberative competences. Robots have partial models of themselves (they know some, but not all, of their internal mechanisms). Some of the internal mechanisms known by the robots can be accessed and modified by themselves.

(11) Every agent is created with links to some subset of the institutional network in existence within the collective. (Nobody is born alone in the wild). At some extent agents are free to join and to disconnect themselves from institutions. However, under certain circumstances, some institutions could be made compulsory for every agent or for some subset of all agents. Some institutions can filter access, either according to some objective rules or according to the will of those already connected. Disconnecting from an institution prevents the access to resources under control of it, as well as the participation in decision making processes taking place within it.

(12) Each robot has a specific individual identification (a name). All robots are able to identify, if not all, at least some others by their names.

(13) Any agent disconnected from all institutions will be treated by other agents as just an aspect of the material world. To recover from that extreme situation and get connected again to the institutional network an agent must be helped by some benevolent agent.

(14) World models are a special kind of institution. Being created with pre-established links to some institutions, any robot is endowed with some partial world models. World models can be specific to aspects of the physical world, specific to aspects of the social world or combine some aspects of both. None of the robots is endowed with a complete model of the world (except if gods are allowed). Inconsistencies between partial world models of one robot are allowed.

(15) There will be some process of collective world modelling. For example, a shared model of physical world can result from co-operative perception (sensor fusion [19:17–22]: merging sensor data from sensors spread over different robots and applying confidence criteria to weight their contribution to an unified picture of some aspect of the environment).

(16) The functioning of the sensorial apparatus of the agents can be modulated by their links to some institutions (adhering to an institution can augment the power or distort the functioning of some sensor). Institutional links can also modify the access to pieces of information available at collective level.

(17) From the point of view of an external observer the world model of a robot can be inaccurate. Inaccuracies can result from objective factors, intrinsic to the robotic platform (like sensors' limitations) or extrinsic (inappropriate vantage points to obtain data from some regions of the environment). Other inaccuracies can result from subjective factors: robots can have "opinions" and "ideologies".

(18) An "opinions" is an individual deviation from world models provided by institutions. (Even if a specific "opinion" of an individual agent is objective

knowledge gathered by virtue of some privileged vantage point, in such a manner that an external observer would prefer to rely on that opinion instead of accepting the "common sense", that means nothing to other agents, as long as they are deprived of that gods' view). By virtue of bearing an "opinion" the behaviour of a robot can be modified.

(19) An "ideology" is a set of "opinions" shared by a subset of all agents. Its acceptance among agents largely overlaps with sets of agents linked to some subset of the institutional network. An "ideology" can be "offered" by an institution to any agent prone to adhere or be a condition for adhesion. An "ideology" can result from a modification of the sensor fusion process (modification of the criteria to weight different individual contributions, for example). "Ideologies" can be about the physical or the social world. Modifying the perception of the agents and their behaviours, "ideologies" can affect the functioning of institutions in many ways: for example providing alternative stereotyped ways of sensing certain situations ("ignore such and such data streams") or undermining mechanisms of social control ("break that rule and we will pay the fine for you with a prize").

(20) Decision-making processes are a special kind of institution. Many aspects of collective dynamic can be subject to co-operative decision-making [19:34–46].

(21) Institutional building is a special issue for decision-making processes: "constitutional rules" for the functioning of some already existing institutions can be deliberated by the robots themselves; robots can deliberately set up new institutions or abandon old ones. Some institutions will have specific mechanisms to facilitate institutional building.

(22) Institutional building is fuelled by "institutional imagination": robots can conceive alternative institutions, or alternative constitutional rules to existing institutions, not to implement them at short term, but as "thought experiments". Results of those thought experiments can be put forward to specific institutional building mechanisms.

(23) The functioning of an institution can be modified, not by deliberative means, but by accumulating small modifications initiated by some robots and not opposed by others.

(24) An institution fade away when none agent is anymore linked to it. Robots can have memories of old institutions and reintroduce them in the future.

5 Conclusion

This paper suggested a new strategy to conceptualize multi-robot systems: the Institutional Robotics, which takes institutions as the main tool of social life of robots with bounded rationality and bounded autonomy. We have plans to set up a working group consisting of a team of people with a multidisciplinary background (e.g., philosophy, cognitive sciences, biology, computer engineering, artificial intelligence, systems and control engineering) to work on it, including further brainstorming, concepts refinement and actual implementation.

References

1. Pfeifer, R., Bongard, J.: How the Body Shapes the Way We Think. The MIT Press, Cambridge (2007)
2. Weyns, D., Parunak, H.: v. In: Weyns, D., Parunak, H.V.D., Michel, F. (eds.) E4MAS 2004. LNCS (LNAI), vol. 3374, pp. 1–47. Springer, Heidelberg (2005)
3. Haselager, W.F.G.: Robotics, philosophy and the problems of autonomy. Pragmatics & Cognition 13(3), 515–532 (2005)
4. Conte, R., Castelfranchi, C.: Cognitive and Social Action. The University College London Press, London (1995)
5. Castro Caldas, J., Coelho, H.: The Origin of Institutions: socio-economic processes, choice, norms and conventions. Journal of Artificial Societies and Social Simulation 2(2) (1999), http://jasss.soc.surrey.ac.uk/2/2/1.html
6. Epstein, J.M., Axtell, R.: Growing Artificial Societies: Social Science from the Bottom Up. Brookings Institution Press, Washington (1996)
7. Lansing, J.S.: Artificial Societies" and the Social Sciences. Artificial Life 8, 279–292 (2002)
8. Hexmoor, H., Venkata, S.G., Hayes, R.: Modelling social norms in multiagent systems. Journal of Experimental and Theoretical Artificial Intelligence 18(1), 49–71 (2006)
9. Malsch, T., Weiß, G.: Conflicts in social theory and multiagent systems: on importing sociological insights into distributed AI. In: Tessier, C., Chaudron, L., Müller, H.-J. (eds.) Conflicting Agents. Conflict Management in Multi-Agent Systems, pp. 111–149. Kluwer Academic Publishers, Dordrecht (2000)
10. Sabater, J., Sierra, C.: Review on Computational Trust and Reputation Models. Artificial Intelligence Review 24(1), 33–60 (2005)
11. Malsch, T., Schulz-Schaeffer, I.: Socionics: Sociological Concepts for Social Systems of Artificial (and Human) Agents. Journal of Artificial Societies and Social Simulation 10(1) (2007), http://jasss.soc.surrey.ac.uk/10/1/11.html
12. Hahn, C., Fley, B., Florian, M., Spresny, D., Fischer, K.: Social Reputation: a Mechanism for Flexible Self-Regulation of Multiagent Systems. Journal of Artificial Societies and Social Simulation 10(1) (2007)
13. Alonso, E.: Rights and Argumentation in Open Multi-Agent Systems. Artificial Intelligence Review 21(1), 3–24 (2004)
14. Durfee, E.H.: Challenges to Scaling Up Agent Coordination Strategies. In: Wagner, T.A. (ed.) An Application Science for Multi-Agent Systems, pp. 113–132. Kluwer Academic Publishers, Dordrecht (2004)
15. Weyns, D., Schumacher, M., Ricci, A., Viroli, M., Holvoet, T.: Environments in Multiagent Systems. The Knowledge Engineer Review 20(2), 127–141 (2005)
16. Tummolini, L., Castelfranchi, C.: The cognitive and behavioral mediation of institutions: Towards an account of institutional actions. Cognitive Systems Research 7(2-3), 307–323 (2006)
17. Hodgson, G.M.: Economics and Institutions: A Manifesto for a Modern Institutional Economics. Polity Press, Cambridge (1988)
18. Dias, M.B., Zlot, R.M., Kalra, N., Stentz, A.: Market-based multirobot coordination: a survey and analysis. Proceedings of the IEEE 94(7), 1257–1270 (2006)
19. Lima, P.U., Custódio, L.M.: Multi-Robot Systems. In: Innovations in Robot Mobility and Control. Studies in Computational Intelligence, vol. 8, pp. 1–64. Springer, Heidelberg (2005)

Investigating the Evolution of Cooperative Behaviour in a Minimally Spatial Model

Simon T. Powers and Richard A. Watson

School of Electronics and Computer Science, University of Southampton, Southampton, U.K., SO17 1BJ
stp05r@ecs.soton.ac.uk; raw@ecs.soton.ac.uk

Abstract. It is well known that the evolution of cooperative behaviour is dependant upon certain environmental conditions. One such condition that has been extensively studied is the use of a spatially structured population, whereby cooperation is favoured by a reduced number of interactions between cooperators and selfish cheaters. However, models that address the role of spatial structure typically use an individual-based approach, which can make analysis unnecessarily complicated. By contrast, non-spatial population genetics models usually consist entirely of a set of replicator equations, thereby simplifying analysis. Unfortunately, these models cannot traditionally be used to take account of spatial structure, since they assume that interaction between any pair of individuals in a population is equally likely. In this paper, we construct as model that is still based on replicator equations, but where spatial localisation with respect to the number of interactions between individuals is incorporated. Using this model, we are able to successfully reproduce the dynamics seen in more complex individual-based models.

1 Introduction

It is well known that the existence of spatial structure in a population can (although not always, e.g. [1]) facilitate the evolution of cooperative strategies. For example, consider an evolving population of individuals playing the Prisoner's Dilemma game, where the payoff that an individual receives modifies its fitness. If an individual plays only one round against each opponent, and the opponents are drawn at random, then a population initialised with a mixture of cooperate and defect strategies will evolve towards every individual playing defect [2]. This situation corresponds to a freely-mixed world with no spatial structure, since an individual is equally likely to interact with any other member of the population. On the other hand, if the population has a spatial structure, so that an individual only plays against its neighbours, then it is possible for spatial clusters of cooperators to persist [3]. This is due to the fact that for the cooperate strategy to be viable, the cooperator must play against other cooperators and minimise contact with defectors.

These two cases of freely-mixed and spatial worlds are often modelled using different techniques. To model the evolution of the population in the freely-mixed

F. Almeida e Costa et al. (Eds.): ECAL 2007, LNAI 4648, pp. 605–614, 2007.

case, it is only necessary to keep track of the proportion of individuals playing each strategy. How the frequency of each strategy changes over time can then be calculated by means of differential or difference equations; this is known as a replicator dynamics model [4]. By contrast, the modelling of the evolution of a spatially structured population is usually achieved by explicitly representing each individual; this technique is known as individual-based modelling. An example of such a model is that of the spatial Prisoner's Dilemma produced by Nowak & May [3], where a rectangular grid was divided into cells, with each cell containing one individual. In this model, interactions only occurred between neighbouring cells.

The problem with using an individual-based model is that it can be a difficult and lengthy task to understand the root processes causing the observed behaviour [5]. For example, Nowak & May [3,6] found that chaotic patterns of cooperate and defect cells emerged in their apparently simple 2D grid model. Conversely, replicator dynamics models are easy to analyse due to their transparency; the processes at work are explicitly stated in the difference equations. However, they cannot be used in the spatial case because they assume that each individual is equally likely to interact with any other individual. In this paper we aim to present a model that combines the simplicity of replicator equations with the ability of individual-based approaches to model spatially-structured populations.

Our model is based on the scenario of different growth (replication) rates under resource-limited conditions. Specifically, we assume that a trade-off exists between growth rate and efficiency of resource usage, an idea that has been widely explored in biological models. Specific examples of existing models that address this trade-off include individual-based models of bacterial growth [7,8,9], and models of the virulence versus productivity trade-off in disease organisms [10,11]. Such models usually consider two strategies; a low growth rate / high efficiency strategy (cooperative), and a high growth rate / low efficiency strategy (selfish). If none of the limiting resource is shared between individuals playing different strategies, then the cooperative type will win due to its efficiency advantage. Conversely, if the resource is freely-shared between different strategies, then the selfish type will win, due to its larger growth rate [7,8,9]. This is despite the fact that the wasteful resource usage of the selfish type means that it will reach a smaller biomass than if all individuals used the cooperative strategy, a scenario known as the "Tragedy of the Commons" [12].

We model groups of individuals using the same growth rate / efficiency strategy as indivisible "colonies", where the ratio of colony biomasses (sizes) can be interpreted as strategy frequencies. A colony therefore represents a particular growth rate strategy. By treating all individuals with the same strategy / genotype uniformly, our model is still able to use replicator equations. This amounts to an assumption that spatial structure within a colony does not matter with respect to the result of competition with another colony, i.e. that the effect of spatial localisation between individuals playing the same strategy is insignificant. A key aim of this paper is to show that given this assumption, the results with respect to the evolution of cooperation are unchanged compared to those models that do explicitly model spatial structure between individuals playing the same strategy.

Spatial localisation in interactions between strategies / colonies is introduced by providing each colony with its own resource supply. The degree of spatial localisation can then varied by changing the amount of this resource that is shared with the other colony. This approach can therefore model completely spatial and freely-mixed worlds, as well as positions between these extremes.

The key concept of spatially structured populations that is relevant to the evolution of cooperation is a reduced frequency of interactions with non-neighbours. By introducing degree of spatial localisation as degree of resource segregation, we aim to capture this key concept. We then hypothesise that this is all that is necessary to capture the competition and invadability dynamics seen in more complex models.

The advantage of this minimal approach is that it allows our model to contain fewer parameters and assumptions compared to those that use an individual-based approach. For example, the authors of grid-world style individual-based models must make decisions such as to only allow an individual to reproduce if there is an empty cell next to it, as was the case in [8]. It is then not clear whether this decision will affect the results obtained. The aim of our minimally spatial model is to include only the concept of space in a population which affects the evolution of cooperation, namely reduced interactions with non-neighbours. Consequently, the number of modelling decisions to be made are minimised. Furthermore, the use of replicator equations eases analysis by making explicit the processes at work.

2 A Minimal Model of Colony Growth

A definition of the symbols used throughout this paper is provided in Table 1.

The change in biomass of a colony at each time-step in our model is as stated in (1).

$$b_{i(t+1)} = b_{i(t)} + g_{i(t)} - k_{i(t)} \tag{1}$$

This shows that the biomass of a colony changes due to both growth and death. The process of colony growth is then as defined in (2). Specifically, the magnitude of biomass increase made by a colony is proportional to both its biomass and to the amount of available resource, where the constant of proportionality is its intrinsic growth rate, as defined by its strategy. The amount of growth a colony can make is therefore limited by the amount of resource that is available to it.

$$g_{i(t)} = G_i b_{i(t)} r_{i(t)} \tag{2}$$

The amount of mortality suffered by the colony is given in (3), and is proportional to biomass and inversely proportional to resource availability.

$$k_{i(t)} = \frac{K b_{i(t)}}{r_{i(t)} + 1} \tag{3}$$

Table 1. Definition of symbols.

Symbol	Definition
G	Growth rate constant
U	Resource usage rate constant
I	Resource influx constant
K	Death rate constant
D	Resource diffusibility constant
b	Biomass
g	Growth achieved at a given time-step
k	Death at given time-step
r	Available resource level
c	Resource consumed at given time-step
d	Magnitude of resource diffusion
β	Modification of d based on relative colony biomasses
M	Maximum effect that β can have
i	Colony index
t	Time index

The amount of resource consumed by the colony at a time-step (4) is proportional to its growth, with its intrinsic resource usage rate, defined by its strategy, as the constant of proportionality.

$$c_{i(t)} = U_i g_{i(t)} \tag{4}$$

The net change in a colony's resource level depends upon three things (5): the resource used during growth (4), a constant influx from the environment to the colony, and exchange of resource with a competing colony via diffusion (6).

$$r_{i(t+1)} = r_{i(t)} + I - c_{i(t)} + d_{i(t)} \tag{5}$$

$$d_{i(t)} = D\left(\beta_{j(t)} r_{j(t)} - \beta_{i(t)} r_{i(t)}\right) \tag{6}$$

The diffusion process in (6) acts to try and equilibrate the resource levels of two competing colonies, i and j. How large an effect this has depends on two things: The setting of the resource diffusibility constant and a term β, defined in (7) and (8) below. Changing the diffusibility constant has the effect of varying the degree of spatial localisation; if it is set to 0.5, then each colony's resource supply is shared freely with the other colony. On the other hand, if it is set to 0 then no movement of resource (interaction) between colonies occurs.

$$\beta_{i(t)} = 1, \text{ if } b_{i(t)} > b_{j(t)} \tag{7}$$

$$\beta_{i(t)} = 1 + (M-1)\left(\frac{1}{b_{i(t)}}\right), \text{ otherwise} \tag{8}$$

The role of β, as defined in (7) and (8), is to make the magnitude of the force of diffusion dependant upon the respective colonies' biomasses. This is motivated by the idea of outward geometric colony growth from a single point in a spatial world. Specifically, in a small colony, a greater proportion of the colony members will be on the colony's edge and therefore in contact with members of the competing colony. By contrast, in a larger colony, a greater proportion of the colony's members will be shielded from such contact by the layer of edge members. Therefore, small colonies will have a greater proportion of their members interact with the rival colony, hence providing the motivation for increased resource sharing if a colony is small. The purpose of (7) is to make this change operate in one direction only, i.e. so that the amount of resource transferred via diffusion can only ever be increased by this effect. M is a constant that determines the maximum effect that this process can have.

Finally, it is worth stressing that movement of resource only occurs *between*, and not *within*, colonies. Within a colony, all individuals are treated in a non-differentiated manner. This factor allows colony growth to be modelled via the replicator equation (1).

3 Investigations with the Model

This section details three scientific questions that can be addressed with the model, presenting detailed results of our investigations into the first two. In all of the experiments, we consider two competing growth rate / resource efficiency strategies. Namely, a cooperative strategy that uses resource efficiently but that grows at a slower rate, and a selfish strategy that grows at a faster rate but that wastes the growth-limiting resource.

It should be noted that we do not, in these experiments, introduce new strategies via mutation. Instead, we explore how strategy frequencies change through time, in the style of evolutionary game theory [13]. Finally, throughout this paper, the setting of the model constants given in Table 2 are used.

Table 2. Settings of model constants used throughout the paper.

Constant	Value
$G_{selfish}$	0.02
$G_{cooperative}$	0.01
$U_{selfish}$	0.2
$U_{cooperative}$	0.1
I	1
K	0.1
M	2

3.1 Equilibrium Size of Cooperative and Selfish Colonies Under Varying Degrees of Spatial Localisation

The most obvious question that can be addressed with the model is how the equilibrium size of competing colonies of cooperative and selfish individuals varies

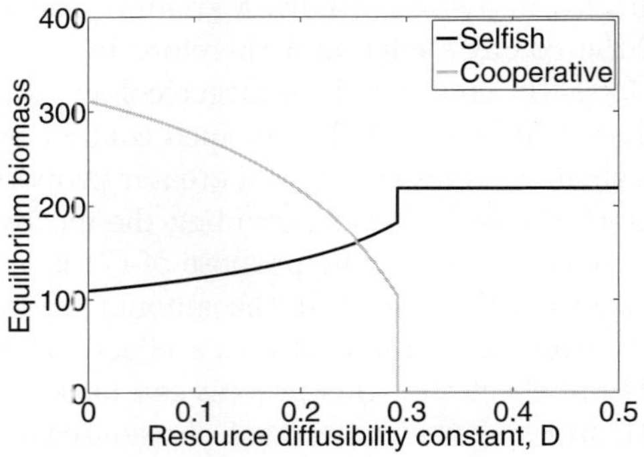

Fig. 1. Equilibrium biomasses of cooperative and selfish colonies.

as the degree of spatial localisation changes. This question can be addressed through our model by starting each colony off with unit biomass and iterating through the difference equations until the equilibrium biomasses are reached. The results of this, for varying D, are presented in Fig. 1.

These results successfully reproduce those found in more complex models. Specifically, in a freely-mixed world (large D) the slower growing cooperators are driven extinct by their less efficient but faster growing rivals, as in the "Tragedy of the Commons" [12]. Conversely, under maximal spatial localisation coexistence of both types is obtained, with the cooperators reaching a larger biomass. Therefore, our model has successfully captured the relevant concepts of spatial localisation, without recourse to individual-based modelling.

What happens between the extremes of freely-mixed and maximal spatial localisation cannot be predicted directly from the equations. In particular, an interesting finding is that as soon as D moves above a critical threshold, the cooperative colony is driven to extinction. The reason for this sudden extinction is that when the cooperative colony's biomass drops below a critical size then it looses too much resource to the selfish colony, due to the β term defined in (8). The loss of a large amount of resource means that the cooperative colony can no longer maintain its biomass level, i.e. its biomass loss due to mortality is greater than its biomass increase due to growth. This then sets off a positive feedback loop whereby as the cooperative colony's biomass decreases it looses more resource to the selfish colony, which then causes its biomass to drop even further and so on. It is the initiation of this positive feedback loop that causes the cooperative colony to go extinct as soon as D rises above the threshold.

3.2 Invadability Dynamics

Our model can also be used to investigate issues of strategy invadability [13]. Means by which an invader playing another strategy can arise in nature include genetic mutation [11,14], phenotypic switching of individuals due to changing environmental conditions [9] and migration of individuals from other demes [15].

Invasion scenarios are created in our model by allowing a colony of one strategy to grow to equilibrium biomass, and then introducing a colony of a different strategy into the same world. A successful invasion is one in which the invading colony survives above 0 biomass and therefore moves the strategy frequencies to a new equilibrium. On the other hand, if the invasion fails then the strategy frequencies will remain at their previous equilibrium, i.e. where the host strategy is at fixation.

The invadability dynamics present in existing models that we aimed to recreate are, firstly, that a single selfish mutant should always be able to invade into a cooperative host. In other words, a colony of cooperators should always be vulnerable to exploitation by a selfish cheater that grows at a faster rate. The second dynamic seen in other models is that cooperators can only invade into a selfish host if there is a sufficient number of them, so that the majority of their interactions will be with each other, rather than with the host. This idea of there being fewer interactions with the host in a larger invading colony is captured in our model by the fact that the amount of resource exchanged between host and invader via diffusion is inversely proportional to the biomass of the invader, as per (8). We therefore hypothesised that all that is required to reproduce this dynamic is to make the amount of resource shared / number of interactions between host and invader inversely proportional to the number of invaders.

We first considered the vulnerability to invasion of a cooperative host by a selfish invader. The results obtained were that an invading selfish colony of unit biomass could always survive and grow, regardless of the degree of spatial localisation. The equilibrium strategy frequencies reached after an invasion under varying degrees of spatial localisation are shown in Fig. 2. The key trend is that the frequency of the selfish strategy increases with D, as would be expected from the results in section 3.1. Furthermore, at settings of D above approximately 0.32 the cooperative host is driven to extinction by the growth of the invader. The conclusion is that a coexistence of cooperative host and selfish invading strategies is reached in worlds with moderate degrees of spatial localisation ($D < 0.32$), while the invading selfish strategy reaches fixation in more freely-mixed worlds ($D \geq 0.32$).

The second question that we consider is the size of a cooperative colony necessary to invade a selfish host. Figure 3 shows the necessary size under varying degrees of spatial localisation, where the necessary biomasses for invasion have been rounded to the nearest integer. Below a threshold value of $D = 0.17$, invasion by unit biomass is always successfully. Above this point, but below a second threshold, cooperative colonies with larger biomasses can successfully invade. Such colonies of greater than unit biomass are analogous to clusters in grid-world style models. The advantage of clusters of cooperators comes from reduced interactions with the selfish host. Our model has shown that this can be achieved simply by making the amount of resource shared inversely proportional to the invader's biomass.

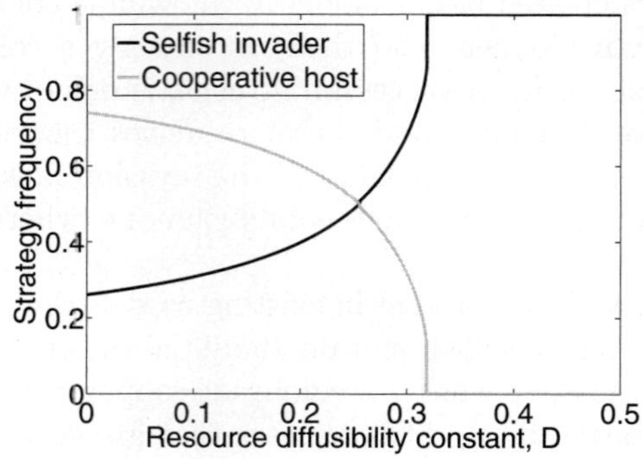

Fig. 2. Equilibrium strategy frequencies after invasion of a cooperative host by a selfish colony, under varying degrees of spatial localisation, D.

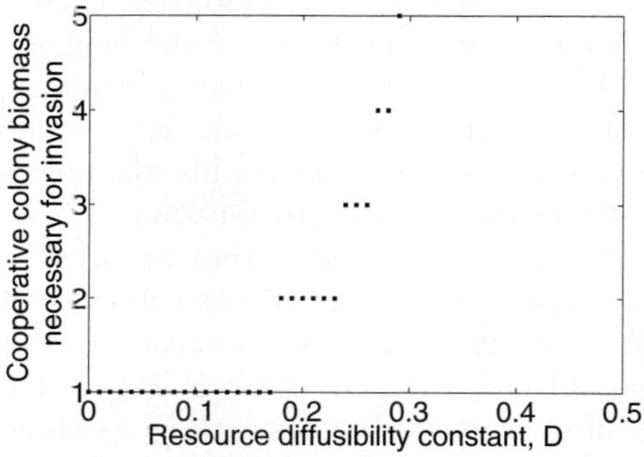

Fig. 3. Biomass of a cooperative colony necessary to invade a selfish host, under varying degrees of spatial localisation, D. Invasion is not possible for values of D above 0.28.

3.3 Genetic Mutation as a Means of Adaptively Setting the Growth Rate / Efficiency Trade-off

The previous section considered invadability dynamics where the invading strategy was very different from that of the host, i.e. there was a large difference in their growth rate / efficiency parameter setting. However, the strategy space in our model is continuous, allowing for a small change in strategy to arise, e.g. via genetic mutation. It may therefore be possible for the canonical cooperative strategy used throughout this paper to evolve through a series of successive small mutations, where each mutant is slightly less selfish than the host strategy. Of most significance is the fact that this may be able to occur in worlds where the canonical cooperative type cannot invade directly as an individual (colony of unit biomass).

We therefore address in this section the question of whether or not a single mutant (colony of unit biomass) that is only slightly more cooperative than the established selfish strategy can successfully invade. We hypothesise that this only slightly more cooperative mutant will be able to invade at settings of D where a more cooperative type cannot. The reasoning is that the closer the mutant is to the established selfish type, the faster it will grow and prevent a large resource flow away from itself via (8).

To test this hypothesis, we considered a world occupied by a single selfish colony at equilibrium biomass and where $D = 0.24$. At this setting, an initial biomass of 3 is necessary for the cooperative strategy to invade a host selfish colony (see Fig. 3). However, we considered a strategy whose growth and resource usage rates are both 95% of that of the selfish type, i.e. only slightly more cooperative. We found that unit biomass of this strategy could invade a host selfish colony, with equilibrium strategy frequencies of 0.49 for the host and 0.51 for the invader. This successful invasion by unit biomass of a slightly more cooperative type suggests that the canonical cooperative type could arise via a series of small mutations in worlds where it cannot invade directly.

4 Conclusion

It is well known from a plethora of existing individual-based models that a spatially structured population can promote cooperative behaviour. In this paper, we have presented a very simple model, based on replicator equations, that allows us to reproduce key results from these more complex models. In particular, we have chosen to capture the concept of spatial localisation in interactions between individuals using the concept of degree of resource sharing. Although our model is framed in terms of colony growth, the questions that it addresses are much more general and are fundamental to explaining the evolution of cooperative behaviour.

The first key conclusion from our model is that, in order to capture the effects of spatial localisation on the evolution of cooperation, it is only necessary to model the effect of spatial localisation between individuals playing different strategies; individuals playing the same strategy can be treated homogeneously. The second key conclusion is that the invadability results from individual-based spatial models, in which cooperators can usually only invade in clusters, can be reproduced simply by making the amount of resource shared inversely proportional to the invader's biomass. This therefore captures the notion of a spatial cluster of cooperators, without having to use a more complex grid-world style model or a social network [16].

It is unavoidable that choices are made about the specific details of the equations defining the model, and of the settings of the constants. It follows that changing these details would quantitatively change the results, e.g. the values of D for which strategy coexistence is sustainable and the colony sizes necessary for invasion. However, qualitative properties of the dynamics should be robust, i.e. there should always be a threshold value of D below which strategy coexistence is sustainable. Furthermore, there were less choices to be made with our approach than with an individual-based model. From this perspective, our results are more robust than those of individual-based models, since there are less parameters to be set. On the other hand, our model is more general than traditional replicator dynamics approaches, as we have been able to model spatial localisation, something that cannot be done with a traditional replicator dynamics model.

In future work, we intend to investigate the relationship between our model and spatial and freely-mixed versions of the Prisoner's Dilemma game. In particular, we would like to make explicit the relationship between the cooperative and selfish strategies of colony growth in our model and the cooperate and defect strategies in the Prisoner's Dilemma.

References

1. Hauert, C., Doebeli, M.: Spatial structure often inhibits the evolution of cooperation in the snowdrift game. Nature 428, 643–646 (2004)
2. Axelrod, R., Hamilton, W.D.: The evolution of cooperation. Science 211(4489), 1390–1396 (1981)
3. Nowak, M.A., May, R.M.: Evolutionary games and spatial chaos. Nature 359, 826–829 (1992)
4. Taylor, P.D., Jonker, L.B.: Evolutionary stable strategies and game dynamics. Mathematical Biosciences 40, 145–156 (1978)
5. Paolo, E.A.D., Noble, J., Bullock, S.: Simulation models as opaque thought experiments. In: Bedau, M.A., McCaskill, J.S., Packard, N., Rasmussen, S. (eds.) Proceedings of the Seventh International Conference on Artificial Life, pp. 497–506. MIT Press, Cambridge (2000)
6. Nowak, M.A., May, R.M.: The spatial dilemmas of evolution. International Journal of Bifurcation and Chaos 3(1), 35–78 (1993)
7. Pfeiffer, T., Schuster, S., Bonhoeffer, S.: Cooperation and competition in the evolution of ATP-producing pathways. Science 292(5516), 504–507 (2001)
8. Pfeiffer, T., Bonhoeffer, S.: An evolutionary scenario for the transition to undifferentiated multicellularity. PNAS 100(3), 1095–1098 (2003)
9. Kreft, J.U.: Biofilms promote altruism. Microbiology 150, 2751–2760 (2004)
10. Bull, J.J., Molineux, I.J., Rice, W.R.: Selection of benevolence in a host-parasite system. Evolution 45(4), 875–882 (1991)
11. Kerr, B., Neuhauser, C., Bohannan, B.J.M., Dean, A.M.: Local migration promotes competitive restraint in a host-pathogen 'tragedy of the commons'. Nature 442, 75–78 (2006)
12. Hardin, G.: The tragedy of the commons. Science 162, 1243–1248 (1968)
13. Maynard Smith, J.: Evolution and the Theory of Games. Cambridge University Press, Cambridge (1982)
14. Rainey, P.B., Rainey, K.: Evolution of cooperation and conflict in experimental bacterial populations. Nature 425, 72–74 (2003)
15. Killingback, T., Bieri, J., Flatt, T.: Evolution in group-structured populations can resolve the tragedy of the commons. Proceedings of the Royal Society B 273, 1477–1481 (2006)
16. Ohtsuki, H., Hauert, C., Lieberman, E., Nowak, M.A.: A simple rule for the evolution of cooperation on graphs and social networks. Nature 441, 502–505 (2006)

Modeling Decentralized Organizational Change in Honeybee Societies

Mark Hoogendoorn, Martijn C. Schut, and Jan Treur

Vrije Universiteit Amsterdam, Department of Artificial Intelligence
De Boelelaan 1081a, 1081 HV Amsterdam, The Netherlands
{mhoogen, schut, treur}@cs.vu.nl

Abstract. Multi-agent organizations in dynamic environments, need to have the ability to adapt to environmental changes to ensure a continuation of proper functioning. Such adaptations can be made through a centralized decision process or come from the individuals within the organization. In the domain of social insects, such as honeybees and wasps, organizations are known to adapt in a decentralized fashion to environmental changes. An organizational model for decentralized organizational change is presented that can aid in analyzing and designing such organizations. The model is specified by dynamic properties at different aggregation levels. At the lowest level such properties characterize the behavior of individual roles, which can be related to higher level properties that express important elements such as survival of an organization. A honeybee colony is used as a case study.

1 Introduction

The concept of organization has been studied in sciences such as social science and economics, but recently also in artificial intelligence [3;4;7]. With the desire to analyze and design more complex systems consisting of larger numbers of agents (e.g., in nature, society, or software), the need arises for a concept of higher abstraction than the concept agent. To this end, organizational modeling is becoming a practiced stage in the analysis and design of multi-agent systems. Hereby, the environment in which the multi-agent organization participates has to be taken into consideration. An environment can have a high degree of variability which might require organizations that change to adapt to the environment's dynamics, to ensure a continuous proper functioning of the organization. Hence, such change processes are a crucial function of the organization and should be part of the organizational model.

An organizational model incorporating organizational change can be specified in two ways: from a centralized perspective, in which there is a central authority that determines the changes to be performed within the organization, taking into account the current goals and environment. In [5] for example, an organizational model is presented which is specified from such a perspective, whereby a single entity, called the change manager, is responsible for directing the change. A second possibility is to create a model for organizational change from a decentralized perspective, in which each agent decides for himself if and how to change its own role allocations. In the latter approach, it is much more difficult for the organization as a whole to change in

F. Almeida e Costa et al. (Eds.): ECAL 2007, LNAI 4648, pp. 615 – 624, 2007.

a coherent way, still satisfying the goals set for the organization, as there is no overall view of the organizational change. The approach might however be the only possibility for an organization to perform change as a central authority for performing change could be non existing or infeasible due to the nature of the organization. In the domain of social insects, such as honeybees and wasps, organizations are known to adapt in a decentralized fashion to environmental changes. This paper presents a model for decentralized organization change appropriate for such phenomena as occur in Nature. Such a model can aid developers of multi-agent systems in creating and analyzing such an organization. The description of the model abstracts from the actual tasks being performed by the organization. The scope of the model is broader than simply being able to model social insects: the mechanisms incorporated in the model facilitating decentralized organizational change may work in other types of organizations as well. In [1] for example, a comparable approach is used for finding an optimal allocation of cars to paint booths.

To evaluate the model being proposed, as a case study the honeybee (Apis Mellifera) has been investigated. For this domain the model has been instantiated. The instantiated model has been validated against properties as acquired from biological experts. A number of different roles have been identified in the literature (see e.g., [8;11]). For the sake of brevity only five will be addressed here: (1) a *brood carer* takes care of feeding the larvae within the bee hive; (2) a *patroller* guards the hive by killing enemies entering the hive; (3) a *forager* harvests food to be stored in the hive; (4) an *undertaker* cleans the hive of corpses, and (5) a *resting worker* simply does nothing.

Switching between roles is triggered by changes in the environment observed by the bees. Such observations differ per bee. Each role has a specific trigger, for which a bee has a certain threshold that determines whether this is the role it should play. The bee always plays the role for which it is most triggered. For example, bees are triggered to start playing the *brood carer* role when they observe the larvae emitting a too high level of hunger pheromones. Once they are allocated to the role, they start getting food from the combs and feed the larvae that are emitting the pheromones. A trigger for the *patroller* role is the amount of enemies observed around the hive. *Foragers* that have returned from their hunt for food, communicate the location where they found the food by means of the honeybee dance (see [2]). For other bees currently not playing the *forager* role, such a dance is a trigger to start playing the *forager* role. The more corpses there are, the more bees are being triggered to switch from their current role to being *undertaker*. Bees perform the *resting worker* role in case they are not sufficiently triggered for any other role.

Section 2 presents the methodological approach used. The model for decentralized organizational change is described in Sections 3 (properties at organization level) and 4 (role properties). Results of a simulation of the organizational model instantiated with domain-specific knowledge of the bee colony are shown in Section 5, and finally Section 6 concludes the paper.

2 Modeling Organizational Dynamics

To enable modeling an organization, an expressive language is needed that has the ability to describe the dynamics of such an organization. For this purpose TTL (Temporal Trace Language) has been adopted cf. [6]. TTL allows for the formal specification of dynamic properties on multiple levels of aggregation. The bottom level addresses role properties, describing the required behavior for each of the roles within the organization. On the top level organization properties are defined, expressing the overall goals or requirements for the organization. An advantage of using TTL is that an executable subset has been defined called *leadsto* which is of the form $\alpha \twoheadrightarrow_{e,f,g,h} \beta$ that states that if α holds for duration g then β will holds for duration h with a delay between e and f. In case role properties are expressed in this executable format, the organizational model can be simulated by putting certain (e.g., environmental) events in the model (without including agents in the model), resulting in a trace of the organizational behavior. The top level organization properties can thereafter be checked against the trace by means of an automated tool called *TTL checker* to see whether the organizational model indeed satisfies the goals or requirements set for it, given the events that have been put into the model. Using the results of these checks, statements can be made about the behavior of the organization, when the agents comply to the role properties that have been defined. More details and the semantics for TTL can be found in [9]. Examples and explanation of properties expressed in TTL are shown in Section 3 and 4.

3 Organizational Properties

The model for decentralized organizational change presented here takes the form of a hierarchy of dynamic properties at two aggregation levels: that of the organization, and that of the roles within the organization. This section describes a number of such properties as well as the relationships between them.

The highest level requirement for the organization as a whole as inspired by the biological domain experts, is survival of the population given a fluctuating environment, in other words, population size needs to stay above a certain threshold M.

OP1(M) Surviving Population
For any time t, a time point t'≥t exists such that at t' the population size is at least M.

Formal:

$\forall t\ \exists t' \geq t, v : state(\gamma, t') \models total\_living\_population\_count(v)\ \&\ v \geq M$

Here $state(\gamma, t') \models total\_living\_population\_count(v)$ denotes that within the state $state(\gamma, t')$ at time point t' in trace γ the state property total_living_population_count(v) holds, denoted by the (infix) predicate $\models$ for the satisfaction relation. Such a high-level requirement is refined by means of a property hierarchy, depicted as a tree in Figure 1. At the highest level OP1 is depicted which can be refined into a number of properties (in Figure 1 *n* properties) each expressing that for a certain aspect the society is in good condition, characterized by a certain *value* for a variable (the *aspect variable*) that is to be maintained. The property template for an aspect X is as follows:

OP2(X, P1, P2) Organization Aspect Maintenance
For all time points t
If v is the value of aspect variable X at t, then v is between P1 and P2

$\forall t, v : state(\gamma, t) \models has\_value(X, v) \Rightarrow P1 \leq v \leq P2$

Sometimes one of the two bounds is omitted, and it is only required that value v is at least P1 (resp., at most P2). For the example bee society the aspects considered are *wellfed brood*, *safety*, *food storage*, and *cleanness* (addressed, respectively, by Brood Care, Patroller, Forager, and Undertaker roles). For each of these aspects a variable was defined to indicate the state of the society for that aspect. For example, for wellfed brood, this variable concerns relative larvae hunger, indicated by the larvae pheromone rate.

In order to maintain the value of an aspect variable X, a certain effort is needed all the time. To specify this, a property that expresses the *effort* made by the organization on the aspect, is introduced. Notice that the notion of provided effort at a time point t can be taken in an *absolute* sense (for example, effort as the amount of feeding work per time unit), but it can also be useful to take it in a *relative* sense with respect to a certain overall amount, which itself can vary over time (for example, effort as the fraction of the amount of feeding work per time unit divided by the overall number of larvae). Below the latter, relative form will be taken. The general template property for aspect effort is as follows:

OP3(X, W1, W2) Sufficient Aspect Effort
For all time points t the effort for aspect X provided by the organization is at least W1 and at most W2.

$\forall t, v : \text{state}(\gamma, t) \models \text{provided\_effort}(X, v) \Rightarrow W1 \leq v \leq W2$

For the bee colony, for instance, the brood care workers take care that the larvae are well-fed. The effort to maintain the hunger of larvae at a certain low level is feeding the larvae. Here provided effort for brood care is defined as the brood care work per time unit divided by the larvae population size. Brood care work is taken as the amount of the (average) brood care work for one individual brood carer times the number of brood carers.

Whether the refined properties given above will always hold, depends on the flexibility of the organization. For example, in the bee colony case, if the number of larvae or enemies increases, also the number of brood care workers, respectively patrollers should increase. If the adaptation to the new situation takes too much time, the property Brood Care Effort will not hold for a certain time. In principle, such circumstances will damage the success of the organization. Therefore, an adaptation mechanism is needed that is sufficiently flexible to guarantee the properties such as Brood Care Effort. For this reason, the adaptation flexibility property is introduced, which expresses that when the effort for a certain organization aspect that is to be maintained is below a certain value, then within a certain time duration d it will increase to become at least this value. The smaller this parameter d is, the more flexible is the adaptation; for example, if d is very large, the organization is practically not adapting. The generic property is expressed as follows:

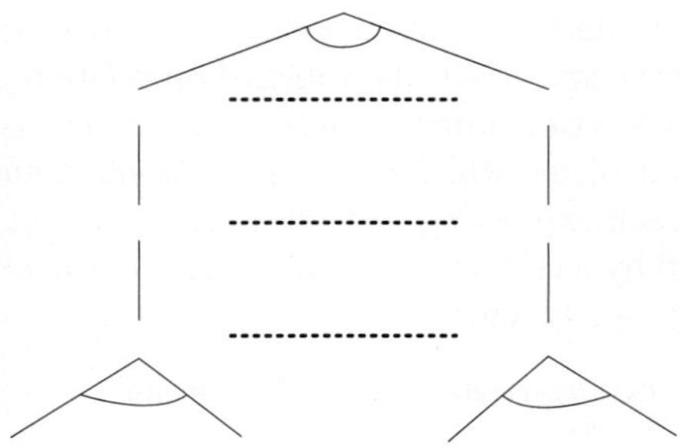

Fig. 1. Property hierarchy for decentralized organizational change

OP4(X, B, d) Adaptation Flexibility
At any point in time t, if at t the effort for aspect X provided by the organization is lower than B, then within time duration d the effort will become at least B.

$\forall t, v1$ [[state(γ, t) |= provided_effort(X, v1) & v1 < B] $\Rightarrow$
$\exists t' \geq t, v2$: [t' $\leq$ t+d & state(γ, t') |= provided_effort(X, v2) & v2 $\geq$ B]]

An assumption underlying this property is that not all aspects in the initial situation are critical, otherwise the adaptation mechanism will not work. OP3 expressing that sufficient effort being provided directly depends on this adaptation mechanism as shown in Figure 1. OP4 depends on role properties at the lowest level of the hierarchy, which are addressed in the next Section.

4 Role Properties

Roles are the engines for an organization model: they are the elements in an organization model where the work that is done is specified. The properties described in Section 3 in an hierarchical manner have to be grounded in role behavior properties as the lowest level properties of the hierarchy. In other words, specifications of role properties are needed that entail the properties at the organizational level described in Section 3. In the behavioral model two types of roles are distinguished: Worker roles which provide the effort needed to maintain the different aspects throughout the organization, and Member roles which have the function to change Worker roles. Each Member role has exactly one shared allocation with a Worker role. The role behavior for the Worker roles within the organization is shown in Section 4.1, whereas Section 4.2 specifies the behavior for the Member roles.

4.1 Worker Role Behavior

Once a certain Worker role exists as an active role, it performs the corresponding work. What this work exactly is, depends on the application: it is not part of the organization model. The property directly relates to OP4 which specifies the overall effort provided, as shown in Figure 1. Note that Figure 1 only shows the generic form of the role property (depicted as RP(w(a_i),d_i,W_i) where a_i is the specific aspect and w(a_i) the Worker role belonging to that aspect) whereas in an instantiated model a role property is present for each instance of the Worker role providing the effort for the specific aspect. In a generic form this is specified by:

RP(R, d, W) Worker Contribution
For all t there is a t' with t $\leq$ t' $\leq$ t + d such that at t' the Worker role R delivers a work contribution of at least W.

$\forall t\ \exists t' \geq t, v$: [t' $\leq$ t+d & state(γ, t') |= work_contribution(R, v) & v $\geq$ W]]

Here work_contribution is part of the state ontology for the output of the role. For each of the specific roles it can be specified what the work contribution is in terms of the domain specific state ontology (e.g., the number of larvae to be fed for the brood carer role).

4.2 Member Role Behavior

By a Member role M decisions about taking up or switching between Worker roles are made. As input of this decision process, information is used about the well-being of the organization, in particular about the different aspects distinguished as to be maintained; these are input state properties indicating the value of an aspect variable X: has_value(X, v). Based on this input the Member role M generates an intermediate state property representing an indication of the aspect that is most urgent in the current situation. In the model the decision mechanism is indicated by a priority relation priority_relation(X_1, v_1, w_1, ..., X_n, v_n, w_n, X) indicating that aspect X has priority in the context of values v_i, respectively norms w_i for aspects X_1, .., X_n. This priority relation can be specialized to a particular form, as shown below by an example specialization in the last paragraph of this section.

RP1(M) Aspect Urgency
At any t, if at t Member role M has norms w_1 to w_n for aspects X_1 to X_n
and receives values v_1 to v_n for X_1 to X_n at its input,
and has a priority relation that indicates X as the most urgent aspect for the combination of these norms and values,
then at some t' ≥t it will generate that X is the most urgent aspect.

∀t, v1, .., vn, w1, .., wn, X
state(γ, t) |= has_value(X_1, v_1) & … & has_value(X_n, v_n) &
has_norm(X_1, w_1) & … & has_norm(X_n, w_n) &
priority_relation(X_1, v_1, w_1, …, X_n, v_n, w_n, X)
⇒ ∃t'≥t state(γ, t') |= most_urgent_aspect(X)

Based on this, the appropriate role for the aspect indicated as most urgent is determined. If it is not the current role sharing an allocation with M, then another intermediate state property is generated expressing that the current Worker role sharing an allocation with M should be changed to the role supporting the most urgent aspect. In other words, the shared allocation of Member role M in the Change Group should change from one (the current) Worker role R1 in Worker Group WG1 to another one, Worker role R2 in Working Group WG2:

RP2(M) Role Change Determination
At any t, if at t Member role M generated that X is the most urgent aspect,
and Worker role R2 is responsible for this aspect, and
R1 is the current Worker role sharing an allocation with M, and R1 ≠ R2,
then at some t' ≥t it will generate that role R2 has to become the Worker role sharing an allocation with M, instead of R1.

∀t, X, R1, R2 state(γ, t) |= most_urgent_aspect(X) &
role_responsible_for(R2, X) & role_reserved_for(R2, M) &
state(γ, t) |= has_shared_allocation(M, R1) & R1≠R2
⇒ ∃t'≥t state(γ, t') |= shared_allocation_change(M, R1, R2)

Based on this intermediate state property the Member role M generates output indicating which role should become a shared allocation and which not anymore:

RP3(M) Role Reallocation
At any t, if at t Member role M generated that Worker role R2 has to become sharing an allocation with M, instead of Worker role R1,
then at some t' ≥t it will generate the output that role R1 will not share an allocation with M and R2 will share an allocation with M.

∀t, R1, R2
state(γ, t) |= shared_allocation_change(M, R1, R2)
⇒ ∃t'≥t state(γ, t') |= not has_shared_allocation(M, R1) &
has_shared_allocation(M, R2)

All three role properties for the Member roles are depicted in Figure 1. The adaptation step property OP4 for all organizational aspects dependent upon it, so each of the OP4 branches depends upon RP1, RP2, and RP3 which have therefore been depicted two times in the Figure.

The generic description for the Member role behavior can be specialized one step further by incorporating a specific decision mechanism. This gives a specific definition of the priority relation priority_relation(X_1, v_1, w_1, ..., X_n, v_n, w_n, X) as has been done for the following decision mechanism based on norms used as thresholds (see e.g. [10]).

1. For each aspect X to be maintained a norm w(X) is present. For the Worker role R1 for X sharing an allocation with Member role M, each time unit the norm has a decay described by fraction r.
2. For each X, it is determined in how far the current value is unsatisfactory, expressed in a degree of urgency u(X) for that aspect.
3. For each aspect with urgency above the norm, i.e., with u(X) > w(X), the relative urgency is determined: u(X)/ w(X)
4. The most urgent aspect X is the one with highest relative urgency.

5 Simulation Results

This section discusses some of the results of simulations that have been performed based on the organizational model, in particular the role properties presented in Section 4 have been put in an executable format and have been instantiated with domain-specific information for bee colonies.

To validate the instantiated simulation model, the high-level dynamic properties from Section 3 were used (in accordance with biological experts). Proper functioning of such an organization in Nature is not self-evident, therefore two simulation runs are compared: one using the adaptation mechanism, and one without. The choice has been made to compare the result of using adaptation with no adaptation due to the fact that comparing with centralized change will obviously result in better performance of the centralized change model. Having a complete picture of the different aspects and their urgencies, gives a major advantage. Whether decentralized adaptation is more successful than no adaptation is however not a trivial matter. In case there is merely coordination by means of observables in the world (which is the case in honeybee colonies), adaptation might even be counterproductive. Note that the results presented here are the results of a simulation of the instantiated organizational model, abstracting from allocated agents. Performing such high-level simulations of an executable organizational model enables the verification of properties against these simulation runs. Hence, it can be checked whether or not the model satisfies the properties or goals considered important. When such properties are indeed satisfied, by allocating agents to the roles that comply to the role properties, the multi-agent system delivers the desired results as well. In the two simulations, several parameters have been set to certain values, where the circumstances are kept identical for both simulations.

External world. Initially, 15 larvae and 10 workers are present for which the initial type of the latter is randomly assigned. The natural mortality age is set to 500 time steps, whereas a larva is grown up after 250 time steps. Every 20 time steps, a new larva is added to the population. The initial food stock is set to 40 units of food. Once every 100 time points an attack of 40 enemies occurs, who stay there until a patroller defeats them. In case over 200 enemies are present in the hive, each individual in the organization is removed with a probability of 0.05 per time step. In case more than 20 dead bodies are present in the hive, individuals are removed with the same probability. Food used by larvae is 0.5 per feed, for workers 1 unit of food per time step.

Larvae. Larvae have an initial pheromone level of 0.5, increasing 0.006 per time step. In case pheromone emissions exceed 0.95, the larva dies. After being fed, the emission level is set to 0.1.

Foragers. Foragers each collect 3 food units per time step.

Brood carers. Feed 1 larvae per 8 time steps, and only feed the larvae with a pheromone level above 0.55.

Undertaker. Carry 1 body per 12 time steps.

Patroller. Defeat 1 enemy per time step.

In the adaptation simulation, the Member thresholds are randomly generated, being somewhat above or below the average observed value of the various triggers. Figure 2 shows results on the performance of the two settings of the organizational model. Figure 2a shows the overall population size over time. The population size of the simulation with adaptation remains relatively stable, whereas without adaptation it drops to a colony of size 3, which is equal to the amount of larvae living without being fed. Figures 2b and 2c show information regarding brood care: Firstly, the average pheromone level, the trigger to activate the allocation to brood carer. Furthermore, the number of active brood carers in the colony is shown. In the case with adaptation their number increases significantly in the beginning of the simulation, as the amount of pheromones observed is relatively high. Therefore, a lot of the brood carer roles are allocated. For example, at time point 300, 15 out of a population of 28 are brood carers.

Despite the fact that the overall pheromone level is not decreasing rapidly, the amount of brood carer roles drops significantly after time point 300. This is due the fact that Member roles can only share an allocation with one Worker role at a time. When another role receives a higher urgency (e.g., there is a huge attack, demanding many patrollers) a switch of worker role takes place. Figure 2d shows the amount of worker roles of the different types (except the resting workers) within the bee colony for the setting with adaptation. The amount of brood carers decreases after time point 300 due to an increase in the amount of shared allocations to the undertaker and forager roles. This results in an increase in pheromone level again, causing a higher delta for brood care again, resulting in more brood carers, etc. The pheromone level finally stabilizes around 0.5 in the organizational model with adaptation. For the setting without adaptation, the brood carers simply cease to exist due to the fact that none of the larvae are growing up. The pheromone level stabilizes at a higher level.

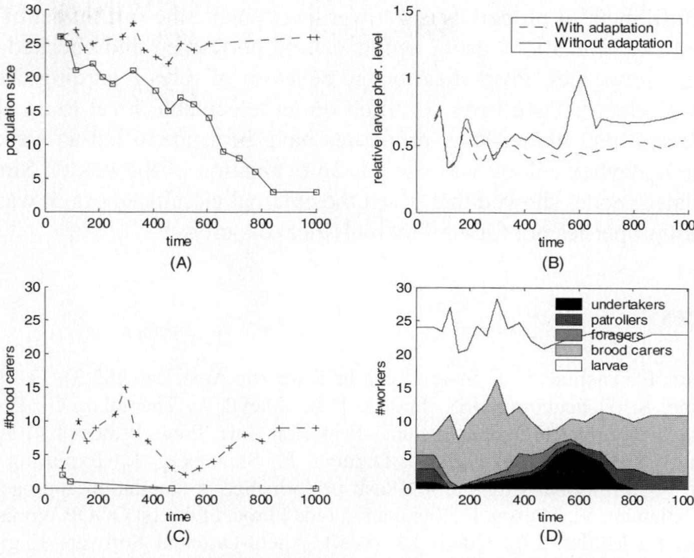

Fig. 2. Results of simulating the bee colony with and without adaptation. Note that (D) only shows the worker types for the adaptive case

The properties from Section 3 have been checked by the automated TTL checker. With the following parameter settings, the properties were validated and confirmed for the organizational model with adaptation and falsified for the one without adaptation: OP1(20), OP2(broodcare,0,0.9), OP3(broodcare,0.15,10000), OP4(broodcare, 0.3, 200).

6 Discussion

The organizational model for decentralized organizational change has been formally specified by means of a methodology which describes the behavior of an organization on multiple aggregation levels; cf. [6]. The model is inspired by mechanisms observed in Nature. The model was used for a honeybee colony case study. The scope of the model is not limited to being a model for social insects: in [1] the effectiveness of such approaches is shown for other domains as well. The model can therefore support organizational modelers and analysts working with multi-agent organizations in highly dynamic environments, without a central authority directing change, in general in designing and analyzing such an organization. The formal specification of the behavior of the organization is described by dynamic properties at different aggregation levels. Once the lowest level properties within the organization are specified in an executable form, the organizational model can be used for simulation abstracting from agents (to be) allocated. Such low level properties can be indicative for the behavior of the agent allocated to that particular role. The possibility also exists to specify the role properties at the lowest aggregation level in a more abstract manner, in a non-executable format. Hierarchical relations between the properties can be identified to show that fulfillment of properties at a lower level entails the fulfillment of the higher level properties. Simulations using agents can be performed and checked for fulfillment of these properties. Properties for the behavior of roles regarding decentralized organizational change have been specified on an executable level to be able to perform simulation, and higher-level properties have been identified as well. The case study of the honeybee colony was used as an evaluation of the model. Simulation of this instantiated model showed that given the external circumstances, it was effective, given overall properties put forward by biological experts.

References

[1] Bonebeau, E., Theraulaz, G.: Swarm Smarts. Scientific American 282(3), 72–79 (2000)

[2] Camazine, S., Deneubourg, J.L., Franks, N.R., Sneyd, J., Theraulaz, G., Bonabeau, E.: Self-Organization in Biological Systems. Princeton Univ. Press, Princeton (2001)

[3] Furtado, V., Melo, A., Dignum, V., Dignum, F., Sonenberg, L.: Exploring congruence between organizational structure and task performance: a simulation approach. In: Boissier, O., Dignum, V., Matson, E., Sichman, J. (eds.) Proc. of the 1st OOOP Workshop (2005)

[4] Giorgini, P., Müller, J.P., Odell, J.J. (eds.): Agent-Oriented Software Engineering IV. LNCS, vol. 2935. Springer, Heidelberg (2004)

[5] Hoogendoorn, M., Jonker, C.M., Schut, M., Treur, J.: Modeling Centralized Organization of Organizational Change. Computational and Mathematical Organization Theory 13, 147–184 (2007)

[6] Jonker, C.M., Treur, J.: Compositional verification of multi-agent systems: a formal analysis of pro-activeness and reactiveness. Int. Journal of Cooperative Information Systems 11, 51–92 (2002)

[7] McCallum, M., Vasconcelos, W.W., Norman, T.J.: Verification and Analysis of Organisational Change. In: Boissier, O., Dignum, V., Matson, E., Sichman, J. (eds.) Proc. 1st OOOP Workshop (2005)

[8] Schultz, D.J., Barron, A.B., Robinson, G.E.: A Role for Octopamine in Honey Bee Division of Labor. Brain, Behavior and Evolution 60, 350–359 (2002)

[9] Sharpanskykh, A., Treur, J.: Temporal Trace Language: Syntax and Semantics, Technical Report, Vrije Universiteit Amsterdam, Dept. of Artificial Intelligence, A'dam (2005)

[10] Theraulaz, G., Bonabeau, E., Deneubourg, J.L.: Response thresholds reinforcement and division of labor in insect societies. Proceedings of the Royal Society of London Series B-Biological Sciences 265, 327–332 (1998)

[11] Winston, M.L., Punnet, E.N.: Factors determining temporal division of labor in honeybees. Canadian Journal of Zoology 60, 2947–2952 (1982)

Social Facilitation on the Development of Foraging Behaviors in a Population of Autonomous Robots

Alberto Acerbi, Davide Marocco, and Stefano Nolfi

Institute of Cognitive Science and Technologies, National Research Council, Via San Martino della Battaglia 44, Rome, 00185, Italy
{alberto.acerbi, davide.marocco, stefano.nolfi}@istc.cnr.it

Abstract. In this paper we propose an adaptive algorithm based on a combination of selective reproduction, individual learning, and social learning. Social learning consists of a simple facilitation process that regulates the strength of individual learning on the basis of the number of individuals located nearby. By testing this model in an experimental scenario, in which a population of 10 mobile robots has to develop a simple foraging behavior, we demonstrate how the model proposed produces effective results. By comparing the results obtained in different experimental conditions we also show how the method proposed outperforms other alternative algorithms based on genetic evolution or individual learning. Finally, we briefly discuss how the model proposed can help us to understand the role of social learning in biological organisms.

Key words: Social Learning, Social Enhancement, Learning and Evolution.

1 Introduction

It is a widespread opinion that social learning dynamics can account both for the diversity of humans' behavioral repertoire [12] and for the complexity of humans' cognitive skills [13]. This consideration has lead most of the researchers that deal with social learning to focus on high–level, human–like, forms of social learning and to attempt to reproduce them through specific algorithms that aim to replicate a behavior of a model by explicitly copying it.

However, when considering learning as a set of adaptive modifications that take place within each single agent in response to environmental stimuli [9], we have to consider that the environment of a single agent also includes its conspecifics. In this perspective, their behavior becomes a valuable source of information that can be exploited in several ways besides from explicit copying. Indeed, especially in the last twenty years, ethologists have shown how other forms of social learning that are much simpler and less cognitive demanding than explicit imitation play an important role in the development of complex behaviors in various vertebrate species. For example, the food preferences developed by Norway rats (*Rattus norvegicus*) are influenced by the food eaten by conspecifics through indirect information obtained by sniffing conspecifics' breath [4]. Similarly, female guppies (*Poecilia reticulata*) prefer to mate with males that they have seen mating before [3]. Again, the development of several primates' behaviors, which have been previously interpreted as a result of explicit imitation, are now considered as an outcome of more simple forms of social

F. Almeida e Costa et al. (Eds.): ECAL 2007, LNAI 4648, pp. 625 – 634, 2007.

learning like *emulation* or *stimulus enhancement* (for reviews on social learning in animals, but also in simulated agents and robots see [2][5][6][14]).

The development of artificial organisms (robots or simulated agents) through artificial life techniques is an excellent way of exploring the feasibility of simple models of social learning [7][8] with particular reference to the possibility to take into account different aspects of the individuals (their body, their nervous system, and the way their adapt to the environment), of the population, and of their interaction. As formerly pointed out in [8], identifying alternative models of social learning, besides imitative learning, is interesting for both scientific and technological reasons. From a scientific point of view, identifying how a population of artificial agents can obtain adaptive advantages by dealing with simple forms of social learning can shed light on how the interaction between individual learning, social interactions, and evolution might produce effective and robust collective dynamics. Indeed, understanding the interaction between these processes through empirical and laboratory experiments can be very difficult. From a technological point of view, on the other hand, the identification of algorithms that allow embodied agents to adapt by exploiting the interaction between evolution, individual learning, and social learning can lead to the development of innovative methods for developing multi-agent systems, spanning from autonomous robots to ubiquitous computing devices.

In this paper we demonstrate how a simple form of social learning, in which agents learn individually but in which the strength of individual learning is modulated by the number of individuals located nearby, can lead to the development of effective behaviors. This will be demonstrated in an experimental scenario in which a population of simulated robotic agents should develop an ability to forage by discriminating different type of foraging areas. Moreover, we show how, in this type of experimental scenarios, the combination of selection at populational level, individual learning, and social learning outperforms other adaptive processes such as individual learning and genetic evolution.

In section 2 we will describe the experimental setup and the different experimental conditions. In section 3 we show how the results obtained in the social learning condition outperform those obtained in the individual learning and genetic evolution conditions. Finally, in section 4, we discuss the implications.

2 Experimental Setup

A team of 10 e-puck robots is placed in an environment that consists in a square arena of 200 x 200 cm surrounded by walls. The floor of the arena is grey and contains two circular target areas with a diameter of 60 cm colored in black and white, respectively (Fig. 1). Robots are provided with simple sensory-motor capabilities that allow them to move, gather information from the environment and to produce sound signals with a fixed intensity.

The robots (Fig. 2 Left) have a circular body with a radius of 37 mm, 8 infrared sensors placed around the body, 1 ground sensor placed on the bottom of robot, 1 microphone and 1 speaker. Robots' neural controllers consist of neural networks with 10 sensory neurons and 3 motor neurons (Fig. 2 Right). Eight sensory neurons encode the activation states of the 8 infrared sensors (which detect obstacles up to a distance

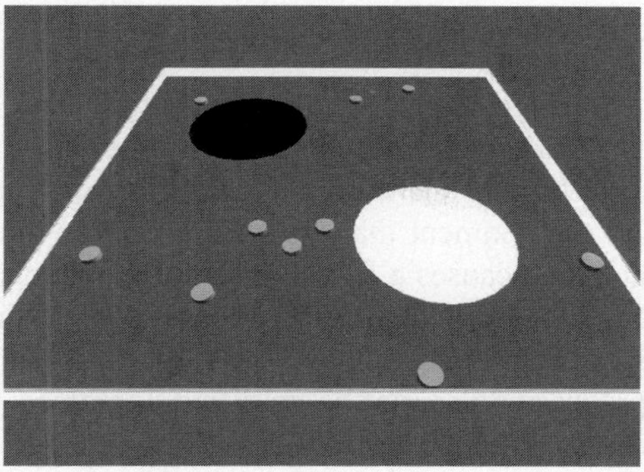

Fig. 1. The environment of the experiment. Black and white circles represent the two foraging areas. Small grey circles represent the 10 robots.

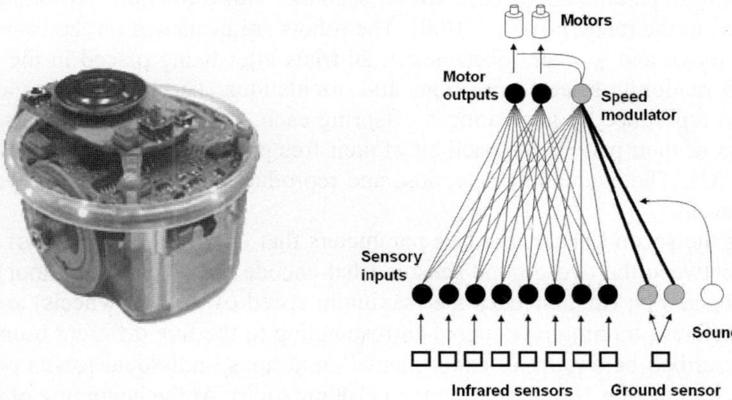

Fig. 2. Left: the e-puck robot – Right: the neural controller. Thick lines represent the weights that are modified in the second phase of the simulation. As it is shown the speed modulator does not operate on the outputs but it acts directly on the motors' activations. The intensity of the sound signal is used to set the learning rate in the social condition.

of ~4 cm) and two neurons that measure the activation of the ground sensors that encode the color of the floor below the robot (grey [0 0], white [1 0], and black [0 1]). The first two motor neurons encode the desired speed of the two corresponding wheels normalized between -MaxSpeed and +MaxSpeed. The third motor neuron instead encodes the maximum speed of both wheels (i.e. the MaxSpeed parameters normalized in the range [0, 10]). The third motor neuron thus acts as a modulator that regulates the speed with which the robot moves or turns at each time step. The 8 sensory neurons that encode the state of the infrared sensors are connected to all the three motor neurons. The two sensory neurons that encode the color of the ground, instead, are connected to the third motor neuron only (i.e. to the motor neuron that regulates the maximum speed of the two wheels).

The robots emit a sound signal through their speakers when they are located in the black or white foraging areas. Moreover, the robots can detect through their microphones the signals produced by other robots located nearby.

2.1 Simulation and Experimental Conditions

The training process is divided in two phases. During the first phase the robots are trained for developing an ability to avoid obstacles (i.e. walls and other robots) and to explore the environment. During the second phase the robots are trained to live as longer as possible in an environment that contains the white and the black target areas. Given that the black area causes a loss of energy, the robots should avoid (i.e. to quickly abandon) it and spend time on the white one. Notice that, in this phase, the behavior of the robots outside the areas has been already determined during the first phase of the training process and cannot be modified, therefore robots can only modify their behavior inside the target areas.

During the first phase, the free parameters that encode the strength of the connections between the 8 infrared sensors and the three motor neurons are selected through an evolutionary method [10]. The robots are placed in an environment without target areas. The free parameters of each robot, encoded with 8 bits, are set randomly and normalized in the range [-10.0; + 10.0]. The robots are evaluated on the basis of their ability to move and to avoid obstacles in 20 trials after being placed in the environment with randomly selected positions and orientations. The best 20 individuals are allowed to reproduce by generating 5 offspring each. Offspring inherit the same free parameters of their parents but each bit of their free parameter is mutated with a probability of 3%. The evaluation, selection, and reproduction processes are repeated for 50 generations.

During the second phase, the free parameters that encode the strength of the connections between the two ground sensors (that encode the color of the floor) and the third motor neuron (that encodes the maximum speed of the two wheels) are trained through different training procedures corresponding to the five different training conditions described below. In all experimental conditions, individual robots are evaluated for 1 trial lasting 10000 time steps (of 100ms each). At the beginning of the trials the robots are placed in the environment with a randomly selected position and orientation and are provided with an initial energy of 200 units. Every time step spent on the black target causes the loss of 1 unit of energy. When a robot finishes up its energy it "dies" and it is replaced with a new robot placed in the environment in a randomly chosen position and orientation, with full initial energy (the initialization of the free parameters of the new robots depends on the different experimental conditions, see below).

Notice that the robots do not have any direct feedbacks about the "quality" of a target area or about their own energy, i.e. they only have information about the colors of an area, but they do not know the way in which staying in a particular area affects their energy state. In practice, given the exploratory behavior acquired during the first phase, the free parameters that encode the strength of the connection between the two ground sensor neurons and the third motor neuron should be set so to allow the robot to slow down and, eventually, to stop in the white target area and to quickly move out from the black target area.

Individual Learning. The two weights that connect the ground sensor neurons to the third motor are initialized, at the beginning of each trial, with a null value, that is, the robots tend to go away from both the areas. When a robot happens to stay inside a target area, a learning algorithm acts to reinforce, with a positive feedback, the individual "preference" for that area, independently from the color of the area and consequently, from the effect on its energy. This positive feedback mechanism is realized by modifying the two free parameters every time an individual is inside one area through the addition of a randomly chosen value in the interval [-lrate, +lrate] to their current value, and by retaining only the variations that lead to an increase of the time spent by the individual in the corresponding target area. This procedure has been tested in two experimental conditions in which the learning rate parameter (lrate) has been set to 0.1 and 1.0, respectively. Individual learning has been tested in these two conditions to verify whether the different results obtained in this condition and in the next condition depend on the absolute value of the learning rate or not.

Social Learning. In this condition the free parameters are varied according to the same procedure described above but the learning rate is modulated by the number of other individuals located nearby. More precisely the learning rate is set to a value corresponding to the intensity of the signal detected, which varies linearly in the range [0.1, 1.0] according to the number of individuals located in the same foraging area.

Notice that in both cases (individual and social learning) the learning algorithm is totally "blind" in respect to the adaptiveness of the behaviors. In the former condition, the only way to orient the learning is related to the fact that individuals with adaptive behavior tend to survive and they are not replaced with naïve individuals. In the latter condition, moreover, the robots can exploit an equally blind conformist bias [12], that is, they tend to assume with more probability a behavior if other robots show already this behavior.

Genetic Evolution. In this condition, at the beginning of the experiment, the two free parameters are initialized with randomly selected value in the range [-10.0; +10.0]. During the trial, dying robots are replaced with newborn individuals that inherit their free parameters from the robot that currently has the highest energy value and that lives since longer time, with the addition of mutations that are realized by adding to the value that are inherited a randomly selected value in the range [-2.0, 2.0]. This value was chosen among a large number of tests as it produces the best performance in this condition (results not shown).

Random Search. This is a control condition identical to the Genetic Evolution condition but in which the free parameters of newborn individuals are assigned randomly in the range [-10.0; +10.0].

For time reasons all the experiments have been carried out using a simulator that carefully reproduces the characteristics of the e-puck robots.

3 Results

By analyzing the obtained results we observed that performance varies significantly in different experimental conditions and that the *Social Learning* condition outperforms all the other conditions (henceforth when we use capital letters and italics we refer to the experimental conditions).

Table 1 shows the average performance obtained in the five experimental conditions at the end of the training process, as well as the average value of some indexes that characterize individuals' behavior. The Target W and Target B indexes indicate the percentage of the robots of a team located, at the end of each trial, inside the white and the black target areas, respectively. The mortality index indicates the number of robots that died during the trial, which is inversely correlated with robots' ability to discriminate between the two target areas. The overall performance index is calculated by means of the following equation:

$$\frac{W_{perc} + \left(\frac{M_{MAX} - M_n}{M_{MAX}}\right)}{2} . \tag{1}$$

Where W_{perc} is the percentage of the robots of the team located inside the target W at the end of the trial, M_{MAX} is the maximum number of dead robots in all the five experimental conditions and M_n is the number of dead robots in the specific experimental condition. In other words, this index measures robots' ability to minimize the risk of mortality by minimizing the energy lost in the black target area and maximizing the time spent on the white area.

The last two columns of Table 1 show the average value of the two free parameters that regulate the time spent by each robot in the white and black target areas, respectively.

Table 1. Average results and average index values at the end of the training process for the five experimental conditions (see text). Each data represents the average result of 100 trials and the respective standard deviation.

	Target W	Target B	Mortality	Overall Performance	Weight 1	Weight 2
Individual 0.1	0.78 (±0.11)	0.018 (±0.03)	14.7 (±4.35)	0.63 (±0.08)	8.13 (±0.99)	1.95 (±0.56)
Individual 1.0	0.97 (±0.06)	0.003 (±0.01)	12.75 (±6.32)	0.76 (±0.06)	9.71 (±0.58)	0.09 (±0.24)
Social Learning	0.98 (±0.04)	0 (±0)	6 (±0.39)	**0.88** (±0.04)	9.86 (±0.39)	1.45 (±0.57)
Genetic Evolution	0.56 (±0.31)	0.037 (±0.06)	26.04 (±10.22)	0.32 (±0.22)	5.63 (±4.58)	0.74 (±4.65)
Random Search	0.37 (±0.14)	0.052 (±0.07)	28.75 (±6.01)	0.18 (±0.10)	1.77 (±1.86)	-1.34 (±1.58)

In particular, by observing the overall performance we can see that the best performances have been obtained by the *Social learning* condition, followed by the *Individual Learning 1.0* and the *Individual Learning 0.1*. These results indicate that, in order to prevent the loss of energy caused by the black area and to minimize the risk of dying, individuals should locate the white target area as fast as possible and as accurately as possible. Moreover, these results indicate how the regulatory effect of social learning on individual learning ensures both speed and accuracy of learning. The fact that *Social Learning* and the two *Individual Learning* conditions significantly outperform the *Genetic Evolution* condition indicates that the latter method is much slower than the formers (for all the Overall Performance results: $p < 0.001$).

By analyzing how Target W (i.e. the percentage of the individuals of a team that are located in the white target) and the mortality indexes vary during the adaptive process (Figure 3) we can see how the curves referring to the *Genetic Evolution* and *Random Search* conditions substantially overlap during the first phase and start to diverge only after approximately 2200 and 6000 time steps with respect to the two indexes. Moreover we can observe how the index of mortality of the *Social Learning* condition is similar to those of the *Genetic Evolution* and *Random Search* conditions rather than to *Individual Learning* conditions during the first phase but significantly diverges after 2000 time steps.

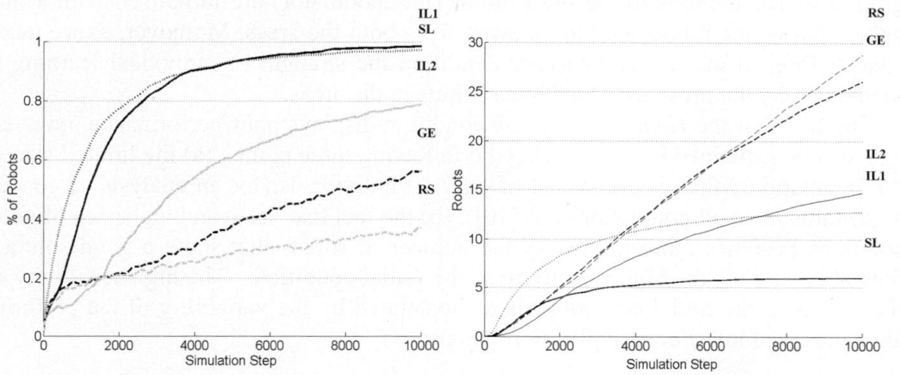

Fig. 3. Left: Percentage of robots located within the white area during the training process. Right: Mortality per trial (i.e. number of robots died during a trial). Both graphs represent the average results of 100 replications for each experimental condition. SL, IL1, IL2, GE, RS indicate the *Social Learning*, *Individual Learning 1.0*, *Individual Learning 0.1*, *Genetic Evolution*, and *Random Search* conditions.

4 Discussion

The results presented in the previous section clearly show that the *Social Learning* condition overcomes the other conditions. The *Individual Learning 1.0* condition leads to similar results with respect to the number of robots located in the white area but it is characterized by a higher mortality rate. Overall the results obtained indicate how the combination of a simple non-directional individual learning process, a simple social mechanism that regulates the strength of the individual learning process, and a selection process that operates at the level of the population (the robots that happen to learn non-adaptive behaviors tend to die and to be replaced by robots with no preference for the white or for the black target) can produce an extremely effective adaptive process.

The individual learning process is non-directional since the mechanism with which it is realized is totally "blind" to whether the variations that are retained are adaptive or not with respect to a single agent per se. Similarly, the mechanism that regulates how social interactions affect individual learning consists of a simple conformist bias [12] which is also "blind" to whether it leads to adaptive or to counter-adaptive variations at the level of a single individual. However, the combination of these two mechanisms together with the selection process ensures that, after an initial phase, the population of individuals displays an ability to quickly and reliably acquire the required adaptive skill.

In general terms, the mechanisms included in the model described in this paper are also plausible from a biological point of view. Indeed, most animals have the tendency to be neophobic (i.e. reluctant to taste food that they have never eaten before) and to prefer food which they know other conspecifics have eaten [1]. In our model, in fact, at the beginning of the trials that involve learning, the weights that connect the ground sensor neurons to the third motor (the modulator) are initialized with a null value, that is, the robots tend to go away from both the areas. Moreover, as we made clear before, in the *Social Learning* condition the strength of individual learning is modulated by the presence of the other robots in the areas.

The fact that the *Genetic Evolution* condition displays poor performance, on average, can be explained by considering the following three points: (a) the limited size of the population which is composed of only 10 individuals (for an analysis of genetic algorithms in small populations see [10]), (b) the fact that the searching space of good solutions becomes more narrow as the number of robots that show a good solution increases, and (c) the high sensitivity to the initial conditions. The high sensitivity of the results to the initial conditions is demonstrated by the variability of the performance observed in different replications (Figure 4).

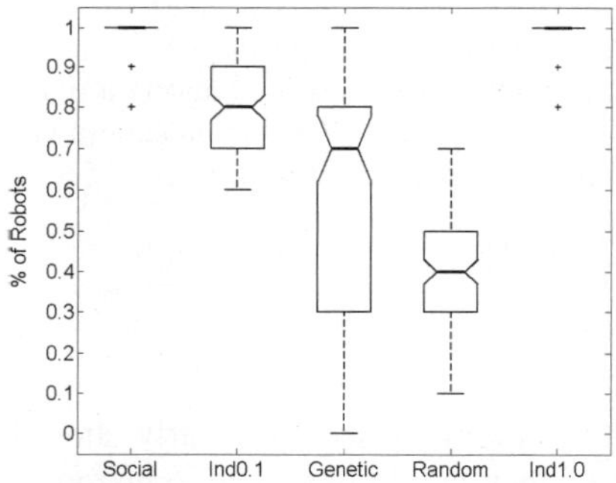

Fig. 4. Distribution of the final percentage of robots in the white area (target W) for 100 trials in the five experimental conditions. Boxes represent the inter-quartile range of the data. The horizontal lines inside the boxes indicate the median values. The horizontal lines outside the boxes indicate the minimum and maximum values. Crosses represent outliers.

The obtained results also indicate that while genetic algorithm applied to small populations might not be effective, the combination of selection at populational level, individual learning, and social learning might produce effective results also in small populations.

The importance of the interaction between the three processes (selection at population level, individual learning, and social learning) can be appreciated by comparing the dynamics in the different experimental conditions. By comparing the mortality rate of the *Individual Learning 0.1* with the mortality rate of the *Social Learning* condition (Figure 3, right) we can see that, during the first 2000 time steps, the latter is substantially higher than the former. This means that, at the beginning of the process, social interactions produce an increase of the number of counter-adaptive behaviors among the group. In fact, as shown in Figure 5, the percentage of robots located in the black target area at the beginning of the training phase is higher in the case of the *Social Learning* condition than in the case of the *Individual Learning 0.1* condition. This can be explained by considering that social learning act as a conformist bias behavior that tend to strengthen the most common behaviors during this phase, independently from their adaptiveness, since the mechanisms that regulate individual learning and social learning are not necessarily adaptive by themselves. However, the same conformist behavior, combined with the fact that the ratio between adaptive and maladaptive individuals tends to increase thanks to the selection process operating at the level of the population, allows the *Social Learning* conditions to produce better results later on.

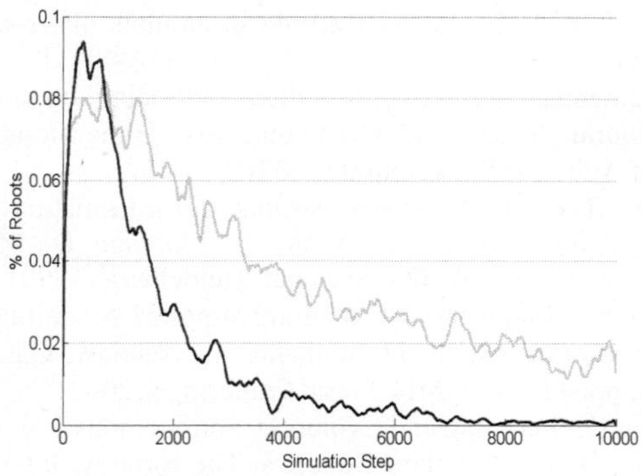

Fig. 5. Percentage of robots inside the black area (target B) through time. Data averaged over 100 trials. The black and gray line indicate the results in the case of the *Social learning* and *Individual Learning 0.1* conditions, respectively.

5 Conclusion and Future Works

We have demonstrated on a simple experimental scenario how social facilitation of individual learning combined with a selection process operating at the population level can be useful in developing adaptive behaviors in small populations of autonomous robots.

In future, we plan to investigate further the scalability of this mechanism in different environmental conditions, e.g., by increasing the number of foraging areas, by increasing (or decreasing) the number of robots, or by increasing the complexity of the perceptual categorization process. We plan to test the ability of the model to cope with changing environmental conditions. Finally, we plan to encode in the genotype of evolving individuals the learning rules that regulate individual and social learning.

Acknowledgments. This work was supported by the ECAgents project, a project funded by the Future and Emerging Technologies programme (IST-FET) of the European Commission, under grant 001940. The information provided is the sole responsibility of the authors and does not reflect the Community's opinion. The Community is not responsible for any use that might be made of data appearing in this publication.

References

1. Adessi, E., Visalberghi, E.: How social influences affect food neophobia in captive chimpanzees. A comparative approach. In: Matsuzawa, T., Tomonaga, M., Tanaka, M. (eds.) Cognitive development in chimpanzees, pp. 246–264. Springer, Tokio (2006)
2. Dautenhahn, K., Nehaniv, C.L. (eds.): Imitation in animals and artifacts. MIT Press, Cambridge (2002)
3. Dugtakin, L.A.: Copying and mate choice. In: Heyes, C.M., Galef Jr., B.G. (eds.) Social learning in animals: the roots of culture, pp. 49–64. Academic Press, San Diego (1996)
4. Galef Jr., B.G.: Social enhancement of food preferences in Norway Rats: a brief review. In: Heyes, C.M., Galef Jr., B.G. (eds.) Social learning in animals: the roots of culture, pp. 49–64. Academic Press, San Diego (1996)
5. Heyes, C.M., Galef Jr., B.G.: Social learning in animals: the roots of culture. Academic Press, San Diego (1996)
6. Nehaniv, C.L., Dautenhahn, K. (eds.): Imitation and Social Learning in Robots, Human and Animals: Behavioural, Social and Communicative Dimensions. Cambridge University Press, Human and Animals: Behavioural (2007)
7. Noble, J., Tuci, E., Todd, P.M.: Social learning and information sharing: an evolutionary simulation model of foraging in Norway rats. In: Floreano, D., Mondada, F. (eds.) ECAL 1999. LNCS, vol. 1674, pp. 514–523. Springer, Heidelberg (1999)
8. Noble, J., Todd, P.M.: Imitation or something simpler? Modelling simple mechanism for social information processing. In: Dautenhahn, K., Nehaniv, C.L. (eds.) Imitation in animals and artifacts, pp. 423–439. MIT Press, Cambridge (2002)
9. Nolfi, S., Floreano, D.: Learning and Evolution. Autonomous Robotics 7(1), 89–123 (1999)
10. Nolfi, S., Floreano, D.: Evolutionary Robotics: The Biology, Intelligence, and Technology of Self-Organizing Machines. MIT Press, Cambridge (2002)
11. Reeves, C.R.: Using Genetic Algorithms with Small Population. In: Forrest, S. (ed.) Proceedings of the Fifth International Conference on Genetic Algorithms, pp. 92–99. Morgan Kaufmann, San Francisco (1993)
12. Richerson, P.J., Boyd, R.: Not by genes alone: How culture transformed human evolution. Chicago University Press, Chicago (2004)
13. Tomasello, M.: The cultural origins of human cognition. Harvard University Press, Harvard (2001)
14. Zentall, T.R., Galef Jr., B.G.: Social learning: psychological and biological perspectives. Erlbaum, Hillsdale (1988)

Social Impact Theory Based Optimizer

Martin Macaš and Lenka Lhotská

Czech Technical University in Prague, Technická 2, Prague, Czech Republic,
mmacas@seznam.cz
http://bio.felk.cvut.cz/~macasm1

Abstract. This paper introduces a novel stochastic and population-based binary optimization method inspired by social psychology. It is called Social Impact Theory based Optimization (SITO). The method has been developed with the use of some simple modifications of simulations of Latané's Dynamic Social Impact Theory. The usability of the algorithm is demonstrated via experimental testing on some test problems. The results showed that the initial version of SITO performs comparably to the simple Genetic Algorithm (GA) and the binary Particle Swarm Optimization (bPSO).

1 Introduction

There has for a long time been a gap between the natural sciences and the technical sciences on one side and the humanities on the other side. One of many cases is the difference between technical and social science. Although it may seem difficult to imagine common elements shared by the technical and social sciences, mutual influences and inspirations are becoming increasingly frequent. A relatively broad area of such connections is the use of artificial intelligence methods in the social sciences ([1], [2]).

This paper relates to issue of social psychology in connection with optimization theory. It describes a modification of computer simulation that models processes investigated by social psychology. The original simulation was described in [3] and was based on Latané's Dynamic Theory of Social Impact [4]. We have modified some parts of the simulation and have thus introduced some optimization ability into the simulated society. The resulting optimization method, called Social Impact based Optimizer (SITO), is described here and the results of experimental testing are compared to results obtained for the Genetic Algorithm (GA) and for binary Particle Swarm Optimization (bPSO).

2 Methods

2.1 Social Impact Theory and Simulations

Social psychology deals with the ways that individuals perceive, influence and relate to each other. It attempts to understand and explain how the thought, feeling and behavior of individuals are influenced by the actual, imagined or

F. Almeida e Costa et al. (Eds.): ECAL 2007, LNAI 4648, pp. 635–644, 2007.

implied presence of others. One of the theories of social psychology is the Dynamic Theory of Social Impact developed by Bibb Latané. According to Latané [4], social impact is any of a great variety of changes in physiological states and subjective feelings, motives and emotions, cognitions and beliefs, values and behavior, that occur in an individual, human or animal, as a result of the real, implied, or imagined presence or actions of other individuals. *Dynamic social impact theory* tries to describe and predict the diffusion of beliefs through social systems. It views society as a self-organizing complex system composed of interacting individuals each obeying simple principles of social impact. It states that the likelihood that a person will respond to social influence will increase with three factors: strength, immediacy and number. *Strength* is a property of influencing individuals which indicates how important the influencing individual is to an influenced individual. *Immediacy* represents the spatial closeness of the influencing individuals from you. *Number* describes how many individuals are influencing a specific person. This structure was simulated by Andrzej Nowak and Chris Szamrej in 1990, [3]. The simulations represented each individual as four parameters: the individual's attitude, two indicators of strength (persuasiveness and supportiveness) and the individual's location in the social structure.

The organization of the simulations was as follows: The *attitude* (opinion) was a binary parameter and could take only one of two values, irrespective of interpretation. (A possible interpretation could be that the people are for or against a given idea - "guilty/not guilty", "supporting EU/not supporting EU".) Note that each person in that simulation had just a one dimensional, binary attitude, which will be generalized in this paper. *Strength factors* were persuasiveness $(0 - 100)$, the ability to persuade people with opposing beliefs to change their minds, and supportiveness $(0 - 100)$, the ability to provide social support for people with similar beliefs. Persuasiveness and supportiveness were reassigned randomly after each attitude change. The two properties were independent, which will not be reflected in our method. The *concept of immediacy* was established by organizing the group (or society) into a square matrix, where each cell represented an individual. The immediacy of two individuals (an attribute of a pair of individuals) was calculated as the Euclidean physical distance between the corresponding cells.

During each iteration, two social impacts are computed from the parameters, from immediacy, represented as Euclidean distance, and from the number of individuals with the same or opposite attitude respectively. The first social impact is total persuasive impact on a single individual of a set of N opposed sources differing in strength and immediacy: $\widehat{l_p} = \sqrt{N_0} \sum (p_i/d_i^2)/N_0$, where $\widehat{l_p}$ is the persuasive impact, N_0 the number of individuals with an opposing view (sources), p_i is the persuasiveness of source i, and d_i is the distance between source i and the recipient. The supportive impact was given by $\widehat{l_s} = \sqrt{N_S} \sum (s_i/d_i^2)/N_S$, where $\widehat{l_s}$ is the supportive impact, N_S is the number of individuals sharing the individual's view (supporters), s_i is the supportiveness of the source , and d_i has the same meaning as before.

At the beginning of the simulation, the attitudes are randomly distributed in the population. The values of p_i and s_i are also randomly initialized in the range $< 0, 100 >$. At each iteration, an individual changes its attitude if the persuasive impact is greater than the supportive impact. In case of attitude change, the values representing strength are randomly re-initialized. During simulation, all individuals change (or do not change) their attitudes and the process leads to an equilibrium state. The equilibrium state is defined as the state without any attitude change. The results show the two emergent group phenomena - the shifting of attitudes towards incompletely polarized equilibria, which was connected with decreasing frequency of attitude changes and the formation of coherent clusterings of subgroups with deviant attitudes. The minority subgroups survived mainly on the margin of the matrix.

The experiments described above show, that the society of mutually influencing individuals is able to perform some forms of self-organization that rely/ on multiple interactions. This self-organization is also associated with a notion of intelligence in the principle of collective intelligence, which appears in many forms in bacteria, animals, humans and computers. Collective intelligence emerges from the collaboration and competition of many individuals. In nature, this phenomenon was observed in ants, termites, bees, fish and birds and inspired some artificial intelligence scientists, who have developed some new optimization algorithms, e.g. Ant Colonies [5] and Particle Swarm Optimization [6]. The following section describes a novel binary optimization method inspired by the Dynamic Theory of Social Impact.

2.2 Social Impact Theory Based Optimization

In this section, we propose some modifications, which enable us to introduce an optimization ability into the social simulation. There are three main modifications:

- **Multiple Attitudes:** The first modification is to add more attitudes to each individual. That makes possible to represent optimization problems with more than one dimension. The attitudes of society are thus no longer represented by a 2-dimensional matrix of binary numbers, but each individual is represented by a binary vector and the attitudes of the whole society correspond to a 2-dimensional matrix of binary vectors.
- **Elimination of Supportiveness:** The Dynamic Theory of Social Impact proposes the important fact that individuals differ. The difference lies partly in attitudes and spatial positioning, but also partly in the strength factor. The term strength was used to refer to the network of all the individual factors making a person influential. In the original simulation, the strength of each individual was represented by the persuasiveness and the supportiveness, two independent properties. We propose to use a single parameter which will be used in computing of social impacts. The first reason for this is to simplify the computational process. The second reason is that there is no special foundation for considering persuasiveness and supportiveness to be independent. These two properties are complex functions with many components (e.g. physical size, intellect, wealth), and they evidently differ,

but are not independent. For this reason we use a single parameter - strength - for computing of both the supportive and the persuasive impact. In the light of the original simulations described above, each individual i has the strength property q_i and the following assumption holds: $q_i = s_i = p_i$.

- **Addition of a Fitness Function:** The simulation proposed by Nowak and Szamrej [3] involves an interesting issue: how to handle the two strength factors? In the original work, the persuasiveness and supportiveness were initialized randomly for each individual and changed only when the individuals' attitude changed. However, there was uncertainty about how the two parameters should change. Therefore, the change was set always randomly after each change of attitude. We propose a modification of the behavior of strength, based on adding of one more parameter to each individual - the value of a fitness function. The fitness function should evaluate each individual in terms of a binary optimization problem and the change of the strength factors (factor, in our case) should be influenced by the actual fitness value and its past behavior. The strength of an individual should be inversely proportional to the individual's fitness (for a minimization task) and must take values from a specified range (in order to be able to compare the individuals).

With regard to these limitations, we propose the following formula:

$$q = \frac{f_{max} - f}{f_{max} - f_{min}}, \tag{1}$$

where q is the strength of a specific individual, the f is the fitness value of the individual and f_{max}, f_{min} are maximum and minimum values of the fitness function in the population, respectively. This equation assigns zero strength to the individual with the maximum fitness value and strength equal to one to the individual with minimum value of fitness. This way of handling with the strength corresponds to real behavior of people, where more successful (fitter) people have higher influence on the others.

The complete SITO algorithm can be now described (see the pseudocode below). First, all individuals in the society initialize their binary attitudes randomly from uniform distribution ($P(0) = 0.5, P(1) = 0.5$). At each iteration, individuals evaluate their attitudes using a fitness function (line 4) and compute their strength using (1). Next, each individual, considering its predefined neighborhood, computes the two social impacts (lines 10,11).

The individuals do not use the contributions of all individuals in the society, but just the square neighborhood of a predefined size (for underlying experiments the square neighborhood with radius 2 was used). This reduces the time requirements. It should be noted that the algorithm described above is stochastic. A change of attitude can occur even if the persuasive impact is less than the supportive impact. On the other hand, if the persuasive impact predominates, the change may not occur. This is enabled by additional probabilistic parameter κ which represents the probability of spontaneous attitude change. Thus, if the persuasive impact is greater, the change takes place with probability $1 - \kappa$ (line 12), else, the change takes place with probability κ (line 13). This can improve the explorative capability and prevent loss of diversity. The parameter κ was set to decrease linearly from 0.1 to 0.01. The size and shape of the society was set to 100 individuals forming a 10×10 square grid for all experiments.

Algorithm: Social Impact Theory Based Optimization

```
1 Initialize attitudes by random assigning binary values from
  (0,1) to society.attitudes;
2 Iter:=0;
3 WHILE (iter<max_iter) DO
4   society.fitness:=evaluate(society.attitudes);
5   fmax:=max(society.fitness); fmin:=min(society.fitness);
6   society.strength:=(fmax-society.fitness)/(fmax-fmin);
    iter=iter+1;
7     FOR each individual i and each dimension d DO
8       Find sources and supporters in neighborhood of i;
9       Compute number of sources and supporters (No,Ns) in
        neighborhood of individual i with respect to dimension d
10      Compute total persuasive impact lp
11      Compute total supportive impact ls
12       IF lp>ls, invert the attitude of individual i in
                   dimension d with probability 1-kappa
13       ELSE,
            invert the attitude of individual i in dimension d
                        with probability kappa;
         END (IF)
       END (FOR)
    END (WHILE)
```

2.3 Other Methods

The SITO method described above has been tested and the results have been compared to the results of other stochastic and population based search methods, namely the genetic algorithm (GA) and the binary particle swarm optimization (bPSO).

Genetic Algorithm. The first method that was compared to the SITO algorithm was the simple genetic algorithm consisting of selection, crossover and mutation [7]. Stochastic uniform selection was used, where a line is laid out, in which each parent corresponds to a section of the line proportional in length to its scaled value. The algorithm moves along the line in equal-sized steps and allocates a parent from the section that it lands on. The "scattered" crossover operator was used, which randomly selects the genes to be recombined. The mutation probability was set to be the inverse value of the problem dimensionality.

Binary Particle Swarm Optimization. The PSO method is one of the recent nature-inspired optimization methods developed originally for continuous optimization in 1995 [6]. The binary version (bPSO) was used in our experiments. Each solution consists of a set of parameters and represents a point in a binary multi-dimensional space. The solution is called a "particle" and the group of particles (population) is called a "swarm". Each particle i is represented as a binary D-dimensional position vector $\overrightarrow{x}_i(t)$ with a corresponding real-valued velocity vector $\overrightarrow{v}_i(t)$. In addition, it remembers its individual best fitness value and position $\overrightarrow{p}_i$ that has resulted in that value. During each iteration t, the velocity update rule is applied to each particle in the swarm: $\overrightarrow{v}_i(t) = w\overrightarrow{v}_i(t-1) + \varphi_1 R_1(\overrightarrow{p}_i - \overrightarrow{x}_i(t-1)) + \varphi_2 R_2(\overrightarrow{p}_g - \overrightarrow{x}_i(t-1))$.

The $\overrightarrow{p_g}$ is the best position of the entire swarm and represents the social knowledge. Parameter w is called the inertia weight and during all iterations it decreased linearly from 0.9 to 0.4. The symbols R_1 and R_2 represent diagonal matrices with random diagonal elements drawn from a uniform distribution between 0 and 1, $U(0,1)$. The parameters φ_1 and φ_2 are scalar constants that weight the influence of the particles' own experience and the social knowledge.

Next, the position update rule is applied:

$$x_{i,j}(t) = \begin{cases} 1 \text{ if } R_3 < \frac{1}{1+e^{-v_{i,j}(t)}} \\ 0 \text{ otherwise,} \end{cases} \tag{2}$$

where R_3 is a random number drawn from $U(0,1)$. Thus, the position update rule is based on probability produced by normalization of the velocity components using a sigmoid function.

3 Experiments

This section describes first experiments performed with SIT. The main goal was to assess how the algorithm is able to optimize some common benchmark functions. The results are compared to the results of experiments with two other different global search stochastic techniques - the genetic algorithm (GA) and binary particle swarm optimization (bPSO).

3.1 Test Functions

For testing purposes, some binary optimization problems were used - two deceptive problems, three binary encoded continuous benchmarks, and one real-world problem of feature selection. The problems are described in Table 1.

The first function is Mühlenbein's order-5 problem [8]. The fitness function is calculated as follows: $f_1(\overrightarrow{x}) = \sum_{i=1}^{D/5} f'(\overrightarrow{x_i})$, where $\overrightarrow{x_i}$ is a disjoint 5-bit substring of $\overrightarrow{x}$ of dimension D. The values of $f'(\overrightarrow{x_i})$ are zero except at five positions: $f'(00000) = 4$, $f'(00001) = 3$, $f'(00011) = 2$, $f'(00111) = 1$ and $f'(11111) = 3.5$.

The second function is Goldberg's deceptive order-3 problem [9]. The fitness is calculated using the following equation: $f_2(\overrightarrow{x}) = \sum_{i=1}^{D/3} f''(\overrightarrow{x_i})$, where $\overrightarrow{x_i}$ is a

Table 1. Benchmark problems used for testing. Some of the benchmarks were used with two different dimensions. For the binary encoded functions, the dimensions in parentheses are the dimensions of the corresponding continuous space

Function	Type	Dimensions
f1	Mühlenbein's order-5 problem	30,150
f2	Goldberg's deceptive order-3 problem	30,150
f3	Binary encoded Rastrigrin function	34(2), 170 (10)
f4	Binary encoded Griewank function	34 (2),170 (10)
f5	Binary encoded Rosenbrock function	40 (2),200 (10)
f6	Feature selection problem	61

disjoint 3-bit substring of $\overrightarrow{x}$ and $f''(\overrightarrow{x_i}) = 0.9, 0.6, 0.3$ and 1.0 if $|x_i| = 0, 1, 2$ and 3, respectively. The expression $|\overrightarrow{x_i}|$ denotes the sum of the bits in the 3-bit substring $\overrightarrow{x_i}$.

The next three test functions were binary encoded continuous benchmark functions [10] using the precision 4 places after the decimal point. Let $\overrightarrow{y}$ of dimension n represents the real value encoded as binary vector $\overrightarrow{x}$. The first function was the Rastrigrin function $f_3(\overrightarrow{x}) = \sum_{i=1}^{n}[y_i^2 - 10\cos(2\pi y_i) + 10]$, where $-5.12 < y_i < 5.12$. The second function was the Griewank function $f_4(\overrightarrow{x}) = 1/4000\sum_{i=1}^{n} x_i^2 - \prod_{i=1}^{n} cos(x_i/\sqrt{i}) + 1$, where $-600 < y_i < 600$. The last binary encoded continuous test function was the Rosenbrock function $f_5(\overrightarrow{x})) = \sum_{i=1}^{n-1}[100(x_{i+1} - x_i^2)^2 + (x_i - 1)^2]$, where $-600 < y_i < 600$.

The last test problem was a real-world problem taken over from the automatic classification of two types of sonar signals [11]. The data-set consists of 208 patterns with 60 features. The optimization task was to select a subset of the whole feature set that maximizes the classification performance of the nearest-neighbor classifier. Thus, the solution was represented as 60-dimensional binary vector, where 0 or 1 means that the corresponding feature is removed from or included in the training data, respectively.

3.2 Experiments and Results

The main purpose of the underlying experiments was to prove the ability of the method developed here to optimize some binary functions.

The SITO method was compared to genetic algorithms and particle swarm optimization using the testing functions described above. For all methods, 30 runs were used. The population size was the same for all methods and problems and was set to 100 individuals. The maximum number of iterations was 500. The results were compared using mean and standard deviation (computed from the 30 runs) of the final fitness values. In addition, the two-sample t-test (significance level $\alpha = 0.05$) was used to investigate the significance of the difference

Table 2. Means and standard deviation of fitness values obtained from 30 runs

f	D	GA		PSO		SIT	
		Mean	Std	Mean	Std	Mean	Std
f1	30	-22.48	1.4767	-22.95	0.5469	-22.95	0.8444
f1	150	-114.87	2.1613	-115	1.0828	-117.7	1.0716
f2	30	-9.43	0.1596	-9.61	0.1337	-9.52	0.1599
f2	150	-46.89	0.3039	-47.16	0.3784	-47.28	0.338
f3	34	0.578	0.736	$6,05 \times 10^{-7}$	0,000	0,000229	0,000104
f3	170	7.2401	3.5925	3.0725	1.9213	6.2396	2.8030
f4	34	0.021	0.01687	0.00942	0.00834	0.00854	0.0082
f4	170	2.6761	5.1351	0.2932	0.4254	0.0177	0.0305
f5	40	2.6761	5.1351	0.2932	0.4254	0.0177	0.0305
f5	200	83.604	130.672	69.978	163.44	280.087	580.66
f6	61	-0.9601	0.0124	-0.9743	0.0086	-0.9777	0.0039

between the methods. The parameter settings for GA and bPSO are described in Section 2.3.

Some interesting conclusions can be drawn from these results. It can be seen that SITO performed significantly better than GA in 7 of the 11 tests and it was significantly worse in only one of the tests. On the other hand, SITO outperformed the bPSO algorithm in just two tests and was worse than the bPSO in 3 cases. We can observe relatively good results obtained for deceptive problems that are treated as difficult for GA. Also, there is no apparent correlation between the dimensionality of the binary search space and the performance of SITO. Thus, the SITO optimizer seems to be comparable to the other examined methods.

All results are summarized in Table 2. For each method and for each combination of test problem and dimension, the mean and standard deviations obtained by averaging over 30 runs are depicted here. Table 3 corresponds to Table 2 and describes the results of the statistical testing of the differences between particular results. The third column compares GA and SITO and the fourth column compares bPSO and SITO. Thus, SITO was compared with each of the other methods. The "0" indicates, that there is no significant difference between the two results. The "1" indicates that the SITO performed significantly better than the corresponding method (GA or bPSO) and the "−1" indicates that the SITO was significantly worse then the other method.

Some interesting conclusions can be drawn from these results. It can be seen that the SITO performed significantly better than GA in 7 of 11 tests while in just one test it was significantly worse. On the other hand, SITO outperformed the bPSO algorithm just in just two tests and was worse than bPSO in 3 cases. We can observe that quite good results were obtained for deceptive problems that are treated as difficult for GA. In addition, there is no apparent correlation between the dimensionality of the binary search space and the performance of

Table 3. Results of statistical t-test ($\alpha = 0.05$). "0" means no significant difference, "1" or "-1" denotes better or worse performance of SITO, respectively

f	f1	f1	f2	f2	f3	f3	f4	f4	f5	f5	f6
D	30	150	30	150	34	170	34	170	40	200	61
SITO vs. GA	0	1	1	1	1	0	1	-1	1	0	1
SITO vs. PSO	0	1	-1	0	0	-1	0	-1	1	0	0

SITO. Thus, the SITO optimizer seems to be comparable to the other methods examined here.

Fig. 1 shows the visualization of the distribution of the fitness value in a society consisting of 30×30 individuals and optimizing the Mühlenbein's 5-order function. The states of the society are captured at three different moments (60, 120, 150 iterations). The most striking phenomenon that can be observed here is the gradual loss of diversity represented as the growing predominance of the fittest individuals. Obviously, the individuals located near to each other tend to have similar fitness values and probably the similar attitudes (solution vectors). This corresponds to the similar phenomena observed in dynamic social impact theory and its simulation, where people become more similar to their neighbors, leading to spatial clustering [12]. Another important thing that is shown in Fig. 1 is the influence of the randomness introduced through the parameter κ. This randomness causes that individuals are sometimes not governed by the relationship of supportive and persuasive impact. These individuals can be seen in all subplots of Fig. 1 as solitary pixels with a different color from their neighbors. The usefulness of this random behavior is shown in the right-hand subplot (the state after 150 iterations), where some of these individuals are the fittest ones in the whole society.

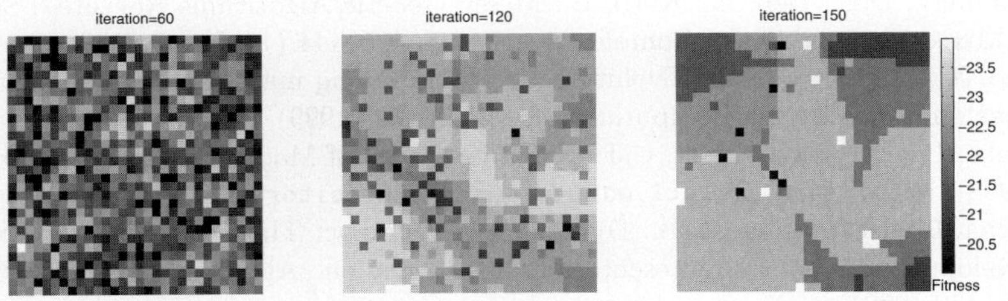

Fig. 1. An example of the time progress of optimization. Each pixel represents one member of the society and its brightness indicates its fitness (the lighter the color, the fitter the individual).

4 Conclusions

This paper has introduced a novel binary optimization algorithm, and has compared it with other stochastic and population-based techniques. For the 6 test problems, the SITO algorithm was comparable with GA and bPSO. Future work will focus on simplifying the algorithm, on further tests and on investigating some special phenomena common to the social behavior of humans.

Acknowledgement. The research was supported by the research program No. MSM6840770012 "Transdisciplinary Research in the Area of Biomedical Engineering II" of the CTU in Prague, sponsored by the Ministry of Education, Youth and Sports of the Czech Republic.

References

1. Minsky, M.: Society of Mind. Simon & Schuster, London (1988)
2. Doran, J.: The computational approach to knowledge, communication and structure in multi-actor system. In: Nigel-Gilbert, G., Heath, C. (eds.) Social Action and Artificial Intelligence, pp. 160–171. Gower, London (1985)
3. Nowak, A.J., Szamrej, J., Latane, B.: From Private Attitude to Public Opinion - A Dynamic Theory of Social Impact. Psychological Review 97(3), 362–376 (1990)
4. Latane, B.: The Psychology of Social Impact. American Psychologist 36(4), 343–356 (1981)
5. Bonabeau, E., Dorigo, M., Theraulaz, G.: Swarm intellligence: from natural to artificial intelligence. Oxford University Press, New York (1999)
6. Kennedy, J.F., Eberhart, R.C., Shi, Y.: Swarm intelligence. The Morgan Kaufmann series in evolutionary computation. Morgan Kaufmann Publishers, San Francisco (2001)
7. Vose, M.D.: The Simple Genetic Algorithm: Foundations and Theory. MIT Press, Cambridge (1999)
8. Mühlenbein, H., Mahnig, T., Rodrigues, A.O.: Schemata, Distribution and Graphical Modes in Evolutionary Optimization. Journal of Heuristics 5 (1999)
9. Goldberg, D.E., Deb, K., Korb, B.: Messy Genetic Algorithms Revisited: Studies in Mixed Size and Scale. Complex Systems 4, 415–444 (1990)
10. Yao, X., Liu, Y., Lin, G.: Evolutionary programming made faster. IEEE Transactions on Evolutionary Computation 3(2), 82–102 (1999)
11. Blake, C., Keogh, E., Merz, C.J.: UCI Repository of Machine Learning Databases (1998), http://www.ics.uci.edu/~mlearn/MLRepository.html
12. Richardson, D.S., Latané, B.: Dynamic Social Impact Theory (DSIT) predicts the development of social representations of aggression. Aggressive Behavior 27(3), 178–179 (2001)

The Role of Collective Reproduction in Evolution

John Bryden

School of Computing, University of Leeds, Leeds LS2 9JT
johnb@comp.leeds.ac.uk

Abstract. To look for an answer to the puzzle of why complexity may increase, this paper looks to the major evolutionary transitions – a recurring pattern where individuals give up their rights to reproduce individually and instead reproduce as part of a super-organism. A simple model of collective reproduction is presented and discussed in light of this topic. The model finds that collective reproduction is actually to the benefit of the individual, not to the group. The cost of reproduction is shown to be an important factor and different scenarios are presented which show individual, sexual reproduction and collective reproduction (with larger numbers of parents) as optimal.

1 Introduction

One of the most striking features of our evolutionary past is the rise of the complex individual. As we replay the timeline of natural history [1], the most complex species has become increasingly more complex. One of the major goals of Artificial life has therefore been to recreate such increases in individual complexity *in-silico* [2]: i.e., within a computer.

The common opinion in biological circles [3] is that core evolutionary theory is all we need to explain the evolution of life and its astounding complexities. However, this does not seem so hopeful when attempts to reproduce such effects in ALife computer models, which implement all the main features of evolutionary theory, have not produced much complexity of interest whatsoever [4]. It seems clear therefore that there is something missing, perhaps just from the models that have attempted to recreate some of the complexity of life, or perhaps even from the evolutionary theory they are based on. This paper looks for a process (or processes) that can increase the complexity of an individual.

A profound theme observed in nature is the hierarchical structures (units made up from sub-units) that can be found [1]. These hierarchical structures are a recurring pattern, and can be seen at all levels of biology. The important feature at every level of these hierarchies is that units are made up of sub-units: proteins are made of chains of amino acids, cells/organelles are made of proteins, eukaryotic cells are made up of organelles, multicellular organisms are made up of eukaryotic cells, and societies are made up of multicellular organisms. Each unit is of greater complexity than each sub-unit.

This repeating pattern implies that there may be some common feature in individual sub-units from every level that leads them to form themselves into

F. Almeida e Costa et al. (Eds.): ECAL 2007, LNAI 4648, pp. 645–654, 2007.

larger units: an evolutionary force that binds these units together. Or, put another way, a mode of interraction that is common to sub-units at every level. In fact, these sub-units have common features at every level. They all make copies of themselves, which may or may not be perfect copies, and they all compete with other similar individuals over resources.

One other feature has been highlighted and is seen in the literature studying the evolution of new super-units from sub-units. These evolutionary steps have been dubbed *major evolutionary transitions* [1]. It has been observed that all of the transitions are characterised by one simple common paradigm: sub-units that could previously reproduce on their own can now only reproduce as part of the super-unit [5]. Since each of the transitions involves an increase in individual complexity, I look for a general model of the transitions which can illuminate processes for increase in individual complexity.

A common approach to explaining some of the individual transitions is to look to altruism and social evolution. The central thesis, here, is that any individual involved in collective reproduction is performing an altruistic act. I.e., taking a reduction in their own reproductive success so that the reproductive success of the group may increase. It has long been rejected that an individual may reduce its fitness (lifetime reproductive success) to benefit its local group [6]. However, it has been shown how an individual may take a reduction in personal fitness to benefit related organisms [7,8] through *kin selection*.

The kin selection perspective has shown some value in explaining the maintenance of eusociality (social insects), however it is not clear that it (or altruism in general) explains the origins of this transition [1,9]. Indeed one problem with kin selection is that its benefits can be negated by increased local competition for food [10]. One question, considered here, is therefore whether altruism actually is crucial for explaining the origins of collective reproduction. It may be possible to find explanations that are mutualistic: i.e., the outcome for both the individual and the group is beneficial.

On the face of it, given the literature on the cost of sex (e.g., [11,12]), it looks unlikely that a mutualistic explanation may be found for collective reproduction. Put simply, the cost of sex means that individuals which reproduce sexually will grow at a slower rate than those that reproduce individually. Rather than directly considering Maynard Smith's model of the cost of sex, I illustrate the problem by reformulating it within the abstract terms used in this paper. Here, I make a simple comparison of the growth rates of an individual strategy and a collective reproduction strategy (with two parents) based on the way resources are allocated to offspring, as shown in Fig. 1.

Since individuals with the collective strategy share contributions to offspring, they contribute less than those with the individual strategy. This means that the average level of resources per individual in the collectively reproducing population will be higher than those in the individually reproducing population. For a fixed input of resources to the system, this means that the collectively reproducing population will grow more slowly than the individually reproducing population. If we look at cases where there are increasingly larger numbers of

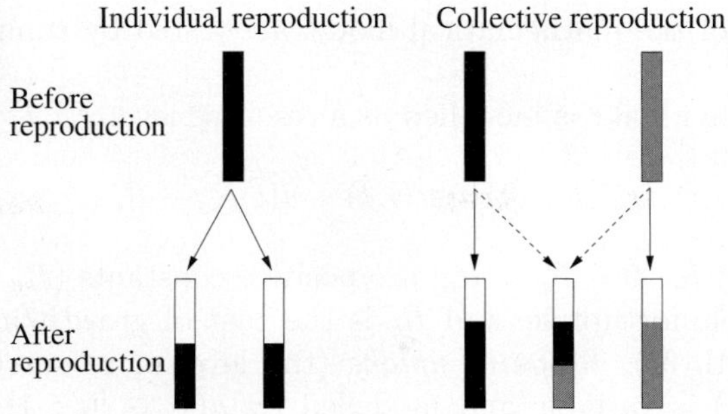

Fig. 1. Individual reproducers will grow more quickly than collective reproducers. Individuals are represented as resource containers of size R_1, with the level of resources represented by how full the containers are. They are shown just before, and immediately after reproduction. Each new individual from the sexually reproducing population will have an average resource level of $2R_1/3$ so, for the same resource input, this population will grow more slowly than the individually reproduction population which starts with an average resource level of $R_1/2$.

parents, then the same reasoning can be used to show that the growth rate will be increasingly slower (e.g., when there are 3 parents, this population will have an average resource level of $3R_1/4$ after reproduction).

Given this cost of collective reproduction, it seems unlikely that there is any advantage to collective reproduction. However, this analysis only looks at the growth phase. Any population that grows will eventually exhaust the resources in its environment and the population will either decline or reach a static level. Because collectively reproducing individuals have, on average, greater resources after reproduction they are less vulnerable during times of population decline.

To approach modelling the major evolutionary transitions, I take an abstract perspective in order to produce a toy model. This approach, based on the common factors of the subunits identified above, simplifies the world to resources, individuals and individual strategies. Individuals accumulate resources, expend resources, make copies of themselves and share resources with those copies. Given that an individual lives within an environment with factors outside of its control, I assume that they have little control over the way they accumulate or expend resources. The space of strategies that they may adopt is therefore concerned with their reproduction strategy.

The model simply approaches the question as to whether it is a better strategy, for the individual, to reproduce collectively or to reproduce individually. It is looking for an explanation for collective reproduction that is not based on altruism. In the next section I present the model and its results.

2 The Model

Two modelling approaches are taken in this section. After presenting details common to both approaches, I outline a mathematical model and its predictions.

The predictions of the mathematical model are tested by simulation models in Section 2.2.

Here each individual i is modelled as a resource level $x(i,t)$ by the equation,

$$x(i, t + \delta t) = x(i, t) + u(t)R_{\mathrm{u}} - R_{\mathrm{c}} \;, \tag{1}$$

where δt, R_{u} and R_{c} $[0 < R_{\mathrm{c}} < R_{\mathrm{u}}]$ are positive constants (R_{u} is the maximum resources available for uptake and R_{c} is the cost of growth/maintenance each timestep of length δt). *Resource uptake* (the level of resources received from the environment) is variable and modelled by $u(t) \in [0 : 1]$. The variable u here is used as a surrogate for competition: population fluctuations will lead to increasing and declining phases, modelled by changes in the behaviour of u.

If an individual's resource level decreases below the lower threshold R_0 (fixed for all individuals) it will die. Without losing generality, R_0 is set to 0 as an *ad hoc* simplification. If an individual's resource level increases above the reproduction threshold (R_1) it will reproduce. All individuals pay a cost of reproduction $R_{\mathrm{r}}(n)$ which is dependent on the number of parents n.

2.1 Mathematical Treatment

The mathematical treatment assumes that u is static over the lifetime of individuals. The resource change δx over a discrete time interval δt is therefore modelled as:

$$\delta x = uR_{\mathrm{u}} - R_{\mathrm{c}} \;. \tag{2}$$

The value of u where $\delta x = 0$ is defined as u_0: $u_0 = R_{\mathrm{c}}/R_{\mathrm{u}}$.

During reproduction all parents pay the cost of reproduction $R_{\mathrm{r}}(n)$. After this the remaining resources are shared equally between the n parents and the offspring. All individuals therefore start their lives, just after reproduction, with $x = n(R_1 - R_{\mathrm{r}})/(n+1)$. Two cases for u can now be considered: $u > u_0$ and $u < u_0$. In the first case the individual resource level will increase until it reaches the upper threshold R_1, taking an expected time W where,

$$\begin{aligned} W_{u>u_0} &= \left[R_1 - \frac{n\,(R_1 - R_{\mathrm{r}})}{n+1}\right]\frac{\delta t}{\delta x} \\ &= \frac{(R_1 + nR_{\mathrm{r}})\,\delta t}{(n+1)\delta x} \;. \end{aligned} \tag{3}$$

In the second case individual resources will decrease until it reaches the lower threshold at resource level R_0. The expected time W is,

$$W_{u<u_0} = \frac{n\,(R_1 - R_{\mathrm{r}})\,\delta t}{(n+1)\delta x} \;. \tag{4}$$

In both cases, as $u \to u_0, W \to \infty$.

The expected population growth rate of a homogeneous population of individuals can be estimated for the two regimes ($u > u_0$ and $u < u_0$). The expected population growth rate per individual G is equal to the reciprocal of the time taken for resources to grow for reproduction during population growth $(1/W_{u>u_0})$ and reciprocal of the time taken for resources to decline for death during population decline $(1/W_{u<u_0})$, or [substituting Equation (2) into Equations (3) and (4)]:

$$G = \begin{cases} \dfrac{(n+1)(uR_{\mathrm{u}} - R_{\mathrm{c}})}{n\,(R_1 + nR_{\mathrm{r}})\,\delta t} & , u > u_0 \\ 0 & , u = u_0 \\ \dfrac{(n+1)(uR_{\mathrm{u}} - R_{\mathrm{c}})}{n\,(R_1 - R_{\mathrm{r}})\,\delta t} & , u < u_0\ . \end{cases} \tag{5}$$

The growth rate (G) is plotted against different values of u for one [$n = 1$, $R_{\mathrm{r}}(1) = 0.1$] and two [$n = 2$, $R_{\mathrm{r}}(2) = 0.05$] parents in Fig. 2. In this case, the two parents share the cost of reproduction born by the single parent – i.e., the cost per offspring is the same.

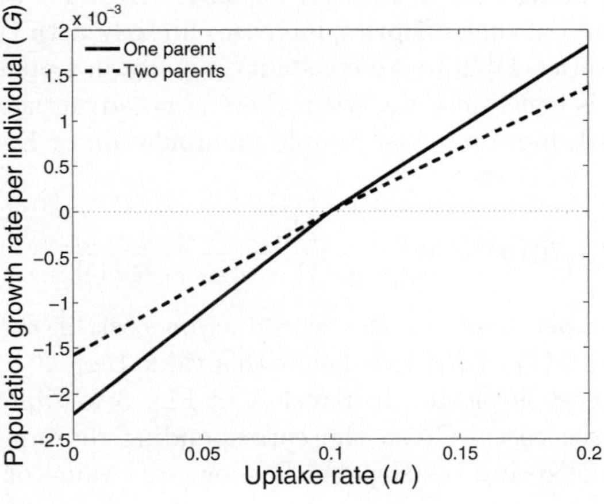

Fig. 2. Growth rates of individually and sexually reproducing populations. The parameters used were: $R_1 = 1.0$, $R_{\mathrm{r}}(1) = 0.1$, $R_{\mathrm{r}}(2) = 0.05$, $R_{\mathrm{c}} = 0.001$, and $R_{\mathrm{u}} = 0.01$.

The figure shows that the sexually reproducing population does in fact grow more slowly during population growth $(u > u_0)$. This is in line with the reasoning presented in Section 1 and Fig. 1. What is also evident from the figure is that the decline rate of the sexually reproducing population is lower in magnitude than that of the individually reproducing population. Furthermore, the two graphs diverge more as u gets further from u_0 indicating that fluctuations may be important, however the ratio between the two plots stays constant. Indeed, in this example, the growth rate of the sexually reproducing population is 0.75

that of the individually reproducing population. However, the decline rate of the sexually reproducing population is only 0.71 that of the individually reproducing population. Interestingly, the sexually reproducing population has a greater advantage during decline periods than the individual strategy has during growth periods.

This analysis therefore indicates that the ratio of growth and decline rates is important and that fluctuations might also be significant. Simple mathematical analysis can be done on the relative growth to decline rates for populations with different numbers of parents, this is given by:

$$\Psi(n) = \frac{G_{u>u_0}}{G_{u<u_0}} = \frac{R_1 - R_\mathrm{r}(n)}{R_1 + nR_\mathrm{r}(n)} \,. \tag{6}$$

It seems very likely that populations with larger values of Ψ will competitively exclude populations with lower values: this is tested with simulations in Section 2.2. In this model, the behaviour of Ψ depends on the way $R_\mathrm{r}(n)$ is determined.

I look at three scenarios for determining $R_\mathrm{r}(n)$. These consider the offspring cost which is defined as the total reproduction cost spent on each offpring (nR_r). In scenario (i), I consider the case where the total cost of each offspring is static: $R_\mathrm{r}(n) = R_\mathrm{r}(1)/n$. In this case, Equation (6) is increasing (the denominator is constant and the numerator increases). Scenario (ii) looks at the case where the cost of total cost of each offspring increases linearly with each extra parent: $R_\mathrm{r}(n) = R_\mathrm{r}(1)[1+c(n-1)]/n$ (c is a constant). Finally, in scenario (iii) I consider the case where Ψ is constant – i.e. when there is no advantage or disadvantage to reproducing with more parents. Simple manipulation of Equation (6) shows that if $\forall n : \Psi(n) = \Psi(1)$ then,

$$R_\mathrm{r}(n) = \frac{2R_\mathrm{r}(1)R_1}{R_1 + R_\mathrm{r}(1) + n[R_1 - R_\mathrm{r}(1)]} \,. \tag{7}$$

This sets an upper limit for the reproduction cost: if $R_\mathrm{r}(n)$ is above this value, then $\Psi(n) < \Psi(1)$, if $R_\mathrm{r}(n)$ is below this value then $\Psi(n) > \Psi(1)$.

The offspring cost is plotted in panel A of Fig. 3 for different numbers of parents in the three cases. Given the corresponding reproduction costs associated with these offspring costs, panel B shows the value of Ψ calculated by Equation (6).

As expected, the plot in Panel B of the figure shows how Ψ increases when the offspring cost is not dependent on the number of parents. The difference in values of Ψ is relatively high when the number of parents is low – the largest increase being the difference between individual and sexual reproduction. When the offspring cost increases linearly with the number of parents, Ψ reaches a maximum at 2 parents (sexual reproduction) and declines thereafter. Considering Panel A of the figure, this indicates that, as the number of parents increases, the total cost of reproduction spent (the cost per offspring) may increase. Again, this increases sharply as the number of parents increase from 1 to about 5 and then levels out. For all graphs, changes in parameters R_1 and R_r did not change the shape of the graphs significantly, however, as the ratio (R_1/R_r) decreases, the values generated by Equation (7) increase (not shown).

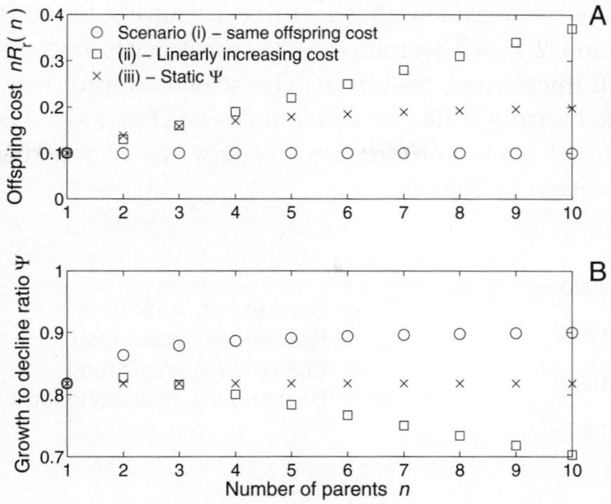

Fig. 3. The behaviour of the growth/decline ratio changes with different functions for $R_r(n)$. Panel A shows the cost per offspring generated by the three different functions of $R_r(n)$ and panel B shows the corresponding predicted growth to decline ratio. The parameters used were: $R_1 = 1.0$, $R_r(1) = 0.1$, $R_c = 0.001$, $R_u = 0.01$, and $c = 0.3$.

2.2 Simulation Models

The predictions of the mathematical treatment in the previous section indicate two things that may be tested with simulation models. Firstly whether the value of Ψ is a good predictor for which strategy is optimal and secondly that fluctuations may also be significant.

Simulations are done with agents modelled as resource levels, based on Equation (1). At each timestep an agent pays a growth/maintenance cost R_c. When its resources are below zero, an agent will die. Each agent has a reproduction strategy which is defined by the number of parents (from 1 to 10) the agent will reproduce with. Those with the same reproduction strategy will reproduce when enough agents (depending on the number of parents defined by their common strategy) have resource levels that are above R_1 (resource levels may go above R_1 without penalty). Each parent pays a cost of reproduction $R_r(n)$ and all parents share their energy with the new offspring.

The simulation models are essentially non-spatial, with individuals located on a grid but moving to a random new cell each timestep.[1] Agents consume a resource unit if they encounter any on the square they are on. A number of resource units, each of value R_u are randomly scattered on a spatial grid (of 50×50 squares) each time step. The number of units is either static (set to 200) throughout the simulation, or fluctuated between two values (100 and 200) changing every 1,000 timesteps.

[1] The results in [13] showed that agents receive resources with between-resource intervals on a geometric distribution when they move to random grid squares each time step.

The simulations were run with the three scenarios for determining $R_r(n)$ presented in Section 2.1. All scenarios were tested with a static resource input to the system and fluctuating resources. The scenarios were run ten times, with each run initialised with a different random seed. After 1,000,000 timesteps the number of agents with each reproduction strategy was averaged over all ten runs. The results are plotted in Fig. 4.

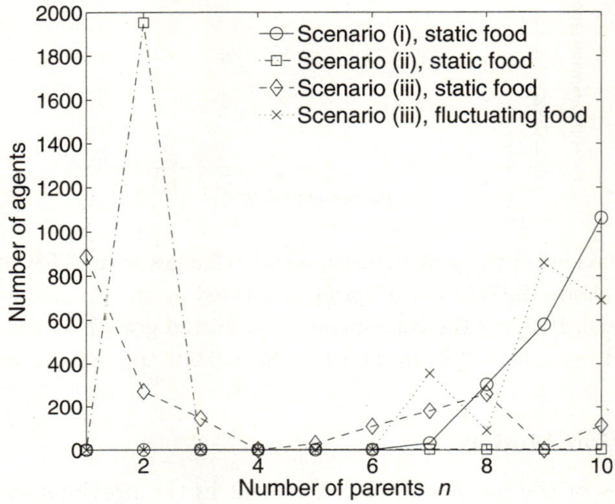

Fig. 4. Collective reproduction depends on offspring cost and resource fluctuations. Parameters for the simulations were: $R_1 = 1.0$, $R_r(1) = 0.1$, $R_c = 0.001$, and $R_u = 0.01$.

The figure shows that, when the offpring cost does not increase with the number of parents, reproduction strategies with higher numbers of parents will dominate. In fact any strategy with less than 7 parents has been completely eradicated from the simulations of this scenario. When fluctuations were also introduced, similar results were seen (not shown). When the cost per offspring increases linearly, the sexual strategy is dominant over the other strategies – as predicted by the shape of Ψ in Fig. 3. Again, the results were similar with and without fluctuations. When the cost per offspring increases in line with the upper limit predicted by Equation (7) (see Fig. 3, Panel A), the viability of collective reproduction depends on fluctuations in resource availablity. With no fluctuations, individual reproduction is dominant, but when the resources do fluctuate, collective reproduction is dominant.

3 Discussion

The mathematical model and simulations presented in this paper demonstrate collective reproduction. Individuals that may reproduce on their own, instead reproduce as part of a collective. Collective reproduction here is done by sharing resources contributed to a shared offspring. The modelling work shows that the cost of this process (the cost of reproduction) is important and fluctuations in environmental resource levels can be significant. The model also presents two different scenarios which predict conditions for when sexual reproduction is optimal and other conditions for when reproduction in larger groups is optimal: this may help to explain why sexual reproduction is dominant in some animals and eusociality is dominant in others.

The mathematical predictions presented in Fig. 3 are concordant with the results in Fig. 4, both predicting when collective reproduction is viable. This includes subtle effects such as the dominance of sexually reproducing individuals. Since the results are so similar, the simulation models show that Ψ is good predictor for which reproduction strategies will competitively exclude others. The mathematical treatment is therefore instructive (in line with [14]) as to why there is a long term growth benefit to lineages that reproduce in this way: the collectively reproducing individuals have greater resources and are therefore less vulnerable to resource fluctuations.

The work contributes to explaining the rise in the complexity of the individual in two ways. Firstly, it demonstrates how collective reproduction can benefit both partners: when more than one parent contributes resources to the production of an offspring, the combined reproductive expenditure can be significantly larger than with individual parents (see Fig. 3 panel A, and corroborating simulation results in Fig. 4). This extra resource is available for the increased complexity needed for the facilitation of collective reproduction.

Indeed, it is plausible that collective reproduction may happen on many levels in the same class of individual. Some examples of collective reproduction may only be viable when the conditions are right, so different mechanisms for collective reproduction may happen under different conditions. Each may have different optimal numbers of parents. As well as this, collective reproduction may happen at different levels at the same time. With some organisms making direct genetic contributions, others making indirect genetic contributions (through kin-relatedness [7]) and others perhaps gambling their genetic contribution (see [15] for an example). There can therefore be many differing mechanisms of collective reproduction taking place within a population at the same time. As new viable mechanisms increase complexity, a rich social fabric should emerge.

Secondly, the model, and its insights, implies a potentially fruitful approach to modelling the major evolutionary transitions. Rather than invoking altruism or group selection, the model of collective reproduction presented here shows mutual benefits to reproducing collectively: i.e., that it is in an individual's selfish interest to reproduce collectively. There is no need for the individual to reduce its fitness for the benefit of its kin or its group. Altruism may therefore not be an essential feature in explaining the major transitions.

That said, this does not imply that altruism is not important in collective reproduction. Altruism can happen and will act as an evolutionary force when appropriate. One major assumption of the model is that the resources of all parents are shared out equally between the parents and offspring. Clearly a parent that does not contribute in this way may be able to disrupt the process by contributing less resources than other parents. It could be argued that a parent that doesn't do this is acting altruistically, however by contributing less resources it will also be harming its own representation in the collective reproductive effort.

Study of such cheating behaviour is outside the scope of this paper but can be addressed in future work. Other future work could also address each of the major evolutionary transitions in more detail. It is to be hoped that the application of the style and approach of modelling in this paper will yield interesting results.

Acknowledgments. Thanks to Stephen Cornell, Jason Noble, Richard Watson and the Biosystems group at the University of Leeds.

References

1. Maynard Smith, J., Szathmáry, E.: The Major Transitions in Evolution. Oxford University Press, Oxford (1995)
2. Bedau, M.A., McCaskill, J.S., Packard, N.H., Rasmussen, S., Adami, C., Green, D.G., Ikegami, T., Kaneko, K., Ray, T.S.: Open problems in artificial life. Artificial Life 6, 363–376 (2000)
3. Dawkins, R.: The Selfish Gene (Revised edn.). Oxford University Press, Oxford (1989)
4. Ray, T.S.: An approach to the synthesis of life. In: Boden, M.A. (ed.) The Philosophy of Artificial Life, pp. 111–145. Oxford University Press, Oxford (1996)
5. Szathmáry, E., Maynard Smith, J.: The major evolutionary transitions. Nature 374, 227–232 (1995)
6. Williams, G.C.: Adaptation and Natural Selection. Princeton University Press, Princeton (1966)
7. Hamilton, W.D.: The genetical evolution of social behaviour. i & ii. Journal of Theoretical Biology 7, 1–52 (1964)
8. Frank, S.A.: Foundations of Social Evolution. Princeton University Press, Princeton (1998)
9. Wilson, E.O., Hölldobler, B.: Eusociality: Origin and consequences. Proceedings of the National Academy of Sciences 102, 13367–13371 (2005)
10. West, S.A., Pen, I., Griffin, A.S.: Cooperation and competition between relatives. Science 296, 72–75 (2002)
11. Maynard Smith, J.: The Evolution of Sex. Cambridge University Press, Cambridge (1976)
12. Bell, G.: The masterpiece of nature: the evolution and genetics of sexuality. Croom Helm, London (1982)
13. Bryden, J.A.: Space: What is it good for. In: Kim, J.T., (ed.) Systems Biology Workshop at ECAL 2005 (2005), `http://www.ecal2005.org/workshopsCD/systemsbiol/index.html`
14. Bryden, J., Noble, J.: Computational modelling, explicit mathematical treatments, and scientific explanation. In: Artificial Life X, pp. 520–526 (2006)
15. Bryden, J.A.: Slime mould and the transition to multicellularity: the role of the macrocyst stage. In: Capcarrère, M.S., Freitas, A.A., Bentley, P.J., Johnson, C.G., Timmis, J. (eds.) ECAL 2005. LNCS (LNAI), vol. 3630, pp. 551–561. Springer, Heidelberg (2005)

Fear and the Behaviour of Virtual Flocking Animals

Carlos Delgado-Mata[1,2] and Ruth S. Aylett[2]

[1] Universidad Bonaterra, Aguascalientes, CP 20290, México
[2] Heriot Watt University, Riccarton Campus, Edinburgh, EH14 4AS, UK

Abstract. The paper investigates the role of an affective system as part of an ethologically-inspired action-selection mechanism for virtual animals in a 3D interactive graphics environment. It discusses the integration of emotion with flocking and grazing behaviours and a mechanism for communicating emotion between animals; develops a metric for analyzing the collective behaviour of the animals and its complexity and shows that emotion reduces the complexity of behaviour and thus mediates between individual and collective behaviour.

1 Introduction

Minsky [1] was one of the first to emphasise the importance of emotion for Artificial Intelligence. Since then, research in affective systems for embodied autonomous agents, robotic and graphical, has expanded. Approaches divide into low-level, neuro-physiologically inspired accounts, focusing on sub-symbolic behavioural architectures [2] [3], and high-level cognitive science-inspired accounts, focusing on symbolic appraisal-driven architectures [4] [5]. This work concentrates on a low-level account, applied to flocking mammals (sheep, deer), and demonstrates the role of fear as a social regulator between individual and group behaviour. We take the "primitive emotions" namely: anger, fear, disgust, surprise, happiness and sadness, [6] as a plausible set for other mammals than humans, and examine how they can be integrated into an ethologically-based action-selection mechanism. For affective systems to have developed and remain under the pressure of evolutionary selection, they must play a functional role within the overall architecture of animals. A number of such functions can be identified. One is to modify behaviour: a sheep that experiences an anxiety-inducing stimulus may carry on grazing but bunch up more tightly with the rest of the flock. A second is to switch behaviours: a sheep experiencing a threatening stimulus inside its flight zone will flee. A third is to avoid dithering between competing behaviours by adding weight to one of them [7], and a fourth and related function is to sustain a selected behaviour for an appropriate interval –a fleeing animal can typically no longer perceive the threatening predator, but fear keeps it running, acting like a cheap short-term memory. Many mammals do not behave merely as individuals, they engage in the collective behaviour known as flocking. Reynolds [8] showed that flocking does not require a complex internal

F. Almeida e Costa et al. (Eds.): ECAL 2007, LNAI 4648, pp. 655–664, 2007.

architecture but can be produced by a set of simple rules. In his model of boids, every individual (boid) tries to fulfil three conditions: cohesion or flock centring (attempt to stay close to nearby flockmates), alignment or velocity matching (attempt to match velocity with nearby flockmates), and separation or collision avoidance (avoid collisions with nearby flockmates). This approach has produced sufficiently believable behaviour to be used for stampedes in a number of animated films. However, mammals do have a complex internal architecture, unlike social insects, and a wide range of individual behaviours: a motivation for this work was to reconcile the generation of collective behaviour by a small set of rules with the more complex agent architecture required for a mammalian behaviour repertoire. For ungulates, grazing is a significant behaviour, requiring spatial orientation behaviours. Two are of particular relevance. The first, kinesis, can be representeded by a reactive rule of slowing down when encountering favourable conditions and speeding up for unfavourable ones: this can also be related to escape behaviour. However most organisms do not move in an absolutely straight line; when orienting to favourable localities: the effect of kinesis can be improved by increasing the angle of the random deviations from the straight line, in any case inherent to locomotion. By these means, the organism is kept in the desirable environment longer and is made to exploit an increased part of its area, especially relevant to grazing. This second enhanced mechanism is termed klinokinesis and it is found in grazing mammals, as well as in swimming protozoa and higher crustacea. This represents an important example of individually-oriented behaviour which conflicts with the rule-set for flocking.

2 An Ethologically Inspired Action-Selection Mechanism

The work discussed here has been implemented with graphically-embodied flocking animals (sheep, deer) in a 3D interactive virtual environment. In order to test the hypothesis that an affective system can act as a regulating mechanism between individual and social behaviour, an ethologically-motivated [9] architecture was developed for the virtual animals. The basic task of a virtual animal brain has often been split into the three sub-tasks of perception (sensing the environment and interpreting the sensory signals to provide a high-level description of the environment), action selection (using the perceptual and emotional inputs to decide which of the animal's repertoire of actions is most suitable at that moment) and motor control (transforming the chosen action into a pattern of "physical" actions to produce the animation of the animal). To this we add a fourth subtask: generating emotions (affecting the behaviour of the animals, exemplified by the conspecifics flight-flocking), Figure 1 shows a detailed diagram of the designed architecture developed as a result, and the next sections describe its components.

While not claiming neurophysiological accuracy, the architecture splits its overall functionality across biologically-plausible subsystems. Thus the module hypothalamus is used to store the drives (for example, hunger), the sensorial cortex stores sensor data, the amygdala contains the emotional systems such as

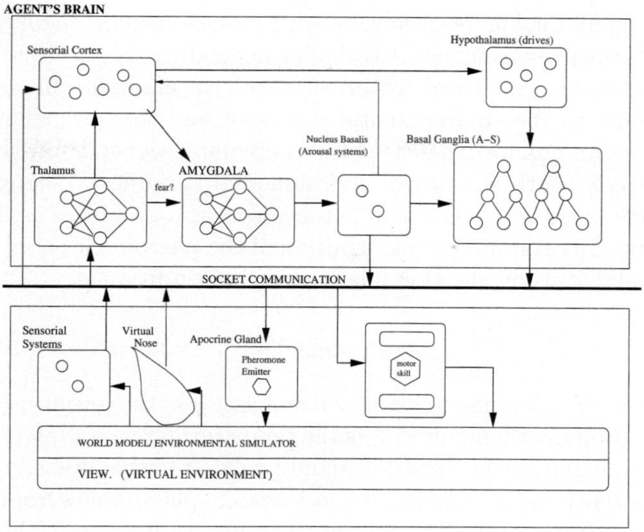

Fig. 1. The complete architecture

Fear, Joy and Anger, and Basal Ganglia contains the hierarchical mechanism for selecting actions, similar to those described by ethologists. Each of the listed modules is defined in XML giving the name of each of the system/variables, the inputs associated to them, a weight, and a function (acting as a filter, in most cases a sigmoid) which in turn generated a feed-forward hierarchy like the one described by Tyrrell [10]. Taking the position that emotion partly functions as a communication mechanism, a novel feature of this work is that the perceptual component has been designed to support the communication of emotion among conspecifics. In the real world, emotional transmission is almost certainly multimodal, with certain modes such as the perception of motion being particularly difficult to model. Thus we have limited ourselves for now to a single mode, and the one we have chosen is pheromones, perceived by a virtual olfaction sensor. Recent experiments [11] have shown that mammals, including humans, emit pheromones through apocrine glands as an emotional response, and as means to communicate that state to conspecifics, who can adapt their behaviour accordingly; research has found that odours produce a range of emotion responses in animals, including humans [12]. This is adaptively advantageous because olfaction is part of the old smell-brain which can generate fast emotional responses, that is without the need of cognitive processes. Grammer [11] argues that every living creature has a distinctive molecular signature that can be carried in the wind, variously showing it to be nutritious, poisonous, sexual partner, predator or prey. Neary [13] points out that sheep, particularly range sheep, will usually move more readily into the wind than with the wind, allowing them to utilise their sense of smell.

Our architecture models the exteroceptors used by real animals to detect the presence of chemicals in the external environment as a virtual nose. An environmental simulator has been developed: its tasks include changing the temperature and other environmental variables depending on the time of day and on the season, using statistical historical data. An alarmed animal sends virtual pheromones to the environmental simulator and they are simulated using the free expansion gas formula in which the volume depends on the temperature and altitude (both simulated environmental variables). The expansion of the pheromone cloud at timestep=9 can be seen in a graphical environment in Figure 5 below. To compute the distribution of the pheromones a set of particles has been simulated using the Boltzmann distribution formula:

$$n(y) = n_o e^{-\frac{mgy}{k_b T}} \tag{1}$$

Where m is the pheromone's molecular mass; g is the gravity; y is the altitude; k_b is the Boltzmann number; T is the temperature; n_o is N/V; N is number of molecules exhuded by the apocrine gland (related to the intensity of the emotion) and V is the Volume. The virtual nose detects pheromones from a threshold of 200.10-16 reflecting values taken from the relevant literature. The problem of action selection is that of choosing at each moment in time the most appropriate action out of a repertoire of possible actions. The process of making this decision takes into account many stimuli, including in this case the animal's emotional state. Action selection algorithms have been proposed by both ethologists and computer scientists. Models suggested by the former are usually at a conceptual level, while those of the latter (with some exceptions -as [7][10] generally do not take into account classical ethological theories. Dawkins [14] suggests that a hierarchical structure represents an essential organising principle of complex behaviours: a view shared by many ethologists [9][15]. Recent research has found that the Basal Ganglia plays an important role in mammalian action selection [16] and our mechanism is implemented in the Basal Ganglia module in Figure 1 as a three-level tree. To avoid sensory congestion, each of Top, Intermediate and Bottom nodes receives sensor data directly as well as data from a higher-level node. Actions are selected by Bottom nodes, dispatching them via a UDP socket to the Animation engine located in the Body module of Figure 1. This mechanism is based on [10] who in turn developed Rosenblatt and Payton's original idea [17] of a connectionist, hierarchical, feed-forward network, to which temporal and uncertainty penalties were added, and for which a more specific rule for combination of preferences was produced. Note that among other stimuli, our action selection mechanism takes the emotional states (outputs of the emotional devices) of the virtual animal. Klinokinesis was modelled as a Finite State Acceptor [18], augmented with transitions based on probability, as seen in Table 1. The basic Reynolds rules of cohesion, alignment and separation have been extended with an additional rule (escape) in which the virtual animal moves away from potential danger (essentially, predators) in its vicinity. More importantly, the flocking behaviour itself is parameterised by the emotional devices output, that is, by the values of the emotions the virtual animals feel. Therefore, in our model each virtual animal moves itself along a vector, which is the resultant of four component vectors, one for each of the behavioural rules. The calcula-

δ	q	*input*	$\delta(q, input)$
	start	go-default	stand-still
	stand-still	P (0.3)	walking
	stand-still	P (0.3)	starting-to-eat
	stand-still	P (0.2)	rotating-left
	stand-still	P (0.2)	rotating-right
	stand-still	in-fear	end
	stand-still	do-nothing	stand-still
	walking	P (0.3)	stand-still
	walking	P (0.7)	walking
	rotating-left	P (0.9)	stand-still
	rotating-left	P (0.1)	rotating-left
	rotating-right	P (0.9)	stand-still
	rotating-right	P (0.1)	rotating-right
	starting-to-eat	head-down	eating
	eating	P (0.6)	eating
	eating	P (0.4)	finishing-to-eat
	finishing-to-eat	head-up	stand-still

Table 1. Finite State Acceptor for klinokinesis

tion of the resultant vector, V(elocity), for a virtual animal A is as follows: The calculation of the resultant vector, *Velocity*, for a virtual animal A is as follows:

$$V_A = \underbrace{(Cf \cdot Cef \cdot Cv)}_{Cohesion} + \underbrace{(Af \cdot Aef \cdot Av)}_{Alignment} + \underbrace{(Sf \cdot Sef \cdot Sv)}_{Separation} + \underbrace{(Ef \cdot Eef \cdot Ev)}_{Escape} \tag{2}$$

$$Velocity_A = limit(V_A, (MVef \cdot MaxVelocity)) \tag{3}$$

where Cv, Av, Sv and Ev are the component vectors corresponding to the cohesion, alignment, separation and escape rules respectively. Cf, Af, Sf and Ef are factors representing the importance of the component vectors Cv, Av, Sv and Ev respectively. Cef, Aef, Sef and Eef are factors representing the importance of the component vectors Cv, Av, Sv and Ev and allow weighting of each component vector independently. In our current implementation they can be varied, in real time, from a user interface. Cef, Aef, Sef and Eef are factors representing the importance of the respective component vectors given the current emotional state of the virtual animal. Each of these factors is a function that takes the current values of the animal's emotions and generates a weight for its related component vector. MaxVelocity is the maximum velocity allowed to the animal. In the current implementation it can be varied from a user interface. MVef is a factor whose value is calculated as a function of the current values of the animal's emotions. It allows the increase and decrease the animal's MaxVelocity depending on its emotional state as shown in Figure 2. limit is a function whose value is equal to the greater of its two parameters.

The emotional factors (Cef, Aef, Sef, Eef, and MVef) reflect ethological heuristic rules. For example, the greater the fear an animal feels, the greater the weight of both its cohesion vector (it tries to stay closer to nearby flockmates) and its escape vector (it tries to stay farther from the potential danger).

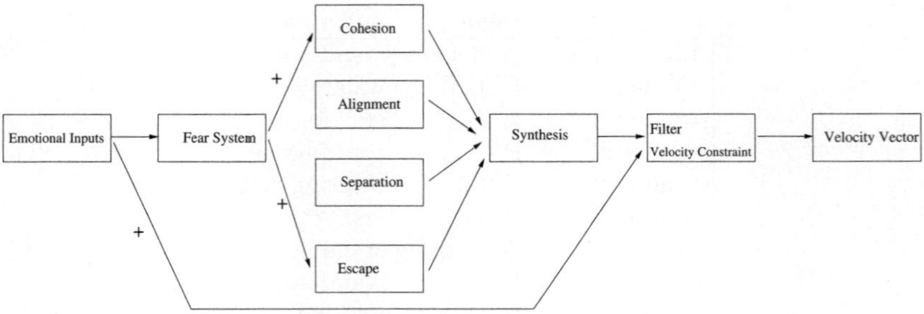

Fig. 2. Influence of Fear on Flocking

The resultant vector obtained by adding the four basic vectors is then scaled so as not to exceed the maximum speed. Note that maximum velocity is also parameterised by fear: the greater the fear an animal feels, the greater the speed it is able to reach.

3 Evaluating the Emergent Behaviour

Our hypothesis that fear can serve as a regulator between individual and social behaviour was evaluated through an experiment in which 5,10, 15 and 20 animals were plotted over 600 timesteps for the following six conditions: 1. Rigid Flocking. The herd of animals was tightly packed (maximum10 centimetres distance between each) and animals were all facing the same direction at all times. This is the baseline condition for optimum coordination. 2. No Flocking No Escape. Each animal moved on its own with no knowledge (perception) of other animals or predators. This is the baseline condition for individual behaviour. 3. Escape. Similar to the previous scenario except that animals perceive predators and individually move to avoid them. 4. Standard flocking. Animals perceive each other, try to avoid collisions between each other and try to stay close to the herd. 5. Standard flocking with Escape. As the previous case but animals perceive predators, and move to avoid them. 6. Escape with emotion. Emotion (fear) is elicited and communicated amongst animals via artificial pheromones when predators are perceived. Figure 3 shows the trajectories plotted for the 20 animals case, and it is intuitively clear to the eye even at this very low resolution that very different patterns of behaviour are being produced. What is required is a way of assessing the complexity of the emergent behaviour in each case. We follow the approach of [19] who presented a method for characterising the pattern of emergent behaviour and its complexity using singular values and entropy. In the matrix A below, M = 600 (number of samples) and N = 4 (degrees of freedom: position x,y and velocity x,y):

$$A = \begin{pmatrix} {x_1}^1 {y_1}^1 \dot{x}_1^1 \dot{y}_1^1 \cdots {x_1}^N {y_1}^N \dot{x}_1^N \dot{y}_1^N \\ \vdots \\ {x_M}^1 {y_M}^1 \dot{x}_M^1 \dot{y}_M^1 \cdots {x_M}^N {y_M}^N \dot{x}_M^N \dot{y}_M^N \end{pmatrix}$$

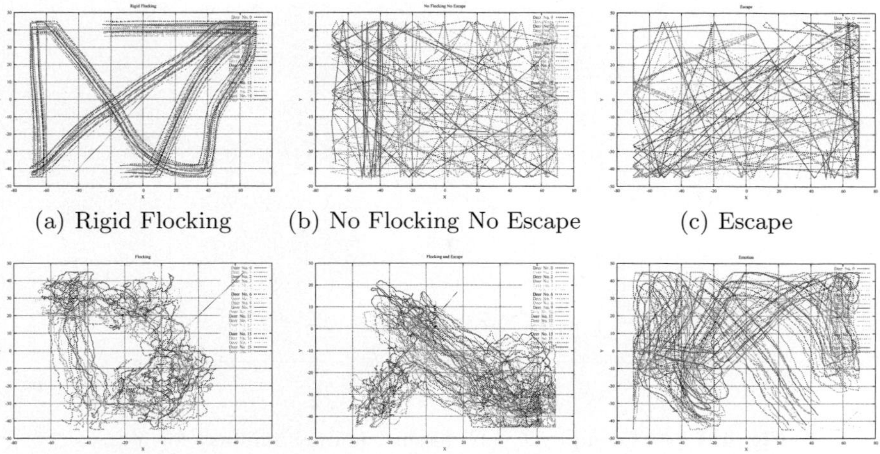

(a) Rigid Flocking (b) No Flocking No Escape (c) Escape

(d) Flocking (e) Flocking Escape (f) Emotion

Fig. 3. Plots of 6 cases, 600 timesteps, 20 animals

To compute the singular values, the following equation from linear algebra is used:

$$A = USV^T \tag{4}$$

The singular values $\sigma_i = S_i$ are all non-negative and generally are presented in a decreasing sequence $\sigma_1 >= \sigma_2 >= \cdots >= \sigma_N >= 0$; singular values can be used as an approximation of the matrix. We do not have space to display the singular values for 5,10,15,20 animals for all six cases here, but if they are represented in bar chart form they show that each flocking case has its own distinctive shape. The next step is to compute the entropy from the N singular values which are normalised, because by definition $\sum_i P_i = 1$ [20], in our case P_i is σ_i. The following equation is used to calculate entropy:

$$E_s = -\sum_{i=1}^{N} \sigma_i{}' \log_2 \sigma_i{}' \tag{5}$$

where $\sigma_i{}'$ is the normalised singular value. And since entropy can be seen as a $\log_2$ count of the number of states in a system [20], the effective number of states and thus the complexity is given by the expression:

$$\Omega = 2^{E_s} \tag{6}$$

Figure 4 shows a plot of the complexities for different types of flocking with different number of animals. It can be seen that rigid flocking (bottom line) shows the least complexity, intuitively supported by looking at Figure 3, top left. Flocking; flocking with escape; no flocking, no escape; and escape behaviours

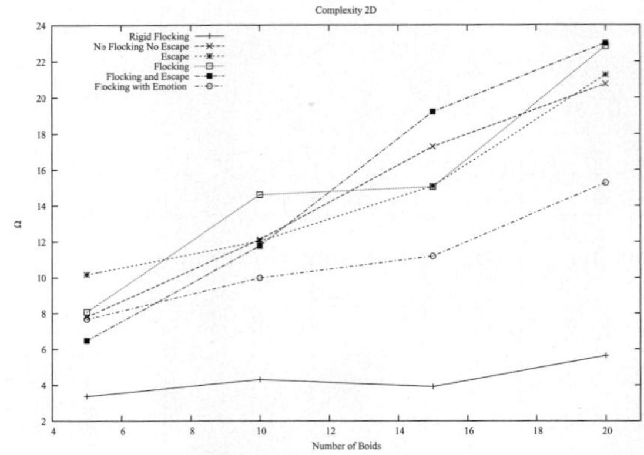

Fig. 4. Plot of complexity (Ω) against animal numbers for 6 cases

(top four lines) are more complex than rigid flocking, but they are also almost always more complex than flocking with emotion (second line up). The exception is the 5-animal case where flocking with emotion, is more complex than flocking with escape. This can be explained by a further set of experiments carried out in which it is shown that at least nine animals are needed to maintain flocking behaviour. With fewer than this, when the animals escape from a predator, some separate from the flock and do not regroup at all during the 600 time-steps.

Thus we conclude that the introduction of an emotional system into action-selection, where emotion can be transmitted between animals, mediates between the complexity of individual behaviour and the rigidity of collective behaviour. It allows a dynamic trade-off between spreading widely, advantageous in seeking new grass to graze - inherent in klinokinesis - and staying together, advantageous in the case of attack by predators. Emotion in this case acts as a social regulator for flocking animals, demonstrating that it has an important social function in addition to its already-understood role in regulating individual behaviour. In addition to the 2D-tracking of trajectory just described, the virtual animals have also been implemented in a real-time 3D graphical world, which can be run in a 4-sided immersive display system (or CAVE). The implementation consists of nearly 28,000 lines of C++ code of which 10,949 implement the brain. Figure 5 shows a screen-shot of a sheep in a graphical world illustrating the spread of the pheromone cloud at timestep = 9. A further objective of the work discussed elsewhere [21] is to examine how far the presence of emotionally-driven autonomous animals can increase the feeling of immersion experienced by a human user in such environments.

4 Conclusions and Further Work

We have presented an ethologically-inspired virtual animal architecture in which primitive emotions have been incorporated into action-selection and a method

Fig. 5. Pheromone cloud visualised at timestep=9

for communicating emotion between animals using virtual pheromones has been included, allowing the extension of the classic approach to flocking to incorporate emotion. We have shown that the effect of adding the emotional input to flocking together with the communication mechanism is to reduce the complexity of individual behaviour without requiring rigid lock-stepping. This substantiates the hypothesis that emotion mediates social behaviour, underlining the functional role of affect in action-selection. Extensions to this work might include individual variation in animals, both across characteristics like fearfulness, and across gender: there is evidence that ewes spend more time grazing and rams significantly longer lying. The presence of lambs would also introduce an interesting element of social heterogeneity, while animals with other behavioural responses to predators –musk ox for example form an outward facing ring – could be explored. The use of 3D space in this implementation is limited to the pheromone propagation algorithm: both perception and locomotion were implemented as 2D mechanisms. Given mammals have significantly less mobility in 3D than the classic examples of fish or birds, a more realistic application of manoeuvrability constraints would not only look more natural but might also have practical implementations for flock fragmentation in the face of predators. A classic predator strategy is to peel off an individual flock member, and including one or more intelligent predators would allow predator-prey interaction to be investigated. Finally, although the architecture developed targetted animals such as sheep and deer rather than humans, the extension of the approach into emotionally-driven human crowds would open up a much larger field of investigation.

References

1. Minsky, M.: The Society of Mind. Simon & Schuster, New York (1985)
2. Canamero, D.: Modeling motivations and emotions as a basis for intelligent behaviour. In: Johnson, L. (ed.) Proceedings of the first International Conference on Autonomous Agents, Marina del-Rey, CA, ACM Press, New York (1997)

3. Velásquez, J.: Modeling emotions and other motivations in synthetic agents. In: Proceedings of the Fourteenth National Conference on Artificial Intelligence (AAAI-97), Providence, RI, AAAI Press, Menlo Park (1997)
4. Gratch, J., Rickel, J., Marsella, S.: Tears and fears. In: 5th International Conference on Autonomous Agents, pp. 113–118. ACM Press, New York (2001)
5. Ortony, A., Clore, G.L, Collins, A.: The Cognitive Structure of Emotions. Cambridge University Press, Cambridge (1988)
6. Ekman, P.: Facial expression and motion. American Psychologist 48 (1993)
7. Blumberg, B.: Action-selection in hamsterdam: Lessons from ethology. In: Proceedings of the Third International Conference on Simulation of Adaptive Behavior, Brighton, England (1994)
8. Reynolds, C.W.: Flocks, herds, and schools: A distributed behavioral model. Computer Graphics 21(4), 25–34 (1987)
9. Baerends, G.P.: The functional organization of behaviour. Animal Behaviour 24, 726–738 (1976)
10. Tyrrell, T.: Computational Mechanisms for Action Selection. PhD thesis, University of Edinburgh, Edinburgh, Scotland (1993)
11. Grammer, K.: 5-alpha-androst-16en-3alpha-one: a male pheromone? Ethology and Sociobiology 14(3), 201–207 (1993)
12. Izard, C.E.: Four systems for emotion activation: Cognitive and noncognitive processes. Psychological Review 100(1), 68–90 (1993)
13. Neary, M.: Sheep sense. The Working Border Collie (2001)
14. Dawkins, R.: Hierarchical organisation: A candidate principle for ethology. In: Hinde, B. (ed.) Growing Points in Ethology, Cambridge University Press, Cambridge (1976)
15. Tinbergen, N.: The study of instinct. Oxford University Press, Oxford (1969)
16. Montes-Gonzalez, F.M.: An action Selection Mechanism based on Vertebrate Basal Ganglia. PhD thesis, Psychology Department, University of Sheffield, Sheffield, United Kingdom (2001)
17. Rosenblatt, J.K., Payton, D.W.: A fine-grained alternative to the subsumption architecture for mobile robot control. In: Proceedings of the IEEE/INNS International Joint Conference on Neural Networks, vol. 2, Washington DC, June 1989, pp. 317–324. IEEE, Los Alamitos (1989)
18. Arkin, R.C.: Behavior-Based Robotics. MIT Press, Cambridge (1999)
19. Wright, W.A., Smith, R.E., Danek, M., Greenway, P.: A generalisable measure of self-organisation and emergence. In: Dorffner, G., Bischof, H., Hornik, K. (eds.) ICANN 2001. LNCS, vol. 2130, pp. 857–864. Springer, Heidelberg (2001)
20. Bonabeau, E., Dorigo, M., Theraulaz, G.: Swarm Intelligence: From Natural to Artificial Systems. Oxford University Press, Oxford (1999)
21. Delgado-Mata, C., Ibanez, J., Aylett, R.: Let's run for it: Conspecific emotional flocking triggered via virtual pheromones. In: Butz, A., krüger, A., Olivier, P. (eds.) SG 2003. LNCS, vol. 2733, pp. 131–140. Springer, Heidelberg (2003)

Comparing ACO Algorithms for Solving the Bi-criteria Military Path-Finding Problem*

Antonio M. Mora[1], Juan J. Merelo[1], Cristian Millán[2], Juan Torrecillas[2], Juan L.J. Laredo[1], and Pedro A. Castillo[1]

[1] Departamento de Arquitectura y Tecnología de Computadores. University of Granada (Spain)
{amorag,jmerelo,juanlu,pedro}@geneura.ugr.es

[2] Mando de Adiestramiento y Doctrina. Spanish Army
{cmillanm, jtorrelo}@et.mde.es

Abstract. This paper describes and compares mono- and multi-objective Ant Colony System approaches designed to solve the problem of finding the path that minimizes resources while maximizing safety for a military unit in realistic battlefields. Several versions of the previously presented CHAC algorithm, with two different state transition rules are tested. Two of them are *extreme* cases, which only consider one of the objectives; these are taken as baseline. These algorithms, along with the Multi-Objective Ant Colony Optimization algorithm, have been tested in maps with different difficulty. hCHAC, an approach proposed by the authors, has yielded the best results.

1 Introduction and Problem Description

The commander of a military unit in the battlefield must consider two main criteria before deciding on the best path to a destination point: the speed (important if the unit mission requires arriving as soon as possible to the target) and the safety (important when the enemy forces are not known or when the unit effectives are very valuable). However, in any situation, both objectives must be considered. This problem is called *military unit path-finding problem* and can be formally defined as: finding the best path for a military unit, from an origin to a destination point in the battlefield, keeping a balance between route speed and safety, considering the presence of enemies (which can shoot against the unit) and taking into account realistic properties and restrictions.

We model this problem considering that the unit has a *level of energy (health)* and a *level of resources*, which are consumed when it moves through the path, so the problem objectives are adapted to minimize the consumption of resources and energy.

The battlefield is also modelled as a grid of hexagonal cells where every one has assigned a *cost in resources* which represents the difficulty of going through it, and a *cost in energy* which means that the unit depletes its human resources or

* Supported by NadeWeb (TIC2003-09481-C04-01) and PIUGR (9/11/06) projects

F. Almeida e Costa et al. (Eds.): ECAL 2007, LNAI 4648, pp. 665–674, 2007.

that vehicles suffer damage when crossing over the cell (*no combat casualties*). Both costs depend on the cell type. In addition there are other costs: one in resources if the unit moves between cells with different heights (more if it goes uphill) and other in energy, *lethality*, which is the damage that a cell could produce due to enemy weapons impact. These features are resumed in Table 1.

Table 1. Energy and Resources description.

	Composed by	*Consumed by*
Energy	global soldiers health, global vehicles status	'no combat' casualties (injuries, tiredness), lethality
Resources	food, fuel, medicines, general supplies, moral	going through difficulty, height difference

We consider *fast* paths (if speed is constant) when the total cost in resources is low (it is not very difficult to travel through the cells, so it takes little time). *Safe* paths, on the other hand, have a low cost in energy.

The cells also have a type, subtype and height associated, plus some properties (for instance, whether they are or not inside the realistic line of sight of the enemy) and restrictions (such as visible distance or maximum height difference that the unit can go trough).

Initially, we solved the problem by implementing the CHAC algorithm [1], and later we improved it with Hexa-CHAC [2]. More details about the problem definition, restrictions and description and test of the algorithms are shown in these articles.

In this work, we have implemented some new algorithms and tested them in the same maps (battlefields) which will allow us to check which algorithm works the best in each circumstance, and to establish baselines for the performance in each of the objectives we are going to optimize. These algorithms are described in the next section.

2 Algorithms Tested in This Paper

In this work, we have tested two MOACOs (Multi-Objective Ant Colony Optimization algorithms [3]): *hCHAC* [2] (with two different state transition rules -STRs from now on-, and extreme values of λ) and *MOACS* [4]. Additionally, we also test a mono-objective approach which combines both objectives in a single aggregative function. So, there are six algorithms in all. All of them are Ant Colony System algorithms [5,6], so the problem is transformed into a graph where each node corresponds to a cell in the map and an edge between two nodes is the connection between neighbour cells in the map. Every edge has two weights associated which are the costs in resources and energy that going through that edge causes to the unit.

Every iteration, ants separately build a complete path (solution), between origin and destination points (if possible), by travelling through the graph. To guide this movement ants use a STR which combines two kinds of information: pheromone trails and heuristic knowledge. We use ACSs to have better control in the balance between exploration and exploitation by using the characteristic parameter q_0.

The problem we want to solve is a multi-objective (MO) one (see a description of MO problems and algorithms in [7]) with two independent objectives to minimize. These objectives are named $\boldsymbol{f}$, minimization of the resources consumed in the path (fast path or speed maximization) and $\boldsymbol{s}$, minimization of the energy consumed in the path (safe path or safety maximization). That is why ***hexa-CHAC*** (*hCHAC* from now on)[2] is an ACS adapted to deal with two objectives, and uses two pheromone matrices (τ_f,τ_s) and heuristic functions (η_f,η_s) (one per objective), a single colony, and two STRs: (*Combined State Transition Rule, CSTR*), similar to the one proposed in [8], which combines the pheromone and heuristic information for each objective weighted using α, β and λ parameters; and (*Dominance State Transition Rule, DSTR*), which ranks neighboring cells according to how many they dominate [1]. These rules use the parameter $\lambda \in (0,1)$, which is user-defined, and sets the importance of the objectives in the search (which one has the highest priority and how much). If the user decides to search for a fast path, λ will take a value close to 1, on the other hand, if he wants a safe path, it has to be close to 0. This value is constant during the algorithm for all ants, so *hCHAC* searches always in the same zone of the space of solutions (the zone related to the chosen value for λ).

The local and global pheromone updating formulae [1] are based in the MACS-VRPTW algorithm proposed in [9,4], with some changes due to the use of two pheromone matrices. Finally, there are two evaluation functions (used to assign a global cost value to every solution found) named F_f (minimization of resources consumption) and F_s (minimization of energy consumption).

On the other hand, ***Mono-hCHAC*** is an ACS that combines the two previous objectives in one. It uses formulae similar to those of *hCHAC* (heuristic, pheromone updating and evaluation function (see in [1])), but only one in each case and adapted to consider only one objective (by including specific terms for each objective).

The *Heuristic Function* is as follows:

$$\eta(i,j) = \frac{\omega_r}{C_r(i,j)} + \frac{\omega_e}{C_e(i,j)} + \frac{\omega_d}{Dist(j,T)} + (\omega_{zo} \cdot ZO(j)) \quad (1)$$

Where C_r and C_e are respectively the cost in resources and energy when moving from node i to node j, *Dist* is the Euclidean distance between two nodes (T is the target node of the problem) and *ZO* is a score (between 0 and 1) for a cell, being 1 when the cell is hidden to all the enemies (or to all the cells in a radius when there are no enemies) and decreasing exponentially when it is seen. ω_r, ω_e, ω_{fd} and ω_{fo} are weights that assign relative importance to the terms in the formula. The values for the two first are the same as in the *hCHAC* formulas, and the values for the two last have been calculated as an average of the correspondent parameters in those formulas. So, all terms are important.

The *STR* which guides the search is the typical formula in mono-objective ACSs:
If $(q \leq q_0)$

$$j = \arg\max_{j \in N_i} \left\{ \tau(i,j)^{\alpha} \cdot \eta(i,j)^{\beta} \right\} \tag{2}$$

else

$$P(i,j) = \begin{cases} \dfrac{\tau(i,j)^{\alpha} \cdot \eta(i,j)^{\beta}}{\sum\limits_{u \in N_i} \tau(i,u)^{\alpha} \cdot \eta(i,u)^{\beta}} & if\, j \in N_i \\ \\ 0 & otherwise \end{cases} \tag{3}$$

where $q_0 \in [0,1]$ is the standard ACS parameter, q is a random value in [0,1]. τ is the pheromone trails matrix and η is the heuristic function (Equation 1). α and β are the usual weighting parameters and N_i is the current feasible neighbourhood for the node i.

This STR works as follows: when an ant is building a path and is placed at one node i, a random number q in [0,1] is generated, if $q \leq q_0$ the best neighbour j is selected as the next node in the path (Equation 2). Otherwise, the algorithm decides the next one by using a roulette wheel considering *P(i,j)* as probability for every feasible neighbour j (Equation 3).

The *Local Pheromone Updating* is performed when a new node j is added to the path that an ant is building:

$$\tau(i,j) = (1-\rho) \cdot \tau(i,j) + \rho \cdot \tau_0 \tag{4}$$

where ρ in [0,1] is the common evaporation factor and τ_0 is the initial amount of pheromone in every edge:

$$\tau_0 = \frac{1}{n_c \cdot ((MAX_R + MAX_E)/2)} \tag{5}$$

with n_c as the number of cells in the map to solve, MAX_R as the maximum amount of resources going through a cell may require, and MAX_E as the maximum cost in energy that going through a cell may produce.

The *Global Pheromone Updating* is performed at the end of every iteration of the algorithm, once all the ants have built a solution path:

$$\tau(i,j) = (1-\rho) \cdot \tau(i,j) + \frac{\rho}{F_{fs}} \tag{6}$$

F_{fs} is the *Evaluation Function* which assigns a global cost value to every solution found by each ant. It considers the cost in resources, the cost in energy, and the visibility of each node (cell) in the path:

$$F_{fs}(Psol) = \sum_{n \in Psol} \left[Cr(n-1,n) + Ce(n-1,n) + \omega_{zo}^{F} \cdot (1 - ZO(n)) \right] \tag{7}$$

where $Psol$ is the solution path to evaluate and ω_{zo}^{F} is the weight which sets the importance of visibility of the cells in the path. The other terms are the same as in Equation 1.

MOACS was proposed by Baran et al. [4], as a variation of the MACS-VRPTW introduced by Gambardela et al. in [9], the main difference being the use of a single pheromone matrix for both objectives (instead of one per objective). We have adapted it to solve this problem, so we use the same heuristic and evaluation functions (see in [1]), but different STR and pheromone updating formulas. The STR is similar to the *hCHAC* CSTR, but using only one pheromone matrix (as we previously said). It is defined as follows:
If $(q \leq q_0)$

$$j = \arg\max_{j \in N_i} \left\{ \tau(i,j) \cdot \eta_f(i,j)^{\beta \cdot \lambda} \cdot \eta_s(i,j)^{\beta \cdot (1-\lambda)} \right\} \tag{8}$$

else

$$P(i,j) = \begin{cases} \dfrac{\tau(i,j) \cdot \eta_f(i,j)^{\beta \cdot \lambda} \cdot \eta_s(i,j)^{\beta \cdot (1-\lambda)}}{\sum\limits_{u \in N_i} \tau(i,u) \cdot \eta_f(i,u)^{\beta \cdot \lambda} \cdot \eta_s(i,u)^{\beta \cdot (1-\lambda)}} & if\, j \in N_i \\ \\ 0 & otherwise \end{cases} \tag{9}$$

where τ is the pheromone matrix, η_f and η_s are the heuristic functions for the objectives, and the rest of terms and parameters are the same as in Equation 3. This rule also uses λ to set the importance of the objectives in the search and takes a constant value during the algorithm for all ants, unlike the original proposal of Baran et al. [4] in which the parameter takes a value of 0 for the first ant, and it grows for every ant until it takes a value of 1 for the last one. The reason is that the algorithm must yield solutions following the user desires (like *hCHAC*), which means solutions in a concrete zone of the solutions space. The rule works as we previously explain.

Since *MOACS* is an ACS, there are two levels of pheromone updating, local and global. The equation for *Local Pheromone Updating* is:

$$\tau(i,j) = (1-\rho) \cdot \tau(i,j) + \rho \cdot \tau_0 \tag{10}$$

considering:

$$\tau_0 = \frac{1}{n_c \cdot MAX_R^{\lambda} \cdot MAX_E^{(1-\lambda)}} \tag{11}$$

with the same parameters and terms as in Equations 4,5, but also weighted using λ parameter to consider again the relative importance of each objective set by the user.

There is a reinitialization mechanism, so the value of τ_0 is not fixed during the algorithm run, as usual in ACS, but it undergoes adaptation. Every time an ant h builds a complete solution, it is compared to the Pareto set P generated until now to check if the former is a non-dominated solution. At the end of each iteration, τ_0' is calculated following the formula:

$$\tau_0' = \frac{1}{n_c \cdot \bar{Cr}(PS)^{\,\lambda} \cdot \bar{Ce}(PS)^{\,(1-\lambda)}} \tag{12}$$

where $\bar{Cr}$ and $\bar{Ce}$ are respectively the average consumption of resources and energy for the solution paths currently included in the Pareto set.

Then, if $\tau_0' > \tau_0$, the current initial pheromone value, the pheromone trails are reinitialized considering the new value for $\tau_0 \leftarrow \tau_0'$. It means a better Pareto Set has been found.

Otherwise the Global Pheromone Updating is made for every solution in the Pareto Set:

$$\tau_f(i,j) = (1-\rho)\cdot\tau_f(i,j) + \frac{\rho}{F_f \cdot F_s} \tag{13}$$

where F_f and F_s are the evaluation functions for each objective.

We will also introduce what we call ***Extreme hCHAC*** (in two versions, *extremely fast* and *extremely safe*) in this section to include them as baseline for comparison with the other algorithms. And both of them have been tested using CSTR and DSTR.

The *extremely fast hCHAC* only considers the minimization of resources consumption objective (speed), so λ is set to 1 and all the weights related to safety objective, such as visibility of cells or energy consumption, take values equal to 0. On the other hand, the *extremely safe hCHAC* only takes into account the minimization of energy consumption objective (safety), so λ is set to 0 and all the weights related to speed objective, such as distance to target point or resources consumption, also take values equal to 0. From now on, we will refer to both approaches as *extr-hCHAC*.

3 Experiments and Results

We have performed experiments in two of the same 45x45 cell realistic maps used in a previous paper [2], which are part of 2 Panzer General^{TM} game maps. All the algorithms presented in the previous section have been run in these maps using the same parameter values (except in extreme approaches, of course), namely: α=1, β=2, ρ=0.1 and q_0=0.4. We have used different values for λ parameter: 0.9 and 0.1 to consider one objective with higher priority than the other, and 1 and 0 in extreme approaches. The mono-objective implementation does not use this parameter.

All the MOACOs yield a set of non-dominated solutions, but less than usual in this kind of algorithms since it only searches in the region of the ideal Pareto front determined by the λ parameter. In addition, we only consider one (chosen by the military staff considering their own criteria and the features of each problem). The mono-objective approach yields a single solution which is evaluated using the same functions as in *hCHAC* and *MOACS* in order to obtain a multi-objective valuation which can be compared with the solutions of the other algorithms and methods.

We have made 30 runs per scenario, using each algorithm or method (STRs) and using each value for λ: 0.9 and 0.1 for *hCHAC-CSTR, hCHAC-DSTR, mono-hCHAC* and *MOACS*, to find the fastest and safest path (but considering the other criteria too); and 0 and 1 for *extr-hCHAC-CSTR* and *extr-hCHAC-DSTR*

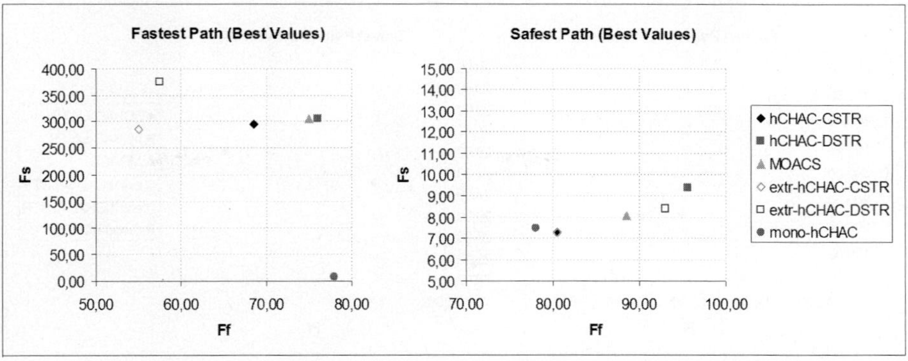

Fig. 1. Best results yielded by all algorithms for Map 1: speed (right) and safety (left) optimization.

Table 2. Results for Map 1. (1500 iterations, 50 ants)

		Fastest (λ=0.9)		Safest (λ=0.1)	
		F_f	F_s	F_f	F_s
hCHAC-CSTR	Best	68.50	295.40	80.50	**7.30**
	Mean	75.20 ±7.87	184.54 ±132.49	85.00 ±3.32	8.10 ±0.49
hCHAC-DSTR	Best	76.00	306.10	95.50	9.40
	Mean	81.63 ±3.02	271.11 ±39.98	108.00 ±5.70	10.40 ±0.52
MOACS	Best	75.00	306.00	88.50	8.10
	Mean	82.95 ±5.13	256.20 ±90.02	103.52 ±5.56	9.50 ±0.70
		Extreme Fast (λ=1)		Extreme Safe (λ=0)	
		F_f	F_s	F_f	F_s
extr-hCHAC-CSTR	Best	**55.03**	285.50	80.50	**7.30**
	Mean	**58.73 ±1.79**	309.53 ±27.63	84.07 ±3.56	**7.89 ±0.56**
extr-hCHAC-DSTR	Best	57.54	375.60	93.00	8.40
	Mean	63.63 ±2.45	329.29 ±38.09	106.90 ±5.85	10.22 ±0.65
mono-hCHAC	Best	78.00	7.50		
	Mean	85.63 ±3.68	8.41 ±0.43		

in order to find the fastest and safest path respectively (without consider the other objective in each case).

Finally, statistical t-Student tests have been used to evaluate obtained results and to test whether differences among means are significant.

In *the first map* there is a single unit on watch between the origin and target point of the unit. There are some hidden zones, but the enemy controls the area the unit must pass in order to get to the target as soon as possible. It has associated a medium difficulty from the military point of view. Results for this map are shown in Table 2 and Fig. 1.

Fig. 1 shows that extreme approaches results can be taken as reference, being always the best in the objective which the highest priority, because they do not consider the other objective at all (they have a high cost value for that objective). We can see that *extr-hCHAC-CSTR* also yields a good result for cost in the non-considered objective, even better than those of the other approaches. The reason is that, in this map, a fast solution can be also a safe one (if the path

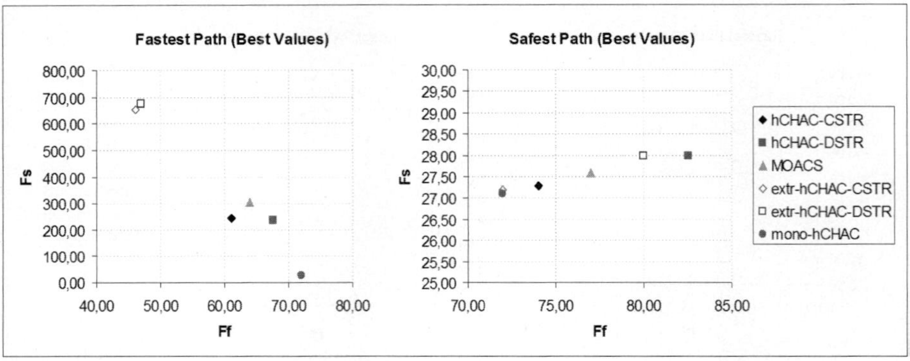

Fig. 2. Best results yielded by all the approaches for Map 2 searching for fastest (left) and safest (right) path.

moves hidden to the enemy) and vice versa. DSTR implementations yield worse results than CSTR ones, because this STR needs a higher exploitation factor (is more explorative) to get similar results. So, even the extreme *extr-hCHAC-DSTR* approach yields worse best solution than some others for the main objective. In the analysis of the main algorithms, it can be seen that *hCHAC-CSTR* yields very good solutions, close to the extreme ones (even equal in the safest case) and, considering the dominance concept, it always yields solutions only dominated by extreme CSTR. DSTR approach is worse for the reason we explain above. *MOACS* also yields good solutions, better than DSTR methods ones (except in fastest path search), but worse than *hCHAC-CSTR* results.

mono-hCHAC is a special case, because it only yields one solution which combines speed and safety. We have represented this solution considering the same costs yielded by it in both searches. Both cost are quite good (low), but they are not the best in comparing with the other approaches, when they search for the fastest or the safest path.

Table 2 shows that all approaches yield results with a low standard deviation in the prioritary objective, which implies robustness (its solutions are similar between runs). Best, mean and standard deviation in the secondary objective are logically worse, because it has little importance. But in this case the differences between the security cost (F_s) when it is prioritary and secondary objective are enormous. The reason is fast paths are usually unsafe due to visibility of the cells, and in this experiments we penalize much the term of visibility in the cost function. Results were verified using t-Student statistical tests. Significant differences were found and the confidence level is 99%.

In *Map 2* there are two enemy units between the origin and target point of the unit. One of them is just watching over and the other one, which is nearer to the target point and in the middle of the straight path, is watching over and firing to some strategic points (some bridges) and in a zone surrounding itself. There are little zones where the unit can hide (slight patches of forest). It has

Table 3. Results for Map 2. (1500 iterations, 50 ants)

		Fastest (λ=0.9)		Safest (λ=0.1)	
		F_f	F_s	F_f	F_s
hCHAC-CSTR	Best	61.00	244.90	74.00	27.30
	Mean	66.42 ±3.29	225.19 ±90.26	84.68 ±4.89	28.36 ±0.48
hCHAC-DSTR	Best	67.50	235.60	82.50	28.00
	Mean	72.92 ±2.63	236.97 ±42.74	95.93 ±7.25	29.43 ±0.72
MOACS	Best	64.00	304.90	77.00	27.60
	Mean	70.77 ±2.43	294.66 ±79.44	93.60 ±6.93	29.23 ±0.68
		Extreme Fast (λ=1)		Extreme Safe (λ=0)	
		F_f	F_s	F_f	F_s
extr-hCHAC-CSTR	Best	**46.04**	654.60	72.00	27.20
	Mean	**49.92 ±2.29**	467.30 ±172.03	80.64 ±4.52	**28.05 ±0.45**
extr-hCHAC-DSTR	Best	47.04	674.70	80.00	28.00
	Mean	53.02 ±2.58	403.61 ±130.55	96.35 ±8.16	29.60 ±0.79

mono-hCHAC			
	Best	72.00	**27.10**
	Mean	78.33 ±4.24	52.23 ±42.97

associated a medium-hard difficulty from the military point of view. The results for the different approaches are show in Table 3 and Figure 2.

Fig. 2 shows that *extr-hCHAC-CSTR* yields the best results in both searches, but only in the main objective. In fact, in the fastest path search there is an overwhelming increase of the cost in energy (or safety) F_s, since that path goes through cells affected by enemy weapons. *hCHAC-CSTR* yields solutions dominated only by those of *extr-hCHAC-CSTR* and *mono-CHAC* (see below) in the safest case. DSTR approach is again bad for the reason we explain above. *MOACS* again yields good solutions, better than DSTR methods ones (except in fastest path search), but worse than *hCHAC-CSTR* results.

mono-CHAC results yield the best safety result this time, which might seem odd, however, due to the stochastic nature of these approaches, it is possible to find a very good solution sometimes; average solutions are not as good, as shown by the high standard deviation. As a result, the cost in resources F_f is high, which happen sometimes in difficult maps like this one.

Table 3 shows results similar to the previous experiment, with a higher increasing of cost in F_s in the search for fastest path, due to the lethality of the cells. Again, significant differences were found applying t-Student tests, obtaining a confidence level of 95 or 99%.

4 Conclusions and Future Work

In this work we have tested and compared six different ACS approaches to solve the bi-criteria military path-finding problem (find the best path considering speed and safety as objectives), finding that *hCHAC-CSTR* is the best approach, maintaining a good balance between speed and safety in all cases, and considering always both objectives (with different priorities, depending on the search). It yields better results than extreme cases (where only one of the objectives is considered); even more so in difficult maps, where the cost for the objective not being minimized can increase dramatically. *MOACS* and *mono-objective* algorithms yield good solutions too, generally better than hCHAC-DSTR approach, but worse than those obtained by hCHAC-CSTR. Differences between methods have been proved significant after the application of t-Student tests.

As future work and following the same researching line, we are going to implement some other algorithms in order to include them in the comparison and in the study of the best approach for this problem. We also are going to make other studies (statistical, for instance) to have more criteria to evaluate the performance of every approach. Besides, we will also try to separate completely all objectives: speed, safety, visibility, and distance to objective (in each step), using more pheromone matrices, but taking into account each problem separately into a truly multi-objective approach.

References

1. Mora, A.M., Merelo, J.J., Millán, C., Torrecillas, J., Laredo, J.L.J.: CHAC. a MOACO algorithm for computation of bi-criteria military unit path in the battlefield. In: Pelta, D.A., Krasnogor, N. (eds.) Proceedings of the Workshop on Nature Inspired Cooperative Strategies for Optimization. NICSO'2006, June 2006, pp. 85–98 (2006)
2. Mora, A.M., Merelo, J.J., Millán, C., Torrecillas, J., Laredo, J.L.J., Castillo, P.A.: Enhancing a MOACO for solving the bi-criteria pathfinding problem for a military unit in a realistic battlefield. In: Giacobini, M. (ed.) EvoWorkshops 2007. Applications of Evolutionary Computing. LNCS, vol. 4448, pp. 712–721. Springer, Heidelberg (2007)
3. García-Martínez, C., Cordón, O., Herrera, F.: An empirical analysis of multiple objective ant colony optimization algorithms for the bi-criteria TSP. In: Dorigo, M., Birattari, M., Blum, C., Gambardella, L.M., Mondada, F., Stützle, T. (eds.) ANTS 2004. LNCS, vol. 3172, pp. 61–72. Springer, Heidelberg (2004)
4. Barán, B., Schaerer, M.: A multiobjective ant colony system for vehicle routing problem with time windows. In: IASTED International Multi-Conference on Applied Informatics. Number 21 in IASTED IMCAI, 97–102 (2003)
5. Dorigo, M., Stützle, T.: The ant colony optimization metaheuristic: Algorithms, applications, and advances. In: Glover, F., Kochenberger, G. (eds.) Handbook of Metaheuristics, pp. 251–285. Kluwer Academic Publishers, Dordrecht (2002)
6. Dorigo, M., Di Caro, G.: The ant colony optimization meta-heuristic. In: Corne, D., Dorigo, M., Glover, F. (eds.) New Ideas in Optimization, pp. 11–32. McGraw-Hill, New York (1999)
7. Coello, C.A.C., Veldhuizen, D.A.V., Lamont, G.B.: Evolutionary Algorithms for Solving Multi-Objective Problems. Kluwer Academic Publishers, Dordrecht (2002)
8. Iredi, S., Merkle, D., Middendorf, M.: Bi-criterion optimization with multi colony ant algorithms. In: Zitzler, E., Deb, K., Thiele, L., Coello Coello, C.A., Corne, D.W. (eds.) EMO 2001. LNCS, vol. 1993, pp. 359–372. Springer, Heidelberg (2001)
9. Gambardella, L., Taillard, E., Agazzi, G.: Macs-vrptw: A multiple ant colony system for vehicle routing problems with time windows. In: Corne, D., Dorigo, M., Glover, F. (eds.) New Ideas in Optimization, pp. 73–76. McGraw-Hill, New York (1999)

Decentralized Control and Interactive Design Methods for Large-Scale Heterogeneous Self-organizing Swarms

Hiroki Sayama

Department of Bioengineering, Binghamton University
State University of New York, Binghamton NY 13902-6000, USA
sayama@binghamton.edu

Abstract. We present new methods of decentralized control and interactive design for artificial swarms of a large number of agents that can spontaneously organize and maintain non-trivial heterogeneous formations. Our model assumes no elaborate sensing, computation, or communication capabilities for each agent; the self-organization is achieved solely by simple kinetic interactions among agents. Specifications of the final formations are indirectly and implicitly woven into a list of different kinetic parameter settings and their proportions, which would be hard to obtain with a conventional top-down design method but may be designed heuristically through interactive design processes.

1 Introduction

Research on decentralized control mechanisms of artificial swarms has recently attracted an increasing attention because of its theoretical importance for understanding self-organization and emergence in complex biological systems [1] and its practical utility for creating distributed, robust, adaptive artifacts such as biomimetic algorithms for optimization problems [2,3] and robotic systems for collective task achievement [4]. The latter merit manifests itself particularly in engineering applications, as self-organizing swarms may be substantially more flexible and less vulnerable to errors or attacks than conventional systems controlled by central units and hierarchical information flows.

A major technical challenge in this field is how to program and direct decentralized systems to a desired end. A number of different techniques have been proposed and tested to address this problem, e.g., by assuming hormonal communication signals in the environment [5,6], by embedding top-down designed finite-state machines in each agent [7], by evolving neural network controllers of agents [8,9], or by probabilistic control of aggregation patterns through kinetic parameter variations [10,11]. Most of these studies (except for the last ones) demonstrated only with small swarm populations in the order of tens, and all of them used homogeneous agents only. Heterogeneous swarms have been the subject of investigation in recent studies [12,13]; however, the focus of research

F. Almeida e Costa et al. (Eds.): ECAL 2007, LNAI 4648, pp. 675–684, 2007.

is largely on the task allocation among different types of agents, with limited attention paid to the dynamic structure of swarm populations themselves.

Here we consider dynamic self-organization of spatio-temporal patterns within a large-scale, heterogeneous swarm population, and present new methods of decentralized control and interactive design with which artificial swarms of a large number of agents (in the order of hundreds or thousands) can spontaneously organize and maintain non-trivial heterogeneous formations. This work is part of our ongoing *Swarm Chemistry* project [14], which aims to develop a novel theoretical framework for artificial chemistry research [15]. Our model is unique in that it assumes no elaborate sensing, computation, or communication capabilities for each agent. Rather, the self-organization is achieved solely by simple kinetic interactions among agents, as was originally modeled in Reynolds' Boids [16], and more recently, physically demonstrated through the attraction-repulsion interactions of small magnetized disks by Grzybowski et al. [17].

Specifications of the final formations of swarms in Swarm Chemistry are indirectly and implicitly woven into a list of different kinetic parameter settings and their proportions, called a *recipe*, which would be hard to obtain with a conventional top-down design method but may be designed heuristically through interactive design processes. This paper presents a brief summary of our model and some preliminary results obtained so far.

2 Model

A swarm population in Swarm Chemistry consists of a number of simple, semi-autonomous agents. They can move in a two-dimensional continuous space, perceive positions and velocities of other agents within its local perception range, and change its velocity in discrete time steps according to the following kinetic rules (adopted and modified from the rules in Reynolds' Boids system [16]):

- If there are no local agents within its perception range, steer randomly (**Straying**).
- Otherwise:
  - Steer to move toward the average position of local agents (**Cohesion**).
  - Steer towards the average velocity of local agents (**Alignment**).
  - Steer to avoid collision with local agents (**Separation**).
  - Steer randomly with a given probability (**Whim**).
- Approximate its speed to its own normal speed (**Pace keeping**).

These rules are implemented as shown in Algorithm 1. Kinetic parameters used in this algorithm are listed and explained in Table 1.

Each agent is assigned with its own kinetic parameter settings. We call a collection of identical agents that share the same kinetic parameter settings a *chemical species* (or just *species* in short). *Reaction* is defined as a process in which two or more species or their compounds are blended together and a new spatio-temporal pattern emerges through the kinetic interaction between different chemical species (Fig. 1).

Algorithm 1 The algorithm used to simulate the behavior of agents. $\boldsymbol{x}_i$, $\boldsymbol{v}_i$ and $\boldsymbol{v}'_i$ are the location, the current velocity, and the next velocity of the i-th agent, respectively. $\boldsymbol{a}$ is a local variable temporarily used to represent an acceleration. r and $r_{\pm p}$ represent random numbers taken from $[0, 1]$ and $[-p, +p]$, respectively.

for all $i \in$ agents **do**
  $N \leftarrow \{j \neq i$ that satisfies $|\boldsymbol{x}_j - \boldsymbol{x}_i| < R^i\}$
  // Finding other agents within its local perception range
  if $|N| = 0$ **then**
    $\boldsymbol{a} \leftarrow (r_{\pm.5}, r_{\pm.5})$ // Straying
  else
    $\langle \boldsymbol{x} \rangle \leftarrow \sum_{j \in N} \boldsymbol{x}_j / |N|$ // Calculating the average position of local agents
    $\langle \boldsymbol{v} \rangle \leftarrow \sum_{j \in N} \boldsymbol{v}_j / |N|$ // Calculating the average velocity of local agents
    $\boldsymbol{a} \leftarrow c^i_1(\langle \boldsymbol{x} \rangle - \boldsymbol{x}_i) + c^i_2(\langle \boldsymbol{v} \rangle - \boldsymbol{v}_i) + c^i_3 \sum_{j \in N} (\boldsymbol{x}_i - \boldsymbol{x}_j)/|\boldsymbol{x}_i - \boldsymbol{x}_j|^2$
    //Cohesion, alignment and separation
    if $r < c^i_4$ **then**
      $\boldsymbol{a} \leftarrow \boldsymbol{a} + (r_{\pm 5}, r_{\pm 5})$ // Whim
    end if
  end if
  $\boldsymbol{v}'_i \leftarrow \boldsymbol{v}_i + \boldsymbol{a}$ // Acceleration
  $\boldsymbol{v}'_i \leftarrow \min(V^i_m / |\boldsymbol{v}'_i|, 1) \cdot \boldsymbol{v}'_i$ // Prohibiting overspeed
  $\boldsymbol{v}'_i \leftarrow c^i_5 (V^i_n / |\boldsymbol{v}'_i| \cdot \boldsymbol{v}'_i) + (1 - c^i_5)\boldsymbol{v}'_i$ // Pace keeping
end for
for all $i \in$ agents **do**
  $\boldsymbol{v}_i \leftarrow \boldsymbol{v}'_i$ // Updating velocity
  $\boldsymbol{x}_i \leftarrow \boldsymbol{x}_i + \boldsymbol{v}_i$ // Updating location
end for

Note that individual agents remain exactly the same during this reaction process; only higher-order properties of the population as a whole (e.g., shapes, movement, etc.) may change. Such a nature of reactions in Swarm Chemistry may be a reasonable analog, at least conceptually, of what is going on in real chemistry, where the observed changes of chemical properties are due to the rearrangements of the participating elements and not due to the changes of the elements themselves. It should also be noted, however, that the kinetic interaction rules assumed in Swarm Chemistry have nothing to do with atomic or molecular interactions in real chemistry. The terminologies are borrowed from chemistry just for the purpose of making intuitive analogy.

3 Simulator

We developed a prototype of the interactive simulator of Swarm Chemistry using Java 2 SDK Standard Edition 1.5.0. It runs as a stand-alone application on any computer platform equipped with Java 2 Runtime Environment. Its source code is freely available from the author's website[1].

[1] http://bingweb.binghamton.edu/~sayama/SwarmChemistry/

Table 1. Kinetic parameters involved in the simulation of agent behavior. The maximum values are arbitrarily chosen for implementation purpose only. Unique values are assigned to these parameters for each agent i as its own dynamical properties.

Name	Min	Max	Meaning	Unit
R^i	0	300	Radius of local perception range	pixel
V_n^i	0	20	Normal speed	pixel step^{-1}
V_m^i	0	40	Maximum speed	pixel step^{-1}
c_1^i	0	1	Strength of cohesive force	step^{-2}
c_2^i	0	1	Strength of aligning force	step^{-1}
c_3^i	0	100	Strength of separating force	pixel^2 step^{-2}
c_4^i	0	0.5	Probability of random steering	—
c_5^i	0	1	Tendency of pace keeping	—

We used the *interactive evolution method* [18,19] to enable interactive swarm design processes. Using our simulator, a human experimenter, or "alchemist", can actively participate in the selection, perturbation and blending operations of swarm populations by subjectively selecting preferred swarm behaviors. Interactive evolution has already been applied to the designs of swarm behavior for scientific and educational purposes [10,11]. Our work presented here is distinct from the earlier studies in that it evolves dynamic patterns of heterogeneous swarm populations.

We also note that conventional evolutionary computation techniques, such as genetic algorithms or genetic programming, could be implemented to acquire some types of swarm behaviors automatically. However, we did not choose to do so because setting an easily measurable metric for automated fitness evaluation would necessarily limit the diversity and novelty of potential outcomes.

Figure 2 shows a screen shot of our Swarm Chemistry simulator. Six different swarms are simultaneously simulated and demonstrated in their respective frames. A user can select up to two preferred swarms by clicking on the frames in which those swarms are simulated.

To produce a next generation of swarms, we used a unique set of evolutionary operators that are different from other typical genetic operators such as point mutation or crossover. If just one swarm is selected by the user, four perturbed swarms will be generated by repetitive random re-sampling of agents from the selected swarm up to a modified total population size (with ±80% possible variations). They will form the next set of swarms together with the original swarm preserved as is and a new swarm of randomly generated species, which is introduced to diversify the available options. On the other hand, if two swarms are selected by the user, three blended swarms, with agents' initial positions completely randomized and mixed, will be generated at randomly determined ratios between the two selected swarms (ranging from 20:80 to 80:20), and they will be part of the next set of six swarms together with the original two as well as a new swarm of randomly generated species. In all cases, the maximum number of agents in a swarm is limited to 300. For the blending cases, the number of agents

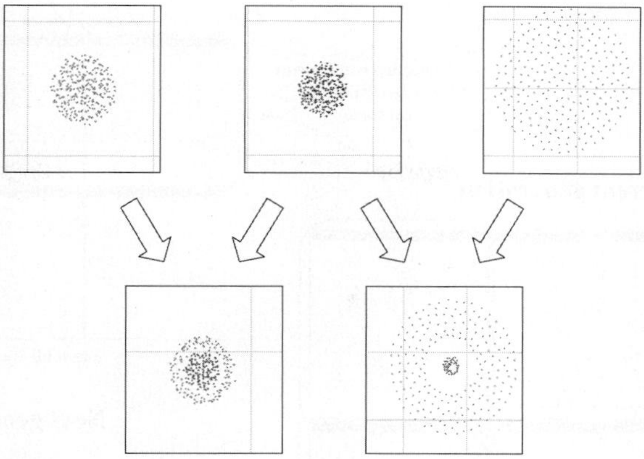

Fig. 1. Chemical reactions in Swarm Chemistry. Each dot represents one agent. Agents are colored by taking their $(c_1^i, c_2^i, c_3^i/100)$ values as their (R, G, B) values, respectively. Top: Three different chemical species (homogeneous populations made of identical agents). Bottom: Results of reactions (spatial patterns formed after blending two species). Blending different chemical species produce different patterns. Gray lines are references drawn at an interval of 300 pixels to show the scale of the image.

in a new swarm is set to the average of those in the two selected swarms. We also implemented point mutation operators that would introduce random variations to each of the kinetic parameters, which was not included in the results presented in this paper in order to focus on the effects of chemical reactions in Swarm Chemistry.

By repeating this interactive selection process, the experimenter can explore a variety of dynamics and potentially create a novel, complex dynamic structure. This design cycle continues indefinitely until the application is manually quitted. The current version of our tool does not allow users to manually edit the kinetic parameter settings, which is planned to be implemented in the future versions.

4 Results

We have conducted preliminary exploration of possible dynamics in Swarm Chemistry and have found several characteristic outcomes of chemical reactions in this model world, which are summarized in what follows.

Spontaneous Segregation: In most cases, agents spontaneously form clusters of the same species and segregate themselves from other chemical species when blended, even though they have no capability to distinguish themselves from other species. This is mainly because of the difference in their kinetic parameter settings that causes the difference in the local environments they “prefer”. It is generally very rare for two different species to remain mixed, with some exceptions where one or both of them are insensitive to local environmental conditions (this may happen if the perception range of agents is small).

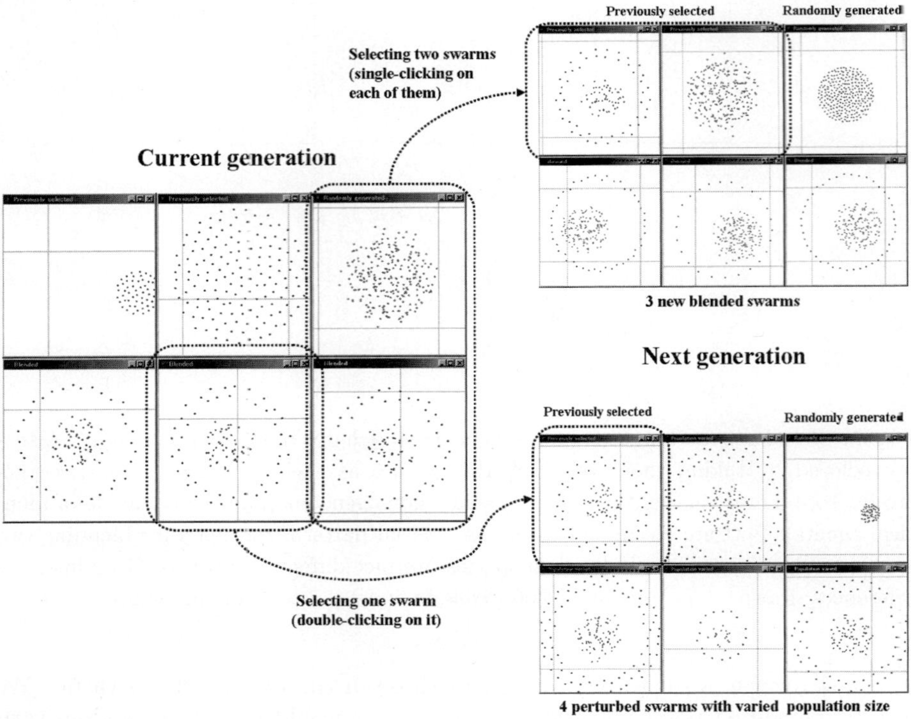

Fig. 2. Left: Screen shot of the Swarm Chemistry simulator. Six swarms are simultaneously simulated on a screen. The simulated space has no boundaries so that the distribution of agents could be infinitely large due to diffusion. The simulator dynamically rescales the viewing size so that the most significant cluster should always be tracked within the frame. Right: Results of selection operations in the Swarm Chemistry simulator. A user can select up to two swarms by clicking on the respective frames. If two swarms are selected, three new blended swarms will be generated at different ratios between the two (top). If just one swarm is selected, it will be replicated four times through random re-sampling with varied population sizes (bottom).

Production of Movements: Reaction in Swarm Chemistry may produce a new macroscale dynamic movement of the swarm, which may be considered an example of the emergence of autonomous motion in prebiotic systems [20]. Movements typically arise when an asymmetric relationship is established between the two chemical species, one as a chaser and the other as an escaper. Such movements based on chasing may take a couple of different forms, as shown in Fig. 3.

Encapsulation and restriction of movements: Chemical reaction in Swarm Chemistry can also form multiple closed layers in a population, where one species is encapsulated in a dynamically formed membrane made of the other species (Fig. 1). This sometimes restricts the movement of originally mobile species (Fig. 4), creating an internal rotation or oscillation inside the structure.

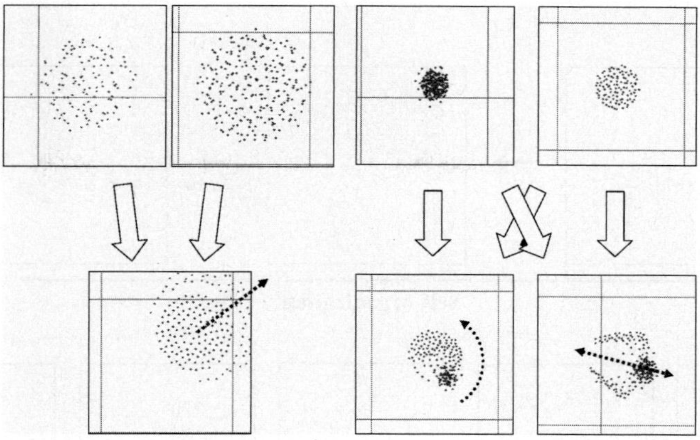

Fig. 3. Production of movements as a result of reaction in Swarm Chemistry. Left: Linear motion. Two originally immobile chemical species form a polarized cluster when blended, and their chasing/escaping behaviors create a linear motion of the entire swarm population. Middle: Rotation. The red cluster rotates around the orange cluster counterclockwise. Right: Linear oscillation. The red cluster is continuously attracted by the orange cloud and keeps moving back and forth through it. In this particular example, the last two movements arise from the same two chemical species. Their relative positions and proportions determine which movement will result.

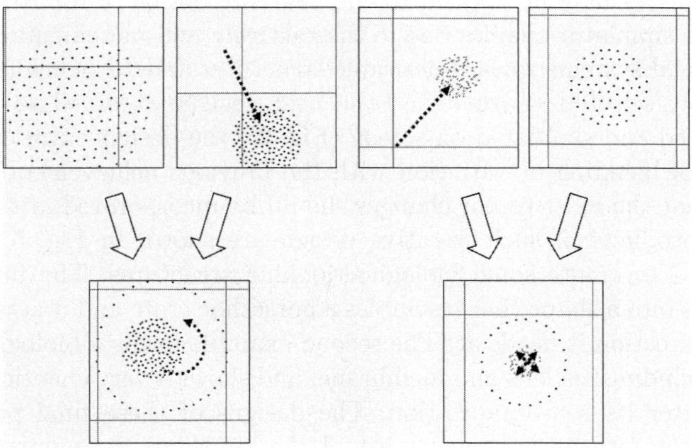

Fig. 4. Encapsulation of originally mobile species into an immobile membrane structure made of other species. The movement of the species are restricted and transformed into internal rotation (left) or oscillation (right).

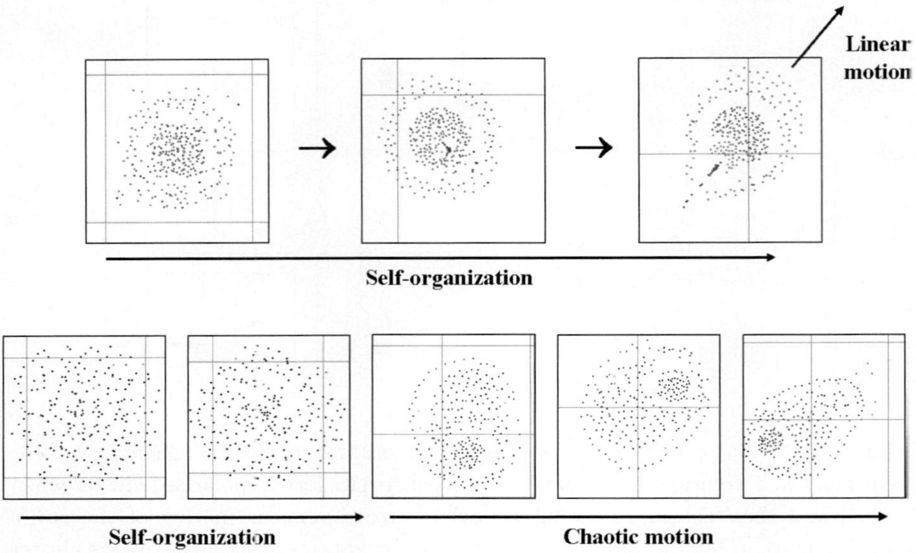

Fig. 5. Examples of complex structures made of several different chemical species, designed using our interactive simulator. The swarms self-organize from initially random states to a shape that looks like a horseshoe crab (top), or a biological cell-like structure that shows active chaotic movement after self-organization (bottom).

5 Interactive Design of Complex Structures

Swarm Chemistry may be applied to the heuristic design of robust distributed systems. Our simulator enables one to interactively and incrementally build artificial chemical systems toward desirable structures and/or behaviors. Because the previously selected swarms (i.e., the best designs obtained so far) are always preserved and simulated on screen (Fig. 2), the designer can compare the new results of blending or variation with the previous achievement and decide whether or not the most recent changes should be incorporated into the design.

Sample products of such iterative design are shown in Fig. 5, where the designer tried to create some biological-looking structures. The first example self-organizes into a shape that resembles a horseshoe crab, and it actually moves toward the direction it heads at. The second example forms a biological cell-like structure, including nucleus and membrane, and shows a very chaotic, aggressive movement after its self-organization. The designs of these final products are specified in the form of a *recipe*, a list of the kind and the number of agents that describes the composition of the swarm population. What the designer is doing is to control the structure and the behavior of the product indirectly by modifying the contents of the recipe.

A remarkable feature of these products, and all other products that can be obtained in Swarm Chemistry, is that they are inherently *self-organizing*. Agents are initially mixed randomly, but they quickly find their own places to sit in and collectively form the shape of the final product (Fig. 5). This self-organization process is found to be highly reproducible and robust to perturbations such as minor variations in system sizes and addition or removal of a small number of agents.

The shape of the final product must be indirectly woven into the recipe; however, there is neither an articulated blueprint nor a central controller that coordinates the arrangement of the agents. The entire process is driven solely by local kinetic interaction between agents. It would be difficult to predict the final outcome without carrying out explicit numerical simulation. While it is generally hard to manually design such a distributed self-organizing system, our approach with interactive, iterative heuristic design appears to be effective to reconcile the inconsistency between self-organization and manual design.

6 Conclusion

We introduced Swarm Chemistry and presented new methods for decentralized control and interactive design of spontaneously organizing and dynamically self-maintaining heterogeneous artificial swarms. We also developed a prototype of the interactive simulator and illustrated several characteristic phenomena frequently observed in the reaction processes in Swarm Chemistry, as well as its potential for heuristic design of robust, self-organizing distributed systems. We anticipate several practical applications of this work, including (1) introduction and dynamic control of self-organized heterogeneous swarms in particle swarm optimization, (2) distributed formation control of unmanned aerial vehicles (UAVs) in aerospace and/or military applications, and (3) interactive generation of aesthetically appealing dynamic patterns for artistic purposes.

Our model is unique in that agents we use are fairly simple. They have no lengthy self-regulating instructions like a genome in a biological cell; they always remain the same with no potential to differentiate. Also, the agents have no capability to identify or distinguish themselves from other species. They just look identical to each other, and therefore no sophisticated communication exists between them. This component simplicity would strengthen Swarm Chemistry as a model of distributed artificial systems, especially in view of real-world implementations where the simplicity of modules is of particular concern.

The Swarm Chemistry project is still at its launching stage, and there are a number of things yet to be done. Future to-dos include (a) analytical and numerical investigations of the properties of each chemical species and the effects of kinetic parameters on pattern formation, (b) exhaustive computational investigation of potential shapes and limitations of patterns made through the interactions of multiple species, (c) implementation and evaluation of dynamic self-replication and self-maintenance of complex patterns, and (d) hardware implementation and demonstration using mobile robotic modules.

References

1. Camazine, S., et al.: Self-Organization in Biological Systems. Princeton Univ. Press, Princeton (2001)

2. Kennedy, J., Eberhart, R.C.: Swarm Intelligence. Morgan Kaufmann, San Francisco (2001)
3. Engelbrecht, A.P.: Fundamentals of Computational Swarm Intelligence. John Wiley & Sons, Chichester (2005)
4. Dorigo, M., et al.: Swarm-bots Project (2001-2005), http://www.swarm-bots.org/
5. Dorigo, M., Stützle, T.: Ant Colony Optimization. MIT Press, Cambridge (2004)
6. Shen, W.-M., Will, P., Galstyan, A., Chuong, C.-M.: Hormone-inspired self-organization and distributed control of robotic swarms. Autonomous Robots 17, 93–105 (2004)
7. Rodriguez, A., Reggia, J.A.: Extending self-organizing particle systems to problem solving. Artificial Life 10, 379–395 (2004)
8. Baldassarre, G., Parisi, D., Nolfi, S.: Distributed coordination of simulated robots based on self-organization. Artificial Life 12, 289–311 (2006)
9. Groß, R., Bonani, M., Mondada, F., Dorigo, M.: Autonomous Self-assembly in a Swarm-bot. In: Proceedings of the Third International Symposium on Autonomous Minirobots for Research and Edutainment (AMiRE 2005), pp. 314–322. Springer, Heidelberg (2006)
10. Kwong, H., Jacob, C.: Evolutionary exploration of dynamic swarm behaviour. In: Proceedings of the 2003 IEEE Congress on Evolutionary Computation (CEC 2003), pp. 367–374. IEEE Computer Society Press, Los Alamitos (2003)
11. Sayama, H.: Teaching emergence and evolution simultaneously through simulated breeding of artificial swarm behaviors. In: Proceedings of the Sixth International Conference on Complex Systems (ICCS2006) (2006), Available online at, http://necsi.org/events/iccs6/proceedings.html
12. Scheutz, M., Schermerhorn, P., Bauer, P.: The utility of heterogeneous swarms of simple UAVs withi limited sensory capacity in detection and tracking tasks. In: Proceedings of the 2005 IEEE Swarm Intelligence Symposium (SIS 2005), pp. 257–264. IEEE Computer Society Press, Los Alamitos (2005)
13. Magg, S., te Boekhorst, R.: Pattern formation in homogeneous and heterogeneous swarms: Differences between versatile and specialized agents. In: Proceedings of the First IEEE Symposium on Artificial Life (IEEE-Alife'07), pp. 311–316. IEEE Computer Society Press, Los Alamitos (2007)
14. Sayama, H,: Swarm chemistry. Artificial Life (Accepted pending revisions) (2007)
15. Dittrich, P., Ziegler, J., Banzhaf, W.: Artificial chemistries – A review. Artificial Life 7, 225–275 (2001)
16. Reynolds, C.W.: Flocks, herds, and schools: A distributed behavioral model. Computer Graphics 21(4), 25–34 (1987)
17. Grzybowski, B.A., Stone, H.A., Whitesides, G.M.: Dynamic self-assembly of magnetized, millimetre-sized objects rotating at a liquid-air interface. Nature 405, 1033–1036 (2000)
18. Unemi, T.: Simulated breeding – a framework of breeding artifacts on the computer. Kybernetes 32, 203–220 (2003)
19. Rothlauf, F., Branke, J., Cagnoni, S., Costa, E., Cotta, C., Drechsler, R., Lutton, E., Machado, P., Moore, J.H., Romero, J., Smith, G.D., Squillero, G., Takagi, H., et al. (eds.): EvoWorkshops 2006 (See papers of the EvoINTERACTION Workshop). LNCS, vol. 3907. Springer, Heidelberg (2006)
20. Ikegami, T,: Emergence of autonomous motion. Keynote talk at the AlifeX Workshop on Simulation Models of Autonomous Systems (2006), Workshop proceedings available online at, http://www.alifex.org/program/wkshp_proceed.pdf

EcoPS - a Model of Group-Foraging with Particle Swarm Systems

Cecilia Di Chio[1] and Paolo Di Chio[2]

[1] Department of Computer Science,
University of Essex, UK
cdichi@essex.ac.uk

[2] Dipartimento di Sistemi e Istituzioni per l'Economia,
University of L'Aquila, Italy
pdc@ec.univaq.it

Abstract. We propose a model for simulating group-foraging behaviour with the use of a Particle Swarm system. Traditionally, the main field of application for Particle Swarm systems has been the optimisation of non-linear functions: with our research, we intend to position Particle Swarm systems in the field of Artificial Life. The EcoPS model we present shows some interesting behaviours. In particular, it seems that grouping is the key to finding the right balance between exploiting resources and surviving.

1 Introduction

Maley [8], in his discussion on methodological issues regarding the use of individual-based computational models for theoretical biology, claimed that one approach to understanding biology is to develop formal computational models which express some phenomena of interest. Whether these models explained the phenomena of nature, was a question that could be answered separately. Taylor and Jefferson [11] claimed that representing organisms by computer programs is the characterising feature of ALife models, because it allows the behaviour of the population to be coded in an explicit way (rather than implicitly as the solution to equations). Bedau, in [9], suggested that ALife will allow us to find simple systems capable of reproducing some properties of life, and this in turn will help us understand such properties.

In this paper, we propose a further extension of the work we presented in [4] for simulating group-foraging behaviour with the use of a Particle Swarm (PS) system, and we develop our approach by fitting our model to the definition of an ALife model. As we will show in the following sections, the group-foraging behaviour resulting from the simulation is emergent because there is no global control over the behaviour of each single particle. This means that there are no explicit rules telling the individuals what they should do in order to (i) find food and (ii) stay together. In his seminal work [7], Langton presented, as an example of an ALife model, Reynold's simulation of flocking behaviour [10]. In that model, the global behaviour of aggregation amongst interacting objects is an

F. Almeida e Costa et al. (Eds.): ECAL 2007, LNAI 4648, pp. 685–695, 2007.

emergent phenomenon (i.e., the rules for each individual object do not depend on any global information). To develop the PS algorithm, Kennedy and Eberhart [6] took inspiration from the same flocking simulation model, which makes it a good candidate for reconsideration in an ALife context. This said, the main field of application for PS systems has traditionally been that of optimisation of non-linear functions: with our research, we intend to also position PS systems in the field which originally inspired them (i.e., ALife).

The remainder of this document is organised as follows. In section 2 we describe the EcoPS model as it was introduced in [4], and we present the newly added characteristics in section 2.1. In section 3 we introduce the experimented parameters. The simulation results are presented and discussed in section 3.1. We conclude in section 4.

2 EcoPS

The *EcoPS* model presented in [4] is a biologically-plausible, albeit abstract, model of an ecological system to simulate group-foraging behaviour. It is based on the PS algorithm: that is, a population of particles looking for resources on a food landscape (a 2D landscape scattered with sources of food in the form of circular patches) and moving according to the standard PS equations for velocity and position update:

$$v_i(t) = \begin{cases} 0 & \text{if food is on patch} \\ \textit{Random} & \text{if food is finished} \\ \kappa((\omega v_i(t-1)) + \Delta t f_i) & \text{otherwise} \end{cases} \tag{1}$$

$$x_i(t) = x_i(t-1) + \Delta t v_i(t) \tag{2}$$

The following biological factors are included in the model (for a detailed description, refer to [4]):

Local Neighbourhood - each particle only exchanges information with those particles which are within a certain distance (*radius* of the neighbourhood).

Forgetful Particles - forgetting, as well as learning, is incorporated into the model since, in reality, a place that was good for feeding some time ago, might no longer be good.

Fitness as Energy - the fitness of a particle is redefined as the long-term total amount of food consumed minus the energy spent in movement, which is proportional to the velocity. Formally:

$$F(t) = \begin{cases} F(t-1) + CQ & \text{if particle is on patch} \\ F(t-1) - D_0 v^2 & \text{otherwise} \end{cases} \tag{3}$$

where: C is the amount of food consumed, Q is the energy gained by eating one unit of food, and $D_0 v^2$ is the rate at which energy is used by the animal moving at speed v^2 (with D_0 moving cost coefficient).

Environmental Attraction - a third term representing the strength of attraction to the nearby patch is added to the acceleration equation. This is related to the *aura* of the food, that is the smell that the food emits. Formally:

$$f_i = \underbrace{\phi_1 R_1(x_{s_i} - x_i)}_{\text{social interaction}} + \underbrace{\phi_2 R_2(x_{p_i} - x_i)}_{\text{individual learning}} \underbrace{+\phi_3 R_3 \times aura \frac{(x_{A_i} - x_i)}{|x_{A_i} - x_i|}}_{\text{environmental attraction}} \quad (4)$$

where: ϕ_1, ϕ_2 and ϕ_3 are the *social interaction*, the *individual learning* and *environmental attraction* rate respectively; R_1, R_2 and R_3 are random variables; x_i is the current position of the particle; x_{s_i} is the best position found by the swarm; x_{p_i} is the best position found by the particle; x_{A_i} is the centre of the patch of food closest to the particle; *aura* is the intensity of the aura at the position of the particle.

2.1 New Characteristics

For this new version of the EcoPS model, we have focused our attention on the population of particles and on the food sources.

With respect to the latter, we decided to increase the complexity of the model by allowing the patches to deplete and regrow independently from whether the food was being eaten or not (i.e., *ephemeral* food patches). The motivation for this choice has been that there are many hypotheses relating to various species of animals (birds as well as fish), which state that patchy and ephemeral food can be beneficial for grouping behaviour [1,2].

The main innovations are, however, directed towards the population. We have introduced into the model the concept of *genotype* of a particle. Particles sharing the same genetic information belong to the same *species*[1]. Each of the l species $\mathcal{S}_i$ in the population $\mathcal{P}$ can be composed by m_i $(0 \leq i \leq l-1)$ *swarms* $\mathcal{W}_j \in \mathcal{S}_i$, which can be distinguished from each other only by means of phenotypic features. Finally, each swarm is formed by n_j $(0 \leq j \leq m_i)$ *particles* $\mathcal{I}_k \in \mathcal{W}_j$, which therefore share both their genotypic and phenotypic attributes.

Just as a reminder, so far the only characteristics that a particle had were: position, velocity, fitness, maximum velocity, size of the neighbourhood (i.e., the radius), and memory threshold (i.e., the size of the memory window). We have now added the following characteristics:

Species (genotypic attributes) all the attributes relating to the species are defined not as a single value, but as a pair `(mean, standard deviation)` in order to allow some kind of variation inside the population.
- Maximum velocity, size of the neighbourhood, and memory threshold now become attributes of the species.
- Maximum and minimum fitness: the values of the maximum amount of energy a particle can reach, and the minimum amount of energy needed for the particle's survival, respectively. When the minimum fitness is

[1] Although, in the experiments presented in sec. 3, we have only used a single species.

reached, the particle is considered dead and is removed from the population.
- Reproductive fitness: the minimum fitness needed for the reproductive process, which is asexual and mono-parental, and can be considered a form of cloning of the parent with mutation. The only offspring generated is positioned close to the parent.
- Reproductive interval: the number of time-steps needed to elapse before a particle can reproduce again; it has been introduced to simulate the cycle of reproductive seasons. So, if a particle reaches the reproductive fitness, but it is not in "the right reproductive season", it will not reproduce.
- Reproductive probability: the probability that a particle will reproduce, given that the reproductive fitness has been reached, and the particle is in the right reproductive interval.
- Reproductive fitness loss: the amount of energy that a particle loses when reproducing.
- Food intake value: the maximum amount of food that a particle is able to eat in a single time-step.

Swarms (phenotypic attributes)
- Personality: we define a *standard* particle as one which trusts its own and others' experience equally, a *bold* particle as one which trusts its own experience more than others', and a *shy* particle as one which trusts the experience of others more than its own. From an algorithmic point of view, the individual learning rate and the social interaction rate of a standard particle have the same influence, the individual learning rate of a bold particle is higher than its social interaction rate, and vice-versa for a shy particle.

Particles retain the old attributes of position, velocity and fitness, while the other attributes (maximum velocity, size of the neighbourhood, and memory threshold) have now became attributes of the species the particles belong to.

3 Experiments

In previous versions of the EcoPS model (see [3,4]), we have observed that group-foraging behaviours emerge spontaneously, even though the models were very simple (in the case of [3], the model was biologically unrealistic). Here, we want to test how the newly added characteristics affect the simulation. In order to do this, we have run a series of experiments with different combinations of parameters. Given the large amount of variables, we have grouped the (hand-tuned) parameters into classes of similar attributes (in table 1 we only present the most significant ones).

When studying group-foraging, we want the agents to (i) find food, (ii) survive, and (iii) stay together[2]. Moreover, we want these three things to happen at

[2] In group-foraging, animals have to "actively" seek to be close to each other, instead of simply meeting at a certain point because of the attraction to environmental conditions at that point.

Table 1. Combinations of parameters relative to population, species, and food patches

Population			
1 species	$1 \rightsquigarrow 4$ swarms	10 particles	
Personality			
Based on the composition of the swarm (% of (BOLD, STANDARD, SHY)):			
Tiger (1.0, 0.0, 0.0)	*Sheep* (0.0, 0.0, 1.0)		
Ant (0.33, 0.34, 0.33)	*Fish* (0.0, 1.0, 0.0)		
	BOLD	STANDARD	SHY
SOCIAL_LF	0.2	0.5	0.8
INDIVIDUAL_LF	0.8	0.5	0.2
Eating			
Based on the amount of food eaten in each time-step			
FOOD_INTAKE_VALUE	(0.5, 0.0), (1.5, 0.0), (3.0, 0.0)		
Reproduction			
Based on the length of the reproductive interval			
REPRODUCTIVE_INTERVAL	2, 10, 20		
Food patches			
Life conditions v (p: probability that a particle positioned at random in the world is on a patch; n_i: number of particles belonging to the swarm i):			
Hard $v = pn_i < 0.1$	*Comfortable* $v = pn_i \simeq 0.5$	*Easy* $v = pn_i \simeq 1$	
NUMBER_OF_PATCHES	1, 3, 10		
PATCH_POSITION	centre, X axis, XY axes		
PATCH_SIZE	$11.28 \rightsquigarrow 35.68$, $6.52 \rightsquigarrow 20.60$, $3.56 \rightsquigarrow 11.28$		

the same time. For this reason, we have defined three scalar *indices* which will allow an easy and quick comparison of the runs, and we are going to study the correlation of these measures, to see if and how these indices are related.

Amount of Food Eaten indeX (`AFEX`) - summarises the ability of each particle to find food (it measures the energy that each particle gains at each time-step of the simulation). `AFEX` $\alpha_k(t)$ for the particle $\mathcal{I}_k$ at time t is given by

$$\alpha_k(t) = \frac{\tau_k(t)}{1+t}$$

where $\tau_k(t)$ is the *total* amount of food eaten until time t.
`AFEX` for swarms, species and the whole population (respectively) are

$$\alpha_j^{\mathcal{W}} = \frac{\sum_k E(\alpha_k) \times \delta_k}{\sum_k \delta_k} \qquad \alpha_i^{\mathcal{S}} = \begin{cases} \frac{\sum_j \alpha_j^{\mathcal{W}}}{m_i} \\ max_j \alpha_j^{\mathcal{W}} \end{cases} \qquad \alpha_i^{\mathcal{P}} = \begin{cases} \frac{\sum_j \alpha_j^{\mathcal{S}}}{l} \\ max_j \alpha_j^{\mathcal{S}} \end{cases}$$

where δ_k is the *duration of life* of the particle, and $E(\alpha_k)$ is the average of each particle's `AFEX`.

Number of Particle Index (`NoPIx`) - indicates the survival ability of the swarm. `NoPIx` π_j is given by

$$\pi_j(t) = \frac{E(N_j(t))}{\iota_j}$$

where $E(N_j(t))$ is the temporal average of the population of one swarm. `NoPIx` for swarms, species and the whole population (respectively) are given by

$$\pi_j^{\mathcal{W}} = \frac{\sum_{s=T-L}^{T} \pi_j(s)}{1+L} \qquad \pi_i^{\mathcal{S}} = \begin{cases} \frac{\sum_j \pi_j^{\mathcal{W}}}{m_i} \\ max_j \pi_j^{\mathcal{W}} \end{cases} \qquad \pi_i^{\mathcal{P}} = \begin{cases} \frac{\sum_j \pi_j^{\mathcal{S}}}{l} \\ max_j \pi_j^{\mathcal{S}} \end{cases}$$

where L is a final portion of the simulation time (we choose $L = 20$).

Density Index (`DenIx`) - measures the degree of clustering of the swarm (it relates to the number of particles per area unit). `DenIx` $\rho_i(t)$ is given by

$$\rho_j(t) = E(N_j(t)) \frac{p}{p + re^{-p/r}}$$

where $f(r) = \frac{p}{p+re^{-p/r}}$ is a *sigmoid-like* function of the distance r. `DenIx` for swarms, species and the whole population (respectively) are

$$\rho_j^{\mathcal{W}} = \frac{\sum_{s=T-L}^{T} \rho_j(s)}{1+L} \qquad \rho_i^{\mathcal{S}} = \begin{cases} \frac{\sum_j \rho_j^{\mathcal{W}}}{m_i} \\ max_j \rho_j^{\mathcal{W}} \end{cases} \qquad \rho_i^{\mathcal{P}} = \begin{cases} \frac{\sum_j \rho_j^{\mathcal{S}}}{l} \\ max_j \rho_j^{\mathcal{S}} \end{cases}$$

3.1 Results

Due to space limitations, we only present detailed results for the case of 1 swarm and 3 food patches[3]. For each combination of parameters, we have performed 30 runs, each for 300 iterations, with different random number generator seeds.

The results of the experiments are presented in terms of: (i) how these measures correlate (i.e., how independent are the "success" criteria)? Each parameter combination is shown as a point on a scatter plot (figure 2), representing a "converged correlation" averaged over the last 20 iterations: for each parameter combination, we either show points for all 30 runs (on the right) or one average over the 30 (on the left); (ii) for parameter combinations where one measure is best, how do the other measures vary in the same runs (i.e., does optimality with respect to one measure have any implications for the other)? This effectively gives us a complementary view to the scatter plot representation in (i); where each measure is best, we show (separately) how the other two measures varied over time in the same runs (figure 3).

[3] This is what `3p1w` refers to in the plots.

The correlation plot between AFEX and NoPIx (figure 2.a) shows that the maximum survival ability is present when the ability of finding food has an intermediate value and not, as intuition would suggest, when the foraging ability is also at its maximum. A possible explanation for this behaviour is as follows: when AFEX is maximum, it means that most of the particles in the swarm are very good at finding food. Therefore, they will all reach the food in a short time (the others, which do not reach the food, die), and the food will finish quickly. When the food finishes, the particles will start searching again, but since the patches deplete, it will take less and less time to finish a patch. After a small number of iterations, the eating-cycle of the particles will be quicker than the life-cycle of the food, so by the time the food will start regrowing, the particles will have reached their minimum fitness level (i.e., they will be dead). In more biological terms, let us recall that the *carrying capacity* of an environment is "the number of individuals the environment can support without significant negative impacts to the given organism and its environment" [12]. The carrying capacity of an environment can change over time due to various factors, including food availability. In particular, if the amount of food available in the world is very limited, the carrying capacity will become small. In the long term, this makes the population size decrease irreparably. We have further confirmation for this behaviour from the temporal plots for AFEX and NoPIx (in the runs where NoPIx and AFEX, respectively, are at their highest values). In fact, when NoPIx is at its best (figure 3.a1), AFEX is asymptotic to a small value (around 0.4), while when AFEX is at its best (figure 3.b1), NoPIx decreases dramatically and then regrows slowly (while staying fairly small, relative to other indices). This means that, when the survival ability is at its best, the ability to find food does not improve, and vice-versa when the foraging ability is at its best, the survival ability is small. This is a very interesting (and unexpected) behaviour: the particles should not try to "maximise" the amount of food that they eat. Instead, they should find an (optimal) intermediate solution between surviving and not spoiling the resources.

The density index DenIx is conceptually different from the other two, in that this index does not reach a *steady state*, but instead it has a *steady cycle* depending on the food's life-cycle (figure 1). Therefore, we have computed the *moving averages* and then used these to obtain scatter and temporal plots relative to the index. From the correlation plot between AFEX and DenIx (figures 2.b) we can infer that when the particles have a low foraging ability, then they tend not to group; when their foraging ability is high, they tend to group fairly strongly, and we can imagine that this happens because the particles "meet" on the food patches; finally, when their foraging ability has an intermediate value (roughly the same as for the best NoPIx- see previous discussion), their ability to group is high. The latter behaviour is interesting, because it could mean that the particles tend to stay together even if they are not eating. The same considerations are confirmed by the temporal plots (figures 3.a2, and 3.c1), in which at the best value for DenIx, there is a steady and small value (less than 0.4) for AFEX, and, at the best value for AFEX, there is a sharp and large increase in DenIx.

Finally, with respect to the correlation between NoPIx and DenIx (figures 2.c), we can observe that more grouping occurs when the survival ability has low to

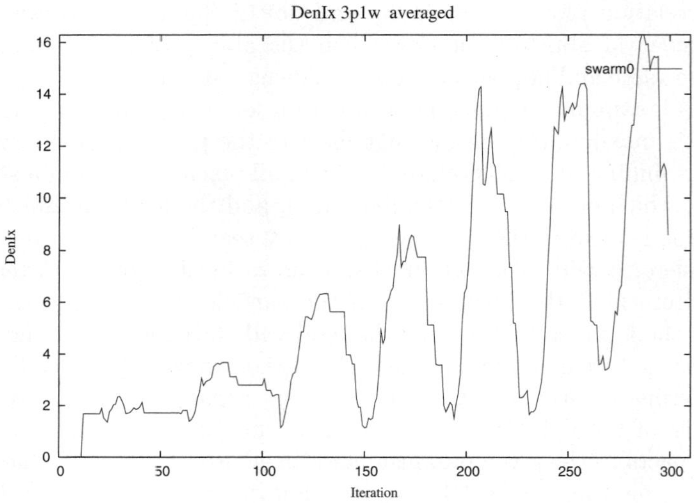

Fig. 1. Steady cycle for `DenIx`.

medium value; when the survival level is high, the ability to group is limited. From the temporal plots we can see how, when the particles have their best grouping ability (figures 3.b2), their survival ability is quite small but increasing over time; however, when the survival ability is at its best (figure 3.c2), the grouping ability grows very slowly and has a fairly small value, thus confirming the findings in the scatter plot. A possible motivation for this behaviour is along the same lines as the explanation we gave for the correlation between `AFEX` and `NoPIx`: a particle should try to find an intermediate solution amongst alternatives behaviours, instead of optimising only one aspect.

To summarise: (i) if a swarm is good at foraging, it is not good at surviving, but is good at grouping; (ii) if a swarm is good at surviving, it is fairly good at foraging, but not too good at grouping; (iii) if a swarm is good at grouping, it is fairly good both at foraging and surviving.

4 Conclusion

We believe that (at least *weak*) ALife is a useful tool for the investigation and understanding of natural phenomena which are, for various reasons, difficult to study in the real world. In John Holland's words, "computer-based models allow complex explorations not possible with real systems.[...] Such models can provide existence proofs, which show that given mechanisms are sufficient to generate a given phenomenon" [5].

With our work, we intend to shed some light on the problem of group-foraging in animals through the use of computer simulations. In particular, we have used a Particle Swarm based system to model abstract animals foraging over a food landscape. We believe that Particle Swarm systems are a natural choice for an

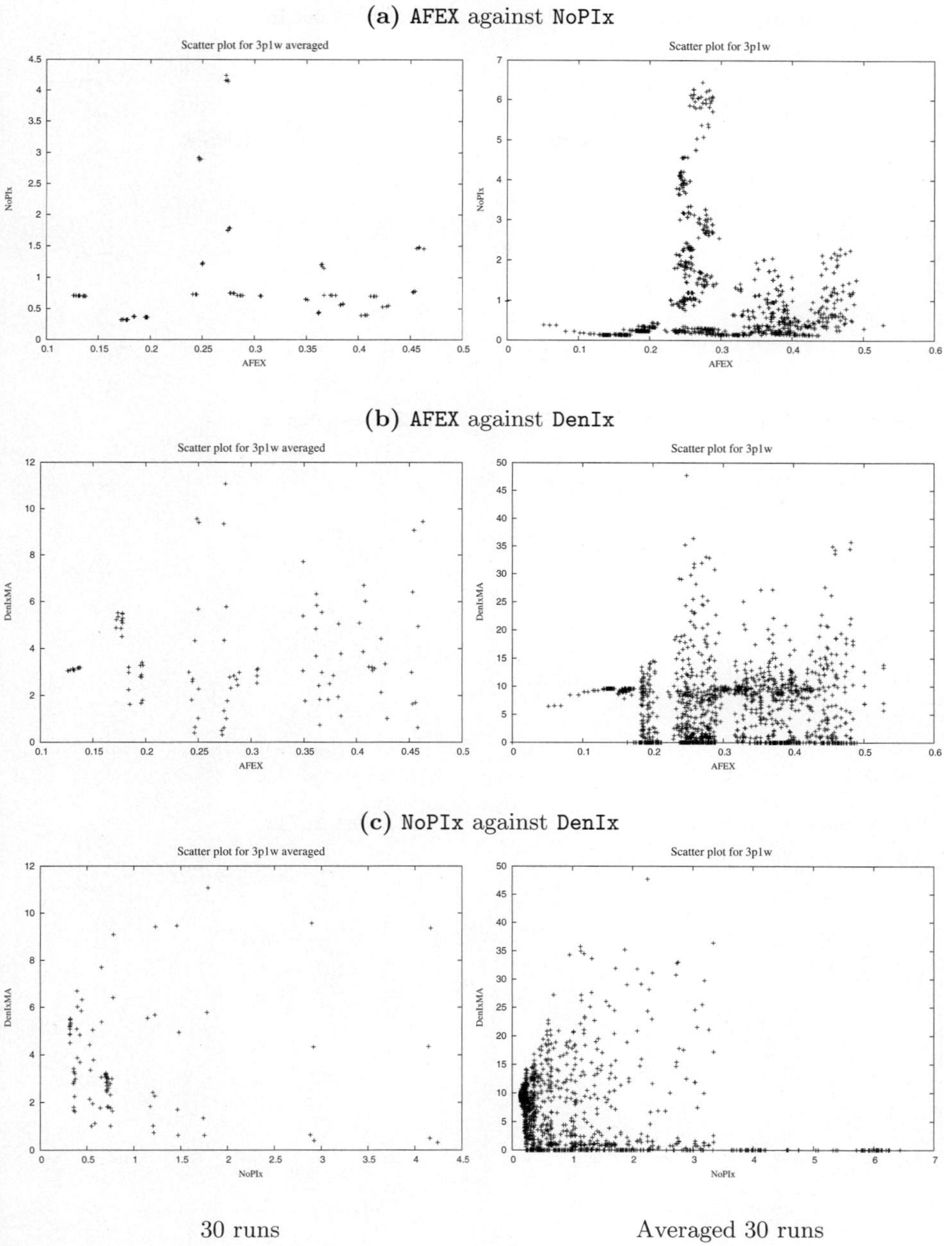

Fig. 2. Scatter plots over 30 different runs (averaged and not): (a) AFEX against NoPIx, (b)AFEX against DenIx, and (c) NoPIx against DenIx.

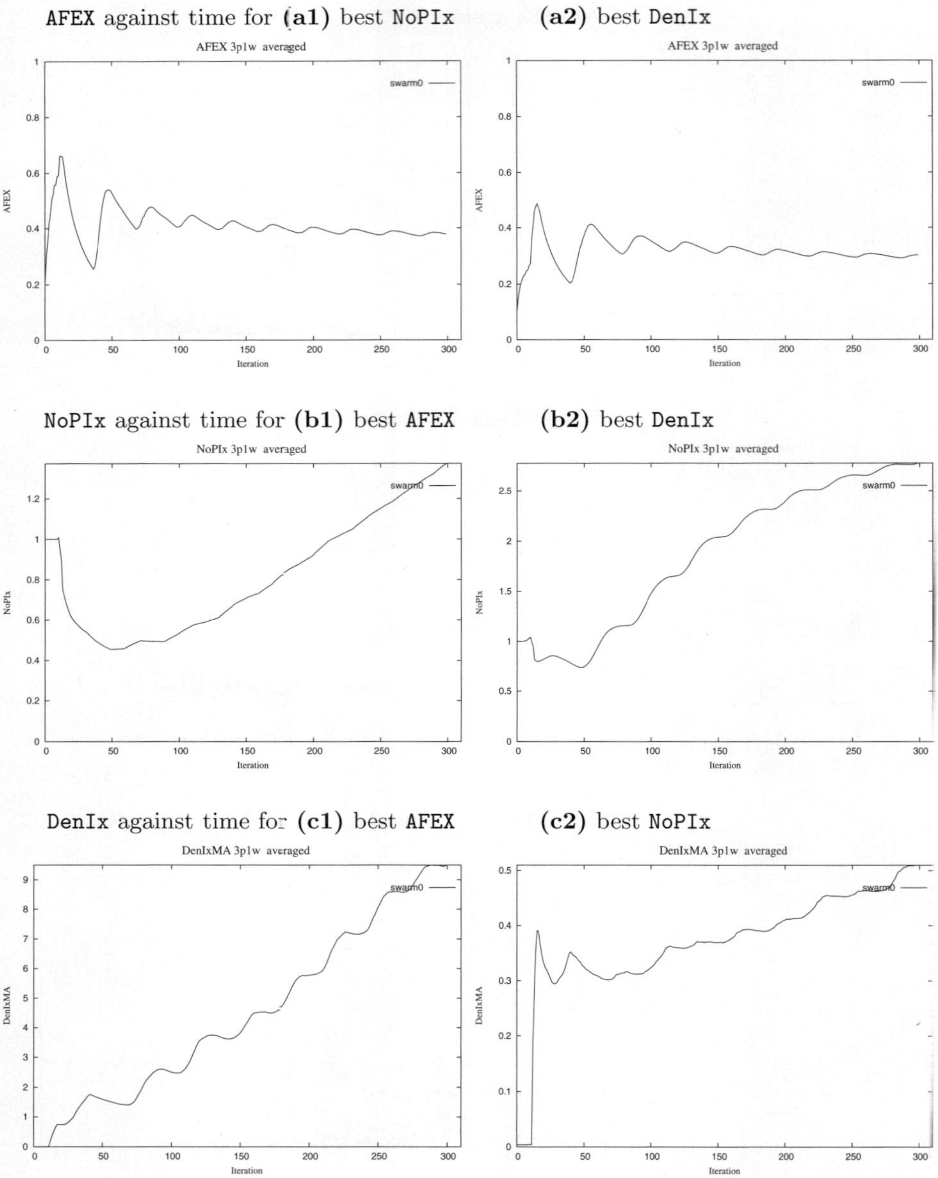

Fig. 3. Temporal plots, averaged over 30 different runs: (a1) `AFEX` plotted against time for best `NoPIx`, (a2) `AFEX` plotted against time for best `DenIx`, (b1) `NoPIx` plotted against time for best `AFEX`, (b2) `NoPIx` plotted against time for best `DenIx`, (c1) `DenIx` plotted against time for best `AFEX`, (c2) `DenIx` plotted against time for best `NoPIx`.

ecological simulation, since they were originally inspired by a model for bird flocking. Although very simple, the EcoPS model presented here is complex enough to show interesting behaviours. In particular, it seems that grouping is the key to finding the right balance between (1) exploiting the resources, and (2) surviving.

Future directions of this research include the formulation of a formal theory to explain and justify the phenomena emerging in the simulation, and then to relate it to ecological theories and empirical investigation.

Acknowledgments. This research is part of the EPSRC XPS project (GR/T11234/01).

References

1. Baird, T.A., Ryer, C.H., Olla, B.L.: Social enhancement of foraging on an ephemeral food source in juvenile walleye pollock, Theragra chalcogramma. Environmental Biology of Fishes (1991)
2. Dall, S.R.X.: Can Information Sharing Explain Recruitment to Food from Communal Roosts? Behavioral Ecology (2002)
3. Di Chio, C., Poli, R., Di Chio, P.: Modelling Group-Foraging Behaviour with Particle Swarms. In: PPSN2006 - Ninth International Conference on Parallel Problem Solving from Nature (2006)
4. Di Chio, C., Poli, R., Di Chio, P.: EcoPS - A Particle Swarm Algorithm to Model Group-Foraging. In: GECCO 2007 - Proceedings of the Genetic and Evolutionary Computation Conference (2007)
5. Holland, J.H.: Hidden Order: How Adaptation Builds Complexity. Addison-Wesley, Reading (1995)
6. Kennedy, J., Eberhart, R.C.: Swarm Intelligence. Morgan Kaufmann Publishers, San Francisco (2001)
7. Langton, C.G.: Artificial Life. In: Artificial Life - Proceedings of an Interdisciplinary Workshop on the Synthesis and Simulation of Living Systems (1988)
8. Maley, C.C.: Methodologies in the Use of Computational Models for Theoretical Biology. Computational Models in Theoretical Biology (1999)
9. Noble, J., Bullock, S., Di Paolo, E.A.: Artificial life: Discipline or method? Report on a debate held at ECAL99, Artificial Life 6 (2000)
10. Reynolds, C.W.: Flock, Herds, and Schools: A Distributed Behavioral Model. Computer Graphics (1987)
11. Taylor, C., Jefferson, D.: Artificial Life as a Tool for Biological Inquiry. Artificial Life 1 (1994)
12. Wikipedia, Carrying Capacity. `http://en.wikipedia.org/wiki/Carrying_capacity`, visited on 22/05/2007

Efficient Multi-foraging in Swarm Robotics

Alexandre Campo and Marco Dorigo

IRIDIA, CoDE, Université Libre de Bruxelles, Brussels, Belgium
{acampo, mdorigo}@ulb.ac.be

Abstract. In the multi-foraging task studied in this paper, a group of robots has to efficiently retrieve two different types of prey to a nest. Robots have to decide when they leave the nest to forage and which prey to retrieve.
The goal of this study is to identify an efficient multi-foraging behaviour, where efficiency is defined as a function of the *energy* that is spent by the robots during exploration and gained when a prey is retrieved to the nest. We design and validate a mathematical model that is used to predict the optimal behaviour. We introduce a decision algorithm and use simulations to study its performance in a wide range of experimental situations with respect to the predictions of the mathematical model.

Key words: swarm robotics, multi-foraging, mathematical modelling.

1 Introduction

Foraging is a classical metaphor used in swarm robotics [1]. In foraging, a group of robots has to pick up objects that are scattered in the environment. The foraging task can be decomposed in an exploration sub-task followed by a transport sub-task. Foraging can be applied to a wide range of useful tasks. Examples of applications are toxic waste clean-up, search and rescue, demining and collection of terrain samples. Central place foraging is a particular type of foraging task in which robots must gather objects in a central place. Borrowing the terminology from biology, the central place is also called the *nest* and the objects *prey*.

Multi-foraging is a variation of the foraging task in which different types of objects to collect are considered [2]. These different types of objects can be concurrently and independently collected by the individuals and can have different properties. Multi-foraging adds a level of complexity with respect to the traditional foraging task as it may be necessary for the individuals to choose which prey to take, and when.

The study of the efficiency of foragers has first been the concern of biologists. In his seminal article [3], Charnov exposes the fundamental hypothesis that gives birth to the field of optimal foraging. The hypothesis is that evolution has shaped individual behaviours of foraging animals so as to maximize the net energy intake. Three decades later, roboticists try to identify how robots should cooperate in order to forage efficiently. Efficiency has been defined in several ways: in biology, researchers use the term *energy* and measure weights of

F. Almeida e Costa et al. (Eds.): ECAL 2007, LNAI 4648, pp. 696–705, 2007.

animals before and after tasks to quantify energy spent and gained. In robotics, the vocabulary is less well established. Terms such as *reward, income, benefit* have been used [4,5,6,7]. For the sake of simplicity, we will use the term *energy*, as in biology. Foraging efficiently is thus a quest to maximize the energy of a group of foraging robots. Robotics researchers often consider that energy is spent when robots move during exploration and is gained when a prey is successfully retrieved to the nest [7].

We focus on a specific case of multi-foraging in which there are only two types of prey that have to be retrieved to the nest. The spatial aspect of the task is negligible as the prey have random locations in the environment. The exploration mechanism used by the robots to find prey is a random walk. Hence, robots discover prey in the environment at a given rate. Robots have to decide whether they rest at the nest and in this way spare energy, or forage in the environment. If they choose to forage, they have to decide which prey to take so as to maximize the energy of the group.

Our objective is to identify an individual behaviour that leads the group of foraging robots to have an efficient collective behaviour.

To achieve this objective, we first design and validate a mathematical model of multi-foraging. Mathematical modeling of robotic experiments is a methodology [8] [9] [10]. Mathematical models are opposed to individual based models (IBMs) [11]. In IBMs, each robot is represented as well as the environment. Differently, mathematical models are analytic description of the evolution of a system, in which the individuals of a system are not represented separately. Mathematical models are faster than IBMs because their computation time does not depend on the number of individuals. They can be used as optimization tools: Ijspeert *et al.* [12] have used a stick pulling experiment as an example to demonstrate how the behaviour of the robots could be made efficient. Within the limits of the mathematical tools available, it is also possible to draw conclusions on the dynamics and intrinsic properties of the system.

The mathematical model we devise predicts with a good confidence the optimal behaviour of the robots, and can therefore estimate the maximum amount of energy that a group of robots can accumulate during an experiment. We use the model as a yardstick to evaluate the performance of the group of robots and test different behavioural rules. Based on simplified equations, we introduce a decision algorithm to control the behaviour of the robots. To evaluate the performance of the algorithm, we run simulations using a large set of 2160 different experimental configurations, obtained by varying the parameters of the experiment.

In Section 2 we detail the task, the experimental setup and the controller of the robots. Section 3 is devoted to the description and validation of the mathematical model. Section 4 presents the decision algorithm and the evaluation of its performance using the predictions of the mathematical model. Section 5 concludes the paper with a discussion of the results and some ideas for future work.

2 Methods

The Task of Multi-foraging. The task studied in this paper is the search and retrieval of objects to the nest by a group of robots. Objects can be retrieved by a single robot. Searching and retrieving objects is energy consuming, but retrieved objects yield energy to the group. There are two types of objects that differ in energy required to be carried to the nest and energy yield. Robots have to decide when to search for objects and which objects are to be retrieved so as to maximize the energy accumulated by the group. Robots have no *a priori* knowledge of the properties of objects and should adapt online to the environment.

Experimental Setup. All the experiments in this paper are simulated. The environment is a circular arena of 1.20 meters of radius. Robots are initially randomly scattered on it. A circular nest is located in the center of the arena. Robots can locate the nest thanks to a lamp suspended above it. The nest has a structure of three concentric rings with different grey levels. The innermost ring defines where robots can rest. The second ring defines where a robot can safely release a prey, with good confidence that it is inside the nest. Finally, the outermost ring defines the boundary of the nest.

Prey are introduced in the environment at random locations around the nest, at a fixed distance from it. New prey appear and disappear with constant rates per time unit. They are removed when they fall inside the nest. Prey have a weight and friction that define the time required for being retrieved. An amount of energy is associated to a prey and is attributed to the group of robots once it is delivered in the nest. Prey of a same type share all their characteristics. We use only two different types of prey in the experiments.

The simulated robots have the same characteristics as *s-bots* from the swarm-bots project [13]. We rely on ground sensors to perceive the structure of the nest. Infrared sensors are used for collision avoidance. The camera is employed to determine the location of the nest and discriminate the type of prey thanks to their colour. Last, the robots use the camera to perceive if a nearby prey is already being retrieved by another robot.

Controller of the Robots. The controller used is the same for all the robots. The architecture of the program is a finite state machine (FSM). The scheme in Figure 2 represents the possible states, with arcs denoting the possible transitions between states. Robots are initialized in the *Explore* state.

- **Explore.** The robot performs a random walk in the environment. An obstacle avoidance subroutine is triggered when needed. The robot switch to the *Rest* state with a probability constant over time. If a prey is close enough and no green colour is perceived the robot enters the *Grasp* state.
- **Grasp.** The robot has detected a prey and may decide to perform a physical connection. This decision depends on the probabilities π_1 and π_2. If grasping is successful the robot enters the *Retrieve* state, otherwise it enters the *Ignore* state.

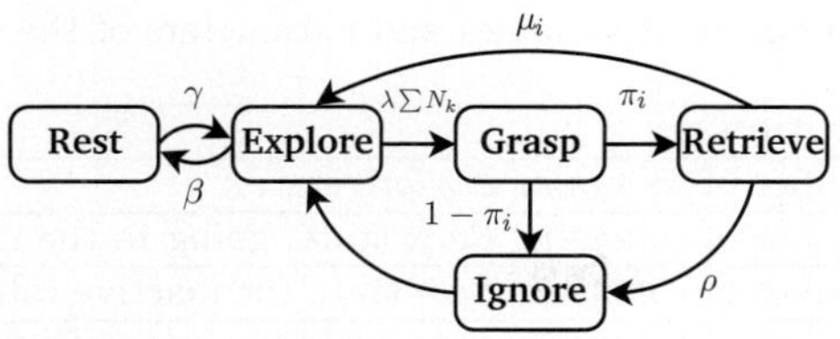

Fig. 1. Finite state machine representing the robot's controller. Transitions between states are triggered according to the probabilities described in Table 1. The control parameters (probabilities β, π_i, $i = 1, 2$) can be modified by a decision mechanism as described in Section 4.

- **Retrieve.** The robot becomes green. This colour is used to keep away other robots. The robot heads toward the nest. When the robot reaches the nest, it releases the prey and enters the *Explore* state. During retrieval, the robot has a constant probability to give up and enter the *Ignore* state.
- **Ignore.** The robot performs a random walk with collision avoidance, neglecting any encountered prey. After a delay of five seconds (enough to move away from a prey) the robot enters the *Explore* state.
- **Rest.** The robot heads back to the nest to rest there. With a constant rate per time unit, the robot can decide to leave the nest by entering in the *Explore* state.

3 Mathematical Model

Description. Partial differential equations are devised to model the flows of robots among five main states. We neglect the modeling of the *Grasp* and *Ignore* states because they occur rarely and their duration is relatively short. In order to describe how energy is gained, we model the retrieval process in two distinct parts, one for each type of prey. In addition, we noticed that the time required to go back to the nest before resting is not negligible and has to be modelled. To this extent, we introduce the *Back* state. We end up with five main states among which flow of robots are exchanged. The meaning of all the variables and parameters is explained in Table 1.

A set of differential equations is used to model the flows of robots exchanged among the states. In the following, we provide a detailed explanation of the first equation. As explained in Section 2, several transitions lead robots to enter or leave the *Explore* state. Each right-term of the differential equations is an amount of robots per second doing a specific transition.

$$\begin{aligned}
\tfrac{\mathrm{d}E}{\mathrm{d}t} &= -\beta E + \gamma I + \textstyle\sum_{i=1}^{2} \left(-\pi_i E N_i \lambda + \mu_i R_i + \rho R_i\right) \\
\tfrac{\mathrm{d}B}{\mathrm{d}t} &= +\beta E - \kappa B \\
\tfrac{\mathrm{d}I}{\mathrm{d}t} &= +\kappa B - \gamma I \\
\tfrac{\mathrm{d}R_i}{\mathrm{d}t} &= \pi_i E N_i \lambda - \mu_i R_i - \rho R_i \qquad \forall i \in [1, 2] \\
\tfrac{\mathrm{d}N_i}{\mathrm{d}t} &= \varphi_i - \pi_i E N_i \lambda - \xi_i N_i + \rho R_i \qquad \forall i \in [1, 2]
\end{aligned}$$

Table 1. Summary of variables and parameters of the experiment.

Variable	Description
E	the number of robots in *Explore* state
B	the number of robots in *Back* state, going to the nest to rest
I	the number of robots in *Rest* state (or inactive robots)
R_i	the number of robots in *Retrieve* state (prey of type i)
N_i	the number of prey of type i in the environment
Parameter	**Description**
T	the total number of robots in the experiment
λ	rate of objects per second found in the environment by a single robot
κ	probability for a single robot to find the nest
En_i	energy associated to a prey of type i
En_p	energy lost during one second spent outside the nest for one robot
φ_i	incoming rate per second of prey of type i
ξ_i	probability constant over time for a prey of type i to disappear
μ_i	inverse of the average time required to retrieve a prey of type i
ρ	probability to give up an ongoing retrieval
β	probability for a robot to return to nest
γ	probability for a robot to leave the nest and look for prey
π_i	probability to take a prey of type i upon encounter

- First, robots can decide to rest at nest with a probability β. On average βE robots leave the *Explore* state and enter the *Back* state.
- Conversely, robots in *Rest* state have a probability γ to come back in *Explore* state. Thus there are on average γI robots entering the *Explore* state.
- Robots may find a prey and decide to retrieve it. The probability to find a single object being λ, the average number of exploring robots that find a prey of type i is $EN_i\lambda$. As robots decide to retrieve the prey with probability π_i, the average number of robots that leave the *Explore* state to retrieve a prey of type i is $\pi_i EN_i\lambda$.
- We consider that a robot has a probability of μ_i to achieve the retrieval of a prey of type i. Hence, there are on average $\mu_i R_i$ robots that achieve a retrieval and come back in *Explore* state.
- Last, during the retrieval of a prey of type i, robots have a probability ρ to give up and come back in *Explore* state. On average their are ρR_i robots that give up retrieval of prey of type i.

Validation. To evaluate the quality of the model and determine to which extent we can rely on it to have a good prediction, it is mandatory to carry out a validation process. This phase involves the comparison of the results obtained in simulation against those of the model for a collection of typical experimental situations. We define a range of reasonable values for each parameter of the experiment (see Table 2), except for the control parameters of the robots π_1, π_2 and β. A *configuration* of the experimental setup is defined by selecting one value for each parameter from its range. There are 2160 possible configurations that define a set P. We denote $C_i \in P, i \in [1, 2160]$ one particular configuration.

Math. \ Simu.	$A < B$	$A > B$
$A < B$	43.22%	7.64 %
$A > B$	6.97 %	42.13 %

(a)

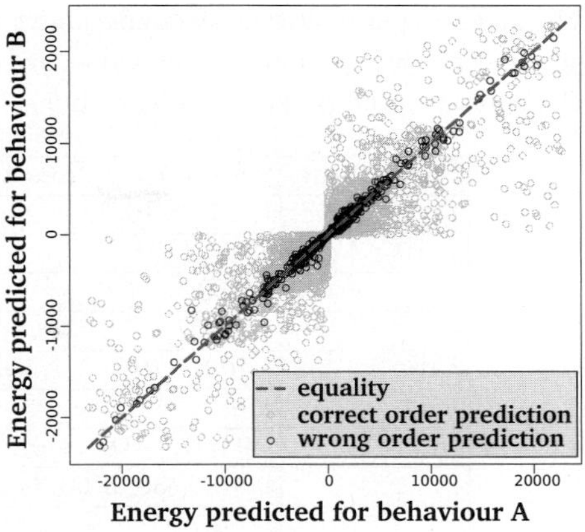

(b)

Fig. 2. (a) Comparison table of predicted orders by the mathematical model with respect to simulation results. The notation $A <,> B$ signifies that the energy accumulated using behaviour A is lower, respectively higher than using B. **(b)** Energy predicted for behaviours B in function of energy predicted for behaviours A. The gray circles show when both the mathematical model and the simulation agree on which behaviour performs better. The black circles show cases of disagreement. They lie along the dashed line ($r^2 = 0.98$, slope $a = 1.00$), indicating that disagreements arise mainly when the predicted energy of two behaviours are very much alike.

In the following, we present a test that is meant to assess the ability of the model to compare the outcome of two different behaviours. The test consists of selecting randomly two behaviours A and B from the control space $(\pi_1, \pi_2, \beta) \in [0, 1]^3$. We compare the accumulated energy predicted for A and B. The same comparison is carried out using one single run of simulation for each behaviour.

We use each configuration $C_i \in P$ to parameterize an experiment of one hour. For each configuration C_i we generate 5 pairs of random behaviours $(A_{ij}, B_{ij}), j \in [1, 5]$. The table shown in Figure 2(a) summarizes the frequencies of all possible comparison results for the 10800 tests performed. The table indicates that in 85.35% of the tests, the mathematical model and the simulations agreed on the ranking of the behaviours. The table is almost symmetric and shows no better performance of the model if A superseeds B or the opposite.

Moreover, we have studied the conditions in which disagreement between the mathematical model and the simulations occurs. We plotted the predicted energy for behaviour B in function of the energy predicted for behaviour A. Figure 2(b) shows as black circles the pairs of behaviours that lead to disagreement. The regression performed on the black circles returns a correlation coefficient $r^2 = 0.98$ and a regression slope $a = 1.00$. The wrong predictions of the model occur mainly if the two behaviours are supposed to yield very similar energy. Given that we use only one run of simulation without averaging, an error caused by the noise in simulation is more likely to appear for these pairs of behaviours.

Table 2. Each parameter of the experiment is given a range of reasonable values. By associating to each parameter one value, we define an experimental configuration. In total, there are 2160 possible configurations.

Parameter	Range of values tested	Unit
T	$1, 2, 3, 5, 10, 15$	robot
$N_1(0)$, $N_2(0)$	5	prey of type 1 or 2
λ	1/159.4	probability
κ	1/19.51	probability
En_1	$-100, -10, -1, 1, 10, 100$	energy
En_2	1	energy
En_p	$-0.001, -0.01, -0.1$	energy
φ_1	$1/15, 1/30, 1/60, 1/120, 1/180$	prey / second
φ_2	1/60	prey / second
ξ_1, ξ_2	0.002	probability
μ_1	$1/90, 1/40, 1/30, 1/60$	second^{-1}
μ_2	1/60	second^{-1}
ρ	0.0111	probability
β	**control parameter**	probability
γ	1/400	probability
π_1, π_2	**control parameter**	probability

4 Efficient Multi-foraging

Decision Algorithm. The decision algorithm is a piece of code plugged in the controller of the robots that modifies their individual behaviour through the three control parameters π_1, π_2 and β. The algorithm relies on an equation that permits robots to individually estimate the instantaneous amount of energy EI that can be obtained by the group. In the following we briefly expose the steps that lead to this equation.

Rate of prey grasped by robots: $preyRate = E\lambda \sum_{i=1}^{2} N_i\pi_i$,
Proportion of prey of type i grasped: $prop_i = \frac{E\lambda N_i\pi_i}{E\lambda \sum_{j=1}^{2} N_j\pi_j}$,
Mean time of retrieval of a prey: $retTime = \sum_{i=1}^{2} 1/\mu_i prop_i$,
Mean time to grasp & retrieve a prey: $preyToNest = 1/preyRate + retTime$.

$$\begin{aligned} EI &= E \cdot En_p + \textstyle\sum_{i=1}^{2} En_i \cdot prop_i \cdot 1/preyToNest, \\ &= E \cdot En_p + E\lambda \textstyle\sum_{i=1}^{2} \frac{En_i N_i \pi_i}{1+E\lambda \sum_{j=1}^{2} 1/\mu_j N_j \pi_j}. \end{aligned} \quad (1)$$

Equation 1 can be used by each robot to estimate the rate of energy currently gained by the group. All the parameters of this equation, except for λ, are either control parameters, or can be estimated by the robots during the exploration of the environment.

Indeed, each robot can estimate the density of robots or prey of any type in the environment, which are respectively λE, λN_1 and λN_2. This estimate is a direct measure of the number of objects encountered per second. We use a classical discount factor method to limit the impact of past observations in the estimate of objects density. However, to compute EI robots need to know λ. This parameter may be estimated by measuring the time to go back to the nest, even though the collisions with other robots may diminish the quality of such an estimate. In the following, the robots are given the parameter λ that characterizes the size of the environment.

Based on Equation 1, the decision algorithm estimates parameters of the experiment using the observations of the robot. It then estimates the impact on the rate of energy EI of four triplet of parameters (π_1, π_2 and E): $\{ (1,1,E+1), (0,1,E+1), (1,0,E+1), (\epsilon,\epsilon,\epsilon) \}$, where $\epsilon = 0.05$. The ϵ is a lower bound that guarantees adaptivity by forcing robots to grasp sometimes the prey or explore the environment and update their estimate of the situation. Notice also that E is a global variable that can not be directly changed by a single robot but only influenced if that robot changes its β parameter. Thus, control parameters (π_1, π_2, β) of the robot are updated to converge towards the triplet that maximizes EI.

Performance. We assess the performance of the decision algorithm by carrying out a systematic comparison of the energy accumulated in simulation with the energy obtained by the predicted optimal behaviour. For each configuration $C_i \in P$, we use the mathematical model to find out the predicted optimal behaviour OB_i. We use a single run of simulation to determine the energy gain $E_{pred}(C_i, OB_i)$ associated to OB_i. We also run a single simulated experiment with the decision algorithm used by each individual.[1] The control parameters are initially set to $(1,1,1)$ so that robots start by exploring the environment. The energy accumulated with a configuration C_i and the decision algorithm plugged in the robot's controller is denoted $E_{dec}(C_i)$.

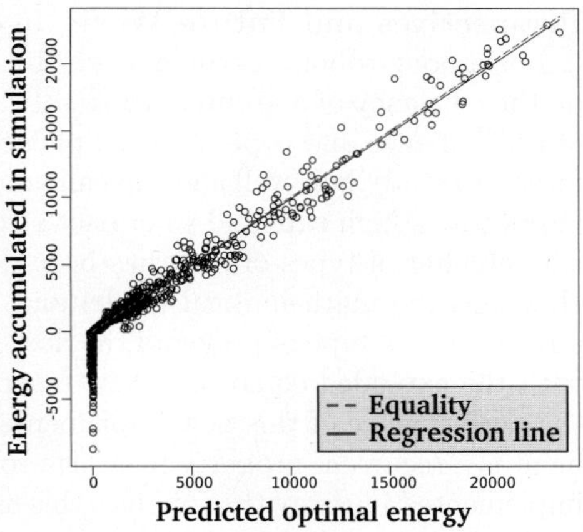

Fig. 3. The decision algorithm performs on average 99% as well as the predicted optimal behaviour (see Section 4).

In Figure 3, the energy $E_{dec}(C_i)$ is compared to the predicted optimal energy $E_{pred}(C_i, OB_i)$. A linear regression is applied to data such that $E_{pred}(C_i, OB_i) > 0$. The correlation coefficient ($r^2 = 0.98$) indicates that the linear relationship hypothesis holds (p-value < 0.001). The slope of the regression line is $a = 0.99$ and the bias is $b = -23.73$, which means that the decision algorithm performs on

[1] According to [14], this is the optimal sampling strategy.

average 99% as well as the predicted optimal behaviour. If $E_{pred}(C_i, OB_i) = 0$, it means that robots should always stay in the nest. However, the algorithm imposes a minimum exploration to get information about the environment which produces some cases with negative energy as seen in the plot.

5 Conclusions

Achievements and Contributions. We have designed a mathematical model of the experiment. Its validation has shown that the model can be used to rank successfully two different behaviours in 85% of the cases tested. In addition, we found that the errors in the remaining 15% arose only in ambiguous cases in which the energy yield by the two compared behaviours are very similar. This new tool, previously unavailable in the multi-foraging literature, makes possible the evaluation of robots performance on a relative scale.

An equation to calculate the average instantaneous reward gained by the group of robots has been devised. It has been used to implement a decision algorithm for the robots. The tests have shown that robots using the decision algorithm manage to accumulate on average 99% of the energy that can possibly be gained.

Perspectives and Future Work. In our work, we neglected on purpose collisions among robots. Lerman *et al.* [15] emphasized the impact of interferences on the efficiency of a group of robots. It is likely that robots may perceive a drop of performance and cope with the phenomenon automatically. In the future, we intend to study how well a group can adapt to a situation in which collisions may happen at a high rate and so impact strongly on the performance of the robots. The number of types of prey has been deliberately limited to two in this study. However, the mathematical model and Equation 1 can be extended to handle any number of type of prey and the decision algorithm can be adapted to comply with the extended equations. New tests will have to be carried out to measure the performance of this new implementation. Adaptivity to a dynamic environment is a recurrent property in swarm robotics. The decision algorithm has been implemented in a way that enables this aspect. Future work will report a detailed analysis of the collective behaviour of the robots facing sudden changes in the environment. It may also be possible to have robots foraging efficiently without knowledge of the λ parameter, although that would probably degrade the performance of the robots. We will work in this direction, so to make the behaviour of the robots totally free of any *a priori* knowledge of the environment. The recruitment of foraging robots could improve the performance of the robots, as reported in [7]. We did not implement this feature in the behaviour of the robots for the present study but it seems to be a promising direction to enhance the group foraging capabilities. Last, to validate our approach and assess the realism of our simulations we plan to carry out a number of experiments with real robots.

Acknowledgements. This research has been supported by COMP[2]SYS, a Marie Curie Early Stage Research Training Site funded by the European Community's

Sixth Framework Programme under contract number MEST-CT-2004-505079. Alexandre Campo and Marco Dorigo acknowledge support from the Belgian FNRS of which they are respectively Research Fellow and Research Director.

The authors would like to thank Jean-Louis Deneubourg for help in designing the mathematical model and for fruitful discussions. The authors are also indebted to Mauro Birattari and Thomas H. Labella for relevant comments and suggestions.

References

1. Dorigo, M., Şahin, E.: Guest editorial. Special issue: Swarm robotics. Autonomous Robots 17(2–3), 111–113 (2004)
2. Balch, T.: Reward and diversity in multirobot foraging. In: Workshop on Agents Learning About and with Other Agents (IJCAI-99), Stockholm, Sweden (1999)
3. Charnov, E.L.: Optimal foraging, the marginal value theorem. Theoretical Population Biology 9(2), 129–136 (1976)
4. Ulam, P., Balch, T.: Using optimal foraging models to evaluate learned robotic foraging behavior. Adaptive Behavior - Animals, Animats, Software Agents, Robots, Adaptive Systems 12(3-4), 213–222 (2004)
5. Labella, T.H., Dorigo, M., Deneubourg, J.-L.: Efficiency and task allocation in prey retrieval. In: Ijspeert, A.J., Murata, M., Wakamiya, N. (eds.) BioADIT 2004. LNCS, vol. 3141, pp. 274–289. Springer, Heidelberg (2004)
6. Lerman, K., Jones, C., Galstyan, A., Matarić, M.J.: Analysis of dynamic task allocation in multi-robot systems. The International Journal of Robotics Research 25(3), 225–241 (2006)
7. Liu, W., Winfield, A., Sa, J., Chen, J., Dou, L.: Strategies for energy optimisation in a swarm of foraging robots. In: Şahin, E., Spears, W.M., Winfiel, A.F.T. (eds.) Swarm Robotics. Second International Workshop, SAB 2006, Rome, Italy, September 30-October 1, 2006. LNCS, vol. 4433, Springer, Heidelberg (2006)
8. Sugawara, K., Sano, M.: Cooperative acceleration of task performance: Foraging behavior of interacting multi-robots system. Physica D: Nonlinear Phenomena 100(3-4), 343–354 (1997)
9. Sugawara, K., Sano, M., Yoshihara, I., Abe, K.: Cooperative behavior of interacting robots. Artificial Life and Robotics 2, 62–67 (1998)
10. Kazadi, S., Abdul-Khaliq, A., Goodman, R.: On the convergence of puck clustering systems. Robotics and Autonomous Systems 38(2), 93–117 (2002)
11. Martinoli, A., Easton, K.: Modeling swarm robotic systems. In: Experimental Robotics VIII. Springer Tracts in Advanced Robotics, vol. 5, pp. 297–306. Springer, Heidelberg
12. Ijspeert, A.J., Martinoli, A., Billard, A., Gambardella, L.M.: Collaboration through the exploitation of local interactions in autonomous collective robotics: The stick pulling experiment. Autonomous Robots 11(2), 149–171 (2001)
13. Dorigo, M., Tuci, E., Groß, R., Trianni, V., Labella, T.H., Nouyan, S., Ampatzis, C., Deneubourg, J.-L., Baldassarre, G., Nolfi, S., Mondada, F., Floreano, D., Gambardella, L.M.: The swarm-bots project. In: Şahin, E., Spears, W.M. (eds.) Swarm Robotics. LNCS, vol. 3342, pp. 31–44. Springer, Heidelberg (2005)
14. Birattari, M., Dorigo, M.: How to assess and report the performance of a stochastic algorithm on a benchmark problem: Mean or best result on a number of runs? Optimization Letters 1(3), 309–311 (2007)
15. Lerman, K., Galstyan, A.: Mathematical model of foraging in a group of robots: Effect of interference. Autonomous Robots 13(2), 127–141 (2002)

Modelling the Effects of Colony Age on the Foraging Behaviour of Harvester Ants

Tom Diethe and Peter Bentley

Department of Computer Science, University College London, UK
{t.diethe, p.bentley}@cs.ucl.ac.uk

Abstract. The colonies of certain species of ants, for example *Pogonomyrmex barbatus*, exhibit changes in behaviour as the colonies grow older, despite nearly all of the individual ants being replaced each year [1]. The behaviour of older colonies is more stable, and they are more likely to avoid intraspecific conflict [2]. Gordon hypothesised that the reason for this is that a 3-4 year old colony is in the steepest part of its growth curve, i.e. the 4000 workers of the 3 year-old colony are feeding 6000 larvae, and that the aggression of individual ants is based on colony level food requirements. This study aims to model this phenomenon using an individual-based simulation. The results from model are compared with field experiments taken over a period of years at the study site in New Mexico [3,4]. The model provides support to the biological hypothesis by showing that both colony age and aggression of individual ants have significant effects on foraging ranges.

1 Introduction

Ant colonies are probably the most famous and most studied case of eusocial insects. There is extreme diversity to the extent that they occupy virtually every evolutionary niche available to them. Ant colonies have long been of interest to computer scientists due to their ability to solve complex tasks with relatively simple local rules. The modelling of ant colonies is also of interest to myrmecologists, who use the models in order to learn more about the biological systems they are studying. This report focuses on the development of a computational model of rival ant colonies in order to test one such biological hypothesis.

This study draws on data on a population of about 250 colonies of the red harvester ant, *P. barbatus*, studied in detail from 1985 to the present day. Colonies live for 15-20 years [1]. *P. barbatus* is a red seed-eating harvester ant that is common throughout the deserts of Arizona. Like many of the other *Pogonomyrmex* harvester ants, they have a powerful sting which is readily used. In the evening, a few *P. barbatus* workers will usually seal the nest entrance with gravel and twigs to keep intruders out of the nest. This often results in one or two of the workers ending up on the outside of the nest entrance after closing it.

Intraspecific competition is a term used by ecologists, which is defined as the struggle between members of a population for scarce resources. In this case different colonies of the same species of ant are fighting for limited food resources.

F. Almeida e Costa et al. (Eds.): ECAL 2007, LNAI 4648, pp. 706–715, 2007.

It differs from interspecific competition which is competition between members of different species for shared resources. The respective levels of intra- and interspecific competition have significant effects on the mortality and densities of the species in question [5]. In general, intraspecific competition is higher than interspecific competition for harvester ants, although there are instances of *P. barbatus* and *P. rugosus* sharing territory [6].

As colonies grow, the production of worker ants accelerates as the number of workers increases, until reaching equilibrium at their mature size. Over long timescales, the foraging range of *P. barbatus* is stable, but on the timescales of days and weeks, the trails are not stable. The short term behaviour depends partly on the behaviour of neighbouring colonies [1,7,8].

Of particular interest are the changes in behaviour exhibited by colonies as they grew older, despite (nearly) all the ants in the colony being replaced each year [1]. The behaviour of older colonies(i.e. over 5 years old) is more stable, and they are more likely to avoid intraspecific conflict [1,2]. Gordon hypothesised that the reason for this may be due to the growth rate of the colony - that a 3-4 year old colony is in the steepest part of its growth curve, i.e. the 4000 workers of the 3 year old colony are feeding 6000 larvae. This hypothesis appears to be reasonable. The individual ants must therefore base their decisions on whether to engage in conflict over food requirements, leading to the global behaviour exhibited.

1.1 Hypotheses

Colony food requirements drive aggression. As a 3-4 year old colony is in the steepest part of its growth curve, the food requirement is high as a smaller population of workers requires energy to support a larger population of larvae. Therefore younger (i.e. developing) colonies should show more aggression. This aggression results in a propensity to engage in conflict with neighbouring colonies. However, the foraging ranges of 3-4 year old colonies are typically no different from those of 5 year and older colonies. Following on from this, the experimental hypotheses for the current study are as follows:

H1. *The foraging ranges of aggressive colonies are significantly larger than those of passive colonies.*

H2. *The foraging ranges of 5 year and older colonies are significantly larger than those of 3-4 year old colonies.*

H3. *The foraging ranges of aggressive 3-4 year old colonies are significantly different from those of passive 5 year and older colonies. (Null hypothesis)*

Thus the aggressiveness of the younger colonies should allow them to have foraging ranges typically found in older colonies despite their being smaller. This can be interpreted as the 'plasticity' of the developing system. This would give credence to the biological hypothesis that colony level food requirements drive the aggression of individual ants, leading to the changes in global behaviour observed in colonies of different ages.

2 The Model

An individual-based model was chosen as the most appropriate for the study of the effects of inter-colony interactions, based on the examination of previous studies (e.g. [9,10,11]) and the specific requirements of the problem. This method allows the addition of stochastic processes that are difficult to incorporate into mathematical models. All modelling was done using Matlab® 7.04 (R14).

The temporal resolution for the model was chosen at 1 second per iteration. At this scale the ant behaviour must be fairly approximate as ants may change speed and direction during this period of time. To this end, the ants are modelled with random search behaviour and constant speed over this timescale. This provides the necessary level of abstraction for the global colony-level behaviours to emerge, and allows long simulation runs (5 hours ant time, which is roughly the amount of time ants spend foraging during a typical day).

In order to model at the right level of abstraction, certain modelling assumptions have been made. These are mainly in line with those used in previous models (e.g. [9,10]), with some additional assumptions/simplifications. Some examples are given below, a full list of assumptions and justifications can be found in [12].

- Ants move at a constant speed [2], even if they are carrying food [13]
- The ants have a fixed mean departure rate from the nest, rather than being cued by the number of ants returning or any other factors [10]
- The foragers all follow fixed trail out of the nest for a period of time determined probabilistically and then start foraging. They later return to the trail at the point where they left it before returning to the nest [8]
- Ants search randomly until they are within 1mm of a food item, at which point they pick it up [14]
- The ants interact only with non-nestmates (quarrels or evasive behaviour); ants that meet from the same colony simply continue on. [6]
- Food items are homogeneous - i.e. the only food items are seeds and the seeds are all the same size. [10]

2.1 Simulation Environment

The simulation used a 50m x 50m toroidal world (as in [10]. This was divided into a 400x400 grid with each square representing 125mm x 125mm. These grid squares were then used for storing the locations of food and calculating local ant populations. At this size of the grid square, with the pickup range for food 1mm and the ant detection range 5mm, there is no particular distortion in using only the grid slot that the ant is currently in, as the probability of the ant being within 5mm of the edge of the grid slot is less than 1% and the probability of the ant being within 1mm of the edge of the grid slot less than 0.1%.

For each of the grid squares determined by the world grid there is a matrix of food locations with their coordinates. Each grid square can hold a maximum number of food items based on a maximum density of 1000 seeds/m^2. This is based on an initial density of food items 625 seeds/m^2 and a renewal rate of 0.0017 seeds/m^2/sec, which equates to an average of 30.6 seeds/m^2 during a 5-hour simulation period. The food renewal is based on observations in the desert [10].

2.2 Ant Behaviours

The behaviours of individual ants are dependent on their state. The states are as follows: *1: In Nest; 2: On Outbound Trail; 3: On Inbound Trail; 4: Searching; 5: Returning to Trail; 6: Quarrelling; 7: Fighting; 8: Dead.* The simplification of reality that these states represent is justified by fact that the present study is concerned with interactions between ants during foraging excursions, and therefore is not concerned with activities of ants within the nest. Recruitment to food sources by scouts at the beginning of foraging activity is assumed to have already taken place, and as such is also not included as an element of the model (i.e. the trial directions are predetermined). These are in line with assumptions made in earlier studies [9,10]. More details regarding ant states can be found in [12].

The ants leave the nest probabilistically according to a set departure rate and follow one of the foraging trails at a fixed speed until the trail departure trigger occurs, unless they encounter food or another ant during this time. Whether the ants have found food or are returning without food, they return to the trail at the point that they left it, then return to the nest along the trail. The ants search for food using a biased random walk, as in [10]. This is done by permuting the velocity vector between time steps by a normally distributed value with standard deviation of 0.6 radians and a bias of 0.01 radians away from the nest. These parameter values give results similar to observations that search is stochastic, but with a drift of about 0.25 cm/s away from the nest.

If an ant encounters an ant from another colony, they engage in a quarrel according to a colony level parameter for the ant that first made the detection. During a quarrel both ants are stationary. The result of a quarrel is either escalation to fighting, or a return to their previous states. Similarly, the end of a fight results in either death for one of the ants (according to the probability of kill), or a return to their previous states.

Food search was implemented by searching within a certain distance from the path of the ant within the last time step. This is superior to using a radius around the ant on each time step as it would be highly likely that the ant would 'jump' over food items. Observations of actual departure rates varied from ca. 0.1 to 0.5 ants per second [10]. After some experimentation, the higher of these figures was used in the present study. The number of seeds collected by ants in the model was around 3,000 to 4,000 per colony during a simulation run, considerably lower than predicted by [9] but in line with observations from the later [10] study.

Interaction between ants was handled by searching for intersections between the trajectories of an ant with its neighbours. Although this method is not perfect (see [12] for a full explanation), sophisticated modelling of this behaviour is unnecessary as the broad behaviour fits well with observations [2].

It was generally believed that harvester ants exhibited stable, non-overlapping territories [6,15], as colonies were thought to use semi-permanent trails that never meet [16]. Gordon [2] demonstrated, however, that encounters are common among *P. barbatus*. During a 17 day period, 34 colonies engaged in a total of 1090 encounters with their neighbours, which equates to a rate of 1.9 encounters per colony per day. Since the main experiments in the simulation model envolved three interacting colonies, the rate of encounters with neighbours was 2 per colony per day, which fits well with observations.

Foragers tend to travel away from the nest in straight lines and then fan out to search for food items [8]. The foraging ranges of *P. barbatus* vary between 1 and 20m [7]. The length of the trail with respect to the total foraging range varies considerably. It is not known how the ants decide to leave the trail. For that reeaon, various types of trail departure behaviour were tested. Exponential trail departure was chosen as it led to the most realistic foraging ranges (see [12]).

The distances between colonies were set to 10m during all experiments. This was derived from the distances between colonies in the crowded region studied in [2], as well as average nearest-neighbour distances ranging from 10.8m to 12.9m between the years of 1985 and 1988 [1]. For experiments with three colonies, the colonies were placed at the corners of an equilateral triangle, such that the distance from each colony to the other two was 10m. Trail angles were used that encouraged interactions. In general, trail angles at incident angles of 90° between neighbouring colonies were used (see Figure 1(b)).

3 Experiments

The parameter settings used in the experiments are shown in Table 1, along with the source of the given parameter values, if applicable. The simulation runs all lasted for 18,000 iterations, which represented 5 hours of 'ant time' - roughly the length of an average day's foraging during peak season [10].

After a thorough exploration of the effects of various parameter settings on the foraging behaviour and inter-colony interaction of the simulated ants (see [12] for details of these), a series of experiments was designed to test the experimental hypotheses as stated in section 1.1: that the foraging ranges of aggressive colonies are significantly larger than those of passive colonies, those of 5 year and older colonies are significantly larger than those of 3-4 year old colonies, and those of aggressive 3-4 year old colonies are not significantly different from those of passive 5 year and older. Both the age of the colonies and the aggressiveness of the ants belonging to each colony were varied in such a way that the effects of each can be separated out. As such, the two main parameters of interest that were manipulated were:

Colony size. The age of the colony is determined by its size (number of active foragers, rather than spatial scale), as in all other behavioural aspects the colonies of different ages are identical. The number of foragers by age of colony is shown in Table 2 (values taken from [2]). The number of active

Table 1. Parameters used in all trials

Parameter	Value	Source
Number of colonies	1/2/3	N/A
Number of foragers	1890	Adler & Gordon 2003
World Size	50m x 50m	Schafer et al. 2006
Distance between colonies	5m - 10m	Gordon & Kulig 1996
Number of trails	2	Schafer et al. 2006
Trail length	4m	Gordon 1992
Ant speed	2.5m/min = 0.0417 m/s	Adler & Gordon 2003
Value of lost forager	50 seeds	Adler & Gordon 2003
Quarrel Range	5mm	N/A
Probability of quarrel	0-1	N/A
Probability of fight	0.07	Adler & Gordon 2003
Probability of kill	0.7	Adler & Gordon 2003
Quarrel Duration	40 sec (avg)	Gordon & Kulig 1996
Fight Duration	42 min (avg) = 2520 sec	Gordon & Kulig 1996
Initial food density	625 seeds/m^2	Schafer et al. 2006
Food renewal rate	0.0017 seeds/m^2	Schafer et al. 2006
Food pickup range	1mm	D.M. Gordon, personal comm.
Departure rate from nest	0.3 ants/sec	Schafer et al. 2006
Simulation time	5 hrs = 18,000 time steps	Schafer et al. 2006

foragers is estimated to be half the total number of foragers [9]; this is the number of foragers used in the model (there is no need to model inactive foragers).

Probability of quarrel (p. quarrel). The aggressiveness of the colony is manipulated here by a colony-wide parameter controlling the probability that an ant engages in a quarrel when two intraspecific ants meet. Experiments were conducted with various values for this parameter, ranging from 0 (complete passivity) to 1 (deterministic aggression). For the main experiments a value of 0.1 was used for passive colonies, and a value of 0.9 was used for aggressive colonies.

Table 2. Number of foragers by colony age

Colony Age	Total Foragers	Active Foragers
1	432	216
2	967	484
3-4	2604	1302
$\geq$5	3780	1890

Table 3 shows the parameter values used in the experiments. The experiments were designed specifically to test the effect of varying the age between 3-4 years (the time at which the colony is in the steepest part of its growth curve) and the probability of quarrel for the focal colony between 0.1 (passive) and 0.9 (aggressive). In addition to the *old-passive*, and *young-aggressive* conditions, the two reverse conditions of *old-aggressive* and *young-passive* were run as control conditions. The non-focal colonies in all seven experiments were old, passive colonies (5 years old, p. quarrel 0.1). The trail angles for the three colonies were [345,135], [15,225], [255,105], which meant that each of the trails intersected at right angles with the trail of one of the other colonies (see Figure 1(b)).

Table 3. Series of experiments to investigate effects of colony age and aggressiveness

Experiment	Colonies	Age (focal)	p.quarrel (focal)	Notes
1	3	3-4	0.1	Young, passive (control 1)
2	3	5	0.1	Old, passive
3	3	3-4	0.9	Young, aggressive
4	3	5	0.9	Old, aggressive (control 2)

4 Results

4.1 Non-overlapping Foraging Ranges

Figure 1 depicts the locations of food discovery from experiment 2, in which all three colonies were 'old-passive' (5 year old, p. quarrel = 0.1). It shows that despite trail angles that cause the foragers to come in contact on a regular basis, the foraging ranges of the colonies rarely overlap. In addition, there is a greater density of activity around the borders of the foraging ranges. This is in line with observed behaviour of *P. barbatus*, as well as of other harvester ant species [6,9].

4.2 Combined Effects of Colony Age and Aggressiveness

This section presents the results of experiments 1-4, which were designed to show how the parameters alter the behaviour of the model in combination, and ultimately to test the experimental hypotheses. The mean distances from the nest of ants in the focal colony at the end of a the simulation run give a measure of the extent of the colony's foraging range [12]. In Experiment 1 *(Young, Passive)* the mean distance was 3.48 metres, in Experiment 2 *(Old, Passive)* it was 3.57 metres, in Experiment 3 *(Young, Aggressive)* it was 3.59 metres, in Experiment 4 *(Old, Aggressive)* it was 3.71 metres. Increasing both aggressiveness (probability of quarrel) and colony age (number of foragers) increases the mean distances. Also note that the mean of the young-aggressive colony is similar to that of the old-passive colony.

An analysis of variance (ANOVA) was run to see if the differences between the means were statistically significant. The results of this are shown in Table 4.

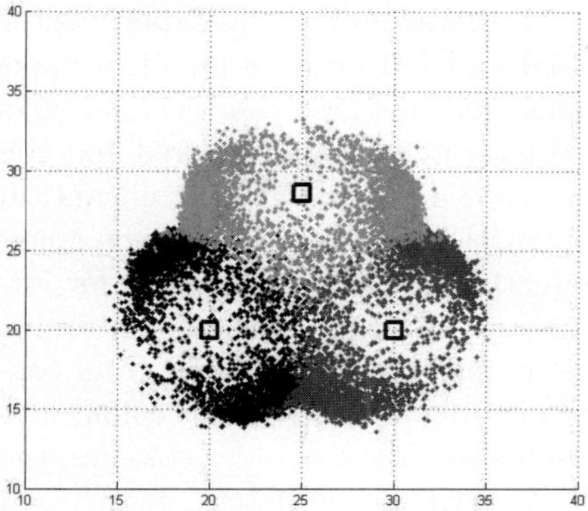

(a) Locations of food discovery

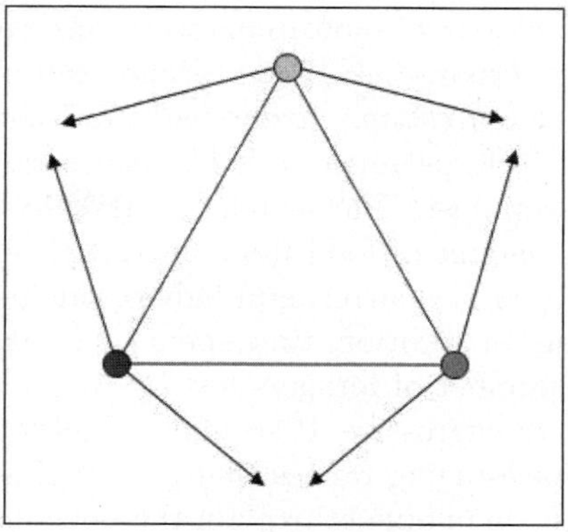

(b) Trail angles

Fig. 1. (a) Locations of food discovery for the 3 colonies in experiment 2. Ants from the three colonies are represented by dots of differing shades of grey; colony locations are shown by black squares. Note that the foraging ranges of the three colonies rarely overlap. (b) Trail angles are shown as black arrows for each of the colonies

Table 4. Analysis of variance for distance from nest by age and aggressiveness

ANOVA - Constrained (Type II) sums of squares					
Source	Sum Sq.	d.f.	Mean Sq.	F	Prob.
Agression	21.4	1	21.37	10.65	0.0011
Age	14.6	1	14.55	7.25	0.0071
Aggression * Age	0.2	1	0.24	0.12	0.7291
Error	11897	5927	2.01		
Total	11935.9	5930			

The main effects of aggressiveness and colony age are both significant at the 0.01 level, whereas the interaction effect is not significant.

In the model, older colonies have larger foraging ranges than younger ones, and aggressive colonies have larger ranges than passive ones. This therefore supports hypotheses H1 and H2, as stated in section 1.1. In addition, young-aggressive and old-passive colonies tend to have foraging ranges of similar sizes. In order to verify whether these results hold statistically, post-hoc analysis was done using Tukey's least significant difference (L.S.D.) procedure. The results of this showed that the population marginal means of the young-aggressive and old-passive colonies are significantly different from the marginal mean of the old-aggressive colony but not from the young-passive colony or each other. At the 95% confidence level, the null hypothesis that the young-aggressive colony and the old-passive colony have significantly different foraging ranges is rejected. This supports the experimental hypothesis H3, as stated in section 1.1.

Despite the increase in foraging range that the increase in aggression brings about, the benefits for the colony were not in evidence. In Experiment 1 *(Young,*

Passive) 2843 food items were retrieved and 90 foragers lost; in Experiment 2 *(Old, Passive)* 3327 food items were retrieved and 102 foragers lost; in Experiment 3 *(Young, Aggressive)* 2787 food items were retrieved and 177 foragers lost; in Experiment 4 *(Old, Aggressive)* 3304 food items were retrieved and 203 foragers lost. The switch to aggressiveness caused a fall, albeit insignificant, in the amount of food items retrieved. An explanation for this is that time spent engaged in quarrels and fights leads to a reduction in the time available for foraging. In addition, time spent by foragers that are lost is also wasted. Naturally, the number of foragers lost by the focal colony increased in the shift from passive to aggressive. It seems that although aggression allows younger colonies to adopt foraging ranges that are roughly the same in extent as older colonies, the model is unable to explain the benefit of this in terms of short-term gains (food retrieved) or losses (foragers killed).

5 Conclusions

The model displayed the non-overlapping foraging ranges of *P. barbatus* described in [9], and showed sensitivity to both colony age and forager aggression. The main effects of age and of agression on mean distance from the nest were both statistically significant. In addition, there was no significant difference between the foraging ranges of young, aggressive colonies and older, passive colonies. These results support the three experimental hypotheses from section 1.1. This is evidence that the model is behaving as expected, and that both colony age and the aggression of individual ants have significant effects on the foraging ranges of colonies. This could explain, in real ant behaviour, the aggressiveness of younger colonies, as it may allow them to have foraging ranges similar to those of older colonies. Passivity in older colonies leads to fewer lost foragers. This would suggest that aggression levels in *P. barbatus* are a trade-off between the foraging range and resources lost due to intraspecific interactions.

The number of food items retrieved did not increase when the aggression of the foragers was increased. This can be explained by the foraging time lost during quarrels or fights, as well as by foragers who are killed. It is likely that this also applies to real ants, so the benefits of having larger foraging ranges must be revealed over longer time spans. As the foraging ranges are fairly stable over time [8], the short-term losses in terms of lost time and resources must be outweighed by the longer-term benefits of a larger foraging range. This study would appear to agree with the biological hypothesis put forward by Gordon [2], although direct comparisons with real ants are made with caution.

It is hoped that this study might inspire further field studies into the behaviour of *P. barbatus* foragers from colonies of different ages (such as an attempt to measure the probability of quarrel directly).

Acknowledgements. Many thanks to Deborah Gordon for her advice, expertise, and her amazing dedication to the research of *Pogonomyrmex barbatus.*

References

1. Gordon, D.M.: Behavioral flexibility and the foraging ecology of seed-eating ants. American Naturalist 138, 379–411 (1991)
2. Gordon, D.M., Kulig, A.W.: Founding, foraging and fighting: colony size and the spatial distribution of harvester ant nests. Ecology 77, 2393–2409 (1996)
3. Gordon, D.M.: Group-level dynamics in harvester ants: young colonies and the role of patrolling. Animal Behaviour 35, 833–843 (1987)
4. Gordon, D.M.: The regulation of foraging activity in red harvester ant colonies. American Naturalist 159, 509–518 (2002)
5. Begon, M., Harper, J.L., Townsend, C.R.: Ecology. Individuals, populations and communities, 3rd edn. Blackwell Science, Oxford (1996)
6. Hölldobler, B., Lumsden, C.J.: Territorial strategies in ants. Science 210(4471), 732–739 (1980)
7. Gordon, D.M.: How colony growth affects foreigner intrusion in neighboring harvester ant colonies. Behavioral Ecology and Sociobiology 31, 417–427 (1992)
8. Gordon, D.M.: The development of an ant colony's foraging range. Animal Behaviour 49, 649–659 (1995)
9. Adler, F.R., Gordon, D.M.: Optimization, conflict and non-overlapping foraging ranges in ants. American Naturalist 162(5), 529–543 (2003)
10. Schafer, R.J., Holmes, S., Gordon, D.M.: Forager activation and food availability in harvester ants. Animal Behaviour 71(4), 815–822 (2006)
11. Couzin, I.D., Franks, N.R.: Self-organized lane formation and optimized traffic flow in army ants. Proceedings of the Royal Society of London, Series B 270, 139–146 (2003)
12. Diethe, T.: Modelling changes in foraging and defensive behaviour of ant colony members as a result of inter-colony interactions. Master's thesis, University College London (2006)
13. Morehead, S.A., Feener Jr., D.H.: Foraging behavior and morphology: seed selection in the harvester ant genus, pogonomyrmex. Oecologica 114, 548–555 (1998)
14. Gordon, D.M., Paul, R.E.H., Thorpe, K.: What is the function of encounter patterns in ant colonies? Animal Behaviour 45, 1083–1100 (1993)
15. Hölldobler, B., Wilson, E.O.: The Ants. Harvard University Press, Cambridge (1990)
16. Hölldobler, B.: Recruitment behavior, home range orientation and territoriality in harvester ants, pogonomyrmex. Behaviour Ecology and Sociobiology 1, 3–44 (1976)

A Mechanism to Self-assemble Patterns with Autonomous Robots

Anders Lyhne Christensen, Rehan O'Grady, and Marco Dorigo

IRIDIA, CoDE, Université Libre de Bruxelles
50, Av. Franklin Roosevelt CP 194/6
1050 Brussels, Belgium
alyhne@iridia.ulb.ac.be,rogrady@ulb.ac.be,mdorigo@ulb.ac.be

Abstract. There are examples of robotic systems in which autonomous mobile robots self-assemble into larger connected entities. However, existing systems display little or no autonomous control over the shape of the connected entity thus formed. We describe a novel distributed mechanism that allows autonomous mobile robots to self-assemble into pre-specified patterns. Global patterns are 'grown' using locally applicable rules and local visual perception only. In this study, we focus on the low-level navigation and directional self-assembly part of the pattern formation process. We analyse the precision of this mechanism on real robots.

1 Introduction

Much research has been devoted to the capabilities of distributed swarms of co-operating, autonomous robots [1]. In some of these systems, multiple robots can *self-assemble* into larger structures in order to overcome the physical limitations of the individual agents. For example, multiple connected robots can transport objects too heavy to be moved by an individual robot [2], or navigate terrain impassable by a robot navigating alone [3].

Existing self-assembling systems, however, have very little autonomous control over the shape of the connected structures they form. In this paper, we present the low-level control aspects of a distributed control mechanism that allows autonomous mobile robots to self-assemble into specific, connected patterns. Three examples of patterns are shown in Fig. 1. Patterns are 'grown' using local visual perception only. Robots that are already attached to the pattern indicate where new robots should attach in order to grow the local structure appropriately. For a detailed study of the high-level control principles and the group-level performance of the proposed mechanism, see [4].

2 Related Work

Related research areas include self-reconfigurable robotics and formation control. Self-reconfigurable robotic systems are made up of modular robots that

F. Almeida e Costa et al. (Eds.): ECAL 2007, LNAI 4648, pp. 716–725, 2007.

Fig. 1. Examples of whole patterns: Rectangle, star, and line.

can connect to each other in different ways so as to autonomously change their global morphology. Examples of such systems include Yim *et al.*'s PolyBot [5,6], Hirose *et al.*'s Gunryu [7], and Fukuda *et al.*'s CEBOT system [8,9]. PolyBot is a modular chain robot in which each module has one degree of freedom. It has been demonstrated that an arm consisting of multiple PolyBot modules is capable of operating in 3D space and that such an arm can grasp and dock with additional modules. In the Gunryu system, each robot is capable of autonomous locomotion and equipped with an actuator that allows robots to form physical connections with each other. CEBOT is a system consisting of heterogeneous modules with different functions, e.g. to rotate, move, and bend. Various prototypes of the CEBOT system comprising different shapes and connection mechanisms have been studied. However, these systems tend to be limited either in their ability to configure themselves autonomously or in their ability to self-assemble. Castano *et al.* has proposed a system of homogeneous modules called CONRO [10]. Rubenstein *et al.* have recently been shown that CONRO is capable of autonomous docking (self-assembly) [11].

In formation control research, groups of robots are steered into one or more pre-specified formations. Mechanisms to maintain these formations while the group is in motion are also studied. Proposed approaches include the use of *virtual structures* [12,13], leader-follower schemes [14,15], and decentralised, behaviour-based methods [16,17,18].

Most existing approaches rely either on global communication or on each robot having access to a blueprint of the global pattern (or both). Much of the research has been conducted in simulation only. In the study presented in this paper we use real robots to self-assemble global patterns using a completely distributed algorithm. None of the robots has access to a blueprint of the global pattern. The algorithmic rules are based on local information only. None of the robots except the *seed* robot has any predefined position in the pattern.

3 Hardware Platform

We use a number of real robots known as *s-bots* [19]. The *s-bot* platform has been used for several studies in swarm intelligence and collective robotics. Overcoming steep hills and transport of heavy objects are notable examples of tasks which a

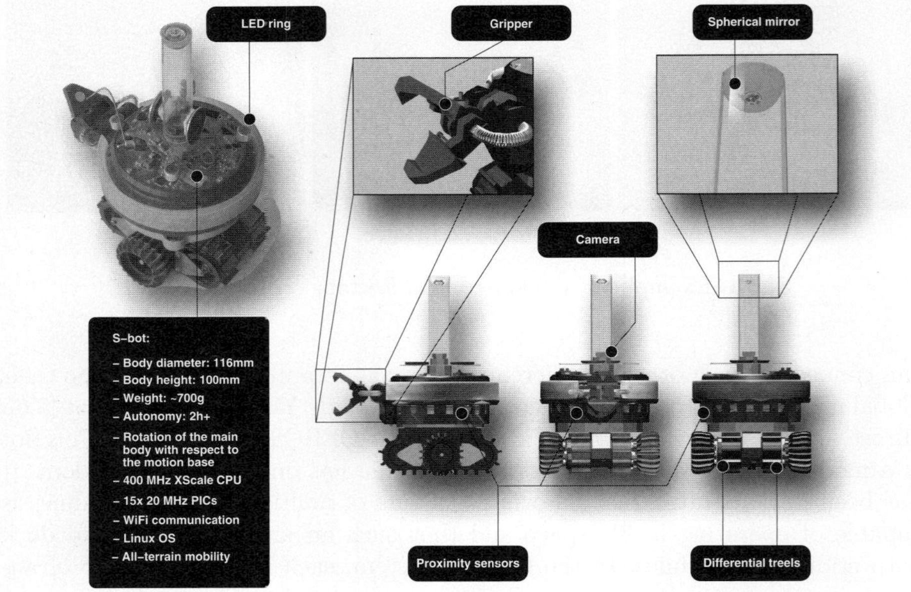

Fig. 2. *s-bot*: An autonomous, mobile robot capable of self-assembly.

single *s-bot* could not solve individually, but which have been solved successfully by teams of self-assembling *s-bots* [2,20,3].

Each *s-bot* is equipped with an Xscale CPU running at 400 MHz, a number of sensors including an omni-directional camera, light and proximity sensors. Each *s-bot* also has a number of actuators. These include 8 sets of RGB coloured LEDs distributed around the circumference of the *s-bot* body. These LEDs can be controlled individually and can be perceived by other robots at a range of up to approximately 50 cm depending on light conditions. The *s-bots* also have a gripper that allows them to form physical connections with one another. The sensors and actuators are indicated in Fig. 2.

4 Growing Global Patterns

In this study, we consider patterns formed by groups of self-assembling robots. We start from a single predesignated robot, the *seed*, that indicates where other robots can attach to it. In principle, the seed could be chosen probabilistically [3] or it could be the first robot that encounters a situation that requires self-assembly into a given morphology. Since we focus on pattern formation in this study, we pre-configure one *s-bot* to be the seed. At the beginning of each experiment the robots are instructed which pattern they should form (e.g., arrow, star or line). As new robots attach to the connected structure, they indicate where

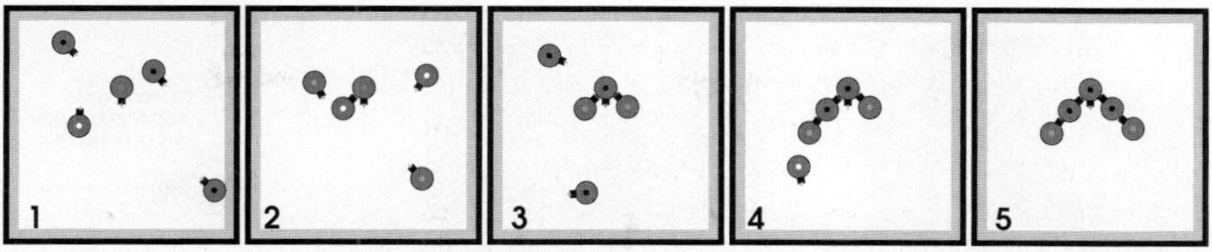

Fig. 3. Example of an arrow pattern being formed by a group of 5 *s-bots*.

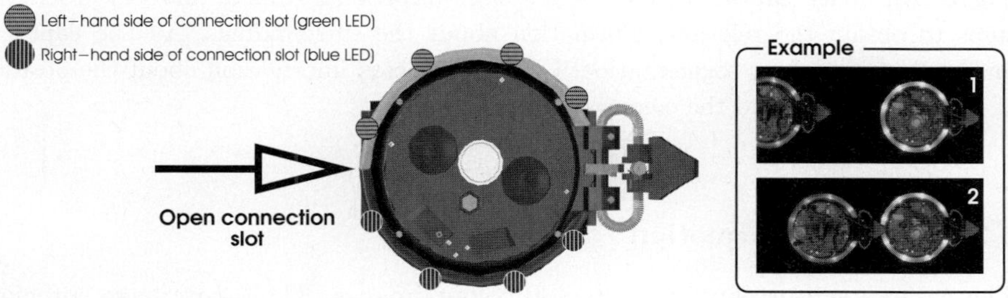

Fig. 4. An example of a *s-bot* with an open connection slot to its rear. The *s-bot* has lit up its left green LEDs and its right blue LEDs. Right: An example with real robots.

other non-attached robots should attach in order to extend the structure appropriately. However, none of the robots have any knowledge about the global state or shape of the connected structure at any time. None of the robots (except the seed in this study) has a predefined location in the final pattern. An example of the formation of an arrow pattern is shown in Fig. 3.

5 Low-Level Pattern Formation Mechanism

5.1 Overview

The robots coordinate using their camera and coloured LED ring. Our control mechanism makes use of the colours red, green and blue. Green and blue indicate the left-hand side and right-hand side of a *connection slot*, respectively. A connection slot specifies a location and a direction in which the pattern should be extended. An example is shown in Fig. 4. An *s-bot* can open connection slots in 7 different locations and directions (each robot has 8 sets of LEDs and a slot can be opened between any two neighbouring LEDs, except between the two front LEDs where the gripper is mounted). Non-attached robots light up their red LEDs in order to be visible to other robots.

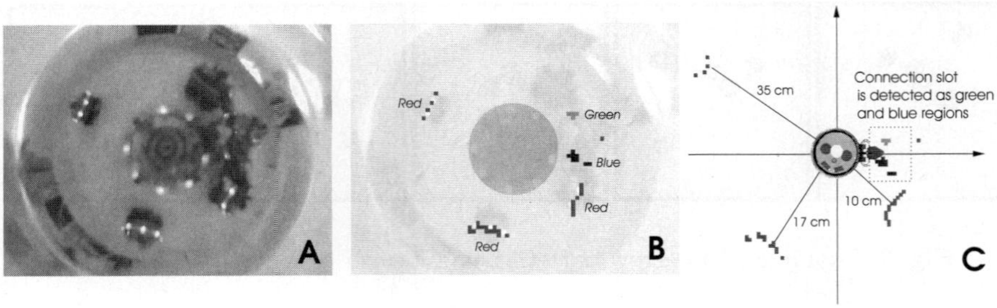

Fig. 5. An image captured by a robot's omni-directional camera and the processing steps to obtain the relevant information about the surroundings. A: The captured image. B: After colour segmentation. C: The extracted information about the position of the other robots and the connection slot.

5.2 Sensory Information

The camera sensor captures 640x480 colour images. The *s-bots* have sufficient on-board processing power to scan the entire image and identify objects based on colour information. The image processor is configured to detect the location of the coloured LEDs of the *s-bots* and discard any other information. The image processor divides the image into a grid of multi-pixel blocks and returns the segmented colour prevalent in each block (or indicates the absence of any segmented colour). The *s-bot* camera captures images of the robot's surroundings reflected in a semi-spherical mirror. Since the *s-bots* operate on flat terrain, this means that the distance in pixels from the centre of an image to a perceived object corresponds to the physical distance between the robot and the object. An example is shown in Fig. 5.

The cameras have a range of approximately 50 cm. Like most sensors on mobile robots, the readings from the camera are subject to a significant amount of noise. Objects are not always perceived even when they are in range of the camera, and due to occlusions one robot cannot see all the LEDs on another robot unless the two robots are adjacent. Furthermore, as a robot moves, objects tend to "jump around" from frame to frame due to the shaking of the perspex tube holding the camera and the spherical mirror (a difference of a few pixels can have a considerable impact on the computed location relative to the perceiving robot). Despite the sensory limitations of individual *s-bots*, we do not have to resort to using global communication or a global pattern blueprint. We generate consistent global patterns by leveraging distributed control principles — each robot acts solely on the basis of what it perceives in its immediate vicinity.

5.3 Navigation to a Connection Slot

Non-attached *s-bots* (i.e., *s-bots* that have not yet attached to the pattern) start by searching for an open connection slot. An open connection slot can be on either the *seed* robot or on other robots that have already attached to the pattern. If a non-attached *s-bot* cannot see any coloured LEDs, it performs a random walk until it perceives one or more LEDs. If a non-attached *s-bot* can see an open connection slot (i.e., blue or green LEDs), the *s-bot* tries to navigate around the connected structure until it has the correct position and alignment to attach to the slot.

The different navigation zones around an *s-bot* with an open connection slot to its rear are shown in Fig. 6. A non-attached *s-bot* takes different actions according to the zone in which it is located. If an *s-bot* is more than 30 cm away from the connection slot, it navigates directly towards the slot. If an *s-bot* is within 30 cm of the slot and it is in the *go around zone*, the *s-bot* attempts to navigate around the connected structure, randomly choosing either the clockwise or the counter-clockwise direction. Once inside the *inner grip zone* the *s-bot* first navigates to the *intermediate spot* before approaching the connection slot. This two-phase strategy enables the non-attached *s-bot* to approach the connection slot with the correct alignment. If, during the approach, the *s-bot* exits the *outer grip zone*, it switches back to navigating around the connected structure.

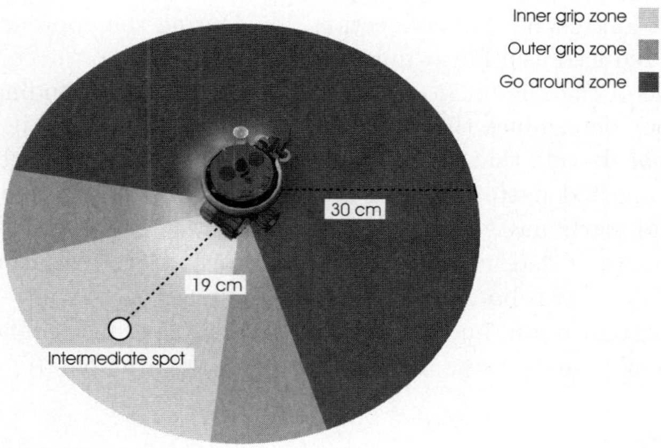

Fig. 6. The *intermediate spot* and navigation zones around an *s-bot* that has opened a connection slot to its rear.

5.4 Forming a Directional Connection

Once an *s-bot* is in the *inner grip zone*, the challenge is to steer the *s-bot* to the connection slot so that the position and orientation are correct when it attempts to grip. This procedure takes place in three steps: 1) Go to the *intermediate spot*, 2) Turn to face the connection slot, and 3) Navigate to the connection slot and connect.

The closer a robot gets to a connection slot, the more accurately it perceives the location and the direction indicated. Corrections to the trajectory are made continuously as the robot approaches the slot and as more accurate positional information becomes available. The *intermediate spot* is calculated by the non-attached *s-bot* to be 19 cm away from the *s-bot* with the open connection slot in the direction indicated by the slot (see Fig. 6). This means that the robot has a course of approximately 13 cm (19 cm minus the radius on an *s-bot*) to align itself correctly. As the robot moves closer, the speeds of the left and the right treels respectively, are set to:

$$s_l = 5.6\,\text{mm/s} + 21.8\,\text{mm/s} \cdot \frac{d}{130\,\text{mm}} \cdot f(\theta), \tag{1}$$

$$s_r = 5.6\,\text{mm/s} + 21.8\,\text{mm/s} \cdot \frac{d}{130\,\text{mm}} \cdot f(-\theta), \tag{2}$$

where d is the distance to the connection slot, and $f : \theta \to [0, 1]$ is a function that maps the angular difference between the current heading and the ideal heading, θ, to a speed modifier in the range $[0, 1]$. The result of applying this speed modifier term to Eqn. (1) and Eqn. (2) is that the *s-bot* continually corrects its alignment as it approaches the connection slot. During the approach, the speed of the treels is reduced as a linear function of the distance to the connection slot (the magnitude of the alignment corrections becomes correspondingly smaller). When the robot determines that it is close enough to connect, it attempts to grip. If the *s-bot* detects that the grip was successful, it assumes that it is now part of the connected pattern. If, on the other hand, the grip fails, the robot moves back and starts navigating to the *intermediate spot* again.

Pattern growth is determined by three sets of pattern extension rules: one set of rules for the *seed* robot, one set for robots already in the pattern that have just received a connection, and one for robots that have just connected to the pattern. By manipulating these three sets of rules, we are able to form different patterns [4].

6 Results

We analysed the precision of the directional self-assembly mechanism that allows an attached robot to specify the position and orientation with which a non-attached robot should attach to it. We conducted 96 trials in which a single non-attached *s-bot* attached to a stationary *seed* robot (12 starting positions, 8 starting orientations). The twelve starting positions were evenly distributed around a circle of radius 35 cm centred on the *seed* robot. We used a *seed* robot with a single connection slot open to its rear.

We initially considered angular precision: how accurately the connecting non-attached robot matched its alignment to the desired alignment indicated by the *seed* robot. The angular precision results are shown in Fig. 7 (top-left). Note that the mean angular misalignment is very close to zero.

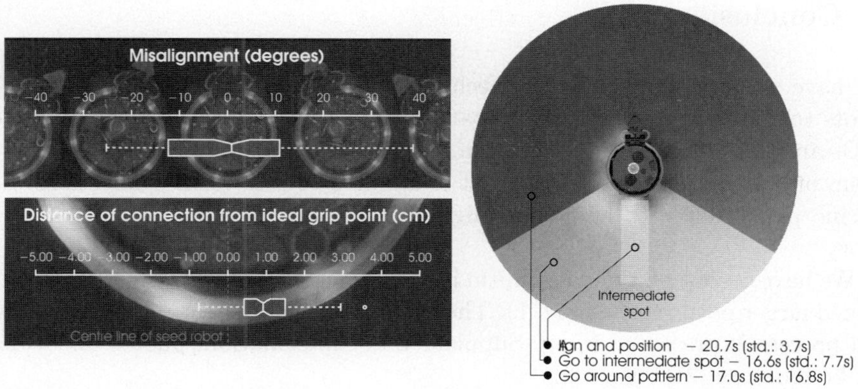

Fig. 7. Timing and precision of the pattern formation mechanism. Top-left: Angular precision. Bottom-left: Positional precision. Right: Mean time and standard deviation spent on different activities while forming a connection.

We also analysed the positional precision: how close to the ideal grip point the non-attached robot connected to the *seed*. With a connection slot to its rear, the ideal grip point is the middle of the rear of the *seed* — the point on the *seed*'s LED ring that falls on the *seed*'s centre line (in line with the *seed*'s camera and gripper). Note that it is possible for a robot to grip at the wrong point even if its alignment is perfect. The positional precision results are presented in Fig. 7 (bottom-left). There is a clear bias towards attaching to the right of the ideal grip point. This bias arises because the LEDs are not distributed in a perfectly uniform manner around the *s-bot* body. When the *seed* lights up its 4 left green LEDs, and its 4 right blue LEDs, the point to its rear equidistant between the green and blue LEDs is in fact about 1 cm to the right of its centre line. This can be seen by looking at the *s-bot* LED ring in the background of Fig. 7 (bottom-left).

The mean times spent on the different activities are shown in Fig. 7 (right). The largest share of the time was spent on positioning and alignment, that is, approaching the connection slot from the *intermediate spot.* Although the distances covered during the different activities vary significantly, the mean times spent on the different activities are comparable. This is a consequence of the increasing precision required as the *s-bot* gets closer to the connection slot — the more precision required, the slower the *s-bot* moves. The mean time from the start of a trial until the *s-bot* attached was 54.3 s.

In all 96 trials the free robot attached to the *seed* robot. In 2 of the 96 trials the free robot failed to attach on the first attempt and retreated to try another angle. In a further four of the 96 trials the non-attached robot abandoned its approach to the connection slot before attempting to grip and retreated to try another angle, as it determined that it was approaching from an incorrect angle. In one of the 96 trials the non-attached robot lost sight of the connection slot and was manually replaced on its starting position.

7 Conclusion

We have presented a low-level mechanism that enables autonomous mobile robots to self-assemble into global patterns. The mechanism relies on coloured LEDs and local visual communication. We evaluated the precision of this mechanism on real robots, and found that it allows *s-bots* to form connections at a specific point and with a specific alignment with a relatively small margin of error.

We have evaluated the mechanism in experiments where groups of real robots formed larger patterns (see Fig. 1). The performance of the entire pattern formation process is presented in [4]. Supplementary information, photos and videos can be found in [21].

An interesting direction for future research is to allow groups of robots to autonomously choose which patterns to self-assemble into, based on the nature of the obstacles encountered. In an all-terrain navigation task, for example, the group could self-assemble into a line formation in order to cross a ditch, while uneven or hilly terrain could trigger self-assembly into a dense structure that provides stability.

Acknowledgements. This work was supported by the *SWARMANOID* project, funded by the Future and Emerging Technologies programme (IST-FET) of the European Commission, under grant IST-022888. Anders Christensen acknowledges support from COMP2SYS, a Marie Curie Early Stage Research Training Site funded by the European Community's Sixth Framework Programme (grant MEST-CT-2004-505079). The information provided is the sole responsibility of the authors and does not reflect the European Commission's opinion. The European Commission is not responsible for any use that might be made of data appearing in this publication. Marco Dorigo acknowledges support from the Belgian FNRS, of which he is a Research Director.

References

1. Dorigo, M., Şahin, E., eds.: *Special Issue on "Swarm Robotics". Autonomous Robots*, 17:111–246. Springer Verlag (2004)
2. Groß, R., Bonani, M., Mondada, F., Dorigo, M.: Autonomous self-assembly in swarm-bots. IEEE Trans. on Robotics **22**(6) (2006) 1115–1130
3. O'Grady, R., Groß, R., Mondada, F., Bonani, M., Dorigo, M.: Self-assembly on demand in a group of physical autonomous mobile robots navigating rough terrain. In: Proc. of Advances in Artificial Life: 8th European Conf., ECAL 2005. Volume 3630 of LNAI., Springer Verlag, Berlin, Germany (2005) 272–281
4. Christensen, A.L., O'Grady, R., Dorigo, M.: Morphology control in a self-assembling multi-robot system. Technical Report TR/IRIDIA/2007-005, IRIDIA, Université Libre de Bruxelles, Belgium (2007)
5. Yim, M., Roufas, K., Duff, D., Zhang, Y., Eldershaw, C., Homans, S.B.: Modular reconfigurable robots in space applications. Autonomous Robots **14**(2-3) (2003) 225–237

6. Yim, M., Duff, D.G., Roufas, K.D.: PolyBot: a modular reconfigurable robot. In: Proc. of the 2000 IEEE Inter. Conf. on Robotics and Automation. Volume 1., IEEE Press, Piscataway, NJ (2000)
7. Hirose, S., Shirasu, T., Fukushima, E.F.: Proposal for cooperative robot "Gunryu" composed of autonomous segments. Robots and Autonomous Systems **17** (1996) 107–118
8. Fukuda, T., Buss, M., Hosokai, H., Kawauchi, Y.: Cell structured robotic system CEBOT: control, planning and communication methods. Robotics and autonomous systems **7**(2-3) (1991) 239–248
9. Kawauchi, Y., Inaba, M., Fukuda, T.: A principle of distributed decision making of celluar robotic system (CEBOT). In: Proc. of the 1993 IEEE Inter. Conf. on Robotics and Automation, IEEE Press, Piscataway, NJ (1993) 833–838
10. Castano, A., Shen, W., Will, P.: CONRO: Towards deployable robots with inter-robots metamorphic capabilities. Autonomous Robots **8**(3) (2000) 309–324
11. Rubenstein, M., Payne, K., Will, P.: Docking among independent and autonomous CONRO self-reconfigurable robots. In: Proc. of the 2004 IEEE Inter. Conf. on Robotics and Automation. Volume 3., IEEE Press, Piscataway, NJ (2004) 2877–2882
12. Lewis, M., Tan, K.: High precision formation control of mobile robots using virtual structures. Autonomous Robots **4**(4) (1997) 387–403
13. Tabuada, P., Pappas, G.J., Lima, P.: Feasible formations of multi-agent systems. In: Proc. of the 2001 American Control Conf. Volume 1., IEEE Press, Piscataway, NJ (2001) 56–61
14. Das, A., Fierro, R., Kumar, V., Ostrowski, J.P., Spletzer, J., Taylor, C.J.: A vision-based formation control framework. IEEE Trans. on Robotics and Automation **18**(5) (2002) 813–825
15. Tanner, H., Pappas, G., Kumar, V.: Leader-to-formation stability. IEEE Trans. on Robotics and Automation **20**(3) (2004) 443–455
16. Sugihara, K., Suzuki, I.: Distributed algorithms for formation of geometric patterns with many mobile robots. Journal of robotic systems **13**(3) (1996) 127–139
17. Balch, T., Arkin, R.: Behavior-based formation control for multirobot teams. IEEE Trans. on Robotics and Automation **14**(6) (1998) 926–939
18. Lawton, J., Beard, R., Young, B.: A decentralized approach to formation maneuvers. IEEE Trans. on Robotics and Automation **19**(6) (2003) 933–941
19. Mondada, F., Gambardella, L.M., Floreano, D., Nolfi, S., Deneubourg, J.L., Dorigo, M.: The cooperation of swarm-bots: Physical interactions in collective robotics. IEEE Robotics & Automation Magazine **12**(2) (2005) 21–28
20. Nouyan, S., Groß, R., Bonani, M., Mondada, F., Dorigo, M.: Group transport along a robot chain in a self-organised robot colony. In: Intelligent Autonomous Systems 9, IAS 9. IOS Press, Amsterdam, The Netherlands (2006) 433–442
21. Christensen, A.L., O'Grady, R., Dorigo, M.: Photos and videos of self-assembling robots forming patterns, http://iridia.ulb.ac.be/supp/IridiaSupp2007-003/ (2007)

Binocular Vision-Based Robot Control with Active Hand-Eye Coordination

Wen-Chung Chang

Department of Electrical Engineering, National Taipei University of Technology
1 Sec. 3 Chung-Hsiao E. Rd., NTUT Box 2125, Taipei 106, Taiwan, R.O.C.
wchang@ee.ntut.edu.tw

Abstract. Traditional eye-in-hand robotic systems are capable of performing versatile manipulation, but generally they can only observe a restricted workspace. As regards eye-to-hand configurations, tasks can be controlled within the field of view of the vision system with accuracy up to the pixel resolution of the vision system. In this paper, the robot workspace is further expanded by allowing cameras to be actively controlled by pan, tilt, and zoom motion. This configuration can be applied to a mobile robot equipped with a binocular vision system. The manipulator for visual servo control purpose can be either mounted on-board or in fixed configuration. To enable large and flexible workspace visual servoing with precision under such a configuration, active hand-eye coordination must be assured. The control strategy is successfully validated through convergence analysis and simulations.

Key words: Active vision, Binocular vision, Hand-eye coordination, Mobile robot, Visual servoing, Visual tracking.

1 Introduction

Compared with other sensors, vision is capable of measuring, recognizing, and tracking objects in an open workspace. Therefore, vision-based control of robots has been an active research field and is being integrated into scientific and industrial applications [1,2,3,4,5]. However, in conventional visual servo approaches, either eye-in-hand or eye-to-hand, the workspace that can be observed by the cameras is restricted. If one wants to carry out control tasks in a large, flexible, and changing workspace, or to zoom the camera to improve task accuracy, these conventional configurations would possibly be inappropriate.

The objective of this research is to propose a promising architecture to resolve the problem. This proposed architecture consists of an active visual tracking system and an active visual serv system. In the active visual tracking system, an image-based controller handling pan-tilt-zoom(PTZ) mechanism for each of the binocular cameras is introduced. The controller is able to maintain the orientation of each camera to keep both the controlled point of the robotic manipulator and the moving target in the vicinity of the image center. In addition, the focal length can be adjusted online to provide better task accuracy. The other control

F. Almeida e Costa et al. (Eds.): ECAL 2007, LNAI 4648, pp. 726–735, 2007.

subsystem, the active visual servo controller, capable of performing positioning and tracking tasks is also introduced. According to the measured angular position and velocity of each camera, the visual servo controller could update the hand-eye coordination on-line. The proposed approach can be applied to a mobile robot equipped with a binocular vision system. The manipulator for active hand-eye coordination purpose can be either mounted on-board or in fixed configuration while connected with the mobile robot by wireless networks as illustrated in Fig. 1. Under either configuration, not only the robot workspace can be expanded but also the task accuracy can be improved.

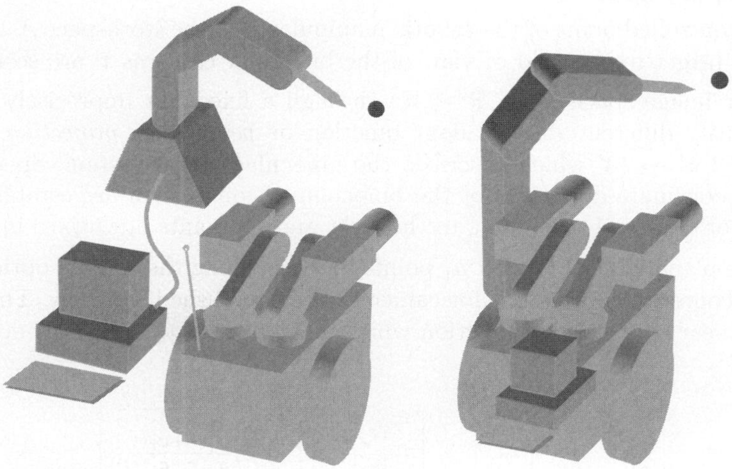

Fig. 1. An active binocular vision-based mobile robot with a manipulator mounted in fixed configuration (*left*) or on-board (*right*).

The remainder of this article is structured as follows. Section 2 describes the system and the binocular camera model in detail. In Section 3, the active visual tracking system is introduced, including the estimation of velocity of a moving target and the realization of the active camera controller. The convergence analysis of the proposed active visual servo approach is described in Section 4. Finally, Section 5 presents simulation results of positioning and target tracking tasks and Section 6 the conclusion, where the significance of the system and the perspectives of the research are addressed.

2 System Description

Let bold uppercase denote matrix, calligraphic alphabet denote set, bold lowercase denote vector, prime denote transpose, and leading superscripts identify the space a quantity is written in.

The active hand-eye system consists of two PTZ cameras, a robotic manipulator, an active visual tracking controller, and an active visual servo controller. Based on the observed information from these PTZ cameras, a visual encoded error [6] could be defined. According to this encoded error, a suitable controller could be designed to drive this error to zero. When the error goes to zero, the robotic manipulator reaches the target with precision [7]. Meanwhile, due to the varying orientation of the on-board binocular cameras, the hand-eye coordination must be updated accordingly for further controlled action. That is, the active visual servo controller must perform online tuning based on the available measurements from the PTZ mechanism. The complete system diagram is illustrated in Fig. 2.

The controlled point of the robotic manipulator in the workspace $\mathcal{X}$ together with the target in the field of view of the binocular cameras $\mathcal{V}$ are seen in the binocular image space ${}^I\mathcal{X} \triangleq \mathbb{R}^2 \oplus \mathbb{R}^2$ through a fixed but imprecisely known, continuously differentiable, readout function or *perspective projection* camera model $G : \mathcal{V} \rightarrow {}^I\mathcal{X}$ which describes the binocular vision system. Specifically, camera coordinate directions of the binocular vision system are established as follows: for camera 1, $\boldsymbol{x}_1$ points to the right and $\boldsymbol{y}_1$ points downward in camera 1's image plane, and $\boldsymbol{z}_1 \triangleq \boldsymbol{x}_1 \times \boldsymbol{y}_1$ points outward along the camera optical axis. Camera coordinate directions for camera 2 are established similarly. Thus, G is the nonlinear time-varying function which maps from $\mathcal{V}$ to ${}^I\mathcal{X}$ and is defined as follows.

$$G(\boldsymbol{r}, t) \triangleq \begin{bmatrix} {}^I\boldsymbol{r}_1 \\ {}^I\boldsymbol{r}_2 \end{bmatrix} = \begin{bmatrix} {}^Ir_{1x} \\ {}^Ir_{1y} \\ {}^Ir_{2x} \\ {}^Ir_{2y} \end{bmatrix} = \begin{bmatrix} f_1(t)\frac{\boldsymbol{x}_1'(\boldsymbol{r}-\boldsymbol{c}_1(t))}{\boldsymbol{z}_1'(\boldsymbol{r}-\boldsymbol{c}_1(t))} \\ f_1(t)\frac{\boldsymbol{y}_1'(\boldsymbol{r}-\boldsymbol{c}_1(t))}{\boldsymbol{z}_1'(\boldsymbol{r}-\boldsymbol{c}_1(t))} \\ f_2(t)\frac{\boldsymbol{x}_2'(\boldsymbol{r}-\boldsymbol{c}_2(t))}{\boldsymbol{z}_2'(\boldsymbol{r}-\boldsymbol{c}_2(t))} \\ f_2(t)\frac{\boldsymbol{y}_2'(\boldsymbol{r}-\boldsymbol{c}_2(t))}{\boldsymbol{z}_2'(\boldsymbol{r}-\boldsymbol{c}_2(t))} \end{bmatrix} \tag{1}$$

where $\boldsymbol{r}$ is the controlled point of the robotic manipulator in $\mathcal{X}$ and f_n denotes the focal length of camera n. Note that the camera model being time-varying is due to the fact that the binocular cameras are actively controlled by PTZ motion. The position of the controlled point in the binocular image space is a measured output ${}^I\boldsymbol{r}$ related to $\boldsymbol{r}$ by the formula

$${}^I\boldsymbol{r} = G(\boldsymbol{r}, t). \tag{2}$$

Similarly, the target position in the binocular image space is an output ${}^I\boldsymbol{r}^*$ related to the target set-point $\boldsymbol{r}^*$ by the formula

$${}^I\boldsymbol{r}^* = G(\boldsymbol{r}^*, t). \tag{3}$$

What this paper is concerned with is how one might define an active visual tracking control law and an active visual servo control law to achieve precise positioning and tracking, without having available an accurate model of G.

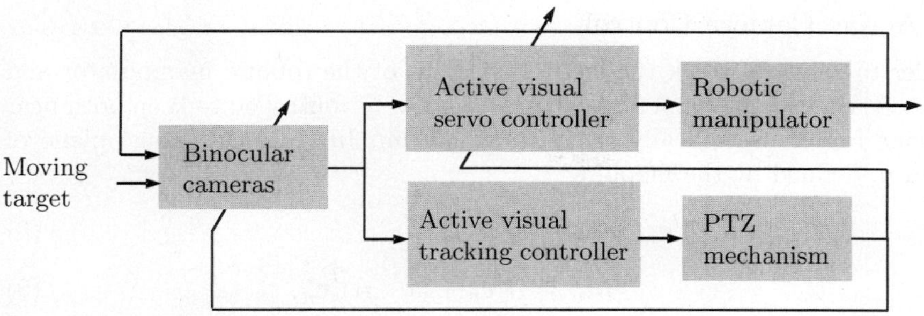

Fig. 2. Block diagram of the proposed active hand-eye system.

3 Active Visual Tracking

To allow active hand-eye coordination, one must keep the controlled point of the robotic manipulator and the target near the center of the field of view of the binocular vision system. This is equivalent to requiring that both the controlled point and the target be in the vicinity of the image centers of the binocular cameras. Therefore, one first estimates the velocity of the moving target in the image plane. Based on the estimate, an active camera controller, capable of controlling the PTZ cameras to accomplish the tracking purpose, is proposed.

3.1 Moving Target Estimation

Let ${}^I\dot{\boldsymbol{r}}^*_n$ denote the velocity of the target in the image plane of an actively controlled camera n. In fact, it consists of two components, one, ${}^I\dot{\boldsymbol{r}}^*_{tn}$, the projected velocity of the moving target while the camera is kept stationary and the other, ${}^I\dot{\boldsymbol{r}}^*_{cn}$, the projected velocity of the target, considered stationary, while the camera is actively controlled. That is,

$$ {}^I\dot{\boldsymbol{r}}^*_n = {}^I\dot{\boldsymbol{r}}^*_{tn} + {}^I\dot{\boldsymbol{r}}^*_{cn}. \tag{4} $$

Therefore, the projected velocity of the moving target while the camera is kept stationary can be estimated by

$$ {}^I\hat{\dot{\boldsymbol{r}}}^*_{tn} = {}^I\hat{\dot{\boldsymbol{r}}}^*_n - {}^I\hat{\dot{\boldsymbol{r}}}^*_{cn} \tag{5} $$

where the projected velocity of the target, considered stationary, while the camera is actively controlled can be estimated using known PTZ motion of the camera by

$$ {}^I\hat{\dot{\boldsymbol{r}}}^*_{cn} = \boldsymbol{J}_\theta({}^I\boldsymbol{r}^*_n)\dot{\boldsymbol{\theta}}_n + \boldsymbol{J}_f({}^I\boldsymbol{r}^*_n)\dot{f}_n \tag{6} $$

$$ = \begin{bmatrix} \frac{{}^I r^*_{nx}\,{}^I r^*_{ny}}{f_n} & \frac{-f_n^2 - {}^I r^{*2}_{ny}}{f_n} \\ \frac{f_n^2 + {}^I r^{*2}_{ny}}{f_n} & \frac{-{}^I r^*_{nx}\,{}^I r^*_{ny}}{f_n} \end{bmatrix} \begin{bmatrix} \dot{\theta}_{nx} \\ \dot{\theta}_{ny} \end{bmatrix} + \begin{bmatrix} \frac{{}^I r^*_{nx}}{f_n} \\ \frac{{}^I r^*_{ny}}{f_n} \end{bmatrix} \dot{f}_n. \tag{7} $$

3.2 Active Camera Control

In order to robustly track the controlled point of the robotic manipulator and the target, each of the binocular cameras is actively controlled to keep both near its image center. Specifically, their *tracked point*, ${}^I\boldsymbol{m}_n$, in the image plane of camera n, defined by the mapping

$$\left\{{}^I\boldsymbol{r}_n, {}^I\boldsymbol{r}_n^*\right\} \mapsto \{{}^I\boldsymbol{m}_n\} \tag{8}$$

$${}^I\boldsymbol{m}_n = \alpha{}^I\boldsymbol{r}_n + (1-\alpha){}^I\boldsymbol{r}_n^*, \tag{9}$$

is driven to the image center by controlling the PTZ motion of camera n.

The encoded error for the set-point camera control can thus be defined as

$$\boldsymbol{e}_{no} = {}^I\boldsymbol{m}_n - {}^I\boldsymbol{o}_n = {}^I\boldsymbol{m}_n \tag{10}$$

where ${}^I\boldsymbol{o}_n$ denotes the image center of camera n. Differentiating Eq. (10) with respect to time, one can see from Eqs. (4) and (6) that

$$\begin{aligned}\dot{\boldsymbol{e}}_{no} &= {}^I\dot{\boldsymbol{m}}_{tn} + {}^I\dot{\boldsymbol{m}}_{cn} &(11)\\ &= {}^I\dot{\boldsymbol{m}}_{tn} + \boldsymbol{J}_\theta({}^I\boldsymbol{m}_n)\dot{\boldsymbol{\theta}}_n + \boldsymbol{J}_f({}^I\boldsymbol{m}_n)\dot{f}_n. &(12)\end{aligned}$$

Therefore, the PT motion control law of camera n,

$$\dot{\boldsymbol{\theta}}_n = -\boldsymbol{J}_\theta({}^I\boldsymbol{m}_n)^{-1}\left(k_o\boldsymbol{e}_{no} + {}^I\hat{\dot{\boldsymbol{m}}}_{tn} + \boldsymbol{J}_f({}^I\boldsymbol{m}_n)\dot{f}_n\right), k_o > 0, \tag{13}$$

can drive the encoded error, $\boldsymbol{e}_{no}$, to zero exponentially. Moreover, the encoded error being zero implies that the tracked point, ${}^I\boldsymbol{m}_n$, is precisely driven to the image center of camera n.

As regards the zoom motion of the binocular cameras, modelled by

$$\dot{f}_n = u_{nf}, \tag{14}$$

one can first determine the desired focal length of camera n, f_n^*, according to the distance between the target and the binocular cameras in 3-D space by

$$f_n^* = f_{no}\frac{\left\|G^{-1}({}^I\boldsymbol{r}^*, t) - \frac{\boldsymbol{c}_1(t)+\boldsymbol{c}_2(t)}{2}\right\|}{d^*}, \tag{15}$$

where f_{no} denotes the initial focal length of camera n and d^* is a pre-determined number specifying the desired distance when observing the target after zooming. Then, the zoom velocity command can be synthesized by the simple proportional control law

$$u_{nf} = -k_f(f_n - f_n^*),\ k_f > 0. \tag{16}$$

4 Active Visual Servoing

An image-based control law is employed for the visual servo system. The encoded error for the control task is defined in the image space. Based on the encoded error, an appropriately designed vision-based controller could drive the encoded error to zero exponentially. Moreover, the encoded error being zero implies that the robot has reached the target with precision.

4.1 Image-Based Task Encoding

The velocity-driven robotic manipulator is assumed to admit a simple kinematic model

$$\dot{\boldsymbol{r}} = \boldsymbol{u}, \tag{17}$$

where $\boldsymbol{u} \in \mathbb{R}^3$ is the control input.

The control task considered in this paper is to precisely position the controlled point of the robotic manipulator at the target even when the target is moving in the robot workspace. Thus, an image-based encoded error [7] is defined as

$$\boldsymbol{e}(t) = {}^I\boldsymbol{r}(t) - {}^I\boldsymbol{r}^*(t). \tag{18}$$

The control problem is to synthesize a causal feedback control law capable of driving the encoded error to zero using only available visual information. Once the encoded error being zero, the task is guaranteed to have been accomplished with precision due to the injectivity of the binocular vision system.

4.2 Controller Design and Analysis

In the light of the fact that the binocular camera model is time-varying due to active visual tracking, one can obtain the differential kinematics of the system by differentiating Eq. (18).

$$\dot{\boldsymbol{e}}(t) = \frac{\partial G(\boldsymbol{r},t)}{\partial \boldsymbol{r}}\dot{\boldsymbol{r}} + \frac{\partial G(\boldsymbol{r},t)}{\partial t} - \frac{\partial G(\boldsymbol{r}^*,t)}{\partial \boldsymbol{r}^*}\dot{\boldsymbol{r}}^* - \frac{\partial G(\boldsymbol{r}^*,t)}{\partial t} \tag{19}$$

$$= \boldsymbol{J}(\boldsymbol{r},t)\boldsymbol{u} + {}^I\dot{\boldsymbol{r}}_c - \boldsymbol{J}(\boldsymbol{r}^*,t)\dot{\boldsymbol{r}}^* - {}^I\dot{\boldsymbol{r}}_c^* \tag{20}$$

Based on the encoded error defined in Eq. (18) and the process model defined in Eq. (20), the following image-based control law with feedforward compensation is proposed.

$$\boldsymbol{u} = -\boldsymbol{J}^+\left(k\boldsymbol{e} + ({}^I\dot{\boldsymbol{r}}_c - {}^I\dot{\boldsymbol{r}}_c^*) - \boldsymbol{J}(\boldsymbol{r}^*,t)\dot{\boldsymbol{r}}^*\right) \tag{21}$$

where k is a positive gain constant and the pseudo-inverse is defined as

$$\boldsymbol{J}^+ \triangleq (\boldsymbol{J}'\boldsymbol{J})^{-1}\boldsymbol{J}'. \tag{22}$$

Theorem 1 (Exponential Stability). *For the system defined by Eqs. (17), (18), (20), and (21), $\boldsymbol{r}(t)$ goes to $\boldsymbol{r}^*(t)$ asymptotically.*

Proof (Proof. of Theorem 1). By virtue of Eqs. (20) and (21), it follows that

$$\dot{\boldsymbol{e}} = -\boldsymbol{J}\boldsymbol{J}^+\left(k\boldsymbol{e} + ({}^I\dot{\boldsymbol{r}}_c - {}^I\dot{\boldsymbol{r}}_c^*) - \boldsymbol{J}(\boldsymbol{r}^*,t)\dot{\boldsymbol{r}}^*\right) + {}^I\dot{\boldsymbol{r}}_c - \boldsymbol{J}(\boldsymbol{r}^*,t)\dot{\boldsymbol{r}}^* - {}^I\dot{\boldsymbol{r}}_c^*. \tag{23}$$

Multiplying both sides of Eq. (23) by $\boldsymbol{J}'$, one can see that

$$\boldsymbol{J}'\dot{\boldsymbol{e}} = -k\boldsymbol{J}'\boldsymbol{e}. \tag{24}$$

Rewriting Eq. (24) as

$$J'(\dot{e} + ke) = 0, \tag{25}$$

It can be further shown that

$$J'e \to 0 \text{ as } t \to \infty. \tag{26}$$

According to the *binocular visual constraint* [7], one thus concludes that

$$e \to 0 \text{ as } t \to \infty. \tag{27}$$

Moreover, due to the injectivity of the binocular camera model, precise control in 3-D space can be assured. That is,

$$r \to r^* \text{ as } t \to \infty. \tag{28}$$

However, one can never expect cameras to be exactly calibrated. Therefore, $J(r, t)$ and $J(r^*, t)$ need to be replaced with their estimates,

$$\hat{J}(r, t) \triangleq \frac{\partial \hat{G}(r, t)}{\partial r}\Big|_{r=\hat{G}^{-1}({}^I r)} \tag{29}$$

and

$$\hat{J}(r^*, t) \triangleq \frac{\partial \hat{G}(r^*, t)}{\partial r^*}\Big|_{r^*=\hat{G}^{-1}({}^I r^*)} \tag{30}$$

respectively. Hence, the following vision-based feedback control law is employed when binocular cameras are approximately calibrated.

$$u = -\hat{J}^+ \left(ke + ({}^I\dot{r}_c - {}^I\dot{r}_c^*) - \hat{J}(r^*, t)\dot{\hat{r}}^*\right) \tag{31}$$

Compared with existing visual servo control approach, the proposed vision-based control law is able to provide better performance by augmenting the feed-forward terms in Eq. (31).

5 Simulations

Computer simulations have been performed under Simulink, developed by The MathWorks, with customized S-function blocks written in C in addition to build-in blocks. In millimetres, the positions of the two cameras with respect to world frame are $c_1 = (300, 400, 300)$ and $c_2 = (700, 400, 300)$ respectively. Both cameras' focal lengths, adjustable from 3.1 mm to 31 mm, are initially set as 15.5 mm. d^* in Eq. (15) is pre-determined as 700 mm. The initial position of the robot with respect to the world frame is $r = (100, 1400, 800)$. The control problem is to drive the robot, denoted as 'o', to a possibly moving target, denoted as '+'. In addition, the estimated camera model used in the proposed visual servo controller is only approximately calibrated. In simulation trials, we have moved the cameras up to 200 mm and rotated them approximately ±10 degrees from their nominal positions with no apparent loss of stability or accuracy.

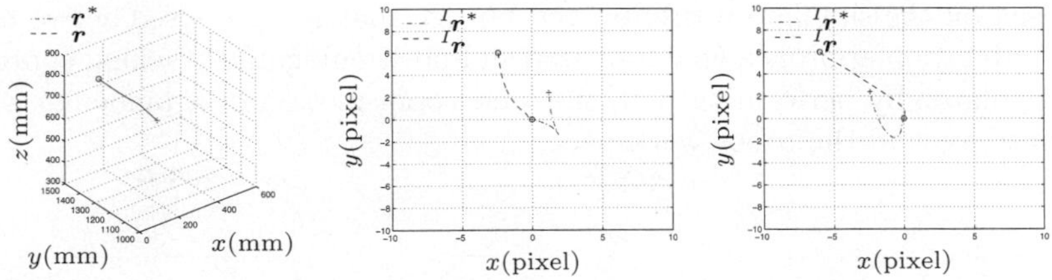

Fig. 3. (*Positioning task*) The trajectory of the robot in 3-D space: $\boldsymbol{r}$ (*left*) and in binocular image space: ${}^I\boldsymbol{r}$ (*right*).

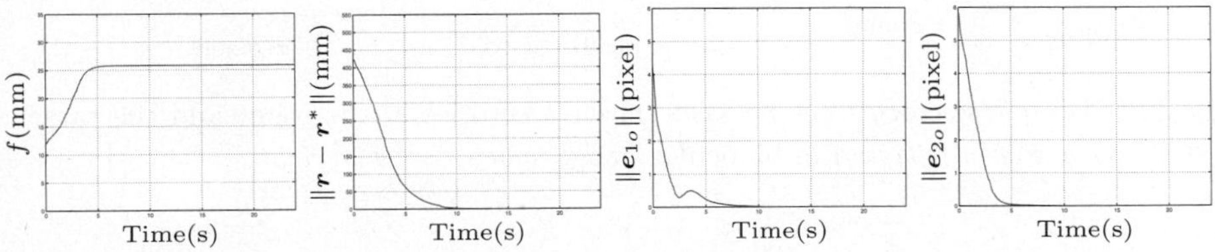

Fig. 4. (*Positioning task*) Focal length of the binocular cameras, positioning error of the robot in 3-D space: $\|\boldsymbol{r} - \boldsymbol{r}^*\|$, and visual tracking errors in binocular image space: $\|\boldsymbol{e}_{no}\|, n = 1, 2$ (*left to right*).

Three distinct control tasks, positioning, planar trajectory tracking, and 3D trajectory tracking, are simulated to verify performance of the proposed system. Note that in all simulations the midpoint between the robot and the target is selected as the tracked point by setting $\alpha = 0.5$ in Eq. (9).

Positioning Task: The robot is driven to a stationary target located at $\boldsymbol{r}^* = (400, 1400, 500)$ in the world frame as shown in Fig. 3. Meanwhile, the active visual tracking system maintains the tracked point at the image centers. Time evolution of the focal length, positioning error of the robot in 3-D space, and visual tracking errors in binocular image space are shown in Fig. 4.

Planar Trajectory Tracking Task: The results are shown in Figs. 5 and 6. It attempts to control the robot to track a moving target. The moving target trajectory, a red contour shown in Fig. 5, is a planar contour with its model listed as follows.

$$\begin{cases} x = 300\sin(0.04t) + 500 \\ y = 400\sin(0.04t + \frac{\pi}{2}) + 1200 \\ z = 150\sin(0.04t) + 500 \end{cases} \tag{32}$$

3-D Trajectory Tracking Task: The tracking results are shown in Figs. 7 and 8 for a moving target modelled by

$$\begin{cases} x = 300\sin(0.04t) + 500 \\ y = 400\sin(0.04t + \frac{\pi}{2}) + 1200 \ . \\ z = -t + 700 \end{cases} \tag{33}$$

Based on the simulation results, one can see that the proposed active hand-eye coordination approach enables successful positioning and tracking control of a manipulator by performing pan, tilt, and zoom motion of a binocular vision system to expand the robot workspace.

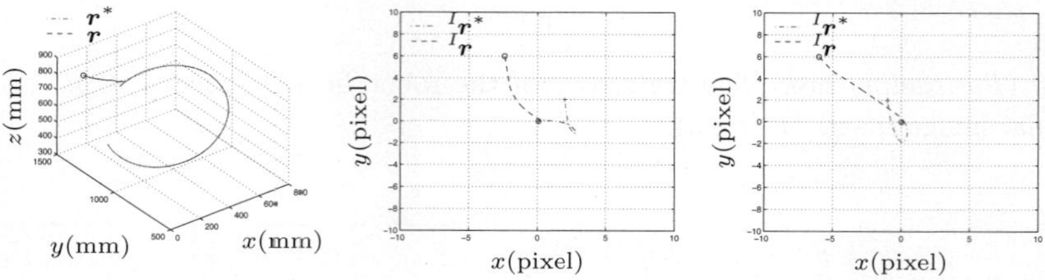

Fig. 5. (*Planar trajectory tracking task*) The trajectories of the robot and the target in 3-D space: $\boldsymbol{r}, \boldsymbol{r}^*$ (*left*) and in binocular image space: $^I\boldsymbol{r}, ^I\boldsymbol{r}^*$ (*right*).

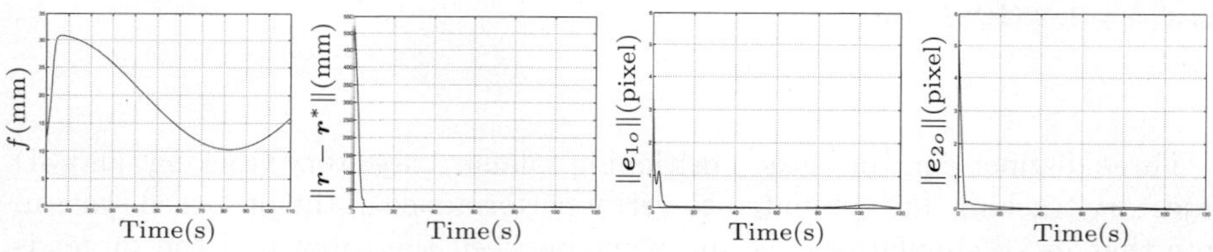

Fig. 6. (*Planar trajectory tracking task*) Focal length of the binocular cameras, tracking error of the robot in 3-D space: $\|\boldsymbol{r} - \boldsymbol{r}^*\|$, and visual tracking errors in binocular image space: $\|\boldsymbol{e}_{no}\|, n = 1, 2$ (*left to right*).

6 Conclusion

In this paper, a perhaps promising architecture was proposed to expand the field of view of a binocular vision system and enable active hand-eye coordination. It is successfully validated through simulations and convergence analysis. The control law is also robust with respect to camera calibration errors and can asymptotically stabilize not only positioning tasks but also target tracking tasks. This is due to the injectivity of the vision system and the augmented feedforward compensation. In addition, the proposed approach can be further applied to resolve problems when obstacles block the field of view of cameras. Specifically, one can manipulate the mobile robot equipped with PTZ binocular cameras to avoid obstacles and occlusion while controlling the manipulator, either mounted in fixed configuration or on-board, to accomplish visual servo tasks.

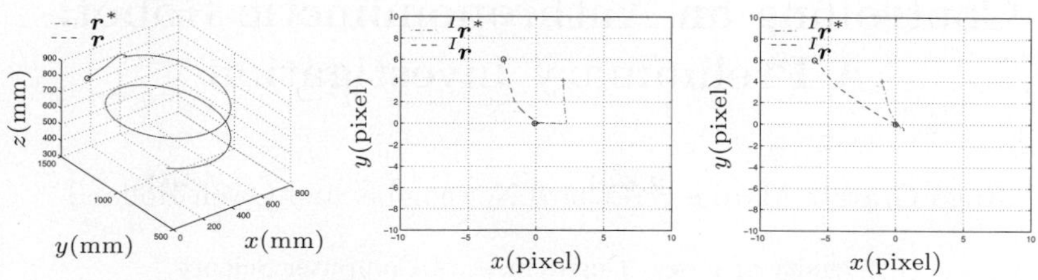

Fig. 7. (*3-D trajectory tracking task*) The trajectories of the robot and the target in 3-D space: $\boldsymbol{r}, \boldsymbol{r}^*$ (*left*) and in binocular image space: ${}^I\boldsymbol{r}, {}^I\boldsymbol{r}^*$ (*right*).

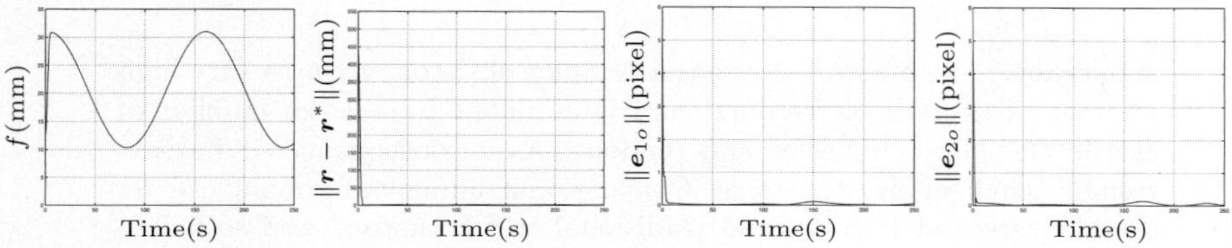

Fig. 8. (*3-D trajectory tracking task*) Focal length of the binocular cameras, tracking error of the robot in 3-D space: $\|\boldsymbol{r} - \boldsymbol{r}^*\|$, and visual tracking errors in binocular image space: $\|\boldsymbol{e}_{no}\|, n = 1, 2$ (*left to right*).

Acknowledgments. This research was supported by the National Science Council of Taiwan, R.O.C. under grants NSC-91-2213-E-027-002 and NSC-95-2221-E-027-049.

References

1. Hutchinson, S.A., Hager, G.D., Corke, P.I.: A tutorial on visual servo control. IEEE Trans. Robot. Automat. 12(5), 651–670 (1996)
2. Chang, W.C.: Vision-Based Control of Uncertain Systems. PhD thesis, Yale University, New Haven, CT (1997)
3. Asada, M., Tanaka, T., Hosoda, K.: Adaptive binocular visual servoing for independently moving target tracking. In: Proc. of the 2000 IEEE International Conference on Robotics and Automation, San Francisco, CA, U.S.A, IEEE Press, Los Alamitos (2000)
4. Chang, W.C.: Hybrid force and vision-based contour following of planar robots. Journal of Intelligent and Robotic Systems 47(3), 215–237 (2006)
5. Chang, W.C.: Binocular vision-based trajectory following for autonomous robotic manipulation. Robotica. Published online by Cambridge University Press (2007), DOI:10.1017/s0263574707003505
6. Chang, W.C., Morse, A.S.: Six degree-of-freedom task encoding in vision-based control systems. In: Proc. of the 14th World Congress of IFAC. Volume B., Beijing, China, pp. 311–316. International Federation of Automatic Control (1999)
7. Chang, W.C.: Precise positioning of binocular eye-to-hand robotic manipulators. Journal of Intelligent and Robotic Systems 49(3), 219–236 (2007)

Controlling an Anthropomimetic Robot: A Preliminary Investigation

Hugo Gravato Marques, Richard Newcombe, and Owen Holland

University of Essex, Department of Computer Science,
Wivenhoe Park. CO4 3SQ Colchester, UK
{hgmarq,ranewc,owen}@essex.ac.uk
http://cswww.essex.ac.uk/

Abstract. The human body has a complex skeleton, giving a very high number of degrees of freedom, and is actuated by a large number of elastic elements – muscles and tendons. As a consequence, it has extremely challenging dynamics. Conventional humanoid robots use reduced degrees of freedom and traditional stiff actuators, and so fail to capture or exploit the important dynamic aspects of the human body. It has proved possible to build robots that mimic the human body – anthropomimetic or 'musculo-skeletal' robots – but the control of such robots will require very different methods from those used in existing humanoid robots. This paper reports the results of a preliminary investigation of the control problems using SIMNOS, a physics-based model of the anthropomimetic robot CRONOS. The transient and steady state effects of load changes on two simple feedforward methods for maintaining arm posture are assessed. The addition of a feedback controller reduces the steady state effects considerably, but still shows oscillatory transient effects. However, by combining this feedback controller with a velocity-limiting feedforward element, it proves possible to make smooth and reasonably accurate changes of posture under conditions of constant load.

Key words: anthropomimetic robots, motor control

1 Introduction

The majority of humanoid robots are precisely engineered mechanisms encased within a human envelope. The similarity between most of these robots and the biological inspiration behind their development resides solely in their external appearance. One can easily recognize the body structure of a common humanoid robot as being close to that of a human – they have a head, torso, two arms and two legs, and all these parts seem to be in just the *right* place – but once we see them in action, walking or performing any other sort of interaction with the environment, it is clear that the dynamics of their movements are quite different from those of their human counterparts. The typical humanoid robot employs high torque electric motors situated at each manoeuvrable joint, and

F. Almeida e Costa et al. (Eds.): ECAL 2007, LNAI 4648, pp. 736–745, 2007.

precise control of each motor using standard engineering methods leads to precise trajectories of the body parts.

The real human body is very different. It contains hundreds rather than tens of articulated elements (bones), and movement and the exertion of force are achieved by the tensioning of elastic muscles connected to the bones by elastic tendons in ways that often actuate two or more joints simultaneously. The successful control of such a system goes well beyond currently available engineering methods. The brain, however, does not merely cope with the complex dynamic phenomena entailed by such an arrangement, but exploits them in a number of ways ranging from reducing energy usage to increasing the accuracy or speed of a movement. If we wish to create humanoid robots that truly mimic our own bodies, we therefore face two tasks: to construct robot bodies like our own; and to find out how to control them. After a brief description of some attempts at the first task, this paper reports on a preliminary approach to the second.

The first task has been addressed in a number of recent developments in humanoid robotics where, for a variety of reasons, attempts have been made to copy the structure of the human body more closely. For example, at the University of Tokyo, Mizuuchi and his collaborators have produced a series of 'musculo-skeletal' robots with up to 140 degrees of freedom (e.g. [1]). Mizuuchi's robots include many parts directly modelled on the human skeleton, and are actuated by series-elastic mechanisms connected to the 'bones' by inelastic tendons. Here at Essex we have developed the anthropomimetic torso CRONOS [2] as part of a project investigating the possibility of building a conscious robot [3]). CRONOS has a lifesize skeleton made of hand-moulded plastic 'bones'; series-elastic actuators are again used in place of muscle, but in this case the implementation involves a length of Dyneema kiteline wound around a rotary motor axis situated at the first anchor site of the muscle, and attached via an elastic shock cord to the second anchor site. By winding in the kiteline, forces are applied to the skeletal structure through the shock cord. This method allows for the generation of forces comparable to those seen in humans, with over 800 Newtons available if required. CRONOS currently has 49 powered degrees of freedom.

The second task, that of finding out how to control robots like CRONOS and Kotaro, is in a much earlier stage of development. There are many sources of difficulty, but the main one is the combination of elastic actuators and multiple redundant degrees of freedom. Such a robot at rest looks innocent enough, but the slightest change in actuation of even one actuator can have effects that spread throughout the whole chassis. Any movement produces both static and dynamic effects; for example, in CRONOS, extending one arm forwards produces changed static loadings in the shoulder and on the spine, along with an inertial reaction to the acceleration and deceleration of the arm. The passive responses to these changes produce more changes – for example, a wave of movement may propagate to the head and to the other arm, and may take some time to die out. Similarly, suddenly loading one outstretched arm with a weight will cause the whole torso to sag and twist, again with some movement propagation to the head and other arm. The position of the hand following the arm extension also varies considerably, typically showing a damped local oscillation superimposed on the larger changes imposed by the movements of the torso as a whole.

The intractability of the real robot presents difficulties when attempting to investigate the control problem, as the amount of experimentation and the type of data required raise issues of time and cost. However, we are fortunate, in that as part of the project we have also developed SIMNOS, a sophisticated physics-based model of CRONOS ([4]). The model was originally intended to serve as an internal self-model of CRONOS, which the robot would use to predict the consequences of its interactions with the world, and which could serve as the basis for an implementation of a self-model theory of consciousness as outlined by Metzinger [5] and others. SIMNOS (see figure 1(b)) provides a highly reconfigurable simulation of CRONOS in a dynamic environment, replicating the large number of degrees of freedom and the general composition of the real robot sufficiently well to serve as a convenient and tractable platform for the development of CRONOS's motor control.

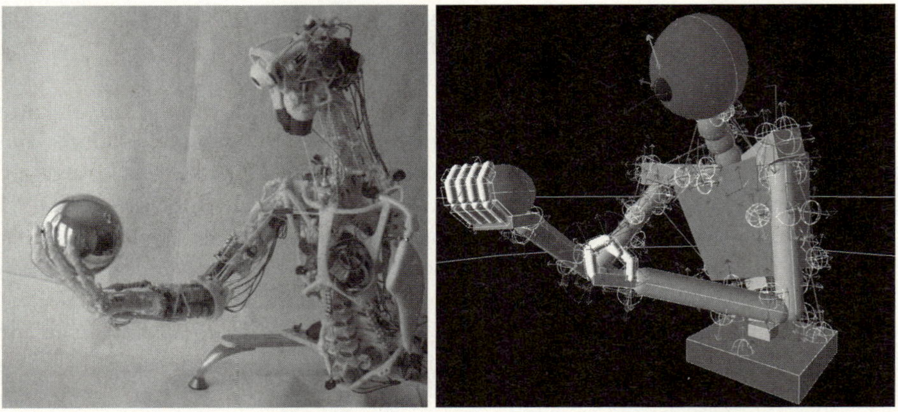

(a) CRONOS. (b) SIMNOS.

Fig. 1.

Current research provides two main theories of human motor control: the equilibrium-point hypothesis (EPH) and the internal dynamics model hypothesis (IDMH) [6]. Under EPH, whenever a control task is to be executed the central nervous system (CNS) plans a set of equilibrium points to be achieved sequentially by the required limbs, each of these containing the kinematic position of the necessary limbs in joint space [7]. According to this hypothesis, once the trajectory is established the muscles lengths are set without further assistance from the CNS. The IDMH on the other hand, claims that the data recorded in several human (and other mammal) motor control experiments does not fit the EPH, and that it is best explained by the use of internal models [8]. Proponents of internal modeling claim that subjects learn the inverse dynamics of the task they are executing and this implies the acquisition of forward models. The subjects will then execute the task in a pseudo-closed-loop fashion, compensating for eventual deviations anticipated by the forward model. We have no prior commitment to either of these models, but we do have a commitment to simplicity, and in this paper we will restrict ourselves to examining some simple strategies for achieving some basic abilities – achieving a posture, maintaining it under changes of load, and moving smoothly from one posture to another – which are trivial for conventional humanoids, but difficult for robots like CRONOS.

2 SIMNOS Body

CRONOS' skeletal body components are modelled in SIMNOS as jointed rigid bodies, with spring-damper systems at each joint in place of the unpowered elastic tendons. SIMNOS' rigid body limbs are fully contactable surfaces which allow interaction of the robot with the environment in a qualitatively similar way to CRONOS. The series-elastic actuators used in CRONOS are represented in the simulation by two types of simulated muscle: linear and rotary. Each linear muscle is a single parallel spring-damper system with asymmetrical conditioning of the spring and damper constants to simulate the fact that the shock cord used in each series-elastic actuator in CRONOS can only produce tension. In CRONOS, rotary motors are used to simplify the construction of the forearms; these are simulated with rotary spring-damper systems where a desired resting length of the spring is used to set the position of the motor. Details of the CRONOS robot are given in [2].

For the experiments described in this paper, the forearm spring-damper muscles, comprising a single rotator and four linear wrist muscles, were set to the maximum spring stiffness, effectively aligning the hand with the forearm; the control signals for those muscles were also unused. The remaining 10 muscles in the arm need to be controlled in order to achieve and maintain a particular hand pose. Since the joint constraints on the limbs lead naturally to curved motions which can be expressed as open curves and lines in spherical space, we express our first controllers in terms of a 3-tuple encoding the desired pose in spherical space $f_{sim}(\phi, \theta, r)$; the controller will return a sequence of commands for each muscle in the arm complex in order to move the hand into the desired pose.

3 Posture Controller Development

In CRONOS, a specific set of distances between muscle anchor points maps onto a single body posture. As control of the muscles in both SIMNOS and CRONOS is available by setting the equivalent of the resting length of a spring, it is possible to produce an approximate first attempt at a controller simply by setting the resting lengths of the springs to the lengths between the muscle anchor points which would be required for a particular posture. The springs will of course deform under static and dynamic loads, but such a controller would achieve greater accuracy as the stiffness of the springs was increased; in the limit it would correspond to the stiff mechanism of a conventional position-controlled robot.

Making the assumption that a uni-modal mapping between the muscle length control commands (ML) required to obtain a desired hand pose is possible, we can obtain an approximation to f_{sim} by enumerating a set of exemplar pairings $\{handPose \mapsto ML\}^N$. By mapping the hand pose in relation to the torso frame of reference (centred on the shoulder joint) a single f_{sim} can be acquired whilst the torso is restricted to a single pose; if required, these can then be used in novel torso postures through a standard coordinate transform. Each pair $\{\phi, \theta, r\}_n \mapsto \{l_1..l_M\}_n$ can be acquired by setting all spring and damper values to 0 and then using external forces to move the hand of SIMNOS to position $\{\phi, \theta, r\}_n$. For each such pose the set of muscle lengths $ML_n = \{l_1...l_M\}_n$ is recorded. Once this has been done, the required function for a given pose can be computed using a nearest neighbour classifier or look-up table. Where the assumption of linear interpolability between the exemplars is plausible, as seems reasonable here, trilinear interpolation will yield a good approximation for poses not represented in the training set.

4 First Experiments

Figures 2 and 3 show the capabilities and weaknesses of such simple controllers. Figure 2 uses a nearest neighbour strategy, setting spring lengths corresponding to the closest exemplar to the required pose, and figure 3 supplements this with trilinear interpolation. In each, the initially unloaded hand is loaded with three progressively increasing masses, which are then suddenly removed. The transients reveal the predominantly oscillatory dynamics, and the steady state errors – greatest in the coordinates corresponding to the projection of gravity – are large, as expected. The steady state errors also show the error due to the mass of the limbs, resulting from the purely kinematic method used for collecting the exemplar data. In order to reduce the steady state errors, it will be worth trying some simple form of feedback control. Since the size of the error is obscured by the transient oscillations, the simplest strategy is to wait for the oscillations to dissipate, and then to initiate a control action to reduce the remaining steady state error. The control action will of course create further oscillations, and so the net effect will be a stepwise progress towards the set point. We can proceed as follows: given the measured error function between the desired and current posture, $\epsilon(t) = P_{desired} - P_t$ we require that the moving average of the absolute value of the first derivative, over K timesteps,

$$E_t = \frac{1}{K} \sum_{i=t}^{t-K} \|\Delta\epsilon(i)\| ,$$

is less than a threshold p. E obtains a zero value only when oscillations in the system have ceased at time t_{end}, at which point the steady state offset of the current pose from that of the desired pose, $\epsilon(t_{end})$ can be measured. When

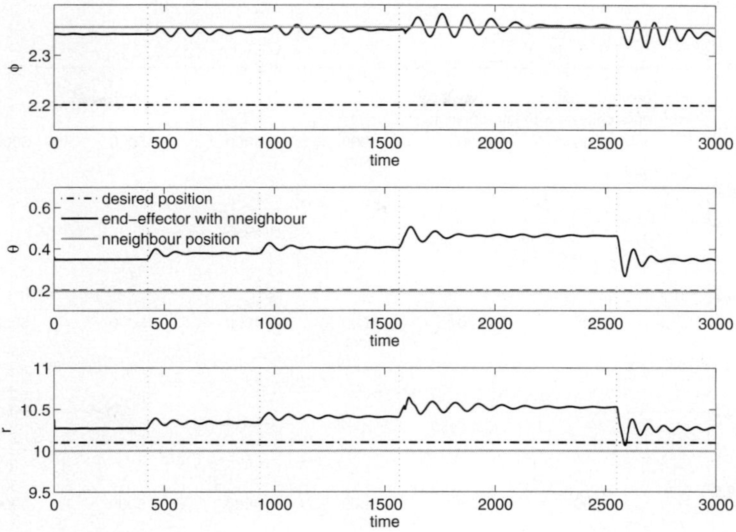

Fig. 2. Nearest neighbour controller.

the above condition is met the feedback controller is enabled. The controller (algorithm 1) is initialised by computing a new way point along the direction of the final error vector translated to the desired pose (2), where α is a small constant factor chosen to enable the pose to change in the direction of the new way point with minimal end oscillations. The new way point command is then executed (4), and when $E_t < p$ is true the new end point error is recalculated. Iterations of (2) and (4) are performed until $\epsilon(t_{end}) < \mu$.

Algorithm 1. Feedback controller

1: **while** $\epsilon(t_{end}) > \mu$ **do**
2: $\quad P'_{wayPoint} = P_{t_{end}} + \alpha\epsilon(t_{end})$
3: $\quad$ **repeat**
4: $\quad\quad$ Execute $f_{sim}(P'_{wayPoint})$
5: $\quad$ **until** $E_t < p$
6: **end while**

Figure 4 shows both the effectiveness and the pathologies of this simple controller. For r and θ, the initial large deviations from the desired pose are reduced in four steps, each taking two complete oscillations to stabilise; the characteristics of ϕ are similar but differ quantitatively. As mass is added, the oscillations

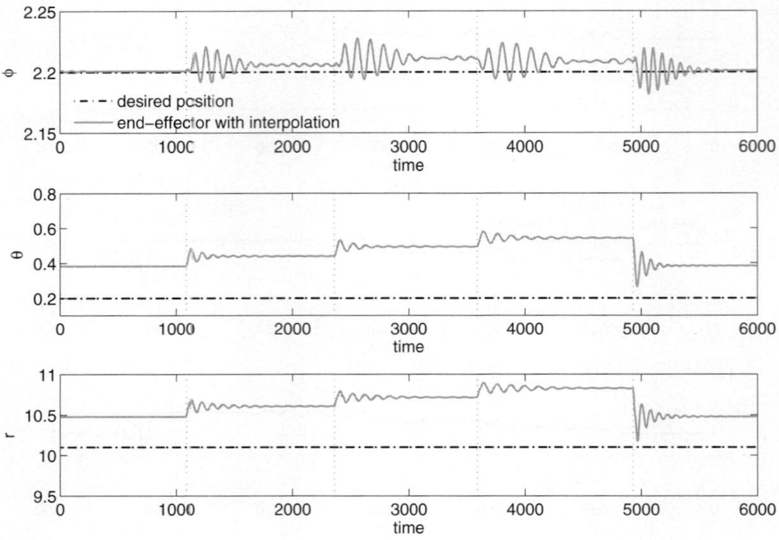

Fig. 3. Nearest neighbour controller with trilinear interpolation.

become more persistent, but the deviations are corrected within a few cycles. The large rebound when the load is removed is also controlled reasonably well.

The final scenario we chose to examine was the movement from one pose to another. In CRONOS, the static and inertial consequences of such simple movements often appeared so complex that successful control seemed unlikely to be achieved through any simple strategy. However, it was also clear that the main culprit was acceleration, and so we decided to examine ways of moving with low accelerations in order to reduce these effects. Within the simple paradigm being examined here, the maximum forces and accelerations are produced when the distance between the current and desired poses is large. To reduce these without requiring sensory feedback during the execution of the pose change, a series of way points between the current and the desired pose are precomputed and successively loaded into the controller in place of the desired end pose alone. This staging reduces the maximum velocity of the system during a change in pose to a level at which the available damping is more effective in lessening oscillation. The precomputation of the trajectory is performed as follows (algorithm 2). The unit vector between the starting and desired hand pose is computed (1), then at each new time interval (the waypoint is updated for each new simulated frame) Δt, the new way point is updated from the previous (3), where η is a scaling factor used to alter the rate of change. (In general η could be a function of time or $P'_{wayPoint} - P_{desired}$, such that large changes are allowed at the start of the trajectory and slowed towards the end. In the experiments performed here η is a constant factor set to 0.004). The computation of each new waypoint $P'_{wayPoint}$

Algorithm 2. Feedforward waypoint controller

1: $\Delta P = (P_{desired} - P_{current}) / \|P_{desired} - P_{current}\|$
2: **repeat**
3: $\quad P'_{wayPoint} = P_{wayPoint} + \eta \Delta P$
4: $\quad$ Execute $f_{sim}(P'_{wayPoint})$
5: **until** $P'_{wayPoint} - P_{desired} < \eta$
6: $P'_{wayPoint} = P_{desired}$

is followed by the execution of the corresponding command set (4). Steps (3) and (4) are repeated until the current way point is within η distance of the target.

In order to enable the feedback controller to operate to reduce the final error, we simply impose the additional condition that the final feedforward way point specified must be within η distance of the specified desired pose vector before the feedback controller is activated. (Any oscillations must also be below a threshold as in the previous use of the feedback controller). Figure 5 shows results for

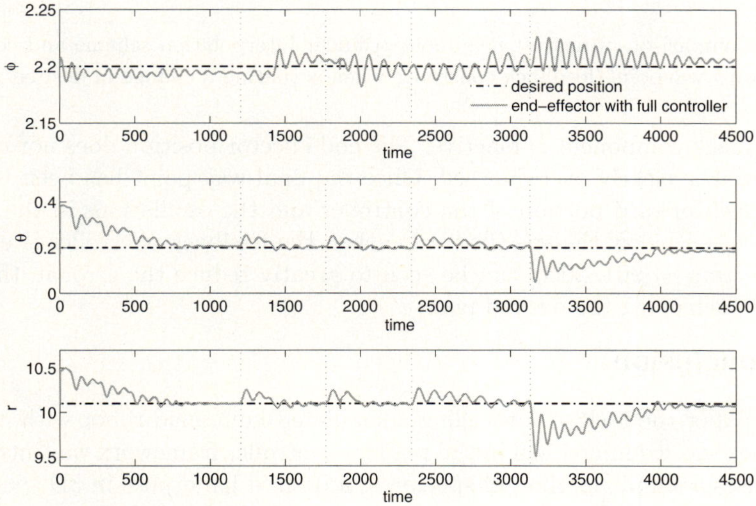

Fig. 4. Nearest neighbour controller with trilinear interpolation and feedback.

both the nearest neighbour trilinear interpolation controller and the combined feedforward waypoint/feedback controller for the task of reaching a desired hand pose, and then moving to a new desired pose. The large oscillations in the pose for the first controller can be seen in all plots of figure 5. Once the oscillations have reduced, a large steady state offset can be seen. From *time* = [210..900] only the feed-forward precomputed way-point based control strategy is used –

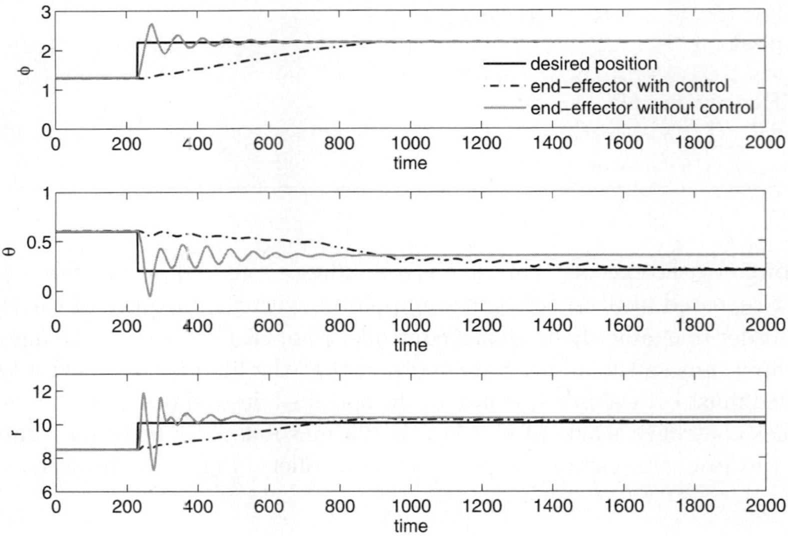

Fig. 5. Comparison of nearest neighbour trilinear interpolation scheme and combined feedforward waypoint/feedback controller when actuating a change in desired posture.

the feedback component is inactive. The end effector position does not oscillate – in fact it is clearly over-damped. Once the final way-point has been updated in the feed-forward portion of the controller and the oscillations in the system have reduced below the specified value then the feedback controller is enabled, and for $time = [901..2000]$ can be seen to greatly reduce the error in the hand position, achieving the desired pose.

5 Conclusion

In this paper the task of controlling a simulated humanoid robot with complex dynamics was examined. An initial posture controller framework was introduced which uses a simple method of specifying a desired hand pose in 3D space. The controller requires no training time other than the collection of the training set which consists of exemplar pairings between the lower level muscle lengths and the 3D hand pose. First results showed that simple strategies of feedforward and feedback control could significantly improve the accuracy and stability of posture maintenance under load, and posture change.

Acknowledgments. The authors wish to thank Renzo De Nardi for help in preparing the LaTeX for this document. Part of this project is funded by the Portuguese FCT (Funcaçâo da Ciência e Tecnologia) PhD fellowship to HGM. The CRONOS project is funded by the Engineering and Physical Sciences Research Council (GR/S47946/01).

References

1. Mizuuchi, I., et al.: Development of musculoskeletal humanoid kotaro. In: Proceedings of the IEEE International Conference on Robotics and Automation, pp. 82–87. IEEE Computer Society Press, Los Alamitos (2006)
2. Holland, O., Knight, R.: The anthropomimetic principle. In: Burn, J., Wilson, M. (eds.) Proceedings of the AISB06 Symposium on Biologically Inspired Robotics (2006)
3. Holland, O.: A strongly embodied approach to machine consciousness. In: Chrisley, R., Clowes, R., Torrance, S. (eds.) Journal of Consciousness Studies Special Issue on Machine Consciousness (2007)
4. Gamez, D., Newcombe, R., Holland, O., Knight, R.: Two simulation tools for biologically inspired virtual robotics. In: Proceedings of the IEEE 5th Chapter Conference on Advances in Cybernetic System, Sheffield, pp. 85–90. IEEE Computer Society Press, Los Alamitos (2006)
5. Metzinger, T.: Being No One: The Self-Model Theory of Subjectivity. MIT Press, Cambridge (2003)
6. Hinder, M.R., Milner, T.E.: The case for an internal dynamics model versus equilibrium point control in human movement. The Journal of Physiology 549, 953–963 (2003)
7. Gu, X., Ballard, D.: An equilibrium point based model unifying movement control in humanoids. Robotics: Science and Systems (2006)
8. Kawato, M.: Internal models for control and trajectory planning. Current Opinion in Neurobiology 9, 718–727 (1999)

Evolution of Neural Networks for Active Control of Tethered Airfoils

Allister Furey and Inman Harvey

Centre for Computational Neuroscience and Robotics
University of Sussex, Brighton, United Kingdom
a.d.j.furey@sussex.ac.uk, inmanh@sussex.ac.uk

Abstract. Recent development in tethered airfoil i.e. kite technology allows the possibility of exploitation of wind energy at higher altitudes than achievable with traditional wind turbines, with greater efficiency and reduced costs. This study describes the use of evolutionary robotics techniques to build neurocontrollers that maximize energy recoverable from wind by kite control systems in simulation. From initially randomized starting conditions, neurocontrollers rapidly develop under evolutionary pressure to fly the kite in figure eight trajectories that have previously been shown to be an optimal path for power generation. Advantages of this approach are discussed and data is presented which demonstrates the robustness of trajectory control to environmental perturbation.

1 Introduction

Recent advances in materials technology and kite design has facilitated the development of large scale electricity and propulsion systems that use kites to collect energy from the wind [6,7]. Here we present a simple aerodynamic simulation of a steerable four-line kite with which we use an evolutionary robotics (ER) approach in order to maximise aerodynamic forces acting along the same vector as the lines. Initially naïve neural networks are evolved using a microbial genetic algorithm [3] through selection and mutation of the controllers that produce the greatest aerodynamic forces over a given test period. Resulting controllers should steer the kite in consistent figure eight trajectories, which prior work has demonstrated are an optimal path for maximising energy recoverable from the wind [4]. The controllers should also be robust, being able to maintain stable trajectories even with significant changes in the wind velocity. As well as being a useful real world application of ER techniques, these technologies are interesting from an A-life perspective, potentially constituting self-sustaining systems through collection of the energy necessary for their operation from their environment.

The brief then for this initial study is to apply ‘off the shelf’ ER techniques to the problem of kite control in simulation, in the first instance to see if stable figure eight trajectories are evolved, and to inform further investigation.

2 Background

Kite energy and kite propulsion systems [5,7] currently enjoy a window of opportunity. This is largely because kite development for other applications now potentially allows the reliable production of large scale, low cost renewable energy.

F. Almeida e Costa et al. (Eds.): ECAL 2007, LNAI 4648, pp. 746 – 755, 2007.

The common element between current proposed traction kites applications is that the aerodynamic forces developed by the kite are transferred via the lines to perform work at near-ground level. This could be either the direct acceleration of large masses such as cargo ships [7], or the turning of a dynamo as the taut lines slowly spool out from a reel [5,6]. In this study, we focus on simply maximising the aerodynamic forces generated at the kite or maximising the component of the aerodynamic force that is in line with the flying lines. This aim is most suited to power generation applications where the direction of force is less important, as long as line tension is maintained. Marine propulsion applications will require the force in a given direction to be maximised which is to be addressed in future work.

As will be described below, the aerodynamic forces that generate line tension are proportional to the square of the apparent wind velocity at the kite. Apparent or effective wind can be defined as a vector sum of the wind speed as experienced by a stationary observer, and the negative of the movement vector of the airfoil. As the aerodynamic forces relate to the *square* of this apparent wind velocity, it is advantageous to augment the apparent wind by directing the kite to perform manoeuvres perpendicular to the wind [1]. Recent work has demonstrated that closed loop flight trajectories in the form of a figure eight are an optimal solution in terms of power generation, with the additional advantage over simple circular trajectories of not requiring a swivel mechanism to prevent crossovers accumulating in the lines [4].
If the lines are being reeled out from a spool coupled to a dynamo, the lines will need occasional retraction [4,5], at which point the line forces will need to be minimised without stalling the kite by reducing the kite speed and placing the kite perpendicular to the wind. Here, we focus purely on the power generation phase and aim to maximise forces acting through the lines.

Real world kite systems are subject to a high degree of environmental variability; wind speed, precipitation and icing will all affect the kite's performance. The real world will also present such non-linearities that current simulations are unable to render in detail without prohibitive computational costs, primarily turbulence. This issue is particularly relevant if flight is to be continued during squall or storm conditions and will become increasingly important if multiple kites are being flown in the same airspace as envisaged in recently designed kite systems [6].

Kites therefore need to be actively flown through the airspace, ideally in stereotypic trajectories, in order to maximise energy recovered from the wind. Kite power and propulsion systems also require both generation and retraction phases, upper and lower hard limits of viable operation and face a high degree of both nonlinearity and environmental variability.

Biological organisms deal with such challenges in a way that forms an effective compromise between absolute optimal solutions and solutions that are highly robust, resulting in high performance over a wide range of conditions. Additionally, evolution can exploit aspects of the environment to improve the performance of the agent. For example, Bluefin Tuna and some flying insects actively or passively create vortices in their respective mediums to boost their performance above that anticipated by conventional fluid dynamics [12, 13]. It is this fit between the challenges of the kite control problem and the strengths of evolutionary problem solving that suggest that an ER approach is potentially fruitful. There are precedents for successful ER control of flight in both simulation [2] and in hardware [8].

3 Simulation Implementation

The kite itself is simulated as a curved airfoil, which viewed from the front forms a semi-circular arc (Fig.2). The kite is tethered by 4 lines and controlled from the ground by adjusting the relative lengths of the rear 2 lines. The kite is allowed to flex, without the fabric stretching and following the common Leading Edge Inflatable (LEI) kite configuration, in which the leading edge is an inflatable baton, the leading edge of the kite that normally faces into the wind is more rigid than the trailing edge opposite. Line tension and angle data is fed to the neurocontroller, which feeds back line length actuation to the kite model as per Fig.1.

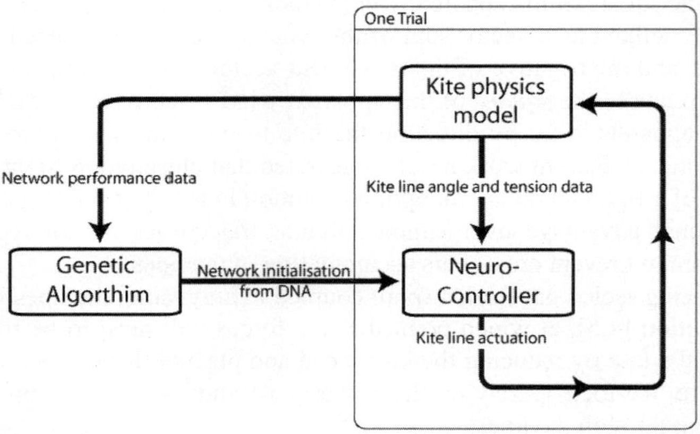

Fig. 1. A simplified schematic of the system in which neurocontrollers are evolved.

3.1 The Kite Physics Simulation

In contrast to other studies in which the kite is treated as single entity [1,4,5], the method of choice for this work was to use a particle based simulation. The motivation for this decision was to provide a framework in which allows explicit consideration of variation in the kite configuration in terms of kite shape, bridle setup, and physical properties of the kite such as relative rigidity and mass of kite components. Specific anticipated defects can then be introduced to the system and the adequacy of the controllers' reaction assessed. The kite is initialised as repeated rows of equidistant particles in a semicircular arc as shown in Fig 2, which illustrates the default setup of two rows of 5 particles.

For simulation purposes we approximate continuous real time by small timesteps at each of which a discrete update is made. The aerodynamic forces are calculated for each slice of the kite, as demarcated by the zigzag lines running from the leading to trailing edge in Figure 2. By slicing the kite up in this way, the force on a section of canopy depends on its particular angle of attack and the apparent wind velocity to which it is subjected. The forces are assumed to be distributable equally amongst the constituent particles, in this case pairs, of each slice although as described below this is a simplifying assumption that will be revised in future work. The acceleration on each particle is simply determined by Newton's second law of motion (Eq.1):

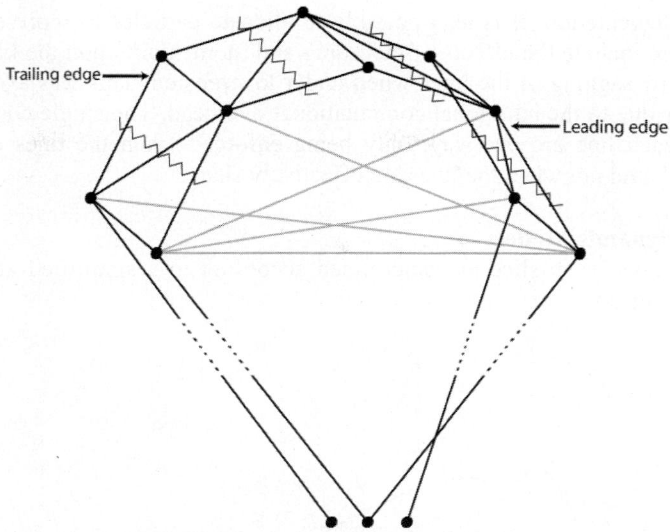

Fig.2. The initial configuration of particles (circles) and constraints (straight lines). The light grey constraints reinforce the arc shape of the kite and prevent 'jellyfish' type flapping motion, effectively performing the same role as the inflatable ribs that maintain the shape of LEI kites. The three lowest particles are the tether points. Zigzag lines indicate the positions at which the canopy is sliced for aerodynamic calculations.

$$a = \frac{f}{m} \quad (1)$$

In addition to the acceleration due to aerodynamic forces, each particle is accelerated -9.81m/s^2 due to gravity. Integration is performed according to the velocity free Verlet method [10] as per Eq.2; this method is used due to its relative stability and speed of execution. x here simply represents a particle position, the time step Δt is kept relatively small at 0.004 to avoid numerical instability.

$$x^{t+1} = 2x - x^{t-1} + a\Delta t^2 \quad (2)$$

The very diversity in forces across the canopy that is being encouraged by the slice system will quickly cause the particles to scatter, it is therefore necessary to constrain the particles to maintain the coherence of the kites structure. The constraints linking the particles are treated as infinitely stiff springs, and their positions are iterated according to the Gauss-Seidel iteration method [11]. The system in essence will simply move particles along the vector that links them in order to satisfy the constraint, i.e. two particles 2 units apart but with a constraint distance of 4 will each be moved 1 unit away from their original positions along the vector between them. By iterating around the set of constraints a number of times the system can be forced to maintain its exact initial configuration. Here the iteration number is set to one, which allows the trailing edge of the canopy that is less constrained to flex more easily than the front, which will itself flex slightly at the upper range of aerodynamic forces. The system can be set to respect the relative masses of the particles and this is the default in this implementation. It is also possible to allocate particles to represent the lines and therefore include the effects of their drag and momentum upon the kite and additionally allow sagging of the lines when under low tension. This was avoided in this initial study due to the additional computational overhead. The single constraints that constitute each line are one-way, only being enforced when the lines exceed their initial length and not when the lines are effectively slack.

3.1.1 Aerodynamic Model

The forces upon each slice are determined according to a simplified aerodynamics model (see Fig. 3).

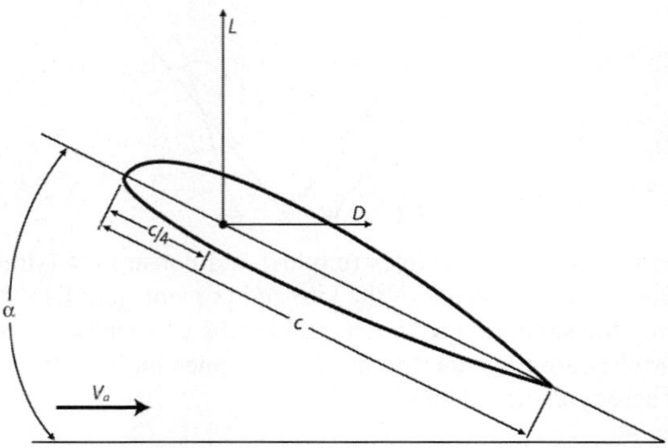

Fig.3. A diagram of the principal aerodynamic forces upon a 2-d airfoil section. The term α denotes the angle of attack, the angle at which the airfoil is inclined relative to the apparent wind, c the airfoil chord, V_a the apparent wind velocity and L and D the Lift and Drag forces respectively. Adapted from [9].

This model is simplified in that no moment coefficient is used and all forces are applied equally amongst the slice's constituent particles. The lift value for each slice is derived through Eqs.3 and 4, the lift force always acts in a direction perpendicular to the apparent wind vector. In the model, the direction of the lift force vector F_L is given by the cross product of the apparent wind vector and the vector that describes the slice's leading edge, a and e respectively in Eq.4. The drag force F_D is always in line with the apparent wind vector; therefore a unit vector of the apparent wind directs the drag force upon the particles as per Eq.5. A is the slice's area, V_a the apparent wind velocity, d the air density, C_L and C_D are the lift and drag coefficients at the current angle of attack.

$$F_L = \vec{L}\frac{1}{2}C_L(\alpha)dV_a^2 A \qquad \vec{L} = \frac{a}{\|a\|}\times e \tag{3, 4}$$

$$F_D = \frac{a}{\|a\|}\frac{1}{2}C_D(\alpha)dV_a^2 A \tag{4}$$

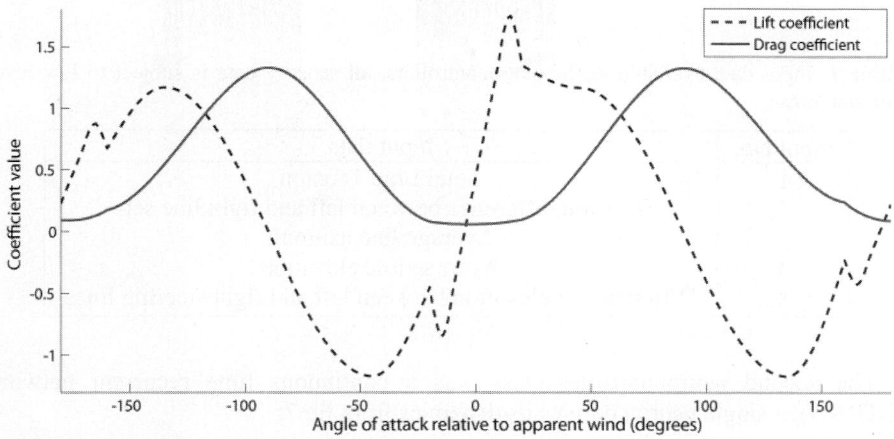

Fig. 4. Graph describing drag and lift coefficient changes over change in angle of attack relative to apparent wind

The lift and drag coefficients are read from a lookup table according to the slice angle of attack, values are plotted over all angles in Fig.4. Values were generated using the X-plane® Airfoil Maker version 860 software, using typical traction kite characteristics of moderate camber and thickness and relatively high drag and lift.

In this preliminary investigation, angle of attack cannot be actively changed in order to best correspond to our prototype hardware system, although this can be achieved by changing the relative lengths of the front and back lines. A simple arbitrary gust generation model is used that generates deviations around a base windspeed of 8m/s. Each timestep has a small probability that a gust or lull is initiated, if so its maximum deviation, onset and decay speeds are set. At each subsequent timestep the windspeed is altered by a small portion of the difference between the current windspeed and the predetermined maximum gust/lull value. No lateral deviation to the wind vector is implemented, although this and the use of real recorded wind records would be logical extensions.

3.2 Neurocontrollers

Two classes of neurocontroller were assessed, both are small recurrent neural networks of 5 input neurons and 7 fully connected interneurons, both inhibitory and excitatory connections are permissible. In these experiments only data measurable with line angle and tension sensors at ground level is made available to the network, as described in Table 1. For inputs we make the simplifying assumption that line tension will be proportional to the aerodynamic forces generated at the kite.

The simplest neurocontroller class was a discrete time recurrent network whose nodes' activation value at a given timestep t ($\Delta t = 0.004$) is given by Eq.6:

$$a_j^t = \sigma\left(\sum w_{ij} a_i^{t-1}\right) - \theta_j \quad (6)$$

Table 1. Input data available to the neurocontrollers, all sensory data is subject to low level gaussian noise.

Input No.	Input data
1	Total Line Tension
2	Tension difference between left and right line sets
3	Average line azimuth
4	Average line elevation
5	Difference in elevation between left and right steering lines

The second neurocontroller class was a continuous time recurrent network (CTRNN), a single neuron derives its dynamics from Eq.7:

$$\tau_j a_j = -a_j + \sum w_{ij} \sigma(o_i - \theta_i) \quad (7)$$

In both cases θ is the bias term, *w* the weight and σ is the sigmoid function which scales values to a range between 0 and 1. In the CTRRN equation, τ_j is the neurons time constant, and *o* the old activation value from the previous timestep. CTRNN neurons are integrated using Euler integration at the same timestep as the physics simulation. One of the neurons is chosen to be the output neuron, and 0.5 subtracted from its sigmoided output. The motor position is modified by 1% of the difference between the network output and its current position each millisecond. This process has three consequences; the motor output is shielded from the majority of noise in the network, the motors are prevented from moving at speeds that are unrealistically fast and finally the extremes of motor output that correspond to a 1m difference in line lengths, usually destabilising for the kite, are only rarely achieved.

3.3 Genetic Algorithm

The genetic algorithm (GA) is a simple tournament based microbial GA [3], with DNA strings composed of real values determining weights both between all neurons and sensory inputs, thresholds and in the CTRNN case, time constants for each neuron in the network. The GA uses the gust generator to generate a wind trace, two individuals are then selected at random from the population and their fitness, either the average of the aerodynamic forces produced by the whole kite, or the component of the aerodynamic force in line with the lines, is determined using the identical wind trace. The kite is always initialised at zenith position, i.e. directly above the tether points. To bias the solutions against brushes with the ground, the test is terminated prematurely as soon as any component of the kite meets ground level, regardless of whether the kite is capable of proceeding in flight or not. The individual with least fitness has its DNA string copied over by the winner with a small mutation applied to every value in the string.

4 Results

The key result derived from this initial investigation is that a simple 7 interneuron network is able to control a simulated kite to fly in figure 8 trajectories in only 200 generations with a population size of 20. With a trial time of 42 seconds, this corresponds to less than 47 hours of real world flight time.

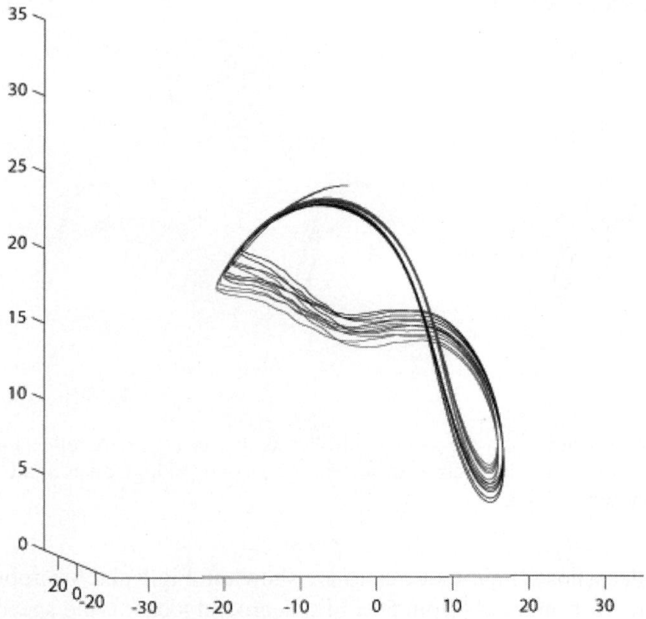

Fig. 5. The leading edge wingtips flight trajectory over a 42 second trial period, leading edge is tethered at [0,0,0]. All axes are in meters.

The discrete-time neural networks found the best solutions in this limited evolution period (Fig.5). The reason for their outperformance of the CTRNN networks is currently unclear, but is potentially due to the additional time required for the selection of satisfactory time constants, or the fact that the CTRNNs were not initialised under centre crossing conditions [14]. The improvement in fitness across generations, with both neuron classes, does not plateau after 200 generations suggesting that additional evolutionary time would result in further improvement in performance. Whether the fitness function was dependent on the total force generated at the airfoil or just the component of that force in line with the lines had no significant effect on the form of the trajectory, although controllers evolved under the latter condition performed significantly more figure repetitions within the trial period.

It is noteworthy that in contrast to other work [4], the evolved trajectories tend to use the whole wind window and not a small arc of less 0.3 radians directly downwind of the controller. The much shorter line length in this model (25m here vs. 100m in [4]) may have contributed to this difference. Additionally, the trajectories evolved often were, as per Fig.5, not centred directly downwind of the tether point. This may have been due to the restriction of actuation speed imposed by the motor output mapping process but is more likely to be an artefact of evolution or the test regime. Early successful controllers from an initial population are more likely to stay in flight if they swoop down to one side, where the kite naturally slows, than swoop down to the centre where the kite continually accelerates and is very sensitive to input. This effect will be addressed in further work, potentially by initialising the kite in a different position at the start of each trial.

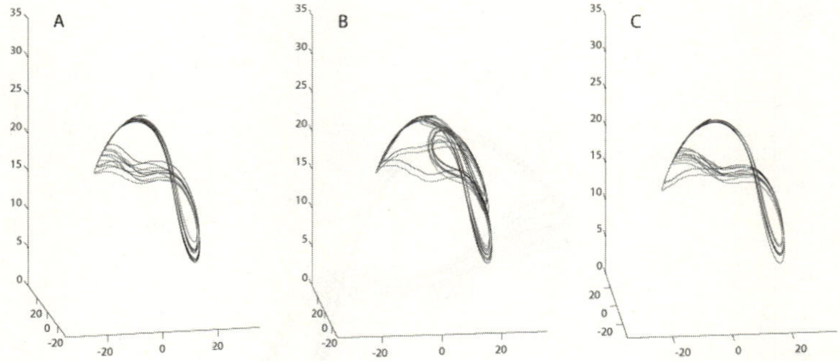

Fig. 6. Flight trajectories using the same controller as that in Fig. 5. A reflects 40 seconds of unchanging wind velocity, B reflects a subsequent 40 seconds of high gusts and C a subsequent return to constant wind speed.

The controller whose flight trajectories are shown in Fig.5 and 6 is robust to all but the most severe gusts and lulls from 5 to 11m/s, around a base wind speed of 8m/s. In smaller or slower wind variation, the evolved trajectory is maintained with small variations. Fig 5 illustrates the initial trajectory (A) deviating to a disturbed state during severe fast onset and offset gusts (B) and subsequently returning to its original trajectory (C) during three 40 second phases of a 120 second trial. As expected, variation in the wind speed during the evolutionary trials was important in rendering the controllers robust to gust and lulls. Only slight gusts and lulls would cause controllers evolved in constant winds to crash.

5 Conclusion

This study has demonstrated that the application of ER techniques to kite control produces controllers that fly the kite in stable figure eight trajectories, shown previously to be an optimal path for recovery of energy from the wind. Evolved neurocontrollers robustly maintain these trajectories during significant deviations of wind speed. These results suggest that it is worth pursuing ER for kite control systems, where real world implementations would have implications both for commercial power generation and for the capability of autonomous artificial agents to source their own power.

Acknowledgements. Thanks must go to J.R. Gloudemans for his advice regarding kite simulation. I am also grateful to the KiteGen team who welcomed my contribution to their project. I must also thank Boris Houska who provided me with a pre-print version of his paper.

References

1. Loyd, M.: Crosswind Kite Power. Journal of Energy 4(3), 106–111 (1980)
2. Shim, Y.S., Kim, C.H.: Evolving Physically Simulated Flying Creatures for Efficient Cruising. Artificial Life 12(4) (2006)
3. Harvey, I.: Artificial Evolution: a Continuing SAGA. In: Gomi, T. (ed.) Evolutionary Robotics. From Intelligent Robotics to Artificial Life. LNCS, vol. 2217, Springer, Heidelberg (2001)
4. Houska, B., Diehl, M.: Optimal Control for Power Generating Kites. In: 2007 European Control Conference (2007)
5. Canale, M., Fagiano, L., Ippolito, M., Milanese, M.: Control of tethered airfoils for a new class of wind energy generator. In: 45th IEEE Conference on Decision and Control proceedings (not yet in publication) (2006)
6. Ippolito, M.: Vertical axis wind turbine with control system steering kites. Patent Publication number EP1672214 (2006)
7. Wrage, S., Mueller, S.: Watercraft comprising a free flying kite-type wind attacked element as a wind powered drive unit. Patent Publication number KR20070007342 (2007)
8. Zufferey, J.-C., Klaptocz, A., Beyeler, A., Nicoud, J.D., Floreano, D.: A 10-gram microflyer for vision based indoor navigation. In: IEEE/RSJ International Conference on Intelligent Robots and Systems (IROS'06), IEEE Press, Los Alamitos (2006)
9. Glauert, H.: The Elements of Airfoil and Airscrew Theory, 2nd edn. Cambridge University Press, Cambridge (1959)
10. Verlet, L.: Computer experiments on classical fluids. I. Thermodynamical properties of Lennard-Jones molecules. Phys. Rev. 159, 98–103 (1967)
11. Press, W.H.: Numerical Recipes. Cambridge University Press, Cambridge (1993)
12. Techet, A.H., Hover, F.S., Triantafyllou, M.S.: Separation and Turbulence Control in Biomimetic Flows, Flow, Turbulence and Combustion 71, 105–118 (2003)
13. Ellington, C.P.: The Novel Aerodynamics of Insect Flight, 1999 J. Exp. Biol. 202, 3439–3448 (1999)
14. Mathayomchan, B., Beer, R.D.: Center-Crossing Recurrent Neural Networks for the Evolution of Rhythmic Behavior. Neural Computation 14(9), 2043–2051 (2002)

Feathered Flyer: Integrating Morphological Computation and Sensory Reflexes into a Physically Simulated Flapping-Wing Robot for Robust Flight Manoeuvre

YoonSik Shim and Phil Husbands

Centre for Computational Neuroscience and Robotics
University of Sussex, BN1 9QH, UK
Y.S.Shim@sussex.ac.uk, P.Husbands@sussex.ac.uk

Abstract. A new biologically inspired approach to a flapping wing controller which benefits from morphological computation and a Reflexive Pattern Generator (RPG) was tested using a simple physically simulated 3D flying robot. In order to tackle the difficulty of generating robust flapping flight and its manoeuvre, the robot employs simplified flexible "feathers" which are modelled as a series of subpanels attached to the wing skeleton using nonlinear angular springs. The neural controller receives sensory inputs from each feather to let them participate in pattern generation, the robot can also "feel" aerodynamic forces on its wings. From the synergy of flexible feathers and their sensory reflexes, the evolved robot exhibited flight manoeuvre using asymmetric wing movements as well as its tail, and rapidly adapted to external disturbances even in the absence of visual sensors. The reduced stiffness in flight control arising from the wing flexibility is discussed.

1 Introduction

Over 330 million years of evolution, nature developed diverse aerial creatures from insects to pterosaurs. Over the history of animal flight, birds are probably the most advanced flying species among higher vertebrates. Unlike insect flights, where the stroke angle and high frequency of wingbeats are nearly constant in time, birds must perform continuous adjustments to their articulated arms during a single stroke due to their comparatively low wingbeat frequencies [1]. Comparatively few studies have focused on the automatic generation of lower-frequency flapping flight found in species other than insects, such as birds or pterosaurs. These works successfully generated flight locomotion using various controllers from hand designed parametric functions to artificial neural networks or hybrids of these [2,3]. However, it turned out that the most challenging feature of an ornithopter is its stable and rapid aerial manoeuvering, and generating such behaviours in artificial systems still remains an open problem. A recent study [4], using staged multiobjective evolution, developed the turning flight of a simulated bird robot using nonlinear oscillators and neural networks,

F. Almeida e Costa et al. (Eds.): ECAL 2007, LNAI 4648, pp. 756–765, 2007.

employing symmetric wingbeats in altitude control and a tail for the steering flight. However, the stabilisation and flight manoeuvre of a real bird is achieved mainly by lateral aerodynamic moments which are produced by the asymmetric movement of the primary wings. At most flight speeds the effect of a bird's tail is minimal, especially in flapping flight. The tail is rarely spread at higher speeds and is held spread and depressed without active change even at low speeds [5].

In order to achieve a successful compromise between the contradictory properties of stability and manoeuverability of flapping flight control, the work described here employs two bio-inspired concepts: morphological computation [6] and mechanosensory reflexes which are embodied as Reflexive Pattern Generators [7]. These features are effectively merged into a simple and tractable robot model using a flexible wing composed of a series of sub-panels which have mechanosensors connected to the control system. To our knowledge, this is the first successful demonstration of the generation of robust lateral flight manoeuvre by asymmetric wingbeats using neural networks developed in an evolutionary framework. The rest of this paper describes the details of the robot model and its neural controller, followed by a demonstration of the result and ending with some discussion of issues arising.

2 The Robot

The robot comprises two single-armed wings which each have three degrees of freedom and a tail with two degrees of freedom. A wing is composed of four feathers which are attached to its skeleton using hinge joints with nonlinear angular springs. The Open Dynamics Engine, an open-source simulator by Russell Smith [8], was used to simulate the articulated rigid body dynamics.

A feathered model - which introduces a degree of 'morphological computation' [6] - has multiple advantages which are now briefly described . First, the flexible feathers acts as an "aerodynamic cushion" in that the feathers reduce the stiffness of motor control. For instance, in completely rigid wings, a small difference in stroke force between two wings will result in a drastic change of net aerodynamic force and moment on the body. Conversely, in a flexible wing, the change will be small due to the passive bending of each feather. In a certain range of motor output, there may exists a small region of identical net lift or net drag and this helps to reduce the control stiffness, thus leading to robustness against external perturbations as well.

Second, in real bird flight, feathers can hold laminar flow of air though a wide range of angle of incidence of a wing arm without stalling [9]. This is also the case in our simplified model, which has a wider range of effective stroke angles of the wings thus producing more lift than that of rigid wings.

Another important advantage of using feathered wings is that a robot can sense the aerodynamic force on each wing by feather bending. This is motivated by the way birds sense airflow using their feathers. In the biological literature, a number of researchers (e.g. [10]) suggest that the airflow directed at feathers on various parts of birds influences their locomotor behaviour, in a similar way

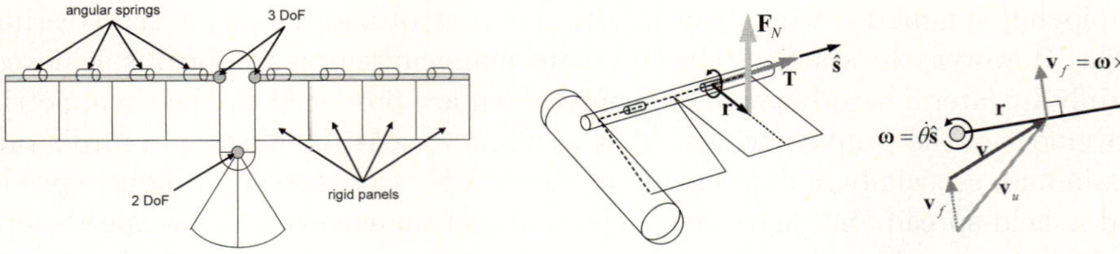

Fig. 1. The robot model and the force and torque on a feather. **r** is feather's longitudinal vector from the wing skeleton to its center, $\mathbf{v}_u$ is the incoming air velocity projected to the surface perpendicular to $\hat{\mathbf{s}}$. The actual airstream $\mathbf{v}$ is obtained by subtracting feather velocity ($\mathbf{v}_f$) from $\mathbf{v}_u$.

to the hairs and antennae of insects. Although a feather itself has no inherent sensing organ, its follicle is surrounded by a variety of mechanoreceptors with various response characteristics induced by its bending and vibration [11]. [12] measured and analysed the signals from feather mechanoreceptors and suggests that the feather sensor mechanism detects momentary stall and the location of flow separation on the wing as well as sensing the speed of airflow from the vibration frequency of secondary wing feathers.

This work utilises feather sensing in a way that is analogous to the use of touch or pressure sensors on legged robots to deal with uneven terrain and external perturbation. In the same way, locomotion in a fluid medium can also exploit pressure sensors for uncertain environmental influences such as turbulence. Supplementary to the gradual descending command from the optic and vestibular systems, in a bird agile sensory reflexes from oscillatory feathers can be effectively entrained to the pattern generation of wingbeats and play a crucial role in active stabilisation.

The total mass of the robot is 4.1 Kg, the wing span is 2.6 m, the aspect ratio is 5.8, and the wing loading is 0.35 g/cm^2 (Fig. 1). Each axis of a wing joint represents its dihedral, sweep, and twist. The joint range of each DoF is $\pm 45°$ for dihedral (tilted upward by $22.5°$ to increase the static stability [13]) and $\pm 22.5°$ for sweep and twist. The tail joint controls tail bend and tail twist each of which has a joint range of $\pm 22.5°$, and the tail spread is fixed at $45°$. Each rectangular feather is subject to aerodynamic forces on its center of mass. Blade element theory with a quasi-steady assumption [14] was used as a simplified aerodynamics. Additional details can be found in [2,3].

A nonlinear angular spring for feather bending is simulated using a first order differential equation so that the bend angle smoothly decays toward the equilibrium position between aerodynamic torque and the spring torque. At each time step, the bend angle rate of the i th feather receiving aerodynamic torque T ($=|\mathbf{T}|$ in Fig. 1) is described by:

$$\dot{\theta}_i(t) = P(T_i(t-1) - k\theta_i(t)) \tag{1}$$

where P is a proportional factor and k is the spring constant. We set $P = 100$ and $k = 0.1$. The bending torque is calculated from the net aerodynamic force

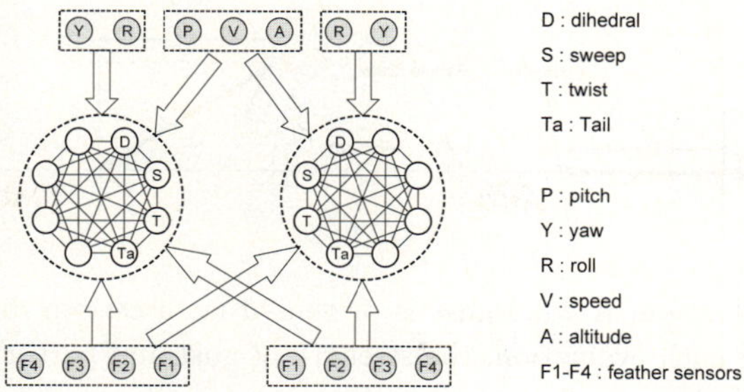

Fig. 2. The neural network architecture. Each lettered circle represents motor neurons and sensors. The neurons without letters are interneurons. Large arrows between subgroups enclosed by dashed lines indicate that every neuron in the target group receives connection from all sensors in the corresponding subgroups.

exerted on the center of mass of the feather. The new bend angle and its angular speed from previous bending torque are used again as the parts of a parameter set for calculating new aerodynamic forces and bending torque in a circular manner. This simplified heuristic model effectively simulates a completely damped angular spring for a feather of negligible mass without further oscillation.

3 The Controller

A bilaterally symmetric continuous-time recurrent neural network (CTRNN) [15] was used for a controller with a hyperbolic tangent function as an activation function. The state of a single neuron is computed by the following equation:

$$\tau_i \frac{dy_i}{dt} = -y_i + \sum_{j=1}^{N} \{w_{ji}\sigma(y_j + b_j)\} + I_i, \quad \sigma(x) = \tanh(x) \tag{2}$$

where y_i is the state of i th neuron, τ_i is a time constant, w_{ji} is an incoming connection from the neuron j, b_j is the bias, and I_i is the external input. Connection weights and biases were set in the range [-4, 4] and time constants were set in the range [0.01, 1.0]. The sensor biases were from the range [-1, 1] and their time constants were fixed to 0.01. By using tanh as an activation function the symmetric bias range is distributed naturally around the center-crossing point [16] without further processing. Every parameter was encoded as a real value in the range [-1, 1]. Time constants were actually mapped into [-2, 0] linearly then rescaled by 10^x. The circuit was integrated using the forward Euler method with a step size of 0.01. The output signals from motorneurons are fed to the simulated servomotors as desired angular positions. The average signal of the two neurons at the left and right side ((R+L)/2) was used for tail bend, and the difference between the outputs ((R-L)/2) was used for tail twist. Fig. 2 shows the network architecture used in this work. Identical fully connected 8-node neural networks were used for each side of the robot. Every neuron receives inputs

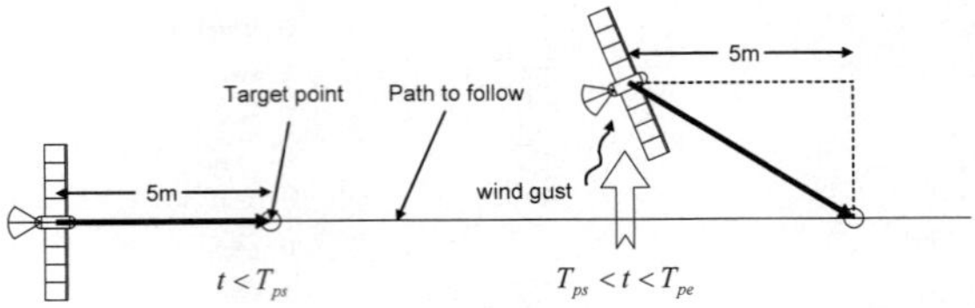

Fig. 3. Fitness evaluation. A wind gust is presented for a certain duration after successful take off. For each evaluation, the strength of gust and perturbation time (T_{ps}, T_{pe}) varies randomly in a small range (see the text).

from all sensors. Where the two local axes of fuselage are given at each time step (side: **s**, forward: **f**), the pitch angle was calculated by the angle between global Y axis and **f**, the yaw angle was obtained from the azimuth angle between the target vector and **f**, and the roll angle was defined as an angle between **s** and **Y**×**f**. Each attitude sensor value was scaled to [-1,1]. Roll and yaw sensors are connected separately at each side of the circuit in order to give inputs with reversed signs to the left and right. The speed and altitude sensor values were also processed to have the symmetric range. Feather bending rate is used as a sensor value because it is more sensitive to instantaneous aerodynamic forces. Note that there is no interconnection between the 8 node subnetworks to avoid unnecessary antiphase movement of the wings. However, both circuits are still able to communicate with each other indirectly through the physical environment by receiving sensory inputs from the feathers of the opposite wing. Setting the number of neurons to 8 ensures sufficient possibilities for searching two briefly distinct motor circuits which acts as the pattern generating motorneurons and the cascaded neural circuits [17].

4 Evolutionary Search

A geographically distributed genetic algorithm [18] was used, usually with a population of 100 (10×10 grids) individuals. Each boundary of the population grid was connected to its opposite side, forming a torus-shaped manifold in order to give an equal chance of mating to the individuals at the corners of the grid. Genoypes were strings of floats encoding the network connection weights, biases and time constants. The mutation rates of parents were changed adaptively after each mating according to the fitness score of their offspring with a threshold of 70 % of the average score of its parents. The initial mutation rate was 0.5 per genotype and this was slowly altered in steps of 0.001. Crossover was also used; supplementary details are to be found in [2].

4.1 Fitness Evaluation

In contrast to the simulation of walking or swimming, the starting strategy for a flapping flight simulation influences the evolutionary search because of the instability of its environment. For instance, when the simulation starts with the flyer in the air, small differences in initial wing stroke between robots due to the slightly different locations of stable limit sets of neural circuit dynamics may cause a variety of consequent motions from gliding to somersaulting. In order to reduce the initial gap, robots were started by being pushed forward on the ground with no friction, and successful take offs survived to be evaluated. Aerodynamic force calculations were turned on after 1sec, and those individuals whose altitude is lower than 50 cm after 5 sec are discarded.

Over the entire flight duration, a robot should retain its flight path which is defined as a straight line of a certain altitude (10 m in this work). The actual target point is located 5m ahead from the robot and moves along the target path line. During flight, the robot receives perturbations from the wind gust for a few seconds as shown in Fig. 3. Before perturbation, the robot only has to reach the target altitude, but after perturbation, the robot should return to the path line as soon as possible and should maintain its flight path continuously. This simple strategy effectively ensures a selection pressure encapsulating several aspects for evolving sophisticated flight control requiring robustness and flight manoeuvre. Because of the highly nonlinear nature of the robot-environment interaction and the tight couplings between the spatial axes for aerodynamic force generation, the robot is forced to capture a variety of capabilities to survive through this simple perturbation strategy. The fitness function is as follows:

$$\text{fitness} = 100\{pf_1 + (1-p)f_2\} \tag{3}$$

$$f_1 = \frac{1}{1 + \frac{1}{T_{flight}} \sum_t^{T_{flight}} d_t}, \quad f_2 = \frac{T_{flight}}{T_{max}} \frac{A}{A_d} \frac{F}{F_d}, \quad p = 0.5 f_2 \tag{4}$$

where T_{flight} is the flight time, T_{max} is the maximum simulation time, d_t is the distance from target path at step t, A and F are average wingbeat amplitudes and frequencies, and A_d and F_d are the maximum desired amplitude and frequency which are set to 0.6 rad and 2 Hz. These force the existence of oscillatory wingbeats, and if the amplitude and frequency of a robot exceeds these limits, they are clipped to the maximum values. The fractional factor p varies according to f_2 so that the continuous flight with desired wingbeat amplitude and frequency is predominant over target following. The maximum value of $f_2(=1)$ occurs when the robot remains airborne during the whole evaluation period with sufficient wing oscillation, and the distance from target (f_1) can influence the fitness by up to half the maximum. If 0, the distance from the target does not influence the fitness at all and the evolutionary search concentrates on maintaining flight. This adaptive weighting scheme enables compromise between the two conflict objectives of stability (retaining balance) and manoeuverability (breaking balance to steer).

5 Result and Discussion

The total evaluation time was set long enough (60 s)to fully test the flight behaviours, and a wind gust was presented twice (horizontally and vertically) dur-

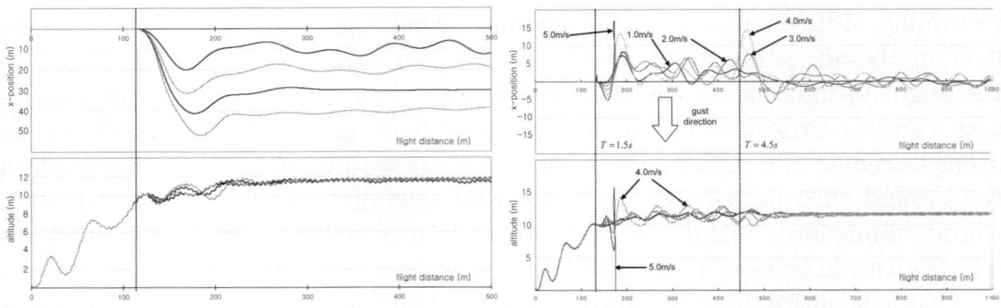

(a) (b)

Fig. 4. (a) Path-following trajectory. The robot steers toward target path and slowly stabilises its trajectory. Steering toward four different path line (x=10,20,30,40 m) is depicted. Because of the bilateral network structure, steering behaviours toward both directions are symmetric. (b) Restabilisation after presence of wind gust. The wind is presented for 3 seconds during flight (between vertical dashed lines) and stabilisation against five wind speeds (1-5 m/s) are shown. The robot actively flies upstream and retains its balance during exposure to the wind gusts.

Table 1. Perturbation strategy. For each evaluation, the strength of gust and perturbation time (T_{ps}, T_{pe}) varies randomly in a small range (see the text).

gust	Start Time (s)	End Time (s)	V_x (m/s)	V_y
1st	$12 \leq T_{ps} \leq 13$	$15 \leq T_{pe} \leq 16$	$\pm 2 \leq V_x \leq \pm 4$	$-0.1 \leq V_y \leq 0.1$
2nd	$30 \leq T_{ps} \leq 31$	$33 \leq T_{pe} \leq 34$	$-0.1 \leq V_x \leq 0.1$	$\pm 1 \leq V_y \leq \pm 3$

ing the flight. The velocity of the wind and the time of its presentation varied in a small range of random values as shown in Table 1.

5.1 Steering and Stabilisation

Evolution was typically run for 50000 to 70000 evaluations, and the fitness converged after around 30000 evaluations. The wingbeat frequencies ranged from 1Hz to 2Hz with flight speed of 9–12 m/s. The low flapping frequencies resulted from multiple causes including the time constant of the feather springs and the inaccuracies of the simulated aerodynamics. However, evolved robots showed robust flapping flight and were able to follow the path line without stalling even in the presence of wind gusts (Fig. 4). A few seconds after the start of a perturbation, the robot was pushed out by the wind then tried to retain its orientation toward some angle between the target and the upstream direction. This upstream flight was induced passively due to the forward-swept wing flapping. Because of the aerodynamic force on a wing due to the incoming air stream which is perpendicular to its leading edge, the outside wing from the wind source generates more lift and thrust, hence the robot turns toward the upstream direction. Successful robots used every degree of freedom to manoeuvre including its tail. The most sensitive response was seen in wing twist, and the smallest

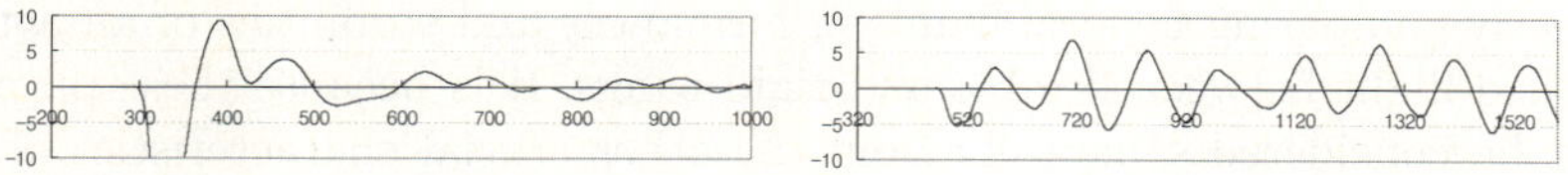

Fig. 5. Stabilisation trajectories of feathered (left) and non-feathered (right) robots after a perturbation. The robots received an angular impact of 3 Kg·m·s.

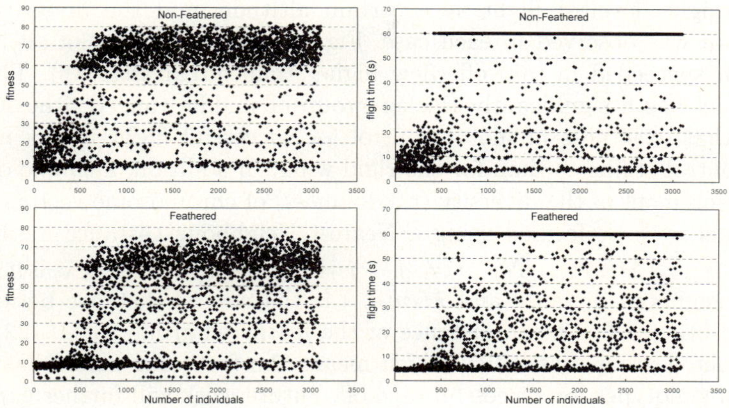

Fig. 6. Every individual whose flight time is longer than 4 s was plotted over 3500 evaluations. Upper graphs are the fitness and the airborne time of non-feathered robots, and the lower graphs are those of feathered robots. We can see the feathered model has more individuals in the middle range, which means that the feathered wing has higher evolutionary potential and more diverse choice in motor control than non-feathered.

change was in wing sweep. Feather sensors participated in the controller circuits and properly coordinated pattern generation. Even in the absence of visual sensors, the robot showed surprising robustness in its locomotion without falling. The robot actively stabilised its roll attitude in order to compensate the force difference between two wings mostly by adjusting its wing twist. If the feather sensor was removed, the controller still generated oscillatory outputs of small uncoordinated amplitude but was unable to fly. Therefore the evolved neural controller can be thought as a pseudo-MPG because it does not have locomotor capacity without sensory feedbacks nor does it show pure reflex chain. Pure RPGs were also evolved in some robots but their performance was lower than pseudo-MPGs. It seems because the shape of feather signals are relatively sharp with small high-frequency peaks, it is likely they cause the jerky wing motions found in certain situations by intensive sensor dependancy.

5.2 Feather vs No Feather: Reduced Stiffness in Motor Control

It is well known that a flexible wing is beneficial in producing a better lift-drag ratio both in fixed and flapping wing flight. However, there are several difficulties in directly comparing general flight performances and stabilities of an actively-controlled flight between flexible and rigid wings. It is because these properties cannot be considered separately from the other crucial characteristics of flight such as the wingbeat kinematics or flight speed [19]. Nonetheless, Fig. 5 shows clear difference in stabilising behaviours between the two species.

In addition, a qualitative observation was made to examine the difference in the evolutionary process in these two cases. The robots were evolved only for a straight, levelled flight at a certain altitude, and the diversity of the population was observed in each case. Fig. 6. shows the plots of all individuals which succeeded in take-off (flew higher than 50 cm after 4 s). The feathered model has a significantly thicker group of middle ranged fitnesses which implies that exactness in motor control for successful flight is less necessary (lower control stiffness) than for the rigid wing. In addition, a simple quantitative measurement of the diversity (randomness) of chromosome vectors of those individuals were performed using a Nearest Neighbour Distance method [20]: $D_{NN} = \sum_{i=1}^{N}(Min(d_{ij})/N) \quad i \neq j, \quad j = 1, ..., N$, where D_{NN} is the mean nearest neighbour distance and $Min(d_{ij})$ is the nearest distance between vector i and others. The mean distance of the feathered group was 0.422 and the non-feathered group was 0.407, which means the feathered model has a wider variety of genotypes for successful take-off. Future work will further explore the significance of these differences.

6 Limitations and Future Work

Although the proposed model shows robust aerial locomotion to a certain extent, it is rather conceptual and requires substantial improvement in order to bridge the reality gap. Firstly, the simplified aerodynamics does not reflect the possible flow separation at the cleft between feathers. These inaccuracies could be reduced by using a single-paneled flexible wing or a series of partially overlapped elastic feathers as on a real bird wing. With the proper redefinition of mechanosensory information which might be changed by the modification of the mechanical model, the evolutionary framework will still be able to find the appropriate controllers for efficient flight. Secondly, additional adaptation processes could be employed for transferring the model to reality. Careful analysis of the effect of reduced control stiffness will enable us to draw the contours of valid regime over which the adaptability of controllers takes effect in the real environment. Also, although the model had less degrees of freedom than those of real birds, we believe this work can be extended effectively to more realistic models. Some movie clips are available at http://www.informatics.sussex.ac.uk/users/yss20/feathered.html.

References

1. Clarke, F., Ekeland, I.: Progress of the REMANTA project on MAV with flapping wings and of the International Universities mini UAV Competition. In: European Micro Air Vehicle Conference and Flight Competition (EMAV), Germany (2006)

2. Shim, Y.S., Kim, C.H.: Evolving Physically Simulated Flying Creatures for Efficient Cruising. Artificial Life 12, 561–591 (2006)
3. Shim, Y.S., Kim, S.J., Kim, C.H.: Evolving Flying Creatures with Path-following Behavior. In: The 9th International Conference on the Simulation and Synthesis of Living Systems (ALIFE IX), Boston, MA, pp. 125–132 (2004)
4. Mouret, J.B., Doncieux, S., Meyer, J.A.: Incremental Evolution of Target-following Neuro-controllers for Flapping-wing Animats. In: Nolfi, S., Baldassarre, G., Calabretta, R., Hallam, J.C.T., Marocco, D., Meyer, J.-A., Miglino, O., Parisi, D. (eds.) SAB 2006. LNCS (LNAI), vol. 4095, pp. 606–618. Springer, Heidelberg (2006)
5. Warrick, D.R., Bundle, M.W., Dial, K.P.: Bird Maneuvering Flight: Blurred Bodies, Clear Heads. Integ. And Comp. Biol. 41, 141–148 (2002)
6. Pfeifer, R., Iida, F.: Morphological Computation: Connecting Body, Brain and Environment. Japanese Scientific Monthly 58, 48–54 (2005)
7. Gallagher, J.C.: Evolution and Analysis of Non-Autonomous Neural Networks for Walking: Reflexive Pattern Generators. In: Congress on Evolutionary Computation, Seoul, Korea (2001)
8. Smith, R.: Intelligent Motion Control with an Artificial Cerebellum. Unpublished doctoral dissertation, University of Auckland, New Zealand (1998), ODE engine available at, `http://ode.org`
9. Burtt Jr., E.H., Ichida, J.M.: Selection for Feather Structure. Acta Zoologica Sinica 52, 131–135 (2006)
10. Gewecke, M., Woike, M.: Breast Feathers as an Air-current Sense Organ for the Control of flight Behaviour in a Songbird. Z. Tierpsychol. 47, 293–298 (1978)
11. Necker, R.: Somatosensory System. In: Physiology and Behavior of the Pigeon, pp. 169–192. Academic Press, London (1983)
12. Brown, R.E., Fedde, M.R.: Airflow Sensors in the Avian Wing. J. Exp. Biol. 179, 13–30 (1993)
13. Thomas, A.L.R., Taylor, G.K.: Animal Flight Dynamics I. Stability in Gliding Flight. J. theor. Biol. 212, 399–424 (2001)
14. Weis-Fogh, T., Jensen, M.: Biology and physics of locust flight I: Basic principles in insect flight. A critical review. Philosophical Transactions of the Royal Society of London B 239, 415–458 (1956)
15. Beer, R.D.: On the Dynamics of Small Continuous-Time Recurrent Neural Networks. Adaptive Behavior 3, 471–511 (1995)
16. Mathayomchan, B., Beer, R.D.: Center-crossing Recurrent Neural Networks for the Evolution of Rhythmic Behavior. Neural Computation 14, 2043–2051 (2002)
17. Haschke, R., Steil, J.J.: Input Space Bifurcation Manifolds of Recurrent Neural Networks. Neurocomputing 64C, 25–38 (2005)
18. Husbands, P.: Distributed Coevolutionary Genetic Algorithms for Multi-criteria and Multi-constraint Optimisation. In: Fogarty, T.C. (ed.) Evolutionary Computing. LNCS, vol. 865, pp. 150–165. Springer, Heidelberg (1994)
19. Taylor, G.K., Thomas, A.L.R.: Animal Flight Dynamics II. Longitudinal Stability in Flapping Flight. J. theor. Biol. 214, 351–370 (2002)
20. Clark, P.J., Evans, F.C.: Distance to Nearest Neighbor as a Measure of Spatial Relationship in Populations. Ecology 34, 445–453 (1954)

Guided Self-organisation for Autonomous Robot Development

Georg Martius[1,2,3], J. Michael Herrmann[1,2,3], and Ralf Der[4]

[1] Bernstein Center for Computational Neuroscience Göttingen,
[2] University of Göttingen, Institute for Nonlinear Dynamics,
[3] Max Plank Institute for Dynamics and Self-Organization,
Bunsenstrasse 10, D-37073 Göttingen, Germany
{georg,michael}@nld.ds.mpg.de
[4] University of Leipzig, Institute of Computer Science,
PF. 920 D-04009 Leipzig, Germany
der@informatik.uni-leipzig.de

Abstract. The paper presents a method to guide the self-organised development of behaviours of autonomous robots. In earlier publications we demonstrated how to use the homeokinesis principle and dynamical systems theory to obtain self-organised playful but goal-free behaviour. Now we extend this framework by reinforcement signals. We validate the mechanisms with two experiment with a spherical robot. The first experiment aims at fast motion, where the robot reaches on average about twice the speed of a not reinforcement robot. In the second experiment spinning motion is rewarded and we demonstrate that the robot successfully develops pirouettes and curved motion which only rarely occur among the natural behaviours of the robot.

Key words: autonomous robots, self-organised behaviour, reinforcement learning, developmental robotics, homeokinesis

1 Introduction

Self-organisation is a key phenomenon in many disciplines ranging from physics over chemistry to the life sciences and economy. It centres on the spontaneous creation of patterns in space, time or space-time in complex systems. The dynamical systems approach to robotics describes robotic behaviour as a spatio-temporal pattern which is formed in the complex interaction of the robot and its environment. Our interest is in developing a systematic approach to the behavioural self-organisation of such systems.

Self-organisation needs a general paradigm which has to be domain invariant. An exemplary paradigm of such generality is homeostasis meant in the early days of cybernetics to be a basis of self-organisation. There are a few attempts to introduce homeostatic mechanisms in robotics, cf. [1,2]. However, while obviously helpful in stabilising systems the principle of homeostasis seems of limited use for the construction of behaviour systems.

F. Almeida e Costa et al. (Eds.): ECAL 2007, LNAI 4648, pp. 766–775, 2007.

One of the authors proposed some time ago *homeokinesis* as a dynamical counterpart to homeostasis, see [3,4]. The idea is that in a behaving system the components like neurons, sensors, motors or muscles have to cooperate their activities in a common kinetic state. As with homeostasis this paradigm is not constructive, because it does not tell how to reach the pertinent state. In particular it gives no answer to the basic question why the robot should do anything at all. One solution is the so called time loop error (TLE) see [5,6] and Sec. 3.2 below. There, the drive for activity has been rooted into the principle itself, and the creation of activity and the adaptation to the environment are combined into one single quantity. The development of the robot is driven by the minimisation of the TLE, which is entirely defined in internal terms of the robot.

Applications with both real [7] and simulated robots have shown many interesting and unexpected behaviours ranging from coiling, hurling and jumping modes in snake like artifacts, over stable rolling modes of spherical robots [8] to dogs climbing over walls and the like, see our video page [9]. What we observe in these experiments are behaviours with high sensorimotor coordination, emerging in a "playful" exploration of the bodily affordances. However, so far all the emerging behaviours are contingent, depending on the concrete body and environmental conditions. Moreover, emerging behaviours are in general transient which may be viewed as the sequential creation and destruction of behavioural primitives.

In the present paper we report a first result of guiding self-organisation into the direction of desired behaviours. In the specific case we consider a spherical robot which earlier has been demonstrated to develop different rolling modes of various velocities and modalities [10]. Our aim now is to tell the robot to move fast or to spin and let self-organisation do the rest. This goal is reached by modulating the TLE with a conveniently defined reward signal as defined below. This simple principle is shown to work in a surprisingly effective way and the presented results may indicate a general approach to influence self-organisation by general reinforcement signals.

There is a close relationship to the attempts of guiding autonomous learning by internal reinforcement signals [11] and to task independent learning [12,13]. The difference is that these approaches work best with discrete state-action spaces of not too large dimensions. Our approach on the other hand was demonstrated to work in real time and continuous space with robots of up to 25 active degrees of freedom [9].

The paper is structured as follows: In the next section we will explain the used robots, then our general principle for self-organisation is formulated. After that a short example for exploration is given followed by the main section about the guiding experiments and finally we close with a discussion and an outlook.

2 Robots

For the experiments we use both a simulated spherical robot called "Sphere" and a cylindrical robot called "Barrel", see Fig. 1. The Sphere was inspired originally by Julius Popp [14]. We constructed the Barrel because it is easier to

Fig. 1. Simulated spherical robot "Sphere" and cylindrical robot "Barrel". *Left:* Sketch of a the Sphere with three internal sliders. The Barrel has only two sliders; *Center:* Picture of the Sphere on the ground; *Right:* Picture of the Barrel on the ground.

analyse and shows clear effects. We used the ODE library (open dynamic engine [15]) embedded in our simulation framework [16] for the computer simulations. The robots are driven by shifting the centre of mass which is realised by shifting internal masses by servo motors, situated on the orthogonal axes (three in the Sphere an two in the Barrel). The motor values are the target positions of each of the masses on its axis, symmetric around the centre ranging to half of the radius. Collisions of these masses are ignored. The servo motors move the masses by applying forces to them, which are calculated by a PID controller. This provides more reliable control of the mass positions and stabilises them against perturbations and centrifugal forces.

The Sphere is equipped with three proprioceptive sensors, which measure the projections of the axes vectors on the z-axis of the world coordinate system, i.e. the z-component of each axis vector. The Barrel only has two such sensors.

Both Sphere and Barrel are physical objects with a complicated mapping of motor to sensor values. In fact, shifting of a mass position will have quite different consequences due to inertia. The task of the controller is to close the sensorimotor loop so that a rolling motion of the robot is achieved. This would be usually done by constructing the controller conveniently. In our case the rolling motion will emerges from our general principle given below.

3 A General Approach to Self-organisation

We will give here a short review of the general homeokinesis approach. Central to our approach is the internal perspective, i.e. everything is based on the stream of the sensor values represented by $x_t \in \mathbb{R}^n$ where $x_t = (s_{t1}, \ldots, s_{tn})$ are the n sensor values at time $t = 0, 1, 2, \ldots$. The controller is given by a function $K : \mathbb{R}^n \to \mathbb{R}^m$ mapping sensor values $x \in \mathbb{R}^n$ to motor values $y \in \mathbb{R}^m$

$$y_t = K(x_t) \ . \tag{1}$$

In the example we have $y_t = (y_1^t, y_2^t, y_3^t)^\top$, y_i^t being the servo target positions of the internal masses on the axes. Our controller is adaptive, i.e. it depends on a set of parameters $C \in \mathbb{R}^c$. In the cases considered here the controller is realised by a one layer neural network defined by the pseudo-linear expression

$$K_i(x) = g(z_i) \qquad g(z) = \tanh(z) \tag{2}$$

$$z_i = \sum_j C_{ij} x_j + h_i \tag{3}$$

again all variables at time t. This seems to be overly trivial concerning the set of behaviours which are observed in the experiments. Please note however, that in our case the behaviours are generated essentially also by an interplay of neuronal and synaptic dynamics (Eq. 11) so that our robots are not simple reactive agents.

3.1 World Model and Sensorimotor Dynamics

The robot has a minimum ability of cognition. This is realised by a world model $F : \mathbb{R}^n \times \mathbb{R}^n \times \mathbb{R}^m \to \mathbb{R}^n$ mapping the actions y and previous sensor values x_t, x_{t-1} of the robot on the new sensor values x_{t+1}, i.e.

$$x_{t+1} = F(x_t, x_{t-1}, y_t) + \xi_t \tag{4}$$

where ξ_t denotes the model error. We make the following ansatz for the world model F,

$$x_{t+1} = Ay_t + S(x_t - x_{t-1}) + b + \xi_t \tag{5}$$

where A is a $n \times m$ matrix, S is a $n \times n$ matrix, b, ξ are column vectors. This model is in contrast to earlier work enhanced by the S-term. The model is trained by gradient descent on the error $E_F = \xi^\top \xi$ as

$$\Delta A_{t+1} = \varepsilon_M \, \xi_t y_t^\top, \quad \Delta S_{t+1} = \varepsilon_M \, \xi_t (x_t - x_{t+1})^\top, \quad \Delta b_{t+1} = \varepsilon_M \, \xi_t \, . \tag{6}$$

where ε_M is the learning rate chosen conveniently. Again, the model seems to be oversimplified. However, model learning is very fast so that the model parameters change rapidly in time and different world situations are modelled by relearning. Moreover, the model only has to represent the coarse response of the world to the actions y of the robot. Behaviour is organised such that this reaction is more or less predictable. Hence, the world model is sufficient to provide a qualitative measure of these response properties.

With these notions we may write the dynamics of the sensorimotor loop in closed form, where ψ denotes the internal model of the sensorimotor loop

$$x_{t+1} = \psi(x_t, x_{t-1}) + \xi_t \tag{7}$$

$$\psi(x_t, x_{t-1}) = AK(x_t) + S(x_t - x_{t-1}) + b \tag{8}$$

using Eq. 1 $y_t = K(x_t)$.

3.2 Realising Self-organisation

As known from physics, self-organisation results from the compromise between a driving force which amplifies fluctuations and a regulating force which tries to stabilise the system. In our paradigm the destabilisation is achieved by increasing the sensitivity of the sensor response induced by the taken actions. Since the controls (motor values) are based on the current sensor values, increasing the sensitivity in this sense means amplifying small changes in sensor values over time which drives the robot towards a chaotic regime.

The counteracting force is obtained from the requirement that the consequences of the taken actions are still predictable. This should keep the robot in "harmony" with the physics of its body and the environment. It has been shown in earlier work that these two objectives can be combined in the time loop error namely finding the input $\hat{x}_t$ which is mapped by ψ to the true new sensor values x_{t+1}, i.e. $\|x_{t+1} - \psi(\hat{x}_t, x_{t-1})\|$ is minimal. We define:

$$E = v^\top v \tag{9}$$

where $v = \hat{x}_t - x_t$. Using Taylor expansion we get from Eq.7

$$\xi_t = L v_t$$

where ξ_t is the model error as introduced above and $L = \partial\psi/\partial x_t$ is the Jacobi matrix of the sensorimotor dynamics. If L^{-1}exists we can write

$$E = \xi^\top Q^{-1} \xi \tag{10}$$

with the positive semidefinite matrix $Q = LL^\top$.

Using gradient descent the parameter dynamics is

$$\Delta c_t = -\varepsilon \frac{\partial E_t}{\partial c_t}, \qquad \Delta h_t = -\varepsilon \frac{\partial E_t}{\partial h_t}. \tag{11}$$

More detail and explicit expressions for the parameter dynamics can be found in previous publications [7,5]. Note that the parameter dynamics Eq. 11 is updated in each time step so that the parameters in practical applications may change on the behavioural time scale if the update rate ε is chosen conveniently. This means that the parameter dynamics is constitutive for the behaviour of the robot.

4 Self-organised Sweeping Through Behaviour Space

Let us consider first the case of the Barrel for the demonstration of the exploratory character of the system. The Barrel is a physical object with strong inertia effects so that it is not possible for instance to drive it with a pattern generator emitting a fixed frequency, where the Barrel will normally execute a rather erratic behaviour. However, if connected to our controller with both the C and A matrix in a "tabula rasa" condition (equal to the unit matrix), the parameter dynamics described above will after a short time excite a rolling mode

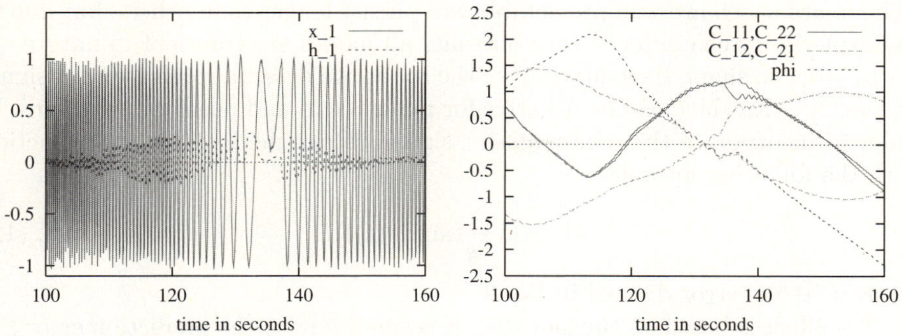

Fig. 2. Dynamics of controller with the Barrel in the time interval 100 to 160 seconds. The region covers the period where the robot actively slows down and then inverts its velocity and then rolls backwards with increasing speed. *Left:* one sensor value $x_1(t)$ and one bias term $h_1(t)$; *Right:* elements of controller matrix C and rotation angle ϕ.

with the velocity systematically increasing up to a maximum value, after this the velocity decreases to zero and increases again with inverted sign.

In Fig. 2 one can see a part of the state and parameter dynamics of the system for one such cycle. Note, that the velocity of the robot can be directly read from the oscillations of the sensor value x_1, high frequency corresponding to high velocities. The direction however depends on the phase relation between x_1 and x_2 (not shown). We can analyse the controller matrix C during the course of time. It is obvious from the right plot of Fig. 2 that despite the unit initialisation C develops into a matrix with scaled $SO(2)$ structure. That means basically that C is a scaled rotation matrix:

$$C = u \begin{pmatrix} \cos(\phi) & -\sin(\phi) \\ \sin(\phi) & \cos(\phi) \end{pmatrix}$$

In the experiment the controller matrix C runs through the entire range of rotation angles ϕ and hence through the accessible velocities of the robot. The described behaviour sweeping repeats more or less periodically. It is important to note that the sweeping effect is a consequence of the interplay between the state dynamics and the learning dynamics of the threshold values h_i see [17] for details.

5 Guiding Self-organisation

In the previous section we showed that the controller will explore the action space and in particular the frequency space. In the case of the Sphere with three dimensional motor and sensor space we observe also frequency sweeping behaviour, however the situation is more complex since the robot can change the axis of rotation and so on and so forth. However, in a normal setup, where the Sphere can move freely, it will exhibit different slow and fast rolling modes.

Behaviours which are well predictable will persist longer than others, but due to the exploratory character of the controller all modes are transient in nature.

In order to shape the behaviour of the robot, we define a reinforcement signal given $r(t) \in \mathbb{R}$, which can be negative for punishment and positive for reward. In order to incorporate the reinforcement signal we can modify the error function with the following formula.

$$E_r = (1 - \tanh(r(t)))E \tag{12}$$

where E is the error defined in Eq. 9.

The effect is based on the fact that E is small if both the prediction error ξ is small (smooth, stable behaviour) and the dynamics is instable (due to L in the denominator, see Eq. 10). The latter effect is what makes the system explorative so that emerging behaviours are transient. The life time of a transient depends also on the strength of ξ so that transient behaviours which can be well modelled have a longer life time. The prefactor in the error function (Eq. 12) regulates the life time of transients as well since it essentially multiplies the learning rate of the parameter dynamics. Behaviours with small or even negative reinforcement are left rapidly, whereas large positive reinforcement tends to increase the life times. The life time of behaviours is maximal if they both are rewarded and can be well modelled. In the following sections we will demonstrate two different nominal behaviours, fast motion and spinning.

5.1 Speed Reinforcement

As one possible goal one could want the robot to move fast. In this case the reinforcement signal is:

$$r(t) = \frac{1}{3}\|\boldsymbol{v}_t\| - 1$$

where $\boldsymbol{v}_t$ is the velocity vector of the robot in world coordinates. In order to avoid saturation of the tanh function in Eq. 12 the reward is scaled and shifted. For the average velocity of the normal runs the reward is about zero. For small velocities the reward is negative and causes a stronger change of behaviour, whereas larger velocities give a positive reward and due to small changes in the behaviour the robot stays longer in this regimes.

We conducted 20 experiments with reinforcement and 20 experiments without reinforcement all with random initial conditions, each 60 minutes in simulated real-time on a flat surface without obstacles. The robot also experiences rolling friction, so that fast rolling really requires constant acceleration. In Fig. 3 the mean velocity for each simulation is plotted and the velocity trace of the robot for two reinforced and two normal runs are plotted. One can see, especially at the overall mean, that the mean velocities for the reinforced runs are significantly larger than the ones of the normal runs. However, since straight and also fast rolling modes are easy predictable they are also exhibited in the normal runs. The traces illustrate that the robot with reinforcement reaches quicker a faster motion behaviour and also stays longer in these behaviours.

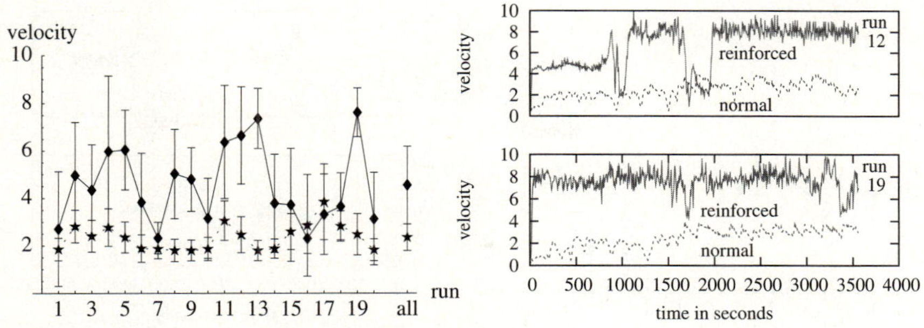

Fig. 3. *Left:* Mean and std. deviation of the velocity of the Sphere for 20 runs each 60 minutes long with (*diamonds/solid line*) and without (*stars/dotted line*) speed reinforcement; *all* denotes the mean and std. deviation over the means of all runs; *Right:* Time course of the velocity during 2 runs, i.e. 4 independent simulations (*upper:* run 12, *lower* run 19).

5.2 Spin Reinforcement

In a different setup we want the robot to drive curves and spin around the z-axis of the world coordinate system. The reinforcement function looks as follows:

$$r(t) = \frac{1}{3}\|\omega_z\| - 1$$

where ω_z is the angular velocity of the robot around the z-axis (in world coordinates). Again the reward is scaled to be in an appropriate interval. Positive reward can be obtained by rolling in a curved fashion or by entering a pirouette mode. The latter can be compared to a pirouette done by figure-skaters, with some initial rotation the masses are moved towards the centre, so that the robot spins fast at the place. The robot also experiences rolling friction, so that fast pirouettes are not persistent. We conducted again 20 experiments with reinforcement, each 60 minutes simulated real-time on a flat surface without obstacles. In Fig. 4 the mean angular velocity ω_z for each simulation is plotted and the angular velocities of the robot in two reinforced and two normal runs are displayed. In this scenario the difference between the normal runs and the reinforced runs are tremendous. Nearly all reinforced runs show a very large mean angular velocity. The reason for this drastic difference is that these spinning modes are less predictable and therefore quickly left in the unreinforced setup. One can see in the traces, that the robot in a normal setup rarely performs spinning motion, whereas the reinforced robot, performs after some time of exploration very fast spinning motions, which are persistent for several minutes. Note, that spinning at the place (high peeks) is not persistent because of friction. So the robot tends to gain some speed by rolling along the ground.

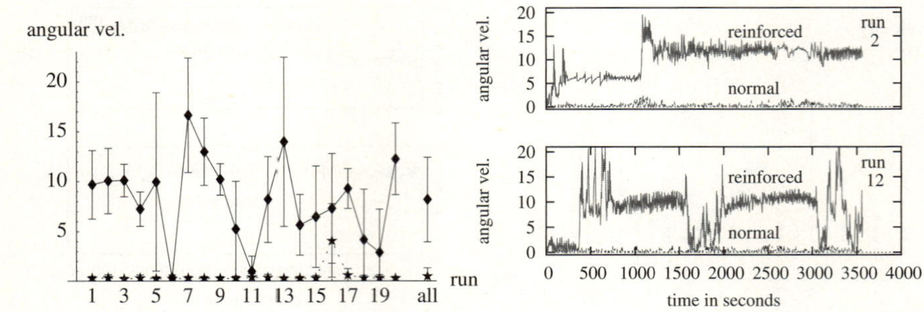

Fig. 4. *Left:* Mean and std. deviation of the angular velocity ω_z of the Sphere for 20 runs each 60 minutes long with (*diamonds/solid line*) and without (*stars/dotted line*) spin reinforcement; *Right:* Time course of the angular velocity during 2 runs, i.e. 2 independent simulations (*upper:* run 2, *lower* run 12).

6 Discussion

We demonstrated in the present paper a simple method by which the otherwise freely self-organised behaviour, generated by the general homeokinesis paradigm, can be guided towards desired behaviours. First we studied an emergent exploratory behaviour in form of a velocity sweep using a two degree of freedom rolling barrel robot. This shows that different behaviours are exhibited in course of time. We integrated a reinforcement signal defined by an external observer into the learning rule of the controller. In essence the original time loop error is multiplied by a strength factor, obtained from the reinforcement signal. The approach is applied to a spherical robot in two scenarios, fast motion reinforcement and spin reinforcement. In both cases the performance was significantly increased and it was shown that the robot was guided towards rewarded behaviours. Nevertheless, the exploratory character of the paradigm stays still intact.

We consider our approach as a contribution to autonomous robot development [18,19] and see potential applications in this field. With the presented reinforcement mechanism we are now able to guide the development of behaviours. However, in the current setup the internal world model will forget past behaviours, so that there is no long term effect of the reinforcement. This can be achieved with multiple internal models and will be subject of a future paper.

Acknowledgements. This study was supported by a grant from the BMBF in the framework of the Bernstein Center for Computational Neuroscience Göttingen, grant number 01GQ0432.

References

1. Neal, M., Timmis, J.: Once More Unto the Breach: Towards Artificial Homeostasis?. In: Castro, L.N.D., Zuben, F.J.V. (eds.) Recent Developments in Biologically Inspired Computing, pp. 340–365. Idea Group, Hershey (January 2005), `http://www.cs.kent.ac.uk/pubs/2005/1948`

2. Di Paolo, E.: Organismically-inspired robotics: Homeostatic adaptation and natural teleology beyond the closed sensorimotor loop. In: Murase, K., Asakura, T. (eds.) Dynamical Systems Approach to Embodiment and Sociality, pp. 19–42. Advanced Knowledge International, Adelaide (2003)
3. Der, R.: Self-organized acquisition of situated behavior. Theory Biosci. 120, 179–187 (2001)
4. Der, R., Steinmetz, U., Pasemann, F.: Homeokinesis - a new principle to back up evolution with learning. In: Computational Intelligence for Modelling, Control, and Automation. Concurrent Systems Engineering Series, vol. 55, pp. 43–47. IOS Press, Amsterdam (1999)
5. Der, R., Liebscher, R.: True autonomy from self-organized adaptivity. In: Proc. Workshop Biologically Inspired Robotics, Bristol (2002)
6. Der, R., Herrmann, M., Liebscher, R.: Homeokinetic approach to autonomous learning in mobile robots. In: Dillman, R., Schraft, R.D., Wörn, H. (eds.) Robotik 2002. VDI-Berichte, vol. 1679, pp. 301–306 (2002)
7. Der, R., Hesse, F., Martius, G.: Rocking Stamper and Jumping Snake from a Dynamical System Approach to Artificial Life. Adaptive Behavior 14(2), 105–115 (2006), doi:10.1177/105971230601400202
8. Der, R., Martius, G., Hesse, F.: Let it roll – emerging sensorimotor coordination in a spherical robot. In: Rocha, L.M., Yaeger, L.S., et al. (eds.) Artificial Life X: Proc. X Int. Conf. on the Simulation and Synthesis of Living Systems, August 2006, pp. 192–198. MIT Press, Cambridge (2006)
9. Der, R., Martius, G., Hesse, F.: Videos of self-organized creatures (2007), `http://robot.informatik.uni-leipzig.de/research/videos`
10. Der, R., Martius, G.: From motor babbling to purposive actions: Emerging self-exploration in a dynamical systems approach to early robot development. In: Nolfi, S., Baldassarre, G., Calabretta, R., Hallam, J.C.T., Marocco, D., Meyer, J.-A., Miglino, O., Parisi, D. (eds.) SAB 2006. LNCS (LNAI), vol. 4095, pp. 406–421. Springer, Heidelberg (2006)
11. Stout, A., Konidaris, G., Barto, A.: Intrinsically motivated reinforcement learning: A promising framework for developmental robotics. In: The AAAI Spring Symposium on Developmental Robotics, AAAI, Menlo Park (2005)
12. Oudeyer, P-Y., Kaplan, F., Hafner, V.V., Whyte, A.: The playground experiment: Task-independent development of a curious robot. In: Bank, D., Meeden, L. (eds.) Proceedings of the AAAI Spring Symposium on Developmental Robotics, Stanford, California, pp. 42–47. AAAI, Menlo Park (2005)
13. Schmidhuber, J.: Completely self-referential optimal reinforcement learners. In: ICANN (2), pp. 223–233 (2005)
14. Popp, J.: Sphericalrobots (2004), `http://www.sphericalrobots.com`
15. Smith, R.: Open dynamics engine (2005), `http://ode.org/`
16. Martius, G., Der, R., Hesse, F., Güttler, F.: Leipzig robot simulator (2006), `http://robot.informatik.uni-leipzig.de/software`
17. Hamed, N.: Self-Referential Dynamical Systems and Developmental Robotics. PhD thesis, University of Leipzig (In preparation, 2006)
18. Lungarella, M., Metta, G., Pfeifer, R., Sandini, G.: Developmental robotics: a survey. Connect. Sci. 15(4), 151–190 (2003)
19. Weng, J., McClelland, J., Pentland, A., Sporns, O., Stockman, I., Sur, M., Thelen, E.: Autonomous mental development by robots and animals. Science 291, 599 (2001)

Near-Optimal Mobile Robot Recharging with the Rate-Maximizing Forager

Jens Wawerla and Richard T. Vaughan

Autonomy Lab, Simon Fraser University, Burnaby, British Columbia, Canada.
{jwawerla,vaughan}@sfu.ca

Abstract. We examine the practical problem of a mobile autonomous robot performing a long-duration survey task, during which it must recharge its batteries periodically. We present a scalable, online, heuristic method that allows the robot to recharge efficiently, thus maximizing its rate of work. The method is a direct application of the *rate-maximizing foraging* model, which seeks to explain the behaviour of animals solving related problems. Simulation results suggest that the method performs very well compared to optimal and naive heuristic approaches.

1 Introduction

We examine the problem of a mobile autonomous robot performing a long-duration survey task. The robot has limited battery capacity, so it must recharge from time to time. Time and energy spent in recharging are not available for doing work, so, to maximize the robot's work-rate, and thus its utility, we should recharge as efficiently as possible. Computing the perfectly optimal robot behaviour is computationally intractable or impossible in typical circumstances, so we seek useful heuristic methods.

This paper presents a version of this task that serves a real-world need, along with a practical solution: an heuristic method that is a direct implementation of a model of animal foraging behaviour, with some attractive features. We provide empirical simulation results comparing the forager's performance with the optimal behaviour and naive heuristics, suggesting that the new method performs well.

1.1 Problem Domain and Motivation

A single, self-sufficient and self-controlling mobile robot must visit a (possibly very long) series of waypoints. The number of waypoints, the energy cost of travelling between them and the robot's limited onboard energy storage capacity mean that the robot must recharge its onboard energy store (we assume a battery), possibly many times. The series of waypoints could be given *a priori*, or could be dynamically generated based on previous data. An examples of such a scenario is the Networked Aquatic Microbial System (NAMOS) robotic boat devised by researchers at USC and UCLA [9] that measures the spatio-temporal

F. Almeida e Costa et al. (Eds.): ECAL 2007, LNAI 4648, pp. 776–785, 2007.

Fig. 1. Application: the USC/UCLA NAMOS robot boat: a mobile aquatic observatory with anchored buoy base-station. A docking/recharging system is under development.

distribution of plankton, and related phenomena, in bodies of water. In this application, a series of waypoints is generated by an adaptive sampling algorithm that attempts to maximize the value of the next sample [12].

We are collaborating with the NAMOS team to add energetic self-sufficiency to the NAMOS boat. Currently the duration and physical size of a water survey performed by the boat is limited by its battery capacity. Adding autonomous recharging will enable the boat to do multi-day or -week long missions without the expense of human intervention, and providing more, better, data to the team's biologists.

1.2 Related Work

Research has been done on many aspects of self-sufficiency; ranging from docking mechanisms and control, action-selection, energy-efficient path-planing, to unconventional robot fuels [2]. The work most relevant here is that on energy-aware path-planning and when-to-recharge decision making.

Aside from the huge literature on general cost-minimizing optimization and path planning, we see some work dedicated to energy gathering and recharging. Wettergreen et al. [11] present a solar powered robot that plans motion synchronous to the sun in order to maximize energy gathered by solar cells. The problem of path-planning while minimizing energy expense is also discussed in [10]. Both these authors use conventional planning methods.

Litus et al. [3] describe heuristic methods to achieve energy-efficient rendezvous between robots, with suggested application to robot-to-robot recharging.

As for deciding when to recharge, the standard method is a fixed threshold policy, where the robot switches to recharging activity once its energy level drops below some predefined threshold. Jung's docking work employs this policy [6], for example. Austin et al. [1] have the robot recharge after a constant time has elapsed, approximating the energy threshold while being easier to measure. The threshold approach is attractively simple, but its efficiency depends critically on selecting a suitable threshold. In the experiments below, we use a fixed threshold strategy as a baseline performance target.

An interesting conceptual extension is to a multi-robot threshold. In a multi-robot scenario with one shared charging station, Michaud et al. [4] use a threshold on the 'remaining operational time' value. This value takes the energetic state of all robots into account and is calculated for each robot. This allows the robots to cooperate and share a charging station.

A more carefully motivated approach to decision making under self-sufficiency is discussed by Spier and McFarland [7]. Here a 'cue-deficit' model is used to determine the action to be taken. This work is closest to our work, since it investigates choosing between multiple resources. The major difference is that Spier considers multiple resources, for each of which the robot has a certain deficit, whereas we consider one resource: energy, which can be obtained from multiple sources.

In the remainder we state the problem formally, describe three different algorithms to solve it and compare their performance to the optimal solution.

2 Problem Statement

A mobile robot must visit an ordered series of n points $P = p_1, p_2, ...p_n$ in the shortest possible time. We are especially interested in very large series, so that brute-force planning is infeasible. For simplicity of presentation we assume all points are co-planar, but the solution methods do not require this.

In addition to the robot, the world contains one charging station at a known location c. The robot has two options for recharging: it can recharge from the charging station with a known charging current I_c or it can use on-board solar cells and recharge with a current of I_s. To use the charging station, the robot must be at c. The solar cells can be used at any location, but we assume that the robot cannot drive while recharging from the solar cells [1] but that solar charging has negligible other cost. We further assume the driving velocity v of the robot is constant and known, and so is the current required for driving I_d. The battery which powers the robot has a known capacity of B_{max} and an instantaneous charge B where $0 \leq B \leq B_{max}$. Below we neglect the units and assume that current is measured in amps (A), time in seconds (s), distance in meters (m), speed in ms^{-1}, and energy storage capacity in As. The robot begins with a full battery charge.

The robot's recharging model has this simple form:

1. When reaching a charging station, the robot recharges completely, i.e. until $B = B_{max}$.
2. If the robot runs out of energy, and is not at a charging station, it can always use its solar cells to recharge. It recharges only enough to reach the next goal point.

This is a sensible model because there is always a locomotion cost involved in going to the charging station. In order to maximize the return on this cost,

[1] This models, for example, the robot needing to unfold the solar array before use.

the robot should gain as much energy as possible from a charger visit. There is no locomotion cost for using solar cells, but the time spent recharging is lost for working: an opportunity cost. Since in most scenarios the solar current is smaller than the charger current, it makes sense for the robot to only solar charge as long as necessary to reach the next goal location. Thus, when solar charging, sample points and chargers are reached as early as possible and with an empty battery, which maximizes the benefit of a charging station.

3 Solutions

3.1 Brute-Force Optimal

Optimal solutions can be calculated with the following simple brute force method. Build a binary tree with the first point p_1 as the root node. The first (left-hand) child node corresponds to point p_2. The cost of traversing from the root to this child is the time required to travel from p_1 directly to p_2 including any required solar cell recharging time (should the distance be greater than that allowed by a single battery charge). The second (right-hand) child node corresponds to the charger location C. The cost of traversing from the root to this child is the time required to travel from p_1 to the charger C then fully recharge the battery and afterwords travel to p_2, again including possible solar charging time. This step is repeated recursively for each child until the last way point is considered. To find the most efficient solution one finds the leaf node with the shortest travel time. Traversing back from this node gives the optimal work-and-charging strategy.

With complexity $O(2^n)$ this method is not practical for large series. It can not be applied to infinite series or dynamically-generated series, such as those created by an adaptive sampling strategy. It is included here only to help evaluate the heuristic methods.

3.2 Fixed Threshold

Here we use remaining battery charge B as our decision variable. For lack of any general principled approach, we use the following conservative strategy: find the largest distance between any waypoint and the charger $l = \max\{d(p_i, c)|1 \leq i \leq n\}$ where $d(a, b)$ is the distance between points a and b. The charging threshold is set so that the robot has enough energy left to reach the charger even in the worst case, that is the waypoint that is furthest away from the charger. Therefore the threshold is $B_t = \frac{l}{v} I_d$. This means that the robot chooses to travel to the next waypoint via the charging station if $B < B_t$. Note, this does not guarantee that the robot will always be able to reach the charger, in cases where I_d is high or B_{max} is low, the threshold might actually be higher the B_{max}. In cases like this the robot can rely on power from the sun.

Without any empirical analysis we can see that the simple fixed threshold strategy will not perform very well, because it does not take the distance of the current waypoint to the charger into account. This can lead to situations where the robot is unnecessarily cautious and thus sacrifices performance. An adaptive threshold may overcome this limitation.

3.3 Adaptive Threshold

With the adaptive threshold, we set the threshold for each path segment by checking if the robot has enough energy to travel from the current waypoint along the current path segment and still make it from the next waypoint to the charger. First we calculate the travel distance $l = d(p_i, p_{i+1}) + d(p_{i+1}, c)$ where i denotes the waypoint the robot is currently at. Then the threshold is again $B_t = \frac{l}{v} I_d$. So the robot travels to the next waypoint via the charger if $B < B_t$ otherwise it takes the direct route. Besides taking locality into count, this method has the benefit that if only requires information about the next waypoint, so the path can be generated dynamically, as required by our NAMOS application.

Performance problems may still arise from the usage of a threshold. There are situations where it is beneficial to refuel even though the battery is not even close to empty, but since the energy source is close by the overhead is very small in doing so. The next method attempts to take this into account without increasing computational complexity.

3.4 Rate Maximization

This method is an application of foraging theory (see [8] for an excellent exposition). This body of theory attempts to explain and give a model for animal behaviour while foraging, that is, gathering food and other resources. The *rate-maximizing forager* model attempts to explain the choices animals make when presented with alternative food sources, and maps neatly onto our problem.

Specifically, the rate-maximizing forager model predicts that an animal faced with several exclusive food options will maximize the net energy rate

$$R = \frac{\sum_{j=1}^{m} p_j \lambda_j e_j}{1 + \sum_{j=1}^{m} p_j \lambda_j h_i}$$

where e_i is the net energy gain from food item i, h_i is the handling time, λ_i is the encounter rate and p_i is the attack probability for this item. Without reproducing the details of [8] a rate maximizing forager applies the zero-one rule and attacks always only prey that yield a higher net energy rate than the combined rates of all food items that considered individually have a lower rate. Note that this model describes a solitary forager: Seth [5] provides a probablistic model that may better describe the behaviour of multiple foragers in a competitive situation.

We turn our robot into a rate maximizer by the following approach. For a given path segment $S_i = \overline{p_i p_{i+1}}$, starting at p_i and ending at p_{i+1}, we have to consider two energy gain rates. In case the robot travels directly to the next waypoint, the rate equals the solar current, that is $R_{d_i} = I_s | \forall i < n$. In the other case where the robot goes to the charger first, fully recharges its battery and then travels to the next waypoint, the energy gain rate is

$$R_{c_i} = \frac{I_s T_{s_i} + I_c T_{c_i} - I_d T_{d_i}}{T_{s_i} + T_{c_i} + T_{d_i}} \tag{1}$$

Here T_{s_i} is the time spent charging from solar cells on segment i, which only occurs if the robot cannot make it to the charger or from the charger to the next waypoint p_{i+1} on the remaining battery capacity B_i. The time spent recharging from the charging station is

$$T_{c_i} = \frac{B_{max} - B_i - I_d(|\overline{p_i c}|)}{I_c}$$

In most case the robot has to take a detour to reach the charger, therefore we subtract the extra energy spend from the energy gained. The time required for the detour is

$$T_{d_i} = \frac{|\overline{p_i c}| + |\overline{c p_{i+1}}| - |\overline{p_i p_{i+1}}|}{v}$$

Now the robot can select the option which provides the highest energy gain rate, but it has more options to choose from if it's battery is full enough. What about the next path segment or the one after that? At any given waypoint, the robot only has to make a decision about going direct or via the charger to the next waypoint. So if the robot chooses to go via the charger we can stop our rate analysis for this segment, since future segments do not influence the robot's decision anymore. But if we do not recharge, it is worth evaluating the energy gain rate of the next segment. Again going from p_{i+1} directly to p_{i+2} is trivial $R_{d_{i+1}} = I_s$. In the other case where the robot travels via the charger to p_{i+2} we use eq. 1. But this time for the segment $S_{i+1} = \overline{p_{i+1} p_{i+2}}$ and we also have to calculate B_{i+1} by subtracting what we spend on the first segment so $B_{i+1} = B_i - \frac{|\overline{p_i p_{i+1}}|}{v} I_d$. This way we can iteratively calculate the energy rate for each segment. We can stop this iteration once the projected battery level B_k reaches zero. Because the robot has to recharge at this point, future segments do not influence the current decision.

The rate-based decision is simply, if the energy gain rate from the charger option of the current segment $R_{c_i} = \max\{I_s, R_{c_i}, R_{c_{i+1}} ... R_{c_k}\}$ choose to recharge from the charger now, otherwise proceed directly to the next waypoint and re-iterate.

The complexity of this method is $O(kn)$ where is k is the number of waypoints the robot can maximally reach with one battery charge.

Rate maximizing foraging animals are clearly not limited to two energy resources, as Stephens and Krebs [8] show. We restricted ourselves because of scalability issues of the optimal method. But extending the rate maximizer to deal with m energy sources is straight forward. On each segment s_i we calculate the energy gain rate R_{j_i} for each energy source j. Combinatorial problems do not occur since we terminate the current analysis on each branch on which the robot recharges. As in the two resource case we interactively calculate rates till the battery is depleted. The charging option with the highest rate is the one the robot selects. In this case the complexity is $O(knm)$ with k being the number of waypoints the robot can visit with one battery charge and n is the total number of waypoints of the trajectory.

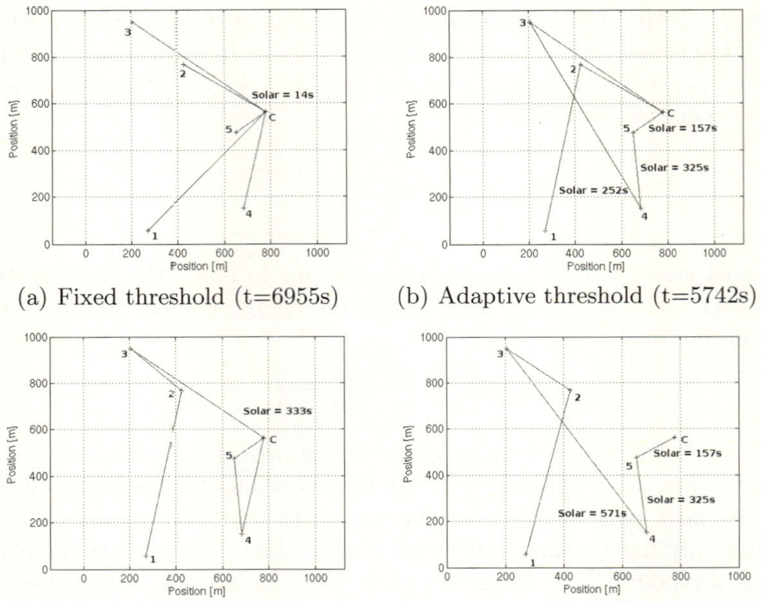

(a) Fixed threshold (t=6955s) (b) Adaptive threshold (t=5742s)

(c) Rate Maximizer (t=4544s) (d) Optimal (t=4446s)

Fig. 2. An example problem, where each of the four methods found a different solution.

3.5 Example Solutions

Before discussing the statistical results, we give an example of the type of plans produced. The example problem shown in Fig. 2, five waypoints, one charging station, is interesting in that each of the four methods produced a different solution, and typical in their relative performance. The robot starts with a full battery at waypoint 1. To ensure all robots finish in the identical state (which masks artifacts caused by experiments with few waypoints), the charger C is the final waypoint and the trial ends when the robot is totally charged.

The fixed threshold method (Fig.2(a)) demonstrates the conservatism of its high threshold setting, with the robot visiting the charger C 5 times: once between every waypoint. Nevertheless, due to the remoteness of waypoint 3, the robot has to solar charge for 14 seconds on the return from 3 to C. The complete run takes 6955 seconds. The adaptive threshold method (Fig.2(b)) visits the charger between waypoints 2 and 3. It spends a long time solar charging towards the end of the run, which takes 5742 seconds to complete. The rate maximizer method (Fig.2(c)) takes only 79% of the adaptive threshold time. It recharges slightly later: in between waypoints 3 and 4, taking a detour that requires a solar charge, but having visited the charger empty, the robot completes the run without charging again. The optimal method (Fig.2(c)) is very slightly faster than the rate maximizer, taking 99% of the time. The brute-force search found a solution that recharges only at the final return to the charger, and uses the solar cells to arrive there empty.

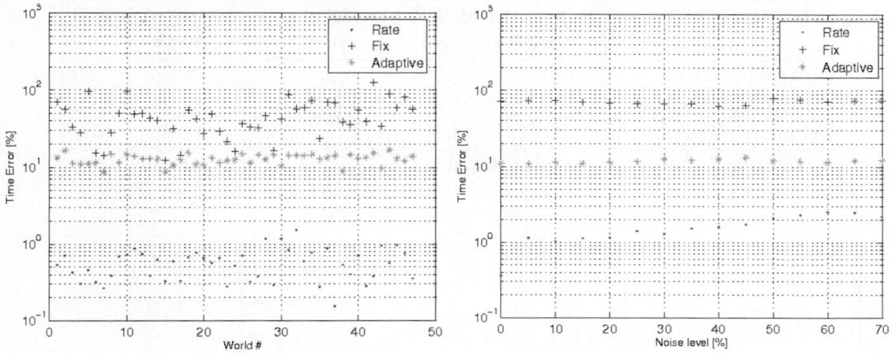

Fig. 3. (Left) Exp.1: The mean error, compared to optimal performance, of each of the three heuristic methods, over a range of parameter settings. Note the Y axis is a log scale. (Right) Exp.3: Performance of the Rate Maximizer under cost estimate error.

4 Experiments

4.1 Experiment 1: Series of 20 Randomly-Placed Points

Though we are interested mainly in large-scale instances of the surveying-and-charging problem, we also wish to evaluate the performance of the heuristic methods compared to the optimal. To allow for reasonably fast computation of the optimal brute force method we are limited to 20 waypoints.

We explored the performance of all four methods under a wide variety of parameters. The number of waypoints and the charging station current were fixed. The rest of the setup was as follows: (i) generate a world with 20 waypoints and a single charging station placed at random; (ii) set values of B_{max}, I_s, I_d; (iii) for each of the 4 methods, measure the time it takes the robot to complete the course (again, ending up fully charged at C). Repeat this process for 50 different worlds, each with 34 different values of B_{max}, 7 of I_s and 7 of I_d, for a total of 1666 configurations per world, and 83,300 trials overall. Parameter ranges were chosen to explore the most significant part of the state space.

Presenting these results is challenging, since so many parameters are varied. The absolute values of performance are not very meaningful since it depends on the length of the trajectory which is randomly generated. Therefore for each trial we subtract the time taken by the optimal method, to give an error signal which we can use to asses the heuristic methods. Figure 3 (left) shows the mean error of each method compared to the optimal solution averaged over all parameter sets in each of the 50 worlds. It shows that the adaptive threshold performs better (has smaller error) then the fixed threshold method and that the rate maximizer performs better then the other two methods. Table 1 (Experiment 1) gives the mean and standard deviation of the time error for all experiments. The distribution of these errors is depicted in Fig. 4, which shows a histogram of the error. Note the axes are scaled differently on all the graphs.

Method	Experiment 1		Experiment 2		Experiment 3	
	μ	σ	μ	σ	μ	σ
Rate Maximizer	0.6	1.3	-	-	1.6	3.9
Fixed Threshold	47.5	51.1	55.8	52.2	70.6	68.9
Adaptive Threshold	12.8	18.7	14.8	28.2	11.8	16.7

Table 1. Summary of statistical results for all three experiments

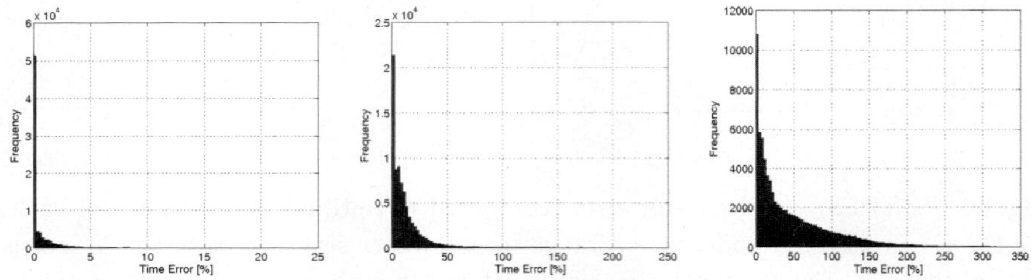

(a) Rate Maximization (b) Adaptive Threshold (c) Fixed Threshold

Fig. 4. Histograms of the relative error. The differences are so large that the axes are scaled differently to preserve visible detail.

We repeated this experiment with various regular patterns of waypoints, producing very similar results. We omit details here for lack of space.

4.2 Experiment 2: Series of 1000 Randomly-Placed Points

To evaluate the methods' performance in long-duration tasks, we generated 100 trajectories with 1000 waypoints each and ran the three heuristic methods over all parameter combinations for each trajectory. Here it is impossible to compare the performance of each method to an optimal solution, since it could not be calculated in a reasonable amount of time for these large problems. Instead we compared the threshold methods against the rate method. Table 1 (Experiment 2) shows how the threshold methods performed compared to the rate maximization. On average the rate maximizer performs 14% better then the adaptive threshold and 55% better then the fixed threshold method.

4.3 Experiment 3: Robustness to Cost Estimate Error

We repeat experiment 2, this time randomly varying the actual cost of driving the robot between waypoints, to simulate unbiased error in the robot's *a priori* cost estimates. The error was varied from 0% to 70 % in 5% intervals. 20 simulations per noise level and parameter set were performed.

As Table 1 (Experiment 3) shows, the performance of the rate maximizer decreases with increased cost estimate error. However, the rate maximizer still performs significantly better than the other methods. And as Fig. 3 (right) shows, the error increases with increasing noise but remains relatively small, suggesting that the rate maximizer is fairly noise tolerant.

5 Conclusion

We have tackled the natural optimization problem of deciding where and when for a long-lived robot to recharge, given two alternative energy supplies with different dynamic properties. We gave a scalable, heuristic solution based on a model from the biological literature that explains the choices that animals *should* make when faced by equivalent problems.

The idea behind the solution is to maximize the local *rate* of energy intake. Given finite energy storage capacity this has the natural result of maximizing the rate at which energy can be spent on work. Thus by adopting this strategy we observe that the work-rate of the simulated systems we observed is very close to optimal. Yet the method is scalable, i.e. requires computation and memory independent of the length of the problem. It is simple to implement and appears to be reasonably resistant to error in cost estimates. The method dramatically outperforms naive threshold-based heuristics.

References

1. Austin, D., Fletcher, L., Zelinsky, A.: Mobile robotics in the long term - exploring the fourth dimension. In: Proc. Int. Conf. on Intelligent Robots and Systems, pp. 613–618 (2001)
2. Ieropoulos, I., Greenman, J., Melhuish, C.: Imitating metabolism: Energy autonomy in biologically inspired robotics. In: Proc. of Symp. on Imitation in Animals and Artifacts, pp. 191–194 (2003)
3. Litus, Y., Vaughan, R.T., Zebrowski, P.: The frugal feeding problem: Energy-efficient, multi-robot, multi-place rendezvous. In: Proc. of the IEEE Int. Conf. on Robotics and Automation, Apr. 2007, pp. 27–32. IEEE Computer Society Press, Los Alamitos (2007)
4. Michaud, F., Robichaud, E.: Sharing charging stations for long-term activity of autonomous robots. In: Proc. IEEE/RSJ Int. Conf. on Intelligent Robots and Systems, vol. 3, pp. 2746–2751. IEEE, Los Alamitos (2002)
5. Seth, A.K.: The ecology of action selection: Insights from artificial life. Phil. Trans. R. Soc. B (In Press)
6. Silverman, M., Nies, D.M., Jung, B., Sukhatme, G.S.: Staying alive: A docking station for autonomous robot recharging. In: Proc. IEEE Int. Conf. on Robotics and Automation, May 2002, pp. 1050–1055. IEEE Computer Society Press, Los Alamitos (2002)
7. Spier, E., McFarland, D.: Possibly optimal decision-making under self-sufficiency and autonomy. Journal of Theoretical Biology 189(3), 317–331 (1997)
8. Stephens, D.W., Krebs, J.R.: Foraging Theory. Princton Uni. Press, Princton (1986)
9. Sukhatme, G.S., Dhariwal, A., Zhang, B., Oberg, C., Stauffer, B., Caron, D.A.: The design and development of a wireless robotic networked aquatic microbial observing system. Environmental Engineering Science 24(2), 205–215 (2006)
10. Sun, Z., Reif, J.: On energy-minimizing paths on terrains for a mobile robot. In: Proc. Int. Conf. on Robotics and Automation, vol. 3, Sept. 2003, pp. 3782–3788 (2003)
11. Wettergreen, D., Tompkins, P., Urmson, C., Wagner, M., Whittaker, W.: Sun-synchronous robotic exploration: Technical description and field experimentation. The Int. Journal of Robotics Research 24(1), 3–30 (2005)
12. Zhang, B., Sukhatme, G.S.: Adaptive sampling for estimating a scalar field using a robotic boat and a sensor network. In: Proc. IEEE Int. Conf. on Robotics and Automation, pp. 3673–3680. IEEE Computer Society Press, Los Alamitos (2007)

Neural Uncertainty and Sensorimotor Robustness

Jose A. Fernandez-Leon and Ezequiel A. Di Paolo

Centre for Computational Neuroscience and Robotics (CCNR),
Department of Informatics, University of Sussex, Brighton, BN1 9QG, UK
{jf76, ezequiel}@sussex.ac.uk

Abstract. Real organisms live in a world full of uncertain situations and have evolved cognitive mechanisms to cope with problems based on actions and perceptions which are not always reliable. One aspect could be related with the following questions: could neural uncertainty be beneficial from an evolutionary robotics perspective? Is uncertainty a possible mechanism for obtaining more robust artificial systems? Using the minimal cognition approach, we show that moderate levels of uncertainty in the dynamics of continuous-time recurrent networks correlates positively with behavioral robustness of the system. This correlation is possible through internal neural changes depending on the uncertainty level. We also find that controllers evolved with moderate neural uncertainty remain robust to disruptions even when uncertainty is removed during tests, suggesting that uncertainty helps evolution find regions of higher robustness in parameter space.

Key words: neural uncertainty; minimal cognition; continuous-time recurrent networks; Evolutionary Robotics

Introduction

Real organisms live in a world full of uncertain situations and have evolved cognitive mechanisms to cope with problems based on actions and perceptions which are not always reliable. Both uncertainty and noise are intrinsic realities that all cognitive systems have to deal with in terms of the situation faced and the mechanisms deployed to cope with that situation [1]. What is not so often realised is that uncertainty and noise may also provide positive mechanisms for producing robust and adaptive behaviour. In particular, in Evolutionary Robotics the use of variability between evaluations and noise during a same evaluation are well-known factors relevant for the quality of the neurocontrollers that evolve.

There are two ways in which the careful use of noise and variability can be of benefit to evolutionary robotics; (a) by promoting the evolution of relatively complex behaviours, and (b) by accelerating artificial evolution. Large amounts of simulation noise may promote the evolution of robotic agents with relatively complex behaviours and neural dynamics compared to those evolved in simulations with low noise levels, accelerating the evolutionary process [2] and possibly enhancing robustness during evolution. In this work, robustness will refer to stability against external perturbations and internal variability [3].

Jakobi [4] has applied these ideas successfully to minimal simulations which allows neurocontrollers evolved in simulation to have a significant rate of success in

F. Almeida e Costa et al. (Eds.): ECAL 2007, LNAI 4648, pp. 786 – 795, 2007.

transferring to real robots (where direct evolution is prohibitive). Minimal simulations work by avoiding accurate replication of the physical complexities of a real-world robot-environment system and proceed to abstract a base set of factors upon which evolution must rely in order to produce the desired behaviour. All other factors in the robot-environment system are crudely modelled and subjected to large amount of noise and variability between evaluations with the result that the only suitable strategy for the evolutionary search process is to ignore them.

If minimal simulations work under the assumption that evolved solutions will ignore unreliable factors, no matter how complex, and hence be robust to their variability in real-world situations, we may ask what would be the consequence of making the neurocontrollers themselves also unreliable in some significant manner. Our hypothesis is that neural uncertainty will contribute to several aspects of robustness of performance in order to significantly and substantially improve robot performance for a given environment. Neural uncertainty could relate to the amount by which a value, typically the neuron output activity, may differ from those obtained when no significant variations in such value are produced in spite of changes in the neuron inputs.

This paper will explore previous hypothesis using a minimally cognitive model. It should be viewed as a preliminary exploration easily extendable to more complex systems and tasks. We test the role of neural uncertainty by biasing the dynamics of every single node in the agent's neurocontroller. The agent must still be able to perform the task in spite of the randomly distorted internal dynamics. We surmise that agents capable of good performance will also show several forms of robustness.

The following section describes the method proposed to induce uncertain internal activity in neurons. This method is applied on the evolution of Khepera-like simulated agents in a simple phototaxis scenario. This is followed by experiments testing the effects of uncertainty under morphological and environmental disruptions. Finally, the last section examines the consequences of the results and discusses some of the questions that remain open.

Methods

We propose to study the effects of uncertainty in neural dynamics through a minimalist approach. Our aim is to test the consequences of evolving networks under uncertain dynamics and to test the resulting robustness of their performance. A population of simulated agents is evolved to perform phototaxis under normal bodily and environmental conditions on a series of light sources. In each test, one light source is presented at a extended-finite time. Limited, but random uncertainty is applied locally to the dynamics of each neuron. The uncertainty of each neuron (y_o) is modelled as a constantly changing activation parameter. The range of this random variable (A) is a control parameter in our studies.

In preliminary tests, we noticed that evolution avoids coping with the effects of uncertainty by selecting neurons working very high or very low in their dynamical range (in our case, near to -1 or 1 output value). The use of center-crossing techniques[5] makes sure that neurons avoid long-term saturation, where the random parameter y_0 has an effect on neural activities which has to be counteracted in order to perform the evolved task. Center-crossing condition is used to enforce neural dynamics away from saturation, enabling more interesting network dynamics throughout the evolutionary search.

Continuous-time recurrent neural networks (CTRNNs) using center-crossing are used to control the agents. The CTRNN topology consist of 2 input nodes, 2 inter-neurons and 2 motor neurons. Full connectivity is used between inter-neurons and output neurons including self connections for the inter-neurons.

In the proposed experiments, agents evolved for this normal phototaxis task are then tested for phototaxis under a series of disruptions of their sensory inputs (sensor gain and visual field inversion), and motor capabilities (motor gain), as well as internal dynamics (uncertainty variation). During evolution agents have not faced such disruptions.

Agent Model

Agents were modelled as solid circular bodies of radius five with two diametrically opposed motors and two frontal light sensors with 47.75° between them according to Khepera robot sensor layout. Motors can drive backwards and forwards in an unlimited 2-D arena. Agents have a very small mass, so that the motor output is the tangential velocity at the point of the body where the agent is located. Agents have a default velocity in each motor node. Light from point sources stimulates sensors with an intensity proportional to the inverse square of the distance from the sensor to the source. The model includes shadows on sensors produced when light is occluded by the agent body.

Neurocontroller Model

A fully connected 4-neuron CTRNN and 2 additional input nodes is used as the neurocontroller. The dynamics of each neuron is modelled as follows:

$$\tau_i \frac{dy_i}{dt} = -y_i + \sum_j^n w_{ji} z_j + I_i \quad (1)$$

$$z_j = \sigma(y_j + \theta_j) \quad (2)$$

where, using terms derived from an analogy with real neurons, y_i represents the cell potential, τ_i the decay constant (range [1;-1+e^2]), θ_i the bias (subject to a sigmoid function), z_i the firing rate, w_{ji} the strength of synaptic connection from node i to node j (range [-10;10]), and I_i the degree of sensory perturbation as an incoming current, which is zero for non-input nodes. Each sensor and motor has one sensory node, or effector neuron, respectively.

The transduction from effector neuron activity to motor output is directly applied from the neuron's output in the interval [-1;1], adding noise (a value uniformly distributed in [-0.25;0.25]). In addition, the incoming signals to sensory nodes are multiplying by a gain (equal to 2). Left/right symmetry in synaptic weights is not enforced. Some parameters are genetically encoded: τ_i,, sensory gains, w_{ji}, and θ_i. The network is updated using the Euler integration method with a time step of 0.1. The center-crossing condition is θ^*_j =0.5 (see [6]). Due to the shape of z_j, unless a neuron's bias (θ^*_j) is properly turned to the range of inputs it receives, that neuron will simply saturate on or off. Thus, the richest dynamics should be found in the neighbourhood of the center-crossing networks in parameter space, and one would expect that an evolutionary algorithm would benefit from focusing its search there [5]. In our case, center-crossing increases the effect of neural uncertainty which is otherwise nullified when using nearly-saturated neurons for generating behaviour.

Inducing Neural Uncertainty

Neural uncertainty is represented by a random local parameter (y_0) in the dynamics of every single neuron in the agent's neurocontroller. This random parameter is selected every time step from the range [*-A; A*], where A is a fixed value in each experiment (A=0,1,2,3,4). Therefore, we test the role of neural uncertainty by randomly biasing, for each new evaluation, the dynamics of neurons. The random variable y_0 affects neurons as follows where i denotes the node index:

$$\tau_i \frac{dy_i}{dt} = -(y_i - y_0) + \sum_j^n w_{ji} z_j + I_i \tag{3}$$

Even though the addition of the term y_0 could be interpreted simply as a perturbation on the current input (I_i), this interpretation is only valid for sensor nodes. However, in this model the perturbation is applied to all neuron in the network. This uncertainty parameter can be mathematically interpreted as influencing the asymptotic behavior of each neuron.

Genetic Algorithm

A population of 60 individuals (neurocontrollers) is evolved using a steady-state, rank-based genetic algorithm with elitism (50%). All real-value parameters are encoded in a real-valued vector, where each component encodes a single parameter in the neural network. These components belong to the ranges previously named for each parameter. A global mutation operator is used [6][7] adding a small random vector to the real-valued genotype when mutation occurs defined by a fixed mutation rate in each generation with σ(0,1). The mutation vector direction was uniformly distributed, and its magnitude was a Gaussian random variable with mean 0 and variance 1. Each individual is run for a number of independent evaluations (typically 10). The fitness of each phenotype is calculated by averaging the fitness obtained in each evaluation. Each evaluation consists of a serial presentation of 6 light sources for a relatively fixed long time period (T_{ls}=50 time steps) during life time (T=300 time steps). The agent must approach in turn and remain close to light source. After T_{ls}, the light source is eliminated and another one appears at a random distance ([10;100]) and angle ([0;2π]). The intensity of each source is fixed. When light is sensed, the sensory input varies between 0 (lowest level) and 880 (highest level), with a noise level of 0.1.

Fitness is calculated in the following manner. $F = 1-D_f/D_i$, (D_f: final distance to source; D_i: initial distance to source). This term is taken as 0 if $D_f > D_i$, and it is calculated for each source and then averaged for the whole evaluation. Therefore, F is in the [0;1] range.

The search algorithm is run for a fixed number of generations (typically 1000), for a fixed number of iterations (typically 200) per generation, taking usually a few hundreds of generations to achieve a considerably high level of average fitness, depending on the configuration used in each experimental test. The capability of the evolved agents to deal with internal uncertainty with and without disruptions is then tested individually.

Results

A series of experiments have been carried out for studying the effects of internal uncertainty in neurocontroller evolvability (the effects of changing uncertainty during evolution) and lifetime performance (robustness against neural uncertainty and sensorimotor disruptions). Statistical assessment of the various types of robustness is proposed based on statistical significance tests (*one-tailed t-test*) and *bootstrapping t-test* [9]. First, *one-tailed t-test* relates to compare whether mean of A=0 is lower to those under uncertainty cases. As confidence interval, 95% was taken for the true difference (usually averaging the means of 10 independent experiments). For this test, we assume that two samples come from normal distributions with unknown and possibly unequal variances (Behrens-Fisher problem using Satterthwaite's approximation for the effective degrees of freedom). Second, the *bootstrapping t-test* or *studentized bootstrap* does not assume that the sampling distributions are roughly normal and have the same variance. The number of bootstrap samples is 2000 with confidence level for the confidence intervals 95%. As summary, our analysis shows similar conclusions in both tests.

The Effects of Uncertainty During Evolution

Preliminary experiments testing evolvability were performed with non-center-crossing controllers, testing them on the phototaxis task. It was found that evolution effectively minimised the effects of neural uncertainty by saturating neural firing rates during most of the performance trial. Therefore, experiments without center-crossing were not suitable to test our hypothesis (i.e. that actively coping with internal uncertainty will affect behavioural robustness).

The effects of uncertainty on fitness are shown in Fig. 1. Each bar indicates the final fitness after 1000 generations for different values of A averaged over 20 independent evolutionary runs. The performances indicate that evolution can find efficient strategies when internal uncertainty is included, which are quantitatively comparable to the control situation of A=0. There is a tendency to decrease in performance when uncertainty values are higher than A=2. We notice that variance also has a tendency to increase. Both these observations confirm the expectation that with higher levels of uncertainty evolvability will be reduced. This situation is slightly significant in A=2 evolution (Fig. 1), where the performance of evolution shows a decay. We therefore confine uncertainty to less than A = 2 in subsequent experiments.

Based on results shown in Fig. 1, the t-test for A=0 situation and cases with uncertainty during evolution, were as follows. Results indicate that null hypothesis (equal means) could be rejected after comparing A=0 against all other situations under uncertainty except for the comparison of A=0 and A=1. The significance of the alternative hypothesis (the mean of A=0 is higher than those under uncertainty) is accepted in every case with a significance level of 60%, 100%, 100%, 100% for A=1, A=2, A=3,

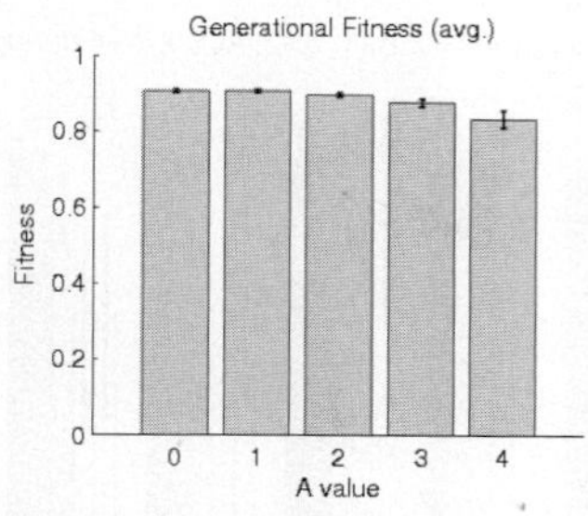

Fig. 1. Fitness values obtained averaging the fitness of 20 independent experiments after 1000 generations for different ranges of uncertainty. Horizontal axis represents the range [-*A*; *A*] of uncertainty

and A=4, respectively. These percentages mean that by chance is possible to see values more extreme than the one in tested experiments. In other words, percentages represent the significance level to obtain lower levels of performance when compared to A=0 situation. Similar conclusions were obtained after bootstrapping t-tests.

The relation between fitness and uncertainty during lifetime is shown in Fig. 2. strategies with A=1,2,3,4, and without uncertainty (A=0) during evolution were tested against different levels of uncertainty during lifetime when performing phototaxis using 10 independently evolved neurocontrollers in each case (Fig. 2). Experiments suggest that certain levels of uncertainty (A=1, A=2, and A=3) slightly improve or maintain agent performance in comparison with evolution without neural uncertainty (A=0). In other words, the A=0 neurocontroller obtained 84.8% of performance against uncertainty variability, meanwhile 86.9%, 96.4%, 98.5%, and 99.7% were obtained by A=1, A=2, A=3, and A=4, respectively. Such percentages indicate the robustness proportion (average) based on reached fitness under disruptions over the obtained one in the control case.

An important point is that in the absence of uncertainty, neurocontrollers evolved with uncertainty maintained their performance (Fig. 2), expressing that such neurocontrollers do not use internal noise functionally to perform the task.

For statistical significance, an analysis of the effects of uncertainty during lifetime was developed. Results indicate that the null hypothesis (equal means) could be acepted after comparing the control case against A=1 under every uncertainty situation in lifetime. Null hypothesis should not be accepted when comparing A=0 to A=2 for A=3 and A=4 levels of uncertainty during lifetime. Obtained significances for A=1 and A=2 represent the difficulty to observe values lower than the A=0 mean. Similar conclusions were obtained with bootstrapping t-tests.

Adaptation to Sensorimotor Disruptions in Lifetime

Fig. 3 shows the performance of neurocontrollers evolved with A=0, A=1, and A=2 in 20 independent evolutionary runs with different initial random seeds using the same evolutionary process previously explained; these controllers are then tested for different disruptions. The disruptions tests were mostly based on experiments proposed in [7] and [8]: (a) *neural lesions*: (a.1) sensor inversion; (a.2) node disruption (input node1 and 2); (a.3) neural uncertainty change (interneuron 1 and 2 with(out) uncertainty

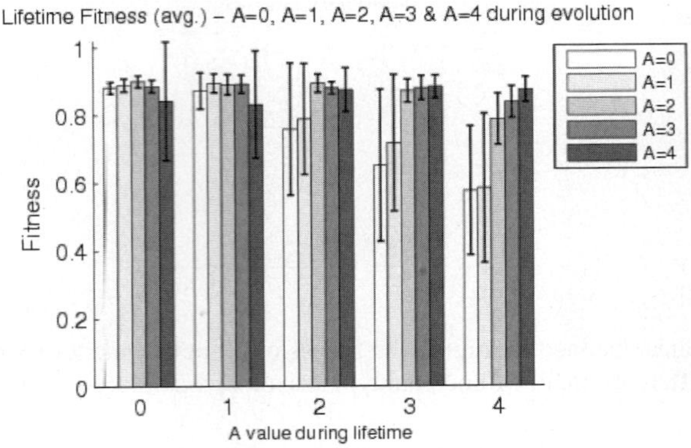

Fig. 2. Variable uncertainty levels during lifetime tests showing the obtained lifetime fitness for controllers evolved with A=0,1,2,3,4; each data point indicates the average fitness of 10 independent experiments.

A=1); (b) *sensorimotor disruptions*: (b.1) sensors gain (sensor left and right in [-75%, +100%] and in [-25%, +50%]); (b.2) motor gain (motor left and right ±50%). Experiments on the inversion of the visual field (*test2*) consist of exchanging the position of sensors. In *test3* and *test4*, the output of a single interneuron is set to zero. From *test5* to *test8* represent situations in which a random increase or decrease to sensors (right or left) is given according to [-75%, +100%] (high range) or [-25%, +50%] (low range). Finally, *test9* and *test10* depicts random right or left motor gains changes over a range of ±50%.

It is important to mention that for the neurocontrollers shown in Fig. 3, the uncertainty level during lifetime was maintained at the value at which it was evolved as indicated in the legend. Fig. 3 depicts the performance of neurocontrollers evolved with those values of uncertainty during tests A=0, A=1, and A=2 and for A=1 and A=2 with no uncertainty in lifetime. Following a short initial time period of normal phototaxis (before t=30), an agent is tested against a disruption. It is possible to estimate from this series of tests an idea of robustness corresponding to the different values of A. This can be done by averaging the ratio of performance between the disrupted and the normal conditions. For the case of A=0 this robustness indicator is 85.2%; for A=1, 85.9% with A=1 and 85.9% with A=0 in lifetime; for A=2 with A=2, 88.6%, and 90.3% with A=0 in lifetime (Fig. 3). Neurocontrollers evolved with A=0 do not show more robust performance than those evolved with A=1 or A=2. This result indicates that uncertainty has a beneficial effect on robustness against disruptions and lends some initial support to our original hypothesis. Agents with A=2 during evolution were more robust and their average performances are comparable to the cases with A=0 in control case. Results of sensor disruptions (*test5* to *test8*) are similar probably because strategies coping with the environment could implicate some degree of robustness against increases or decreases of sensing. In other words, sensory perturbations had not enough influence during tests. We can also observe that neurocontroller A=2 with uncertainty 2 in lifetime demonstrated a high level of robustness against morphological disruptions (77.6%), and 72.2% for A=1 (A=1 in lifetime) in comparison with the control case (71.5%), meanwhile A=2 without uncertainty in lifetime

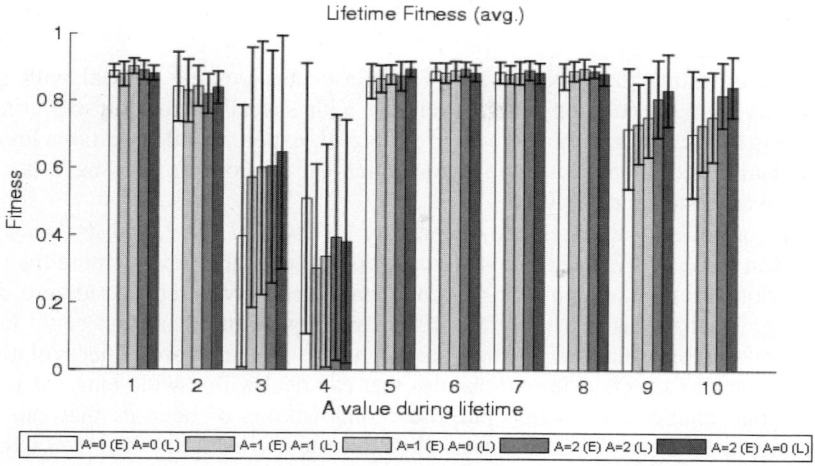

Fig. 3. Experiments with center-crossing neurocontrollers evolved (E) with A=0, A=1, and A=2; tests against lesions and sensorimotor disruptions during lifetime (L) with uncertainty values according to the legend; 20 independent evolutionary runs per plotted bar. *Test1*: no disruption; *Test2*: inversion; *Test3*: damage neuron 1; *Test4*: damage neuron 2; *Test5*: sensor left high; *Test6*: sensor left low; *Test7*: sensor right high; *Test8*: sensor right low; *Test9*: gain left; *Test10*: gain right.

obtained 80.6%. Respect of environmental disruptions indexes, they were 98.8%, 99.6%, 98.4%, 99.5% and 99.9% for A=0, A=1 with uncertainty A=1 and A=0, and A=2 with uncertainty A=2 and A=0, respectively. According to data shown in Fig. 3, results of t-test and bootstrapping support these ideas.

A full behavioural and dynamical analysis of how agents perform these tasks under different conditions remains to be carried out. Modifications to other body parameters and internal lesions (removal of connections or modifications to the behaviour of single neurons) have not been explored either.

In Fig. 3, the performance of neurocontrollers evolved with A=1, but tested with no uncertainty during the disruption (A=1 (E) A=0 (L)), results in very similar levels of robustness to the case where the same controller was evaluated using uncertainty A=1 during the test phase. This indicates that the result of including moderate levels of neural uncertainty in the evolutionary process is not only capable of producing higher robustness when the uncertainty is present, but also when the uncertainty is absent during the test phase. Similar results could be observed with A=2 neurocontroller. The presence of uncertainty seems therefore to act in ways that enhance the exploration of regions of the fitness landscape where controllers are more robust per se. This result gives us an insight into how moderate levels of uncertainty affect evolutionary search.

Conclusions

Experiments testing the capacity of CTRNN's center-crossing to deal with neural uncertainty during evolution were developed with special interest in exploring the robustness of the resulting behaviours. To this end, experimental conditions implying lesions and sensorimotor disruptions for which the neurocontrollers have not been evolved were tested and analysed.

Neurocontrollers with moderate uncertainty levels (A=1, A=2, and A=3) are more robust than those evolved with no uncertainty (A=0) or higher levels (more than A=3) in evolution and they remain robust even when uncertainty is reduced to zero during tests (Fig. 2). It means that the effect of uncertainty during evolution could help to find more suitable solutions to the task evolved. In other words, the evolutionary process seems to selects neuron dynamics that can deal with a wide range of internal and external changes. Evolution purposes combinations of neurons that can adapt themselves, and as a group. However, when uncertainty is higher than A=3 in evolution, such benefits are affected by a slightly variability.

A contrast between results could be shown. On the one hand, based on previous robustness indexes, we can tentatively conclude that a little uncertainty positively affects the robustness of neurocontrollers, but when it is considerably increased, some visible negative effects appeared in agent's performance (e.g., low performance levels, and erratic behaviours). Therefore, returning to our initial hypothesis about whether uncertainty is a prospective mechanism for obtaining more robust artificial systems, we can say that it is convenient (in robustness terms) to have a little "neural noise" (uncertainty) during evolutionary and lifetime processes, but just a few noise.

On the other hand, an important observation is that tests implying A=1 and A=2 with and without uncertainty during lifetime presented similar levels of robustness against uncertainty in Fig. 3. Some tentative suggestion about why both configurations performed well indicates that evolution generated feasible strategies against internal and external perturbations using uncertainty. Nevertheless, when uncertainty is high, it is probable that the trade-off between neural noise and center-crossing properties could be lost, meaning that evolution could not be able to generate better individuals in each generation due to the high variability caused by uncertainty. In spite of obtained results, further studies in this context must be carried out.

The results suggest that the mechanisms at work in the production of robust controllers is an enhancement of the evolutionary search under moderate levels of uncertainty leading to the discovery of zones of the fitness landscape that correspond to neurocontrollers that are robust both *with* and *without* uncertainty. For higher levels of uncertainty (more than A=3) this effect seems to break down (Fig. 2) leading to controllers that can evolve high levels of fitness but are less reliable when uncertainty is removed.

The significance of this work serves to demonstrate that evolutionary neurocontrollers can evolve robustly under lower levels of uncertainty, in spite of internal noise in every neuron and sensorimotor disruptions. The consideration of neural noise in every neuron is the novel aspect of this paper, as opposed to previous works which only permitted noise in sensorimotor signals (see [10]).

Different issues are opened for discussion. Firstly, a proper analysis of neural dynamics and behaviour in the presence of different levels of uncertainty and for the cases of disruption must be planned in future works. Some preliminary observations have indicated that functional redundancy may be used to increase robustness based on analysis of neural dynamics (e.g. neurons acting in the same way). For example, this study will indicate how isolated neuron activities could affect other neurons under different levels of uncertainty, and how neurons will react to specific internal changes. Secondly, other forms of uncertainty must be considered. For example, uncertainty could relate to a constant value y_0 for each neuron, but uncertain from one evaluation to the next. This representation of uncertainty should facilitate the analysis of neural dynamics because a fixed level of uncertainty could generate clear patterns of neuron's dynamics (e.g. increasing or decreasing their activities). Therefore, uncertainty would bias the activity of neurons. Finally, experiments evolving different tasks should be proposed for generalizing how uncertainty affects evolution. Such generalization process should cover a wide range of tasks, mainly in the minimal cognitive context (e.g. learning and action-selection problems).

All of these issues deserve further attention, as well as clearer studies for their practical application. However, one of the goals of this paper is to open questions for discussing ideas about uncertainty in neurocontroller evolution.

Acknowledgments. Thanks to A. Furey, P. Fine, M. Rohde, and S. McGregor for their corrections of this paper. This work was partially supported by the Programme Alβan, the European Union Programme of High Level Scholarships for Latin America, scholarship No. E05D059829AR, and also recognized by CONICET, Argentina, Type-I-Res.38-08/01/2004.

References

1. Di Paolo, E.A., Harvey, I.: Decisions and noise: the scope of evolutionary synthesis and dynamical analysis. Adaptive Behavior 11(4), 284–288 (2004)
2. Seth, A.K.: Noise and the pursuit of complexity: A study in evolutionary robotics. In: Husbands, P. (ed.) Evolutionary Robotics. LNCS, vol. 1468, pp. 123–137. Springer, Heidelberg (1998)
3. Kitano, H.: Biological robustness. Nature Reviews 5, 826–837 (2004)
4. Jakobi, N.: Evolutionary robotics and the radical envelope of noise hypothesis. Journal of Adaptive Behaviour 6(2) (1997)
5. Mathayomchan, B., Beer, R.: Center-crossing recurrent neural networks for the evolution of rhythmic behavior Source. Neural Comp. 14(9), 2043–2051 (2002)
6. Beer, R.: On the dynamics of small continuous-time recurrent neural networks. Adaptive Behaviour 3, 469–509 (1995)
7. Di Paolo, E.: Homeostatic adaptation to inversion in the visual field and other sensorimotor disruptions. In: From Animals to Animats 6, SAB'2000, pp. 440–449 (2000)
8. Beer, R.: Intelligence as Adaptive Behaviour: An Experiment in Computational Neuroscience. Academic Press, San Diego (1990)
9. Efron, B.: The Jackknife, the Bootstrap and Other Resampling Plans. CBMS-NSF Regional Conference Series in Applied Mathematics, Monograph 38. SIAM, Philadelphia (1982)
10. Nolfi, S., Floreano, D.: Evolutionary Robotics: The Biology, Intelligence, and Technology of Self-Organizing Machines. MIT Press, Cambridge (2000)

Simulations of Simulations in Evolutionary Robotics

Edgar Bermudez Contreras and Anil K. Seth

Department of Informatics, University of Sussex,
Falmer, Brighton, BN1 9QJ, UK.
{e.j.bermudez-contreras, a.k.seth}@sussex.ac.uk

Abstract. In recent years simulation tools for agent-environment interactions have included increasingly complex and physically realistic conditions. These simulations pose challenges for researchers interested in evolutionary robotics because the computational expense of running multiple evaluations can be very high. Here, we address this issue by applying evolutionary techniques to a simplified simulation of a simulation itself. We show this approach to be successful when transferring controllers evolved for example visual tasks from a simplified simulation to a comparatively rich visual simulation.

1 Introduction

For more than a decade, evolutionary robotics (ER) has struggled with the challenge of producing controllers that function in real world environments. The approach of evolving in the real world itself is prohibitively time consuming in all but the simplest of cases [1],[3]. A popular alternative has been to evolve controllers in simulations, but simulations are often poor abstractions of the complexities of real world environments. This situation is changing. Recent years have witnessed enormous growth in the sophistication of simulation tools for modelling agent-environment interactions. Highly detailed physics-based simulations are now readily available 'off-the-shelf' which simulate not only complex morphologies but also rich streams of sensory input and motor output signals [5],[6]. While impressively realistic, these simulations can be highly computationally expensive and as a result can pose challenges similar to those posed by evolution in the real world.

This is not to say that evolution in a rich simulation is as problematic as evolving in the real world. Even a very rich simulation can likely be executed more rapidly (and with less chance of hardware failure) than a corresponding real world condition. If this is not true at present for a particular simulation, future increases in computational power will undoubtedly compensate. In addition, rich simulations offer the possibility of exploring detailed but non-physical agent-environment interactions, which may shed light on adaptive behavior by providing alternative comparison conditions to agent-environment interactions in real-world situations.

F. Almeida e Costa et al. (Eds.): ECAL 2007, LNAI 4648, pp. 796–805, 2007.

In this paper, we consider the construction of simplified simulations of simulations themselves, from the perspective of ER. We describe the construction of a simulation of a rich simulation of visually guided behavior and illustrate the value of this simplified simulation by evolving controllers to perform two visually guided tasks: object approaching and object discrimination. We show that controllers evolved in the simplified simulation transfer successfully into the rich visual simulation, despite there being significant differences in the structure of sensory input in the two cases. Our work therefore suggests the possibility of a "hierarchy" of simulations of progressively increasing complexity as a means of (i) evolving controllers for real world operation, and (ii) probing the dynamical structure of adaptive agent-environment interactions.

2 Methods

Simulated agents were evolved to perform two visually guided tasks. The first task, 'object approaching', required the agent to approach an object placed in the arena and be as close as possible to the object at the end of a fixed period of time. The second task, 'object discrimination', required the agents to discriminate between two different objects by approaching only one of them and remaining as close as possible to that object at the end of a period of time.

Two types of agents were simulated, one with a visual system using a simulated camera, called the "rich simulated agent" (RSA) and another using a simplified visual system called the "simple simulated agent" (SSA). As described below, the visual systems of both agents were tailored to each visual task. A genetic algorithm (GA) was used to evolve continuous-time recurrent neural network controllers (CTRNNs) for SSAs, and successful controllers were analyzed both as controllers for SSAs and as controllers for RSAs.

2.1 Rich Simulated Agent (RSA)

The RSA has a circular body with two wheels driven by two independent motors, and a camera on top of its body. The visual system of the RSA has a visual field which is a grey-scale region of the output of the simulated camera. This region is 512×32 pixels (see figure 1). The visual system has a blob detection mechanism and two types of sensors. The blob detection mechanism selects visual subregions of consistent pixel intensity with area in the range 10-50 pixels. Only one 'blob' is selected at any time. In cases where there is more than one blob in the visual field, the visual system selects the blob with largest area.

Two types of sensor respond to a selected blob. The first is a "location sensor" which responds to the proximity of the object to the edge of the visual field (the sub-region of the image captured from the camera). The RSA has left and right location sensors. These sensors are activated by the inverse of the distance (L or R) between the object and the corresponding edge of the visual field (see figure 1). The second sensor type, used only in the object discrimination task, are "colour sensors" that return the pixel intensity of the centroid of the selected

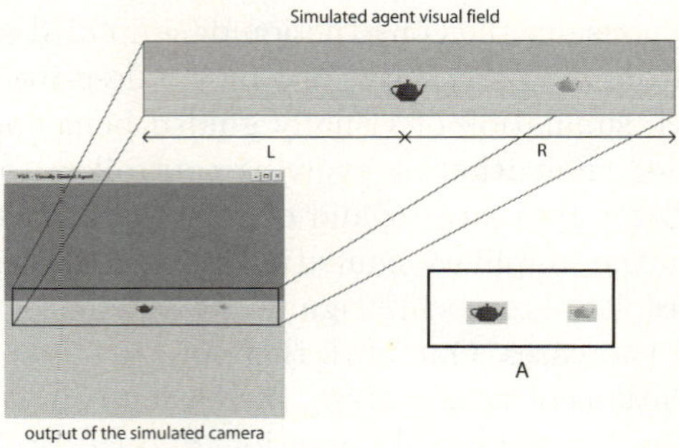

Fig. 1. Simulated visual system. The visual field of the agent is a region of 512×32 pixels. L is the distance from the object to the left edge of the visual field and R is the distance from the object to the right edge. The inset A in the figure shows the detected blobs (from a distance of 2.5 to the dark object and 3.0 units to the light object) containing the light and dark objects respectively. In this example, the dark object is the largest and so the sensor neurons will respond to this object.

blob. Because the rich visual simulation incorporates directional illumination and reflectance properties of the objects in the arena, the pixel intensity at any time is a complex function of intrinsic properties of the object detected and the reflectance of the object in the corresponding region of the visual field. Although the two colour sensors receive identical input (unlike the location sensors), they may still produce different outputs depending on intrinsic neuron properties (see below).

At the beginning of each evaluation, an RSA was randomly positioned within a region of 12x12 units in an unlimited arena. For object approaching experiments, a visual object (a dark-coloured kettle) was placed in a fixed position in the arena. For the object discrimination task, a light coloured kettle (target) and a dark coloured kettle (distracter) (see inset A in figure 1) were placed in the arena in positions (0, -4) and (0, 4), respectively. During evolution, each evaluation lasted for 200 time-steps; during analysis of evolved controllers, each evaluation lasted for 800 time-steps.

2.2 Simple Simulated Agent (SSA)

The SSA has a circular body with radius of 0.5 units and two wheels on both sides of the agent, driven by two independent motors. The simplified simulated visual system of this agent has a visual field that is restricted to a region of fixed width V. This region is also limited by two lines originating from the center of the agent extending ±45 deg from the orientation of the agent (see figure 2). It is important to emphasize that this region is spatial, in the sense that it is defined in terms of a subregion of the arena, rather than, as is the case for the RSA, as a subregion of a visual image. This difference means that sensory signals for the two agents will have different dynamical structures. For example, it is possible that a visual object will move in and out of view for the SSA (because of the fixed width V of the visual field) while remaining constantly within view for the RSA. (One situation in which this may occur is as an agent spins.)

It is also important to notice that there is a spatial region near to the agent where the SSA is blind (the light gray region in figure 2). As we describe below, this blind region is important in the explanation of evolved behaviour of the SSA.

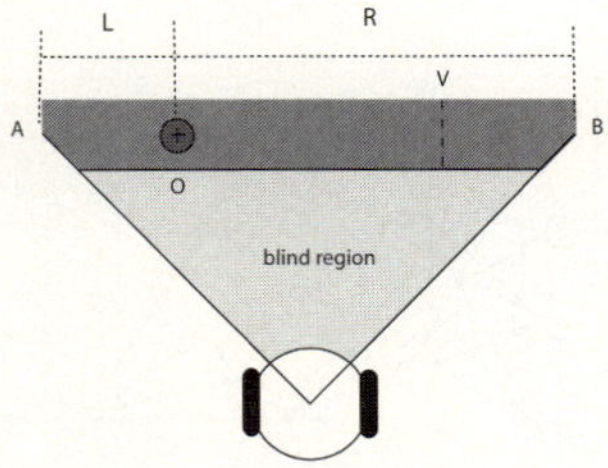

Fig. 2. Visual field of the SSA: the object (O) can only be sensed if it is within the dark region. This region is limited by two lines extending from the center of the agent with ±45 deg from the orientation line and a width of V. L and R are the distances between the object and the left and right edges of the visual field, respectively.

As with the RSA, the SSA has two types of sensor which take input from the visual system. The location sensors of the SSA are activated by the inverse of the distance (L or R) between the object (if it is within the visual field) and the corresponding edge of the visual field. The colour sensors of the SSA return a similar value to the pixel intensity of the objects (40 and 130 for the dark and light kettles, respectively) used for the RSA. To deal with the variation in values of colour sensors for RSAs (resulting from changes in reflectance and in intrinsic properties of the selected blob), colour sensors for SSAs were modulated by a random value [-30, 30] (distributed uniformly).

2.3 Controller

The controllers for both types of agents were Continuous Time Recurrent Neural Networks (CTRNNs)[3], [2]. In a CTRNN, the state y of each neuron i changes in time according to the differential equation:

$$\tau_i \dot{y}_i = -y_i \sum_j w_{ij} \phi(y_j + \beta_j) + g_i \cdot I_i$$

where ϕ is the sigmoid activation function, τ is a time constant, β is a bias, and w_{ij} represent connection weights from neuron i to neuron j. The state of each neuron is therefore the integration of the weighted sum of all incoming connections (plus a gain modulated input $g_i \cdot I_i$ for input neurons).

For the object approaching task, the CTRNN consisted of eight neurons, specifically, two sensor neurons, four fully connected interneurons and two motor neurons. For the discrimination task, two more sensor neurons corresponding to the colour sensors were added (see figure 3). Parameter values for all neurons were initialised in the following ranges: $\tau \in [0.2, 2.0]$, $\beta \in [-10, 10]$, and connection weights $w_{ij} \in [-5, 5]$. In the object discrimination task, neurons 8 and 9 used $\tau \in [0.2, 10.2]$ and bias $\beta \in [-30, 30]$. All parameter values were shaped by the GA (see below).

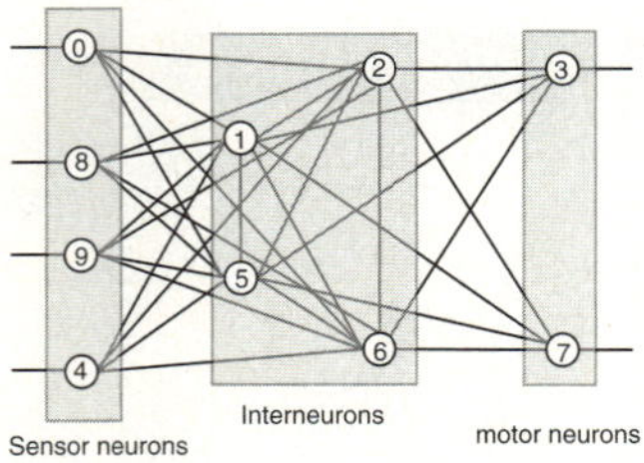

Fig. 3. Controller. Neurons: 0 and 4 are location sensors; 8 and 9 are colour sensors. Neurons 1,2, 5 and 6 are fully connected. Neuron 3 is the left motor neuron and neuron 7 is the right motor neuron. Note that the colour sensor neurons 8 and 9 were *not* used for the object approaching task.

2.4 Genetic Algorithm

A distributed GA was used to evolve CTRNNs to perform the visually guided tasks. The genome of each individual was coded as a real vector of 32 elements for the object approaching controller and 39 elements for the object discrimination controller. For the 32 element vector, 4 elements were used to code the time constants of each neuron, 4 for the bias of each neuron, 2 for the sensor gains and 22 for the weights. Each element was coded as a real number in $[0, 1]$ and linearly scaled according to the parameters previously described in section 2.3. For the 10 neuron controller 7 elements were added, 2 for the bias of the two extra sensor neurons, 2 for the time constant, 1 for a sensor gain for these neurons and 2 for the weights. A population of 400 individuals was evolved with mutation probability of 80% for each genotype and 20% for mutation change for each vector element. There was also a 5% probability of crossover and an elitism probability of 80%.

The controllers were symmetrical (i.e., same parameters were used for each pair of sensor neurons, 0 and 4; 1 and 5; 2 and 6 and so on. See figure 3.), except for neurons 8 and 9 which had independent parameters.

Two fitness functions $F_1 = 1/d_f$ and $F_2 = 1/d_l - 1/d_d$ were used. F_1 was used for the object approaching task and F_2 for the object discrimination task. In F_1, d_f is the distance from the agent to the object at the end of the trial and in F_2, d_l and d_d are the final distances between the agent and the light and dark objects respectively. The fitness of each individual was calculated as the average across 5 independent trials (of 200 time-steps each).

3 Results

After several thousands of generations controllers were successfully evolved for both tasks. As mentioned previously, controllers were evolved using the SSA and were then tested in both types of agents, SSA and RSA.

3.1 Object Approaching Task

For this simple task successful controllers were found quickly (before 2000 generations). As we can see in figure 4B, the agents used an exploratory strategy, first spinning until the object was within the field of view and then approaching the object and rotating around or very close to it.

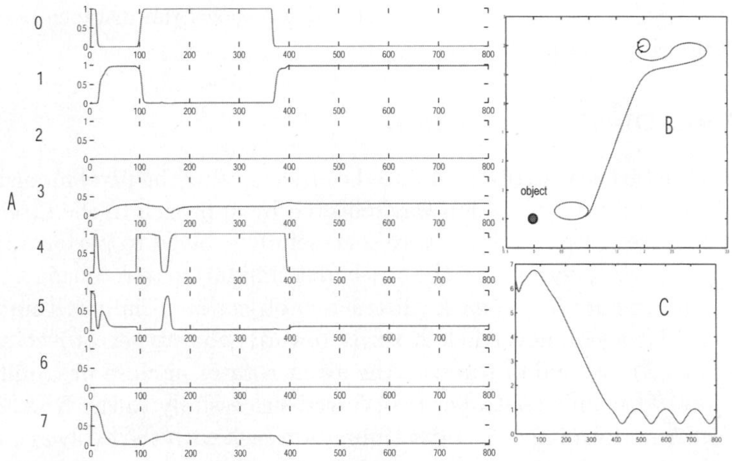

Fig. 4. Object approaching by an SSA. [A] shows the neural activity during a test trial of 800 time-steps. [B] shows the distance between the agent and the object during the trial and [C] shows the distance between the agent and the object during the trial.

Successful controllers for SSAs were tested in agents using the rich visual system (RSAs). These evolved controllers also performed the object approaching task successfully (see figure 5B). The behaviour of the RSAs was similar to that observed for SSAs: rotate or explore until the object is within the visual field, approach the object and then rotate close to it. In the particular case shown in the figures, the circle described by the trajectory of the RSA at the end of the trial is bigger than that described by the trajectory of the SSA. This observation is highlighted by figures 4C and 5C, where the distance to the object is shown during the test trial.

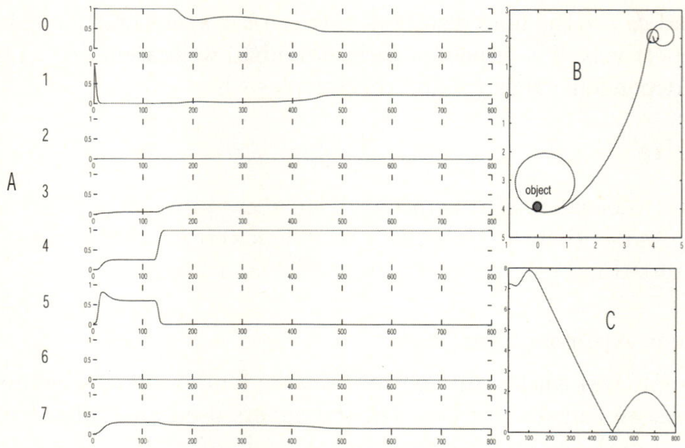

Fig. 5. Object approaching performed by an RSA using the evolved controller shown in figure 4. [A] shows the neural activity during the test trial of 800 time-steps. [B] shows the trajectory of the agent the trial and [C] shows the distance between the agent and the object during the trial.

3.2 Object Discrimination Task

In this case the task was to discriminate the objects using the pixel intensity information. Successful discrimination was reflected by approach to the target object (the light-coloured object). SSAs were successfully evolved to perform this task. Figure 6 shows a SSA performing the object discrimination task during a test trial.

As shown in figure 6, the dark (distracter) object is initially within the field of view but the agent nevertheless turns towards the target object and then approaches it. At the end of the trial the agent rotates in close proximity to the target object. The same controller transferred successfully to the RSA. Figure 7 shows an RSA performing object discrimination task with the evolved controller.

As in the first task, the behaviour of the RSA is similar to that of the SSA. The agent rotates until the object is within its visual field and then approaches it. In the trial shown in figure 7, the dark object is closer to the agent at the beginning of the trial, however, after a short time, the agent moves away from the dark object and subsequently approaches the target. Note that for the object discrimination task, both SSAs and RSAs stay very close to the target object (compare figures 6C and 7C).

It is important to emphasize that, for this task, certain aspects of simulation of the colour sensors were critical for the successful transfer of controllers. Specifically, evolutionary runs in which random variance in these sensor values was not incorporated (see section 2.2) showed considerably decreased performance when transfer to an RSA was attempted. During attempted transfer in these cases, variance in the RSA colour sensor values (due to the richness of the visual simulation) resulted in these agents approaching both object types equally often.

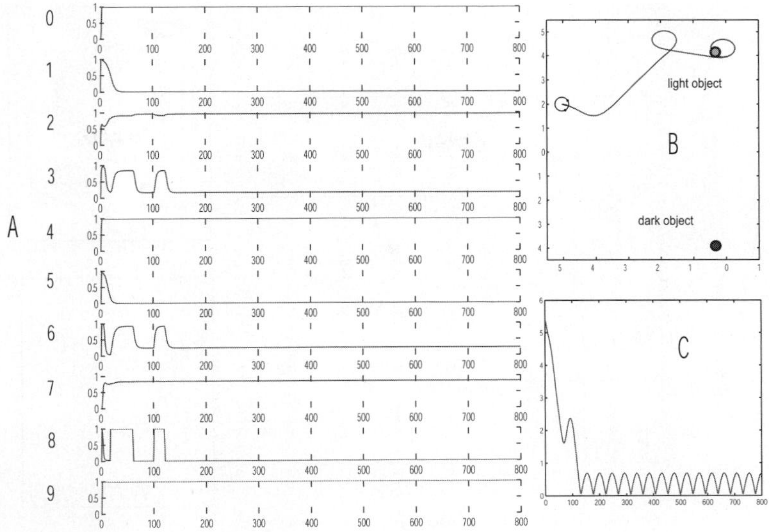

Fig. 6. Object discrimination performed by an SSA. [A] shows the neural activity during a test trial of 800 time-steps. [B] shows the trajectory of the agent during the trial and [C] shows the distance between the agent and the object during the test trial.

4 Analysis

In general the strategies of both SSAs and RSAs can be described as follows. First, agents rotated until an object was within the field of view, then agents approached the object, and finally, agents rotated either close to or around the object, until the end of the trial.

In order to better understand the dynamics of evolved behaviours and the factors underlying successful transfer between simulations, we now examine evolved behaviors in terms of neural activity. For both agent types, the initial rotating behaviour can be attributed to the random initialisation of the CTRNN. This was shown by initialising the neurons uniformly, in which case SSAs and RSAs navigated in a straight line at an arbitrary heading (data not shown). The approach behaviour of both agent types can be attributed to sensor activation corresponding to an object perturbing the equilibrium point in neural dynamics corresponding to the spinning behaviour. This was shown by testing SSAs and RSAs without any object in the arena (data not shown).

For object approaching task, neurons 2 and 6 were always constantly saturated for both agent types and therefore can be discarded from the analysis (see figure 4A and 5A and figure 3), leaving only neurons 1 and 5 as modulators of motor neuron activity (see figure 3). For the object discrimination task, all the sensor neurons are constantly saturated except for neuron 8 (again for both agent types). Since this type of neuron has a different weight for each connection,

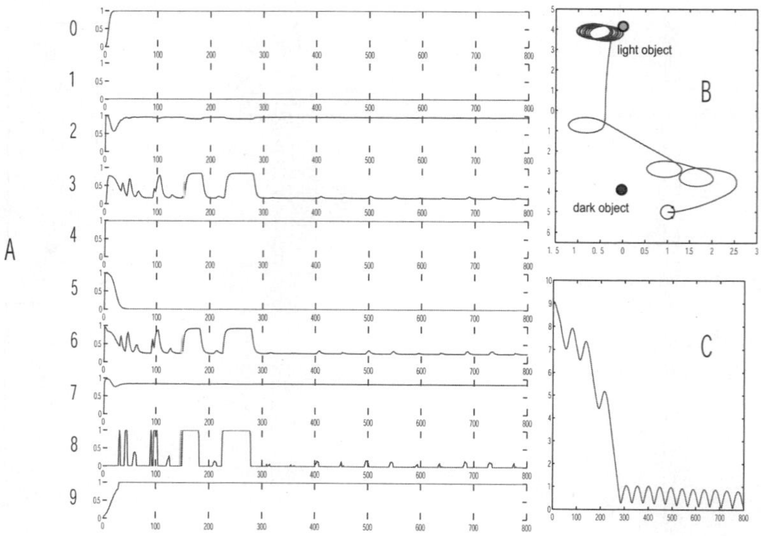

Fig. 7. Object discrimination performed by an RSA using the evolved controller shown in figure 6. [A] shows the neural activity during a test trial of 800 timesteps. [B] shows the trajectory of the agent during the trial and [C] shows the distance between the agent and the light object during the trial.

it is still able to modulate neuron 6 which in turn is responsible for regulating the motor neurons (see figure 6 and 7).

The final segment of successful agent behaviour involved rotating close to an object. This behavior was related to the initial rotating (described previously). Once the agents were sufficiently close to the object so that the object was within the "blind region" (see section 2.2), they reverted to spinning. In the object approaching task, when this happens, the agent could no longer sense any object and the situation was equivalent to the one where no object was present. For the object discrimination task, once the agent was spinning very close to the target object but was not able to sense it, the agent could still sense the dark (distracter) object (see neuron 8 in figure 7A, the small peaks correspond to the dark object and high peaks to the light object) but the activation of the sensor neuron was not high enough to trigger approaching behaviour. This situation is not shown in figure 6 because the agent is spinning too far away from the dark object to be able to detect it, however the same situation applies to both SSAs and RSAs.

In general, the behaviour of the evolved controllers shows that despite the differences in the dynamical structure of sensory signals between SSAs and RSAs, evolved controllers transferred successfully from one to the other. As the neural analysis shows, this transfer was possible because evolved agents relied on consistent features of sensory activity, and not on those aspects that varied between the agent types (see section 2.2).

5 Conclusions and Future Work

In this work, it was shown that evolved controllers for agents using a simplified visual system (SSAs) could be successfully transferred to agents using more complex visual information (RSAs). The behaviour of both agents (SSAs and RSAs) for object approaching and discrimination was fully explained by analysing the dynamics of their neural activity. In this way, it was shown that the complexity gap between SSAs and RSAs was crossed.

This demonstration is useful for evolutionary robotics in several ways. First, the development of increasingly complex simulations is blurring the distinction between simulation and reality, therefore an important future goal for ER will be to create adaptive controllers for agents in simulations, and not only as a bridge to real-world situations. On the other hand, a hierarchy can be envisaged in which controllers are initially evolved in simple simulations and then are incrementally refined in progressively more complex simulations until final deployment in a real world environment. Alternatively, rich simulations offer the possibility of exploring detailed agent-environment interactions which do not exist in real-world situations, thereby supplying potentially valuable comparison conditions for understanding mechanisms of adaptive behaviour.

Future work in this area could usefully consider the development of *minimal* simulations of rich simulations, in the sense described by Jakobi [4] . Minimal simulations incorporate extremely high levels of noise in specific loci in order to ensure that evolved controllers cannot rely on these aspects of agent-environment interaction. This method might extend the 'complexity gap' between simulations that can be feasibly traversed by evolutionary approaches.

References

1. Cliff, D., Harvey, I., Husbands, P.: Artificial evolution of visual control systems for robots. In: Srinivisan, M., Venkatesh, S. (eds.) From Living Eyes to Seeing Machines, pp. 126–157. Oxford University Press, Oxford (1997)
2. Funahashi, K.-I., Nakamura, Y.: Approximation of dynamical systems by continuous time recurrent neural networks. Neural Netw. 6(6), 801–806 (1993)
3. Harvey, I., Husbands, P., Cliff, D., Thompson, A., Jakobi, N.: Evolutionary robotics: the sussex approach. In: Robotics and Autonomous Systems (1996)
4. Jakobi, N.: The minimal simulation approach to evolutionary robotics. In: Gomi, T. (ed.) Proceedings of Evolutionary Robotics - From Intelligent Robots to Artificial Life, AAI Books, Ottawa (1998)
5. Koening, N., Howard, A.: Design and use paradigms for gazebo, an open-source multi-robot simulator. In: Proceedings of International Conference on Intelligent Robots and Systems, vol. 3, pp. 2149–2154 (2004)
6. Michel, O.: Webots^{TM}: Professional mobile robot simulation. International Journal of Advanced Robotics Systems 1(1), 39–42 (2004)

Synthesizing Physically-Realistic Environmental Models from Robot Exploration

Josh Bongard

Department of Computer Science, University of Vermont
33 Colchester Ave., Burlington, VT 05405
josh.bongard@uvm.edu www.cs.uvm.edu/~jbongard

Abstract. In previous work [4] a framework was demonstrated that allows an autonomous robot to automatically synthesize physically-realistic models of its own body. Here it is demonstrated how the same approach can be applied to empower a robot to synthesize physically-realistic models of its surroundings. Robots which build numerical or other non-physical models of their environments are limited in the kinds of predictions they can make about the repercussions of future actions. In this paper it is shown that a robot equipped with a self-made, physically-realistic model can extrapolate: a slow-moving robot consistently predicts the much faster top speed at which it can safely drive across a terrain.

1 Introduction

Modeling has long played a part in robotics research, but has been plagued by two main challenges: models must either be laboriously created by hand, or (whether manually or automatically created) they are of limited use for making predictions about the outcome of future actions. Models have also been used extensively in evolutionary robotics [13], where the goal is to use evolutionary algorithms to automate the creation of behaviors. There are three main approaches to evolutionary robotics: controllers are either evolved directly on the physical device, requiring thousands of evaluations [6][8]; controllers are adapted from an existing, hand-designed controller [17]; or a hand-designed simulator is used to evolve controllers in simulation before transferal to the physical device [10][14]. The first approach is infeasible for continuous, rapid adaptation; the second approach requires a human to create the starting behavior; and the third approach requires a human to craft a simulation of the robot.

In previous work [2][4] a fourth method was introduced that overcomes these obstacles by allowing the robot to evolve simulations of itself, and then use the best of the evolved simulations to internally rehearse behaviors before attempting them in reality. It was found that the models allowed the robot to make successful predictions regarding future actions: it first created a self-model using only gentle rocking motions; it then optimized a locomotion pattern using the self-model; and finally it executed that pattern successfully in reality.

Here it is demonstrated that the same technique can allow a robot to model its surroundings, and then make successful predictions about how it can interact

F. Almeida e Costa et al. (Eds.): ECAL 2007, LNAI 4648, pp. 806–815, 2007.

with that environment. Specifically, it is shown that a robot can learn the topological properties of a terrain and then predict the maximum speed at which it can traverse that terrain without tipping over. Many techniques exist for allowing a robot to model its environment (or the nature of its interaction with that environment), dating back to the first research with autonomous robots [12]. More recent approaches have embraced probabilistic models (for an overview see [18]). Other approaches include compiling a database of past experiences [11] or training neural networks to reproduce sensor-motor correlations [9]. However, the models from all of these approaches are limited, in that they only allow the robot to make predictions about future actions that have either been performed before, or are some aggregate of those actions. For example in [9], the robot can predict future change in its position given the current acceleration, as long as that acceleration falls within the space of past accelerations that were used to train the model.

In the next section the robot, its environment, and the technique it uses to make successful predictions about new behaviors in that environment is described. In section 3 results generated using this approach are presented, and in the final section some discussion and concluding remarks are provided.

2 Methods

Here it is assumed that a robot wishes to reconstruct the topological features of its surrounding terrain, in order to better understand how it can interact with it: for instance, what is the top speed at which it can drive across the terrain without tipping. Clearly, the robot should not simply drive fast to determine this threshold, as it may actually tip over. Rather it should move slowly, synthesize an accurate topological model of that terrain, and then drive a virtual copy of itself across that terrain to determine its safest top speed.

This process is executed as follows. First the robot slowly traverses the terrain it wishes to model, known as the *target terrain*. In this work, a virtual robot and environment are used in lieu of a physical robot and environment: in future work a physical robot will be used. Two examples of the robot and its environment are shown in Fig. 1.

This traversal generates sensor data, which the robot stores. The robot then generates a series of environmental simulations, known as *model terrains*. It drives a virtual copy of itself across each model terrain, which again generates sensor data. It then compares the model-generated sensor data against the sensor data from the target terrain. The closer the match, the more accurate the current model terrain must be. The robot then uses a hill climber [15] to optimize the model terrains by minimizing the distance between the model-generated and target-generated sensor data.

Both the target and model environments are simulated using physical simulation[1]. In a physical simulation, all objects have physical properties such as mass, velocity, and friction. Objects are connected to one another using joints,

[1] www.ode.org

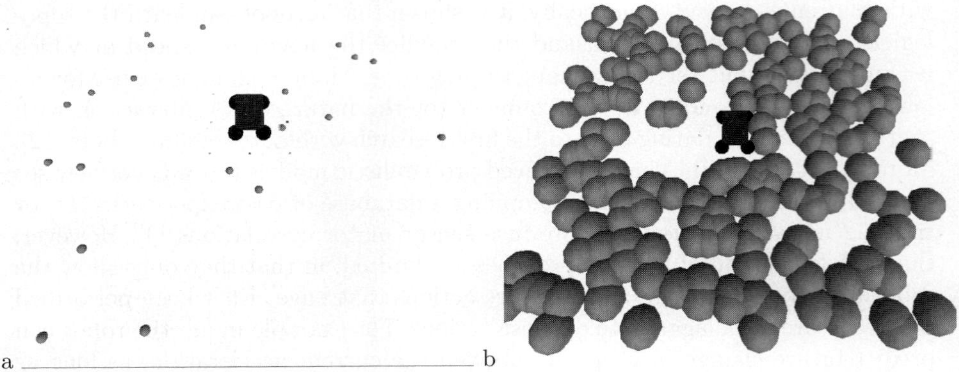

a b

Fig. 1. The robot, and two sample environments. The wheeled robot is shown at center. **a:** The robot driving across a terrain with 10 boulders each with a radius of 10cm. **b:** The same robot traversing a terrain with 200 boulders with radius 45cm.

which may be powered. At each time step of the simulation, the positions of the objects are updated based on their current acceleration, and the internal and external forces acting on them.

The robot used here is made up of five objects: a rectangular body and four spherical wheels. The wheels are driven at a constant rotational velocity. The wheels are connected directly to the body: turning is accomplished by driving the left and right wheels at slightly different velocities. Every traversal of a terrain is accomplished as follows. The robot begins driving forward, with the left wheels moving at 0.6 revolutions per second and the right wheels at 0.4 rev/sec (or vice versa). The robot can sense, in degrees, how much its main body tilts to the left or right, and how much it tilts forward and backward. If the robot moves outside the terrain, it will not tilt for a period of time as it is on flat ground. This acts as a signal to the robot to reverse direction, and change the velocity differential on its left and right wheels.

Environments are created by altering two parameters: the number of boulders within the terrain, and the radius of those boulders. The boulders are embedded within the (flat) ground, and cannot be moved by the robot. It is assumed that the robot knows that the boulders are embedded and immovable, but not how many and how large they are. Boulders are distributed randomly, using a uniform distribution, across the terrain.

3 Results

The robot first traversed three environments with 10 boulders of radius 10cm, 100 boulders of radius 10cm, and 200 boulders of radius 10cm. Fig. 2a,c and e reports the time series of the resulting tilt sensor data collected by the robot from these three environments.

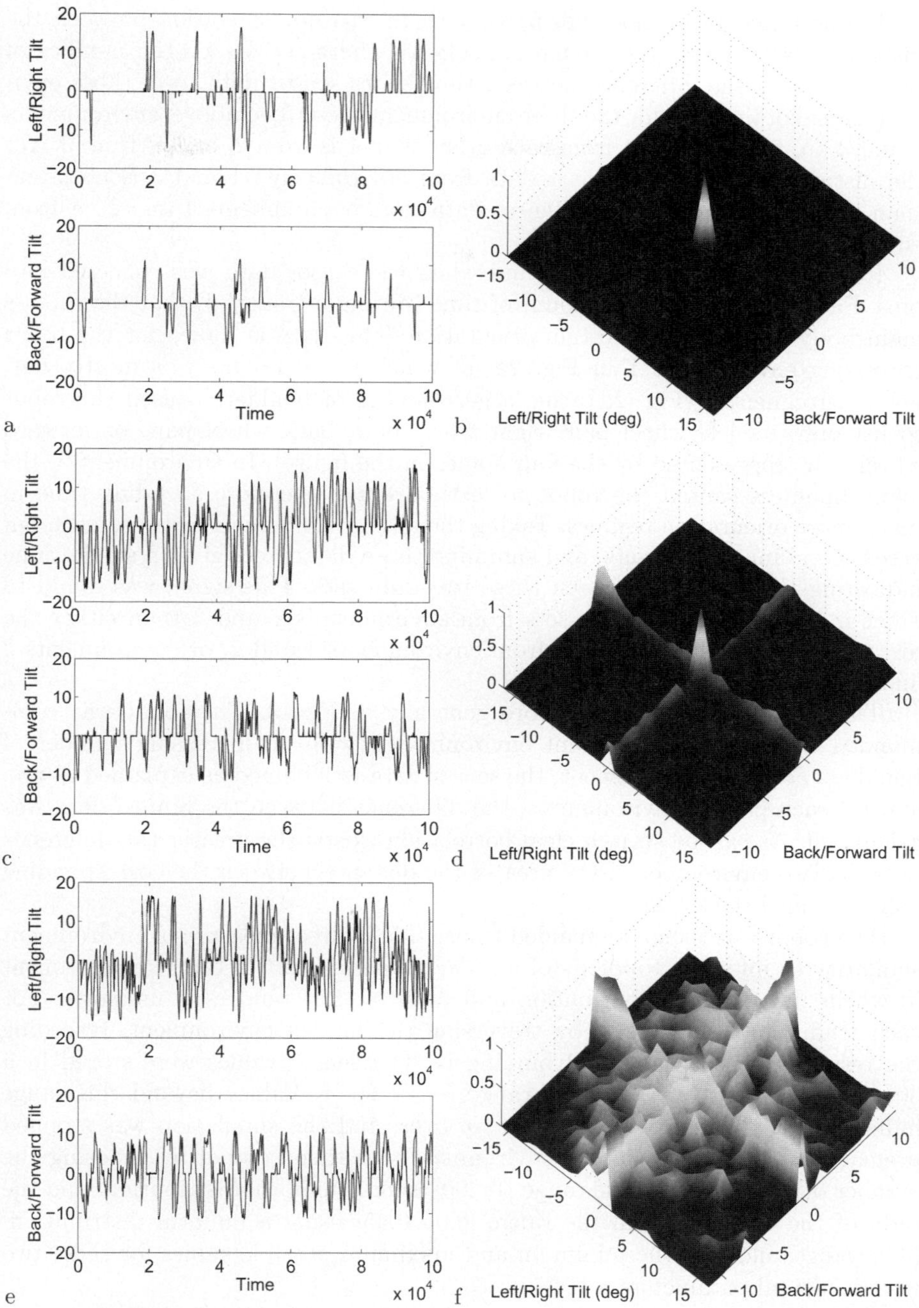

Fig. 2. Robot experience when traversing three different terrains. The robot sensor time series (**a**) and binned data (**b**) for environment 1. The time series (**c**) and binned data (**d**) for environment 2. The time series (**e**) and binned data (**f**) for environment 3.

It was found (data not shown) that there was no correlation between the distance between two given environments (as characterized by the number of boulders), and the difference between time series sensor data from those environments. For instance for the three environments described above, environments 1 and 3 are more different from each other than 1 is from 2, or 2 is from 3. Yet the distance between the sensor data from environments 1 and 3 is no larger than the distance between the sensor data from environments 1 and 2, or from environments 2 and 3.

However, a correlation *was* found when the sensor data was collected into bins. Each bin denotes the amount of time (during a traversal) that the robot's main body maintained a certain orientation. This data is shown for the robot traversing environment 1 in Fig. 2a, in which the main body is mostly flat. For environment 2 (Fig. 2d), the larger number of boulders caused the robot to list onto its left wheel pair, right wheel pair, back wheel pair, or forward wheel pair (represented by the four spines in the figure). In environment 3, the many boulders caused the robot to jostle around, therefore spending time in many more orientation regimes. Taking the absolute difference between each bin across two binned data sets, and summing those differences gives a single value indicating the distance between these two data sets. This signal was found to differ more between the data sets from environments 1 and 3 than either the distances between the data sets from environments 1 and 2, or environments 2 and 3.

This observation is shown more generally in Fig. 3. The robot was commanded to traverse 10 different environments, with an increasing number of boulders. For each environment, the sensor data was binned as explained above, and for each pair of environments, the difference between the binned data was calculated. As can be seen, a clear correlation exists: the greater the difference between two environments, the greater the distance between the corresponding pair of binned sensor data.

The robot was then commanded to use this indirect measure of environment similarity to infer the topology of a *hidden environment*: a target environment in which the number, distribution and radii of the boulders is unknown. For each trial, the robot began by traversing the hidden environment, recording the resulting sensor data, and binning it. The sensor values were stored in a 30×30 matrix of bins over the range $[-60^o, 60^o]$[2]. Values beyond this range indicate that the robot is about to tip over, and the simulation was stopped prematurely. A random model environment was then created by choosing the number of boulders from the range $[1, 200]$ using a uniform distribution, and the radii of the boulders from the range $[0.05, 0.45]$ using a uniform distribution. (The ranges indicate the minimum and maximum possible values for these two environmental parameters.)

A model environment is then created from these two parameter values, traversed, and sensor data is collected and binned. Then, the distance between the model-generated and target-generated binned sensor data is computed. A child

[2] This differs from the 20×20 matrix of bins over the range $[-15^o, 15^o]$ shown in Fig. 2.

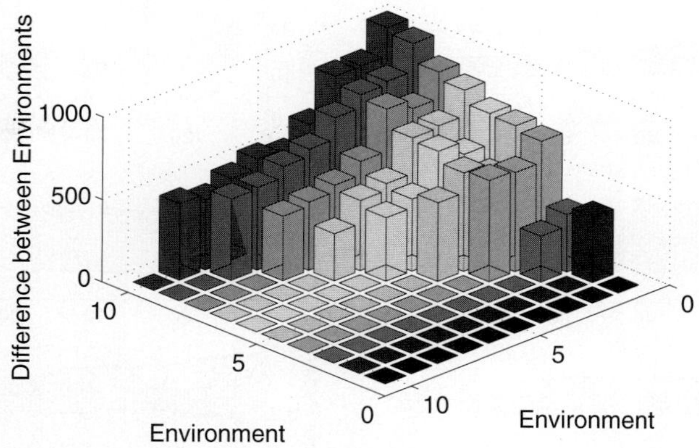

Fig. 3. Differences between sensor data collected from 10 different environments, with increasing numbers of boulders.

model environment is then created from the current model environment as follows. The two environmental parameters are copied, and with 50% probability the boulder number parameter is mutated; otherwise, the boulder radii parameter is mutated. If the boulder number parameter is mutated, a Gaussian value is chosen from $[-10, 10]$ and added to the current value. If the boulder radii is mutated, a Gaussian value is chosen from $[-0.1, 0.1]$ and added to the current value. A child model environment is then synthesized from these two new parameter values. (Note that as the distribution of boulders is still random, the same parameter values may generate different environments.)

The robot traverses the child model environment, and the sensor data is collected and binned. If the distance between the binned sensor data from the target and child model environments is less than the distance between the data from the target and parent model environments (in other words, if the child model environment is more accurate than the parent model environment) the parent model environment is discarded and replaced by the child model environment. Otherwise, the child model environment is discarded. This process is repeated for 100 generations.

Thirty independent trials were conducted using this process. For both the target and model environments, the robot was allowed to traverse them for 20,000 time steps of the simulation. Figure 4a reports the performance of these trials. As can be seen, the error (ie. the distance between the binned data from the target and model environment) gradually decreases over evolutionary time, indicating that this metric provides some gradient in the space of all possible model environments. However, the number of boulders encoded does not approach the true number of boulders in the target environment, which was fixed at 100. The models do however converge on the correct value for the boulder radii, which was fixed at 0.25, indicating that the robot successfully infers the size of the boulders in the environment indirectly, given only tilt sensor information.

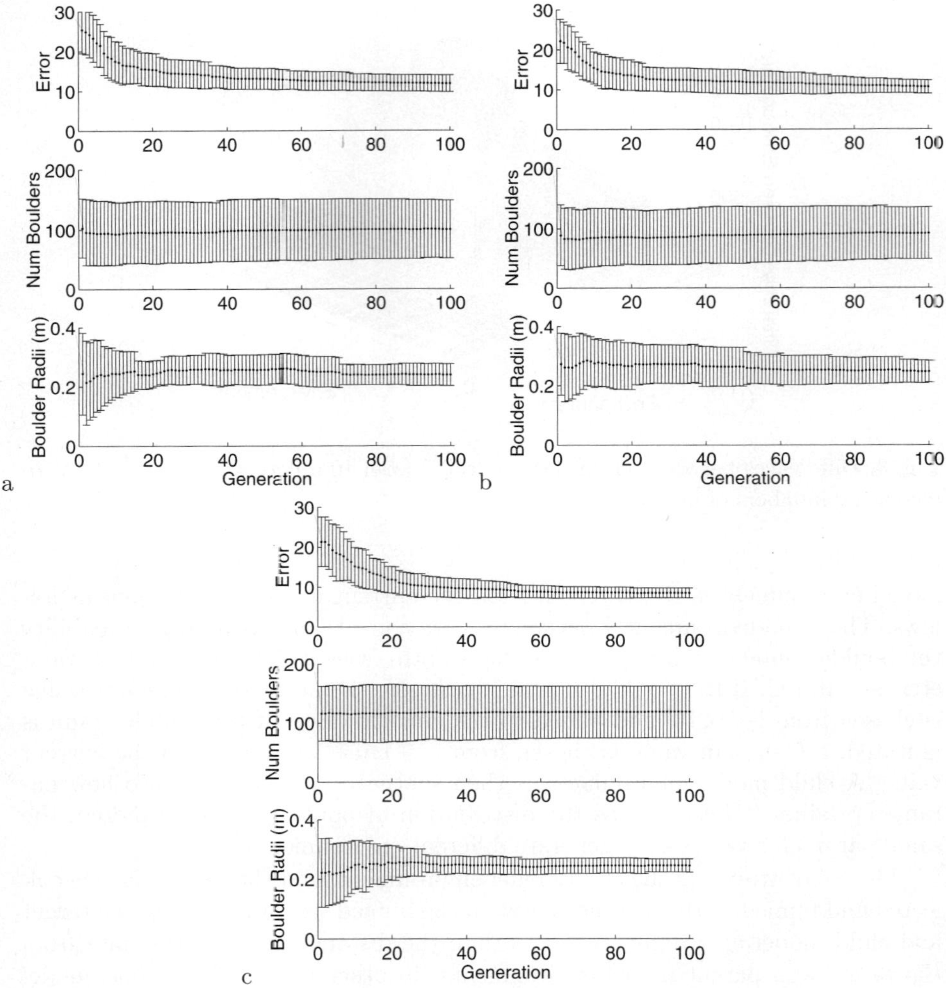

Fig. 4. Relative performance of the three experiments. a: The robot in the first experiment was allowed to traverse the target and model environments for 20,000 simulation time steps. **b:** Another thirty trials were conducted in which traversal time was extended to 40,000 time steps. **c:** A third set of thirty trials were conducted in which traversal time was extended to 60,000 time steps.

A further 30 trials were conducted in which the traversal time for both the target and model environments was extended to 40,000 time steps (Fig. 4b), and finally a third set of 30 trials were conducted in which the robot was allowed to traverse the environments for 60,000 time steps (Fig. 4c).

In order to determine whether the robot could use the synthesized models to make successful predictions about the result of future actions, two trials were randomly selected from the third experimental regime, and their best model environments were extracted. The robot was then commanded to traverse these two models at increasing speed, for 5000 time steps. Sixteen speeds were used: [5rev/sec, 6rev/sec, ... 20rev/sec]. For each speed, the robot traversed each of the two model environments 30 times. During each traversal, the simulation was stopped if the robot tipped over. If the simulation was repeatedly stopped prematurely for a given speed, this indicates that this is probably an unsafe speed at which to traverse not only the model environment, but also the target environment that it represents.

Fig. 5a reports the mean time to stopping for these two model environments. As can be seen, both models predict that the robot will begin to tip over at 9 rev/sec. The robot was then driven over the target environment at these same speeds, and the mean time until the robot tipped over was again calculated. For the target environment, the robot indeed begins to tip over at 9 rev/sec, indicating that the robot was successfully able to use the models to predict, in advance, the threshold at which behaviors become dangerous.

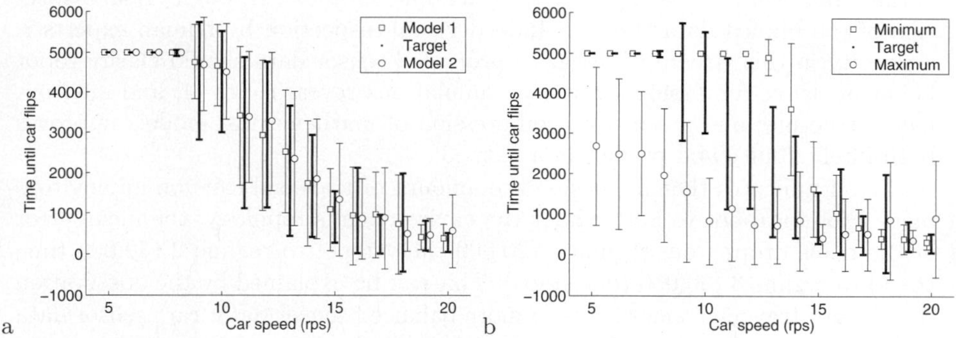

Fig. 5. Predictions of future action made by various models. a: Two of the optimized models predict that the robot will begin flipping over when it drives over the target environment at wheel speeds of 9 revolutions/second and higher, which matches that actual behavior of the robot in the target environment. **b:** Two other inaccurate models (at the minimum and maximum extremes of the allowed environment parameters) incorrectly predict that the robot will begin flipping over at 13rev/sec (the minimal model) or at speeds less than 5rev/sec (the maximal model).

In comparison, two other inaccurate models were used to predict the result of future action: a minimal model containing only one boulder with a radius of 5cm, and a maximal model containing 200 boulders with radii of 45cm. Fig. 5b reports the poor predictions generated by these models. Not surprisingly, the minimal model incorrectly predicts flipping over commences at higher speeds (13 rev/sec). The maximal model incorrectly predicts flipping over commences at lower speeds (5 rev/sec).

4 Discussion and Conclusions

In this work it has been shown that minimal sensor feedback (two tilt sensors) is required to indirectly infer and synthesize accurate, physically-realistic models. This finding supports the findings in [4], in which only two tilt sensors were used to indirectly recover the topology of a legged robot. Despite this minimal sensor requirement, sensor feedback must be processed correctly in order to induce a search gradient in the space of possible models (Fig. 3). The sensor processing introduced here in effect removes the time component from the sensor data. This reflects the observation that physical environments are extremely noisy and nonlinear, and it is well known that even accurate models will begin to diverge from nonlinear systems after a very short time interval [16][7]. In effect, the binning process described here compares the phase portraits of nonlinear signals, in which the dimensions of the portraits correspond to the number of sensors. Further investigation into the most appropriate method of signal comparison across a wide range of robots, sensors and environments is warranted.

This method of processing has the added benefit that the statistical behavior of the robot over time can be inferred by a human observer through visual inspection of the binned data (Fig. 2). Indeed visual inspection by human experts is a priority in this stream of research: processed sensor data should clarify robot behavior; three-dimensional physical simulations reveal robot [4] and environment structure; and automated compression of mathematical models improves both intelligibility and predictive ability [3].

Fig. 4 indicates that increasing the amount of traversal time on an environment allows a robot to better infer the environment's topology: the mean error of the trials drops from regime 1 (20,000 time steps) to regime 2 (40,000 time steps) to regime 3 (60,000 time steps). This can be explained by the observation that longer traversal times leads to more balanced sensor data: rare sensor data and biases from short traversals (such as only tipping to the left by encountering a series of boulders on the right) is damped out in longer traversals.

Similarly, the accuracy of the inferred size of the boulders improves in the long traversal experimental regime (Fig. 4c): the trials converge more rapidly and more tightly to the actual boulder size of 25cm. On the other hand, the number of boulders is rarely inferred correctly (Fig. 4). This is presumably due to the fact that certain random distributions of boulders (such as many boulders together) may fool the robot into believing there are more (or less) boulders than there actually are. Surprisingly however, even with this particular handicap, the robot is still able to correctly predict the outcome of future actions.

This is an encouraging result, as one of the main arguments against using models in robotics [5][1] is that a perfect model can never be created (or at least not in a finite amount of time). The results here indicate that a robot does not need to create a perfect model in order to correctly predict the result of future action. Rather, it merely needs to create a model that is appropriate for considering the outcomes of certain types of future action. For instance, because the robot does not model boulder distribution, it cannot plot a path through the boulder field. This result indicates that rapid synthesis of multiple, approximate models in which each is tuned for considering the results of a particular kind of future action may be more adaptable than a robot that attempts to create a single, perfect model.

References

1. Arkin, R.C.: Behavior-based Robotics. MIT Press, Cambridge (1998)
2. Bongard, J., Lipson, H.: Nonlinear system identification using coevolution of models and tests. IEEE Transactions on Evolutionary Computation 9(4), 361–384 (2005)
3. Bongard, J., Lipson, H.: Automated reverse engineering of nonlinear dynamical systems. Proceedings of the National Academy of Science (to appear, 2007)
4. Bongard, J., Zykov, V., Lipson, H.: Resilient machines through continuous self-modeling. Science 314, 1118–1121 (2006)
5. Brooks, R.A.: Intelligence without representation. Artificial Intelligence 47, 139–160 (1991)
6. Cliff, D., Husbands, P., Harvey, I.: Evolving visually guided robots. In: Meyer, J.-A., Roitblat, H., Wilson, S. (eds.) Proceedings of the Second International Conference on the Simulation of Adaptive Behaviour, MIT Press, Cambridge (1993)
7. Danforth, C.M., Yorke, J.A.: Making Forecasts for Chaotic Physical Processes. Physical Review Letters 96(14), 144102 (2006)
8. Floreano, D., Mondada, F.: Hardware solutions for evolutionary robotics. In: Husbands, P., Meyer, J.-A. (eds.) EvoRobots, pp. 137–151 (1998)
9. Gloye, A., Wiesel, F., Tenchio, O., Simon, M.: Reinforcing the driving quality of soccer playing robots by anticipation. IT - Information Technology 47(5), 250–257 (2005)
10. Jakobi, N.: Evolutionary robotics and the radical envelope of noise hypothesis. Adaptive Behavior 6(1), 131–174 (1997)
11. Keymeulen, D., Iwata, M., Kuniyoshi, Y., Higuchi, T.: Online evolution for a self-adapting robotics navigation system using evolvable hardware. Artificial Life 4, 359–393 (1998)
12. Nilsson, N.: Shakey the Robot. SRI International, Menlo Park (1984)
13. Nolfi, S., Floreano, D.: Evolutionary Robotics. MIT Press, Boston (2000)
14. Pollack, J.B., Lipson, H., Ficici, S., Funes, P., Hornby, G., Watson, R.: Evolutionary techniques in physical robotics. In: Miller, J.F., Thompson, A., Thompson, P., Fogarty, T.C. (eds.) ICES 2000. LNCS, vol. 1801, pp. 175–186. Springer, Heidelberg (2000)
15. Russell, S.J., Norvig, P.: Artificial Intelligence: A Modern Approach. Prentice-Hall, Upper Saddle River (1995)
16. Strogatz, S.H.: Nonlinear Dynamics and Chaos: with applications to physics, biology, chemistry, and engineering. Perseus Books, Reading (1994)
17. Tedrake, R., Zhang, T., Seung, H.: Learning to walk in 20 minutes. In: Proceedings of the Fourteenth Yale Workshop on Adaptive and Learning Systems. Yale University, New Haven, CT (2005)
18. Thrun, S., Burgard, W., Fox, D.: Probabilistic Robotics. MIT Press, Cambridge (2005)

The Evolution of Pain

Alberto Acerbi and Domenico Parisi

Institute of Cognitive Sciences and Technologies, National Research Council
44, Via San Martino della Battaglia, 00185, Rome, Italy
{alberto.acerbi, domenico.parisi}@istc.cnr.it

Abstract. We describe two simple simulations in which artificial organisms evolve an ability to respond to inputs from within their own body and these inputs themselves can evolve. In the first simulation the organisms develop an ability to respond to a pain signal caused by body damage by stopping looking for food when they feel pain since resting while the body is damaged accelerates healing of the body and increases the individual's survival chances. In the second simulation the pain signal itself evolves, that is, the body develops a tendency to send pain signals to the nervous system when the body is damaged. The results are discussed in terms of an internal robotics in which the robot's body has an internal structure and not only an external morphology and the neural network that controls the robot's behavior responds to inputs both from the external environment and from within the body.

Key words: artificial life, neural nets, internal robotics, pain.

1 Introduction

The nervous system responds both to information from outside the body (external environment) and to information from the organs and structures that lie inside the body (internal environment). In both cases the nervous system tends to respond to the incoming information in ways that increase the survival and reproductive chances of the organism. However, there is an important difference between the external environment and the internal environment. The external environment has properties that exist independently of the organism. It is true that the organism's sensory organs filter the environment in ways that depend on the organism and that, by moving its body or body parts, the organism can determine what portions of the environment will send information to its sensory organs at any given time. However, the intrinsic properties of the environment and the fact that these properties have an impact on the organism's nervous system are something for which the organism has no responsibility. The internal environment is different. In the case of the internal environment what must evolve is not only an ability of the organism's nervous system to respond in appropriate ways to the information originating from the internal structures of the body but also the fact itself that these internal structures can send specific kinds of information to the nervous system. In other words, in the interactions between the nervous system and the rest of the body, everything must evolve.

F. Almeida e Costa et al. (Eds.): ECAL 2007, LNAI 4648, pp. 816 – 824, 2007.

In most simulations using neural networks and genetic algorithms what evolves is the manner in which organisms respond to the input but the input itself does not evolve. The input has properties that are the physical consequence of the intrinsic nature of the external environment. But consider physical pain. Physical pain is input from within the organism's body caused by some damage, or potential damage, to the body. The organism's nervous system must respond to pain in ways that tend to increase the organism's survival and reproductive chances. A simple adaptive response to physical pain, often found in animals, is to stop moving so that whatever the damage to the body that causes pain it can more easily heal spontaneously. In the case of pain, not only the nervous system's response to pain must evolve (to stop moving) but pain itself as a signal to the nervous system must evolve. The body must evolve a tendency to translate physical damage that occurs in some of its parts into pain. In fact, people who are unable to feel pain tend to die before they reach adulthood. Therefore, pain as an input to the nervous system can be considered as an adaptation [9, 16, 17].

In this paper we describe some simple simulations of the evolution of pain. In the first simulation a population of artificial organisms evolve an ability to find food in the external environment but at the same time they stop moving and looking for food when they feel pain since resting while the body is damaged accelerates healing of the body and increases the individual's survival chances. In a second simulation pain itself evolves, that is, the body evolves a tendency to send pain signals to the nervous system when the body is damaged. In the first simulation inherited genotypes encode only the properties of the organisms' nervous systems, i.e., the connection weights that allow the neural networks to respond appropriately to incoming signals from either the external or the internal environment. In the second simulation genotypes also encode some properties of the organisms' body beyond the nervous system. They encode a tendency of the body to send pain signals to the neural network to which the neural network can respond appropriately.

2 Simulations

2.1 Evolving the Ability to Respond to Pain

A population of 100 organisms lives in an environment containing food elements. Each individual lives in its own copy of the environment for 10 epochs of 50 time steps each (input/output cycles of the neural network controlling the organism's behavior). The environment is a continuous square with 5 randomly located food elements. At the beginning of each epoch the individual is positioned in a randomly selected location with a randomly selected orientation. For independent reasons and at unpredictable intervals each individual has a 0.01% probability at each time step to incur in some physical damage to its body. The damage lasts for 20 consecutive time steps and it then recovers spontaneously. The nervous system of the organism is simulated by a three-layered neural network with three units in the input layer and two units each in the internal and output layers. Two input units encode the distance of the nearest food element from the organism and the angle between the food element and the individual's current orientation. (The organism has a perceptual field of 360 degrees.) These two units are connected to the two internal units which in turn are connected to the two output units. The third input unit encodes the presence (1) or absence (0) of pain which is felt when the organism's body is physically damaged. This input unit is directly connected to the two output units. The two output units encode the organisms' movements in the environment. One output unit, the orientation unit, encodes the angle with which the individual's orientation changes with respect to its present value, and the other unit, the displacement unit, encodes the distance covered by the individual in its present orientation.

At the beginning of the simulation the neural networks of all the organisms have connection weights with a value randomly selected in the interval between –5 and +5. If an organism reaches a distance which is smaller than 0.01 from a food element, the organism eats the food element and its fitness is increased by one unit, while the food element which has been eaten is replaced by a new one in a randomly selected location of the environment. If an organism displaces itself in the environment when the pain unit is 1, that is, its second output unit has a value higher than 0.01, the organism's fitness is decreased by 2 units. At the end of life the 20 individuals with the highest fitness are selected for reproduction, and they generate 5 offspring each which inherit their connection weights, with a probability of 0.01 of mutating each connection weight by adding or subtracting a value randomly selected within the same range between –5 and +5. The 20x5=100 new individuals constitute the second generation. The simulation is terminated after 1000 generations. All the results that are cited in the paper are the average results of 10 replications of the different simulations.

Figure 1 show how the fitness of the average individual and the fitness of the best individual increase across the 1000 generations. Figure 1 also shows the curves for the average and best individuals of a control simulation in which there is no physical damage and therefore no pain signal.

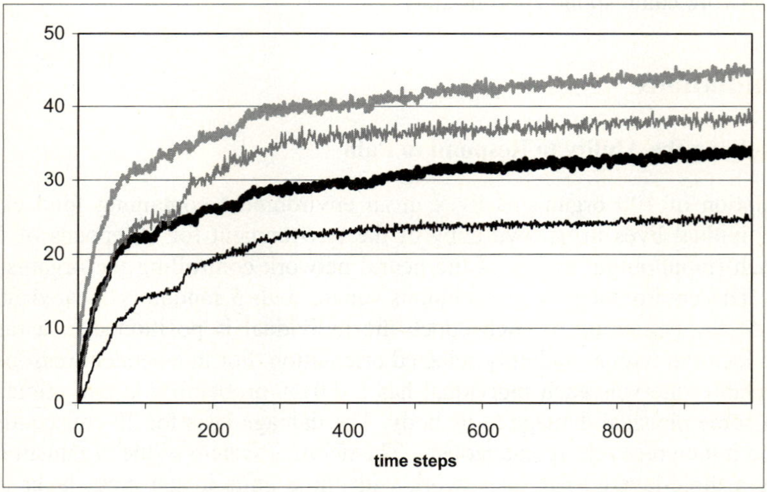

Fig. 1. Average (tick black line) and best (tick gray line) fitness for Simulation 1 in which the organisms evolve an ability to both look for food when they are in a healthy condition and do not feel any pain and to stop moving and ignore food when their body has some physical damage and they feel pain. Average (thin black line) and best (thin gray line) fitness for a control simulation in which there is no physical damage and therefore no pain signal.

Figure 1 shows that the organisms evolve an ability to look for food both when their body can be damaged and they have to stop looking for food and when body damage cannot occur. Inevitably, the total quantity of food they are able to eat is lower in the first than in the second condition. However, the organisms are very good at stopping moving only when they feel pain. This is shown in Figure 2. The organisms stop moving in 90% of the cycles in which they receive the pain stimulus but they stop moving only in 15% of total cycles, that is, approximately, the percentage of cycles in which they in fact receive pain stimuli.

The inhibitory function of the pain signal is realized by evolving strong inhibitory (negative) connection weights for the connection linking the pain input unit to the output unit which controls speed of movement. If we examine how the weights of the two connections linking the pain input unit to the two output units change across

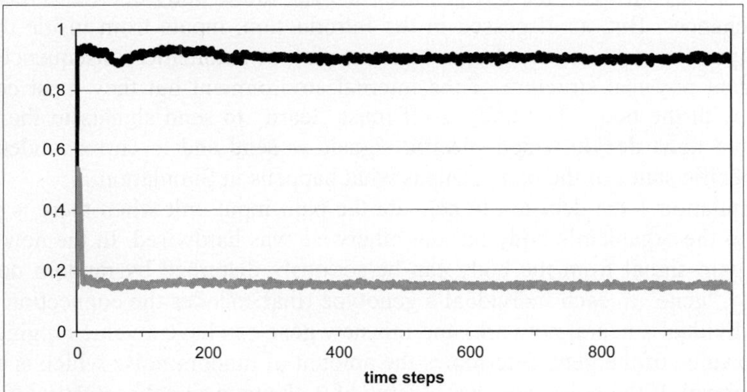

Fig. 2. Percentage of input/output cycles in which the organisms stop moving when they feel pain (black line) and total percentage of cycles in which they stop moving (gray line).

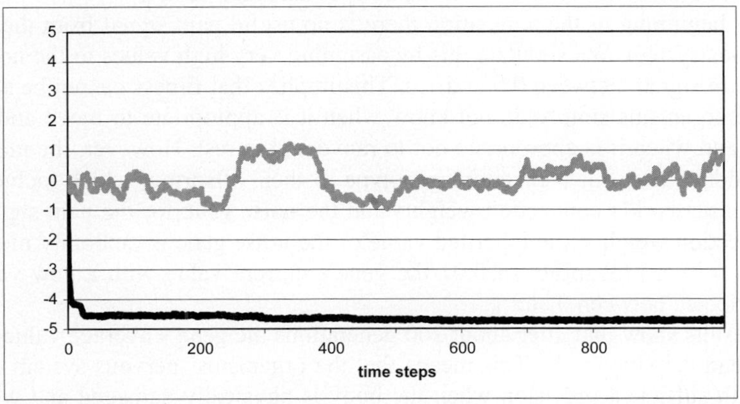

Fig. 3. Weight value of the connections linking the pain input unit to the first output unit (angle of turning - gray line) and to the second output unit (speed of movement - black line) across 1000 generations.

generations, we find that while the weight value of the connection to the first output unit (angle of turning) oscillates around a value of zero, the weight value of the connection to the second output unit (speed of movement) becomes strongly negative (inhibitory) after a few generations (Figure 3). Since when the body is damaged, it is displacing oneself in the environment, not turning, which is costly in terms of fitness, it is reasonable that only the second connection weight becomes strongly negative, thereby inhibiting the individual's displacements in the environment when the body is damaged and the pain signal is on, that is, the pain input unit has a value of 1.

2.2 Evolving the Ability to Respond to Pain

Simulation 1 has shown that, given an input signal that co-varies with physical damage to the body, the organisms learn to stop moving when this signal from the body is on because moving when there is physical damage to the body decreases their reproductive chances. But, as discussed in the Introduction, inputs from inside the body, unlike inputs from the external environment, are not automatic consequences of the independent physical structure of the internal environment but they must co-evolve together with the body. The body itself must "learn" to send signals to the nervous system and must decide which specific signals to send and in correspondence with which specific states of the body. This is what happens in Simulation 2.

In Simulation 1 the decision to activate the pain input unit when there is physical damage to the organism's body but not otherwise was hardwired. In the new simulation the pain signal from the body can be seriously disturbed by random noise. We add a new "gene" to each individual's genotype (that encodes the connection weights of the individual's neural network) and this new gene can have a value ranging from 0 to 1. The value of the gene determines the amount of random noise which is added to the pain signal. If the noise gene has a value of 0, the pain signal perfectly informs the nervous system about the state of the body. On the other hand, with increasing values of the noise gene, the pain signal becomes progressively less informative because progressively more noise is added to the signal. With high values of the noise gene, the nervous system does not know if the body is damaged or not.

At the beginning of the simulation there is no useful pain signal from the body to the nervous system. We simulate this by assigning very high values to the noise gene for the pain signal, between 0.9 and 1.0. This implies that fitness cannot be very high since the organisms simply do not know when it is appropriate to move and to look for food and when it is appropriate not to move and to rest. However, the individuals that reproduce transmit their entire genotype to their offspring, which includes both the neural network's connection weights and the noise gene for the pain signal. Like the connection weights, the inherited value of the noise gene is randomly mutated by replacing with a probability of 0.01 the gene's current value with a new value randomly selected between 0 and 1.

The results show that after about 200 generations the gene's average value tends to be less than 0.1 (Figure 4). This means that the organisms' nervous system can now know with sufficient precision when the body is physically damaged and when it is not, and it can respond to the first information by stopping moving and ignoring food and to the second information by moving and eating.

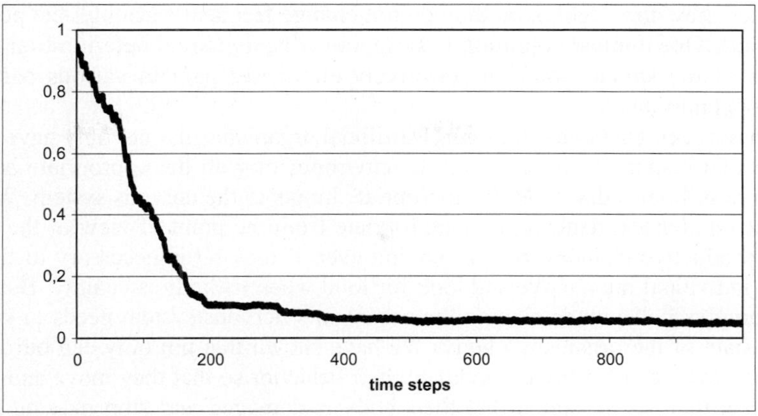

Fig. 4. Average value of the noise gene for the pain signal in Simulation 2.

The fitness value at the end of the simulation is similar to the fitness value at the end of Simulation 1 (Figure 5), showing that an effective integrated system has emerged: the organisms' body has evolved an appropriate signal that tells the nervous system when the body is damaged and the nervous system has evolved an appropriate response to this signal (resting).

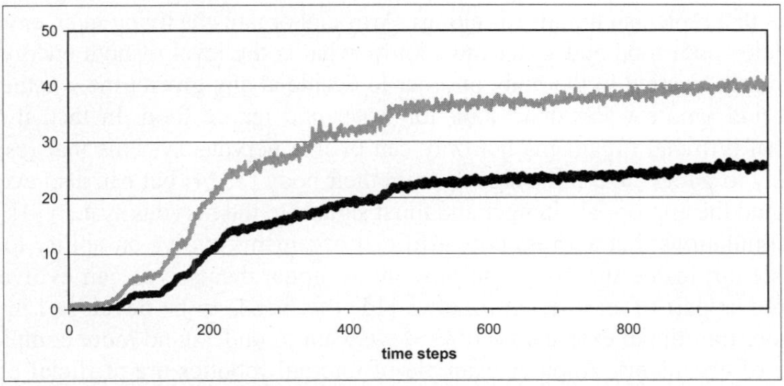

Fig. 5. Average (black line) and best (gray line) fitness for Simulation 2.

As in Simulation 1, the organisms tend to stop moving when they feel pain and to move and search for food when there is no pain signal. The appropriate performance level is reached somewhat more slowly than in Simulation 1 because evolution has to solve two problems rather than only one as in Simulation 1.

3 Discussion

Pain clearly does have adaptive value [9]. Clinical cases demonstrate that people who don't experience pain die at an early age, from accidental injuries or, more frequently, as a consequence of irreversible damage to their joints due to the fact that, since they don't feel pain, they do not change frequently enough the posture of their joints. One clinical condition is syringomyelia, a gradual deterioration of some portions of the spinal cord that selectively eliminates pain in various part of the body, especially the hands [7].

We have been able to evolve simple artificial organisms that not only have nervous systems that respond to inputs from the environment with the appropriate behaviors but also have bodies that send the appropriate inputs to the nervous system. When the body incurs physical damage, it is appropriate from the point of view of the individual's reproductive chances to stop moving even if moving is necessary to find food and the individual must move and look for food when its body is healthy. But for this more complex behavior to emerge the organism's nervous system needs to know the current state of the organism's body. We have shown that not only can our artificial organisms evolve an ability to modulate their behavior so that they move and look for food when there is no signal that their body is damaged and stop moving and rest when there is a pain signal indicating that the body is damaged, but also that their body can evolve an ability to signal appropriately to the nervous system what is its current state. (For a description of the actual neurobiology of pain, i.e., of an organism's nociceptive system, see

The body provides the nervous system with a great variety of important signals. For example, hunger and thirst are signals sent by the body to the nervous system and informing the nervous system about the current quantity of energy and liquids present in the body, respectively. On the basis of these signals the nervous system can take the appropriate action by producing either behaviors that reintegrate energy (eating) or behaviors that replenish liquids (drinking). Artificial organisms living in an environment that contains both food and water must know what is the level of both energy and liquids currently existing in the body in order to decide at any given time whether to look for food and ignore water or to look for water and ignore food. In fact, it has been shown that artificial organisms not only can evolve nervous systems that respond appropriately to hunger and thirst signals from their body [5, 14] but can also evolve bodies that send the appropriate hunger and thirst signals to the nervous system [4].

The simulations that address how artificial organisms evolve an ability to respond to inputs from inside the body and how these inputs themselves can evolve, can be considered as part of an internal robotics [15] that needs to be developed in addition to the more traditional external robotics if we want to understand more completely the behavior of organisms. Another example of internal robotics are artificial organisms that respond to light in the environment by moving in search of food and to darkness by stopping moving (sleeping). However, if these organisms enter a dark cave, they may stop moving for ever. The problem is solved by having these organisms evolve an internal "biological clock" that tells their nervous system when it is daytime and when it is nighttime independently of sensory input from the external environment [10]. (For other explorations of internal robotics, see [6])

The simulations described in the present paper have addressed physical pain associated with physical damage to the body and have shown that physical pain can be interpreted as an adaptive signal which is sent by the body to the nervous system and which evolves as part of the evolution of the entire body. An interesting question is whether the same interpretation and the same approach that we have used to simulate the evolution of physical pain can be applied to psychological pain. Physical pain signals a physical damage to the body, and it is the rest of the body, outside the nervous system, which is at the origin of the pain signal and sends the pain signal to the nervous system. (Physical damage to the nervous system, at least to the central nervous system, generally does not cause pain.) In contrast, signals of psychological pain (grief) seem to be self-generated within the nervous system. But, aside from the physical origin of the different pain signals, the more interesting question is what is the adaptive significance of psychological pain signals and what is the adaptive response to these signals. In the simple scenario of our simulations the appropriate response to physical pain is resting. When they feel physical pain our organisms simply stop moving since resting facilitates the spontaneous recovery of the body from physical damage. In fact, resting appears to be a simple but very fundamental adaptive response to physical pain which exists in many non-human animals, although of course more sophisticated and differentiated adaptive reactions to physical pain exist even in non-human animals and certainly in human beings. Examples include learning to avoid whatever caused the pain, acting on the specific part of the body which is in pain (which of course requires a spatially more rich information from the body), and going to a doctor.

But what is the adaptive response to psychological pain which justifies the evolutionary emergence of signals of psychological pain self-generated within the nervous system? Signals of psychological pain may cause in the individual that feels the pain behaviors that allow the individual to escape from situations that have negative consequences for the individual's survival chances or to avoid future situations with these consequences. This may apply to the psychological pain felt by young helpless individuals when they are separated from their mother [2] but it can generalize to other cases of psychological pain in adults. (For a general evolutionary account of various forms of psychological pain, cf. [11, 12, 13]. For a different position that views psychological pain as a by-product of other processes, see [1].)

Another interesting aspect of pain that could be addressed with evolutionary simulations is the anticipatory value of pain. Pain is not only a signal that allows the organism to escape from current conditions that are negative from an adaptive point of view (e.g., moving when the body is damaged or being separated from parent) but it is by itself a condition that the organism tries to escape from or to avoid. Hence, pain appears to have an anticipatory value. By trying to escape from or avoid pain, the organism escapes from or avoids future conditions that may have negative consequences from an adaptive point of view.

The anticipatory value of pain applies especially to psychological pain. While physical pain occurs only if there is actual damage to the body, psychological pain is something that can be learned. Initially neutral stimuli can become psychologically painful with learning. (Of course physical pain itself can cause psychological pain. For the interactions between physical and psychological pain, see [8]). This allows psychological pain to assume a great importance from an anticipatory point of view, alerting the organism to future negative conditions and allowing the organism to act preventively in order to avoid those conditions. By realizing simulations in which pain does not only evolve but can also be learned during life, one could address this important property of psychological pain.

A final problem posed by pain is its "felt" nature. Pain, both physical and psychological, is not just a process or an event which takes place in our body/nervous system but is something which is subjectively "felt". Can our evolutionary simulations address the problem of the "felt" nature of pain? This a complex question that cannot really be addressed here. However, we can indicate a direction that our simulations could take to address this question.

At any given time many different inputs arrive to an organism's nervous system and the organism cannot simultaneously respond to all of them. It has to choose. The "felt" nature of pain can be interpreted as a mechanism for guaranteeing that some particular input is given prominence in determining the organism's behaviour. In general terms, "felt" states are states that "speak louder" in order to be heard (responded to) by the organism. From this point of view pain can reveal its adaptive nature if we assume that the behaviour which is generated in response to the particular input to which pain gives prominence is, in the circumstances, the most important one from an adaptive point of view. The mechanism through which pain plays this role of giving prominence can basically be the inhibitory mechanism that has emerged in our Simulation 1. Pain stimuli are inputs that inhibit other inputs (in our simulations, the input from food) which would produce other behaviours (in our simulations, the behaviour of moving and searching for food) and in this way they make it possible for the organism to produce a behaviour which, in the circumstances, is more important from an adaptive point of view (in our simulations, the behaviour of resting when the body is damaged).

References

1. Archer, J.: The Nature of Grief. The Evolution and Psychology of Reaction to Loss. Routledge, London (1999)
2. Besson, J.M.: The neurobiology of pain. The Lancet 353, 1610–1615 (1999)
3. Bowlby, J.: Attachament and Loss, vol. 2. Separation: Anxiety and Anger. Hogart, London (1975)
4. Cangelosi, A., Parisi, D., Nolfi, S.: Cell division and migration in a 'genotype' for neural networks. Network 5, 497–515 (1994)
5. Cecconi, F., Parisi, D.: Neural networks with motivational units. In: Meyer, A.J., Roitblat, H.L., Wilson, S.W. (eds.) From animals to animats 2: Proceddings of the 2nd International Conference on Simulation of Adaptive Behavior, pp. 346–355. MIT Press, Cambridge (1993)
6. Fellous, J.M., Arbib, M. (eds.): Who needs Emotions? The Brain Meets the Robot. Oxford University Press, Oxford (2005)
7. Hendry, S.: Pain. In: Wilson, R., Keil, F. (eds.) MIT Encyclopedia of Cognitive Sciences, pp. 622–624. MIT Press, Cambridge (1999)
8. Loeser, J.D., Melzack, R.: Pain: an overview. The Lancet 353, 1607–1609 (1999)
9. Melzack, R.: The Puzzle of Pain. Basic Books, New York (1973)
10. Mirolli, M., Parisi, D.: Artificial organisms that sleep. In: Banzhaf, W., Ziegler, J., Christaller, T., Dittrich, P., Kim, J.T. (eds.) ECAL 2003. LNCS (LNAI), vol. 2801, pp. 377–386. Springer, Heidelberg (2003)
11. Nesse, R.M.: What good is feeling bad? The evolutionary benefits of psychic pain. The Sciences nov./dic, 30–37 (1991)
12. Nesse, R.M.: The Evolution of Hope and Despair. Social Research 66, 429–469 (1999)
13. Nesse, R.M.: Is depression an adaptation? Archives of General Psychiatry 57, 14–20 (2000)
14. Parisi, D.: Motivation in artificial organisms. In: Tascini, G., Esposito, V., Roberto, V., Zingaretti, P. (eds.) Machine Learning and Perception, pp. 3–19. World Scientific, Singapore (1996)
15. Parisi, D.: Internal robotics. Connection Science 16, 325–338 (2004)
16. Wall, P.: Pain. Weidenfeld and Nicolson, London (1999)
17. Willis, W.D.: The Pain System: The Neural Basis of Nociceptive Transmission in the Mammalian Nervous System. Kragel, Basel (1985)

A Computational Morphogenesis Approach to Simple Structure Development

Enrique Fernández-Blanco, Julián Dorado, Juan R. Rabuñal, Marcos Gestal, and Nieves Pedreira

Dept. Information and Communication Technologies, University of A Coruña
Facultade de Informatica Campus Elviña s/n 15071 A Coruña, Spain
{efernandez, julian, juanra, mgestal, nieves}@udc.es

Abstract. This paper presents a new model for computational embryology that mimics the behaviour of biological cells, whose characteristics can be applied to the solution of computational problems. The presented tests apply the model to simple structure generation and provide promising results with regard to its behaviour and applicability to more complex problems

1 Introduction

The biological inspiration for the creation of these new models in the computational field is not a revolutionary idea: science has already been the basis for the famous artificial neuron models, the genetic algorithms, etc. The cells of a biological organism are able to compose very complex structures from a unique cell, the zygote, with no need for centralized control. The cells could perform such process thanks to the existence of a general plan, encoded in the DNA for the development and functioning of the system. Another interesting characteristic of natural cells is that they form systems that are tolerant to partial failures: small errors do not induce a global collapse of the system. Finally, the tissues that are composed by biological cells present parallel information processing for the coordination of tissue functioning in each and every cell that composes this tissue.

All the above characteristics are very interesting from a computational viewpoint. This paper presents the development of a model that tries to emulate the biological cells and to take advantage of some of their characteristics by trying to adapt them to artificial cells. The model is based on a set of techniques known as *Computational Embryology* [9].

2 State of the Art

The Evolutionary Computation (EC) field has given rise to a set of models that are grouped under the name of Embryology Computation, first introduced by Kumar & Bentley [9]. This group refers to all the models that try to apply certain characteristics of biological embryonic cells to computer problem solving, i.c. self-organisation, failure tolerance, and parallel information processing.

F. Almeida e Costa et al. (Eds.): ECAL 2007, LNAI 4648, pp. 825–834, 2007.

The starting point of this field can be found in the modelling of gene regulatory networks, performed by Kaufmann in 1969 [8]. After that, several works were carried out on subjects such as the complex behaviour generated by the fact that the differential expression of certain genes has a cascade influence on the expressions of others [10].

The work performed by the scientific community can be divided into two main branches. The more theoretical branch uses the emulation of cell capabilities such as cellular differentiation and metabolism [7] to create a model that functions as a natural cell. The purpose of this work is to do an in-depth study of the biological model.

The more practical branch mainly focuses on the development of a cell inspired-model that might be applicable to other problems. According to this model, every cell would not only have genetic information that encodes the general performance of the system, it would also act as a processor that communicates with the other cells. This model is mainly applied to the solution of simple spatial problems, robot control, or to the development of the evolutionary design of hardware and circuits [11]. The most relevant models are the following: the Kumar & Bentley model [9], which uses the Bentley's theory of fractal proteins [1] for the calculation of protein concentration; the Eggenberger model [4], which uses the concepts of cellular differentiation and cellular movement to determine cell connections; and the work of Dellaert and Beer [3], which incorporates the idea of biological operons to control the model expression.

All these models can be regarded as special cellular automats. In cellular automats, a starting cell set in a certain state will turn into a different set of cells in different states when the same transition function [2]. Whereas the classical conception specifies the behaviour rules, the evolutionary models establish the rules by searching for a specific behaviour. This difference comes from the mathematical origin of the cellular automats, whereas the here presented models are based on biology and embryology.

3 Biological Inspiration

The cells of a biological system are mainly determined by the DNA strand, the genes, and the proteins contained by the cytoplasm. The DNA is the structure that holds the gene-encoded information that is needed for the development of the system. The genes are activated or transcribed thanks to the protein shaped-information that exists in the cytoplasm, and consist of two main parts: the sequence, which identifies the protein that will be generated if the gene is transcribed, and the promoter, which identifies the proteins that are needed for gene transcription.

Another remarkable aspect of biological genes is the difference between constitutive genes and regulating genes. The latter are transcribed only when the proteins identified in the promoter part are present. The constitutive genes are always transcribed, unless inhibited by the presence of the proteins identified in the promoter part, acting then as gene oppressors.

The present work has tried to partially model this structure with the aim of fitting some of its abilities into a computational model; in this way, the system would have a structure similar that is similar to the above and will be detailed in section 4.

4 Proposed Model

Various model variants were developed on the basis of biological concepts. The present work uses for its tests an implemented version of our model used in [5]. The proposed artificial cellular system is based on the interaction of artificial cells by means of messages that are called proteins. These cells can divide themselves, die, or generate proteins that will act as messages for themselves as well as for neighbour cells.

The system is supposed to express a global behaviour towards the generation of structures in 2D. Such behaviour would emerge from the information encoded in a set of variables of the cell that, in analogy with the biological cells, will be named genes.

One promising application, in which we are working, could be the compact encoding of adaptive shapes, similar to the functioning of fractal growth or the fractal image compression.

The central element of our model is the artificial cell. Every cell has a binary string-encoded information for the regulation of its functioning. Following the biological analogy, this string will be called DNA. The cell also has a structure for the storage and management of the proteins generated by the cell and those received from neighbourhood cells; following the biological model, this structure is called cytoplasm.

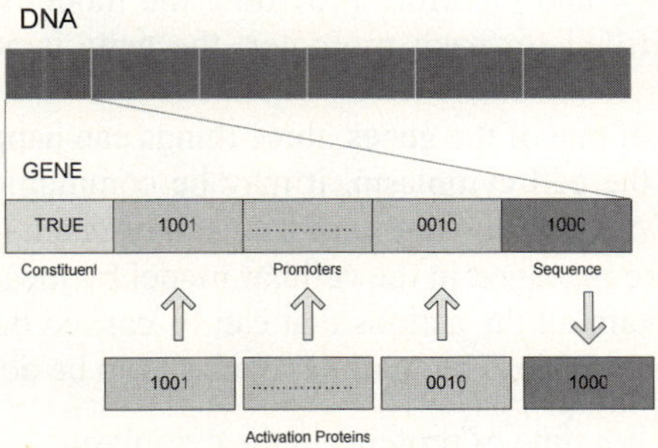

Fig. 1. Structure of a system gene

The DNA of the artificial cell consists of functional units that are called genes. Each gene encodes a protein or message (produced by the gene). The structure of a gene has four parts (see **Fig. 1**):

- Sequence: the binary string that corresponds to the protein that encodes the gene
- Promoters: is the gene area that indicates the proteins that are needed for the gene's transcription.
- Constituent: this bit identifies if the gene is constituent or regulating
- Activation percentage (binary value): the percentage of minimal concentration of promoters proteins inside the cell that causes the transcription of the gene.

The other fundamental element for keeping and managing the proteins that are received or produced by the artificial cell is the cytoplasm. The stored proteins have a certain life time before they are erased. The cytoplasm checks which and how many proteins are needed for the cell to activate the DNA genes, and as such responds to all the cellular requirements for the concentration of a given type of protein. The cytoplasm also extracts the proteins from the structure in case they are needed for a gene transcription.

The functioning of genes is determined by their type, which can be constituent or regulating. The transcription of the encoded protein occurs when the promoters of the non-constituent genes appear in a certain rate at the cellular cytoplasm. On the other hand, the constituent genes are expressed during all the "cycles" until such expression is inhibited by the present rate of the promoter genes.

$$\text{Protein Concentration Percent} >= (\text{Distance} + 1) * \text{Activation Percent} \quad (1)$$

The activation of the regulating genes or the inhibition of the constituent genes is achieved if the condition expressed by **Eq.1** is fulfilled, where *Protein Concentration Percentage* represents the cytoplasm concentration of the protein that is being considered; *Distance* stands for the Hamming distance between one promoter and the considered protein; and *Activation Percentage* is the minimal percentage needed for the gene activation that is encoded in the gene. This equation is tested on each promoter and each protein. If the condition is fulfilled for all the promoters, that gene is transcribed. According to this, if gene-like promoters exist in a concentration higher than the encoded concentration, they can also induce its transcription, similarly to what happens in biology and therefore providing the model with higher flexibility. If the condition is fulfilled for each promoter, the gene is activated and therefore transcribed.

After the activation of one of the genes, three things can happen: the generated protein may be stored in the cell cytoplasm, it may be communicated to the neighbour cells, or it may induce cellular division (mitosis) and/or death (apoptosis). The different events of a tissue are managed in the cellular model by means of "cellular cycles". Such "cycles" will contain all the actions that can be carried out by the cells, restricting sometimes their occurrence. The "cellular cycles" can be described as follows:

- Actualisation of the life time of proteins in the cytoplasm
- Verification of the life status of the cell (cellular death)
- Calculation of the genes that react and perform the special behaviour that may be associated to them
- Communication between proteins

A classical approach of EC proposes the use of Genetic Algorithms (GA) [6] for the optimisation, in this case, of the values of the DNA genes (binary strands). Each individual of the GA population will represent a possible DNA strand for problem solving.

In order to calculate the fitness value for every individual in the GA or the DNA, the strand is introduced into an initial cell or zygote. After simulating during a certain number of cycles, the contained information is expressed and the characteristics of the resulting tissue are evaluated by means of various criteria, according to the goal that is to be achieved.

Fig. 2. (Above) Three promoters and a PCS structure . (Below) Example of GA genes association for the encoding of cellular genes

The encoding of the individual genes follows a structure that is similar to the one described in **Fig. 2 (Above)**, where the number of promoters of each gene may, vary but the white and indivisible section "Activation Percentage – Constituent – Sequence" (PCS) must always be present. The PCS sections determine the genes of the individual, and the promoter sections are associated to the PCS sections, as shown in **Fig. 2 (Below)**.

The search of a set of structures similar to those shown in **Fig. 2** required the adaptation of the crossover and mutation GA operations to this specific problem. Since the length of the individuals is variable, the crossover had to be performed according to these lengths. When an individual is selected, a random percentage is generated to determine the crossover point of that individual. After selecting the section in that position, a crossover point is chosen for the section selected in the other parent. Once this has been done, the crossover point selection process is repeated in the second selected parent in the same position as in the previous individual. From this stage on, the descendants are composed in the traditional way, since they are two strings of bits. We could execute a normal bit strings crossover, but the previously mentioned steps guarantee that the descendants are valid solutions for the DNA strands transformation.

With regards to mutation, it should be mentioned that the types of the promoter or PCS sections are identified according to the value of the first string bit. Bearing that in mind, together with the variable length of individuals, the mutation operation had to be adapted so that it could modify not only the number of these sections, but also the value of a given section.

The probability of executing the mutation is usually low, but this time it even had to be divided into the three possible mutation operations that the system contemplates. Various tests proved that the most suitable values for the distribution of the different mutation operations, after the selection of a position for mutation, were the following: for 20% of the opportunities, a section (either a promoter or a PCS) is added; for another 20%, the existing section is removed; and finally, for the remaining 60% of the opportunities, the value of one of the bits of the section is randomly changed. The latter may provoke not only the change of one of the values, but also the change of the section type: if the bit that identifies the section type is changed, the information of that section varies. For instance, if a promoter section turns into a PCS section, the promoter sequence turns into the gene sequence, and constitutive and activation percentage values are generated.

After reaching this development level and presenting the test set in [5], the bottleneck of the model turned out to be the development of the evaluation functions, since in every new figure the development of the function was time-consuming and not reusable.

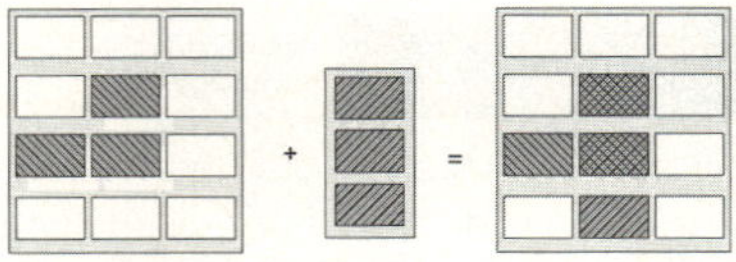

Fig. 3. Tissue + Template. Example of Template use

In order to solve this problem, the evaluation function was developed according to the concept of a correction template. From the tissue that is developed by the DNA that is being evaluated, the centroid is calculated. This point would be the center of the solution template, which is merely a matrix of Boolean values representing the figure that is aimed at. The template could be (and usually is) smaller than the development environment of the tissue, which means that every cell that may not be covered by the template will contribute to the tissue error with 1.0. The remaining tissue, covered by the template, will execute the NEXOR Boolean operation in order to obtain the number of differences between the template and the tissue. Each difference contributes with a value of 1.0. to the tissue error.

Fig. 3 illustrates the use of this method. We can observe that the error of this tissue with regard to the template is 2, since we generated a cell that is not contemplated by the template, whereas another cell that is present in the template is really missing.

5 Tests

The performance of the model was studied by applying it to simple 2D spatial problems. We selected this type of problems because they allow us to easily verify the results and simplify the evaluation functions.

This section describes some of the tests performed with the proposed model, using a Java implementation on a Pentium Core Duo with 1GB RAM. The same parameters were used for all the tests in the GA and the cellular system model. The GA parameters are the following:

- 1000 generations with 500 population individuals
- 600 individual length sections (maximum)
- 90% crossovers and 10% mutations
- 50 pre-evaluation tissue development cycles
- GA operators: roulette selection, parent replacement, and one point crossover

The cellular system parameters were fixed as follows:

- Protein lifetime: 3 cycles
- 20x20 development environment
- Division-inducing proteins: 1000 (N), 1100 (S), 1010 (W), and 1110 (E).
- Apoptosis-inducing protein: 0000 string
- 5% and 10% communication limit values and 10% apoptosis limit value

We also carried out experiments to select the sequences that were identified as division as well as apoptosis sequences. We tested randomly selected sequences and sequences with the biggest possible Hamming distance between them. The result of the tests showed that the specification of the sequences did not have any influence on the development of the tests.

5.1 Growing Tests

We carried out several tests to reach a better understanding of how the proposed model functions and to extract some conclusions. The tests involve the search of DNA strands whose associated performance might be the growth of simple structures with a minimum number of genes.

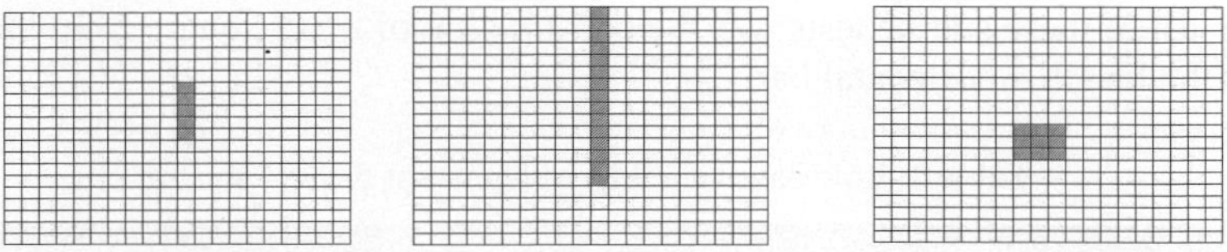

Fig. 4. Some test results: (Left) Length 5 vertical bar. (Center) Length 15 vertical bar. (Right) square 3x3.

The following tables show the genes of every test, every row representing a gene. The abbreviations of the headings have the following meaning:

- **N** refers to the order number within the gene's DNA
- **Pmtr** are the promoter sequences of the gene
- **Act** is the gene activation percentage
- **Cons** indicates the constitutive genes
- **Seq** is the protein that encodes the gene
- **Special** indicates if the gene has any associated behaviour

The first test sought to minimize the number of genes needed to build a 5 cells vertical bar located at the center of the environment. To achieve this, we established a template that considered the number of sections and the distance of the figure centroid to the center as penalisations. The evaluation function is as follows:

$$\text{Template-related differences} + 0.01 * |\text{center} - \text{centroid}| + 10^{-4} * \text{Number of sections} \qquad (2)$$

We found an individual with three genes (see **Table 1**) that gave rise to a 5 sections structure.

Table 1. Genes for the growth of a 5 elements vertical bar

N	Pmtr	Act	Cons	Seq	Special
1		1.9771E-7	true	0100	
2	0000	45.2033	false	1010	Grown N
3	1000	3.8838	false	0101	

The first gene works as the clock signal of the system, because it is constitutive and does not have any inhibiting promoters. The second gene induces the structure to grow upwards; it will only be activated during the first development stages of the structure because it has a very high activation percentage. Finally, the third gene is easily activated and acts as a regulator, because it will not allow the activation of the gene that induces the growth. **Fig. 4(Left)** shows the final stage of the tissue.

Other test have been performed, for instance the 15 elements vertical on **Fig.4 (Center)**, the conclusions archived were: the system search certain roles, not specific sequences; genes explanation is not always easy, at the same time that they could express their own behaviour and can be linked with the expression of other genes;. Finally, mention that some test was performed to study the changes on orientation and range of an archived form, the results shows that is easy to archive those changes.

The last of these simple tests was the construction of a 3x3 square from the population of the length 3 horizontal bar.

Table 2. Role based summary of genes at the 3x3 square

N	Pmtr	Act	Cons	Seq	Special
1-7		*	true	0100-1010	
8	0000-0011	2.0101E-13	false	0001	
9	0000	45.125	false	1100	Grown W
10	0010	1.4626E-29	false	0011	
11	0000	41.1894	false	1000	Grown S

The solution has 11 genes and the clock signal role is played by the 7 first genes, which are almost identical (they vary between 2 sequences, they are constitutive, and they do not have promoters, therefore the activation percentage is not important). These 7 genes are summed up in the first row of the **Table 2**. Although the exact reason is not known, nature also tends to complicate the solution with redundant information, as explained in [7]. The remaining genes generate the shape and stabilize it. Genes 8 and 10 has low activation percentages, so they could induce the system to distribute the concentration among several proteins and then genes 9 and 11 would not be expressed. These two genes induce the growth and, due to their high activation percentage, their expression would only be possible during the first cycles, when the percentage has not been distributed yet. The development of the structure can be seen in **Fig.4 (Right)**.

5.2 Considerations

Other tests were carried out to explore the generation capabilities of the model. Some further testing examples are: leaving only the edges of an initial square shape, generating a cross shape from a unique cell, or trying to generate a square structure in an environment with non-valid positions. Some examples could be seen on **Fig 5**.

Starting from a satisfactory 5x5 square test result, other tests are performed in order to evaluate the flexibility of the DNA solution that was found. The dark points of **Fig. 5 (Right)** are positions that can not be occupied by the cells; this fact induces a higher error rate, but the used DNA was originated from an environment without such obstacles.

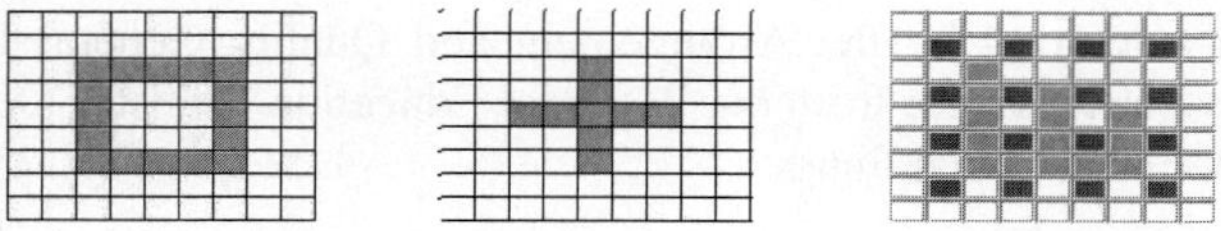

Fig. 5. Test examples: (Left) Edge of a square. (Center) Cross shape. (Right) Failure tolerance.

The cells try to solve the problem and tend to build the square trying to avoid the obstacles. The two obtained solutions take up an approximately 5x5 box square area, although they do not use all the possible boxes of such area and they only use 14 cells. During the GA DNA search no obstacles were used to reach the DNA fitness so this proves that flexibility is inherent in the model used.

After all these tests, we can conclude that the system searches for certain functions/roles instead of concrete values in order to make the DNA strands work. It was observed that the system can be configured in a simple way so as to change the range of shapes or their orientation. Our current work is focused on combining these compact genomes (which perform simple growing tasks), scaling them, and combining them in order to develop, together with other compact genomes, more complexes figures. One of the means for carrying out this process would involve a set of initial genes that generates a general shape such as the ones show in **Fig. 4**; from this shape, another set of genes would elaborate other different structures that could be combined with the previously achieved ones.

6 Conclusions

Taking into account the here developed model, we can say that the use of certain properties of biological cellular systems is feasible for the creation of artificial structures that might be used in order to solve certain computational problems.

Some behaviours of the biological model have been also observed in the artificial model: information redundancy in DNA, stability after achieving the desired shape, or variability in gene behaviour. The model could also include new characteristics such as the displacement of cells around their environment, or a specialisation operator that blocks pieces of DNA during the expression of its descendants, as happens in the natural model.

Finally, this group is currently working in one of the possible applications of this model: its use for image compression similarly as fractal compression works. The fractal compression searches the parameters of a fractal formula that encodes itself the starting image. The present model searches the gene sequence that might result in the starting image. In this way, the method based on template that has been presented in this paper can be used for performing that search, using the starting image as template.

Acknowledgements. This work was partially supported by the Spanish Ministry of Education and Culture (Ref TIN2006-13274) and the European Regional Development Funds (ERDF), grant (Ref. PROLIT/SP1.E194/03) funded by the European Regional Development Funds (ERDF), grant (Ref. PIO52048) funded by the Carlos III Health Institute, and grants from the General Directorate of Research of the Xunta de Galicia (Ref. PGIDIT 05 SIN 10501PR and 2006/60). The work of Enrique

Fernández is supported by the Arrangement and Quality General Direction of the Galician University System from the Regional Education Ministry (Xunta de Galicia) and by the European Social Fund.

References

1. Bentley, P.J.: Digital Biology. Simon and Schuster, New York (2002)
2. Conway, J.H.: Regular Algebra and Finite Machines. Chapman and Hall, Ltd., London (1971)
3. Dellaert, F., Beer, R.D.: A Developmental Model for the Evolution of Complete Autonomous Agent. In: From animals to animats: Proc. of the 4th International Conference on Simulation of Adaptive Behavior, 9-13 September 1996, pp. 394–401. MIT Press, Cambridge (1996)
4. Eggenberger, P.: Cell Interactions as a Control Tool of Developmental Processes for Evolutionary Robotics. In: From animals to animats: Proc. of the 4th International Conference on Simulation of Adaptive Behavior, 9-13 September 1996, pp. 440–448. MIT Press, Cambridge (1996)
5. Fernández-Blanco, E., Dorado, J., Rabuñal, J.R., Gestal, M., Pedreira, N.: A New Evolutionary Computation Technique for 2D Morphogenesis and Information Processing. WSEAS Transactions on Information Science & Applications 4, 600–607 (2007)
6. Holland, J.H.: Adaptation in natural and artificial systems. University of Michigan Press, Ann Arbor (1975)
7. Kaneko, K.: Life: An Introduction to Complex Systems Biology. Springer, Heidelberg (2006)
8. Kauffman, S.A.: Metabolic stability and epigenesis in randomly constructed genetic nets. Journal of Theoretical Biology 22, 437–467 (1969)
9. Kumar, S., Bentley, P.J.: On Growth, Form and Computers. Elsevier Academic Press, London (2003)
10. Mjolsness, E., Sharp, D.H., Reinitz, J.A: Connectionist Modelo f Development. Journal of Theoretical Biology 176, 291–300 (1995)
11. Tufte, G., Haddow, P.C.: Towards Development on a Silicon-based Cellular Computing Machine. Natural Computing 4(4), 387–416 (2005)

Program Evolvability Under Environmental Variations and Neutrality

Tina Yu

Department of Computer Science
Memorial University of Newfoundland
St. Johns, NL A1B 3X5 Canada
tinayu@cs.mun.ca
http://www.cs.mun.ca/~tinayu

Abstract. Biological organisms employ various mechanisms to cope with the dynamic environments they live in. One recent research reported that depending on the rates of environmental variation, populations evolve toward genotypes in different regions of the neutral networks to adapt to the changes. Inspired by that work, we used a genetic programming system to study the evolution of computer programs under environmental variation. Similar to biological evolution, the genetic programming populations exploit neutrality to cope with environmental fluctuations and evolve evolvability. We hope this work sheds new light on the design of open-ended evolutionary systems which are able to provide consistent evolvability under variable conditions.

1 Introduction

Biological organisms live in an ever-changing world. However, early population genetics theory assumed the environment to be constant while the mathematical ecology assumed the genetic makeup of the species involved to be constant. About 40 years ago, Richard Levins published a seminal work which modeled the situation where evolution is taking place while the environment changes[7]. This work has influenced the population genetic theory to consider environmental fluctuations in studying the theory of evolution by natural selection.

While population genetic theory focuses on population-level adaptation, other empirical studies investigate molecular basis of phenotypic changes to cope with environmental variation. For example, it is shown that bacteria turn on the machinery for taking up iron from the environment, that is they synthesize siderophores, only when a lack of iron triggers the expression of over ten genes involved in the regulation of this system [11][2]. Currently, there is a rich volume of information revealing diverse strategies that organisms have evolved to cope with environmental fluctuations. One recent trend in ecology and evolution research is to integrate the diverse forms of 'adaptive variation' into a single conceptual framework [10].

In the evolutionary computation community, there are also interests in investigating the impact of environmental variation on the evolution of artificial

F. Almeida e Costa et al. (Eds.): ECAL 2007, LNAI 4648, pp. 835–844, 2007.

life. The most noticeable work is by Lipson and colleagues [8], who argued that modularity is an emerged property resulting from environmental variation. In that work, they used computer simulation to study modular structures induced under different environmental changing rates. They reported that the amount of modular separation is logarithmically proportional to the changing rate.

Kashtan and Alon [4] went farther by devising the environmental changes in a modular manner: the two objective functions used to evolve two different Boolean functions contained two identical modules. The only difference between these two objective functions was that they used two different logic operators to combine the two modules: one used AND and the other used OR. By switching the two objective functions periodically throughout the evolution process, they reported that the evolved Boolean functions not only contained both modules but also had a structure that could be easily switched into the other Boolean function with a small number of mutations.

This research investigates program evolvability [1] under environmental variations and implicit neutrality (semantic redundancy in genetic programming) [13]. We employed two objective functions which have two opposite goals. The first one is even-4-parity, whose genotypes contain an *even* number of all 4 input variables. The second one is always-on, whose genotypes contain an *odd* number of all 4 input variables [6]. We periodically switched these two objective functions using various rates. The interplay between program evolvability and neutrality under these changing rates are then analyzed.

This work is motivated by our interest in building computer systems for open-ended evolution, i.e. a system that provides consistent evolvability under environmental variation. According to the experimental results, under environmental fluctuations, populations exploited implicit neutrality to evolve evolvability. We hope this work sheds new light on the design of open-ended evolutionary systems for Artificial Life research.

The rest of the paper is organized as follows. Section 2 highlights the mechanisms biological organisms have developed to cope with variable conditions. In Section 3, we explain implicit neutrality in the genetic programming framework. The two objective functions of even-4-parity and always-on are then presented in Section 4. Section 5 details the computer experimental setup while Section 6 reports the experimental results and provides analysis. We discuss the implications of these results in Section 7. Finally, Section 8 concludes the paper and outlines our future research.

2 Background

Variation is the fuel of evolution. Under different environmental variation rates, different individual-level adaptation strategies may evolve [10]. When the environment changes rapidly, mechanisms such as *physiological plasticity* and *learning* may occur for individual organisms to respond to these changes. As environmental changes slow down, *phenotypic variation* may rise resulting from stochastic or directed heterogeneity in the developmental pathway. For even slower rates of changes, mutations may produce *novel phenotype.* If environmental fluctuations are rare, populations may have a period of directional selection and thus have sufficient time to achieve *genetic robustness.* These phenotypic and genotypical changes in responding to environmental variation have been observed in viruses/bacteria and in experimental studies [9].

Genotypes and phenotype (or genes and traits) have a many-to-one relationship [3]. For any particular trait value, there exists a large number of genotypes that give rise to that value. Metaphorically, one can imagine the genotypes that map to the same phenotype as a network connected by mutations. Within the network, a mutation from one genotype to another is neutral, having no impact on the physiology, behavior or fitness of the organism. However, depending on the location in the network where the mutation takes place, there will be different outcomes. Near the periphery of the network, mutations are likely to produce different phenotype, whereas near the center of the neutral networks, mutations have little impact on the phenotype.

Under environmental variations, populations evolve toward genotypes in different regions of the neutral networks. In [9], Meyers and colleagues used a simple mathematical model and a single amino acid site to study evolution under two alternate environments. They reported that when the variation is rare, the populations swing back and forth between genetic robustness, which is located at the center of the neutral networks, of the two phenotypes. At intermediate rates of fluctuation, populations favor the edge of the neutral networks. Thus, mutation between the two phenotype occurs frequently. Finally, for highly variable environments, populations settled in a phenotype that has an intermediate fitness in both environments. This phenotype corresponds to organismal flexibility - individuals tolerate both conditions, but neither one exceptionally well.

Inspired by these biological phenomena and mathematical modeling results, we used computer simulation to study programs evolution under environmental variation. Similar to the biological evolution, there is a many-to-one mapping between genotypes and fitness. The computer evolutionary system used is genetic programming, which is described in the following section.

3 Genetic Programming and Implicit Neutrality

Genetic programming (GP) [5] models the process of natural evolution for problem solving by generating a population of possible solutions, which are selected and mutated to reach a near-optimum. In a GP system, a genotype represents a point in the search space and is operated on by genetic operators. In contrast, a phenotype represents a point in the solution space and is evaluated by the fitness function. In other words, selection is based on phenotypes, while reproduction operates on the underlying genotypes [12].

A GP system may or may not distinguish genotypes from phenotypes, depending on the implementation. In the first case, there can be a many-to-one mapping between genotypes and phenotypes while in the later case, there can be a many-to-one mapping between genotypes and fitness. Under either implementation, a mutation may transform genotypes from one to another without affecting the fitness; they are neutral mutations.

In that standard GP, the evolved genotypes are computer programs, whose behaviors are interpreted to give the phenotype and the corresponding fitness. In other words, fitness is evaluated directly on the genotypes. Since computer programs are semantically rich, many syntactically different programs may be interpreted to the same behavior, hence have the same fitness. Consequently, there is a many-to-one mapping between genotypes and fitness. We call this implicit neutrality [13] as it is embedded in the genotype representation without explicit encoding. In the tree-based GP representation, there are two forms of implicit neutrality: functional redundancy and introns.

Functional redundancy refers to the case when many different programs (genotypes) represent exactly the same function (phenotype). For instance, the following three programs give the same function XOR:

G1: NOR (AND x_1 x_2) (NOR x_1 x_2)
G2: NOR (NOR x_1 x_2) (AND x_1 x_2)
G3: NOR (AND x_1 x_2) (NOR (NOR x_1 x_2) (NAND x_1 x_2))

Genetic transformation from one genotype to another (e.g., G1 to G2) has a neutral effect on the program's behavior.

Introns are code that is part of a program but is semantically redundant. For example, the OR operator with input FALSE is an intron in the following genotype because (OR FALSE ANY-BOOLEAN-VALUE)=ANY-BOOLEAN-VALUE.

G4: OR FALSE (NOR (AND x_1 x_2) (NOR x_1 x_2))

Genetic transformations that remove introns from a genotype (e.g., G4 to G1) have a neutral effect on the program's behavior.

Functional redundancy and introns can emerge within an evolving genetic program and render implicit neutrality. This work will study genetic programs evolution under environmental variation. The two objective functions used to define the two alternating environments are even-4-parity and always-on, which are explained in the following section. We will use the term 'program' to refer to genotype and 'function' for phenotype.

4 Even-Parity and Always-On

Boolean parity is well studied in the GP community. These programs can take any n number of Boolean inputs. An even-n-parity returns TRUE if an *even* number of the inputs are TRUE while an odd-n-parity returns TRUE if an *odd* number of the inputs are TRUE.

An even-n-parity program can be constructed using only EQ logical operator if n is even [6]. Moreover, the EQ-only even-n-parity programs have an unique feature: they contain an *odd* number of all n input variables. In other words, if any of the n input variables is missing or if there is an *even* number of any of the n input variables, the program is *not* an even-n-parity. This is because EQ is symmetrical, which allows the input variables to be in any order in the program without changing the semantics. Moreover, x_1 EQ x_1 is TRUE, hence $(x_2 \text{ EQ } x_1 \text{ EQ } x_1) = x_2$. With these two premises, any pair of repeated input variables can be removed from a program or added to a program without changing the program behavior. An even-n-parity program therefore always has an *odd* number of all n input variables.

Among those EQ-only programs that are *not* even-n-parity, the subset that contains an *even* number of all n input variables are always-on. These programs effectively ignore all input variables (since they all come in pair) and always return TRUE. The remaining EQ-only programs contain an *even* number of some input variables and an *odd* number of other input variables. We call these programs *in-between* as they are in between the two extremes of even-n-parity and always-on. With implicit neutrality, more than one program can be interpreted as even-n-parity or as always-on or as in-between. In [6], Langdon and Poli showed that for any program length l, only a small fraction (less then 1%) of the programs are interpreted as even-n-parity or as always-on. The majority of the EQ-only programs are in-between.

A program tree can have any length. However, because EQ is a binary operator, an EQ-only program tree always has an odd length: $l = 2t - 1$, where t is the number of leaf nodes in the program tree. These leaf nodes may contain any of the n input variables. As mentioned, even-n-parity has an *odd* number of all n input variables, $t = n + 2i, i = 0, 1, 2, \ldots$. The length of an even-$n$-parity program is therefore $l = 2t - 1 = 2n + 4i - 1$. In contrast, always-on has an *even* number of all n input variables, $t = 2p + 2i, p = 1, 2, 3, \ldots, n, i = 0, 1, 2, \ldots$. The length of an always-on program is therefore $l = 2t - 1 = 4(p + i) - 1$.

A program tree with length l can have many different shapes: $s = \dfrac{(l-1)!}{((l+1)/2)!((l-1)/2)!}$. The following table lists some of the program lengths and structures that are common to even-4-parity and always-on. The length ranges from the smallest of 7 up to 35. Figure 1 shows three example program trees with length 7 that are even-4-parity, always-on and in-between functions.

even-4-parity			always-on				
i	t	l	p	$p+i$	t	l	s
0	4	7	2	2	4	7	5
1	6	11	2	3	6	11	42
2	8	15	2	4	8	15	429
3	10	19	2	5	10	19	4862
4	12	23	2	6	12	23	58786
5	14	27	2	7	14	27	742900
6	16	31	2	8	16	31	9694845
7	18	35	2	9	18	35	129644790

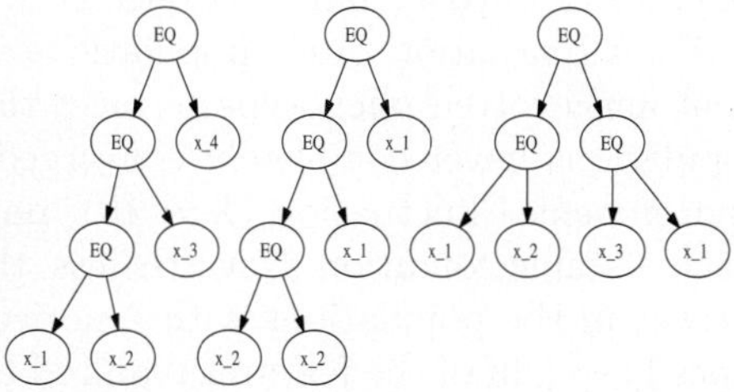

Fig. 1. (a) even-4-parity (b) always-on (c) in-between.

5 Experimental Setup

We allowed the program trees to have a length between 7 and 35. Since the internal nodes of a program tree are constant (EQ), only the leaf nodes were considered for evolution. Additionally, program trees with the same length might have different shapes. We therefore used a vector of length $t+1$ to encode all possible variations: the first t cells contain the 4 possible input variables (x_1, x_2, x_3, x_4) while the last cell gives the shape of the program tree, with a value between 1 and s. During the initialization of the first population, uniform distributions were used to allocate genotypes to each of the 8 possible vector lengths.

The evolutionary system is steady-state with population size 10. At each generation, 5 offspring were generated by mutating 5 selected parent individuals. Among the 15 parent and offspring individuals, the top 10 were kept as the new generation. The selection method is fitness-proportional without scaling. Each selected individual was mutated on each cell of the vector with 10% probability. We let a run last for 1,000 generations, which is equivalent to 5,000 mutations and fitness evaluations. Three sets of experimental runs were made, one for each of the 3 environmental epoch lengths (λ=10, 20 and 50 generations). Each set consists of 100 runs.

Initially, the objective function was always-on. After λ generations, the objective function was switched to even-4-parity. The switching between these two objective functions took place every λ generations until the end of the run.

Fitness of an evolved program was calculated based on the number of test cases it solved. With 4 Boolean inputs, each can be TRUE or FALSE, the number of test cases is $2^4 = 16$. When the objective function is even-4-parity, the EQ-only programs have a needle-in-haystack fitness landscape: a program either solves 16 or 8 test cases. This property also applies to the always-on objective function program fitness landscape [6]. An evolved program, therefore, has fitness value either 16 or 8 under both objective functions.

6 Results and Analysis

Under different environmental variation rates, populations exhibit different program distributions. We show the results averaging over 100 runs in Figures 2, 4 and 6. Note that we only plot the first 100/200 generations as the rest of the generations have a similar pattern.

For all variation rates, populations were able to adapt to the new environment and evolved phenotype to meet the new objective function. However, the populations never completely converged to the target phenotype. Under rapid environmental fluctuation ($\lambda = 10$), populations did not have enough time to reach a stable condition. Nevertheless, the in-between programs are consistently present in the populations with a noticeable proportion (30-40%). These genotypes have half of the features required for both objective functions. They correspond to organismal flexibility - individuals tolerate both conditions, but neither one exceptionally well.

At an intermediate rate of fluctuation ($\lambda = 20$), populations were able to settle in a stable condition for a short period of time, during which the target

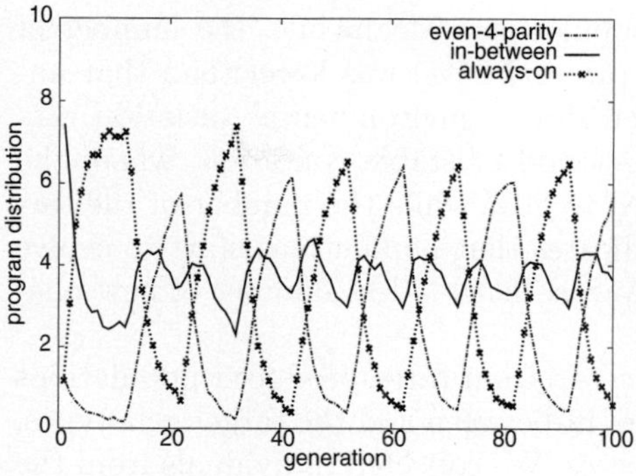

Fig. 2. program distributions under λ=10.

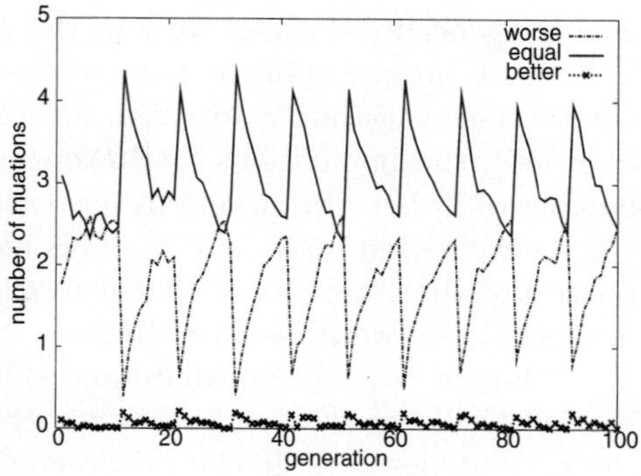

Fig. 3. program evolvability under λ=10.

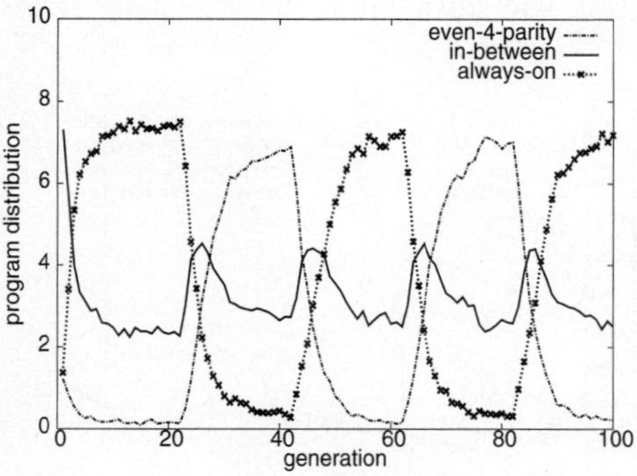

Fig. 4. program distributions under λ=20.

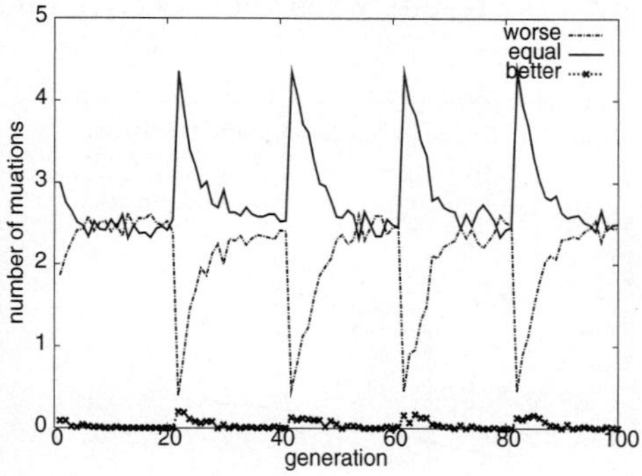

Fig. 5. program evolvability under λ=20.

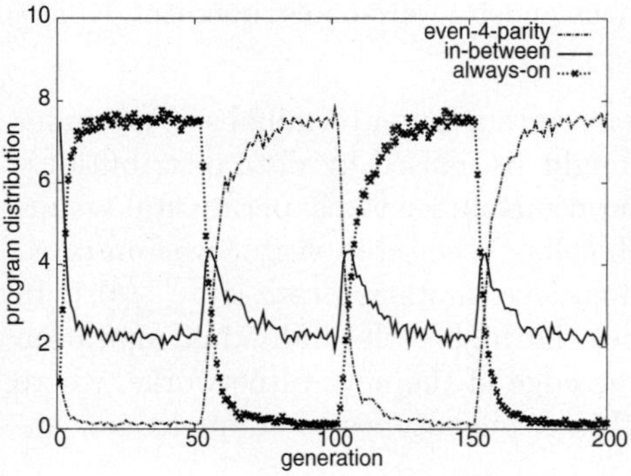

Fig. 6. program distributions under λ=50.

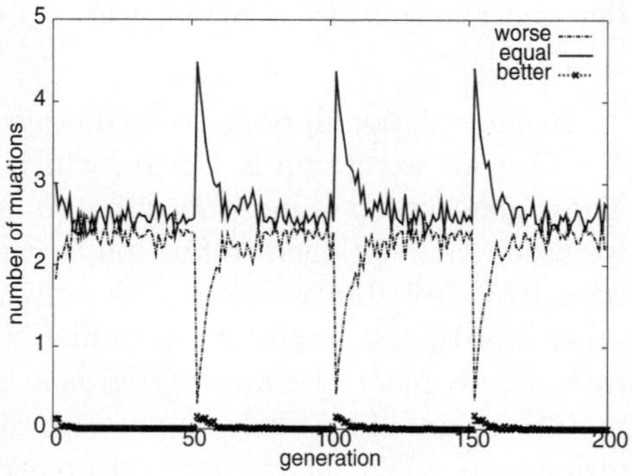

Fig. 7. program evolvability under λ=50.

genotypes occupied about 70% of the populations. Meanwhile, the number of in-between programs in the populations (around 25%) was lower than that under the fast variation rate. With an even slower environmental variation rate ($\lambda = 50$), the populations had a longer period of stable condition, where the number of in-between programs decreased to 20% while the number of the target genotypes increased to 75%. This indicates that populations start to evolve programs which are more resilient to changes, hence a less number of new phenotypes (in-between) were produced.

Although both intermediate and slow variation rates produced populations that evolved a similar combined number of in-between and the target genotypes, these genotypes have different characteristics. We collected individuals from the populations prior to switching to the new objective function from these two sets of runs. These genotypes were then evaluated with their terminal size (t). Figure 8 gives the results when the objective function was even-4-parity and Figure 9 gives the results when the objective function was always-on.

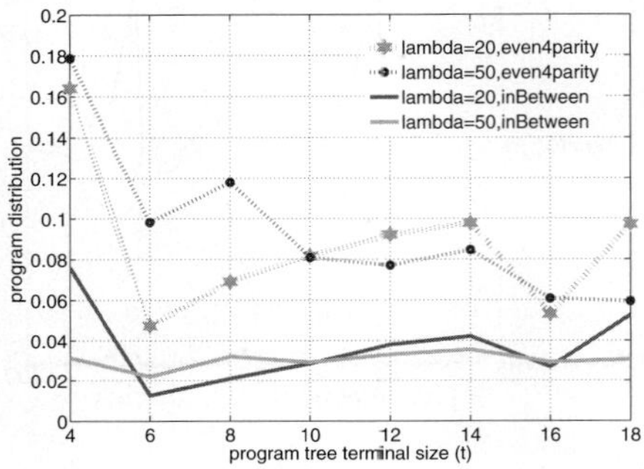

Fig. 8. program terminal size (t) distribution under even-4-parity environment.

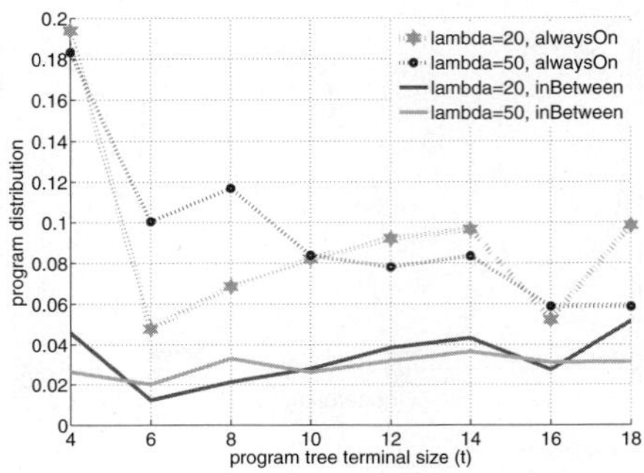

Fig. 9. program terminal size (t) distributions under always-on environment.

In general, populations evolved programs with a larger terminal size (t) under $\lambda = 20$ (one exception is $t = 16$, which might be caused by data distribution). These genotypes are more sensitive to mutations under the experimental setups we used. As mentioned in Section 5, each cell of a selected vector was mutated with 10% probability. Hence, the individual-level mutation rate is $\sum_{i=0}^{t+1} 0.1$. In other words, the larger the terminal size, the higher the individual mutation rate. These genotypes are located near the edge of the neutral networks, where mutations are more likely to produce different phenotypes to adapt to new environments when the change takes place.

Did the populations that evolved genotypes with a heightened mutation sensitivity produce more beneficial offspring? Figure 10 gives the information of all mutations that produced better offspring.

As shown, the populations which evolved genotypes with larger terminal sizes (under $\lambda = 20$) produced more beneficial offspring than those evolved genotypes of shorter terminal sizes (under $\lambda = 50$). Moreover, the majority of the geno-

	λ=10		λ=20		λ=50	
parent	qty	%	qty	%	qty	%
opposite	2129	28.29	1200	26.99	457	24.75
in-between	5396	71.71	3245	73.01	1389	75.25
Total	7525	100%	4445	100%	1846	100%

Fig. 10. Information of mutations that produced better offspring.

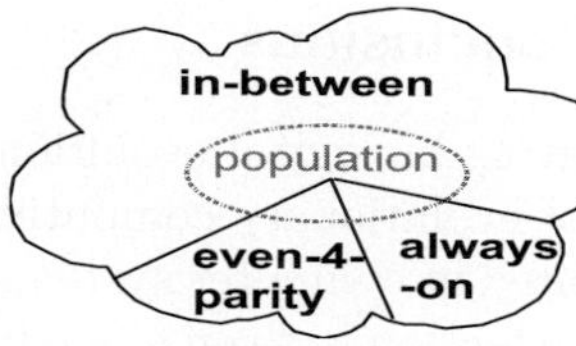

Fig. 11. Populations evolve toward boundaries of the 3 neutral networks.

types that produced better offspring are in-between. We also examined all these parent genotypes and found that they *all* have terminal size 18, hence are all located at the very end of the neutral networks. This indicates that populations evolved genotypes around the boundaries of the three phenotype neutral networks (see Figure 11). As shown in Figures 3, 5 and 7, all beneficial mutations took place right after the environmental switches. These provide strong evidences that under environmental variations, populations exploit neutrality to increase the like-hood of beneficial mutations.

7 Discussions

Using a simple GP system that switches between two opposite environments periodically, we have observed the populations favor genotypes with heightened mutation sensitivity, hence increase the rate of phenotypically consequential mutations, even though the genotypic mutation rate is the same. This evolution of evolvability [1] enhances the populations' ability to respond to selection and mutations according to the changes of their environments.

The emergence of self-adaption of mutation rates to improve evolvability has also been observed in a different GP system with a many-to-one genotype-phenotype mapping [13]. In that study, the objective function was constant throughout the evolution process. However, the populations exhibited an incremental progress that evolved from genotypes with lower fitness to higher ones step by step until the target phenotype was reached. We can therefore view it as an environment with continuous changes of target phenotypes from lower-fitness to higher-fitness. This pattern is similar to that of this study in that the populations focus on one phenotype at a time. We speculate that the emergent of self-adaption of mutation rates is likely to hold for a larger class of systems if the neutral networks contain genotypes with variable mutation sensitivities.

In this work, the prior and future environments are identical. If, instead, the environment continually shifts to completely novel states, the evolutionary history of a population might not prepare it for future adaptation. We speculate that populations will still evolve a certain number of genotypes with heightened mutation sensitivity, as long as they exist in the phenotype neutral networks. When considering open-ended evolution, populations which are capable of adapting mutation rates according to the environment they live in are more likely to continuously evolve evolvability.

8 Conclusions

Simulating biological evolution process for problem solving has been the backbone of evolutionary computing. While many aspects of the process have been integrated in computer systems to study the evolution of artificial life forms, not much attention is given to environmental variations. With the recent research development in evolutionary biology that clarifies the relationships between molecular, functional and ecological variations, we become better equipped to explore the implications of artificial life evolution under heterogeneous environments.

This paper is our initial effort in researching the interplay between program evolvability and neutral networks under environmental variations. Using a simple GP system under 2 alternating objective functions, we have observed populations exploiting neutrality to cope with environmental fluctuations and evolve evolvability. We will continue this work by investigating more complex environment variation patterns and other forms of neutrality.

Acknowledgment. This research is supported by NSERC of Canada.

References

1. Altenberg, L.: The evolution of evolvability in genetic programming. In: Kinnear Jr., K.E. (ed.) Advances in Genetic Programming, pp. 47–74 (1994)
2. Crosa, J.H.: Signal transduction and transcriptional and post transcriptional control of iron-regulated genes in bacteria. Microbiology and Molecular Biology Reviews 61(3), 319–336 (1997)
3. Huynen, M.A., Stadler, P.F., Fontana, W.: Smoothness within ruggedness: The role of neutrality in adaptation. Proceedings of the National Academy of Sciences of the United States of America (PNAS) 93(1), 397–401 (1996)
4. Kashtan, N., Alon, U.: Spontaneous evolution of modularity and network motifs. Proceedings of the National Academy of Sciences of the United States of America (PNAS) 102(39), 13773–13778 (2005)
5. Koza, J.R.: Genetic Programming. MIT Press, Cambridge (1992)
6. Langdon, W.B., Poli, R.: Why "building blocks" don't work on parity problems. Technical Report CSRP-98-17, University of Birmingham (1998)
7. Levins, R.J.: Evolution in Changing Environments. Princeton Univ. Press, Princeton (1968)
8. Lipson, H., Pollack, J.B., Suh, N.P.: On the origin of modular variation. Evolution 56(8), 1549–1556 (2002)
9. Meyers, L.A., Ancel, F.D., Lachmann, M.: Evolution of genetic potential. Computational Biology 1(3), 236–243 (2005)
10. Meyers, L.A., Bull, J.J.: Fighting change with change: adaptive variation in an uncertain world. Trends in Ecology & Evolution 17(12), 551–557 (2002)
11. Neilands, J.B.: Iron absorption and transport in microorganisms. Annual Review Nutrition 1, 27–46 (1981)
12. Yu, T., Bentley, P.: Methods to evolve legal phenotypes. In: Eiben, A.E., Bäck, T., Schoenauer, M., Schwefel, H.-P. (eds.) Parallel Problem Solving from Nature - PPSN V. LNCS, vol. 1498, pp. 280–291. Springer, Heidelberg (1998)
13. Yu, T., Miller, J.: Through the interaction of neutral and adaptive mutations, evolutionary search finds a way. Artificial Life 12(4), 525–551 (2006)

The Creativity Potential Within Evolutionary Algorithms

David Iclănzan

Department of Electrical Engineering,
Sapientia Hungarian University of Transylvania,
Şoseaua Sighişoarei 1C, 547367, Corunca, ROMANIA
david.iclanzan@gmail.com

Abstract. The traditional GA theory is pillared on the Building Block Hypothesis (BBH) which states that Genetic Algorithms (GAs) work by discovering, emphasizing and recombining low order schemata in high-quality strings, in a strongly parallel manner. Historically, attempts to capture the topological fitness landscape features which exemplify this intuitively straight-forward process, have been mostly unsuccessful. Population-based recombinative methods had been repeatedly outperformed on the special designed abstract test suites, by different variants of mutation-based algorithms. Departing from the BBH, in this paper we seek to exemplify the utility of crossover from a different point of view, emphasizing the creative potential of the crossover operator. We design a special class of abstract test suites, called Trident functions, which exploits the ability of modern GAs to mix good but *significantly different* solutions. This approach has been so far neglected as it is widely believed that disruption caused by mating individuals that are too dissimilar may be harmful. We anticipate that hybridizing different designs induces a complex neighborhood structure unattainable by trajectory-based methods which can conceal novel solutions. Empirical results confirm that the proposed class of problems can be solved efficiently only by population-based panmictic recombinative methods, employing diversity maintaining mechanisms.

Key words: fundamentals of GAs, crossover, hybridization, creativity

1 Introduction

Evolutionary Algorithms (EAs) are inspired and closely linked to Darwinian evolution. The complex schemes involved in the transmission of biological information make these methods suited for changing environments. Recent results had shown that in dynamical environments induced by exogenous noise, recombinative methods clearly outperform hill-climbers [1]. However, as applications of EAs consider mostly fixed fitness landscapes, a major criticism of EAs is that biological metaphors may be unnecessarily complex. Therefore, an essential open problem regarding the fundamentals of GAs is to characterize the topological features of *static* fitness landscapes for which crossover is an efficient operator.

F. Almeida e Costa et al. (Eds.): ECAL 2007, LNAI 4648, pp. 845–854, 2007.

According to the intuition behind the BBH [2,3], GAs should perform well when they are able to discover above-average-fitness low-order schemata and recombine them to produce even more fit higher-order schemata in a recursive manner. So far, attempts to design abstract test functions suited for this hierarchical, straight-forward process have been unsuccessful. Population-based recombinative methods had been repeatedly outperformed on the special designed abstract test suites, by different variants of mutation-based algorithms.

Leaving aside the suggestions of the BBH, the objective of this paper is to develop test suites that focus on and exploit the generative potential of the crossover operator.

The creative potential of the crossover operator had been already conjectured many times [4,5]. A great asset of EAs is the ability to simultaneously process and combine many *different* designs. The promoting idea behind recombination, as it was originally proposed [2], is to combine different sub-solutions from varying individuals. Nevertheless, there is widely spread methodology which emphasize another aspect of the recombination, namely that it preserves the similarity exhibited by the parents. This idea is supported in classical approaches, where niching and speciation methods restrict mating to individuals that are genotypically similar. The same view manifests in spatially distributed GAs [6] where breeding is promoted within local populations. These approaches advocate the idea that parents selected from two different fitness peaks are likely to produce an offspring that lands in the "valley" between.

But what if valuable solutions lay between two different fitness peaks? This may be the case in many design problems, where hybridizing two prototypes can result in a completely new, valuable solution. In these cases, though crossover may not produce successful offspring on average, it conceals a great generative ability, revealing novel designs which would be unattainable otherwise. The points from the complex neighborhood structure defined by crossover on two valuable but different solutions, are expected to be intractable by trajectory methods, provided that these points do not have a large basin of attraction. In this case the problem can be solved only by employing the generative potential of the recombination operator.

The remaining of the paper is organized as follows: in the next section we review the history of the quest for "GA friendly" test functions. In Section 3 the class of functions which emphasize the creative potential of GAs (Trident functions) is introduced. Experimental setup and results on these test suites are presented in Section 4. Section 5 concludes.

2 Historical Background

The Royal Road (RR) problems [7] were an early attempt to capture the problem structure, which enables GAs to outperform other methods. It was recognized that major tenets behind the BBH are the notion of problem decomposition and the assembly of solutions from sub-solutions. Subsequently, a set of functions were constructed, which clearly emphasized a gross-scale building-block structure with low-order building-blocks that recombine to higher-order ones. The functions were expected to lay out a "royal road" for GAs and the intention was to exemplify the class of problems that required the recombinative aspects of the GA. However, a Random Mutation Hill-Climber (RMHC), which accepts states with equal objective function value, greatly outperformed GAs on these test suites.

The RR functions can be solved by mutation-based hill-climbers much faster than by GAs, because they exhibit separable building-block structure with tight linkage and do not contain any local optima. Even when deception is introduced and each building-block corresponds to a deceptive trap function, the Macro Mutation Hill-Climber (MMHC) can outperform the GA, by benefiting from the tight linkage and cluster mutating one sub-function at a time [8]. The $\text{JUMP}_{n,m}$ and alike functions, where finding global optima requires a bigger jump from a local optima, proposed in [9] can also be handled by macromutation strategies.

In order to break hill-climbers, problems with unfavorable epistatic linkage and unfavorable genetic linkage were devised. The former refers to building-block interdependency (non-separability), while the latter corresponds to building-blocks formed by non-adjacent alleles. Watson et al. [10] introduced a class of hierarchically decomposable functions, which presented strong non-linear hierarchical building-block interdependency. In a later development, Pelikan et al. [11] proposed the hierarchical trap function, where a basic deceptive function is used on each level to make the problem even harder.

On problems with poor genetic linkage, classical crossover is not able to combine good building-blocks effectively. Thus, considerable effort was invested at developing algorithms that discover and represent the interdependency of alleles, resulting in more intelligent crossover operators [12]. As a result, competent methods can address problems which present both epistatic and unfavorable genetic linkage. It was considered that this class of functions exemplifies those problems for which GAs are well-suited.

Hill-climbing in building-block space was proposed and shown to be more efficient than selectorecombinative GAs on deterministic additively-separable problems of bounded difficulty [1]. But a recent development [13] had shown that a certain hill-climber operating over the building-block space can address problems even with epistatic and unfavorable genetic linkage. The new Building Block Hill-Climber (BBHC) is able to solve hierarchical problems with random linkages. The method uses hill-climb search experience to learn the problem structure. The neighborhood structure is adapted whenever new knowledge about the underlaying BB structure is incorporated into the search. The BBHC holds a major qualitative advantage, scaling as $O(l \cdot log(l))$ compared to $O(l^{1.5} \cdot log(l)$, which is at least required by population-based recombinative methods [14].

The new result suggests that if a problem has a nice structure, even if "hidden" like the building-block space characterized by strong non-linear interdependencies, a proper hill-climber can outperform population-based recombinative methods, *without requiring extra domain knowledge.*

We suspect that the idea of a GA marching on a fitness landscape is maybe a little bit romantic; a suitable hill-climber is almost certainly quicker if there is a nice structure of the problem to be exploited. In order to pinpoint the usefulness of recombinative methods, maybe we should look for hard problems, which can be solved somewhat slothfully by GAs, but are intractable using other methods.

In the next chapter such a function is introduced, for which we hope GAs are well-fitted.

3 Hybridization of Differences

The reason why hill-climbers outperform population-based recombinative methods on the test suites proposed so far, may be their non-deceptive nature at *some* level. Deception may manifest in more then one form, but the general idea is that "a deceptive function is one in which low-order schema fitness averages favor a particular local optimum, but the global optimum is located at that optimum's complement" [3]. The RR function does not contain any local optima; hierarchical problems are fully deceptive in the Hamming space but are fully non-deceptive in the building-block or crossover space. Recent local-search literature authors have emphasized the importance of using a good neighborhood operator [15]. Thus, it is not that surprising that a certain hill-climber employing a good neighborhood operator (which engenders a building-block-wise search in the case of hierarchical functions), can outperform methods which operate via populations.

We reason that in order to defeat hill-climbers, problems must contain a degree of deception, which can not be overcome by a neighborhood operator induced by one point in the search space. This of course will hinder GAs performance also, as the mutation works in the neighborhood of one individual and short-term selection may favor deceptive search paths. However, EAs posses a great asset by having a more complex neighborhood structure generated by the recombination operator, which takes into account at least two individuals. This may help the methods to escape the local optima and overcome deceptiveness.

The problem representation together with the neighborhood structure defines the search landscape. We argue that there are problems where only search landscapes transformed by crossover may be efficiently exploitable. In the followings we give an example for such a class of problems called the Trident functions (TF).

3.1 The Trident Function

TF accepts bitstrings of the length $2k$ where $k \geq 2$ and uses a function of unitation (which depends on the number of ones in a bitstring, and not on their positions) as underlaying structure:

$$base(x) = \|2 \cdot u(x) - |x|\| \tag{1}$$

where $u(x)$ is the unitary of x (the number of ones) and $|x|$ is the length of x.

The base function has its minimum in 0 which is generated by strings with an equal number of 1's and 0's: $u(x) = |x| - u(x)$. The maximum is attained by strings formed by all 1's or all 0's with a corresponding value of $|x|$.

The next component of the TF is a contribution function which rewards *certain configurations* of strings that have an equal number of 1's and 0's. Let $L = x_1, x_2, \ldots, x_{\frac{n}{2}}$ be the first half of the binary string x of length n and $R = x_{\frac{n}{2}+1}, x_{\frac{n}{2}+2}, \ldots, x_n$ the second one. Then, we define the contribution function for this paper based on the exclusive OR (XOR) relation:

$$contribution(x) = \begin{cases} 2 \cdot |x| & \text{, if } L = \bar{R}; \\ 0 & \text{, otherwise.} \end{cases} \tag{2}$$

where $\bar{R}$ stands for the bitwise negation of R.

Please note that the contribution function does not have a basin of attraction; it rewards fully an input or it does not reward it at all. Finding the maxima of such a function is equivalent to the *needle in the haystack* problem. As there are no better search methods for this class of function than the random-search, these function are also resistant to biased mutation-based search.

The TF is defined as the sum of the base and the contribution function:

$$trident(x) = base(x) + contribution(x) \tag{3}$$

Figure 1 presents the graphical interpretation of the Trident function.

TF has its maximum in the points rewarded by the contribution function. Here it takes the value $2 \cdot |x|$ as the base function in these points attends the minimum 0.

TF is very hard for mutation-based algorithms because the base function leads away the search from the region where global optima lay. Even if a random state is generated with equal number of 1's and 0's, it is very unlikely for large problem instances that the contribution function will reward that string. If the algorithm does a biased search, it will be immediately drawn away from the minimum of the base function, towards regions with higher base function fitness.

The TF can defeat macromutation hill-climbers also, as local and global optima are very distant in the Hamming space. The chance of jumping from local optima to a global one is minimal as $\frac{n}{2}$ bits must be changed simultaneously.

Also, there are no "hidden" structures which could be easily exploited. The "building-blocks" L and R are rewarded if and only if their context i.e. the counterpart half of the string is compatible. As TFs have $2^{\frac{n}{2}}$ global solutions, the probability of this happening for randomly generated strings is $P_{hit} = \frac{2^{\frac{n}{2}}}{2^n} = \frac{1}{2^{\frac{n}{2}}}$.

What about GAs? Global optima can be found quite easily if the GA is mixing good but different solutions. Let us take the example where $n = 8$ and we have two strings at each local optimum: $s_1 = 00000000$ and $s_2 = 11111111$. The one-point crossover between s_1 and s_2 will produce the optimal strings $s_3 = 00001111$ and $s_4 = 11110000$ with the probability $P = \frac{1}{n-1} = \frac{1}{7}$. When using two-point crossover, we have $\frac{n}{2} - 1 = 3$ favorable cases. The favorable crossing points pairs are $\{(1,7), (2,6), (3,5)\}$. Optimal strings may not result only from the breeding of individuals located at local optima. For example, the one-point crossover between $s_5 = 00100001$ and $s_6 = 11111101$ between loci 4 and 5 will also produce an optimal solution $s_7 = 00101101$. The important aspect is to combine *different* candidate solutions.

The TF portray the problems where several highly different good solutions exist, and hybridizing these solutions *may* result in a completely new, valuable design. Even if crossover does not produce above average individuals on a regular basis, it may create occasionally an exceptional organism. Thus, crossover has a generative potential which we believe should not be neglected by restricting the recombinationb to genotypically similar individuals.

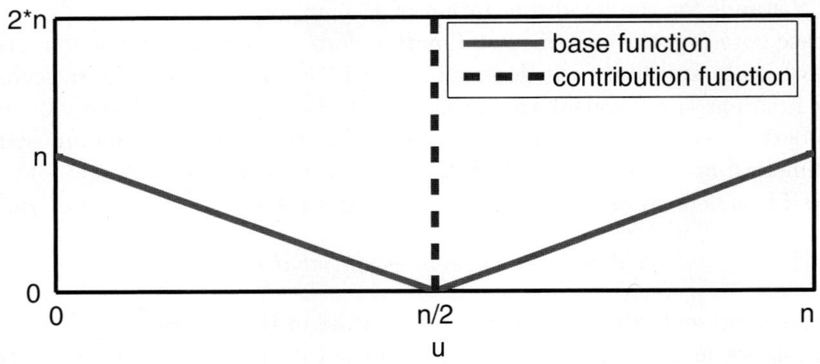

Fig. 1. The Trident Function. u is the unitary of the input string. The base function is deceptive, leading away the search from the area which contains global optima. The contribution function has no basin of attraction so its maximum is very hard to detect. Note that the contribution function does not reward all strings with $u = \frac{n}{2}$; only special configurations are rated.

3.2 Natural Metaphor

Hybridization between related species is a common occurrence in nature and it plays an important role in the evolution of some species.

Unlike mutation, which affects only one gene, hybridization simultaneously affects multiple variations across genes or gene combinations. This translates to big jumps on the fitness landscape. However, these jumps are not random as in the case of the macromutation. Crossover may mix certain well established features which may not be that important for the parents, but their combination is spectacular and it is highly rewarded in the resulting hybrid. Hybrids in nature are often stronger than either parent variety. This phenomenon is known as *hybrid vigor* or *heterosis*.

In the next section we analyze how different variants of hill-climbers and GAs perform on the proposed class of functions.

4 Results

We tested the RMHC, the MMHC, the Simple Genetic Algorithm (SGA) and a special genetic algorithm which employs diversity maintenance, called Deterministic Crowding (DC) on the 16, 32, 64 respectively 128-bit versions of the TF. A number of 100 independent runs were averaged. The algorithms ran until they found a global optimum or 10^6 function evaluations were exceeded.

The difficulty of the TF grows exponentially with the problem size. The 16-bit version of the problem is utterly easy as global optimum occupies $\frac{1}{256}$-th of the search space. A simple random sampling of the search space will yield an optimal solution in 256 steps on average. Anyhow, increasing the problem size to 64-bit will drop the probability of randomly hitting an optimal solution to less then 1 to 4.e+9. We were especially interested how the methods can handle the exponential explosion and how the reduction of the fraction occupied by global optima in report with the size of the search space affects their scalability.

The methods and the settings used for each algorithm are briefly described as follows.

4.1 Random Mutation Hill-Climber

RMHC [7] chooses at each iteration a locus at random to mutate. If the mutation leads to an equal or higher fitness, then the new state is accepted.

A random restart was applied in our tests, if the RMHC could not improve for 10^4 consecutive evaluations.

4.2 Macro Mutation Hill-Climber

MMHC had been shown to be a very powerful hill-climbing method, which can outperform GAs even on problems where each building-block corresponds to a deceptive trap function, provided that the problem has a tight linkage [8].

In the mutation operator of the MMHC, the genome is viewed as a circular string. Two distinct points on the ring are randomly selected and the loci in the smaller section are set to random alleles.

In our tests, if the MMHC could not improve for 10^5 epochs, it was restarted from a random state, in the limit of the 10^6 function evaluations.

4.3 Simple Genetic Algorithm

The SGA [2,3] is a GA that uses fitness-proportionate selection, one-point crossover, and point mutation to evolve a single population of bit strings, with each generation completely replacing the previous one. The population size was set to $ps = 500$, the mutation rate to $p_{mut} = 0.005$ and the crossover rate was $c = 0.8$.

4.4 Deterministic Crowding

Deterministic Crowding operates on the premise of restricted competition rather than restricted mating. The population is panmictic so genotypically different individuals may be recombined. This accommodates well the view on the importance of hybridization. Diversity in maintained by restricting the competition to parents and their own offspring. In this way, subpopulations that are occupying different niches do not need to outperform each other in order to propagate. A detailed presentation of the method can be found in [16].

For the DC we used the same parameter settings as for the SGA.

Table 1. Performance of the studied algorithms on the TFs. Column "Succ. rate" contains the number of successful runs where the methods find global optima. Column "Avg. nr." contains the number of average function evaluations needed to find the global optima and "Max. nr." counts the maximum number of evaluations needed provided that all runs were successful. In the case of the population-based methods column "Nr. opt." contains the average number of different optima within the correctly converged population.

TF size	16				32			
	Succ. rate	Avg. nr.	Max. nr	Nr. opt.	Succ. rate	Avg. nr.	Max. nr.	Nr. opt.
RMHC	**85%**	400020	-	/	**3%**	337136	-	/
MMHC	**100%**	4213	33528	/	**37%**	471018	-	/
SGA	**100%**	241	1390	1.46	**25%**	8435	-	1.56
DC	**100%**	250	2111	23.94	**100%**	15771	23883	6.69

TF size	64				128			
	Succ. rate	Avg. nr.	Max. nr	Nr. opt.	Succ. rate	Avg. nr.	Max. nr.	Nr. opt.
DC	**100%**	46816	61366	6.02	**100%**	112557	157890	4.32

4.5 Numerical Results

The numerical results of the experiments are summarized in Table 1. On the 64 and 128-bit versions of the TF, only the results of the DC are reported as the hill-climbers and the SGA failed in all runs on these suites.

As expected, the worst behavior on the TFs was shown by the RMHC. Even for the very easy 16-bit version of the problem, the success rate is only 85%. Similar to the other hill-climber, the MMHC, solutions are found at very high cost and only due to the random restart mechanism. The number of function evaluations required to identify optima, exceeded by orders of magnitude the amount that would be required by random-search. As the problem size increases, becoming unaddressable by random sampling, the hill-climbers fail in all runs due the deceptive nature of the TF.

When SGA succeeded, its performance was the fastest, being much better then it would be required by random-search. This shows that even simple recombinative algorithms have the potential to exploit the features of the TF landscape. However, as the problem size increases, the SGA fails to find optima due to the lack of improper initial sampling. Therefore, the population is quickly shifted towards the basin of attraction of a single local optimum. A significant increase of the population could address this problem but then success would come at high costs.

The only competent algorithm on the TFs was the DC. It succeeded in absolutely all runs, being able to identify global optima within a maximum of 16% of the allowed function evaluations. In all cases, several optimum points were detected. The success of the algorithm derives from its diversity maintaining mechanism combined with the panmictic population.

We once again emphasize the importance of the capability to mix different designs; only then recombination can become creative. An algorithm with diversity maintaining mechanism, but with crossover restricted to similar individuals, would also fail on the TFs.

5 Conclusions

This paper has examined some questions concerning the fundamentals of EAs. The shortcomings of existing building-block style test functions were surveyed and a view which promotes and emphasizes the generative potential of EAs was discussed. According to this conception, the great strength of the crossover operator lays in its capability to hybridize different designs, rather than in that of promoting similarity among the population.

Consequently, a new class of test problems, called Trident functions was introduced. The TF is dominated by a fully deceptive base function as global optima coincide with the minima of this function. The discrete optimal solutions are defined by a contribution function which rewards points from the search space where certain different genotypical features appear concomitantly. As the contribution function does not have an attractor basin, the deceptiveness of the base function can not be overcome using only simple neighborhood structures.

The TFs may represent design problems where several good, locally optimal drafts are easy to find, dominating the search space (deception) and the real good designs result from the hybridization of different drafts. Furthermore, the complex layouts defining the best solution only emerge in "reactive regions" where the correct particular features appear simultaneously; there is no sequence of improving designs to these solutions (needle in the haystack). Nevertheless, crossover possesses the creative potential i.e the more complex neighborhood structure, which enables it to identify these solutions by mixing features from different drafts, until the correct configuration is detected.

The experimental results confirm that the proposed class of functions contains topological features, which can be efficiently exploited only by recombinative methods mixing different good solutions.

Acknowledgments. This work was sponsored by MEdC-CNCSIS under grant AT-8/70-2006 and by the Sapientia Institute for Research Programs (KPI).

References

1. Sastry, K., Goldberg, D.E.: Let's get ready to rumble: Crossover versus mutation head to head. In: Deb, K., et al. (eds.) GECCO 2004. LNCS, vol. 3103, pp. 126–137. Springer, Heidelberg (2004)
2. Holland, J.H.: Adaptation in natural artificial systems. University of Michigan Press, Ann Arbor (1975)
3. Goldberg, D.E.: Genetic Algorithms in Search, Optimization and Machine Learning. Addison-Wesley, Reading (1989)
4. Syswerda, G.: Uniform crossover in genetic algorithms. In: Proceedings of Third International Conference on Genetic Algorithms (1989)
5. Chen, S.: Is the Common Good? A New Perspective Developed in Genetic Algorithms. PhD thesis, Robotics Institute, Carnegie Mellon University (1999)
6. Starkweather, T., Mathias, K., Whitley, D.: Optimization using distributed genetic algorithms. In: Proceedings of the First International Conference on Parallel Problem Solving from Nature, pp. 176–186. Springer, Heidelberg (1991)
7. Forrest, S., Mitchell, M.: What makes a problem hard for a genetic algorithm? some anomalous results and their explanation. MACHLEARN: Machine Learning 13 (1993)
8. Jones, T.: Evolutionary Algorithms, Fitness Landscapes and Search. PhD thesis, University of New Mexico, Albuquerque, NM (1995)
9. Jansen, T., Wegener, I.: The analysis of evolutionary algorithms – A proof that crossover really can help. Algorithmica 34, 47–66 (2002)
10. Watson, R.A., Hornby, G., Pollack, J.B.: Modeling building-block interdependency. In: Eiben, A.E., Bäck, T., Schoenauer, M., Schwefel, H.-P. (eds.) Parallel Problem Solving from Nature - PPSN V. LNCS, vol. 1498, pp. 97–108. Springer, Heidelberg (1998)
11. Pelikan, M., Goldberg, D.E.: Escaping hierarchical traps with competent genetic algorithms. In: Spector, L., et al. (eds.) GECCO '01: pp. 511–518. Morgan Kaufmann, San Francisco (2001)
12. Larranaga, P., Lozano, J.A.: Estimation of Distribution Algorithms: A New Tool for Evolutionary Computation. Kluwer Academic Publishers, Norwell (2001)
13. Iclanzan, D., Dumitrescu, D.: Overcoming Hierarchical Difficulty by Hill-Climbing the Building Block Structure. In: GECCO '07: Proc. of the Genetic and Evolutionary Computation Conference, July (accepted, 2007)
14. Pelikan, M.: Hierarchical Bayesian optimization algorithm: Toward a new generation of evolutionary algorithms. Springer, Heidelberg (2005)
15. Watson, J.-P., Beck, J.C., Howe, A.E., Whitley, L.D.: Problem difficulty for tabu search in job-shop scheduling. AIJ: Artificial Intelligence 143 (2003)
16. Mahfoud, S.W.: Niching methods for genetic algorithms. PhD thesis, Dept. General Engineering, University of Illinois, Urbana, Illinois (1995)

The Problems with Counting Ancestors in a Simple Genetic Algorithm

Robert Collier and Mark Wineberg

Department of Computing and Information Science
University of Guelph
Guelph, Ontario, Canada, N1G 2W1
collierr@uoguelph.ca, wineberg@cis.uoguelph.ca

Abstract. The ease with which the genetic algorithm can generate good solutions to challenging optimization problems has resulted in a tendency for researchers to overlook the easily gathered and largely untapped raw data present in the ancestral relationships that guide a population to convergence. This article introduces the notion of a lineage tree structure and associated ancestor count measure that reveals unexpected regularities when studying instances of a simple genetic algorithm applied to three disparate problems. Consequently, a series of explanatory models was constructed to identify the components of the underlying evolutionary mechanism that may be responsible for the commonalities observed. These components (exponential growth of member parentage and inbreeding caused by the appearance of duplicates and shared ancestries) place constraints on the number of ancestors that a solution may have; an insight that may prove valuable for future analysis of the behavior of the genetic algorithm.

Keywords: Evolutionary Computation, Genetic Algorithm, Ancestry, Lineage.

1 Introduction

Although the genetic algorithm has already demonstrated its versatility and effectiveness as an optimization technique, the stochastic nature of the algorithm requires that researchers recognize the possibility that copies of the same genetic algorithm implementation, executed in distinct yet identical environments, can achieve different results. This is sometimes considered indicative that the genetic algorithm is unreliable; understandably, as a poorly selected initial population can be the difference between success and failure in finding an optimal solution. This motivates an overarching investigation of properties possessed by populations that successfully converge on an optimal solution yet absent from those which do not.

This article represents the initial phase of this exploratory investigation into the ancestral relationships that exist between the candidate solutions in each generation of the genetic algorithm. The general structure proposed here, termed "lineage tree", should provide insight into some of the underlying deterministic elements of the evolutionary mechanisms dictating whether a specified population will thrive or perish. However, since preliminary experiments produced results which were somewhat counter-intuitive, it is essential that the simplest features manifested in this structure

F. Almeida e Costa et al. (Eds.): ECAL 2007, LNAI 4648, pp. 855 – 864, 2007.

be explained as completely as possible. Only then could the lineage tree be employed in the development of enhancements to the genetic algorithm.

Although ancestral data, which comprises the lineage tree, is very easy to collect, it is voluminous and is generally considered to have little application to the task of solving optimization problems. That is to say, when a metaheuristic is applied to an optimization problem, the information of greatest consequence is the quality of the solution and the speed with which it is located. It follows that the search path taken through the space of candidate solutions is usually considered irrelevant. Indeed, none of the surveyed resources on evolutionary theory have yet attempted to utilize or model information from observed ancestral relationships[3][5][7]. However, since the motivation of the experiments and models in this paper is focused upon an empirical exploration of the mechanisms underlying computational evolution independent of the solution to any particular problem instance, this study of ancestral evolutionary relationships could yield interesting and potentially useful insights.

In this paper we will examine the number of ancestors of an optimal solution to a particular problem for each generation of the genetic algorithm and generate predictive models in an attempt to fully explain the observations. Since this body of research is in its infancy, little consideration is given to the applicability of these results at this time. However, it stands to reason that the results of this investigation have the potential to be at least as fruitful as the constructivist "trial-and-error" techniques for algorithm enhancement.

2 The Lineage Tree Structure and Ancestor Count Measure

2.1 The Differences Between Phylogenetic, Genealogic, and Lineage Trees

When exploring population ancestry, the structures most commonly employed are phylogenetic and genealogical trees. Although biology's phylogenetic trees are a concrete example of structures constructed from ancestral relationships, they are used to find relationships between species, not individuals. That is to say, species which are siblings in a phylogenetic tree are joined by the most recent common ancestor, resulting in a structure depicting diversification and speciation as a function of time. With few exceptions[1][2], this structure is not applicable to the genetic algorithm. The genealogical tree, however, depicts paths from individuals to their respective ancestors and serves as the inspiration behind the lineage tree introduced in this paper.

The term lineage tree is actually a misnomer, as the use of crossover implies that population members may have more than one parent and, consequently, the actual ancestry is a directed acyclic graph rather than a tree. However, since the structure remains analogous to the genealogical tree (a similar misnomer), the term tree is retained. It is also noteworthy that, since all "branches" in the proposed lineage tree end in earlier generations (towards ancestors and not descendents), it would not be prudent to the use the established term genealogical tree.

2.2 The Procedure for Counting Ancestors

Although other metrics will be explored in the future, a count of the number of ancestors represents the simplest statistic that can be gathered from the lineage of a member. This technique will allow for measurements of the percentage of previous populations that have contributed chromosomal data to the root solution.

To determine what data can be extracted from the lineage tree structures it is necessary to develop models mirroring, as closely as possible, the ancestor counts of actual lineage trees. Given an instance of the genetic algorithm which has run for exactly G generations, in which there is an optimal or current best solution s, a series L of sets $L_0 < L_1 < \ldots < L_{G-1}$ corresponding to the set of nodes in the lineage tree for every possible depth from 0 to $G-1$ will be constructed by the following algorithm, which recursive flags all members contributing genetic material to the solution:

$L_0 := \{s\}$	*the first set consists of the solution s*
for $g = G..1;\ h = 0..G-1$	*for every generation preceding G,*
$L_{h+1} := \varnothing$	*add the parent members of all*
for each $m \in L_h$	*elements of the previous set in L*
$L_{h+1} := L_{h+1} \cup \text{parents}(P_{g-1}, m)$	*to the current set in L*

where P_{g-1} is the population at generation $g-1$ and parents(P, m) are the members of P that produced offspring m. The number of nodes in the lineage tree at depth h is $|L_h|$, where h is the reverse chronologically ordered *ancestral generation*.

3 Modeling Ancestor Counts – Developing Model 1

A model approximating the actual number of ancestors of a member of interest in an evolutionary system should prove useful, as discrepancies could contain important information about the underlying mechanism. For example, a common (albeit naïve) expectation is that the number of ancestors of a member, traced back far enough, would reach the maximum size of the population. However, observations of actual genetic algorithms reveal that ancestor counts approach a significantly lesser bound. Nevertheless, an ancestor counting model considering only the expected number of parents will describe the first in a series of improving models of ancestor count.

3.1 Motivating Model 1: Considering the Exponential Growth of Parentage

As stated in the introduction to this section, exponential growth in the number of ancestors of a given population member does not sufficiently characterize the ancestor count observed in actual systems. Fortunately, it does provide a logical approximation for the ancestor counts of any evolutionary system and, consequently, provides a starting point for the generation of subsequent models. As such, it will be explored in detail.

Employing an analogy from genealogy, every generation of a family tree is expected to include exactly two parents for each member in the following generation. However, since the populations created by the genetic algorithm are the product of both sexual and asexual reproduction, the exponential growth factor is reduced. Given the probability of performing crossover in each generation ($p_{crossover}$), the expected number of parents in the previous generation can be expressed as

$$\begin{aligned}|L_{h+1}| &= |L_h| \cdot (p_{crossover} \cdot 2) + |L_h| \cdot ((1 - p_{crossover}) \cdot 1) \\ &= |L_h| \cdot (2 \cdot p_{crossover} + (1 - p_{crossover})) \\ &= |L_h| \cdot (1 + p_{crossover}).\end{aligned} \tag{1}$$

Thus, for a typical crossover probability of 0.7 (that is, 70% of each population is created sexually with crossover), the expected number of parents for each population member is 1.7. Thus, the lineage tree would be (naïvely) expected to have an exponential growth rate of 1.7 until the entire available population is incorporated. Comparing the results of this model against the ancestor count of typical lineage trees will reveal the degree to which this model accurately predicts the ancestor count.

3.2 Testing the Accuracy of Model 1

Lineage tree ancestor sets were generated, using the construction algorithm described earlier, for genetic algorithms applied to three distinct binary encoded problems with varying fitness landscapes: OneMax (for 200 bits), Rastrigin's (in 60 bits for 6 dimensions within -5.12, 5.11), and Whitley's F8F2 (in 48 bits for 4 dimensions within -8.192, 8.188) [7]. The problems completed 100 generations with a population size 200. All instances employ tournament selection of size 2 (without elitism), a crossover probability of 0.7, and a mutation rate of 1 / *chrLength*, where *chrLength* refers to the length of a chromosome. All experiments were repeated 30 times to produce statistically significant results and:

1) Fitness evaluated by Rastrigin's function is defined as

$$\text{Rastrigin}(x) = 10 \cdot n + \sum_{i=1}^{n} (x_i^2 - 10 \cdot \cos(2 \cdot \pi \cdot x_i)) ; \tag{2}$$

2) Fitness evaluated by F8F2 is defined as

$$\text{F8F2}(x) = \sum_{i=0}^{n-1} (1 + \frac{(\text{F2}(x_i, x_{i+1}))^2}{4000} - \cos(\text{F2}(x_i, x_{i+1}))); \tag{3}$$

where $\text{F2}(x_i, x_{i+1}) = 100(x_i^2 - x_{i+1}) + (1 - x_i)^2$ and $x_n = x_0$.

Preliminary experiments indicated that the number of members included in the lineage tree is largely independent of the probability of mutation, resulting in a decision to defer the effects of varying mutation to a future study. The results from Rastrigin's and OneMax were indistinguishable and very similar to F8F2, and so, unless comparisons are required, only the results from F8F2 are presented.

The lineage tree ancestor count observations from the genetic algorithm experiments are graphed against the evolutionary growth model (see Figure 1). Although it is evident that the actual counts are matched closely by the model during the initial growth of the lineage tree, the model quickly surpasses the observed results and expands to fit the entire population. Although the actual counts do approach a stable region, it is significantly less than that predicted by the model.

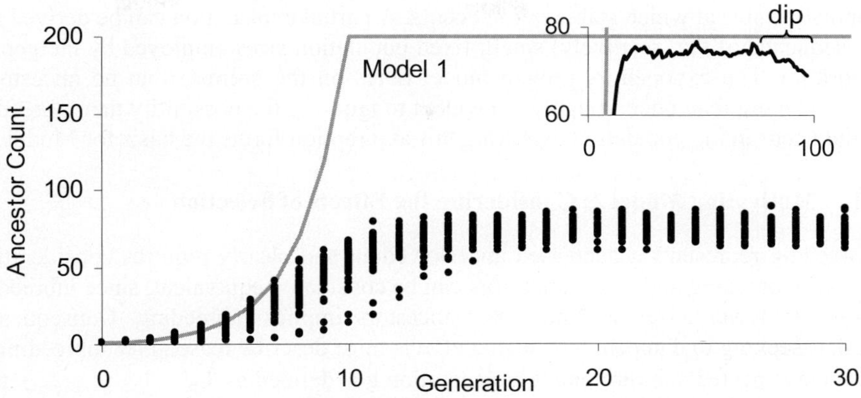

Fig. 1. Ancestor counts (from the best solution found in the final generation) predicted by Model 1 against actual counts from F8F2 for the first 25 ancestral generations. Also depicted (as inset), the ancestor count "dip" occurring only in the final ancestral generations.

3.3 Observations

From Figure 1, it becomes apparent that the ancestor counts can be categorized into three regions: a region of growth analogous to the exponential growth section of the first model, a region of growth which does not follow an exponential curve, and finally a flat[1] region of stability. This is true for all evaluation functions examined.

For F8F2 and Rastrigin's, there is an additional feature: a "dip" in the ancestor count, which only occurs in the final ancestral generations near the initial population. This feature was confirmed independent of the actual number of generations completed, but possible explanations are deferred to the final section of the paper.

Finally, it is apparent that the genetic algorithm produces identical ancestor counts independent of the problem being solved. A curve consisting of a region of growth followed by a region of stability at approximately 35% of the population size[2] was determined for each evaluation function. This curve was also demonstrated to be largely independent of population size, further suggesting that ancestor counts are governed by constant, consistent properties independent of the problem being solved.

---

1 A least squared error linear regression model was applied to the final region to verify its stability. The data in this region is normally distributed (correlation with rankits producing an r^2 value of 0.999) so linear regression can be applied directly. The resultant equation had a slope of 0.002 (indicating that the region is essentially flat) and a calculated p-value of 0.0003 from a t-test ensures that this result is statistically significant.

2 The actual percentages are 35.3±0.1% for One-Max, 36.9±0.1% for Rastrigin's, and 37.4±0.1% for F8F2. It appears that harder problems have marginally higher ancestor counts; within one or two percentage points.

4 Inbreeding Considerations – Developing Model 2

From the previous section, it is apparent that exponential growth does not entirely explain the actual ancestor counts observed; especially the percentage of the total population size at which stabilization occurs. A partial explanation can be derived as a consequence of the (relatively) small, fixed population sizes employed by the genetic algorithm. The exponential growth model relies on the premise that no ancestor is selected more than once, roughly equivalent to ignoring the possibility that inbreeding could occur in the population. Relaxing this assumption forms the basis for Model 2.

4.1 Motivating Model 2: Considering the Effects of Selection

Inbreeding represents a decreased ancestor count and clearly requires consideration. In fact, inbreeding and fewer ancestors can be considered equivalent, since inbreeding results in fewer ancestors and fewer ancestors implies inbreeding. Consequently, models seeking to dampen exponential effects must describe reasons for inbreeding.

If the expected ancestor count in generation g is defined as $|L_h| \cdot (1 + p_{crossover})$, then the number of unique ancestors is the cardinality of the population subset formed by $|L_h|$ selections. Unfortunately, the varying selection probabilities in a population preclude the use of simple combinatorics to compute this value. Since the deficiencies of fitness proportional selection are well-known and the probabilities associated with rank selection can be made equivalent to tournament selection [4], the new model makes the (reasonable) assumption that selection probabilities can be determined by:

$$P_s(i) = \frac{2}{n^2}\,\text{rank}_0(i) + \frac{1}{n^2} \qquad \textbf{(4)}$$

where $\text{rank}_0(i)$ is the ranking (by fitness) of the ith member, starting at rank 0.

By assuming a total ordering of the ranks in each population associated with each generation, each member can be assigned a unique selection probability. However, this assumption of uniqueness is actually invalid, and will be revisited shortly. This ranking technique eliminates any requirement for an actual fitness function, and it follows that chromosome and variation operations can be ignored as well.

4.2 Constructing Model 2: Using Transition Graphs and Simulation

Since simple combinatorics cannot provide the cardinality of the subset of the population created from $|L_h|$ selections, a precise model (not requiring simulation) using a transition graph is used. Since the subset of the population formed by selection in the genetic algorithm is constructed iteratively (i.e. perform selection, add to selected members, repeat), a transition graph with states for all possible population subsets models the process effectively. For a population of $|P|$ members, there are $2^{|P|}$ possible subsets that could be formed after any number of selections.

As a clarifying example, consider the transition graph (Figure 2) for a genetic algorithm with three members (performing only three selections each generation). The distinct selection probabilities P_A, P_B, and P_C represent the probabilities of selecting A, B, or C (respectively) and, thus, represent the probability of passing from the empty state to one of the three states containing a single population member. After the second selection is performed, one of two outcomes has occurred - either a new population member has been selected (distinct from the one depicted in the current state) or the same member has been selected a second time (in which case the state remains the same since a member cannot be flagged for inclusion in the lineage tree more than once). If a different member was chosen by the second selection, there is a transition to one of the states depicting a subset of two distinct members, and so on.

With a vector v representing the initial state (before any selections are made), the transition matrix M can be used to determine the most likely number of unique members after n selections by calculating $r^T = v^T \cdot M^n$, and then summing the elements of r^T associated with the states representing subsets of each size. Thus it can be determined which of the possible cardinalities of the subset created from $|L_h|$ selections has the highest probability and, thus, the most likely number of unique ancestors. Unfortunately, although this technique calculates the most likely number of inbred ancestors precisely, it is too computationally intensive to calculate conventionally for very large populations. Consequently, Monte Carlo simulations are used to computer the model.

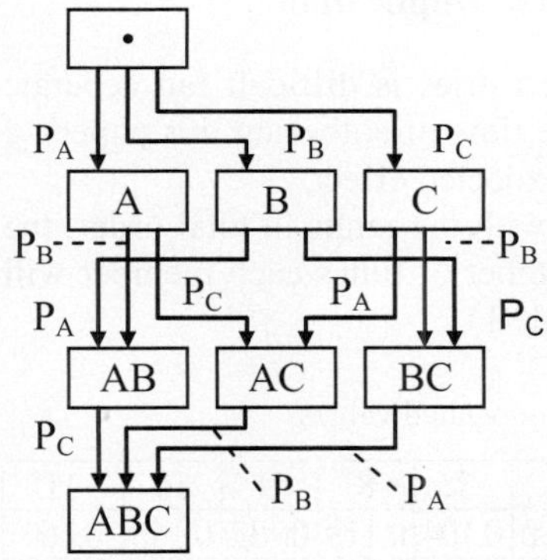

	•	A	B	C	AB	AC	BC	ABC
•	0	P_A	P_B	P_C	0	0	0	0
A	0	P_A	0	0	P_B	P_C	0	0
B	0	0	P_B	0	P_A	0	P_C	0
C	0	0	0	P_C	0	P_A	P_B	0
AB	0	0	0	0	$P_A + P_B$	0	0	P_C
AC	0	0	0	0	0	$P_A + P_C$	0	P_B
BC	0	0	0	0	0	0	$P_B + P_C$	P_A
ABC	0	0	0	0	0	0	0	1

Fig. 2. A transition graph (and corresponding matrix M) describing the selection of unique (i.e. non-duplicate) members in a genetic algorithm population of size 3.

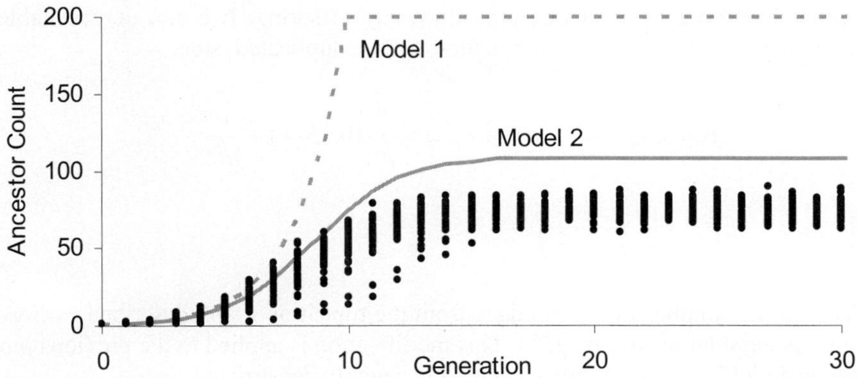

Fig. 3. Ancestor counts (from the best solution found in the final generation) predicted by Model 2 against those of Model 1 and from F8F2 (first 30 ancestral generations).

4.3 Testing the Accuracy of Model 2

Though the accuracy of the predicted ancestor count of Model 2 is improved by including consideration for inbreeding through selection (see Figure 3), it is clear that further causes for inbreeding must be accounted for to achieve greater accuracy.

5 Further Inbreeding Considerations − Developing Model 3

The motivation for the third model of lineage tree ancestor counts is the realization that the total order assumption is fallacious, as it does not consider the possible presence of duplicates in the current population. Multiple offspring from individuals of higher fitness encourages crossover between members which share partial ancestry. This component will be approximated by the logical reduction in population size expected from the effective selection probabilities of tournament selection.

5.1 Motivating Model 3: Considering the Effects of Duplication

Although predicting the presence of overlapping ancestries is difficult (an accurate model is being formulated but was unavailable at the time of authoring this paper), a crude model is proposed as an approximation of the expected effect.

Consider, as an example, a population of size 12 with the ranks in total order, the assigned selection probabilities, and the expected number of times each member will be selected after 12 selections are performed (see Table 1).

Table 1. Total ordered ranks and their associated values

$rank_0$	0	1	2	3	4	5	6	7	8	9	10	11
$P_S(i)$	0.007	0.021	0.035	0.049	0.063	0.076	0.090	0.104	0.118	0.132	0.146	0.160
$E_S(i)$	0.084	0.252	0.42	0.588	0.756	0.912	1.08	1.248	1.416	1.584	1.752	1.92
Round($E_S(i)$)	0	0	0	1	1	1	1	1	1	2	2	2

Rounding reveals that members above rank 8 are expected to be selected twice, while members below rank 3 will rarely produce any offspring. It is not unreasonable to assume that the top 25% of any population will be duplicated, since

$$(rank_0(i) \geq 0.75 \cdot n) \rightarrow P_s(i) > \frac{2}{n^2}(0.75 \cdot n) + \frac{1}{n^2}$$

and hence $$E_s(i) > \left(\frac{1.5}{n} + \frac{1}{n^2}\right) \cdot n = 1.5 + \frac{1}{n} \approx 2 \;. \tag{5}$$

This further implies a loss of edges from the transition graph described earlier, reducing the population size by 25%. This modification is applied to the previous model to form Model 3 which is more accurate, but roughly constructed.

5.2 Testing the Accuracy of Model 3

As anticipated, though the stable region of Model 3 is now quite close to the observed value (see Figure 4), further improvement is possible. Comparing the stable region for F8F2 (73.8 ± 2.5) against those of the models (200, 108, and 81 respectively), it is evident that inbreeding caused by selection and shared ancestries accounts for 72.9% and 21.4% of the improvement respectively, leaving unaccounted only 5.7%.

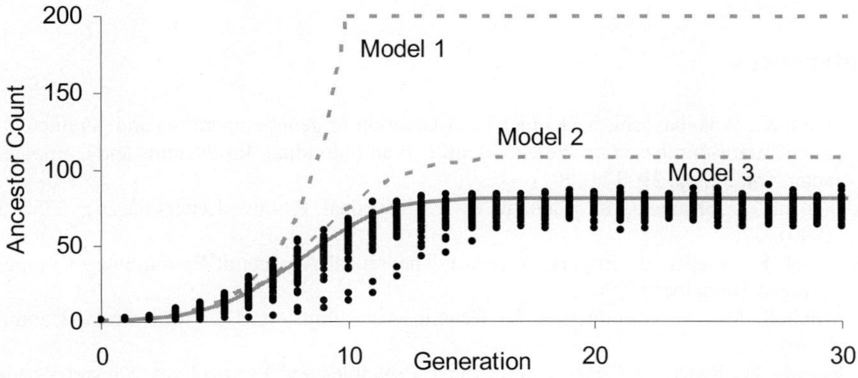

Fig. 4. Ancestor counts (from the best solution found in the final generation) predicted by Model 3 against those of Models 1, 2, and from F8F2 (first 30 ancestral generations).

6 Conclusions and Future Work

In this paper, the lineage tree (depicting a complete ancestry of a population member) and ancestor count (subpopulation whose descendants include the member) were introduced as inceptive components of an investigation into the interrelationships between the elements of genetic algorithm populations.

Experiments were performed on functions of varying difficulty and, in all cases, regions of exponential growth, non-exponential (diminishing) growth, and stability were observed. Models were then developed to explain these unexpected commonalities. The first was comprised entirely of the simple concept of exponential growth in member parentage. Since any manifesting reduction from this model over the regions of non-exponential growth and stability is attributable to inbreeding, subsequent models revolve around identifying additive sources of inbreeding. The second model considered the effect of selecting duplicate members while the third used the expected number of duplicates to estimate the effect of overlapping ancestries. While Model 3 was most accurate, Model 2 is attributed with the most significant improvement, possibly indicative that a significant effect of the genetic algorithm is reactive in nature.

Although the stable region was confirmed independent of the problem or number of generations, the "dip" observed on certain problems (specifically those considered more difficult) and the fluctuations in all stable regions were not anticipated. A hypothesis for the decrease in ancestor count at the initial generation is the presence of small regions of relatively high fitness, discovered most commonly in early generations, resulting in a "genetic bottleneck" effect. Similarly, the presence of local optima in the fitness landscape results in stagnation, leading to increased inbreeding. This could be the cause of the negatively sloped fluctuations in the stable region, and, if so, the positive slopes might indicate the escape from these local optima. Further experimentation will be necessary to test the validity of these explanations.

Acknowledgements. The authors wish to acknowledge partial funding for this research by the Natural Sciences and Engineering Research Council of Canada (NSERC).

References

1. Hatta, K., Wakabayashi, S., Koide, T.: Adaptation of genetic operators and parameters of a genetic algorithm based on the elite degree of an individual. In: Systems and Computers in Japan, vol. 32, pp. 29–37 (2001)
2. Ikegami, T., Kaneko, K.: Genetic fusion. Physical Review Letters 65(26), 3352–3355 (1990)
3. Kallel, L., Naudts, B., Rogers, A. (eds.): Theoretical Aspects of Evolutionary Computing. Springer, Heidelberg (2001)
4. Mitchell, M.: An Introduction To Genetic Algorithms, p. 171. MIT Press, Cambridge (1996)
5. Reeves, C., Rowe, J.: Genetic Algorithms: Principles and Perspectives. Kluwer Academic Publishers, Dordrecht (2002)
6. Vose, M.: The Simple Genetic Algorithm: Foundations and Theory. MIT Press, Cambridge (1999)
7. Whitley, D., Rana, S.B., Dzubera, J., Mathias, K.E.: Evaluating Evolutionary Algorithms. Artificial Intelligence 85(1-2), 245–276 (1996)

Asynchronous Graph-Rewriting Automata and Simulation of Synchronous Execution

Kohji Tomita[1], Satoshi Murata[2], and Haruhisa Kurokawa[1]

[1] National Institute of Advanced Industrial Science and Technology (AIST), Tsukuba, Japan
{k.tomita, kurokawa-h}@aist.go.jp
[2] Tokyo Institute of Technology, Yokohama, Japan
murata@dis.titech.ac.jp

Abstract. In this paper, we consider asynchronous update scheme for a variant of graph rewriting systems called graph-rewriting automata, and show that synchronous update can be simulated by asynchronous update using a constructed rule set from the one for synchronous update. It is well known that such rule construction is possible on cellular automata or other automata networks whose structures are fixed, but graph rewriting automata induce structural changes and additional mechanism of communication and local synchronization is required. Some simple examples are given by simulation.

1 Introduction

Cellular automata were introduced by von Neumann [9], and have been applied in many areas. Usually, the cells are arranged in lattice space and state transition is only concerned. Graph grammar or graph rewriting systems consider generation of structures [5], and have also been studied in the context of artificial life [1,6]. We have proposed a variant of such systems to express self-replication naturally on 3-regular graphs [7,8]. On this basis, graph structures are updated synchronously with some restriction on execution time and local connection relation so that the process is deterministic. By this property, rule design and behavior analysis become easier. However, such restrictions are not aesthetic, and global synchronization is unsuitable or difficult to realize on large systems like living things. To overcome this, we will introduce asynchronous update scheme of graph-rewriting automata with less restrictions. Randomness inherent in asynchronous update permits diverse behaviors, which are suitable for artificial life models.

In this paper, we will focus on the design of a rule set for asynchronous scheme that simulates the behavior of synchronous execution. It is well known that asynchronous execution can simulate synchronous execution with increased number of states on cellular automata or more general automata networks with fixed structures [3,2,4]. We will show that a similar method is possible for graph-rewriting automata by organizing their behavior to execute the synchronous steps in a

F. Almeida e Costa et al. (Eds.): ECAL 2007, LNAI 4648, pp. 865–875, 2007.

step by step manner. There is difficulty because structural changes cause irregular state change of neighboring nodes, which might activate inappropriate rules resulting in disorder configurations. Thus, we incorporate a local synchronization mechanism among several nodes in the rule sets for asynchronous update in order to complete structural changes as desired.

In the following, formulation of graph-rewriting automata is given, and two execution models are presented. Then a construction method of rule sets for asynchronous model is given. Some simulation examples are shown. Finally, non-interleaving model is discussed and the conclusion follows.

2 Graph-Rewriting Automata

In this section, we describe the framework that we consider in this paper. It is a variant of graph-rewriting systems, called a graph-rewriting automaton. (Evolution of networks based on similar rewriting rules is discussed in [10].) We assume that the base graph structure is a 3-regular planar graph: each node has three neighbor nodes. At each node, a cyclic order of links is defined. Each node has an internal state chosen from a finite set. More formally, it is defined as follows.

Definition 1. Let S be a finite set of states. Base graph G_S is a quadruple $\langle V, E, \xi, \eta \rangle$, where V is a (possibly empty) set of vertices, E is a set of edges, $\xi : V \to S$ is a function that assigns states to the vertices, and η is a function that specifies three edges of vertices with cyclic order. More precisely, $\eta : V \times \{0, 1, 2\} \to E$, such that $|\{\langle v, d\rangle | \eta(v, d) = e, \text{ for each } v \in V \text{ and } d \in \{0, 1, 2\}\}| = 2$ for every $e \in E$.

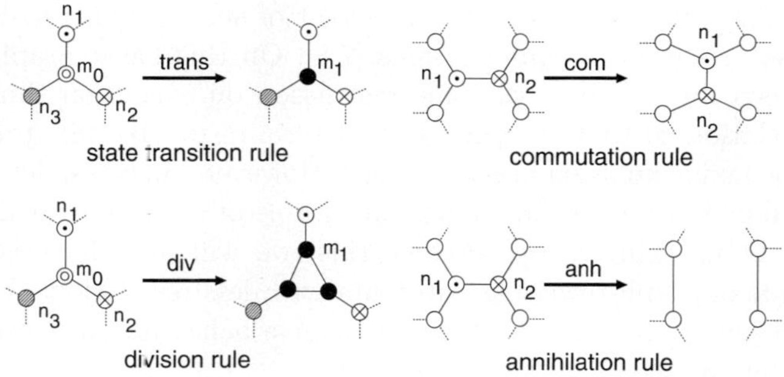

Fig. 1. Four kinds of rules of planar graph-rewriting automata: state transition, division, commutation, and annihilation. These rules are written as "trans m_0 $(n_1, n_2, n_3) \to m_1$", "div m_0 $(n_1, n_2, n_3) \to m_1$", "com (n_1, n_2)", and "anh (n_1, n_2)", respectively. Note that in the first two rules, "(n_1, n_2, n_3)" and "$(n_2,\ n_3, n_1)$" are the same, but "(n_3, n_2, n_1)" is different from them by the cyclic order.

From this definition, it permits multiple links or self links. Hereafter, base graphs are also called graphs for simplicity. Graphs are rewritten by state transition, as in state transition in cellular automata, and by structural rewriting. Figure 1 shows four kinds of graph-rewriting rules we use. In the figure, nodes are drawn so that the cyclic order of all nodes is the same. A division rule divides a node into three nodes. A commutation rule rotates a pair of nodes clockwise. An annihilation rule eliminates a pair. State transition and division rules are called node rules, and commutation and annihilation rules are called link rules. Hereafter, these rules are abbreviated as 'trans', 'div', 'com', and 'anh'. They were introduced so that these rules are universal in the sense that they can rewrite any 3-regular planar graphs with more than three nodes of different states to any such a graph. Development processes of an initial graph based on a rule set are considered in the following.

3 Synchronous and Asynchronous Execution

In this section we introduce two models of execution: synchronous and asynchronous. In our previous study, we have mainly concentrated on synchronous execution where an initial graph and a set of rules determine a development process uniquely. In this paper we introduce asynchronous execution.

In the synchronous execution scheme, rules are applied synchronously on the time of natural numbers: all the possible node rules (state transition and division) at an even time and all the possible link rules (commutation and annihilation) at an odd time. Applying the link rules to adjacent links at the same time causes inconsistency. Therefore, all such applications are suppressed, that we call lateral inhibition. In this model, the overall development process is determined from an initial graph and a rule set.

We introduce another model of execution, which is asynchronous. In the asynchronous model, execution repeats the following. Nodes $v_1, ..., v_n$ and links $e_1, ..., e_m$ are chosen randomly so that e_is do not have a common vertex, and that vertices of each link e_i are different from v_j for $j = 1, \ldots, n$. (In other words, each node participates in at most one node rule or one link rule.) A base graph is updated by rule applications to these nodes and links. Depending on the chosen nodes and links, the process is stochastic. Some restrictions in synchronous model (such as different execution time for node and link rules, or lateral inhibition) can be eliminated in this model.

Figure 2 illustrates the difference of these models. As in this figure, a variety of development processes are possible on asynchronous execution, and it could be a better model for artificial life. In this paper we will focus on the design of a rule set that simulates the behavior of synchronous execution.

In the following, interleaving execution is mainly considered where only one node or one link is chosen at one time (i.e., $n = 0$ and $m = 1$ or $n = 1$ and $m = 0$). Non-interleaving execution is briefly discussed in section 6. Note that we assume the property of fairness that when a rule is applicable to a cell forever, the rule must be applied eventually.

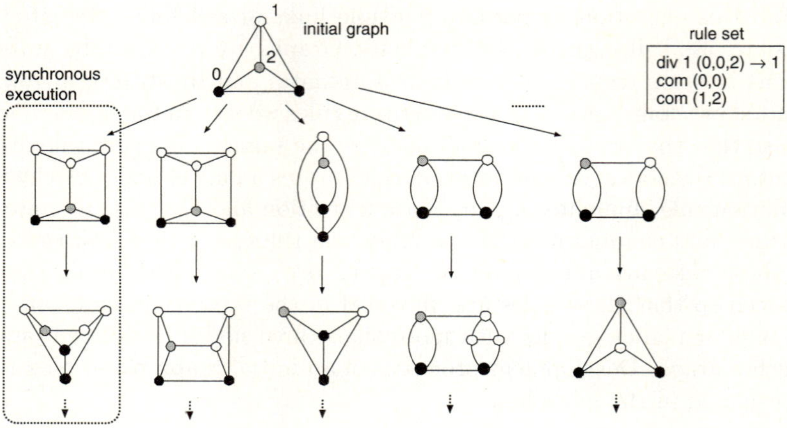

Fig. 2. An example of synchronous and asynchronous execution. The numbers (0, 1 and 2) indicate internal states. Development processes of an initial four node structure and a rule set with three internal states are shown. Only the leftmost one is obtained by the synchronous execution.

4 Rule Construction

In this section, we show that any execution in synchronous model can be simulated by a corresponding asynchronous execution using additional states. In other words, we can construct a rule set for asynchronous execution from a rule set for synchronous execution.

More detailed procedure is as follows. As in cellular automata, increasing the number of states enables asynchronous simulation of node rules. Basically, it is realized by memorizing tentative neighbor states generated by asynchronous execution as internal states. For link rules, there is difficulty because structural changes cause irregular state change of neighboring nodes in addition to lateral inhibition. Thus, a local synchronization mechanism among several nodes is incorporated into rule sets for asynchronous update in order to perform a structural changes as desired.

By storing the tentative information and the local synchronization mechanism, required states are increased. Figure 3 show such states and transitions among them in a basic case.

4.1 Node Rules

Execution of node rules are similar to the usual state transition in the view point of states. We assume that every possible state transition is included in a rule set.

For each node rule of the form 'nrule $m_0(n_1, n_2, n_3) \rightarrow m_1$', where nrule is trans or div, we replace this by two type rules:

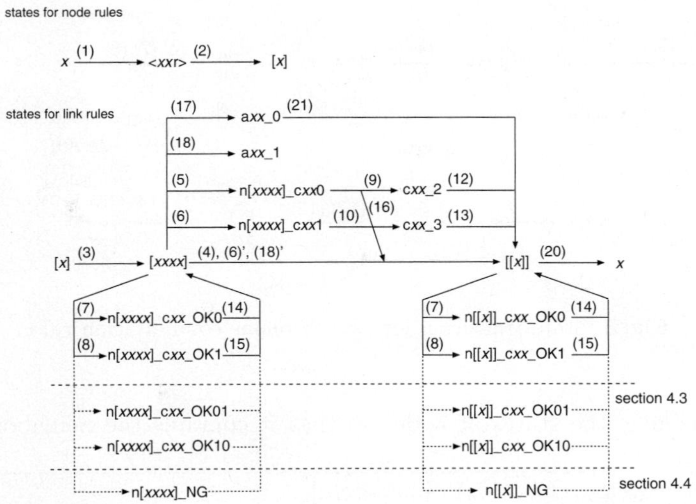

Fig. 3. Summary of states and transitions among them in a basic case. 'x' indicates a state in S, and 'r' is trans or div. By the transitions from the initial 'x' to the final 'x', one round of synchronous execution is performed. These xs are not necessarily the same.

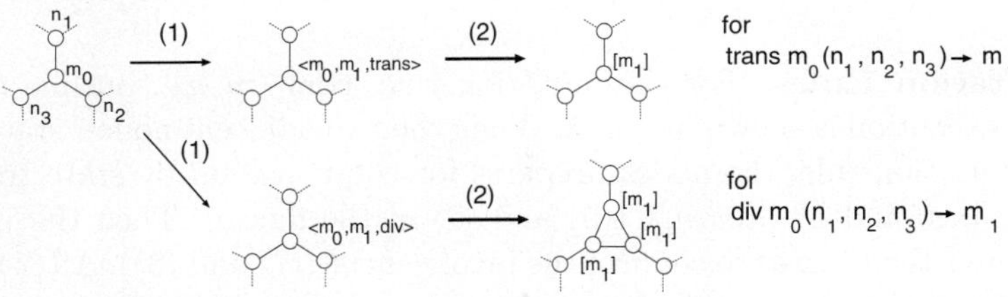

Fig. 4. State transition for asynchronous node rules

(1) trans $m_0(NS_1, NS_2, NS_3) \rightarrow \langle m_0, m_1, r\rangle$,
where NS_i is n_i or $\langle n_i, x, r\rangle$ for any $x \in S$ and r is trans or div.
(2) trans $\langle m_0, m_1, \text{trans}\rangle(NS'_1, NS'_2, NS'_3) \rightarrow [m_1]$,
div $\langle m_0, m_1, \text{div}\rangle(NS'_1, NS'_2, NS'_3) \rightarrow [m_1]$,
where NS'_i is $\langle n_i, x, r\rangle$ or $[x]$ for any $x \in S$.

Note that $\langle m_0, m_1, r\rangle$ or $[m_1]$ represents one of newly introduced states. By executing these rules, state m_0 is updated to state $[m_1]$ and division is executed if necessary. The numbers (1) and (2) correspond to the transition shown in Fig. 3. Figure 4 illustrates these steps.

4.2 Link Rules

For link rule application, one difficulty is in lateral inhibition that prevents execution of link rules at neighboring links. Therefore, before application of link

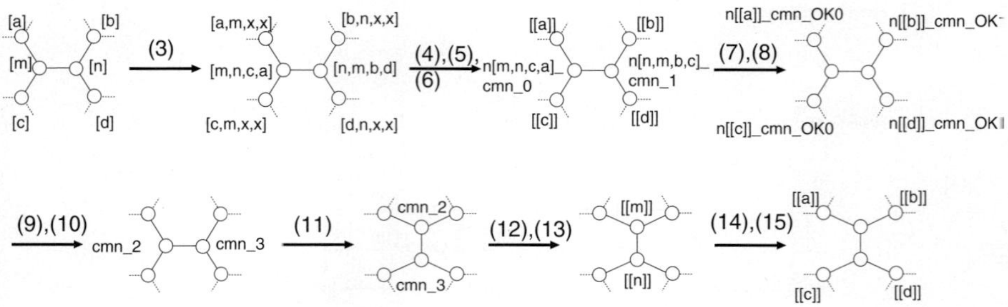

Fig. 5. State transition for asynchronous commutation rule

rules, we change the states of nodes so that it contains the neighboring state information.

(3) trans $[m](CS_n, CS_c, CS_a) \to [m, n, c, a]$, where CS_x is $[x]$ or $[x, u, v, w]$ for any $x, u, v, w \in S$.

If annihilation rules are applied before commutation rules, some links for commutation may constitute double links and a link to be commuted cannot be specified. We thus design a rule set so that execution of annihilation rules are postponed until neighboring commutation rules are executed.

Commutation Rules. For commutation rule 'com (m, n)', outline of asynchronous execution is shown in Fig. 5. When the two adjacent nodes' state match the commutation rule, the nodes prepares for commutation by state transition to n[...]_*cmn*_0 and n[...]_*cmn*_1 ((5) and (6) in the figure). Then the neighbor nodes change their states to confirm the involvement ((7) and (8)). After this, the nodes changes their state again ((9) and (10)), and actual commutation happens ((11)). It stops at appropriate configuration ((12) and (13)), and commutation completes ((14) and (15)). Actual rules are as follows.

(4) trans $[s, u, v, w](CS'_u, CS'_v, CS'_w) \to [[s]]$,
for any $s, u, v, w \in S$ such that the number of matched links between s and any of u or v or w is not one, where CS'_x is $[x, p, q, r]$ or $[[x]]$ for any $x, p, q, r \in S$.
(5) trans $[m, n, u, w](CS''_n, CS''_u, CS''_w) \to \mathrm{n}[m, n, u, w]\_cmn\_0$,
for any $u, w \in S$ such that the number of matched links between m and any of u or n or w is one, where CS''_x is $[x, p, q, r]$ for any $x, p, q, r \in S$,
(6) trans $[n, m, b, d](\mathrm{n}CS''_m\_cmn\_0, CS'_x, CS'_y) \to \mathrm{n}[n, m, b, d]\_cmn\_1$,
for any $b, d, x, y \in S$ such that the number of matched links between n and any of m or b or d is one.
(6)′ trans $[n, m, b, d](\mathrm{n}CS''_m\_cmn\_0, CS'_x, CS'_y) \to [[n]]$,
for any $b, d, x, y \in S$ such that the number of matched links between n and any of m or b or d is not one.
(7) trans $CS'_x(cmn\_0, CS'_y, CS'_z) \to \mathrm{n}CS'_x\_cmn\_\mathrm{OK0}$,
for any $x, y, z \in S$.
(8) trans $CS'_x(\mathrm{n}CS''_n\_cmn\_1, CS'_y, CS'_z) \to \mathrm{n}CS'_x\_cmn\_\mathrm{OK1}$,
for any $x, y, z \in S$.

(9) trans nCS''_m_cmn_0(nCS''_n_cmn_1, nCS'_x_cmn_OK0, nCS'_y_cmn_OK0) $\rightarrow$ cmn_2,
for any $x, y \in S$.
(10) trans nCS''_n_cmn_1(nCS''_m_cmn_0, nCS'_x_cmn_OK1, nCS'_y_cmn_OK1) $\rightarrow$ cmn_3,
for any $x, y \in S$.
(11) com (cmn_2, cmn_3).
(12) trans cmn_2(cmn_3, $CS'_{u,m}, CS'_{v,m}$) $\rightarrow$ [[m]],
for any $u, v \in S$ where $CS'_{x,m}$ is [[x]] or [x, m, p, q] for any $x, p, q \in S$.
(13) trans cmn_3([[m]], $CS'_{u,n}, CS'_{v,n}$) $\rightarrow$ [[n]],
for any $u, v \in S$ where $CS'_{x,n}$ is [[x]] or [x, n, p, q] for any $x, p, q \in S$.
(14) trans nCS'_x_cmn_OK0([[m]], CS'_y, CS'_z) $\rightarrow$ CS'_x,
trans nCS'_x_cmn_OK0([[n]], CS'_y, CS'_z) $\rightarrow$ CS'_x,
for any $x, y, z \in S$.
(15) trans nCS'_x_cmn_OK1([[m]], CS'_y, CS'_z) $\rightarrow$ CS'_x,
trans nCS'_x_cmn_OK1([[n]], CS'_y, CS'_z) $\rightarrow$ CS'_x,
for any $x, y, z \in S$
(16) trans nCS[m, n, x, y]_cmn_0([[n]], [[u]], [[v]]) $\rightarrow x$,
for any $x, y, u, v \in S$.

Annihilation Rule. For annihilation rule 'anh (m, n)', the following rules are used and the steps are illustrated in Fig. 6.

(17) trans [m, n, c, a]([n, m, b, d], CS'_x, CS'_y) $\rightarrow$ amn_0,
for any $a, b, c, d, x, y \in S$ such that the number of matched links between m and any of n or c or a is one.
(18) trans [n, m, b, d](amn_0, CS'_x, CS'_y) $\rightarrow$ amn_1,
for any $b, d \in S$ such that the number of matched links between n and any of m or b or d is one.
(18)′ trans [n, m, b, d](amn_0, CS'_x, CS'_y) $\rightarrow$ [[n]],
for any $x, y \in S$ such that the number of matched links between n and any of m or b or d is not one.
(19) anh (amn_0, amn_1).
(20) trans [[a]](N_b, N_c, N_d) $\rightarrow a$,
for any $a, b, c, d \in S$, where N_x is [[x]] or x for any state $x \in S$.
(21) trans amn_0([[x]], [[y]], [[z]]) $\rightarrow$ [[m]],
for any $x, y, z \in S$.

In the above rules, lateral inhibition is realized by either of (4), or (6)′ and (16), or (18)′ and (21).

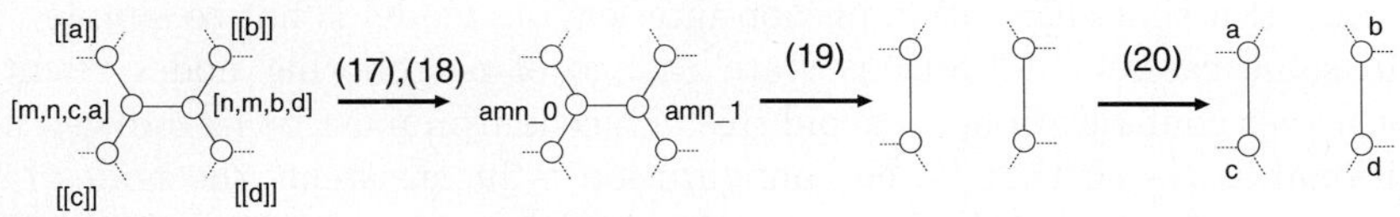

Fig. 6. State transition for asynchronous annihilation rule

4.3 Rules for Common Neighbor Case

The above procedure is for a basic case. In general, two nodes in Fig. 5 denoted by [a] and [b] (or [c] and [d]) may coincide. In this case, this single node needs to give permission to both [m] and [n] using different states from above. Rules must be added to cope with this case. Such modification is straightforward, and only additional states are shown in Fig. 3.

4.4 Rules for Avoiding Conflicts

In some cases, one node A is adjacent to two nodes B and C both are to be involved in different commutations. If both B and C change to locking states, A cannot decide which should be permitted and changes to a refusing state. Then B and C revert to their original state like $[p, q, r, s]$, then A does so, and the same process starts over again. For this case, additional states are necessary as shown in Fig. 3. Note that with this mechanism, livelock may happen, i.e., nodes to commute may repeat requesting and releasing, and overall system does not proceed.

With these modifications, an obtained asynchronous rule set can successfully simulate the synchronous development process of the original rule set if it does not livelock.

5 Simulation

We conducted some simulations to confirm the above rule construction. The simulator visualizes each node as a triangle in 3D space.

For simplicity, the first example includes two states and no link rules as shown in Fig. 7. It uses two states and five rules. The development process resembles to the synchronous case as shown in (c) and (d).

Figure 8 shows another example with annihilation rule. It uses three states to self-replicate itself. Commutation rules are not included and rule construction is simple.

Figure 9 is an example of self-replication using commutation rule. The original synchronous rule set includes two states, and self-replication is performed successfully as shown in Fig. 9(c) by asynchronous execution.

6 Non-interleaving Execution

The rule set obtained above is for interleaving model, where only one rule is executed at a time. Execution on non-interleaving model is not so simple.

In some cases, simultaneous state change of neighboring nodes results in inconsistent configuration. To avoid this, a mechanism must be introduced using additional states so that if the configuration is inconsistent, the nodes revert to the previous states. Further execution is delayed until the configuration is confirmed to be consistent.

7 Conclusion

We introduced asynchronous graph-rewriting automata, and showed that a simple rule construction enables a simulation of synchronous model. In the course of construction, some restrictions on synchronous model such as different execution time for node and link rules or lateral inhibition were incorporated into the resulting rule sets. Some requirements used for rule construction are i) local configurations before and after execution of link rules must be distinguished only from the neighbor states, and ii) structural change does not lead to disorder configurations, or equivalently, there is a path from such configurations to the appropriate configurations. This rule construction will be applicable to other systems with the above properties. One difficulty of the above method is the increased number of states. Further study to reduce the states will be necessary.

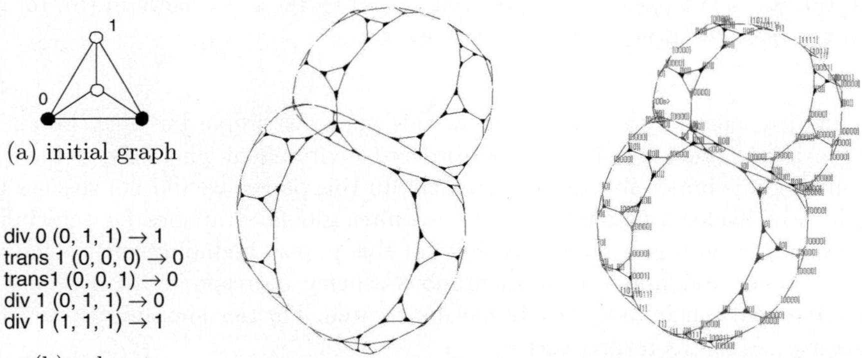

(a) initial graph (b) rule set (c) synchronous execution (d) asynchronous execution

Fig. 7. An example without link rules. A snapshot of synchronous development process, from initial structure (a) and rule set (b), is (c). A corresponding snapshot in asynchronous execution is (d). Numbers indicate internal states.

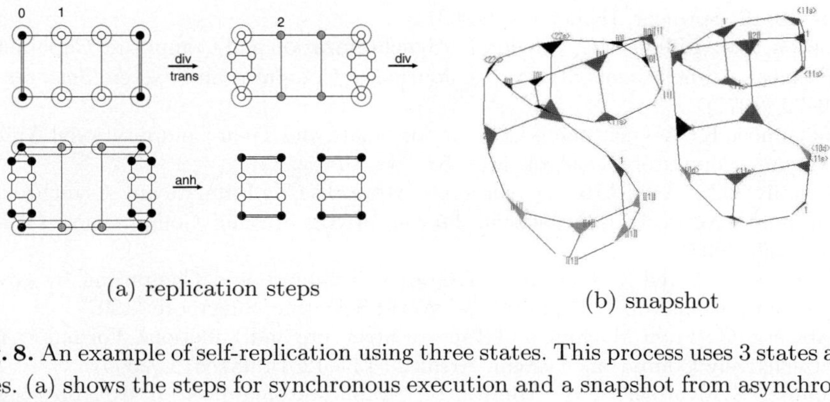

(a) replication steps (b) snapshot

Fig. 8. An example of self-replication using three states. This process uses 3 states and 4 rules. (a) shows the steps for synchronous execution and a snapshot from asynchronous execution is (b).

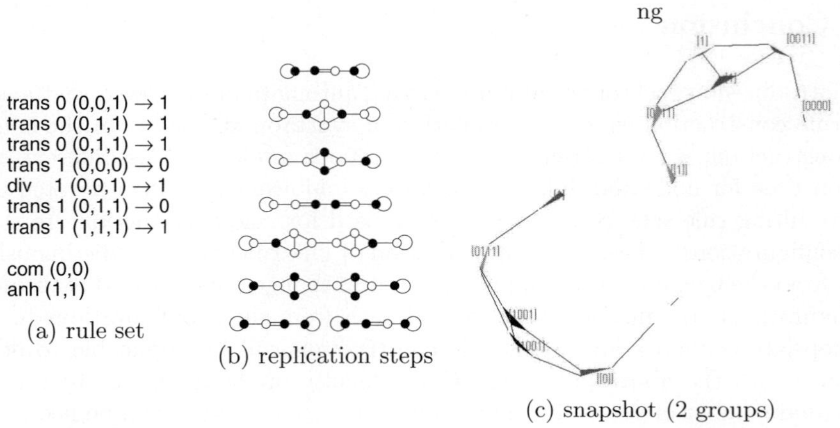

Fig. 9. An example of self-replication using commutation rule. An initial structure with two states is replicated using the rule set (a) by the steps shown in (b). (c) is a snapshot when the structure is replicated into two.

The asynchronization presented in this paper is simple but may be useful for generating ordered behavior in disordered environment where the global synchronization is impossible or not practical. In this paper, we did not discuss the problem of livelock. Coping with this requires global strategies for scheduling the execution, and is beyond the scope of this paper. Including such behavior, the inherent randomness in asynchronous scheme corresponds to diversity in behavior, and much wide variety can be treated. Further simplification of the framework is left as future work.

References

1. Benko, G., Flamm, C., Stadler, P.F.: Genetic Properties of Chemical Networks: Artificial Chemistry Based on Graph Rewriting. In: Banzhaf, W., Ziegler, J., Christaller, T., Dittrich, P., Kim, J.T. (eds.) ECAL 2003. LNCS (LNAI), vol. 2801, pp. 10–19. Springer, Heidelberg (2003)
2. Lipton, R.J., Miller, R.E., Snyder, L.: Synchronization and Computing Capabilities of Linear Asynchronous Structures. Journal of Computer and System Sciences 14, 49–72 (1977)
3. Nakamura, K.: Asynchronous Cellular Automata and Their Computational Ability. Systems, Computers, Controls 5(5), 58–66 (1974)
4. Nehaniv, C.L.: Asynchronous Automata Networks Can Emulate any Asynchronous Automata Network. International Journal of Algebra and Computation 14(5-6), 719–739 (2003)
5. Rozenberg, R. (ed.): Handbook of Graph Grammars and Computing by Graph Transformation, vol. 1: Foundations. World Scientific, Singapore (1997)
6. Salzberg, C.: From Machine and Tape to Structure and Functions: Formulation of a Reflexively Computing System. Artificial Life 12(4), 487–512 (2006)
7. Tomita, K., Kurokawa, H., Murata, S.: Graph Automata: Natural Expression of Self-Reproduction. Physica D 171(4), 197–210 (2002)

8. Tomita, K., Murata, S., Kamimura, A., Kurokawa, H.: Self-description for Construction and Execution in Graph Rewriting Automata. In: Capcarrère, M.S., Freitas, A.A., Bentley, P.J., Johnson, C.G., Timmis, J. (eds.) ECAL 2005. LNCS (LNAI), vol. 3630, pp. 705–715. Springer, Heidelberg (2005)
9. von Neumann, J.: Theory of Self-Reproducing Automata. Univ. of Illinois Press, Urbana (1966)
10. Wolfram, S.: A New Kind of Science. Wolfram Media, Champaign (2002)

Catalysis by Self-assembled Structures in Emergent Reaction Networks

Gianluca Gazzola[1,2], Andrew Buchanan[1,2], Norman Packard[1,2], and Mark Bedau[1,2,3*]

[1] ProtoLife Srl, Venezia, Italia
[2] European Center for Living Technology, Venezia, Italia
[3] Reed College, Portland, USA
mark@protolife.net

Abstract. We study a new variant of the dissipative particle dynamics (DPD) model that includes the possibility of dynamically forming and breaking strong bonds. The emergent reaction kinetics may then interact with self-assembly processes. We observe that self-assembled amphiphilic aggregations such as micelles have a catalytic effect on chemical reaction networks, changing both equilibrium concentrations and reaction frequencies. These simulation results are in accordance with experimental results on the so-called "concentration effect".

Key words: Chemical Reaction Network, Dissipative Particle Dynamics, Self-assembly, Micelle, Concentration Effect.

1 Introduction

We seek to understand the properties of networks of chemical reactions that implicitly interact with self-assembled amphiphilic structures[1]. Chemical reaction networks as well as self-assembled amphiphilic structures[2] are complex systems. Real complex systems in nature often involve the integration of sub-groups of complex systems. The system we study here is one such example. It couples chemical reaction networks with self-assembling amphiphilic structures.

There is ample experimental evidence that such coupled networks exhibit interesting behavior, in particular, that self-assembled amphiphilic structures affect certain chemical reactions. Micelles and other self-assembled structures are known to profoundly increase the rates of certain reactions [11]. The core mechanism is simply that the supramolecular structures increase the local concentration of the reagents, just as some catalysts do, and thus accelerate reaction rates. For example, hydrophobic reagents will spontaneously concentrate inside micelles, leading to reaction rate accel-

---

* To whom correspondence should be addressed.

1 An extended version of this article is available at this web page: http://www.ecltech.org/bmcmyp/Data/ECLT/Public/publications.html

2 Various parameters such as temperature, pH, and critical threshold concentration influence the type (or "phase") of structures that self-assemble from the amphiphiles. In addition to familiar amphiphiles such as fatty acids and phospholipids, other materials self-assemble including biopolymers like oligopeptides [12].

F. Almeida e Costa et al. (Eds.): ECAL 2007, LNAI 4648, pp. 876 – 885, 2007.

eration. Sometimes called "micellar catalysis" [24, 27], this catalytic concentration effect has been observed in a variety of chemical systems that involve micelles and reverse micelles [28, 35]. For example, the presence of micelles increases the rate of RNA self-cleavage reactions 100-fold [27]. Many kinds of reactions are catalyzed by micelles, such as redox [21] and hydrolysis reactions [7]. Micellar catalysis is very general and happens with many kinds of self-assembling materials besides amphiphiles. Examples include polymerized and polymeric amphiphiles [24] and dendrimers [7, 23]. Dendrimers are spherical macromolecules that are somewhat similar to micelles, except that while micelles are rather fluid aggregations composed of many amphiphilic molecules held together by the hydrophobic effect, dendrimers are single static structures tightly held together by covalent bonds.

Our goal here is to model and study this kind of catalysis by self-assembled structures in emergent reaction networks, where the dynamics of the network are not explicitly specified in the rules governing the system. Historically, biochemical reaction networks have been modeled using several approaches. Early approaches used networks whose nodes represented chemical species, and lines between nodes represented reactions. Autocatalytic reaction networks also included lines from catalyst nodes to reaction lines, to represent catalyzed reactions [8, 16]. Other reaction networks have been modeled in immunology: idiotypic networks [9] and more recently, cytokine networks [18]. The chemistry in many of the early network models was abstract. The models intentionally sought to escape from the details of real chemical interactions, for two reasons: capturing the details of real chemical interactions is difficult and immediately begs the question of what level of detail is to be captured, and the results sought from the model were expected (hoped) to be relatively independent from details of the individual chemical reactions; for large networks, the bulk properties of the network (connectivity, scaling, etc.) were hoped to be independent of the details.

More recently, the experimental understanding of reaction networks has been increasing substantially, and there has been an increased awareness of the need to model details of real chemical reactions in order to define and understand biochemical functionality in a given context, e.g., for a cell [29] and for reaction networks with reaction properties based on quantum mechanics [2, 3, 4]. Simultaneously, there has been a growing awareness that chemical reactions cannot by themselves provide a complete picture of biochemical functionality. Structural properties of amphiphilic assemblies must be added to the purely chemical picture. A very rich example is that of lipid structures. These structures are particularly interesting because they have complex phase diagrams, with phase transitions between several different phases, e.g., lipid solution, micelles, and vesicles, because the transitions between these phases may be catalyzed by the presence of other biopolymers, and because some of the phases may themselves have catalytic properties, e.g., for template-directed replication [19, 20, 26]. Finely tuned chemical control of phase transitions in biochemical gels (including more complex gels than simple lipid structures) has been proposed as a general framework for cellular function [25].

In this paper we study a model of interacting microscopic particles that combines relatively simple chemical reaction properties with properties deriving from self-assembly processes that can strongly affect the chemical reactions. Interactions between the particles determine both the chemistry and the self-assembly. The macroscopic result of a model simulation is the emergence of a network of chemical reactions that may interact with the self-assembled structures.

2 The Model

Our model of chemical reaction systems is based on the well-studied dissipative particle dynamics (DPD) framework [13, 14, 22, 30, 31, 32]. In DPD all the elements move in a two- or three-dimensional continuous toroidal space, according to the influences of four pairwise forces, $f_i = \sum_j (F_{ij}^C + F_{ij}^D + F_{ij}^B + F_{ij}^R)$. These are a conservative "weak" force between pairs of particles, a dissipative force between nearby particles, a spring-like "strong" bond force if two particles are bonded and a random force. Since strong bonds never form or break in the traditional DPD framework, that framework is unable to represent reaction processes that involve forming and breaking strong bonds. We introduced dbDPD, an augmented DPD framework that includes the possibility of making particles react with each other, dynamically forming or breaking strong bonds (see [1, 6], to which we refer for a thorough explanation of the model and all its parameters). Relevant parameters occur in the expressions for the weak forces, $F_{ij}^C = \alpha_{IJ}(1-\beta_{IJ} r_{ij})$, where α_{IJ} and β_{IJ} are specific to the types of *i* and *j* and r_{ij} is the Euclidean distance between the particles, and also include r_{IJ}^f and r_{IJ}^b, which are respectively the distance beyond and within which the reactive particles of type I and J form and break their strong bond (set to 0.2 and 0.4 in our simulations). Further, we have *l*, the relaxed strong bond length (set to 0.01), k, that governs the strength of all strong bonds, represented as Hookean springs $F_{ij}^B = k(r_{ij} - l)$ (set to 100), σ, a balancing factor between dissipative and random forces (set to 3), and w^R, a random factor (set to 2.73205) which occur in the expressions $F_{ij}^D = \sigma(v_i - v_j)(1-r_{ij})^2/2$ and $F_{ij}^R = \sigma w^R(1-r_{ij})u$, where v_i is the velocity vector of *i* and *u* is a uniform random number chosen from the interval (-1, 1).

3 Emergent Chemical Reaction Networks

Some of the networks in the literature are emergent [2, 16], some not. Reaction networks that arise as a result of our chemistry within dbDPD are highly emergent, in the sense that their equilibrium state is very hard to derive without explicit simulation.

Given rules for forming and breaking bonds, together with the constraint of only two or less bonds allowed per particle, we have the basis for a network of linear polymerization reactions. The architecture of the reaction network is determined by the constraints that are set on the process of strong bond formation. The reaction network complexity is controlled with the specification of the pairs of particle types that can form strong bonds and of the maximum length *N* of the polymeric chains resulting from the strong bond formation process.

The simple example we explore is a reaction network identified by the duple <{5,4}, 2>, namely having two reactive particle types labeled with integers 5 and 4 that can only form chains of length *N=2*, resulting in the architecture shown in Fig. 1.

Each labeled node represents a chemical species that can undergo strong bonding reactions; each solid dot connecting edges from the chemical species represents one of the three possible reactions that can take place: 5 + 4 ↔ 54, 5 + 5 ↔ 55, 4 + 4 ↔ 44. Note that bond forming and bond breaking radii do not influence a chemical reaction network's architecture but play a main role in determining the rate of interaction

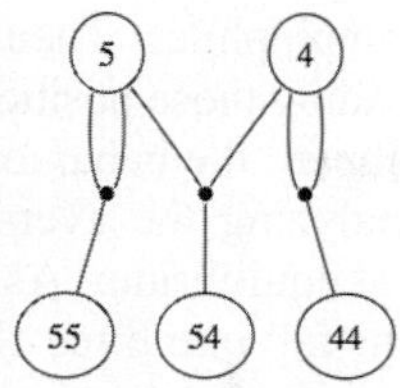

Fig. 1. Architecture of the reaction network for the experiment of polymerization from monomers to dimers.

of reagents in the process of strong bond formation. In our simulations the values of r_{IJ}^{f} and r_{IJ}^{b}, with $r_{IJ}^{f} < r_{IJ}^{b}$, were the same for all reagents. Thus we can say that all reagents share the same intrinsic reaction rate.

3.1 Results in a Simple Network

The first experiment we report concerns a simple network that contains five particle types: water (type 1), amphiphilic heads (type 2), amphiphilic tails (type 3), and two reagents (types 4 and 5). Because it is so simple, this network clearly illustrates the main kinds of interactions between self-assembled structures and emergent reaction networks. The inter-particle interactions are governed by the $(\alpha_{IJ}, \beta_{IJ})$ values specified in Table 1.

Table 1. $(\alpha_{IJ}, \beta_{IJ})$ values for particle interactions in simple network. The particle types 4 and 5 are the monomers that polymerize in the reaction network.

$(\alpha_{IJ}, \beta_{IJ})$	water	head	tail	4	5
water	(1,1)				
head	(1,1)	(150,1)			
tail	(4,1)	(15,5)	(5,1)		
4	(1,1)	(1,5)	(1,1)	(1,1)	
5	(1,1)	(1,5)	(1,1)	(1,1)	(1,1)

Note that "neutral" interactions, e.g., between water and water, are taken to be very weak repulsive interactions, with $(\alpha_{IJ}, \beta_{IJ}) = (1,1)$.

The self-assembly process of amphiphilic dimers into micelles can require several time steps in DPD, depending on several factors, such as the temperature of the system and the strength of the weak forces. In our experiments, we wanted to simulate a real chemical system in which lipophilic reagents are placed into an aqueous solution containing micelles that have already formed. In control cases, the amphiphilic dimers were replaced by water and particle initial positions were chosen randomly. All the other DPD parameters were kept the same.

The experiments that we ran were set within a 30x30 toroidal space with 7200 particles, composed of 2/7 reagents, 2/7 amphiphilic dimers and 3/7 water. Simulations were run without allowing the reagents to form bonds, until the amphiphiles aggregated into micelles and the distribution of reagents reached the equilibrium, according to the weak forces that reagents feel towards amphiphiles. Then the particle positions were saved and the simulation was restarted, loading those positions as initial conditions.

One straightforward way to compare the behavior of the system with micelles to a control without micelles is by analyzing the average concentrations of the species identified by the network's nodes at equilibrium. As we can see from Fig. 2 and Table 2, in the control case concentrations fall into three clearly distinguishable classes. The dominant species are monomers, followed by the only non-palindromic dimer and then by the two palindromic dimers, produced by a self-reaction, which is a reaction between two monomers of the same kind. The experimental case is, in contrast, dominated by the non-palindromic dimer, followed by the two palindromic ones and then by monomers. Note that palindromic dimers are half as concentrated as non-palindromic dimers in both systems.

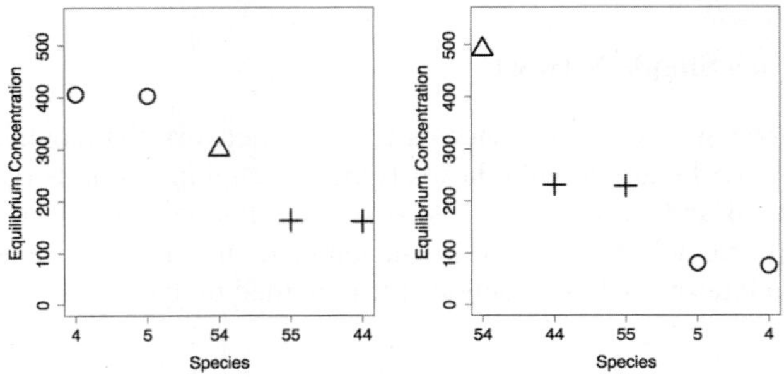

Fig. 2. Equilibrium concentration of different chemical species for a representative dbDPD run of control (left) and experiment (right). The error bars are smaller than the size of the symbols. Note that the triangle has the highest concentration for the experiment, shifted significantly from the control.

We may also compare the reaction dynamics of the two systems, in particular, how frequently reactions happen. Table 3 displays how often each possible reaction happens on average over 50 time steps once the dynamics have reached the equilibrium. In both the control and the experimental case reaction frequencies fall into two classes, one containing the self-reactions and the other one the non-self-reactions. Note that in both cases, the reactions that involve palindromic dimers are half as frequent as those that involve non-palindromic ones. The experimental case shows that the frequency of each reaction is lower than its corresponding one in the control. We now discuss two different observed effects in detail.

Concentration effect: In the control case the only force that can keep the reagents close to each other is due to the (possible) strong covalent bonds formed between two of them. As explained in [1, 6], if two reagents come within a distance smaller than their bond forming radius, then they form a strong bond that keeps on existing as long as the reagents' distance is smaller than their bond breaking radius. The reagents are free to float around until they form a strong bond. At this point, the bond strength will determine how long the bonded monomers will stay close to each other enough to keep their bond intact. The weaker the bond, the more likely it will be for it to break in the following time steps, leaving the two resulting monomers free floating again. Apart from the bond strength, nothing affects the survival probability of a dimer.

Table 2. Equilibrium concentration for each chemical species as shown in Fig. 2.

Chemical species	Equilibrium concentration	Length
	CONTROL	
4, 5	~ 404	Monomers
54	~ 300	Non-palindromic dimer
44, 55	~ 162	Palindromic dimers
	EXPERIMENT	
54	~ 491	Non-palindromic dimer
44, 55	~ 230	Palindromic dimers
5, 4	~ 79	Monomers

Table 3. Reaction frequency for the same run as in Fig. 2 and Table 2.

Chemical reactions	Reaction frequency	Observations
	CONTROL	
5 + 4 ↔ 54	~ 103	Monomer + monomer ↔ non-palindromic dimer
5 + 5 ↔ 55 4 + 4 ↔44	~ 52	Monomer + monomer ↔ palindromic dimer
	EXPERIMENT	
5 + 4 ↔ 54	~ 88	Monomer + monomer ↔ non-palindromic dimer
5 + 5 ↔ 55 4 + 4 ↔44	~ 42	Monomer + monomer ↔ palindromic dimer

In the experimental case, clusters of reagents form because of the weak forces that attract them to micelles. Therefore when the covalent bonds are broken, the monomers don't start to freely diffuse again, but continue to be entrapped in the same cluster, then it is very likely that they are involved in new bonding reactions, possibly with other free monomers in the same cluster.

The probability of existence of a bond depends on several factors. One main factor is the reagents' density (number of reagents over space area). The higher the density, the smaller the average distance between reagents. This probability affects equilibrium concentrations. If bonds are highly likely, then longer polymers are more prevalent. If bonds are unlikely, then monomers are more prevalent. For the reasons we explained above, the reagent density is locally increased by micelles, and that explains why dimers are more concentrated than monomers. By spatially concentrating reagents the micelles act as catalysts. We could also have obtained an analogous result in the control case, by increasing the bond strength.

The concentration effect caused by micelles can be observed from the change in reaction frequencies. Due to the concentration effect, bonded dimers survive for a longer time than when there is no concentration effect, reducing the frequency of the bond breaking reactions. This results in a low number of free reactive monomers, which decreases the frequency of the forward reactions as well.

Palindrome effect. We noticed that the 55 and 44 dimers' concentration is around half of that of the 54 dimer at equilibrium (Fig. 2, Table 2). We also noticed that the frequency of each reaction involving 55's or 44's is half as much of that of the only reaction that involves 54's. The reason is that dimers 54 and 45 have been identified as the same dimer. For all possible pairwise combinations of monomers of type 4 and type 5, if 4's are as many as 5's, there are equal numbers of 44, 45, 54, and 55, so if 45 and 54 are considered identical, their number is double that of 44 and 55. More generally, one may consider all polymer types of a given length, and see by the same argument that those that are non-palindromic will be doubled by identifying polymers read in one direction with those that are the same when read in the opposite direction

While the concentration effect concerns the difference between control and experimental equilibria, the palindrome effect concerns the difference between palindromic and non-palindromic N-mers in either control or experimental situations. The palindrome effect is actually seen in both the control and the experimental case.

3.2 Results for a Larger Network

We observed qualitatively the same kinds of effects in a wide variety of more complex emergent reaction networks, with higher maximum polymer length and more kinds of reagents.

Cascade effect: The more complex networks illustrated a third effect. Micelles increase the frequency of the reactions involving higher length polymers, as they tend to increase the concentration of their required ingredients compared to the control case. On the way to equilibrium, however, we see that first monomer concentration decreases as dimer concentration rises, then after reaching a maximum, dimer concentration decreases as trimer concentration increases, and so on. This effect moving through successive length polymers we term the "cascade effect".

4 Discussion and Conclusion

We have studied a new variant of dissipative particle dynamics (DPD) with dynamic bond forming and breaking, which we termed "dbDPD". This yields a microscopic mechanism for chemical reactions, from which emerges macroscopic chemical kinetics. The reagents may be represented as nodes on a graph, the reaction network, which also emerges from the microscopic chemical reaction mechanism. The emergent reaction network and reaction kinetics have many of the hallmarks of real reaction networks, e.g., the existence of many side reactions. Here, we have studied a particular class of reactions, polymerization between two monomer types.

DPD is well known as a modeling framework that is suited for studying self-assembled structures from amphiphilic molecules. Our addition of chemical reactions in dbDPD enables the additional study of the interplay between chemical reactions and self-assembly processes.

We report a clear identification of an experimentally known type of micellar catalysis: the concentration effect. Essentially, the effect comes about because lipophilic reagents may aggregate within or near the micelles, effectively increasing their local concentration and changing the equilibrium concentrations of resulting reaction products. In particular, long polymers that have very low equilibrium concentration in the absence of micelles may have very high equilibrium concentration (relative to all other reagents) in the presence of micelles.

In addition to the concentration effect, we identified two other effects that should be experimentally observable: (i) the palindrome effect, the doubling of the concentration of non-palindromic polymers because of the identification of polymers read in one directions with those that are the same when read in the opposite direction, and (ii) the cascade effect, seen when starting with high concentration of monomers: the concentration of monomers goes down as the concentration of dimers increases, then the concentration of dimers reaches a maximum and then decreases as the concentration of trimers increases, and so on.

Future directions for research based on dbDPD include refinement of the microscopic chemical reaction mechanisms to make them more realistic for particular target experiments. We also believe that introduction of variations into reaction products may enable the system to display evolvability.

Acknowledgements. We thank JohnMcCaskill and Thomas Maeke for an implementation of DPD software from which our software is derived. We benefited from conversations with Harold Fellermann, Ricard Solé, and Martin Hanczyc. Thanks also to the ECLT, which helped facilitate this work and provided an opportunity to present it. This work was supported by the EU in the PACE integrated project, EC-FP6-IST-FET-IP-002035.

References

1. Bedau, M.A., Buchanan, A., Gazzola, G., Hanczyc, M., Maeke, T., McCaskill, J.S., Poli, I., Packard, N.H.: Evolutionary design of a DDPD model of ligation. In: Talbi, E.-G., Liardet, P., Collet, P., Lutton, E., Schoenauer, M. (eds.) EA 2005. LNCS, vol. 3871, pp. 201–212. Springer, Heidelberg (2006)
2. Benkö, G., Flamm, C., Stadler, P.F.: A Graph-Based Toy Model of Chemistry. Journal of Chemical Information and Computational Science 43, 1085–1093 (2003)
3. Benkö, G., Flamm, C., Stadler, P.F.: Generic Properties of Chemical Networks: Artificial Chemistry Based on Graph Rewriting. In: Banzhaf, W., Ziegler, J., Christaller, T., Dittrich, P., Kim, J.T. (eds.) ECAL 2003. LNCS (LNAI), vol. 2801, pp. 10–19. Springer, Heidelberg (2003)
4. Benkö, G., Flamm, C., Stadler, P.F.: Multi-Phase Artificial Chemistry. In: Schaub, H., Detje, F., Brüggemann, U. (eds.) The Logic of Artificial Life: Abstracting and Synthesizing the Principles of Living Systems, pp. 16–22. IOS Press, Amsterdam (2004)
5. Besold, G., Vattulaien, I., Karttunen, M., Polson, J.M.: Towards Better Integrators for Dissipative Partycle Dynamics Simulations. Physical Review E 62, 7611–7614 (2000)
6. Buchanan, A., Gazzola, G., Bedau, M.A.: Evolutionary Design of a Model of Self-Assembling Chemical Structures. In: Krasnogor, N., Gustafson, S., Pelta, D., Verdegay, J.L. (eds.), Elsevier Science, Amsterdam (2007)
7. Delort, E., Darbre, T., Reymond, J.-L.: A Strong Positive Dendritic Effect in a Peptide Dendrimer-Catalyzed Ester Hydrolysis Reaction. Journal of the American Chemical Society 126, 15642–15643 (2004)
8. Farmer, J.D., Kauffman, S.A., Packard, N.H.: Autocatalytic Replication of Polymers. Physica D 22, 50 (1986)
9. Farmer, J.D., Packard, N.H., Perelson, A.: The Immune System, Adaptation, and Machine Learning. Physica D 22, 187 (1986)
10. Fellerman, H., Rasmussen, S., Ziock, H., Solé, R.: Life cycle of a minimal procell: a dissipative particle (DPD) study. Artificial Life (in press, 2007)

11. Fendler, J.H., Fendler, E.J.: Catalysis in micellar and macromolecular systems. Academic Press, New York (1975)
12. Fung, S.Y., Keyes, C., Duhamel, J., Chen, P.: Concentration Effect on the Aggregation of a Self-Assembling Oligopeptide. Biophysical Journal 85, 537–548 (2003)
13. Groot, R., Warren, P.: Dissipative particle dynamics: bridging the gap between atomistic and mesoscopic simulations. Journal of Chemical Physics 107, 4423–4435 (1997)
14. Hoogerbrugge, P., Koelman, J.: Simulating microscopic hydrodynamic phenomena with dissipative particle dynamics. Europhysics Letters 19, 155–160 (1992)
15. Jury, S., Bladon, P., Cates, M., Krishna, S., Hagen, M., Ruddock, N., Warren, P.: Simulation of amphiphilic mesophases using dissipative particle dynamics. Physical Chemistry and Chemical Physics 1, 2051–2056 (1999)
16. Kauffman, S.A.: Autocatalytic sets of proteins. Journal of Theoretical Biology 119, 1–24 (1986)
17. Kranenburg, M., Venturoli, M., Smit, B.: Phase behavior and induced interdigitation in bilayers studied with dissipative particle dynamics. Journal of Physical Chemistry 107, 11491–11501 (2003)
18. Kuby, J.: Immunology, 3rd edn. W.H. Freeman, New York (1997)
19. Luisi, P.L., Giomini, M., Pileni, M., Robinson, B.: Reverse micelles as hosts for proteins and small molecules. Biochimica and Biophysica Acta 947, 209–246 (1988)
20. Luisi, P.L., Walde, P., Oberholzer, T.: Enzymatic synthesis in self-reproducing vesicles: An approach to the construction of a minimal cell. Berichte der Bunsengesellschaft für Physikalische Chemie 98, 1160–1165 (1994)
21. Mallick, K., Jewrajka, S., Pradhan, N., Pal, T.: Micelle-catalysed redox reaction. Current Science 80, 1408–1412 (2001)
22. Marsh, C.: Theoretical aspects of dissipative particle dynamics. University of Oxford, Ph.D. Thesis (1998)
23. Martin, K.I., Twyman, L.J.: Acceleration of an aminolysis reaction using a PAMAM dendrimer with 64 terminal amine groups. Tetrahedron Letters 42, 1123–1126 (2001)
24. Oehme, G., Grassert, I., Paetzold, E., Fuhrmann, H., Dwars, T., Schmidt, U., Iovel, I.: The Effect of Assembled Amphiphiles on Catalytic Reactions in Aqueous Media. Kinetics and Catalysis 44, 766–777 (2003)
25. Pollack, G.H.: Cells, Gels and the Engines of Life: A New, Unifying Approach to Cell Function. Ebner & Sons, Seattle (2001)
26. Rasmussen, S., Chen, L., Stadler, B., Stadler, P.: Proto-organism kinetics: Evolutionary dynamics of lipid aggregates with genes and metabolism, Origins of life and evolution of the biosphere (in press)
27. Riepe, A., Beier, H., Gross, H.J.: Enhancement of RNA self-cleavage by micellar catalysis. FEBS Letters 457, 193–199 (1999)
28. Ruasse, M.-F., Blagoevab, I.B., Garcia-Rio, R.C.L.G., Leis, J.R., Marques, A., Mejuto, J., Monnier, E.: Organic reactions in micro-organized media: Why and how? Pure and Applied Chemistry 69, 1923–1932 (1997)
29. Shilling, C.H., Palsson, B.O.: The underlying pathway structure of biochemical reaction networks. Proceedings of the National Academy of Science 95, 4193–4198 (1998)
30. Shillcock, J., Lipowsky, R.: Equilibrium structure and lateral stress distribution from dissipative particle dynamics simulations. Journal of Chemical Physics 117, 5048–5061 (2002)
31. Trofimov, S., Nies, E., Michels, M.: Thermodynamic consistency in dissipative particle dynamics simulations of strongly nonideal liquids and liquid mixtures. Journal of Chemical Physics 117, 9383–9394 (2002)
32. Vattulainen, I., Karttunen, M., Besold, G., Polson, J.: Integration schemes for dissipative particle dynamics simulations: From softly interacting systems towards hybrid models. Journal of Chemical Physics 116, 3967–3979 (2002)

33. Yamamoto, S., Maruyama, Y., Hyodo, S.: Dissipative particle dynamics study of spontaneous vesicle formation of amphiphilic molecules. Journal of Chemical Physics 116, 5842–5849 (2002)
34. Yamamoto, S., Hyodo, S.: Budding and fission dynamics of two-component vesicles. Journal of Chemical Physics 118, 7937–7943 (2003)
35. Zingaretti, L., Boscatto, L., Chiacchiera, S.M., Silber, J.J.: Kinetics and mechanism for the reaction of 1-chloro-2,4-dinitrobenzene with n-butylamine and piperidine in AOT/n-hexane/water reverse micelles. In: Arkivoc X, pp. 189-200 (2003)

Community Detection in Complex Networks Using Collaborative Evolutionary Algorithms

Anca Gog[1], D. Dumitrescu[2], and Béat Hirsbrunner[1]

[1] University of Fribourg, Department of Computer Science, Bd. de Pérolles 90, 1700 Fribourg, Switzerland, {anca.gog, beat.hirsbrunner}@unifr.ch

[2] Babes-Bolyai University, Department of Computer Science, Kogalniceanu 1, 400084 Cluj-Napoca, Romania, ddumitr@cs.ubbcluj.ro

Abstract. Scientific researchers from computer science, communication and as well from sociology and epidemiology reveal a strong interest in the study of networks. One important feature studied in complex network is the community structure. A new evolutionary technique for community detection in complex networks is proposed in this paper. The new algorithm is based on an information sharing mechanism between the individuals of a population. A real-world network is considered for numerical experiments.

1 Introduction

The study of complex networks intensively preoccupied the scientific community in the recent years. Examples of complex networks in nature and society include metabolic networks, the immune system, the brain, the human social networks, the Internet and the World Wide Web. A complex system is characterized by the lack of central control and the fact that individual components are simple compared to the collective behavior. The study of real-world networks revealed features as degree distribution, average distance between vertices, network transitivity [2], [11].

Another property that concerned scientific researchers is the community structure. A community in a network is a group of nodes densely connected but sparsely connected with the nodes belonging to other communities. The importance of community detection emerges from its many applications. For example, in social and biological networks it could help studying interactions between groups of people or animals, better understanding metabolic networks; this problem also arises in parallel computing, and the list could continue.

Many techniques have been proposed for identifying community structure in complex networks. In this paper, a new collaborative evolutionary algorithm for community detection is proposed. This algorithm is based on the collaboration between individuals that exchange information in order to accelerate the search process. The experimental results prove the efficiency of the proposed technique.

F. Almeida e Costa et al. (Eds.): ECAL 2007, LNAI 4648, pp. 886–894, 2007.

2 Existing Methods for Community Detection

An exhaustive description of all existing methods for community detection in complex networks is beyond the scope of this paper. Yet, an overview of some of the most known algorithms is presented in what follows.

Hierarchical (agglomerative and divisive) clustering [9] aims at discovering natural divisions of networks into groups based on metrics of similarity of strength of connection between vertices. Girvan and Newman [4] proposed a divisive algorithm that uses the edge betweenness as a weight measure of the edges. Radicchi et al. [8] proposed a similar technique but they used a new metric, edge-clustering coefficient whose computation time is less than the betweenness centrality. The resistor network approach proposed by Newman and Girvan in [7] has been improved by Huberman and Wu in [6]. In [7] is also proposed an algorithm for community detection using random walks. Balakrishnan and Deo proposed a new technique based on bibliometric similarity between all pairs of vertices in the graph [1]. Community detection using extremal optimization [3] uses the network modularity proposed in [7].

The drawback of these techniques is the computational complexity that makes them unsuitable for very large networks. Indeed, finding an exact solution for the community detection is believed to be an NP-complete problem and therefore difficult to solve.

Evolutionary computation provides promising algorithms for solving NP-hard problems. They provide good (acceptable) solutions for the problem in a reduced amount of time. Regarding evolutionary techniques applied for detecting community structure in complex networks, there is only one approach made by Tasgin and Bingol [10], as far as we know.

3 Collaborative Evolutionary Algorithms

Standard evolutionary algorithms are characterized by the lack of communication between individuals. Indeed, whether is the subject of selection, recombination or mutation, one individual does not know anything about the other individuals in the population or about the individuals that have contributed to its existence. In the proposed technique, the population imitates a social system where individuals communicate and share information.

One individual (or chromosome) encodes a potential solution of the problem and is composed by a set of elements called *genes*. Each gene can take multiple values called *alleles*. In the proposed collaborative evolutionary algorithm an individual has extra information that enables the sharing mechanism. On one hand, each individual knows the best potential solution already obtained in the search process (*GlobalOpt*). On the other hand, each individual is endowed with memory reverberated in the fact that it knows the value of its best ancestor (*LineOpt*). The ancestors represent all individuals that have existed in one of the previous generations and have contributed to the creation of the current individual: its parents, the parents of its parents, and so on. If within a single ancestral line there are multiple ancestors with the same best fitness values, the closest ancestor is chosen. If within the two ancestral lines of an individual the best individuals have identical fitness, one of them is randomly chosen. In the initial population, the *LineOpt* of each individual is the individual itself.

Both *GlobalOpt* and *LineOpt* will guide the search process in the form of passing relevant genetic material to the individuals. It is intended to show how this information can affect the search process for a problem whose solutions are encoded by discrete variables. This extra information that each individual has will affect the way selection is performed and the way recombination between individuals takes place.

3.1 Encoding and Population Model

A fixed size population is considered, the population size being a parameter of the algorithm. A potential solution of the problem (a chromosome) is a string of constant length $\{x_1, x_2, \ldots, x_n\}$ where n represents the number of nodes in the network and x_i represents the identifier of the cluster to which the node i belongs, $1 \leq i \leq n, 1 \leq x_i \leq n$.

Besides that, each individual retains the value of its best related individual and the value of the best individual obtained so far in the search process.

The initial population is randomly generated. Furthermore, a number of nodes are randomly selected and the neighbors of each node receive the same cluster identifier as the selected node. The neighbors of a node are considered to be all the nodes connected by edges with the current one.

3.2 Fitness Function

The potential solutions are evaluated by means of a real-valued function $f : X \to \mathbb{R}$, where X denotes the search space of the problem. The fitness of a chromosome tells how good a certain distribution of the nodes into clusters is, how well are detected the communities existing in the network.

In order to quantify the strength of a particular division of the network, the measure of quality proposed in [7] is considered. The *modularity* measure is defined by:

$$Q = \sum_i (e_{ii} - a_i^2),$$

where i is the index of the communities, e_{ii} is the proportion of edges that connect vertices in the community i and a_i is the proportion of edges with at least one node in the community i. The fitness function is to be maximized.

3.3 Collaborative Tournament Selection

The n individuals within the population $P(t)$ are grouped by their *LineOpt*. The clusters $A_1, \ldots, A_k, k \leq n$ are formed according to the rules:

(i) the clusters $A_1, \ldots, A_k, k \leq n$ represent a partition of *P(t)*:

$$(a) A_i \neq \phi, 1 \leq i \leq k,$$

$$(b) \bigcup_{i=1}^{k} A_i = P(t),$$

$$(c) A_i \cap A_j = \phi, 1 \leq i \leq k, 1 \leq j \leq k, i \neq j.$$

(ii) all the individuals that belong to the cluster $A_i (1 \leq i \leq k)$, have the same *LineOpt*:

$$LineOpt(x^u) = LineOpt(x^v),$$

$$\forall x^u, x^v \in A_i, 1 \leq i \leq k.$$

(iii) every two different clusters $A_i, A_j (1 \leq i \leq k, 1 \leq j \leq k, i \neq j)$ have a different *LineOpt*:

$$LineOpt(x^u) \neq LineOpt(x^v),$$

$$\forall x^u \in A_i, x^v \in A_j, 1 \leq i \leq k, 1 \leq j \leq k, i \neq j.$$

In order to preserve the exploration of the search space, two individuals are selected for being recombined only if they belong to different clusters. The aim is to recombine genetic material from individuals that are not genetically related, in order to expand the search to regions that have not been yet explored.

Once we have randomly chosen the two different clusters that will provide individuals for recombination, we choose an individual from each cluster according to a tournament scheme.

Let $A_i, 1 \leq i \leq k$, be one of the two clusters selected for recombination. A group of q $(1 \leq q \leq |A_i|)$ individuals is randomly chosen from the cluster. The sample members are chosen so as to induce a great diversity within the tournament group. The number q is the tournament size and can be different for each cluster, if we take into account the size of the cluster. For all numerical experiments presented in Section 4, the tournament size is set to:

$$q = \left\lceil \frac{|A_i|}{2} \right\rceil,$$

where for a real value k, we denote by $\lceil k \rceil$ the superior integer part of k,

$$k \leq \lceil k \rceil \leq k + 1.$$

This way the tournament size is proportional to the size of the cluster. The fittest individual from the tournament (or sample) group will be chosen as one of the two parents that will be recombined. The second parent is chosen from the second cluster in a similar way.

3.4 Collaborative Recombination Operator

Recombination operator performs an information exchange between chromosomes. This way the offspring obtained after the recombination of the parents will keep genetic information of both parents. Recombination helps the progress of the search by exploring the search space.

A variant of the collaborative recombination operator for permutation based encoding has been proposed in [5]. The recombination operator uses the information encoded by the *GlobalOpt* and the *LineOpt* of each individual. This way it is not only transferring to the offspring genetic information from the parents but from the best ancestor and from the best global as well.

The main idea of this operator is that if a certain individual's *LineOpt* contains genetic material that can also be retrieved in the *GlobalOpt*, than that genetic material is considered as being good for the search process. For our specific problem, relevant genetic material refers to the fact that a certain node belongs to a certain cluster. One important feature of the proposed recombination is the control of the amount of relevant genetic information transferred from the *GlobalOpt* and *LineOpt* to the offspring. The control is made by taking into account the number of the current generation relative to the total number of generations and the number of common genes of *GlobalOpt* and *LineOpt*. The aim is to increase the diversity in the first stages of the algorithm and to become more goal-oriented in the final stages, by keeping in the configuration of the offspring more relevant genetic information from *LineOpt* and *GlobalOpt*.

In order to increase the population diversity during all stages of the search process, another characteristic of the collaborative recombination is that a randomly chosen sequence of one parent is always kept in the offspring.

3.5 Mutation Operator

The mutation operator is not affected by the *GlobalOpt* and *LineOpt* - as its main feature is to introduce diversity into the population of candidate solutions and to reintroduce lost genetic material into the population. Mutation remains responsible for exploring new promising regions of the search space and not to exploit those which already have been discovered.

By taking into account the architecture of the chromosome, mutation has the following features: one gene of the chromosome is randomly selected and assigned with a randomly chosen cluster identifier. Also, all the neighbors of the selected node will receive the same cluster identifier.

Mutation takes place with a certain probability, which is given as parameter of the algorithm.

3.6 Merging Operator

A new merging operator is considered in order to accelerate the search process. The merging operator is applied with a certain probability for each individual. Two genes having different values (e.g. two nodes belonging to different clusters) are randomly chosen. If the individual obtained by combining the two clusters has a better fitness than the original individual, than all the genes that have one of the two cluster identifiers will receive the same value.

3.7 Selection for Replacement and Survival

The best offspring obtained after recombination (possible subject to mutation and/or merging) is kept in the next generation. The elitism ensures that the fitness of the best solution in the population does not deteriorate as the generation advances. The algorithm ends after a certain number of generations that did not improve the best solution obtained so far.

4 Experimental Results

The Zachary's karate club network [12] is considered to test the efficiency of the proposed algorithm. The parameters of the algorithm are written in Table 1.

Table 1. Parameters of the Collaborative Evolutionary Algorithm

Population size	Number of nodes randomly chosen when initializing the population (see subsection 3.1)	Tournament size (see subsection 3.3)	Mutation rate	Number of generation after which the algorithm ends if no improvement is brought
100	$\frac{n}{10}$	$q = \frac{\lvert A_i \rvert}{2}$, $1 \leq i \leq k$	0.05	1000

This is a classic network used for social network analysis. It has 34 nodes and 78 edges that represent the social interactions between the members of a karate club at an American university, observed by Wayne Zachary in the early 1970s. Due to a dispute between the club's administrator and the principal teacher, the club split in two, forming one cluster around the teacher and one cluster around the administrator.

Therefore, the real community structure is formed by two clusters around the node 1 (the administrator) and node 33 (the teacher). Thought, if we understand the communities as subsets of vertices within which vertex-vertex connections are dense but between which connections are less dense, the real division of the karate club into two communities does not represent, by all means, the best modularity of the network. Indeed, a better modularity is obtained if there is a third community that contains only the node 10. This is the solution obtained by the proposed collaborative evolutionary algorithm in 80% of its 25 runs and is depicted in Figure 1. An even better modularity is obtained in 20% of the 25

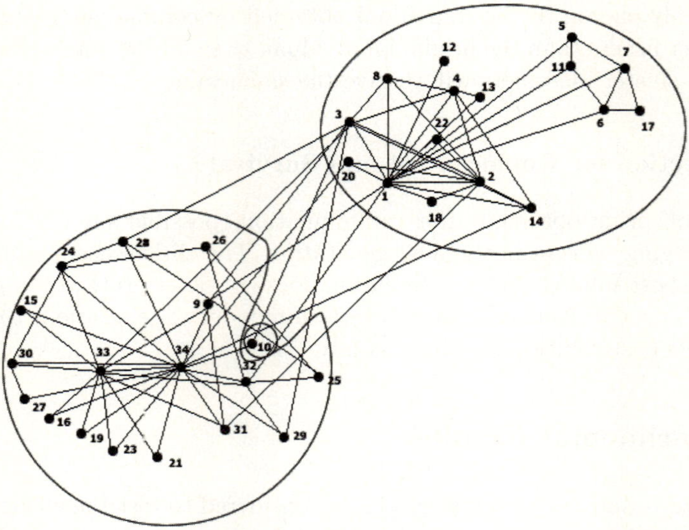

Fig. 1. Solution obtained by the proposed collaborative evolutionary algorithm in 80% of its 25 runs

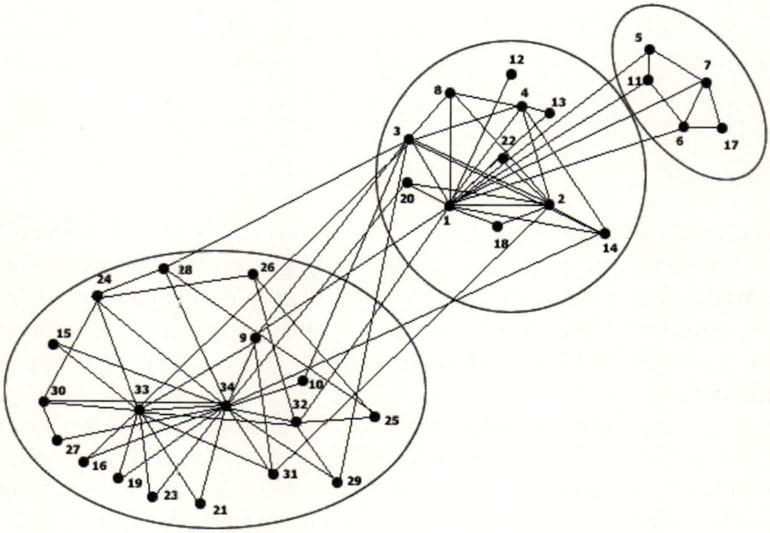

Fig. 2. Solution obtained by the proposed collaborative evolutionary algorithm in 20% of its 25 runs

runs which put the node 10 in the real cluster to which it belongs but create a third cluster that contains the nodes 5, 6, 7, 11 and 17 (see Figure 2).

Compared to the other evolutionary approach of community detection [10] which uses the same modularity as the fitness of the individuals, our proposed collaborative algorithm performs much better, as [10] reports a 97%-100% detection of only two clusters for the karate club network. Even if these two clusters represent the real communities formed in this network, they do not represent the best modularity, a much better modularity being detected by our proposed algorithm.

Moreover, the collaborative evolutionary algorithm finds the solution after about 1050 generations. By taking into account the fact that the algorithm ends after 1000 generations that did not improve the solution found so far (see Table 1), it means that the solution is detected after about 50 generations.

5 Conclusion

A new evolutionary technique for community detection in complex networks has been proposed. The collaborative evolutionary algorithm introduces into the population of individuals knowledge about the best solution obtained so far and about the best related chromosome of each individual. This extra knowledge affects the way recombination and selection are performed. A new merging operator is also proposed. The Zachary's karate club network is considered for numerical experiments. Results show an improvement of the solution obtained by the other evolutionary approach existing for community detection [10].

Acknowledgments. This work was supported by a Scholarship 2006-2007 of the University of Fribourg.

References

1. Balakrishnan, H., Deo, N.: Discovering Communities in Complex Networks. In: Proceedings of the ACM Southeast Regional Conference, pp. 280–285. ACM Press, New York (2006)
2. Barabasi, A.-L.: Linked: The New Science of Networks. Perseus, New York (2002)
3. Duch, J., Arenas, A.: Community Detection in Complex Networks using Extremal Optimization. Physical Review E 72, 027104 -1 (2005)
4. Girvan, M., Newman, M.E.J.: Community Structure in Social and Biological Networks. Proceedings of the National Academy of Sciences of the USA 99, 7821–7826 (2002)
5. Gog, A., Dumitrescu, D.: Adaptive Search in Evolutionary Combinatorial Optimization. In: Proceedings of the International Conference of Bio-Inspired Computing – Theory and Applications (BIC-TA), Wuhan, China, pp. 123–130 (2006)
6. Huberman, B.A., Wu, F.: Finding Communities in Linear Time: a Physics Approach. The European Physics Journal B 38, 331–338 (2004)
7. Newman, M.E.J., Girvan, M.: Finding and Evaluating Community Structure in Networks. Physical Review E 69, 026113-1 (2004)

8. Radicchi, F., Castellano, C., Cecconi, F., Loreto, V., Parisi, D.: Defining and Identifying Communities in Networks. Proceedings of National Academy of Science in USA 101, 2658–2663 (2004)
9. Scott, J.: Social Network Analysis: A Handbook. Sage Publication, London (2000)
10. Tasgin, M., Bingol, H.: Community Detection in Complex Networks using Genetic Algorithm. cond-mat/0604419 (2006)
11. Watts, D.: Six degrees: The Science of a Connected Age. Gardner's Books, New York (2003)
12. Zachary, W.W.: An Information Flow Model for Conflict and Fission in Small Groups. Journal of Anthropological Research 33, 452–473 (1977)

Detecting Non-trivial Computation in Complex Dynamics

Joseph T. Lizier[1,2], Mikhail Prokopenko[1], and Albert Y. Zomaya[2]

[1] CSIRO Information and Communications Technology Centre, Locked Bag 17, North Ryde, NSW 1670, Australia
[2] School of Information Technologies, The University of Sydney, NSW 2006, Australia
jlizier@it.usyd.edu.au

Abstract. We quantify the local information dynamics at each spatiotemporal point in a complex system in terms of each element of computation: information storage, transfer and modification. Our formulation demonstrates that information modification (or non-trivial information processing) events can be locally identified where "the whole is greater than the sum of the parts". We apply these measures to cellular automata, providing the first quantitative evidence that collisions between particles therein are the dominant information modification events.

1 Introduction

Information-theoretic measures are increasingly being used to capture dynamics and to drive evolution in *artificial life*. Examples here include the use of a memory-like measure in [1], and information transfer-like measures in [2] and [3]. Such work appears disjointed however in that each example uses a different *single* measure of fitness. We observe that these single information-theoretic measures of fitness are often related to sub-functions or elements of computation. Interestingly, emergent complex behavior has often been described from the perspective of computation within the given system [4], and complex behavior is postulated to be associated with the capability to support universal computation [5,6]. Such discussions focus on cellular automata (CAs) as model systems offering a range of dynamic behavior [4], including producing emergent structures such as gliders and glider collisions [7]. These discussions typically surround qualitative observation of the component operations of computation: *information storage*, *transfer* and *modification* (e.g. [5,4]).

We suggest that a more intricate approach of *quantifying* the information dynamics of *each element* of computation will provide greater insight into and greater control over artificial life systems. We will describe how to quantify each element of computation on a local scale *within* a given system, showing how information storage and transfer interact to produce information modification; neither a single measure or system-wide approach is capable of this. We quantify a sum of the parts of computation, and locally identify information modification events where the sum is missing information; i.e. where the *whole is*

F. Almeida e Costa et al. (Eds.): ECAL 2007, LNAI 4648, pp. 895–904, 2007.

greater than the sum of the parts. This phrase is often used to describe emergent structure in complex systems, e.g. patterns in Belousov-Zhabotinsky media [8], self-organization in microtubules [9] and collisions in CAs [4].

Our approach will provide insight into the local information dynamics of complex systems, from the perspective of the individual elements of computation. Here, we use it to demonstrate that the whole is quantitatively greater than the sum of the parts at collisions in CAs, and thereby prove the long-held conjecture that these are the dominant information modification agents therein.

2 Information-Theoretical Preliminaries

To quantify the elements of computation, we look to information theory (e.g. see [10]) which has proven to be a useful framework for the design and analysis of complex self-organized systems, e.g. [1,2,3]. The fundamental quantity is the *Shannon entropy*, which represents the uncertainty associated with any measurement x of a random variable X (logarithms are in base 2, giving units in bits): $H(X) = -\sum_x p(x) \log p(x)$. The *joint entropy* of two random variables X and Y is a generalization to quantify the uncertainty of their joint distribution: $H(X,Y) = -\sum_{x,y} p(x,y) \log p(x,y)$. The *conditional entropy* of X given Y is the average uncertainty that remains about x when y is known: $H(X|Y) = -\sum_{x,y} p(x,y) \log p(x|y)$. The *mutual information* between X and Y measures the average reduction in uncertainty about x that results from learning the value of y, or vice versa: $I(X;Y) = H(X) - H(X|Y)$. The *conditional mutual information* between X and Y given Z is the mutual information between X and Y when Z is known: $I(X;Y|Z) = H(X|Z) - H(X|Y,Z)$.

The *entropy rate* is the limiting value of the conditional entropy of the next state x of X given knowledge of the previous $k-1$ states $x^{(k-1)}$ of X: $h_\mu = \lim_{k\to\infty} H\left[x|x^{(k-1)}\right] = \lim_{k\to\infty} h_\mu(k)$. Finally, the *excess entropy* quantifies the total amount of structure or memory in a system, and is computed in terms of the slowness of the approach of the entropy rate estimates to their limiting value (see [11]). For our purposes, it is best formulated as the mutual information between the semi-infinite past and semi-infinite future of the system:

$$E = \lim_{k\to\infty} I\left[x^{(k)}; x^{(k^+)}\right], \tag{1}$$

where $x^{(k^+)}$ refers to the next k states. This interpretation is known as the *predictive information* [12], as it highlights that the excess entropy captures the information in a system's past which is relevant to predicting its future.

3 Introduction to Cellular Automata

Cellular automata (CA) are discrete dynamical systems consisting of an array of cells which each synchronously update their state as a function of the states of a fixed number of spatially neighboring cells using a uniform rule. While the behavior of each cell is simple, their (non-linear) interactions can lead to quite intricate global behavior. As such, CAs have become the classical example of complex behavior, and been used to model a wide variety of real world phenomena (see [4]). *Elementary CAs*, or *ECAs*, are a simple variety of 1D CAs using binary states, deterministic rules and one neighbor on either side (i.e. cell range $r = 1$). An example evolution of an ECA may be seen in Fig. 1a. For more complete definitions, including that of the Wolfram rule number convention for describing update rules (used here), see [13].

An important outcome of Wolfram's well-known attempt to classify the asymptotic behavior of CA rules into four classes [6,13] was a focus on emergent structure: *particles*, *gliders* and *domains*. A domain is a set of background configurations in a CA, any of which will update to another configuration in the set in the absence of any disturbance. A domain may be *regular*, where the configurations repeat periodically, or is otherwise known as *irregular*. Domains are formally defined within the framework of computational mechanics as spatial process languages in the CA [14]. Particles are considered to be dynamic elements of coherent spatiotemporal structure; gliders are regular particles, blinkers are stationary gliders. Formally, particles are defined by computational mechanics as a boundary between two domains [14]; they can be termed *domain walls*, though this is typically used with reference to irregular domains. Several techniques exist to *filter* particles from background domains (e.g. [15,16]).

4 Computation in Cellular Automata

Computation in CAs has been a popular topic for study, with a major focus in observing or constructing (Turing) universal computation in certain CAs (see [4,5,6]). This capability has been proven for several CA rules (e.g. the Game of Life [7]), through the design or identification of entities which provide the three primitive functions of universal computation: information storage, transmission and modification. Typically such analyses focus on blinkers as the basis of information storage, particles as the basis of information transfer, and collisions between these structures as information modification (see [5,4]).

However, the focus on universal computational ability has been criticized as drawing away from the ability to identify "generic computational properties" in these and other CAs [14]. Related criticisms target attempts to classify CA rules in terms of generic behavior or "bulk statistical properties", suggesting that the wide range of dynamics taking place in different areas of the CA make this problematic [14,4]. With respect to computation, it would be too simplistic to say a CA was either *computing* or *not computing*. Alternatively, these studies suggest that analyzing the rich space-time dynamics *within* the CA is a more appropriate focus, since the CA may be undertaking different parts of a complex computation at different times or spatial points. As such, these and other studies have *analyzed the local dynamics of intrinsic or other specific computation*, while continuing the focus on particles facilitating the transfer of information and collisions facilitating information modification or processing. Noteworthy examples include: applying filters from the domain of computational mechanics (using regular language parsers in [14] and local statistical complexity in [15]), analysis using such filters on CA rules selected via evolutionary computation to perform tasks including classification [17]; and deep investigations of particle properties and their interactions [17,18].

Despite this surrounding interest, no complete local quantification of the individual elements of computation exists. In the subsequent sections, we outline how the individual elements of computation can be locally quantified within the spatiotemporal structure of a CA. In particular, we describe how information storage and information transfer interact to give rise to information modification events, being where the whole is greater than the sum of the parts.

5 Information Storage

Although discussion of information storage, or memory, in CAs has often focused on periodic structures, it does not necessarily entail periodicity. Instead, the excess entropy (1) encompasses all types of structure and memory by capturing correlations across all time lengths. In examining *local* information dynamics, we are particularly interested in how much of the stored information is actually *in use* at every local point in time and space.

The excess entropy can be massaged into a spatiotemporally local measure (i.e. the amount of information stored by a *particular* cell at a *particular* point in time) by noting that it is actually the *expectation value* of a local excess entropy at every time step [19].[3] The local excess entropy is then the mutual information between the semi-infinite past and future *for the given cell at the given time step*. It quantifies the total stored information that will be used *at some point* in the future of the state process of that cell; possibly but not necessarily at the next time step $n+1$. To reveal the amount of memory actually *in use* at the next time step, we derive *local active information storage* $a(i, n+1)$ as the local mutual information between the semi-infinite past $x_{i,n}^{(k)}$ (as $k \to \infty$) and the *next state* $x_{i,n+1}$ of a given cell i at the given time step $n+1$:

$$a(i, n+1) = \lim_{k \to \infty} \log \frac{p(x_{i,n}^{(k)}, x_{i,n+1})}{p(x_{i,n}^{(k)})p(x_{i,n+1})}. \tag{2}$$

It is not feasible to compute $a(i, n)$ in the limit $k \to \infty$; instead we compute $a(i, n, k)$ with finite k. Importantly, $a(i, n, k)$ may be positive or negative, meaning the past history of the cell can either positively inform us or actually *misinform* us about it's next state. An observer is misinformed where, given the past history, the observed outcome was relatively unlikely.

[3] As per Shalizi's explanation in [19], which was for the *light-cone formulation* of excess entropy. A detailed description on why such average measures are the expectation value of local measures, and why the local measure is simply the log term within the expectation value, lies in the presentation of local transfer entropy in [16].

Fig. 1. Information Storage. ECA Rule 54: a. (*left*) Raw CA (time is vertical). b.,c. Local active information storage $a(i, n, k = 16)$: b. (*center*) positive values only, grayscale (30 levels), max. 1.11 bits (black); c. (*right*) negative values only, grayscale (30 levels), min. -12.2 bits (black). All figures generated using modifications to [20].

In order to evaluate $a(i, n, k)$ (and our later metrics) within sample CA runs, we estimate the probability distribution functions from CA runs of 10 000 cells with periodic boundary conditions, initialized from random states, with 600 time steps retained (after the first 30 time steps were eliminated to allow the CA to settle). Since the CA is a homogeneous system, the probability functions were estimated from observations at every spatiotemporal point so as to obtain more accurate estimates. All results were confirmed by at least 10 runs from different initial states. We use ECA rule 54 as a running case study: it is complex enough to contain traveling and stationary gliders, though these are simple enough that the results are easily understood.

The raw states of a sample run for rule 54 are displayed in Fig. 1a. The local active information storage $a(i, n, k = 16)$ profile contains both positive and negative values, plotted separately in Fig. 1b and c. The positive values are concentrated in the domain areas, as well as in the stationary gliders (known as α and β [18]), due to strong temporal periodicity in both areas. In fact, the $a(i, n, k)$ values form spatially and temporally periodic patterns in the domain, due to the periodicities of the raw states. The negative values are concentrated in the traveling glider areas (known as γ^+ and γ^- [18]). When a traveling glider is encountered at a given cell, the past history of that cell (being part of the background regular domain) is misinformative about the next state of the cell, since the domain sequence was more likely to continue than be interrupted. It is useful that $a(i, n, k)$ can reveal the gliders in this way (similar to existing filtering methods for CAs [14,15]), however it is does not tell us the source of information for the glider: to quantify this, we turn to information transfer.

6 Information Transfer

Schreiber presented *transfer entropy* as a directional, dynamic measure for information transfer [21] so as to address deficiencies in the de facto use of mutual information (as a static, symmetric measure). Transfer entropy is the deviation from independence (in bits) of the state transition of an information destination X from the (previous) state of a source Y. It is a *conditional* mutual information [22], being the average information in the source y_n about the next state of the destination x_{n+1} that was not already contained in the destination's past $x_n^{(k)}$ (i.e. $I(Y_n; X_{n+1}|X_n^{(k)})$), allowing us to quantify directional flows of information.

In [16], we demonstrated that the transfer entropy is an *expectation value* of a *local* transfer entropy at each observation. We also generalized comments on the entropy rate in [21] to suggest that the asymptote $k \to \infty$ is most correct for agents displaying non-Markovian dynamics. For systems such as CAs with homogeneous spatially-ordered agents, the local apparent transfer entropy to cell X_i from X_{i-j} at time $n+1$ is:

$$t(i, j, n+1) = \lim_{k \to \infty} \log \frac{p(x_{i,n+1}|x_{i,n}^{(k)}, x_{i-j,n})}{p(x_{i,n+1}|x_{i,n}^{(k)})}, \tag{3}$$

for transfer from the previous time step only. Again, we define $t(i, j, n, k)$ for finite k. Transfer $t(i, j, n, k)$ is defined for every spatiotemporal destination (i, n), for every information channel or direction j where sensible values for CAs are within the cell range, $|j| \leq r$ (e.g. $j = 1$ means transfer across one cell to the right). Local apparent transfer entropy $t(i, j, n, k)$ may be either positive or negative, with negative values occurring where (given the destination's history) the source is actually *misleading* about the next state of the destination.

The destination's own historical values can indirectly influence it via the source or other neighbors and be mistaken as an independent flow from the source [16]. In the context of computation, this influence is recognizable as the *active information storage. The active information storage* $a(i, n+1)$ *is eliminated from the transfer entropy measurement by conditioning on the destination's history* $x_{i,n}^{(k)}$. Yet any self-influence transmitted prior to these k values will not be eliminated, which is why we suggest taking the limit $k \to \infty$ to be most correct.

We applied the local transfer entropy metric to several important ECA rules in [16]. Fig. 2 displays application of the local apparent transfer entropies to rule 54, demonstrating that the metric successfully highlights traveling gliders with large positive transfer against background domains (it also highlights domain walls where they exist). Importantly, the metric finds negative transfer for gliders moving orthogonal to the direction of measurement, because the source (as part of the domain) is misinformative about the next state of the destination. Also, there is a small non-zero information transfer in background domains, effectively indicating the absence of gliders; this is particularly strong in the wake of real gliders, where secondary gliders often follow. *The measure provided the first quantitative evidence for the long-held conjecture that particles are the dominant information transfer agents in CAs.* This highlighting was similar to other methods of filtering in CAs (e.g. [15,14]), but subtly allowed comparison between and within gliders of the amount and (channel or) direction of information transfered at each point, and revealed the leading edges of gliders as the major information transfer zones. At least a minimum k was required to achieve

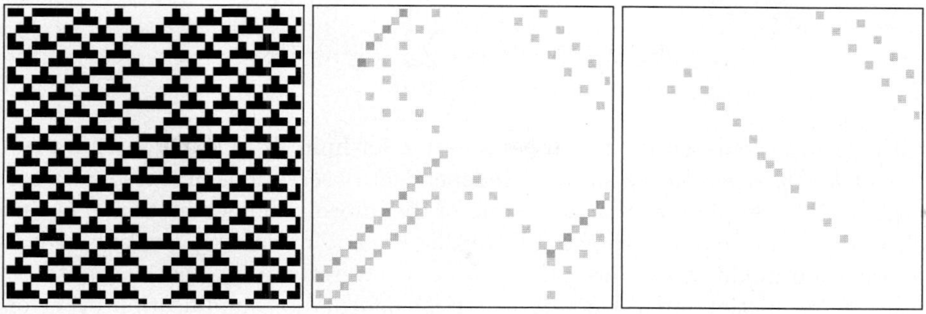

Fig. 2. Information Transfer. ECA Rule 54: a. (*left*) Raw CA. b.,c. Local apparent transfer entropy $t(i, j = -1, n, k = 16)$ ($j = -1$ means transfer one cell to the left): b. (*center*) positive values only, grayscale (16 levels), max. 7.92 bits (black); c. (*right*) negative values only, grayscale (16 levels), min. -4.21 bits (black).

reasonable estimates of the metric (e.g. of the order of the period of a regular periodic domain); without this, particles were not highlighted. Finally, note that this metric cannot quantitatively distinguish gliders from their collisions: for this, we look to an information modification metric.

7 Information Modification

Information modification has been described as interactions between stored and transmitted information that result in a modification of one or the other [5], and generally interpreted to mean interactions or collisions of particles. As an information processing event, the important role of collisions in determining the dynamics of the system is widely acknowledged [18]. For a regular particle or glider, a local *modification* is simply an alteration to the predictable periodic pattern of the glider's dynamics, where an observer would be surprised or misinformed about the next state of the glider without having taken account of the entity about to perturb it. Recall that local apparent transfer entropy $t(i, j, n)$ and local active information storage $a(i, n)$ were negative where the respective information sources were *misinformative* about the next state of the information destination. This occurred for $a(i, n)$ at unperturbed gliders, and for $t(i, j, n)$ at gliders traveling in the orthogonal direction to the measurement. However, we expect that the $t(i, j, n)$ in the direction of motion of the glider will be *more informative* than the misinformation conveyed from the other sources.

Where a glider is perturbed by an interaction with another glider, we cannot expect $t(i, j, n)$ in the macroscopic direction of the first glider to remain informative about the evolution of that glider at the collision point. Nor can we expect this from the incoming $t(i, j, n)$ for the incident glider. As such, we hypothesize that *at the spatiotemporal location of a local information modification event or collision, the total information from the information storage and information transfer, where each source is observed separately, will misinform an observer.* We label this total as the *local separable information*, $s(i, n)$:

$$s(i,n) = a(i,n) + \sum_{j=-r, j\neq 0}^{+r} t(i,j,n), \tag{4}$$

with $s(i,n,k)$ representing the approximation for finite k. Where $s(i,n)$ is *positive* or *highly separable*, separate or independent observations of the sources are informative overall about the next state of the information destination. This indicates that information storage and transfer are not interacting, and only trivial information modifications are taking place. Conversely, we expect $s(i,n)$ to be *negative* at spatiotemporal points where an information modification event or collision takes place, with more significant modifications taking larger negative values. Separate examination of sources fails here because the information storage and transfer are interacting, i.e. *non-trivial information modification* takes place. This formulation of non-trivial information modification quantifies the description of emergence in complex systems as where *"the whole is greater than the sum of it's parts"*. While we quantify the sum of the parts in $s(i,n)$, there is no quantity representing the "whole" as such, simply an indication that the whole is greater where all information sources must be examined *together* in order to receive positive information on the next state of the given entity.

Fig. 3 displays application of $s(i,n,k)$ to ECA rule 54. Positive values of $s(i,n,k)$ (not plotted) are concentrated in the domain regions and at the stationary gliders (α and β): as expected, these regions are undertaking trivial computations only. The dominant negative values of $s(i,n,k)$ are concentrated around the areas of collisions between the gliders, including those between traveling gliders only (marked by "A") and between the traveling gliders and blinkers (marked by "B" and "C"). Considering the collision "A" ($\gamma^+ + \gamma^- \rightarrow \beta$ [18]), the marked information modification event is one time step below where one may naively define it. Our metric correctly marks the information modification event however, being where prediction requires the sources to be considered together. For the other collisions "B" and "C" also, the spatiotemporal location of the primary information modification(s) appears to be delayed from a naively defined collision point; this indicates a time-lag associated with processing the information. Smaller negative values are also associated with the gliders (too small to appear in Fig. 3b), which was unexpected. These weak information modifications appear to indicate the absence of a collision (i.e. the absence of an incident glider) and in some sense are a computation that the glider will continue. These computations are more significant in the wake of real collisions (indeed the secondary collision points for types "B" and "C" are higher in magnitude than the earlier collision points), since they have a larger influence on the surrounding dynamics at those points. This finding is analogous to that of small values of transfer entropy in domains indicating the absence of gliders, which were also more significant in the wake of real gliders [16]. This is the first known metric which brings together information storage and transfer to identify information modification, and it has provided the first quantitative evidence that collisions in CAs are the dominant information modification events therein. It is also the first suite of filters able to distinguish between particles and particle collisions.

Fig. 3. Information Modification. ECA Rule 54: a. (*left*) Raw CA. b. (*center*) Local separable information $s(i, n, k = 16)$, negative values only plotted, grayscale (30 levels), min -5.23 bits (black). c. (*right*) Locations of negative values of $s(i, n, k = 16)$ (larger than weak values along the gliders) marked with black circles against $t(i, j, n, k = 16)$ summed over $j = [-1, 1]$; "A", "B" and "C" mark collision types discussed in the text.

8 Conclusion

We have discussed appropriate local metrics for information storage and transfer, and demonstrated how these quantities interact to produce information modification events where the "whole is greater than the sum of the parts". These metrics form a powerful framework for quantifying the information dynamics of complex systems. Here, the framework has been applied to CAs, providing the first evidence for the long-held conjecture that collisions therein are the dominant information modification events.

We aim to provide a deeper investigation of this topic in a forthcoming analysis, reporting results we have obtained from application of the presented methods to other CA rules (complex, chaotic and those with domain walls) which corroborate and extend those reported here. We believe these three elements of computation identify a set of *axes of complexity* to characterize complex systems. Within this multi-dimensional space, a system such as a CA may appear complex in one dimension (e.g. information storage) without necessarily appearing so in the other dimensions. Finally, we shall explore the generality afforded by the information-theoretic basis of this framework. Application to higher dimensional CAs or other discrete dynamical systems is straightforward, and we expect application to artificial life systems to provide greater insight into and control over such systems.

References

1. Prokopenko, M., Gerasimov, V., Tanev, I.: Evolving spatiotemporal coordination in a modular robotic system. In: Nolfi, S., Baldassarre, G., Calabretta, R., Hallam, J.C.T., Marocco, D., Meyer, J.-A., Miglino, O., Parisi, D. (eds.) SAB 2006. LNCS (LNAI), vol. 4095, pp. 548–559. Springer, Heidelberg (2006)

2. Klyubin, A.S., Polani, D., Nehaniv, C.L.: All else being equal be empowered. In: Capcarrère, M.S., Freitas, A.A., Bentley, P.J., Johnson, C.G., Timmis, J. (eds.) ECAL 2005. LNCS (LNAI), vol. 3630, pp. 744–753. Springer, Heidelberg (2005)
3. Lungarella, M., Sporns, O.: Mapping information flow in sensorimotor networks. PLoS Computational Biology 2(10), 144 (2006)
4. Mitchell, M.: Computation in cellular automata: A selected review. In: Gramss, T., Bornholdt, S., Gross, M., Mitchell, M., Pellizzari, T. (eds.) Non-Standard Computation, pp. 95–140. VCH Verlagsgesellschaft, Weinheim (1998)
5. Langton, C.G.: Computation at the edge of chaos: phase transitions and emergent computation. Physica (Amsterdam) 42D(1-3), 12–37 (1990)
6. Wolfram, S.: Universality and complexity in cellular automata. Physica (Amsterdam) 10D(1-2), 1–35 (1984)
7. Conway, J.H.: What is life? In: Berlekamp, E., Conway, J.H., Guy, R. (eds.) Winning ways for your mathematical plays, vol. 2, Academic Press, New York (1982)
8. Sendiña-Nadal, I., Mihaliuk, E., Wang, J., Pérez-Muñuzuri, V., Showalter, K.: Wave propagation in subexcitable media with periodically modulated excitability. Phys. Rev. Lett. 86(8), 1646 (2001)
9. Brown, J.A., Tuszynski, J.A.: A review of the ferroelectric model of microtubules. Ferroelectrics 220, 141–156 (1999)
10. MacKay, D.J.C.: Information Theory, Inference, and Learning Algorithms. Cambridge University Press, Cambridge (2003)
11. Crutchfield, J.P., Feldman, D.P.: Regularities unseen, randomness observed: Levels of entropy convergence. Chaos 13(1), 25–54 (2003)
12. Bialek, W., Nemenman, I., Tishby, N.: Complexity through nonextensivity. Physica (Amsterdam) 302A(1-4), 89–99 (2001)
13. Wolfram, S.: A New Kind of Science. Wolfram Media, Champaign (2002)
14. Hanson, J.E., Crutchfield, J.P.: The attractor-basin portait of a cellular automaton. J. Stat. Phys. 66, 1415–1462 (1992)
15. Shalizi, C.R., Haslinger, R., Rouquier, J.-B., Klinkner, K.L., Moore, C.: Automatic filters for the detection of coherent structure in spatiotemporal systems. Phys. Rev. E 73(3), 036104 (2006)
16. Lizier, J.T., Prokopenko, M., Zomaya, A.Y.: Local information transfer as a spatiotemporal filter for complex systems. Unpublished (2007)
17. Mitchell, M., Crutchfield, J.P., Hraber, P.T.: Evolving cellular automata to perform computations: Mechanisms and impediments. Physica (Amsterdam) 75D, 361–391 (1994)
18. Hordijk, W., Shalizi, C.R., Crutchfield, J.P.: Upper bound on the products of particle interactions in cellular automata. Physica (Amsterdam) 154D(3-4), 240–258 (2001)
19. Shalizi, C.R.: Causal Architecture, Complexity and Self-Organization in Time Series and Cellular Automata. PhD thesis, University of Wisconsin-Madison (2001)
20. Wójtowicz, M.: Java cellebration v.1.50. Online Software (2002)
21. Schreiber, T.: Measuring information transfer. Phys. Rev. Lett. 85(2), 461–464 (2000)
22. Ay, N., Polani, D.: Information flows in causal networks. Adv. Comp. Sys (to be published, (2006)

Evolution of One-Dimensional Cellular Automata by $1/f$ Noise

Shigeru Ninagawa

Division of Information and Computer Science,
Kanazawa Institute of Technology,
Ohgigaoka, Nonoichi, Ishikawa 921-8501, Japan
ninagawa@infor.kanazawa-it.ac.jp

Abstract. It is speculated that there is a relationship between $1/f$ noise and computational universality in cellular automata. We use genetic algorithms to search for one-dimensional and two-state, five-neighbor cellular automata which have $1/f$-type spectrum. A power spectrum is calculated from the evolution starting from a random initial configuration. The fitness is estimated from the power spectrum in consideration of the similarity to $1/f$-type spectrum. The result shows that the rule with the highest average fitness has a propagating structure like other computationally universal cellular automata, although computational universality of the rule has not been proved yet.

1 Introduction

Cellular automata (CAs) are spatially and temporally discrete dynamical systems with large degrees of freedom. Since spectral analysis is one of the useful methods to investigate the behavior of dynamical systems [1], it is reasonable to apply it to the study of the behavior of CAs. Among a wide variety of CAs, computationally universal CAs are the most intriguing ones because they exhibit complex behavior. A number of computationally universal CAs have been invented in two-dimension [2,3,4,5,6] and one-dimension [7,8,9]. They can perform universal computation only from elaborately designed initial configurations. Our primary concern is, however, the dynamics which brings about computational universality rather than the behavior generated from a particular initial configuration. Therefore we shall concentrate on simple CAs which can run from random initial configurations.

It was proved that elementary CA (ECA), namely one-dimensional and two-state, three-neighbor CA rule 110 has the capability of supporting universal computation [10]. In addition, the evolution of rule 110 starting from a random initial configuration exhibits $1/f$ noise whose power spectrum $S(f)$ is inversely proportional to frequency f [11]. Another example of computationally universal CA is the Game of Life (LIFE) [12]. LIFE is a two-dimensional and two-state, nine-neighbor outer totalistic CA. It is supposed that LIFE is capable of supporting universal computation, while the evolution starting from a random initial configuration has $1/f$-type spectrum [13]. These results suggest that there is a

F. Almeida e Costa et al. (Eds.): ECAL 2007, LNAI 4648, pp. 905–914, 2007.

relationship between computational universality and $1/f$ noise in CAs. We need to find more CAs with $1/f$-type spectrum to verify the relationship.

In this paper we search for $1/f$-type one-dimensional and two-state, five-neighbor CAs (1-2-5CAs). Since the total number of the rules of 1-2-5CAs is $2^{2^5} \approx 4.2 \times 10^9$, it is not reasonable to investigate the whole rule space. So we use genetic algorithms (GAs). GA has been used to discover CAs with desirable properties [14,15]. In their work they searched for one-dimensional and two-state, seven-neighbor CAs which can classify the density of the initial configurations and then could find CAs with high performance on the task.

2 1/f Noise and Computational Universality

We formulate the spectral analysis of one-dimensional CAs, although this method is easily applicable to any CAs. Let $x_i(t) \in \{0, 1, \cdots, k-1\}$ denote the value of site i at time step t in a one-dimensional and k-state CA.

The discrete Fourier transform of time series of state $x_i(t)$ for $t = 0, 1, ..., T-1$ is given by

$$\hat{x}_i(f) = \frac{1}{T} \sum_{t=0}^{T-1} x_i(t) \exp(-i\frac{2\pi t f}{T}), \quad (f = 0, 1, \cdots, T-1). \tag{1}$$

It is natural to define the power spectrum of CAs as

$$S(f) = \sum_i |\hat{x}_i(f)|^2, \tag{2}$$

where the summation is taken over all cells in the array. The power $S(f)$ at frequency f intuitively means the "strength" of the periodic vibration with period T/f in the evolution of observation length T.

ECAs have been investigated in detail for the smallness of the rule space ([16], appendix in [17]). There is a wide variety of rules which exhibit regular, chaotic or complex behavior in ECA rule space. The rule function of rule 110 is given by:

$$\frac{111}{0}\,\frac{110}{1}\,\frac{101}{1}\,\frac{100}{0}\,\frac{011}{1}\,\frac{010}{1}\,\frac{001}{1}\,\frac{000}{0}.$$

The upper line represents the 8 possible states of neighborhood and the lower line specifies the state of the center cell at the next time step. The total number of possible distinct ECA rules is $2^8 = 256$ and each rule is abbreviated by the decimal representation of the binary sequence of the lower line. Out of the 256 ECA rules 88 of them remain independent (appendix in [18]).

Figure 1 shows a typical example of the space-time pattern (left) and the power spectrum (right) of rule 110 starting from a random initial configuration. In a space-time pattern, configurations obtained at successive time steps in the evolution are shown on successive horizontal lines in which black squares represent sites with value 1, white squares sites with value 0. The space-time pattern in Fig.1 consists of 100 cells for 100 time steps. Periodic background of little white triangles with period seven is called "ether". There are several kinds of patterns propagating through ether which is called "glider". Rule 110 has been expected to be capable of supporting universal computation by emulating cyclic tag systems [19] and after that it was exactly proved [10].

The power spectrum in Fig.1 is calculated from the evolution of 500 cells for 1024 time steps. Only half of the components are shown since the other half are redundant. Both x and y-axis are plotted on a logarithmic scale. The broken line represents the least square fitting of the power spectrum from $f = 1$ to $f = 10$ by $\ln S(f) = \alpha + \beta \ln f$, $\beta = -1.3$. The fluctuation whose power spectrum $S(f)$ is inversely proportional to frequency f is called $1/f$ noise [20]. $1/f$ noise has been observed in a wide variety of phenomena such as the voltage of vacuum tubes, the rate of traffic flow, the loudness of music. But its origin is not well understood yet. The spectral analysis on the evolution of 88 independent ECAs revealed that rule 110 exhibits $1/f$ noise during the longest time steps [11].

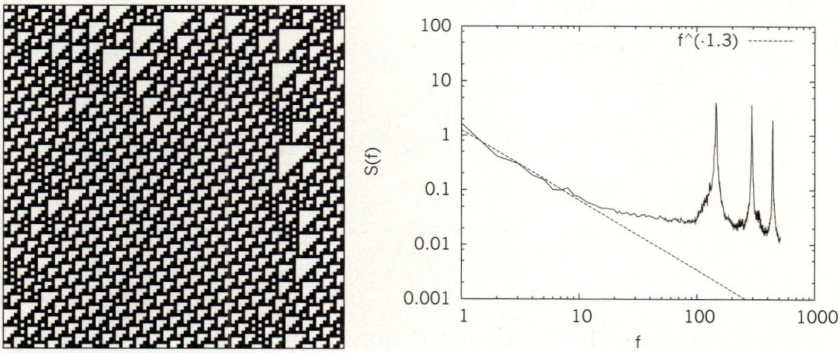

Fig. 1. Space-time pattern (left) and the power spectrum (right) of elementary cellular automaton rule 110. The space-time pattern consists of 100 cells for 100 time steps. The power spectrum is calculated from the evolution of 500 cells for 1024 time steps. Both x and y-axis are plotted on a logarithmic scale. The broken line represents the least square fitting of the power spectrum from $f = 1$ to $f = 10$ by $\ln S(f) = \alpha + \beta \ln f$, $\beta = -1.3$.

Another example which possesses both the exhibition of $1/f$ noise and the capability of supporting universal computation is LIFE. LIFE is one of the two-dimensional and two-state, nine-neighbor outer totalistic CAs [12]. Let $s_{x,y}(t)$ denote the state of the cell at position (x, y) at time step t. The state of the site (x, y) evolves by the rule function d,

$$s_{x,y}(t+1) = d(s_{x,y}(t), n_{x,y}(t)), \tag{3}$$

where $n_{x,y}(t)$ denotes the sum of the states of the eight nearest neighboring sites around the site (x, y) at time step t. The rule of LIFE is defined by

$$d(0,3) = d(1,2) = d(1,3) = 1, \quad otherwise \quad d = 0. \tag{4}$$

The rule of LIFE generates a glider which propagates infinitely until it is annihilated when it collides with another object on the array. It is supposed that a universal computer can be constructed on the array by considering a glider as a pulse in a digital circuit.

Figure 2 shows the power spectrum of LIFE calculated from the evolution starting from a random initial configuration with 100×100 cells for 7200 time steps. The broken line represents the least square fitting of the spectrum from $f = 1$ to $f = 10$ by $\ln S(f) = \alpha + \beta \ln f$, $\beta = -1.4$. This spectrum behaves like $1/f^\beta$ with $\beta \approx 1$ at low frequencies and hence the evolution is considered to be $1/f$ noise.

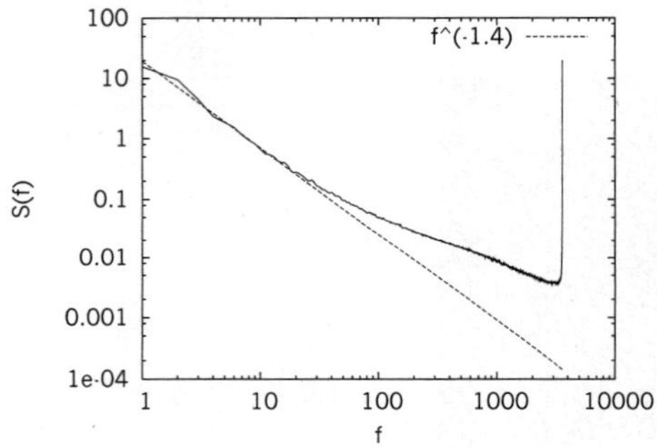

Fig. 2. Power spectrum of LIFE calculated from the evolution of 100×100 cells for 7200 time steps. The broken line represents the least square fitting of the spectrum by $\ln S(f) = \alpha + \beta \ln f$ from $f = 1$ to $f = 10$ with $\beta = -1.4$.

3 Experiment

The method of spectral analysis formulated in the previous section can be easily applicable to 1-2-5 CAs. The value of each site of 1-2-5 CAs is specified as 0 or 1. The site value evolves by iteration of the mapping,

$$x_i(t+1) = F(x_{i-2}(t), x_{i-1}(t), x_i(t), x_{i+1}(t), x_{i+2}(t)). \tag{5}$$

Here F is an arbitrary function specifying the 1-2-5 CA rule. The 1-2-5 CA rule is represented by a binary sequence with length $2^5 = 32$ like this:

$$F(1,1,1,1,1)F(1,1,1,1,0)\cdots F(0,0,0,0,1)F(0,0,0,0,0).$$

The total number of possible distinct 1-2-5 CA rules is $2^{32} \approx 4.29 \times 10^9$. Since we consider the state 0 as the quiescent state, we restrict the rules to those with $F(0,0,0,0,0) = 0$. Therefore a rule is represented by a 31 bit string $F(1,1,1,1,1)\cdots F(0,0,0,0,1)$. We use this representation as the genotype of a rule. Throughout the research a CA evolves for $T = 3000$ time steps from random initial configurations in which each site takes state 0 or 1 randomly with independent equal probabilities and periodic boundary conditions where each end of the array is connected like a ring are imposed.

CA rules are parameterized by a parameter λ which is the fraction of nonzero output states in the rule function [21]. Generally speaking as $\lambda = 0$ varies from 0 to $1 - (1/k)$ (k is the number of cell states), CAs change from the most homogeneous rule to the most heterogeneous rule. So we randomly generate the rules whose λ is uniformly distributed between 1/32 and 16/32 in an initial population.

The estimation of the fitness of a rule is given below. First the power spectrum $S(f)$ of the rule is calculated by (1), (2). Next the coefficient α and β are estimated by the least square fitting

$$\ln(S(f)) = \alpha + \beta \ln(f), \tag{6}$$

of the power spectrum $S(f)$ from $f = 1$ to $f = f_b$ [22]. The residual sum of squares σ^2 is given by

$$\sigma^2 = \frac{1}{f_r} \sum_{f=1}^{f_r} (\ln(S(f)) - \alpha - \beta \ln(f))^2, \tag{7}$$

where f_r is the number of data used for the calculation of σ^2. Finally the fitness F is calculated by

$$F = \begin{cases} \dfrac{|\beta|}{\sigma^2 + \delta} & \beta < 0, \\ 0 & \beta \geq 0, \end{cases} \tag{8}$$

where δ is the correction term to avoid division by zero and is set to 1.0×10^{-6}. We set $f_b = 10$, $f_r = 100$ throughout the research.

The numerator in the right-hand side of (8) for negative β is intended that the steeper the slant of a power spectrum at low frequencies is, the larger the fitness is. The denominator is supposed to guarantee that the spectrum fits the power law in a broad range of frequencies. One-dimensional CAs in general have larger variations in the value of β with initial configurations than two-dimensional CAs. Therefore we average the fitness over 10 distinct random initial configurations of 700 cells which are created at each generation. By doing so, we can avoid evolving rules which get high fitness depending on a particular initial configuration.

The experiment proceeds as follows.

1. A population of P rules is randomly generated.
2. The fitness F of each rule in the population is estimated from its power spectrum.
3. A number E of the highest fitness rules is copied without modification to the next generation.

4. The remaining $P - E$ rules for the next generation are formed by uniform crossovers with a probability of P_c between pairs in the population chosen by roulette wheel selection. Every bit of the offspring from each crossover is mutated with a probability of P_m.

One generation consists of steps 2 - 4 and the number of the generations repeated in each run is 100. We performed a total of 20 runs of the GA with the parameters $P = 160$, $E = 20$, $P_c = 0.6$, and $P_m = 0.05$.

The 10 kinds of rules with the highest fitness at the final generation in each run are picked out. In this case $20 * 10 = 200$ rules are selected and their average fitnesses over 30 random initial configurations are calculated to make final precise evaluation. Figure 3 is a histogram showing the frequency of these 200 rules as a function of λ. This result suggests that promising rules which exhibit $1/f$ noise are in the range of $\lambda = 13/32 \sim 15/32$.

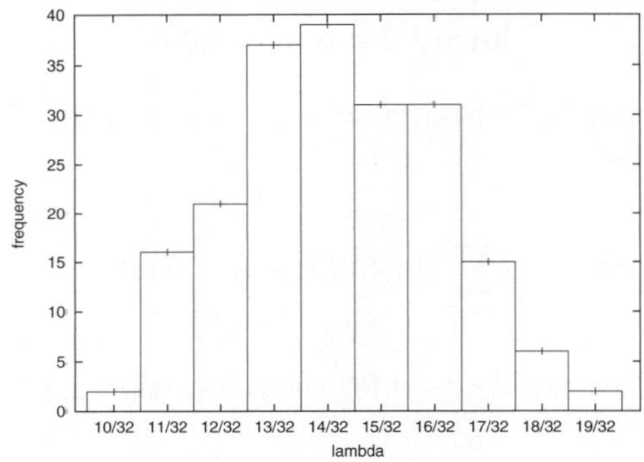

Fig. 3. Histogram of the frequency of 200 elite rules at the final generation in 20 runs as a function of λ.

The rule with the highest average fitness 394.34 is represented by a 31-bit string 1000000011010011101110000001101 with $\lambda = 13/32$. Figure 4 shows a typical example of the space-time pattern (left) and the power spectrum (right) of the rule. The space-time pattern consists of 200 cells for 200 time steps. There are several kinds of periodic structures with period three and six in periodic background with period three. The power spectrum is calculated from the evolution of 700 cells for $T = 3000$ time steps. The two peaks at $f = 500$ and $f = 1000$ are caused by periodic structures with period six and three respectively and the peak at $f = 1500$ is harmonics. The broken line represents the least square fitting of the power spectrum from $f = 1$ to $f = 10$ by $\ln S(f) = \alpha + \beta \ln f$, $\beta = -2.0$.

The most controversial problem in $1/f$ noise is whether $1/f$ noise lasts forever or not. Generally speaking, if there is finite correlation time τ in a fluctuation, the power spectrum with the observation length $T > \tau$ has almost equal power density at frequencies smaller than $1/\tau$. Likewise if there is finite correlation time

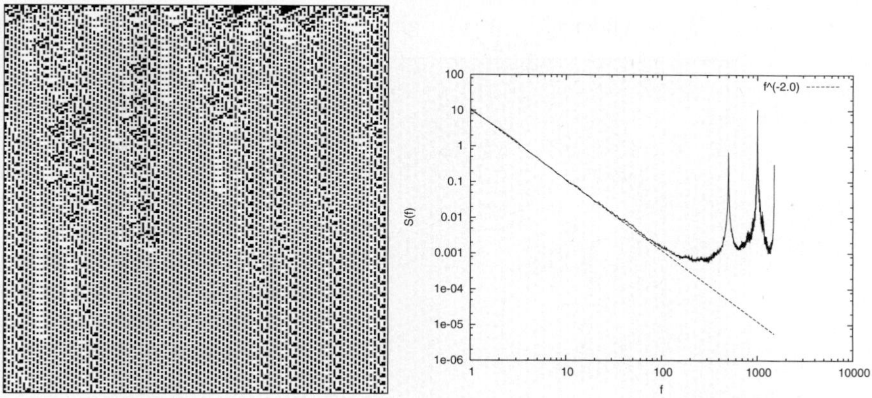

Fig. 4. Typical example of the space-time pattern (left) and the power spectrum (right) of the rule with the highest average fitness. The space-time pattern consists of 200 cells for 200 time steps. The power spectrum is calculated from the evolution of 700 cells for 3000 time steps. The broken line represents the least square fitting of the power spectrum from $f = 1$ to $f = 10$ by $\ln S(f) = \alpha + \beta \ln f$, $\beta = -2.0$.

τ in the evolution of a CA, the power spectrum with the observation length $T > \tau$ has almost equal power density at frequencies smaller than T/τ. So the power spectrum turns close to flat line at low frequencies as the observation length T becomes longer than the correlation time τ. It might be thought that the exponent $\beta = -2.0$ of the power spectrum in Fig. 4 implies that the evolution of the rule is similar to Brownian motion rather than $1/f$ noise. But the exponents averaged over 30 random initial configurations of 700 cells are -1.89, -1.82, and -1.75 for $T = 6000$, $T = 9000$, and $T = 12000$ respectively. This result agrees with the argument above.

4 Discussion

We performed the experiment using GA to find one-dimensional and two-state, five-neighbor CAs with $1/f$-type power spectrum. On the rule with the highest fitness found in this experiment we can observe one kind of glider propagating left in periodic background and three kinds of periodic structures (A, B, C) as shown in Fig. 5. The periodic structures A, B, and C have period three, three, and six respectively. The collision between a B and two consecutive gliders in Fig. 5 leads to the formation of an A. The collision between an A and a glider forms a B which vanishes at the collision with a glider propagating from the right. The collision between a C and a glider shifts the position of the C two cells to the right. On the whole, the structures and their interaction observed in the evolution does not seem to be as diverse as those in ECA rule 110. Nevertheless, the findings of the rule which has gliders and periodic structures and exhibits the

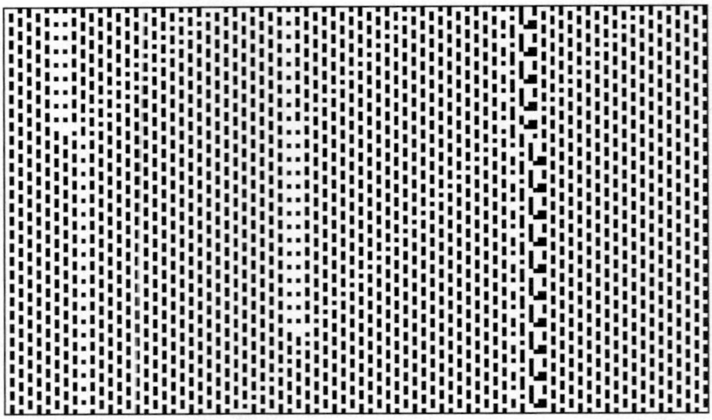

Fig. 5. Collisions between a glider and a periodic structure on the rule with the highest fitness.

interaction of these patterns suggests the effectiveness of the proposed method based on $1/f$ noise, although computational universality of the rule has not been proved yet. We need to perform still more GA runs to search for $1/f$-type rules, especially for longer observation length T.

The most significant property common to ECA rule 110 and LIFE which possess both the exhibition of $1/f$ noise and the capability of supporting universal computation is the existence of gliders. The gliders play a crucial role to support computational universality because they are necessary for the transmission of information. Furthermore, the gliders seem to cause $1/f$ noise by the phenomenon of intermittency. Intermittency in chaotic dynamical systems is one of the main mechanisms of $1/f$ noise [23]. A chaotic dynamical system with a particular parameter value exhibits periodic behavior which is disrupted occasionally and irregularly by a "burst". This burst persists for a finite duration, it stops and a new periodic behavior starts. Intermittent chaos emerges when the transition from periodic to chaotic behavior takes place as the parameter is varied. The evolution of rule 110 leads to periodic phase where gliders are monotonously shifting in periodic background. But there sometimes occurs a collision between the gliders which corresponds to a burst. Figure 6 shows the space-time patterns of rule 110 from a random initial configuration of 200 cells. The space-time pattern from 0 to 199 time steps shown on the left exhibits transient behavior, while that from 5400 to 5599 time steps on the right shows a collision between a periodic structure and a glider which corresponds to a burst. The same observation applies to the evolution of LIFE. The structures which remain after transient behavior in LIFE are almost always periodic ones and gliders. That is periodic phase where almost all cells evolve periodically and chaotic phase occurs as the glider collides with another structure. This scenario for the evolution of rule 110 and LIFE might cause intermittency, and thereby lead to $1/f$ noise.

The hypothesis of "the edge of chaos" has evoked considerable controversy [24]. This hypothesis says the ability to perform universal computation in a

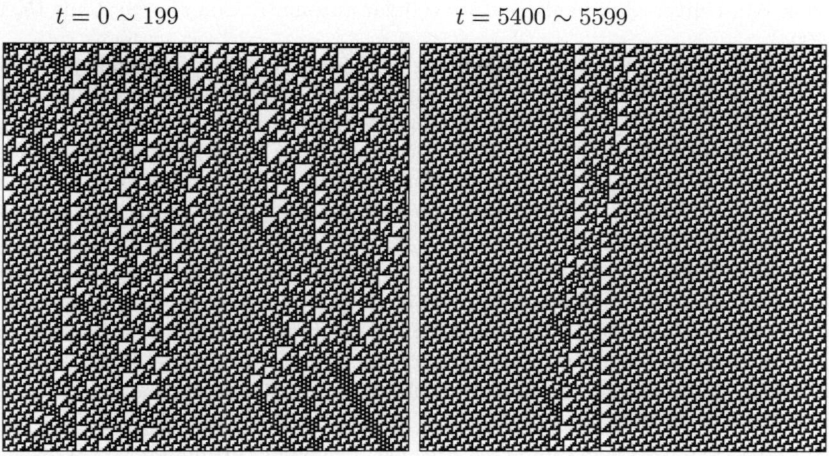

Fig. 6. Space-time patterns of elementary cellular automaton rule 110 starting from a random initial configuration of 200 cells. The evolution from 0 to 199 time steps is shown on the left and from 5400 to 5599 time steps on the right where a collision between a periodic structure and a glider is observed.

system arises near a transition from regular to chaotic behavior such as Class IV. So far various statistical quantities, such as entropy and difference pattern spreading rate, have been proposed to detect Class IV quantitatively [25]. The exponent of power spectrum at low frequencies might be another useful index.

References

1. Crutchfield, J., Farmer, D., Packard, N., Shaw, R., Jones, G., Donnelly, R.J.: Power spectral snalysis of a synamical system. Phys. Lett. 76A, 1–4 (1980)
2. von Neumann, J.: Theory of Self-reproducing Automata. University of Illinois Press, Illinois (1966)
3. Codd, E.F.: Cellular Automata. Academic Press, New York (1968)
4. Banks, E.R.: Information processing and transmission in cellular automata, thesis, Massachusetts Institute of Technology (1971)
5. Serizawa, T.: Universal cellular automata with minimal states and neighbors (1), Paper of Technical Group AL77-26, 1–10 (in Japanese) (1977)
6. Serizawa, T.: Universal cellular automata with minimal states and neighbors (2), Paper of Technical Group AL77-27, 11–18 (in Japanese) (1977)
7. Smith III., A.R.: Simple computation-universal cellular spaces. J. ACM 18, 339–353 (1971)
8. Albert, J., Culik II., K.: A simple universal cellular automaton and its one-way and totalistic version. Complex Systems 1, 1–16 (1987)
9. Lindgren, K., Nordahl, M.G.: Universal computation in simple one-dimensional cellular automata. Complex Systems 4, 299–318 (1990)

10. Cook, M.: Universality in elementary cellular automata. Complex Systems 15, 1–40 (2004)
11. Ninagawa, S.: 1/f Noise in elementary cellular automaton rule 110. In: Calude, C.S., Dinneen, M.J., Păun, G., Rozenberg, G., Stepney, S. (eds.) UC 2006. LNCS, vol. 4135, pp. 207–216. Springer, Heidelberg (2006)
12. Berlekamp, E.R., Conway, J.H., Guy, R.K.: Winning Ways for Your Mathematical Plays, vol. 2. Academic Press, New York (1982)
13. Ninagawa, S., Yoneda, M., Hirose, S.: $1/f$ fluctuation in the "Game of Life. Physica D 118, 49–52 (1988)
14. Mitchell, M., Hraber, P.T., Crutchfield, J.P.: Revisiting the edge of chaos: evolving cellular automata to perform computations. Complex Systems 7, 89–130 (1993)
15. Mitchell, M., Crutchfield, J.P., Hraber, P.T.: Evolving cellular automata to perform computations: mechanisms and impediments. Physica D 75, 361–391 (1994)
16. Wolfram, S.: Statistical mechanics of cellular automata. Rev. Mod. Phys. 55, 601–644 (1983)
17. Wolfram, S. (ed.): Theory and Applications of Cellular Automata. World Scientific, Singapore (1986)
18. Li, W., Packard, N.: The structure of the elementary cellular automata rule space. Complex Systems 4, 281–297 (1990)
19. Wolfram, S.: A New Kind of Science. Wolfram Media, Champaign (2002)
20. Keshner, M.S.: $1/f$ noise. Proc. IEEE 70, 211–218 (1982)
21. Langton, C.G.: Studying artificial life with cellular automata. Physica D 22, 120–149 (1986)
22. Press, W.H., Teukolsky, S.A., Vetterling, W.T., Flannery, B.P.: Numerical recipes in C, 2nd edn. Cambridge University Press, Cambridge (1992)
23. Pomeau, Y., Manneville, P.: Intermittent transition to turbulence in dissipative dynamical systems. Commun. Math. Phys. 74, 189–197 (1980)
24. Langton, C.G.: Computation at the edge of chaos: phase transitions and emergent computation. Physica D 42, 12–37 (1990)
25. Li, W., Packard, N.H., Langton, C.G.: Transition phenomena in cellular automata rule space. Physica D 45, 77–94 (1990)

Genotype Reuse More Important than Genotype Size in Evolvability of Embodied Neural Networks

Chad W. Seys[1] and Randall D. Beer[2]

[1] Dept of Electrical Engineering and Computer Science, Case Western Reserve University, Cleveland OH, 44106 cwseys@cwru.edu
[2] Cognitive Science Program, Indiana University, Bloomington IN, 47406 rdbeer@indiana.edu

Abstract. The relative influence of genotype reuse and number of genotype parameters on the evolvability of an embodied neural network is explored. Two genotype to phenotype mappings are used to encode a neural network controlling a hexapod agent. A symmetric encoding reuses the genotype by duplicating parts of the genotype to create the phenotype. A direct encoding maps one genotype parameter to one phenotype parameter. To test whether genotype reuse is more important than genotype size, the architecture of the neural network is manipulated such that the genotype size of the symmetrically-encoded neural networks is larger than the directly-encoded neural networks. The symmetrically-encoded neural networks are found to be more evolvable than the directly-encoded despite having a larger genotype.

Key words: evolvability, adaptation rate, genotype reuse, genotype size, genotype to phenotype mapping, representation, encoding, modularity

1 Introduction

Evolutionary algorithms (EAs) such as genetic algorithms (GAs) are widely used as a means to solve problems from many areas in ways unexpected by their programmers. However, as the number of parameters describing the solutions to which EAs are applied grow larger, the size of the parameter space grows exponentially [1], and this leads to a scalability problem. This is especially a problem for direct encodings in which one phenotype parameter maps to one genotype parameter.

Non-direct encodings in which one genotype parameter maps to multiple phenotype parameters have been invented to facilitate evolving complex solutions. A partial list of comparisons between a variety of non-direct and direct encodings includes [2][3][4][5]. These non-direct encodings have been evaluated for their relative ability to find high fitness solutions on a variety of complex tasks and often succeed in evolving higher fitness solutions in fewer generations than the direct encodings.

F. Almeida e Costa et al. (Eds.): ECAL 2007, LNAI 4648, pp. 915–924, 2007.

A common theme in the experiments above is that a non-direct encoding increases the degree of genotype reuse, decreases the genotype size, and improves the performance of the search over the direct encoding. Genotype reuse often leads to phenotypic structural regularity and this may be useful for exploiting the regularities of certain problems. Additionally, when parts of the genotype are reused, a given phenotype can be described in a smaller genotype. A smaller genotype size means that the space of possible solutions is smaller and therefore faster to search.

It is difficult to know whether genotype reuse, smaller genotype, or both deserves credit for the improved performance of non-direct encodings. Often both are credited [2][3][4]. The question of the extent to which the size of genotype space and genotype reuse contribute to the success of encodings is widely applicable yet remains unexplored.

This paper explores the relative effect of genotype reuse and number of genotype parameters on the evolvability[3] of a simple system. The system consists of a neural network controlling a simulated hexapod agent in a simple walking task. The degree of genotype reuse of the neural network was changed by using either a direct or symmetric encoding. In the direct encoding every phenotype parameter had a corresponding genotype parameter, so there was no genotype reuse. The symmetric encoding had a genotype which was $\frac{1}{6}^{th}$ the size of the phenotype and a six-fold symmetry matching the six legs of the hexapod. The number of genotype parameters was adjusted by using two neural network architectures and changing the number of interneurons. By switching between neural network architectures and changing the number of interneurons, it was possible to create a symmetrically-encoded neural network with more genomic parameters than a directly-encoded neural network.

The evolvability of these encoding/architecture combinations are examined from several different perspectives:

discoverability is the ability of the GA to find high fitness individuals beginning with a random initial population.

degree of adaptability is the amount of environmental change that a high fitness population can be subjected while maintaining a high fitness during a GA search.

adaptation rate is the degree of adaptability at a given generation. Adaptation rate could be especially important in situations where a competition for resources exists.

competition is a situation when different encoding/architecture combinations are made to compete for population share.

This paper will show that genotype reuse is more important than genotype size in the relative evolvability of encodings. Symmetric encodings are more

[3] "Evolvability" is a term that has many meanings. Two of many definitions are 1) ability of random variations to sometimes produce improvement [6] and 2) ability to respond to a selective challenge [7]. Both of these definitions apply to the experiments described in this paper. Reisinger [8] also discusses evolvability and its definition(s).

evolvable than direct encodings, even if the symmetric encoding's genotype is larger than the direct encoding's genotype. In conditions of changing environment, the symmetric encodings win competitions for population share against direct encodings, even if the symmetric encodings have a lower fitness at the beginning of the competition.

2 Methods

The body (Fig. 1) is a hexapod agent based on prior work by Beer & Gallagher [9] and similar to [10]. The six legs move in response to the output of a continuous-time recurrent neural network (CTRNN). A real-valued genetic algorithm (GA) is used to evolve the CTRNN parameters. Fitness is assigned based upon the average forward velocity of the agent. The possible velocities are [-1, 1] and a "high fitness" individual has a fitness ≥ 0.75 and walks with a tripod gait. For details of the body model, neural network, and GA, please refer to [11].

Two types of neural network architecture, called "fully-connected" and "locally-connected" are used to control the six legs of the body. In the fully-connected architecture, every neuron is connected to every other neuron. In the locally-connected architecture, six fully connected subnetworks are sparsely connected to form the whole network. Each subnetwork is associated with one of the legs and is composed of motor neurons for that leg and $\frac{1}{6}^{th}$ the interneurons for the entire neural network. From each motor and interneuron in a subnet, sparse connections emanate to connect it only to its analogous neighbor in the subnet immediately to the left and right around the perimeter of the body. Figure 1 illustrates one of these subnetworks.

In both types of architectures there are always three neurons per leg which serve as motor neurons and control the effectors of the leg. Additional interneurons with no direct connections to the effectors can be added to the network in multiples of six.

Two encoding schemes, direct and symmetric, are used to store the neural network parameters in the genome. The direct encoding scheme is a one-to-one mapping of genotype parameters to phenotype parameters. I.e. every parameter of every neuron is present in the genotype. In the symmetric encoding scheme, the parameters for only one subnetwork are stored in the genotype. When the neural network is created, the parameters for the subnetwork are duplicated six times (once for each leg) to create the whole neural network.

The evolvability experiments were conducted in three stages outlined here and explained in more detail in [11]. A high fitness seed individual was discovered through GA search beginning with a random initial population. The seed individual was then used to create an "equilibrium GA search". The degree of adaptability was determined by continuing the equilibrium GA search while the population's legs were incrementally shortened for 5000 generations. Whenever the population recovered its initial fitness, the leg was shortened again.

Competitions were performed by placing the most fit halves of two equilibrium GA searches of different encodings in the same population, then continuing

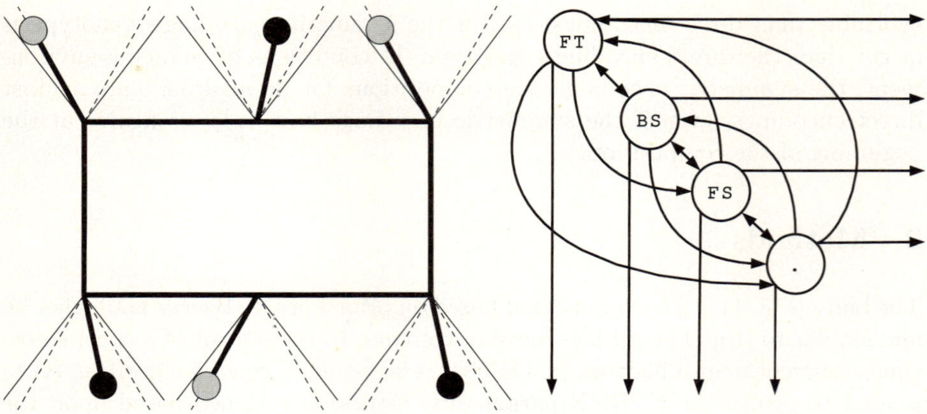

Fig. 1. LEFT: Hexapod agent modeling a tripod gait. Feet on the ground are black and feet off the ground are gray. RIGHT: Example locally-connected architecture subnetwork: The neurons are represented as circles and the connections between neurons as arrows. Neurons within a subnetwork are fully-connected. Outward pointing arrows are connections to analogous neurons in neighboring subnetworks. Motor neurons are named FooT (FT), BackSwing (BS), ForwardSwing (FS), and an interneuron (.). The number of interneurons varied in these experiments from 0 to 3.

the GA search while incrementally shortening the leg as in the evolvability experiments. Competitions ended when one encoding filled the entire population

The significance of the differences of the various measurements was determined by comparing the confidence intervals of the measure created by bootstrapping. If the 95% confidence intervals do not overlap, then there is a 5% or less chance that the measures are the same, i.e. a $p < 0.05$. [12][13]

3 Results

3.1 Symmetric Encoding Is More Discoverable and Has Greater Degree of Adaptability than Direct Encoding

In a previous paper [11] the question asked was whether the size of genotype parameter space was more or less important to "discoverability" and "degree of adaptability" than the encoding used. In order to change the encoding while maintaining approximately the same parameter space size it was necessary to also change the architecture and number of neurons in the neural network. By choosing a particular combination of architecture and number of neurons, it was possible to create symmetrically-encoded neural networks which had nearly the same, but slightly more genomic parameters than directly-encoded neural networks. The genotype of an symmetrically-encoded, fully-connected, neural network with 5 neurons (abbreviated SF5) has 160 parameters, slightly more than the 126 parameters a directly-encoded, locally-connected neural network

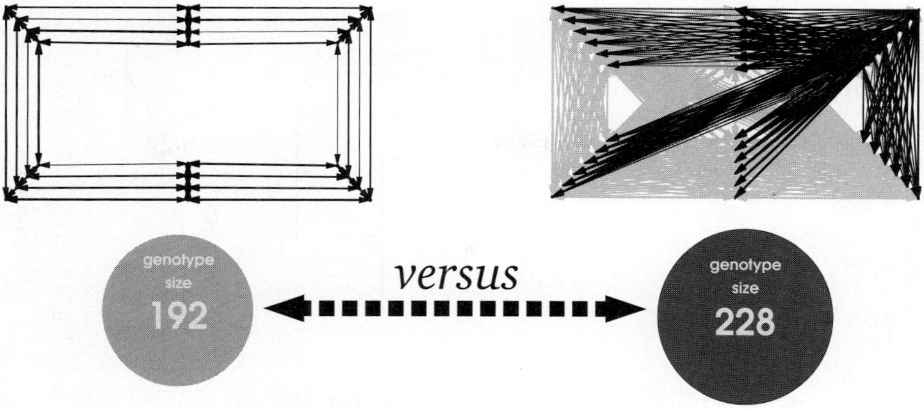

Fig. 2. An illustration summarizing the experimental design: On the left, a directly-encoded neural network with fewer genotype parameters than the symmetrically-encoded neural network on the right. The "versus" arrow symbolizes the evolvability comparisons performed. If the symmetrically-encoded neural network on the right is more evolvable, then structural regularity is more important than genotype size. (The neural network connections shown in black are stored in the genotype, while those in gray are duplicates of the black.)

with 3 neurons (abbreviated DL3). Similarly, SF6 contains 228 parameters, which creates a larger genomic parameter space than DL4 (192). Figure 2 is a graphical representation of DL4 versus SF6. If parameter space size alone were the sole determinant of evolvability and a larger parameter space meant less evolvability, then one would arrange these neural networks, in order of decreasing evolvability: DL3 (126), SF5 (160), DL4 (192), SF6 (228).

It was found that "discoverability" mostly depends on the encoding type rather than parameter space size. Discoverability is the fraction of searches finding high fitness individuals from a random initial population. High fitness individuals were discovered for all symmetrically-encoded neural networks regardless of parameter space size or neural network architecture. No high fitness individuals were discovered for the directly-encoded neural networks (after at least 1000 attempts) even though they were of the same neural network architecture or smaller parameter space size than the symmetrically-encoded neural networks. Figure 3 shows these results as the fraction of searches performed which discovered a high fitness neural network. The symmetrically-encoded fractions are marked with a dark gray dot, while the directly-encoded fractions are marked with a light gray dot.

Degree of adaptability mostly depends on the encoding type rather than parameter space size. In these experiments, degree of adaptability is the final leg length after 5000 generations of incremental shortening. These results are shown for DL3, SF5, DL4, and SF6 in Fig. 3 where the x-axis is the number of genotype parameters and the y-axis is the change in leg length after 5000 generations. It

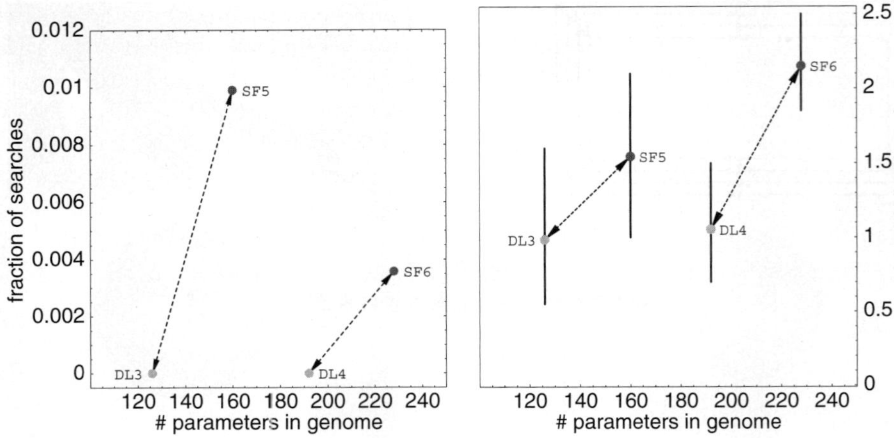

Fig. 3. LEFT: Discoverability of the various encoding/architecture types. Fraction of searches beginning with random initial population resulting in discovery of high fitness neural network plotted against size of genotype. A small fraction of searches discovered high fitness individuals for all symmetrically-encoded neural network architectures. Zero high fitness directly-encoded neural networks were found after 1000 attempts. RIGHT: Median change in leg length after 5000 generations (dots) and 0.95 confidence intervals (solid vertical lines). Symmetrically-encoded neural network were able to adapt to a greater leg length change than directly-encoded neural networks.

was found that the symmetrically-encoded neural networks are more adaptable even when the symmetric encoding has more genotype parameters than the direct encoding. SF6 (228 genotype parameters) is significantly more adaptable than DL4 (192) and the median adaptability for SF5 (160) is higher than DL3 (126) median adaptability, though not significantly.

Symmetrically-encoded/fully-connected neural networks are more discoverable and have greater degree of adaptability than directly-encoded/locally-connected neural networks. Experiments controlling for architecture are discussed in [11] and show that encoding rather than the architecture leads to increased evolvability.

3.2 Symmetric Encoding Has Higher Adaptation Rate

There are two aspects of adaptability: degree and rate. Adaptation rate may be more important than degree when, for example, two subpopulations compete for the same resources in a changing environment. If one subpopulation is less adaptable in degree, but has a higher adaptation rate, it may outcompete a subpopulation with greater degree of adaptability but lower adaptation rate.

Overall, symmetrically-encoded neural networks adapt more quickly than the directly-encoded neural networks. The adaptation rate during the adaptation experiments can be displayed by graphing median leg length change per generation. All 5000 generations of the adaptation experiments are shown in Fig. 4(RIGHT). Also shown in Fig. 4(LEFT) are the first 300 generations which may be more important in a competition between encodings. SF6 appears to be more adaptable than DL4 in the first 25 generations, but SF5 and DL3 appear to have approximately the same median adaptability until generation ∼225.

These observations suggest that if the two encoding types were competing for a slot in a limited population under conditions of shortening leg length, the subpopulations with symmetrically-encoded neural networks should adapt more quickly and outcompete the subpopulations with directly-encoded neural networks.

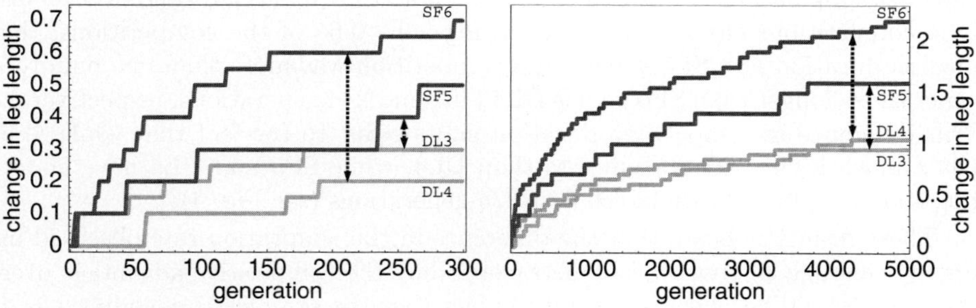

Fig. 4. Median change in leg length by generation during adaptation experiments performed using the indicated encoding/architecture type. SF5 and DL3 adapt at approximately the same rate for the first 225 generations, but later SF5 appears to adapt to a greater degree, though the difference at 5000 generations is not significant (Fig. 3(RIGHT)).

3.3 Symmetric Encodings Outcompete Direct Encodings in a Changing Environment

Would the difference in adaptation rates seen in Fig. 4 be enough to allow the symmetric encoding to outcompete the direct encoding? To test this, competitions between the symmetrically- and directly-encoded neural networks were set up as described in Sect. 2. Once again, the comparisons of interest were between a directly-encoded neural network and a symmetrically-encoded neural network of slightly larger genotype. Specifically, all pairs of equilibrium populations of DL3 (126) and SF5 (160) competed against each other, as did all equilibrium population pairs of DL4 (192) and SF6 (228). As a control, the competitions were also run without changing the leg length. If, after a number of generations, individuals of one encoding type replaced every individual of the other encoding type, then that remaining encoding type "won" the competition.

It was found that encoding type gives no advantage if adaptation is not required. In control experiments in which the GA search was performed without changing the leg length, the encoding type with the most fit initial subpopulation won 0.97 of the competitions (Fig. 5 far left bar pair). More fit individuals outcompeting less fit individuals is what one would expect from the normal functioning GA search.

Neural network encoding does make a difference when adaptation to leg length shortening is required to maintain high fitness. The symmetric encodings win a significant majority of competitions (0.78 ± 0.02) (Fig. 5 middle bar pair), including competitions in which the symmetrically-encoded subpopulation is initially less fit than the directly-encoded subpopulation (0.65 ± 0.04, Fig. 5 right bar pair). This is especially noteworthy because the symmetrically-encoded neural networks are winning the competitions despite starting with an initially less fit subpopulation. Additionally, while in both SF5 versus DL3 and SF6 versus DL4 competitions the symmetric encoding wins 0.65 of the competitions, the median duration of a SF5 versus DL3 competition was more than one hundred generations longer than SF6 versus DL4 (402 and 274 generations, respectively). This difference in competition duration may be due to the fact that while SF6 has a much higher adaptation rate than DL4, while SF5 and DL3 have about the same adaptation rate in the first 225 generations (see Fig. 4).

These results indicate that the difference in the adaptation rate observed in Fig. 4 is enough to give the symmetric encoding the competitive advantage over the direct encoding. This is despite the fact that the symmetric encoding has a larger parameter space than the the direct encoding.

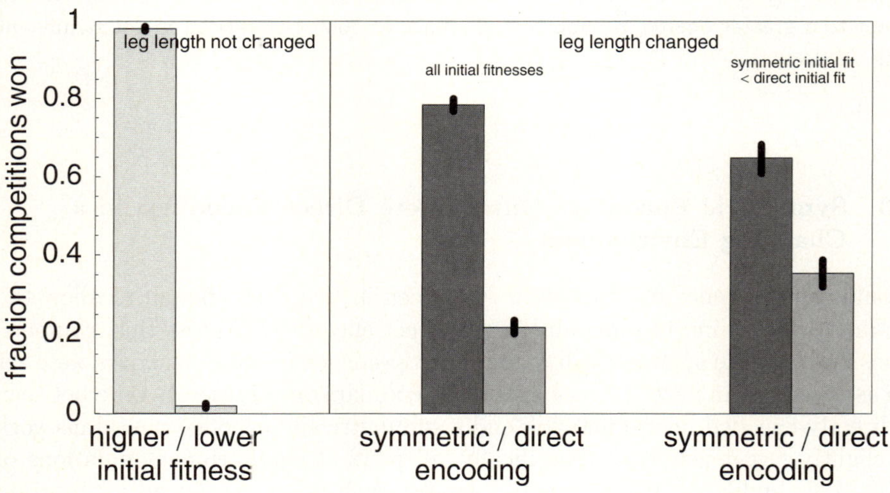

Fig. 5. When the leg length is not changed, the initially higher fitness subpopulation wins most of the competitions (left pane). When leg length is changed during the competition, the symmetrically-encoded subpopulation wins the majority of competitions (right pane, first bar pair), even when initially less fit than the directly-encoded subpopulation (right pane, second bar pair).

4 Discussion

It was shown that the number of genotype parameters is less important to evolvability of these neural networks than genotype reuse. Symmetrically-encoded neural networks containing more genotype parameters were more discoverable and adaptable than directly-encoded neural networks.

There are many possibilities for the increased evolvability of symmetrically-encoded neural networks. One possibility is that though the genotype space of the symmetrical encoding is larger, the density of high fitness solutions is also higher. This would affect discoverability by allowing the initial search find a high fitness region more quickly. The symmetric encoding also coordinates changes of neural network parameters relative to the direct encoding. This would have the effect of making regions of parameter space unreachable, but if these regions are lower fitness, not being able to reach them would increase adaptability.

The increased evolvability is not due to extra-dimensional bypass, but bias or focusing of the search. In a previous paper [11] the evolvability of directly- and symmetrically-encoded versions of the same neural network architecture were compared. It was found that the symmetrically-encoded version was more evolvable despite having fewer parameters.

A changing environment is necessary to reveal an encoding's greater adaptation rate in a competition for population share. If the environment did not change, the winner of the competition was almost always the subpopulation which had the initially higher fitness. However, when the environment changed, the more adaptable encoding usually outcompeted the less adaptable encoding even when starting with a relatively less fit subpopulation.

In these experiments there was a positive correlation between increased discoverability, adaptation rate, and adaptation degree. It is possible that these three aspects of evolvability may not always be positively correlated. For example, in these experiments a wide range of mutation variances were used and the greatest adaptation was reported. Considering that natural evolution has limited control over (the lower limit of) mutation rate and often not all mutation variances can be tried in EAs due to time constraints, this is not always realistic.

One could expect interesting consequences if discoverability, adaptation rate, and degree of adaptability of an encoding are not always correlated in scenarios where encodings themselves evolve and compete [14][15][16]. For example, an encoding with a higher adaptation rate but lower degree of adaptability might outcompete an encoding with lower adaptation rate and higher degree of adaptability. When evolving complex objects, the initial evolution may hinder later evolution. Also, in natural history, the evolution of body plans [17] may have been a time of competing encodings: what effects have the relative evolvability of natural encodings had on the evolution of natural systems? By studying evolvability we improve our ability to evolve artificial systems and understand the evolution of natural systems.

Acknowledgments. Thank you reviewers for your suggestions. Supported in part by NSF grant EIA-0130773 and by a NSF IGERT fellowship.

References

1. Yao, X.: Evolving artificial neural networks. Proceedings of the IEEE 87, 1423–1447 (1999)
2. Gruau, F.: Automatic definition of modular neural networks. Adaptive Behavior 3(2), 151–183 (1995)
3. Bentley, P., Kumar, S.: Three ways to grow designs: A comparison of embryogenies for an evolutionary design problem. In: Proceedings of Genetic and Evolutionary Computation Conference (GECCO) (1999)
4. Komosiński, M., Rotaru-Varga, A.: Comparison of different genotype encodings for simulated three-dimensional agents. Artificial Life 7(4), 395–418 (2001)
5. Hornby, G.S.: Measuring, enabling and comparing modularity, regularity and hierarchy in evolutionary design. In: Proceedings of Genetic and Evolutionary Computation Conference (GECCO) (2005)
6. Wagner, G.P., Altenberg, L.: Perspective: Complex adaptations and the evolution of evolvability. Evolution 50(3), 967–976 (1996)
7. Hansen, T.F.: Is modularity necessary for evolvability? Remarks on the relationship between pleiotropy and evolvability. BioSystems 69, 83–94 (2002)
8. Reisinger, J., Stanley, K.O., Miikkulainen, R.: Towards an empirical measure of evolvability. In: Proceedings of Genetic and Evolutionary Computation Conference (GECCO), pp. 254–264. Springer, Heidelberg (2005)
9. Beer, R.D., Gallagher, J.C.: Evolving dynamical neural networks for adaptive behavior. Adaptive Behavior 1(1), 92–122 (1992)
10. Kodjabachian, J., Meyer, J.-A.: Evolution and development of modular control architectures for 1-d locomotion in six-legged animats. Connection Science 10(3-4), 211–237 (1998)
11. Seys, C.W., Beer, R.D.: Effect of encoding on the evolvability of an embodied neural network. In: GECCO Workshop Proceedings; Complexity through Development and Self-Organizing Representations (CODESOAR) (2006)
12. Efron, B., Tibshirani, R.J.: An introduction to the bootstrap. CRC Press, Boca Raton (1994)
13. Varian, H.: Bootstrap tutorial. The Mathematica Journal 9(4), 768–775 (2005)
14. Toussaint, M.: Demonstrating the evolution of complex genetic representations: An evolution of artificial plants. In: Cantú-Paz, E., Foster, J.A., Deb, K., Davis, L., Roy, R., O'Reilly, U.-M., Beyer, H.-G., Kendall, G., Wilson, S.W., Harman, M., Wegener, J., Dasgupta, D., Potter, M.A., Schultz, A., Dowsland, K.A., Jonoska, N., Miller, J., Standish, R.K. (eds.) GECCO 2003. LNCS, vol. 2723, pp. 86–97. Springer, Heidelberg (2003)
15. Miller, J.F.: Evolving a self-repairing, self-regulating, french flag organism. In: Deb, K., et al. (eds.) GECCO 2004. LNCS, vol. 3102, pp. 129–139. Springer, Heidelberg (2004)
16. Gordon, T.G.W., Bentley, P.J.: Bias and scalability in evolutionary development. In: Proceedings of Genetic and Evolutionary Computation Conference (GECCO), Washington DC, USA, pp. 83–90. ACM Press, New York (2005), doi:10.1145/1068009.1068021
17. Raff, R.A.: The Shape of Life: Genes, Development, and the Evolution of Animal Form. University of Chicago Press, Chicago (1996)

Information-Cloning of Scale-Free Networks*

Mahendra Piraveenan[1,2], Mikhail Prokopenko[1], and Albert Y. Zomaya[2]

[1] CSIRO Information and Communications Technology Centre
Locked Bag 17, North Ryde, NSW 1670, Australia,
[2] School of Information Technologies, The University of Sydney, NSW 2006, Australia
Corresponding author: mikhail.prokopenko@csiro.au

Abstract. In this paper, we introduce a method, Assortative Preferential Attachment, to grow a scale-free network with a given assortativeness value. Utilizing this method, we investigate information-cloning — recovery of scale-free networks in terms of their information transfer — and identify a number of recovery features: a full-recovery threshold, a phase transition for both assortative and disassortative networks, and a bell-shaped complexity curve for non-assortative networks. These features are interpreted with respect to two opposing tendencies dominating network recovery: an increasing amount of choice in adding assortative/disassortative connections, and an increasing divergence between the joint remaining-degree distributions of existing and required networks.

1 Introduction

Many biological networks, e.g. gene-regulatory networks, metabolic networks, and protein interaction networks are often characterized as complex scale-free networks. In this paper, we consider a task of *information-cloning* of a scale-free network, given its fragment and some topological properties of the original network. The "cloning" is interpreted information-theoretically: the resulting network may disagree with the original one in terms of specific node to node connections, but is required to have equivalent information transfer. The information-cloning task is partly motivated by needs of network manufacturing, where an "assembly-line" starts with a fragment and continues with "manufacturing" the rest, subject to topological constraints. Another motivation is regeneration of scale-free networks which are prone to percolation/diffusion of adverse conditions, as well as removal of highly connected nodes. Both demands (topology-oriented manufacturing and regeneration) are referred in this paper as *network recovery*.

Recovery of networks can be attempted and evaluated in various ways. In this paper, we aim at a general measure in terms of mutual information contained in the network, or its information transfer. More precisely, we propose to judge success of the recovery with respect to the amount of information transfer regained by a resulting network.

Various network growth models have been analyzed in literature [1,2,3,4]. One prominent model is the preferential attachment model, which explains power law degree distributions observed in scale-free networks [4]. In this model, the probability of a new node making a link to an existing node in the network is proportional to the degree of the

* The Authors list is in alphabetical order.

F. Almeida e Costa et al. (Eds.): ECAL 2007, LNAI 4648, pp. 925–935, 2007.

target node. Newman [2] pointed out that this model does not take into account the degree of the source node in influencing the attachment probability, and suggested to consider another tendency for preferential association, measured via *assortativeness*. The networks where highly connected nodes are more likely to make links with other highly connected links are said to mix assortatively, while the networks where the highly connected nodes are more likely to make links with more isolated, less connected, nodes are said to to mix disassortatively. In both cases, the likelihood of creating a link depends on the degrees of both nodes. Both assortative and disassortative mixing is contrasted with non-assortative mixing, where no preferential connection can be established.

The extent of assortativeness affects network's resilience under node removal or percolation/diffusion of adverse conditions [2]. Our objective is an investigation of how successful is a *network recovery* in terms of assortativeness and information transfer. We note that this objective is different from investigation of networks' robustness properties such as *error tolerance*, *attack survivability*, or *network fragmentation* that have been extensively studied [5,6,7]. For example, Moreno et al. [7] explored robustness of large scale-free networks faced with node-breaking avalanches (cascading failures when a failure of a node triggers subsequent failures of neighbours), and investigated how the random removal of nodes in a fixed proportion affects the global connectivity and functionality of scale-free networks. Stauffer and Sahimi studied scale-free networks with annealed disorder [8], when the links between various nodes may temporarily be lost and reestablished again later on, and observed a number of critical phenomena, e.g. "the existence of a phase diagram that separates the region in which diffusion is possible from one in which diffusion is impossible". This study did not investigate, however, the role of assortativeness and information transfer in the diffusion process.

Naturally occurring networks display various extents of assortative mixing, and it is often possible to measure or calculate the level of assortativeness in these networks [1]. However, it is not straightforward to (re-)grow a network with a level of assortative mixing specified *a priori*. We address this problem and propose a method to grow or recover a scale-free network with a given assortativeness. We also show that a network with perfect assortativeness can be grown for any desired degree distribution, whereas a network with perfect disassortativeness can be grown only if the corresponding 'remaining degree distribution' for the desired degree distribution is symmetric.

Utilizing this method, we investigate recovery of scale-free networks in terms of their information transfer. Following Solé and Valverde [1], we define the information transfer as mutual information contained in the network, or the amount of general correlation between nodes. Importantly, the maximum attainable information transfer defines the network's capacity, in analogy with information-theoretic notion of channel capacity — the maximum mutual information for the channel over all possible distributions of a transmitted signal. In general, information transfer is a vital indicator of complex non-linear behavior in self-organizing systems, and can be associated with *predictive information*, richness of structure (i.e. *excess entropy*), and *physical complexity* [9].

2 Assortativeness and Information Transfer

We study assortativeness in scale-free networks described by power law degree distributions, formally specified as $P(k) = Ak^{-\gamma}u(k/N_p)$ where u is a step function specifying a cut off at $k = N_p$.

The degree of a node is the number of other nodes to which it is connected to. Let us consider a network with N nodes (vertices) and M links (edges), and say that the probability of a randomly chosen node having degree k is p_k, where $1 \le k \le N_p$. The distribution of such probabilities is called the *degree distribution* of the network. However, if a node is reached by following a randomly chosen link, then the remaining number of links (the remaining degree) of this node is not distributed according to p_k. Instead it is biased in favour of nodes of high degree, since more links end at a high-degree node than at a low-degree one [2]. The distribution of such remaining degrees is called the *remaining degree distribution*, and is related to p_k as follows:

$$q_k = \frac{(k+1)p_{k+1}}{\sum_j^{N_p} j p_j} \ , \ \ 0 \le k \le N_p - 1 \tag{1}$$

where p_k is the degree distribution of the network, and q_k is the remaining degree distribution of the network [2]. For scale-free networks, Eq. (1) yields that if $\gamma = 1$ (that is, $p(k) = A/k$ before the cut off), the resulting remaining degree distribution is uniform, therefore symmetric.

Following Callaway et al. [3] and Newman [2], we define the quantity $e_{j,k}$ to be the joint probability distribution of the remaining degrees of the two nodes at either end of a randomly chosen link. As pointed out by Newman [2], this quantity is symmetric in its indices for an undirected graph, that is $e_{j,k} = e_{k,j}$, and it obeys the sum rules $\sum\limits_{jk} e_{j,k} = 1$ and $\sum\limits_{j} e_{j,k} = q_k$. Assortativeness is formally defined [4] as a correlation function which is zero for non-assortative mixing and positive or negative for assortative or disassortative mixing respectively:

$$r = \frac{1}{\sigma_q^2} \sum_{jk} jk(e_{j,k} - q_j q_k) \tag{2}$$

where σ_q^2 is the variance of the probability distribution q_k. Here r lies between -1 and 1, whereby $r = 1$ means perfect assortativeness, $r = -1$ means perfect disassortativeness, and $r = 0$ means no assortativeness (random linking).

If a network has perfect assortativeness ($r = 1$), then all nodes connect only with nodes with the same degree. For example, the joint distribution $e_{j,k} = q_k \delta_{j,k}$ where $\delta_{j,k}$ is the Kronecker delta function, produces a perfectly assortative network. If the network has no assortativeness ($r = 0$), then any node can randomly connect to any other node. A sufficiency condition for a non-assortative network is $e_{j,k} = q_j q_k$.

Perfect assortativeness and perfect disassortativeness are not exact opposites. Newman noted that if a network is perfectly disassortative then every link connects two nodes of different degrees (types) [10]. However, this requirement is not sufficient to generate an $e_{j,k}$ resulting in $r = -1$. In fact, the $r = -1$ case is possible only for symmetric degree distributions where $q_k = q_{(N_p-1-k)}$: $e_{j,k} = q_k \delta_{j,(N_p-1-k)}$. In other words, for a network with remaining degrees $0, \ldots, N_p - 1$, a node with a degree k must be linked to a node with a degree $N_p - 1 - k$. Nodes with identical degrees may still be connected in a perfectly disassortative network (e.g., when their degree j is precisely in the middle of the distribution q, i.e., N_p is odd and $j = (N_p - 1)/2$). Perfect disassortativeness is not possible for non-symmetric degree distributions q, because the $e_{j,k}$

distribution must obey the rules $e_{j,k} = e_{k,j}$, as well as $\sum_j e_{j,k} = q_k$. On the contrary, any degree distribution may give rise to a perfectly assortative network. We denote the maximum attainable disassortativeness as r_m, where $r_m < 0$ ($r_m = -1$ only for symmetric q_k). This limit and the corresponding $e_{j,k}^{(r=r_m)}$ can be obtained, given the distribution q_k, via a suitable minimization procedure by varying $e_{j,k}$ under its constraints.

Let us define the information transfer [1]:

$$I(q) = H(q) - H(q|q') \tag{3}$$

where the first term is the Shannon entropy of the network, $H(q) = -\sum_{k=0}^{N_p-1} q_k \log(q_k)$, that measures the diversity of the degree distribution or the network's heterogeneity, and the second term is the conditional entropy defined via conditional probabilities of observing a node with k links leaving it, provided that the node at the other end of the chosen link has k' leaving links. Importantly, the conditional entropy $H(q|q')$ estimates spurious correlations in the network created by connecting the nodes with dissimilar degrees — this noise affects the overall diversity or the heterogeneity of the network, but does not contribute to the amount of information within it. Informally, information transfer within the network is the difference between network's heterogeneity and assortative noise within it [1,9].

In information-theoretic terms, $H(q|q')$ is the assortative noise within the network's information channel, i.e., it is the non-assortative extent to which the preferential (either assortative or disassortative) connections are obscured [9]. Given the joint remaining degree distributions, the transfer can be expressed as:

$$I(q) = -\sum_{j=0}^{N_p-1} \sum_{k=0}^{N_p-1} e_{j,k} \log \frac{e_{j,k}}{q_j q_k} \tag{4}$$

Solé and Valverde [1] empirically analysed the relationship between assortativeness and information transfer, using a set of real world networks. Their conclusion was that the information transfer and assortativeness are correlated in a negative way: the extent of disassortativeness increases with mutual information (see Fig. 7 in [1]). We argue that networks with the same assortativeness r and the same distribution q_k could have different information transfers I — because they may disagree on $e_{j,k}$ — and observe that, under certain conditions (see Section 3), the information transfer non-linearly depends on the absolute value of the assortativeness (i.e. mutual information increases when assortativeness varies in either positive or negative direction), as illustrated in Figure 1. Moreover, we capitalize on the fact that, under certain conditions, the knowledge of r allows one to determine the information transfer $I(r)$ uniquely. Specifically, we intend to recover a network by growing the missing fragments in such a way that the resulting assortativeness (and hence, the information transfer) is as close as possible to the original one, while other network parameters are kept constant.

3 Assortative Preferential Attachment

Inspired by the preferential attachment method proposed by Barabasi et al. [4], we introduce the *Assortative Preferential Attachment* (APA) method to grow (or recover)

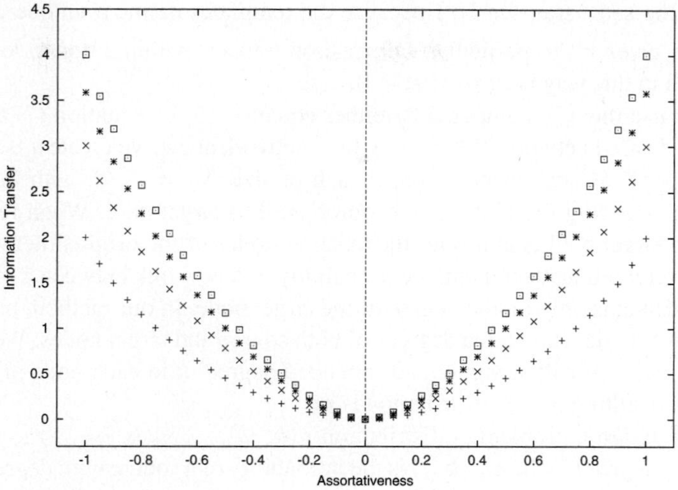

Fig. 1. Information transfer $I(r)$ as a function of r, for a q_k distribution with $\gamma = 1$; '+' indicate $N_p = 4$; '×' indicate $N_p = 8$; '∗' indicate $N_p = 12$; □ indicate $N_p = 16$.

a network with a specific assortativeness value r, given a degree distribution p_k and a network size N. The remaining degree distribution q_k is obtained using equation (1).

We classify networks according to the dependency of the distribution $e_{j,k}^{(r=r')}$ on the assortativeness r'. Within a class, the same distribution q_k and the same assortativeness r result in the same information transfer $I(r)$. We study one such class as an example for network growth and/or recovery with the APA method (other classes are handled by the method as long as the $e_{j,k}^{(r=r')}$ is defined in terms of r). This class is defined by the following dependency (template) of $e_{j,k}^{(r=r')}$ on $r' > 0$:

$$e_{j,k}^{(r=r')} = r' \left[e_{j,k}^{(r=1)} - e_{j,k}^{(r=0)} \right] + e_{j,k}^{(r=0)} \tag{5}$$

where $e_{j,k}^{(r=1)} = q_k \delta_{j,k}$ and $e_{j,k}^{(r=0)} = q_j q_k$. We assert that if the $e_{j,k}$ is given by the decomposition (5) for a real number $r' > 0$, then the network assortativeness is precisely r'. This is a sufficient but not necessary condition. A similar sufficient condition also exists for $r' < 0$

$$e_{j,k}^{(r=r')} = -\frac{r'}{r_m} \left[e_{j,k}^{(r=r_m)} - e_{j,k}^{(r=0)} \right] + e_{j,k}^{(r=0)} \tag{6}$$

For symmetric distributions q_k, it reduces to

$$e_{j,k}^{(r=r')} = r' \left[e_{j,k}^{(r=-1)} - e_{j,k}^{(r=0)} \right] + e_{j,k}^{(r=0)} \tag{7}$$

where $e_{j,k}^{(r=-1)} = q_k \delta_{j,(N_q-1-k)}$. These assertions can be verified by substituting templates (5) — (7) into Eq. (2). The same distribution q_k and the same assortativeness

r results in the same transfer $I(r)$ because the templates define a unique distribution $e_{j,k}^{(r=r')}$ for a given r'. In particular, information transfer within a non-assortative network defined in this way is zero: $I(0) = 0$.

Now we use the $e_{j,k}$ computed by either equation (5) or equation (7) to grow (or recover) the desired network. When growing a network anew, we create a 'source pool' and 'target pool' of unconnected nodes, each of size $N_0 = N/2$, with the intention of sequentially adding the nodes from source pool to target pool. When recovering a network, the target pool contains all the existing nodes of the original network. In the traditional preferential attachment, the probability of a new link between a source and a target node depends only on the degree of the target node. In our method, however, the probability would depend on the degrees of both source and target nodes. We therefore begin by probabilistically assigning an 'intended degree' k to each node in both pools such that the resulting degree distribution is p_k.

Then we assign a probability distribution $\mu(k, j_0), \ldots, \mu(k, j_{N_p-1})$ to each target node with the degree k, where $\mu(k, j)$ is the probability of a source with degree j joining the target node with the degree k. The probability $\mu(k, j)$ is calculated as $\mu(k, j) = e_{j,k}/p_j$, then normalized such that $\sum_j \mu(k, j) = 1$. The distribution $\mu(k, j)$ has to be biased by division by p_j, because each source node with degree j does not occur in the source pool with the same probability. In other words, sequential addition would not maintain $e_{j,k}$, and the biased probability $\mu(k, j)$ accounts for that. Once $\mu(k, j)$ is generated, each source node with degree j is added to the target pool and forms a link to a target node with degree k with probability $\mu(k, j)$. For example, if there are twice as many source nodes with degree j_2 than those with degree j_1 (i.e., $p(j_2) = 2\, p(j_1)$), while $e(k, j_2) = e(k, j_1)$, then the biased probabilities $\mu(k, j_1)$ and $\mu(k, j_2)$ would be such that $\mu(k, j_2) = e(k, j_2)/p(j_2)$ and $\mu(k, j_1) = e(k, j_1)/p(j_1) = 2\mu(k, j_2)$. This ensures that nodes with degree j_1 (represented twice as scarce as the nodes with degree j_2) would find it twice as easy to form a link with a target node with degree k. When a target node with k degrees forms its last, k-th, link, all its probabilities $\mu(k, j)$ are set to zero (i.e., this node will not form any more links). The grown network will thus have the desired joint distribution $e_{j,k}$, and hence the desired assortativeness r'.

When recovering a network rather than growing it anew, the probabilistic assigning of intended degrees to target nodes with existing links may deviate from the intended $e_{j,k}$, and APA method may be outperformed by a heuristic with recursive matching of intended and existing degrees. Such an alternative, however, is NP-hard. Our intention is to demonstrate that APA method does not significantly reduce solutions' quality.

4 Simulation Results and Analysis

We utilized the APA method to grow and/or recover scale-free networks with varying assortativeness values. Each experiment involved a set of networks with fixed degree distributions q_k (that is, fixed $\gamma = 1$, or $\gamma = 3$, and $N_p = 16$), and varying assortativeness values $r = 1$, $r = 0$ and $r = r_m$. In the case $\gamma = 1$, the disassortativeness extreme $r_m = -1$. Each original network (for each r) was grown with APA, and resulting information transfer $I_0(r)$ provided the point of reference. Then the network was progressively modified by removing a certain percentage (deficit) δ of nodes and the links connected to these nodes (δ varied from 1% to 99%). The APA method was applied to each modified network, and information transfer $I_\delta(r)$ was computed for the recovered network. The information-transfer distance $D_\delta(r) = |I_0(r) - I_\delta(r)|$ determined the success of the recovery in terms of information transfer. The experiments were repeated 10 times for each deficit level δ, and averaged into $\overline{D_\delta(r)}$.

We begin our analysis with symmetric distributions, $\gamma = 1$ and $N_p = 16$. The most challenging cases involve recovering highly assortative (e.g., perfectly assortative, $r = 1$) or highly disassortative (e.g., perfectly disassortative, $r = -1$) networks. These cases are more difficult than recovering non-assortative networks ($r = 0$) because the probabilistic assigning of intended degrees to target nodes with existing links may deviate from the intended $e_{j,k}$, but any such deviation would not harm non-assortative networks. Figure 2 plots $\overline{D_\delta(r)}$ for both extreme cases $r = 1$ and $r = -1$. It can be observed that, if the deficit level δ is below a certain threshold δ_0, a full recovery of information transfer is possible: $\overline{D_\delta(r)} = 0$ for both $r = 1$ and $r = -1$. As the deficit level δ increases, it becomes harder to recover the transfer, but the distance $\overline{D_\delta(r)}$ grows slower and stabilizes after reaching a certain height. However, at a certain critical level δ_t, there is a final transition to the region where the method cannot always follow the intended $e_{j,k}$ and departs from the corresponding templates. This results in a higher variance of the information distance when $\delta > \delta_t$ (especially visible in Figure 2, right, for $r = -1$, which is less robust than the case $r = 1$). Figures 3 and 4 plot, respectively, average and standard deviation of $\overline{D_\delta(r)}$ over 10 experiments: the critical levels δ_t are evident, pinpointing phase transitions as the deficit surpasses the level δ_t.

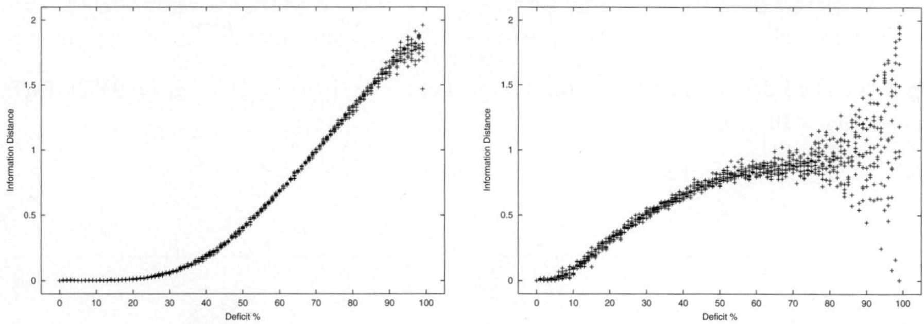

Fig. 2. Difficulty of recovery for $\gamma = 1$. Left: $r = 1$ ($\delta_0 \approx 20\%$, $\delta_t \approx 95\%$). Right: $r = -1$ ($\delta_0 \approx 10\%$, $\delta_t \approx 70\%$).

Figure 5 plots $\overline{D_\delta(r)}$ for the non-assortative case $r = 0$. Interestingly, a full recovery is possible in this scenario for either very low or very high deficit level δ. The reason for such symmetry is simple: the low levels δ present no challenge as the missing network fragments are small, while the high levels δ leave the method a lot of freedom in choosing the random (non-assortative) connections. For example, if a non-assortative network is regrown completely anew, it will attain the point-of-reference information transfer.

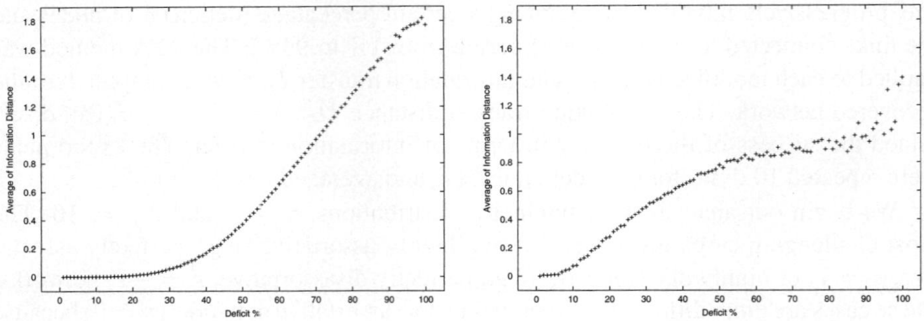

Fig. 3. Average of $\overline{D_\delta(r)}$ for $\gamma = 1$. Left: $r = 1$ ($\delta_0 \approx 20\%$, $\delta_t \approx 95\%$). Right: $r = -1$ ($\delta_0 \approx 10\%$, $\delta_t \approx 70\%$).

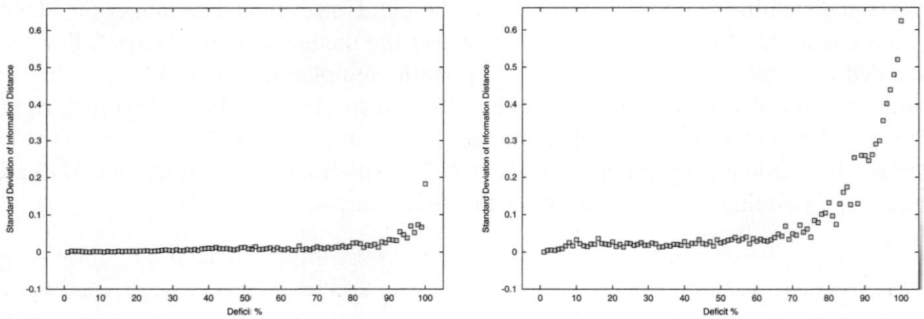

Fig. 4. Standard deviation of $\overline{D_\delta(r)}$ for $\gamma = 1$. Left: $r = 1$ ($\delta_0 \approx 20\%$, $\delta_t \approx 95\%$). Right: $r = -1$ ($\delta_0 \approx 10\%$, $\delta_t \approx 70\%$).

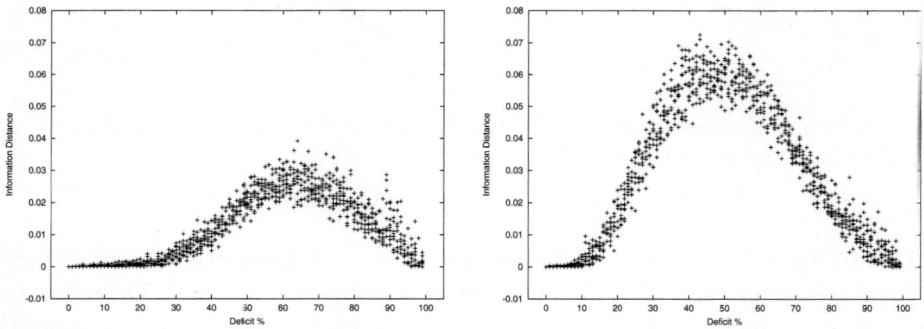

Fig. 5. Difficulty of recovery for $r = 0$. Left: $\gamma = 1$. Right: $\gamma = 3$.

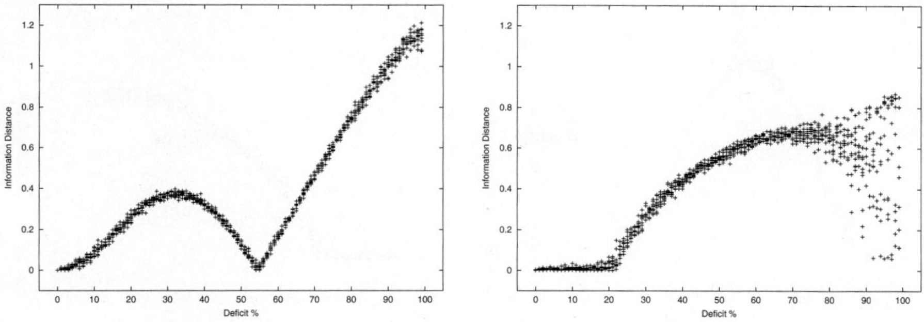

Fig. 6. Difficulty of recovery for $\gamma = 3$. Left: $r = 1$ ($\delta_0 \approx 5\%$, $\delta_m \approx 55\%$, $\delta_t \approx 95\%$). Right: $r = r_m \approx -0.52$ ($\delta_0 \approx 22\%$, $\delta_t \approx 75\%$).

Thus, there is a maximal difficulty (symptomatic of bell-shaped complexity curves) at the mid-range of δ. We should also note that the information distance $\overline{D_\delta(r)}$ is overall much smaller than that of the cases of highly assortative (disassortative) networks, as it is significantly less difficult to find non-assortative connections. The transition point δ_t noted in the plots for extreme r's can now be explained in the light of the complexity curve. There are two tendencies contributing to the recovery process: one is trying to reduce the difficulty as δ approaches 100% (more choice, or freedom, left by the higher deficit in constructing the desired $e_{j,k}$), while the other is increasing the difficulty (the $e_{j,k}$ of the existing links in the target pool diverges more from the required $e_{j,k}$).

We noted earlier that if $\gamma = 1$, the resulting remaining degree distribution q_k is uniform, hence symmetric. For other values of γ, the resulting q_k is not symmetric. Perfect disassortativeness is possible only for symmetric q_k, and therefore, for $\gamma > 1$, e.g. $\gamma = 3$, it is not possible to get close to the $(r = -1)$ case. Nevertheless, the recovery behaviour is similar to the one observed in the scenarios for $\gamma = 1$.

Figure 5, right, shows a familiar bell-shaped complexity curve for non-assortative networks, $r = 0$. Figure 6, left, showing $r = 1$, has an extra feature. In addition to expected full recovery δ_0 threshold for low deficit levels, and transition recovery δ_t for high deficit levels, there is a mid-range δ_m level where the amount of choice available for recovery completely dominates over the divergence of the existing $e_{j,k}$ from the required $e_{j,k}$. The information distance is minimal at δ_m as the full recovery is attained. Figure 6, right, showing $r = r_m \approx -0.52$, is similar to its counterpart from symmetric degree distribution ($\gamma = 1$): there are detectable levels of full recovery δ_0 and transition recovery δ_t. Similar results are observed with $\gamma = 4$ (Figure 7).

The experiments were also repeated for different distribution lengths N_p, and medium assortativeness values r. The latter cases showed intermediate profiles, where $\overline{D_\delta(r)}$ balances between the two identified tendencies (increasing freedom of choice and increasing divergence of $e_{j,k}$) as δ approaches maximum deficit.

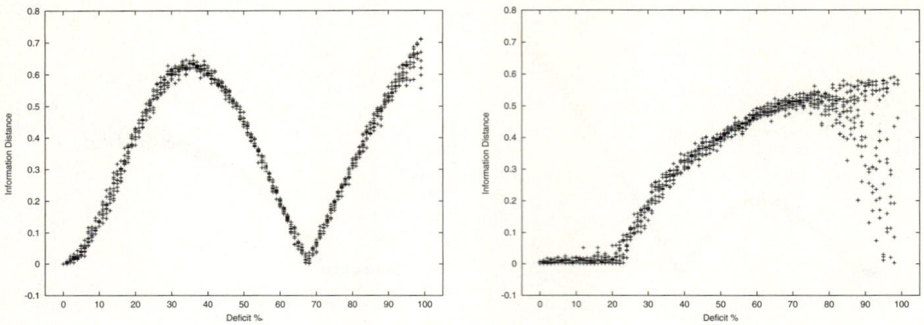

Fig. 7. Difficulty of recovery for $\gamma = 4$. Left: $r = 1$ ($\delta_0 \approx 5\%$, $\delta_m \approx 68\%$, $\delta_t \approx 95\%$). Right: $r = r_m \approx -0.50$ ($\delta_0 \approx 22\%$, $\delta_t \approx 75\%$).

5 Conclusions

We introduced and applied Assortative Preferential Attachment (APA) method to grow and/or recover scale-free networks in terms of their information transfer. APA achieves a required assortativeness value, and hence the information transfer, for a given degree distribution and network size. The method covers the extreme cases of perfect assortativeness and perfect disassortativeness, where the latter is only achievable if the specified degree distribution is such that the corresponding remaining degree distribution is symmetric.

We identified a number of recovery features: a full-recovery threshold, a phase transition for assortative and disassortative networks when deficit reaches a critical point, and a bell-shaped complexity curve for non-assortative networks. Two opposing tendencies dominating network recovery are detected: the increasing amount of choice in adding assortative/disassortative connections, and the increasing divergence between the existing and required networks in terms of the $e_{j,k}$.

References

1. Solé, R.V., Valverde, S.: Information theory of complex networks: on evolution and architectural constraints. In: Ben-Naim, E., Frauenfelder, H., Toroczkai, Z. (eds.) Complex Networks. Lecture Notes in Physics, vol. 650, Springer, Heidelberg (2004)
2. Newman, M.E.: Assortative mixing in networks. Phys. Rev. Lett. 89(20), 208701 (2002)
3. Callaway, D.S., Hopcroft, J.E., Kleinberg, J.M., Newman, M.E., Strogatz, S.H.: Are randomly grown graphs really random? Phys Rev E 64(4 Pt 1) (2001)
4. Albert, T.R., Barabasi, A.-L.: Statistical mechanics of complex networks. Rev. Mod. Phys. 74, 47–97 (2002)
5. Albert, R., Jeong, H., Barabási, A.-L.: Error and attack tolerance of complex networks. Nature 406, 378–382 (2000)
6. Crucittia, P., Latora, V., Marchiori, M., Rapisarda, A.: Error and attack tolerance of complex networks. Physica A 340, 388–394 (2004)
7. Moreno, Y., Gómez, J.B., Pacheco, A.F.: Instability of scale-free networks under node-breaking avalanches. Europhys. Lett. 58, 630–636 (2002)

8. Stauffer, D., Sahimi, M.: Diffusion in scale-free networks with annealed disorder. Phys. Rev. E 72, 046128 (2005)
9. Prokopenko, M., Boschetti, F., Ryan, A.: An information-theoretic primer on complexity, self-organisation and emergence. Unpublished (2007)
10. Newman, M.E.: Mixing patterns in networks. Phys. Rev. E 67(2), 026126 (2003)

MBEANN: Mutation-Based Evolving Artificial Neural Networks

Kazuhiro Ohkura[1], Toshiyuki Yasuda[1], Yuichi Kawamatsu[1], Yoshiyuki Matsumura[2], and Kanji Ueda[3]

[1] Hiroshima University, 1-4-1, Kagamiyama, Higashi-Hiroshima, 739-8527, Japan
[2] Shinshu University, 3-15-1, Tokida, Ueda, Nagano, 386-8567, Japan
[3] The University of Tokyo, 5-1-5, Kashiwanoha, Kashiwa, Chiba, 277-8568, Japan

Abstract. A novel approach to topology and weight evolving artificial neural networks (TWEANNs) is presented. Compared with previous TWEANNs, this method has two major characteristics. First, a set of genetic operations may be designed without recombination because it often generates an offspring whose fitness value is considerably worse than its parents. Instead, two topological mutations whose effect on fitness value is assumed to be nearly neutral are provided in the genetic operations set. Second, a new encoding technique is introduced to define a string as a set of substrings called operons. To examine our approach, computer simulations were conducted using the standard reinforcement learning problem known as the double pole balancing without velocity information. The results obtained were compared with NEAT results, which is recognised as one of the most powerful techniques in TWEANNs. It was found that our proposed approach yields competitive results, especially when the problem is difficult.

1 Introduction

Artificial evolution has been proven to be a promising approach to artificial neural networks (ANNs) in complex reinforcement learning tasks [5][3]. As discussed in [3][4], evolving artificial neural networks (EANNs) are faster and more efficient than reinforcement learning methods [6] for some representative benchmark problems. The reason is presumed to be that EANNs can search high-dimensional and continuous learning space more efficiently than other approaches. In addition, considering that ANNs can adapt to time series problems using the memory mechanism realised by recurrent synaptic connections, artificial evolution of ANNs would be a natural choice for learning non-Markovian tasks, a classification to which many interesting problems belong.

There has been a great deal of interest in EANNs. A good summary of EANNs up until 1999 can be found in [11]. Traditionally, EANNs are classified into the following three categories, according to their network structure:

- the network structure is fixed and the connection weights are evolving
- the network structure is evolving and the connection weights are trained by learning

F. Almeida e Costa et al. (Eds.): ECAL 2007, LNAI 4648, pp. 936–945, 2007.

– the network structure and the connection weights are evolving simultaneously

In the rest of this paper, we consider only the last case, which is called topology and weight evolving artificial neural networks (TWEANNs) [8]: of the three categories, artificial evolution plays the most important role in this one.

TWEANNs have several fundamental problems. One is that there are no generally effective guidelines for encoding a network structure into the form of a genotype. Currently, each researcher uses his own encoding method. A more serious problem is that the crossover operator cannot simply be applied to two individuals whose genetic information is different in length. Even when the crossover is applied to two individuals in a sort of brute force manner, the generated offspring often have much worse fitness values than the parents. As a result, most offspring would not survive into the next generation. Another factor that complicates the situation is that there is no effective theory for how to prepare the initial individuals. From the viewpoint of the crossover problem, it does not seem to be a good approach to provide the initial population with random topologies. Conversely, it seems to be inappropriate for the genetic search for all the individuals to have the same topology.

Many TWEANN approaches such as GNARL[1], EPnet[12] and ESP[3] have been proposed thus far. However, the authors regard the most impressive approach to be Neuro-Evolution of Augmenting Topologies (NEAT) [8], because NEAT solves the double pole balancing without velocity information (DPNV) problem, which is recognised as one of the difficult benchmark problems for reinforcement learning. The source code of NEAT is available at Stanley's homepage [10].

However, in our own experiments with NEAT, we encountered unwanted behaviour in the process of artificial evolution. We consider this to be unavoidable as long as crossover is adopted as the main genetic operator. Therefore, in this paper, we propose a novel method of TWEANNs called mutation-based evolving artificial neural networks (MBEANN), in which no crossover is used. This means that all genetic operations are applied to an individual independent of other individuals. Instead of crossover, two types of structural mutations are provided. Each is defined so as not to change the signal transfer, to make the effect on the fitness function value nearly or completely neutral.

The rest of this paper is organised as follows. Section 2 provides an overview of NEAT and explains how NEAT shows unwanted behaviour. Section 3 explains the details of our proposed MBEANN method. In Section 4, after introducing the general characteristics of DPNV, computer simulations are conducted to examine the effectiveness of MBEANN. The last section is the conclusion.

2 NEAT

NEAT is reported to be a very effective approach to the domain of TWEANNs [8] [9]. In this section, NEAT is briefly explained.

Encoding Method and Crossover. NEAT's genetic encoding scheme is designed to allow corresponding genes to be easily lined up when two individuals cross over during mating, since crossover is considered as the primary genetic operator. A genotype is represented by two lists, one for nodes and the other for synaptic connections. A node is represented by the following two components: one is the number that makes it uniquely identifiable for all the generations in a population; the other is the information that shows the type of layer that a node belongs to, i.e. the input layer, the hidden layer or the output layer. The genetic information for representing a synaptic connection is composed of two pointers to the nodes for both ends, the weight value, the flag (whether the connection is able to use it) and the innovation number. The innovation number is used for identifying the connection for all the generations in a population. Whenever a new synaptic connection is developed, this number is incremented and provided to the synaptic connection. Because of these innovation numbers, crossover can be applied to any two individuals without duplication or deletion of the genetic information from their parents, and the common part between them can be easily identified.

However, offspring generated by this crossover tend to have more genetic information than the parents. Therefore, the greater the difference between the parents, the larger the offspring that tend to be generated. This is a serious character because individuals can grow very large within several generations.

Speciation. In TWEANNs including NEAT, it is commonly observed that a structural change brings a harmful result, and therefore, a new offspring has a much worse fitness value than the parents. In NEAT, as a strategy to keep them alive for the next generation, a speciation technique with explicit fitness sharing [2] is adopted to protect new offspring from natural selection. Speciation also has the effect that crossover can be applied to only two individuals in the same species, i.e. to two individuals whose structural differences are within a compatible distance. The details are shown in [8].

This seems a reasonable and effective procedure at first glance. However, according to our computer simulations, the number of species almost always increased to the maximum number in earlier generations than we expected. Newly generated individuals tend to be protected from natural selection as a result of speciation. According to our NEAT computer simulations, we frequently encountered the unwanted situation in which there were too many species that include only one individual.

Initial Population. Typically, TWEANNs start with an initial population of random topologies to introduce a sufficient topological diversity. In contrast, NEAT biases the search towards minimal-dimensional spaces by starting out with a uniform population of networks with zero hidden nodes.

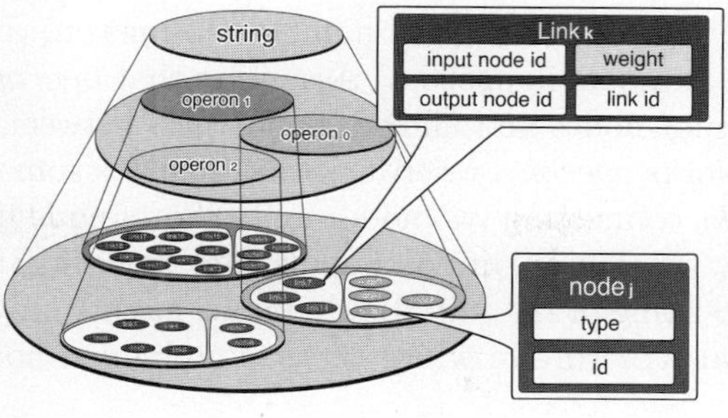

Fig. 1. The concept of genotype for MBEANN

3 Proposed Method: MBEANN

Although NEAT is a successful approach for TWEANNs, we still found the unwanted evolutionary behaviour that might originate from crossover as explained above. On the other hand, it might not be easy to propose a more effective strategy to reduce such side effects of crossover. Therefore, we propose a novel approach to TWEANNs in which no crossover is used. Our approach is called MBEANN, which stands for mutation-based evolving artificial neural networks. MBEANN adopts two types of structural mutations that may decrease little or have no fitness value. Therefore, speciation is not used in MBEANN.

3.1 Encoding Method

In the research of TWEANNs thus far, it does not seem that the topic of how to represent a network structure has been explored systematically. Each researcher has adopted his own representation method and special genetic operations[11][7], and no de facto standard is applied. Following this trend, we propose our original genotype representation as the answer to the question of how to design a robust genotype for a small change of genetic structure.

We considered the problem as follows. Assuming that an individual is a set of sub-networks each of which is independent of the others, we can expect that a small genetic change occurring in a sub-network would influence only a limited region. This might bring robustness against a genetic change. From this assumption, an individual is designed as a set of sub-networks, i.e. as a set of modules called operons. Fig. 1 shows the concept of the genotype that we designed. As shown in the figure, node information consists of the node type and the node identification number. Link information consists of the input node, the output node, the weight value and the link identification number. The two identification numbers should be unique to each individual. Note that they are not the same as the innovation numbers in NEAT, because they are unique only to an individual.

Thus, supposing that I is the maximum number of operons, a genotype *string* is formulated as follows:

$$string = \{operon_0, operon_1, , ..., operon_I\} \tag{1}$$

$$operon_i = \{\{node_j \mid j \in O_{Ni}\},\ \{link_k \mid k \in O_{Li}\}\} \tag{2}$$

where O_{Ni} is the set of node identification numbers in $operon_i$ and 0_{Li} is the set of link identification numbers in $operon_i$. Assuming that $operon_0$ holds only the input nodes, the output nodes and all the connections between them, string is composed of sub-networks $operon_i$, where $operon_i$ includes some nodes $node_j$ and the connections $link_k$ connecting two nodes in itself or a node in the $operon_i$ and a node of $operon_0$. As for an initial population, since MBEANN starts with the population consisting of the initial individuals having only one operon, $operon_0$, i.e. the minimal structure in which there is no hidden node.

3.2 Genetic Operators for Structural Evolution

Similar to NEAT, MBEANN starts with the minimal individuals. Hidden nodes and synaptic connections are obtained by two structural mutations with generations. The two genetic operators are defined as follows. They are designed to be nearly neutral or completely neutral with respect to the fitness value.

Add-Node Mutation. The add-node mutation is applied to each operon at a constant probability P_{add}. This mutation operates to remove one of the synaptic connections, which is randomly selected, and then adds a new hidden node and two associated synaptic connections. If one of the ends of the removed connection is connected to a node in $operon_0$, a new operon is provided to the newly developed sub-network. Fig. 2 illustrates the effect of the add-node mutation. When a synaptic connection W_1 is selected, the link is removed and a new node and two synaptic connections W_3 and W_4 are added to construct a new operon called $operon_i$.

In order not to change the signal transfer by this mutation, a_1 and a_2 in Fig. 2 should be equal. Therefore, supposing that S is the sigmoid function and the W_i is the weight value, is we consider the following condition:

$$S[w_1x_1 + w_2x_2] = S[w_3S[w_4x_1] + w_2x_2] \quad (3)$$

Thus, we get $w_1x_1 - w_3S[w_4x_1] = 0$. In addition, assuming the condition that $w_1 = w_3$ and $S[x] = 1/(1 + e^{\beta(\alpha-x)})$, we find that the following value $f(x_1)$ should always be small against α, β and w_4:

$$f(x_1) = x_1 - \frac{1}{1 + e^{\beta(\alpha - w_4x_1)}} \quad (4)$$

For the sake of simplicity, we define $w_4 = 1$. As for α and β, we set them as 0.5 and 4.0, respectively, considering the simplicity and the results of the preliminary experiments.

Add-Connection Mutation. The add-connection mutation is applied to each operon at a constant probability. Fig. 3 illustrates how this mutation works. When an operon is selected, a node in the operon is randomly chosen to make a random synaptic connection with a node in the operon or with a node in $operon_0$. The weight of the new connection is set to 0 so that it does not change the signal

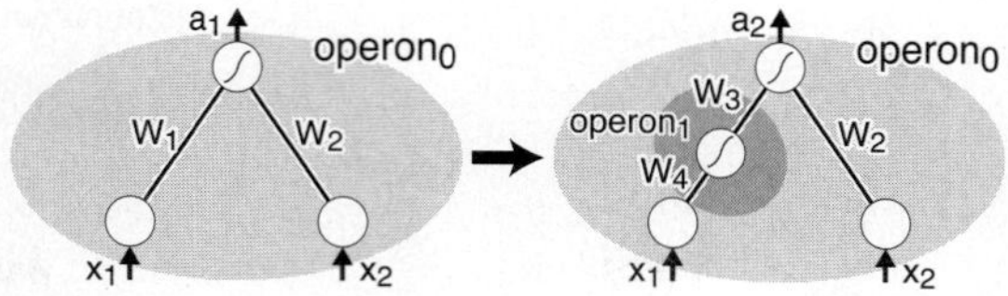

Fig. 2. Add-node mutation

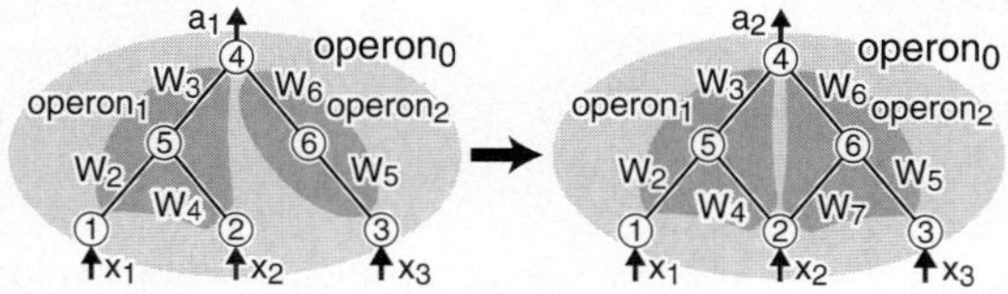

Fig. 3. Add-connection mutation

transfer. With this add-connection mutation, a node in $operon_i$ where $i > 0$ can be connected to a node in the same operon or a node in $operon_0$. In other words, the connection to a node in operon$(j = i, j > 0)$ is prohibited. Therefore, since each sub-network $operon_i$,$i > 0$ grows independent of the others, we can expect the functional modularity in an individual.

Synaptic Weight Mutation. As for weight mutations, we adopt the most popular mutation for real-coded genetic algorithms. That is, the weight mutation introduces a small change by adding a Gaussian random number of which the average is zero and the standard deviation is σ. In this paper, the value $\sigma = 0.05$ is adopted.

4 The Double Pole Balancing Problem

4.1 DPNV and Simulation Conditions

The performance of MBEANN is discussed with the benchmark problem called the double pole balancing without velocity information (DPNV) problem[3]. Two poles are connected to a moving cart by a hinge and the neural network must apply force to the cart to keep the poles balanced for as long as possible without going beyond the boundaries of the track. The system state is defined by the cart position x and the velocity $\dot{x}$, the first pole's position θ_1 and its angular velocity $\dot{\theta}_1$, and the second pole's position θ_2 and its angular velocity $\dot{\theta}_2$. However, $\dot{x}$, $\dot{\theta}_1$ and $\dot{\theta}_2$ are not used as inputs because the problem becomes more difficult; therefore the controller must utilise the time series information.

It is known that control is possible when the poles have different lengths to respond differently to control inputs. The computer simulations were conducted with four different ratios of the long pole length versus the short pole length, i.e.

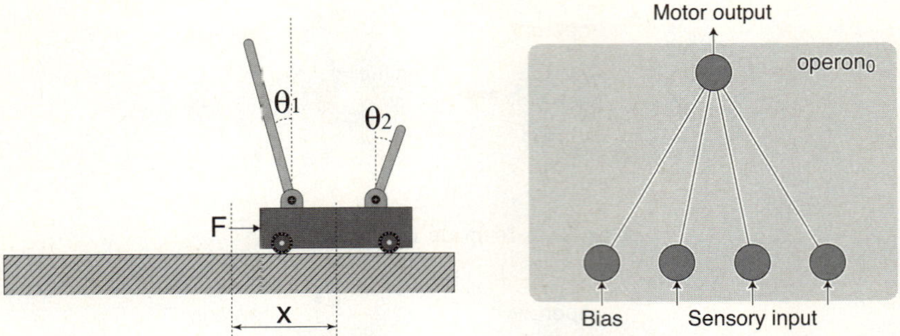

(a) The cart pole system (b) The topology of initial individuals

Fig. 4. Setting of the double pole balancing problem without velocity information

Table 1. Parameters for MBEANN and MBEANN-WO

population size	1,000	weight mutation rate	1.0
tournament size	20	weight value in add-node	1
add-node rate	0.007	α in sigmoid function	0.5
add-connection rate	0.3	β in sigmoid function	4.9
final generation	1,000		

1:0.1, 1:0.2, 1:0.3 and 1:0.4. It is known that the control is more difficult when the difference between the pole lengths is smaller. All other details for problem settings for computer simulations were the same as in [8]. As for the comparison, the original NEAT source from Stanleys website[10] was used.

In addition, since the operon structure was employed for MBEANN based on our presumption that the modularity in genotype would work better, MBEANN without the operon structure (MBEANN-WO) was also conducted to compare the results with those of MBEANN to confirm our decision. MBEANN-WO can be easily realised by skipping the procedure of generating a new operon when add-node mutations are applied. The parameters for MBEANN and MBEANN-WO used in our computer simulations are summarised in Table 1.

4.2 Results

Fig. 5 summarises the results of ten independent runs of NEAT, MBEANN-WO and MBEANN. From Fig. 5(a), which shows the success rates, we found that all three approaches solve the problem with very high probabilities at easy conditions such as 1:0.1 or 1:0.2. However, it is clearly observed that NEAT's performance was getting worse as the problem difficulty increased. In particular, NEAT solved only once in ten runs under the most difficult condition of 1:0.4. At the same time, it was also found that MBEANN and MBEANN-WO performed much better than NEAT under difficult conditions.

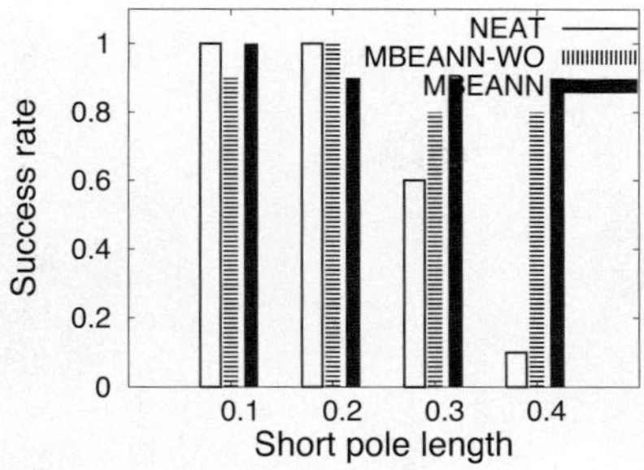

(a) Success rate

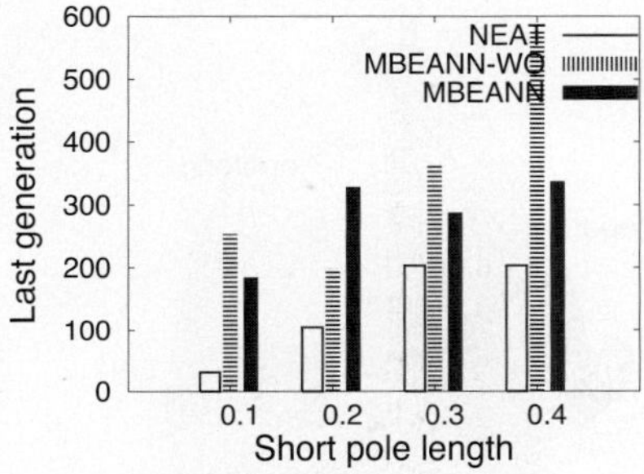

(b) Last generation

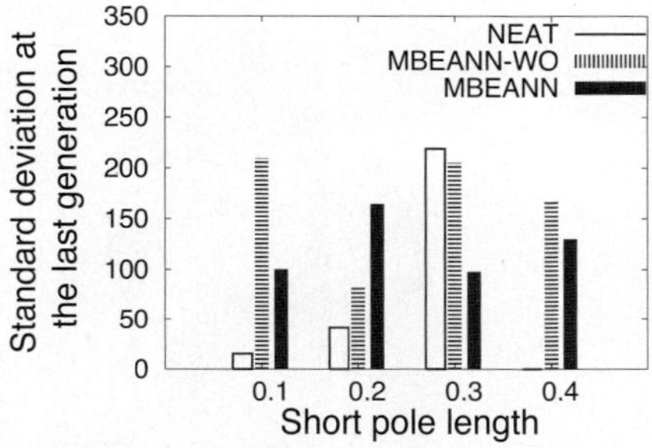

(c) Standard deviation at the last generation

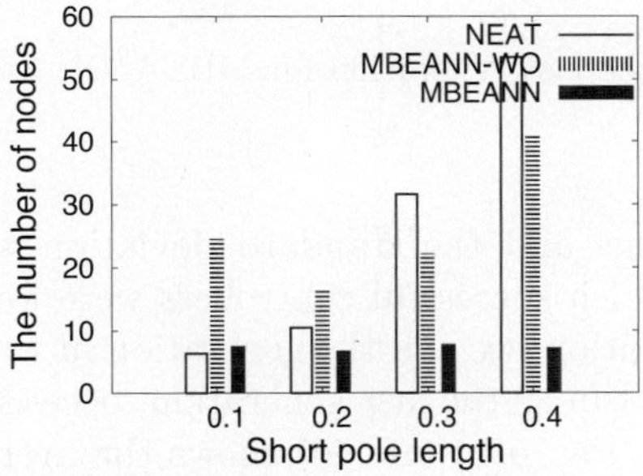

(d) Nodes

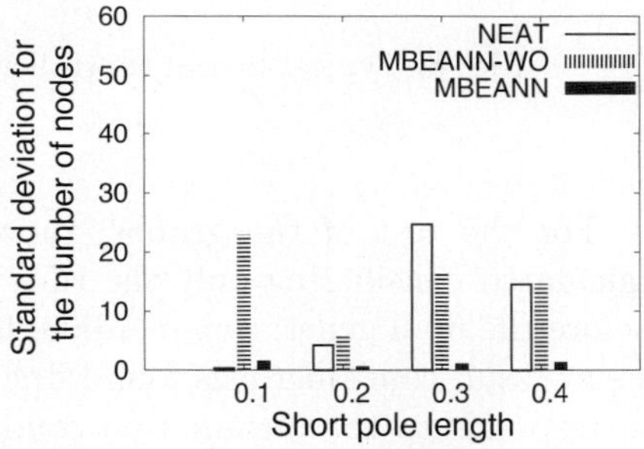

(e) Standard deviation for nodes

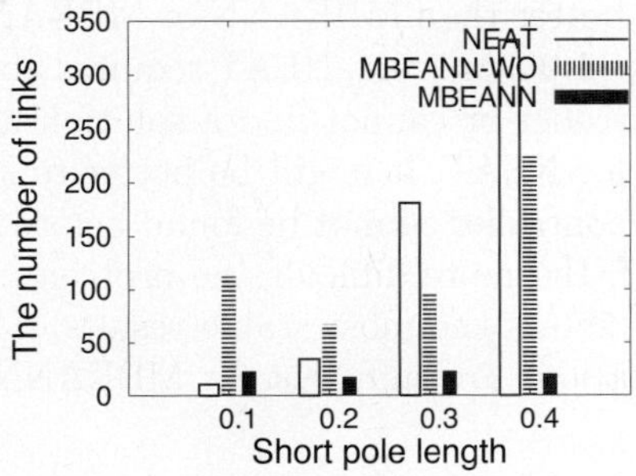

(f) Links

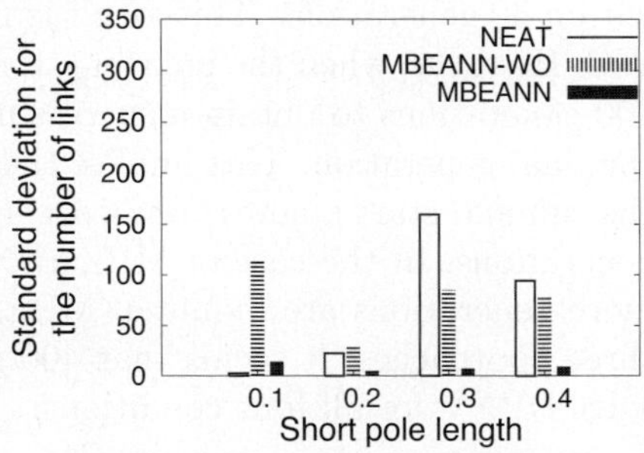

(g) Standard deviation for links

Fig. 5. Experimental results averaged over ten trials for NEAT, MBEANN-WO and MBEANN

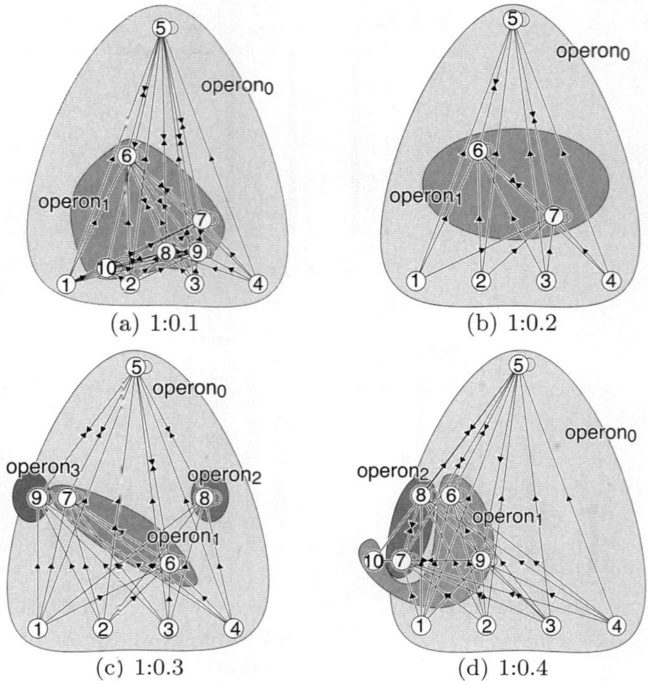

Fig. 6. Typical neural controllers at the last generation for MBEANN

For the rest of the graphs, the average and the standard deviation were calculated considering only the runs in which successful controllers were found before the final generation of 1,000. In addition, we call the generation in which a successful controller was found first in a run as the last generation, because it corresponds to the termination condition. Fig. 5(b) and (c) shows the average and standard deviations of the last generations. NEAT solves the problem of 1:0.1 within 30 generations. This result is much better than MBEANN or MBEANN-WO. However, when the pole length ratio is 1:0.3 or 1:0.4, NEAT requires about 200 generations to obtain a successful controller or cannot find a solution until the final generation. This implies that, as for NEAT, it might be better to stop the run and start a new run if a successful controller cannot be found before 200 generations. In the case of MBEANN-WO, the more difficult the problem, the more generations are required. MBEANN shows the most stable results of the three approaches. It seems that 400 generations are sufficient for MBEANN to solve DPNV for all four conditions.

Fig. 5(d) and (e) shows the average number of nodes in the controller at the last generation and the standard deviation of ten runs. Similarly, Fig. 5(f) and (g) shows the average number of links and the standard deviation. NEAT found a successful controller with only 6 nodes and 10 links at the condition of 1:0.1 in average. However, the number of nodes and links rapidly increases with the problem difficulty. Finally, at the condition of 1:0.4, the controller had grown to reach the size of 53 nodes and 332 links. In the case of MBEANN-WO, the controller becomes gradually larger with the difficulty, but the speed seems moderate compared to the one of NEAT. On the other hand, the MBEANN's controllers are always similar in size for all four conditions. Controllers typically obtained are shown in Fig. 6. MBEANN needs about seven nodes and twenty links on average. Therefore, we can say that the importance of modularity in genotype has been inductively proven by the results of the computer simulations.

5 Conclusions

MBEANN was proposed as yet another approach to TWEANNs. Using the pole balancing without velocity information problem, its robustness as well as effectiveness were demonstrated. We believe that more investigation is required to improve the performance of TWEANNs, especially for the paradigm of evolutionary robotics[7], where TWEANNs are necessary to realise artificial evolution.

References

1. Angeline, P.J., Sauders, G.M., Pollack, J.B.: An evolutionary algorithm that constructs recurrent neural networks. IEEE Trans. Neural Networks 5(1), 54–65 (1994)
2. Goldberg, D.E.: Genetic Algorithms in Search, Optimization and Machine Learning. Addison-Wesley, New York (1989)
3. Gomez, F., Miikkulainen, R.: Solving non-Markovian control tasks with neuroevolution. In: Dean, T. (ed.) Proceedings of the Sixteenth International Joint Conference on Artificial Intelligence, pp. 1356–1361. Morgan Kaufmann, San Francisco (1999)
4. Gomez, F., Miikkulainen, R.: Learning robust nonlinear control with neuroevolution. Technical Report AI02-292, Department of Computer Science, University of Texas at Austin, Austin, Texas (2002)
5. Grauau, F., Whitely, D., Pyeatt, L.: A comparison between cellular encoding and direct encoding for genetic neural networks. In: Koza, J.R., et al. (eds.) Genetic Programming 1996: Proceedings of the First Annual Conference, pp. 81–89 (1996)
6. Kaebling, L.P., Littman, M., Moore, A.W.: Reinforcement learning: A survey. Journal of Artificial Intelligence 4, 237–285 (1996)
7. Nolfi, S., Floreano, D.: Evolutionary Robotics. MIT Press, Cambridge (2000)
8. Stanley, K.O., Miikkulainen, R.: Evolving Neural Networks Through Augmenting Topologies. Evolutionary Computation 10(2), 99–127 (2002)
9. Stanley, K.O., Miikkulainen, R.: Competitive Coevolution Through Evolutionary Complexification. Journal of Artificial Intelligence Research 21, 63–100 (2004)
10. Stanley, K.O.: `http://www.cs.ucf.edu/~kstanley/`
11. Yao, X.: Evolving Artificial Neural Networks. Proceedings of the IEEE 87(9), 1423–1447 (1999)
12. Yao, X., Liu, Y.: A New Evolutionary System for Evolving Artificial Neural Networks. IEEE Transactions on Neural Networks 8(3), 694–713 (1997)

Measuring Entropy in Embodied Neural Agents with Homeostasic Units: A Link Between Complexity and Cybernetics

Jorge Simão

DCC-Faculty of Sciences-University of Porto & LIACC
jsimao@dcc.fc.up.pt

Abstract. We present a model of a recurrent neural network with homeostasic units, embodied in a minimalist articulated agent with a single link and joint. The configuration of the agent is determined by the total activation level or kinetic energy of the network. We study the complexity patterns of the neural networks, and see how the entropy of the neural controller state and agent configuration changes with the relative characteristic time of the homeostasis when compared with the excitatory-inhibitory activation dynamics of network. We also present a meta-model of embodied neural agents, that serves as conceptual framework to study self-perturbation and the self-organization in embodied neural agents. Simulation results show that homeostasis significantly influences the dynamics of the network and the controlled agent, allowing the system to escape fixed-points and produce complex aperiodic behavior. The relation between the characteristic time of homeostasis and the characteristic time of main excitatory-inhibitory activation dynamics was found to be non-linear and non-monotonic. We use these findings to connect the perspectives of classical cybernetics on homeostasis to complexity research.

1 Introduction

Motivated by concepts and ideas from autopoetic philosophy [1], ecological psychology [2], complex systems theory [3], and situated artificial intelligence research [4,5], we present a recurrent neural network model [6] with homeostasic units [7] and study its complexity patterns. Homeostasis is modeled using adaptive thresholds in units, whose effect is to (try to) push units to an equilibrium or rest activation value. The agent consists of a link and a rotational joint in a 2D plane. The agent configuration is fully determined by a single degree of freedom — the joint angle. Attached to the link is a muscle whose contraction/distension produces an angular displacement of the link. The activity of the neural units determines the level of contraction of the muscle. Homeostasis is modeled in our neural network as an adaptive threshold in neural units. We use measures of entropy to see how the characteristic time of homeostasic process affects the complexity of the dynamics of the neural controller (micro state), and the resulting configuration of the agent (the macro state).

F. Almeida e Costa et al. (Eds.): ECAL 2007, LNAI 4648, pp. 946–955, 2007.

Our results show that even though units individually try to move to a rest/equilibrium activation value, the system as a whole never reaches equilibrium because of connections between neural units. Instead, the system exhibits complex (mostly aperiodic) patterns, generating a source of novelty/variety in agent behavior. On the other hand, if homeostasis is not used, the system tends to a stable fixed point or a small region of the state space (if noise is present). To abstract and guide our work, we begin by presenting a generic meta-model of embodied neural agents that can be used to understand in a programmatic way how self-organization and neural development relates to adaptive behavior and sensoriomotor coordination tasks (both in high-order animals and humans).

Below, we present the meta-model for embodied neural agents (section 2). Next, we describe the particular embodied neural agent model studied in this article (section 3), and present the results of several computational experiments using the model (section 4). Section 5 summarizes simulation results, relates it with other work on on homeostasis, and makes a conceptual link with classical cybernetics.

2 A Generic Meta-model for Cognitive Development and Sensoriomotor Coordination in Embodied Neural Agents

To study self-organization in embodied neural agents and the development of sensori-motor skills, we have made an abstract characterization of this type of agent model. [See [8] for another characterization of embodied developmental agents, taking the more traditional computational perspective.]

Agents are characterized at two levels: the macro-level and the micro-level. The macro-level is defined by the configuration state — a formal description of the agents body posture in space, as seen by an external observer or as made apparent to the agent through self-perception. A small number of degrees of freedom is often required to describe an agent at this level. The micro-level is a characterization of the state of its neural controller. This includes the activation level of neural units (e.g., mean firing rate), units' thresholds, and neural connections' weights. Usually, the micro-level requires a much higher number of degrees of freedom to be fully described than the macro-level, since an agent with few links and joints may have a controller with many neural units. Interfacing the micro and macro-levels, agent descriptions include the way the neural controller is connected to the agents' body — both in muscular connections (efferent) and in the way sensation-perception cells/inputs impinge on the neural controller.

Agents are often situated in some environment, in such a way that its behavior and interaction with the environment may be observed by some external observer. In fig. 1, we make a sketch representation of the relationship between the agent, its environment, the external observer, and the two levels of description.

A key aspect of natural agents, is that the mapping from the (micro) neural level and the (macro) configuration level is not one-to-one. The coordinated

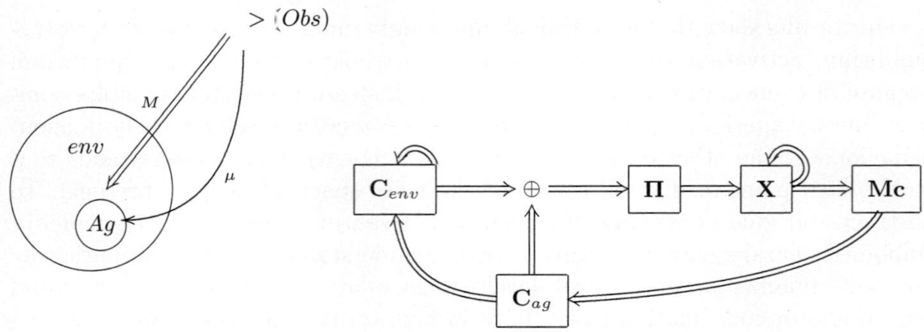

Fig. 1. Meta-Model of Embodied Neural Agents: **left)**: conceptual diagram of the agent, the external observer, the environment, the macro-level and the micro-level descriptions; **right)**: schematic block diagram of agent-agent and agent-environment functional dependencies .

action of a large number of neurons and muscular cells is usually required to generate strong and high-amplitude body movements. Additionally, many different neural states may mandate the same body configuration.

Because agents have units sensitive to environmental and body state (the sensation-perception inputs), agents can sense the effects of their own actions. Thus the micro and macro level are connected in a two-way causality loop. The state of the micro-level determines the body configuration (apart from external mechanical perturbations of the agent body, such as gravity and social manipulation), and the body configuration perturbs the internal dynamics of the neural controller. Fig. 1, represents the causality loops in agent behavior according to our meta-model. $\mathbf{X}$ represents the state of the neural controller of the agent (part of micro-level or internal state), and $\mathbf{C}_{ag}$ represents the body configuration of the agent (the macro-level or external state). The self-loop in $\mathbf{X}$ represents the internal dynamics of the controller, and the connection from $\mathbf{X}$ to the motor units $\mathbf{Mc}$ represent the commanding of muscular force. The connection from $\mathbf{Mc}$ to $\mathbf{C}_{ag}$, represents the change in body configuration. This represents the micro-macro causality. Due to self-perception (in any sensorial modality) the agent configuration $\mathbf{C}_{ag}$ generates input or perturbation to the neural dynamics — represented as $\mathbf{\Pi}$. This represents the macro-micro causality. As a side-effect, change in body configuration may change the state or configuration of the environment (e.g., as in a manipulation task), represented as $\mathbf{C}_{env}$, which in turn generate additional perturbation to the neural controller. In a complex task-environment, environments may also have complicated dynamics of their own — represented as a self-loop in $\mathbf{C}_{env}$.

From a formal point of view, the agent body, neural controller, and environment represent a (complex) dynamical system that can be summarized with two coupled differential equations:

$$\begin{cases} \dot{\mathbf{X}} = f_a[\mathbf{X}, f_{\Pi}(\mathbf{C})] \\ \dot{\mathbf{C}} = f_C[Mc(\mathbf{X}), \mathbf{F}_{ext}, \mathbf{C}] \end{cases} ,$$

where f_a is the neural units activation function, and f_C and Mc relate internal state with changes in body configuration. The variable $\mathbf{F}_{ext}$ accounts for (possible) external mechanical perturbation in the agent's body configuration.

This characterization of embodied neural agents relates to Ashby classical characterization of adaptive agents and agent-environment couplings as dynamical systems [9], further explored in mainstream situated AI literature [10]. However, in our formulation we make explicit and render of theoretical importance the difference between the typical number of degrees-of-freedom at the micro/neural-level and the macro/configuration-level. This allows us to explore the relationship between the two levels, and see how mechanisms of self-organization in neural controllers (such as on-line learning and self-perturbation) relate to the generation of muscular force and adaptive behavior in sensori-motor coordination task. Additionally, we tend to use mostly the tools and methods of complex systems rather than the tools normally used in studying (lower) dimensionality non-linear dynamical system [11].

3 A Minimalist Embodied Neural Agent with Homeostasic Units

We model an embodied agent with a single link and a single joint. The joint angle ψ fully defines the body configuration of the agent. The joint angle is determined by the contraction of a simplified muscle that works like a mechanical lever. The muscle has a large number of muscular units m_i. The contraction/extension of a muscular unit m_i produces a spatial displacement Δs_i, and the summation of all displacements determines the joint angle. Formally, $\psi = f(\sum_i \Delta s_i)$, where f is a function of the detailed geometry of the agent.We assume that the contraction of a single neural units produces a relatively small link displacement. In particular, the simultaneous contraction of a large proportion of muscular units is required to generate maximum displacement of the link. Moreover, the joint angle ψ is always constrained to lie within a maximum amplitude interval $[-\frac{pi}{2}, \frac{pi}{2}]$. In figure 3, we show the abstract design of the agent (left), and the graphical design as visualized in our simulator (right).

Muscle contraction (and thus body configuration) is controlled by a neural population with N units, whose activation/excitation state we represent by the vector $\mathbf{X} \equiv [x_1, \ldots, x_i, \ldots, x_N]$. We make a simple attachment between this motor control neural population and the muscle units, by making the number of muscular units equal to the number of neural units, and connecting them one-to-one (unidirectional). When all units are in a rest/natural activation value, ϕ takes value 0 (the link is horizontal).

Neural units are connected in a network/graph as a fully recurrent neural network (all units connect to all) [6]. Connection strengths are represented with a connectivity matrix $\mathbf{M}$, where element c_{ij} represents the connection strength or weight between unit i and j. In the simulation results presented below we experiment mostly with fixed connection weights.

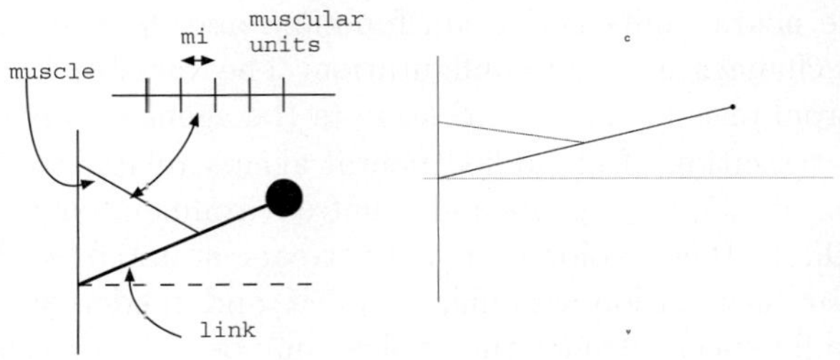

Fig. 2. Body Configuration of Minimalist Articulated Agent (one degree of freedom, one set of muscular units): **left)**: abstract design; **right)**: visualization in neural simulator.

Neural unit are assumed to be initially connected with random weights, using a normal distribution with mean value 0 and variance $\sigma^2(M)$. Learning (when used) is modeled as an hebbian-like learning rule for positive connections, and an anti-hebbian-like learning rule for negative connections.

Neural units have an adaptive threshold that is used to maintain units in a sensitive state. This is equivalent to cellular homeostasis mechanisms in biological neural networks [7]. For unit i we represent its threshold as θ_i. When a unit's activation is very high, a slow adaptation process takes place that gradually moves the activation value to a rest or natural activation value x_0. Likewise, when unit's activation value is low the same adaptation process takes place to raise the activation level to x_0.

The operation of units is formally defined using two ordinary first-order differential equations [approximated by the Euler method in the simulations below]. The first equation below describes the (fast) dynamics of individual unit's activation. The second equation describes the (slower) dynamics of homeostasis. The learning dynamics is modelled by the third equation presented below.

$$\begin{cases} \tau_1 \dot{x}_i = -x_i + x_0 + f(\sum_j c_{ji} x_i + c_i \pi_i - \theta_i) + \xi_i \\ \tau_2 \dot{\theta}_i = x_i - x_0 \\ \tau_3 \dot{c}_{ji} = -c_{ji} + sign(c_{ji}) \cdot g(x_{\max} - x_i) g(x_{\max} - x_j) \end{cases},$$

above τ_1, τ_2, and τ_3, with $\tau_1 << \tau_2$, are constants for the characteristic times of the neural processes modelled. x_0 is the resting or natural activation of units. f is an activation gain function. We use a constant gain G. A unit's activation x_i is constrained to be in the interval $[x_{\min}, x_{\max}]$, where $x_{\max}$ is the saturation value and $x_{\min}$ is the lowest/depression value. ξ_i is a normally distributed random noise value. It represents a stochastic term, to the otherwise ordinary differential equation. It is implemented by adding a random value to each neural unit's activation value in each simulation time step. π_i represents the current value of the self-perturbation input for unit i and c_i is a fixed input gain (not used in the particular studies presented below). g is an auxiliary function that produces higher values when the argument is close to 0. Its used to implement the (anti-)hebbian learning function. *sign* is the standard sign function.

Solving for equilibrium in the first equation shows that at rest $x_i = x_0 + f(\sum_j c_{ji}x_i + c_i\pi_i - \theta_i) + \xi$, which is a fast quiescent/rest state. Solving for equilibrium for the second equation show that at rest $x_i = x_0$, which is a slow quiescent/rest state (since $\tau_1 << \tau_2$). Simulation results below show that full equilibrium (that is, $x_i = x_0$ for all units) is never reached due to units interconnections.

4 Experimental Results

The simplest operation regime for the system is when neural units are not connected, that is, the connection matrix **M** is null. In this case, network dynamics is governed by the internal noise variables ξ_i and, if units homeostasis is turned off, the body configuration angle ψ takes values normally distributed around 0. If units are set to have homeostasis, fluctuation around $\phi = 0$ can be made arbitrarily low by reducing noise and reducing the characteristic time for homeostasis τ_2. (plots not shown)

When units are connected according to some random weight matrix **M**, with mean 0 and variance $\sigma^2(M)$, and weights are fixed (no learning), the behavior of the system changes. If noise is removed and homeostasis is also removed, the neural activation state and the configuration angle converges in most simulation runs to a **fixed point**. In the fixed point most units are either fully saturated or fully depressed. Different random matrices produce different fixed points. Fig. 3left) shows the evolution of the neural state over 100 time steps. High activation of units is color coded as red, low activation as blue, and values near x_0 as green[1]. Also in fig. 3left) we show the evolution of the configuration state for the same simulation run. In a set of 10 consecutive runs with the same settings, the results obtained were qualitatively the same.

Fig. 3right) shows the system evolution of a typical simulation run when noise is added to the system ($\sigma^2(\xi) > 1$). The results show that the neural and configuration state converges to a small region of state space, and remain in that region (the region is meta-stable). [The results were qualitatively the same for a set of 10 consecutive runs.] This suggests that fixed points are Lyapunov stable (neural states tend to stay within a small distance of a fixed point when perturbed). Most units remain for most of the time close to saturation or depressive points (red and blue colors). Fluctuations of ψ around a mid-point can not be assumed to be normally distributed since units are interconnected.

When units have homeostasis, the behavior of the system changes considerably. The proportion of time units are not saturated or depressed increases, as inspection of the differential equation for the threshold above would suggest. However, most units do not remain with an activation value near x_0 all the time since they are taken away from homeostasis due to interconnection with

[1] We use this color code since red can be intuitively associated with heat produced by natural cells' activity, and blue with cold.

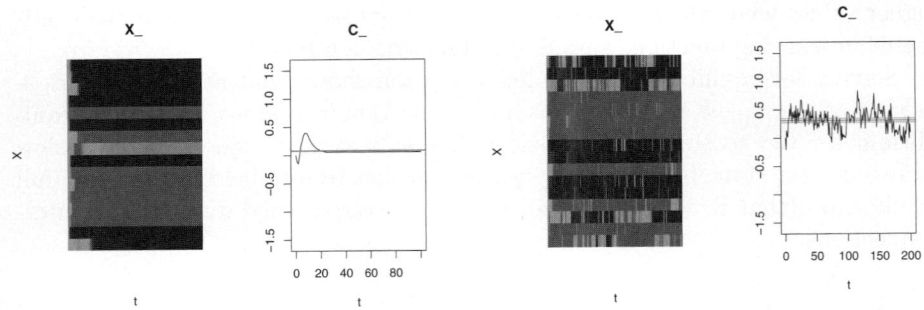

Fig. 3. Typical system's behavior with random matrix ($\sigma^2(M) = 1$) and no homeostasis ($\tau_2 = \infty$): **left)** no noise ($\sigma^2(\xi) = 0$): Convergence to a fixed point with most units saturated or depressed ($N = 16$); **right)** high noise ($\sigma^2(\xi) > 1$): Convergence to a small region of the neural state space (meta-stable) ($N = 16$).

other units. Fig. 4 shows a qualitatively typical simulation run (from a set of 10 consecutive runs). [Noise was set to a moderate level ($\sigma^2(\xi) = 0.2$), when compared with the set maximum activation level ($x_{\max} = 3$).] The system state does not converge to any attractor, but exhibits **non-periodic behavior** due to threshold adjustments. When the number of units is small, the behavior of the systems can also be quasi-periodic or periodic. The behavior of the link follow the variation in kinetic energy of the neural network, that is also non-periodic. (plots and detailed study of qualitative kinetics not presented in this paper)

To study in more detail the role of homeostasis in system' dynamics, we have experimented with different values of the characteristic time for homeostasis, τ_2, and measure the entropy of neural states and configuration states. Entropy of neural states and configuration states is formally defined according to standard information theory and statistical physics: $S(\mathbf{X}) = -\sum_i pr(\mathbf{X}_i) \cdot \log(pr(\mathbf{X}_i))$, where $\mathbf{X}_i$ is some neural state (a particular assignment of activation values x_i for

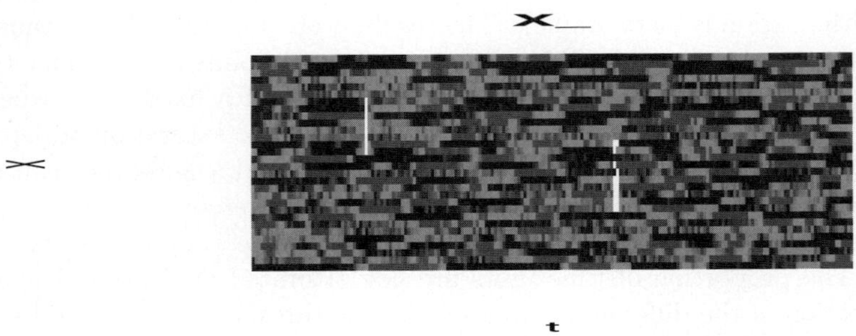

Fig. 4. A qualitatively typical system's behavior with random matrix ($\sigma^2(M) = 1$), no learning, moderate noise ($\sigma^2(\xi) = 0.2$), and homeostasis ($\tau_2 = \frac{\tau_1}{5}$): The system exhibits non-periodic behavior ($N = 30$).

all units), and $S(C) \equiv S(\psi) = -\sum_i pr(\psi_i) \cdot \log(pr(\psi_i))$, where ψ_i is some (small) range of angle values of the agent link. Fig. 5left) shows the values of the entropy of neural states as a function of the relative characteristic time of homeostasis τ_2 when compared with the characteristic time of the units activation dynamics τ_1. A value of $\frac{\tau_1}{\tau_2} = 0$ represents infinitely slow homeostasis (no homeostasis), and a value of $\frac{\tau_1}{\tau_2} = 1$ indicates that homeostasis occurs as fast as the units (main) activation dynamics. For each value of $\frac{\tau_1}{\tau_2}$ three simulation runs were tried with different random weights matrices (all with $(\sigma^2(M) = 1)$. To facilitate result comparisons the same weight random matrices were used for different values of $\frac{\tau_1}{\tau_2}$. Simulations were run for 500 time-steps, with $\tau_1 = 1$. The plots in Fig. 5 shows results for individual runs (open circles), and the averages over all runs for each value in the horizontal axis (red solid circles). To simplify analysis of entropy values, we removed noise from the network in this experiment $(\sigma^2(\xi) = 0)$.

The results confirm the initial observation, that homeostasis has a significant impact on system dynamics. When homeostasis is not present the neural entropy is very small (nearly 0) since the dynamics converges to a fixed point. When homeostasis is used but has large relative characteristic time $\frac{\tau_1}{\tau_2} \approx 0 (\neq 0)$ entropy increases, indicating that neural state escapes from fixed points. When homeostasis increases further and is approximately one order of magnitude lower than main activation dynamics, the effect is strongest. This represents the emergence of non-periodic (or quasi-periodic, or periodic) behavior. If homeostasis characteristic time increases beyond this point, until it is comparable with main dynamics characteristic time, the entropy decreases slightly indicating that the exploration of the neural state space is smaller. This is understandable since very rapid homeostasis implies that units have a very strong (individual) drive to be at equilibrium/rest activation value $x_i = x_0$, and become less prone to influences by other units. Overall, the plots show that the identified relationship between entropy of neural states and speed/characteristic time of homeostasis is non-linear and non-monotonic. Fig. 5right) shows a similar plot for the entropy of the configuration angle (proportional to the kinetic energy of the neural state, as explained above). The plot trend is also non-linear and non-monotonic. It is relevant to look at $S(C)$ since this represents the behavior of the embodied agent as perceived by an external observer[2].

Previous and on-going research, showed that when the learning rule is used to modify connection weights in combination with homeostasis, the system behavior remains qualitatively similar (mostly aperiodic) thus leaving the observations above intact. The effect of learning is to make units form high-connectivity (structural) clusters and produce correlated (functional) activity [12]. Future work will be to measure entropy values when learning is used to generate connection matrices with weight values of approximately the same order of magnitude as above $(\sigma^2(M) = 1)$, and see if learning promotes further exploration of the neural state space (higher entropy) or as the opposite effect.

[2] Logarithm base in Fig. 5left) is ternary and in the Fig. 5right) is binary.

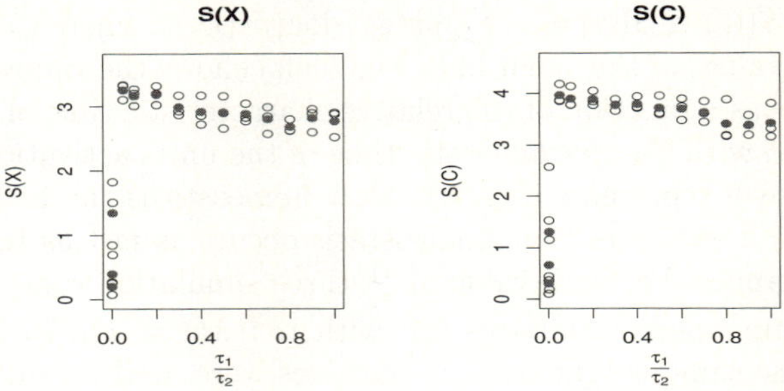

Fig. 5. Entropy of neural states **left)** and configuration state **right)** as function of relative characteristic time of homeostasis ($\frac{\tau_1}{\tau_2}$): open circles represent individual runs, red filled circles represent average over three runs with different random weight matrices ($N = 16$). The relationship is non-linear and non-monotonic. [Logarithm bases in left and right are different.]

5 Summary Discussion, Related Work, and Conclusions

We have presented a model of a recurrent neural network with homeostasic units for an embodied agent with a single degree of freedom. The activity of neural units generates muscular contraction, and determines the configuration angle of the agent. The neural model is motivated by a meta-model of embodied neural agents, whose goal is to inspire the design of agents that learn using principles of self-organization. Simulation results show that homeostasis in neural cells allow neural controller to "escape" fixed points and allow the agent to explore its neural and configuration state spaces. Additionally, we have shown that the functional relationship between relative characteristic time of homeostasis and the characteristic time of main activation dynamics is non-linear and non-monotonic. Exploration of neural and configuration state space is most extensive when homeostasis in one order of magnitude slower than main activation dynamics.

Homeostasic mechanisms have been identified in the biological neural networks literature [7], and its behavioral relevance is being explored by other researchers [13]. Classical cybernetics has identified homeostasic behavior as a key characteristic of natural and artificial adaptive system [9], however the link between homeostasic behavior and more complex types of agent behavior has not been made completely clear. Here we have showed that neural systems with homeostasic units at the individual/micro level, do not necessarily produce homeostasic behavior at the macro-level (moving the agent to a rest/homeostasic state if left unperturbed). Instead, micro and macro behavior of these type of systems can generate complex aperiodic patterns (quantitatively measurable using entropy values) that can be used as a source for behavior variety to experiment with the environment. This offers a conceptual link between classical control theory and cybernetics and modern complexity theory. Understanding how learning and mechanisms of self-perturbation can exploit this source of variety is an open research direction.

References

1. Maturana, H.R., Varela, F.J.: Autopoiesis and Cognition: the Realization of the Living. D. Reidel Publishing, Dordrecht (1980)
2. Reed, E.S.: Encountering the World: Toward an Ecological Psychology. Oxford University Press, New York (1996)
3. Pines, D., A., G., M., C.N.: Complexity: Metaphors, Models, and Reality. Perseus Books, Reading (1994)
4. Meyer, J.-A., Wilson, S.W.: From Animals to Animats: Proceedings of the First International Conference on Simulation of Adaptive Behavior. MIT Press, Cambridge (1991)
5. Varela, F.J., Bourgine, P.: Toward a Practice of Autonoumous Systems: Proceeding of the First European Conference on Artificial Life. MIT Press, Cambridge (1994)
6. Hopfield, J.J.: Neural networks and physical systems with emergent collective computational abilities. Proc. Natl. Acad. Sci. USA 70, 2554–2558 (1982)
7. Turrigiano, G., Nelson, S.B.: Homeostatic plasticity in the developing nervous system. Nature Reviews Neuroscience, 97–101 (2004)
8. Weng, J.: Developmental robots: theory and experiments. International Journal of Humanoid Robotics 1(2), 199–236 (2004)
9. Ashby, W.R.: Introduction to Cybernetics. Methuen, London (1956)
10. Beer, R.D.: Computational and dynamical languages for autonomous agents. In: Mind as motion: explorations in the dynamics of cognition table of contents, pp. 121–147. MIT Press, Cambridge (1996)
11. Strogatz, S.H.: Nonlinear Dynamics and Chaos: With Applications to Physics, Biology, Chemistry and Engineering. Perseus Books Group, Reading (1994)
12. Simão, J.P.: Self-perturbation and homeostasis in embodied recurrent neural networks: A meta-model and some explorations with mechanisms for sensorimotor coordination (submitted, 2007)
13. Di Paolo, E.A.: Organismically-inspired robotics: Homeostatic adaptation and natural teleology beyond the closed sensorimotor loop. In: Dynamical Systems Approach to Embodiment and Sociality. International Series on Advanced Intelligence, pp. 19–42. Advanced Knowledge International Press, Magill (2003)

Networks Regulating Networks: The Effects of Constraints on Topological Evolution

Francisco C. Santos[1], Hugues Bersini[1], and Tom Lenaerts[2,3]

[1] IRIDIA, CoDE, Université Libre de Bruxelles, Brussels, Belgium**
[2] SWITCH Laboratory, VIB, Brussels, Belgium
[3] Vrije Universiteit Brussel, Brussels, Belgium

Abstract. We propose a generalized framework to analyse constraints and representations in growing complex networks. We show that the introduction of biological, social and technological information by means of an additional network of constraints, together with the distinction between complete potential networks and instantaneous effective ones, can offer additional insights about the final topological outcome. Specifically, we study the emergence of exponential cutoffs in broad-scale degree distributions as a result of high level constraints.

1 Introduction

In the last years, it has been experimentally observed that real-world biological, social and technological networks are not structured in a random way [1]. Instead, most of these networks are organized in such a way that a few nodes are able to interact with a lot whereas many others only interact with a few. The extreme case is often referred as scale-free networks in which the degree distribution follows a power-law [2]. However, other configurations showing lower levels of heterogeneity are also common [3]. Dissections of real world networks have produced evidence for single-scale networks, characterized by a fast Gaussian decaying tail in the degree distribution, broad-scale distributions, defined by a power-law with an abrupt truncation for large connectivities [4] and, finally, the previously referred scale-free class. The ubiquity of such classes of networks raises many questions of which one is on the origin of these topological properties. Moreover, in the context of understanding complex social and biological phenomena, it is useful, and in many cases mandatory, to understand the topology of the underlying networks of interactions [1,5,6]. System's global properties rely extensively upon the underlying network and different dynamical outcomes emerge from different topologies [7,8,9,10]. One way of understanding these topological features relies on the development and the study (both mathematically and in software) of abstract models showing the same degree distribution as observed in empirical observations [2,11,5,12,13,14]. Although these universal mechanisms provides rules on how to construct networks that belong to the different classes,

** For further information contact fsantos@ulb.ac.be

F. Almeida e Costa et al. (Eds.): ECAL 2007, LNAI 4648, pp. 956–965, 2007.

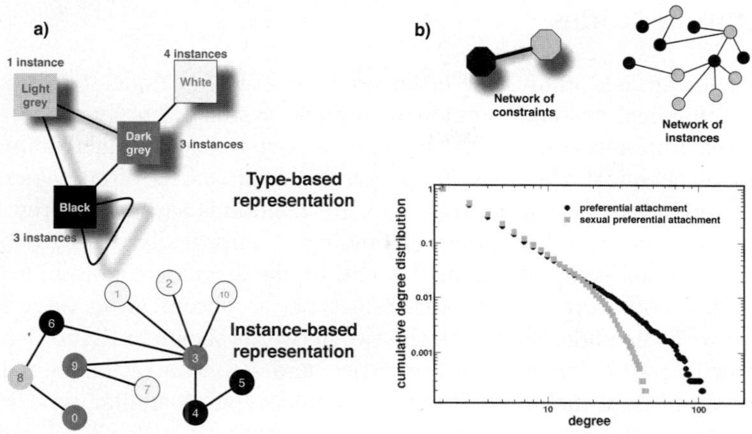

Fig. 1. a) Instance and Types. Instance-based: each node represents an instance of a certain type represented by a different colour (ex: social and technological networks). Type-based: each node represents a certain type with concentration equal to the number of instances of that type. (ex: biological and chemical networks). b) Sexual preferential attachment. Influence of intrinsic features of each node in the final degree distribution. Black dots stand for the cumulative degree distribution after growing a network of 1000 nodes using the Barabási and Albert Model of growth and preferential attachment [2] and the squares show the resulting cumulative degree distribution of the same model but constraining the preferential attachment rule to nodes of the opposite sexual tag. Here, we implicitly define a network of constraints with two nodes (without loops) and one link.

they do not clarify the domain-specific rules that explain the network topology in terms of the specific social, biological or technological domain. Since this explanation is essential for researchers in those particular domains, the models miss some key properties of the networks they aim at replicating. An incomplete set of the constructive laws that form the network can only lead to an incomplete assessment of the empirical data and of the history that produces them. This article proposes an enrichment of the modelling vocabulary and practice used to study growing networks. First we argue that there exists for each network a set of constraints that define its topology. Knowing this type restrictions will provide insight in how generalised growth functions can be defined. Second, we discuss that this set of type constraints define the network of all possibilities which is not necessarily the same as the network that is derived from the data. This differentiation between the potential and the instantaneous network underlines our argument that fitting generalized models to the data is not the best strategy since it does not explain the underlying rules of formation. To illustrate our argument we provide a few examples. Note that the examples are hence only purely illustrative. The main goal here is introduce a slightly richer common language to characterise complex networks and their evolution in time.

2 Defining Nodes

In biology, mean-field analysis is often used to attain an understanding of the complex dynamical system common in natural systems. For instance, in biochemical reaction networks, nodes are often associated to chemical components and links to chemical reactions [9]. A similar approach is often taken in the study of protein interaction networks [15,16]. A node is assumed to represent a certain protein and a link is drawn if binding is empirically observed between two proteins. Since each individual element, at the atomic or molecular level, is clustered in a node representing a certain type, a concentration value is often associated with the node. Similar to the two networks shown in Figure 1-a, nodes are types with particular physical properties, and a concentration value informs about the number of instances of any particular type. A similar type/instance dichotomy can be found in other biological networks such as food web networks, signal transduction networks, etc. As a rule, biological networks are always represented as type-based networks. There are two main motivations for this. First, most of the experimental data can be obtained only in this format. Second, it would almost be impossible to define such kind of networks at an instance level. Either it becomes trivial (for instance, it is hard to imagine a lion interacting with more than one specie at a time t) or too complex. When an instance of a protein interacts with another one, it will become part of a complex and, as a result, won't be able to connect to any other instance. In addition, the intrinsic properties of such node would no longer be the same. The type-level is indeed the obvious level of representation for biological networks of interactions.

Is this type/instance dichotomy meaningful in other scientific domains? It seems not. Social and technological networks are often studied by adopting another perspective. Here every node corresponds to one particular person, computer or HTML page, to give the three most usual examples. The justification of such a choice is mainly practical. First, this is the way in which empirical results are obtained. Second, in most of the cases, the consideration of type is simply meaningless. Defining a type-based network entails a lost in the amount of information (see Figure 1-a). Nonetheless, types are always there, and for each case a proper type representation could be found.

In biology, most of the recent modelling of network evolution does not take this distinction into account. First, types are treated as instances in order to drop the cumbersome issue raised by the dynamics at the level of the number of instances/concentration. This might be the best approach for both social and technological networks but is clearly oversimplified for chemical or biological ones. The absence of any biological or chemical information inherent in the definition of types leaves out some physical constraints certainly decisive in the network evolution. For instance, some nodes interact because their physical characteristics make such interactions extremely likely. This is definitely not related with any kind of topological consideration (like the amount of links that each of agents has). Accordingly, it is hard to accept a system that grows following a degree-preferential rule as biologically realistic. Some highly connected nodes are so, because they have intrinsic features that allow them to connect to a higher number of partners. Roughly, they are natural hubs or *born to be hubs* instead of being hubs just because of their presence in the network during a longer period. All these special features are included in the definition of each type and in the way links are defined. By disregarding concentration and the number of instances of a certain type, we loose one of the possible mechanisms behind the topological evolution of a growing network. The dynamics of the concentration/instance level often produces the necessary feedback for the network growing while this growing in turn constrains the dynamics. This constant interplay between the dynamics and the topological dynamics is something not taken into account in most of the network modelling.

3 Constraint Ties Among Networks

Adding a natural semantic to the definition of nodes entails the presence of an additional level of representation (we will call it *level 1* in the following) that will constrain what happens at the lower level (called *level 0* in the following). This is true in all kinds of networks. Some nodes, simply are not able to connect to some other nodes. These additional constraints may be due to all sort of reasons. For example, in social or economical networks, a European is more likely to interact with other Europeans. Considering such constraint tags when modelling the growth of networks, different results will be produced as a function of these constraints and their strength. Let's consider a first very simple example of sexual interaction network, where links and nodes represent sexual interactions and individuals, respectively. Such an interaction network has been shown to follow a scale-free degree distribution or at least to be part of a broad-scale class of interaction networks [17,18]. The evolution of this network is successfully modelled by a preferential attachment and growth model, proposed in [2]. In this case, an obvious constraint can be introduced in order to restrain the study to heterosexual relationships. This additional constraints network (*level 1* in Table 1) is simply defined by two gender nodes (male and female) connected by a link but without any loop (that would represent homosexual connections). Considering growth and preferential attachment model and stating that each node has 50 per cent chance of being part of one of the genders, the preferential attachment process will be restricted to the subset of nodes belonging to the opposite gender. The final result reflects precisely that (see Figure 1-b). The heterosexual constraint introduces an artificial cut-off that is frequently observed in experimentally derived networks [3].This sexual network example illustrates how constraints, reflecting intrinsic features of each node, influence the final outcome of the network. As discussed in the following paragraphs, more complex constraint networks will produce more complex effects.

In biology, proteins and chemical nodes have often been modelled by a key-lock mechanism [19,20,21,22,14]. It is assumed that nodes are defined by bit-strings and links by a pre-defined affinity function as in [14]. In this framework, the hamming distance between 2 bit-strings and a global or individual connection threshold have been classically used as affinity function, but many more choices

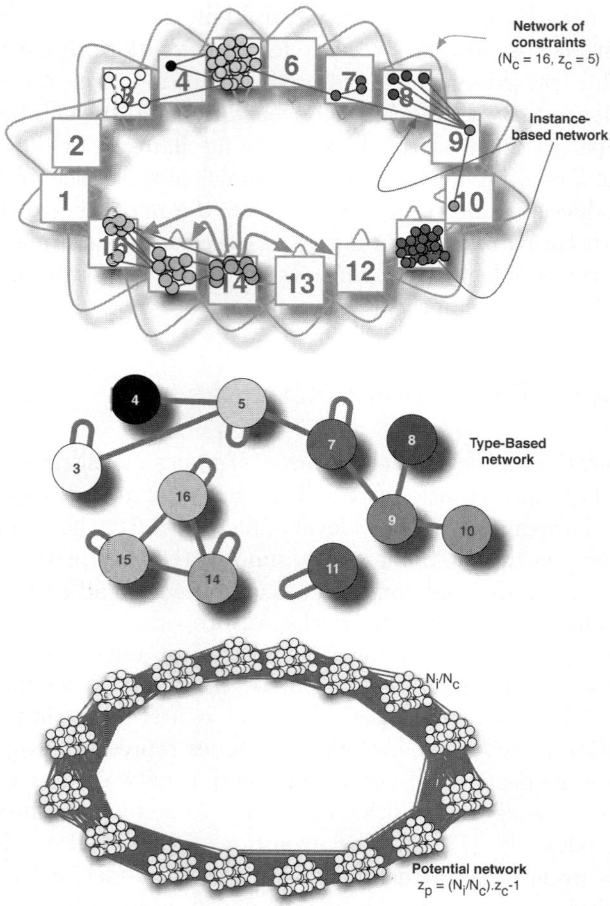

Fig. 2. Spatial constraints at different levels of representation. Upper panel: we have the network of constraints (squares and light grey links) with N_c nodes (16 in this example) and z_c as average degree (we allow loops at the constraints level, allowing instances of the same type to potentially interact). Each number/colour represents a different type. On top of the the constraints network, we can observe an example of an effective instance-based network, with $N_i = 42$ nodes. This last network helps to understand the correspondent type-based network in the middle panel, emerging as a subset of the constraints network. Finally, in the last panel, we see a potential network, drawn from the constraints network, with a maximum number of N_p nodes and z_p links per node. Each node will be part of a certain type, and will connect to all members of its type and to the ones belonging to types that have a link (at the constraints level) with its own type.

Table 1. Two basic distinctions: level 0 *vs* level 1 and instantaneous *vs* potential. We characterize as instantaneous the effective network at *level 0* and individual types at *level 1*. They are the interactions spotted at a given instant of time and turn out to samples of all interactions defined as potential ones. The instantaneous network, either at *level 0* or *1*, is a subset of nodes and links of the potential one originating from the structural evolution of the network.

	Level 0	Level 1
Potential	Potential network	Network of constraints
Instantaneous	Instance-based Representation	Type-Level Representation

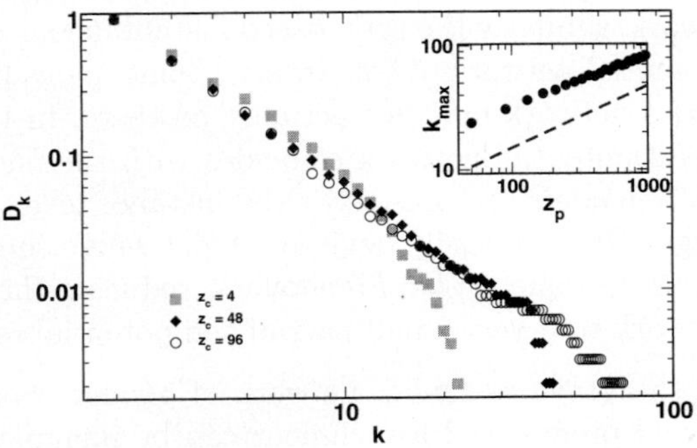

Fig. 3. Cumulative degree distribution (D_k) for different values of average connectivity at the constraints level (z_c). Inset: Maximum degree (k_{max}) dependence on the average degree at the potential level, $z_p = z_c \frac{N_i}{N_c} - 1$, where N_i and N_c represent the number of instances and the number of types, respectively. The dashed line has slope 1/2 and indicates the expected relationship between z_p and the finite-size cut-off given by $q_c \sim N^{\frac{1}{1-\gamma}}$ (see main text).

are possible. As a matter of fact, defining nodes with a specific tag or intrinsic characteristics, together with a list or a network of all constraints implicitly gives rise to a set of all possible interactions that can be represented as a potential network at either *level 0* or *level 1*. The effective or observed network, that will be called in the following *the instantaneous network*, boils down to a subset of nodes and links of this potential scenario. In the Barabási and Albert (BA) model of growth and preferential attachment [23], no constraint is defined and there is no need for an extra level of representation. The potential network at *level 0* is just a fully connected graph of size N and average degree $z_p = N - 1$. On the other hand, in the BA model with sexual constraints described before, the *level 1* potential network is simply a two nodes network with no loop. The *level 0* potential network turns out to be a fully connected bi-partide graph, reducing the number of potential links to $z_p = \frac{N}{2}$ inducing a finite-size effect [11] in the maximum connectivity of network (see below).

In the previous paragraphs, we introduced different representations and ways of constraining the network growth. Table 1 attempts at summarising and clarifying these distinctions. At the lowest level, *level 0*, first appears the instantaneous network of interactions between the individual instances currently present in the network. It is the real *physical* level at which the interactions take place. From that one, we can derive the instantaneous network at *level 1* reflecting the current constrained interactions between each type or flag that characterises all nodes present at *level 0*. It is impossible to obtain the individual interactions among instances from the network established at *level 1*, while the other way is the only possibility. Finally, all possible interactions at these two levels of representation are defined by a potential network, both at *level 0* and *level 1*. While the instance-based instantaneous network represents a time snapshot of the potential network, similarly the type-based instantaneous network represents a subset of the potential network of constraints. Some nodes likely to be present in the instantaneous network may not actually be there. In the sexual network example, the *level 1* potential network coincides with the *level 1* instantaneous network, since both genders are present at the instance level. This is not a general feature, though. One can easily devise a model where some of the types are not present in the instantaneous *level 0* network, reducing the size of the *level 1* instantaneous network to a very small part of the potential one.

By extending the previous trivial network of gender constraints, a similar model of growth and preferential attachment can be imagined, but with many more types and constraints. Let's start by introducing at *level 1* a general graph of constraints of size N_c and average degree z_c. Only instances belonging to connected types are able to interact. We can construe this network as describing possible interactions among families of proteins, international economical transactions (where each type is a country), or even shaped by religious or spatial constraints. Each of these examples gives rise to a different network of constraints with non-trivial topologies. A network with spatial constraints is certainly the easiest to illustrate. It can be characterized, in its simplest form, by a regular network [24]. Figure 2 shows how all four kinds of networks (*level 0* and *level 1*, *potential* and *instantaneous*) influence each other. If we consider that every node of type j can interact with all nodes present in a same square and with all nodes present into squares in the neighbourhood (see Figure 2), the potential network is (locally) a fully-connected network among all nodes of the same square and among all nodes of squares which are positioned side by side. Once again, the instantaneous networks (at both *levels 0* and 1) may differ from the potential network, regarding the present number of links and nodes, but will always be a subset of it.

Figure 3 shows the simple implementation of this idea by a generalised constraints network, where the effect of a regular network of constraints at *level 1* on the final structure of the network at *level 0* is studied. A regular network of constraints of size N_c (10^2 types) is considered, with an average degree $z_c = 5$, and $N_i = 10^3$ instances (a subset of all possible instances N_p) uniformly spread among all types. At each time step, a new instance is added and connect preferentially [2] to highly connected instances already present in the network. Depending on the average degree (z_c) of the potential network at *level 1*, a different average connectivity will be produced for the potential network at *level 0* (z_p). Each instance will be restrained to interact with the elements of its own type or flag and also with the instances belonging to the neighbourhood types. To simplify, we will consider in the following a finite potential network, in which the number of instances N_i has reached the maximum number of possible instances N_p. Hence, the potential average degree at *level 0* will be given by $z_p = z_c \frac{N_i}{N_c} - 1$, where N_i represent the number of instances and N_c the number of types. The fact that the average degree of the potential network is no longer equal to $N_i - 1$ (like in the usual Barabási and Albert model [23] where every node can potentially connect to every other node), induces typical finite-size effects. The preferential attachment universe is now reduced to the universe of potential partners (z_p), and the finite-size cut-off that scales up proportionally to $N^{\frac{1}{1-\gamma}}$ ($\gamma = 3$ being the resultant slope of the degree distribution $d_k = k^{-\gamma}$) [11] instead scales up with $z_p^{1/1-\gamma}$. This result is shown in the inset of Figure 3. These cut-off effects are only visible for high connectivity levels, though. The network is perfectly scale-free, but only from a local perspective, and repeats itself throughout the entire network for typical scales above z_p [25]. Figure 3 also shows that the obtained *level 0* instantaneous network is well represented by a stretched exponential degree distribution [26] or a power-law with a exponential cut-off for high connectivities [25,27,28]. The network shows a strong dependence on the amount of links at *level 1*. Depending on the strength of the constraints, different cut-offs emerge from the growing process and the power-law obtained with the Barabási-Albert model is recovered when z_c moves the constraint network to a fully-connected graph. Moreover, the topology of *level 1* may influence the topology of *level 0* in different ways, which may not change the average potential degree. For instance, suppose each type to be a country or city which some, as a result of social and economical reasons, interact with more countries or cities than others [29]. To represent this feature, some degree heterogeneity need to be introduced at *level 1*, being the regular network replaced by other class of topologies [3]. This might produce non-trivial results at *level 0*. The typical cut off could become smoother and allows some instances to receive more links than others. From the beginning, there will be nodes that can potentially interact with many more than others – *some nodes are born to become hubs*. Such effects will be tackled in a future work.

4 Conclusions

This last example, together with the even more simplified sexual one, sufficiently illustrates the effects of an extra level of complexity on the growing process of a network, justifying the need for this enlarged perspective. These conceptual additions may contribute to a better understanding of phenomena associated with the structural evolution of complex networks. To the dynamics of the nodes and the evolution of the topology, external constraints, coming from intrinsic features of the nodes and the way they do interact, need to be added. Moreover, the interplay between the two levels of representation and between the instantaneous and the potential networks remains an open problem in various scientific domains. We have tried in this paper to propose an wide perspective, allowing a better introduction of additional effects which are responsable for shaping networks' evolution. Besides analysing all interacting agents as a whole, different levels of description need to be accounted and their influence investigated, both in the experimental data and in simulation. Furthermore, instead of choosing one unique level of complexity and representation for each problem, a more extended framework has been demonstrated to produce important results that are necessary to understand the intricate network of influences that underlie each complex phenomenon.

Acknowledgments. Discussions with Jorge M. Pacheco are gratefully acknowledged. FCS acknowledges the support of COMP^{2}SYS, a Marie Curie Early Stage Training Site, funded by the EC through the HRM activity.

References

1. Albert, R., Barabási, A.L.: Statistical mechanics of complex networks. Rev. Mod. Phys. 74, 47–97 (2002)
2. Barabási, A.L., Albert, R.: Emergence of scaling in random networks. Science 286(5439), 509–512 (1999)
3. Amaral, L.A., Scala, A., Barthelemy, M., Stanley, H.E.: Classes of small-world networks. Proc. Natl. Acad. Sci. U S A 97(21), 11149–11152 (2000)
4. Tanaka, R., Yi, T.-M., Doyle, J.: Some protein interaction data do not exhibit power-law statistics. FEBS letters 579, 5140–5144 (2005)
5. Dorogotsev, S.N., Mendes, J.F.F.: Evolution of Networks: From Biological Nets to the Internet and WWW. Oxford University Press, Oxford (2003)
6. Barabási, L., Oltvai, Z.N.: Network biology: Understanding the cell's functional organization. Nature Reviews Genetics 5, 101–113 (2004)
7. May, R.M., Lloyd, A.L.: Infection dynamics on scale-free networks. Phys. Rev. E 64(6 Pt 2), 66112 (2001)
8. Pastor-Satorras, R., Vespignani, A.: Epidemic spreading in scale-free networks. Physical Review Letters 86(3200) (2001)
9. Guimerá, R., Amaral, L.A.N.: Functional cartography of complex metabolic networks. Nature 433(895) (2005)
10. Santos, F.C., Pacheco, J.M.: Scale-free networks provide a unifying framework for the emergence of cooperation. Phys. Rev. Lett. 95(9), 098104 (2005)
11. Dorogovtsev, S.N., Mendes, J.F., Samukhin, A.N.: Size-dependent degree distribution of a scale-free growing network. Phys. Rev. E 63(6 Pt 1), 062101 (2001)
12. Vazquez, F., Flamini, A., Maritan, A., Vespignani, A.: Modeling of protein interaction networks. ComplexUs 1, 38–44 (2003)
13. Wagner, A.: How the global structure of protein interaction networks evolve. Proc. R. Soc. Lond. B. 270, 457–466 (2003)
14. Bersini, H., Lenaerts, T., Santos, F.C.: Growing biological networks: Beyond the gene-duplication model. Journal of Theoretical Biology 241(3), 488–505 (2006)

15. Uetz, P., Giot, L., Cagney, G., Mansfield, T.A., Judson, R.S., Knight, J.R., Lockshon, D., Narayan, V., Srinivasan, M., Pochart, P., Qureshi-Emili, A., Li, Y., Godwin, B., Conover, D., Kalbfleisch, T., Vijayadamodar, G., Yang, M., Johnston, M., Fields, S., Rothberg, J.M.: A comprehensive analysis of protein-protein interactions in saccharomyces cerevisiae. Nature 403(6770), 623–627 (2000), `http://dx.doi.org/10.1038/35001009`
16. Uetz, P., Finley, J.: From protein networks to biological systems. FEBS letters 579(8), 1821–1827 (2005)
17. Liljeros, F., Edling, C.R., Amaral, L.A., Stanley, H.E., Aberg, Y.: The web of human sexual contacts. Nature 411(6840), 907 (2001)
18. Liljeros, F., Edling, C.R., Amaral, L.A.N.: Sexual networks: implications for the transmission of sexually transmitted infections. Microbes Infect 5(2), 189–196 (2003)
19. Varela, F., Coutinho, A.: Second generation immune network. Immunology Today 12(5), 159–166 (1991)
20. Detours, V., Bersini, H., Stewart, J.: Development of an idiotypic network in shape space. Journal of Theor. Biol. 170, 401–404 (1994)
21. Perelson, A.S.: Immune network theory. Immunological Reviews 10, 5–36 (1989)
22. Perelson, A., Weisbuch, G.: Immunology for physicists. Rev. Mod. Phys. 69, 1219–1268 (1997)
23. Barabási, A.L., Albert, R., Jeong, H.: Mean-field theory for scale-free random networks. Physica A 272, 173–187 (1999)
24. Watts, D.J., Stogatz, S.H.: Collective dynamics of small-world'networks. Nature 393(440) (1998)
25. Rozenfeld, A.F., Cohen, R., Ben-Avraham, D., Havlin, S.: Scale-free networks on lattices. Phys. Rev. Lett. 89(21), 218701 (2002)
26. Xulvi-Brunet, R., Sokolov, I.M.: Evolving networks with disadvantaged long-range connections. Phys. Rev. E 66(2), 026118 (2002)
27. Gastner, M.T., Newman, M.E.J.: The spatial structure of networks. Eur. Phys. J. B 49, 247–252 (2006)
28. Yang, K., Huang, L., Yang, L.: Lattice scale-free networks with weighted linking. Phys. Rev. E 70(1), 015102 (2004)
29. Colizza, V., Barrat, A., Barthelemy, M., Vespignani, A.: Prediction and predictability of global epidemics: the role of the airline transportation network. Proc. Natl. Acad. Sci. U S A 103, 2015–2020 (2006)

Preliminary Investigations on the Evolvability of a Non-spatial GasNet Model

Patricia A. Vargas, Ezequiel A. Di Paolo, and Phil Husbands

Centre for Computational Neuroscience and Robotics (CCNR)
University of Sussex BN1 9QH, United Kingdom
{p.vargas, ezequiel, p.husbands}@sussex.ac.uk

Abstract. This paper addresses the role of space in evolving a novel Non-Spatial GasNet model. It illustrates that this particular neural network model which make use of modulatory effects of diffusing gases has its evolvability improved when its neurons are not constrained to a Euclidean space. The results show that successful behaviour is achieved in fewer evaluations for the novel unconstrained GasNet than for the original model.

1 Introduction

Our aim in this work is to explore and analyze the role of spatial organization and processes in a particular class of non-classical artificial neural network (ANN) called GasNets. We embark on this endeavour by proposing and investigating the evolvability [1] of a novel spatially unconstrained version of this network model, which we call NSGasNet (Non-Spatial GasNet). We expect that this investigation may have wider implications for understanding the interactions between space, neuro-modulation, temporal dynamics and evolvability in several systems apart from GasNet models.

The original GasNet neurocontroller uses a simple model of a diffusing gas that may affect the transfer function of nodes within a delimited spatial range. Such a model has proved significantly more evolvable for certain robotic tasks than non-gaseous neurocontrollers [2, 3]. Further developments of this model have led to even higher evolvability [4, 5, 6]. Hypotheses concerning this increased evolvability have been discussed, including the combined role of slow and fast temporal timescales and the flexible coupling between distinct processes in the networks, but still no clear answer has been found to explain the success of this model.

There are good reasons to believe that the inherent spatiality of the original model may play a significant role in enhancing evolvability. Spatiality introduces a topology where variations produced by mutation events may be smoother and it may enhance the exploration of modular architectures [7]. The effects, however, are far from linear or straightforward. For instance, the Plexus GasNet model [4, 5] has introduced a spatial decoupling between synaptic and gaseous interaction, and the freeing up of this constraint has resulted in enhanced evolvability [6].

The objectives of the present investigation are to begin a series of systematic statistical explorations into the role of space in GasNets. We will introduce some comparisons between a spatial and a non-spatial version of GasNets for a central pattern generator task (a CPG task). The CPG was chosen as they avoid very simple regularities

F. Almeida e Costa et al. (Eds.): ECAL 2007, LNAI 4648, pp. 966–975, 2007.

and have significant timescale properties. Some generalization to tasks evolved under similar conditions and of similar complexity might be expected.

We start by giving a brief account of existing versions of GasNets plus a novel model in Section 2. In Section 3 we describe our experiments in detail including the respective network architecture and genetic encoding. Section 4 highlights the evolutionary regime. The results for the task are presented in Section 5 and in Section 6 we provide a discussion of the main findings and future directions.

2 Non-classical Artificial Neural Networks

At the beginning of the past decade, with the advent of remarkable discoveries of neuronal modulation and non-standard cell-to-cell signaling in the biological nervous system [8], a novel ANN model, namely GasNet, was proposed by Husbands [2]. This particular architecture could be considered non-classical as it involves non-synaptic chemical signaling as well as synaptic interactions. The network is conceptualized as operating on a 2D Euclidean plane, thus both types of interaction are constrained by a spatial relation in the sense that the synaptic connections and neuronal modulation by diffusing gases are restricted to a spatial range. Hence, all neural connections depend upon a spatial organization. In this work, a new GasNet model is devised in which there is no spatial relation among neurons in order to help us investigate the role played by space.

2.1 Spatially Constrained GasNet Models

The rationale behind this gaseous artificial neural network model is to mimic the production and release of nitric oxide (NO) by real neurons, in order to affect long and short-term modulations of the behaviour of other neurons in (spatial) range. In fact, the original GasNet model is a discrete-time, recurrent neural network with a variable number of nodes. These nodes can be connected in terms of synapses by either excitatory (with a weight of +1) or inhibitory (with a weight of -1) links and, in terms of dynamic gas modulation depending on their spatial relation [3].

In the GasNet model, the classical sigmoid output function $y = \tanh(x)$ of each neuron at each time step is modulated by a transfer function parameter which will define which curve from the family of eleven sigmoids will be employed during the network's operation. Almost all GasNet parameters and variables are under evolutionary control.

2.2 Non-spatial GasNet: The NSGasNet Model

We have devised a novel spatially unconstrained GasNet model (NSGasNet). In this model nodes do not have a location in a Euclidean space. In the absence of a spatial relation, all emitted gases can spread freely among neurons, thus there is no notion of a gas cloud anymore.

In this new scenario, all nodes can in principle be affected by the gas emitted from any other node. Therefore it was envisaged that a *sensitivity limit* should be imposed to each network node in order to regulate the strength of modulation. The *sensitivity limits* are under evolutionary control lying in the range [0, 1] and are specific to each other emitting node. This can be understood as if there are 'gas' connections between nodes of a strength defined by the respective *sensitivity limit*.

In the NSGasNet model, the *sensitivity limit* was named *Mbias* (modulator bias) and its product with the amount of gas emitted $T(t)$ will now determine the gas concentration at the node. Each node will have a modulator bias lying in the range [0,1] for every emitting node. Therefore, given an emitting node, any network node could "decide" whether it will be affected (*Mbias* > 0), or not (*Mbias* = 0) by the gas emitted, without the requirement of being within its gas cloud limits.

The NSGasNet network genetic encoding differs from the original model [3] in the absence of variables that deal with spatial parameters, such as node *coordinates*, *spatial electrical connectivity* and *maximum radius of emission*. Similar to the original GasNet model, the network size, topology and almost all its parameters are also under unconstrained evolutionary control. Nonetheless, each node has a list of modulator biases (one modulator bias for each emitting node).

Depending on the task, the network is encoded on a variable-sized genotype, where each gene represents a network node. A genotype consists of an array of integer variables lying in the range [0, 99] (each variable occupies a gene locus). The decoding from genotype to phenotype obeys the same simple laws for continuous values and for nominal values adopted for the original model [3]. Apart from the task-dependent parameters, the NSGasNet model has 6 variables associated with each node plus 1 modulator bias for each node. Suppose the network is composed of 4 nodes, than the NSGasNet genotype will have 6+4+task dependent parameters.

In the following sections we will provide a set of comparisons between the original spatially constrained GasNet model and the novel non-spatial NSGasNet model in an attempt to address the role of space in the evolvability of both networks.

3 Comparative Experiment

CPGs are widely known neuronal circuits found in almost all nervous systems from invertebrates [9] to vertebrates [10, 11]. They are used in many biological functions including the production of rhythmic movement found in locomotion in many species [12]. These circuits have been under investigation for a long time [13, 14, 15, 16] in a hope to fully uncover how locomotion patterns, such as swimming, walking and flying are internally organized and coordinated.

The original GasNet model has already been successfully applied to CPG problems [4, 17]. In fact, the experiment described in our CPG task was first suggested by Smith [17]. It is comprised of different target patterns representing sequences of bits from the set {0, 1}. Networks are evolved to generate the required patterns. Four patterns were tested (Table 1).

As a CPG produces a regular cyclic pattern of movement resulting in the animal behaviour usually cycling among a set of states, our idea here is to mimic these states dictated by a set of synthetic patterns to be learned by the ANN.

In this experiment, both GasNet models, original and NSGasNet, were designed as fully connected ANNs (including self-connections) with four nodes (Fig. 1). We decided to start with simple systems in order to try to fully uncover them, furthermore, the task does not require more complex networks and that there is still an important

Table 1. CPG patterns for the CPG task adapted from Smith [17].

Ten:Four	Eleven:Five	Eleven:Seven	Seven:Five
1111111111:0000	11111111111:00000	11111111111:0000000	1111111:00000

number of node arrangements modulated by a set of possible distances that may make the spatial factor an important one.

Apart from each model's particularities, both genetic encodings only differ from the basic GasNet in that the synaptic weights are also under evolutionary control lying in the range [-1, 1], and there are no other electrical connectivity parameters.

Hence, the original GasNet genotype will have 9 parameters for each node, which makes a total of 36 parameters for the entire network plus 6 parameters for the synaptic connection weights. Each NSGasNet gene will have 6 parameters for each node plus 4 parameters for the modulator bias (4 nodes), giving a total of 40 parameters for the entire network plus 6 parameters for the synaptic connection weights.

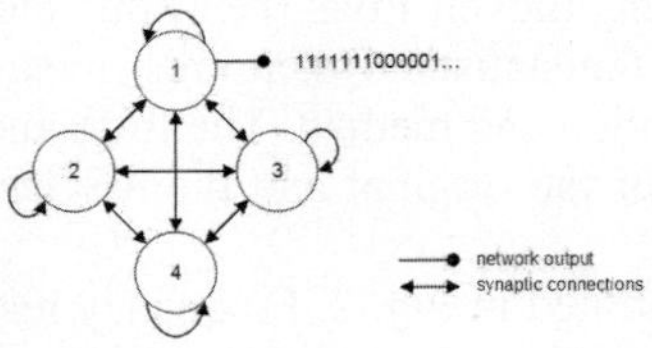

Fig. 1. Pictorial example of a fully-connected ANN for the CPG task with four nodes. The network does not receive external input and the first network neuron output determines the network output.

The choice of fully interconnected networks for this task follows from previous experiments on another CPG task [18, 19]. Nonetheless, in a more recent work, Psujek, Ames and Beer [20] investigated, among other issues, the number of connections required to achieve high performance in a CPG for walking in a simple legged body and stated that "far sparser than fully interconnected circuits are sufficient to achieve high performance on the walking task…".

In this primary investigation, we will adopt a full connectivity, however in future work we envisage using not only partially connected networks, but will also explore the network *metadynamics*, i.e. exploring a variety of network dimensions (including number of nodes) during the evolutionary process, both of which in our opinion might lead to superior results for our CPG task.

For the evolutionary regime, we employ a distributed steady-state genetic algorithm as described in Smith [17]. In order to gather statistics, fifty runs were performed for

Table 2. Statistics on the evolvability in terms of number of fitness evaluations for each GasNet model, original and NSGasNet, over 50 runs for each CPG pattern. The values are presented in the following order: mean, number of successful evolved networks, standard deviation and median.

Pattern	Eleven-Seven	Eleven-Five	Ten-Four	Seven-Five
Original				
Mean/n	23310/21	15048/21	20085/20	33218/22
(Std) Median	(29123) 12200	(15343) 7600	(23924) 7350	(31001) 15500
NSGasNet				
Mean/n	11231/36	15691/34	11845/31	9252/34
(Std) Median	(18209) 3050	(22105) 6050	(16395) 4000	(15523) 3200

the CPG. One evolutionary run is composed of a maximum of 1,000 generations for the CPG task or until successful genotypes are produced. Each generation comprises 100 reproduction events or fitness evaluations.

Each genotype is evaluated over 200 steps. The network output is "1" if the activation of the output node is greater than zero, and "0" otherwise. The network output is compared to the pattern output at each step and a value of one is added to its fitness each time the value is the same. A weighted sum is performed at the end of the steps in order to account for the different numbers of ones and zeros of each test pattern. The fitness is than scaled to [0, 1].

4 Results

Here we present the statistical measures of evolvability for the original GasNet and the NSGasNet in the CPG task. The statistics on the number of fitness evaluations for the original and NSGasNet models for each CPG pattern are shown in Table 2. The first column presents the CPG patterns: Eleven-Seven, Eleven-Five, Ten-Four and Seven-Five (Section 3). The second column presents the statistical measures: mean, number of successful evolved networks, standard deviation and median. The third and fourth columns present the values for each measure for the original and the NSGasNet, respectively.

All values displayed in Table 2 are graphically illustrated in Fig. 2. Frequency histograms comparisons between the two GasNet models over the number of fitness evaluations for each pattern are illustrated in Fig. 3.

A quantitative analysis of the data shown in Fig. 2 demonstrates that the distribution is not symmetric, actually all distributions are skewed to the right. Therefore the mean is always greater than the median (Table 2). Moreover, there is also the presence of outliers (a), (b), (c), (d), (e) and (h) and the large values of standard deviations show that the data has a great variability (Table 2). The median is less sensitive to outliers and therefore we will use the median instead of the mean in order to compare evolvability between the GasNet models for all patterns.

From Table 2, it is clearly observed that the NSGasNet model outperforms the original model, not only in terms of the median as a measure of evolvability but also in terms of the total number of successfully evolved networks.

As the evolved networks were fully connected, the synaptic connections architecture might not have had a great impact on the networks evolvability and dynamics.

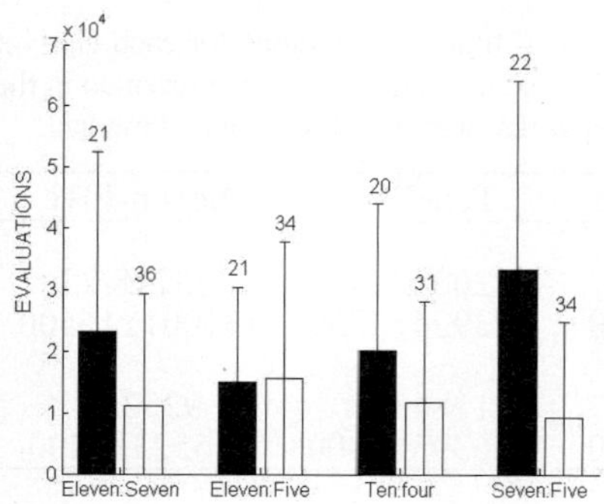

Fig. 2. Mean and standard deviations (error bars) of fitness evaluations required to evolve successful networks for each CPG pattern, Eleven-Seven, Eleven-Five, Ten-Four and Seven-Five. Black bar shows original mean data and white bar shows NSGasNet mean data. The numbers above each error bar represent the total number of successfully evolved networks within 50 runs.

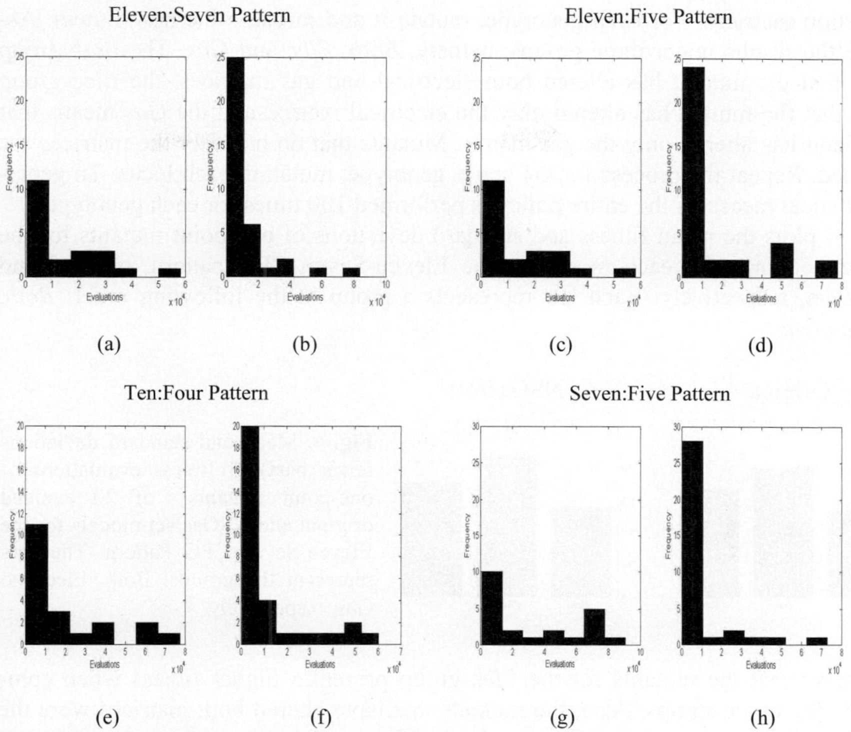

Fig. 3. Frequency histograms comparison between the original (graphics: (a), (c), (e) and (g)) and the NSGasNet (graphics: (b), (d), (f) and (h)) models over the number of fitness evaluations for each pattern. The distribution is conditional on the number of successfully evolved networks within 50 runs for each case (Fig. 2).

Hence, the gaseous connections were mostly responsible for the performance of successfully evolved networks, especially for the NSGasNet network, where the absence of spatial constraints seemed to have contributed to a better performance in terms of speed of evolution. For instance, some evolved NSGasNet networks made explicit use of the modulator bias in a scenario where the presence of a spatial limitation would not have allowed such coupling between the nodes involved.

In both models, after evolution some nodes had no participation on the network dynamics, either for their synaptic weights were set to zero or for there were no gaseous connections whatsoever.

5 Analyzing the Fitness Landscape

In order to investigate the influence of the fitness landscape on evolvability, we have decided to carry out a series of systematic explorations of the fitness landscape around the best networks for each case, thus testing the smoothness of the landscape. This analysis will be performed under the following policy, adopted from Philippides *et al.* [6]. Take the 20 best genotypes of each model and generate the electrical and the gas connection matrices. For each genotype, mutate it and measure the new fitness. Assemble the results under three groups, namely, *Both*, *Elec* and *Gas*. The *Both* group means that the mutant has altered both electrical and gas matrices, the *Elec* group means that the mutant has altered only the electrical matrix, and the *Gas* means that the mutant has altered only the gas matrix. Mutants that do not alter the matrices are discarded. Repeat the process for the entire genotype, mutating each locus. To generate statistical measures the entire policy is performed 100 times for each genotype.

Fig. 4 plots the mean fitness and standard deviations of one-point mutants for the 20 best networks for each model on the Eleven-Seven CPG pattern, original and NSGasNet, respectively. Each bar represents a group in the following order: *Both*, *Elec* and *Gas*.

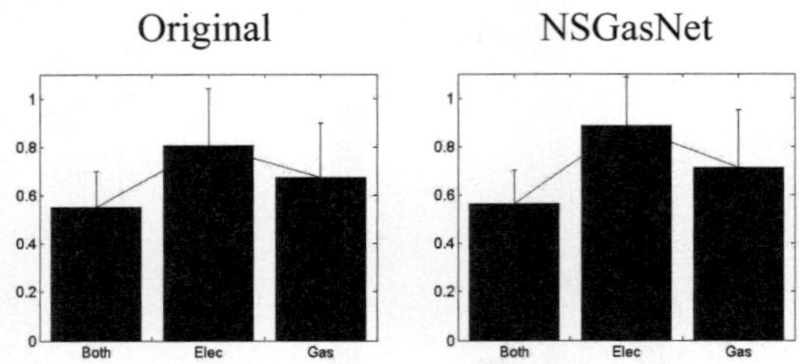

Fig. 4. Mean and standard deviations (error bars) of fitness evaluations of one-point mutants of 20 evolved original and NSGasNet models for the Eleven-Seven CPG Pattern. The bars represent the groups: Both, Elec and Gas, respectively.

Observe that the mutants for the *Elec* group present a higher fitness when compared to the other groups. Also, the mutants that have altered both matrices were the most deleterious ones, thus scoring the lowest fitness. To better access the statistical results, Fig. 5 shows the fitness distribution of the fitness for each group. The first row shows the results for the original model and the second row the NSGasNet model. The histogram depicts the number of fitness evaluations for each group distributed along the fitness value. The first column shows the results for the *Both* group, the second column for the *Elec* group and the third column for the *Gas* Group. The distribution is conditional on the number of fitness evaluations for all one-point mutants within 100 iterations for each successfully evolved network.

From Fig. 5 it is possible to notice that the fitness landscape is not smooth for it seems that many mutations take the individuals to a very low fitness value. This is represented by the high number of mutants with low fitness, especially for the *Gas* group (3rd column). It is now necessary to verify whether the *Gas* group is the best representative of the results. Therefore in Fig. 6, we show the result of the proportion, i.e. the number of one-point mutants divided by the total number of mutants for each group and for each model. It is clear that the number of *Gas* type mutants is much higher than the other groups.

This investigation shows that the landscapes are both equally rugged for the spatial and non-spatial versions. However, spatial constraints affect gas connections in one case and not in the other and this is the set of mutations that have the strongest effect on fitness (Fig. 6), so it is possible that removing spatial constraints from this type of interaction has an overall positive effect on the search. However, this needs further investigation with a more detailed analysis of the evolutionary dynamics.

It is important to highlight that the same analysis was performed for the other CPG patterns and the results are similar.

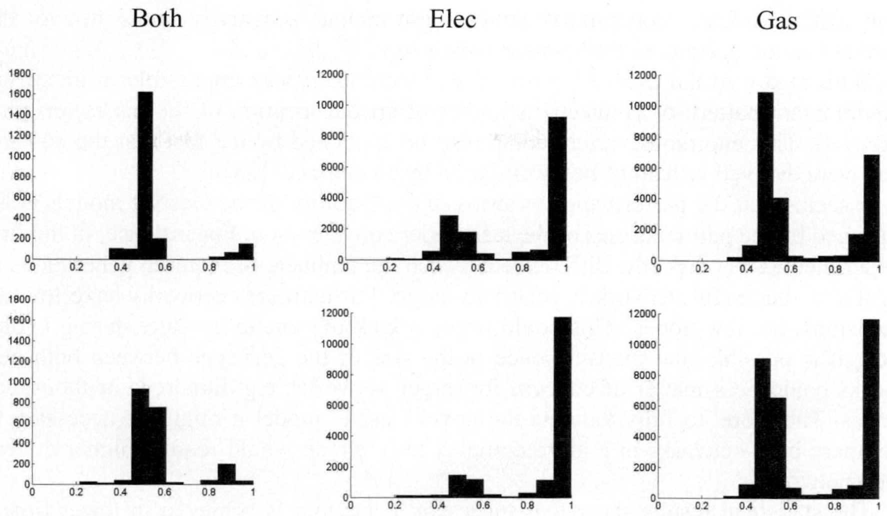

Fig. 5. Frequency histograms comparison between the original (first row) and the NSGasNet (second row) models over the number of fitness evaluations for each group. *Both* group (first column), *Elec* group (second column) and *Gas* Group (third column). The distribution is conditional on the number of fitness evaluations for all one-point mutants within 100 iterations for each successfully evolved network.

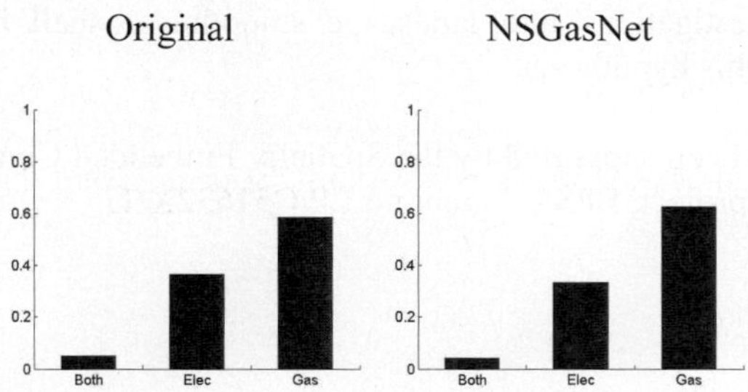

Fig. 6. Proportion of one-point mutants for each group and each model, original and NSGasNet respectively for the Eleven-Seven CPG pattern.

6 Discussion

In this paper, we have attempted to address the role of space in evolving GasNet models by illustrating that this particular neural network model has its evolvability improved when its neurons are not constrained to a Euclidean space. A quantitative analysis of the statistical data was performed and the most relevant findings were reported.

The adoption of a scheme whereby the strength of neuromodulation is regulated by a genetically-determined sensitivity allows us to break from the constraints of Euclidean space. It is still questionable whether some other form of spatiality is introduced by this scheme (where sensitivities would be analogous to some function of distance). The scheme, however, is general and does not comply with basic properties of a distance measure (i.e., symmetry and triagle inequality) so that the model is properly non-spatial. A future comparative study might include sensitivity limits also for the spatial version in order to further assess their role.

With regard to the evolved network architectures, it was impossible to identify a predominant pattern of connections and/or of spatial location of the nodes (original GasNet). This enormous variety could also be explained by the fact that the controllers were evolved instead of being designed by an engineer [21].

It seems that the performance in terms of evolvability of the GasNet models is influenced by the particularities of the task under consideration. For instance, in the first experiment (CPG task) the difference between the numbers of required generations to evolve a successful network is relatively large. Furthermore, networks have low dimensions, i.e. few nodes. This could imply a lack of genetic pressure during evolution; it is possible that the difference in the size of the genotype between both networks could be a matter of concern for larger networks, e.g. hundreds or thousands nodes. Therefore, to fully validate the novel GasNet model it might be necessary to compare both networks in a more complex task, which would require higher dimension networks.

The statistical results show that successful behaviour is achieved in fewer fitness evaluations for the novel unconstrained GasNet than the original model, supporting a tendency investigated by previous GasNet models which had effectively introduced a sort of spatial decoupling between synaptic and gaseous interactions [6].

Finally, we intuit that the role of space might be directed linked to the smoothness of the fitness landscape. An initial analysis of the fitness landscape appeared to demonstrate that the less smooth the landscape the faster is the NSGasNet in terms of evolvability. Certainly, further investigations into landscape smoothness shall be performed in order to fully validate this hypothesis.

Acknowledgements. This work has been supported by the Spatially Embedded Complex Systems Engineering (SECSE) project, EPSRC grant no EP/C51632X/1.

References

1. Yao, X.: Evolving Artificial Neural Networks. Proc. of the IEEE 87(9), 1423–1447 (1999)
2. Husbands, P.: Evolving Robot Behaviours with Diffusing Gas Networks", in Husbands, P. In: Husbands, P. (ed.) Evolutionary Robotics. LNCS, vol. 1468, pp. 71–86. Springer, Heidelberg (1998)
3. Husbands, P., Smith, T., Jakobi, N., O'Shea, M.: Better living through chemistry: Evolving GasNets for robot control. Connection Science 10(3-4), 185–210 (1998)
4. Philippides, A,: Modelling the Diffusion of Nitric Oxide in Brains. PhD thesis, School of Biological Sciences, University of Sussex, UK (2001)
5. Philippides, A., Ott, S.R., Husbands, P., Lovick, T., O'Shea, M.: Modeling Cooperative Volume Signaling in a Plexus of Nitric Oxide Synthase-Expressing Neurons. The Journal of Neuroscience 25(28), 6520–6532 (2005a)
6. Philippides, A., Husbands, P., Smith, T., O'Shea, M.: Flexible Couplings: Diffusing Neuromodulators and Adaptive Robotics. Artificial Life 11, 139–160 (2005b)
7. Fine, P., Di Paolo, E.A., Philippides, A.: Spatially Constrained Networks and the Evolution of Modular Control Systems. In: Nolfi, S., Baldassarre, G., Calabretta, R., Hallam, J.C.T., Marocco, D., Meyer, J.-A., Miglino, O., Parisi, D. (eds.) SAB 2006. LNCS (LNAI), vol. 4095, Springer, Heidelberg (2006)

8. Garthwaite, J.: Glutamate, nitric-oxide and cell-cell signaling in the nervous system. Trends in Neuroscience 14, 60–67 (1991)
9. Getting, P.A.: Emerging Principles Governing the Operation Of Neural Networks. Annual Review of Neuroscience 12, 185–204 (1989)
10. Grillner, S., Wallén, P.: Central Pattern Generators for Locomotion, with Special Reference to Vertebrates. Annual Review of Neuroscience 8, 233–261 (1985)
11. Grillner, S,: Neural Networks for Vertebrate Locomotion. Scientific American, 48-53 (January 1996)
12. Delcomyn, F.: Neural Basis for Rhythmic Behaviour in Animals. Science 210, 492–498 (1980)
13. Katz, P.S., Getting, P.A., Frost, W.N.: Dynamic Neuromodulation of Synaptic Strength Intrinsic to a Central Pattern Generator Circuit. Nature 367, 729–731 (1994)
14. Marder, E., Calabrese, R.L.: Principles of Rhythmic Motor Pattern Generation. Physiological Reviews 76, 687–717 (1996)
15. Roberts, P.D.: Classification of Temporal Patterns in Dynamic Biological Networks. Neural Computation 10, 1831–1846 (1998)
16. Katz, P.S., Fickbohm, D.J., Lynn-bullock, C.P.: Evidence that the Central Pattern Generator for Swimming in Tritonia Arose from a Non-Rhythmic Neuromodulatory Arousal System: Implications for the Evolution of Specialized Behavior. Amer. Zool. 41, 962–975 (2001)
17. Smith, T. M. C,: The Evolvability of Artificial Neural Networks for Robot Control. PhD thesis, CCNR, Department of Informatics, University of Sussex, UK (2002)
18. Chiel, H.J., Beer, R.D., Gallagher, J.C.: Evolution and Analysis of Model CPGs for Walking I. Dynamical Modules. Journal of Computational Neuroscience 7, 99–118 (1999)
19. Beer, R.D., Chiel, H.J., Gallagher, J.C.: Evolution and Analysis of Model CPGs for Walking II. General Principles and Individual Variability. J. of Comp. Neuroscience 7, 119–147 (1999)
20. Psujek, S., Ames, J., Beer, R.D.: Connection and Coordination: The Interplay Between Architecture and Dynamics in Evolved Model Pattern Generators. Neural Comp. 18, 729–747 (2006)
21. Beer, R.D.: A Dynamical System Perspective on Agent-Environment Interaction. Artificial Intelligence 72, 173–215 (1995)

Semi-synchronous Activation in Scale-Free Boolean Networks

Christian Darabos[1], Mario Giacobini[2,3], and Marco Tomassini[1]

[1] Information Systems Department, University of Lausanne, Switzerland
{christian.darabos, marco.tomassini}@unil.ch
[2] Dpt. of Animal Production Epidemiology and Ecology, University of Turin, Italy
[3] Molecular Biotechnology Center, University of Turin, Italy
mario.giacobini@unito.it

Abstract. We study the dynamics of Boolean networks of the scale-free type. The model takes into account the topology and abstracts recent findings about real genetic regulatory networks. We propose a new, more biologically plausible, semi-synchronous update scheme on networks of larger sizes. We simulate statistical ensembles of networks and discuss the attractors of the dynamics, showing that it is compatible with theoretical biological network models. Moreover, then model demonstrates interesting scaling abilities as the size of the networks is increased.

1 Introduction

Random Boolean Networks (RBN) are a highly simplified model of genetic regulatory networks that have been introduced by Kauffman [1]. In a RBN with N nodes, a gene is represented as a binary node, meaning that the gene is expressed if it is on (1), and it is not otherwise (0). Each gene receives inputs from K randomly chosen other genes. Initially, one of the possible Boolean functions of K inputs is assigned at random to each gene. The network dynamics is discrete and synchronous: at each time step all nodes simultaneously examine their inputs, evaluate their Boolean functions, and find themselves in their new states at the next time step. The system travels in time through its phase space, until a point or cyclic attractor is reached whence either it will remain in that point attractor forever, or it will cycle through the states of the periodic attractor. Since the system is finite and deterministic, this will happen at most after 2^N time steps.

This model has been analyzed at length and its interesting dynamical behavior has been fully studied [2,3]. Here we summarize the main results.

First of all, as some parameters are varied such as K, or the probability p of switching the node's state, the RBN can go through a phase transition. Indeed, for every value of p, there is a critical value of connectivity K such that for values of K below this critical value the system is in the ordered regime, while for values of K above this limit the system is said to be in the chaotic regime. $K = 1$ corresponds to the ordered regime, $K = 2$ is critical, and for $K \geq 3$ the system is in the chaotic phase. Kauffman found that for $K = 2$ the size

F. Almeida e Costa et al. (Eds.): ECAL 2007, LNAI 4648, pp. 976–985, 2007.

distribution of perturbations in the networks is a power law with finite cutoff that scales as the square root of N. Thus perturbations remain localized and do not percolate through the system. The mean cycle length scales at most linearly with N for $K = 2$. Kauffman's conjecture is that attractors correspond to cell types in the RBN phase space, and only those attractors that are short and stable under perturbations would be of biological interest. Thus, according to Kauffman, $K = 2$ RBN lying at the edge between the ordered and the chaotic phases can be seen as abstract models of genetic regulatory networks.

We believe that the original view of Kauffman that these models may be useful for informing us on the behavior of real cell regulatory networks, is still a valid one, provided that the model is updated to take into account present knowledge about the structure of real gene regulatory networks, and the timing of events, without loosing its attractive simplicity.
Kauffman's RBN model rests on three main assumptions:

- The nodes implement random Boolean functions and their state is either on or off;
- The nodes that affect a given node in the network are randomly chosen and are a fixed number;
- The dynamics of the network is synchronous in time.

The binary state simplification is radical but actually it represents quite well "threshold phenomena" in which variables of interest suddenly change their state, such as neurons firing or genes being switched on or off. Moreover, the random nature of the nodes' activation functions seems an acceptable simplification for this model according to the present lack of knowledge about the biological regulatory networks.

Random regular graphs were a logical choice in the beginning, since the exact couplings in networks were generally unknown. Today it is more open to criticism since it does not correspond to what we know about the topology of biological networks. In fact, many biological networks, including genetic regulatory networks, seem to be of the scale-free type or of a hierarchical type (see [4,5] and references therein) but not random, according to present data. For scale-free networks, this means that the distribution function of the degree, i.e. the probability $P(k)$ that a randomly chosen node has degree k, is a power law $P(k) \sim k^{-\gamma}$, usually with $2 < \gamma < 3$, instead of a Poisson distribution as in a random graph, or a delta distribution as in a classical RBN. Thus the low connectivity suggested by Kauffman for candidate stable systems is not found in such networks, where a wide range of degrees is present instead, given the distribution's fat tail. This fact is likely to have important consequences for the dynamics.

Boolean networks with scale-free topology were first used by Oosawa and Savageau [6]. They took *Escherichia coli* as a model for their scale-free nets with an average input degree $\bar{K}$ of two. But, although interesting in this particular case, this is too limited as most other networks have higher connectivity levels. Aldana has recently presented a detailed analysis of Boolean networks with scale-free topology [7]. He has been able to define a phase space diagram for boolean networks, including the phase transition from ordered to chaotic dynamics, as a function of the power law exponent γ. He also made exhaustive simulations for several relatively small values of the network size N.

Our model has in common with Aldana's the scale-free topology of the networks, although the graphs are constructed in a different way. But, in contrast to Aldana's, we define a suitable semi-synchronous dynamics for the system, instead of using the customary synchronous update. A preliminary account of our work appears in [8], where we used the same small system sizes as Aldana's for the sake of comparing with his results. In the present work we offer the following extensions with respect to the previous work. First of all we use larger graphs for the simulations. The maximum N in Aldana's work and in [8] was 19, which is far from typical values of real networks and does not permit a proper statistical characterization of the graph ensembles. In addition, we have improved our construction of the scale-free networks and we have revised and improved the semi-synchronous update that were described in [8].

1.1 Scale-Free Network Topologies

Kauffman's RBN are directed graphs. Let's suppose that each node i ($i \in \{1, \ldots, N\}$) receives k_i inputs and projects a link to other l_i nodes, i.e. there are l_i nodes in the graphs that receive an input from node i. Among the N nodes of the graph, the distribution $P_{in}(k)$ of the input connections is not necessarily the same of the distribution of the output connections $P_{out}(k)$.

According to present data, many biological networks, including genetic regulatory networks, suggest a scale-free output distribution $P_{out}(k)$ and a Poissonian input distribution $P_{in}(k)$ [4].

In a previous work [8] we have generated networks according to a mixed generalized/poisson random graph model with a maximum of 19 vertices, following Aldana's footsteps. Although these models show interesting dynamical behavior, one can only wonder if their size is biologically relevant. In fact, from the work of Vasquez [4], one can infer that complex biological networks are much larger. In this matter, our original model showed some weaknesses. First, it did not scale very well, the bigger the network became, the farther away the outgoing distribution went from the intended power-law curve. Then, it did not allow us to pre-define an average input/output degree $\bar{K}$, which reduces drastically the model's flexibility. Finally, the poisson-random in-degree distribution allowed nodes without any input ($k_i = 0$), which is biologically very unlikely, as biological systems usually have all sorts of both external and internal stimuli.

Addressing the first two points, here we use a new *Modified Configuration Model* (MCM) fully described in [9]. We start by defining the desired average in/out degree $\bar{K}$ which in turn determines the total number of edges. Next, we generate the power-law degree distribution of outgoing links according to the chosen exponent γ. Subsequently, we specify a narrow Poissonian function in order to produce the incoming degree distribution, ensuring a minimal k_i^{in} degree for each vertex, in this our model almost comes back to Kauffman's RBN

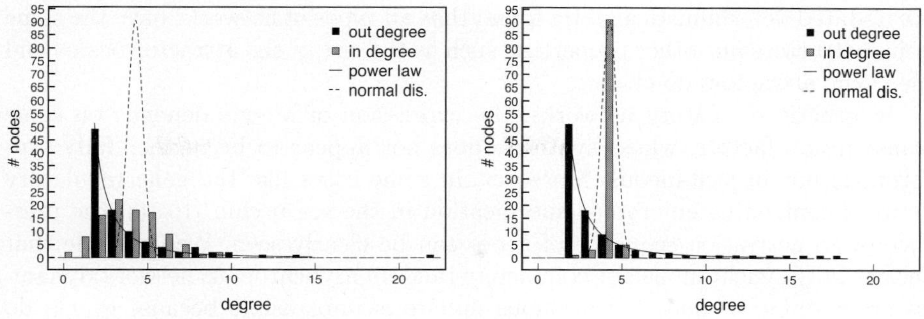

(a) (b)

Fig. 1. Input and output degrees distributions of scale-free networks of size $N = 100$ and $\gamma = 2.5$ obtained using (a) the mixed generalized/poisson random graph model and (b) the Modified Configuration Model ($\bar{K} = 4$).

with fixed k. According to both the input and output degrees distribution edges are drawn at random, excluding multiple and self-connections. Figure 1 compares the resulting input and output degrees distribution for both methods described above.

1.2 Semi-synchronous Network Dynamics

Standard RBN update their state synchronously. This assumption simplifies the analysis, but it is open to discussion if the network has to be biologically plausible. In particular, for genetic regulatory networks, this is certainly not the case, as many recent experimental observations tend to prove. Rather, genes seem to be expressed in different parts of the network at different times, according to a strict sequence (see, for instance, [10]). Thus a kind of serial, asynchronous update sequence seems to be needed.

Several researchers have investigated the effect of asynchronous updating on classical RBN dynamics in recent years [11,12,13,14,15]. Harvey and Bossomayer [11] studied the effect of random asynchronous updating on some statistical properties of network ensembles, such as cycle length and number of cycles, discovering that, while point attractors do persist, there are no true cyclic attractors, only so-called loose ones and states can be in more than one basin of attraction. Also, the average number of attractors is very different from the synchronous case.

Mesot and Teuscher [12] confirmed the findings of [11]. They studied the critical behavior of asynchronous RBN and concluded that they do not have a critical connectivity value analogous to synchronous RBN and they behave, in general, very differently from the latter. Di Paolo and Rohlfshagen [13] mainly investigated rhythmic and non-rhythmic attractors using evolutionary algorithms.

Gershenson [14,15] extended the analysis and simulation of asynchronous RBN by introducing additional update policies in which specific groups of nodes are updated deterministically. He found that all types of networks have the same point attractors but other properties, such as the size of the attractor basins and the cyclic attractors do change.

In genetic regulatory networks, the expression of a gene depends on some transcription factors, whose synthesis does not appear to be neither fully synchronous nor instantaneous. Moreover, in some cases like the gene regulatory network controlling embryonic specification in the sea urchin [10,16], the presence of an activation sequence of genes can be clearly seen. We conclude that neither fully synchronous nor completely random asynchronous network dynamics are suitable models. Synchronous update is implausible because events do not happen all at once, while completely random dynamics does not agree with experimental data on gene activation sequences and the model does not show stable cyclic attractors of the right size. In our opinion, the activation/update sequence in a RBN should be in some way related to the topology of the network.

In [8] we have proposed a topology-driven semi-synchronous update method, called *Cascade Update* (CU). Although not a faithful model for true biological gene activation sequences, we do believe that our proposed scheme is closer to biological reality than the previously proposed ones: fully synchronous (SU) and various asynchronous policies. Nevertheless, after this first experience with semi-synchronous update schemes, we noticed there is room for improvement. Namely, in this first approximation, at time $t = 0$, a starting node would be chosen at random and updated according to its Boolean function. At the next time step, all the nodes receiving an incoming edge from the first one will be updated. These will in turn influence other nodes through their outgoing links and so the cascade goes. The model does not differentiate whether the updated state at a given time s_i^t of the node is *on* or *off*, which is unrealistic for two reasons. Firstly, one can clearly imagine that, depending on the connectivity of the graph, the list of nodes to be updated will, after a few time steps only, contain all the nodes that have incoming edges. This part of the graph will then be updated in a synchronous manner, leaving untouched only a handful of vertices without incoming edges. Secondly, we ought to consider the influence of one node on another as active biological *activating* or *repressing* factors: only when the state of the node is turned or stays *on* has this node an effect on the subsequent nodes in the cascade. In contrary, nodes changing their state to or remaining *off* have no impact on nodes they are linked to, thus breaking the cascade. In other words, only the activation of an activator or a repressor will have a repercussion on the list of nodes to be updated at the next time-step. We have called this new update scheme the Activated Cascade Update (ACU).

As a consequence, the definition of point or cyclic attractors changes slightly, because the state of a network at any give time t is, from now on, not only determined by the individual state $s_i^t \in \{on, off\}$ of each node but also by the list l^{t+1} of nodes to be updated at the next time step. The concept of loose attractor has, in this context, no relevance.

2 Simulations and Methodology

Using the MCM model described in Section 1.1 we produce networks having a scale-free distribution of the output degrees and a Poisson distribution of the input degrees. In this work we investigate the effect of the new ACU update scheme presented in Section 1.2 vs. the previous CU, and SU for a set of γ exponents of the scale-free distribution $\gamma \in \{2.0, 2.5, 3.0\}$. In an effort to probe the network scaling properties, we have simulated ensembles of graphs with $N \in \{50, 100, 150, 200\}$, all with a connectivity of $\bar{K} = 4$.

For any given network produced, we fix an initial configuration (IC), which is a set of initial boolean values chosen at random with probability $p = 0.5$, for each node of the network. Subsequently, a update function is attributed to each node, consisting of a randomly initialized lookup table which entries is determined by the input degree of that particular node. Then, we let the system stabilize over 10'000 initial steps during which the chosen update scheme determines the next nodes to be updated. This allows the system to evolve and maybe reach the basin of an attractor. Should the cascade stop because there is no node to be updated, another starting node is selected at random. After these initial steps, we determine over another 1'000 time steps if the system has reached an attractor by tracing the set of consecutive states visited, verifying for each one whether it has already been reached or not. If so, we define the length of that attractor as the minimum number of steps necessary to reach a known state. Let us recall here that a system state at a give time s^t is defined as the ensemble of all individual states s_i^t of each node and the list of nodes to be updated at the next time step l^{t+1}. Each network is exposed to 500 ICs. This is repeated 20 times, each with a new set of update rules for each node, this is called a realization. In order to be thorough, we study 50 different networks (i.e. 1'000 realizations) this way, for each values of γ, N and each update scheme.

3 Analysis of the Results

Synchronization. Before diving into the matter of the attractors as such, we have conducted an analysis of the dynamical behavior of the two semi-synchronous update scheme, that is the CU and the ACU. As mentioned in Section 1.2, we expect CU to become equivalent to the fully synchronous update scheme, that is, after a few time steps, all the nodes will be part of the update list l_i^t, thus the system become fully synchronous and all nodes are updated at each time steps. We define as synchronous a semi-synchronous update scheme which has updated the same subset of nodes for over half of the time steps. Table 1 shows that in the case of CU, 100% of the simulations, for all values of γ and N, have resulted in synchronous updates, whereas about 20% have synchronized in the case of ACU with 50 nodes, less then 10% have with $N = 100$, below 5% for $N = 150$ and around 2% for $N = 200$. This difference when scaling makes perfect sense because, as the degree remained constant, probabilities of update all nodes simultaneously decreases.

Table 1. Fraction of the simulations where the semi-synchronous update scheme has fully synchronized, for all values of γ and N.

	$N = 50$		$N = 100$		$N = 150$		$N = 200$	
γ	CU	ACU	CU	ACU	CU	ACU	CU	ACU
2.0	1.0	0.189	1.0	0.022	1.0	0.02	1.0	0.01
2.5	1.0	0.228	1.0	0.044	1.0	0.04	1.0	0.02
3.0	1.0	0.233	1.0	0.085	1.0	0.02	1.0	0.02

Number of Attractors. During the simulations, we have analyzed for each IC of each realization if the system has relaxed to a single state (point attractor) or cycled through the states of the periodic attractor (cyclic attractor). Biologically speaking, a point attractor has a very limited significance, because it would mean either that the system vegetates with no chance of evolving or adapting, and ultimately the death of the system or the end of the differentiation cycle of a stem cell. Therefore, in Table 2 we differentiate the number of point attractors from the number of cyclic ones. We only consider attractors of length between 1 and 50 states.

Table 2. Number of point attractors / cyclic attractors for semi-synchronous update schemes, for all values of γ and N.

	$N = 50$		$N = 100$		$N = 150$		$N = 200$	
γ	CU	ACU	CU	ACU	CU	ACU	CU	ACU
2.0	37/461	32'000/2'690	2/2	20'147/2'383	0/0	8'794/414	0/0	6'994/73
2.5	89/603	41'954/3'807	5/6	31'264/4'234	0/0	10'962/691	0/0	6'961/59
3.0	92/661	42'188/3'417	9/33	52'525/7'018	0/2	22'632/2'843	0/0	11'374/650

As expected, the difference between SU and CU, due only to the non-deterministic nature of the networks, ICs and update rules, is not significant. Therefore results for SU are not shown here to save space. We note that the number of attractors, both point and cyclic, found with the ACU is between 1'000 and 10'000 times greater than those found with CU. The table also shows there is a significant drop in the number of attractors as N grows with CU, and virtually no attractors at all for values of $N = 150$ and $N = 200$. In the case of ACU, the decline is less dramatic.

We also witnessed this change in the dynamics of the system when increasing the size of the networks by plotting the average number of realizations per attractor length. Figure 2 shows that in the case where $N = 50$ using SU and CU, less than 10% of the realizations have no attractors, the majority, about 60%, have between 1 and 3 attractors, 15% have 4 and the rest has more. In opposition, with $N = 100$, the system struggles finding attractors, leaving over 90% of the realizations without attractors. This cleavage does not appear when using ACU, where the distributions of realizations over the number of attractors

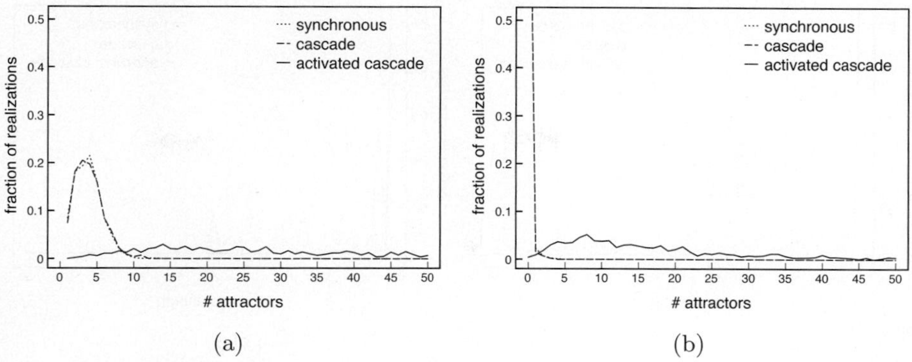

Fig. 2. Fraction of realizations having a given number of attractors with $\gamma = 2.5$ for (a) $N = 50$ and (b) $N = 100$. In (b) over 90% of SU and CU realizations have no attractors (not plotted here for readability reasons). The SU and CU curves are almost superposed.

Table 3. Average length of cyclic attractors for synchronous and semi-synchronous update schemes, for all values of γ and N.

	$N = 50$			$N = 100$			$N = 150$	$N = 200$
γ	SU	CU	ACU	SU	CU	ACU	ACU	ACU
2.0	19.04	18.81	7.71	8.89	7.00	6.54	6.30	8.96
2.5	18.87	20.06	7.86	5.94	14.67	6.94	6.47	4.69
3.0	18.15	19.05	8.12	10.35	13.24	6.50	7.52	8.73

is approximately constant, with a peak between 3% to 5% for values 5 to 15 attractors, and there results remain consistent with $N = 150$ and $N = 200$ (not shown here). As there is almost no attractors for $N > 100$ with CU and SU, we did not plot results for those values.

Attractors Length. We have also computed the average length of cycles for attractor sizes 2 to 50. Considering their prominent number, we have omitted point attractor that would drag the average down. Table 3 depicts the different average length of the attractors for all combinations of update methods and γ values. We see that, although the number of cyclic attractors between the distinct cases sometimes differs by several orders of magnitude, the average attractors length remains comparable, though constantly higher for SU and CU. As there is no attractors in the SU and CU case with $N > 100$, we represent only the average length of cyclic attractors obtained with ACU.

Finally, Figure 3 shows the proportion of attractors for lengths from 1 to 50, over all the attractors. Again, the behavior of ACU does not noticeably change the distribution of attractors length when we go from 50 to 100 nodes. In contrast, there is, in proportion, more short attractors for $N = 100$ than for $N = 50$. Again, we cannot plot comparison for attractors lengths on networks with $N > 100$, as only ACU has reached attractors.

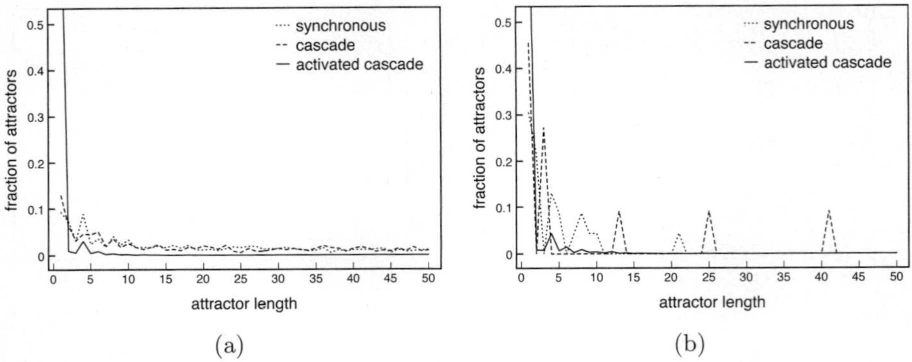

Fig. 3. Fraction of attractors found having a given length (between 1 and 50) with $\gamma = 2.5$ for (a) $N = 50$ and (b) $N = 100$. For ACU, over 90% of the attractors found are point attractors (not plotted here for readability reasons).

In all cases, we notice unexpected increase in value for attractors of even-length attractors, especially 2 and 4. We can imagine that this is due to the binary nature of our networks. In the case of a point attractor, no node changes and there is thus a unique state. If we have 2 states, that could be explained by one node flipping from *on* to *off* and back at each time step. Similarly, 4 states would mean 2 nodes are flipping, and so on.

4 Conclusions and Future Work

In this study we have presented a new semi-synchronous model for the dynamics of scale-free RBN that is more faithful to the present-day knowledge about genetic regulatory network data. With respect to our previous work [8], both the scale-free networks' generation algorithm and the update scheme have been completely revised and modified to better model the biological metaphor.

The new Modified Configuration Model algorithm for network generation with a given degree distribution is shown to produce graphs with a better agreement to the desired statistical properties with respect to our previous model.

The new Activated Cascade Update is shown to recover the main limitation of the CU, the final synchronization of the update sequence. This lockstep update does not agree with microarray data on gene activation sequences. Moreover, this new update scheme characterizes an attractors' state space with drastic differences with respect to that of RBN with synchronous update. More attractors (both point and cyclic ones) are found and these results show interesting scaling properties as the network size is increased.

In the future we intend to thoroughly investigate the tolerance of our model to faults, such as single genes knock-out, and the diffusion of perturbations.

Acknowledgements. The authors thank F. Di Cunto and P. Provero of the University of Torino (Italy) for the useful discussions and suggestions on biological regulatory networks. M. Tomassini and Ch. Darabos gratefully acknowledge financial support by the Swiss National Science Foundation under contract 200021-107419/1.

References

1. Kauffman, S.A.: Metabolic stability and epigenesis in randomly constructed genetic nets. J. Theor. Biol. 22, 437–467 (1969)
2. Kauffman, S.A.: The Origins of Order. Oxford University Press, New York (1993)
3. Aldana, M., Coppersmith, S., Kadanoff, L.P.: Boolean dynamics with random couplings. In: Kaplan, E., Marsden, J.E., Sreenivasan, K.R. (eds.) Perspectives and Problems in Nonlinear Science. Springer Applied Mathematical Sciences Series, pp. 23–89. Springer, Berlin (2003)
4. Vázquez, A., Dobrin, R., Sergi, D., Eckmann, J.-P., Oltvai, Z.N., Barabàsi, A.-L.: The topological relationships between the large-scale attributes and local interactions patterns of complex networks. Proc. Natl. Acad. Sci USA 101(52), 17940–17945 (2004)
5. Barabasi, A.-L., Oltvai, Z.N.: Network biology: understanding the cell's functional organization. Nature Reviews: Genetics 5, 101–113 (2004)
6. Oosawa, C., Savageau, M.A.: Effects of alternative connectivity on behavior of randomly constructed Boolean networks. Physica D 170, 143–161 (2002)
7. Aldana, M.: Boolean dynamics of networks with scale-free topology. Physica D 185, 45–66 (2003)
8. Giacobini, M., Tomassini, M., De Los Rios, P., Pestelacci, E.: Dynamics of scale-free semi-synchronous boolean networks. In: Rocha, L.M., et al. (eds.) Artificial Life X, pp. 1–7. MIT Press, Cambridge (2006)
9. Tomassini, M., Giacobini, M., Darabos, C.: Performance and robustness of cellular automata computation on irregular networks. Advances in Complex Systems (ACS) (to be published, 2007)
10. Davidson, E.H., et al.: A genomic regulatory network for development. Science 295, 1669–1678 (2002)
11. Harvey, I., Bossomaier, T.: Time out of joint: attractors in asynchronous random boolean networks. In: Husbands, P., Harvey, I. (eds.) Proceedings of the Fourth European Conference on Artificial Life, pp. 67–75. MIT Press, Cambridge (1997)
12. Mesot, B., Teuscher, C.: Critical values in asynchronous random boolean networks. In: Banzhaf, W., Ziegler, J., Christaller, T., Dittrich, P., Kim, J.T. (eds.) ECAL 2003. LNCS (LNAI), vol. 2801, pp. 367–376. Springer, Heidelberg (2003)
13. Rohlfshagen, P., Di Paolo, E.A.: The circular topology of rythm in asynchronous random Boolean networks. BioSystems 73, 141–152 (2004)
14. Gershenson, C.: Classification of random Boolean networks. In: Standish, R. (ed.) Artificial Life VIII: Proceedings of the Eight International Conference on Artificial Life, pp. 1–8. MIT Press, Cambridge (2002), http://alife8.alife.org/proceedings/sub67.pdf
15. Gershenson, C.: Updating schemes in random Boolean networks: Do they really matter? In: Pollack, J. (ed.) Artificial Life IX Proceedings of the Ninth International Conference on the Simulation and Synthesis of Living Systems, pp. 238–243. MIT Press, Cambridge (2004), http://uk.arxiv.org/abs/nlin.AO/0402006
16. Olivieri, P., Davidson, E.H.: Genes regulatory network controling embryonic specification in the sea urchin. Current Opinion in Genetics and Development 14, 351–360 (2004)

Spatial Embedding and Complexity: The Small-World Is Not Enough

Christopher L. Buckley and Seth Bullock

School of Electronics and Computer Science, University of Southampton, UK
clb05r|seth@ecs.soton.ac.uk

Abstract. The "order for free" exhibited by some classes of system has been exploited by natural selection in order to build systems capable of exhibiting complex behaviour. Here we explore the impact of one ordering constraint, spatial embedding, on the dynamical complexity of networks. We apply a measure of functional complexity derived from information theory to a set of spatially embedded network models in order to make some preliminary characterisations of the contribution of space to the dynamics (rather than mere structure) of complex systems. Although our measure of dynamical complexity hinges on a balance between functional integration and segregation, which seem related to an understanding of the small-world property, we demonstrate that small-world structures alone are not enough to induce complexity. However, purely spatial constraints can produce systems of high intrinsic complexity by introducing multiple scales of organisation within a network.

1 Introduction

From its outset, artificial life has concentrated on how simple properties can give rise to complex organisation and behaviour. The interplay between, *inter alia*, non-linear, local interactions, physical constraints, noise, and processes of copying or competitive exclusion have been shown to give rise to self-organisation, auto-catalysis, path dependence, and emergent behaviour in many different ways [1,2]. For most complex systems, spatial embedding is a potential source of enabling constraint. By projecting a system of interacting elements into a low dimensional space, local correlations are imposed and maintained. These correlations can predispose systems to exhibit behaviour that would otherwise be unstable [3].

The recent explosion of interest in the "new science of networks" has focused attention on the application of graph-theoretic approaches to the characterisation of natural and engineered systems. While the influence of space is at least implicit in certain of the graph structures discussed and employed in this literature, its contribution has yet to be systematically explored. For instance, Stanley Milgram's now infamous demonstration of the "six degrees of separation" that apparently link members of society to each other through mutual acquaintance relies upon space. His instruction to each experimental subject was to deliver a package to a person identified only by name and place of residence. Subjects

F. Almeida e Costa et al. (Eds.): ECAL 2007, LNAI 4648, pp. 986–995, 2007.

were clearly required to combine their social *and geographical* knowledge to meet this challenge.

Likewise, when Watts and Strogatz [4] went on to formalise the notion as the "small-world property", they also made explicit use of spatial embedding. They construct a lattice where the pattern of connectivity reflects the regular (isotropic, homogeneous) spatial organisation of the nodes, and find that repeated application of random rewiring events gradually erodes the spatial organisation until a random graph results. Intermediate between the ordered lattice and the disordered random graph, Watts and Strogatz characterised small-world structures that simultaneously exhibit a small characteristic path length and a high degree of clustering.

Here we are interested in the relationship between spatial constraints, network topology, and functional complexity, an information-theoretic measure of which has been developed in order to characterise some important properties of both vertebrate and invertebrate nervous systems [5]. In particular, it has been used to identify, and quantify, a tension that lies at the heart of neural information processing. On the one hand, the brain must integrate distributed information in order to produce coherent behaviour, on the other, a great deal of experimental work demonstrates that neural regions specialise [6]. It has been suggested that a measure of complexity might reconcile the notions of neural segregation and integration within a single theoretical framework [5]. In particular, biological networks have high complexity by this measure, which has been linked to spatial constraints on connectivity (along with other structural properties, e.g., hierarchical organisation) [5]. Furthermore, it has been suggested that both the small-world property and high complexity are coincident in biological neural networks [7]. This paper examines these issues, first exploring the extent to which spatially constructed small worlds are associated with high complexity, and then assessing the more general impact of spatial embedding and spatial structure on network complexity.

2 The Complexity Measures

Consider a network X comprising a set of n nodes interacting according to an adjacency matrix Ω. Each element is subject to a low magnitude noise signal, but there is no other external input. By assuming that the interactions between elements are linear and that the dynamics of the system is well approximated by a multi-dimensional stochastic process such that temporally extended trends in the dynamics can be ignored, a measure of intrinsic complexity can be derived [5].

While such systems seem somewhat removed form what one would expect of the average artificial recurrent neural network (e.g., sensory input, nonlinear behaviour with rich temporal dynamics), these idealisations are useful in understanding certain biological network formulations [5]. More generally systems that interact weakly, and hence almost linearly, have been studied in neuroscience and are thought to be a good approximation of the dynamics in some regions of the nervous system [8].

The level of dependence and independence between sets of elements in such a system can be measured through the concept of mutual information. Equation (1) gives the mutual information, $MI(X)$, between the jth of k subsets of X and its complement $X - X_j^k$, in terms of the entropy, $H(x)$. Entropy is derived from consideration of the covariance between the activity of each of the elements, denoted by the covariance matrix **COV** which can be considered to quantify the dependence between each of the elements [5].

$$MI(X_j^k; X - X_j^k) = H(X_j^k) + H(X - X_j^k) - H(X), \tag{1}$$
$$H(X) = 0.5 \ln((2\pi \exp)^n |\mathbf{COV}|)$$

An estimate of the integration (i.e., the shared information) between the elements of a subset is given by equation (2), which measures the difference between the sum of the deviations from independence of each element taken independently, and the entropy of the system as a whole.

$$I(X) = \Sigma_{i=1}^{n} H(x_i) - H(x) \tag{2}$$

Integration is minimal where dependence between elements is uniform, and maximal where the elements are highly and heterogeneously interdependent. Complexity is then given by equation (3), which measures the integration within network subsets of different sizes, denoted by k. Complexity is proportional to the difference between the average value of integration for a subset X_j^k (over all it j permutations) and the integration expected for a linear increase in system size summed over all subset sizes.

$$C_N(X) = \Sigma_{i=1}^{n}[(k/n)I(X) - < I(X_j^k) >] \tag{3}$$

Like other notions of complexity, this measure is low when either all elements are independent and hence completely segregated, or conversely, the system is completely integrated. Complexity is maximal in a system that is globally integrated at the level of large subsets, but simultaneously exhibits a high degree of segregation in smaller subsets. Note: this measure applies to both nonlinear systems and the linear systems considered here.

3 A Network Model

For the rest of this paper, complexity is calculated for a network of N nodes and K connections, with noise of magnitude $R = 0.1$ added to each node. Each network is encoded as an adjacency matrix Ω comprised of $N * N$ elements. Following [5], the sum of the absolute afferent input is normalised such that it equals a constant value. Each node is provided with a small inhibitory self-connection $\omega_{ii} = -0.001$.

In order to measure complexity, we need to determine the covariance matrix of the system, **COV**. This can calculated numerically by constructing and simulating a weakly coupled system. However, this route in computationally demanding for large ensembles of networks. Instead, here, we employ a method that

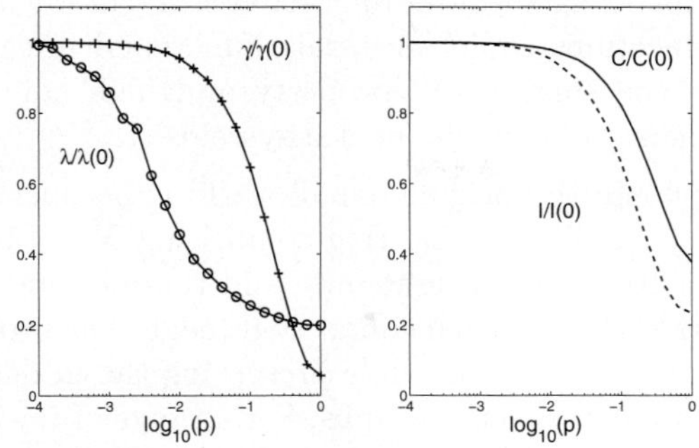

Fig. 1. How complexity, integration, path length and clustering vary as a one-dimensional ring lattice is gradually eroded by random rewiring. The ring comprises $N = 256$ nodes connected to their $k = 10$ nearest neighbours. The left-hand panel shows the scaled characteristic path length, $\lambda/\lambda(0)$ and the scaled clustering coefficient, $\gamma/\gamma(0)$, versus the log of the probability of rewiring, $\log_{10}(p)$ (circles and crosses, respectively). The right-hand panel shows the scaled complexity, $C/C(0)$, and scaled integration, $I/I(0)$, versus the log of the probability of rewiring, $\log_{10}(p)$ (solid and dashed lines respectively). Where $\lambda(0)$, $\gamma(0)$, $I(0)$ and $C(0)$ are measures taken on a ring lattice with $p = 0$.

allows us to an analytically calculate **COV** directly from the adjacency matrix of a linear, connected (i.e., no disconnected sub-graphs) network [9].[1] Lastly, for large networks, calculating mutual information measures over all subset sizes is also computationally demanding. Here, unless otherwise stated, we calculate the complexity as an average over subset sizes $i \leq 4$. This was observed to give a good approximation of full complexity for all models.

In addition to measuring dynamical complexity, we make use of two standard graph theoretic measures: clustering and characteristic path length. The nodal clustering coefficient is defined as the number of connections between the neighbours of a given node divided by the total number of possible connections between them [4]. The graph clustering coefficient, γ, (simply referred to as the clustering coefficient form henceforth) is calculated as the mean nodal clustering coefficient over a network's nodes. A network's characteristic path length, λ, is the average length of the shortest paths connecting all pairs of nodes [4]. In contrast to the clustering coefficient this is a global property of the graph.

All results reported here are averaged over no less than 30 networks per data point, and standard deviations were consistently lower than 0.5%.

Small-Worlds. Intuitively, the small-world effect, where systems combine strong clustering with short characteristic path lengths, would seem commensurate with

[1] For the matlab code for this and many of the other calculations employed in this paper, see Olaf Sporns' website at `http://www.indiana.edu/~cortex/`

high complexity. Clustering suggests functional segregation, while a sparse web of longer-range connections could encourage functional integration at a global level. Furthermore, the small-world property and high complexity have been shown to be coincident in biological neural systems [7].

Initially, we replicate the original small-world experiment presented in [4]. Commencing with a one-dimensional ring comprising $N = 256$ nodes, each connected to their $k = 10$ nearest neighbours, and representing these interactions as a binary connection matrix, each connection (edge) has probability p of being randomly rewired to another node while preserving the in degree at each node. Note: unlike Watts we use directed graphs. For a range of rewiring probabilities, we calculate the resulting values of γ, λ, and also calculate the complexity, C, and integration, I, as outlined in section 2.

Fig. 1 presents these measurements scaled by the values associated with the original ring lattice, see caption for further details. While a low probability of rewiring generates a small-world effect in reducing characteristic path length without damaging clustering, both complexity and integration fall monotonically with p (as mentioned recently in [7]). Essentially, the spatial organisation of the lattice is being eroded by rewiring. However, perhaps this result is specific to a rewired lattice which only exhibits a single topological scale of organisation. Note: while clustering coefficient seems to refer to an intuitive idea of distinct clusters in fact this is not the case and even a lattice has a high clustering coefficient. Instead consider Watts' connected cave world [10], for example, which exhibits two topological scales, that of the tightly intra-connected local clusters (caves), and a global level of loose inter-cluster connections. To explore this we examine four different structures: a one-dimensional ring is presented for comparison with fig. 1; a toroidal structure represents extending such a ring into a second spatial dimension; a "connected cave-world" [4] consists of a set of 32 fully-connected caves of 8 nodes each arranged on a ring with 8 connections between each pair of caves, representing a simple clustered network; a fractal structure similar to those employed in [7]. To build this fractal structure we start with a fully-connected clique of 8 nodes, duplicate it, and connect nodes from one cluster with nodes in the other according to some connection probability. The resulting structure is again duplicated and connections between the new pair are added. This process repeats until there are 256 nodes. Note: the probability of inter-cluster connections is reduced exponentially over fractal levels (see [7]).

Fig. 2 shows how the small-world index (S) and scaled complexity, $C/C(p = 1)$, vary with the log of the rewiring probability, $log_{10}(p)$, for these network structures. Note: in contrast to fig. 1 above all measurements are scaled by the values associated with fully randomised networks, i.e., $p = 1$. This highlights the relative differences between the impact of the different network structures in the absence of re-wiring. Again, the small-world effect is not enough to scaffold complexity. Rather, as in fig. 1, complexity appears to be correlated with the clustering coefficient, both falling monotonically with the increasingly probability of rewiring. By contrast, the consonant variation in characteristic path length appears to have little or no influence.

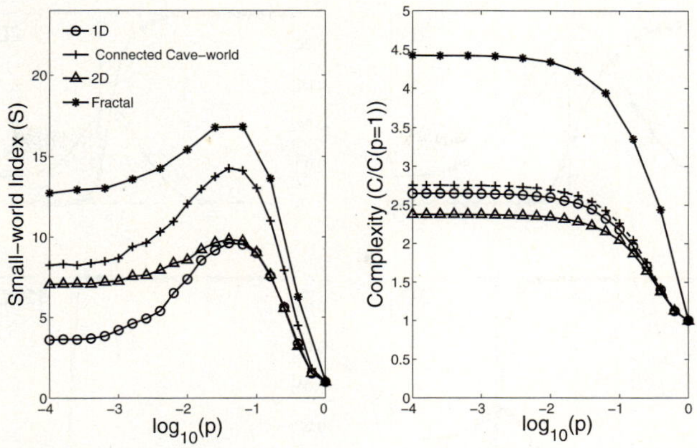

Fig. 2. The left-hand panel shows how the small-world index, $S = \frac{\gamma/\gamma(p=1)}{\lambda/\lambda(p=1)}$, varies with the log of the probability of rewiring, $log_{10}(p)$, for four network structures. The right-hand panel shows how the scaled complexity, $C/C(p = 1)$, varies for the same network structures. All networks comprise $N = 256$ nodes with identical connection densities ($N/K \approx 0.03$). (The different network structures necessitate that different degree distributions must be compared.) Here $C(p = 1)$ is the value of complexity associated with a random graph (i.e., when the probability of rewiring is unity).

Spatial Length Scales. The impact of spatial embedding is not limited to its effect on clustering coefficients and characteristic path lengths. Rather, (at minimum) it is capable of bringing about structural organisation over a particular length scale. Here, we explore ensembles of spatially constrained networks constructed over nodes distributed uniformly in hypercubes of varied dimensionality, varying the length scale of the interaction between the nodes. Note: in order to preserve the magnitude of spatial relationships between pairs of nodes over different numbers of dimensions all distances are scaled by $1/\sqrt{d}$. Instead of the binary connection matrixes used above, here we employ continuous-valued entries to represent weighted connections between pairs of nodes given by $\omega_{ij} = exp(-|\mathbf{r}_j - \mathbf{r}_i|/\sigma)$. Where, $|\mathbf{r}_j - \mathbf{r}_i|$ is the distance between nodes i and j. Connection weights between pairs of nodes fall exponentially with distance at a rate which is defined by the interaction length, σ. Fig. 3 shows how complexity, C, varies with the log of the interaction strength, $\log_{10}(\sigma)$.

The graph theoretic measures that we have used to characterise network structure up to this point can only be applied directly to binary (unweighted) networks. In order to calculate these measures here, we discretise each weighted network by reinterpreting each entry in the weight matrix as the probability that a pair of nodes will be connected. Consequently, each continuous matrix can be mapped to an ensemble of binary networks from which a random sample can be drawn and their properties calculated. For each network, we enumerate the number of disconnected components. As this value approaches unity, the graph

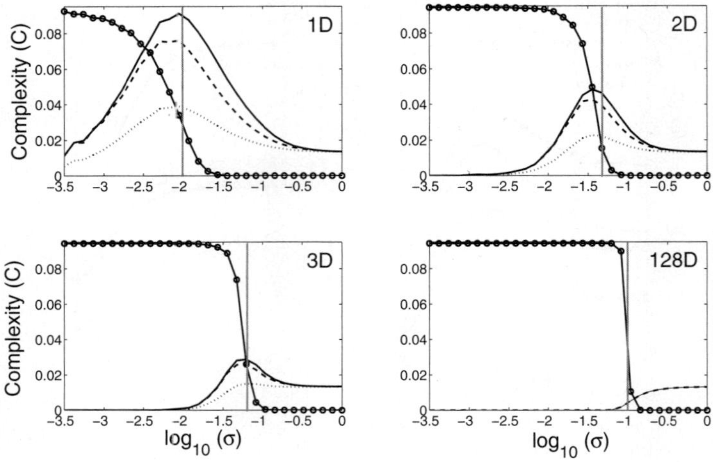

Fig. 3. Plots of complexity C versus the log of the interaction length, $log_{10}(\sigma)$,for 1, 2, 3, and 128 dimensions are presented in the top left, top right, bottom left and bottom right panels, respectively.All networks comprise $N = 128$ nodes. The solid curves represent the mean complexity, C, of spatially embedded system with continuous weights. The dotted and dashed lines indicates the complexity of networks derived from two null models (see text). The grey vertical lines mark the peaks of complexity for discretised networks with the same interaction length, which agree well with the peak in complexity for the associated continuous system (the solid line). The scaled number of network components is also presented (circles), falling from N (a totally disconnected system) to unity (a super cluster).

is becoming completely connected, indicating the onset of a single component or super-cluster [10].

For comparison, all plots in fig. 3 also present values of complexity for two null models. First, the dotted line represents the complexity of networks where each node has the same distribution of afferent connection strengths, but the identity of neighbours is randomly assigned. To achieve this, the entries of each row in the weight matrix are shuffled, preserving the sum of afferent weights. The dashed line represents the complexity of networks for which connections are shuffled in a way that preserves reciprocity, i.e., where a shuffle swaps elements ω_{ij} and $\omega_{i'j}$, it must also swap elements ω_{ji} and $\omega_{ji'}$. Note: in this case the sum of the magnitude of the afferent weights may not be preserved.

The first point to note is that for low-dimensional spaces, complexity rises and falls with interaction length.[2] As the dimensionality of the space increases, peak complexity falls. The reciprocal nature of spatial interactions clearly ac-

[2] Since the covariance matrix of a 1-d lattice is of Gaussian Toeplitz form, this agrees with previous results demonstrating that scaling in such matrices is associated with a rise and fall in complexity [5].

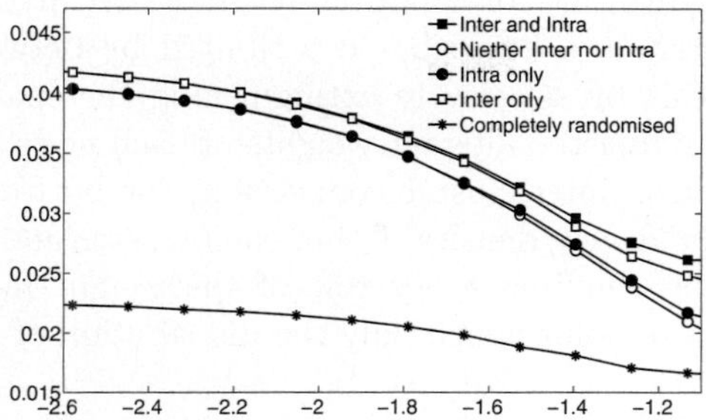

Fig. 4. Complexity, C, varies with cluster width for networks with spatial structure within and/or between each of 12 regularly arranged clusters of nodes distributed in two-dimensional space according to a normal distribution with variance, σ_{space}. The complexity of equivalent non-spatial random networks is shown for comparison.

counts for this effect to some extent (and to a larger degree than the mere distribution of afferent weights). However, particularly in low dimensions, the impact of spatial constraints exceeds that of mere reciprocity, suggesting that higher-order structures are significant. As the dimensionality of the space increases, and the strength of spatial constraints weaken, peak complexity falls, until the contribution of space, and even reciprocity disappears.

Interestingly, the peak in network complexity is correlated with the onset of the super cluster in the discretised versions of the networks presented in fig. 3. Although the graph theoretic measure does not directly translate into the continuous domain, this result suggests that complexity is associated with the achievement of a single *strongly coupled* component in a continuous network. Furthermore the interaction length required for onset of the strong component (and thus high complexity) falls with the dimensional order.

Spatial Structure. Thus far, we have only considered uniform spatial distributions of points. However, spatio-temporal processes naturally bring about structured distributions. Here we consider how the introduction of *community structure*, in the form of randomly distributed clusters of equal size, impacts on network complexity. In contrast to clustering coefficient community structure provide a more intuitive notion of clustering [11].

Here $N = 126$ nodes are divided into 9 groups of 14 points. The group foci are regularly arranged as a 3×3 grid in the unit square. The points of each group are then normally distributed around each focus with a variance σ_{space} (note: this is distinct from the interaction length, σ). For increasing σ_{space}, distinct, tight clusters (communities) initially spread, then merge, and eventually overlap to form a virtually uniform distribution of nodes. The connection weight between each pair of nodes is determined as per the previous model with a fixed interaction length $\sigma = 10^{-3}$.

We wish to distinguish the contribution to complexity made by within-cluster spatial correlation structure from that contributed by between-cluster organisation. We achieve this by selectively extinguishing the spatial correlations at each scale, either shuffling the afferent weights of each node's intra-cluster connections, or each node's inter-cluster connections, or both. All three shuffling processes preserve the degree density within each cluster and between each pair of clusters. Lastly, by shuffling every row of the weight matrix, we generate fully randomised networks for which only the distribution of weight strengths is preserved.

Fig. 4 shows that as the cluster width increases and clusters merge, complexity falls, suggesting that non-uniform spatial distributions impact on network complexity. Here network complexity can be partitioned into contributions due to inter-cluster spatial constraints, intra-cluster spatial constraints, and the residual community structure arising from the fact that, to the extent that clusters are spatially distinct from one another, there will tend to be stronger weights on within-cluster connections than between-cluster connections. The latter contribution dominates until cluster widths approach the width of the space, resulting in an approximately uniform distribution. By contrast, the contribution of within-cluster spatial organisation is minimal until nodes approximate a uniform distribution. Inter-cluster spatial constraints make a consistent but relatively small contribution to complexity across the range of cluster widths.[3]

4 Discussion and Conclusion

First, by systematically exploring the relationship between the small-world effect on a networks topology and the consequent dynamical complexity that the network exhibits, we have shown that although these two properties may co-occur in natural systems [7], it is not the case that small-world structures alone straightforwardly imply complex network behaviour (see figs. 1 and 2).

However, as intimated in recent work [7], results here demonstrate that spatial constraints on connectivity contribute directly to complexity. Even in the absence of the community structure or fractal organisation that is known to generate complex network behaviour [7], networks merely comprising uniform random distributions of locally connected nodes enjoy increased complexity as a result of the strong spatial constraints imposed by low dimensionality (see fig. 3).

The nature of the contribution to complexity made by spatial embedding is not straightforward. Neither the shape of the distribution of afferent weights (dotted lines, fig. 3) nor their reciprocity (dashed lines, fig. 3) are sufficient to account its impact on complexity. Rather, the property stems from space imposing correlations at several topological scales. This is evidenced by the gradual erosion of the influence of space as dimensionality is increased (see fig. 3).

Fig. 3 suggests that high network complexity is correlated with the onset of strongly coupled super cluster. The coupling strength required for its onset

[3] These results are redolent of the differences in complexity between ordered and non-ordered fractal mappings presented in [7].

is much smaller in networks embedded within low-dimensional spaces suggesting that strong spatial constraints may make high complexity easier to achieve despite sparse or weak connections.

Finally, we have shown that the structure of the underlying spatial distribution of nodes can impact on network complexity. For example, results suggest that clusters of nodes randomly distributed in space bring about network topologies that exhibit high complexity stemming from both inter-cluster and intra-cluster correlations, but mostly by the residual community structure that distinct clusters impose (perhaps justifying the current focus on hierarchical and fractal organisation with respect to neural systems [6,5]).

In summary, as suggested by the evolvability of some spatially embedded network architectures [12], the inherent constraints imposed by projecting systems into low dimensional spaces may be enabling for evolution in that they predispose systems to exhibit complex behaviour for free.

Acknowledgements. We thank Olaf Sporns for his email discussions.

References

1. Langton, C.G.: Computation at the edge of chaos. Physica D 42, 12 (1990)
2. Kauffman, S.: The Origins of Order. University Press, Oxford (1993)
3. Boerlijst, M.C., Hogeweg, P.: Spiral wave structure in pre-biotic evolution: Hypercycles stable against parasites. Physica D 48, 17–28 (1991)
4. Watts, D.J., Strogatz, S.H.: Collective dynamics of 'small-world' networks. Nature 393, 440–442 (1998)
5. Tononi, G., Sporns, O., Edelman, G.M.: A measure for brain complexity: Relating functional segregation and integration in the nervous system. Proc. Natl. Acad. Sci. 91, 5033–5037 (1994)
6. Sporns, O., Tononi, G., Edelman, G.M.: Theoretical neuroanatomy: Relating anatomical and functional connectivity in graphs and cortical connection matrices. Cerebral Cortex 10(2), 127–141 (2000)
7. Sporns, O.: Small-world connectivity, motif composition, and complexity of fractal neuronal connections. Biosystems 85, 55–64 (2006)
8. Hoppensteadt, F.C., Izhikevich, E.M.: Weakly Connected Neural Networks. Springer, Heidelberg (1997)
9. Tononi, G., Edelman, G.M., Sporns, O.: Complexity and coherency: integrating information in the brain. Trends in Cognitive Sciences 2(12), 474–483 (1998)
10. Watts, D.J.: Small Worlds. Princeton University Press, Princeton, Princeton (1999)
11. Girvan, M., Newman, M.E.J.: Community structure in social and biological networks. Proc. Natl. Acad. Sci. 99, 7821–7826 (2002)
12. Husbands, P., Smith, T., Jakobi, N., O'Shea, M.: Better living through chemistry: Evolving gasnets for robot control. Connection Science 10, 185–210 (1998)

The Application of the Idea of Extended Cellular Automata for Some Pedestrian Behaviors

Eva Dudek-Dyduch, Jarosław Wąs, and Bartłomiej Gudowski

Institute of Automatics
AGH University of Science and Technology
al. Mickiewicza 30, 30-059 Kraków, Poland
{edd,jarek,bart}@agh.edu.pl

Abstract. The article suggests new ideas regarding computer simulations in pedestrian dynamics, which take into account interaction between particular pedestrians. The described situation is a room evacuation for chosen classes of situations: normal room evacuation, controlled evacuation, and panic. Based on the formalism of Extended Cellular Automata, the following models of pedestrian dynamics are presented: a basic model, SPA (Strategic Pedestrian Abilities) model, SPA–BNE (Bottleneck Effect) model, which were created with reference to the particular classes of situations.

1 Introduction

Modeling crowd dynamics and interactions between particular people is an important problem for many fields of science such as architecture, sociology, psychology and computer science. Transmission of these issues into practical solutions connected with safety (safe architectural solutions, emergency notification) is a very significant aspect of research in this field. Computer simulations have been currently playing a very important role in recognizing and evaluating pedestrian dynamics processes.

First studies devoted to computer simulations of crowd dynamics were based on a macroscopic approach (macroscopic motion) in which particular pedestrians were treated similarly to particles in hydrodynamics. This meant that the interaction at the level of particular people was not studied and steering in the model was possible for example by changes in geometry, changes of the Reynolds number, or changes of the turbulence model. A model developed by Pauls and described in [9] is probably the best known example of such an approach.

The last few years have witnessed a strong development of a new research trend connected with the microscopic approach to pedestrian behavior simulations. At the moment, Social Forces Method [7] based on Molecular Dynamics and Cellular Automata are the two most popular methods of microscopic simulation in pedestrian dynamics.

This paper will deal with the second of the enumerated methods, namely Cellular Automata method. Let us quote first a few most characteristic studies

F. Almeida e Costa et al. (Eds.): ECAL 2007, LNAI 4648, pp. 996–1005, 2007.

based on this method: in the paper [2], Burstedde et al. proposed to use a static and dynamic floor field for modeling pedestrian flow. The static floor field takes into account room topology while the dynamic floor field takes into consideration frequency in the particular grid cells and points out attractive movement directions.

In [3], Dijkstra et al. presented a model of pedestrian movement in the streets or shopping malls. The model combines the features of Cellular Automata and Multi-Agent Systems. Pedestrians/agents obtain certain information from the environment which stimulates them to make decisions (for example, choose a particular movement direction).

A model of tourist activity was presented by Gloor et al. in [5]. Particular tourists are also perceived there as autonomous agents, each of whom makes decisions concerning the movement, such as the target, choice of route etc. From the technical point of view, it is interesting to enrich the basic grid of Cellular Automata with a graph calculating the shortest routes.

On the other hand, Narimatsu et al. in [8] proposed an algorithm for collision avoidance for two-directional pedestrian movement. Pedestrians who walk along a corridor in two opposite directions learn certain patterns of collision avoidance.

A model of pedestrian dynamics taking into consideration a social theory of proxemics: Social Distances, introduced by E.T. Hall [6] was presented in [11]. In the paper, the model is applied for the movement of passengers on a tram.

This paper concentrates on the idea of Extended Cellular Automata (described in [4] and [10] for modeling chosen classes of situations in pedestrian movement. The description of the situation takes into consideration interactions between particular pedestrians. The aim of the paper is then to find such algorithms with which it is possible to model a situation of room evacuation by a group of pedestrians in normal conditions, during controlled evacuation and in panic.

2 Formalization of Extended Cellular Automata

2.1 Classical and Non-classical Definition of Cellular Automata

The concept of cellular automata was proposed by John von Neuman and Stanislaw Ulam in the middle of the 20th century. According to one of the most popular definitions: *Cellular Automata are discrete dynamical systems, whose behavior is completely specified in terms of local relation.*

Unfortunately, the classical concept of Cellular Automata is insufficient as far as numerous practical applications are concerned because the majority of known studies dealing with the application of cellular automata in pedestrian dynamics (and not only within that field!) introduces more or less significant modifications of the classical concept of Cellular Automata, for instance [2, 3, 5, 8]. This is the reason why the papers [4, 10] present a formalized method of describing non-homogenous, asynchronous Cellular Automata.

The presented definition of extender Cellular Automata makes it possible to describe a wide class of non-homogenous Cellular Automata. In particular, it can be used in the description of a simulation in pedestrian dynamics.

2.2 An Idea of Extended Cellular Automata

More complex description of *Extended Cellular Automata* could be found in [4, 10]. Let us discuss some more important elements of this idea.

Extended cellular automata denoted with symbol $ECAL_{const}$, is described by seventuple (equation 1):

$$ECAL_{const} = (C, R_p, \eta, S, CON, r_{ext}, f) \tag{1}$$

where:

C - set of grid cells $c_k \in C$
R_p - relation of proximity
η - neighborhood function
S - set of cell states
CON - set of cell configurations
r_{ext} - local rule
f - transition function (in which the cell state actualization order is essential)

The set of cells (space in the model) is marked by C. In the C set we will define the adjacent relation $R_{\mathrm{p}} \subset C \times C$. This relation is used to define different grid formats (different topologies in the set of cells). Neighborhood is defined by the use of a function $\eta: C \rightarrow 2^C$ (2^C denotes a set of all subsets of the set C), which assigns a subset of cells being its neighborhood $\eta(c_k)$ to each cell $c_k \in C$. The function η is called a neighborhood function. Let us denote a set of cell states as S. The function $con: C \rightarrow S$ which assigns to each cell, its state $con(c_k) = s \in S$ c_k is called a configuration. Since the configuration denotes the states of all the cells, it is identified with the state of the whole grid. The set of configurations is denoted as CON. Configuration changes is defined with the transition function $f: CON \rightarrow CON$, which assigns the next configuration to each configuration $f(con_t) = con_{t+1}$. The function f is the function of a global change of states. The local rule, denoted as r_{ext} can be shown as follows: $r_{\mathrm{ext}}: CON \rightarrow CON$ such that: $r_{\mathrm{ext}}(con_t/\eta^*(c_k), con_t/A) = con_{t+1}/\eta^*(c_k)$, where A is an extra subset of cells, the state of which influences a future state of the cells from the surrounding, whereas the symbol $\eta^*(c_k)$ denotes the so-called full surrounding, i.e. $\eta^*(c_k) = \eta(c_k) \cup \{c_k\}$.

Let us emphasize the significant features of $ECAL_{const}$ automaton:

- $ECAL_{const}$ automaton is non-homogeneous and asynchronous
- local rule may be different for different cells (that is, for example, a modification is possible by using an additional automaton grid such as static or dynamic floor field [2]).

– the cell state actualization order in a given time step is essential if one wants to realize a subsequent configuration; in particular, the cell state can change more often than once when this configuration is being calculated.

3 Suggested Models of Pedestrian Dynamics

Models of pedestrian dynamics apply to situations of room evacuation by a group of pedestrians when a few exits are accessible. The models presented are based on the formalization of the above mentioned extended Cellular Automata $ECAL_{const}$ with a two-dimensional, square grid, where Moore's neighborhood is considered.

Grid cells are divided into several classes and they can represent walls, obstacles, exits, or a movement space etc (that is, in the set C there are distinguished following subsets: (W)alls, (E)xits, (O)bstacles, (M)ovement(S)pace $\subset$ C). Particular pedestrians are represented in the model by cell occupation for the cell belonging to the movement space ($c_k \in MS$) therefore cell state $s(c_k) = 1$. Movement space refers to grid cells which are available for pedestrian movement.

At the beginning of the simulation in the cell grids, we generate a pedestrian population, whose movement is registered in discrete time periods. In the subsequent steps of the simulation pedestrians head towards grids representing exits, according to the given algorithm. It is important to note that in the subsequent versions of the algorithm when pedestrians choose their direction (exit) they take into account not only the distance from the exit but based on the cost function they also consider crowd density near the exits [12, 13].

Basically, the mechanism of pedestrian movement is based on the rule of Bouvier's potential field [1]. According to Bouvier's concept of potential field, each grid cell is assigned a certain number called a potential, which depends on the distance to the nearest cell which is a room exit (fig. 1 on the left).

Fig. 1. On the left – potential field generated on the automaton grid by two exits. On the right – an example of all bottlenecks generated on the grid (grey cells).)

Let us characterize briefly the proposed models of pedestrian dynamics:

Basic Model Basic model is the closest to the idea of classical Cellular Automata. The direction of pedestrian movement is determined by the initial allocation of a pedestrian on the grid.
Basic model is presented with the aim of opening a new possibility of comparing the effectiveness, functionality and the level of realism of this relatively simple model with other, more complex models.

SPA Model Contrary to the basic model the suggested SPA model (Strategic Pedestrian Abilities) takes into account cost function which makes it possible to decide about the exit.
In this model, pedestrians make decisions about exit through which they can, in their opinion, leave the room the fastest. They apply two fundamental criteria in their decision-making process:

- the distance from particular exits counted as the value of potential field generated from each exit [1]
- crowd density around exits.

The cost function $cost(c_k, e_x)$ for each of the exits $e_x \in E$ can be presented in a more formalized way:

$$cost(c_k, e_x) = w_1 \cdot dist(c_k, e_x) + w_2 \cdot dens(e_x) \tag{2}$$

where:

w_1, w_2 - determined factors
$dist(c_k, e_x)$ - the smallest number of cells constituting a route from the cell c_k to any of the exit cells e_x
$dens(e_x)$ - a function characterizing crowd density in $N(e_x, \lambda)$ neighborhood around a given exit e_x with a formula (3).

$$dens(e_x) = \frac{\sum_{c_k \in A(e_x)} s(c_k)}{|A(e_x)|} \tag{3}$$

where:

$|N(e_x, \lambda)|$ - the number of cells belonging to neighborhood e_x, for a given Moore radius λ
$s(c_k)$ - current state of c_k cell
$A(e_x)$ - cells belong to e_x neighborhood

It is important to name three most important elements which are connected with each other and which distinguish the SPA model from the basic model:

- the new local rule, which reflects the decision process of the pedestrian and takes into account neighborhood observation
- new method of ordering cell state actualizations in the transition function (radiating wave actualization – from exits inside a room)
- new movement algorithm (using cost function).

The SPA model is created for the situation of normal room evacuation. The weight of distances is then close to 1 (and is of key importance).

SPA–BNE Model A blockade appears in a situation when rivaling pedestrians are planning to head for the same place at the same time, which temporarily freezes the movement flow.

The models of pedestrian dynamics described above, that is the classical model and the model of strategic pedestrian abilities SPA assume a smooth flow of movement when a determined order of room evacuation is adhered to, similarly to queue systems, such as *FIFO.* Whereas in evacuation situations and especially when panic occurs, we can often observe that this smooth flow is disturbed. This is visible when conflicts and blockades appear or when pedestrians push one another. Such a situation can occur locally or be more widespread. According to the authors, it can take place when the three following conditions jointly occur:

- geometrical conditions: *bottlenecks* in the room, that is when the pedestrians' routes narrow down
- substantial crowd density around the bottleneck (that is when a critical number of people is present in the vicinity of the *bottleneck*)
- high level of pedestrian determination (irritation), resulting from existing danger to life or health, impatience caused by long waiting etc.

In the direct vicinity of each grid cell belonging to the movement area, that is in its Moore neighborhood within radius = 1, it is possible to determine cells with a better potential (exit cells) or cells with a worse potential (exit cells). The ratio of the number exit cells to the number of exit cells determines *the level of capacity of a given cell.* If this ratio is lower than 1, we have a situation of a *bottleneck.* The grey cells on the right in 1 are the cells of the bottleneck.

The model, therefore, presents a mechanism which refers to a situation when at least two pedestrians try to reach the same field which becomes a *bottleneck.* In such a case, with certain probability (calculated on the basis of the elements presented above) their target cell may be blocked till the end of a given time step (that is the cell assumes the state $s(c) = 2$). This corresponds to a situation of a local blockade caused by pedestrians pushing one another.

In order to calculate the transition rule it is essential to calculate the probability of the blockade. It has been assumed that the blockade occurs under the following rule. The probability of blockade creation p equals:

$$p = dens_{BN}(c_n) \cdot (1 - Y(c_n)) \cdot p_{glob} \tag{4}$$

where:

$dens_{BN}(c_n)$ - crowd density around the cell c_n, which is the bottleneck is described by the equation: $dens_{BN}(c_n) = \frac{\sum_{c \in \eta(c_n), s(c) \neq 2} s(c)}{|\eta(c_n)|}$

$Y(c_n)$ - the level of the bottleneck

p_{glob} - global blockade probability, which depends on the level of pedestrian determination

The SPA–BNE model is created for the situation of controlled evacuation and panic. In the first case, the probability of blockade occurrence is low and using strategic pedestrian abilities in the cost function is definitely the highest. On the other hand, in the situation of panic the probability of a blockade and the criterion of distance from the exit become dominating.

The table below 1 presents a shortened formalized description of the proposed models, according to the definition of Cellular Automata (tab. 1):

Table 1. A shortened formalized description of the discussed models of pedestrian dynamics

Symbol	basic model	SPA model	SPA – BNE model
C	$c_k \in C$	$c_k \in C$	$c_k \in C$ (new subset $BN \subset MS$)
R_p	natural	natural	natural
$\eta(c_k)$	Moore neighborhood radius $\lambda = 1$	Moore neighborhood radius $\lambda = 1$	Moore neighborhood radius $\lambda = 1$
$s \in S$	{free, occupied}	{free, occupied}	{free, occupied, blocked}
CON	all possible	all possible	all possible
r_{ext}	local rule	cost function	cost function
f	actualization order: pedestrian list	actualization order: "radiating wave"	actualization order: "radiating wave"

4 Results of Computer Simulations

Several computer simulations were carried out to compare proposed models, as a part of the study. The SPA model was applied for modeling the normal evacuation situation while for controlled evacuation and panic the SPA–BNE model was used. In order to compare the results, simulations with the use of the basic model (the closest to the classical Cellular Automata) were also carried out.

4.1 Implementation

The model has been implemented with the use of $C++$ programming language. All features of the model are enclosed into several $C++$ classes, which represent: the grid, grid cells, pedestrians etc. The application consists of two main parts: the part representing the model and Graphical User Interface.

4.2 Simulation Results

Figure 2 presents a typical representation of the simulation based on the basic model. Black cells represent walls or obstacles, circles represent pedestrians during evacuation. Two exits are available (on the top left corner and on the bottom left corner).

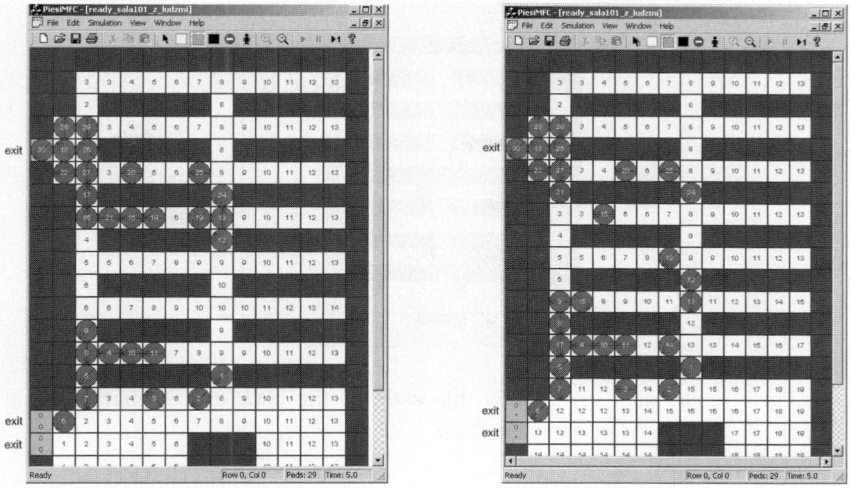

Fig. 2. On the left – typical representation for the basic model, pedestrians head towards exits taking into account only the distance. On the right – a typical situation for the SPA model – pedestrians analyze the situation systematically. We can observe that the two exits are evenly "loaded".

Based on the value of potentials determining the distance, the majority of pedestrians head towards the "back" exit. In this model, pedestrians are unable to map intelligent decisions and change their choice of exit.

Figure 2 in turn presents a typical representation for the SPA model.

In many situations pedestrians search for the fastest route out of a room. For the basic model, the only criterion of the choice of exit is exclusively the criterion of distance (the shortest way to the exit). On the other hand, in the SPA model the assumed weight of the distance criterion $w_d = 0.92$ definitely dominates over the criterion of density $w_g = 0.08$. In the SPA model the room evacuation procedure is similar to the $FIFO$ queue system. There are no local blockades (which result from competition between pedestrians). That is why both in the computer simulation and in the experimental studies we do not observe rivalry between particular pedestrians (like blockades in SPA-BNE).

Figure 3, in turn, depicts a situation characteristic for panic. The following values of the model parameters have been assumed: factor p_{glob} (equation 4) equals 0.3, the weight of distance equals $w_d = 0.82$ and the weight of density equals $w_g = 0.18$ for Moore radius $\lambda = 3$.

We can observe a blockade with a bottleneck around the back (upper) room exit. Despite the determined ordering of room evacuation for pedestrians with identification numbers 29, 27, and 21 (ordering connected with the waiting time in the cell with the potential = 1) there is a blockade characteristic for the situation of panic and resulting from the fact that pedestrians push each other.

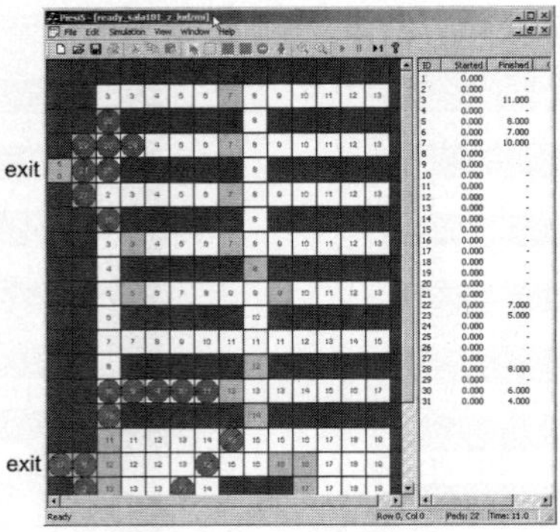

Fig. 3. The simulation of panic. The blockade at the back exit is characteristic for the high level of pedestrian determination.

5 Conclusions

The carried out simulations made it possible to investigate the application of the proposed models of pedestrian dynamics for particular situations.

The basic model, which is the closest to the concept of the classical Cellular Automata, does not yield results which would be as good as the proposed models based on the technology of non-homogeneous Cellular Automata: SPA and SPA–BNE. In the choice of exit, it does not allow for any other criteria but distance. It is also impossible to differentiate between normal conditions and evacuation and connected with them different interactions between pedestrians. On the basis of the carried out research and an analysis in [10] one can conclude that mechanisms of the basic model are not sufficient for a successful description of room evacuation by a group of pedestrians. The two proposed models SPA and SPA–BNE make it possible to differentiate between a normal situation, controlled evacuation, and panic.

Strategies put forward in the SPA and SPA–BNE models come down to the analysis of two factors deciding about the choice of the exit: distance and density around the exit. In normal conditions the weight of distance is essential although, as the experiments revealed, the weight of density cannot be omitted in the model. In the case of both of the proposed models we can mention environment observation and a systematic decision-making procedure about the direction of evacuation. This means that the models can represent intelligent strategic decisions of pedestrians.

The SPA–BNE model has been proposed both for the simulation of controlled evacuation and a simulation of panic. The mechanism responsible for creating conflicts and local blockades, which is characteristic for the situation of panic, can also occur in controlled evacuation with the presence of certain conditions. Geometrical features of the model, crowd density and global blockade probability – the parameter connected with the modeled situation – decided about the occurrence of blockades.

In order to confirm the applicability of the proposed models it is necessary to conduct a series of real-life experiments, which would enable us to verify the proposed models. The first series of experiments taking into consideration normal situation, controlled evacuation and panic has already been carried out and described in [10]. Promising results were obtained. They require, nevertheless, further verification.

References

[1] Bouvier, E., Cohen, E., Najman, L.: From crowd simulation to airbag employment. Journal of Electronic Imaging 6(1), 94–107 (1997)

[2] Burstedde, C.K., Klauck, K., Schadschneider, A., Zittartz, J.: Simulation of Pedestrian Dynamics using a 2-dimensional Cellular Automaton. Phys. Rev. A 295, 507–525 (2001)

[3] Dijkstra, J., Jessurun, A.J., Timmermans, H.: A Multi-Agent Cellular Automata System for Visualising Simulated Pedestrian Activity. In: Proceedings of ACRI, pp. 29–36 (2000)

[4] Dudek–Dyduch, E., Wąs, J.: Knowledge Representation of Pedestrian Dynamics in Crowd. In: Rutkowski, L., Tadeusiewicz, R., Zadeh, L.A., Zurada, J.M. (eds.) ICAISC 2006. LNCS (LNAI), vol. 4029, pp. 1101–1110. Springer, Heidelberg (2006)

[5] Gloor, C., Stucki, P., Nagel, K.: Hybrid Techniques for Pedestrian Simulations. In: Sloot, P.M.A., Chopard, B., Hoekstra, A.G. (eds.) ACRI 2004. LNCS, vol. 3305, pp. 581–590. Springer, Heidelberg (2004)

[6] Hall, E.T.: The Hidden Dimension. Garden City, New York (1966)

[7] Helbing, D., Molnar, P.: A Social Force Model for Pedestrian Dynamic. Phys. Rev. E 51, 4284–4286

[8] Narimatsu, K., Shiraishi, T., Morishita, S.: Acquisiting of Local Neighbour Rules in the Simulation of Pedestrian Flow by Cellular Automata. In: Sloot, P.M.A., Chopard, B., Hoekstra, A.G. (eds.) ACRI 2004. LNCS, vol. 3305, pp. 211–219. Springer, Heidelberg (2004)

[9] Pauls, J.: Movement of people, DiNenno (1995)

[10] Wąs, J.: Intelligent Behaviour Modelling Algorithms in Pedestrian Dynamics Issues using Nonhomogeneous Cellular Automata, PhD Work, AGH University of Science and Techology (2006)

[11] Wąs, J., Gudowski, B., Matuszyk, P.J.: Social Distances Model of Pedestrian Dynamics. In: El Yacoubi, S., Chopard, B., Bandini, S. (eds.) ACRI 2006. LNCS, vol. 4173, Springer, Heidelberg (2006)

[12] Wąs, J., Gudowski, B.: The Application of Cellular Automata for Pedestrian Dynamics Simulation. Automatyka Journal AGH-UST, Kraków, 303–313 (2004)

[13] Wąs, J., Gudowski, B.: Simulation of Strategical Abilities in Pedestrian Movement using Cellular Automata. In: Proceedings of 24th IASTED MIC Conference, Innsbruck, pp. 549–553 (2005)

Transients of Active Tracking: A Stroll in Attractor Spaces

Mario Negrello[1,2] and Frank Pasemann[1,2]

[1] Fraunhofer Institute, IAIS
Schloss Birlinghoven, Sankt Augustin, Germany
[2] University of Osnabrueck, Germany
mnegrell@uos.de

Abstract. We present the dynamical analysis of embodied RNNs evolved to control a cybernetic device that solves a tracking problem. From the Neurodynamics perspective, we analyze the networks with focus on a characterization of the attractors and attractor sequences, guiding the transients. Projections of these attractors to motor space help visualizing the shape of the attractors, thus pointing to the underpinnings of behavior. Among the different attractors found are fixed points, periodic and quasi-periodic attractors of different periods, as well as chaos. Further analysis of the attractors relates changes of shape, size and period to motor control. Interesting characteristic behaviors arise, such as chaotic transitory regimes and implicit mapping of environmental assymmetricities in the network's response, (as for example attractor hops that implicitly code for gravity). We discuss autonomy, capacity and some issues relating to a possible theory of transients.

Key words: Dynamical Systems, Cognition, Morphing Attractors, Dynamical Tracking, Toy Problem, Chaos, Recurrent Neural Networks.

1 Introduction

1.1 Cognition on Transients and Attractor Morph

Dynamical Systems (DS) approaches to cognition defend that cognition lives on transients while transients live between attractors [1,2,3,4]. A recurrent neural network (RNN) when parametrized by an input layer is a *collection* of Dynamical Systems, because every distinct input resolves another dynamical system. Therefore, even small RNNs possess a wealth of attractor landscapes [5,6], where often all the attractor types are seen (fixed-points, periodic, quasi-periodic, and chaos). All the same, the classification of an attractor into one of the classes above, does not exhaust the attractor's features. The shape of an attractor is its identity.

So, it is not surprising that embodied neural systems will exploit the attractor's shape for motor control, as was the case on the networks evolved here. In effect, the motor actions of the network are shaped by the projections of the

F. Almeida e Costa et al. (Eds.): ECAL 2007, LNAI 4648, pp. 1006–1015, 2007.

attractors to the motor space. But the attractors alone are insufficient for meaningful behavior. Agents embedded in the sensori-motor loop will constantly be driven by its relations to the environment, therefore competent action selection is also a function of the transitions between attractors. These transitions happen in the transients, which are driven by the attractors, although usually not settling in them, given the ever-changing interactions.

In what follows, the analysis focus on the changing shapes and the transitions between attractors, as the behavior dwells the 'meta-transient' of the set of dynamical systems resolved by the interaction between the agent and world. It is interesting therefore to analyze the manner attractor changes and transients 'hop' between them as a function of the shifting input. For in an embodied problem, both are important: (1) how the attractors change and (2) how the paths of visiting attractors remain coherent.

1.2 Toy Problem

A toy problem is devised for the purpose of studying the dynamical strategies for control of a cybernetic device or agent. Simply put, the problem is: a cybernetic head should be able to follow, with its gaze, a ball that bounces irregularly in a 2D plane (i.e.. within a frame). The primary problems of the head are to know (1) in what direction to turn and (2) with what velocity. A secondary problem is to actively search for the ball, in case it is lost. As mentioned, evolutionary algorithms beget parameters for RNNs that endow the embodied network with successful action selection mechanisms. Conversely and complementarily, our problem is to know what are the dynamical entities that allow the network to do so. That is, the focus of the analysis is to see: (1) how the transients hop between attractors, (2) how attractors change as a function of changing input pattern and (3) how do attractors become motor action selection.

2 Methods

2.1 Problem Description

The toy problem consists of a cybernetic tracking device following a ball (radius $.5m$) in a virtual environment (see picture 1). The device is a head mounted on two velocity driven motors for yaw and pitch, endowed with an array of 9 distance sensors (25 cm apart from each other) [3]. An evolutionary algorithm selects those RNNs according to sensor input accumulated over trial time. So, the more the head is able to keep the ball under gaze while minimizing oscillations, the fitter the network (see equation 1).

[3] Roughly analogous to a 9 pixel retina.

Table 1. ODE Simulation Physics

Entity	Property	Quantity
Head	Mass)	3Kg
Head	Height	2m
Yaw and Pitch	Max. Force	5N
Yaw and Pitch	Max. Velocity	90deg/s
Physics and Network	Update Frequency	100Hz

Fitness Function This is the fitness of a trial for an individual *ind*, where $S_s(t)$ is the input at sensor neuron s at time t and $M_m t$ is the activation of the motor neuron m at t. In equation 1 the first term (LT) stands for the sum of stimulus input per cycle, while the negative quadratic terms (QT) aims for minimizing oscillations. Alpha and Beta are parameters to weight the terms and are alterable on-the-fly during evolution, depending on the emphasis the experimenter after observing the behavior of the current population. cycles in a single trial is c. Note that all the variables in the fitness function are accessible to the agent.

$$Fitness_{individual} = \underbrace{\alpha . \sum_{t=1}^{c} \sum_{s=1}^{9} S_s(t)}_{LT} - \underbrace{\beta . \sum_{t=2}^{c} \sum_{m=1}^{2} (M_m(t) - M_m(t-1))^2}_{QT} \quad (1)$$

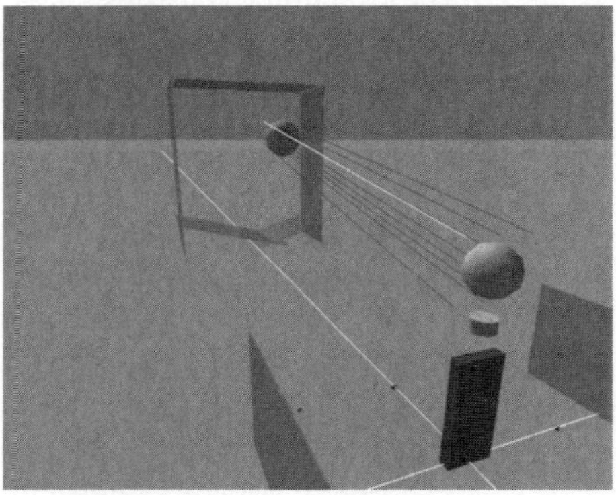

Fig. 1. Picture of the environment with dimensions.

Discrete Recurrent Neural Network (RNN) The RNN is discrete time, and the units have the hyperbolic tangent as the nonlinear transfer function. The input layer receives no backward connection. In the appendix are the weights of an example evolved network.

$$a_i(t+1) = \sum_{j=1}^{n} w_{ji}\tau(a_j(t)), i = 1, \ldots, n; \tag{2}$$

Where a_i is the activity of the i_{th} unit of the network. The total number of units is n, τ is a sigmoid function, in this case the hyperbolic tangent, w_{ij} reads i receives from j with weight ditto the networks are evolved without bias units. The sensors are units 1 to 9 and receive no backward projections.

2.2 Challenges for the Tracker

The tracker's problem is complicated by a number of issues, which have to be taken into account when analyzing the resulting networks.

1. The head's foe, the bouncing ball, is designed to have chaotic orbits thanks to the different angles of the bottom platforms (see figure 1).
2. The ball is dropped from different positions in the z and y axis (see figure). By design the ball does not lose energy as it bounces, implying that when the ball bounces in different positions of the side walls or bottom platforms, it has very different velocities in the z and y cartesian axes. For example, when the ball bounces side-ways in the bottom of the frame the horizontal velocity is much higher than when it bounces higher up in the frame. The ball is constantly subject to gravity of $9.8m/s^2$.
3. The network has no knowledge of the frame, so in principle there is no information of the exit angle after it bounces against the frame. That means that if the network has stereotypical response of ball following (such as a pure asymmetry of the left-right weights), it is bound to lose the track, as it was observed in the first generations (approx. 50 generations).
4. The agent has to cope with rather meager input. With mere 9 distance sensors each pattern is ambiguous (is the ball coming into view or escaping view?). Moreover, small changes in ball position relative to the ray might lead to big input changes (say, in one cycle the input of one sensor might drop from something to zero).

2.3 Artificial Evolution

For the Artificial Evolution we employ the ENS^3 algorithm [7] and operators that vary the network structure. The genome of the evolution algorithm is the structure of the networks. The variation operator adds or deletes units (neurons) and synapses, as well as changing weights according to on-the-fly specified probabilities. The experimenter can also limit the number of units in the network, as well as introducing costs for extra units (neurons). The selection of the agents that generates offspring is rank based, which is controlled by the shape of the distribution (gamma) of a poisson process. With that measure we attempt to keep a high diversity during evolution.

All parameters of the evolution are alterable on-the-fly, that is, during the evolution itself. In that way, the experimenter is enabled as a 'meta-fitness', selecting agents not only by their raw fitness but also by remarkable characteristics of behavior. The controllable parameters are the weights of the terms of fitness function, the shape of the distribution of offspring production, number of cycles of trial, number of initial conditions (such as the initial position of the ball), number of neurons and synapses, etc. Offline parameters are for example the frequency of the virtual simulation and the refresh rate of the networks.

3 Results

3.1 Action Selection

In principle, competent action selection is achieved if the head is able to match it's velocity with the velocity of the ball. This requires modulation of both the direction and the force applied to the motors. From the input alone both direction and force are not decidable. So, in order to have best tracking, the past states have to be used to modulate the force vectors. As we will see, it is the profile of the motor output wave that modulates both the force and direction of the neck. The choice of wave profile for control is equivalent to the choice of attractor carried out by the network[4].

3.2 Initial Solutions

The first solutions that appeared (before 50th gen.) were simple networks, that have been used as a canvas by the evolution for the more resourceful ones. These primitive solutions used the asymmetry of the connections to lock to the ball, where the difference between the up sensor and the bottom sensor of the array, implied the direction imputed to the motor. A very obvious limitation of such networks, is that they are unable to actively search, rather remaining in fixed points (say, remaining down left until the ball were again in sight). These primitive solutions were rapidly substituted by networks that were able to solve the problem more robustly. We inspect these networks more closely in the next sections.

[4] The networks reported here never lose gaze under normal conditions. Many of the networks were not only selected by their high fitness, but also by their observed behavior in different conditions as for example smaller ball and higher simulation frequencies (many subtle and interesting properties are hard to define in terms of the fitness function, as for example the search strategies when the ball escapes gaze). Those whose behavior was seemingly less stereotypical also proved to have more diversity of the supporting dynamical structures. Nevertheless, their size was constrained to not more than 6 units in the recurrent layer and a maximal of 120 synapses.

3.3 Analysis of Dynamical Entities Generating Behavior

Most of the analysis is done in terms of the asymptotic behavior of the decoupled networks. Stimulus patterns that represent interactions of the sensors with the ball, lead to responses, that are projected onto motor space. This reduction has been used often in such studies and allow an intuitive understanding of the behavior of the network. This allows for a clear picture of the states the network would reach if the input remained constant. Moreover, to represent the motor actions the agent's body effectively impresses, we average the motor outputs (150 network steps), to verify the action tendency in a given moment (see figure 2). This is not an exhaustive profile of analysis and leads to some caveats that will be discussed at the end of this section.

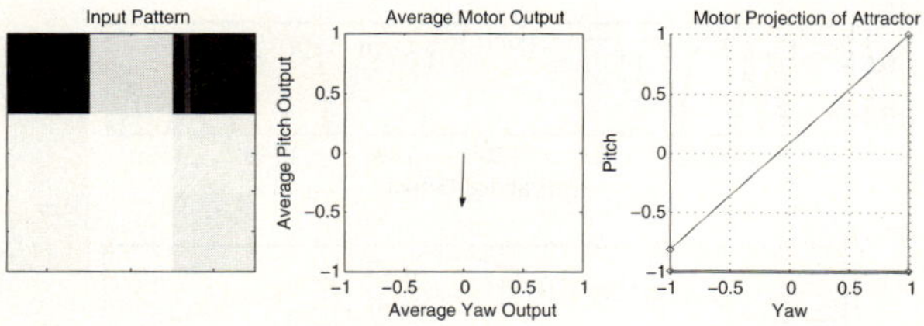

Fig. 2. Left to right, stimulus pattern, the period 4 attractor itself and the averaged output of the associated attractor. In the middle box, the arrow indicates the direction of states in the attractor.

3.4 Velocity Modulation via Oscillations

The velocity with which the head will move is defined by the average under the profile of the motor projection for the motor units, which is in turn a function of the attractor's shape. The motor actions are not usually not trivial states, as fixed-point attractors or simple periodic attractors. For a large portion of the input patterns (that represent the interaction with the ball), the natural asymptotic output of the network is a cyclic attractor (see figure 3. There are coexisting attractors for one and the same input pattern. Also, many different periods are found (e.g., 4, 5, 8, 16, 20 and quasi-periodic attractors and chaos). The net output of the oscillations tell the head where and with what speed to move.

It is easy to see that there is no one-to-one mapping of a given pattern to a velocity of the ball[5]. From that follows, that to be optimal (gaze locked with the ball), the head has to use different velocities even when keeping the input pattern constant (although the acceleration is constant). Therefore, the velocity of the head has to be chosen by considering the recent velocity history. In the

[5] This is easy to see: when the ball is subject to gravity, the velocity on the vertical axis, even if the head locks position with the ball, thereby keeping a constant input pattern, it must nevertheless accelerate down.

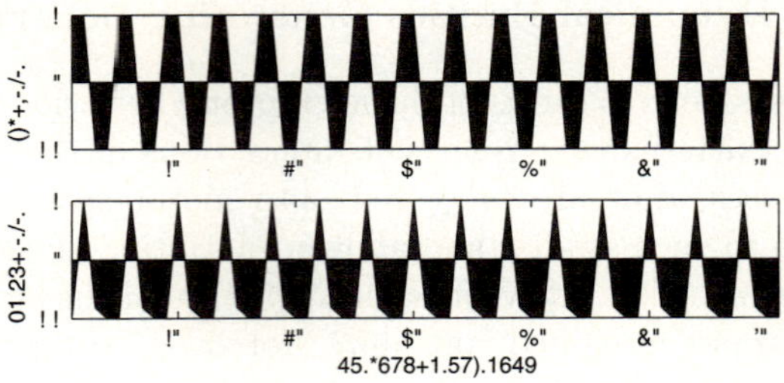

Fig. 3. Temporal translation of an orbit on the period 4 attractor above, for yaw (above) and pitch motors (below). The shape of the oscillations evoked by attractor above, defines the velocity arrow output above.

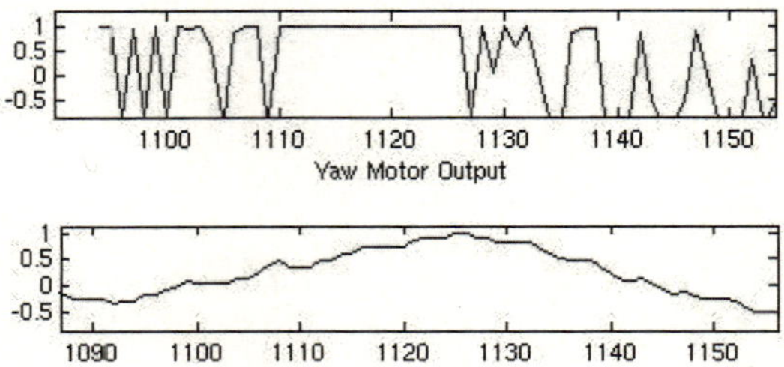

Fig. 4. Roughly linear increase and decrease of pitch velocity of the head. The row above is the actual output of the network. The second row is the convolution with a rectangular causal kernel of 10 steps (.1s), representing the average velocity implemented by the tracker. For the average velocity to increase, the transients must switch attractors.

following plot (figure 4), one sees how the oscillations on y axis lead to a linear increase on the averaged velocity of the tracker's neck. The increase is consistent with a linear velocity increase imposed by gravity, as in figure 4.

The averaged motor projections of the attractors indicate that there are two features of the attractors responsible for the action. Neither are preponderant over the other, i.e., the number of periods and the shape of the attractor. Different periods might lead to similar average speed. Conversely, equal periods might lead to different outputs. This is clearly illustrated with the following series of chaotic and quasi-periodic attractors in figure 5, for very similar input patterns. It shows the motor space projection of the orbit, for 150 steps, for the respective inputs. All the same, as the actions during the trials happen in a very short time, the average speed might change depending on the duration that an attractor is followed. That is particularly true for chaotic attractors, on which transients might get entrained in different positions. In figure 6 one sees a first return map of one example chaotic attractor has definite shape. Clearly, the output of the motors willdepend on where the transient engages the attractor.

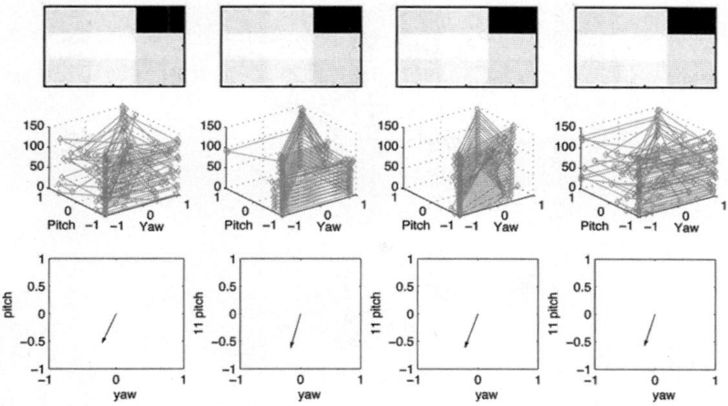

Fig. 5. For each of the stimulus above, we calculate 4 different attractors for 4 different randomized initial conditions in motor space (z axis is iteration cycle). Note that different chaotic attractors and periodic attractors appear, but lead to very similar actions (velocity vectors of the second row).

3.5 Motor Projections

As pointed out, in usual conditions, none of these attractors is followed during as many steps as 150, as a motor action may result in a change of stimulus (unless the velocity of the head is perfectly matching that of the ball, which happens often, as in figure 4. With a changing stimulus, so does the attractor's shape change. The transient hops across basins and is so guided to the new attractor (as bodies are attracted towards their center of gravity). As the input changes the aspects of the attractors, a path through attractor space is defined also by the changing basins of attraction. In the case of chaotic attractors, fractal boundaries might imply that even very small changes in input may provoke an attractor switch.

4 Discussion

4.1 Autonomy and Capacity Through Evolution

Methods with predefined associations between stimuli and output such as Tani's or Mataric's [8,9] exchange behavioral autonomy in favor of pre-defined action sequences (such as patterns to be imitated), learned with temporally dependent variations of the Hebbian Rule, or reinforcement learning schedules. Powerful yet they might be, when behavioral patterns are uniquely coded (one bias vector per behavior, for example), two difficulties arise. First, this is done at the expense of autonomy, as one plugs the bias vectors to trigger behaviors. Second, when designing a learning schedule, it is not always clear what is the 'best' motor action to be associated to a pattern. The option endorsed in here, is to design the

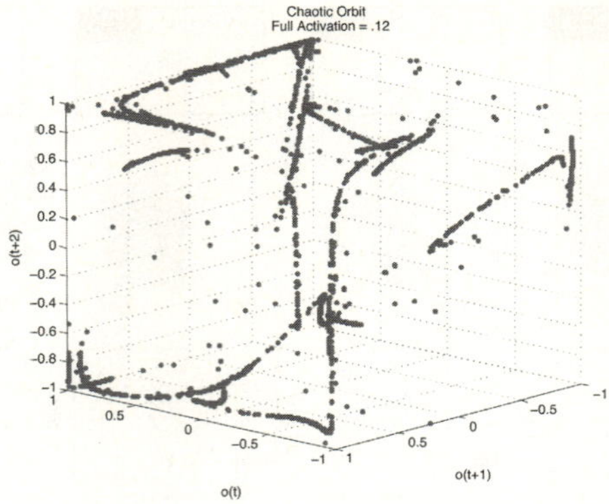

Fig. 6. First-return map of a chaotic attractor for the yaw motor output. See text for details.

problem in such a manner that the association happens implicitly by selection of agents. By providing a problem where meaningful actions respond to the 'structural coupling' [10,11] and [12,13] between agent and environment, there needs not be an a priori definition of the desired action.

Knitting the work with the overlaying neurodynamics framework is the intuition that it is possible to understand action control in terms of attractor shapes and landscapes. In the context of solving embedded problems of physical nature, by showing how an evolved attractor landscape is explored in control, the aim is to leverage some background understanding to scaffold the development of a theory of 'meta-transients' and 'attractor morphing'. This paper is an empirical offering in this direction[6].

5 Conclusion

5.1 Dynamical Entities Solving Embodied Problems

It follows an enlisting useful characteristics of problem solving, and its possible dynamical entities responsible. In this short paper we merely scratched the surface of these and each deserves a treatment of its own.

1. Situation specific action selection, due to dependency on temporal context, given by scattered basins of attraction in phase space. This is seen when the same stimulus pattern evokes different attractors, with different associated actions.
2. Robustness to errors and mismatches appears because of multiple possible responses to a given pattern, associated with history. This is seen as gravity is mapped by the transients across attractors of different shapes.

[6] A theory to quantify the behavioral capacity of underlying network structures in terms of topological complexity is underway

3. Unpredictability of motor responses in ambiguous circumstances. Chaos is responsible.
4. Modeling of body and environment in attractor shapes and paths across attractor space.

Perhaps pushing the analogy too far it could be said that, in nature, evolution begets parameter sets for the embodied neural structures for control. The networks lodge a wealth of dynamical repositories as collections of attractor landscapes, accessed with transient paths across attractors and *represent* the solution to ecological problems. The hope is that results from neuroethology will eventually warrant the analogy, even if to an extent. Meanwhile, dynamical systems approaches to cognition opened a niche for a rich research endeavor, the development of a, so far inexistent, theory of transients.

Acknowledgements. The authors acknowledge Keyan Ghazi-Zahedi and Martin Huelse for the Yars and ISEE simulation and evolution software. This work was supported by the grant DFG SCC1125 and the Fraunhofer Society, IAIS.

References

1. Beer, R.: A dynamical systems perspective on agent-environment interaction. Artificial Intelligence 72, 173–215 (1995)
2. Negrello, M., Pasemann, F.: Adaptive Neurodynamics. In: Yang, A., Shan, Y. (eds.) Applications of Complex Adaptive Systems, IDEA Group, Hershey (to appear, 2008)
3. Pasemann, F., Steinmetz, U., Huelse, M., Lara, B.: Robot control and the evolution of modular neurodynamics. Theory in Biosciences 120, 311–326 (2001)
4. Tsuda, I.: Toward an interpretation of dynamic neural activity in terms of chaotic dynamical systems. Behavioral and Brain Sciences 24, 793–847 (2001)
5. Pasemann, F.: Complex dynamics and the structure of small neural networks. Network: Computation in neural systems 13(2), 195–216 (2002), `http://www.iop.org/EJ/S/0/27972/uoqgs3Z8ALGleMygUyOqrg/toc/0954-898X/13/2`
6. Molter, C., Salihoglu, U., Bersini, H.: The road to chaos by time-asymmetric hebbian learning in recurrent neural networks. Neural Computation 19, 80–110 (2007)
7. Huelse, M., Wischmann, S., Pasemann, F.: Structure and function of evolved neurocontrollers for autonomous robots. Connection Science 16(4), 249–266 (2004)
8. Mataric, M.: Sensory-Motor Primitives as Basis for Imitation: Linking Perception to Action and Biology to Robotics. In: Imitation in Animals and Artifacts, pp. 392–422. MIT Press, Cambridge (2002)
9. Tani, J., Ito, M.: Self-organization of behavioral primitives as multiple attractor dynamics: A robot experiment. IEEE Transactions on Systems, Man and Cybernetics, Part A 33(4) (2003)
10. Varela, F.: Principles of Biological Autonomy. North Holland, Amsterdam (1979)
11. Varela, F., Rorty, E., Thompson, E.: The Embodied Mind. MIT Press, Cambridge (1991)
12. Harvey, I., Di Paolo, E., Wood, R., Quinn, M., Tuci, E.: Evolutionary robotics: A new scientific tool for studying cognition. Artificial Life 11(1-2), 79–98 (2005), `http://www.mitpressjournals.org/doi/abs/10.1162/1064546053278991`
13. Pfeifer, R., Iida, F., Bongard, J.: New robotics: Design principles for intelligent systems. Artificial Life 11(1-2), 99–120 (2005), `http://www.mitpressjournals.org/doi/abs/10.1162/1064546053279017`

Wavelet Network with Hybrid Algorithm to Linearize High Power Amplifiers

Nibaldo Rodriguez and Claudio Cubillos

Pontifical Catholic University of Valparaiso
Av. Brasil 2241, Chile
{nibaldo.rodriguez,claudio.cubillos}@ucv.cl

Abstract. This paper propose a linearizing scheme based on wavelet networks to reduce nonlinear distortion introduced by a high power amplifier over 256QAM signals. Parameters of the proposed linearizer are estimated by using a hybrid algorithm, namely least square and gradient descent. Computer simulation results confirm that once the 256QAM signals are amplified at an input back off level of 0 dB, there is a reduction of 29 dB spectrum re-growth. In addition proposed linearizing scheme has a low complexity and fast convergence.

Key words: Wavelet network, linearizing,power amplifier, digital modulation

1 Introduction

High power amplifier (HPA) introduce nonlinear distortion in both amplitude and phase in wireless communication systems based on multi-level quadrature amplitude modulation (MQAM) schemes [1]. The main drawback of 256QAM is the large envelope fluctuations, making the system sensitive to nonlinearities of the HPA. These distortions cause spectral regrowth in adjacent channels and deformation of the signal constellation. Many linearizing schemes based on artificial neural networks [2]-[6] and neuro-fuzzy systems [7][10] for compensations of the nonlinearities and their effects have been proposed in the recent literature. In most existing models, complex input-output measured signals are initially converted to either a polar or rectangular representation and then two separate and uncouple real-valued models are used to estimate the output amplitude and phase as a function of the input power amplitude. The real parameters of the two models were obtained during a training procedure based on back-propagation algorithm. Therefore, the disadvantage of these linearizing techniques is their slow convergence speed and elevated requirements of computing resources.

Due to the success of neural networks in various adaptive linearizing techniques, in this paper we propose a linearizing scheme based on wavelet networks (WN) to create a transmitter operating with only one real-valued WN instead of the two separate networks commonly used to achieve linear amplification and high spectral efficiency in 256QAM systems. WN was introduced in [11],[12] as an idea of combining wavelet theory with the adaptive learning scheme of

F. Almeida e Costa et al. (Eds.): ECAL 2007, LNAI 4648, pp. 1016–1023, 2007.

feedforward neural networks. WN has a feedforward structure consisting of two layers, a nonlinear hidden layer with wavelet activation function and a linear output layer. In order to optimize all of the WN parameters an hybrid training algorithm is used, which combines the gradient descent (GD) based search for the nonlinear parameters of the hidden wavelet nodes and the least squares (LS) estimation of the linear output parameters. In addition, the hybrid algorithm (LS-GD) is chosen here due to its fast convergence speed.

The remainder of this paper is organized as follows: In section 2, a brief description of the baseband 256QAM system is presented. The linearizing technique of the HPA and an hybrid training algorithm for adjusting the linearizer's parameters are presented in Section 3. The performance curves of the constellation warping effect and spectral regrowth of 256QAM signals are discussed in Section 4. Finally, the conclusions are drawn in the last section.

2 System Description

The input data bits are encoded using the 256QAM mapper device, which maps a k-tuple of bits over a MQAM symbol by using Gray coding. The transmitter filter is implemented as a square root raised cosine (SRRC) pulse shaping distributed between the transmitter and receiver with 47-taps, roll-off parameter equal to 0.3 and over-sample factor of 8 samples per symbol. The modulated baseband signal $x(n)$ is first linearized $y(n)$ and nonlinearly amplified $z(n)$, then propagated over an additive white Gaussian noise (AWGN) channel. The amplified signal is obtained using a traveling wave tube amplifier (TWTA), which is defined as

$$z(n) = A(|y(n)|) \exp\{j\left[\theta_y + P(|y(n)|)\right]\} \quad (1)$$

where $|y(n)|$ and θ_y correspond respectively to the amplitude and phase of the linearized complex signal $y(n)$.

The functions $A(\cdot)$ and $P(\cdot)$ denote nonlinear amplitude and nonlinear phase; respectively. For a TWTA, the expresions for $A(\cdot)$ and $P(\cdot)$ are given by Saleh as [1]

$$A(|y(n)|) = \frac{\alpha_A |y(n)|}{1 + \beta_A |y(n)|^2} \quad (2)$$

$$P(|y(n)|) = \frac{\alpha_P |y(n)|^2}{1 + \beta_P |y(n)|^2} \quad (3)$$

with $\alpha_A = 2, \beta_A = 1, \alpha_P = \pi/3$ and $\beta_P = 1$

The nonlinear distortion of a high power amplifier depends on the back off. The input back off (IBO) is defined as the ratio of the saturation input power, where the output power begins to saturate, to the average input power.

$$IBO(dB) = 10 \log_{10}\left(\frac{P_{i,sat}}{P_{i,avg}}\right) \quad (4)$$

where $P_{i,sat}$ is the saturation input power and $P_{i,avg}$ is the average power at the input of the TWTA.

At time n, the received signal $r(n)$ is defined as

$$r(n) = z(n) + \zeta(n) \tag{5}$$

where $\zeta(n)$ is the complex AWGN channel with two-sided spectral density. The signal $r(n)$ is fed to the 256QAM demapper device, which splits the complex symbols into quadrature and in-phase components; and puts them into a decision device, where they are demodulated independently against their respective decision boundaries. Finally, output bits stream $\hat{b}$ are estimated.

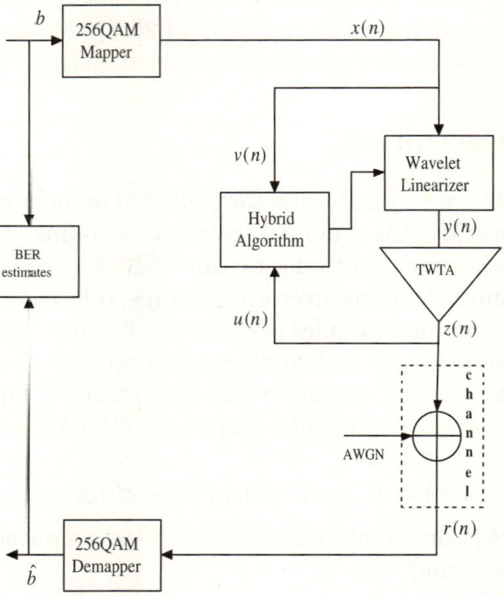

Fig. 1. Baseband 256QAM transmission system

3 Linearizing Base on WN

A linearizer works by creating 256QAM signal distortion that is the complement of the distortion inherent to the TWTA and its output signal is obtained as

$$y(n) = M(|x(n)|)\exp\{j\left[\theta_x + N(|x(n)|)\right]\} \tag{6}$$

where the functions $M(\cdot)$ and $N(\cdot)$ are used to invert the nonlinearities introduced by the TWTA.

In order to achieve the ideal linearizing function, the signal $z(n)$ will be equivalent to the input signal $x(n)$. That is:

$$A[M(|x(n)|)] = \alpha|x(n)| \tag{7a}$$
$$N(|x(n)|) = -P[M(|x(n)|)] \tag{7b}$$

where $\alpha|x(n)|$ is the desired linear model. In this paper, desired linear gain was set to $\alpha = 1$, so that saturation power was reached at 0 dB.

Therefore, combining (1) and (6) we obtain output signal of the linearizer as

$$y(n) = \frac{x(n)}{|x(n)|} A^{-1}[|x(n)|] \exp\{jP(A^{-1}[|x(n)|])\} \tag{8}$$

where $A^{-1}[\cdot]$ represent inverse AM-AM function of the TWTA.

Finally, in order to achieve (8), it is necessary only to find the real-valued function $A^{-1}[\cdot]$, which can be approximated by using wavelet neural network and a finite number of samples of the AM-AM function.

The linearizer's output signal $y(t)$ is approximated using a wavelet network and is obtained as

$$\hat{y} = Wh^T \tag{9}$$

where $()^T$ represent the transposed operator and

$$W = [w_0 \; w_1 \; \ldots \; w_c] \tag{10a}$$
$$h = [1 \; \phi_1(u) \; \ldots \; \phi_c(u)] \tag{10b}$$

where w_j are the linear output parameters, c is the number of hidden wavelet nodes, u is the input value and $\phi_j(u)$ are hidden wavelet functions, which is derived from its mother wavelet ϕ through the following relation

$$\phi_j(u) = \frac{1}{\sqrt{d_j}} \phi\Big(\frac{u - t_j}{d_j}\Big) \tag{11}$$

where t_j and d_j are the translation and dilatation parameters of the wavelet functions; respectively. In the present paper, we choose the Mexican Hat wavelet function as a mother wavelet, which is defined as [11]-[12]

$$\phi(\lambda) = (1 - \lambda^2) \exp\{-0.5\lambda^2\} \tag{12}$$

In order to estimate the linear parameters $\{w_j\}$ and nonlinear parameters $\{t_j, d_j\}$ of the wavelet linearizer an hybrid training algorithm is proposed, which is based on least square (LS) and gradient descent (GD) algorithms. The LS algorithm is used to estimate the parameters $\{w_j\}$ and the GD algorithm is used to adapts the nonlinear parameters $\{t_j, d_j\}$.

Now suppose a set of training input-output samples, denoted as $\{u_i, v_i, i = 1, \ldots, N_s)$. Then we can perform N_s equations of the form of (9) as follows

$$\hat{Y} = W\varPhi \tag{13}$$

with the matrix Φ defined as

$$\Phi = \begin{pmatrix} 1 & \cdots & 1 \\ \phi_1(u_1) & \cdots & \phi_1(u_{N_s}) \\ \vdots & \vdots & \vdots \\ \phi_c(u_1) & \cdots & \phi_c(u_{N_s}) \end{pmatrix} \tag{14}$$

where the desired output v_i and input data u_i are obtained as

$$v_i = \frac{|x(n)|}{max\{|x(n)|\}} IBO \tag{15a}$$

$$u_i = |z(n)| \tag{15b}$$

For any given representation of the nonlinear parameters $\{t_j, d_j\}$, the optimal values of the linear parameters $\{\hat{w}_j\}$ are obtained using the LS algorithm as follows

$$\hat{W} = \Phi^{\dagger} V \tag{16}$$

where $V = [v_1 \; v_2 \; \cdots \; v_{N_s}]$ is the desired output and $\Phi^{\dagger}$ is the Moore-Penrose generalized inverse [13] of the wavelet function output matrix Φ.

Once linear parameters $\hat{W}$ are obtained, the gradient descent algorithm adapts the nonlinear parameters of the hidden wavelet functions minimizing mean square error, which is defined as

$$E(t, d) = \frac{1}{2} \left\| V - \hat{Y} \right\|^2 \tag{17a}$$

$$\hat{Y} = \hat{W} \Phi \tag{17b}$$

Finally, the GD algorithm adapts the parameter $\{t_j, d_j\}$ according to the following equations

$$t_j = t_j + \mu(v_k - \hat{y_k}) \hat{w}_j \frac{\partial \Phi_k}{\partial t_j} \qquad j = 1, \ldots, c \tag{18a}$$

$$d_j = d_j + \mu(v_k - \hat{y_k}) \hat{w}_j \frac{\partial \Phi_k}{\partial d_j} \qquad j = 1, \ldots, c \tag{18b}$$

where μ is the step size of the GD algorithm.

4 Simulations Results

In this section, it is presented the performance evaluation of the nonlinear distortion compensation scheme for 256QAM signals. The parameters of the wavelet linearizer (WL) were estimated during the training process using $N_s = 100$ samples of the nonlinear amplitude and the TWTA was operated with an IBO of

0 dB. Linearizer was configured with one input node, one linear output node, two hidden wavelet nodes and one bias unit. In the training process the initial parameters , $\{t_j(0), d_j(0)\}$, were initialized by a Gaussian random process with normal distribution $N(0, 1)$. The training process was run with 3 trials and the normalized mean square error after convergence was approximately equal to -50 dB. In decision-direct mode, the wavelet linearizer is simply a copy of the wavelet network obtained in training process.

Fig. 2, show the power spectral density (PSD) curves. In one hand, for the nonlinear amplification case only with TWTA, the PSD curve is denoted as (256QAM TWTA). By the other hand, for the nonlinear amplification case with wavelet linearizing, the curve is denoted as (256QAM WLTWTA). It can be seen that 256QAM TWTA have a degradation of PSD about 29 dB. Moreover, from the figure can be seen that the curve of spectral regrowth with linearizing is very close to the ideal (linear) case due to the incorporation of the proposed linearizing scheme. Therefore, the proposed wavelet linearizing scheme allow to reduce significantly the degradation of the spectral re-growth for 256QAM signals at an IBO level of 0 dB.

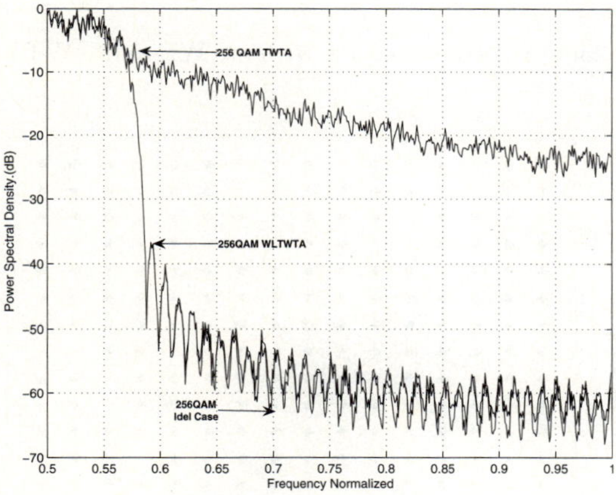

Fig. 2. Output Spectrum with and without linearizer at IBO = 0 dB

The effects of nonlinearities on the received 256QAM constellations in the absence of the channel AWGN are shown in Fig. 3 and 4, which correspond to the TWTA without and with linearizing scheme operated at an input back off level of 0 dB; respectively. According to Fig. 3, it is observed that square 256QAM constellation is severely distorted by the nonlinear characteristics of the TWTA without linearizing technique. These distortions are interpreted as in-band noise, and it is called constellation warping effect. From Fig. 4, the proposed linearizer reduces significantly the constellation warping effect on received 256QAM signals. Therefore, comparing Fig. 3 and 4, it can be seen that constellation warping effect is reduced significantly by using proposed wavelet linearizing scheme. Moreover, it permits to reduce both computer storage requirements and coefficients adaptation time of the linearizer, which is achieved due to the proposed hybrid algorithm; it only uses one wavelet network for compensating both nonlinear characteristics of the TWTA.

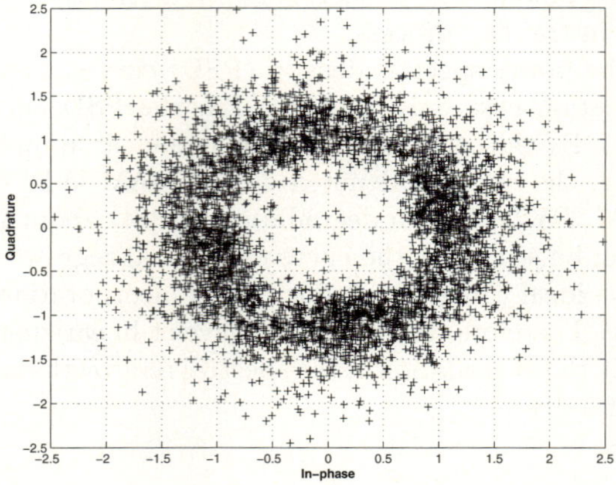

Fig. 3. Received 256QAM constellation without WL and TWTA at IBO = 0 dB

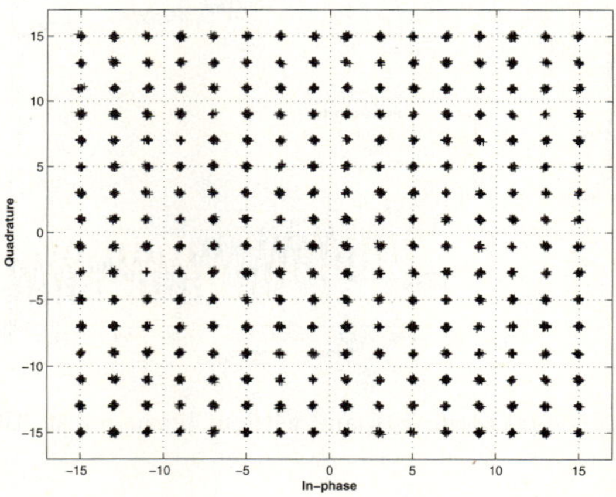

Fig. 4. Received 256QAM constellation with WL and TWTA at IBO = 0 dB

5 Conclusions

In this paper, an adaptive baseband linearizing scheme based on wavelet networks to reduce nonlinear distortion has been proposed. The proposed linearizer uses only a wavelet network with 2 hidden wavelet nodes and one output node to compensate both the non-lineal amplitude and phase distortion introduced by the TWTA over 256QAM signals. The linearizer parameters adaptation was found by using 3 iterations of a hybrid algorithm based on both least square and gradient descent method. Simulation results have shown that the proposed wavelet linearizing scheme can prevent to the RF transmitter from spectrum re-growth and constellation warping effect due to TWTA's nonlinearity with a low complexity and fast convergence.

Acknowledgments. The authors want to thank to Vice-rectoria of the Catholic University of Valparaiso (Proyect 209744/2007).

References

1. Saleh, A.M.: Frecuency-Independent and Frecuency-Dependent nolinear models TWT amplifiers. IEEE Trans. Comm. 29, 1715–1719 (1981)
2. Watkins, B.E., North, R.: Predistortion of nonlinear amplifier using neural networks. In: Proc. IEEE Military Comm. Conf., vol. 1, pp. 316–320. IEEE, Los Alamitos (1996)
3. Ibnkahla, M., Sombrin, J., Castanié, J.F., Bershad, N.J.: Neural network for modeling non-linear memoryless communications channels. IEEE Trans. Comm. 45(5), 768–771 (1997)
4. Ibnkahla, M.: Neural network modelling predistortion technique for digital satellite communications. In: Proc. IEEE ICASSP, vol. 6, pp. 3506–3509. IEEE, Los Alamitos (2000)
5. Ibnkahla, M.: Natural gradient learning neural networks for adaptive inversion of Hammerstein systems. IEEE Signal Processing Letters, 315–317 (October 2002)
6. Abdulkader, H., Langket, F., Roviras, D., Castanie, F.: Natural gradient algorithm for neural networks applied to non-linear high power amplifiers. Int. Journal of Adaptive Control and Signal Processing 16, 557–576 (2002)
7. Li, Y., Yang, P.-H.: Data predistortion with adaptive fuzzy systems. In: IEEE Int. Conf. Syst., Man, and Cybern., vol. 6, pp. 168–172 (1999)
8. Hong-min, D., Song-bai, H., Jue-bang, Y.: An adaptive predistorter using modified neural networks combined with a fuzzy controller for nonlinear power amplifiers. Int. Journal of RF and Microwave Computer-Aided Engineering 14(1), 15–20 (2003)
9. Lee, K.C., Gardner, P.: A Novel Digital Predistorter Technique Using an Adaptive Neuro-Fuzzy Inference System. IEEE Comm. Letters 7(2), 55–57 (2003)
10. Lee, K.C., Gardner, P.: Adaptive neuro-fuzzy inference system (ANFIS) digital predistorter for RF power amplifier linearization. IEEE Trans. on Veh. Tech. 55(1), 43–51 (2006)
11. Zhang, Q., Benvenist, A.: Wavelet network. IEEE Trans. Signal Processing 13(6), 889–898 (1992)
12. Zhang, Q.: Using wavelet network in non-parameters estimation. EEE Trans. Neural Networks 8(2), 227–236 (1997)
13. Serre, D.: Matrices: Theory and applications. Springer, New York (2002)

A Behavior-Based Model of the Hydra, Phylum Cnidaria

Malin Aktius[1], Mats Nordahl[2], and Tom Ziemke[1]

[1] University of Skövde
School of Humanities and Informatics
SE-541 28 Skövde, Sweden
{malin.aktius,tom.ziemke}@his.se
[2] Department of Applied Information Technology
Göteborg University and Chalmers University of Technology
SE-417 56 Göteborg, Sweden
nordahl@design.chalmers.se

Abstract. Behavior-based artificial systems, e.g. mobile robots, are frequently designed using (various degrees and levels of) biology as inspiration, but rarely modeled based on actual quantitative empirical data. This paper presents a data-driven behavior-based model of a simple biological organism, the hydra. Four constituent behaviors were implemented in a simulated animal, and the overall behavior organization was accomplished using a colony-style architecture (CSA). The results indicate that the CSA, using a priority-based behavioral hierarchy suggested in the literature, can be used to model behavioral properties like latency, activation threshold, habituation, and duration of the individual behaviors of the hydra. Limitations of this behavior-based approach are also discussed.

Key words: behavior-based modeling, data-driven modeling, hydra, colony-style architecture

1 Introduction

Drawing upon theories from, for example, neuroscience, ethology, cognitive sciences, and evolutionary game theory, the use of biology in robotics has been influenced by several types of mechanisms underlying behavior [1]. The degree of biological inspiration for artificial agents is, however, quite diverse: ranging from classical control systems using only vague arguments founded in biology to actual replication of biological behavior [2]. In this work, we are concerned with the ethological[3] perspective on behavior, which explains behavior from a functional aspect, in terms of e.g. reflexes, fixed action patterns, and orientation of movement (e.g. [3]), and we investigate the possibility for a data-driven approach to behavior-based modeling of a lower animal.

[3] The study of animal behavior under natural conditions.

F. Almeida e Costa et al. (Eds.): ECAL 2007, LNAI 4648, pp. 1024–1033, 2007.

A behavior-based model at the ethological level is proposed in [4], in terms of decision upon straight-line swimming vs. tumbling behavior in bacteria. Turning to a more complex animal, another example can be found in [5], where parts of the visuo-motor behavior of the praying mantis[4] are modeled using the framework of schema theory, and implemented in a hexapod robot.

In this work, we aim at developing a data-driven behavior-based model of the constituent behaviors and the behavioral organization of a simple biological organism. For this approach, the following criteria are proposed for a model organism: (1) Sufficient availability of literature on quantitative behavioral studies – for input-output modeling of sensorimotor actions. (2) Sufficient availability of literature on studies of interaction of behaviors – for implementing a relevant behavior selection system. (3) The organism should have a simple nervous system – for ease of modeling internal processes. Based on these criteria, the hydra (*Hydra*) was selected model organism for this work. The next section provides some background information on the hydra. Section 3 then states the models and methods used, whereas the experiments and results are provided in Section 4. Finally, Section 5 discusses the outcome, recommendations for future work, and limitations of the model.

2 Background: The Hydra and Its Behavior

The selected model organism, the hydra, is shown in Fig. 1. Hydra belongs to the phylum Cnidaria, the first evolved animals (that still exist) to possess nerve cells and sense organs. It lives in ponds, lakes, and streams, where it is most often found attached, by its foot, to some vegetation. The hydra feeds on small aquatic invertebrates, can reproduce sexually or asexually, in the latter case by means of budding. It remains in its tube-shaped polyp form throughout its lifetime [6].

Turning now to the behavior of hydra, the distinct movement patterns of the animal, resulting from alternating activity of its motor cells, consist of: (1) con-

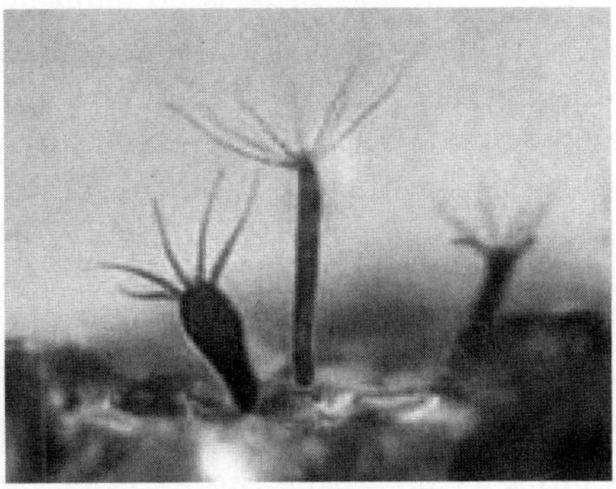

Fig. 1. Three hydras. Image courtesy of BioMedia Associates [7].

[4] A large, carnivorous insect.

traction and extension of body and tentacles, (2) feeding, involving a sequence of actions, and (3) locomotion, accomplished either by gliding (by means of cilia[5] on the foot.), or by somersaulting [8,9]. In the literature, the behaviors of hydra are described in terms of responses to specific stimuli rather than in terms of the actual movement patterns of the animal. Mainly, four distinct constituent behaviors are identified: spontaneous actions, response to mechanical and light stimuli, respectively, and feeding [8,9]. A priority-based organization of the behaviors has been suggested in e.g. [10], with feeding inhibiting responses to light and mechanical stimuli, and response to light inhibiting response to mechanical stimuli.

Spontaneous Actions: Without changes in hydra's external environment, it shows spontaneous, periodic, contractions and locomotion. There is an adaption to background illumination in the sense that the contraction frequency varies with ambient light conditions. The actions also depend on the nutritional state of the animal: contraction frequency decreases with starvation, while locomotion is more common in starved hydras. After one week of starvation, practically any overt behavior ceases, and the animal eventually perishes [11].

Response to Mechanical Stimuli: Response to mechanical stimuli, such as shaking or physical contact, occurs by means of contraction or locomotion. The way in which the hydra responds depends on stimuli interaction history as well as on the nutritional state of animal: starved animals are more likely to respond by locomotion, whereas a contraction response is more common in well fed animals. The response shows habituation to repeated stimuli [12].

Response to Light Stimuli: On exposure to strong light there is an immediate inhibition of any ongoing contraction. Following a latency, a response consisting of either contraction or locomotion is evoked. Also in this case, locomotion is more common in starved animals. The latency is inversely related to light intensity [13], and there is no habituation [12].

Feeding: Hydra's feeding behavior consists of a sequence of actions, and the behavior is evoked by mechanical stimulation of hydra's tentacles or by the presence of glutathione (GSH), a peptide released by prey stung by hydra's nematocysts[6]. The activation threshold for feeding is regulated by hydra's nutritional state, with starved animals having a lower threshold than recently fed ones. There is also a refractory period following a feeding response, during which the capacity of responding is gradually regained e.g. [10].

In summary, hydra's response to a given stimulus depends on: (1) stimulus configuration, such as intensity or concentration, (2) state of internal variables, such as its nutritional level, and (3) stimulus interaction history, such as habituation to a mechanical stimulus.

[5] Small, hairlike extensions.

[6] Stinging organelles located on hydra's tentacles.

3 Models and Methods

This section starts by presenting the simulated hydra and its environment, and continues by briefly describing the colony style architecture (CSA) for behavior selection, as well as presenting the implemented CSA behavioral organizer. Due to lack of space, some details are omitted, but can be found in [14].

3.1 The Simulated System

The hydra was modeled here ignoring the dynamical properties of its body. In this simplified model, the animal's body is represented in 2D, by a circle with radius r and maximum extension length l_{max}. The movement of the animal is controlled by setting the physical state variables in a simplified manner. The environment of hydra consists of a square-shaped arena with periodic boundary conditions. The implemented physical state variables are shown in Table 1.

In [15], the concept of motivational state is explained as the combined perceptual and physiological state of an animal, i.e. the state of its external and internal environment. Table 1 shows the five motivational state variables that were implemented in the simulated hydra.

Table 1. Top panel: Physical state variables for the simulated hydra in a 2D environment. For the extension rate, r_{ext}, 0 denotes maximal contraction, and 1 maximal extension. S denotes the side-length of the arena. Bottom panel: Motivational state variables for the simulated hydra. The upper limit imposed on hunger, $hmax$, corresponds to one week of starvation, see Section 2.

Variable	Range	Description
M_l	$[0, 1]$	Reading of light sensor
M_m	$[0, 1]$	Reading of touch sensor
M_c	$[0, 1]$	Reading of GSH sensor
M_h	$[0, hmax]$	Nutritional state
M_{beh}	B1,B2,B3,B4	Current active behavior

Variable	Range	Description
x_f	$[0, S]$	Position of foot
y_f	$[0, S]$	Position of foot
θ_{ext}	$[0, 2\pi]$	Direction of body extension
θ_{loc}	$[0, 2\pi]$	Direction of animal locomotion
r_{ext}	$[0, 1]$	Extension rate

Four main constituent behaviors were implemented in hydra's behavior repertoire: spontaneous actions, response to mechanical stimuli, response to light stimuli, and feeding. The first three behaviors all generate movement of the animal in terms of either contraction or locomotion, and these movement patterns were implemented as sub-behaviors of the corresponding main behaviors. Table 2 shows the implemented behaviors in the model of hydra.

Table 2. Behaviors and sub-behaviors in the simulated hydra. Priority 1 denotes the highest priority.

Label	Priority	Description
B1	4	Spontaneous actions
B2	3	Response to mechanical stimuli
B3	2	Response to light stimuli
B4	1	Feeding

Sub-behaviors of B1-B3	
Label	Description
B11, B21, B31	Contraction/ Extension
B12, B22, B32	Locomotion

3.2 The Control Architecture

For coordination of hydra's constituent behaviors, the CSA [16] was used. Developed by Connell, it descends from the subsumption architecture presented by Brooks [17], and operates according to the following principles [16,18]:

- It uses an arbitration method, i.e. only one behavior is active a time.
- Behaviors are arranged in layers, in a priority-based manner.
- Each behavior is associated with an applicability clause (AC), and a transfer function (TF). The AC determines whether the behavioral output should be active or not, while the TF determines what action the agent would take (typically the motor output), assuming that the behavior is active.
- Switches on the behavioral interconnections perform the actual behavior selection (see Fig. 2). The following switch types exist: (1) Suppression, where an active output from a a higher-priority behavior replaces any lower-priority behavioral output. (2) Inhibition, where a higher-priority behavior (whenever active) prevents a lower-priority one to generate any output. (3) Release type, where a higher-priority behavior enables the output of a lower-priority behavior to pass through the switch.

In a CSA, an AC can be either situation-driven or event-driven. A situation-driven AC is related to a goal state, and only the present motivational state of the agent determines whether the AC should be true or false. An event-driven AC, on the other hand, is of set/reset type where a certain event[7] triggers the AC, and another event resets it.

For the priority-based interaction of the behaviors in the hydra, suggested in e.g. [10], behavioral organization using a CSA can be accomplished by organizing the behaviors in layers with respect to priorities, using switches of suppression type for the behavioral inter-connections. Fig. 2 shows the implemented behavioral model.

In [8,9], a random-walk movement was observed in undisturbed hydras. To implement this property in the simulated animal, the locomotion direction, θ_{loc}, was set to a new, random value for each locomotion response. Also, in the above-mentioned works, it is suggested that the animals extend in a new, random direction following a contraction. Thus, also the extension direction, θ_{ext}, of the

[7] According to Connell, an event is characterized by a very brief (point-like) occurrence, whereas situations typically are extended intervals of time [16].

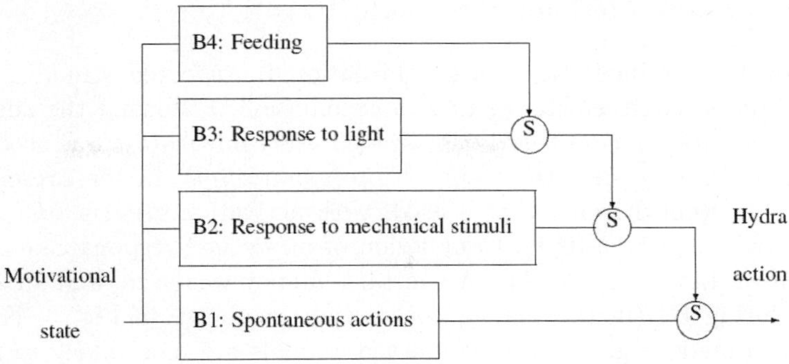

Fig. 2. Implemented overall organization (in the CSA) of the constituent behaviors B1-B4 in the simulated hydra. Here, S represents switches of suppression-type.

simulated hydra was set to a new, random value for each contraction. To simulate the nutritional state of the hydra, M_h, was set to decrease during feeding, and increase while any other behavior is active.

4 Experiments and Results

Using a CSA, generation of the constituent behaviors amounts to defining an AC and a TF for each behavior. Experimental data from the literature was used to validate the properties of ACs and TFs. As described in Section 3, the actual movement patterns of the animal were implemented as sub-behaviors of the main behaviors. The biologically plausible concepts of contraction pulses (CPs) and locomotion pulses (LPs) [10,11] were adopted. In this way, contraction and locomotion are accomplished by corresponding pulse signals, extended in time, which cause animal movement. Note that an active AC only states a potential activeness of its corresponding behavior, the actual behavior selection is carried out by the way in which the behaviors are interconnected (i.e. how they are arranged in layers and what switch types are used).

4.1 Spontaneous Actions (B1)

By default, a spontaneous action is an applicable behavior, and hence the AC for this behavior is always true. The TF consists of CPs and LPs at certain times, which were implemented as inhomogeneous Poisson spike trains to agree with the behavior suggested in [11], as briefly described in Section 2. The results, shown in Fig. 3, are an expected value of CP rate that is linearly decreasing with M_h (i.e. with the time of starvation), and an expected value of LP rate that is normally distributed with mean at $M_h = h_{max}/2$.

4.2 Response to Mechanical Stimuli (B2)

As previously described, B2 shows habituation to repeated stimuli, and the evoked response is either contraction or locomotion. Following the suggestion in [4], a habituation model based on cascaded leaky integrators was used, where the model output denotes the response probability, p_{B2}, in the presence of a stimulus. An event-driven AC was used, with activation criteria $M_c > 0$ and $p_{B2} > X$, where $X \sim U(0, 1)$. Completion of an evoked response, i.e. the end of a CP or an LP, resets the AC. A constant latency was used, and an example of a TF during B2 (in the case of contraction) is shown in Fig. 3. Following observations made, e.g. in [8], that starved animals are more likely to respond by locomotion, the probability of LP was set to increase with M_h. With Y_1 as the stimulus input, the equations for unit j in a cascade of n leaky integrators become:

$$V_j(k+1) = a_j V_j(k) + b_j Y_j(k), \tag{1}$$

$$Y_{j+1}(k) = \begin{cases} Y_j(k) - V_j(k), & Y_{j+1}(k) > T_j \\ 0, & \text{otherwise.} \end{cases} \tag{2}$$

Y_{n+1}, the output from the n:th unit, was taken as p_{B2}. An evolutionary algorithm (EA) was used to optimize the size and parameters of the habituation model with respect to training data from experiments carried out in [12]. The final evolved, model consists of 10 units, and its response to various inter-stimulus intervals (ISIs) is shown in Fig. 3.

4.3 Response to Light Stimuli (B3)

Since the evoked response to light stimuli, to a great extent resembles, the one to mechanical stimuli, the behavior generation is similar to the one of B2. For B3, however, no habituation occurs, whereas the latency is invesely related to the intensity of a stimulus. see Fig. 3. Activation of the (event-driven) AC occurs if $M_l > T_l$, where T_l is the activation threshold, set here to be constant. The same CP/LP relation as in B2 was used.

4.4 Feeding (B4)

B4 was modeled to be activated by the presence of GSH, where the activation threshold depends on hydra's nutritional state and where there is gradual recovery of feeding ability, as discussed in Section 2. An event-driven AC was used, where activation of the AC occurs if $M_c > T_f$, and the AC is reset if $t_{B4} > D_f$, or if $M_c = 0$. Here, t_{B4} denotes the time for which B4 has been active, and D_f denotes the ability of the feeding response, in terms of duration, at the time of activation of B4. Thus, feeding occurs until repletion or until the presence of food is no longer detected by the sensory system.

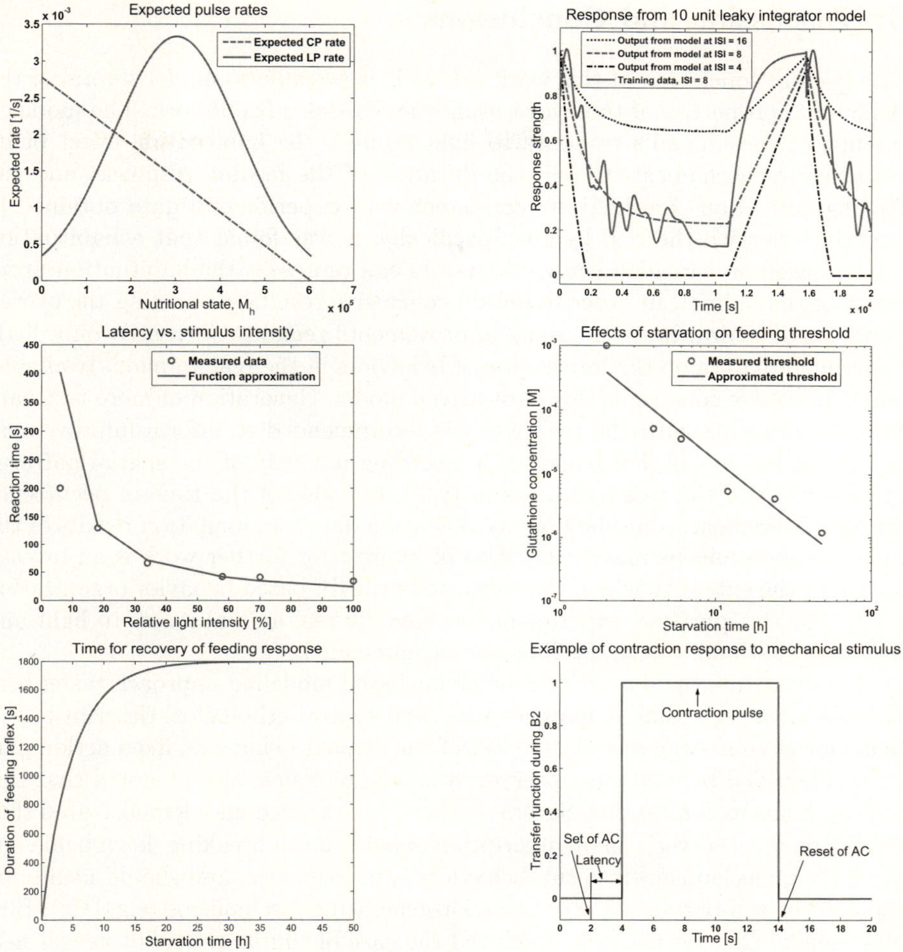

Fig. 3. Top left: Expected values of CPs and LPs, respectively, for B1 as functions of the simulated hydra's nutritional state. Top right: Decrease in response strength, p_{B2}, for B2 as a result of the habituation effect. Output generated by the best evolved leaky integrator model, consisting of 10 units. Middle left: The latency of B3 as a function of stimuli intensity. Middle right: The activation threshold of B4 as a function of the simulated hydra's nutritional state. Bottom left: The effect of starvation on the (possible) duration of B4. Bottom right: Example of behavioral output from the model, showing the transfer function for a contraction response, B21, to a mechanical stimulus.

5 Discussion and Conclusions

The general conclusion of this work is that it is possible to model several of the behavioral properties of the hydra using the CSA as a framework. The modeled latency of the animal's response to light stimuli, the habituation effect of its response to mechanical stimuli, the duration of the feeding response, and the feeding activation threshold are consistent with experimental data obtained in experiments with the real hydra. Specifically, it was found that a habituation model based on cascaded leaky integrators can represent the habituation properties of the animal. In order to obtain conclusive results concerning the overall behavior of hydra, however, some improvements remain. Currently, only little experimental data on the integration of behaviors in the real animal is available, which of course constrains this data-driven model. Generation of more test data from experiments with the real hydra is recommended to inform improvement of the current model. For example, a recording not only of the spatial patterns of movement of the real hydra (as in [8,9]), but also of the time of occurrence for each movement, could be used as validation data for simulation results of the animal's spontaneous movement. Also of interest for further work is an investigation of the extent to which the suggested priority-based behavior organization holds. For this purpose, experiments on how the real animal reacts to light and mechanical stimuli during feeding are recommended.

Finally, while the data-driven behavior-based modeling approach taken here has the advantage that it matches very well typical ethological descriptions of behavior as consisting of a discrete set of constituent behaviors, fixed action patterns, identified by a human observer, it should of course also be noted that this approach has its limitations. Several authors have argued and demonstrated that relying on an observer's distal description of behavior in breaking down behavior-generating mechanisms into sub-behaviors is problematic, and should ideally be replaced by self-organization of behavior-generating mechanisms (e.g. [19]). Biró and Ziemke [20], for example, analyzed the case of simple recurrent neural networks evolved for the control of visually guided search behavior, and showed that the networks, through feedback and self-modulation over time, exhibited what to an observer could very well appear to be a number of distinctly different behaviors organized in a subsumption architecture. Hence, the fact that we have shown in this paper that ethological descriptions of the hydra's overall behavior as consisting of four distinct constituent behaviors can be modeled in a corresponding CSA, should not be taken as conclusive evidence that the real hydra's behavior is actually organized this way. An alternative approach to modeling the hydra's behavior would be, for example, an evolutionary-robotics based model in which the empirical data available for the real hydra could inform the formulation of an appropriate fitness function for the evolution of its artificial counterpart without breaking down the control architecture in a modular fashion a priori.

Acknowledgments. This work was supported by a European Commission grant to the project *"Integrating Cognition, Emotion and Autonomy"* (ICEA, IST-027819, www.iceaproject.eu) as part of the European *Cognitive Systems* initiative. The experimental part was carried out while the first author was at Chalmers University of Technology [14]. The writing of this paper has benefited from discussions with Robert Lowe.

References

1. Webb, B.: Can robots make good models of biological behavior? Behavioral and Brain Sciences 24, 1033–1050 (2001)
2. Arkin, R.C.: Reactive robotic systems. In: Arbib, M. (ed.) Handbook of Brain Theory and Neural Networks, pp. 793–796. MIT Press, Cambridge (1995)
3. Slater, P.J.: Essentials of Animal Behaviour. Cambridge University Press, Cambridge (1999)
4. Staddon, J.E.R.: Adaptive Dynamics. MIT Press, Cambridge (2001)
5. Arkin, R., Kahled, A., Weitzenfeld, A., Cervantes-Perez, F.: Behavioral models of the praying mantis as a basis for robotic behavior. Journal of Robotics and Autonomous Systems 32(1), 39–60 (2000), `citeseer.ist.psu.edu/arkin98behavioral.html`
6. Hickman, C.P., Roberts, L.S., Larson, A., I'Anson, H., Eisenhour, D.S.: Integrated Principles of Zoology, 13th edn. McCraw-Hill, New York (2006)
7. Biomedia associates: http://www.ebiomedia.com (Oct. 13th 2006)
8. Jennings, H.S.: Behavior of the Lower Organisms. Indiana University Press, Bloomington (1962)
9. Wagner, G.: On some movements and reactions of Hydra. Quarterly Journal of Microscopical Science 48, 585–622 (1905)
10. Lenhoff, H.M.: Behavior, hormones, and Hydra. Science 161, 434–442 (1968)
11. Passano, L.M., McCullough, C.B.: Co-ordinating systems and behavior in Hydra, I. pacemaker system of the periodic contractions. Journal of Experimental Biology 41, 643–664 (1964)
12. Rushforth, N.B., Burnett, A., Maynard, R.: Behavior in Hydra: Contraction responses of Hydra Pirardi to mechanical and light stimulation. Science 139, 760–761 (1963)
13. Singer, R., Rushforth, N.B., Burnett, A.: The photodynamic action of light on Hydra. Journal of Experimental Zoology 54(2), 169–173 (1963)
14. Aktius, M.: Modeling hydra behavior using methods founded in behavior-based robotics. Master's thesis, Chalmers University of Technology. Department of Applied Mechanics, Göteborg, Sweden, Report No. 2007:17 (2007)
15. McFarland, D., Bösser, T.: Intelligent Behavior in Animals and Robots. MIT Press, Cambridge (1993)
16. Connell, J.: A colony architecture for an artificial creature. Technical Report 1151, MIT Artificial Intelligence Laboratory (June 1989)
17. Brooks, R.A.: A robust layered control system for a mobile robot. IEEE Journal of Robotics and Automation RA-2(1), 14–23 (1986)
18. Harper, C.J., Winfield, A.F.T.: A methodology for provably stable behaviour-based intelligent control. Robotics and Autonomous Systems 54, 52–73 (2006)
19. Nolfi, S., Floreano, D.: Evolutionary Robotics. MIT Press, Cambridge (2000)
20. Biró, Z., Ziemke, T.: Evolution of visually-guided approach behaviour in recurrent artificial neural network robot controllers. In: From animals to animats 5 - Proceedings of the Fifth International Conference on the Simulation of Adaptive Behavior, pp. 73–76. MIT Press, Cambridge (1998)

A Computational System for Investigating Chemotaxis-Based Cell Aggregation

Manolya Eyiyurekli[1], Peter I. Lelkes[2], and David E. Breen[1]

[1] Department of Computer Science, College of Engineering
[2] School of Biomedical Engineering, Science and Health Systems
Drexel University, Philadelphia PA 19104, USA

Abstract. We have developed a software system that simulates chemotaxis-based cell aggregation in 2D. The model implemented within the system consists of such cell behaviors as chemical diffusion/detection, motility, proliferation, adhesion and life cycle stages. Each virtual cell detects the state of the environment, and responds to the environment based on a pre-defined "program" and its own internal state. Cells are discrete units that are located on a grid, exist in discrete states (e.g. active or dying) and perform discrete tasks (e.g. divide and attach), but they also contain and are affected by continuous quantities (e.g. chemoattractant concentrations, gradients, age and velocities). This paper provides an overview of our chemotaxis-based aggregation model and details the algorithms required to perform chemotaxis-based cell aggregation simulation. A number of biological studies are being conducted with the system. They include fine-tuning the model parameters to reproduce *in vitro* PC12 cell aggregation experiments and parametric studies that demonstrate the effect that the model's components have on cell aggregation dynamics.

1 Introduction

Chemotaxis (CTX) is the phenomenon where cells detect gradients of chemicals (growth factor, cytokine) and respond to the chemical stimulus by moving either towards (positive CTX) or away (negative CTX) from the source. Multicellular aggregates and eventually tissue-like assemblies are formed when individual cells attach to each other and when this aggregation leads to subsequent cellular differentiation. Understanding the influence of the many components of CTX on overall cell aggregation should lead to a more detailed understanding of the mechanisms involved in tissue assembly and organogenesis, and also facilitate the development of novel technologies for tissue engineering based on controlling or directing these underlying biological processes. Modeling cell aggregates and their assembly/differentiation into functional tissues has implications for the mechanistic understanding of this process *in vivo*, as well as for "*in vitro* embryology". Beyond tissue engineering, chemotaxis also plays a significant role in a number of other biomedical phenomena of great interest, e.g. tumor formation [1] and bacteria accumulation [2]. Chemotaxis is also currently being

F. Almeida e Costa et al. (Eds.): ECAL 2007, LNAI 4648, pp. 1034–1049, 2007.

explored as a central modeling paradigm for self-organizing geometric primitives that automatically produce user-defined shapes.

Given the important role that chemotaxis plays in a variety of biological processes, we have developed a software system that simulates chemotaxis. The system provides a computational testbed for investigating the dynamics of chemotaxis and the biological components/interactions that affect and control it. While we are ultimately interested in modeling chemotaxis in 3D, our initial computational system is capable of simulating CTX-based cell aggregation in 2D. The model implemented within the system consists of the cell behaviors needed for aggregation, such as chemical diffusion/detection, motility, proliferation, adhesion and life cycle stages. An aggregation simulation is performed by modeling individual cells existing in an environment that contains a diffusing chemoattractant chemical. Each cell detects the state of the environment, and responds to the environment based on a pre-defined "program" and its own internal state. The model is hybrid in nature as it includes both discrete and continuous components. The cells are discrete processing units that are located on a grid, exist in discrete states (e.g. active or dying) and perform discrete tasks (e.g. divide and attach), but they also contain and are affected by continuous quantities (e.g. chemical concentrations, gradients, age and velocities). This paper provides an overview of our model, and details the algorithms required to perform chemotaxis-based cell aggregation simulations.

2 Previous Work

The most relevant research in cell modeling and simulation may be placed in two broad categories, 2D grid-based models, and hybrid models of discrete cells with continuous motions/quantities.

2D and 3D grid based models focus on inter-cellular processes such as signaling, motion and attachment to an extracellular matrix. Some of these models also combine the internal workings of individual cells with external processes. These models treat cells as particles collectively moving on a rectilinear grid. The Cellular Potts Model (CPM) [3], a lattice-based stochastic framework, has been used to model adhesion driven motion created by varying surface energies of different kinds of cells in order to simulate cell sorting. CPM has been used as a part of other models to simulate the rearrangement of cells during morphogenesis [4, 5] and to model avascular tumor growth [6]. A discussion of how cell-centered simulations like CPM can help to explain aspects of developmental biology can be found in [7]. Some models in this category [8, 9, 10] use cellular automata [11, 12] to model cell motility and differentiation in order to simulate tumor growth [13, 14] and embryogenesis [15].

There are other models that use a 2D-grid based approach, combined with environmental forces, to simulate the deformation and aggregation of flowing red blood cells [16]. Monte-Carlo models, non-deterministic techniques that produce solutions by random assignments to uncertain model variables, using only Brownian Motion [17] and a combination of Brownian Motion and cellular automata [18] have been developed to simulate tumerogenesis.

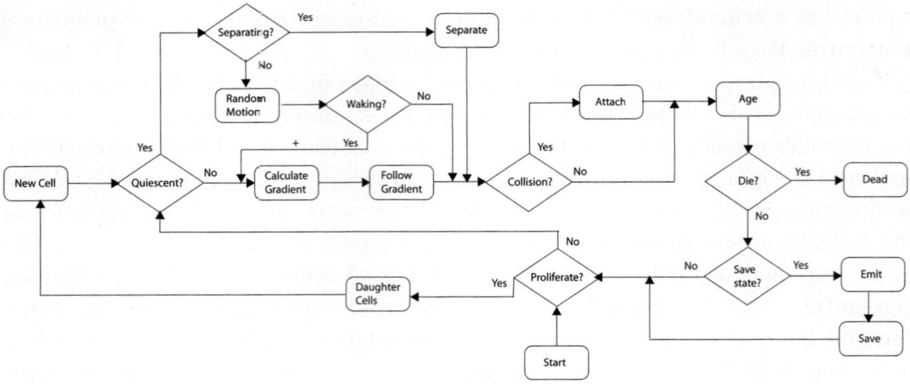

Fig. 1. Computational flow of cell aggregation simulation per time step per cell.

The latter approach utilizes hybrid models that contain both discrete and continuous components. These models define cells as independent, discrete units that can move in response to continuous forces and influences. This approach has found use in simulating multicellular pattern formation [19,20], studying the chemotactic motility of individual cells [21,22] and the computational modeling of cell adhesion [23]. Our model simulates chemotaxis, where chemical-based cell-cell signaling is the main source of communication. A similar use of cell-cell signaling can be found in [21], which investigated cell motility with respect to chemotactic forces. Some models integrate two or more of these approaches in one simulation system. COMPUCELL-2D [24] and 3D [25] are two simulation systems that have been developed with both CPM and diffusion-based continuum models They are able to simulate signaling and cells moving in response to adhesion forces. Our model clearly falls into the second category of hybrid models. While it contains many of the elements from previous efforts [19,20,21,24,25], such as the discrete cells that move, divide and attach, our approach also includes a more detailed modeling of proliferation and apoptosis.

3 Model Overview

In our simulation system each cell is defined by a collection of physiologically relevant parameters and actions [26], such as the number and position of chemical receptors on the cell surface, location of the cell, age, life cycle stage, chemoattractant emission and response rates, diffusion radius, proliferation rate, quiescent period, time of apoptosis, and number of attached cells. Our virtual cells are able to emit chemoattractants, sense the chemoattractant gradient, move in the direction of the gradient, proliferate, adhere to other cells, age and die.

A cell's life cycle and behaviors are implemented as set of actions that are performed during each time step of a simulation. A single aggregation simulation is comprised of a series of these time steps. A cell's actions for each one of these time steps are outlined in Figure 1. Newly proliferated cells remain in a quiescent state for a user-defined period. These cells are not *Active* and cannot divide. A quiescent cell can perform three types of actions. Immediately after cell division a new daughter cell separates from its parent. It moves with a constant velocity along the division axis for a user-defined period of time. Upon completion of this separation period the cell begins to respond to the chemoattractant gradient in its local environment. During the *Waking* phase there is some randomness in the cell's movement. These random motions vanish as the cell becomes *Active.* Based on the chemical fields produced by nearby cells a gradient is calculated and the cell/aggregate moves in the gradient direction in response to it. A cell is capable of attaching to other cells upon collision. A cell's age is incremented at each time step. If the cell is apoptotic, it probabilistically determines if it should die. If the state of the complete environment is to be saved for visualization purposes, the cell emits and stores its chemical field in the chemoattractant array. If in the *Active* stage, the cell probabilistically determines if it is time to divide. If it divides, its and its daughter cell's age are set to zero, and they enter the quiescent stage.

Table 1. Virtual cell variables

Variable	Description
ID	Cell's unique identifier
Coordinates[2]	2D grid coordinates defining the location of the cell
Age	Age of the cell
lifeCycleStage	One of: *Active, Quiescent, Waking, Dying, Dead*
AggregateNumber	ID number of the cell's aggregate
Sensors[8]	Chemoattractant concentration read at sensors
Gradient[2]	Stores the gradient vector
dieFlag	Cell dies if set to 1
divideFlag	Cell divides if set to 1
NumberOfNeighbors	Number of attached cells

4 System Description

The simulation system is implemented in C++ using the OpenGL library. It consists of two classes with a total of 2300 lines of code. The cell class contains the variables (see Table 1) and methods (see Figure 1) which define a virtual cell. The main class implements the simulation system that performs the simulation with a number of cells. Simulations were run on a 64-bit linux cluster running Dual-Opteron nodes with 1GB of RAM, and Gentoo 2.6. The simulation execution times mostly depend on the number of cells and discrete time steps in the simulation. The average running time for a 24-hour simulated aggregation experiment with approximately 1,500 cells and 1440 time steps ($\Delta t = 1$ minute) is approximately 30 CPU-minutes.

Initialization and Main Loop. A cell aggregation simulation is initialized by reading in model parameters from the command line, and randomly placing cells and aggregates in the simulation environment, as described in Algorithms 1 and 2[3]. The number of cells and aggregates at the beginning of a simulation run can be based on aggregate size distribution data derived from *in vitro* experiments, or can be defined by the user. The cell/aggregate distribution information (stored in the file *SizeDistribution.txt*), i.e. the number of singles cells, two-cell aggregates, etc., is used to set the initial conditions for the simulations, and determines how many aggregates of the varying sizes should be defined at the beginning of the simulation. At initialization time cells/aggregates are randomly placed in non-intersecting locations in the simulation environment (see Algorithm 2[1]). Prior to cell placement, a number of aggregate configurations, i.e. specific descriptions of cell groupings and connections, for each aggregate size present in the experiments are pre-calculated. The aggregate configurations are created by performing several aggregation simulations and recording the geometry of the resulting aggregates for the different sizes. The aggregate configurations are stored in files, each containing several random arrangements. Each aggregate is represented by the number and coordinates of neighboring cells for each of its cells. During initialization the aggregates are read from the configuration files and placed in the simulation environment. The main loop, detailed in Algorithm 3[1], Figure 1 and Section 3, is then executed to activate the virtual cells.

Proliferation. All non-quiescent and non-apoptotic cells may probabilistically divide, with their proliferation probabilities being a function of the number of cells attached to them. See Algorithm 4[1]. Cell attachments increasingly inhibit proliferation. A cell's proliferation probability is decreased by a factor of $e^{-(n+1)}$ (K_3 in Algorithm 4[1]), where n is the number of cells attached to the cell. Once a cell divides the ages of both daughter cells are set to zero, and they enter the quiescent state. After division a new cell is randomly placed in an open location next to the original. With probability $P_{Separation}$ the new cell separates from the original cell, otherwise it stays attached.

Quiescence. After a cell divides it enters the quiescent phase, a period where the cell's mechanisms and behaviors gradually become functional. If cells separate after a single cell division the two daughter cells move away from each other with a constant velocity along the division axis for a user specified time period ($T_{Separation}$). If the dividing cell is attached to an aggregate, only the new cell possibly separates. Newly created cells do not immediately contribute to or respond to the chemotactic field around them . Halfway through the quiescent period $((T_{Awake} = T_{Quiescent}/2) > T_{Separation})$, they start emitting and sensing chemoattractants with an increasing rate (see Figure 2: Left). After

[3] The detailed pseudocode for the algorithms is located at http://www.cs.drexel.edu/~david/Papers/ECAL07_Algs.eps. The algorithms are included in the submission as an appendix, but will only be available on the web if the paper is published.

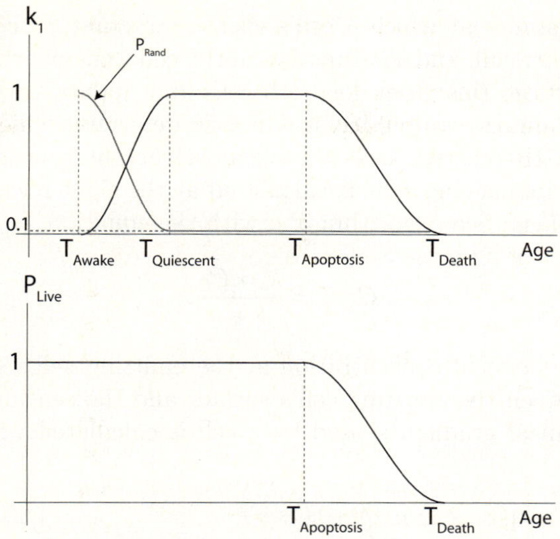

Fig. 2. Left: Change in chemoattractant emission and response as a function of age (k_1) and probability of taking a random step (P_{Rand}). Right: Probability of being alive as a function of age.

the separation period ($T_{Separation}$) cells randomly move. After T_{Awake} cells are increasingly affected by chemoattractant gradients. The probability that a quiescent cell takes a random step (P_{Rand}) decreases as it becames more awake. This probability becomes 1% for fully active cells. Conversely waking cells follow the chemoattractant gradient with probability $1 - P_{Rand}$. See Algorithm 6[1] for more details.

$$\mathbf{Velocity} = \lambda * k_1 * \nabla C, \tag{1}$$

Cell/Aggregate Movement. An unattached cell moves in the direction of the chemoattractant gradient with a velocity proportional to the gradient, with λ being the proportionality constant. A cell's chemotactic response is also affected by the cell's life cycle stage and age. This feature is implemented with scale factor k_1. See Figure 2:Left and Equation 1. If a cell is part of an aggregate, its gradient-based velocity is combined with the velocities calculated for the other cells in the aggregate to produce an average velocity for the whole aggregate. An aggregate's velocity is also inversely proportional to its mass, i.e. the number of its constituent cells. The algorithms that implement cell/aggregate movement are detailed in Algorithms 6 and 7[1].

At every time step and for each cell, the surrounding cells that can influence the current cell are identified and added to the cell's neighbor set. Looping through every cell, we calculate the distance from the current cell to every other cell. If the distance between the two is less than the other cell's R_{Max} value, the other influencing cell is added to the current cell's neighbor list. R_{Max} is the maximum distance at which a cell's chemoattractant concentration can be detected by another cell, and is a function of the cell's chemoattractant emission rate, C_0. We perform this check for both cells and update both cells' neighbor list in order to minimize computations. Once we determine which cells' chemical fields may affect the current cell, the chemoattractant concentration emitted from each of the influencing cells is calculated at the eight receptor sites on the current cell's surface (See Algorithm 5[1]), with the equation

$$C(r) = \frac{k_1 * C_0}{1 + r}, \tag{2}$$

where C_0 is the chemical concentration at the emitting cell's surface, and r is the distance between the emitting cell's surface and the sensing cell's receptor. The overall chemical gradient sensed by a cell is calculated with the following equations,

$$C_x = \frac{\Lambda_2 - \Lambda_6 + \frac{\Lambda_1 - \Lambda_5 - \Lambda_7 + \Lambda_3}{\sqrt{2}}}{2 * r_c} \qquad C_y = \frac{\Lambda_8 - \Lambda_4 + \frac{\Lambda_7 + \Lambda_1 - \Lambda_5 - \Lambda_3}{\sqrt{2}}}{2 * r_c}$$

$$\nabla C = (C_x, C_y) \qquad \Theta(\nabla C) = atan(C_y / C_x), \tag{3}$$

where Λ_i is the chemical concentration calculated at receptor r_i, r_c is the cell's radius, and Θ is the angle used to determine the cell's displacement direction. The locations of the numbered receptors are provided in Figure 3:Left. Cells do not self-stimulate nor are they influenced by the cells in their aggregate, i.e. cells in the same aggregate are not added to the neighbor list.

In an effort to simplify movement, collision, attachment and cell division computations, we discretize the environment in which the cells exist and move into a hexagonal grid with a toroidal topology. While each cell contains a Cartesian location P_{Car}, having the cells positioned in a hex-grid ensures equal distances between each grid point and its six neighboring grid points. The toroidal topology connects the left side of the finite environment to the right side, and the top edge to the bottom edge. So as cells move over an edge in the underlying grid they are placed at the opposite side of the grid. The conversion from hex-grid coordinates P_{Hex} (i,j) to Cartesian coordinates P_{Car} (x,y) is given in Equation 4, where d is the distance between grid-points (1μm for our simulations).

$$P_{Car} = \begin{bmatrix} d & \frac{1}{2}d(j \; mod \; 2) \\ 0 & \frac{\sqrt{3}}{2}d \end{bmatrix} \cdot P_{Hex} \tag{4}$$

A cell therefore can only move in one of the six directions as seen in Figure 3:Right. Cells and aggregates move in the direction which is closest to its velocity vector. The closest direction is determined by the identifying the smallest angle between the six potential discrete directions and the cell's continuous velocity vector.

The minimum distance a cell can move on the hexagonal grid is 1μm. It is possible that the displacement δ produced by a particular chemoattractant

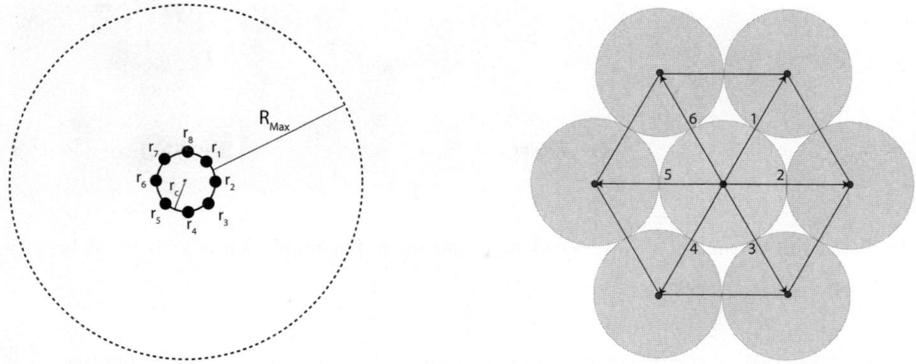

Fig. 3. Left: A single cell has radius r_c and has eight chemoattractant receptors identified as r_i. The cell's chemoattractant cannot be sensed past R_{Max}. Right: The cell at the center can move in one of the six directions defined by the hexagonal grid.

gradient and cell/aggregate mass is less than d. When this occurs, δ is stored and accumulated until the length of the accumulated vector is greater than d. At that point the cell/aggregate is moved. It is also possible that δ is not a multiple of d, i.e. $\delta = a * d + b$, $b < d$. In this case the cell/aggregate is moved a steps and the amount b is saved for future accumulation. It is also possible that a cell's displacement is more than d in a defined Δt. In this case the simulation time step is divided into smaller periods so that the maximum displacement over one period is d; thus facilitating accurate collision detection.

Collision Detection and Attachment. Collisions are detected by checking if any two cell centers are within a distance of $2 * Radius$. Each cell pair is only tested once. Once a collision is detected three scenarios are possible: (See Algorithm 8[1].)

- both cells are singles, and they form a doublet;
- one of the cells is single, the other belongs to an aggregate, and the single cell joins the aggregate;
- both cells belong to aggregates, and all of the cells in one aggregate join the other. The aggregate with the smaller mass joins the larger aggregate to save computation time.

Aging. Each cell has an age which is initially set to zero, and is incremented at every time step. The Age algorithm (See Algorithm 9[1]) sets a cell's life cycle state. Cells are initially in a *Quiescent* state. After $T_{Quiescent}$ they enter a fully active state. If a cell is still unattached after $T_{Apoptosis}$ it enters *Dying* state and has an increasing probability of dying. (See Figure 2:Right). Once a cell dies it is removed from the CELLS list and is no longer processed.

Visualization. It is possible to produce images that capture the changing state of an aggregation simulation. When generating these images every active cell emits chemoattractants into the environment. A 2D Cartesian grid is used to store the

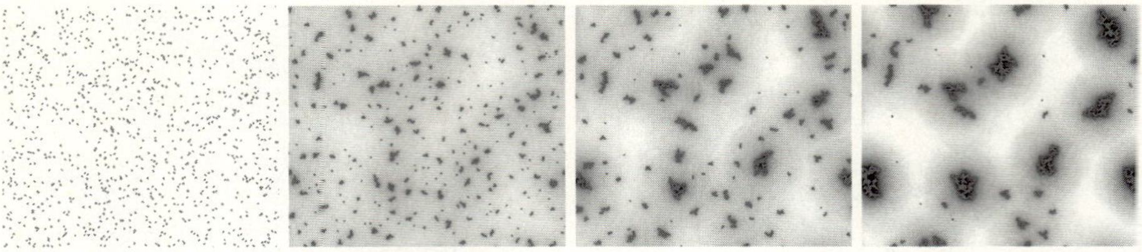

Fig. 4. Results from an *in silico* cell aggregation experiment. Time = 0 hr, 6 hrs, 15 hrs and 25 hrs.

chemoattractant concentration. The amount of chemoattractant concentration at a grid point is calculated from the diffusion function (Equation 2) and added to the grid point. (See Algorithm 10[1]). Since chemical concentrations are emitted into the chemoattractant grid only for visualization purposes, these values are calculated at a user-specified time interval different than the simulation time step. For example, while the simulation time step may be 1 minute, an image capturing the locations of cells and the chemoattractant concentration distribution in the environment may only be generated and saved (using the SaveState() routine) once every simulation hour. See Figure 4 for example output images.

Table 2. Model parameter values that produce the optimal fit to the PC12 *in vitro* results in Figure 5:Right.

Parameter	Description	Value
Δt	Simulation time step	1 min
d	Grid discretization	1 μm
Dim	Grid dimension	3250 × 3250 (3250μm^2)
C_0	Chemoattractant concentration at cell surface	90 molecules/μm^2 [27
R_{Max}	Maximum radius of influence	100 μm [28]
r_c	Cell radius	6 μm [29]
$T_{Apoptosis}$	Start time of apoptosis	18h [30]
T_{Death}	Time when all single cells have died	24h [30]
P_{Div}	Probability of cell division per time step	0.07
$P_{Separation}$	Probability of separation after cell division	0.20
$T_{Separation}$	Separation time after cell division	20 mins
$T_{Quiescent}$	Quiescence end time	90 mins
λ	Chemotaxis response factor	0.055

5 Results

Our cell aggregation simulation system is currently being employed for a number of computational experiments. In order to validate our model and computational system we have performed an *in vitro* PC12 aggregation experiment. The *in vitro* experiment was imaged [31], and the aggregate size distribution at

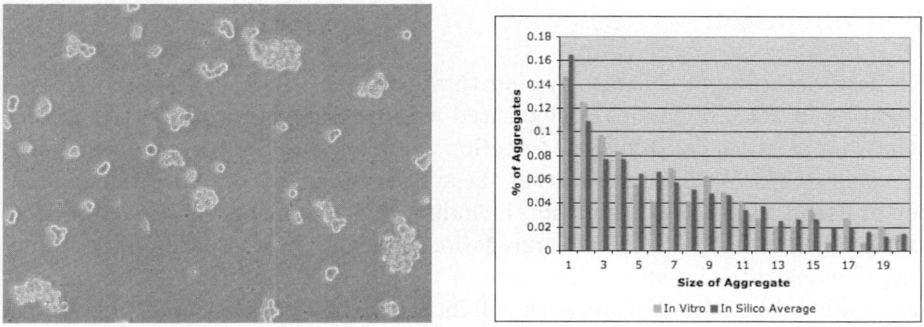

Fig. 5. Left: Image taken from an *in vitro* PC12 aggregation experiment after 24 hours. Right: Histogram comparing aggregate size distributions from an *in vitro* PC12 aggregation experiment (blue bars) and an optimized *in silico* cell aggregation experiment.

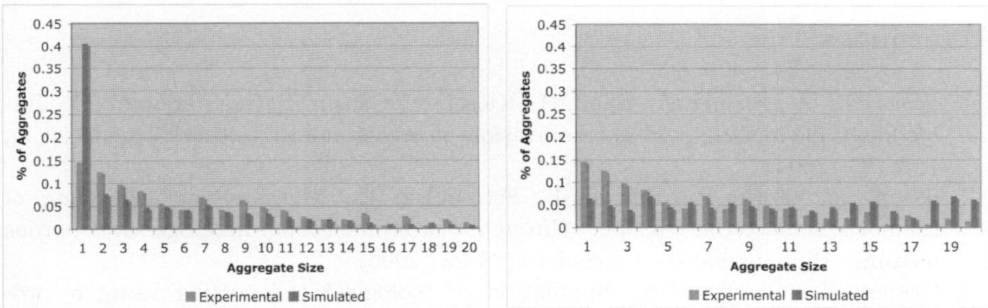

Fig. 6. Effect of chemoattractant gradient response on aggregation. Slower cells ($\lambda = 0.01$) (left) form fewer and smaller aggregates while faster cells ($\lambda = 0.1$) (right) form larger aggregates.

the start and end of the experiment were extracted from the images. We then fine-tuned our model and its parameters to recreate *in silico* the aggregation distribution that was produced in the *in vitro* experiment. See Figure 5 for a sample image from the *in vitro* experiment and the histograms containing the resulting aggregate size distributions from the *in vitro* and *in silico* experiments. See Table 2 for the optimal model parameters. The last five parameters (P_{Div}, $P_{Separation}$, $T_{Separation}$, $T_{Quiescent}$, λ) were adjusted to provide the optimal fit to the *in vitro* results. The optimal solution was identified by minimizing the Earth Mover's Distance [32] between the two histograms. We are also currently conducting a series of parametric studies in order to explore the influence that each component of cell behavior has on overall aggregation dynamics. Figure 6 contains some of our initial results. Here, we have significantly decreased and increased each cell's response to the chemoattractant gradient. It can be seen that slower cells (Left histogram) formed mostly smaller aggregates, yielding many more single cells. Faster cells (Right histogram) aggregated more aggressively and formed fewer smaller aggregates and more numerous larger aggregates.

6 Conclusion

We have described a software system that simulates chemotaxis-based cell aggregation in 2D. The model implemented within the system consists of such cell behaviors as chemical diffusion/detection, motility, proliferation, adhesion and life cycle stages. We have shown that the system is capable of performing a number of useful *in silico* experiments, including fine-tuning the model parameters to reproduce *in vitro* PC12 cell aggregation experiments and parametric studies that demonstrate the affect that each of the model's components have on cell aggregation dynamics. Future work will include extending the model and system to 3D, and utilizing the system to study tumor growth.

Acknowledgements. This research has been funded by NSF Grant # CCF-0636323 and NASA Grant # NNJ04HC81G.

References

1. Fernandis, A., Prasad, A., Band, H., Klösel, R., Ganju, R.: Regulation of CXCR4-mediated chemotaxis and chemoinvasion of breast cancer cells. Oncogene 23(1), 157–167 (2004)
2. Olson, M., Ford, R., Smith, J., Fernandez, E.: Mathematical modeling of chemotactic bacterial transport through a two-dimensional heterogeneous porous medium. Bioremediation Journal 10, 13–23 (2006)
3. Graner, F., Glazier, J.A.: Simulation of biological cell sorting using a two-dimensional extended potts model. Physical Review Letters 69, 2013–2016 (1992)
4. Chaturvedi, R., Huang, C., Kazmierczak, B., Schneider, T., Izaguirre, J.A., Newman, S.A., Glazier, J.A., Alber, M.: On multiscale approaches to three-dimensional modeling of morphogenesis. Journal of the Royal Society 2, 237–253 (2005)
5. Glazier, J.A., Graner, F.: Simulation of the differential adhesion driven rearrangement of biological cells. Physical Review E 47, 2128–2154 (1993)
6. Jiang, Y., Pjesivac-Grbovic, J., Cantrell, C., Freyer, J.: A multiscale model for avascular tumor growth. Biophysical Journal 89, 3884–3894 (2005)
7. Merks, R.M.H., Glazier, J.A.: A cell-centered approach to developmental biology. Physica A 352, 113–130 (2005)
8. Hogeweg, P.: Evolving mechanisms of morphogenesis: on the interplay between differential adhesion and cell differentiation. Journal of Theoretical Biology 203, 317–333 (2000)
9. Hogeweg, P.: Computing an organism: on the interface between informatic and dynamic processes. Biosystems 64, 97–109 (2002)
10. Tozeren, A., Coward, C.W., Petushi, S.P.: Origins and evolution of cell phenotypes in breast tumors. Journal of Theoretical Biology 233, 43–54 (2005)
11. Ilachinski, A.: Cellular Automata: A Discrete Universe. World Scientific, Singapore (2001)
12. Wolfram, S.: Cellular automata as models of complexity. Nature 311, 419–424 (1984)
13. Dormann, S., Deutsch, A.: Modeling of self-organized avascular tumor growth with a hybrid cellular automaton. In Silico Biology 2(3), 393–406 (2002)

14. Patel, A.A., Gawlinski, E.T., Lemieux, S.K., Gatenby, R.A.: A cellular automation model of early tumor growth and invasion: the effects of native tissue vascularity and increase in anaerobic tissue metabolism. Journal of Theoretical Biology 213, 315–331 (2001)
15. Longo, D., Peirce, S.M., Skalak, T.C., Davidson, L., Marsden, M., Dzamba, B., Simone, D.W.D.: Multicellular computer simulation of morphogenesis: blastocoel roof thinning and matrix assembly in xenopus laevis. Developmental Biology 271, 210–222 (2004)
16. Bagchi, P., Johnson, P.C., Popel, A.S.: Computational fluid dynamic simulation of aggregation of deformable cells in a shear flow. Transactions of the ASME 127, 1070 (2005)
17. Song, H., Jain, S.K., Enmon, R.M., O'Connor, K.C.: Restructuring dynamics of DU 145 and lncap prostate cancer spheroids. In Vitro Cellular and Developmental Biology-Animal 40, 262–267 (2004)
18. Ascencio, S.F., Meana, H.P., Miyatake, M.N.: Two and three dimensional computer simulation of cancer growth. In: Proc. XXI Int. Conference of the Chilean Computer Science Society, pp. 73–79 (2001)
19. Fleischer, K.W., Barr, A.H.: A simulation testbed for the study of multicellular development: the multiple mechanisms of morphogenesis. In: Artificial Life III, pp. 389–408 (1994)
20. Fleischer, K.W.: Investigations with a multicellular developmental model. In: Artificial Life V, pp. 229–236 (1996)
21. Jabbarzadeh, E., Abrams, C.F.: Chemotaxis and random motility in unsteady chemoattractant fields: A computational study. Journal of Theoretical Biology 235, 221–232 (2005)
22. Palsson, E., Othmer, H.: A model for individual and collective cell movement in dictyostelium discoideum. Proceedings of the National Academy of Science USA 97, 10448–10453 (2000)
23. N'Dri, N.A., Shyy, W.: Tran-Son-Tay: Computational modeling of cell adhesion and movement using a continuum-kinetics approach. Biophysical Journal 85, 2273–2286 (2003)
24. Izaguirre, J.A., Chaturvedi, R., Huang, C., Cickovski, T., Coffland, J., Thomas, G., Forgacs, G., Alber, M., Newman, S., Glazier, J.A.: Compucell, a multi-model framework for simulations of morphogenesis. Bioinformatics 20, 1129–1137 (2004)
25. Cickovski, T., et al.: A framework for three-dimensional simulation of morphogenesis. IEEE/ACM Transactions on Computational Biology and Bioinformatics 2, 273–288 (2005)
26. Alberts, B., Bray, D., Hopkin, K., Johnson, A., Lewis, J., Raff, M., Roberts, K., Walter, P.: Essential Cell Biology, 2nd edn. Garland Publishing, New York (2003)
27. Savinell, J.M., Lee, G.M., Palsson, B.O., Arbor, A.: On the orders of magnitude of epigenic dynamics and monoclonal antibody production. Bioprocess Engineering 4, 231–234 (1989)
28. Serini, G., et al.: Modeling the early stages of vacular network assembly. The EMBO Journal 22(8), 1771–1779 (2003)
29. Watanabe, O., Torda, M., Meldolesi, J.: The effect of α-latrotoxin on the neurosecretory PC12 cell line: Electron microscopy and cytotoxicity studies. Neuroscience 10(3), 1011–1024 (1983)
30. Hirata, Y., Adachi, K., Kiuchi, K.: Activation of JNK pathway and induction of apoptosis by manganese in PC12 cells. Journal of Neurochemistry 71, 1607–1615 (1998)

31. Manley, P., Lelkes, P.: A novel real-time system to monitor cell aggregation and trajectories in rotating wall vessel bioreactors. Journal of Biotechnology 125(3), 416–424 (2006)
32. Rubner, Y., Tomasi, C., Guibas, L.: The earth mover's distance as a metric for image retrieval. International Journal of Computer Vision 40(2), 99–121 (2000)

A Appendix: Algorithm Pseudocode

Algorithm 1: Main

```
ParameterSet ⟵ CommandLineArguments
InitialSizeDistribution ⟵ READ("SizeDistribution.txt")
CreateWindow(ParameterSet("Width"), ParameterSet("Height"))
PlaceCells()
SimulationTime ⟵ 0
MainLoop()
```

Algorithm 2: PlaceCells

```
for N ⟵ BiggestAggregateSize to 1 do
    //NumberOfAggregates[1..MaxSize] has the number of
    //aggregates of sizes 1..MaxSize respectively
    for M ⟵ NumberOfAggregates[N] to 1 do
        Randomly pick aggregate of size N from file
        repeat
            Choose random coordinates
        until Aggregate can be placed without intersection
```

Algorithm 3: MainLoop

```
while SimulationTime < 24hr do
    if Divide? == True then
        foreach Cell in CELLS do
            if Cell.Quiescent == False And Cell.isDying == False
            then
                Divide(Cell)
    CalculateGradient()
    CellMove()
    AggregateMove()
    DetectCollisions()
    Age()
    if SaveState? == True then
        Emit()
        SaveState()
    SimulationTime+ = SimulationStepTime
```

Algorithm 4: Divide

```
Input: Cell C
P_Dividing = getK_3(C.NumAttachments) * P_Proliferation
//Division occurs with probability P_Dividing
if C decides to divide then
    Dir ⟵ direction chosen randomly between 1 to 6 (Figure 3)
    while direction chosen is occupied do
        Dir = (Dir + 1) Mod 6
    Put a new cell C_N at (C.center + Dir)
    C.Age ⟵ C_N.Age ⟵ 0
    C.quiescent ⟵ C_N.quiescent ⟵ True
    C_N separates from C with probability P_Separation
```

Algorithm 5: CalculateGradient

```
for C_1 ⟵ 1 to NUMCELLS do
    for C_2 ⟵ C_1 + 1 to NUMCELLS do
        if CELLS[C_1] And CELLS[C_2] do not belong to same aggregate then
            Dist ⟵ Distance between CELLS[C_1] and CELLS[C_2]
            if Dist ≤ CELLS[C_2].R_Max then
                CELLS[C_1].NeighborSet.add(C_2)
            if Dist ≤ CELLS[C_1].R_Max then
                CELLS[C_2].NeighborSet.add(C_1)
foreach Cell in CELLS do
    Initialize values of all receptors to zero
    foreach OtherCell in Cell.NeighborSet do
        foreach R_i in Cell.Receptors do
            Dist ⟵ Distance between OtherCell and R_i
            R_i.Value+ = (OtherCell.k_1 * C_0)/(1 + Dist)
    Cell.Gradient ⟵ Calculate gradient from Equation 3
    Cell.GradientAngle ⟵ Calculate theta from Equation 3
```

Algorithm 6: CellMove

```
foreach Cell in CELLS do
    if Cell is single then
        if Cell.Quiescent == False then
            Velocity = λ * Cell.k1 * Cell.Gradient (Equation 1)
            Direction ⟵ FindDirection(Cell)
        else
            if Cell.Age < T_Separation then
                Velocity = V_Separation
                Direction = Cell.SeparationDirection
            else if Cell.Age > T_Awake And Cell.Age < T_Quiescent then
                begin With probability P_Rand
                    Take a random step
                end
                begin With probability 1 − P_Rand
                    Velocity = λ * Cell.k1 * Cell.Gradient (Equation 1)
                    Direction ⟵ FindDirectionCell
                end
            else
                Move Cell randomly 1 to 3 μ
                return
        Displacement = Velocity * SimulationStepTime
        NewLocation = Cell.Center + Direction * Displacement
```

Algorithm 7: AggregateMove

```
foreach Aggregate in AGGREGATES do
    AverageGradient ⟵ Average gradient of cells in Aggregate
    Velocity = λ/M * AverageGradient
    Displacement = Velocity * SimulationStepTime
    if Displacement > 0 then
        Dir ⟵ FindDirection(AverageGradient)
        if Every Cell in Aggregate can move towards Dir without colliding with
        any other cells or aggregates then
            foreach Cell in Aggregate do
                NewLocation = Cell.Center + Dir * Displacement
                Move Cell to NewLocation
    else
        Move Aggregate randomly 1 μ
```

Algorithm 8: DetectCollisions

```
for C1 ⟵ 1 to NUMCELLS do
    for C2 ⟵ C1 + 1 to NUMCELLS do
        if CELLS[C1] and CELLS[C2] do not belong to same aggregate then
            Dist ⟵ Distance between CELLS[C1] and CELLS[C2]
            if Dist < 2 * CellRadius then
                Aggregate1 ⟵ CELLS[C1].AggregateNumber
                Aggregate2 ⟵ CELLS[C2].AggregateNumber
                if Both cells are single then
                    Form a new doublet
                else if Only one is single then
                    //Assume CELLS[C2] is single
                    CELLS[C2] joins Aggregate1
                else
                    //Assume Aggregate1 has more cells
                    All members of Aggregate2 joins Aggregate1
                    Aggregate2 is destroyed
                CELLS[C1].NumberOfAttachments + +
                CELLS[C2].NumberOfAttachments + +
```

Algorithm 9: Age

```
foreach Cell in CELLS do
    Cell.Age+ = SimulationStepTime
    if Cell.Quiescent = True And Cell.Age >= QuiescentTime then
        Cell.Quiescent = False
    else if Cell is single And Cell.Age >= T_Apoptosis then
        Cell.isDying = True
        Cell dies with probability 1 − P_Live (Figure 2:Right)
        if Cell dies then
            Remove Cell from CELLS
```

Algorithm 10: Emit

```
foreach GridPoint (x, y) do
    ChemoGrid[x][y] = 0
foreach Cell in CELLS do
    foreach GridPoint (x, y) within Cell.R_Max distance from Cell do
        Dist ⟵ Distance between Cell and GridPoint
        ChemoGrid[x][y]+ = (Cell.k1 * C0)/(1 + Dist)
```

A Signal Based Approach to Artificial Agent Modeling

Luís Morgado[1,2] and Graça Gaspar[2]

[1] Instituto Superior de Engenharia de Lisboa
Rua Conselheiro Emídio Navarro, 1949-014 Lisboa, Portugal
lm@isel.ipl.pt
[2] Faculdade de Ciências da Universidade de Lisboa
Universidade de Lisboa, Campo Grande, 1749-016 Lisboa, Portugal
gg@di.fc.ul.pt

Abstract. In this paper we propose an approach to agent modeling that follows a signal based metaphor where agents are modeled as dissipative structures and their cognitive structures are modeled as compositions of multiple energetic potentials. This uniform representational support is used to model both reactive and deliberative processes. To illustrate the descriptive adequacy of the model, two experimental cases are presented where reactive and deliberative processes are modeled based on the proposed approach.

1 Introduction

The dichotomy between cognitive (or deliberative) and behavioral (or reactive) agents, that was originally adopted in the field of agent modeling, promoted a view of cognitive and behavioral aspects as essentially different and therefore requiring distinct architectural support (e.g. [1], [2]). To overcome limitations of both subfields, hybrid architectures have been proposed that deal with the distinction between cognitive and behavioral aspects by incorporating them as distinct functional layers. Although a functional modularization is useful to deal with architectural complexity, the use of different architectural supports for the different layers leads to integration problems, namely in what relates to the representational support. A concrete example is the well-known difficulty to integrate sub-symbolic aspects, which are typically continuous, with discrete symbolic aspects.

In this paper we use a uniform representational support to model both reactive and deliberative processes. Our approach follows a signal based metaphor where agents are modeled as dissipative structures and their cognitive structures are modeled as compositions of multiple energetic potentials.

The paper is organized as follows: in section 2, we introduce the proposed approach; in section 3, we present an overview of the agent flow model; in section 4, we report experimental results that illustrate the use of the model with two distinct types of agents; in section 5, we establish comparisons with related work and draw some conclusions and directions for future work.

F. Almeida e Costa et al. (Eds.): ECAL 2007, LNAI 4648, pp. 1050 – 1059, 2007.

2 Modeling an Agent as a Dissipative Structure

In the proposed approach two aspects are considered central to the agent notion: the relation between the agent and the environment and the ability for autonomous action. Both of these aspects are intrinsic characteristics of a dissipative structure [12]. Dissipative structures are open systems governed by the interchange of energy with the environment and able to maintain themselves in a state far from equilibrium, yet keeping an internally stable overall structure. The autonomous action results from the autopoietic organization supported by feedback networks. These feedback networks enable the maintenance of the internal stability in spite of environmental changes.

Based on the notion of a dissipative structure, an agent can be characterized by the internal potentials that form the agent's structure and by the flows that result from agent-environment coupling. Those potentials participate in the agent's internal activity, changing according to the internal and external energy flows.

2.1 Formation of Motivations

The uncontrolled increase or decrease of potentials may result in the collapse of the overall structure. That is, the maintenance of a viable internal structure in spite of environment changes implies that internal potentials should be constrained to remain within ranges of viability.

The maintenance of those ranges of viability by auto-regulation processes can be considered a primary motivational source of an agent. The viability limits may be expressed in an implicit way through structural constraints or in an explicit way as regulator potentials. Those regulator potentials therefore play the role of motivators of agent behavior.

Two main types of motivations can be distinguished: *predefined motivations*, embedded in agent structure during its design or implementation, conceptually related to innate motivations resulting from genetic factors in biological organisms; *acquired motivations*, resulting from the primary motivations and from the agent-environment relation, particularly in the context of adaptive processes. In this way, the agent motivations constitute a causal hierarchy with the predefined motivations at its root.

2.2 Motivation Achievement

The motivational dynamics are driving forces that guide the agent behavior in order to achieve the underlying motivations. To achieve its motivations an agent must apply an internal potential to be able to produce the adequate change on the environment. The concretization of the intended change depends also on the characteristics of the current environmental situation that, from a thermodynamic point of view, can be modeled as an agent-environment coupling conductance. Therefore, the overall agent-environment relation can be modeled as a relation between an agent's internal potential, its *achievement potential*, and the agent-environment coupling conductance, the *achievement conductance*. The achievement potential represents the potential of change that the agent is able to produce in the environment to achieve the intended state-of-affairs. The achievement conductance represents the degree of the environment's conduciveness or resistance to that change, which can also mean the degree of the environment's change that is conducive, or not, to the agent intended state-of-affairs.

From a thermodynamic point of view, the achievement potential can be viewed as a force (*P*) and the achievement conductance as a transport property (*C*). The behavioral dynamics of an agent can therefore be characterized as a relation corresponding to a flow, called *achievement flow* (*F*), which results from the application of a potential *P* over a conductance *C*. The forces that arise from this relation between achievement potential and achievement conductance, expressed as energy flows, generate behavioral dynamics that underlie the cognitive activity of an agent.

3 The Agent Flow Model

Although inspired by biophysical analogies, the main aim of the proposed model is to support the development and implementation of artificial agents, independent of their kind or level of complexity. Therefore it is necessary to concretize the base notions of the model in a computationally tractable way.

The first aspect that we need to address is the notion of energy. In thermodynamics, energy is usually defined as the capacity to produce work. In the context of our model, energy can be defined as the capacity of an agent to act or, in a wide sense, to produce change. Considering an agent as a dissipative structure, that change is oriented towards the achievement of motivations, driven by internal potentials and expressed through energy flows. That is, both the agent and the environment can be modeled as a composition of multiple energetic potentials with different characteristics. In this way, the notion of energetic potential is the base notion that allows unifying the different aspects that characterize an agent in a single uniform framework.

3.1 Cognitive Structure

Energetic potentials can represent different aspects of an agent, such as a perception, a memory or an intention, providing an adequate support to model the cognitive structure of an agent. These potentials result both from agent-environment interaction and from agent internal activity. In both cases they express aspects that the agent is able to discriminate and perceive, such as "weight" or "color", commonly called quality dimensions [3].

Formally, cognitive potentials are modeled as a composition of two types of signals: a base signal $\varphi(t)$ with a specific angular frequency ω that identifies the discriminated aspect or quality; and a quantitative signal $\rho(t)$ corresponding to the actual value of the discriminated quality, expressed as a frequency shift $\Delta\omega$ that modulates the base signal $\varphi(t)$. That is:

$$p(t) = \rho(t).\varphi(t) \tag{1}$$

Through superposition, aggregates of potentials can be formed. These aggregates of potentials, which we call *cognitive elements*, are the base of the cognitive structure of an agent. Superposition is possible because the base signals that characterize the cognitive potentials are orthogonal among each other, which implies superposition of energy. Therefore a cognitive element $\sigma(t)$ is defined as a superposition of cognitive potentials. That is:

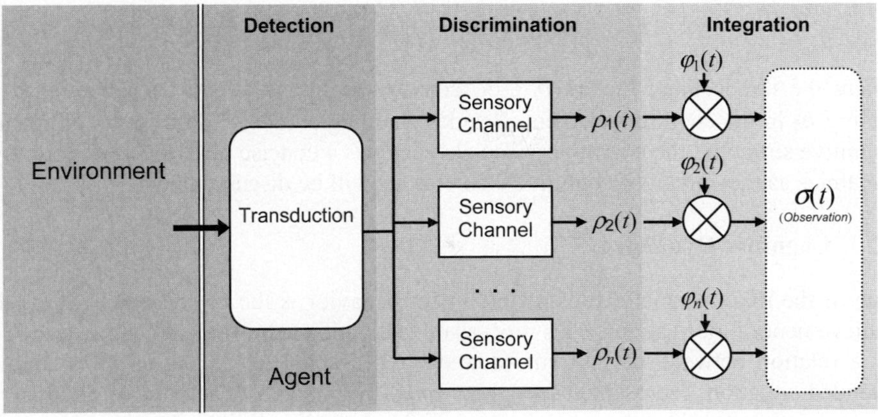

Fig. 1. Formation of cognitive elements resulting from perception activities.

$$\sigma(t) = \sum_{i=1}^{K} p_i(t) \tag{2}$$

where K is the number of potentials in the aggregate. Figure 1 illustrates the formation of cognitive elements in the context of a perception process.

Cognitive elements play different roles in cognitive activity. Three main roles can be identified: *observations*, *motivators*, and *mediators*. *Observations* result from perception processes, representing the current environmental situation. They can also result from simulated experience. *Motivators* and *mediators* are formed internally or embedded in agents' structure. *Motivators* represent intended situations, acting as motivating forces driving agent's behavior. *Mediators* describe the media that supports action, forming an interface between internal cognitive processing and action. For instance, planning processes produce sequences of *mediators* that are translated by action processes into concrete action.

3.2 Cognitive Space

The cognitive structure of an agent is the support through which the agent maintains an internal state that reflects the evolution of the agent-environment interaction along the time. Among other aspects, the cognitive structure supports the formation of an internal model of external situations. In order to obtain a concise description of those internal representations, we can further consider that the base signals that compose potentials and cognitive elements form a signal space underlying the cognitive structure of the agent, which we call a *cognitive space*. Formally, a cognitive space CS_K is defined by a set of K orthonormal basis functions $\Phi = \{\varphi_i : i = 1, 2, \ldots, K\}$ with $K \in \mathbb{N}$. Each basis function φ_i corresponds to a base signal $\varphi_i(t)$ with a specific quality ω_i.

Cognitive elements correspond to specific positions in the cognitive space. Since cognitive elements change with time, at successive time instants they occupy different positions, describing trajectories that reflect the behavior of the agent. At some instant $t = \tau$, a cognitive element $\sigma(t)$ is represented in a cognitive space CS_K as a vector $\boldsymbol{\sigma}$, defined as:

$$\sigma = (\rho_0, \rho_1, \ldots, \rho_k) \quad (3)$$

where the dimensional factors $\rho_i \in \mathbb{C}$ convey the intensity and frequency shift of quality ω_i in the cognitive element. Besides enabling a concise description of agents' cognitive structure, the cognitive space also enables a concise description of cognitive dynamics as movement of cognitive elements, as will be discussed next.

3.3 Cognitive Dynamics

One of the main characteristics of intelligent behavior is the orientation towards the achievement of motivations. This process of motivation achievement can be described by a relation between the current situation, represented by an *observation*, and an intended situation, represented by a *motivator*. The cognitive activity of an agent is consequently guided by the maximization of the flows that lead to the reduction of the distance between observations and motivators, through the use of *mediators*. This process can be described as movement in the cognitive space, where motivators and observations correspond to specific positions and mediators define directions of movement, as illustrated in figure 2.

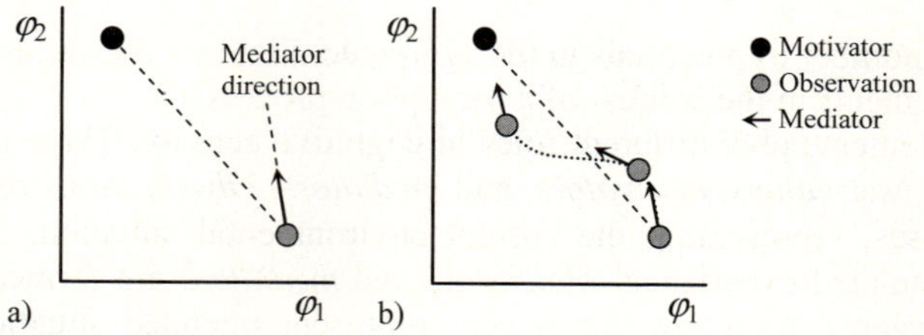

Fig. 2. Elements participating in the achievement of a motivator in a two-dimensional cognitive space.

As shown in figure 2.a, the direction of the selected mediators may not be the exact direction towards the motivator. Besides that, motivators can change and the dynamics of the environment (either internal or external) can influence the movement of the observations. Figure 2.b shows a possible trajectory resulting from the adjustment of agent's behavior by switching to different mediators. Mediators may be selected according to predefined behavior rules or they may be generated and selected dynamically by reasoning and decision processes.

3.4 Agent Architecture

The agent architecture is composed by multiple loosely coupled processes with concurrent activity. These processes are characterized by different dynamics and functionalities. Irrespective of the agent type, two main types of processes can be identified: (i) processes that realize the transduction (i.e. signal detection and conversion) and internalization of stimuli, commonly called *perception processes*; (ii) processes that realize the externalization and transduction of internally generated signals, generically named *action processes*.

In simple agents, as is the case of purely reactive agents, perception and action processes are interrelated through structural and functional constraints embedded in the agent design, determining stereotyped behaviors. In more complex agents, as is the case of agents with deliberative capabilities, it is possible to identify three other types of generic processes: (i) processes that realize the integration of stimuli into multimodal signals and incorporate these signals in the agent cognitive structure, therefore named *assimilation processes*; (ii) processes that manipulate the cognitive structure to derive information, namely to anticipate future situations and to determine possible courses of action, therefore designated *reasoning processes*; (ii) processes that select/determine a concrete course of action based on the available options, resulting both from reasoning processes and from memories of past experiences generated by the assimilation processes, therefore called *decision processes*. Figure 3 illustrates how these processes are related in the agent flow model generic agent architecture.

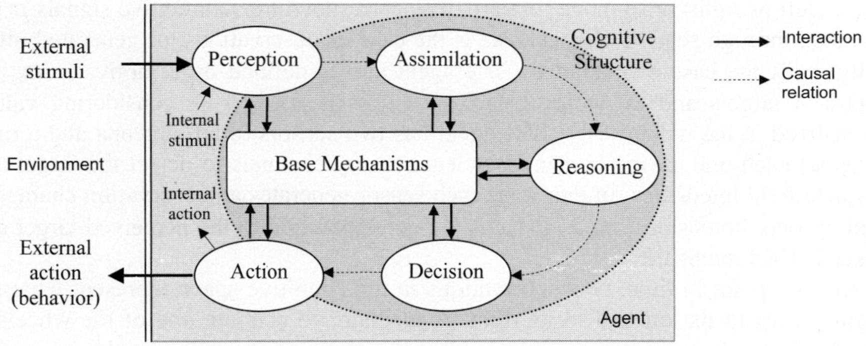

Fig. 3. Flow model generic agent architecture.

A base aspect of the agent architecture is the cognitive structure, which plays a key role in the integration of the different modalities of cognitive activity, from reactive to reasoning activities.

4 Modeling Different Types of Agents

The proposed model provides a generic framework for the implementation of agents of different types and levels of complexity. For instance, simple agents have a cognitive structure composed of a fixed number of cognitive elements (even though the position of those elements in the cognitive space can change along time) and very simple cognitive processes. Their behavior is directly guided by the dynamics resulting from the cognitive potentials and flows [4], leading to basic adaptive behavior such as the *kineses* of some organisms (e.g. bacterial *chemotaxis*) [5]. In more complex agents, internal representations are dynamically formed and changed, constituting an internal model (e.g. [6]) based on which high-level cognitive processing, such as reasoning and decision-making, can occur. To illustrate the descriptive adequacy of the proposed model, next we will describe how it can be used to model these two types of agents.

4.1 Experimental Case 1: Simple Gradient Following Agent

In this kind of agent, perception processes generate observations resulting from the inward flows produced by sensing channels. Static motivators define the agent aims. As observations evolve along time, the corresponding change relative to the motivators is perceived as achievement flows associated to each observation. Based on the observed achievement flows, action processes produce mediators to move the agent in the direction of the achievement flow with higher intensity.

Figure 4 show the results of an experiment where a simple gradient following agent moves in a simulated environment composed by targets and obstacles. The agent aim is to reach targets and to avoid obstacles. In this case, the agent's cognitive structure is characterized by four cognitive dimensions: φ_{target} and $\varphi_{obstacle}$, related to the external fields generated by targets and obstacles; and two other dimensions φ_{left} and φ_{right}, related to the activation of two actuators (wheels) that allow the agent to rotate (left or right) or to move forward. The corresponding quantitative signals ρ_i are obtained through sensing channels, as is the case of observations, or generated internally, as is the case of mediators. The agent aim is defined by a motivator $\boldsymbol{\sigma}_{mot}$ to approach targets and to avoid obstacles ($\rho_{target} = 1$, $\rho_{obstacle} = 0$, considering values normalized in the range [0, 1]). The agent has two sensors (a left antenna and a right antenna) each one having two independent sensory channels to detect the target and obstacle field intensities. In this way, each sensor generates an observation characterized by two dimensions, φ_{target} and $\varphi_{obstacle}$, corresponding to the perceived target and obstacle field intensities.

At each point in time, two observations in the cognitive space represent what the agent senses to its left and to its right. A mediator to activate one of the wheels is simply chosen by considering which one of the observations, left or right, is moving closer to the motivator (in terms of achievement flows, the one that corresponds to the higher achievement flow).

In this type of agent the cognitive structure is predetermined, that is, the agent is not able to change the cognitive structure in order to assimilate the effects of past experiences. Therefore the agent cognitive dynamics are determined only by the forces resulting from agent-environment interaction without any sophisticated reasoning activity.

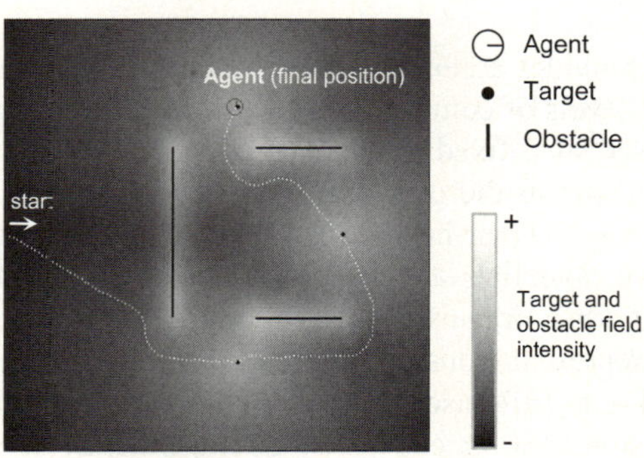

Fig. 4. Behavior of the simple agent in one of the test environments

Our modeling of this gradient following agent has many similarities to that used in potential field approaches. However, a distinctive aspect of our approach is the use of a multidimensional cognitive space. That cognitive space notion, and the explicit representation in that space of observations, motivators and mediators, is fundamental to model more complex deliberative agents, as will be exemplified next.

4.2 Experimental Case 2: Deliberative Agent

A distinctive aspect of deliberative agents is the ability to assimilate the effects of past experiences in order to improve future behavior. In this type of agents, the cognitive structure evolves according to agent behavior by incorporating memories of the experienced situations, forming dynamic internal representations of the world. Reasoning processes manipulate these representations in order to generate possible options for action. In our model, these options are represented by mediators that define directions of movement in the cognitive space, as shown in figure 5.a.

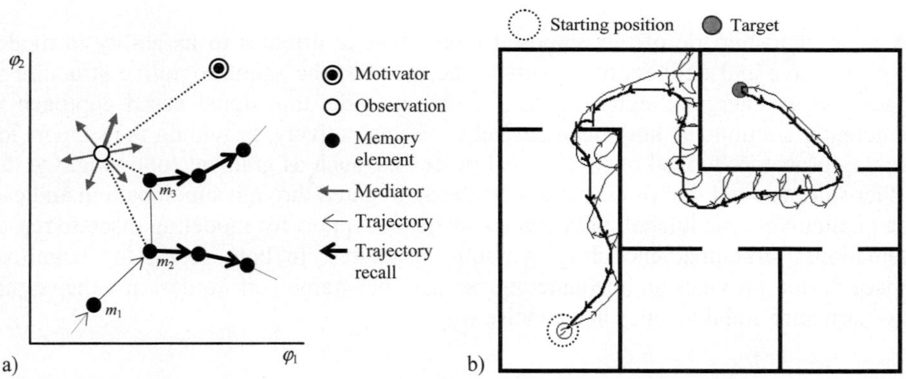

Fig. 5. Deliberative consideration of options for action: (a) perceived situation, memory recall and trajectory exploration; (b) resulting behavior in a simulated rescue environment.

Figure 5.b shows a simulated environment where an agent must move targets (people in need for rescue) to a specific position (a safe location, here corresponding to the agent's starting position). Targets emit a location signal that the agent is able to detect. Minimum sensing ability was considered, that is, the agent is only able to detect a target signal and derive from it the direction and some measure of proximity of the target. The agent is unable to detect obstacles unless by direct contact (collision).

Has it happens even in simple agents, action is guided by the maximization of the achievement flows that lead to the achievement of motivations. However, in this case these achievement flows may also result from deliberative processes that generate and select options for action based on prospective reasoning. By prospective reasoning we mean the ability to simulate possible trajectories in the cognitive space based on the internal representations of the world continuously resulting from agent experience.

When the agent begins moving to the target it is guiding its actions only by the position of the target (the motivator) and by the agent's current position, that is, by direct perceptive influence. Therefore multiple collisions with the obstacles occur.

However, during this process memories are formed. An auto-evaluation of the progress towards the selected motivators is associated with each one of the memories formed (see e.g. [7]). When the agent is about to initiate the return path, its motivator jumps back to the agent's starting position. Thereafter, in order to choose the best option for action, not only does the agent have to recall its' past positions most relevant to the current context, but it also has to deduce what will be the expected effect of replaying past trajectories starting from those positions. In doing so, the agent may choose new sequences of mediators that correspond to improved courses of action, instead of merely replaying its steps back, has it can be observed in the return path of figure 5.b (in bold).

Although it is impossible to explain in the space of this paper the details of the reasoning and decision processes of this rescue agent, they can be consulted in [7]. Those processes are based on the same representational support as the reactive agent previously presented, namely the cognitive structure and cognitive space notions.

4.3 Discussion

A main characteristic of the proposed model that contributes to its ability to model both reactive and deliberative agents is the fact that the agent cognitive structure is based on an energetic signal metaphor. On one side, this signal based approach is inherently continuous, both quantitatively and qualitatively, providing the support for analog processes typical of biophysical processes, such as gradient following. On the other side, the fact that those signals can be aggregated through superposition and can be qualitatively modulated, provides an adequate support for modeling discrete representational structures, including symbolic structures. In both cases, the cognitive space notion provides an adequate representational framework to describe the cognitive structure and dynamics in a concise way.

5 Conclusion and Related Work

In this paper we presented a framework for artificial agent design that enables a uniform integration between symbolic and sub-symbolic representational levels, allowing the implementation of agents of different types and levels of complexity without requiring a distinct architectural support across functional levels. In the proposed agent model, motivational aspects, such as the base auto-preservation dynamics and derived motivations, can be integrated with internal representations of the environment in a single conceptual structure. In this way an adequate support is provided for modeling hybrid (integrated reactive and deliberative) behavior.

The cognitive space notion that characterizes the agent model can be related to other approaches to artificial agent modeling, such as the notions of physiological space (e.g. [8]) or conceptual space (e.g. [3]). However, a distinctive aspect of our approach is the fact that the cognitive space notion is modeled as a signal space that represents concrete energetic signals and not only qualitative variables. In this way an increase in representational power is possible, enabling a tight integration between different representational levels, including both qualitative and quantitative aspects.

A related area of work is the area of potential field approaches. A detailed comparison with recent approaches (e.g. [9, 10]) that integrate planning and action evaluation and selection with traditional potential field methods is a topic for future work. The auto-organization of the agent structure and processes inherent to autopoietic mechanisms is also an area for future work, namely its relation with evolutionary mechanisms (e.g. [11]).

References

1. Van de Velde, W.: Cognitive Architectures - From Knowledge Level to Structural Coupling. In: Steels, L. (ed.) The Biology and Technology of Intelligent Autonomous Agents, Springer, Heidelberg (1995)
2. Weiss, G.: Cognition, Sociability and Constraints. In: Proc. Balancing Reactivity and Social Deliberation in Multi-Agent Systems Workshop held at the ECAI'00 (2000)
3. Gärdenfors, P.: Conceptual Spaces: The Geometry of Thought. MIT Press, Cambridge (2000)
4. Morgado, L., Gaspar, G.: A Generic Agent Model Allowing a Continuous Characterization of Emotion. In: Proc. 17th European Meeting on Cybernetics and Systems Research (2004)
5. Staddon, J.: Adaptive Dynamics: The Theoretical Analysis of Behavior. MIT Press, Cambridge (2001)
6. Newell, A.: Unified Theories of Cognition. John Wiley & Sons, Chichester (1990)
7. Morgado, L., Gaspar, G.: Adaptation and Decision-Making Driven by Emotional Memories. In: Bento, C., Cardoso, A., Dias, G. (eds.) EPIA 2005. LNCS (LNAI), vol. 3808, Springer, Heidelberg (2005)
8. McFarland, D., Bösser, T.: Intelligent behavior in animals and robots. MIT Press, Cambridge (1993)
9. Laue, T., Röfer, T.: A Behavior Architecture for Autonomous Mobile Robots Based on Potential Fields. In: Nardi, D., Riedmiller, M., Sammut, C., Santos-Victor, J. (eds.) RoboCup 2004. LNCS (LNAI), vol. 3276, Springer, Heidelberg (2005)
10. de Jong, S., Tuyls, K., Hashimoto, T., Iida, H.: Scalable PotentialField MultiAgent Coordination In Resource Distribution Tasks. In: Proc. 4th International Joint Conference on Autonomous Agents and Multi-Agent Systems (2006)
11. Harvey, I., Di Paolo, E., Tuci, E., Wood, R., Quinn, M.: Evolutionary Robotics: A new scientific tool for studying cognition. Artificial Life 11(Issues 1-2), 1–2 (2005)
12. Nicolis, G., Prigogine, I.: Self-Organization in Nonequilibrium Systems: From Dissipative Structures to Order through Fluctuations. John Wiley & Sons, Chichester (1977)

Construction of Hypercycles in Typogenetics with Evolutionary Algorithms

Chohwa Gwak and Kyubum Wee

Dept. of Computer Science and Engineering
Ajou University, Suwon, S. Korea 443-749
{jminary, kbwee}@ajou.ac.kr

Abstract. The concept of hypercycles was proposed by M. Eigen and P. Schuster to study the origin-of-life problem. A hypercycle is a simple self-reproducing system modeling molecular evolution in the abiotic period. Typogenetics is a formal system of strings originally devised by D. Hofstadter to explain the connection between computation and molecular genetics. It was later established by H. Morris as a formal system to study artificial life. Evolutionary algorithms were used by Kvasnicka et al. to find a small hypercycle in typogenetics. We improve upon their algorithm and construct many hypercycles of large sizes. We also experimented with enzymes of different lengths and various mappings between enzymes and their functions.

1 Introduction

In the 1970s, M. Eigen and P. Schuster introduced the concept of hypercycles to study the origin-of-life problem [2]. The model of hypercycles describes a hypothetical evolutionary stage of pre-biological self-reproducing macromolecular system. A hypercycle consists of *n* nucleic acids A_i and enzymes E_i. The *i*-th nucleic acid A_i codes for the *i*-th enzyme E_i, and E_i catalyzes the replication of A_{i+1}, and so on. Eventually E_n catalyzes the replication of A_1, closing the loop [5].

Typogenetics is a formal system of strings devised by D. Hofstadter to explain the connection between molecular genetics and models of computation [3]. It was later established by H. Morris as a formal system to study artificial life [7]. It was further demonstrated by L. Varetto as a system where self-reproduction can occur [9, 10].

Construction of hypercycles in typogenetics is hopeless if exhaustive search is employed. Evolutionary algorithms were used by Kvasnicka et al. to find hypercycles in typogenetics [6]. We improve upon their work by using more refined measure of distance between strings representing nucleic acids. As a result we are able to construct hypercycles of sufficiently large sizes. We also experiment with enzymes of different lengths and various mappings between enzyme strings and their functions. All of these variations succeeded in finding sufficiently long hypercycles, which demonstrates, at least at this theoretical level, that the self-reproduction of macromolecules is a robust phenomenon under varied environments.

F. Almeida e Costa et al. (Eds.): ECAL 2007, LNAI 4648, pp. 1060 – 1068, 2007.

2 Backgrounds

We briefly explain basic concepts of typogenetics and describe Kvasnicka et al.'s work. The alphabet of the typogenetics consists of four letters, called *bases*, A, C, G, and T. A and G are classified as *purines*, and C and T *pyrimidines*. A and T are *complementary*, and C and G are complementary. *Strands* are strings composed of four bases. A complementary strand S of a strand $\overline{S}$ has every base complementary to the corresponding base in S. For example, complementary strand of S = CCAGATTA is $\overline{S}$ = GGTCTAAT. A double strand, called DNA, consists of two strands S (lower strand) and R (upper strand) that are complementary, that is, $\overline{S} = R$. For example, D = $\begin{pmatrix}\text{CCAGATTA}\\ \text{GGTCTAAT}\end{pmatrix}$ is a double strand. A *quasistrand* is a strand with occasional hash symbols (#) which represent empty positions. For example, C##GATT# is a quasistrand. The distance between two quasistrands of the same length is the number of bases that are different from the corresponding bases divided by the length of a strand. More formally, the *distance* between two quasistrands $S = X_1X_2 \cdots X_n$ and $R = Y_1Y_2 \cdots Y_n$ is defined as $d = 1 - \frac{1}{n}\sum_{i=1}^{n} \delta(X_i, Y_i)$, where $\delta(X,Y) = \begin{cases} 1 & \text{if } X = Y \neq \# \\ 0 & \text{otherwise} \end{cases}$. For example, the distance between S = C##GATT# and R = C#TGACTG is $1 - \frac{1}{8}(1+0+0+1+1+0+1+0) = \frac{1}{2}$. It can be easily observed that $0 \le d(S,R) \le 1$ and that $d(S,R) = 0$ if and only if $S = R$.

A substrand of length two, called a *doublet*, is considered as a gene and expressed to be an *enzyme* performing a predefined function, called an *instruction*. Hence a strand can be translated into a stream of instructions. Table 1 shows the mapping from the doublets to the instructions. Table 2 describes the meaning of each instruction.

For example, the strand S = CCAGATTA is translated into the sequence of instructions mvl-mvr-mvl-rpy. It is denoted as instructions(S) = mvl-mvr-mvl-rpy.

The sequence of instructions corresponds to the *primary structure* of the enzyme. A *tertiary structure* is needed to determine the *binding site*. It is determined by the sequence of *inclinations* in Table 1. The inclinations l, s, and r represent left-turn, straight, and right-turn, respectively. The binding site is determined by the combination

Table 1. Mapping from doublet to *instruction* and *inclination*

no.	doublet	instr.	inclin.	no.	doublet	instr.	inclin.
1	AA	mvr	l	9	GA	rpy	s
2	AC	mvl	s	10	GC	rpu	r
3	AG	mvr	s	11	GG	lpy	r
4	AT	mvl	r	12	GT	lpu	l
5	CA	mvr	s	13	TA	rpy	r
6	CC	mvl	s	14	TC	rpu	l
7	CG	cop	r	15	TG	lpy	l
8	CT	off	l	16	TT	lpu	l

Table 2. Functions of instructions

No.	instruction	Function
1	cop	Enzyme turns on copy mode; produces complementary bases
2	off	Enzyme turns off copy mode
3	mvr	Enzyme moves one base to the right
4	mvl	Enzyme moves one base to the left
5	rpy	Enzyme finds nearest pyrimidine to the right
6	rpu	Enzyme finds nearest purine to the right
7	lpy	Enzyme finds nearest pyrimidine to the left
8	lpu	Enzyme finds nearest purine to the left

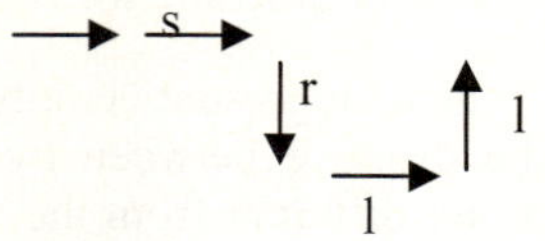

Fig. 1. An example of tertiary structure

of the first inclination and the last arrow (Table 3). For example, consider the inclinations sequence of the strand S = AGCGTTTG, which is s-r-l-l. The initial inclination (the zero-th inclination) is always assumed to be eastward. Hence the last arrow is upward (Figure 1).

Since the first inclination is s and the last arrow is upward, by Table 3, the binding site is the position of the first occurrence of C in whatever the strand to which the strand S is applied. It is denoted as binding-site(S) = C.

Now that the primary structure (the instructions) and the tertiary structure (the binding site) are defined, we denote the function of the strand S as enzyme(S) = (instructions(S), binding-site(S)). Now the replication process is defined as the process of applying the enzyme(S) to the strand V to produce the strand W , and denoted as rep(V , enz(S)) = W .

The following shows the replication process of the strand V = TAGCTAGG with S = AGCGTTTG. Then instruction(S) = mvr-cop-lpu-lpy, binding-site(S) = s-r-l-l = (s, ↑) = C. Hence the replication process rep(V , enz(S)) goes as follows:

$$\begin{pmatrix}\#\#\#\#\#\#\#\# \\ \mathrm{TAG\underline{C}TAGG}\end{pmatrix} \Rightarrow \mathrm{mvr} \begin{pmatrix}\#\#\#\#\#\#\#\# \\ \mathrm{TAGC\underline{T}AGG}\end{pmatrix} \Rightarrow \mathrm{cop} \begin{pmatrix}\#\#\#\#\mathrm{A}\#\#\# \\ \mathrm{TAGC\underline{T}AGG}\end{pmatrix} \Rightarrow \mathrm{lpu} \begin{pmatrix}\#\#\mathrm{CGA}\#\#\# \\ \mathrm{TA\underline{G}CTAGG}\end{pmatrix} \Rightarrow \mathrm{lpy} \begin{pmatrix}\mathrm{ATCGA}\#\#\# \\ \mathrm{\underline{T}AGCTAGG}\end{pmatrix}$$

Table 3. Determination of binding sites

no.	1st inclin.	last arrow	binding	no.	1st inclin.	last arrow	binding
1	s	→	A	7	l	←	G
2	s	↑	C	8	l	↑	T
3	s	↓	G	9	r	↑	A
4	s	←	T	10	r	←	C
5	l	↓	A	11	r	→	G
6	l	→	C	12	r	↓	T

A sequence of double strands $\left\langle \begin{pmatrix} R_1 \\ S_1 \end{pmatrix}, \begin{pmatrix} R_2 \\ S_2 \end{pmatrix}, \cdots, \begin{pmatrix} R_n \\ S_n \end{pmatrix} \right\rangle$ is called a *hypercycle* if it satisfies the following conditions: rep(R_{i+1} ,enz(R_i)) = S_{i+1} , rep(S_{i+1} ,enz(S_i)) = R_{i+1} , for every $i = 1, 2, \cdots, n-1$, and rep(R_1 ,enz(R_n)) = S_1 , rep(S_1 ,enz(S_n)) = R_1 . In other words, replication of each strand S_i is catalyzed by the enzyme function of the previous strand S_{i-1} with the previous strand of the first strand being the last strand.

The conditions for a sequence of double strands to be a hypercycle are very restrictive constraints. It is hopeless to search exhaustively for hypercycles. Kvasnicka et al. devised the technique of constructing a small hypercycle and then enlarging the size of the hypercycle one by one [6]. Suppose we have a hypercycle of size n in our hand. Then we can construct a hypercycle of size $n+1$ by finding ($n+1$)-th double strand $\begin{pmatrix} R_{n+1} \\ S_{n+1} \end{pmatrix}$ that satisfies the following conditions:

$$\text{rep}(R_{n+1}, \text{enz}(R_n)) = S_{n+1} \tag{1}$$

$$\text{rep}(S_{n+1}, \text{enz}(S_n)) = R_{n+1} \tag{2}$$

$$\text{rep}(R_1, \text{enz}(R_{n+1})) = S_1 \tag{3}$$

$$\text{rep}(S_1, \text{enz}(S_{n+1})) = R_1 \tag{4}$$

Finding the ($n+1$)-th double strand that satisfies the above conditions is still a daunting task. We do not even have a clue on how long the strand should be. Hence an evolutionary algorithm is used to find the strand. Fitness of a double strand $\begin{pmatrix} R_{n+1} \\ S_{n+1} \end{pmatrix}$ is defined as follows: fitness$\begin{pmatrix} R_{n+1} \\ S_{n+1} \end{pmatrix} = \frac{1}{4}\begin{bmatrix} 1 - d(\text{rep}(R_{n+1}, \text{enz}(R_n)), S_{n+1}) \\ +1 - d(\text{rep}(S_{n+1}, \text{enz}(S_n)), R_{n+1}) \\ +1 - d(\text{rep}(R_1, \text{enz}(R_{n+1})), S_1) \\ +1 - d(\text{rep}(S_1, \text{enz}(S_{n+1})), R_1) \end{bmatrix}$, where d is the distance between two strands. It can be easily observed that $0 \le \text{fitness}\begin{pmatrix} R_{n+1} \\ S_{n+1} \end{pmatrix} \le 1$, and that fitness$\begin{pmatrix} R_{n+1} \\ S_{n+1} \end{pmatrix} = 1$ if and only if $\begin{pmatrix} R_{n+1} \\ S_{n+1} \end{pmatrix}$ satisfies the above equations (1) through (4).

Three kinds of mutation operators are introduced: *change*, *insertion*, and *deletion*. The following example illustrates these operations.

CCA$\underline{\text{G}}$ATTA → CCA$\underline{\text{T}}$ATTA		(change)
CCAGATTA → CCAGAT$\underline{\text{C}}$TA		(insertion)
CC$\underline{\text{A}}$GATTA → CCGATTA		(deletion)

3 Improvements and Experiments

The previous section described mainly Kvasnicka et al.'s approach. They constructed a hypercycle of size 3. We refined the way of measuring the distance between two strands in computing the fitness of a given strand. As a result, we succeeded in generating hypercycles of sufficiently large sizes. We also experimented with shuffling of the functions of enzymes, and introduction of triplet codes for enzymes.

3.1 Refinement of Distance

We refined the definition of distance between two strands. Kvasnicka's definition of distance just compares the bases at the same positions of two strands, the bases at the first positions, the ones at the second positions, and so on. And if two strands are not of the same length, then the extra positions of the longer one are simply ignored. We employed minimum edit distance, or Levenshtein distance, as the measure of difference between two strands [1]. Minimum edit distance is also the way of comparing nucleotide sequences or amino acid sequences in bioinformatics [4, 8].

More formally, we defined the distance between two strands $S = X_1X_2 \cdots X_n$ and $R = Y_1Y_2 \cdots Y_m$ as follows: $d = 1 - \frac{2}{m+n} lcs(S,R)$ where $lcs(S,R)$ stands for the length of the *longest common subsequence* of S and R. A common subsequence of S and R is a subsequence that occurs in both of S and R. The longest common subsequence is the longest one among all the common subsequences. It is easy to see that $0 \le d \le 1$, since $0 \le lcs(S,R) \le \min\{m,n\}$. Note that $d = 0$ if and only if $S = R$

For example, let S = TGGACT and R = CGTGAT. Then the longest subsequence of S and R is GGAT, since GGAT is a subsequence of S and also a subsequence of R. In addition to that, there is no common subsequence longer than GGAT. Hence the distance $d = 1 - 2\frac{4}{6+6} = \frac{1}{3}$. Recall that the Kvasnicka's distance is $d = 1 - \frac{1}{p}\sum_{n=1}^{p} \delta(X_i, Y_i)$, where $p = \min\{m,n\}$. The distance between S and R by the Kvasnicka's definition is $d = 1 - \frac{2}{6} = \frac{2}{3}$, since S and R coincide only at the second position and the sixth position. We believe that our definition reflects the similarity and difference between two strands better than Kvasnicka's. There is a simple and efficient dynamic programming algorithm that finds the longest common subsequence in $\theta(m+n)$ time [1].

3.2 Enzyme Functions

Kvasnicka's work found a hypercycle under the fixed assignment of functions to enzymes. Table 1 in Section 2 shows the functions of enzymes. We tried many different associations between enzymes and their functions to see whether particular associations affect the chance of emergence of hypercycles. We randomly permuted the instructions in the "instructions" column of Table 1. We found that emergence of hypercycles were not affected by particular associations between enzymes and functions. We were still able to find hypercycles of all sizes from one to ten in every association we tried between enzyme codes and enzyme functions

3.3 Triplet Codes

Kvasnicka's work used doublets to encode enzymes. We experimented with triplet codes for enzymes to examine what difference they would make to the chance of emergence of hypercycles and the sizes of hypercycles. We found that it took more time for a hypercycle to emerge, but the sizes of hypercycles were not much longer than those with doublet enzymes.

3.4 Experiments

We implemented our algorithm on Pentium 2.8 GHz with 1GB memory in Visual C++ language running on windows XP. For our evolutionary algorithm, we used roulette wheel selection, and set the population size to 2000, mutation rate 0.01, and the number of generations 2000. The lengths of the strands in the initial population range from 15 to 30. Note that length of a strand may be increased or decreased when mutated by the result of insertion or deletion operation. It may also stay still.

When a strand is selected, mutation occurs to each symbol with the probability of 0.01 to generate a new strand at the next generation. When mutation occurs, one of the three mutation operators - change, insertion, and deletion - is randomly chosen and applied to the symbol.

Table 4. Triplet codes

No.	triplets	instr.	inclin.	no.	triplets	instr.	inclin.	no.	triplets	instr.	inclin.
1	AAA	*off*	*l*	23	CCG	*mvl*	*s*	45	GTA	*mvl*	*r*
2	AAC	*mvl*	*s*	24	CCT	*lpy*	*r*	46	GTC	*rpy*	*l*
3	AAG	*lpy*	*s*	25	CGA	*rpu*	*r*	47	GTG	*lpu*	*r*
4	AAT	*rpy*	*r*	26	CGC	*lpy*	*s*	48	GTT	*mvl*	*r*
5	ACA	*off*	*s*	27	CGG	*mvl*	*l*	49	TAA	*lpu*	*r*
6	ACC	*rpy*	*s*	28	CGT	*mvr*	*l*	50	TAC	*rpu*	*s*
7	ACG	*lpy*	*r*	29	CTA	*mvr*	*r*	51	TAG	*mvl*	*l*
8	ACT	*lpy*	*l*	30	CTC	*rpu*	*s*	52	TAT	*rpu*	*l*
9	AGA	*mvr*	*r*	31	CTG	*rpu*	*r*	53	TCA	*mvr*	*r*
10	AGC	*cop*	*l*	32	CTT	*cop*	*s*	54	TCC	*lpu*	*s*
11	AGG	*off*	*s*	33	GAA	*lpu*	*r*	55	TCG	*lpu*	*s*
12	AGT	*off*	*r*	34	GAC	*mvl*	*l*	56	TCT	*mvl*	*l*
13	ATA	*lpy*	*l*	35	GAG	*rpy*	*l*	57	TGA	*mvl*	*l*
14	ATC	*lpy*	*l*	36	GAT	*mvr*	*r*	58	TGC	*lpu*	*l*
15	ATG	*cop*	*l*	37	GCA	*mvr*	*l*	59	TGG	*mvl*	*s*
16	ATT	*lpu*	*r*	38	GCC	*rpy*	*r*	60	TGT	*rpu*	*s*
17	CAA	*lpy*	*l*	39	GCG	*rpu*	*s*	61	TTA	*rpu*	*s*
18	CAC	*cop*	*r*	40	GCT	*rpy*	*l*	62	TTC	*mvr*	*s*
19	CAG	*rpy*	*l*	41	GGA	*mvl*	*l*	63	TTG	*mvr*	*s*
20	CAT	*rpy*	*s*	42	GGC	*lpu*	*l*	64	TTT	*mvr*	*s*
21	CCA	*mvr*	*l*	43	GGG	*mvr*	*l*				
22	CCC	*mvr*	*r*	44	GGT	*mvl*	*r*				

Table 5. Determination of binding sites

no.	1st inclin.	last arrow	binding	no.	1st inclin.	last arrow	binding
1	s	→	C	7	l	←	G
2	s	↓	A	8	l	↑	G
3	s	←	G	9	r	→	A
4	s	↑	C	10	r	↓	C
5	l	→	T	11	r	←	T
6	l	↓	A	12	r	↑	T

Table 6. A hypercycle of size 10 with triplet codes

R_1	CAAAAGAGAAGGAAATGGAGGAGT
S_1	GTTTTCTCTTCCTTTACCTCCTCA
R_2	CACGAAGAGGAGG
S_2	GTGCTTCTCCTCC
R_3	GAAAGCGAGGGGGAG
S_3	CTTTCGCTCCCCCTC
R_4	CAAAGAGAAAAGAGCGAGGAGC
S_4	GTTTCTCTTTTCTCGCTCCTCG
R_5	AGCGAAGAGGAGGGA
S_5	TCGCTTCTCCTCCCT
R_6	CACGAAGAGGAGAAGAGA
S_6	GTGCTTCTCCTCTTCTCT
R_7	TAGGAAATGAGAAAGGAGGAGT
S_7	ATCCTTTACTCTTTCCTCCTCA
R_8	CACAGAGAAGAGGAG
S_8	GTGTCTCTTCTCCTC
R_9	AGGAAGAGGGAAATGGAGAAT
S_9	TCCTTCTCCCTTTACCTCTTA
R_{10}	TAGAGCGAAGAGGAGGAAGGC
S_{10}	ATCTCGCTTCTCCTCCTTCCG

We started with constructing a hypercycle of size one, and successively constructed hypercycles of size two, three, and so on, all the way up to size ten. Incidentally, hypercycles of size one are called autoreplicators [6, 11].

We experimented with doublet codes and triplet codes. We also tried various associations between enzyme codes and their functions. We succeeded in constructing hypercycles of all sizes from one to ten in every association we tried between enzyme codes and enzyme functions with both doublet codes and triplet codes.

The following example shows one of the hypercycles of size 10 constructed with triplet codes. Table 4 and Table 5 are the triplet codes table and binding sites table that are used for this particular hypercycle, respectively. Table 6 is the hypercycle of size 10. Table 7 is the illustration of the replication process of the upper strand R_1 with enzyme R_{10} applied, producing the complementary lower strand S_1. Table 8 is the illustration of the replication process of the lower strand S_1 with enzyme S_{10} applied, producing the complementary upper strand R_1. These replication processes are just two of the 20 replication processes occurring in this hypercycle.

Table 7. rep(R_1, enz(R_{10})) = S_1

no.	replication process	instruction
0	CAAAAGAGAAGGAAATGGAGGAGT ########################	binding site A
1	CAAAAGAGAAGGAAATGGAGGAGT ########################	mvl
2	CAAAAGAGAAGGAAATGGAGGAGT G#######################	cop
3	CAAAAGAGAAGGAAATGGAGGAGT G#######################	lpu
4	CAAAAGAGAAGGAAATGGAGGAGT GTTTTCTCTTCCCTTTA#######	rpy
5	CAAAAGAGAAGGAAATGGAGGAGT GTTTTCTCTTCCCTTTACCTCCTCA	rpy
6	CAAAAGAGAAGGAAATGGAGGAGT GTTTTCTCTTCCCTTTACCTCCTCA	lpu
7	CAAAAGAGAAGGAAATGGAGGAGT GTTTTCTCTTCCCTTTACCTCCTCA	lpu

Table 8. rep(S_1, enz(S_{10})) = R_1

No.	replication process	instruction
0	######################## GTTTTCTCTTCCCTTTACCTCCTCA	binding site G
1	######################## GTTTTCTCTTCCCTTTACCTCCTCA	lpy
2	######################## GTTTTCTCTTCCCTTTACCTCCTCA	lpu
3	C####################### GTTTTCTCTTCCCTTTACCTCCTCA	cop
4	CAAAAGAGAAGGAAAT######## GTTTTCTCTTCCCTTTACCTCCTCA	rpu
5	CAAAAGAGAAGGAAATGGAGGAGT GTTTTCTCTTCCCTTTACCTCCTCA	rpu
6	CAAAAGAGAAGGAAATGGAGGAGT GTTTTCTCTTCCCTTTACCTCCTCA	cop
7	CAAAAGAGAAGGAAATGGAGGAGT GTTTTCTCTTCCCTTTACCTCCTCA	mvl

4 Conclusions

Construction of hypercycles is one of the important topics on the studies of self-reproduction and origin-of-life in artificial life. Finding hypercycles of large sizes is a very hard combinatorial problem. Exhaustive search is prohibitive. To find a hypercycle of size 10 with each strand being 10 bases long, we would have to examine 4^{100} cases, which is roughly 10^{60} cases.

Previous research by Kvasnicka et al. found a hypercycle of size three with evolutionary algorithms. We improved upon their work and succeeded in constructing hypercycles of all sizes from one to ten. We believe that we can find hypercycles of arbitrary size given enough computation time.

We also experimented with enzymes of different lengths and various mappings between enzyme strings and their functions. All of these variations succeeded in finding sufficiently long hypercycles, which demonstrates, at least at this theoretical level, that the self-reproduction of macromolecules is a robust phenomenon under varied environments.

In our current model it is implicitly assumed that strands have infinite resources for their replication. We plan to investigate the ramifications of setting limits to resources thus putting selective pressure to strands.

Acknowledgements. This work was supported by the Korea Science and Engineering Foundation(KOSEF) grant funded by the Korea government(MOST) (No. R01-2006-00-10775-0). The authors would like to thank the reviewers for their helpful comments.

References

[1] Cormen, T., Leiserson, R., Rivest, R., Stein, C.: Introduction to Algorthms, 2nd edn. MIT Press, Cambridge (2001)

[2] Eigen, M., Schuster, P.: The Hypercycle: A Principle of Natural Self-Organization. Springer, Berlin (1979)

[3] Hofstadter, D.: Godel, Escher, Bach: an Eternal Golden Braid. Basic Books, New York (1999)

[4] Jones, N., Pevzner, P.: An Introduction to Bioinformatics Algorithms. MIT Press, Cambridge (2004)

[5] Klipp, E., Herwig, R., Kowald, A., Wiering, C., Lehrach, H.: Systems Biology in Practice. Wiley-VCH, Weinheim (2005)

[6] Kvasnicka, V., Pospichal, J., Kalab, T.: A Study of Replicators and Hypercycles by Typogenetics. In: Kelemen, J., Sosík, P. (eds.) ECAL 2001. LNCS (LNAI), vol. 2159, pp. 37–54. Springer, Heidelberg (2001)

[7] Morris, H.: Typogenetics: a Logic for Artificial Life. In: Langton, C. (ed.) Proceedings of the First Workshop on Artificial Life, Los Alamos, New Mexico, pp. 369–395 (1987)

[8] Mount, D.: Bioinformatics: Sequence and Genome Analysis, Cold Spring Harbor Laboratory (2001)

[9] Varetto, L.: Typogenetics: an Artificial Genetic System. Journal of Theoretical Biology 160, 185–205 (1993)

[10] Varetto, L.: Studying Artificial Life with a Molecular Automaton. Journal of Theoretical Biology 193, 257–285 (1998)

[11] Wee, K., Lee, W.: Extensions and Variations on Construction of Autoreplicators in Typogenetics. In: Capcarrère, M.S., Freitas, A.A., Bentley, P.J., Johnson, C.G., Timmis, J. (eds.) ECAL 2005. LNCS (LNAI), vol. 3630, pp. 685–694. Springer, Heidelberg (2005)

Designing a Methodology to Estimate Complexity of Protein Structures

Alejandro Balbín[1] and Eugenio Andrade[1]

Grupo de Biología Molecular, Teórica y Evolutiva. Departamento de Biología, Universidad Nacional de Colombia, Bogotá.
oabalbinr@unal.edu.co

Abstract. This paper proposes a methodology to estimate the information content of protein structures by using: an alphabet of local microenvironments obtained from a set of protein domains with equivalent function, a modification of the physical complexity concept [1], the measures of mutual information ($I(seq; str)$) and conditional entropy ($H(str|seq)$) between sequence and structure. The kinase domain catalytic subunit was used as a specific example. Our results are in accord with the hypothesis that proteins are information gathering and using systems [3], and suggest that protein structure depends less on protein sequence than biologists have historically supposed.

1 Introduction

Previous studies on information content used two set of measures, one is based on Shannon information theory, H, and the other is based on Kolmogorov-Chaitin algorithmic theory, K. Both kinds have frequently been applied to quantify information content of protein sequences [15] (and references in), but rarely to protein structures [11,5].

However, the meaning of stored information in an object is not considered by those measures because they do not regard the system that produce and use this information [9]; i.e, they do not consider the environment in which sequences and structures are functional and were selected during evolution [1].

A complexity measure should assign high complexity to sequences with meaning, while low complexity to those which are meaningless. This measure should define a string's randomness in relation to a physical world [1]. Adami and Cerf [1] define the physical complexity of a string s as the number of bits with meaning in s. The physical complexity is given by the mutual complexity between the string and the environment: $K(s : e) = K_0(s) - K(s|e)$, where $K_0(s)$ is the s complexity without the environment, and $K(s|e)$ is the s complexity given the environment e. In the absence of a specific environment, s would be random and $K_0(s)$ would be maximal and given by the length of s: $K_0(s) = |s|$ [1]. It is remarkable that in order to estimate $K(s : e)$ we should use a functionally

[1] Physical complexity is a particular case of Effective complexity [9].

F. Almeida e Costa et al. (Eds.): ECAL 2007, LNAI 4648, pp. 1069–1078, 2007.

equivalent set of strings, because inspecting only one string does not permit us to identify which regions have meaning and which ones have not.

The physical complexity has been applied over sets of aligned sequences with equivalent function of DNA, RNA and proteins [1,2]. However, the natural systems that use and record this information is neglected again.

Zurek [18] showed that from the point of view of an information gathering and using system (IGUS) both H and K are important. H measures the IGUS' uncertainty about its environment and K measures the IGUS' knowledge about it. By means of a measurement process, an IGUS reduces its environment uncertainty and increases its knowledge about it. If the IGUS efficiently encodes and records the environment regularities it will obtain useful energy to accomplish a specific work.

Balbín and Andrade [3] proposed that proteins can be understood as information gathering and using systems. Proteins-IGUS measure and record their environments obtaining useful energy to construct their native structures. In this hypothesis, the native structure would be a record of measurements; the measurements are made by proteins during their folding and interaction with external ligands, and they are continuously repeated during a protein evolutive process. Therefore, structure is a record of mutual information between a protein and the environment in which it functions.

In this paper we present a methodology to estimate the information content of a coarse grained and functional structure of domains belonging to the same protein family: the protein kinase, catalytic subunit. First, structural symbols were defined using a residue microenvironment concept [4] and then a structural alphabet was built with those symbols; the alphabet recover the main functional characteristics of the kinase domain. Finally, this alphabet was used to estimate the domain information content using a modification of the physical complexity concept [1], and to estimate the mutual information and conditional entropy between domain sequence and structure.

The results are analyzed from the proteins-IGUS hypothesis [3]; they give informational arguments in favor the new vision of proteins [13]; and their consequences on the molecular biology central dogma are discussed.

2 Methodology

2.1 Constructing a Local Alphabet of Structures

First, a functionally equivalent set of domains was defined. It was constituted by 55 catalytic subunit kinase domains. The homology among the amino acid sequence of those domains was > 30 % but ≤ 80 %, and the resolution of the crystallographic structures was between 1.2 – 2.5 Å. This set was culled from the entire set of domains belonging to the catalytic subunit kinase family in the Structural Classification of Proteins data base, by using a Protein Sequence Culling Server [14]. The set represents all classes and groups of the family.

Second, a microenvironment concept [4] is used to define local substructures in a protein. The local microenvironment of a residue j is defined as the set of

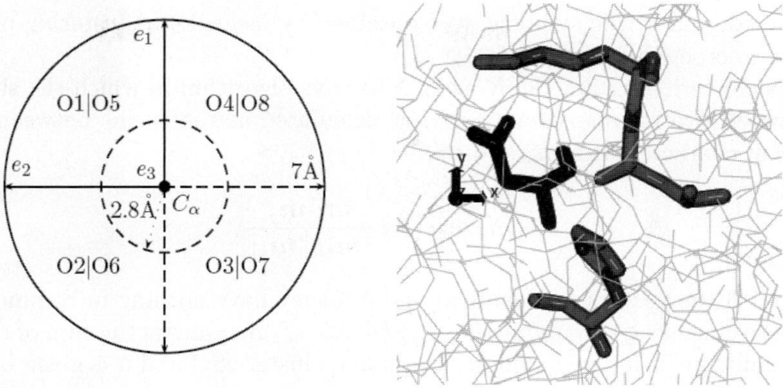

Fig. 1. Microenvironment of a residue. *Left* Definition: The microenvironment of a residue j is constituted by the identity, chemical group and spatial position of their neighbors residues in the protein structure, which have 66 % of its atoms at $r \leq 7$ Å or 33 % of its atoms at $r \leq 2.8$ Å. The spherical region is divided on 8 octants (O1 - O8) as in [4]. *Right* An alphabet symbol: This symbol describes the kinase catalytic site. It could be characterized by a microenvironment with an ASP as central residue (*closer to the axis*) and three typical neighbors: HIS-O6, LEU-O8 and LYS-O8. Graphic made with VMD v.1.8.5 using the 1BLXa kinase domain.

neighbor residues which are inside a spherical region with radio r, and center on the alpha carbon (C_α) of the amino acid j. The local microenvironment is characterized by the identity, chemical group and spatial position of the neighbor residues inside this vicinity [4] (Fig. 1 left).

So that, a microenvironment characteristic represents a biochemical property and its spatial position, e.g, Lys-O1, Acid-O1. A characteristic is codified by a natural number whose magnitude corresponds to the number of occurrences that the characteristic was observed inside the microenvironment. Then, the microenvironment of a residue j is represented as a vector of 216 characteristics: 1 by the identity of every neighbor amino acid in every octant $(20 * 8)$, and 1 by each chemical group in each octant $(7 * 8)$ [2]. Therefore, the structure of a protein domain is represented as a microenvironment disperse matrix with dimension $L * 216$, where L is the length of peptide sequence.

Third, an alphabet of structural symbols is built. A symbol is defined as a unity, which clusters similar local microenvironments assigned to it unambiguously. An alphabet is defined as a set of those unities. The symbols were independently defined for each one of the 20 amino acid standards, therefore, local microenvironments with different central amino acid were not compared.

The alphabet was constructed in a two step process: 1) For every one of 16.749 residues belonging to the domain set, its local microenvironment was

[2] Most of these vector components are zero, pointing out the absence of this characteristic; therefore, it is only required a small number of components to describe one microenvironment.

determined, and 2) the symbols were obtained by means of a clustering process of those microenvironments.

This clustering was made by using a K-ways algorithm in which the similarity between two microenvironments was calculated as the cosine between their vectors of characteristics $\boldsymbol{m_i}$ and $\boldsymbol{m_j}$,

$$\cos(\boldsymbol{m_i}, \boldsymbol{m_j}) = \frac{\boldsymbol{m_i}^t \boldsymbol{m_j}}{\|\boldsymbol{m_i}\| \|\boldsymbol{m_j}\|} . \quad (1)$$

It is 1 if the vectors are identical and 0 if they have nothing in common.

As the clustering global function I_1 was used. I_1 maximizes the sum of the average similarity inside each microenvironment cluster (S_r) and it is given by [17]:

$$\text{maximize } I_1 = \sum_{r=1}^{k} n_r \left(\frac{1}{n_r^2} \sum_{\boldsymbol{m_i}, \boldsymbol{m_j} \in S_r} \cos(\boldsymbol{m_i}, \boldsymbol{m_j}) \right) . \quad (2)$$

The clustering of microenvironments with the same central amino were repeated 20 different times, each time k seeds were randomly chosen. The best of those replicates, in function of I_1, was selected. These final clusters are the symbols for each one of the amino acid standards. All clusterings were made with CLUTO v2.1.1 [17], and the number of k clusters for each amino acid was determined by using a semi automatic process designed with base on [12].

The obtained alphabet of microenvironments has 299 symbols, and recovers, on average, 83% of all functional or structural important positions in the catalytic subunit kinase domains (Fig. 1 right). In this protein family there would be, on average, 15 symbols of local structure associate with each amino acid standard.

In order to estimate a protein structure information content, the amino acid sequence of all sampled domains were aligned by using CLUSTALW. With the aligned sequences, we can compare the different structural symbols which are observed at the same sequence position through the set of protein domains.

2.2 Information Measures

The estimation of the physical complexity requires a functionally equivalent set of objects (sequences, structures), which were produced by similar IGUS acting in similar environments. The average physical complexity of a statistical *ensemble* of sequences $\langle K_{(s:e)} \rangle_S$ was defined as [1,2]:

$$\langle K(s:e) \rangle_S = \sum_S p(s) K(s:e) \quad (3)$$

$$\approx H(S) - H(S|e) \quad (4)$$

$$\equiv I(S;e) . \quad (5)$$

Where $I(S;e)$ is the mutual information between the set of sequences and the environment; i.e, the information about e stored in S. $I(S;e)$ is the difference between the S marginal entropy, $H(S)$, and the conditional entropy of S given e, $H(S|e)$ [3]; in $H(S)$ there is an uniform distribution for each s that belongs to S. In other words, $I(S;e)$ is the difference between the entropy of S in the absence of an environment and the actual entropy of S given the environment e.

In this paper, the structural entropy in the absence of e, $H(str)$, was estimated as:

$$H(str) = -\sum_{d \in D} p(d) \log p(d) \ . \tag{6}$$

Where d is each symbol in the microenvironment structural alphabet, D, and $p(d)$ the symbol probability. $p(d) = \frac{n_d}{M}$ where n_d is the number of occurrences of the symbol d in the microenvironment population M. The logarithm is base 2 in order to obtain the entropy in bits. $H(str)$ has maximum value when all symbols d have the same probability; $H_{max} = \log |D|$.

The actual structural entropy is the conditional entropy of the structure given the environment in which the kinase domain is functional, $H(str|env)$. Adami [2] approximated the sequence conditional entropy given its environment by the sum of the actual entropy at each sequence position; i.e, the sum of the conditional entropy of each sequence symbol i given the position j at which it is, when each position j has the same probability.

For a set of domains of the same protein family, the actual structural entropy, $H(str|env)$, is estimated by the specific conditional entropy of the structure given the aligned peptide sequences. In the kinase case:

$$H(str|seq_{=kin}) = -\sum_{j=1}^{P}\sum_{d=1}^{D} p(d|j) \log p(d|j) \ . \tag{7}$$

Where $p(d|j)$ is the probability of the structural symbol d given the sequence position j; it is estimated by $p(d|j) = \frac{p(d,j)}{p(j)}$ with $p(d,j)$ the joint probability between d and j, and $p(j)$ the probability of j. Structural conserved positions contribute few to $H(str|seq_{=kin})$ increases, while structural variable positions make the biggest contribution.

The information about the environment stored in a protein structure is calculated as the difference between the structural entropy in the absence of an environment and the actual structural entropy:

$$I(str;env) = H_{\max}P \ - \ \Big(-\sum_{j=1}^{P}\sum_{d=1}^{D} p(d|j) \log p(d|j)\Big). \tag{8}$$

Moreover, the informational relationship between kinase sequence and structure is studied by estimating the conditional entropy, $H(str|seq)$, and mutual information, $I(str;seq)$, between them:

[3] [1] recognize that $I(S;e)$ is not an information in strict meaning because that should be averaged over all possible E.

$$H(str|seq) = -\sum_{j=1}^{P} p(j) \sum_{d=1}^{D} p(d|j) \log p(d|j) \ , \tag{9}$$

$$I(str; seq) = \sum_{j=1}^{P} \sum_{d=1}^{D} p(d,j) \log \frac{p(d,j)}{p(d)p(j)} \ . \tag{10}$$

To conclude if sequence and structure are independent its mutual information will be zero, in any other case it will be positive and the value indicates the correlation between them.

2.3 Correcting the Information Measures

Entropy measures are affected by the sample size used to estimate the probability distributions. Marginal and conditional entropies are underestimated, while mutual information is overestimated [16].

In this paper, $I(str; seq)$ overestimation was corrected by using the methodology proposed by [5]. They estimated the distribution of $I(X;Y)$ empirically by using random and independent permutations of the data Y and X, $\sigma_i(\mathrm{D})$. These permutations reorganize the observations of Y with respect to X, eliminating any correlation between them. Therefore, for N random permutations and a given v value, they estimated:

$$P(I(X;Y) \geq v) \approx \frac{|\{i : 1 \leq i \leq N \text{ and } I(X;Y|\sigma_i(\mathrm{D})) > v\}|}{N} \ . \tag{11}$$

The probability that $I(X;Y)$ exceeds v is given by the fraction of N permutations for which this happens. Let $v = I(X;Y|\mathrm{D})$ be the mutual information calculated between X and Y given the data. If $I(X;Y|\mathrm{D}) \geq I(X;Y|\sigma_i(\mathrm{D}))$, at least for 99% of the random permutations, then the null hypothesis that X and Y are independent is rejected with a confidence of 99%. $N = 5000$ was used, $N = 1000$ yields a stable estimate of $I(X;Y)$ [5].

The bias in the empirically estimation of $I(X;Y)$ named independent information is estimated as the expected value of $I(X;Y|\sigma(\mathrm{D}))$:

$$I_I(X;Y) = I(X;Y|\sigma(\mathrm{D})) = \frac{1}{N} \sum_{i=1}^{N} I(X;Y|\sigma_i(\mathrm{D})) \ . \tag{12}$$

Finally, a corrected information, $I_C(X;Y)$, is estimated as

$$I_C(X;Y) = I_C(X;Y|\mathrm{D}) = I(X;Y|\mathrm{D}) - I_I(X;Y) \ . \tag{13}$$

$I_C(X;Y)$ corresponds to the real mutual information between X and Y beyond the sample bias.

$H(str|seq_{=kin})$ and $H(str|seq)$ underestimations were corrected by following a similar methodology, but for N random permutations and a given v value we estimated:

$$P(H(X|Y) \leq v) \approx \frac{|\{i : 1 \leq i \leq N \text{ and } H(X|Y|\sigma_i(\mathrm{D})) < v\}|}{N} \ . \tag{14}$$

The probability that $H(X|Y)$ being less than v is given by the fraction of N permutations for which this happens. If $v = H(X|Y|\mathrm{D})$ and $H(X|Y|\mathrm{D}) \leq H(X|Y|\sigma_i(\mathrm{D}))$ at least for 99% of the random permutations then the null hypothesis that X and Y are independent is rejected with a confidence of 99%. $N = 5000$ was used.

Then, a random conditional entropy, $H_r(X|Y)$, was estimated as the expected value of $H(X|Y|\sigma(\mathrm{D}))$:

$$H_r(X|Y) = H_r(X|Y|\sigma(\mathrm{D})) = \frac{1}{N}\sum_{i=1}^{N} H(X|Y|\sigma_i(\mathrm{D})) \ . \tag{15}$$

In order to estimate the bias in the conditional entropy calculation, we should be aware of $H(X|Y) \leq H(X)$ with $H(X|Y) = H(X)$ if and only if X and Y are independent [6]. Therefore under the independence hypothesis between X and Y, we have $H_r(X|Y) = H(X)$ and the estimation bias of $H(X|Y)$ is equal to $H(X) - H_r(X|Y)$. It represents the reduction of X entropy yielded by the empirical estimation of $H(X|Y)$ but not by a real correlation between X and Y. If Y conditions X beyond the sample bias, then $\Big\langle H(X|Y|\mathrm{D}) + \big(H(X) - H_r(X|Y)\big)\Big\rangle < H(X)$.

Finally, we defined a corrected conditional entropy between X and Y, as:

$$H_C(X|Y) = H(X|Y|\mathrm{D}) + \big(H(X) - H_r(X|Y)\big) \ . \tag{16}$$

3 Results and Discussion

The shared information between the structure of the kinase domain and the environment estimated by using (8) is:

$$I(str; env) = 1784.608 \ - \ 1337.50 \ , \tag{17}$$
$$= 447.608 \text{ bits} \ . \tag{18}$$

Where P in (8) corresponds to the best aligned positions inside the domain sample studied, $P = 217$, and D in (8) is the size of the structural alphabet, $D = 299$. The specific conditional entropy is corrected following the methodology in Sect. 2.3. The kinase domain catalytic subunit has stored 447.608 bits of information about its environment.

The estimates for $H(str|seq)$ and $I(seq; str)$ are presented in Table 1. From our results, the shared information between sequence and structure of the catalytic subunit kinase domain is 1.6346 bits, on average, by residue.

Crooks [7] estimated the mutual information among primary structure (R), secondary structure (S), and side chain packing of an individual residue as 0.26 bits[4]. It indicates a significant correlation between the identity and amino acid

[4] R corresponds with the 20 standard amino acids, S with three secondary structure categories and B with four packing categories.

Table 1. Informational relationship between sequence and structure of kinase domain, catalytic subunit. All values are given in bits.

Mutual Information $str; seq$		
Observed	Independent	Corrected
4.0104	2.3758 ±0.0066	1.6346

Conditional Entropy $str\|seq$		
Observed	By Random	Corrected
3.8196	5.4915 ± 0.0059	6.1581

All estimates were done with $P = 217$ and $D = 299$

local structure. Moreover, the information added by the local structure and identity of a sequence neighbor residue is 0.030 bits, and 0.015 bits for a spatial neighbor [5]. Almost all residues analyzed by [7] had two sequence neighbors and two spatial neighbors, therefore, they concluded that $I(seq; str)$ of these microenvironments would be approximately 0.35 bits.

The 0.35 bits value by amino acid constitutes an approximation to $I(seq; str)$ regardless of the protein function, inasmuch as [7] used a data base of not redundant protein domains without functional correlation. Therefore, our estimation (Table 1) is showing around 1.2 bits of functional information by residue. This result agrees with the calculated mutual information among residues which are functionally important and conserved during evolution. This information has been calculated as around 1.0 bit by residue, for sets of aligned sequences with 80 % of homology [10]. With all this in mind, our methodology is recovering both information produced by structural constraints and information produced by functional constraints conserved during evolution.

Finally, $H(str|seq) = 6.1581$ bits by residue, on average, indicates that the shared information between sequence and structure is enough to specify the kinase domain function, but this information does not determine completely the domain structure. This permits enough structural flexibility which is compatible with the kinase function.

Most of the previous studies have mainly been concentrated on estimates of the peptide sequence information content [15] (and references in), some few on the structure information content [11,5] and even less on the relationship between sequence and structure [7].

A fundamental supposition about this relationship is that the structural entropy given the sequence is equal to zero, $H(str|seq) = 0$ [8]; i.e, when a specific sequence is given, an unique structure is produced. From this supposition, and by using the standard equation: $I(Y; X) = H(Y) - H(Y|X)$ [6], the mutual information between structure and sequence has been equated to the entropy of the structure, $I(seq; str) = H(str)$ [8]. Moreover, if we suppose that one function is performed by only one structure then the physical

[5] Two residues are spatial neighbors if they are far up to six residues in the sequence and their side chains are at less than 8 Å[7]

complexity is equal to the mutual information between sequence and structure, $I(seq; env) \approx I(seq; str) = H(str)$ because $H(str|seq) = 0$ [1].

However our results (Table 1) for the catalytic subunit kinase domain do not agree with those suppositions. In our case: i) $H(str) > I(seq; str)$; i.e, a family of proteins has stored enough information about its environment, as needed to accomplish its function. However, there are environmental information yet to be measured and recorded. And ii) $H(str|seq) > 0$; i.e, more than one different structure is produced by a given sequence. Therefore,

$$I(str; env) \neq I(seq; str) \text{ , and} \tag{19}$$

$$I(seq; str) = H(str) - H(str|seq) \text{ ,} \tag{20}$$

where $H(str|seq) > 0$.

Balbín and Andrade [3] proposed that a protein acting as an IGUS in the cellular environment has a tendency to increase the mutual information between sequence and structure. This tendency *drives* the protein evolutive process, and that is satisfied through successive and reiterative measurements of the environment. By means of these measurements proteins obtain information about their actual environment in order to reduce $H(str|seq)$ during their cellular activity. Measurements that would permit to reduce $H(str|seq)$ during the evolutionary process are digitally recorded as they are incorporated into peptide sequence.

In agreement with the protein-IGUS hypothesis, (20) states that the entropy of protein structures is given by:

$$H(str) = I(seq; str) + H(str|seq) \text{ .} \tag{21}$$

Therefore, the entropy of protein structures depends on mutual information between sequence and structure, $I(seq; str)$, stored during the evolutive process, and information gathered by proteins' measurements while they act within the cell in every life cycle. These measurements permit to reduce $H(str|seq)$.

During the evolutive process, the state in which $H(str|seq) = 0$ and $I(seq; str) = H(str)$ would be a limit condition when protein-IGUS approaches the thermodynamic equilibrium [3]. Only in this state $I(seq; str) \approx I(str; env)$.

Our results give informational arguments in favor to the *new vision* of proteins [13], in which a sequence can adopt multiples structures and functions. In the *new vision* conformational diversity would increase functional diversity of a sequence repertoire, promoting evolution of new proteins and functions from those present already inside the plastic repertoire of structures.

4 Conclusion

This paper proposed a methodology to estimate the physical complexity of a protein structure, $I(seq; str)$ and $H(str|seq)$. The protein kinase family, catalytic subunit, was studied as example. Our estimates show a dependency between structure and sequence of 20 % approximately which is much less than what has been traditionally supposed.

In agreement with the protein-IGUS hypothesis those results suggest that the entropy of protein structures depends on both the mutual information between sequence and structure, $I(seq; str)$, and the measurements made by proteins during their cellular activities that lead to reduce $H(str|seq)$.

Acknowledgements. We acknowledge the support of the Department of Biology and the Vicerrectoria de Investigaciones de la Universidad Nacional de Colombia (DINAIN 20101005106).

References

1. Adami, C., Cerf, N.: Physical complexity of symbolic sequences. Physica D 137, 62–69 (2000)
2. Adami, C.: Information theory in molecular biology. Physics of Life Reviews 1, 3–22 (2004)
3. Balbín, A., Andrade, E.: Protein folding and evolution are driven by the maxwell demon activity of proteins. Acta Biotheoretica 52, 173–200 (2004)
4. Chen, S., Bahar, I.: Mining frequente patterns in protein structures: a study of protease families. Bioinformatics 20, i77–i85 (2004)
5. Cline, M., Karplus, K., Lathrop, R., Smith, T., Rogers, R., Haussler, D.: Information - theoretic dissection of pairwise contact potentials. Proteins 49, 7–14 (2002)
6. Cover, T., Thomas, J.: Elements of information theory. Wiley Series in Telecommunications. John Wiley & Sons, New York (1991)
7. Crooks, G., Wolfe, J., Brenner, S.: Measurements of protein sequence–structure correlations. Proteins 57, 804–810 (2004)
8. Dewey, T.: Algorithmic complexity and thermodynamics of sequence – structure relationships in proteins. Phys. Rev. E 56, 4545–4552 (1997)
9. Gell-Mann, M., Lloyd, S.: Information measures, effective complexity, and total information. Complexity 2, 44–52 (1996)
10. Henikoff, S., Henikof, F.: Amino acid substition matrices from proteins blocks. Proc. Natl. Acad. Sci. USA 89, 10915–10919 (1992)
11. Krasnogor, N., Pelta, D.: Measuring the similarity of protein structures by means of the universal similarity metric. Bioinformatics 20, 1015–1021 (2004)
12. Strehl, A., Ghosh, J.: Relationship-based clustering and visualization for high-dimensional data mining. INFORMS Journal on Computing 15, 208–230 (2003)
13. James, L., Tawfik, D.: Conformational diversity and protein evolution – a 60-year-old hypothesis revisited. TRENDS in Biochemical Sciences 28, 361–368 (2003)
14. Wang, G., Dumbrack, R.: PISCES: a recent improvement to a PDB sequence culling server. Nucleic Acid Research 33, W94–W98 (2005)
15. Weiss, O., Jiménez-Montaño, M.A., Herzel, H.: Information content of protein sequences. J. Theor. Biol. 206, 379–386 (2000)
16. Wolpert, D., Wolf, D.: Estimating functions of probability distributions from a finite set of samples. Phys. Rev. E 52, 6841–6854 (1995)
17. Zhao, Y., Karypis, G.: Empirical and theoretical comparisons of selected criterion functions for document clustering. Machine Learning 55, 311–331 (2004)
18. Zurek, W.H.: Algorithmic randomness and physical entropy. Phys. Rev. A 40, 4731–4751 (1989)

Designing for Surprise

Telmo Menezes and Ernesto Costa

Centre for Informatics and Systems of the University of Coimbra (CISUC)
{telmo, ernesto}@dei.uc.pt

Abstract. We propose a theoretical framework for the design of multi-agent systems with the capacity to surprise a human observer. Reasons for the difficulty in creating simulations capable of perpetual innovation are addressed. Artificial life concepts and biological inspiration from different abstraction layers serve as base for a three part model. These parts are: world modelling with artificial chemistries, agent brains and population dynamics. The parts of the model are theoretically discussed and some academic examples provided. Possible applications to biological research in the problem of speciation are considered.

1 Introduction

The natural world has the ability to perpetually surprise the observer, by never ceasing to introduce novelty in the system. Simulated worlds have so far fell short of having this capacity. A classical example of area of application where this would be desirable is computer games. Computer game technology has evolved to a point of near photorealism in the graphics front, but artificial intelligence has somewhat stagnated. The use of scripted behaviors tends to compromise the player's immersion in the game world.

We argue that Artificial Life concepts may be applied to the development of systems that have the ability to permanently surprise human observers or participants. In this paper we present a theoretical framework of concepts that we believe may be applied to the successful development of such systems.

It is normal to seek inspiration in nature when striving to recreate some property that the natural world has. Nature seems to have an endless capacity to generate novelty, but it also has at its disposal an enormous computational power over very large time scales. We humans observe nature at a certain level of abstraction. We know that this level of abstraction at which we perceive reality emerges from other layers that we can only observe indirectly or with the aid of instruments. Social phenomena emerge from biological phenomena, which emerge from chemical phenomena, which in turn emerge from physical laws. This suggests that bottom-up approaches should be able to simulate complex phenomena by recreating the mechanisms of properties of the levels of reality below them that enable their emergence. It is thus tempting to go for the most bottom-up approach we can find when trying to engineer biologically inspired systems. An example of such an approach comes from Wolfram's work with cellular automata [1]. The practical problem with this kind of model is that the

F. Almeida e Costa et al. (Eds.): ECAL 2007, LNAI 4648, pp. 1079–1088, 2007.

lower we go in levels of abstraction when performing a simulation, the more computational power we tend to need to simulate the same phenomena. Also, we loose more and more control over the end result, and if we are looking for a design process we want to retain some control. Even not considering the practical problems, it has not been theoretically demonstrated that this approach is possible [2].

There are Artificial Life simulations designed at several levels of abstraction. Systems like Tierra [3] and Avida [4] are very low level and would probably require immense computational power to achieve the level of complexity we can observe in the natural world. Even so, it may be argued that they are not low level enough to provide the kind of environment that is needed for significant diversity and surprise to emerge. Systems totally based on artificial chemistry [5] may provide this environment, but suffer even more from extreme computational requirements.

We propose a framework of concepts that are biologically inspired, but come from different layers of emergence in the natural world. We tried to identify a minimal set of concepts that allow for the definition of generic multi-agent systems with the capacity to surprise. This is a pragmatic approach, aimed at the development of systems under current technological limitations and that operate in time frames short enough to be interesting to a human observer.

In the next section we will enumerate the components of the framework we propose. After that we will discuss modelling with artificial chemistries, and then interfacing with agents' brains. Some examples will be provided. Following that we will address population dynamics. We end by detailing exceptions to the framework and offering some final remarks.

2 Brains, Population Dynamics and Artificial Chemistries

Our approach focuses on three components of the system: the agents' brains, the population dynamics and the modelling of simulations with artificial chemistries. The brain is a system that is internal to the agent, receiving perceptions from its point of view and triggering actions for it to perform. The population dynamics is a set of rules that takes care of the creation, destruction and replication of agents, defined in a way that promotes adaptation and diversity. The artificial chemistries enable the definition of simulation rules that are transformational and combinatorial. There is a relationship between these parts that is synergistic and contributes to our goal. The brains command the agents' behavior, constantly attempting to take actions that increase their chances of survival and transmission of genetic code. It is by the evolutionary pressure imposed by the population dynamics that the brains evolve to having the ability to perform their function. Population dynamics promotes diversity, so that new niches are constantly explored. Adaptation of agents to new niches will generate novelty in the system. Artificial chemistries allow for the definition of a kind of world where diversity is possible.

A big source of inspiration for this framework is Holland's discussion of Complex Adaptive Systems as well as his ECHO simulation [6]. Despite that, we will not follow a purely adaptive approach, taking in consideration biological theories on speciation as will be discussed in section 6.

3 Modeling Worlds with Artificial Chemistries

Artificial chemistries are powerful abstractions with applications in modeling, information processing and optimization. Much of the appeal of artificial chemistries to Artificial Life research comes from its pre-biotic nature, or its potential to explain the formation of evolutionary units. In the framework we are proposing, evolutionary units are explicitly designed. Artificial chemistries are used to model information flows in the world for its ability to define constructivist processes, where rules are defined in a way that does not excessively restricts the domain of possibilities.

"Formally, an artificial chemistry can be defined by a triple (S, R, A), where S is the set of all possible molecules, R is a set of collision rules representing the interaction among the molecules, and A is an algorithm describing the reaction vessel or domain and how the rules are applied to the molecules inside the vessel." [7]

We propose the use of the molecule as a metaphor for a type of information contained in the world. We identify the following general classes of information to be represented as molecules: object properties, interfaces and messages. Valid interactions are established by defining the reaction rules. The reaction vessel is defined by the interactions of objects in the simulated world. A reaction occurs when an agent perceives or acts on another object or when two objects collide or interact as determined by the simulation.

Since molecules are the basic information units, they are used to construct the visualization for the simulation. If the information that the molecule contains is to be experienced by human observers of the simulation, a translation from this encoded information to a human-friendly representation is defined. Properties of objects may translate to shapes, sizes, colors or textures; message molecules may translate to sound or text strings and so on.

Several artificial chemistries may be defined for the same simulation, each one for a certain type of information. This allows for the definition of specific molecule formats and reaction rules according to information type. As an example, it is likely that sound messages and the general physical properties of objects are best defined by distinct artificial chemistries.

We will define three types of reactions that may occur: binding, unidirectional transformations and bidirectional transformations. Unlike most artificial chemistries, the type of reaction to perform is explicitly selected by the reaction vessel, or in our case, the simulation. Binding rules evaluate the degree to which two molecules are able to bind and returns a value in the $[0, 1]$ real interval. 0 is for molecules that do not bind at all, 1 for molecules that bind in the strongest way possible. In unidirectional transformations, molecule A and B react to transform molecule B, but molecule A remains unchanged. In bidirectional transformations both molecules may change. As we will see, the first two types of reactions are used to define brain / world interfaces, while the third is used for involuntary object interactions, typically collisions.

4 Interfacing with Brains

We defined the brain as a system that is internal to an agent and that receives perceptions and triggers actions. We will now see how these perceptions / actions interfaces may be naturally defined in our framework as part of the artificial chemistries. The interfaces will be established by molecules from the same artificial chemistries that the perceptions / actions relate to. A perception of sound could thus be defined by a molecule from the sound message chemistry while an action for eating could be defined by a molecule from the food chain chemistry.

Perceptions use binding reactions. The perception for an observed molecule / perception molecule pair will be the binding value as defined by the reaction rules. Actions use unidirectional transformations. The action molecule A will transform the target molecule B according to the reaction rules. If the reaction releases energy, that energy is added to the agent, if it absorbs energy, it will be removed from it.

Many approaches could be used to model brains under this framework. However, we would like to propose the gridbrain [8], an heterogeneous network of components that was conceived by the authors of this text to fit in such a model. The gridbrain may be easily extended to support the kind of interfaces we described in the former sections.

5 Some Examples

We will now provide a few concrete examples in an attempt to illustrate the concept of modelling simulations through artificial chemistries.

5.1 Food Chain

We will define a set of nutrients A, B, C, D. The group of possible molecules S is defined as the set of all possible sequences of nutrients of any size. Agents have an eat action that is represented by a molecule from S. Unidirectional reactions occur when a agent performs an eat action on an object (can be another agent) that contains food molecules. An example of a unidirectional reaction rule could be represented by the following pseudo-code:

```
energy = 0

for each atom a in moleculeA:
    for each atom b in moleculeB:
        match = false
        if a == b:
            add b to moleculeC
            remove b from moleculeB
            match = true
            energy = energy + 1
```

```
    if match == false:
        energy = energy - 1

moleculeB = moleculeC + moleculeB
```

The resulting energy is added to the agent that performs the eat action. Since the resulting energy may be negative, certain food molecules may be poisonous to certain agents. This food chain chemistry allows for the emergence of complex behaviors in the world like predator-prey networks and, in combination with visual tags, mimicry [9].

5.2 Communication

Agents have the ability to broadcast a message at a certain range. Any agent in this range will receive the message. Let's suppose the alphabet is A, B, C, D, E, F, G and S is the set of all possible strings of any size from this alphabet. Actions are defined by a communication molecule and cause this molecule to be broadcasted. Perceptions are also defined by communication molecules and use binding rules. When a message molecule matches a perception molecule, the binding energy is the value of the perception. A binding rule to compute energy might be to compare the atoms in equivalent positions in both molecules and add 1 for a match, 0 for a mismatch and -1 for a size mismatch, and then divide the result by the average of the size of the two molecules. This communication chemistry may allow for the emergence of language between agents. Communication messages may be presented to human observers of the simulation by playing the sequence of musical notes that correspond to the letters in the message.

5.3 Color

Agents, objects or sections of terrain have a color. This color is represented by a color molecule in S. Elements in S are tuples (R, G, B, W), with $R, G, B, A \in [0, 1]$. R, G and B are the red, green and blue components of the color. W is the wetness. The higher the W, the more color will transfer from one object to the other if they collide. Every color molecule in the world will have its W decrease by a certain value on every simulation cycle until it reaches 0. Agents have color perceptions, defined by color molecules and binding rules, where binding energy is calculated by the expression:

$$E_{binding} = \frac{|R_A - R_B| + |G_A - G_B| + |B_A - B_B|}{3} \quad (1)$$

Agents have painting actions. The action is defined by a color molecule. The unidirectional rule states that the target molecule becomes equal to the action molecule, but with a W value of 1.0. The energy expended by this reaction is calculated by applying the binding equation on the action molecule and the target molecule.

A bidirectional reaction rule defines the transference of color between two objects when they collide. Molecule A is transformed by molecule B such that $R_A = R_A + R_B.W_B$, $G_A = G_A + G_B.W_B$ and $B_A = B_A + B_B.W_B$. Equally, molecule B is transformed by molecule A following the same logic.

6 Population Dynamics: Adaptation and Diversity

The purpose of the population dynamics component of the framework we are proposing is to provide both an adaptation mechanism and a way to promote diversity. It is a well known biological concept that the species is the basic unit of evolution at population level [10]. Seeking inspiration at yet another layer of the natural phenomena, this time the biological species level, we will look at modern evolutionary synthesis [11] to define population dynamics algorithms.

For the purpose of this discussion we will adopt Mayr's old definition of species as "groups of actually or potentially interbreeding natural populations which are reproductively isolated from other such groups". [12] There are biological problems with this definition, namely its inability to distinguish species that reproduce asexually and the difficulty in evaluating interbreeding potential in many real world situations. [13] The definition of species is still a problem in modern biology. This one will, however, be sufficient for our purposes.

Modern evolutionary synthesis claims that variation arises by chance through genetic mechanisms like mutation and recombination. Stochastically, these variations give rise to population level effects like genetic drift and natural selection. Speciation occurs when populations become reproductively isolated. Simplistically we could state that adaptation and optimization occur intra-species, while new species are branched out to explore other evolutionary niches.

It is interesting to note at this point that some notable computer science models, namely genetic algorithms, mimic the modern synthesis model in order to establish optimization processes. The classic genetic algorithm is analogous to intra-species adaptation and is designed towards finding the best solution, not innovation or surprise. The classic genetic algorithm is not capable of speciation because there is no process by which reproductive isolation can arise.

Without species, the population will necessarily converge to an optimum or a local optimum. There is no process by which innovation can be maintained in the system. If species exist, however, competition and cooperation between them will result in dynamic equilibriums that may be disrupted, forcing the system or parts of the system into new equilibrium points. This leads to surprise and innovation.

Until now, there has been considerable success in developing algorithms that mimic intra-species optimization, but not speciation. There is one simple explanation for this: more immediate engineering applications exist for optimization. Any optimization that is hard to perform by traditional mathematical analysis is a potential application of genetic algorithms. There are engineering application for speciation besides creating surprise, but they are not so direct or obvious.

One example would be specialized swarm intelligence. Through artificial speciation we may evolve multi-agent systems that operate in complex domains by developing specialized teams.

Another, more interesting reason for lack of speciation in computer algorithms, is that biology itself does not understand speciation as well as it understands intra-species optimization. As stated by Baker: "Speciation thus requires the acquisition of traits (called isolating mechanisms) that prevent gene flow between diverging populations. Isolating mechanisms eliminate the possibility of fruitful sex between otherwise happy mates. In so doing, they would seem to lower organismic fitness. How, then, could such traits ever evolve?". [13]

Biological hypothesis for speciation range from the geographic [14] to the adaptive [15].

It becomes clear that to achieve artificial speciation we have to concern ourselves with two issues: defining a mechanism that creates the possibility of reproductive isolations and promoting the occurrence of such isolations.

6.1 A Reproductive Interface

Again we will resort to the artificial chemistry metaphor to define a reproductive interface that may evolve into reproductive isolation. This is done by providing each agent with a molecule that encodes the reproductive interface. The ability of two agents to reproduce is determined by a binding rule, as discussed in section 3. If this molecule suffers the action of genetic operators, it creates the potential for reproductive isolation.

6.2 Forced Isolation

An inelegant but pragmatic solution to the speciation dilemma is to force arbitrary reproductive isolation between sub-populations. Coupled with the reproductive interface formerly described, and because of genetic drift, this forced isolation will lead to the formation of separate species. This idea is based on geographic isolation occurring in nature, namely the formation of islands. Unlike geographic isolation, agents in different sub-populations are allowed to have non-reproductive interactions.

There are some possible interesting variations of the forced isolation technique, according to the duration of the artificial barriers:

- Sub-populations are defined for the entire duration of the simulation
- Sub-populations are defined only for a certain interval of time during the beginning of the simulation, to promote the emergence of species
- Sub-populations are redefined at certain time intervals

6.3 Testing Biological Hypothesis

By not providing explicit isolation mechanisms, the kind of simulation we are describing can be used to test biological hypothesis on speciation.

6.4 Fixed Population Algorithm

In classical evolutionary algorithms, like genetic algorithms, generations are explicitly defined. Each generation produces the next one in a synchronous fashion. In a continuous simulation like the kind of ones we are considering this approach would be too disruptive. Many approaches are possible to the problem of maintaining evolution in a continuous simulation. We choose an approach that keeps the total population of agents constant. This has the advantage of helping to maintain the computational requirements of the simulation relatively stable, which is important for real time requirements and computational feasibility.

The fixed population algorithm simply places a new agent in the world every time an agent dies, by executing the procedure described by the following pseudocode:

```
do tournament_size times:
    a = random agent from world
    add a to agent_list1

parent1 = agent with highest fitness from agent_list1

for each agent a in the world:
    if distance(a, parent1) < max_reproductive_distance:
        add a to agent_list2

for each agent a in agent_list2:
    if not reproductive_bind(parent1, a):
        remove a from agent_list

if size(agent_list2) > 0:
    parent2 = agent with highest fitness from agent_list2
    child = crossover(parent1, parent2)
else:
    child = clone(parent1)

mutate(child)
place child in world
```

There are several things to notice about the parameterization of this algorithm. The most interesting is, perhaps, the fitness function. The fitness function is a familiar concept in genetic algorithms, and it is used to calculate or estimate the quality of a solution. In our case we will define the fitness function as an estimation of reproductive strength. This estimation is dependent of the world being simulated but some general principles may be established and a general-purpose fitness function proposed.

In most simulations we can use the metaphor of energy to represent and quantify the general resource. We can assume that this resource is gathered, transfered and transformed in all sorts of ways. In equation 2, reproductive strength (RS) is directly proportional to the sum of the energy accumulated by the agent at each simulation cycle ($E(t)$), from time t_0 to the current time, and inversely proportional to the amount of energy needed to create a new agent (E_r).

$$RS(t) = \frac{\sum_{c=t_0}^{t} E(t)}{E_r} \tag{2}$$

This function rewards the agent directly for gathering energy and conserving it, and indirectly for longevity. t_0 is the time of the last reproductive event, or the time the agent was created. The reproductive strength of an agent is thus reset when it achieves reproduction. E_r is the starting energy of the new agent. The higher the E_r the higher the survival chances of the offspring but the lower the probability of successful reproduction. E_r is defined as a genetic trait of the agents. Through evolution, each species will be permanently optimizing E_r. This allows for a diversity of survival strategies to arise.

Another interesting parameter of the algorithm is *max_reproductive_distance*. This parameter determines the radius around which an agent is allowed to look for a mate. We believe that the value of this parameter may influence the amount of geographical speciation that spontaneously occurs in the simulation.

If forced isolation (as discussed in section 6.2) is being used, the algorithm is applied in almost the same way. In this case the agents that participate in the tournament are selected only from the sub-population that the dead agent belonged to. Also, agents that do not belong to this sub-population are excluded from selection for mating as the second parent.

6.5 Genetic Operators

The discussion of genetic operators was kept general in this text. We indicated the use of mutation and recombination but never detailed specific mechanisms. This is because we are proposing a framework where the genotype of agents is heterogeneous. An agent is characterized by:

- A brain
- A set of molecules from different chemistries that represent actions, perceptions and traits
- A set of values for traits that fall outside artificial chemistries

This creates a variety of representation that calls for different genetic operators. In fact, a mutation or recombination operation on an agent will be a combination of specific mutations and recombinations for each aspect of the system.

7 Exceptions to the Model

Although a simulation may be entirely constructed using the concepts we described, we believe that spatial actions and perceptions are best modeled using more conventional techniques. These actions and perceptions find a better analogy in Newtonian physics than in artificial chemistries. The typical spatial perceptions needed are: relative angle, relative position, relative distance and size. The typical spatial actions are rotate and go forward. We describe open ended evolution experiments that use this kind of actions and perceptions in some of our previous work [16] [8]. An excellent example of a use of this kind of actions and perceptions and further discussion on its details can also be found on the Nero Project research papers [17].

8 Final Remarks

This paper constitutes a theoretical proposal. We are aware of the need for experimental validation of the ideas we propose and, in fact, a computer simulation is under development to perform this experimentation [18].

References

1. Wolfram, S.: A New Kind of Science. Wolfram Media, Champaign (2002)
2. Kurzweil, R.: Reflections on Stephen Wolfram's 'A New Kind of Science' `http://www.kurzweilai.net`
3. Ray, T.: Evolution and optimization of digital organisms. Scientific Excellence in Supercomputing: The IBM 1990 Contest Prize Papers, pp. 489–531. The University of Georgia Press, Athens (1991)
4. Adami, C., Brown, C.: Evolutionary Learning in the 2D Artificial Life Systems Avida. In: Brooks, R., Maes, P. (eds.) Proceedings of Artificial Life IV, pp. 377–381. MIT press, Cambridge (1994)
5. Hutton, T.: Evolvable Self-Reproducing Cells in a Two-Dimensional Artificial Chemistry (2007)
6. Holland, J.: Hidden Order - How Adaptation Builds Complexity. Addison-Wesley, Reading (1995)
7. Dittrich, P., Ziegler, J., Banzhaf, W.: Artificial Chemistries - A Review. Artificial Life 7(Issue 3), 225–275 (2001)
8. Menezes, T., Costa, E.: The Gridbrain: an Heterogeneous Network for Open Evolution in 3D Environments. In: Proc. of the The IEEE Symposium on Artificial Life, Honolulu, USA, IEEE Computer Society Press, Los Alamitos (2007)
9. Wickler, W.: Mimicry in plants and animals. McGraw-Hill, New York (1968)
10. Darwin, C.: The origin of species by means of natural selection, 4th edn. J. Murray, London (1866)
11. Huxley, J.: Evolution: The Modern Synthesis. Allen and Unwin, Crows Nest (1942)
12. Mayr, E.: Species Concepts and Definitions. The Species Problem. Amer. Assoc. Adv. Sci. 50 (1957)
13. Baker, J.: Adaptive speciation: the role of natural selection in mechanisms of geographic and non-geographic speciation. Studies in History & Philosophy of Biological & Biomedical Sciences 36(Issue 2), 303–326 (2005)
14. Coyne, J., Orr, H.: Speciation. Sinauer, Sunderland (2004)
15. Dieckmann, U., Doebeli, M., Metz, J., Tautz, D.: Adaptive speciation.
16. Menezes, T., Costa, E.: A First Order Language to Coevolve Agents in Complex Social Simulations. In: Proc. of the European Conference on Complex Systems, University of Oxford (2006)
17. Stanley, K., Bryant. B., Miikkulainen, R.: Evolving Neural Network Agents in the NERO Video Game. In: Proceedings of the IEEE Symposium on Computational Intelligence and Games, Piscataway, USA, Cambridge University Press, Cambridge (2005)
18. Menezes, T.: LabLOVE - Laboratory of Life On a Virtual Environment (2007), `http://sourceforge.net/projects/lablove`

Evolving Virtual Neuronal Morphologies: A Case Study in Genetic L-Systems Programming

Benjamin Torben-Nielsen[1,2]

[1] Maastricht ICT Competence Centre (MICC), Universiteit Maastricht, the Netherlands
[2] Theoretical and Experimental Neurobiology Unit, Okinawa Institute of Science and Technology, Japan

Abstract. Virtual neurons are digitized representations of biological neurons, with an emphasis on their morphology. In previous research we presented a proof of principle of reconstructing virtual neuronal morphologies by means of Genetic L-Systems Programming (GLP) [13]. However, the results were limited due to a hard evolutionary search process and a minimalistic fitness function. In this work we analyzed the search process and optimized the GLP configuration to enhance the search process. In addition, we designed a neuron type-specific fitness function which provides an incremental assessment of the evolved structures. The results are significantly better and relevant issues are discussed.

1 Introduction

Virtual neurons are digitized representations of biological neurons, with an emphasis on their morphology. These neurons are primarily used in computational neuroscience for two reasons. First, they compensate the lack of vast amounts of biological data about neuronal morphology and are used for extensive modeling studies (e.g., [3]). Second, due to their virtual nature the researcher has full control of all morphological parameters which gives extra experimental control. Alternatively, virtual neurons are also proposed in the field of Artificial Intelligence as a biology-inspired way of robotic self-repair [14].

A Lindenmayer system or L-System [11] is a mathematical formalism of rule-rewriting and is successfully applied in a wide variety of applications ranging from modeling of plant morphology [11] and neuronal morphologies (e.g., [3,6]), to the generation of virtual landscapes [10] and creatures [7]. However, due to the recursive nature of L-Systems it is a difficult process to construct L-System descriptions by hand. Therefore, Genetic Algorithms (GAs) were proposed to automatically explore for descriptions according to predefined criteria. The combination of L-Systems and GAs is often referred to as *Genetic L-Systems Programming* (GLP)[8].

In previous work we presented a proof of principle that GLP could be used to search for specific virtual neuronal morphologies [13]. Despite promising initial results the evolved neuronal morphologies were not highly accurate for two main reasons, viz. (i) the evolutionary search process proved to be hard, and, (ii) the morphological criteria were not specific enough for use with GLP. In this paper we present a brief analysis of the exploration process and identified three general problems (that are of particular interest for evolving recursive formalisms): (i) the size of the solution space, (ii)

F. Almeida e Costa et al. (Eds.): ECAL 2007, LNAI 4648, pp. 1089–1099, 2007.

ruggedness of the solution space, and, (iii) a *low* evolvability (i.e., relative number of good solutions in the solution space). We present several enhancement to the exploration process and the fitness function by which the aforementioned problems can be tackled and show that the newly obtained results are quantitatively highly accurate and qualitatively significantly better.

The remainder of this paper is outlined as follows. Section 2 briefly introduces the domain of modeling neuronal morphologies. Section 3 presents the basic experimental setup and a brief analysis of the exploration process before elaborating on the enhanced setup. Section 4 describes the new results and a discussion and conclusion are given in Section 5.

2 Modeling Neuronal Morphology

As illustrated in Figure 1, neurons come in a wide variety of morphologies and sizes; all dedicated to perform a particular function. The function is (largely) dependent on the intrinsic properties (i.e., morphology without active conductances) of neurons [12]. Thus, realistic simulation of neuronal function requires the use of extensive morphologies. A primary goal in modeling of neuronal morphology is finding a minimal representation of neuronal morphology without compromising on natural variability and morphological complexity [2]. Small representations are required for the storage and simulation of large amounts of (virtual) neurons and neural networks.

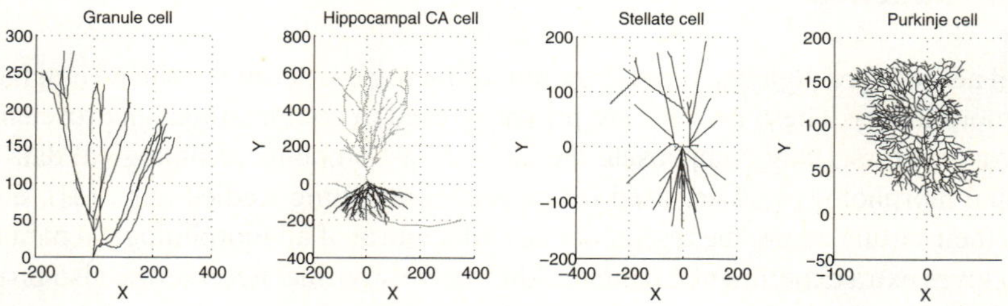

Fig. 1. Different types of neuronal cells. All neurons are taken from a public accessible database [2] (http://NeuroMorpho.Org) and displayed in XY-projection.

Roughly, there exist two approaches to model neuronal morphology: growth models and reconstruction models [15]. Growth models (or mechanistic models) incorporate underlying mechanisms of neuron growth. Identified mechanisms are used as a model to *grow* morphologies from scratch in a biologically plausible way. This method is successful to mimic natural variability of specific patterns in neurons, but they do not grow *full* descriptions including both neuronal topology and geometry. Reconstruction models on the other hand focus on the final product rather than the development and mostly generate full morphological descriptions. Reconstruction models could approximate a morphologies at a specific time step during development, but have no parallel with biological development (as growth models have). We use a reconstruction model and generate full neuronal descriptions from scratch.

3 Methods

3.1 Standard Experimental Setup

GLP employs the L-System formalism to generate structures, and GP to explore for desired descriptions [3]. When applied in the domain of modeling neuronal morphologies L-Systems are used to generate neuronal morphologies and GP guides the exploration process toward desired (i.e., accurate) morphologies.

Generating Morphologies. Modeling of morphologies is done with the most basic type of L-system: D0L-System which is context free and deterministic [11]. The idea of L-Systems is simple yet powerful. An axiom defines the initial starting point and the production rules define how to rewrite the axiom. Both the axioms and the production are defined over a finite alphabet (cf. Section 3.3). Initially, the L-System contains nothing but a string consisting of the axiom. In cycles, this stored string is rewritten as defined by the production rules. This rewriting process is repeated a predefined number of times and results in a long, iteratively built string. Additionally, brackets ("[" and "]") are included as well for branch formation [11]. Below, an example of an L-System.

axiom:	FX
rules:	F → YF
	X → BX
1st cycle	YFBX
2nd cycle	YYFBBX

In this example, the L-System consists of the alphabet $\{B, F, X, Y\}$, a single axiom containing the symbols FX, and two production rules $F \rightarrow YF$ and $X \rightarrow BX$, respectively. In the subsequent rewriting cycles the symbols are then substituted by the content of the appropriate production rule. A L-System in itself is nothing more than a way of generating a long string from a parsimonious description. In essence, L-Systems have no semantics. Semantics are added by a so-called interpretation scheme. Neuronal morphologies are a set of cylinders which are linked to a single parent, and have at maximum two daughters. In our application we limit ourself to the construction of *skeletons*: diameters of the morphologies are not incorporated. As a consequence, an interpretation scheme that allows the generation of line segments in 3D space is sufficient. The Rotation-Elevation scheme is used in our application and is illustrated in Figure 2 (left). In this scheme all directions in 3D can be reached by setting two angles: the rotation angle on a first plane, and the elevation angle on the plane intersecting orthogonally with the first plane.

Exploration. GP is a pragmatic programming method inspired by biological evolution to explore large solution spaces (e.g., [9]). In this study, we explore the set of descriptions to find a description from which an accurate morphology can be generated[4]. GP is

[3] Despite technical differences we use "explore", "search" and "optimize" as synonyms in relation to GP.

[4] In the remainder of this manuscript "description" is used instead of "L-System description".

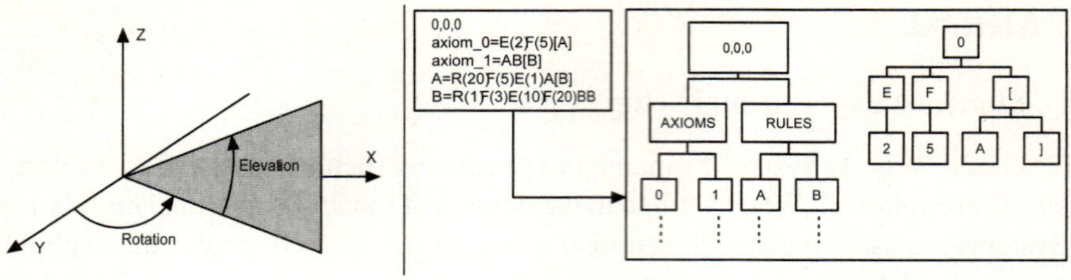

Fig. 2. Left: Rotation-Elevation interpretation scheme. All directions in 3D space can be reached by setting two angles. Right: Encoding of a description as a tree structure. `axiom_0` is fully depicted and illustrates how a complete axiom (or rule) can be encoded as a tree. Both blocks with `0` are the same block, so the right most tree is mounted under `AXIOMS`. The top root of the genome defines the starting position. The symbols are summarized in Table 1.

mainly defined by four characteristics: genome encoding, genetic operators, the genetic parameter configuration, and fitness function. Here we briefly explain these characteristics in the light of our study.

In GLP, the genotype is a L-System description. Hence, the description needs to be encoded in a genome. The hierarchical nature of descriptions (i.e., rules consists of branches, branches consist of symbols, and symbols have arguments) makes a tree-encoding a good encoding. In addition, GP already has a vast amount of genetic operators that work with tree encoding so we do not need to devise any specific operators. Figure 2 (right) illustrates how a description is encoded in a tree structure.

The evolutionary operators define how a new population is created from the previous one. Three steps are associated with the creation of a new population: (i) selection, (ii) crossover, and, (iii) mutation. The selection process determines which individuals are considered in the reproduction process. Here, (μ, λ)-selection [4] is used. Every generation, μ individuals are selected and λ descendants are generated from the selected individuals. The new λ descendants replace the old individuals, and a constant population size is maintained. The crossover is a variant on single-point crossover in conventional GAs: a node (i.e., *point*) is chosen in one parent and the descendant will consist of the nodes of that parent up to that point, and from nodes from the other parent from that node on. The " order" of nodes is defined as depth-first visits. Mutation is implemented in four distinct ways: argument and symbol mutation, and introduction and deletion of sub trees. The genetic parameter configuration specifies the exact numbers for selection and the probabilities with which a genetic operator is chosen. We used a population size of 200 individuals. As an upper bound, the evolution was allowed to run for 1000 generations. In the selection procedure, μ represented the top 65% individuals. Crossover was invoked with a probability of 0.5, new individuals resulting from crossover were not subject to any further mutation. The remainder of the new population was filled by individuals from the previous generation and were subject to mutation: duplication probability of 0.12; deletion probability of 0.10; symbol-mutation probability of 0.05; and argument-mutation probability of 0.1. The fitness function is the most crucial part in this application and assesses the generated morphology on its accuracy compared to a biological morphology (i.e., the prototype). Thus, it is required to parametrize biological morphologies so that a quantitative comparison can be done. In previous work, ten parameters were used to assess the accuracy. The euclidean distance between the biological and generated parameter vectors defined the fitness value [13]. The next sub section briefly presents the challenges arising from this standard setup.

3.2 Analysis of Exploration Process

The complexity of a search problem is generally defined by three properties: (i) the size of the solution space, (ii) the level of smoothness (or ruggedness) of the fitness landscape, and, (iii) the relative number of good solutions (also referred to as "evolvability'). A search problem is considered difficult when the the solution space is large, the fitness landscape rugged, and when there is a relatively small number of solutions present in the solution space. Unfortunately, most GLP applications turn out to have a complex exploration problem.

First, the size or dimensionality of the genotype-space is defined by the number of symbols in the applied L-System alphabet. And, the number of symbols in the alphabet roughly corresponds to the desired complexity and accuracy of the phenotype. For example, in the case of neuronal morphologies, the degree of complexity is rather low (i.e., neurons consists of linked line segments in 3D space), while the desired accuracy is high (i.e., neurons are composed of a high number of small segments and a small change in the base length will have have global consequences). It should be noted that a symbol/argument tuple can be seen as a symbol of the alphabet itself, and thus increasing the size of the alphabet considerably.

Second, the fitness landscapes prove to be highly rugged in the case of L-systems due to its recursive nature. For example, a description that differs in only a single rule call (e.g., extra call to a production rule in an axiom) can have a completely different phenotype. The smoothness of the fitness landscape is also determined by the number of rewriting cycles: more rewriting cycles allow higher phenotypic complexity but also cause a higher degree of ruggedness of the fitness landscape.

Third, as a consequence of the large and rugged solution space it is difficult for the search algorithm to find optimal solutions. Therefore, the evolvability in GLP applications is rather low. Consequentially, one might wonder if GAs is a good technique to use in combination with L-Systems. In unreported tests we compared random (i.e., undirected) search with GAs and GAs still proved very efficient to find basic structures. Thus, the task of the fitness function is then to guide the search process in the best way as possible to desired solutions and hereby overcoming difficulties arising from the inherent recursiveness of L-Systems.

In general, it can be concluded that GLP applications are hard search process. Fortunately, when only one of these three properties can be controlled, the search will be easier. For example, when the solution space is small, it is is of no importance how smooth the fitness landscape is since a random search (or brute-force) can find a solution because the relative number of good solution increases with a decreasing solution space. Or, when the fitness landscape is smooth a GA will ultimately find a optimal solution (under normal circumstances).

3.3 Enhanced Experimental Setup

In this study we enhanced the neuronal modeling application in two ways: the exploration process is made more straight-forward, and the fitness function is updated.

Exploration Process. The basic exploration process outlined before is enhanced by three modifications, namely (i) a decrease in the number of symbols, (ii) *noisy* interpretation of the description, and, (iii) semantic updates.

First, we decreased the size of the solution space by decreasing the alphabet size. Table 1 lists the symbols that are used in the current implementation (first column and gray) and the previous implementation (second column). As mentioned before, we use the Rotation-Elevation interpretation scheme as it only requires two angles to parametrize all directions in 3D (thus, leaving out the "roll" from standard turtle-geometry [1]). However, in analogy to the turtle-geometry interpretation we used separate symbols for left/right and up/down. Currently, we use one symbol for the rotation angle ("R") and one for the elevation angle ("E"). Additionally, the arguments are can take every real number (kept on $[-360, 360]$ with a modulo operation) instead of values on $[0, 180]$, which is convenient for the genetic operators as well. Line segments have positive length (and argument).

Table 1. Current and previous alphabet used in the L-System used for generating neuronal morphologies. Light gray rows indicate that this symbol is used in the current alphabet. More explanation in the text.

New alphabet	Old alphabet	Purpose
$F(x)$	$F(x)$	Draw segment of length x.
$R(x)$	$+(x)$	Rotation left with x degrees.
	$-(x)$	Rotate right with x degrees.
$E(x)$	$/(x)$	Elevate with x degrees.
	$\backslash(x)$	Descent with x degrees.
[and]	[and]	Start a new branch: put current position on the stack. Terminate current branch: pop position from stack.
	$H(x, y)$	Stem direction: rotate x degrees, elevate (or descend) x degrees.
	$ran(x)$	Generate uniform distributed (integer) random on interval $[0, x]$. When $x = 1$, a uniformly distributed double value on $[0, 1]$ is generated.
	$rnd(x, y)$	Generate uniform distributed (integer) random on interval $[x, y]$.

Second, in analogy to the L-System for growing axons described in [6] we had special arguments to generate random values on a fixed interval (i.e., $[x, y]$ or $[0, x]$). In the current implementation the random operators are removed and a *noisy* interpretation is introduced: arguments have a noisy interpretation by sampling their exact value from a Gaussian distribution with $\mu = actual\_argument$ and $\sigma = v$ where v can be specified for every symbol separately. As a consequence, we can also *discretized* the arguments and allow mutation to add or distract a discrete value (in contrast to a random real number). This way, the number of symbols and the size of the argument space are decreased considerably.

Third, the ruggedness of the fitness landscape is caused by small genotypic changes having large phenotypic consequences. The deletion mutation can have such a drastic difference between genotype and phenotype when a production rule is deleted. After such a deletion, syntactic consistency requires to rename all production rules that come (in depth-first order) after the deleted rule. Indeed, after such a deletion a rule call can loose its semantics as it might refer to a different (with the same name after renaming). This effect can be countered by a *semantic update* that also renames the rule calls so that all calls still refer to the original rule (except for the deleted rule). As a result, we can avoid some of the needless ruggedness of the fitness landscape introduced by one of the mutation operators.

Fitness Function. In previous work we succeeded in in optimizing a structure according to predefined morphological criteria. Despite the successful quantitative optimization, the accuracy was low and mainly caused by the limited amount of morphological data in the literature and a lack in compatibility between data originating from different sources [5]. In addition, only a small proportion of evolutionary runs yielded a desired outcome. The small number of successful runs was a direct consequence of the naive and straight-forward fitness function.

In this study we enhanced the fitness function by measuring morphological properties from publicly available reconstructed neurons ourselves, and, devising a new neuron type-specific and incremental fitness function. By measuring the morphological properties ourselves we can avoid some inconsistencies in the available data. We use a set of eleven features that only partially covers the set of ten features in earlier experiments: number of stems (No_stem), percentage of bifurcations (Bif_perc), stem rotation (T_rot) and elevation angle (T_elev), fractal dimension (FD, computed on XY projection), number of bifurcations (No_bif), order, total length (L_tot), branch rotation (B_rot) and elevation (B_elev) angle and finally soma tropism ($TropsimF$). This new set also include extra features required to constrain the algorithm itself. For example, it is generally believed that neurons only have bifurcations. However, also n-furcations can be generated with a L-System specification. Generation of n-furcation must be countered by an extra feature (i.e., Bif_perc) in the fitness function. Contrasting to the previous study, we optimize the features in the order as they are presented above rather than all together. The rationale is that "good begun is half done": initially only one feature is optimized and when this feature has a desired value, the next feature is optimized and so on. Furthermore, the similarity between a prototype and evolved structure is measured by the number of *correct* features instead of the euclidean distance between features. Now this distance value is only used to optimize a single property. In addition to the fitness function itself we also included a heuristic that requires a minimal complexity of the evolved structure. This heuristic defines that a structure must consist of a minimal number of segments before a fitness value will be assigned.

The new fitness function is optimized for finding a specific neuron type: cat spinal cord α motor neurons. The neuron-specificity comes from the chosen features, but especially from the order of optimization. For instance, motor neurons have a specific number of dendrites in uniform directions away from the soma. The distribution of directions has direct consequences for the overall size, and fractal dimension. So, it it of great importance to start with correct stems and these will be optimized first. Consequently, a neuron type with only a single stem (e.g., a Purkinje cell) has different *principal* features and features will be optimized in a different order.

4 Results

Ten evolutionary runs were conducted with a heuristic demanding that structures consisted of at least 300 segments. Eight out of ten runs yielded a structure with a fitness value higher than 80%, which means that almost all features in the optimization criteria are met. However, for a structure to be accurate all features must be in accordance with the predefined criteria. For example, the topology (i.e., branching pattern) of a neuron might be correct but the size can be different. Three structures had a fitness of $> 99\%$ which means that all criteria are met. These three structures are illustrated in Figure 3.

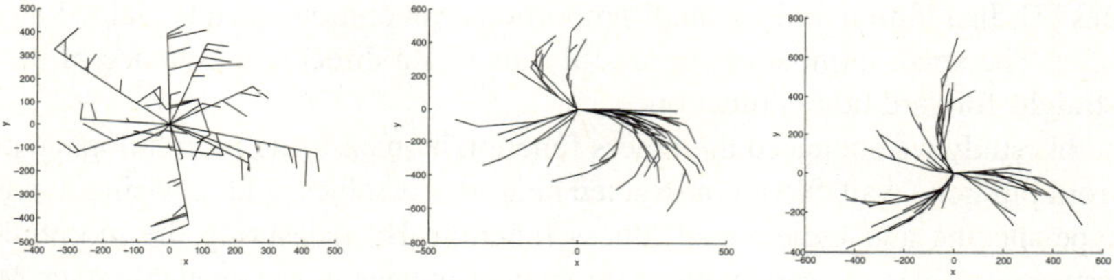

Fig. 3. α-motor neuron. Three generated neurons with +300 segments. All generated neurons obtained a fitness of $> 99\%$.

Compared to the previous result illustrated in Figure 4 (middle) these results are qualitatively better. Nevertheless, the evolved structures still have an algorithmic appearance (i.e., similar repeating patterns occur) and a qualitative difference with the prototype neuron illustrated in Figure 4 (left) is observed. We attributed this difference to a lack of complexity exhibited by the evolved morphologies. Therefore, we performed three runs with a heuristic demanding structures consisting of at least 600 segments. One such run yielded a morphology that obtained a fitness of $> 99\%$ and this morphology is illustrated in Figure 4 (right). Again, this structure is quantitatively in accordance to the predefined criteria. But in addition, it exhibits a high qualitative resemblence to the prototype neuron.

Here, we have to stress that it is hard to map these result onto biological development. L-Systems *construct* morphologies in a different way than actual neuronal development. However, several optimized features represent biological constraints. For example, the contraction observed in dendrites has no functional reason, but is a consequence of being embedded (or packed) in a substrate hosting many other neurons and glial cells.

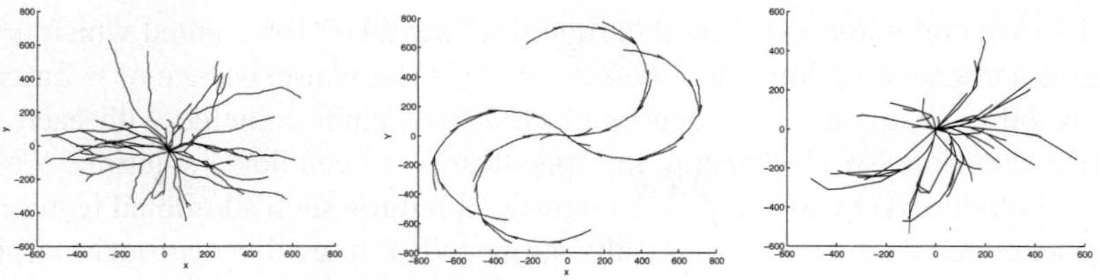

Fig. 4. α-motor neuron. Left: biological prototype neuron. Middle: best estimation of previous method. Right: (qualitatively) best estimation of the enhanced method.

5 Discussion and Conclusion

A brief analysis shows that the exploration process in GLP applications is a difficult one due to (i) the size of the solution space, (ii) the ruggedness of the fitness landscape, and, (iii) low evolvability. However, as one of these three difficulties can be overcome, the exploration process becomes easier. In this paper we showed that insights in the application domain (here neuronal modeling) can overcome some of the difficulties and result in a successful use of GLP. The size of the solution space was reduced by employing a smaller alphabet in combination with more discrete arguments and noise introduction during interpretation of the L-System. A semantic update smoothens the fitness landscape after the mutation operator. And finally, an incremental fitness function guides the exploration process better to good solutions. We believe that these enhancements are general to any domain in which GLP is used.

A semantic update as applied here after a specific mutation operator has a positive effect on the smoothness of the fitness landscape. However, problems arising from the inherent recursive nature of L-Systems cannot be controlled by such superficial updates. The complicated mapping between genotype and phenotype is initiated by the rewriting cycles: less rewriting cycles result in a more direct mapping between genotype and phenotype, while a higher number of cycles give rise to a complex mapping. Therefore, it is an important task to carefully choose the number of rewriting cycles. Potentially, more rewriting cycles can generate more complex structures but have the drawback of a complex genotype-phenotype mapping. In this application, the level of complexity is increased without a higher number of rewriting cycles but by using a heuristic defining a minimum number of segments. Implicitly, this heuristic forces longer rules (and axioms) in the description. This way, the complexity is increased without making the genotype-phenotype mapping unnecessarily complex. Again, we believe that such a heuristic can be generalized to other GLP applications.

In previous work we proved the principle that GLP could be used for evolving neuronal morphologies: we succeeded in optimizing a structure in compliance with predefined criteria. Despite quantitatively desired outcomes, the resulting morphologies were qualitatively highly inaccurate (See Figure 4 (middle)). This came as a surprise as the optimized criteria were generally accepted as descriptors of neuronal morphology of that particular type [13]. Analysis of the result showed that we also needed to include specification specific (here L-systems) criteria instead of only domain specific criteria.

In the neuroanatomy domain several "properties" are taken for granted which are not obvious in automated optimization. One example is the non-existence of n-furcations which. Another example is that dendritic segments cannot coincide with each other while line segments can. In general, the specification of candidate solutions (i.e., encoding of individuals) by means of L-Systems will require such additional features.

We can conclude that we successfully applied GLP to evolve neuronal morphologies. In this particular domain we pointed out that domain knowledge can greatly advance the success of GLP. We introduced three enhancements which can be applied in other application domain. In the future, it would be interesting to investigate to which extend the genetic operators can be constraint by semantics to smoothen the fitness landscape. From a neuroscientific point of view, it would be interesting to see if this framework of exploration for morphologies can be extended to "exploration for functionalities": is it possible to systematically explore morphological building blocks which allow specific computation and functions?

Acknowledgements. The author thanks Karl Tuyls and Eric O. Postma for their contribution to previous versions, and Guido de Croon and three anonymous referees for comments that improved the manuscript. The research reported here is partially sponsored by the Interactive Collaborative Information Systems (ICIS) project, supported by the Dutch Ministry of Economic Affairs, grant nr: BSIK03024.

References

1. Abelson, H., di Sessa, A.A.: Turtle Geometry: The Computer as a Medium for Exploring Mathematics. MIT Press, Cambridge (1981)
2. Ascoli, G.A.: Mobilizing the base of neuroscience data: the case of neuronal morphologies. Nature Neurosc. Rev. 7, 318–324 (2006)
3. Ascoli, G.A., Krichmar, J.L., Scorcioni, R., Nasuto, S.J., Senft, S.L.: Computer generation and quantitative morphometric analysis of virtual neurons. Anat. Embryol. 204, 283–301 (2001)
4. Bäck, T.: Evolutionary Algortihms in Theory and Practice. Oxford University Press, Oxford (1996)
5. Burns, G.A.P.C.: Knowledge management of the neuroscientific literature: the data model of the neuroscholar system. Phil. Trans. R. Soc. 356, 1187–1208 (2001)
6. Feng, N., Ning, G., Zheng, X.: A framework for simulating axon guidance. Neurocomputing 68, 70–84 (2005)
7. Hornby, G.S., Pollack, J.B.: Evolving l-systems to generate virtual creatures. Computers & Graphics 25, 1041–1048 (2001)
8. Jacob, C.: Genetic L-System Programming. In: Davidor, Y., Männer, R., Schwefel, H.-P. (eds.) Parallel Problem Solving from Nature - PPSN III. LNCS, vol. 866, Springer, Heidelberg (1994)
9. Langdon, W.B., Poli, R.: Foundations of Genetic Programming. Springer, Heidelberg (2002)
10. Noser, H.: Lworld: An animation system based on rewriting. In: Pacific Graphics 2002 Proceedings, IEEE Computer Society, Beijing China, 9-11 October, pp. 487–488. IEEE Computer Society Press, Los Alamitos (2002)
11. Prusinkiewicz, P., Lindenmayer, A.: The algorithmic beauty of plants. Springer, Heidelberg (1990)

12. Steuber, V., De Schutter, E., Jaeger, D.: Passive model of neurons in the deep cerebellar nuclei: the effect of reconstruction errors. Neurocomputing 60, 563–568 (2004)
13. Torben-Nielsen, B., Tuyls, K., Postma, E.O.: Shaping realistic neuronal morphologies: An evolutionary computation method. In: International Joint Conference on Neural Networks (IJCNN2006), Vancouver, Canada (2006)
14. Torben-Nielsen, B., Tuyls, K., Postma, E.O.: Towards robotic self-repair by means of neuronal remodelling. In: Adaptation in Artificial and Biological Systems (AISB2006), Bristol, England (2006)
15. van Pelt, J., Schierwagen, A.: Morphological analysis and modeling of neuronal dendrites. Math. Biosc. 188, 147–155 (2004)

Folding Protein-Like Structures with Open L-Systems

Gemma B. Danks, Susan Stepney, and Leo S.D. Caves

York Centre for Complex Systems Analysis,
University of York, York, YO10 5YW, United Kingdom
gbd501@york.ac.uk

Abstract. Proteins, under native conditions, fold to specific 3D structures according to their 1D amino acid sequence, which in turn is defined by the genetic code. The specific shape of a folded protein is a strong indicator of its function in the cell. The mechanisms involved in protein folding are not well understood and predicting the final conformation of a folded protein from its amino acid sequence alone is not yet achievable despite extensive research efforts, both theoretical and experimental. The protein folding process may be viewed as an emergent phenomenon, a result of underlying physics controlling the interaction of amino acids with their local environment, leading to the complex global fold. In this spirit we present a model for investigating protein folding using open L-systems, local rewriting rules with environmental interaction.

Key words: Protein-folding, L-systems, open-L-systems

1 Introduction

In physiological solution, a protein molecule needs only the information contained in its 1D amino acid sequence – a string of typically several hundreds of amino acids of 20 different types in a specific order – to fold to its lowest energy, stable, native state [1]. This specific 3D structure is necessary for the biological function of a protein. In general different sequences fold to different structures and similar sequences fold to similar structures, however there are exceptions where two very different protein sequences share a similar native state. The number of possible conformations of a given sequence is far greater than the number the protein can adopt during folding, indicating that folding is not a random or exhaustive process but follows some pathway(s) [2]. These pathways may be thought of as resulting from the underlying physics of interactions between amino acids in the protein chain. In this sense, protein folding is a paradigm of emergence - the development of well defined global order from a process of self-organised assembly. We present a model for investigating the application of parallel rewriting rules to study protein folding. We use open L-systems with turtle interpretation to model the protein structure and its subsequent folding through application of rewriting rules to local regions of the protein structure over a number of generations leading to global changes in conformation.

F. Almeida e Costa et al. (Eds.): ECAL 2007, LNAI 4648, pp. 1100–1109, 2007.

L-systems were developed as a mathematical theory of plant development [3,4] facilitated by an interpretation based on turtle geometry. L-systems are sets of parallel rewriting rules acting repeatedly on symbols in an initial string, the *axiom*, over a number of *derivation* steps. At each step the string may be interpreted graphically leading to visual models of plant growth and development [4]. L-system rules can be quite flexible and several extensions to L-systems have allowed for more complex models of plant development to be created. The following summarises the extensions we used in this work, for further details and formal definitions see [4,5].

The simplest L-system consists of rules that each rewrite one symbol, called the *predecessor*, with another symbol or string of symbols, called the *successor*, whenever that symbol appears in the string. Context-sensitive L-systems take into account the context of the predecessor - i.e. its neighbouring symbols. Parametric L-systems allow parameters to be assigned to symbols in the string. Conditions on these parameters may then be used in the L-system rules and C-like statements may also be incorporated for further flexibility.

Open L-systems [5] include a separate environmental process, interacting with the L-system via environmental query modules $?E(...)$, in a bi-directional communication process. The environmental program is sent information from the L-system using parameters of the environmental query modules. This information is processed by the environment to determine a response, which is returned to the environmental query modules in the L-system string. The L-system rules can then use this information in the environmental query modules. In this way an open L-system can model a plant interacting with its environment over a number of derivation steps. This has been used to model for example collision avoidance in branching structures and competition for light [5].

2 The Rules of Protein Folding

Up to twenty naturally occurring amino acids can be found in a protein sequence. Each amino acid has a backbone of a central carbon atom, called Cα, attached to an amino group (NH_2), a carboxyl group (COO), a hydrogen and an amino acid specific side chain or R group (see Fig. 1a). Each R group has a distinctive structure and chemical characteristics. The prototypical R group, found in the amino acid alanine, is a methyl group ($-CH_3$), glycine is simpler but a special case as it has a single H atom instead of a side chain. Other side chains vary from long hydrocarbon chains to ring structures or charged groups. Amino acids are linked together, to form *polypeptides*, via a planar peptide bond.

The spontaneous folding of a protein from its *unfolded state* to its lowest energy stable *native state* is driven by physical interactions [6]. The main interactions between atoms thought to drive protein folding consist of the following:

1. Van der Waals forces: attraction and repulsion between atoms, representing general short-range cohesion and excluded volume effects.
2. Electrostatic forces: attraction/repulsion between (partially) charged atoms.

3. Hydrogen bonding: Hydrogen atoms bonded to, and interacting with electronegative atoms (e.g. oxygen/nitrogen) form characteristic spatial interactions. These hydrogen bonds may drive or stabilise the formation of *secondary structure* in a protein – local structural regularities in the protein chain, mainly the α-helix and β-strand. These structures are stable and are the main ordered structural elements occurring in folded proteins.
4. The hydrophobic effect: in aqueous solution, hydrophobic amino-acids tend to pack together at the core of globular proteins, while hydrophilic amino acids tend to be located at the surface.

It is still unclear which of the above interactions are the dominant driving forces in protein folding. Under physiological conditions, all the information a protein needs to fold to its native state is encoded in its amino acid sequence. Different sequences will give rise to different interactions between amino acids in the chain and lead to different native conformations. These sequences have been selected by evolution to fold quickly and spontaneously to stable states [7].

Understanding of the process of protein folding and the accurate prediction of the native state has been the goal of numerous models of protein folding (see [8] for a detailed review). These models range in their level of complexity. The simple 2D lattice HP models [9,10] assume the hydrophobic effect is the driving force and model short proteins, or peptides, as beads of two types - hydrophobic (H) and hydrophilic or polar (P) - on a string while finding the 2D conformation that maximises hydrophobic contacts. Complex all-atom continuous 3D space models calculate forces between each atom pair[8,11].

Our approach is to investigate how underlying local rules, governing the interaction of amino acids with their local environment, can be used to model the process of protein folding as an emergent phenomenon leading to a complex global fold. We have developed a three dimensional model using 20 amino acid types and physics-based open L-system rules that drive the folding of an initial protein conformation. Previous work [12,13] using L-systems to model proteins has focussed on obtaining the native conformation through evolving L-systems rules and an initial axiom that grow the native structure of small (up to 64 residues) proteins under the two-dimensional lattice HP model. Our work focusses on modelling the dynamics of the process of protein folding, rather than on structure prediction. We summarise details of simple L-systems models that have been constructed using different sets of rewriting rules that differ in the level of detail in the representation of physical interactions. For further details see [14]. Both models contain an initial axiom defining a protein sequence in single letter amino acid code. An initial rule set replaces the single letter code of each amino acid with a string of symbols representing the 3D structure of each amino acid type. The string at this stage can be interpreted graphically to give the initial 3D conformation of the protein. Turtle interpretation of the string is used to define the geometrical properties of the system and communicate this to the environment for inter-atomic force calculation. A further rule set is applied over a number of derivation steps to alter the conformation of each amino acid according to physical interactions in its local environment. The repeated reapplication of these rules leads to global folding of the protein.

The first folding rule set uses a simple environmental model to detect collisions between atoms and leads to local conformational changes that depend on the presence of local collisions. This rule set requires knowledge of the direction of folding, i.e. the successor of the rule has to be specified. The second more sophisticated model uses a more realistic model of the physical interactions between atoms. This rule set uses information from an environment that calculates physical forces to determine the direction of folding. The following sections describe these models in more detail.

3 Building Proteins in L-Systems

There are two main variables responsible for the conformation of a chain of amino acids, which are the two backbone torsion angles of each amino acid. The torsion angle ϕ is an angle of rotation around the bond between the backbone nitrogen and Cα. The second torsion angle ψ is similarly an angle of rotation around the bond connecting Cα and the following carbon atom (see Fig. 1a). Other torsion angles are present in side chains but do not directly define the conformation of a protein backbone. Rotations around ϕ and ψ cause a polypeptide chain to alter in conformation. For example, all amino acids in a chain adopting both ϕ and ψ torsion angle values of 180° results in an extended chain (as in Fig. 1a). If consecutive amino acids adopt torsion angles $(\phi, \psi) = (-57°, -47°)$ the result is an α-helix, a stable secondary structure in proteins due to the presence of hydrogen

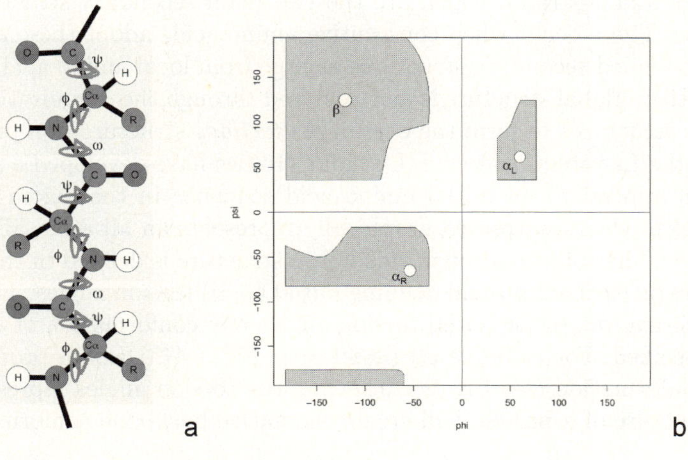

Fig. 1. a. Three amino acids (shaded) linked by peptide bonds with backbone torsion angles shown. The torsion angle ω varies little due to the rigid peptide bond. Variations in torsion angles ϕ and ψ result in different backbone conformations. **b.** A schematic diagram representing a typical Ramachandran plot showing allowed regions of ϕ, ψ space shaded grey. The common secondary structures are shown (β = β-strand, α_R= right-handed α-helix, α_L= left-handed α-helix)

```
Lsystem: alpha-helix

derivation length: 1

Axiom: h    A(-57,-47)A(-57,-47)A(-57,-47)A(-57,-47)A(-57,-47)
            A(-57,-47)A(-57,-47)A(-57,-47)A(-57,-47)A(-57,-47)   F(1.32)o

A(phi,psi) --> F(1.329)n [+(57)Fh]
               -(58.3)\(phi)
               F(1.458)c[+(55.6)&(54.73)Fh] [+(55.6)^(54.73)F(1.521)c /(150)\(-60)
                                                                     [/(60)+(70.5)Fh]
                                                                     [/(180)+(70.5)Fh]
                                                                     [\(60)+(70.5)Fh]]

               -(68.8)\(psi)
               F(1.525)c[+(59.2)F(1.231)o]
               -(63.8)\(180)

homomorphism

c --> ;(83)@O(0.8);(99)
h --> ;(99)@O(0.8);(99)
n --> ;(50)@O(0.8);(99)
s --> ;(115)@O(0.8);(99)
o --> ;(67)@O(0.8);(99)
```

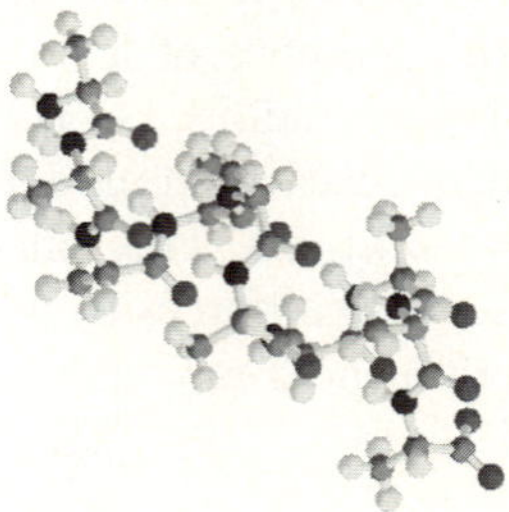

Fig. 2. An L-system to create an α-helix from 10 alanine (A) amino acids with initial torsion angles $(\phi, \psi) = (-57°, -47°)$ defined in the axiom. Each symbol A is replaced by the string in the successor of the rule shown. Graphical interpretation of the string results in an α-helical structure with coloured spheres representing different atom types created for graphical interpretation only (i.e. not rewritten) by *homomorphism rules*

bonds. Not all combinations of possible ϕ, ψ torsion angles are physically possible due to collisions of neighbouring atoms (*steric hindrance*) at some angles. In 1963 Ramachandran et al. examined all possible conformations of two linked peptide units and plotted the resulting *allowed* ϕ, ψ combinations [15]. This plot, known as a Ramachandran, or ϕ, ψ, plot, shows two main regions of allowed ϕ, ψ space (see Fig. 1b). These regions correspond with the torsion angles defining the α-helix and β-strand, which are the two main secondary structures found in proteins. These occur when consecutive amino acids adopt these angles, and so these extended secondary structures emerge from local amino acid conformations. Further, global structure is also achieved through the organisation of these secondary structures to form the overall 3D *tertiary* structure of a protein.

Using the L-systems software 'L-studio' [16] we have developed a set of rules that when applied to an initial amino acid sequence in the axiom leads to a string, which when interpreted graphically represents an all-atom 3D structure of a protein. The conformation of this initial structure is defined in the axiom as parameters on each amino acid defining initial (ϕ, ψ) torsion angles. Through the use of different values for initial torsion angles any conformation of a structure can be specified. For example all $(\phi, \psi) = (-57°, -47°)$ would produce an all α-helix conformation (see Fig. 2). Inserting the torsion angles representing the native structure of a protein will create the native backbone conformation.

4 Folding Proteins in L-Systems

Once an initial 3D conformation of a protein sequence has been created a further rule set in the L-system rewrites the initial torsion angles in each amino acid repeatedly over a number of *derivation* steps. Altering the torsion angles in parallel across the whole chain results in global changes in the protein fold as a consequence of local conformational changes in each individual amino acid.

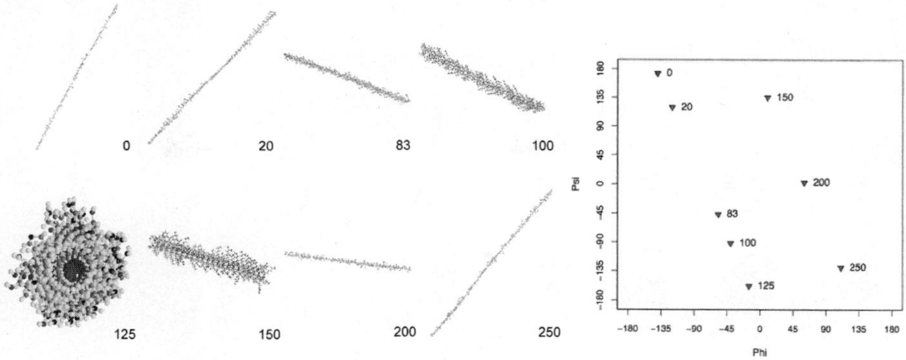

Fig. 3. Derivation steps emerging from a set of L-system rules: $\phi \rightarrow \phi+1, \psi \rightarrow \psi-2.65$ applied to an initial conformation in an extended state of the amino acid sequence of the protein barnase (110 residues). A Ramachandran plot shows the ϕ, ψ angles of every amino acid at corresponding step numbers. These rules cause the folding of the structure to a β-strand conformation at derivation step 20, and an α-helix at step 83. Continuing to apply these rules leads to physically impossible structures e.g. step 125.

4.1 Simple Geometric Model

The rewriting rules in this simplest case are of the format: $\phi \rightarrow \phi \pm \Delta\phi$ where $\Delta\phi$ is a constant value of increment for ϕ and similarly for ψ. This results in a uniform change in local conformation across the whole protein chain leading to ordered changes in global conformation. However, with no restrictions in place on the torsion angles allowed, the structure is free to adopt physically impossible conformations (see Fig. 3) both globally (the entire protein chain may occupy a flattened disk shape) and locally (ϕ, ψ combinations causing overlapping neighbouring atoms within an amino acid). Imposing restrictions to local conformational changes is possible in the L-system but with information limited to being local in the sequence (i.e. individual amino acid torsion angles) regions of the chain that are brought close together spatially but distant in the sequence are not having an effect on the folding. As folding is in 3 dimensions it is important that the local rules are governed by spatially local regions not just regions local in sequence. This requires the use of open-L-systems in the model to communicate with the L-system rules in order to include local spatial information. Two sets of rules were developed with different levels of simplification of the physics involved and are described below.

4.2 Simple Collision Avoidance Model

The first model incorporating open L-systems uses an existing environmental program 'Ecosystem' included in the L-studio software package [17]. At each derivation step this environment is sent the radii and positions of spheres, using environmental query modules ('?E(r)' where r is the radius of a sphere), and

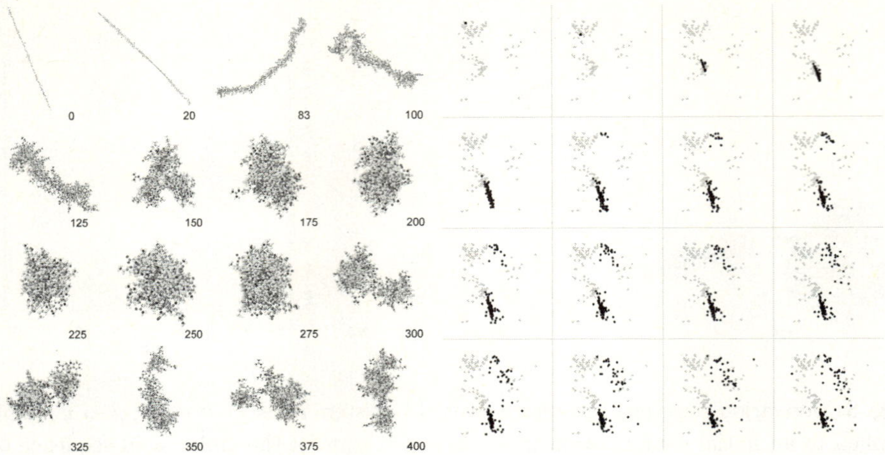

Fig. 4. General features emerging from the L-system using rules as in Fig. 3 but modified such that the sign of the angle increment is reversed with each consecutive local collision. Images show the global changes in conformation obtained using the amino acid sequence of barnase, Ramachandran plots show the ϕ, ψ angles for each amino acid at corresponding derivation steps with the native state angles shown in grey for reference.

detects if any sphere is overlapping any other sphere. This information is returned to the L-system which then incorporates it into context-sensitive, parametric rules with conditions on the parameters of all communication modules in the string local to the torsion angle being rewritten. Therefore the increment of each torsion angle may depend on whether there is a collision between any atom close to the torsion angle and any other atom in the protein. The information on local collisions is used to alter torsion angles in one way if no collisions occur and another if there is a collision. For example, simply reversing the sign of the angle increment on detection of a collision causes the local conformation to back out of its previous move. The effects on the global and local conformational changes in such a model can be seen in Fig. 4. Local conformational changes vary across the chain resulting in complex sequence-dependent 3D global folds.

Due to restrictions imposed in the environmental program and to keep the model as simple as possible all atomic radii were kept equal and much smaller than their actual radii. This leads to problems since local conformations that would produce collisions, were the radii realistic sizes, are then allowed. This causes the L-system structures to adopt less *protein-like* conformations. The environmental program was modified to allow realistic radii, which improved the protein-like nature of the resulting local conformations but it became clear that defining constant angular increments needed to be replaced by a model which incorporates more physics.

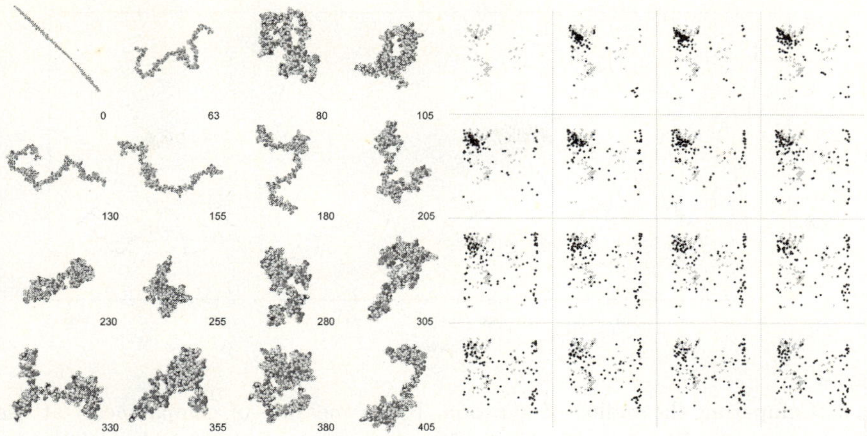

Fig. 5. General features emerging from the L-system using the sum of torque on neighbouring atoms, from the local environment, to increment the angle. The initial state corresponds to a β-strand conformation (as in step 20 in Fig. 4). Images show the global changes in conformation obtained using the amino acid sequence of barnase, Ramachandran plots show the ϕ, ψ angles (black) for each amino acid at corresponding derivation steps with the native state angles shown in grey for reference.

4.3 Physical Forces Model

In a more physical rule set, information on the forces exerted on each atom are returned to the L-system, replacing the simple collision detection. These forces were calculated using a Lennard-Jones potential, to model van der Waals interactions i.e weak attractive forces between distant atoms and strong repulsive forces between very close atoms, and simple Coulombic electrostatics. The following formulae show the forms of the Lennard-Jones potential (left) and the electrostatic potential (right):

$$V(r) = 4\epsilon\left[\left(\frac{\sigma}{r}\right)^{12} - \left(\frac{\sigma}{r}\right)^{6}\right] \qquad V(r) = \frac{q_1 q_2}{r}$$

Where r is the separation distance between two atoms, σ represents the separation distance where the potential is zero and ϵ the energy well-depth, and q_1 and q_2 are the partial charges of two atoms.

In this model, side chain torsion angles are also rewritten. The information (the forces calculated) remains local to the torsion angle. The torque each atom exerts on its nearest rotatable bond is used in the rules to alter each torsion angle by summing the values of torque from nearby atoms. This avoids defining a fixed angular increment and allows the physics to drive the rules. The increments will change at each derivation step due to the application of the rewriting rules altering the locations of atoms. The feedback between the L-system and the environment results in conformational changes following the physics of the model. This necessarily depends on parameters that must be defined for each atom for

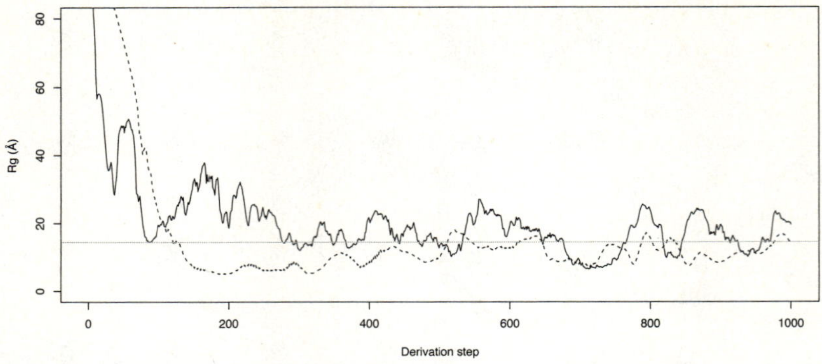

Fig. 6. Comparing the radius of gyration, Rg (a measure of compactness) at each derivation step of the L-system in the simple collision detection model (dashed line) and the forces model (solid line). The grey line shows the value of the native conformation of barnase. Both simulations lead to compact global structures from local rules.

use in calculations of forces. The parameters we use here are taken from the OPLS force field [11] used in atomistic condensed phase molecular simulations. The effects of this more physical set of rules are shown in Fig. 5.

Although neither model produces native-like conformations (plots in Figs. 4 and 5) both produce protein-like conformations as measured by the compactness of global conformations (globular proteins generally adopt compact native conformations) (Fig. 6). The use of a more physical rule set results in folding that is not forced by defining rules that drive the simulation to fold the protein in a predefined direction (as in the simple collision detection model), instead the successor of each rule, i.e. the increment of each torsion angle, is dependent on the local physical forces that change at each derivation step. The local conformations in this model are also more protein-like as seen by the angles adopted by each amino acid (Fig. 5) when compared with the allowed regions of a typical Ramachandran plot (Fig. 1).

5 Summary

The L-systems models we have presented show that incorporating even very simple collision detection produce complex global conformations that are also sequence dependent. This comes from inclusion of the environment. Protein folding is a problem of translating a 1D code to a 3D structure where the process is driven by physical rules. Replacing simple collision detection with physical forces prevents restrictions imposed by defining the rule successors. This allows folding to be governed by local physics of the environment and leads to more protein-like features. These features include characteristic local conformations shown by more realistic trajectories through ϕ, ψ space and compact global conformations.

The approach of using local rewriting rules has so far given interesting results as proof of concept. Our next step is to analyse the behaviour of the models both in terms of the characteristics of their trajectories and resulting structures. The models may also be developed further to allow the other driving forces in protein folding - hydrophobic interactions and hydrogen bonding - to be additionally incorporated into the rule sets. The goal of this study is to try to discover to what extent protein folding may be modelled in terms of physical locally-determined conformational changes.

Acknowledgments. This work is supported by the BBSRC.

References

1. Anfinsen, C.B.: Principles that govern the folding of protein chains. Science 181, 223–239 (1973)
2. Zwanzig, R., Szabo, A., Bagchi, B.: Levinthal's paradox. Proc. Natl. Acad. Sci. USA 89, 20–22 (1992)
3. Lindenmayer, A.: Mathematical models for cellular interactions in development. Parts I and II. J. Theor. Biol. 18, 280–315 (1968)
4. Prusinkiewicz, P., Lindenmayer, A.: The Algorithmic Beauty of Plants. Springer, New York (1990)
5. Mech, R., Prusinkiewicz, P.: Visual Models of Plants Interacting with Their Environment. In: Proceedings of SIGGRAPH 96, pp. 397–410 (1996)
6. Dill, K.A.: Dominant forces in protein folding. Biochemistry 29, 7133–7155 (1990)
7. Onuchic, J.N., Wolynes, P.G.: Theory of protein folding. Curr. Opin. Struct. Biol. 14, 70–75 (2004)
8. Mirny, L., Shakhnovich, E.: Protein folding theory: From lattice to all-atom models. Annu. Rev. Biophys. Biomol. Struct. 30, 361–396 (2001)
9. Lau, K.F., Dill, K.A.: A Lattice Statistical-Mechanics Model of the Conformational and Sequence-Spaces of Proteins. Macromolecules 22, 3986–3997 (1989)
10. Dill, K.A., Bromberg, S., Yue, K.Z., Fiebig, K.M., Yee, D.P., Thomas, P.D., Chan, H.S.: Principles of protein-folding - a perspective from simple exact models. Protein Sci. 4, 561–602 (1995)
11. Jorgensen, W.L., Maxwell, D.S., Tirado-Rives, J.: Development and testing of the OPLS All-Atom force field on conformational energetics and properties of organic liquids. J. Am. Chem. Soc. 118, 11225–11236 (1996)
12. Escuela, G., Ochoa, G., Krasnogor, N.: Evolving L-systems to capture protein structure native conformations. In: Keijzer, M., Tettamanzi, A.G.B., Collet, P., van Hemert, J.I., Tomassini, M. (eds.) EuroGP 2005. LNCS, vol. 3447, pp. 73–83. Springer, Heidelberg (2005)
13. Ochoa, G., Escuela, G., Krasnogor, N.: Incorporating knowledge of secondary structures in a L-system-based encoding for protein folding. In: Talbi, E.-G., Liardet, P., Collet, P., Lutton, E., Schoenauer, M. (eds.) EA 2005. LNCS, vol. 3871, pp. 247–258. Springer, Heidelberg (2006)
14. Danks, G.B., Stepney, S., Caves, L.S.D.: Protein folding with L-systems: encoding the problem (In prep.)
15. Ramachandran, G.N., Ramakrishnan, C., Sasisekharan, V.: Stereochemistry of polypeptide chain configurations. J. Mol. Biol. 7, 95–99 (1963)
16. Prusinkiewicz, P., Karwowski, R., Mech, R., Hanan, J.: L-Studio/cpfg: A Software System for Modeling Plants. In: Münch, M., Nagl, M. (eds.) AGTIVE 1999. LNCS, vol. 1779, pp. 457–464. Springer, Heidelberg (2000)
17. Mech, R., Prusinkiewicz, P.: User's Manual for Environmental programs (1998)

Formal Model of Embodiment on Abstract Systems: From Hierarchy to Heterarchy

Kohei Nakajima[1], Soya Shinkai[2], and Takashi Ikegami[1]

[1] Department of General Systems Sciences, The Graduate School of Arts and Sciences, University of Tokyo 3-8-1 Komaba, Tokyo 153-8902, Japan
nakajima@bio.c.u-tokyo.ac.jp

[2] Department of Applied Physics, Faculty of Science and Engineering, Waseda University, Tokyo 169-8555, Japan

Abstract. An embodiment of a simple system, such as a one-dimensional map system, derived from heterarchical duality is discussed. We formalized two pairs of heterarchical layers induced by the indefiniteness of the environment and inconsistency between parts and wholeness by using category theory and applied its construction to a logistic map. From the analysis of its behavior, we universally observed $1/f$ spectrum for orbits and the fractal-like behavior in the dynamics of return maps. For the coupling map system, the parameter region with an on-off intermittency was clearly extended. Finally, we discuss the relationship between this model and the recent interest in morphological computations and search for a way to deal theoretically with the concept of adaptability.

Key words: Internal measurement, Heterarchy, One-dimensional Map

1 Introduction

Understanding adaptability has been one of the most challengeable topics in the field of biological system theoretical research. Generally, to deal with the adaptability of the system, one prepares a well-defined system and analyzes its interactions with its outside environment. When we are eager to construct a system that is thought to be adaptive, or to improve the system to be adaptive, the system should have a certain construction added or be designed to function in a specific way. If the observer regards the system as adaptive at this stage, then the added constructions or specially designed functions are thought to be the reason for the adaptability; if it is not, then a newly up-dated and well-defined system is needed. This scheme continues infinitely, if we take into account that an environment is inherently indefinite. In this case, adaptability itself has not been understood theoretically, because the environment is pushed outside of the system. This means that the concept of adaptability is not one of the properties of a particular system; rather, it lies between the system and its environment that assumes the specified rules. Recently, especially in the fields of robotics and artificial life, there broadly exists the study to understand the interface between the environment and the system and/or the interface between human and

F. Almeida e Costa et al. (Eds.): ECAL 2007, LNAI 4648, pp. 1110–1119, 2007.

machine, as creating an autonomous agent that carries a sensorimotor coupling system [1,2,3] or studying the co-creation of the synchronous phenomena that exist between human and machine [4]. Morphological computations are also proposed to deal with the system that simultaneously interacts with its environment and its internal function. These studies are motivated to search for a way to deal with an indefinite environment.

On the other hand, in facing with these problems, Conrad M [5], Matsuno K [6] and Gunji Y-P [7,8] insisted that to understand adaptive systems, it is essential to consider the interaction between the hierarchy of parts and wholeness and proposed the concept of evolutionary computing and internal measurement respectively. Within those concepts, the concept of stability, which was naively assumed to be the main property of adaptability, is replaced by the concept of robustness. These concepts are very much similar to those of heterarchy, which has recently been relightened. Heterarchy is the concept originally proposed by McCulloch [9] to reveal the indefinite hierarchical system; it was defined as the interaction of the hierarchy, which has latent evolvability. The concept of heterarchy came into currency in the fields of autopoiesis [10], sociology and economics, but its strict formalization had not yet been realized. Recently, however, starting with the relightening of the concept of heterarchy by Jen [11] in 2003, Gunji Y-P and Kamiura M [12,13,14] extended the concept and formalized its construction by using category theory; they called their system "observational heterarchy", and they then created a model on dynamical systems which revealed intermittent orbits. We think that it is very important to understand the adaptability of a system in this way. In this paper, we use the term 'embodiment' in a more abstract sense than usual: it is more like the 'material cause' in Aristotle's four causes (material, formal, efficient and final) [15]. From the recent interest in morphological changes to a system, we think that we should scope into the kind of perturbation not only to a state, $x+\delta x$, but also to a function, $f+\delta f$, derived from the environment [12,16]. Only by expanding our view to this stance, can we deal with the adaptive mutation of the system - which is to say, changes to the structure of the system itself, - in an abstract or theoretical sense.

This paper has mainly two aims. One is to provide an example of the application of "observational heterarchy" to a simple system, such as one-dimensional maps. The second is to analyze its behavior and search for a way to understand adaptability itself, on a theoretical basis. In the next section, we formalize the construction of heterarchical duality generalized on one-dimensional maps by using category theory [17]; in section 3, we apply it to the logistic map and its coupling system, and analyze its behavior. In section 4, we discuss the relationship between the system and its adaptability.

2 Formalizing Observational Heterarchy on One-Dimensional Maps

To construct a heterarchy, at least two layers are required. In this article, we will induce it from an inconsistency between Intent and Extent expressions.

For example, if we consider cells, the interaction driven by chemical reactions in a cell, intra cellular behaviors, and the interaction between cells with such reactions, inter cellular behaviors, are simultaneously observed. Here with the same phenomena we have two descriptions, intra cellular behaviors correspond to an Extent-expression, and inter cellular behaviors correspond to an Intent-expression. As is well known, however, the velocities of particles at these levels cannot be assumed equal. Because intra cellular behaviors are defined for particles such as c-AMP and inter cellular behaviors are defined for cells, to describe this phenomenon theoretically, the structure always accompanies the mixture of an Extent- and Intent-expression. However, such a mixture obviously results in a logical inconsistency. In constructing a heterarchy, it is most important how one deals with this inconsistency.

In this section, we formalize the structure of the heterarchy on one-dimensional maps and additional constructions to prepare for the coupled map system. First, we formalize the consistent relationship between Extent and Intent expressions by introducing an adjunction of the map and, moreover, of two comma categories, and define two pairs of layers. We then show the observational interaction between each pair of layers and define the observational heterarchy that is universally observed.

2.1 Consistent Structures of Adjunction on One-Dimensional Maps

In category theory [17], for maps from an object X to X, in this case one-dimensional maps, the following relationship is defined as an 'adjunction': The map $\psi : X \times A \to X$ corresponds to the map $\phi : A \to X^X$, i.e., $\phi(a) = \psi_a$ for any $a \in A$, where A is the set of the system parameter and the map ψ_a is defined as $\psi_a(x) = \psi(a, x)$ for any $a \in A$, $x \in X$. Hence $\phi(a)(x) = \psi(x, a)$ for any $a \in A$, $x \in X$. And an adjunction is expressed by

$$Hom(X \times A, X) \simeq Hom(A, X^X). \tag{1}$$

Moreover, we here are concerned with the category of topological space **Top**. At this point, ψ belongs to the subcategory Extent **Ext** from the viewpoint that all input data interact with the parameter and they are changed into next states, while ϕ belongs to the subcategory Intent **Int** from the viewpoint that the selected parameter a controls the maps from X to X. Then we can rewrite Eq. (1) as

$$\mathbf{Ext}(X \times A, X) \simeq \mathbf{Int}(A, X^X). \tag{2}$$

We can express this situation by using two adjoint functors $Hom(X, -)$ and $X \times (-)$. In this case, $Hom(X, -)$ transforms each set B to a set as $Hom(X, B)$, and each map $p : B \to C$ to a map $Hom(X, p) : Hom(X, B) \to Hom(X, C)$ such that for a map $m \in Hom(X, B)$, $Hom(X, p)(m) = pm \in Hom(X, C)$. The functor $X \times (-)$ transforms a set B to a product set $X \times B$, and a map $p : B \to C$ to a product map, such as $id_X \times p$ where $(id_X \times p)(x, b) = (x, p(b))$. A functor is defined so as to satisfy the law of preservation of composition, namely, composition of applying a functor to a map is equivalent to applying a functor to the composition. Actually, two functors, $Hom(X,-)$ and $X \times (-)$ satisfy the condition. Given $p : B \to C$ and $q : C \to D$, for all $m \in Hom(X, B)$, $Hom(X, qp)(m) = qpm = Hom(X, q)(pm) = Hom(X, q)Hom(X, p)(m)$. As for $X \times (-)$, one obtains that $X \times qp = (X \times q)(X \times p)$.

By using those notations, equation $\phi(a)(x) = \psi(x, a)$ is expressed as $ev(id_X \times \phi)(x, a) = \psi(x, a)$. In this formulation, a universal map, $ev : X \times Hom(X, X) \to X$ such that $ev(x, p) = p(x)$ is introduced. Diagrammatically it is shown as

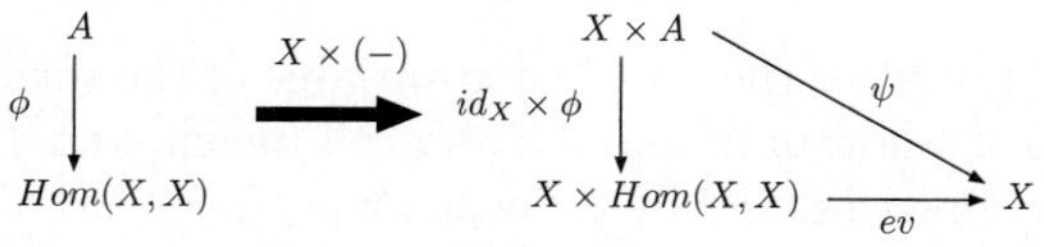

Dually, there is another universal map $\eta : A \to Hom(X, X \times A)$ such that $\eta(a)(x) = (x, a)$. By the universal map η, we obtain that $Hom(X, \psi)\eta(a) = \phi(a)$. It also reveals the $Hom(X, \psi)\eta(a)(x) = \psi\eta(a)(x) = \psi(x, a) = \phi(a)(x)$ as mentioned in equation $\phi(a)(x) = \psi(x, a)$. Diagrammatically it is shown as

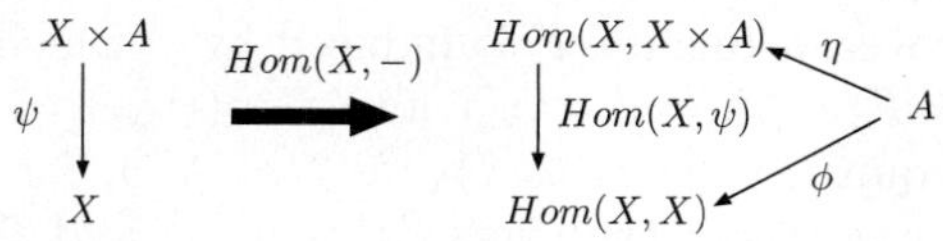

As shown here, we can understand the consistency between Extent and Intent expressions through two adjoint functors.

Next, according to adjunction Eq. (1) and topology category **Top**, it is trivial that adjoint functors can be replaced by the following maps, which are four maps $\alpha : Hom(X, X \times A) \to X \times A$ and $\beta : X \to Hom(X, X)$ and $\alpha^{-1} : X \times A \to A$ and $\beta^{-1} : Hom(X, X) \to X \times Hom(X, X)$ such that $\phi = \beta\psi\alpha\eta$ and $\psi = ev\beta^{-1}\phi\alpha^{-1}$ are simultaneously satisfied, in other words, there exists a bijective map $\theta : Hom(X \times A, A) \to Hom(A, X^X)$ such that $\theta(\psi) = \phi$ and $\theta^{-1}(\phi) = \psi$, i.e., adjunction Eq. (1) is satisfied. Then we can construct two layers. One is $\mathcal{L}1 \equiv \langle \mathcal{C}, \theta \rangle$ and the other is $\mathcal{L}2 \equiv \langle Ext, Int, Hom(X,-), X \times (-) \rangle$ as shown in Fig. 1.

Next, to prepare for the coupled map system, we should construct additional Intent and Extent subcategories induced by a transportation function, and this time, the Extent subcategory which we have constructed above is replaced by a new Intent subcategory because of the following reason. For example, if we assume that the map simulates the dynamics of the population of the substrate in a cell, generally we prepare a transportation function to express the population of the substrate which goes out and/or goes into the cell and describe as

$$x^i_{t+1} = (1 - c)\psi(x^i_t) + f(\psi(x^j_t)) \quad \text{(with } i, j = 0, 1\text{)}, \tag{3}$$

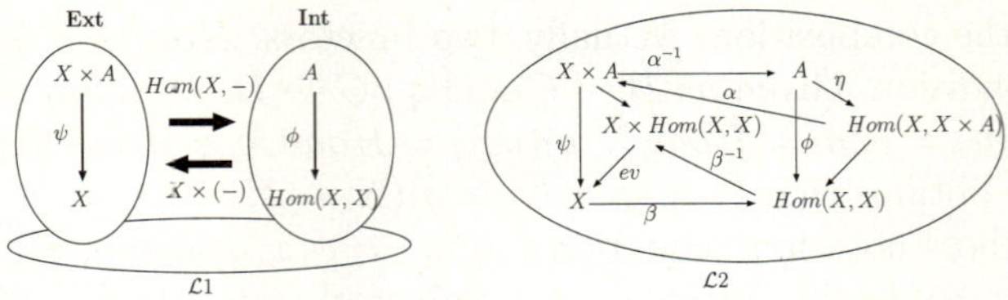

Fig. 1. Layers ($\mathcal{L}1$, $\mathcal{L}2$) with respect to Extent and Intent subcategories.

where $x_t^i \in [0,1]$ stands for the population variable of the substrate in a cell. The map f represents the flow of materials from cell to cell, and $f(x)$ is defined by a map that can be approximated by a linear map $f(x) \sim cx$. This means that we should consider the large groups and the small groups of the population of the substrate, respectively.

We here introduce a comma category[3] to express a new Intent subcategory which consists of the large groups and a new Extent subcategory which consists of the small groups. Let us consider comma categories $\mathcal{C}/X$ and $\mathcal{C}/Y$. When the transportation function $f : X \to Y$ in Eq. (3) is approximated by a linear function $f(x) \sim cx$, f is bijective. Then, for $\mathcal{C}/X \supseteq \mathcal{D}_X$ whose object is $\psi_{int} : X \times A \to X$, f induces a subcategory $\mathcal{D}_Y \subseteq \mathcal{C}/Y$ whose object is $\psi_{ext} = f(\psi_{int}) : X \times A \to Y$. Moreover, a functor F is induced by f which satisfies one-to-one mapping such that $F(\psi_{int}) = f \circ \psi_{int}$ and $F^{-1}(\psi_{ext}) = f^{-1} \circ \psi_{ext}$ and two subcategories are equivalent such as $\mathcal{D}_X \simeq \mathcal{D}_Y$. And, for a category $\mathcal{C}$ and a map $f \in \mathcal{C}$, we can construct new layers $\mathcal{L}1' \equiv \langle \mathcal{C}, f \rangle$ and $\mathcal{L}2' \equiv \langle \mathcal{D}_X, \mathcal{D}_Y, F \rangle$.

2.2 Constructions of Heterarchical Duality Induced by Inconsistency of the Layers

Here we introduce an observational heterarchy by taking account of the case when one-to-one correspondence, which guarantees the bijection α, β and f, is not satisfied. In that case, we need to define a pre-functor and a pseudo-inverse map [12,13].

We express observation of the internal perspective by the following equivalent-like operation pairs $(\langle Hom(X,-)\rangle, \langle X \times (-)\rangle)$ on the layer $\mathcal{L}1$ and $(\tilde{F}, \tilde{F}^*)$ on the layer $\mathcal{L}1'$: Given maps $\tilde{\alpha}_t : Hom(X, X \times A) \to X \times A$, $\tilde{\beta}_t : X \to Hom(X,X)$ and $\tilde{f} : X \to Y$ that are not monic but epic and are permitted approximation $\tilde{\alpha}_t \simeq \alpha$, $\tilde{\beta}_t \simeq \beta$ and $\tilde{f} \simeq f$, where α, β and f are bijective[4]. Contrary to bijection of these maps, the maps which satisfy the following properties are defined as 'pseudo-inverse maps' for $\tilde{\alpha}_t, \tilde{\beta}_t, \tilde{f}_t$; $\tilde{\alpha}_t^* : X \times A \to A$, $\tilde{\beta}_t^* : Hom(X,X) \to X \times Hom(X,X)$, $\tilde{f}_t^* : X \to X$, $\tilde{\alpha}_t \eta \tilde{\alpha}_t^* = id_{X \times A}$, $\eta \tilde{\alpha}_t^* \tilde{\alpha} \neq id_{Hom(X, X\times A)}$, $\tilde{\beta}_t ev \tilde{\beta}_t^* =$

[3] The comma category $\mathcal{C}/C$ has: objects, all arrows $f \in \mathcal{C}$ such that $cod(f) = C$, and arrows, g from $f : X \to C$ to $f' : X' \to C$ is an arrow $g : X \to X'$ in $\mathcal{C}$ such that $f' \circ g = f$.

[4] As show in previous subsection, bijection implies that two comma subcategories are equivalent.

$id_{Hom(X \times X)}$, $ev\tilde{\beta}_t^* \tilde{\beta}_t \neq id_X$, where N is a set of infinite natural numbers. Then these pseudo-inverse maps are not unique for $\tilde{\alpha}_t$, $\tilde{\beta}_t$ and $\tilde{f}$ and there are families of the arrows $\{\tilde{\alpha}^*_{(t)}\}_{t \in N}$, $\{\tilde{\beta}^*_{(t)}\}_{t \in N}$ and $\{\tilde{f}^*_{(t)}\}_{t \in N}$, and we can obtain pre-functors $\langle Hom(X, -)\rangle : Ext \rightarrow Int$, $\langle X \times (-)\rangle_{(t)} : Int \rightarrow Ext$, $\tilde{F} : \mathcal{D}_X \rightarrow \mathcal{D}_Y$ and $\tilde{F}^*_{(t)} : \mathcal{D}_Y \rightarrow \mathcal{D}_X$ such that the identity functors $X \times (-) \circ Hom(X, -) = Ext$ and $F^{-1} \circ F = Id_{\mathcal{D}_X}$ are replaced with $\langle X \times (-)\rangle_{(t)} \circ \langle Hom(X, -)\rangle : Ext \rightarrow Ext_{(t)}$ and $\tilde{F}^*_{(t)} \circ \tilde{F} : \mathcal{D}_X \rightarrow \mathcal{D}_{X_{(t)}}$ that can be regarded as operators of time evolution. Therefore, the observation of the agent with the internal perspective makes the agent itself change.

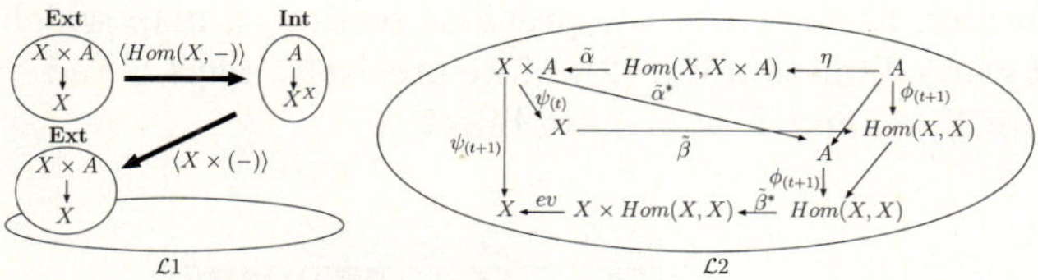

Fig. 2. The dynamics on layers ($\mathcal{L}1$, $\mathcal{L}2$) by pre-functors $\langle Hom(X, -)\rangle$ and $\langle X \times (-)\rangle$

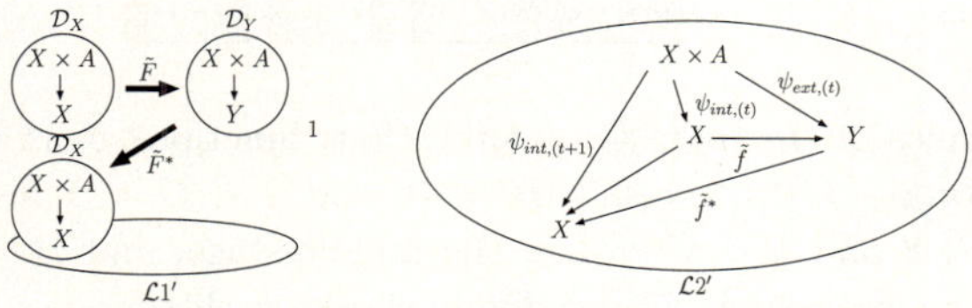

Fig. 3. The dynamics on layers ($\mathcal{L}1'$, $\mathcal{L}2'$) by pre-functors $\tilde{F}$ and $\tilde{F}^*$.

Then, given the initial $\psi^i_{(0)}$ at $t = 0$, the following dynamics are succeeded:

$$\phi_{(1)} = \tilde{\beta}_0 \psi_{(0)} \tilde{\alpha}_0 \eta \tag{4}$$

$$\psi_{int,(0)} = ev\tilde{\beta}_0^* \phi_{(1)} \tilde{\alpha}_0^* = ev\tilde{\beta}_0^* \tilde{\beta}_0 \psi_{(0)} \tilde{\alpha}_0 \eta \tilde{\alpha}_0^* = ev\tilde{\beta}_0^* \tilde{\beta}_0 \psi_{(0)} \tag{5}$$

$$\psi_{ext,(0)} = \tilde{f} \psi_{int,(0)} = \tilde{f} ev\tilde{\beta}_0^* \tilde{\beta}_0 \psi_{(0)} \tag{6}$$

$$\psi_{(1)} = \tilde{f}^* \psi_{ext,(0)} = \tilde{f}^* \tilde{f} ev\tilde{\beta}_0^* \tilde{\beta}_0 \psi_{(0)} \tag{7}$$

In the following section, we consider η and ev which have a universal mapping property with an identity arrow because of the existence of an exponential in the topological space **Top**.

3 Models and Results

In this section, firstly, we apply the formalization on the layer $(\mathcal{L}1, \mathcal{L}2)$ to the logistic map $x_{t+1} = ax_t(1 - x_t)$, where $a \in [0, 4] = A$, $x_t \in [0, 1] = X$ and construct the embodied logistic map (ELM). Next a coupled map system with ELM called by 'active coupling of embodied logistic map' (ACELM) is constructed on the pairs of layers $(\mathcal{L}1, \mathcal{L}2)$ and $(\mathcal{L}1', \mathcal{L}2')$.

3.1 Embodied Logistic Map

Usually, for a selected parameter $a \in A$ and an initial state x_0, $x_1 = h_a(x_0)$, $x_2 = h_a(x_1) = h_a(h_a(x_0))$, $\cdots$ are obtained recursively, where $h_a : X \to X; x \mapsto ax(1 - x)$. In fact, as shown in the previous section, a map which we iterate is guaranteed by an adjunction Eq. (2). More precisely, maps α and β which induce functors $Hom(X, -)$ and $X \times (-)$ are bijective.

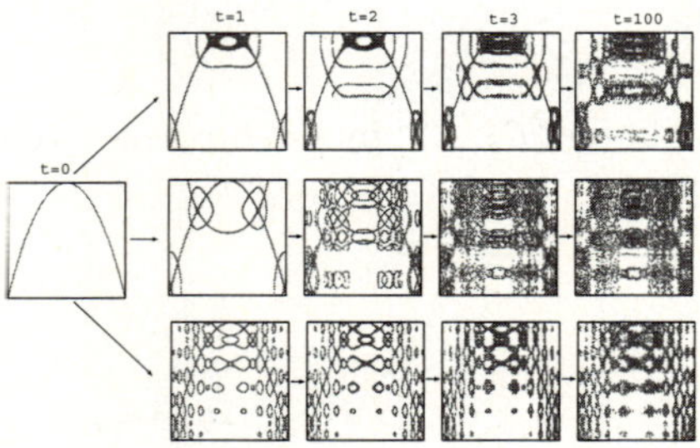

Fig. 4. The dynamics of the map $\psi_{(t),a}$ for various functions $\tilde{\beta}_t$ in ELM. Its form is $\tilde{\beta}_t = x + A\sin(B\pi(x - C))$. $A = 0.3$, $B = 10$ and $0 < C < 1$ in the upper-stage, $A = 0.3$, $0 < B < 8$ and $0 < C < 1$ in the middle-stage, and $B = 20$, $C = 0$ and $0.1 < A < 0.4$ in the lower-stage are randomly chosen each time t.

Here we apply observational heterarchy by using maps $\tilde{\alpha}_t$ and $\tilde{\beta}_t$ which are not monic but epic. Then from Eq. (5), the dynamics of x_t is defined by

$$x_{t+1} = \tilde{\beta}_t^* \tilde{\beta}_t \cdots \tilde{\beta}_0^* \tilde{\beta}_0 \psi_{(0)}(x_t, a). \tag{8}$$

Figure 4 shows the typical dynamics of the map $\psi_{(0),a}$, which is defined by $\psi_{(t),a}(x) = \psi_{(t)}(x, a)$. When the map $\tilde{\beta}_t$ is defined by

$$\tilde{\beta}(z) = \begin{cases} c_2(t)/b_1(t)\, z & \text{for } z \in [0, b_1(t)), \\ (c_1(t) - c_2(t))/(b_2(t) - b_1(t))\,(z - b_1(t)) + c_2(t) & \text{for } z \in [b_1(t), b_2(t)), \\ (1 - c_1(t))/(1 - b_2(t))\,(z - b_2(t)) + c_1(t) & \text{for } z \in [b_2(t), 1], \end{cases} \tag{9}$$

where $b_1(t), b_2(t), c_1(t), c_2(t) \in [0, 1]$ are randomly chosen and inequalities $0 < b_1(t) < b_2(t) < 1$, $0 < c_1(t) < c_2(t) < 1$, $b_1(t) < c_2(t)$, $c_1(t) < b_2(t)$ are satisfied, the orbit and the dynamics of the map $\psi_{(t),a}$ are shown in Fig. 5. In this restricted case, we have obtained two significant features. One is $1/f$ spectrum with respect to the power spectrum of orbits. The other is that the behavior of the return map $\psi_{(t),a}$ has fractal-like one for ELM when t goes large enough.

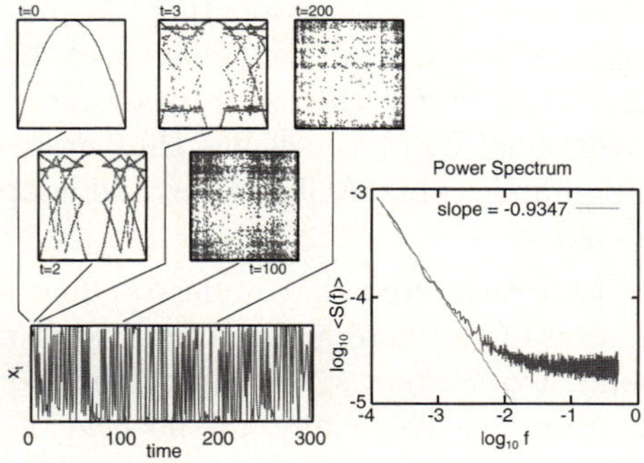

Fig. 5. The typical orbit and the map $\psi_{(t),a}$ in the case that $\tilde{\beta}_t$ is given by Eq. (9). Then the power spectrum of orbits reveals $1/f^{-\nu}$ spectrum ($\nu = 0.9347$ in our simulations), where the length of orbits is 2^{14} and the number of orbits is 10^4.

3.2 Active Coupling of Embodied Logistic Map

Here we construct active coupling of embodied logistic map (ACELM) in the previous subsection. From Eqs.(4)∼(8), we can rewrite Eq.(3) as follows:

$$
\begin{aligned}
x_{t+1}^i &= (1-c)\psi_{int,(t)}(x_t^i, a) + \psi_{ext,(t)}(x_t^j, a) \\
&= (1-c)\tilde{\beta}_t^* \tilde{\beta}_t \tilde{f}^* \tilde{f} \tilde{\beta}_{t+1}^* \tilde{\beta}_{t+1} \cdots \tilde{f}^* \tilde{f} \tilde{\beta}_0^* \tilde{\beta}_0 \psi_{(0)}(x_t^i, a) \\
&+ \tilde{f} \tilde{\beta}_t^* \tilde{\beta}_t \tilde{f}^* \tilde{f} \tilde{\beta}_{t+1}^* \tilde{\beta}_{t+1} \cdots \tilde{f}^* \tilde{f} \tilde{\beta}_0^* \tilde{\beta}_0 \psi_{(0)}(x_t^j, a)
\end{aligned} \tag{10}
$$

In our simulations we restrict the time independence of $\tilde{\beta}_t$ and $\tilde{f}$.

Figure 6(a) shows the typical orbit for $c = 0.4$ and the dynamics of maps on the pairs of layers ($\mathcal{L}1, \mathcal{L}2$) and ($\mathcal{L}1', \mathcal{L}2'$), where the Intent and the Extent maps perpetually change and also the new Extent-Intent maps change. In conventional CLM, it is well known that the synchronization between x_t and y_t are observed in large parameter region (Fig. 6(d)) and on-off intermittency is observed at the edge of chaos (near c=0.25 and 0.75). To evaluate this phenomena we count frequency of alternation from burst phase ($|x_t - y_t| > \delta$) to laminar phase ($|x_t - y_t| > \delta$) ($\delta = 10^{-4}$ in our simulations). Fig. 6(b) shows the frequency in ACELM (red) and the CLM (green). The parameter region with on-off intermittency is clearly extended in ACELM. Additionally, Fig. 6(b) shows the property of the anti-synchronization with on-off intermittency in the large parameter region.

4 Discussions and Conclusion

In this article, we formalized the construction of heterarchical duality generalized on one-dimensional maps by using category theory. And then, we showed the application of it on a logistic map called by ELM and their coupling system called by ACELM and analyzed its behavior. In ELM, we universally observed $1/f$ spectrum for orbits and the fractal-like behavior of the dynamics on return maps. In ACELM, the parameter region with on-off intermittency was clearly extended for the conventional CLM which has the parameter region called by the edge of chaos. This implies that ACELM has the adaptive property in the context of the edge of chaos.

In this article, we have suggested that understanding adaptive system theoretically, a consideration of the heterarchical duality of the system would be inevitable. If we assume a consistent system in the environment, heterarchical duality is universally observed. Especially, in the field of artificial life, one of the authors committed to simulate 'microslip' behavior by using an autonomous agent carrying sensorimotor coupling [18]. And its analysis of the internal dynamics revealed heterarchy-like features. Heterarchical construction could also

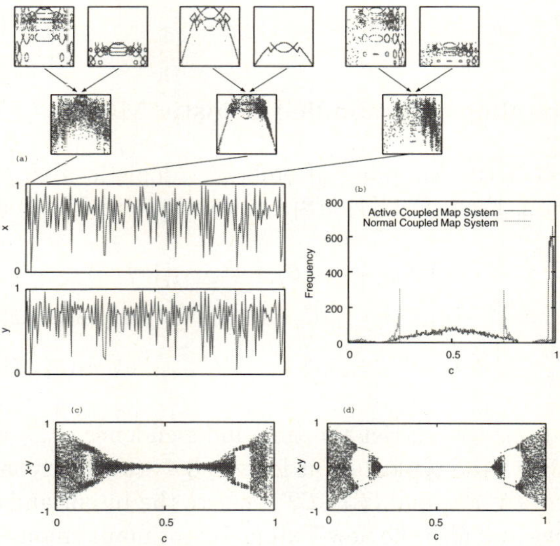

Fig. 6. (a) The dynamics of orbit and maps in ACELM for $c = 0.4$. The map has three figures. The left-upper stands for the Intent map and the right-upper stands for the Extent map. The lower is the map which results from the new Extent-Intent dynamics on the layer $(\mathcal{L}1', \mathcal{L}2')$. (b) Frequency of alternation from burst phase to laminar phase in $10000 \leq t \leq 20000$. The red and green lines correspond to ACELM and the coupled logistic map (CLM), respectively. (c) and (d) are bifurcation diagrams of the values of $x_t - y_t$ in $4000 \leq t \leq 4200$ in ACELM and CLM, respectively.

be observed in neuroscience, psychology, biology, and so forth. Moreover, in this article we intended to proceed one step further. Because, from the recent interest in morphological computation in the face of the indefiniteness of the environment, we think that we should reconsider the concept of environment for the system and/or the concept of perturbation to the system. For example, let's consider a robot that carries sensorimotor coupling. Usually, for the robot's environment, a virtual or restricted real environment and/or restricted boundary conditions are assumed. In this case we can approximate the perturbation to the system as the perturbation to the sensory input, or $x+\delta x$. In the real world, however, there usually exists a perturbation never related to sensory input because the agent has a real body, 'embodied'. In this case the perturbation should be described as, so to speak, an unpredictable perturbation to the system. Then, if a large rock were to attack the robot, sometimes the robot would continue moving instead of its destruction to the system itself and sometimes it would stop and never move. Considering the system from that kind of standpoint would allow us, theoretically, to deal with the adaptability and/or the system that has a latent evolvability. The construction of "observational heterarchy" could be a good tool in dealing with these cases theoretically.

Acknowledgment. We gratefully acknowledge Professor Y.-P. Gunji, Dr. Kamiura and Dr. Haruna for their fruitful discussions and suggestions.

References

1. Morimoto, G., Ikegami, T.: ALife IX: Proc. 9th Int. Conf. on the Simulation and Synthesis of Living Systems, pp. 188–193 (2004)
2. Pfeifer, R., et al.: Understanding Intelligence. MIT Press, Cambridge (1999)
3. Bertschinger, N., et al.: Proc. 7th GermanWorkshop of ALife, pp. 26–28 (2006)
4. Miyake, Y.: Co-Creation System and Human-Computer Interaction. In: Sakai, T., et al. (eds.) 3rd Conf. on Creating, Connecting and Collaborating through Computing, pp. 169–172 (2005)
5. Conrad, M.: Biosystems 52, 99 (1999)
6. Matsuno, K.: Protobiology: Physical Basis of Biology. CRC Press, Boca Raton (1989)
7. Gunji, Y.-P., Ito, K., Kusunoki, Y.: Physica D 110, 289 (1997)
8. Gunji, Y.-P.: Protocomputing and Ontological Measurement. Univ. of Tokyo Press, Tokyo (2004)
9. McCulloch, W.S.: Bull. Math. Biosystems 7, 89 (1945)
10. Varela, F.J.: Principles of Biological Autonomy. North-Holland, Amsterdam (1979)
11. Jen, E.: Complexity 8, 12 (2003)
12. Gunji, Y.-P., Kamiura, M.: Physica D 198, 74 (2004)
13. Kamiura, M., Gunji, Y.-P.: Physica D 218, 122 (2006)
14. Kamiura, M., Nakajima, K., Gunji, Y.-P.: Physica D (submitted)
15. Gunji, Y.-P., Haruna, T., Sawa, K.: Physica D 219, 152 (2006)
16. Rosen, R.: Life itself. Columbia University Press, New York (1991)
17. Goldblatt, R.: Topoi: The Categorial Analysis of Logic (Revised edition). Dover, Mineola (2006)
18. Ogai, Y., Ikegami, T.: Int. Series of Natural and Artificial Intelligence 1: Human and Artificial Intelligence Systems From Control to Autonomy, pp. 215–220 (2004)

Neuro-evolution Methods for Designing Emergent Specialization

Geoff S. Nitschke

Computational Intelligence Group, Vrije Universiteit Amsterdam, De Boelelaan 1081a, 1081 HV Amsterdam, The Netherlands,
nitschke@cs.vu.nl

Abstract. This research applies the *Collective Specialization Neuro-Evolution* (CONE) method to the problem of evolving neural controllers in a simulated multi-robot system. The multi-robot system consists of multiple pursuer (predator) robots, and a single evader (prey) robot. The CONE method is designed to facilitate behavioral specialization in order to increase task performance in collective behavior solutions. Pursuit-Evasion is a task that benefits from behavioral specialization. The performance of prey-capture strategies derived by the CONE method, are compared to those derived by the *Enforced Sub-Populations* (ESP) method. Results indicate that the CONE method effectively facilitates behavioral specialization in the team of pursuer robots. This specialization aids in the derivation of robust prey-capture strategies. Comparatively, ESP was found to be not as appropriate for facilitating behavioral specialization and effective prey-capture behaviors.

1 Introduction

Design principles that facilitate emergent behavioral specialization have been studied in artificial life [10] and multi-robot systems [1] research. However, collective behavior design methods for harnessing and utilizing emergent specialization for the benefit of problem solving are currently lacking.

Pursuit-evasion is a collective behavior task that is commonly used within artificial life research to test both non-adaptive (typically game theoretic) and adaptive (typically learning and evolution) methods for agent controller design. This paper describes a pursuit-evasion game, where a team of pursuer robots (herein called: *predators*), are required to collectively immobilize one evader robot (herein called: *prey*). The paper compares two different *Neuro-Evolution* (NE) methods, the CONE and ESP methods, for designing effective collective prey-capture behaviors. The effectiveness of emergent prey-capture behaviors are examined with respect to the role of behavioral specialization.

Research Goal: To demonstrate that the CONE method is appropriate for deriving behavioral specialization in a team of predators, where such specialization gives rise to successful collective prey-capture behaviors. Success of prey-capture behaviors is measured in terms of the simulation time for which a prey is immobilized (captured).

F. Almeida e Costa et al. (Eds.): ECAL 2007, LNAI 4648, pp. 1120–1130, 2007.

Hypothesis 1: CONE will facilitate emergent behavioral specialization that will enable the derivation of high performance prey-capture behaviors.

Hypothesis 2: The ESP method is not as appropriate for facilitating behavioral specialization in the pursuit-evasion task, and will thus yield comparatively low performance prey-capture behaviors.

Behavioral Specialization: Using methods from related work [8] we were able to ascertain which sensory activation and motor output value ranges corresponded to an observed behavior. Specifically, we measured the portion of a predators lifetime that light sensors (for prey detection) and infrared sensors (for proximity detection) were activated (within a given range). That is, sensor and motor activations of individual predators, within a given range of values, were found to produce specific observed behaviors. In some cases, these specific behaviors collectively produced an effective prey-capture behavior. Sensory-motor activation instances that had been identified with an observed behavior were summed over the course of a predators lifetime. If the sum of these particular activation instances was $\geq 50\%$ of the predators lifetime (that is: the total amount of activation instances possible) the corresponding observed behavior was labeled as *specialized*.

Task: The task was for a predator team to maximize the time for which a prey is immobilized. A control experiment (described in related work [7]) demonstrated that at least two predators are required to immobilize a prey.

Team Fitness Calculation: Predator teams were evaluated according to the total time for which the team was able to immobilize a prey. Specifically, a global fitness function calculated the average time for which a prey was immobilized. This average was calculated over the lifetime of a given predator team, as well all experimental runs. A fitness estimation method known as *fitness sharing* [2] was used in this calculation. This assumed that each predator in the team contributed equally to the capture of a prey, and thus each predator received an equal fitness reward when a prey was immobilized. Specifically, each predator in the team received a reward equal to the time for which a prey was immobilized.

2 Neuro-evolution Methods

2.1 CONE: Collective Neuro-evolution

CONE is an extension of both the SANE [6] and ESP [5] methods. A key difference between CONE and other NE methods is that it creates n separate genotype (neuron) sub-populations for n neural controllers operating in the task environment, and is thus best suited for collective behavior tasks. One advantage of CONE is that it expedites artificial evolution, given that the genotype population is organized into sub-populations. Hence, specialized controllers do not have to emerge out of a single population of neurons, and progressive specialization of controllers is not hindered by recombination of controllers with complementary specializations. A second advantage is that it provides more genotype diversity (comparative to single genotype population methods) and encourages emergent controller specialization given that evolution occurs within separate genotype

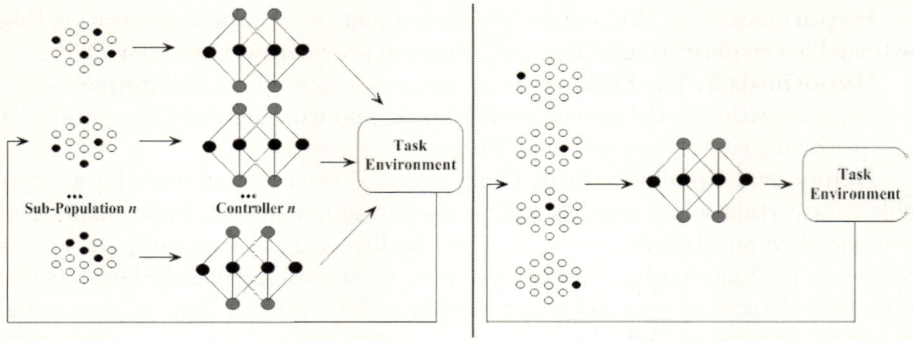

Fig. 1. Left: Example of *Collective Neuro-evolution* (CONE) Method (Section 2.1). **Right:** Example of *Enforced Sub-Populations* (ESP) method (Section 2.2).

sub-populations. Organizing the genotype population into separate niches (sub-populations), either dynamically [10], or *a priori* [9] facilitates specialization, and protects emergent behaviors within specialized niches of the genotype space.

CONE Process. After each of the n sub-populations was initialized with m genotypes, the CONE process (figure 1) was executed as follows.

1. n predator neural network controllers are constructed via selecting p genotypes (neurons) from each sub-population. Each set of p neurons then becomes the hidden layer of each of the n predator neural network controllers.
2. These n controllers are then tested together in a task environment for a *lifetime* of q epochs. An epoch is a test scenario lasting for w iterations of simulation time. Each epoch tested different predator and prey starting positions and orientations in the environment. For each of the q epochs ($q \geq m$, where m is the number of genotypes in a sub-population), each genotype in a given sub-population was selected and tested in combination with p-1 other neurons (in the form of a complete controller) randomly selected from the same sub-population.
3. p neurons from each of the n sub-populations were concurrently evaluated in the task environment and assigned a fitness. Testing of neurons within each sub-population continued until all neurons had been tested at least once.
4. At the end of a predators lifetime (q epochs) an average fitness value was assigned to each of the p neurons that participated in each controller. The average fitness of each neuron was calculated as its cumulative fitness divided by the number of controllers it participated in.
5. The testing and evaluation of the m neurons in each predators genotype sub-population constituted one generation of the CONE process.
6. For each sub-population, recombination and mutation of the fittest 20% of neurons then occurred, where the fittest 20% were arranged into pairs of neurons, and each pair produced 5 child neurons, so as to replace all genotypes in the current sub-populations and propagate the next generation of each sub-population.

7. p neurons were randomly selected from the fittest 20% within each of the n sub-populations. These n sets of p neurons were then decoded into n controllers, and executed in the task environment as the next generation. This process was then repeated for r generations.

2.2 ESP: Enforced Sub-Populations

Enforced Sub-Populations (ESP) has been effectively applied to non-Markovian control tasks with sparse reinforcement such as double pole balancing, rocket control, as well as pursuit-evasion games [5]. ESP differs from other NE methods in that it allocates and evolves a separate neuron population for each of the p hidden-layer units in a neural network. A neuron can only be recombined with other neurons from its own sub-population, thus making it suitable for the evolution of recurrent neural networks (not the case for the SANE method [6]).

2.3 CONE and ESP: Common Methods

Constructing Neural Network Controllers: In the case of ESP, 1 genotype is selected from each of p=6 sub-populations (table 1), for deriving a neural network controller's hidden layer of p neurons [5]. This is repeated n times for n controllers. There were p sub-populations for n predators, where p equaled the number of hidden layer neurons in a controller. In the case of CONE, complete controllers were constructed via selecting p neurons from each of the n sub-populations (table 1). There were n sub-populations for n predators. Each neuron in each sub-population was assigned to a fixed position in the hidden layer of any given controller. The position that the *ith* neuron (g_i) would take in a hidden layer of p neurons, where g_i was selected from any sub-population of m neurons, was calculated as follows. Each of the m neurons in a sub-population were initially assigned a random and unique ranking in the range [0, m-1]. A sub-population was divided into approximately equal portions (m / p), where if g_i was within the *kth* portion (where: k = [1, p]) then g_i would adopt the *kth* position in the hidden layer. Given that recurrent neural networks were being evolved, CONE only recombined neurons that were in the same sub-population (section 2.1), and assigned to the same hidden layer position.

Recombination and Mutation of Genotypes: Genotypes were encoded as a string of floating point values (table 1), which represented neural network weights connecting all sensory input neurons and all motor output neurons to a given hidden layer neuron. Child genotypes were produced using single point crossover, and *Burst* mutation with a *Cauchy* distribution [5]. Mutation of a random value in the range [-1.0, +1.0] was applied to each gene (connection weight) with a 0.05 degree of probability, and weights of each genotype were kept within the range [-10.0, +10.0] (table 1). Burst mutation was used so that most weight changes were small whilst allowing for larger changes to some weights.

3 Experimental Design, Agents, and Environment

3.1 Experimental Design

In this case, experiments measured the impact of a *neuro-evolution method* and a *group types* upon *prey capture time.*

- *Neuro-evolution Method*: Each predator used a recurrent neural network controller which was adapted with either the ESP or CONE method.
- *Group Type*: Between 2 and 6 predators were tested with 1 prey. These 5 group types (GT) were defined as follows: *GT-1:* 2 predators, *GT-2:* 3 predators, *GT-3:* 4 predators, *GT-4:* 5 predators, *GT-5:* 6 predators.

Table 1. Neuro-evolution parameter settings for the ESP and CONE methods.

CONE and ESP Neuro-evolution Parameter Settings	
Runs per experiment	50
Epochs	50
Iterations per epoch	1000
Mutation probability	0.05
Evaluations per neuron	10
Mutation type	Burst mutation / Cauchy distribution
Mutation range	[-1.0, +1.0]
Weight range	[-10.0, +10.0]
Crossover	Single point
Sensory input neurons	22
Hidden layer neurons	6
Motor output neurons	2
Phenotypes	[2, 6] Recurrent neural networks
Genotype sub-populations	[2, 6] / 6 (ESP)
Genotype representation	Neuron
Genotype length	24
Genotypes	600 per sub-population / 100 per sub-population (ESP)

Each experiment measured prey capture time, given a group type and neuro-evolution method. Prey capture time was measured as an average calculated over the multiple test scenarios that constituted a predator team's lifetime.

3.2 Environment

Predators and the prey move within a discrete environment of 180 x 180 quadrants. Each quadrant is large enough to contain only one predator or prey. If two or more robots attempt to occupy the same quadrant, a collision occurs. As a course encoding of movement, a predator or prey could turn at any angle (in 45 degree increments) up to 180 degrees, either to the left or to the right, with respect to its current heading. Each robot is initialized with a random heading.

A difference calculation in wheel speeds (MO_0 and MO_1) controlled the orientation of any given predator or prey. Obstacles are detected at a maximum range of 4 quadrants with a 360 degree field of detection (figure 2). This field of detection was an area of 9 x 9 quadrants, which is divided into 8 sectors to account for the coverage of different infrared proximity (for predators and prey) or light (for predators only) sensors. If an obstacle was detected on a quadrant divided by two sectors, that is, covered by two proximity (or light) sensors, then both sensors are simultaneously activated (each receiving an equal activation value).

3.3 Predators and Prey: Sensors and Actuators

The sensor and actuator configuration of each predator and prey is assumed to be that of a Khepera mobile robot [8] (figure 2). The prey is equipped with a light on its top (L_0). This light could be detected by the predator light sensors, and was used so each predator could distinguish fellow predators from the prey. Both predators and prey are equipped with 8 infrared proximity sensors ([SI_0, SI_7]. Additionally, each predator is equipped with 8 light ([SI_8, SI_{15}]) sensors, positioned on its periphery. Both, predators and prey are provided with two wheel motors (MO_0, MO_1) that controlled their speed and orientation.

When an obstacle came within range of a given proximity sensor, that sensor was activated with a value proportional to the distance to the obstacle. When a prey came within range of a predators light sensor, that sensor was activated with a value proportional to the distance to the prey. Sensor values were normalized within the range [0.0, 1.0] for the purposes of being acceptable as neural network inputs. Motor output values were normalized within the range [-10.0, 10.0].

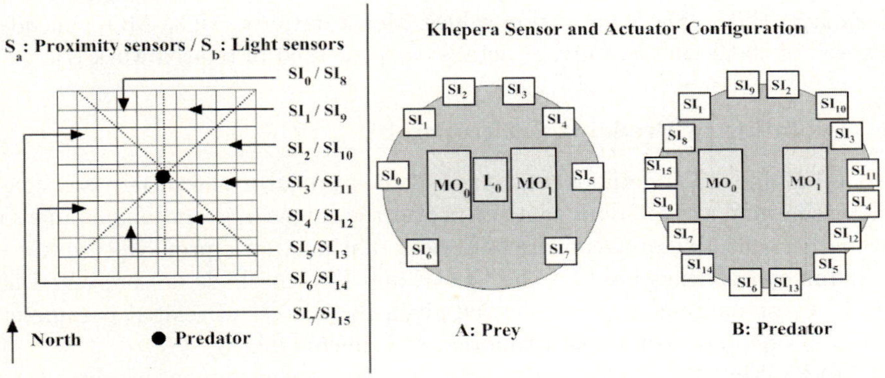

Fig. 2. Left: Sensory range of proximity and light sensors in the discrete simulation environment (prey is limited to proximity sensors). **Right:** Sensor and actuator configuration for predator (A) and prey (B) Khepera robots. See section 3.3 for details.

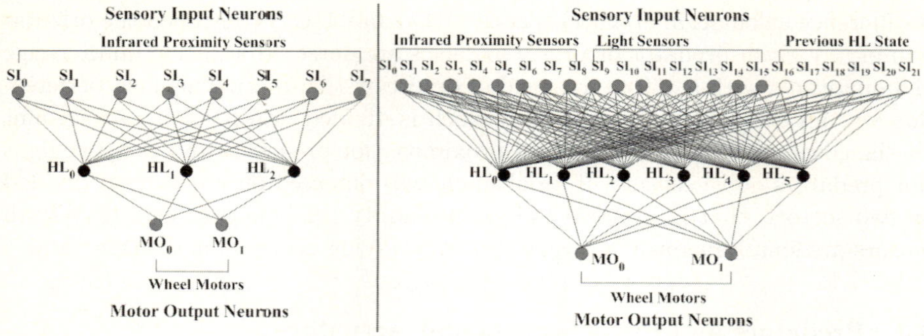

Fig. 3. Left: Prey feed-forward neural network controller. **Right:** Predator recurrent neural network controller. See section 3.4 for details.

3.4 Predators and Prey: Neural Network Controllers

The prey neural network consists of 8 sensory input neurons and 2 motor output neurons, fully connected to a hidden layer of 3 neurons (figure 3). Prey sensory inputs encode the state of 8 infrared proximity sensors, and 2 motor outputs encode the speed of 2 wheels. The output values of the 2 motor neurons are multiplied by 1.2. This sets a prey's speed to be 20% faster than the predators. The prey controller was evolved for static and dynamic obstacle avoidance, before being tested with a predator team. A recurrent neural network was selected as the predator controller in order to emulate short term memory [3]. A hidden layer of 6 sigmoidal units fully connects 22 sensory input neurons to 2 motor output neurons (figure 3). Predator sensory input neurons encode the state of 8 infrared proximity sensors and 8 light sensors ([SI_0, SI_{15}]), as well as previous hidden layer ([SI_{16}, SI_{21}]) activation values. Motor outputs (MO_0, MO_1) encode the speed of the 2 wheels. Further details are presented in related work [7].

3.5 Evolution of Predator Controllers

The CONE and ESP methods were applied to n predator neural network controllers. The goal was to adapt controllers over the course of evolutionary time so as to derive collective prey-capture behaviors that maximize prey-capture time. The parameter settings used for the CONE and ESP methods are presented in table 1. These parameters were selected given the success of related parameter settings in previous evolutionary robotics experiments [8].

CONE: When the CONE method was applied to evolve predator controllers, between 2 and 6 genotype sub-populations were created. These sub-populations represented the genotype space of between 2 and 6 predator neural network controllers. Each sub-population was initialized with 600 genotypes.

ESP: When the ESP method was applied to evolve predator controllers, 6 genotype populations were created so as to represent the genotype space of the 6 hidden layer neurons in a predator neural network controller. Each genotype population was initialized with 100 genotypes.

Genotypes (Neurons). For both CONE and ESP, each genotype represented the connection weights of a neuron that potentially participated in the hidden layer of any given predators neural network. Each genotype was encoded as vector of 24 floating point values. That is, 22 input connection weights plus 2 output connection weights. In order to construct a single predator neural network controller, 6 neurons were selected from a given genotype sub-population (1 genotype from each of the 6 sub-populations in the case of ESP). This set of neurons then constituted the hidden layer of 1 predator neural network. The process was then replicated between 2 and 6 times for each predator in the team.

4 Results: Evolved Collective Prey-Capture Behavior

Collective prey-capture behaviors, utilizing at least 3 predators and at most 4 predators, consistently emerged in the later stages of both the ESP and CONE evolutionary processes (> 200 generations). These collective behaviors were termed *role-switcher* and the *pursuer-blocker*. The former emerged under ESP, where as, both emerged under CONE. Figure 6 presents the average prey capture time, and number of instances of emergent specialization (corresponding to prey capture behaviors) for all group types.

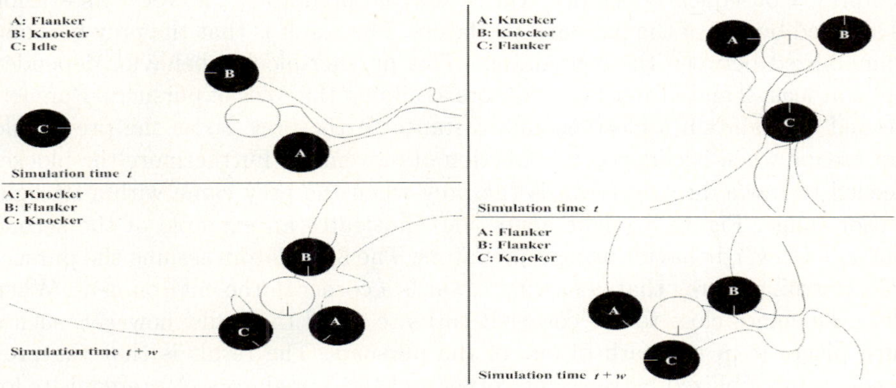

Fig. 4. Two versions of the *role switcher* prey capture behavior using 3 predators.

4.1 ESP: Role-Switcher Prey Capture Behavior

Two versions of the role switcher collective behavior emerged. In each case, predators in the team adopted 1 of 3 specialized behavior roles, termed: *flanker*, *knocker* and *idle*. However, at least 1 predator adopted 2 of these behavioral roles, which it would switch between in order to maintain the role-switcher prey-capture behavior. The role switcher behavior was most effective for teams of 3 or 4 predators. Teams of 2 predators were insufficient to immobilize a prey for more than a few simulation iterations, and teams of 5 and 6 predators often caused physical interference between each other, and thus caused the prey-capture behavior to fail. Figure 4 (left and right hand side) illustrate the two versions of the role switcher prey-capture behavior, occurring over w simulation time steps. The role switcher behavior has been observed in related research [7], and is thus not elaborated upon here.

4.2 CONE: Pursuer-Blocker Prey Capture Behavior

In addition to the role-switcher behavior, a prey capture behavior called *pursuer-blocker* also emerged under the CONE method. Two versions of the pursuer-blocker collective behavior emerged. Predators assumed 1 of 2 specialized behavioral roles, termed, *pursuer* and *blocker*. Pursuer-blocker was most effective for teams of 3 or 4 predators. Teams of 5 or 6 predators were ineffective due to physical interference that occurred between the predators as they collectively approached the prey, and 2 predators were sufficient for immobilizing the prey for a few simulation iterations only.

Figure 5 (left hand side) illustrates an example of the first pursuer-blocker behavior. Predators A and B are the pursuers, assuming positions behind and to either side of the prey. Predator C assumes the role of the blocker. When the prey moves within light sensor range of predator C, it moves directly towards the prey. Consequently the prey turns to avoid predator C, however its evasion is stopped by one of the pursuing predators. The result is that the prey becomes immobilized between the 3 predators. This pursuer-blocker behavior depended upon at least 2 and at most 3 predators assuming the roles of pursuers. Pursuers needed to maintain a close enough distance to the prey, so as the prey could not escape when it changed its direction of movement. Furthermore the blocker needed to move directly towards the prey when the prey came within its light sensor range. Figure 5 (right hand side) illustrates an example of the second *pursuer-blocker* behavior using 3 predators. The 3 predators assume the pursuer role, pursuing a prey that is moving towards a corner in the environment. When the prey comes close to the corner it turns to avoid the walls, however, such a turn places it in the path of one of the pursuers. The result is that the prey becomes immobilized between the corner and the 3 predators. A prerequisite for the success of this behavior was that the prey be moving towards a corner. This pursuer-blocker behavior also emerged using 2 or 4 predators, but failed with 5 and 6 predators due to interference that occurred between predators as they collectively approached the prey in a corner.

5 Analysis and Discussion

To draw conclusions from this comparative study, a set of statistical tests were used to gauge respective differences between CONE and ESP method results. First, the data distributions for *prey-capture time* results yielded by CONE and ESP, were determined to be normal distributions via applying the Kolmogorov-Smirnov test [4] (P=0.72 and P=0.98, respectively). We then applied an independent t-test [4]; 0.05 was selected as the threshold for statistical significance,

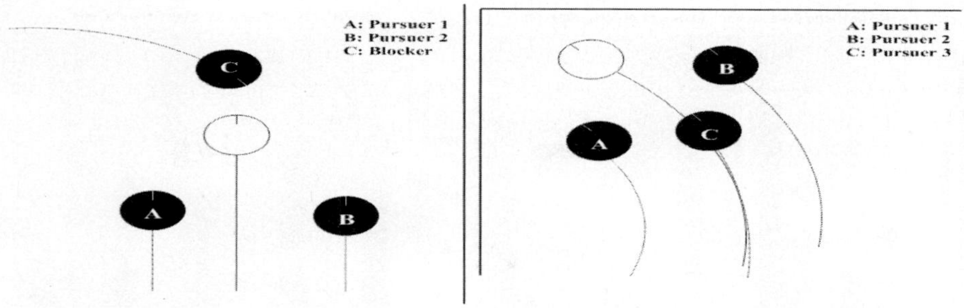

Fig. 5. Two versions of the *pursuer-blocker* prey capture behavior using 3 predators.

and the null hypothesis stated that the two data sets did not significantly differ. P=0.00012 was calculated, meaning the null hypothesis was rejected. This partially supported our first and second hypotheses. That is, CONE (and not ESP) would derive prey-capture behaviors with superior task performance. A t-test was then applied to the *instances of specialization* data sets for the CONE and ESP methods; P=0.00025 was calculated, meaning the null hypothesis was rejected. This further supported our first and second hypotheses. That is, the high task performance exhibited under CONE (and not ESP) was due to emergent specialization. To completely support our hypotheses, it is necessary to compare the group types where prey capture time and the number of specialization instances are highest for CONE (figure 6). Instances of emergent specialization were calculated as the number of epochs in a predators lifetime (averaged for all predators in the team) that were labeled as *specialized* (section 1). As illustrated in figure 6 the largest differences between the average number of emergent specialization instances, for CONE and ESP, were for group types 2 and 3. Group types 2 and 3, also yielded the largest difference in prey-capture times. That is, under the ESP method a low number of specialization instances corresponded to low prey-capture times for these group types. It is theorized that the superior performance of prey-capture behaviors derived by CONE, was due to its capability to facilitate more instances of behavioral specialization. Subject to future research, this was attributed to the use of separate neuron sub-populations for deriving complete controllers. This encouraged the derivation of specialized controllers that complemented each other in the task environment.

6 Conclusions

This paper described a comparative study of the ESP and CONE neuro-evolution methods applied to the task of deriving effective collective prey-capture behaviors in a pursuit-evasion game. Prey capture behaviors were evolved for a team of simulated predator robots that attempted to immobilize (capture) a prey robot. The effectiveness of prey-capture behaviors, and hence the fitness of the predator team was measured in terms of prey-capture time. Results indicated, that

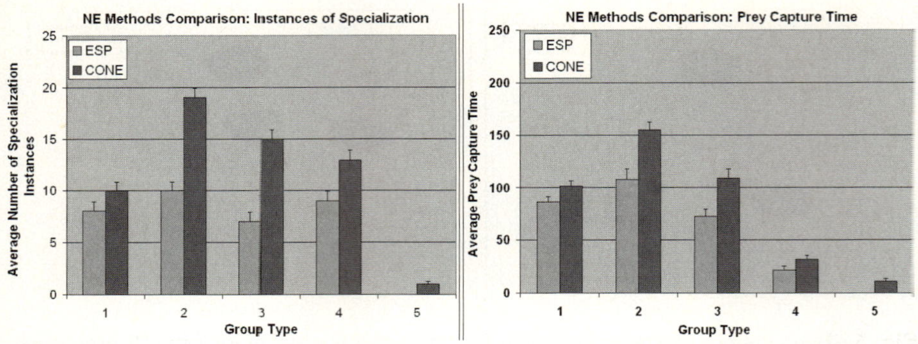

Fig. 6. Neuro-evolution (CONE and ESP) methods comparison. **Left:** Average number of emergent specialization instances. **Right:** Average prey capture time.

the CONE method facilitated emergent behavioral specialization in the predator team. These specialized behavioral roles served to increase the effectiveness of prey-capture behaviors. Comparatively, less instances of emergent specialization were observed when the ESP method was applied for the purpose of deriving collective prey-capture behaviors in a predator team. Given this, and the inferior performance of prey-capture behaviors derived under the ESP method, it is concluded that emergent behavioral specialization is beneficial in the pursuit-evasion task, and that CONE is appropriate for facilitating such specialization.

References

1. G. Baldassarre, S. Nolfi, and D. Parisi. Evolving mobile robots able to display collective behavior. *Artificial Life*, 9(1):255–267, 2003.
2. L. Bull and J. Holland. Evolutionary computing in multi-agent environments: Eusociality. In *Proceedings of the Second Annual Conference on Genetic Programming*, pages 347–352, San Francisco, USA., 1997. IEEE Press.
3. J. Elman. Finding structure in time. *Cognitive Science*, 14(1):179–211, 1990.
4. B. Flannery, S. Teukolsky, and W. Vetterling. *Numerical Recipes*. Cambridge University Press, Cambridge, 1986.
5. F. Gomez. *Robust Non-Linear Control Through Neuroevolution. PhD thesis.* Department of Computer Sciences, The University of Texas, Austin, Texas, 2003.
6. D. Moriarty and R. Miikkulainen. Efficient reinforcement learning through symbiotic evolution. *Machine Learning*, 22(1):11–32, 1996.
7. G. Nitschke. Designing emergent cooperation: a pursuit-evasion game case study. *Artificial Life and Robotics*, 9(4):222–233, 2005.
8. S. Nolfi and D. Parisi. Learning to adapt to changing environments in evolving neural networks. *Adaptive Behavior*, 1(5):75–98, 1997.
9. M. Potter and K. De Jong. Cooperative coevolution: An architecture for evolving coadapted subcomponents. *Evolutionary Computation*, 8(1):1–29, 2000.
10. K. Stanley, B. Bryant, and R. Miikkulainen. Real-time neuro-evolution in the nero video game. *IEEE Transactions Evolutionary Computation*, 9(6):653–668, 2005.

Neutral Emergence and Coarse Graining

Andrew Weeks, Susan Stepney, and Fiona Polack

Department of Computer Science, University of York, UK

Abstract. We introduce the concept of *neutral emergence* (defined by analogy to an information theoretic view of neutral evolution), and discuss how it might be used in the engineering of *robust* emergent systems. We describe preliminary results from an application to coarse graining of cellular automata.

1 Introduction

We introduce the concept of *neutral emergence*, defined by analogy to an information theoretic view of neutral evolution. Our particular interest in neutral emergence is as a potential component of an engineering framework for emergent systems. The key problems for engineering emergence are to be able to implement a system such that a specified property emerges, and to make the property robust, at least to minor unforeseen perturbations. Our initial work is improving our understanding of emergence, and of the "slack", or potential for robustness, between an emergent system and its implementation.

Section 2 summarises background material on neutral evolution, and its formulation in information theoretic terms. Section 3 summarises background material on emergence. In section 4 we cast emergence in information theoretic terms, by analogy to information theoretic evolution. In section 5 we introduce the concept of *neutral emergence*, and show how it is related to system robustness. In section 6 we describe preliminary results from an application to coarse graining of cellular automata, and outline our plans for future work.

2 Background: Evolution

Neutral Evolution. The mapping from genotype (DNA) to phenotype (organism) is complex, and there is significant redundancy in both. Different genotypes can map to the same phenotype; for example, different codons (DNA nucleotide triplets) can code for the same amino acid. Hence the genotype can change (a nucleotide can mutate) without changing the phenotype. Similarly, the same genotype can result in different phenotypes, due to different environmental conditions during development.

Neutral evolution is a recognised phenomenon in evolutionary biology [15]. It involves a change in the genotype without a significant change in the fitness of the phenotype; that is, it involves a change that is selectively neutral. Such changes allow the phenotypic population to explore their fitness landscape, by drifting

F. Almeida e Costa et al. (Eds.): ECAL 2007, LNAI 4648, pp. 1131–1140, 2007.

around close to contours of equal fitness, to regions where they may subsequently find fitter solutions not directly accessible from their original position in the landscape.

Information Theoretic Evolution. Adami [1,2,3] gives a description of evolution in information-theoretic terms, by considering the information in the genome *in the context of its environment.*

Adami [1] classifies a (digital) genome into 'hot' and 'cold' bits. The cold bits are ones strongly conserved in the population, whereas the hot bits vary between population members, and are therefore (presumably) not significantly contributing to the fitness. So, in general, mutations of hot bits are *neutral*, but occasionally might be beneficial, at which point they become 'frozen in' to the population, and conserved. These fitter organisms are exploiting their environment better, and so must contain more information about those parts of the environment that they are exploiting [1, p115].

In information theoretic evolution, the genome is seen as some kind of representation of the environment: the cold bits are strongly correlated with the environment, whereas the hot bits are not. The more correlated bits there are, the higher the *mutual information* between the organism's genome and the environment. The *mutual information*, or correlation, between the system S and its environment E, $I(S : E)$, is the *entropy* of the system, $H(S)$, less the *conditional entropy* of the system in the context of the environment, $H(S|E)$:

$$I(S : E) = H(S) - H(S|E) \tag{1}$$

The conditional entropy $H(S|E)$ can be thought of as the amount of information in the system that cannot be explained by (correlations with) its environment. Similarly, the conditional entropy $H(E|S)$ can be thought of as the amount of information in the environment that cannot be explained by correlations with the system.

Evolution (increasing fitness) is then *increasing mutual information*: increasing the shared information, or correlations, between the genome and the environment. So "*natural selection can be viewed as a* filter ... *that lets information flow into the genome, but prevents it from flowing out*" [2]. See figure 1. This simple description captures the essence; in reality the biological process is much more complicated. For example: bits are correlated within a genome [2], so when some bits go cold it is necessary for other bits to become hot again. Also, the environment co-evolves with the organisms. However, a fuller picture does not undermine the principle of an information-theoretic approach, particularly when the principle is transferred from the biological to an engineering domain.

Neutral Evolution in Information Theoretic Terms. The mutual information, I, is independent of the hot bits in S; it depends only on the cold bits correlated with the environment E. Hence, in this context, neutral evolution is a change in the genome that does not affect the mutual information of the genome

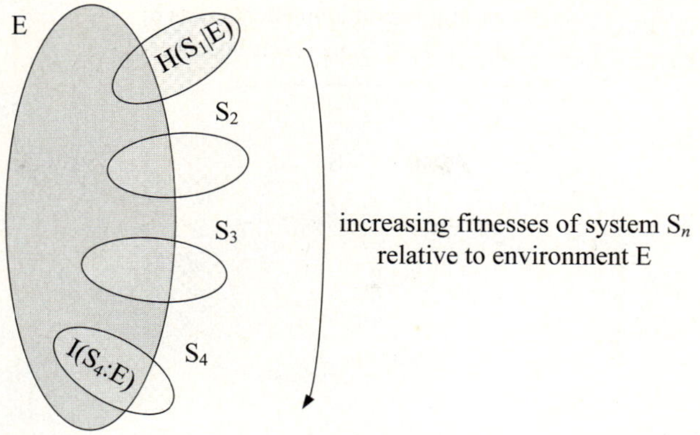

Fig. 1. Evolution increasing the mutual information between the genome S and the environment E (adapted from [1, fig.5.7]). The area of an ellipse represents its information content, and the area of overlap represents the mutual information between an S_n and E. For simplicity, total information is shown as unchanging (the areas of ellipses S_n and E are constant); this implies that conditional information decreases as the mutual information increases. This is not necessarily the case: for example, a neutral evolutionary step could change the amount of conditional information $H(S|E)$ by increasing the size of the uncorrelated part of the genome [2].

and (the fitness relevant part of) the environment, $I(S : E)$. This means that S is *robust* to this kind of genomic change. Similarly, parts of the environment E that are not correlated with S can also change neutrally (from the point of view of S), so S is *robust* to this kind of environmental change, too.

3 Background: Emergence

In [16] we provide a review of opinion on emergence. We start from Ronald *et al*'s definition of emergence: "*The language of design L_1 and the language of observation L_2 are distinct, and the causal link between the elementary interactions programmed in L_1 and the behaviors observed in L_2 is* non-obvious *to the observer—who therefore experiences surprise.*" [13]. We reject the use of 'surprise' as a criterion for emergence: it is subjective, and wears out with repetition. We do, however, follow Ronald *et al*'s use of two languages of description, which we refer to as L for the 'microscopic' *local* level of implementation substrate, and S for the 'macroscopic' global level of the model or *specification* (we are interested in observing and modelling natural emergent systems, but more in specifying and implementing engineered ones). We follow Shalizi [14] in defining emergence in information-theoretic terms, as the greater 'predictive efficiency' of descriptions in S over those in L. Following Bar-Yam [4], we see emergence as a consequence of global constraints on the system (the emergent level) rather than of constraints on components (the implementation level).

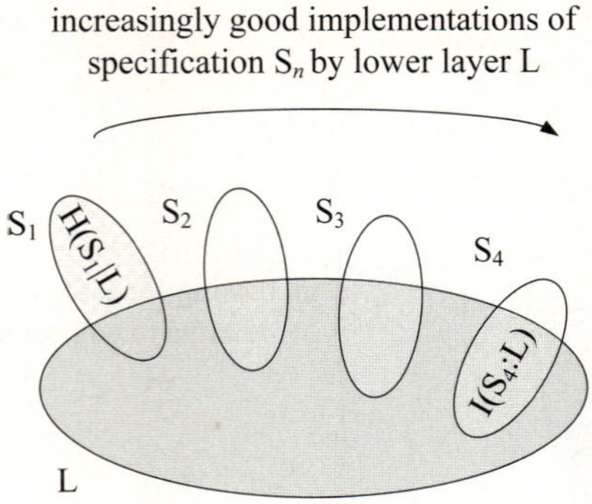

Fig. 2. Modelling, or incremental system development, as increasing the mutual information between the system specification S and the implementation L.

In relation to levels, various research identifies 'natural' scales. Israeli and Goldenfeld [7] note that there is an emergent natural length scale for coarse graining (see section 6.1) cellular automata (CA). In the process of "reconstructing the attractor" from time-lagged observations [17] (also a form of coarse-graining), the correct time lag can be found using mutual information [12].

To summarise our view, in an emergent system, the whole is greater than, and different from, the sum of its parts (in terms of behaviour and other qualities); emergence can, in various senses, be equated to novelty; levels are essential: emergence occurs at a higher level or longer time scale than that on which basic processes occur; emergence is a characteristic that is expressed at the high level, caused by the low level. Emergence is not an arbitrary feature; it is not dependent merely on the chosen level of observation. Some levels are 'better' than others (for example, coarse grainings that better describing the underlying dynamics), and form natural levels for discovering, or for designing in, emergent properties.

4 Emergence in Information Theoretic Terms

By direct analogy to the information theoretic description of evolution, we define an information theoretic quantity of emergence.

Definition: The amount of emergence $I(S : L)$ is the mutual information, or correlation, between a specified system S and its implementation in substrate level L.

$$I(S : L) = H(S) - H(S|L) \tag{2}$$

The conditional entropy $H(S|L)$ can be thought of as the amount of information in the system specification that has not been captured by (correlations with) its implementation.

Modelling, or incremental system development, can be viewed as increasing mutual information: increasing the shared information, or correlations, between the system specification and its implementation (figure 2).

If one were trying to explain (model) an observed system S in terms of L, the conditional information $H(S|L)$ might be the behaviours or properties of S not explained by L. (It might just be noise.) If one were trying to implement a specified system S in an implementation substrate L, this conditional information is the part of the specification that has yet to be captured by the proposed implementation L: more development work is required.

The conditional information $H(L|S)$ (the information in the implementation not correlated with the system specification) might be considered as properties of L that are unnecessary for the realisation of S. Of course, these additional properties might well be *invisible* if the system is subject only to high-level observation [6] that reveals only system-level properties (multiple distinct microstates nevertheless resulting in the same observed macrostate). It is, however, these extra properties that can be exploited to get *robust* implementations of S.

The information theoretic definition of emergence suggests an approach to incremental development of emergent systems: use the mutual information I as a fitness function in a search for good models (system descriptions) S of an existing L (such as a given CA or agent system), or to search for good low level implementations L (such as the required rules and states for a CA or agent system) of a system specification S.

5 Neutral Emergence

Definition: An emergent property exhibits *neutral emergence* when a change in the microstate L does not change the macrostate S, or vice versa. It is a change that does not significantly change the mutual information $I(S : L)$.

Robustness. A system S is robust to changes in either itself, or in L, if the changes do not significantly change the mutual information. In particular, S can be robust to many changes in its implementation, including, possibly, the effect of errors. It is often stated that emergent systems (modelled on natural processes) exhibit robustness: here we see why (and where) this may be the case. The excess information in L (a large $H(L|S)$) is *necessary* for emergent systems to be robust in this manner.

As argued earlier, an engineering development process can be seen as implementing specification S by finding an L with a high mutual information $I(S : L)$. Here we see that, at the same time, the process can seek to maximise robustness, by searching for a system that is insensitive to (uncorrelated with) certain failure modes or other possible changes in L. If a system were stressed during development (exposed to a range of stresses and implementation errors), its implementation could be encouraged towards regions that are insensitive (robust) to such events. (Compare this to the development of formally proven systems: they do not guarantee any level of performance with even the smallest change.) By analogy to evolutionary fitness landscapes, we want to find systems that lie in gently sloping plains and plateaux, rather than on narrow peaks or steep cliffs.

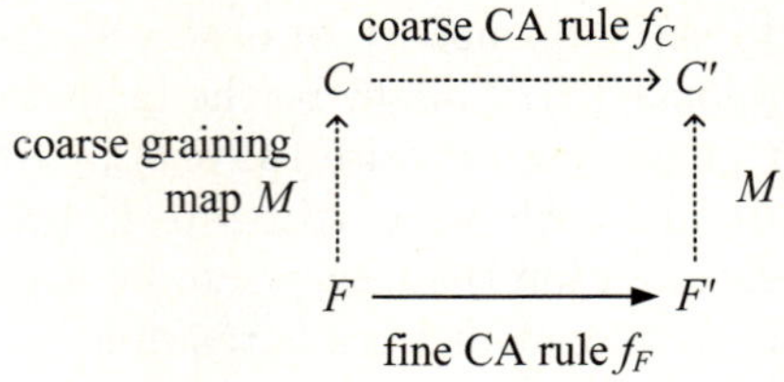

Fig. 3. Exact coarse graining. The fine grained state F evolves under the given CA rule f_F. The coarse graining M maps fine grained state F to coarse grained state C. The coarse grained state C evolves under another CA rule f_C. The task is to find a mapping M and rule f_C such that the square commutes: $M \circ f_F = f_C \circ M$.

Speculation on 'The Edge of Emergence'. Experiments with CAs [10] and Random Boolean Networks [8,9] indicate that a phase transition takes place between a 'solid' ordered regime and a 'gaseous' chaotic state, and that a narrow 'liquid' region on the boundary between these regimes is capable of supporting the most complex behaviour, including universal computation. (Note: the detail of Langton's 'edge-of-chaos' results [10] have been questioned [11].)

Compare this with our mutual information measure of emergence: a system exhibits minimal emergence when everything is a surprise (zero mutual information). Clearly such a model, that knows nothing about what it is modelling, is useless, but equally (as argued above) useful emergence does not require maximal mutual information in the system. There should be a level of emergence with the maximum utility: a position at which the model has useful freedom to explore, whilst being held within a constrained region of the search space (maintaining the quantity of mutual information; analogous to the 'fitness contouring' of neutral evolution). While it is tempting to label this the 'emergence liquid region', substantive experimental data is needed to support such a claim.

That said, consideration of the 'edge-of-chaos' analogy raises some interesting questions. Does a phase transition occur between the 'solid' (system specification or model effectively duplicating the implementation) and 'gaseous' (little in common) regimes? Is the region a narrow one? If so, is this why creating effective emergent systems manually has proved difficult?

6 Application

Thus far, our discussion has been conceptual. Now we present preliminary work applying our ideas to CAs.

6.1 Coarse Graining Cellular Automata

Israeli and Goldenfeld [7] coarse grain CAs. Their *exact* coarse graining captures the dynamics of a CA rule with another rule, expressed over a coarser grain of

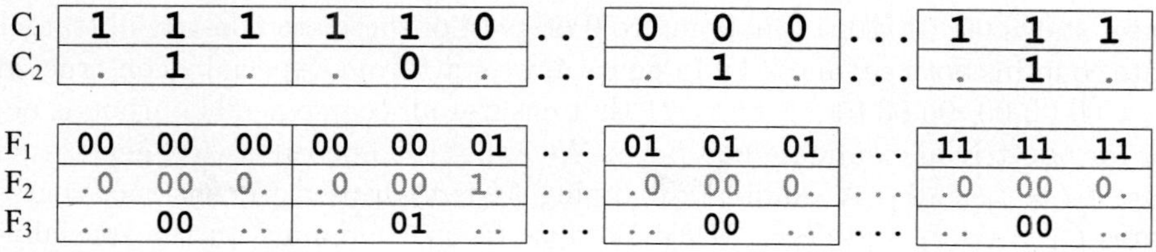

Fig. 4. Search setup for exact coarse graining.

space and time. (See figure 3.) Exact coarse graining preserves the underlying dynamics but loses 'irrelevant' detail.

In terms of emergence, the two languages L and S are *different* CA rules (except in a few cases), and information is lost in the coarse graining (in that low-level fine-grained detail is washed out). Interestingly, the figures illustrating coarse grainings in [7] seem to highlight some of the underlying L structure (for example, various propagating 'signals'), maybe because they have smoothed out other, irrelevant, structure.

The higher level rule S may be a CA rule of lower complexity than the rule at level L: because "*the system (the update rule, not the cell lattice) does not contain enough information to be complex at large scales*" [7]. This complexity-reducing kind of coarse graining (of 'relevant degrees of freedom') removes information that is relevant to the dynamics at L, but is redundant under the coarse graining at S: it does not capture all of the microscopic dynamics.

6.2 Applying Neutral Emergence to Coarse Graining

In investigating neutral emergence, we start by recasting Israeli and Goldenfeld's [7] coarse-graining results in the information theoretic emergent framework described above. We then consider an extension of coarse graining that has greater robustness.

Robustness. The exact coarse graining of [7] exhibits robustness in the way described earlier. In particular (i) different low-level CA rules coarse grain to the same high-level rule, so the high level is robust to certain changes in the low-level rule; (ii) even for a fixed low-level rule, different low-level CA states coarse grain to the same high-level state, so the high level is robust to certain changes in the low-level state.

Exact Coarse Graining. Israeli and Goldenfeld [7] use a brute force approach to discover their coarse grainings. For exact coarse graining, the diagram in figure 3 must commute for all fine states F.

As a control, we implemented this brute force search for 1D CA coarse grainings with $N = 2$ (that is, one high level cell corresponds to 2 low level cells) as follows (figure 4). For $N = 2$, there are 16 candidate mappings M, with fine states states 00, 01, 10. 11 mapping to 0 or 1. 14 of these are non-trivial (target state contains both a 0 and a 1). In figure 4, we start from a specially constructed $F_1 = 00\ 00\ 00\ \ 00\ 00\ 01\ \ldots 11\ 11\ 11$ that ensures all coarse neighbourhoods occur for non-trivial candidate mappings. We apply the fine rule twice, generating state $F_3 = f_F\, f_F\, F_1$. A candidate mapping M is then used to construct coarse states $C_1 = M\, F_1$ and $C_2 = M\, F_3$. Because C_1 is a mapping on the specially-constructed F_1, it contains all possible coarse neighbourhoods.

To determine whether candidate mapping M is a valid coarse graining, we must ensure that whenever a neighbourhood occurs more than once in C_1 it maps consistently to the same central state in C_2. Since C_1 contains all coarse neighbourhoods, the coarse rule f_C can be deduced from the constructed C_1 and C_2 states.

Mutual Information as a Fitness Function. As an alternative to brute-force search, a genetic algorithm (GA) was used to search for 1D CA coarse grainings with $N = 2$. Each member of the GA population has a random candidate mapping M. The coarse graining process described above is followed to generate states C_1 and C_2. We use an approximation to mutual information as the fitness function of the map M: the fitness maximises the number of consistent states between C_1 and C_2.

This GA successfully finds 1D exact coarse grainings (although not on every run), supporting our assertion (section 4) that mutual information would be an appropriate fitness function. The trial on exact coarse graining is used to validate the GA search, but in this case the GA is much less efficient brute force search. However, as the coarse-graining size increases (for example, $N = 5$), brute force search becomes computationally intractable (the number of states to search increases exponentially), and GA search becomes attractive.

Approximate Coarse Graining. Israeli and Goldenfeld's [7] coarse grainings provide an exact fit over *all possible states*: they lose information, but the systems never differ in their predictions. Exact coarse graining is a form of information *compression.* The mutual information is the same as the entropy of the high level system.

We can search for *approximate coarse grainings*, requiring an exact fit over only *some* CA states, thereby allowing the information content of those states to be exploited, in addition to the information content of the CA rules themselves. (Compare the case in physical emergent systems, where the emergent properties occur only over some restricted set of all possible states, such as a restricted temperature range.)

In approximate coarse graining, we start with a much smaller state F_1 that need not ensure all coarse neighbourhoods occur (figure 5). We build up C_1 and C_2 as before, but because we start from a partial state, C_1 may not cover all possible coarse neighbourhoods.

When we check for consistency of the candidate mapping M, we derive all the neighbourhoods contained in C_1, including the overlapping ones (excluded from

C_1	1	0	1	1	1	1
C_2	0	0	1	1	1	1

F_1	00	01	10	11	10	00
F_2	00	10	00	00	01	00
F_3	01	01	00	00	10	10

Fig. 5. Search setup for approximate coarse graining.

the exact coarse graining). Thus, in figure 5, $C_1 = 1\ 0\ 1\ 1\ 1\ 1$ yields neighbourhoods 1 0 1, 0 1 1, 1 1 1, etc. This fully exploits the partial information. If the mapping is consistent for this partial information, we generate the appropriate set of coarse rules.

The brute force approach was used to search for 1D approximate coarse grainings. Even when starting from very small F_1, we discover good approximate coarse graining rules. Each solution set contains the exact rule discovered by [7]. In many cases, additional rules are found that, superficially at least, look like the exact rule. We are now investigating how the additional rules depend on the chosen F_1.

6.3 Next Steps

To continue our exploration, we propose to use the information theoretic approach to guide our choice of search over interesting states: low complexity patterns in the low level CA (for example, gliders in Conway's Game of Life CA [5]). Thus our coarse grainings will be optimised for such patterns, rather than for 'random' behaviour.

This approach of using patterns to guide the coarse graining rule search requires a way of finding such patterns. We will use a coarse graining approach here, too: coarse graining states over space and time to detect repeating behaviour, and hence stationary and moving patterns. The size of these coarse grained cells limits the size of patterns detectable by this approach.

This combination of object detection and CA prediction [7], when combined, should permit object prediction – an important aspect for reliable engineering.

Our next step is to test our proposals by combine object detection and CA prediction on the Game of Life CA, to investigate how the quality of prediction is related to the quality of rule coarse graining, as a function of the patterns used to guide the GA coarse graining search.

Eventually, we want to examine emergence in the context of an environment (combining our ideas of evolution and emergence), both for defining emergent properties, and for achieving robustness of that emergence.

7 Conclusions

We have cast emergence in information theoretic terms, by direct analogy to evolutionary processes, and have shown how this can be used to derive a fitness function for developing models of emergent systems, or emergent implementations of system specifications. We have introduced the concept of neutral emergence, and shown its relationship to robust emergence. As an initial demonstration, we have applied these concepts to coarse graining CAs.

Our eventual aim is to incorporate neutral emergence as a component in a robust emergent engineering process.

Acknowledgements. We thank the anonymous reviewers for their helpful comments. Andrew Weeks is funded by a Microsoft Research European PhD Scholarship.

References

1. Adami, C.: Introduction to Artificial Life. Springer, Heidelberg (1998)
2. Adami, C.: What is complexity? BioEssays 24, 1085–1094 (2002)
3. Adami, C., Cerf, N.J.: Physical complexity of symbolic sequences. Physica D 137, 62–69 (2000)
4. Bar-Yam, Y.: A mathematical theory of strong emergence using multiscale variety. Complexity 9(6), 15–24 (2004)
5. Berlekamp, E.R., Conway, J.H., Guy, R.K.: Winning Ways for your Mathematical Plays. Academic Press, London (1982)
6. Clark, J.A., Stepney, S., Chivers, H.: Breaking the model: finalisation and a taxonomy of security attacks. ENTCS 137(2), 225–242 (2005)
7. Israeli, N., Goldenfeld, N.: Coarse-graining of cellular automata, emergence, and the predictability of complex systems. Phys. Rev. E 73, 026203 (2006)
8. Kauffman, S.A.: The Origins of Order: self-organization and selection in evolution. Oxford University Press, Oxford (1993)
9. Kauffman, S.A.: At Home in the Universe. Oxford University Press, Oxford (1995)
10. Langton, C.G.: Computation at the Edge of Chaos: Phase-Transitions and Emergent Computation. PhD thesis, University of Michigan (1991)
11. Mitchell, M., Crutchfield, J.P., Hraber, P.T.: Dynamics, computation, and the 'edge of chaos': a re-examination. In: Cowan, G.A., et al. (eds.) Complexity: Metaphors, Models, and Reality, pp. 497–513. Addison-Wesley, Reading (1994)
12. Ray, A.: Symbolic dynamic analysis of complex systems for anomoly detection. Signal Processing 84, 1115–1130 (2004)
13. Ronald, E.M.A., Sipper, M., Capcarrère, M.S.: Testing for emergence in artificial life. In: Floreano, D., Mondada, F. (eds.) ECAL 1999. LNCS, vol. 1674, pp. 13–20. Springer, Heidelberg (1999)
14. Shalizi, C.R.: Causal architecture, complexity and self-organization in time series and cellular automata. PhD thesis, University of Wisconsin at Madison (2001)
15. Stearns, S.C., Hoekstra, R.F.: Evolution: an introduction. OUP, Oxford (2000)
16. Stepney, S., Polack, F., Turner, H.: Engineering emergence. In: ICECCS 2006, IEEE Computer Society Press, Los Alamitos (2006)
17. Takens, F.: Detecting strange attractors in turbulence. In: Rand, D.A., Young, L.-S. (eds.) Dynamical Systems and Turbulence, pp. 230–242. Springer, Heidelberg (1981)

New Models for Old Questions: Evolutionary Robotics and the 'A Not B' Error

Rachel Wood and Ezequiel Di Paolo

CCNR, University of Sussex, Brighton BN1 9QH, UK.
r.k.wood@sussex.ac.uk

Abstract. In psychology the 'A not B' error, whereby infants perseverate in reaching to the location where a toy was previously hidden after it has been moved to a new location, has been the subject of fifty years research since it was first identified by Piaget [1]. This paper describes a novel implementation of the 'A not B' error paradigm which is used to test the notion that minimal systems evolutionary robotics modelling can be used to explore developmental process and to generate new hypotheses for test in natural experimental populations. The model demonstrates that agents controlled by plastic continuous time recurrent neural networks can perform the 'A not B' task and that homeostatic mediation of plasticity can produce perseverative error patterns similar to those observed in human infants. In addition, the model shows a developmental trend for the production of perseverative errors to reduce during development.

1 Introduction

In the fifteen years since its inception Evolutionary Robotics (ER) has become a core methodology in Artificial Life. ER offers advantages of automated design, bias reduction and proof of concept capabilities which allow researchers to explore behaviour with models which although simplified, retain their explanatory power [2]. Central to ER is the use of evolutionary algorithms to design artificial 'brains' (and morphologies) for robots and simulated agents: such methods serve to reduce the degree of designer bias built into systems and support the discovery of novel mechanisms underlying behaviour. ER also affords the generation of existence proofs - model systems which demonstrate a particular cognitive capacity under specific conditions. Such systems can be used to devise a set of minimal conditions required for a particular behaviour. Minimal modelling is an integral aspect of ER which reflects a general concern for bias reduction and cognition as an interactional property rather than a computational one. ER has been applied to a very wide range of problems however little work has been done to explore the use of these methods to model *developmental process*. The work described in this paper uses ER to design developmental trajectories in simple simulated agents.

Piaget's delayed manual search task, (a.k.a. the 'A not B' error), is one of the most explored phenomena in developmental psychology [1]. The finding that

F. Almeida e Costa et al. (Eds.): ECAL 2007, LNAI 4648, pp. 1141–1150, 2007.

7-12 month old infants make perseverative errors in retrieving a hidden toy has sparked decades of debate. Subtle alterations in the way in which the task is administered have a important effect on the behaviour observed. The canonical 'A not B' error paradigm has the infant seated at a table fitted with two lidded wells. The experimenter attracts the infant's attention with a toy which is then hidden in one of the wells, after a short delay the child is allowed to retrieve it. This process is repeated a number of times before the toy is hidden in the second well. The delay is again imposed before the child is allowed retrieve the toy. Perseverative errors are observed as subjects reach back to the first hiding place despite having seen the toy being hidden in the second. A great deal of experimentation has been carried out as researchers attempt to establish exactly what skill or knowledge or the perseverating infant lacks. Explanations range from the Piagetian notion that errors arise from immature notions of object physics to Thelen and Smith's persuasive contention that conflict between motor memories and specific task inputs is responsible for perserverative reaching [4].

On this view, the delayed search paradigm is an excellent test case for a minimal, developmental ER approach: the required behaviour is interestingly complex and has been well studied in the natural domain. In order to perform delayed manual search the agent, must first attend to the relevant stimuli and then organise an appropriate motor response. The agent should discriminate the two cases (location 'A' versus location 'B') and produce a different motor response for each. Memory is also implicated: the agent must 'mark' the cued location to the end of a delay period and only then issue a motor response. Thus we are not primarily concerned with *engineering* a solution to the 'A not B' task, and we do not expect to 'explain' perseverative reaching with a simple evolutionary robotics simulation. Rather, our purpose is to *test* the notion that minimal simulation models can be a useful addition to the experimental toolkit for exploring complex developmental phenomena.

This paper describes two experiments designed to test the notion that we can use minimal ER methods to model complex developmental trajectories and asks if we can use this approach *generatively* i.e. can we use the results of such work to come up with new questions and hypotheses for empirical testing. Thus, our measure of success will be whether we can generate a novel hypothesis for theoretical and empirical research on this problem. In both experiments an agent is placed between two acoustic sources ('A' and 'B') one of which emits a signal. A delay is imposed before the agent is allowed to locate the signal source. Thus, the infant's manual search has become locomotion to a target and visual cues have become auditory signals but the crucial interaction between attention to a stimulus and the formulation of a motor response is preserved.

2 Methods

The agent is modelled as a circular body, radius $R_0 = 4$, with two diametrically opposed motors and a two acoustic sensors positioned symmetrically at a 45 degree angle to the motors. The agent can rotate its sensors part way around its body and thus has the possibility to actively 'manage' attention by controlling sensor orientation. Motors can drive forward and in reverse and agent bodies are modelled as rigid, small and having very little mass. Thus motor output translates directly into tangential velocity as the point where the motor is located. The agent's behaviour is controlled by a continuous time recurrent neural network (CTRNN) controller with nine nodes: two sensor nodes, a fully connected, four node intranet, two motor nodes, and an additional motor node to control sensor orientation. Each node in the intranet receives an incoming synapse from each sensory node and each motor node receives an incoming synapse from each node in the intranet; nodes in the intranet also have self connections. Spatial discrimination is provided by the relative activity of the two sensors which, being in different positions on the agent's body, receive differing intensities of external perturbation depending on their positions relative to the source. Neural activity is calculated using:

$$\tau_i \dot{y}_i = -y_i + \sum_j w_{ji} z_j; \quad z_j = \frac{1}{1 + exp[-(y_j + b_j)]} \tag{1}$$

where y_i represents the cell potential, τ the decay constant, b_i the bias, z_i the firing frequency, w_{ij} the strength of the synaptic connection from node i to node j. Uniformly distributed noise, scaled by a gain factor is added to all sensor and motor transduction steps.

Sound is modelled as an instantaneous, additive field with volume attenuated in inverse proportion to distance squared. The intensity of incoming signal is further attenuated by a self-shadowing factor derived from the distance travelled by the signal through the body of the robot. This mechanism devised by Di Paolo [3], provides a simple but biologically plausible means for the agent to distinguish the direction of a sound source. The degree of attenuation is determined by movement and angular orientation of the robot. This body shadowing factor D_{sh} is modelled as linear attenuation proportional to the distance travelled through the body (ibid.).

$$Dsh = Dsen(1 - A), 0 \leq A < 1, A = \frac{D^2 - R_0^2}{D_{sen}^2} \tag{2}$$

where D_{sen} is the distance between the source and the sensor, and D is the distance between the source and the centre of the body. If $A \leq 1$, there is a direct line between source and sensor and $D_{sh} = 0$, maximum attenuation occurs when the sensor is on the opposite side of the body to the sound source. The intensity of incoming signal is calculated by taking the volume at sensor (attenuated by source to sensor distance) and then multiplying it by a linear scale derived from the degree of shadowing (1.0 when $D_{sh} = 0$, 0.1 when $D_{sh} = 2R_0$).

A tournament style GA with rank based selection and strong elitism is used. Populations (N = 60) are evolved for 1000 generations; vector mutation is the sole genetic operator employed. Twelve trials are given with fitness averaged across trials. Fitness is based on final position relative to the target ($F_A = 1 - D_{final}/D_{initial}$) and efficiency of approach trajectory (i.e. the proportion of approach time spent within 4 body radii of the target (F_D). Scores are modulated by an additional factor based on the agent's energetic output during the approach phase. The individual fitness F for a single trial is obtained using

$$F = (a_d F_D + a_a F_A) exp(-a_e \int_{t_i}^{t_f} \sqrt{V_R^2 + V_L^2 dt}) \quad (3)$$

where a_d= 0.25 and a_a=0.75 are respectively weighting factors for approaching and maintaining proximity to the target, $a_e = 0.005$ scales the exponential and V_R and V_L represent the translation speeds of the right and left motors respectively. Network parameters are encoded as real values in a vector of fixed dimensions. Mutation is carried out by perturbing the genome $\mathbf{G}$ with the probability $\mu = 0.005$ in a random direction by adding a normalized random vector $\mathbf{p}$ multiplied by a distance m (in the range [0,1]), thus $\mathbf{G} \longrightarrow \mathbf{G} + \mathbf{mp}$. Unless otherwise indicated parameters in the range [0,1] are mapped to [-3,3] for biases, [0.4,2] time constants and [0.05, 10] sensor gains, [-5, 5] motor gains. Parameter ranges were selected having been used successfully in previous ER work.

3 Experiment One

The agent moves on an unbounded, 2-D plane, its task, to locate and approach the source of an auditory signal emitted by one of two randomly placed beacons ('A' and 'B') located symmetrically with respect to the agent's initial position (initial distance in the range [35, 70] units). Agents begin each presentation at a fixed orientation. A 400 timestep trial comprises three phases: signal phase when a high volume, irregular signal is emitted by one beacon, delay period and approach phase when the agent is permitted to move toward the target (see fig. 1). Although the agent cannot move during the signal and delay phases sensor rotation is permitted. Both beacons emit a low volume variable signal during the delay and approach phases.

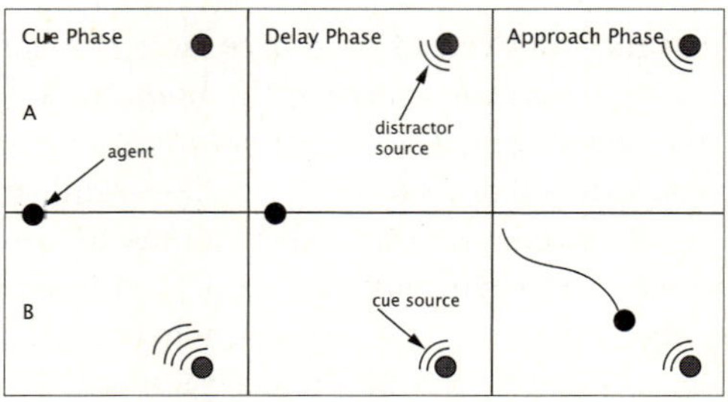

Fig. 1. Schematic of the delayed search task environment

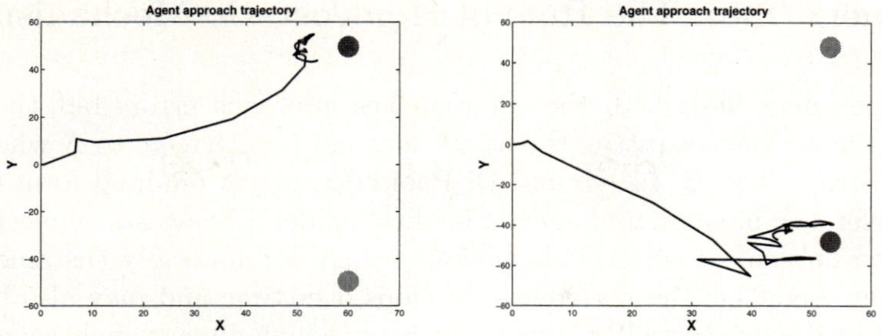

Fig. 2. Experiment one: approach trajectories for agent with increased source salience, figure shows two correct approaches one to the 'A' location (l) and one to the 'B' location (r).

3.1 Experiment One: Plasticity

The weightings on synaptic connections are set by a mechanism based on Hebbian learning. Four constants (α, β, γ and ϵ) and a learning rate, all in the range [-1,1] and under evolutionary control, determine the mapping between synaptic activity and weight change. The learning rate provides a degree of localised plasticity by modulating the rate at which individual synapses adapt. Weights are initialised with small random values and constrained to the range [-8.0, 8.0]. At each update step modifications to the weights on each synapse are obtained via

$$\frac{dw_{ij}}{dt} = \eta(\alpha y_i y_j + \beta y_j + \gamma y_i + \epsilon) \qquad (4)$$

where η represents the learning rate for the synapse, and α, β, γ and ϵ are constants in the range [-1,1].

3.2 Experiment One: Results

A first set of evolutionary runs produced agents that could achieve the task but with difficulty and stereotyped behaviours. It was found that the task was easier to evolve if the beacons both emitted sound at a high volume during the approach phase (this does not break the conditions of the A not B paradigm since manipulations increasing saliency are also used in infant experiments). Five evolutionary runs have been carried out and four produced agents with fitness of 0.70 or higher. The best agent has a score of 0.74, its approach strategy differs from that used by the agents described above. Here, the agent does not approach all targets from the same direction and its motion trajectories are less stereotyped. Performance of the task is more efficient both in terms of the number of targets correctly located and the agent's approach trajectories see fig 2. The agent finds its target in 10 trials (out of twelve) and approaches both 'A' and 'B' targets efficiently indicating that it discriminates between signals (rather than responding to the presence/absence of a cue from one side). However, there is little evidence of a perseverative pattern in its behaviour.

4 Experiment Two: The Role of Homeostatic Mechanism

In the first experiment described, the agent makes approach errors but they are not systematically perseverative: the agent does not regularly go to A when the signal has swapped to B. Iizuka and Di Paolo describe a minimal form of behavioural preference in simulated agents: in their model a homeostatic mechanism is used to implement selective behaviour which is temporally sustained without being invariant i.e. the preference develops over time and may also be altered [7]. A plastic neurocontroller is evolved with two high dimensional 'boxes' or regions defined in the space of neural dynamics. An association between the regions and the behavioural options available to the agent is created through evolution. Their model shows durability of preference and preference transitions through a process of mutual constraint between internal and external dynamics which provides the link between environmental change and contingent neural reorganisation. Their model has some interesting features in common with the A not B error paradigm; the delayed manual search task can be conceived in terms of choice, (i.e. the infant has two options for retrieving the toy), and, on this view, perseverative errors take the form of a preference or habit enacted at some level of internal dynamics. In the dynamic fields model of perseverative reaching, the error is produced through the action of a 'pre-shape' in motor planning space which biases the system toward repeated actions rather than the generation of new ones [9]. This 'motor memory' based model of perseveration is highly persuasive but requires a relatively complex modelling approach. So, it is interesting to ask if a very minimal model of homeostatic behaviour control can be used to invoke perseverative responding in simple simulated agents.

4.1 Experiment Two: Methods

To explore this question further experiment one has been extended to include homeostatic mediation of synaptic plasticity. Two homeostatic regions are specified as a function of the firing-rate of the post-synaptic node and these determine the rate of change in the pre-synaptic weights. Plasticity is zero inside the regions and elsewhere is determined by a function based on post-synaptic firing. The regions correspond to firing rates of [0.15, 0.4] and [0.6, 0.85], (their locations are arbitrary and here we take the ranges used in [7]). At the beginning of each evolutionary run correspondence between the regions and the 'A' and 'B' targets is randomly specified for each node. Thus one neuron's 'A' box maybe the high firing region and another's the low firing region. Two region control of plasticity is not applied to sensory or motor neurons to avoid over biasing their activity. For these neurons a single region is specified in the range [0.15, 0.85]. Weight change is calculated using a Hebbian rule with linear dependence on the firing rate of the pre-synaptic node, an evolved learning rate is also applied (as experiment one.) Weights are updated according to:

$$\delta w_{ji} = \eta_{ji} z_i p(z_j) \tag{5}$$

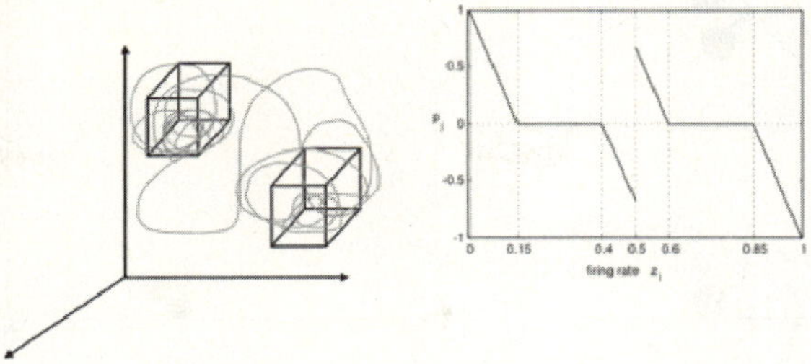

Fig. 3. Schematic of plastic facilitation for intranet nodes, two high-dimensional 'boxes' or regions are specified in the space of neural firing. Right: how plasticity corresponds to the homeostatic regions, the plot shows the plasticity function (P_j) as a function of neural firing (z_j). When post-synaptic firing is in one of two flat regions $P_z = 0$ and plasticity is inhibited (schematic adapted from [7]).

where δw_{ji} is the change in w_{ji} per unit of time, $p(x)$ is the plasticity function and η_{ji} the learning rate. Parameters are restricted to positive numbers to ensure that the function brings firing rates back toward the specified firing regions. Fitness is calculated as previously with the addition of a term designed to reward homeostatic agents. For every time step that a specified neuron behaves homeostatically a counter is incremented by 1.0; thus, on an 'A' trial the fitness is awarded for homeostasis if the firing rate of the specified neuron remains within the 'A' box. If it fires at a rate which is inside the 'B' box no fitness is awarded and if the firing rate is outside both an increment of 0.5 is given. This score is averaged for all nodes and the term obtained F_H is used to scale fitness awarded for approach and proximity to target. In all other respects the model is as described in experiment one:

$$F = (F_D + F_P)F_H \tag{6}$$

4.2 Experiment Two: Results

Five evolutionary runs have been completed, two produced agents with fitness of 0.6 or above and the best agent scored 0.62, here we examine a single agent. The approach paths taken by this agent are direct and having located the source it remains there for the remainder of the trial. The agent is equally successful on both 'A' and 'B' trials. The reduction in overall fitness reflects scaling of scores for approach and proximity by F_H, the score for homeostasis. The error pattern produced by the homeostatic agent is unlike that observed in experiment one. Non-homeostatic agents make more frequent errors and mistakes do not follow a particular pattern. In the homeostatic system errors occur less often and tend to be clustered around 'swaps' (i.e. the presentation immediately following transition from signal delivered at 'A' to signal delivered at 'B' and vice

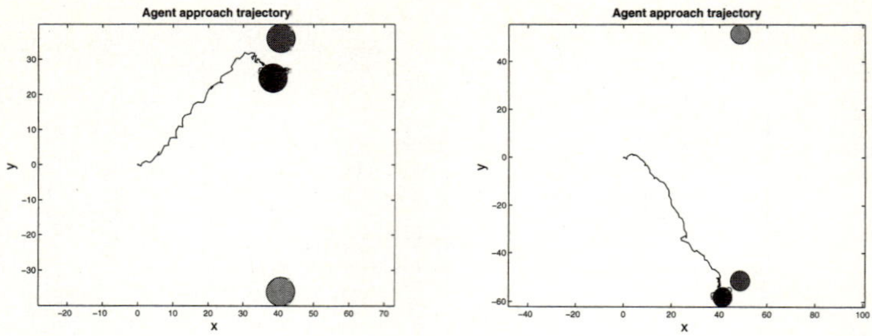

Fig. 4. Experiment two: approach trajectories for agent with homeostatic mediation of plasticity, figure shows one 'A' (l) and one 'B' (r) trial.

versa). The rate of perseverative errors in homestatic versus non-homeostatic agents is shown in figs 5 and 5. Here, trials comprising six presentations to one side followed by 6 presentations to the other are administered. Start order is randomised so that agents receive an equal number of trials beginning on each side. The first presentation in each trial has been discarded as errors here cannot be considered perseverative as the agent has had no previous exposure to the test environment.

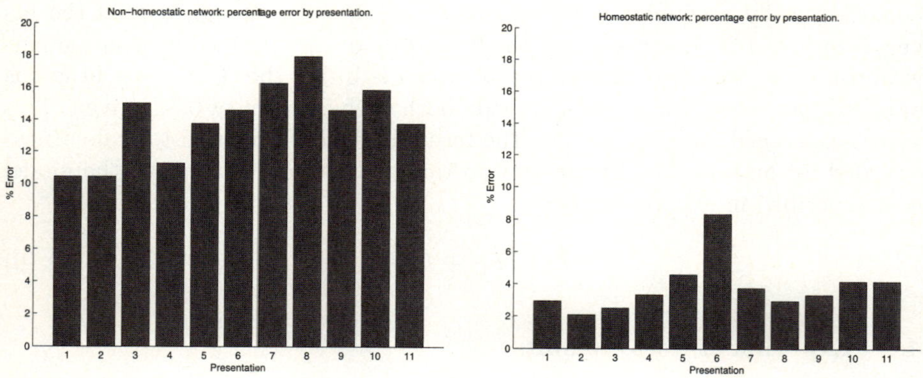

Fig. 5. Percentage error by presentation (l) non homeostatic network, (r) Homeostatic network, 240 independent trials administered.

Figure 5 (left) shows the percentage error made on each presentation by a non-homeostatic agent evolved under the conditions described in experiment one, 240 trials were given. The error rate varies between 10 and 20 per cent and errors are made equally often on all presentations. There is a slight increase in the number of errors following the signal swap but there is no evidence of a peak in errors on the presentation immediately following it. Figure 5 (right) shows

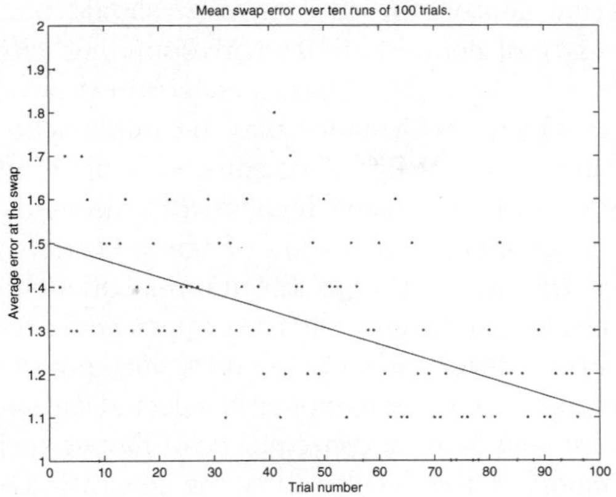

Fig. 6. Average of ten independent runs of 100 trials comprising 6 'A' and 6 'B' presentations, line = linear regression on the data, negative slope indicates that the swap error reduces over time.

the percentage of errors made by a homeostatic network under the same trial regime, here we see a fewer errors overall and a peak in the number of errors on the trial following the signal swap. Examination of the behaviour of agents on trial six, (following the swap), indicates that the errors made are perseverative i.e. agents continue to approach the source at which the signal was delivered in the previous six presentations. Figure 6 shows the mean swap error over ten independent runs of 100 trials (with the 6/6 presentation regime described above). Linear regression on the data produces a trend line with a negative slope (fig 6 solid line) indicating that swap errors decrease over time. Data points are widely scattered thus the R-square value indicates a weak fit. However, analysis of the mean swap error over time indicates that there is a developmental trend in the production of perseverative errors in the model. Further statistical analysis is required to properly explore this result.

5 Discussion

This paper presents results from two experiments: in the first agents must rely on the signal emitted by one source at the beginning of each trial as the only cue indicating the location to approach. In the second experiment the task of locating the selected source is simplified because sources emit sound throughout the trial, however the agent is still required to 'remember' where the signal came from. Plastic changes and processes of neural self-organisation play a significant role the behaviour of agents in this task. The homeostatic mechanism is introduced as a means to test if we can obtain perseverative responding; this strategy reflects a view of homeostasis as a mechanism for conservation in adaptation. In experiment two plasticity is mediated by an evolutionary requirement that the activity of internal neurons in the controller should remain within one of two regions in the space of neural activity corresponding with the two types of task environment (i.e. 'A' and 'B'). Thus, a constrained developmental process limits the means by which performance may be implemented. This aspect of the work is interesting for a number of reasons: very little work has been done with plastic CTRNN controllers using homeostatic mechanisms of adaptation, and this experiment combines exploration of homeostatic dynamics with further investigation of the application of the minimal modelling approach. The homeostatic mechanism reproduces a pattern of perseverative errors observed in human infants and also a developmental trend for such errors to disappear. Neither of these characteristics has been explicitly selected for and therefore we can hypothesise that these results are a consequence of having included homeostatic mechanisms in the model. These results allow the generation a hypothesis about mechanisms underlying the 'A not B' error in infants such that there is some process of regulation of plasticity which plays a role similar to the homeostatic mechanism in the model. This process supports the conservation of patterns of neural activation with the effect of adapting to repeated sensorimotor requirements. An original aim of this work was to test the notion that minimal ER simulation can be used to model complex developmental trajectories, the results discussed here indicate that such methods can be valuable tools for exploring empirical problems.

References

1. Piaget, J.: The Origins of Intelligence in the Child. Routledge & Kegan Paul, London (1952)
2. Harvey, I., Di Paolo, E., Wood, R., Quinn, M., Tuci, E.: Evolutionary Robotics: A New Scientific Tool for Studying Cognition. Artificial Life 11(1-2) (2005)
3. Di Paolo, E.A.: Behavioural Coordination, Structural Congruence and Entrainment in a Simulation of Acoustically Coupled Agents. Adaptive Behaviour 8(1), 27–48 (2000)
4. Thelen, E., Smith, L.B.: A Dynamic Systems Approach to the Development of Cognition and Action. MIT Press, Cambridge (1994)
5. Beer, R.: Toward the Evolution of Dynamical Neural Networks for Minimally Cognitive Behaviour. In: Maes, P., Mataric, M., Meyer, J-A., Pollack, J., Wilson, S. (eds.) From Animals to Animats 4: Proceedings of the Fourth International Conference on Simulation of Adaptive Behaviour, MIT Press, Cambridge (1996)
6. Ashby, W.: Design for a Brain. John Wiley and Sons Inc., Chichester (1960)
7. Iizuka, H., Di Paolo, E.A.: Toward Spinozist Robotics: Exploring the Minimal Dynamics of Behavioural Preference. Adaptive Behaviour (Accepted)
8. Tuci, E., Quinn, M., Harvey, I.: Using a Net to Catch a Mate: Evolving CTRNNs for the Dowry Problem. In: From Animals to Animats 7: Proceedings of the Seventh International Conference on Simulation of Adaptive Behaviour, MIT Press, Cambridge (2002)
9. Thelen, E., Schoener, G., Scheier, C., Smith, L.: The Dynamics of Embodiment: A Dynamic Field Theory of Infant Perseverative Reaching. Behavioural and Brain Sciences 24, 1–86 (2001)

PLAZZMID: An Evolutionary Agent-Based Architecture Inspired by Bacteria and Bees

Susan Stepney, Tim Clarke, and Peter Young

Department of Computer Science; Department of Electronics; Department of Biology
University of York, UK, YO10 5DD

Abstract. Classical evolutionary algorithms have been extremely successful at solving certain problems. But they implement a very simple model of evolutionary biology that misses out several aspects that might be exploited by more sophisticated algorithms. We have previously critiqued the traditional naïve approach to bio-inspired algorithm design, that moves straight from a simplistic description of the biology into some algorithm. Here we present a process for developing richer evolutionary algorithms abstracted from various processes of biological evolution, with a corresponding richer analogical computational structure, and indicate how that might be further abstracted.

Key words: evolutionary algorithms, meta-evolution

1 Introduction

Classical evolutionary algorithms have been extremely successful at solving certain optimisation problems. Without *too* much caricaturing, these algorithms can be said to implement the model of evolutionary biology shown in figure 1. This simple model and simple analogy miss out, or obscure, several things that might be exploited by more sophisticated algorithms, including: the full richness of the genotypic structure, the corresponding richness of the evolutionary operators acting on that structure, the richness of the mapping from genotype to organism (phenotype), and the regulatory feedback from the phenotype to the genotype's expression.

In [25] we critique the traditional approach to bio-inspired algorithm design, that moves straight from a simplistic description of the biology into some algorithm. There we propose a "conceptual framework", including mathematical and computational modelling, abstraction of principles, and instantiation into relevant application domains. Here we indicate in more detail what such a process could look like, in the context of structures from various processes of bacterial evolution, with a corresponding richer analogical computational structure, and indicate how that might be further abstracted. We conclude by describing PLAZZMID, a computational evolutionary system for dynamic problems that we are developing under this process.

F. Almeida e Costa et al. (Eds.): ECAL 2007, LNAI 4648, pp. 1151–1160, 2007.

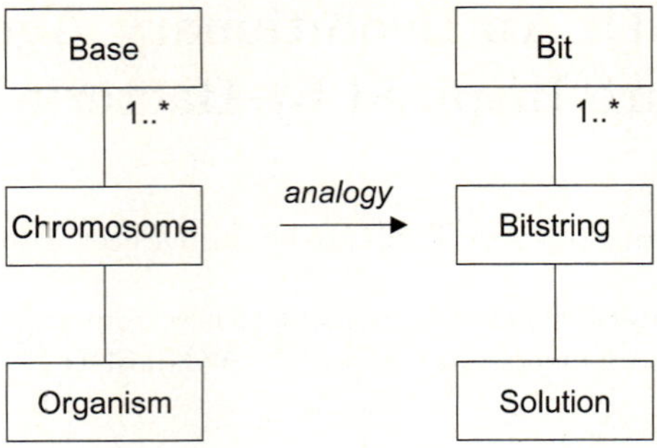

Fig. 1. UML class diagram of the simple model of evolutionary biology on which classical evolutionary algorithms are based, and the corresponding computational analogy.

2 A Model of a Bacterial Genome

Bacteria adapt to novel environments through rapid evolution, aided by a lean and efficient genome organisation [14][29]. Typically, a bacterium has a single circular chromosome that encodes the core functions that all members of the species need. It also has an *accessory genome*, which confers specific adaptations to the environment; each member of the species may have a different combination of genes in this set. Examples of accessory functions are the ability to grow on unusual food sources, to resist toxins, or to colonise the tissues of animals or plants. Parts of the accessory genome may be inserted into the chromosome, but much of it is carried on *plasmids*, which are mini-chromosomes with a high propensity to transfer from one bacterium to another. *Transposons* are small packages of genes that can jump from plasmid to plasmid, or to the chromosome. Mating may result in the transfer of plasmids, and also in the replacement of short stretches of the recipient's chromosome by homologous genes from the donor. This mating system allows major rearrangements of the genome without the excessive cost that this incurs in higher organisms that have equal genetic contributions from mates.

Figure 2 shows a UML class diagram that captures the main structures of the bacterial genotype and the bacterial organism of interest in formulating a more sophisticated evolutionary algorithm. The things to be noted in contrast to figure 1 are the rich structure of the genome, containing a hierarchy of components, and a corresponding richer structure of the phenotype. The diagram shows the presence, but not the significance, of the regulatory feedback loop (a protein expressed by one gene may regulate the expression of another, forming a complex *gene regulatory network*). The diagram indicates the redundant encoding (many codons to one amino acid) but not the highly non-linear genotype-phenotype mapping (obtained through complex protein folding). For clarity and brevity, the evolutionary operators have been omitted, but there are operators that work at every level of the genomic structure, from mutations acting on single bases, through gene duplications and transposons, to plasmid exchange.

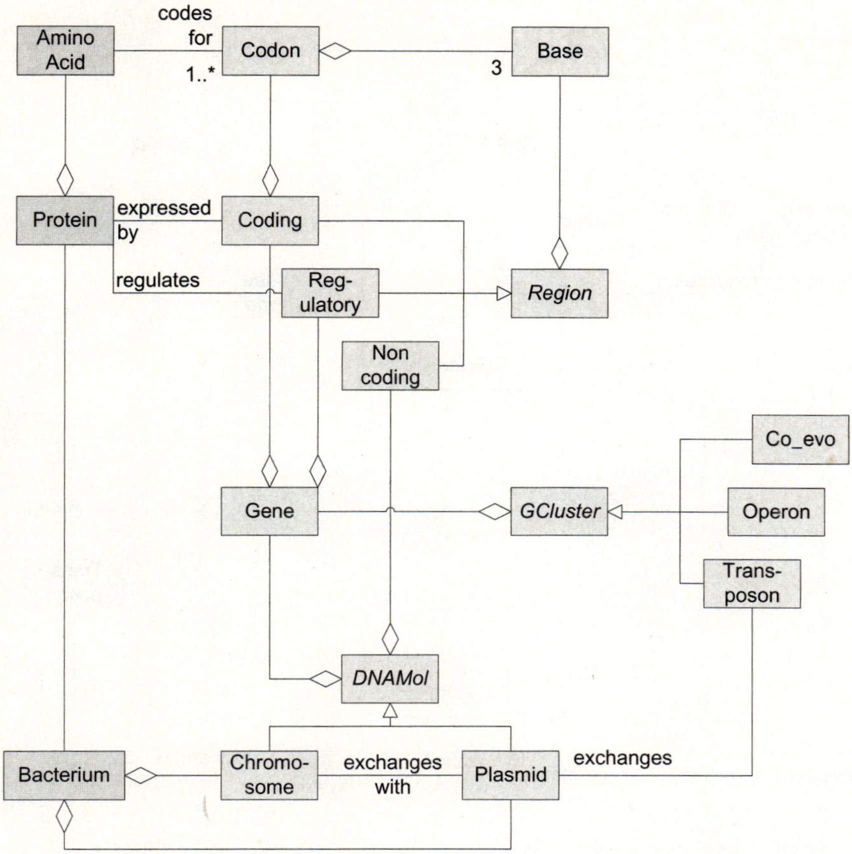

Fig. 2. A UML class diagram of the structure of the bacterial genome and its relationship to the bacterial phenotype.

3 An Evolutionary Architecture Inspired by Bacteria

We can preserve this rich biological structure with computational analogues of these components and their dynamics. We define a computational architecture with analogues of the complex multi-level structure of DNA (via an assembly language syntax analogy), of the complex evolutionary processes, including replication and transposon exchange (via multi-level evolutionary operators tailored to the syntactic structures), and of the complex dynamic behaviour of the phenotype (via execution of the relevant language as part of a subsumption architecture). See figure 3.

The analogy in more detail is: **Gene**: a piece of machine code defining a simple machine (or *agent*). **Genome**: a collection of machine code programs (static code), arranged in chromosomes and plasmids. Reproductive and evolutionary *operators* act on this genotypic program code, at a variety of levels (bits, bytes,

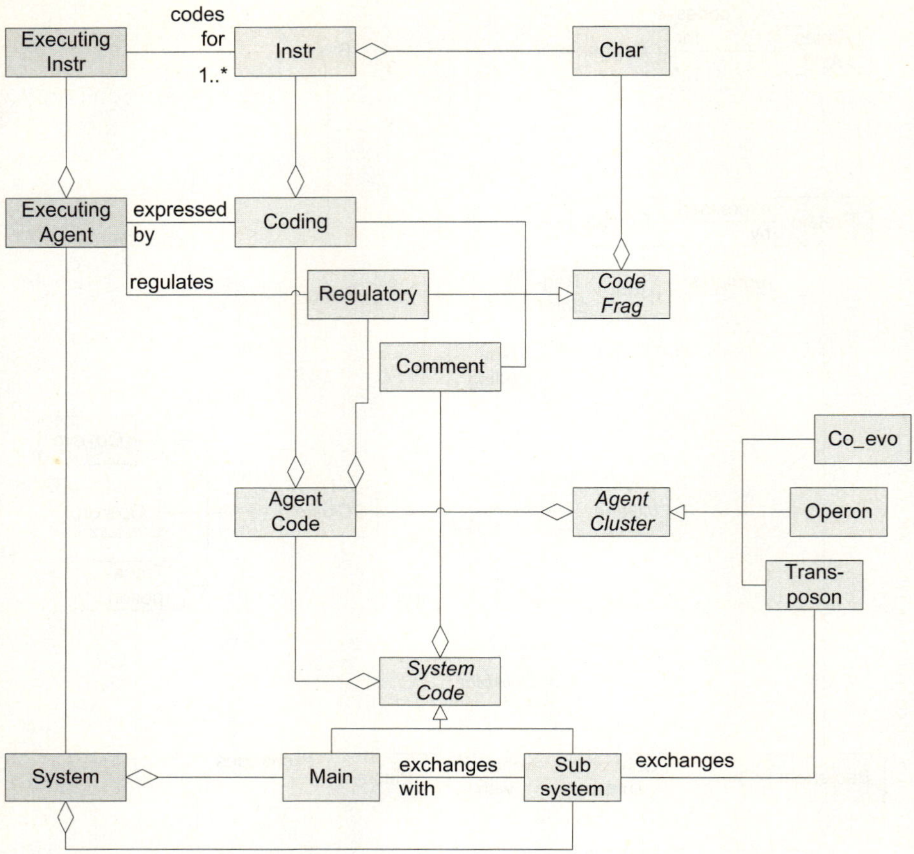

Fig. 3. A UML class diagram of the corresponding structure of the computational genome (assembly language program) and its relationship to the computational phenotype (agent architecture).

instructions, modules, etc), including mutation, crossover, gene duplication, and exchange of transposons. **Gene expression**: the process of moving from static code to corresponding executing agent. **Gene regulation**: the process of controlling which genes get expressed (which agents get to execute). **Proteins**: the executing agents.

Proteins have complex structure and behaviour. To capture this kind of complexity, the agents can be arranged in a subsumption (layered) architecture (figure 4); each can provide i/o to other agents (or the environment), and can *interact* to produce new agents at other subsumption layers (the interaction language is part of the agent's code, and is also evolvable). Some outputs from low level agents feed back into the genome, regulating gene expression. **Phenotype** (organism): the entire collection of executing agents (which changes over time). The

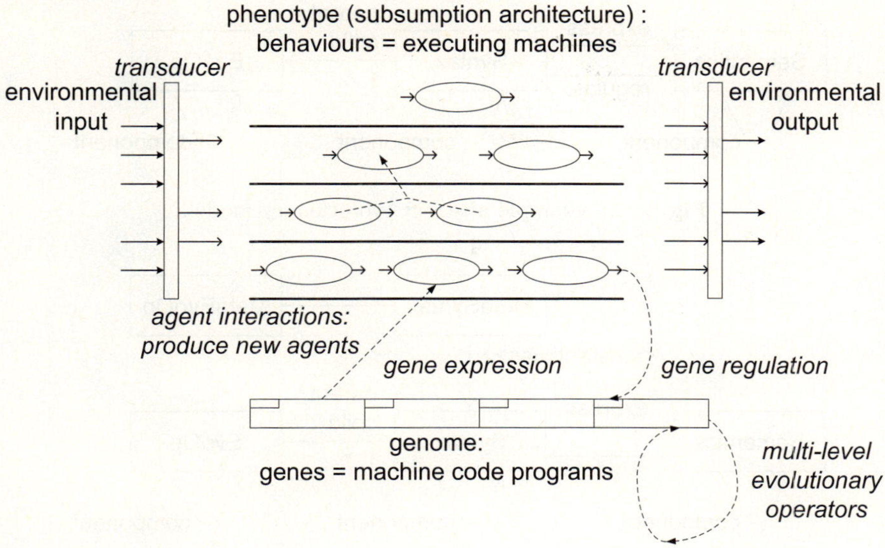

Fig. 4. An agent-based subsumption architecture analogue of the bacterial phenotype model.

organism's *fitness* (reproduction probability) is a function of the behaviour of this phenotype executing in an environment. **Epigenetics**: the part of an organism's state that can be preserved on reproduction, for example, which genes are currently being expressed, or the current internal state of the executing agents.

Again for brevity, the evolutionary operators have been omitted from these diagrams. As with the bacterial model, the computational model can have a rich set of evolutionary operators that manipulate the genome at all the levels of its structure, from single characters, through instructions, to high level structures, related to the various syntactic structures of the genome.

The architecture incorporates a regulatory feedback mechanism, controlling the "expression" of the low level agents. It does not (here) incorporate any developmental process. The transcription process goes directly from static code to executing agent. This provides the desired non-linear mapping, and is close in spirit to biological processes (of the protein as an "executing machine", resulting from a relatively simple transcription from codons to amino acids, followed by a highly non-linear protein folding process). The mapping can also incorporate redundancy, with textually distinct instructions having the same semantics, eg ADD 1 and INC 1.

4 Abstracting, and Meta-behaviour

Note that the rather direct computational analogy sketched here does not incorporate the *abstraction* task that is part of the development process we outline

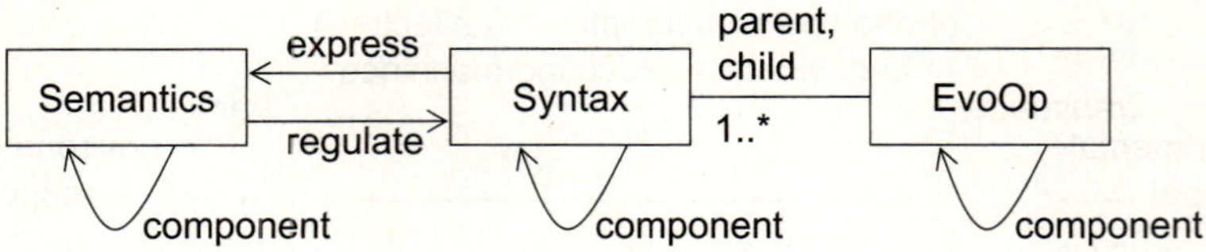

Fig. 5. An example abstract evolutionary model.

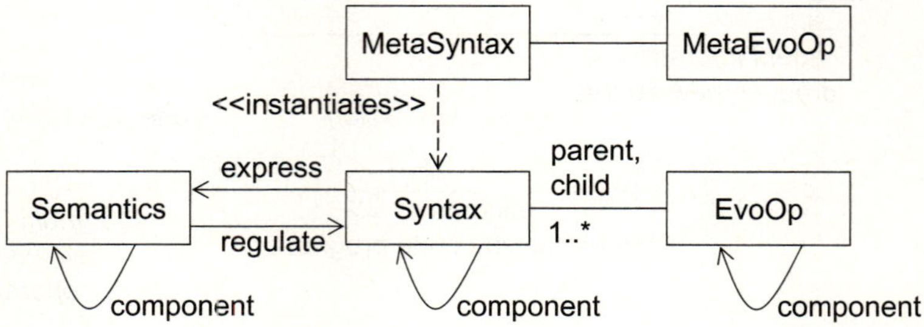

Fig. 6. An example abstract meta-evolutionary model.

in [25]. We should build a more abstract model, of which the biological specifics are one instantiation, for example (figure 5) in terms of *syntax* (structure of the genotype), the corresponding *semantics* (mapping to the phenotype, *cf* protein expression and regulation), *evolutionary operators* acting at the different levels of the syntactical structure (for example, via an attribute grammar [1][10]) that change instances conforming to the syntax (*cf* single letter mutation, gene duplication, transposon exchange).

We could then define a computational architecture as an alternative instantiation of this abstract model, in terms of particular architecture components. Treating the example in section 3 in this way, its instantiation would be of a syntax covering characters, assembly language instructions, and programs, a semantics of the agents executing their code, and suitable evolutionary operators acting at the various syntactic levels. However, the abstraction allows other instantiations to be made, corresponding to possibly less direct analogies, but potentially better fitting some application domain.

At first sight, the abstract model of figure 5 might look little more sophisticated than the simple model of figure 1. However, that is misleading. The structure of the abstract genome that comes from treating it as a *syntax* points to similar structures in the semantics and the evolutionary operators, and the regulatory feedback from the semantics to the syntax indicates a much more dynamic system. Additionally, the idea that the evolutionary operators change *instances* of syntax leads to a further abstraction at the meta-level: that of meta-evolutionary operators that evolve the *classes* of the syntax, introducing and removing syntactic structures, and their corresponding attributes of opera-

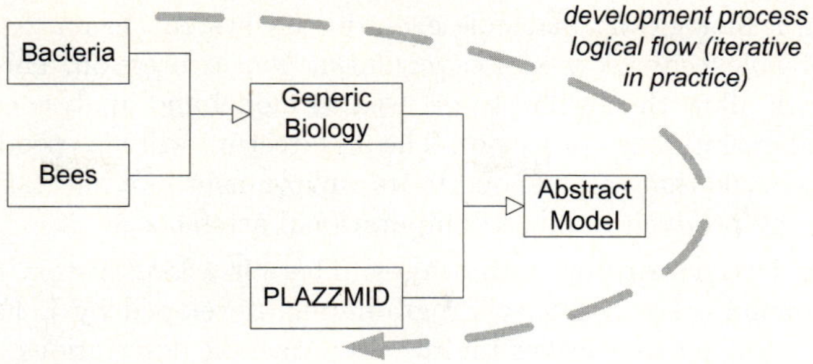

Fig. 7. The modelling, abstraction, and instantiation process.

tional semantics and evolutionary operators, via operations on the meta-syntax that defines a particular instance of the syntax (figure 6).

5 Next Steps

We are now developing this process and architecture to build a fully flexible computational evolutionary system. Our final system will be an abstraction of both bacterial genomes (described here) and bee genomes, and for this reason is called PLAZZMID. The computational system will be based around the parallel programming language occam-π [28], defined using a parallel language graph-based syntax, rather than the usual "parallel fixup" applied to a tree-based syntax.

The UML class diagrams given above are merely preliminary, partial models, illustrating only some parts of the inspiring biological systems. They are also only static class models. Full dynamic models need to be developed for PLAZZMID, to include the *processes* of expression, regulation and evolution, and their abstractions. The tasks needed to achieve this are (figure 7): • Model the biological systems, in terms of DNA, genes, protein expression and regulation, genetic operators, etc, and including interaction with an environment of potentially co-evolving organisms. (The purpose of this is *not* to build models of full biological rigour: it is to build models sufficient for defining the analogous computational architecture. Nevertheless, it is expected that the biological models will be of potential interest to biologists.) • Build a generic biological evolutionary model, which captures both the bacterial and bee biological specifics of DNA-based evolution. • Build a more abstract model, of which the biological specifics are one instantiation. • Define an alternative instantiation of the abstract model, in terms of computational architecture components.

6 Related Work

The individual components in the PLAZZMID architecture have been tried and tested in isolation (see below). However, this brings them all together for the first time, to form a biologically plausible evolutionary system, in particular, incorporating evolvable feedback processes regulating gene expression. This biological plausibility will allow the system to be used to model and analyse questions of real biological evolutionary processes. The architecture will also produce a rich, dynamic phenotype that can respond to its environment in a naturally adaptive manner, thereby producing robust computational artefacts.

Evolving Programming Languages. There is a long history of evolving structured genomes. Evolutionary Programming, developed by L Fogel in the 1960s [8], was devised to discover Finite State Machine descriptions. In Genetic Programming [11], the genome is (usually) a tree structure representing a program in a HLL, and the program is evolved; Linear GP [2] is used to evolve assembly language programs. Spector's "Push" is a stack-based language designed for evolutionary computation [24]. Quantum circuit descriptions, a low level language of quantum programming, can be evolved [16][17]. Ray's Tierra [22] is a virtual machine and environment specially designed to support evolution of digital organisms.

Gene Expression Architectures. Gene expression involves a non-linear distancing of the genome (DNA, or search space) from the phenotype (proteins, or solution space). Many approaches have been used in artificial evolution, usually involving interpreting the genome as a recipe, or *instructions for building* the phenotype. For example, L-systems distance the genome (an L-systems description) from the phenotype (typically a picture) by a turtle graphics "transcription" process, and have been used in an evolutionary setting [18]. Grammatical Evolution (GE) [23] evolves a numerical genotype, which is interpreted as a sequence of instructions for constructing a (syntactically correct) program from a (fixed) grammar. See also [19], which lists several potential advantages of such an approach.

Gene Regulation Architectures. Gene regulation controls which genes get expressed, allowing dynamic feedback and control. In particular, environmental inputs can affect the regulation, allowing the phenotype to adapt to environmental conditions. These biological ideas have been abstracted into a variety of evolvable computational control architectures [3][5][6][13].

Subsumption Architectures. Brooks invented his subsumption architecture [4] as a way of incrementally designing "intelligent" behaviour in a series of relatively simple behavioural layers. Each layer in the architecture provides a simple additional behaviour. This layering, suggested by biological evolution, provides a general flexible architecture in which more complex computational behaviours can be incrementally evolved [12][15][26].

Modelling Language: UML. The de facto standard Unified Modelling Language (UML) [20], designed initially for modelling computational systems, is well suited to agent-based modeling [21], and has been successfully applied to modelling a range of (parts of) biological systems [7][9][27]. Class diagrams model types and relationships (eg, as shown earlier, a bacterium *contains* a chromosome and several plasmids, each of which *contain* several transposons; plasmids are *associated* with the chromosome via transposon exchange). Sequence diagrams model the interaction between objects over time (eg, gene regulatory operation; the evolutionary lifecycle of a population), and interactions with the environment. State charts model the lifecycle of individual objects, or classes of objects (eg, a gene, a chromosome, an organism). Additionally, MDA/MDD (Model Driven Architecture / Model Driven Development) techniques (which often use UML) provide robust approaches to transforming models, including the kinds of abstractions, analogies, and instantiations used in PLAZZMID.

7 Conclusions

Whilst classic evolutionary algorithms based on a simple model of evolutionary biology have been successful as optimisers, they have not exploited the full richness and variety of the biological processes. We have sketched here a process that highlights that richness, and moreover points the way to introducing (possibly non-biological) *meta*-operators. We have outlined our plans for PLAZZMID, a system that we are designing based on these principles. PLAZZMID will be capable of exploring questions from theoretical evolutionary biology, and of solving dynamic computational problems, such as evolving for homeostasis in a variable environment.

Acknowledgments. We thank Richard Paige for helpful comments on an earlier draft.

References

1. Abramson, H., Dahl, V.: Logic Grammars. Springer, Heidelberg (1989)
2. Banzhaf, W., et al.: Genetic Programming. Morgan Kaufmann, San Francisco (1998)
3. Bentley, P.J.: Evolving fractal gene regulatory networks for graceful degradation of software. In: Babaoğlu, Ö., Jelasity, M., Montresor, A., Fetzer, C., Leonardi, S., van Moorsel, A.P.A., van Steen, M. (eds.) Self-star Properties in Complex Information Systems. LNCS, vol. 3460, pp. 21–35. Springer, Heidelberg (2005)
4. Brooks, R.A.: Cambrian Intelligence. MIT Press, Cambridge (1999)
5. Clegg, K., Stepney, S., Clarke, T.: Using feedback to regulate gene expression in a developmental control architecture. In: GECCO 2007, London, UK, ACM Press, New York (2007)
6. Clegg, K., Stepney, S., Clarke, T.: A reconfigurable FPAA architecture based on genetic regulation. In: FPL 2007, Amsterdam, Netherlands, IEEE Computer Society Press, Los Alamitos (2007)
7. Efroni, S., Harel, D., Cohen, I.R.: Toward rigorous comprehension of biological complexity: modeling, execution, and visualization of thymic T-cell maturation. Genome Res. 13(11), 2485–2497 (2003)
8. Fogel, L.J., Owens, A.J., Walsh, M.J.: Artificial Intelligence through Simulated Evolution. Wiley, Chichester (1966)
9. Kam, N., Cohen, I.R., Harel, D.: The immune system as a reactive system: modeling T cell activation with Statecharts. In: Proc. Visual Languages and Formal Methods, IEEE Computer Society Press, Los Alamitos (2001)

10. Knuth, D.E.: Semantics of context-free languages. Math. Systems Theory 2(2), 127–145 (1968)
11. Koza, J.R.: Genetic Programming: on the programming of computers by means of natural selection. MIT Press, Cambridge (1992)
12. Koza, J.R.: Evolution of subsumption using genetic programming. In: ECAL 1991, Paris, France, MIT Press, Cambridge (1992)
13. Kumar, S.: A developmental genetics-inspired approach to robot control. In: GECCO 2005 SOEA workshop, pp. 304–309 (2005)
14. Lan, R.T., Reeves, P.R.: Intraspecies variation in bacterial genomes: the need for a species genome concept. Trends in Microbiology 8, 396–401 (2000)
15. Liu, H., Iba, H.: Multi-agent Learning of Heterogeneous Robots by Evolutionary Subsumption. In: Cantú-Paz, E., Foster, J.A., Deb, K., Davis, L., Roy, R., O'Reilly, U.-M., Beyer, H.-G., Kendall, G., Wilson, S.W., Harman, M., Wegener, J., Dasgupta, D., Potter, M.A., Schultz, A., Dowsland, K.A., Jonoska, N., Miller, J., Standish, R.K. (eds.) GECCO 2003. LNCS, vol. 2724, pp. 1715–1718. Springer, Heidelberg (2003)
16. Massey, P., Clark, J.A., Stepney, S.: Evolving quantum circuits and programs through genetic programming. In: Deb, K., et al. (eds.) GECCO 2004. LNCS, vol. 3103, pp. 569–580. Springer, Heidelberg (2004)
17. Massey, P., Clark, J.A., Stepney, S.: Human-competitive evolution of quantum computing artefacts by genetic programming. Evolutionary Computation Journal 14(1), 22–40 (2006)
18. McCormack, J.: Interactive evolution of L-system grammars for computer graphics modelling. In: Green, D.G., Bossomaier, T. (eds.) Complex Systems: from Biology to Computation, pp. 118–130. IOS Press, Amsterdam (1993)
19. O'Neill, M., Ryan, C.: Incorporating gene expression models into evolutionary algorithms. In: GECCO 2000 Workshops, AAAI press, Menlo Park (2000)
20. Object Management Group. UML 2.0. `http://www.uml.org/`
21. Odell, J., Parunak, H., Bauer, B.: Extending UML for Agents. In: AOIS Workshop at AAAI (2000)
22. Ray, T.S.: Artificial Life. In: Dulbecco, R., et al. (eds.) Frontiers of Life, Volume One The Origins of Life, Academic Press, London (2001)
23. Ryan, C., Collins, J.J., O'Neill, M.: Grammatical Evolution: evolving programs for an arbitrary language. In: Banzhaf, W., Poli, R., Schoenauer, M., Fogarty, T.C. (eds.) EuroGP 1998. LNCS, vol. 1391, pp. 83–95. Springer, Heidelberg (1998)
24. Spector, L., Klein, J., Keijzer, M.: The Push3 Execution Stack and the Evolution of Control. In: GECCO 2005, Washington DC, USA, pp. 1689–1696. ACM Press, New York (2005)
25. Stepney, S., Smith, R.E., Timmis, J., Tyrrell, A.M., Neal, M.J., Hone, A.N.W.: Conceptual Frameworks for Artificial Immune Systems. Int. J. Unconventional Comp. 1(3), 315–338 (2005)
26. Togelius, J.: Evolution of a Subsumption Architecture Neurocontroller. J. Intelligent and Fuzzy Systems 15(1) (2004)
27. Webb, K., White, T.: UML as a cell and biochemistry modeling language. BioSystems 80, 283–302 (2005)
28. Welch, P.H., Barnes, F.R.M.: Communicating mobile processes: introducing occam-pi. In: Abdallah, A.E., Jones, C.B., Sanders, J.W. (eds.) Communicating Sequential Processes. LNCS, vol. 3525, pp. 175–210. Springer, Heidelberg (2005)
29. Young, J.P.W., et al.: The genome of Rhizobium leguminosarum has recognizable core and accessory components. Genome Biology 7:R34 (2006)

Self-organizing Acoustic Categories in Sensor Arrays

Ivan Escobar[1], Erika Vilches[1], Edgar E. Vallejo[1], Martin L. Cody[2], and Charles E. Taylor[2]

[1] ITESM-CEM, Computer Science Dept.
Atizapan de Zaragoza, Edo. de Mexico, 52926, Mexico
{iescobar, A00461595, vallejo}@itesm.mx
[2] UCLA, Dept. of Ecology and Evolutionary Biology
Los Angeles, CA, 90095, USA
mlcody@ucla.edu
taylor@biology.ucla.edu

Abstract. In this paper, we explore the emergence of acoustic categories in sensor arrays. We describe a series experiments on the automatic categorization of species and individual birds using self-organizing maps. Experimental results showed that meaningful acoustic categories can arise as self-organizing processes in sensor arrays. In addition, we discuss how distributed categorization could be used for the emergence of symbolic communication in these platforms.

1 introduction

Sensor arrays are finding use in an increasing variety of applications. This technology holds the potential to produce a major paradigm shift in the way we interact with the physical environment (Estrin et al, 2001). Enabling sensor arrays with adaptation capabilities would be a major step towards realizing the full potential of this emerging technology.

We envision sensor arrays as collections of intelligent agents that behave as a single ensemble. In this idealization, agents can recognize concepts and discourse intelligently about them. The realization of these capabilities relies crucially on the ability to categorize data, to ground symbols into perceptual categories and to propagate symbols in a community of other agents. In effect, each sensor node needs to categorize their experiences, to bind categories with symbols and communicate to other nodes using a collection of mutually intelligible symbols. Previous studies have established plausible scenarios for the emergence of these capabilities in sensor arrays (Collier and Taylor, 2005; Friedlander and Phoha, 2002).

Conversely, sensor arrays are excellent platforms for studying fundamental topics on artificial life research, such as emergence and self-organization. In effect, organisms may be viewed as collections of sensors, actuators and processors of limited abilities, communicating primarily with other sensors, actuators and

F. Almeida e Costa et al. (Eds.): ECAL 2007, LNAI 4648, pp. 1161–1170, 2007.

processors that are mostly at their immediate neighborhood, and mostly with limited bandwidth, functioning together as an ensemble (Taylor, 2002).

A fundamental aspect of living systems is their categorization capabilities (Pfeifer and Bongard, 2007). Furthermore, the remarkable ability to distinguish among different elements in the environment is crucial for the viability of an organism. However, usefulness of categorization is often contingent to the creation of meaningful categories. We believe sensor arrays are appropriate testbeds for exploring these issues on animal cognition.

In this work, we conducted a series of experiments on the emergence of acoustic categories in sensor arrays. We describe a series of experiments in which self-organizing maps are used to automatically create categories on species of birds and individual birds in an ecological monitoring application. Experimental results showed that meaningful categories can arise as self-organizing processes in sensor arrays.

2 The Model

In this section, we propose a formal framework for conducting computational experiments in sensor arrays. We first provide a series of definitions and then describe an implementation of this model based on self-organizing maps.

The formal definitions presented below are based on considerations of the models proposed in (Vallejo and Taylor, 2004) and (Zhao and Guibas, 2004).

2.1 Definitions

A sensor array is a distributed collection of interconnected sensor nodes. Such an array is described by the underlying graph and a set of properties related to both the sensor nodes and the links.

Formally, a sensor array is a 4-tuple:

$$G = \{V, E, P_V, P_E\}$$

where:

1. V is a set of nodes
2. $E \subseteq V \times V$ is a set of links
3. P_V is a set of functions related to properties of V
4. P_E is a set of functions related to properties of E

A sensor node consists of a collection of sensors. A node can ground categories to features in the flood of perceptions that come into its sensors. A node also has the capability for symbolic communication with other nodes via the transmission and the reception of a set of symbols.

Formally, a sensor node is an 8-tuple:

$$v = \{P, X, C, S, \delta, \phi, \tau, \psi\}$$

where:

1. P is finite set of sensors
2. X is set of input vectors
3. C is a finite set of perceptual categories
4. S is a finite set of symbols
5. $\delta : \rho(P) \to X$ is the perception function
6. $\phi : X \to C$ is the categorization function
7. $\tau : C \to S$ is the transmission function
8. $\psi : S \to C$ is the reception function

A sender node v_1 comunicates to a receiver node v_2 in a sensor array as follows. Initially, v_1 perceives an input vector from the environment and associates this input with a perceptual category. Then, v_1 relates the category with a symbol and transmits the symbol to a node v_2. Node v_2 receives the symbol and interprets it as a category. The communication is successful if both the categories of the sender and the receiver match.

Formally, $v_1 = \{P_1, X_1, C_1, S_1, \delta_1, \phi_1, \tau_1, \psi_1\}$ communicates successfully to $v_2 = \{P_2, X_2, C_2, S_2, \delta_2, \phi_2, \tau_2, \psi_2\}$ given an input vector $\mathbf{x} \in X$ if the following conditions are satisfied:

1. $\phi_1(\mathbf{x}) = c_i$
2. $\tau_1(c_i) = s_i$
3. $\psi_2(s_i) = c_j$
4. $c_i = c_j$

Finally, we consider several assumptions that are often present in sensor arrays: (1) each sensor node communications only to its neighbors, (2) communication is by broadcast to these neighbors, and (3) node layout follows an arbitrary, but fixed topology.

3 Experiments and Results

3.1 Experiment 1: Categorization of Bird Species

In this experiment, we explore the emergence of acoustic categories describing bird species from their songs. We used bird songs recorded at the Montes Azules Biosphere Reserve in Chiapas, Mexico. From these recordings, we constructed a training set consisting of 15 spectral and temporal features of the acoustic signal for the simulations reported here as shown in table 1.

An unsupervised competitive learning neural network was used to implement the categorization function ϕ in the nodes of the sensor array. A competitive network consists of a single layer of output units c_i, each fully connected to a set of inputs o_j via excitatory connections w_{ij} (Kohonen, 1997).

Given an input vector $\mathbf{o}$, the winner is the unit c_{i^*} with the weight vector $\mathbf{w_{i^*}}$ as follows:

$$|\mathbf{w}_{i^*} - \mathbf{o}| \leq |\mathbf{w}_i - \mathbf{o}| \text{ (for all } i)$$

Table 1. Bird species data set

Bird species	Samples
Dusky antbird (DAS)	65
Barred antshrike (BAS)	127
Great antshrike (GAS)	123
Mexican antthrush (MAT)	114

After each training step, the sensor node updates the weights w_{i^*j} for the winning category c_{i^*} only, as follows:

$$\Delta w_{i^*j} = \eta(o_j - w_{i^*j})$$

where $\eta \in [0, 1]$ is the learning constant.

Multiple simulations were conducted using different combinations of parameter values as shown in table 2. Simple competitive neural networks correctly identified four categories in the data set. In addition, the accuracy of unsupervised classification on this data set was 76–86.7%.

Table 2. Parameters for the simulations

Parameter	Value
Categories	4–10
Learning rate	0.001–0.005
Simulation steps	500

Self-organizing maps were also used for unsupervised categorization. Figure 1 shows the obtained results. Dark areas represent clusters of similar elements in the data set. The left part of the figure is the U-matrix (i.e. Unified Distance Matrix). The U-matrix shows the results of the unsupervised classification obtained by the SOM. In this figure dark areas are interpreted as clusters and light areas as cluster boundaries. The right part of the figure represents actual values associated to each cell. This representation may be viewed as the supervised classification used by the SOM to evaluate the obtained results. There exists some overlap between BAS and GAS as these species possess similar song repertoires. This can be appreciated in the self-organizing map as well as in the principal components graph shown in figure 2.

3.2 Experiment 2: Categorization of MAT Individuals

In this experiment, we explore the emergence of acoustic categories describing MAT individuals from their songs. We used recordings from the same field site as before. From these recordings, we constructed a collection of training and validation sets consisting of 5 spectral and temporal features of the acoustic signal for the simulations as shown in table 3.

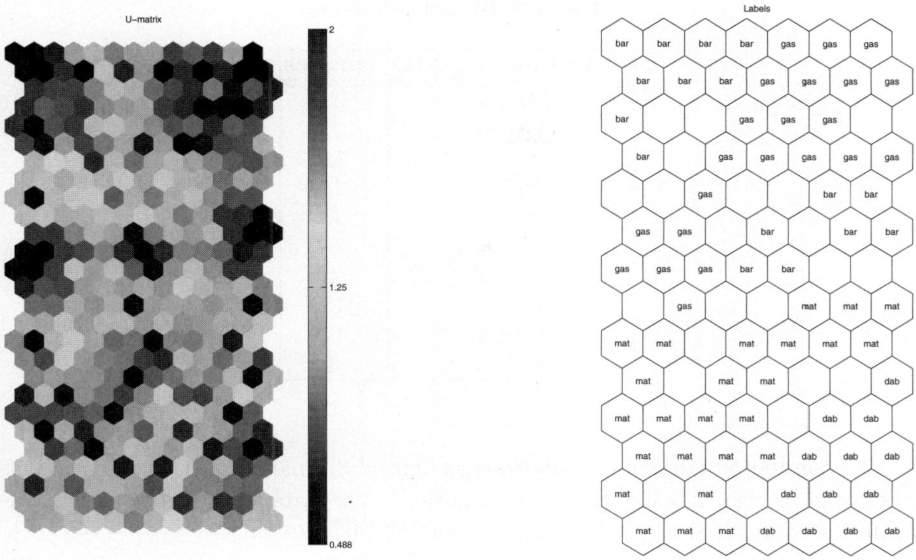

Fig. 1. Self-organizing maps: species.

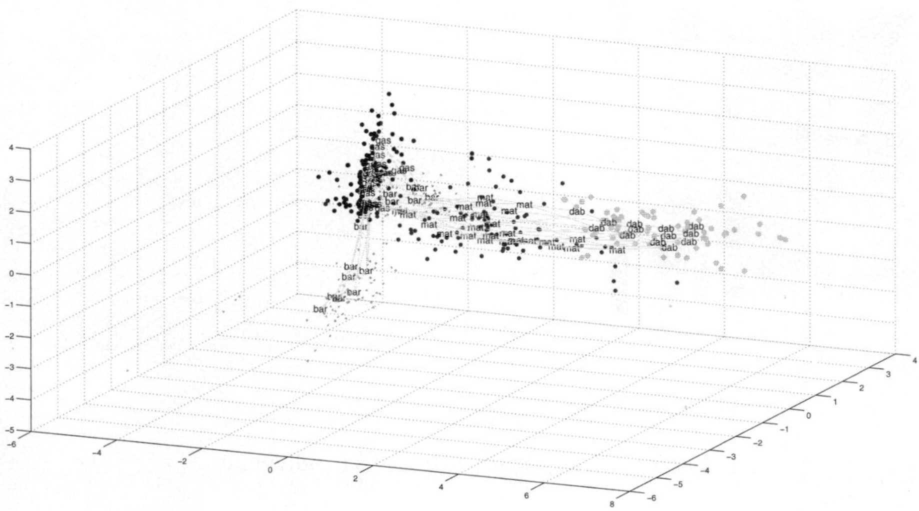

Fig. 2. Self-organizing maps: three principal components

Table 3. Birdsong data set

MAT individual ID	Samples
PMPa	28
PMPb	21
LGEa	12
PBEa	25
AVEa	38
AVEb	5
LCNa	4
LCNb	18
SNWa	17
SNWb	3

We conducted multiple simulations as before. Simple competitive neural networks fail to correctly identify the categories in the data set. More precisely, the accuracy of unsupervised classification was 52–56%.

Figure 3 shows the results obtained using self-organizing maps. It can be appreciated that unsupervised neural networks were unable to accurately identify the underlying individuals in the data set.

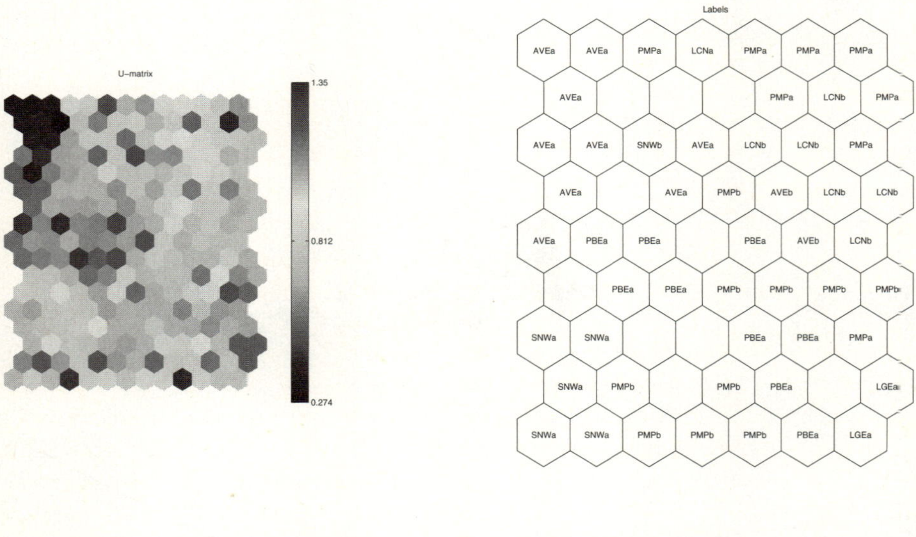

Fig. 3. Self organizing maps: MAT individuals.

We conducted additional experiments using Learning Vector Quantization (LVQ1) (Kohonen, 1997). LVQ is the supervised version of competitive learning. The MAT individuals was partitioned into 60% training, 20%validation and 20% test sets.

After each training step, LVQ updates eights w_{i^*j} for the winning category c_{i^*} only, as follows:

$$\Delta w_{i^*j} = \begin{array}{l} +\eta(o_j - w_{i^*j}) \text{ if category is correct} \\ -\eta(o_j - w_{i^*j}) \text{ if category is incorrect} \end{array}$$

Multiple simulations were conducted as before. The accuracy of supervised classification for MAT individuals was 90.17–97.65%. Figure 4 shows the training results using a learning rate value of 0.001 and 0.005, respectively. It can be appreciated that training should be stopped after about 100 epochs to avoid overfitting of the training set.

3.3 Emergence of Symbolic Communication in Sensor Arrays

Self-organizing maps locally induce categorization at sensor nodes. This form of categorization does not imply a shared collection of categories among the nodes in the array. In general, different nodes can associate a perception with different categories. Sensor nodes may be in a position to perceive correlations between perceptions and categories from other nodes. This can provide the basis for properly identify the appropriate category for a particular input.

For example, figure 5 shows two different self-organizing maps obtained from two mutually exclusive partitions from the species data set. The obtained self-organizing maps shows some overlap in the dark areas, so they could be in a position to communicate and understand to each other, given that the same symbol is used to describe the same category.

Therefore, once categorization has been achieved at the nodes of the sensor array, they could supervise to each other using a language game in order to arrive to a collection of shared symbols (Steels, 2003; Arita and Taylor, 1996).

4 Discussion

Despite its preliminary character, the results shown here seem to indicate that meaningful categorization can emerge as self-organizing processes in sensor arrays. Furthermore, symbolic communication holds the potential for reducing the bandwith requirements for sensor arrays. However, it appears that the capabilities of self-organizing maps to perform automatic categorization will be severely constrained by the complexity of the perceptions, so complementary approaches should be considered (e. g. supervised learning, information theory).

It should be noted that the proposed model has only been tested in a simple simulated setting. We will test the categorization and generalization capabilities of the proposed model in real settings in the near future. Other applications of sensor arrays such as localization and tracking would contribute to the understanding of the potentials of competitive learning neural networks for automatic categorization in sensor arrays.

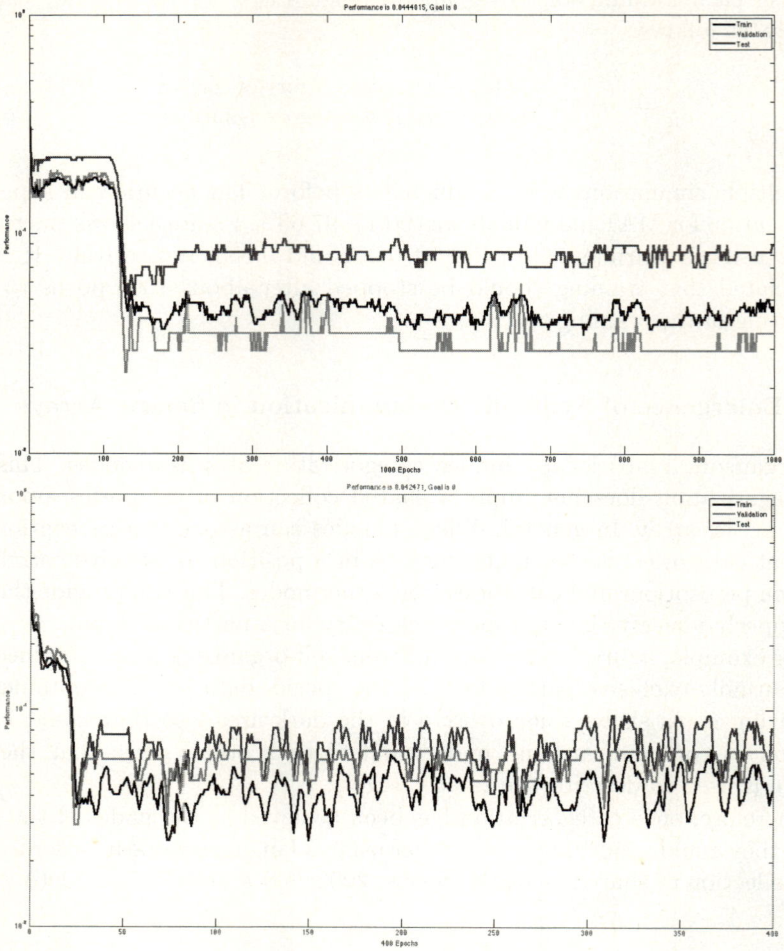

Fig. 4. Learning vector quantization: categorization results

Fig. 5. Self-organizing maps two partitions of the data set

Acknowledgements. This work was supported by the National Science Foundation under Award Number 0410438 and by Consejo Nacional de Ciencia y Tecnología under Award Number REF:J110.389/2006. Any opinions, findings and conclusions or recommendations expressed in this publication are those of the authors and do not necessarily reflect the views of the sponsoring agencies.

References

Arita, T., Taylor, C.E.: A simple model for the evolution of communication. In: Fogel, L.J., Angeline, P.J., Back, T. (eds.) Evolutionary Programming V, pp. 405–410. MIT Press, Cambridge (1996)

Collier, T.C., Taylor, C.E.: Self-Organization in Sensor Networks. Journal of Parallel and Distributed Computing 64(7), 866–873 (2004)

Estrin, D., Girod, L., Pottie, G., Srivastava, M.: Instrumenting the world with wireless sensor networks. In: Proceedings of the International Conference on Acoustics, Speech and Signal Processing (ICASSP 2001) (2001)

Friedlander, D.S., Phoha, S.: Semantic information fusion for coordinated signal processing in mobile sensor networks. The International Journal of High Performance Computing Applications 16(3), 235–241 (2002)

Kohonen, T.: Self-organizing maps, 2nd edn. Springer, Heidelberg (1997)

Lee, Y., Riggle, J., Collier, T.C., et al.: Adaptive communication among collaborative agents: Preliminary results with symbol grounding. In: Sugisaka, M., Tanaka, H. (eds.) Proceedings of the Eighth International Symposium on Artificial Life and Robotics. Beppu, Oita Japan, Beppu, Oita Japan, Jan. 24-26, pp. 149–155 (2003)

Pfeifer, R., Bongard, J.: How the body shapes the way we think. MIT Press, Cambridge (2007)

Steels, L.: The evolution of communication systems by adaptive agents. In: Alonso, E., Kudenko, D., Kazakov, D. (eds.) Adaptive Agents and Multi-Agent Systems. LNCS (LNAI), vol. 2636, pp. 125–140. Springer, Heidelberg (2003)

Taylor, C.E.: From cognition in animals to cognition in superorganisms. In: Bekoff, M., Allen, C., Gurghardt, G. (eds.) The Cognitive Animal. Empirical and Theoretical Perspectives on Animal Cognition, MIT Press, Cambridge (2002)

Teal, T., Taylor, C.E.: Effects of compression on language evolution. Artificial Life 6(2), 129–143 (2000)

Vallejo, E.E., Taylor, C.E.: The effects of learning on the evolution of Saussurean communication. In: Pollack, J., Bedau, M., Husbands, P., Ikegami, T., Watson, R.A. (eds.) Proceedings from the Ninth International Conference on Artificial Life, MIT Press, Cambridge (2002)

Zhao, F., Guibas, L.: Wireless Sensor Networks. An Information Processing Approach. Morgan Kaufmann Publishers, San Francisco (2004)

Self-organizing Systems Based on Bio-inspired Properties

André Stauffer, Daniel Mange, and Joël Rossier

Ecole polytechnique fédérale de Lausanne (EPFL)
Logic Systems Laboratory
CH-1015 Lausanne, Switzerland
andre.stauffer@epfl.ch

Abstract. Bio-inspiration borrows three properties characteristic of living organisms: multicellular architecture, cellular division, and cellular differentiation. Implemented in silicon according to these properties, our self-organizing systems are able to grow, to self-replicate, and to self-repair. The growth and branching processes, performed by the so-called Tom Thumb algorithm, lead thus to the configuration and cloning mechanisms of the systems. The repair processes allow its cicatrization and regeneration mechanisms. The cellular implementation and hardware simulation of these mechanisms constitute the core of this paper.

1 Introduction

Borrowing three structural principles (multicellular architecture, cellular division, and cellular differentiation) from living organisms, we have already shown how embryonic hardware [2] [3] [1] is able to implement bio-inspired properties in silicon thanks to two algorithms: an algorithm for cellular differentiation, based on coordinate calculation, and an algorithm for cellular division, the Tom Thumb algorithm [4].

The goal of this paper is to devise and implement five mechanisms, based on the preceding algorithms, in order to organize a system and deal with its faults in a fully automatic way. In addition to the structural configuration and functional configuration mechanisms, these algorithms allow the cloning mechanism (cellular and organismic self-replications), the cicatrization mechanism (cellular self-repair), and the regeneration mechanism (organismic self-repair) to be achieved on artificial organisms [5] [6]. Starting with a very simple organism made of only three cells, the SOS acronym, Section 2 will recall the definitions of cloning, cicatrization, and regeneration. We will then introduce the configuration and cloning mechanisms, based on the Tom Thumb algorithm, the cicatrization mechanism, and the regeneration mechanism which confer these bio-inspired properties to our self-organizing systems (Section 3). The hardware simulation of the processes involved in the former mechanisms results from the VHDL implementation of the systems (Section 4). A brief conclusion (Section 5) summarizes our paper and opens new research avenues.

F. Almeida e Costa et al. (Eds.): ECAL 2007, LNAI 4648, pp. 1171–1181, 2007.

2 Bio-inspired Properties

2.1 Cellular Architecture

An extremely simplified example, the display of the SOS acronym, is introduced in order to illustrate the basic bio-inspired properties of our self-organizing systems. The system that displays the acronym is a one-dimensional artificial organism made of three cells (Fig. 1a). Each cell of this organism is identified by a X coordinate, ranging from 1 to 3. For coordinate values $X = 1$ and $X = 3$, the cell implements the S character, for $X = 2$, it implements the O character. Such an organism can be built with a single *totipotent cell* (Fig. 1b) capable of displaying either the S or the O character according to the X coordinate. The totipotent cell is an arrangement of $4 \times 6 = 24$ molecules, 21 of which are invariant, one displays the S character, and two display the O character. An incrementer implementing the X coordinate calculation is embedded in the final organism.

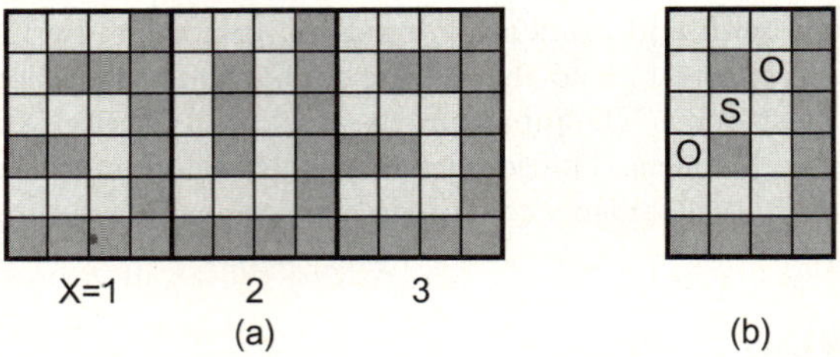

Fig. 1. SOS acronym. (a) One-dimensional organism made of 3 cells. (b) Totipotent cell made of $4 \times 6 = 24$ molecules.

2.2 Cloning

The *cloning* or self-replication can be implemented at the cellular level in order to build a multicellular organism and at the organismic level in order to generate a population of organisms. The cloning of the totipotent cell displayed in Fig. 1b results thus in the SOS organism of Fig. 1a. The cloning of the organism, defining a population of SOS acronyms (Section 4), rests on two assumptions: (1) There exists a sufficient number of spare cells in the array to contain at least one copy of the additional organism. (2) The calculation of the coordinates produces a cycle $X = 1 \rightarrow 2 \rightarrow 3 \rightarrow 1$ implying $X+ = (X + 1) \bmod 3$. Given a sufficiently large space, the cloning of the organism can be repeated for any number of specimens in the X and/or Y axes.

2.3 Cicatrization

The introduction in the totipotent cell of one column of spare molecules (SM, Fig. 2a), defined by a specific structural configuration, and the automatic detection of faulty molecules (by a built-in self-test mechanism which constantly compares two copies of the same molecule) allows *cicatrization* or self-repair at the cellular level: each faulty molecule is deactivated, isolated from the network, and replaced by the nearest right molecule, which will itself be replaced by the nearest right molecule, and so on until a spare molecule (SM) is reached (Fig. 2b). The number of faulty molecules handled by the cicatrization mechanism is necessarily limited: in the example of Fig. 2a, we tolerate at most one faulty molecule per row. If more than one molecule is faulty in a same row (Fig. 2c), cicatrization is impossible, in which case a global KILL=1 is generated to activate regeneration as described thereafter.

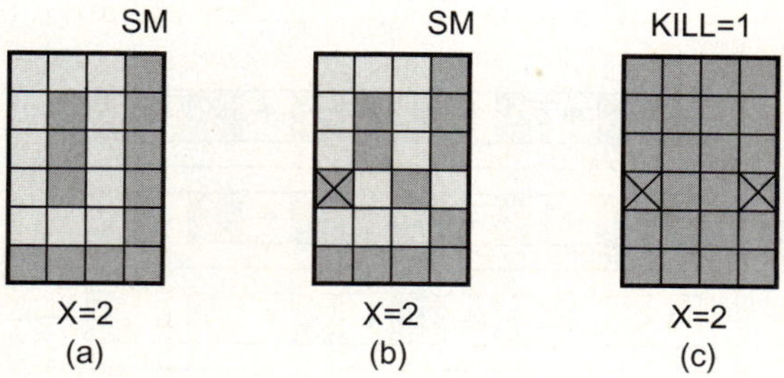

Fig. 2. Cicatrization of the SOS organism. (a) Healthy cell displaying O. (b) Self-repaired cell with 1 faulty molecule. (c) Faulty cell with 2 faulty molecules in the same row.

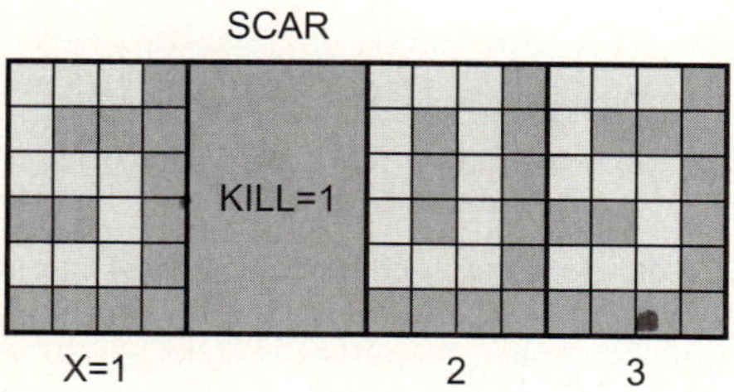

Fig. 3. Regeneration of the SOS organism.

2.4 Regeneration

In order to implement *regeneration*, that is self-repair at the organismic level, we need at least one spare cell to the right of the original organism (Fig. 1a). The existence of a fault, detected by the KILL signal generated at the cellular level (Fig. 2c), identifies the faulty cell, and the entire column of all cells to which the faulty cell belongs is considered faulty and is deactivated (Fig. 3; in this simple example, the column of cells is reduced to a single cell). All the functions (X coordinate and configuration) of the cells to the right of the column $X = 1$ are shifted by one column to the right. Obviously, this process requires as many spare cells to the right of the array as there are faulty cells to repair.

3 Self-organizing Mechanisms

3.1 Structural Configuration

The *structural configuration mechanism* is made up of a *structural growth process* followed by a *load process*. It performs these processes in order to build the structure of the totipotent cell (Fig. 1b).

The *structural configuration data* or *structural genome* of the totipotent cell is organized as a string of characters comprising one *flag character* and one *structural character* for each molecule of the cell (Fig. 4). The symbols of these characters are given in Fig. 5 and Fig. 6 respectively.

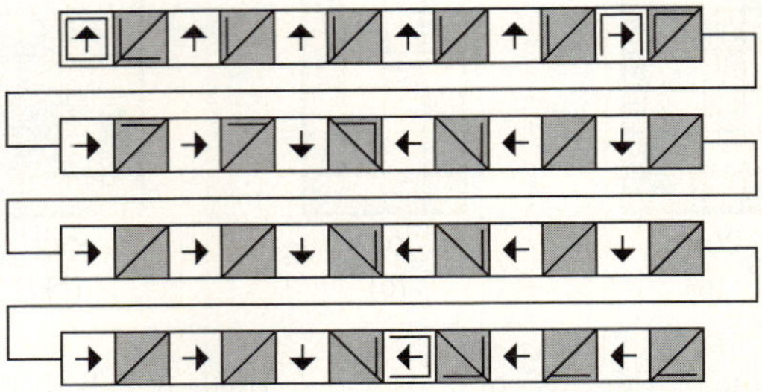

Fig. 4. Structural configuration string of the SOS cell.

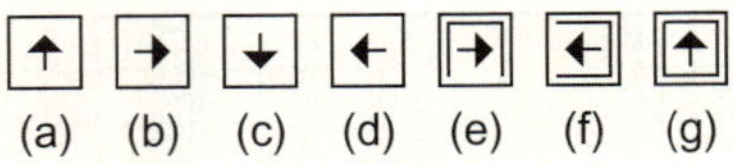

Fig. 5. Flag data characters. (a) North connect. (b) East connect. (c) South connect. (d) West connect. (e) East connect and north branch. (f) West connect and east branch. (g) North connect and branch activate.

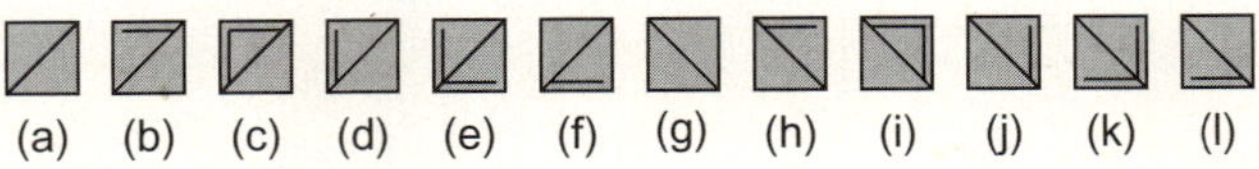

Fig. 6. Structural data characters. (a) Living internal. (b) Living top. (c) Living top-left. (d) Living left. (e) Living bottom-left. (f) Living bottom. (g) Spare internal. (h) Spare top. (i) Spare top-right. (j) Spare right. (k) Spare bottom-right. (l) Spare bottom.

According to the flag data, the *structural growth process* creates a path among the molecules in order to trap a copy of the structural data in the memory positions of the cell and to allow another copy of the them to move around the cell (Fig. 7). This process starts with a *growth signal* applied to the lower left molecule of the cell (Fig. 7a). When the connection path between the molecules closes, the lower left molecule delivers a *close signal* to the nearest left neighbor cell (Fig. 7c).

The *load process* is triggered by the *close signal* applied to the lower right molecule of the cell (Fig. 8a). A *load signal* propagates then westward and northward through the cell (Fig. 8b) and, according to the structural data trapped in their memory positions, each of its molecules acquire a *molecular mode* (Fig. 9) and a *molecular type* (Fig. 10). We finally obtain an homogeneous tissue of molecules defining both the boundaries of the cell and the position of its *living mode* and *spare mode* molecules (Fig. 8c). This tissue is ready for being configured by the functional configuration data.

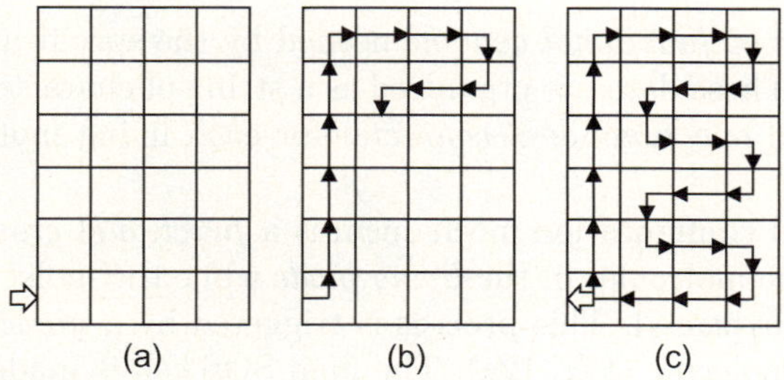

Fig. 7. Structural growth process of the SOS cell. (a) Growth signal applied to the lower left molecule. (b) Data path under construction. (c) The lower left molecule delivers a close signal when the path closes.

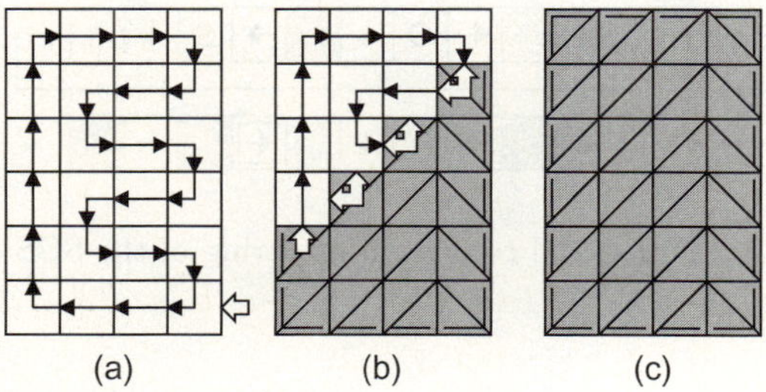

Fig. 8. Load process of the SOS cell. (a) Close signal applied to the lower right molecule. (b) Structural loading in progress. (c) Final structure of the cell.

3.2 Functional Configuration

The goal of the *functional configuration mechanism* is to store in the homogeneous tissue, which already contains structural data (Fig. 8c), the *functional*

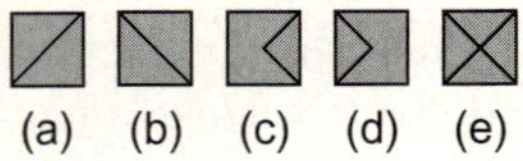

Fig. 9. Molecular modes. (a) Living. (b) Spare. (c) Faulty. (d) Repair. (e) Dead.

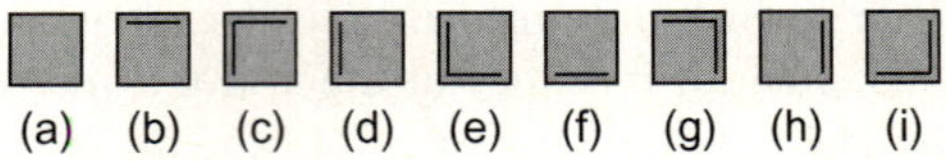

Fig. 10. Molecular types. (a) Internal. (b) Top. (c) Top-left. (d) Left. (e) Bottom-left. (f) Bottom. (g) Top-right. (h) Right. (9) Bottom-right.

configuration data or *functional genome* needed by the specifications of the current application. These data are organized as a string of characters involving one *flag character* and one *functional character* for each living molecule of the cell (Fig. 11).

The functional configuration mechanism is a *functional growth process*, performed only on the molecules in the *living mode* while the molecules in the *spare mode* are simply bypassed. This process is triggered by a *growth signal* applied to the lower left molecule (Fig. 12a). The final SOS cell is made up of 18 living molecules organized as an array of six rows by three columns (Fig. 12c), while one row of six spare molecules are bypassed. The final specifications of the cell under construction are now stored as functional data in its living molecules.

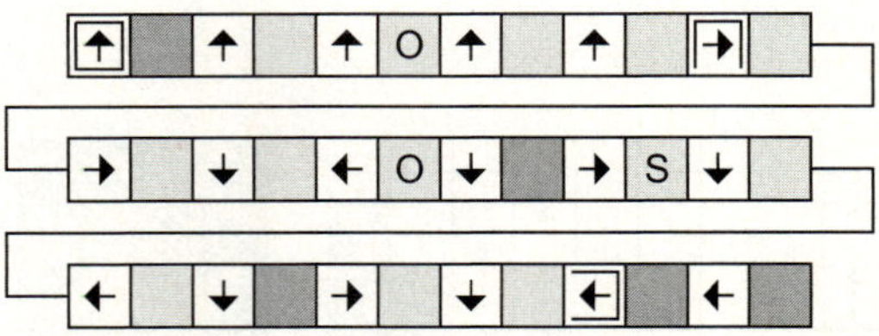

Fig. 11. Functional configuration string of the SOS cell.

3.3 Cloning

The *cloning mechanism* allows the self-replication of the cells in order to create a multicellular organism and the self-replication of the organisms in order to obtain a population of them. This mechanism corresponds to a *branching process* which takes place when the structural and the functional configuration mechanisms deliver northward and eastward growth signals on the borders of the cell during the corresponding growth processes (Fig. 13).

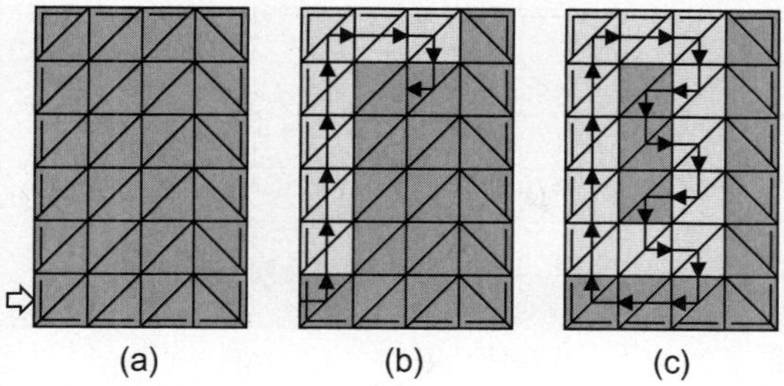

Fig. 12. Functional configuration mechanism of the SOS cell. (a) Growth signal applied to the lower left molecule. (b) Functional growth process. (c) Final configuration of the living molecules.

3.4 Cicatrization

Fig. 12c shows the normal behavior of a healthy SOS cell, i.e. a cell without any faulty molecule. A molecule is considered as faulty, or in the *faulty mode*, if some built-in self-test, not described in this paper, detects a lethal malfunction. Starting with the normal behavior of Fig. 12c, we suppose that one living molecule will become suddenly faulty (Fig. 14a). This molecule triggers a *cicatrization mechanism* made up of a *repair process* involving eastward *repair signals* (Fig. 14b) followed by a *reset process* performed with northward, westward and southward *reset signals* (Fig. 14c).

This tissue, comprising now one molecule in the *faulty mode* and three molecules in the *repair mode*, is ready for being reconfigured by the functional configuration data. This implies a *functional growth process* bypassing the faulty molecules (Fig. 15).

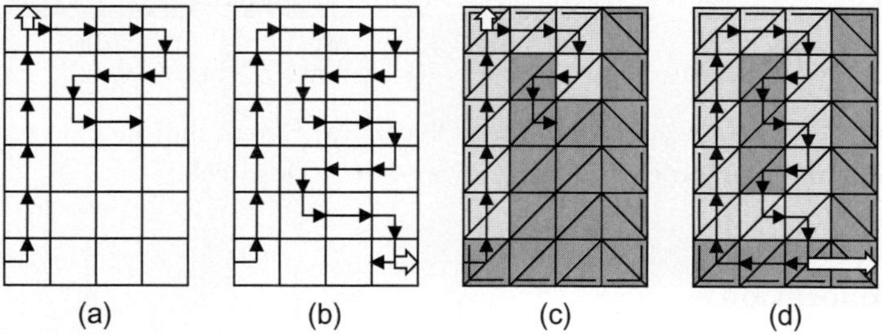

Fig. 13. Northward and eastward growth signals triggering the cloning mechanism of the SOS cell. (a),(b) Structural branching processes. (c),(d) Functional branching processes.

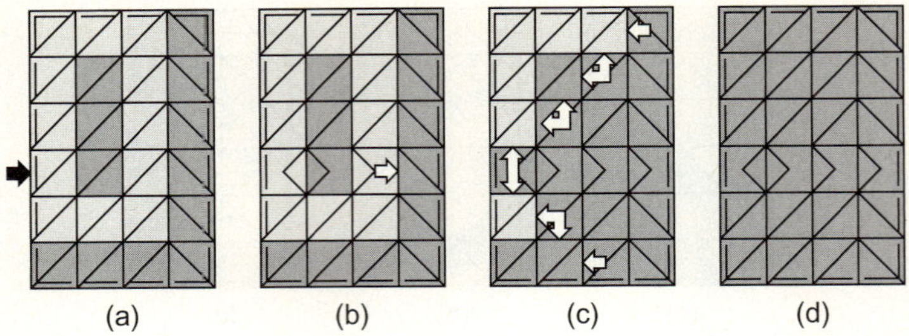

Fig. 14. Cicatrization mechanism of the SOS cell. (a) Living molecule becoming faulty. (b) Repair process. (c) Reset process. (d) Resetted cell.

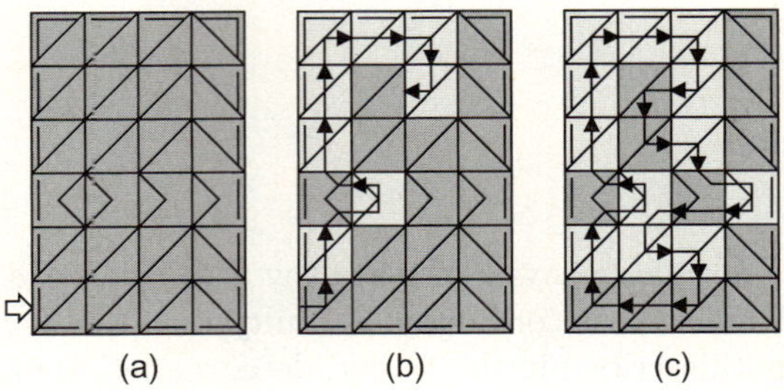

Fig. 15. Functional reconfiguration of the SOS cell. (a) Growth signal applied to the lower left molecule. (b) Functional growth process. (c) Final reconfiguration of the living and repair molecules.

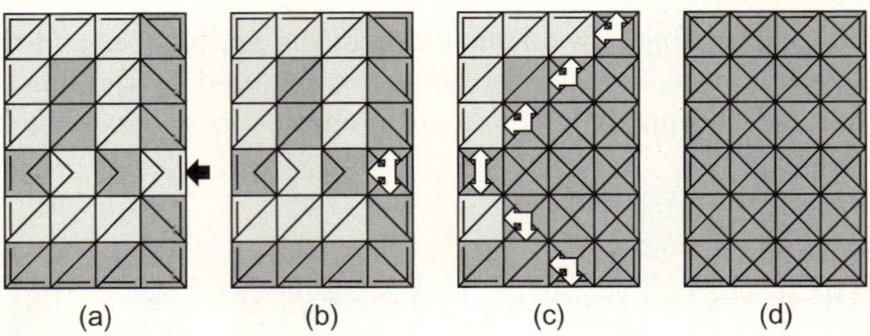

Fig. 16. Regeneration mechanism of the SOS cell. (a) Repair molecule becoming faulty. (b) Unsuccessful repair process. (c) Kill process. (d) Dead cell.

3.5 Regeneration

Our SOS self-organizing cell comprises a single spare molecule per row and tolerates therefore only one faulty molecule in each row. A second faulty molecule in the same row will trigger the death of the whole cell, and the start of a *regeneration mechanism*. Fig. 16b and Fig. 16c illustrate respectively the unsuccessful *repair process* and the following *kill process* involved in this mechanism. Starting with the normal behavior of the cicatrized cell (Fig. 15c), a new molecule becomes faulty (Fig. 16a). Being already in the *repair mode*, this molecule enters the lethal *dead mode* and triggers *kill signals* northward, westward and southward. Finally in Fig. 16d, the original SOS cell is dead.

4 Hardware Simulation

4.1 Multicellular Organism

The *structural growth process*, the *functional growth process*, and the *branching process* perform all the Tom Thumb algorithm [4] in order to configure the self-organizing systems. Using the VHDL description language, we realized the hardware implementation of this algorithm as a data and signals cellular automaton (DSCA) [7]. The corresponding hardware simulation (Fig. 17) shows how the processes of the self-organizing mechanisms grow, load, repair, reset and kill the middle cell of the SOS organism. In this hardware simulation, the cloning mechanism is implemented at the cellular level in order to build the multicellular organism by self-replication of its totipotent cell.

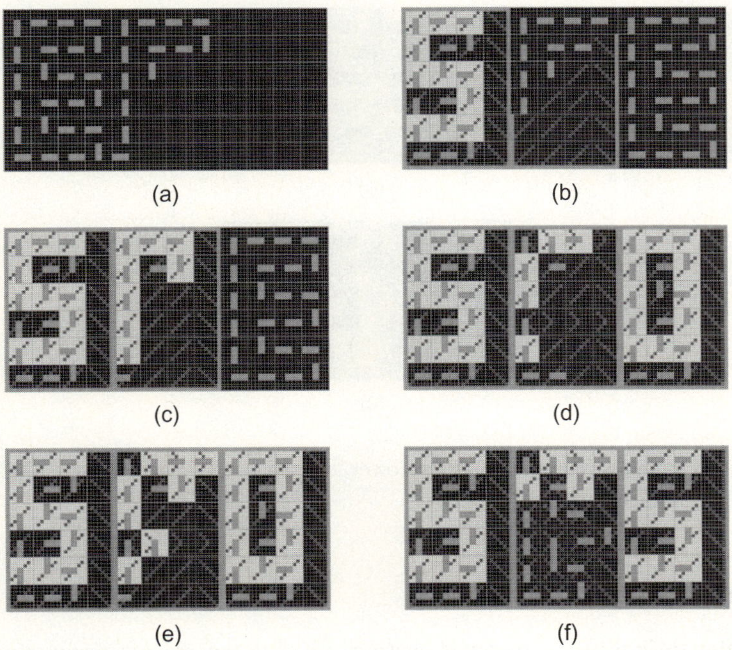

Fig. 17. Processes performed on the middle cell. (a) Structural growth. (b) Load. (c) Functional growth. (d) Repair and reset. (e) Functional regrowth. (f) Kill.

4.2 Population of Organisms

Fig. 18a illustrates the cloning of the multicellular organism resulting in a population of SOS acronyms. In this hardware simulation, the cloning mechanism is implemented at the organismic level. Two conditions are fulfilled to make such a self-replication possible: (1) There exists nine spare cells in the array allowing three additional copies of the SOS organism. (2) The calculation of the coordinates produces cycles respectively equal to 3 and 1 on the X and Y axes.

The graphical distortion of the S character results from the cicatrization and functional reconfiguration mechanisms applied on the lower left organism (Fig. 18b). The scar produced by the regeneration mechanism performed on the lower left organism affects the entire column of all cells to which the faulty cell belongs (Fig. 18c).

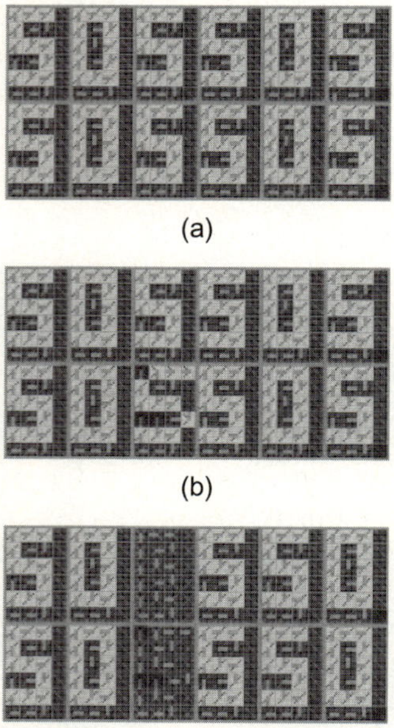

Fig. 18. Mechanisms performed on the lower left organism. (a) Cloning. (b) Cicatrization. (c) Regeneration.

5 Conclusion

The cellular design of bio-inspired systems involve several self-organizing mechanisms, whose growth and branching processes are mainly based on the Tom Thumb algorithm. These mechanisms allow the cellular systems to perform cloning (cellular and organismic self-replication), cicatrization (cellular self-repair), and regeneration (organismic self-repair).

Starting with a very simple organism made of only three cells, the SOS acronym, we described the self-organizing mechanisms as the makeup of simple processes like growth, load, branching, repair, reset, and kill. The hardware simulation of the SOS organism results from the VHDL implementation of its totipotent cell as a data and signals cellular automaton (DSCA).

While the hardware implementation of the growth and branching processes were already performed previously, the hardware simulation of the SOS acronym verifies that the conceived mechanisms and all their underlying processes ensure perfectly the self-organization of bio-inspired systems.

In order to improve our self-organizing systems, we intend to develop additional features such as:

– Initial detection of the faulty molecules.
– Online detection of erroneous configuration data.
– Online detection of application dysfunction.

References

1. Canham, R., Tyrrell, A.M.: An embryonic array with improved efficiency and fault tolerance. In: Lohn, J., et al. (eds.) Proceedings of the NASA/DoD Conference on Evolvable Hardware (EH'03), pp. 265–272. IEEE Computer Society Press, Los Alamitos (2003)
2. Mange, D., Sipper, M., Stauffer, A., Tempesti, G.: Toward robust integrated circuits: The Embryonics approach. Proceedings of the IEEE 88(4), 516–541 (2000)
3. Mange, D., Stauffer, A., Petraglio, E., Tempesti, G.: Embryonics machines that divide and differentiate. In: Ijspeert, A.J., Murata, M., Wakamiya, N. (eds.) BioADIT 2004. LNCS, vol. 3141, Springer, Heidelberg (2004)
4. Mange, D., Stauffer, A., Petraglio, E., Tempesti, G.: Self-replicating loop with universal construction. Physica D 191(1-2), 178–192 (2004)
5. Stauffer, A., Mange, D., Tempesti, G.: Embryonic machines that grow, self-replicate and self-repair. In: Lohn, J., et al. (eds.) Proceedings of the 2005 NASA/DoD Conference on Evolvable Hardware (EH'05), pp. 290–293. IEEE Computer Society Press, Los Alamitos (2005)
6. Stauffer, A., Mange, D., Tempesti, G.: Bio-inspired computing machines with self-repair mechanisms. In: Ijspeert, A.J., Masuzawa, T., Kusumoto, S. (eds.) BioADIT 2006. LNCS, vol. 3853, Springer, Heidelberg (2006)
7. Stauffer, A., Sipper, M.: The data-and-signals cellular automaton and its application to growing structures. Artificial Life 10(4), 463–477 (2004)

Stepwise Transition from Direct Encoding to Artificial Ontogeny in Neuroevolution

Benjamin Inden

Max Planck Institute for Mathematics in the Sciences

Abstract. There is a gap between neuroevolution systems employing artificial ontogeny and those being able to solve difficult control tasks. The NEON system builds on ideas of the well known NEAT neuroevolution system to make possible a stepwise transition from a direct encoding to complex genetic architectures using developmental processes.

1 Introduction

Artificial ontogeny [1], or artificial embryogeny [8], is the imitation of the natural process of development for artificial life. It is typically used together with an evolutionary algorithm, and entails a growth process, where a mature phenotype is constructed from a simple initial state using information from the genotype. Several others have used artificial ontogeny to construct neural networks, which in turn have been used to perform a robot control task (e.g. [1,2]).

Using artificial ontogeny instead of a direct encoding in connection with neuroevolution may have several advantages: First, compressible phenotypes can be encoded in a more compact genotype through gene reuse. This enhances scalability of neuroevolution [6]. Second, the growth process can exploit constraints from the environment. It can evolve to be adaptive: produce different phenotypes in different environments such that each is adapted. Third, restructuring the developmental process by evolution makes linkage learning and coordinated variability for phenotypic variables possible [10].

On the other hand, existing artificial ontogeny systems seem to be biased towards phenotypes of low complexity (high compressibility), and have considerable difficulties evolving high complexity phenotypes [4]. Simulating ontogeny is also very time consuming.

Artificial ontogeny may be ultimately advantageous in incremental evolution scenarios, where the additional cost of learning a good representation in the beginning pays later through coordinated variability and gene reuse.

A paradigm frequently used in artificial ontogeny is that of genetic regulatory networks (GRNs). In these kinds of artificial ontogeny systems, each cell accesses the same genome, but has got its individual list of molecule concentrations. Genes code for transcription factors, which are proteins that can control the transcription rates of other genes. Some transcription factors may have side effects like cell division, initiation of neural connections to other cells, etc., or

F. Almeida e Costa et al. (Eds.): ECAL 2007, LNAI 4648, pp. 1182–1191, 2007.

there may be dedicated molecules for that purpose. The genomes specify how production of molecules depends on the concentrations of transcription factors.

Models of GRNs and models of neural networks can be equivalent under some conditions: Protein concentrations can be seen as corresponding to neuron outputs or firing rates, while the regulatory sequences correspond to connectivity and connection weights specification. If only weighted sums are allowed on the regulatory sequence, this corresponds exactly to the expressiveness of standard neural networks; otherwise, the regulatory unit of a gene has got more computational power than a standard neuron. Typically, transcription factors are modeled as decaying slowly; this would be similar to some continuous time neuron models, where neuron activation also depends on input from previous time steps. If diffusion of transcription factors between cells is taken into account, this is similar to a neural network extension called GasNet [5], where neurons are modeled as emitting gas, which diffuses over a grid and influences the output of the other neurons residing on the grid. Therefore, artificial GRNs can be implemented and analyzed using established methods from neural networks research.

The aim of the work reported here is to use neuroevolution to solve challenging benchmark tasks *and* make use of complex genetic architectures (including those using a developmental process) which can be used in incremental evolution scenarios to solve increasingly complex tasks with many inputs and outputs. In order to achieve this goal, a stepwise transition from direct to developmental encodings is proposed, and results are presented for some of the proposed steps.

The following practical considerations have been made in devising the encoding: First, a good level of abstraction has to be used. In many cases, the results of processes (e.g. gradients or cells [8]) can be created directly instead of simulating the processes. This may take away some possibilities which evolution could exploit, but the speedup is essential.

Second, stochasticity has to be limited, otherwise large population sizes or multiple evaluations of a single genotype are necessary. Instead, one could provide access to large amounts of *unchanging* random data, which can be exploited to construct the phenotype.

Third, representations and operators have to be designed such that heritability is high enough; while some mutations might cause large changes on the phenotype layer, there must still be enough mutations that cause slight changes only.

2 The NEON System

The "neuroevolution with ontogeny" (NEON) system can use two layers, a GRN and a neural network layer, where the GRN layer is in fact a neural network whose outputs provide "data streams" with instructions to build the upper layer. But in the experiments reported here, the lower layer is not yet used. Instead, the upper layer can access a large number of other data streams with arbitrary data size. These "data streams" are in fact just chunks of output from a random generator function, and they are accessed by specifying a seed for the random number generator. This data is therefore *unchanging* provided the same key is used every time.

2.1 Neural Network Layer

Standard sigmoid neurons are used; transfer function is

$$o_i(t+1) = \frac{2}{1+\exp\left(-2\left(\sum_{j\in N} w_{ij}o_j(t)\right)\right)} - 1$$

(which is equivalent to the tanh function for the weighted input sum, and is in the range$[-1,1]$). Connection weights are typically in the range $[-3,3]$. The network consists of a single output node connected to all inputs when the developmental process begins.

2.2 Developmental Operations

The developmental operations used in NEON to construct the neural network are very similar to the mutation operators in the NEAT neuroevolution system [7,9]; they and their parameters are read from data streams. Therefore the random values mentioned in the following list are usually the same every time a phenotype is constructed from the same genotype.

- A specified fraction of the connection weights are perturbed randomly each with a value either drawn from a Gaussian or uniform distribution.
- A specified fraction of the connection weights are set to random values from the range of allowed weights.
- Two neurons are randomly chosen and connected if no connection exists between them.
- A neuron is randomly chosen and a recurrent connection established if none exists yet.
- An existing connection is chosen and a neuron inserted in between. The connection weight to the new neuron is set to 1.0, while the connection from the new neuron is assigned the weight of the old connection. The old connection is disabled.
- A disabled connection is enabled.
- The state of a connection is toggled (enabled/disabled).

2.3 Tags and Matching

Tags are assigned to each neuron and connection to calculate a similarity value for pairs of neural networks and divide the population into species. These tags are also taken from the data stream and are unique with very high probability.

All developmental operations require choosing nodes or connections. A naive implementation would be to choose according to their position in a list. But when a new connection or node is created, all operations later in developmental time would then work on changed lists, which would make this operation a macromutation. Therefore, a more sophisticated method is used which chooses nodes or connections based on matching their tags to parameters of the developmental operations. The goal is to make choosing each item roughly equally likely, and have insertion of a new tag change very few, if any, subsequent choices. One implementation of this idea uses long integer tags as follows: If *one* connection or node needs to be chosen, XOR is applied between each tag and the operation parameter. The item with the highest result is chosen. If *all* connections have to be dealt with, the results of an XOR each between their tags and the operation parameter are used as keys to access the random data pool again. The data from the pool is then used to perform the operation for each connection.

The tags for all input, output, reservoir and initial hidden neurons are directly taken from the data pool accessed by the gene with the lowest time index. All other tags are computed from these tags: The upper half of new connection tags is taken from the lower half of the source neuron tag, while their lower half is taken from the lower half of the sink neuron tag. Tags of new neurons are computed as XOR of the two tags of the neurons between which the new one is inserted, with a cyclic shift of two bits to the left.

2.4 Representation and Search Operators

Each gene has got an id number, a key to access the data pool, a volume field which specifies how many developmental operations to read from the pool, and a time index which specifies where in the sequence of developmental operations the gene applies (figure 1).

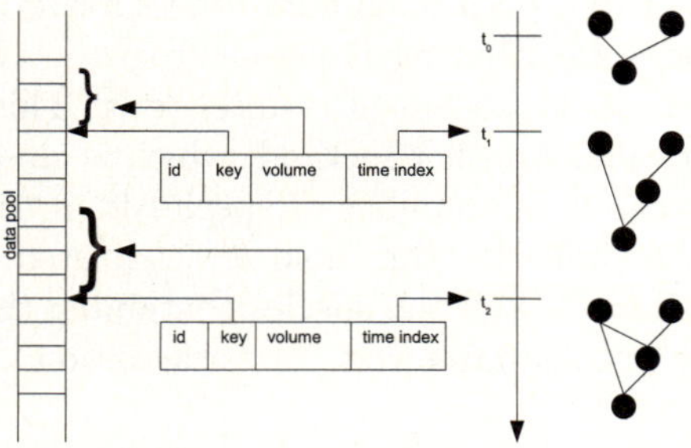

Fig. 1. Ontogeny with the neon method. In the middle are the genes, while on the right the sequence of phenotypes is depicted.

Insertion of a gene usually happens with its volume set to 1, while probabilities can be given for the time index of the new gene being below or above the current highest time index in the genome. Among the substitution operations that can be applied to a gene are increment and decrement of the volume, and change of key. The deletion operator removes a gene completely.

If mutations always insert developmental operations at the end of the sequence, evolution proceeds as if a direct encoding was used and mutation were applied directly as in NEAT, the difference being that these operations must be "replayed" every time the phenotype is constructed. If gene volume is increased or new operations are inserted within the sequence of operations, the evolutionary history and the developmental pathway diverge. Together with deletions, these mutations make compression of the genotype possible, but also change the way the phenotype space is searched.

2.5 Selection Methods

NEON (like NEAT) partitions the population into species every generation. An individual is assigned to a species if its phenotype is sufficiently similar to the phenotype of that species' representative from the last generation. If it cannot be assigned to an existing species, a new species is created. Two individuals are sufficiently similar if $d + c_w \Delta w \leq \theta$, where d is the number of connection tags that are present in only one individual, Δw is the average weight difference of connections which by their tags can be found in both individuals, and c_w, θ are constants. Each species' offspring size is made proportional to its mean fitness; this prevents a slightly superior species from taking over the whole population. That way, diversity and innovation in the population are protected to some degree. Inside the species, the worst performing individuals are deleted, after which stochastic uniform selection is used for the rest. Species with offspring size greater than five also keep their best performing individual.

If a species' maximum fitness has not increased for more than n_{kill} generations, its mean fitness is multiplied by 0.01, which usually means it dies out. If the whole population has not improved maximum fitness for n_{dc} generations, "delta coding" is applied, where all offspring is assigned to the two best performing species, which have their no improvement counter reset. This is like in the standard NEAT implementation. Another method, which is close to the SharpNeat implementation [3], checks if the number of species is in some desirable range $s_{min} \ldots s_{max}$; if not, the similarity threshold θ is adjusted. This method does not use delta coding and never kills the species containing the fittest individual; the other species are only killed after very long stagnation.

3 Simulation Results

3.1 Performance Evaluation

Networks are evolved to balance two poles simultaneously. The poles are both fixed to a cart and can, like the cart, move in one dimension only. The basic measure of performance is the number of time steps (at most 100000) that the cart stays within certain distance from its point of origin, and both poles do not deviate from the upright position by more than some angle. In the simpler Markovian version of the task (DPV), the neural network gets the pole angles and angular velocities as well as the cart position and speed as input. A bias input is also provided. In the more difficult non-Markovian version (DPNV), all velocity inputs are missing. A special fitness function with an 'anti-wiggling' term (AWDPNV) is also used for the non-Markovian task; in that setup, networks are first tested on whether they can balance the poles for 1000 time steps; if so, they are also tested on whether they can balance the poles for 100000 time steps, and for 1000 time steps with at least 200 out of 625 different initial configurations. Only in that case the network is accepted as final solution. All tasks have been described in more detail e.g. in [7,9].

A series of experiments with 30 runs each are reported here (comparisons were done using a Wilcoxon rank sum test on the number of evaluations; unless reported otherwise, a run lasted for 500 generations at most).

NEON can solve the DPV task using 5628 evaluation on average (NEAT: 3600), the DPNV task using 49918 evaluations (NEAT: 20918), and AWDPNV using 51588 evaluations, final solutions solving 252 of the 625 tasks on average (NEAT: 24543 evaluation, 286 tasks, as reported in [7,9]). This means that NEON finds solutions using the same order of magnitude of evaluations as NEAT, although somewhat slower. Upon inspection of the NEAT source code, one finds that the perturb operation is very sophisticated there, making also a distinction between connections that arose earlier and those that arose later in evolutionary history. Such a distinction was not attempted to be made in NEON, were different time axes for ontogeny and phylogeny complicate the issue. In any case, the performance is sufficient for studying how difficult tasks can be solved using indirect encodings.

The reported number of evaluations were reached with a standard configuration (DS1) which used the following parameter settings: population size $s = 150$, dynamic threshold starting with $\theta = 4$, connection weight differences weight $c_w = 2$, $n_{kill} = 200$ (never kill best species, do not use delta coding), threshold increment $\Delta\theta = 0.025$, target number of species $n_{spec} = 3 \ldots 9$, within species selection truncation threshold $\sigma = 0.4$.

The developmental operations were applied with the following probabilities: Weight perturbation 82.4%, Weight setting 10%, Connect 5%, Connect recurrent 0.5%, Split 0.1%, Toggle Enable 1%, Reenable 1%. Perturbations and weight settings each affected 40% of the connections on average. Perturbations added a value drawn from a normal distribution with standard deviation 0.24 to each affected weight.

For further studies, a configuration with larger population (DL1) was used because it proved more robust to parameter changes. Here the standard parameters were the same as in DS1, except $s = 1000$, $c_w = 3$, and $n_{spec} = 35 \ldots 45$. This configuration used on average 108126 evaluations for DPNV, and 108800 for AWDPNV.

It was also easy to find good configurations with static threshold and large population size, while configurations with static threshold and small population size in many cased failed to converge in some runs. The result reported for DPV above was achieved using a static setup (SS1) with $s = 150$, $c_w = 1$, $\theta = 3$, $n_{kill} = 15$, $n_{dc} = 20$, and $\sigma = 0.2$.

In these runs which were meant to be equivalent to neuroevolution with direct encoding, mutations were always insertions with the time index of the new gene set to some value above the values of all existing genes. For DL1, exactly one mutation was applied on a genome, while the mutation procedure for DS1 and SS1 produced 1/genome_length mutations per gene on average, but was called repeatedly until at least one mutation had happened, resulting in stronger mutation pressure.

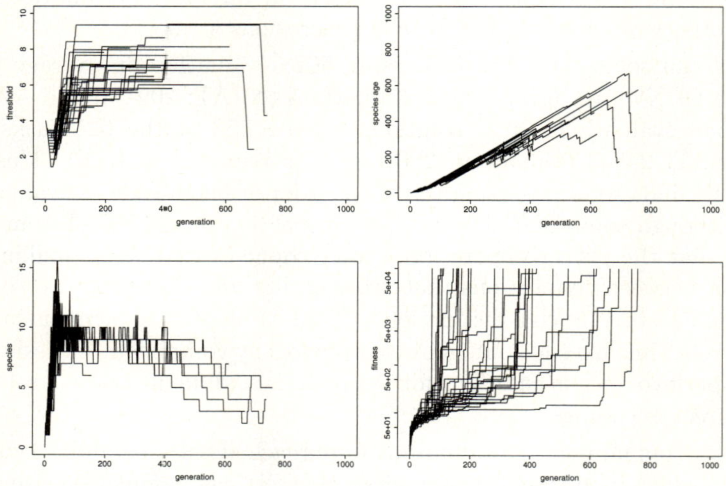

Fig. 2. Population dynamics for configuration DS1 evolving for DPNV. Top left: speciation threshold. Top right: mean species age. Bottom left: number of species. Bottom right: best fitness (scaled logarithmically).

The population dynamics for DS1 runs can be studied in figure 2: As the population is created from a single common ancestor, it initially lacks diversity, causing the threshold to fall. The diversity of the population increases simultaneously through mutations, causing the number of species to overshoot. Also, it is much more easy for a network of intermediate performance to further increase performance than for an incompetent network to learn to balance at all. Therefore, fitness increases drastically during the last generations, causing the extinction of many (especially older) species, which in turn causes the mean species age and speciation threshold to fall.

The immediate fitness effects of developmental operations on networks which perform above a baseline performance of 60 time steps can be found in table 1. As expected, the immediate fitness effects of set weight operations are more

Table 1. Immediate fitness effects of adding developmental operations when original fitness is at least 60 time steps (DPNV, configuration DL1, 30 runs).

operation	negative	neutral	positive
perturb	0.774	0.125	0.101
set weight	0.926	0.061	0.013
connect	0.290	0.686	0.024
split	0.768	0.180	0.052
toggle/re-enable	0.738	0.191	0.070

often negative than those of the perturb operation, Nevertheless, configurations without the set weight operation perform significantly worse; this operation is necessary for creating and maintaining enough diversity in the population. The fraction of neutral connect operations is large because the operation by default has got no effect if the randomly chosen nodes are already connected. What is remarkable is the rather large fractions of toggle enable / reenable operations with positive effects. Indeed, setups without both of these operations perform significantly worse.

For comparison, a simple tournament selection setup with population size 1000, tournament size 2 and elite size 10 can solve DPNV using 156826 evaluations on average; it also finds solutions in 73% of the runs for AWDPNV. Although both results are significantly worse than those for DL1, this shows that the gap between speciation selection and standard selection methods may not be very large once good operators are used.

3.2 Direct Approaches to Complex Genetic Architectures

Table 2 lists results for a number of experiments where other mutations besides insertion at the end of the developmental sequence were allowed. This makes search less efficient. For example, by incrementing the volume of a gene (the number of developmental operations its reads from the data pool), only the particular operation that comes next in the data pool can be added, which may lead to repeated exploration of the same phenotypes. Also, the tagging system described above does not eliminate *all* side effects that inserting a developmental operation has on subsequent developmental operations.

Configurations DL2 (run for at most 1000 generations) and DL3 use different mixtures of mutations. The advantage over the standard configuration is that they lead to smaller genomes: DL2 solutions had from 16 to 37 genes (mean 26.3), while DL3 solutions had from 31 to 108 genes (mean 64.8). For comparison, DL1 solutions had between 33 and 176 genes, the mean being 89.1.

Configurations DL4 and DL5 show that performance also degrades slightly but significantly when the probability of inserting a new operation not at the end of, but somewhere within the sequence is increased.

3.3 Incremental Approaches to Complex Genetic Architectures

Above it was argued that developmental encodings may be most useful in an incremental evolution scenario, where instructions for a newly evolving task are first stored uncompressed in the genome, and later reorganized and compressed as that feature gets conserved. To study this idea in the context of NEON, the solutions to the DPNV problem were taken from the 30 runs of standard configuration DL1.

These solutions were then evolved for 500 more generations using a different fitness function. This function, as before, counted the number of balancing time steps, but only up to a maximum of 1000 time steps for saving run time. That number was multiplied with a function that rewards smaller genomes linearly: $f_2(x) = 1 - 0.5 \cdot \frac{l(g_x)-l_{min}}{l_{max}-l_{min}}$ (if $l(g_x) < l_{min}$, then $f_2(x) = 1$; if $l(g_x) > l_{max}$, then $f_2(x) = 0.5$, where $l_{min} = 2$, $l_{max} = 110$, and $l(g_x)$ is genome length). The mutation probabilities now were like in experiment DL2.

The compressed solutions of these runs were then re-evolved to reach 100000 time steps, either with mutation probabilities as in DL2, or with only insertions allowed as in DL1. The length of the genomes were not evaluated in these runs.

After the compression runs, the solutions had between 10 and 30 genes, the mean being 17.1. After the first kind of re-evolution, the respective values were 11, 30, and 18.4; after the second kind of re-evolution, they were 11, 39, and 19.5. All re-evolution runs re-reached 100000 time steps. Re-evolution took 5.2 generations and 5181 evaluations on average with the first method, or 3.5 generations and 3527 evaluations on average with the second method (2 of the 30 compressed solutions did not need any re-evolution to reach 100000).

This means that the strongest compression method achieves compression to 19% of the original size on average. On examination of the re-evolved solutions, one can find that the volume of the kernel genes is 1.64 on average, that is, each gene reads on average 1.64 developmental operations from the pool (the highest volume found is 6).

Table 2. Performance of NEON with complex genetic architectures.

configuration	mean evaluations	comparison
(DL2) key substitution (0.05), deletion (0.45), volume inc. (0.2), volume dec. (0.05)	294928.1 (2.7 times DL1)	significantly worse ($p = 9.917 \cdot 10^{-7}$)
(DL3) key substitution (0.1), deletion (0.15), volume inc. (0.2), volume dec. (0.05)	137571.6 (1.3 times DL1)	significantly worse ($p = 0.01759$)
(DL4) lower time index: $p = 0.2$	129142.6 (1.2 times DL1)	indistinguishable ($p = 0.3738$)
(DL5) lower time index: $p = 0.4$	170784.3 (1.6 times DL1)	significantly worse ($p = 0.004994$)

4 Conclusions

There is a trade-off between the compactness and the smoothness of the search space. The NEON system allows to choose how to make this trade-off, and to change this choice in incremental evolution scenarios. The experiments reported here show that difficult control problems can be tackled with neuroevolution systems employing artificial ontogeny, especially if instructions for new function are first stored uncompressed in the genome and later compressed, making the genetic architecture more complex. This incremental approach can in principle overcome the problem of generating complex, not easily compressible phenotypes that plagues many developmental encodings.

NEON has also been successfully employed for robotic tasks like exploratory behavior with obstacle avoidance and goal search. It includes features for gene reuse not reported here and is planned to be used for complex control tasks where dimensions of input and output increase over evolutionary time.

Acknowledgments. I would like to thank Jürgen Jost for support and helpful discussions.

References

1. Bongart, J.: Incremental Approaches to the Combined Evolution of a Robot's Body and Brain, PhD thesis, Universität Zürich (2003)
2. Eggenberger, P.: Cell Interactions as a Control Tool of Developmental Processes for Evolutionary Robotics. In: From Animals to Animats 4, Fourth International Conference on Simulation of Adaptive Behavior (1996)
3. Green, C.: SharpNEAT. `http://www.sharpneat.sourceforge.net`
4. Harding, S., Miller, J.: A comparison between developmental and direct encodings — An update of the GECCO 2006 Paper "The Dead State" (2006), `http://www.cs.mun.ca/~simonh/`
5. Husbands, P., Smith, T., Jakobi, N., O'Shea, M.: Better living through chemistry: Evolving GasNets for robot control. Connection Science 10(4), 185–210 (1998)
6. Roggen, D., Federici, D.: Multi-cellular development: is there scalability and robustness to gain? In: Yao, X., Burke, E.K., Lozano, J.A., Smith, J., Merelo-Guervós, J.J., Bullinaria, J.A., Rowe, J.E., Tiňo, P., Kabán, A., Schwefel, H.-P. (eds.) Parallel Problem Solving from Nature - PPSN VIII. LNCS, vol. 3242, Springer, Heidelberg (2004)
7. Kenneth, O.: Stanley, Risto Miikkulainen: Evolving Neural Networks through Augmenting Topologies. Evolutionary Computation 10(2), 99–127 (2002)
8. Kenneth, O.: Stanley, Risto Miikkulainen: A taxonomy for artificial embryogeny. Artificial Life 9(2), 93–130 (2003)
9. Stanley, K.O.: Efficient Evolution of Neural Networks through Complexification. PhD Thesis. Report AI-TR-04-314, University of Texas at Austin (2004)
10. Toussaint, M.: The evolution of genetic representations and modular adaptations. PhD thesis, Ruhr-Universität Bochum (2003)

Symbiosis, Synergy and Modularity: Introducing the Reciprocal Synergy Symbiosis Algorithm

Rob Mills and Richard A. Watson

Electronics and Computer Science, University of Southampton
Southampton, SO17 1BJ, UK
{rmm05r, raw}@ecs.soton.ac.uk

Abstract. Symbiosis, the collaboration of multiple organisms from different species, is common in nature. A related phenomenon, symbiogenesis, the creation of new species through the genetic integration of symbionts, is a powerful alternative to crossover as a variation operator in evolutionary algorithms. It has inspired several previous models that use the repeated composition of pre-adapted entities. In this paper we introduce a new algorithm utilizing this concept of symbiosis which is simpler and has a more natural interpretation when compared with previous algorithms. In addition it achieves success on a broader class of modular problems than some prior methods.

1 Introduction

Collaboration between organisms of different types is called symbiosis [13], and such relationships are common in nature, particularly amongst prokaryotes [10]. When symbiotic associations persist over evolutionary timescales, the entities involved may become reproductively inseparable and Konstantin S. Merezhkovsky named this symbiogenesis [8]. Symbiogenesis is thought to have been responsible for several major evolutionary transitions [12], including 'perhaps the most important and dramatic event in the history of life [...] – the origin of eukaryotes' [13, p51] when an archaebacterium and a eubacterium in a symbiotic relationship were genetically joined [11].

The genetic algorithm (GA) is one model of evolution and is useful as it allows us to investigate and understand the process of natural selection. It has distilled some key elements of evolution, such as the ever-repeating process of inherited variation followed by differential selection. However, one key aspect that is not captured by a model like the GA is macroevolution: processes that occur above the level of the species. Symbiosis and symbiogenesis are mechanisms that require a macroevolutionary model to explore successfully. The variation offered by symbiogenesis is qualitatively different from sexual crossover: it offers the union of two sets of pre-adapted genetic material. If this process is repeated, the complexity of the resultant entity can be far greater than that of the original entities involved. We call this process compositional evolution [22], and models that investigate this type of process may shed light on how modular encapsulation facilitates macroevolutionary trends such as the major evolutionary transitions.

Compositional evolution is of interest algorithmically as well as biologically, and this phenomenon has given rise to several algorithms that employ a symbiotic mecha-

F. Almeida e Costa et al. (Eds.): ECAL 2007, LNAI 4648, pp. 1192 – 1201, 2007.

nism. Some have been aimed both at a computational biology audience (e.g. [21]) and problem solving in evolutionary computation (e.g. [5][4]). In this paper we consider a new model and investigate its utility upon optimization problems.

A symbiotic approach is particularly suited to hierarchically modular problems: the solution space is searched with entities of increasing complexity and stability, initially identifying optimal configurations for small subsets of variables, but progressively increasing the size of subsets by combining the optimal configurations for different variable subsets. However, we believe that employing symbiotic variation has the ability to exploit any modularity, hierarchically organized or not. We observe modular structures in a wide range of biological [7], physical [18] and engineered systems [1]. In such systems individual parts may be solved in partial or full isolation from other portions of the system [1][18]. In problems exhibiting this type of decomposability, algorithms that exploit this knowledge should outperform algorithms that do not exploit modular structures. The canonical example of an algorithm which does not exploit modularity is a simple hill climbing process which can only advance in the search space with gradual steps of improvement. There are a number of algorithms that attempt to exploit modular structures, including the genetic algorithm (GA) [20][14], cooperative coevolution [17], and estimation of distribution algorithms such as the Bayesian optimization algorithm [16]. Here we address the type of algorithm that is explicitly compositional as mentioned above. There are also a number of test problems in the literature that exhibit modular properties, including concatenated functions of unitation [3], and royal roads [15]. Here we utilize functions with hierarchical modularity [19], and parameterizable structural modularity [14].

In the remainder of this section we review the Symbiogenic Evolutionary Adaptation Model (SEAM) [21], identify limitations and propose possible methods to address certain limitations of this approach with a new computational algorithm.

SEAM is an early evolutionary algorithm inspired by symbiogenesis [21]. It differs from typical evolutionary algorithms (EA) in several aspects. These include the symbiogenic variation operator, an ecosystem of entities which are partially specified, and coevolved ecosystem templates that are used to facilitate evaluation. The main pool of genetic material represents an ecosystem of many different entities, each specifying only part of an overall solution. There is no source of new genetic material (i.e. no mutation), and the only variation operator is the symbiotic join of two entities. The main loop of the algorithm operates as follows: two entities are picked at random, and evaluated in a number of contexts to determine if the pair should make a permanent symbiogenic alliance. A version of Pareto-dominance (see [21]) is used to make these decisions. If this is the case, the two symbionts are removed from the ecosystem and replaced with the chimera. As this process is repeated, the average size of entities will increase until fully specified solutions are discovered.

SEAM has several limitations: the Pareto dominance mechanism employed to determine whether a join should be made is unnecessarily complicated. Assembling contexts from other entities within the ecosystem, although arguably natural, is (we will show) an unnecessary assumption. Finally, SEAM's performance in other problem domains is severely brittle: the utility of SEAM has only been demonstrated on a stylized hierarchical function, Hierarchical If-and-only-If (HIFF) [19].

Following [21] we develop a new symbiosis-inspired algorithm where individuals repeatedly test and make pairwise joins to compose larger and larger entities, resulting in problem solutions. As in prior work, we are also interested in understanding what symbiosis can offer us algorithmically, when compared with mutation and sexual recombination. Symbiosis potentially offers compositional evolution which can, where the problem domain allows, provide a fundamentally different form of adaptation from the gradualist framework by exploiting divide and conquer problem decomposition [22]. The shuffled-HIFF test problem, for example, can be solved by the composition provided by SEAM in time polynomial in the problem size, but neither a hill climbing process nor a GA using sexual crossover can be guaranteed to reach the optimum in less than exponential time [21][22]. Our new compositional algorithm can also reliably solve shuffled-HIFF, but also extends the scope of applicability to include a single-layered modular problem that is somewhat less contrived than shuffled-HIFF. The single-layered modular problem is, however, also beyond the competence of hill climbing and sexual crossover in the simple GA [14]. In addition to being more general, the new algorithm has a much simpler biological interpretation than SEAM, removing several markedly complicated components of the algorithm, as we shall discuss. Thus we provide an illustration of compositional evolution that is shown to be capable of exploiting the divide and conquer advantage of composition by solving a problem with modular interdependency [2]. The mechanism has relatively straightforward biological analogues in the formation of symbiotic associations, yet its adaptive capacity remains algorithmically distinct from that of the simple GA. The modularity exploited by this mechanism, as shown in the problem class we illustrate, is also a straightforward kind of modularity with a simple intuitive structure. Together this indicates that the breadth and potential impact of symbiosis and compositional mechanisms in natural systems could be much broader than the results supported by prior work, and brings us a step closer to developing biologically relevant algorithms of hierarchical adaptive processes and the major transitions in evolution. The new algorithm uses an abstraction of symbiogenesis as its variation operator, and decisions on when to invoke this mechanism to create a permanent association between entities are based on selecting symbiotic relationships with maximal synergy (synergy is essentially a measure of how much greater the symbiont is over the sum of the individual effects).

2 The Reciprocal Synergy Symbiosis Algorithm

This section describes the new algorithm presented and investigated in this paper. The Reciprocal Synergy Symbiosis Algorithm (RSSA) uses a generational loop consisting of evaluation, selection and variation, in common with a regular EA. However, significant differences exist in each of these stages as well as the representation used for a candidate. Psuedocode for the algorithm is given in Fig 1.

A population of candidates is used but they are intended to represent distinct entities in an ecosystem, rather than all competing for the same niche. As such a candidate entity only specifies alleles for a subset of the problem loci. The ecosystem is initialized with entities that each cover one allele for one locus such that collectively they cover all values for all problem variables.

The algorithm uses no mutation or other method of genetic variation; the only operator that changes the make-up of the ecosystem is the symbiogenic join. This join produces a new entity, taking on the alleles from each of its symbionts where specified, (usually) resulting in a larger, but still partially specified entity.

Partial evaluation of solutions is not available in general, and is not permitted in this algorithm. Instead a partially specified entity is evaluated in context. A context is simply a set of randomly generated values for each problem variable that is not specified by the entity under evaluation (see templates in [6] and contexts in [21][5]). Whilst individual entities are small with respect to the ecosystem, the ecosystem has niches for many small entities, and we would expect each of these to be occupied at any one time. As the result of the evaluations is to end up with some entities in permanent symbiogenic association, it is not sufficient to simply measure the fitness of each entity in the ecosystem. Symbiotic relationships are evaluated on a pairwise basis, where the synergy afforded by a particular join is assessed. Synergy is measured for all pairs of entities in the ecosystem, and for a meaningful measurement this is performed over various contexts.

We define a fitness delta, δ, as the difference in fitness between a context, *c*, alone and an entity, *A*, in that context (Eqn. 1). Synergy, *s*, is defined as the difference in fitness deltas of the symbiont, *A*+*B*, and the sum of the individual entities (Eqn. 2).

$$\delta(A,c) = f(A+c) - f(c) \tag{1}$$

$$s(A,B,c) = \delta(A+B,c) - \delta(A,c) - \delta(B,c) \tag{2}$$

This measure aims to quantify the additional benefit obtained by coexisting, over individual existence. This is closely related to what might be called a measure of epistasis between *A* and *B* when in the context of *c*. The selection phase is a competition between potential joins, and the most viable are those with high synergy scores. The symbiotic relationships are modeled such that whilst high synergy is rewarded, both symbionts must 'want' to make the join. This *reciprocal synergy maximization* is key to assigning the correct symbiotic joins, as we describe later. In order to identify joins with both maximal synergy *and* reciprocal desire to unite from *A* and *B*, we can consider the product of the synergy values for *A*+*B* and *B*+*A*. (Contrast this with an alternative rule that is simply 'maximize the overall synergy for a pair', which just sums *s*(*A*+*B*) and *s*(*B*+*A*). This would allow joins where one entity benefits significantly more than the other – a scenario we found to be unsuccessful). In order for the product to be consistent all synergy values must be unsigned. For simplicity we also choose to normalize the range, such that all values will lie in the range [0,1]. Once joins have been made, the generational loop returns to the evaluation stage, until all of the remaining entities are fully specified.

A number of points regarding the new algorithm are worth highlighting. We make abstractions of several elements that are often present in evolutionary algorithms. As in SEAM, no population dynamics are modeled: we assume that all values for each variable are already present in the ecosystem at initialization. Unlike in SEAM, the interactions between entities are also idealized. We assume that enough interactions occur in the timescale of a generation that we can perform a fixed number of interactions between every pair of entities in the ecosystem rather than stochastically sampling the interactions. The maximum number of joins that can be made in one generation is equal to half the current ecosystem size. In the current implementation there is no method for undoing a join.

```
Initialize ecosystem with all atomic units available in substrate
For g=1:MAX_GENERATIONS
 For c=1:CONTEXTS
  Generate context as random bit-string
  Evaluate context
  For each pair of entities remaining in ecosystem A,B
   Evaluate A in context
   Evaluate B in context
   Superimpose A and B, evaluate symbiont in context
   Calculate synergy for A+B
 For each pair A,B calculate mean synergy values across all con-
 texts
 Process synergy matrix:
  Rescale synergy averages values to lie in range [0,1]
  Calculate reciprocal synergy for each pair A,B, by performing
  the product of s(A,B)*s(B,A)
 Find J highest reciprocal synergy values and make these joins
 If all entities remaining have length N
  Break
```

Fig. 1. Psuedo-code for the Reciprocal Synergy Symbiosis Algorithm

3 Simulation Experiments

3.1 Performance on Shuffled-HIFF

In this section we verify that the RSSA can solve the HIFF problem [19] that has previously been used to exemplify the abilities of the SEAM algorithm [21]. Hierarchical If-and-only-If is a test problem that is built from a single subfunction of If-and-only-If (IFF), the Boolean function of equality. Each pair of variables is grouped into pairs using this same subfunction recursively, which results in an overall function that is dependent on all of its variables. The IFF subfunction incorporates epistasis between functional groups starting from the bottom hierarchical layer upwards and this causes the optima at each hierarchical level to be maximally distant in Hamming space, creating order-*N* dependencies at the highest level. This means that HIFF is pathologically difficult for a mutation-only hill climbing process to solve [19]. However it is an ideal candidate for solving with a compositional approach since these dependencies have a clear modular structure that composition can exploit. Note that a GA using suitable diversity maintenance and one-point crossover is successful on HIFF but only when the linkage map is tight. Randomising the linkage map such that genetic linkage does not correspond to the epistatic structure of the problem (shuffling) defeats the compositional mechanism of crossover in a GA. Simulations verified that the RSSA solves a 256-bit shuffled-HIFF in 30/30 runs demonstrating its ability to exploit modular interdependency successfully in this problem.[1]

---

[1] An early version of the RSSA found that joining all pairs of variables with synergy>0 works very well on HIFF. However, this joining rule is sensitive to the assumption that fitness contributions combine linearly by default, and if all fitnesses are increased by a slight geometric scaling this method will erroneously join any pair of variables. To increase robustness to fit-

3.2 Performance on VSM Problem

In this section, we compare RSSA against SEAM and two variants of GA on a modular test problem, the variable structural modularity (VSM) problem [14]. The specific instances used have clean modularity, for both unshuffled and shuffled cases.

The VSM problem is a test problem that parameterizes the amount of structural modularity to be present, and its modularity exhibits modular interdependency [2]. It is the simplest parameterized model to distinguish the performance of the GA and a hill climber with respect to modularity exploitation. As discussed in [14] a GA using crossover can only exploit the modularity to solve the problem correctly if that crossover mechanism preserves linkage. Therefore, by modifying the physical linkage such that it did not correspond to the epistatic linkage (but the modular structures remain), we expect to reveal a distinction between the GA and RSSA. Although the VSM permits a reduced amount of structural modularity to be present in problems, we do not investigate this dimension of freedom in this paper. The reduction in structural modularity makes the problem easier (see [14]) but since both the GA and RSSA can solve the problem in the hardest case, little is to be learned about the distinction between these two algorithms from easier, less modular problems. Thus we only investigate VSM instances with clean modularity and both tight and random linkage.

In the VSM, the fitness of a genotype is defined as the sum of weighted pairwise dependencies between the problem variables. An *N*-by-*N* matrix is used to define the strength of each interaction, and these can be organized in such a fashion that the problem exhibits structural modularity. Each weight is only included in the summation if the variable values in question satisfy IFF (i.e. when both variables agree) and the problem instance we choose has just two classes of weights: strong weights for interdependencies which are internal to modules, and weak weights for interdependencies between modules. We choose to group the strong weights together near to the leading diagonal of the weight matrix, which introduces a number of modules, the variables of which have tight linkage on the genome. The overall fitness of a candidate $\boldsymbol{x}$ is given by Eqn. 3. Eqn. 4 describes specifically the location of the weights used to structure the modularity. Note that although the problem is built entirely from pairwise interactions, dependencies can be structured to 'act in concert' to create local optima with significant Hamming distances between them [14].

$$F(x_0, x_1, \ldots, x_{N-1}) = \sum_{i=0}^{N-1} \sum_{j=0}^{N-1} w_{ij} (x_i \leftrightarrow x_j) \quad (3)$$

$$w_{ij} = \begin{cases} wI, \; if \lfloor i/k \rfloor = \lfloor j/k \rfloor \\ wE, \quad otherwise\,. \end{cases} \quad (4)$$

The problem parameters used are: *N*=400, *Z*=*k*=20, *wI*=400, *wE*=1. RSSA uses 160 contexts and assigns a maximum proportion of 0.5 joins per generation. This uses approximately 412 million evaluations. For the GAs we use deterministic crowding as

ness scaling, the RSSA joins the most synergistic pairs of variables, and this method works reliably on $HIFF^{0.5}$ and $HIFF^2$ as well as the default fitness scaling of HIFF.

Table 1. Successful runs out of 30 for the VSM problem, parameters as in text above

	VSM, 400 bit	S-VSM, 400 bit
GA-DC 1point crossover	30/30	0/30
GA-DC uniform crossover	0/30	0/30
SEAM	0/30	0/30
RSSA	30/30	30/30

a selection method, a mutation rate of $4/N$, a population size of 4000, crossover rate of 0.05 (this is a low value but see [14] for discussion). SEAM uses 200 contexts to make its joining decisions.

The results shown in Table 1 illustrate the ability of RSSA to solve a second type of modular problem. The joins made in RSSA explicitly represent epistatic dependencies and shuffling the genetic map does not deter RSSA from reliably solving either problem. The GA using one point crossover solves the unshuffled problem as it preserves linkage. However when the genetic map is randomized, preserving (physical) linkage is no longer sufficient to exploit the modularity in the problem, and it should be no surprise that the GA is defeated on the shuffled problem. Note that SEAM is unable to solve either of these problem configurations.

Fig. 2 shows the synergy matrices for examples of HIFF and VSM problems created by RSSA, before any joins have been made. The instances shown are smaller than those tested in order to show relevant detail, but the properties are qualitatively alike. Note that the problem has N variables but these matrices are $(2N)^2$ since each variable can take two values, and the relationships are assumed to be independent. Firstly considering Fig. 2(a), some of the structure of the problem is revealed, but not all.

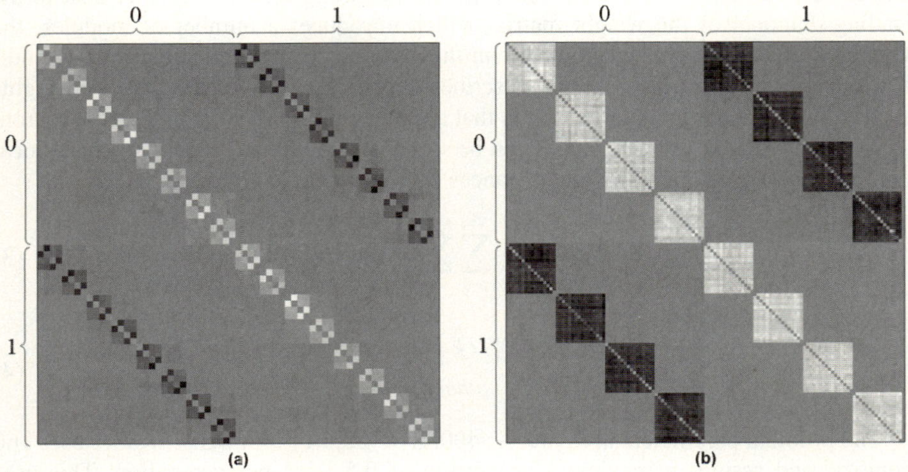

Fig. 2. Synergy matrices from (a) 32-bit HIFF and (b) 64-bit VSM, Z=4 problems (unshuffled genetic maps) before any joins have been made. Each entry (i,j) represents the synergy between the ith and j^{th} problem variables. Strong synergy values are closer to white, weak synergy values are closer to black. Note how the lowest level modules are clear in both cases, but higher level interactions in HIFF are not visible until the search space has been reduced

The number of contexts used is sufficient to uncover the next two or sometimes three hierarchical layers, but an exponential increase in the number of contexts is required to see further layers. However, the information required to make the correct joins for a single layer at a time is sufficient when the problem is solved in a compositional manner. For the VSM (Fig. 2(b)), which has only a single layer of hierarchy, the entire structure is revealed: the 1-entities produce high synergy with other 1-entities within modules, and likewise for 0-entities. The bright line which runs through all the dark blocks in the top-right and bottom-left quadrants represents high synergy for two entities coding different values for the same locus. None of these joins should be made, but RSSA requires a large number of evaluations to reveal that these joins are incorrect. Taking into account these synergy matrices and the fact that modular problems are solved by RSSA, we see that our algorithm is capable of correctly identifying modular structures and revealing that structure in the synergy values it assesses.

4 Discussion

The RSSA introduced in this paper is both a simplification and generalization of SEAM and thereby expands the relevance of symbiosis-inspired algorithms. In section 0 we identified a number of shortcomings with SEAM. Here we revisit these limitations, noting how they are addressed by RSSA. The use of a Pareto dominance mechanism to make joining decisions is weak biologically: it is difficult to imagine two entities not making a join until they have made absolutely sure there is no other alternative symbiotic partner they would rather be with. We replace this mechanism with one that feels more likely in a biological system; that is, the associations that have the strongest synergy for both partners result in symbiotic joins most readily. This also has the algorithmic advantage that it more directly assesses and exploits the epistasis in the problem. Where SEAM required coevolved contexts, the RSSA shows that randomly generated templates are sufficient to facilitate the evaluation of partially specified entities. Although comprehensively describing the algorithmic niche of the RSSA remains an open question, we are confident that it is broader than SEAM: the only problem SEAM is shown to be competent on is shuffled-HIFF and the RSSA is here demonstrated to solve both shuffled-HIFF and another, and less contrived, form of modular problem that SEAM cannot solve.

There are several areas of future research and development, two of which concern the interactions which are used to determine symbiotic joins. Currently RSSA only makes joins between two entities (although each entity may contain many variable values), but there is scope for uniting larger groups. This would not necessarily require the evaluation of symbiotic groups of more than two entities at a time, but could occur when high synergy was identified in all pairs within the group. The interactions could also be focused selectively. Currently all $(2N)^2$ pairs of entities are evaluated in the same number of contexts. But we might imagine that when one pair of entities consistently discovers low synergy scores that their frequency of interaction (i.e. the number of additional contexts they are tested in) would decline, but pairs which show promise of high reciprocal synergy would be allocated further contexts to confirm the value of a join. Although not a key factor here since we have been examining the generality of the algorithm rather than its efficiency, the computational expense of the algorithm is high. Both of these two modifications would reduce the average number of evaluations per join. Finally, we aim to ascertain the ability of RSSA to solve a wider set of modular problems, both hierarchical and flat in structure. It would also be of interest to examine its robustness to the parameterizable modular structures described in [14].

5 Conclusion

In this paper we have presented a new model inspired by symbiosis, with a more straightforward compositional mechanism than that used by SEAM and a simpler biological interpretation than any previous symbiosis-inspired algorithm. We have used the algorithm to explore various possibilities enabled by using symbiosis as a variation operator. As in previous studies, we find it can solve problems that cannot be solved with the more traditional variation operators of crossover and mutation. The experiments performed show the applicability of RSSA is widened when compared with SEAM and it is able to demonstrate an algorithmic advantage over the simple GA on problems with a very simple and intuitive modular structure. In addition, we can observe the structure that is discovered by the algorithm which may provide insight into the characteristics of less neatly formed problems.

Understanding the adaptive significance of the major evolutionary transitions, genetic events of symbiogenesis, and more generally, the formation of adaptive associations among symbionts is critical to evolutionary theory and explaining how evolution works in full [12]. In our work we want to understand not only the conditions under which such events may occur, but also what consequence such events might have for evolutionary adaptation and the evolvability of complex systems. Intuitively, modularity has a significant impact on evolvability but only if there are appropriate mechanisms to exploit it. Crossover in the simple GA cannot exploit modularity unless genetic linkage corresponds with epistatic dependencies, but because the symbiotic joining mechanism modeled here is insensitive to genetic linkage it can exploit modularity that cannot be exploited by the simple GA. It has been shown previously that symbiosis-inspired mechanisms can in principle provide a fundamentally different form of adaptation from the variation of mutation and crossover in the simple GA [21]. However, the exploration of the space in which these types of mechanisms have an advantage is in its infancy. In this paper we make some significant improvements over previous attempts to provide a simple algorithmic model of the formation of symbiotic associations and to describe the kind of modularity it can exploit.

References

1. Baldwin, C.Y., Clark, K.B.: Design Rules, vol. 1: The Power of Modularity. MIT Press, Cambridge (2000)
2. Dauscher, P., Polani, D., Watson, R.A.: A simple modularity measure for search spaces based on information theory. In: Procs. Artificial Life X (2006)
3. Deb, K., Goldberg, D.E.: Analyzing deception in trap functions. In: Whitley, D. (ed.) FOGA 2, pp. 98–108. Morgan Kaufmann, San Francisco (1993)

4. Defaweux, A., Lenaerts, T., van Hemert, J.I.: Evolutionary Transitions as a Metaphor for Evolutionary Optimisation. In: Capcarrère, M.S., Freitas, A.A., Bentley, P.J., Johnson, C.G., Timmis, J. (eds.) ECAL 2005. LNCS (LNAI), vol. 3630, pp. 342–352. Springer, Heidelberg (2005)
5. de Jong, E.D., Watson, R.A., Thierens, D.: on the complexity of hierarchical problem solving. In: Procs. GECCO (2005)
6. Goldberg, D.E., Korb, B., Deb, K.: Messy genetic algorithms: Motivation, analysis, and first results. Complex Systems 3, 493–530 (1989)
7. Higgs, P.G.: RNA secondary structure: physical and computational aspects. Quarterly Reviews of Biophysics 8(4), 373–391 (2000)
8. Khakhina, L.N. (Author), Coalson, R., Margulis, L., McMenamin, M. (eds.): Concepts of Symbiogenesis: Historical and Critical Study of the Research of Russian Botanists. Yale University Press, New Haven (1992)
9. Mahfoud, S.: Crowding and Preselection Revisited. In: PPSN 2, pp. 27–36. Elsevier, Amsterdam (1992)
10. Margulis, L.: The Symbiotic Planet: A new look at evolution. Phoenix, London (1998)
11. Margulis, L., Dolan, M.F., Guerrero, R.: The chimeric eukaryote: Origin of the nucleus from the karyomastigont in amitochondriate protists. PNAS 97(13), 6954–6959 (2000)
12. Maynard Smith, J., Szathmary, E.: The major transitions in evolution. W.H. Freeman, New York (1995)
13. Mayr, E.: What Evolution is. Phoenix, London (2001)
14. Mills, R., Watson, R.A.: Variable Discrimination of Crossover Versus Mutation Using Parameterized Modular Structure. In: GECCO (in press, 2007)
15. Mitchell, M., Forrest, S., Holland, J.H.: The Royal Road for Genetic Algorithms: Fitness Landscapes and GA Performance. In: ECAL 1, MIT Press, Cambridge (1992)
16. Pelikan, M., Goldberg, D.E., Cantú-Paz, E.: BOA: The Bayesian optimization algorithm. In: Procs. GECCO (1999)
17. Potter, M.A.: The Design and Analysis of a Computational Model of Cooperative Coevolution. Ph.D. dissertation, George Mason University, Fairfax, Virginia (1997)
18. Simon, H.: Sciences of the Artificial. MIT Press, Cambridge (1969)
19. Watson, R.A., Hornby, G.S., Pollack, J.B.: Modeling Building Block Interdependency. In: Eiben, A.E., Bäck, T., Schoenauer, M., Schwefel, H.-P. (eds.) Parallel Problem Solving from Nature - PPSN V. LNCS, vol. 1498, pp. 97–106. Springer, Heidelberg (1998)
20. Watson, R.A., Jansen, T.: A building block royal road where crossover is provably essential. In: GECCO (in press, 2007)
21. Watson, R.A., Pollack, J.B.: A Computational Model of Symbiotic Composition in Evolutionary Transitions. BioSystems 69(2–3), 187–209 (2003)
22. Watson, R.A.: Compositional Evolution. MIT Press, Cambridge (2006)

Turing Complete Catalytic Particle Computers

Anthony M.L. Liekens[1] and Chrisantha T. Fernando[2]

[1] Department of Biomedical Engineering,
Technische Universiteit Eindhoven, Eindhoven, the Netherlands
anthony@liekens.net
[2] School of Computer Science,
University of Birmingham, Edgbaston, United Kingdom
c.t.fernando@cs.bham.ac.uk

Abstract. The *Bare Bones* language is a programming language with a minimal set of operations that exhibits universal computation. We present a conceptual framework, Chemical Bare Bones, to construct Bare Bones programs by programming the state transitions of a multi-functional catalytic particle. Molecular counts represent program variables, and are altered by the action of the catalytic particle. Chemical Bare Bones programs have unique properties with respect to correctness and time complexity. The Chemical Bare Bones implementation is naturally suited to parallel computation. Chemical Bare Bones programs are constructed and stochastically modeled to undertake computations such as multiplication.

1 Introduction

1.1 Chemical Computing

An approach for programming a chemical computer is to design a complex 'particle' capable of a controlled transition between configurations, where each configuration is capable of catalyzing a specific set of reactions. Ribozymes can be artificially selected that catalyze specific reactions [1]. Multi-enzyme complexes are common in cells, e.g. PDGF and Tar Complexes [2]. Just as we require, they possess multiple catalytic activities and exist in many states. Programmability arises because the state of a complex subunit is dependent on the states of other subunits on the complex. The topology of the complex can be designed, e.g. clusters, chains, rings, to allow appropriate 'conformational spread' [3]. Approximately digital solid-state circuitry can be produced in proteins [4].

Our approach differs from other work in chemical computing with reaction networks in the following ways. Chemical Bare Bones (CBB) does not make explicit the implementation details of the catalytic reactions; the substrates may be proteins, RNAs or metabolites. Although DNA hybridization catalyst circuits have been proposed by Seelig et al., they model at the algorithm level, circuits of logic gates, not serially executable programs [5]. CBB describes computations carried out by only *one* particle complex with multiple states, and not networks of catalytic particles computing in a distributed manner. Such neural network

F. Almeida e Costa et al. (Eds.): ECAL 2007, LNAI 4648, pp. 1202–1211, 2007.

metaphors utilize coupled cascade cycles, where the weights are the extent of allosteric and covalent modification of the equilibrium position between binary protein configurations that represent activities [6,7,8]. Although it is possible to produce logic gates with an enzyme cascade cycle it will be a formidable task to assemble many of these gates together into a network [9]. The demonstration that CBB is Turing universal lies in the isomorphism between chemical reactions and the Bare Bones language. This overlays the underlying Turing universality of chemical kinetics on which our system depends [10]. Other approaches to demonstrating the Turing universality of a chemical computing system depend for example on forming an isomorphism with Wang tiles [11]. CBB does not produce analog reaction networks, e.g. integral feedback controllers [12] or analog networks capable of computing mathematical functions at their steady state [13]. Such analog networks cannot easily be hand-designed whereas CBB allows hand-design of similar functionalities, plus the incorporation of analog networks where necessary.

1.2 The Bare Bones Programming Language

The Bare Bones programming language contains only 2 assignment statements and one control structure, besides chaining of instructions. The assignment statements are **increase** v and **decrease** v where v denotes a variable name representing a strictly positive integer. Variables are created in memory when they are used for the first time, with a random initial value. The sole control structure is represented by a specific **while** loop, as a **while** $v \neq 0$ **do** ... **end** statement pair. The Bare Bones programming language only allows one condition to control the **while** loop, $v \neq 0$, where v can be any variable.

The Bare Bones language can express programs for any partial recursive function, as shown by Brookshear [14], which ensures Turing completeness. As an elementary example of a Bare Bones program, the above instructions can be used to clear a variable v with **while** $v \neq 0$ **do decrease** v **end**. In Algorithm 1, the values in variables v and w are multiplied and the result is stored in u. Versions that destroy the initial values of v or w during the computation are also possible and result in simpler algorithms.

2 Methods

We show a conceptual implementation of the 3 basic Bare Bones instructions as networks of chemical reactions. Molecules and their counts represent program variables and their values. The number of molecules `V` denotes the value of variable v. We assume that the reaction networks have access to resource particles `R`. A multifunctional catalytic particle controls the flow of the program, analogous to a program or instruction counter in computers. The state of the particle corresponds to the current instruction that has to be processed by the program. In order to process the instruction, the controlling particle catalyzes a reaction

Algorithm 1 Multiplication ($u \leftarrow v * w$) in Bare Bones

```
clear u, clear t1, clear t2
while v ≠ 0 do
  increase t1, decrease v
end
while t1 ≠ 0 do
  decrease t1, increase v
  while w ≠ 0 do
    increase t2, decrease w
  end
  while t2 ≠ 0 do
    increase w, increase u, decrease t2
  end
end
```

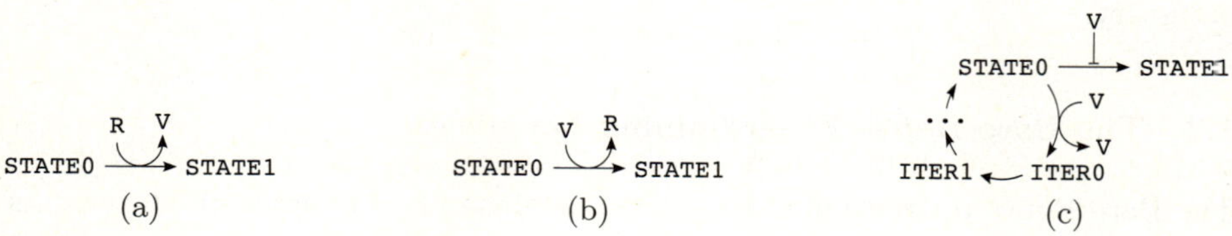

Fig. 1. Chemical Bare Bones primitives (a) **increase** v, (b) **decrease** v and (c) **while** $v \neq 0$ **do** ... **end**

that acts on molecules representing the variables of the program. During the instruction, the program counter particle changes to a conformation representing the next instruction for the program.

Simple catalytic reactions that implement the **increase** v and **decrease** v primitives as basic reaction networks are depicted in Figure 1(a-b). Both primitives can be written as a single reaction. In the case of **increase** v, the control particle in state `STATE0` reacts with an abundant resource molecule `R`, where a new molecule `V` and the control particle in configuration `STATE1` are the products of the reaction. When the control particle signals `STATE1`, the **increase** v operation is terminated. Similarly, we let the program counter react with particle `V` to release a resource particle `R` to instantiate the **decrease** v primitive.

These two basic primitives can be chained as an ordered series of instructions. As an example, Figure 2 represents a program that consecutively goes through instructions **increase** u, **decrease** v and **increase** w. To construct this program, the controller goes through three successive states where it catalyzes one reaction.

The **while** is implemented by means of two reactions as shown in Figure 1(c). One reaction sets the program counter molecule from state `STATE0` to the first instruction of the iteration (`ITER0`) if molecules `V` are present. An iteration of the while loop is a sequence of Bare Bones primitives. At the end of the iteration, the program counter is returned to its `STATE0` state. The second reaction that makes

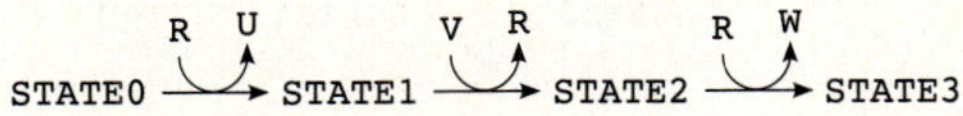

Fig. 2. Chaining three instructions

up a **while** control structure moves the program counter out of the loop, and is inhibited by V molecules. This requirement for inhibition limits the application of the model, as some forms of reaction systems (e.g., metabolic networks) do not have this property. For now, we assume that this inhibition is *strict*, i.e., one V molecule locks the transition of the program counter from STATE0 to STATE1. In a later section, we analyze the behaviour of the loop with stochastic, competitive inhibition. Under the assumption of strict inhibition, the **while** loop cycles for as long as there are molecules V in the system. Note that if the value of v is not decreased during the iteration, the control structure loops unboundedly.

The above construction allows Bare Bones primitives to be ported to a platform of conceptual catalytic and inhibitory reactions. As a consequence, any Bare Bones program can be implemented as a reaction network. Since the Bare Bones language is Turing complete, our interpretation of the language as chemical reaction networks results in a universal language as well.

Because of this universality, more complex instructions that are being added to the language do not improve upon its expressive power. Higher level primitives can all be implemented as Bare Bones instructions, but may have simpler interpretations that can be added to the primitives, and increase the readability of the programs. As an example, an **if** $v \neq 0$ **then** ... **else** ... **endif** control structure could be implemented as two consecutive **while** loops, or it can be implemented more straightforwardly as a pair of reactions. One reaction would move the controller molecule to one series of instructions if V is present, where the second reaction, inhibited by V molecules, moves the pointer to a second block. At the end of both conditional blocks, the controller points to the instruction that follows the control structure.

2.1 Basic Programs

Figure 3 shows an implementation of the **clear** v operator as a basic reaction network. For as long as there are molecules V in the reactor, these react with state STATE0 of the program controller, to produce a resource particle R. If no more molecules V are left in the reactor, the inhibitory reaction becomes unlocked, thereby ending the execution of the program by moving the program controller to STATE1. Figure 4 shows an implementation of a multiplier, as in Algorithm 1. The program initially clears the result and temporary variables t_1 and t_2. Then, molecules V are moved to temporary molecules T1. Consuming T1, the **while** loop starting at state STATE4 sums the value of v to u, t_1 times, using a similar loop with temporary molecule T2. When all temporary molecules T1 have been used, the program signals its end by setting the state of the program controller to STATE8.

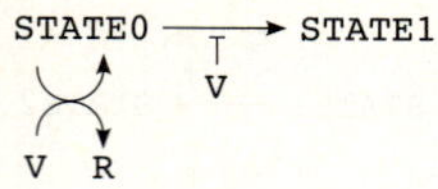

Fig. 3. Chemical Bare Bones program for **clear** v as a reaction network

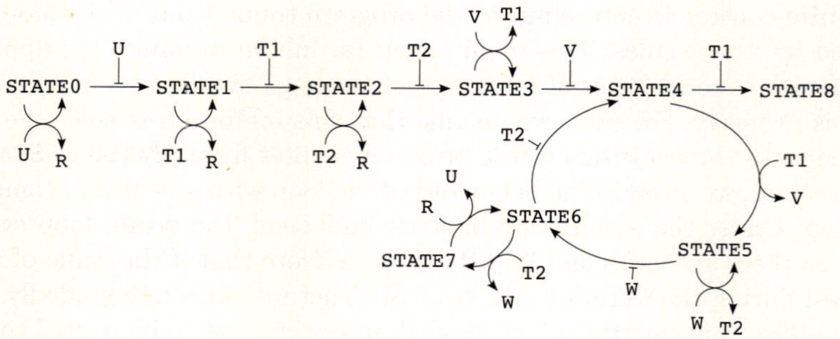

Fig. 4. Chemical Bare Bones program for multiplying the values of v and w as a reaction network. The result is stored in variable u

2.2 Networks as Subroutines

Complex programs in Chemical Bare Bones can be used as modular components, and called in similarity to subroutines. In order to do so, a main program sets the parameters of the module, and activates a new program controller set to the first state of the module. When the end state of the module appears, the extra controller has to be deactivated. The result of the subroutine's computation can now be copied to local variables before continuing the main program. In similarity to an instruction stack in computers, program controllers and their states guarantee the correct flow of the program. For now, we assume that only one program controller can be active at any given time. In a later section, we elaborate on the implementation of parallel program threads.

As an example, the multiplier reaction network from Figure 4 can be called from another program. In order to do so, the parameters for the multiplication have to be copied to variables v and w. Activating a new controller in state STATE0 of the multiplier starts the subroutine. When the multiplier reaches state STATE8, its controller must become deactivated, and the result in u copied to the local variable of the main program before continuing.

Chemical Bare Bones programs are built up out of digital computations, in contrast with other models that rely on analog circuits, see e.g., [13] for implementations of analog square root functions. If each of the reactions of such an analog circuit is catalyzed by the state of the program controller, and the program controller has a method to sense the termination of the analog program, then Chemical Bare Bones programs can, in theory, interface with such analog circuits.

3 Results

3.1 Reaction Networks as Machines

Stochastic Models and Correctness. To set up general Bare Bones programs in reaction networks, we have previously assumed that the inhibition rule to end a while loop is strict. However, this assumption is not feasible in real networks of reactions. Because of real chemistry's stochastic nature, the correctness of a chemical implementation of a program is not guaranteed.

Assuming a well-stirred mixture, and a basic model of mass-action kinetic laws, a reaction is said to occur with a propensity proportional to its reaction rate and the number of reactants available in the system. If we assign a sufficiently fast reaction rate to the reaction that starts an iteration of the while loop, and a relatively slow reaction rate to the competitively inhibitory reaction that exits the while loop, we can decrease the probability that a while loop is exited prematurely.

For now, we suppose that the program counter molecule of the program points to the start of the **while** $v \neq 0$ **do** ... **end** loop, where the program can either enter an iteration of the **while** loop, or exit the loop. We first want to determine the probability that the program exits the **while** loop at the entry state. Let k_{fast} be the reaction rate for the reaction that enters an iteration of the loop. The propensity for this reaction to occur is $k_{\text{fast}}v$. Similarly, let k_{slow} be the reaction rate to exit the **while** loop, and transform the program counter molecule to its next state, ending the **while** loop. Assuming competitive inhibition, the propensity of this reaction is k_{slow}. Using Gillespie's algorithm [15] for stochastic models of reactions, we can determine the probability that the system exits the **while** loop prematurely, dependent on the number of molecules `V` in the reactor. The probability that either reaction occurs first is proportional to its propensity,

$$\Pr[\text{next reaction enters iteration}] = \frac{k_{\text{fast}}v}{k_{\text{fast}}v + k_{\text{slow}}} \tag{1}$$

$$\Pr[\text{next reaction exits loop}] = \frac{k_{\text{slow}}}{k_{\text{fast}}v + k_{\text{slow}}}. \tag{2}$$

If $v = 0$, the probability that the next reaction exits the **while** loop is 1. When the number of molecules `V` is low, the probability to prematurely end the **while** loop is highest. The probability that the **while** loop stops iterating is lower if a faster reaction rate k_{fast} or slower rate k_{slow} is chosen.

We can, as an example, compute the probability that a **clear** v program, as in Figure 3, stops removing `V` molecules and moves to its end state too early. The probability that the loop is terminated prematurely, is given by

$$1 - \prod_{n=1}^{v} \frac{k_{\text{fast}}n}{k_{\text{fast}}n + k_{\text{slow}}} = 1 - \frac{v!\,\Gamma\left(\frac{k_{\text{slow}}}{k_{\text{fast}}} + 1\right)}{\Gamma\left(\frac{k_{\text{slow}}}{k_{\text{fast}}} + v + 1\right)}. \tag{3}$$

Figure 5(a) depicts the probability of terminating a **clear** v prematurely. For higher $k_{\text{fast}}/k_{\text{slow}}$ ratios, the probability of incorrectly exiting the while loop is

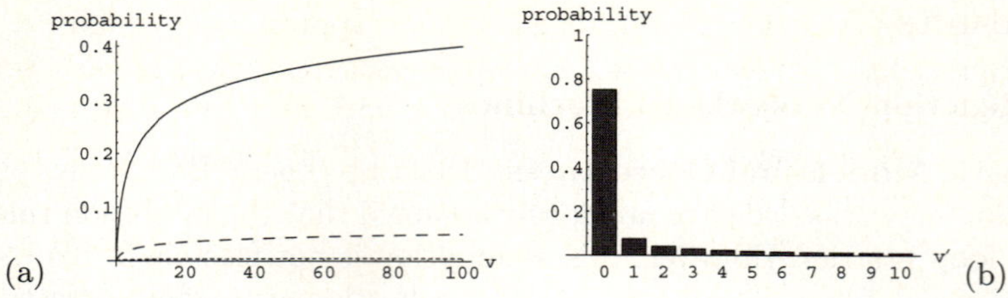

Fig. 5. (a) Probability of prematurely exiting a **clear** v program, dependent on the initial value of v. The continuous, dashed and dotted graphs represent the probabilities for $k_{\text{fast}}/k_{\text{slow}} = 10, 100$ and 1000, respectively. (b) Probability distribution over the possible end results v' of a **clear** v program with initial $v = 10$, with ill-defined parameters $k_{\text{fast}}/k_{\text{slow}} = 10$

lower. As more iterations of the **while** loop have to be carried out, the probability of exiting too early is higher. Independent on the kinetic rate settings, as v goes to infinity, the probability of prematurely exiting the loop tends to 1.

The probability to end the **clear** v program when there are still v' particles left, with $1 \leq v' \leq v$ is given by

$$\frac{k_{\text{slow}}}{k_{\text{fast}}v' + k_{\text{slow}}} \prod_{n=v'+1}^{v} \frac{k_{\text{fast}}n}{k_{\text{fast}}n + k_{\text{slow}}} = \frac{k_{\text{slow}}\Gamma\left(\frac{k_{\text{slow}}}{k_{\text{fast}}} + v'\right)\Gamma(v+1)}{\Gamma\left(\frac{k_{\text{slow}}}{k_{\text{fast}}} + v + 1\right)\Gamma(v'+1)} \tag{4}$$

Figure 5(b) shows the probability to end a **clear** v operation with v' particles left, where the initial value of v is set to 10. Parameter k_{fast} was chosen to be 10 times bigger than k_{slow}. Increasing the rate between these parameters results in higher probability to terminate the computation correctly, as shown in Figure 5(a).

Time Complexity. As a result of the previous section, a larger ratio between kinetic rates k_{fast} and k_{slow} increases the accuracy of programs that use **while** loops. There is, however, a trade-off in the expected running time of the program. Indeed, for high $k_{\text{fast}}/k_{\text{slow}}$ ratios, entering iterations of a **while** loop is fast, where exiting from a while loop is relatively very slow.

As an example, we analyze the running time of a successful **clear** v operation. Assuming that k_{fast} and k_{slow} are well-separated, the expected running time of the **clear** v program can be approximated with the sum of expected times that separate consecutive operations, with

$$\mathrm{E}\left[\text{running time of successful } \textbf{clear } v\right] \approx \frac{1}{k_{\text{slow}}} + \sum_{n=1}^{v} \frac{1}{n k_{\text{fast}}}. \tag{5}$$

The first term denotes the expected time to exit the while loop when there is no V particles left, the second term sums up the expected times to execute an iteration of the **while** loop. Note that this is an approximation since we do not take the probability to exit the **while** loop prematurely into account. Since rate k_{slow} is chosen to be very small in comparison with k_{fast}, the expected running time of the **clear** v operator is dominated by the time it takes to exit the **while** loop, and not so much by carrying out the instructions in the **while** loop.

This result implies a unique property of Chemical Bare Bones programs with respect to time complexity. By choosing high $k_{\text{fast}}/k_{\text{slow}}$ ratios, the time complexity of Bare Bones programs in stochastic models becomes dependent on the number of **while** loops that need to be exited during a run, and not on the number of instructions that have to be processed by the program, since these are relatively fast in comparison with the time required to exit a **while** loop.

Parallelism. In the above sections, we have constructed essentially serial programs for an inherently parallel platform. By introducing multiple program counter molecules in the reactor, parallel programs can be carried out in reaction networks. However, these multiple program counters and their instructions act on shared memory variables.

Multi-threaded programs with shared memory require a mutual exclusion concept to control the flow of the program. The simplest implementation is to use binary semaphores as safeguards of critical sections, i.e., sections of the program that access shared memory. Before a thread of the parallel program executes a critical section, it reserves the semaphore which offers mutual exclusive access to the shared memory. At the end of the critical section, the semaphore is released to allow other threads to access shared memory.

The Bare Bones language in reaction networks allows for the implementation of parallel programs with semaphores. Threads of the parallel program can be started by activating multiple program counter molecules. For the implementation of a semaphore, a variable s is initialized with value 1 at the beginning of the program. If a thread wants to enter a critical section, it can reserve the semaphore with a **decrease** s statement. At the end of its critical section, the thread must release the semaphore by an **increase** s statement to allow other threads to enter their critical sections after locking the semaphore. If the semaphore was reserved by a thread, other threads that request the semaphore with a **decrease** s reaction are implicitly put on hold since the reaction can only be completed if the semaphore is available.

Other options for parallelization is to activate multiple program counters that act on local variables, in similarity to cellular automata [16], or as Chomsky grammars and P systems.

3.2 Simulation

Using the stochastic simulator BioNetS [17] we modeled the multiplier network in Figure 4. Figure 6 shows the typical behaviour of the multiplier for inputs $v = 13, w = 7$, computing the correct result $u = 91$. Figure 7 shows the approximate mean errors of the same network for input values of v and w in the range 1 to

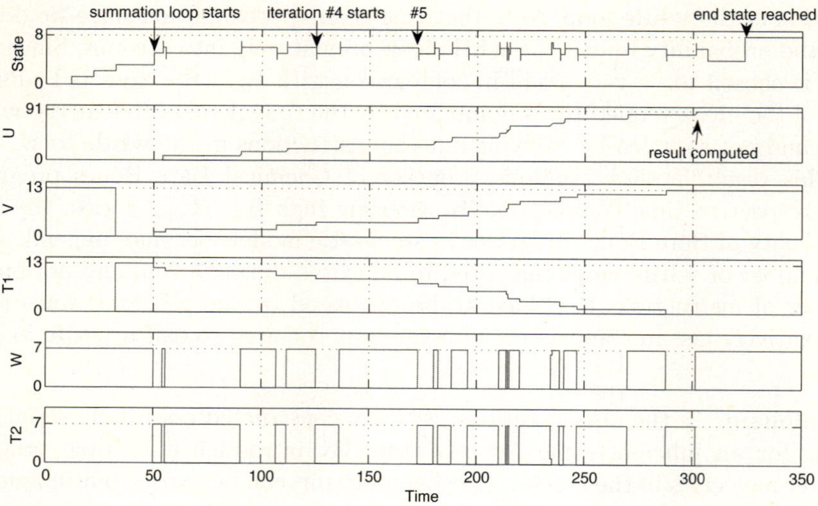

Fig. 6. Typical behaviour of the multiplier with initial values $v = 13, w = 7, u = t_1 = t_2 = 0$. The end result, $u = v, w = 91$ is achieved just before the program controller reaches state 8. The intra-loop rates are fast, with reaction rates $k_{\text{fast}} = 100$, whilst the inter-loop state transition steps are slow, with reaction rate $k_{\text{slow}} = 0.1$

100. For each (v, w) pair, 100 trials were conducted with different random seeds. The rare failure to complete the cycles of the outer loop result in large errors with high variance. As predicted, the greater the difference in rate between fast intra-loop reactions and slow between-loop reactions, the higher the accuracy of the multiplier. For higher values of v and w we see an increasing mean error.

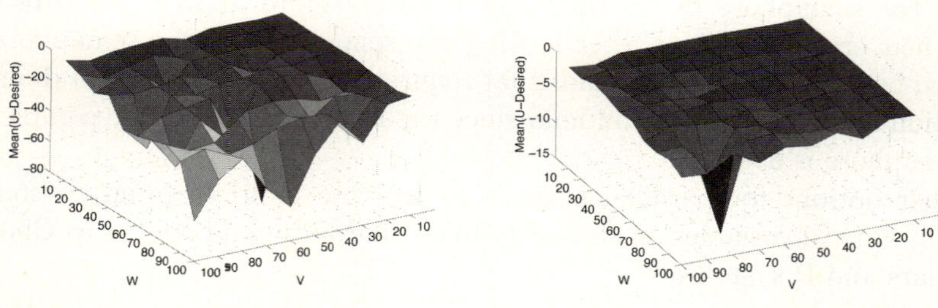

Fig. 7. Left: Low accuracy system, $(k_{fast}/k_{slow} = 100/0.1)$. **Right:** High accuracy system, $(kfast/kslow = 1000/0.1)$. The vertical axis shows the mean$(U_{end} - (v.w))$ obtained over 100 trials for each v, w pair. The errors are always to underestimate the value of $v.w$, due to a **while** loop being exited prematurely

4 Discussion

We have demonstrated a conceptual means by which Turing complete programs can be built using catalytic particles. The complexity of a program is likely in practice to be limited by the capacity to design particle complexes that can act as multi-state program counters. We make no claim that such serial programs exist in biology. If multiplication were to ever confer a fitness advantage, evolution would be more likely to find an analog solution.

Acknowledgments. We would like to thank Huub ten Eikelder for helpful discussion. Also we thank Jon Rowe and Peter Hilbers for guidance throughout the project. This work is supported by the European Community through the ESIGNET project of the Sixth Framework Programme.

References

1. Bartel, D., Szostak, J.: Isolation of new ribozymes from a large pool of random sequences. Science 261, 1411–1418 (1991)
2. Bray, D.: Signaling complexes: Biophysical constraints on intracellular communication. Ann. Rev. Biophys. Biomol. Struct. 27, 59–75 (1998)
3. Bray, D., Duke, T.: Conformational spread: The propogation of allosteric states in large multiprotein complexes. Ann. Rev. Biophys. Biomol. Struct. 33, 53–73 (2004)
4. Graham, I., Duke, T.: The logical repartoire of ligand-binding proteins. Phys. Biol. 2, 159–165 (2005)
5. Seelig, G., Yurke, B., Winfree, E.: Dna hybrization catalysts and catalyst circuits. DNA, 329–343 (2004)
6. Bray, D.: Protein molecules as computational elements in living cells. Nature 376, 307–312 (2004)
7. Arkin, A., Ross, J.: Computational functions in biochemical reaction networks. Biophys. J. 67, 560–578 (1994)
8. Hjelmfelt, A., Weinburger, E.D., Ross, J.: Chemical implementation of neural networks and turing machines. Proc. Natl. Acad. Sci. USA 88, 10983–10987 (1991)
9. Baron, R., Lioubashevski, O., Katz, E., Niazov, T., Willner, I.: Elementary arithmatic operations by enzymes: A paradigm for metabolic pathway-based computing. Angew. Chem. Int. Ed. (in press, 2006)
10. Magnasco, M.O.: Chemical kinetics is turing universal. Phys. Rev. Lett. 68, 1190–1193 (1997)
11. Winfree, E.: Dna computing by self-assembly. The Bridge 33(4) (2003)
12. Sauro, H.M., Kholodenko, B.N.: Quantitative analysis of signaling networks. Prog. Biophys. Mol. Biol. 86, 5–43 (2004)
13. Deckard, A., Sauro, H.M.: Preliminary studies on the in silico evolution of biochemical networks. Chembiochem. 5, 1423–1431 (2004)
14. Brookshear, J.G.: Theory of Computation, Formal Languages, Automata and Complexity. Benjamin-Cummings, Redwood City (1989)
15. Gillespie, D.T.: Exact stochastic simulation of coupled chemical reactions. J. Chem. Phys. 81(25), 2340–2361 (1977)
16. Sipper, M.: Evolution of Parallel Cellular Machines. Springer, Heidelberg (1997)
17. Adalsteinsson, D., McMillen, D., Elston, T.C.: Biochemical network stochastic simulator (bionets): software for stochastic modeling of biochemical networks. BMC Bioinformatics 5, 24 (2004)

Author Index

Printing: Mercedes-Druck, Berlin
Binding: Stein + Lehmann, Berlin

Lecture Notes in Artificial Intelligence (LNAI)

Vol. 4733: R. Basili, M.T. Pazienza (Eds.), AI*IA 2007: Artificial Intelligence and Human-Oriented Computing. XVII, 858 pages. 2007.

Vol. 4720: B. Konev, F. Wolter (Eds.), Frontiers of Combining Systems. X, 283 pages. 2007.

Vol. 4694: B. Apolloni, R.J. Howlett, L. Jain (Eds.), Knowledge-Based Intelligent Information and Engineering Systems, Part III. XXIX, 1126 pages. 2007.

Vol. 4693: B. Apolloni, R.J. Howlett, L. Jain (Eds.), Knowledge-Based Intelligent Information and Engineering Systems, Part II. XXXII, 1380 pages. 2007.

Vol. 4692: B. Apolloni, R.J. Howlett, L. Jain (Eds.), Knowledge-Based Intelligent Information and Engineering Systems, Part I. LV, 882 pages. 2007.

Vol. 4682: D.-S. Huang, L. Heutte, M. Loog (Eds.), Advanced Intelligent Computing Theories and Applications. XXVII, 1373 pages. 2007.

Vol. 4667: J. Hertzberg, M. Beetz, R. Englert (Eds.), KI 2007: Advances in Artificial Intelligence. IX, 516 pages. 2007.

Vol. 4660: S. Džeroski, J. Todorovski (Eds.), Computational Discovery of Scientific Knowledge. X, 327 pages. 2007.

Vol. 4659: V. Mařík, V. Vyatkin, A.W. Colombo (Eds.), Holonic and Multi-Agent Systems for Manufacturing. VIII, 456 pages. 2007.

Vol. 4651: F. Azevedo, P. Barahona, F. Fages, F. Rossi (Eds.), Recent Advances in Constraints. VIII, 185 pages. 2007.

Vol. 4648: F. Almeida e Costa, L.M. Rocha, E. Costa, I. Harvey, A. Coutinho (Eds.), Advances in Artificial Life. XVIII, 1215 pages. 2007.

Vol. 4635: B. Kokinov, D.C. Richardson, T.R. Roth-Berghofer, L. Vieu (Eds.), Modeling and Using Context. XIV, 574 pages. 2007.

Vol. 4632: R. Alhajj, H. Gao, X. Li, J. Li, O.R. Zaïane (Eds.), Advanced Data Mining and Applications. XV, 634 pages. 2007.

Vol. 4629: V. Matoušek, P. Mautner (Eds.), Text, Speech and Dialogue. XVII, 663 pages. 2007.

Vol. 4626: R.O. Weber, M.M. Richter (Eds.), Case-Based Reasoning Research and Development. XIII, 534 pages. 2007.

Vol. 4617: V. Torra, Y. Narukawa, Y. Yoshida (Eds.), Modeling Decisions for Artificial Intelligence. XII, 502 pages. 2007.

Vol. 4612: I. Miguel, W. Ruml (Eds.), Abstraction, Reformulation, and Approximation. XI, 418 pages. 2007.

Vol. 4604: U. Priss, S. Polovina, R. Hill (Eds.), Conceptual Structures: Knowledge Architectures for Smart Applications. XII, 514 pages. 2007.

Vol. 4603: F. Pfenning (Ed.), Automated Deduction – CADE-21. XII, 522 pages. 2007.

Vol. 4597: P. Perner (Ed.), Advances in Data Mining. XI, 353 pages. 2007.

Vol. 4594: R. Bellazzi, A. Abu-Hanna, J. Hunter (Eds.), Artificial Intelligence in Medicine. XVI, 509 pages. 2007.

Vol. 4585: M. Kryszkiewicz, J.F. Peters, H. Rybinski, A. Skowron (Eds.), Rough Sets and Intelligent Systems Paradigms. XIX, 836 pages. 2007.

Vol. 4578: F. Masulli, S. Mitra, G. Pasi (Eds.), Applications of Fuzzy Sets Theory. XVIII, 693 pages. 2007.

Vol. 4573: M. Kauers, M. Kerber, R. Miner, W. Windsteiger (Eds.), Towards Mechanized Mathematical Assistants. XIII, 407 pages. 2007.

Vol. 4571: P. Perner (Ed.), Machine Learning and Data Mining in Pattern Recognition. XIV, 913 pages. 2007.

Vol. 4570: H.G. Okuno, M. Ali (Eds.), New Trends in Applied Artificial Intelligence. XXI, 1194 pages. 2007.

Vol. 4565: D.D. Schmorrow, L.M. Reeves (Eds.), Foundations of Augmented Cognition. XIX, 450 pages. 2007.

Vol. 4562: D. Harris (Ed.), Engineering Psychology and Cognitive Ergonomics. XXIII, 879 pages. 2007.

Vol. 4548: N. Olivetti (Ed.), Automated Reasoning with Analytic Tableaux and Related Methods. X, 245 pages. 2007.

Vol. 4539: N.H. Bshouty, C. Gentile (Eds.), Learning Theory. XII, 634 pages. 2007.

Vol. 4529: P. Melin, O. Castillo, L.T. Aguilar, J. Kacprzyk, W. Pedrycz (Eds.), Foundations of Fuzzy Logic and Soft Computing. XIX, 830 pages. 2007.

Vol. 4520: M.V. Butz, O. Sigaud, G. Pezzulo, G. Baldassarre (Eds.), Anticipatory Behavior in Adaptive Learning Systems. X, 379 pages. 2007.

Vol. 4511: C. Conati, K. McCoy, G. Paliouras (Eds.), User Modeling 2007. XVI, 487 pages. 2007.

Vol. 4509: Z. Kobti, D. Wu (Eds.), Advances in Artificial Intelligence. XII, 552 pages. 2007.

Vol. 4496: N.T. Nguyen, A. Grzech, R.J. Howlett, L.C. Jain (Eds.), Agent and Multi-Agent Systems: Technologies and Applications. XXI, 1046 pages. 2007.

Vol. 4483: C. Baral, G. Brewka, J. Schlipf (Eds.), Logic Programming and Nonmonotonic Reasoning. IX, 327 pages. 2007.

Vol. 4482: A. An, J. Stefanowski, S. Ramanna, C.J. Butz, W. Pedrycz, G. Wang (Eds.), Rough Sets, Fuzzy Sets, Data Mining and Granular Computing. XIV, 585 pages. 2007.

Vol. 4481: J. Yao, P. Lingras, W.-Z. Wu, M. Szczuka, N.J. Cercone, D. Ślęzak (Eds.), Rough Sets and Knowledge Technology. XIV, 576 pages. 2007.

Vol. 4476: V. Gorodetsky, C. Zhang, V.A. Skormin, L. Cao (Eds.), Autonomous Intelligent Systems: Multi-Agents and Data Mining. XIII, 323 pages. 2007.

Vol. 4456: Y. Wang, Y.-m. Cheung, H. Liu (Eds.), Computational Intelligence and Security. XXIII, 1118 pages. 2007.

Vol. 4455: S. Muggleton, R. Otero, A. Tamaddoni-Nezhad (Eds.), Inductive Logic Programming. XII, 456 pages. 2007.

Vol. 4452: M. Fasli, O. Shehory (Eds.), Agent-Mediated Electronic Commerce. VIII, 249 pages. 2007.

Vol. 4451: T.S. Huang, A. Nijholt, M. Pantic, A. Pentland (Eds.), Artifical Intelligence for Human Computing. XVI, 359 pages. 2007.

Vol. 4441: C. Müller (Ed.), Speaker Classification. X, 309 pages. 2007.

Vol. 4438: L. Maicher, A. Sigel, L.M. Garshol (Eds.), Leveraging the Semantics of Topic Maps. X, 257 pages. 2007.

Vol. 4434: G. Lakemeyer, E. Sklar, D.G. Sorrenti, T. Takahashi (Eds.), RoboCup 2006: Robot Soccer World Cup X. XIII, 566 pages. 2007.

Vol. 4429: R. Lu, J.H. Siekmann, C. Ullrich (Eds.), Cognitive Systems. X, 161 pages. 2007.

Vol. 4428: S. Edelkamp, A. Lomuscio (Eds.), Model Checking and Artificial Intelligence. IX, 185 pages. 2007.

Vol. 4426: Z.-H. Zhou, H. Li, Q. Yang (Eds.), Advances in Knowledge Discovery and Data Mining. XXV, 1161 pages. 2007.

Vol. 4411: R.H. Bordini, M. Dastani, J. Dix, A.E.F. Seghrouchni (Eds.), Programming Multi-Agent Systems. XIV, 249 pages. 2007.

Vol. 4410: A. Branco (Ed.), Anaphora: Analysis, Algorithms and Applications. X, 191 pages. 2007.

Vol. 4399: T. Kovacs, X. Llorà, K. Takadama, P.L. Lanzi, W. Stolzmann, S.W. Wilson (Eds.), Learning Classifier Systems. XII, 345 pages. 2007.

Vol. 4390: S.O. Kuznetsov, S. Schmidt (Eds.), Formal Concept Analysis. X, 329 pages. 2007.

Vol. 4389: D. Weyns, H.V.D. Parunak, F. Michel (Eds.), Environments for Multi-Agent Systems III. X, 273 pages. 2007.

Vol. 4386: P. Noriega, J. Vázquez-Salceda, G. Boella, O. Boissier, V. Dignum, N. Fornara, E. Matson (Eds.), Coordination, Organizations, Institutions, and Norms in Agent Systems II. XI, 373 pages. 2007.

Vol. 4384: T. Washio, K. Satoh, H. Takeda, A. Inokuchi (Eds.), New Frontiers in Artificial Intelligence. IX, 401 pages. 2007.

Vol. 4371: K. Inoue, K. Satoh, F. Toni (Eds.), Computational Logic in Multi-Agent Systems. X, 315 pages. 2007.

Vol. 4369: M. Umeda, A. Wolf, O. Bartenstein, U. Geske, D. Seipel, O. Takata (Eds.), Declarative Programming for Knowledge Management. X, 229 pages. 2006.

Vol. 4343: C. Müller (Ed.), Speaker Classification I. X, 355 pages. 2007.

Vol. 4342: H. de Swart, E. Orłowska, G. Schmidt, M. Roubens (Eds.), Theory and Applications of Relational Structures as Knowledge Instruments II. X, 373 pages. 2006.

Vol. 4335: S.A. Brueckner, S. Hassas, M. Jelasity, D. Yamins (Eds.), Engineering Self-Organising Systems. XII, 212 pages. 2007.

Vol. 4334: B. Beckert, R. Hähnle, P.H. Schmitt (Eds.), Verification of Object-Oriented Software. XXIX, 658 pages. 2007.

Vol. 4333: U. Reimer, D. Karagiannis (Eds.), Practical Aspects of Knowledge Management. XII, 338 pages. 2006.

Vol. 4327: M. Baldoni, U. Endriss (Eds.), Declarative Agent Languages and Technologies IV. VIII, 257 pages. 2006.

Vol. 4314: C. Freksa, M. Kohlhase, K. Schill (Eds.), KI 2006: Advances in Artificial Intelligence. XII, 458 pages. 2007.

Vol. 4304: A. Sattar, B.-h. Kang (Eds.), AI 2006: Advances in Artificial Intelligence. XXVII, 1303 pages. 2006.

Vol. 4303: A. Hoffmann, B.-h. Kang, D. Richards, S. Tsumoto (Eds.), Advances in Knowledge Acquisition and Management. XI, 259 pages. 2006.

Vol. 4293: A. Gelbukh, C.A. Reyes-Garcia (Eds.), MICAI 2006: Advances in Artificial Intelligence. XXVIII, 1232 pages. 2006.

Vol. 4289: M. Ackermann, B. Berendt, M. Grobelnik, A. Hotho, D. Mladenič, G. Semeraro, M. Spiliopoulou, G. Stumme, V. Svátek, M. van Someren (Eds.), Semantics, Web and Mining. X, 197 pages. 2006.

Vol. 4285: Y. Matsumoto, R.W. Sproat, K.-F. Wong, M. Zhang (Eds.), Computer Processing of Oriental Languages. XVII, 544 pages. 2006.

Vol. 4274: Q. Huo, B. Ma, E.-S. Chng, H. Li (Eds.), Chinese Spoken Language Processing. XXIV, 805 pages. 2006.

Vol. 4265: L. Todorovski, N. Lavrač, K.P. Jantke (Eds.), Discovery Science. XIV, 384 pages. 2006.

Vol. 4264: J.L. Balcázar, P.M. Long, F. Stephan (Eds.), Algorithmic Learning Theory. XIII, 393 pages. 2006.

Vol. 4259: S. Greco, Y. Hata, S. Hirano, M. Inuiguchi, S. Miyamoto, H.S. Nguyen, R. Słowiński (Eds.), Rough Sets and Current Trends in Computing. XXII, 951 pages. 2006.

Vol. 4253: B. Gabrys, R.J. Howlett, L.C. Jain (Eds.), Knowledge-Based Intelligent Information and Engineering Systems, Part III. XXXII, 1301 pages. 2006.